THE AMERICAN RAILROAD FREIGHT CAR

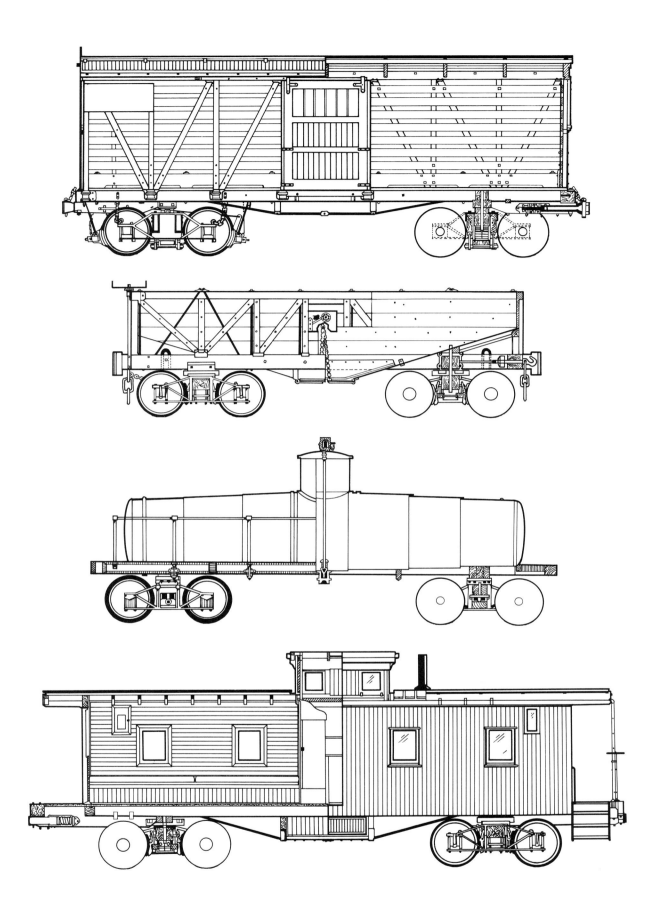

From the Wood-Car Era to the Coming of Steel

The American Railroad Freight Car

JOHN H. WHITE, JR.

THE JOHNS HOPKINS UNIVERSITY PRESS BALTIMORE AND LONDON

© 1993 The Johns Hopkins University Press
All rights reserved
Printed in the United States of America on acid-free paper

The Johns Hopkins University Press
2715 North Charles Street
Baltimore, Maryland 21218-4319
The Johns Hopkins Press Ltd., London

LIBRARY OF CONGRESS CATALOGING-IN-PUBLICATION DATA
White, John H., 1933–
    The American railroad freight car ; from the wood-car
era to the coming of steel / John H. White, Jr.
        p.   cm.
    Includes bibliographical references and index.
    ISBN 0-8018-4404-5
    1. Railroads—United States—Freight cars—History.
I. Title.
TF470.W45   1993
625.2'4'0973—dc20   92-24255

A catalog record for this book is available from the British
Library.

Facing title page: From top to bottom, a stock car (1880),
a coal car (c. 1861), a tank car (1870), and a caboose (1894).
Drawings by Allyn Massey.

This book is dedicated to my brother ROBERT J. WHITE,
who, like the master car builders of the past,
has served the engineering community long and well.

CONTENTS

PREFACE

The idea for this book was born almost twenty-five years before serious work began on the text. During my first months at the Smithsonian I was directed to begin work on an exhibit script for the Railroad Hall in the projected Museum of History and Technology. My instructions were rather vague, but it was clear that a broad and inclusive exhibit was envisioned by my superior. "Just make sure nothing important is left out" was the general message, but when I asked for more particulars, my questions were dismissed with the assurance that the railroad literature was substantial and just a little reading on my part would answer any query. As a very junior employee, I accepted the counsel of my seniors and set to work.

This was in the fall of 1958. I spent every hour not consumed by other duties in the stacks of the Library of Congress. A stack pass allowed direct access to the abundant books available on railways. The quantity of material was sizable, as predicted. But after a few weeks it became clear that most of the volumes were concerned with short lines or the finance and regulation of class one roads. In scanning the pages of these thick and thin volumes, I found a few nuggets, but in the main I uncovered little that proved useful in outlining the growth and development of the American railways' physical plant, except perhaps for the steam locomotives. The volumes by Alfred W. Bruce (*The Steam Locomotive in America*, 1952) and Angus Sinclair (*Development of the Locomotive Engine*, 1907) offered the kind of guidance I was seeking. The library staff sent me on to the Association of American Railroads, then located in the Transportation Building just a few blocks from the White House. It contained the finest railway library in the nation, now sadly dispersed. It would surely hold the sources I sought. The staff there was informed and helpful, but after a few visits it seemed clear that no one had compiled the studies I needed, outside of the locomotive field. Someone, I think it was a colleague at the museum, suggested that I try the Patent Office. There

I would find experts in just about every subject and field.

My informant was correct. There was an expert on railway engineering. My audience with this person started out on an optimistic note. He surely knew railway technology and expressed over and over again his willingness to help the Smithsonian in any way possible. But it also became clear after a few minutes that my expert understood only the most recent developments in the field and had little or no grasp of matters before his time. He went on for some time explaining the recent developments in cushion draft-gear apparatus, then the hottest topic in the industry. But when I began to press for information on nineteenth-century car construction, he admitted his ignorance about the "old times," as he called them. "Why," he said, "I thought the Smithsonian kept files on the ancient history of railroading and so could answer any question on the subject." At this point I declined to explain the limitations of our supposedly infinite information repositories, but it was clear that there was no quick or easy route to follow. The Northwest Passage definitely did not exist, and if I wished to proceed to India, it would be over a long and arduous road.

The information did exist, but it was scattered about like so much low-grade ore. Mining out the information was more labor-intensive than I had imagined. The task goes on, and I have even now hardly scratched the surface. Yet by late 1958 my mission was obvious. The sources were all around me or anyone else, for that matter, who cared to pursue the subject. Principally, they consisted of the trade journals and technical society transactions that recorded in remarkable detail the day-by-day progress or lack thereof in the engineering field. There are many gaps and omissions, of course; this is particularly true for the early period (pre-1870), but the later literature reflects back on the pioneers just enough to help us assemble a fairly complete picture of what transpired throughout the railway age. Naturally, it all took far longer than I had anticipated back in the last years of the Eisenhower presidency. My thoughts for a series of short outline volumes were discouraged by the same cadre of wise old men who told me that there was already an abundance of literature on the subject. But now I was advised that should I think to add anything between hard covers, it should be a more formal study. And so the first book, a volume on early locomotive design, took seven years to write and another year or two to see through print. The passenger car volume took even longer, and now the freight car study, which barely enters the twentieth century, has taken longer yet. Time is fleeting, and the energy of youth has long passed. Yet with enthusiasm for railway preservation at an all-time high, surely others will carry on the work.

How and why I came to write this book seems obvious to me as just explained, but it is hardly so to someone outside the history of technology, which is after all a very narrow field. About the time I began serious work on this book in 1980, I was introduced to a formidable matron at a garden party. This lady understood that I worked at the Smithsonian and wondered in which of the art departments I was stationed. After explaining that I worked in the transportation history field, she asked about my latest project. I then briefly described the projected freight car study. But before I could finish, she exclaimed, "Oh, they don't make you work on things like *that*, do they?" The perceived lack of glamour or intellectual worth was neatly summarized by this lady's rejoinder. It also identified the disdain of this late industrial age for the mundane yet practical necessities of everyday life. Cheap, dependable haulage of goods is essential to civilized life, at least as we know it in the West. Yet the importance of a subject will never make it a popular one. Indeed, within the field of railroad history, the least important topics tend to be the most popular ones.

I suppose it might be worth dedicating this volume to everyone who believes that railroads carry only passengers, and that once that service has been suspended, the line is abandoned. Or perhaps it would be better to recognize the people who never receive book dedications, the common workers who ran American rail freight service for so many generations. This would involve not just the expected train crews but yard clerks, freight agents, inspectors, car repairers, and the like. We are speaking now of millions of individuals over a century and a half. Some were surely intelligent, dedicated, and sober. Many, however, were not exceptional; they were just ordinary people trying to earn a living. The skilled and the unskilled, the energetic and the lazy, the mean-spirited and the good-hearted—they all worked together, somehow, to keep the cars moving. Individually they are forgotten except by their immediate families,

and their biographies can never be reassembled. But as an army of workers, they should be remembered for service that never wins medals or public acclaim but is nonetheless essential to the nation's daily well-being. Something about their working days and the equipment they fabricated and operated is given in the following pages. The trades pursued by many of these workers have become as obscure as certain occupations of the Middle Ages. For example, a colleague asked me to help with his family genealogy. Relatives had told him that his grandfather was a journal inspector at the C&O shops, Huntington, West Virginia, but none could explain what journals were nor indeed why it was advisable to inspect them.

I must make it clear that this is a hardware study. Its central mission is to explain the principal types of freight cars used on American railways from 1830 to about 1910. The materials and construction techniques employed are important secondary themes. The basic arrangement of the book is a simple breakdown of cars by type of service. Coal cars are included in the bulk carrier section (Chap. 5), for example. However, to emphasize the more specialized nature of rolling stock during the first forty years of American railroading, all early cars, including those used to carry coal, have been treated in Chapter 2. The subject matter does not always allow for neat packaging, however, for the industry, when backed up with traffic, would use just about any type of car to move cargo. Boxcars were accordingly often used to move coal even though they were not well suited for such service. Some cars do not fit neatly into just one category, and so they were placed in what appeared to be the most logical position. The index has been constructed to alert readers to the more obvious crossovers.

Many important subjects—traffic, rates, operations—are introduced in the first chapter, but the reader must understand that these topics are not the focus of the book. Once again, the subject is the cars themselves. The weighty subjects touched upon in Chapter 1 should be taken up and more fully treated by a historian more conversant with economics and business matters. My intention is only to outline the essential reasons that hundreds of thousands of cars ambled across America in such a restless manner.

This study is intended as a reference work. It has few surprises to offer other than the early appearance of many ideas now considered very modern. Containers can be traced back to the pre-railroad era, and covered hopper cars date from the 1830s, though the latter scheme did not achieve much popularity until around 1950. The railroads' resistance in adopting specialized cars for exceptional cargoes, like cattle or produce, may surprise many readers. The investment costs appear to have been the primary source of this attitude. It took powerful forces outside the industry, primarily big shippers like Swift and Armour, to break down this prejudice. The industry was in fact continually being picked away at by outsiders like Westinghouse, Janney, Schoen, and many others who wanted the railroads to adopt their apparatus. Important improvements such as power brakes, semiautomatic couplers, and steel cars were the result of the persistence of outsiders who would, in time, convince the railroads of the value of their products.

Regrettably, I have not been able to devise a thesis or single overall focus for this book. That railroad cars were an important factor in the development of America in the industrial age is obvious. Just how to draw anything more profound out of this study I must leave to more creative historians.

The record for the American freight car is like a great heap of broken tiles, and it was my job to sort through these unrelated shards and reassemble an intelligible mosaic. My task was made more difficult because so little serious historical work has been done in this area. This is particularly true in the interpretive side of the subject. The trade journals and car dictionaries contain a great deal of data that tells what happened and when it happened, but never why it happened. Why, for example, did the arch-bar truck become the standard in about 1870? Every car builder of the time surely understood why this was so, but not one of them seems to have recorded it.

In my assembling and shifting, I have attempted to give some regional coverage in the selection of the cars shown to make this volume a national survey. Some areas or railroads have received more coverage than others because they were more important, or simply because more information was available. No exact balance was attempted and some lines have been ignored, but just about every trunk road has received at least some attention. At the same time, an effort was made to represent the work of major designers and commercial car builders.

In the course of writing this book I was helped by many individuals and institutions and it is a pleasure to acknowledge them at this time. Wil-

liam D. Edson, George W. Hilton, Robert C. Post, and Thomas T. Taber III offered helpful criticisms on portions of the text in its earliest, draft stages. The handwritten manuscript was ably typed by Mary E. Braunagle, Johanna Kinley, and Edwin B. Wynn. Many experts and collectors gave freely of their information and illustrations. These include Gerald W. Best, Michael A. Collins, Donald Duke, Stephen Drew, Arthur D. Dubin, Richard Dole, Edward T. Francis, Walter P. Gray, George Hart, John Hankey, Herbert H. Harwood, Jr., Anthony Hall-Patch, John W. Lozier, Benjamin Kline, Bob Richardson, Ellen Swartz, John N. Stine, Robert M. Vogel, Susan Tolbert, and William L. Withuhn. Others who offered help and encouragement during the long years this work was in preparation include Peter B. Bell, Arthur P. Mollela, Arnold Menke, Dick Dennison, Robert L. Emerson, Albert Eggerton, Paul Moore, Jackie Pryor, Joyce Palmer, and Marilyn Wandrus. Betty Tone deserves praise for her fine art work, especially the dust jacket illustration. I also owe a special word of thanks for the meticulous care in copy editing by Mary V. Yates.

Many libraries, museums and corporations proved helpful in the creation of this study. The Allegheny Foundation provided a generous grant that assured a high-quality publication. The Johns Hopkins University Press staff was more than cooperative at every stage of the book's preparation, and special thanks are due Henry Tom, Lee Sioles, Martha Farlow, and Penny Moudrianakis.

The Museum of American History Library staff—Rhoda Ratner, James P. Roan, Helen Holley, Bridget J. Burke, Lindsey Ealy, and Stephanie Thomas—was always ready to help locate any book or serial needed for my study. Other institutions that were helpful include the California State Railroad Museum, the Colorado Railroad Museum, the Maine Historical Society, the Denver Public Library, the De Golyer Foundation Library, the State of Pennsylvania Railroad Museum, the National Railway Museum (England), the Railway and Locomotive Historical Society, the Science Museum, and the Southern Pacific Transportation Company.

THE AMERICAN RAILROAD FREIGHT CAR

1 PARADE OF THE CARS

Cheap transportation is the essence of railroad freight service. The ability simply to move great masses of goods long distances is not enough. Capacity must be coupled with the ability to transport goods economically. With a few exceptions, speed was not an important factor in nineteenth-century American transport. Ships had provided inexpensive, long-distance haulage since ancient times, but much of the United States was landlocked. Overland wagon rates were too high, except for short distances or very valuable goods. Canals, rivers, and lakes offered inland highways useful to the development of some regions. But many areas were not traversed by rivers, and most lakes helped only on a small part of the journey. Canals, of course, were useless in rugged topography. Too many locks boosted construction costs to unrealistic levels and slowed the already languid pace of boats as they transferred from one lock to the next. Worse yet, most inland waterways were open only in warm-weather months—hardly a total service system. Then in the spring, floods crippled most canals and at least some rivers. If the summer was especially dry, the rivers and canal reservoirs dried up. The railroad, conversely, was an all-weather, every-season route. It could be counted on to operate every day except during the more capricious acts of nature. The railroad was soon recognized as a handmaiden to commerce. It proved a reliable avenue for industry that could cross the loftiest mountain ranges, the driest desert, the wettest swamp, the widest river, and the loneliest prairie.

Put very simply, railroads were, and are, a service industry. They have nothing to sell but transportation. That service must be economical, orderly, reliable, and reasonably prompt. It is true that a few lines have mined coal or harvested lumber or even rolled a little rail, but by and large they are selling only one product—transportation. Their chief customer has always been industry, and this was true even at the peak of rail passenger travel. In the middle 1850s freight receipts already exceeded passenger revenues by 20 percent.[1] At the time of the 1880 Census it was reckoned that freight brought in 64.5 percent of all railroads' earnings. By 1903 this figure had jumped to 73 percent.[2] A few years earlier L. F. Loree had insisted that practically all the profits of the industry came from freight service because of the high cost associated with passenger operations.[3] Profits were surely more important to industry than gross revenues, be the source passenger or freight.

Comparing freight and passenger transportation might be likened to comparing a necessity and an extravagance. The cartage of fuel, food, clothing, and building materials pertains to the necessities of life, at least for civilized life as we have come to know it. Most travel, however, is largely frivolous, whimsical, or impulsive. Holidays, wedding trips, reunions, anniversary excursions, or clan meetings are pleasant but hardly necessary. If no fuel arrives at the power plant, no grain at the mill, no cotton at the gin, we are all likely to suffer. Yet if most passenger travel were aborted, it would be more an annoyance than a debacle. Even in Victorian times most business communications could be carried on with relative ease via the mails and telegraph. Personal meetings were only occasionally a necessity. But cut off the freight cars' ceaseless rolling anytime after 1860, and the nation would be in ruins.

From small and uncertain beginnings in 1830, the railroad network grew until by 1860 it had come to dominate the inland commerce of the nation. Mass transportation had come to the United States. The flow of goods over long distances was established. George Rogers Taylor, in his 1951 book *The Transportation Revolution*, considered railroads to be one of the most revolutionary inventions of all time. Taylor based this judgment not on the sophistication of the technology, but on the economic impact of affordable transportation. The consequences are impressive: the creation of cities, the settlement of western lands, the introduction of dry farming into the Great Plains, and the general shift of inland traffic from the North and South to the East and West. The railroad, this new

FIGURE 1.1 *(opposite page)*
A freight train from the wood-car era is seen rushing out of the past in this scene dating from the 1880s on the Delaware & Hudson Railroad. (Delaware & Hudson Photo 17655)

FIGURE 1.2

agency of transportation, was fully justified in real economic terms of net earnings and transport cost savings. Albert Fishlow concludes that this saving amounted to 15 percent per year between 1830 and 1860.[4] The nation benefited from low shipping rates, which stimulated the exchange of goods between all states. The agricultural sector probably benefited most from the miracle of cheap overland transit. This breaking down of barriers to the distribution of goods meant an end to local markets. Of more significance, it meant an end to national markets, for American grain was now competitive on the world market. In effect, the railroads had increased the size of the market.

The monumental nature of the railroad freight traffic, the sheer bigness of the operation, the tonnage handled, the distances traveled, the climates encountered—all these make it a difficult subject to explain. It is easier to handle, of course, when presented on a more human scale. What did it mean to average citizens? Potentially it offered an improvement in the quality of everyday life. It meant that city dwellers might have cheaper farm products. It meant that they could expect a greater variety of foods year-round, though this was not much in evidence until after 1880. Luxuries like oranges and lemons became everyday fare. It meant that farmers benefited from the sale of these products, and that unskilled workers might move out into the Great Plains to establish a new life, secure in the knowledge that there was an economical way to ship grain or hogs to market. And this was true despite Granger agitation to the contrary. Conversely, those who were weary of life on the

land might now go to the cities and find employment in industry.

The railroad became a pipeline between the factory and the consumer. Cheap brass clocks from Connecticut or buggies from Cincinnati went at low rates from the maker to the buyer anywhere in the land. Even the poor benefited from low freight charges. The poorest of the poor could count on cheap bread, because the railroads routinely moved wheat at extremely low rates, ensuring a ready supply of the staff of life at a price all could afford. James Ward tells us in his book *Railroads and the Character of America* that in 1842 cheap coal was made available to the poor of Philadelphia who could not afford firewood.[5] In times of drought or flood, the railroad could rush in relief supplies. The misery of Chicago citizens after the Great Fire of 1871 was lessened considerably by such emergency service. And so the average citizen did benefit directly and daily from the humble freight cars that rattled and lumbered constantly about them. But it is likely that few of the beneficiaries thought much about the role of boxcars in their lives. Indeed, the average citizen rarely thinks much about any public service until it ceases to function.

Romance and beauty are almost never associated with freight cars. Plain, ponderous, and work-a-day, these faithful servants, which function so efficiently as breadwinners for the railroad industry, receive little attention and less compassion. Their cheerless and dowdy appearance—for most look old long before their time—has won them few admirers. Shippers see them as a free warehouse but

THE AMERICAN RAILROAD FREIGHT CAR

FIGURE 1.3

Wrecked and battered cars were a normal part of railroad operations. In the rear-end collision depicted here, one lucky boxcar, sans trucks, landed on top of an errant locomotive. Its cargo of grain is seen spilled alongside the locomotive. (*Locomotive Engineering*, Jan. 1892)

FIGURE 1.4

Hard use and wrecks meant an early end to most boxcar careers. A wrecking crew cleans up after a cornfield meet on the Kansas City, Fort Scott & Memphis around 1890. (Collection of J. O. Riley)

PARADE OF THE CARS

little more. Tramps, near the bottom of the social ladder, appreciate the boxcar more than most. It was, after all, their palace car. If anyone could understand the romantic side of the stalwart boxcar, it was the hobo, himself a wanderer who shared his daily travels with the slogging freights that moved unnoticed across the landscape of America.

The roving cars were always in motion as their ranks rolled across the plains, over mountain summits, down into valleys, from there to follow a river, disappear into a tunnel, pass through a woods, later to cross a desert or follow some ancient path like the Santa Fe Trail, the trail of the Holy Faith. Battered by snows while crossing the Continental Divide in some high pass, or sunbaked on a dusty desert plateau, the sides of the cars took on the weatherbeaten look of the nomad. There was an imperative mobility to these wayfarers whose broad, dull red sides were like a chain of billboards moving slowly across the land. Most were a thousand miles away from home; grinding on seemed to bring them nearer only because they were not standing still.

Actually, the cars stood dead still much of the time, but their public image is one of ceaseless movement. Cars stored in a yard do not attract our attention, but at least a few will notice their ponderous passage at night as they rumble across an overpass, their bulky shapes dimly illuminated by a street light. They look like a great line of elephants hooked together tail to trunk, weaving slowly along a track toward some distant graveyard. They move along obediently, but with the heavy resignation of chattel. One is reminded of the enforced wanderings of Captain Vanderdecker in the legend of the Flying Dutchman. Travel becomes torture to those condemned to move on forever. Vanderdecker remains a romantic figure in western culture; is it too much to expect the same for the lowly boxcar?

The downtrodden gain a certain romantic cachet as objects of pity. Freight cars might be likened to galley slaves or migrant workers—used roughly and treated hard, unappreciated and underpraised. Conversely, they can be seen as a symbol of strength and endurance, for they surely could take it. Here was a brawny, toil-hardened vehicle ready for the long haul. It was of the be-strong school, ready for a life of lifting and working. It in fact demonstrated nicely Victorian notions about

FIGURE 1.5

Beauty is not a factor commonly associated with railroad freight equipment, but there is something handsome in a robust or rustic way about most of these cars. This sturdy 40-footer was built late in the wooden era (1913) by Haskell & Barker of Michigan City, Ind.

THE AMERICAN RAILROAD FREIGHT CAR

fitness and success belonging to the strong. In a world red in tooth and claw, the boxcar was the perfect model of a rough and tough survivor. In such a world only the fit survive and reproduce. The weak fail and are replaced by the fit. Surely, these primitive laws seemed right for this most Darwinian of vehicles.

Some of the ideas just recounted inspired writers in the wood-car era to speak of these stalwarts of freight service. In 1902 *Railway Age* published the story of an old freight car written by an anonymous scribe who signed himself The Inspector.[6] The old car described by The Inspector was reaching the end of its career and worked only slow freight in long, tiresome trips. Its floor was dirty, and its roof leaked so much that it carried only low-grade goods, everything from lumber to pig iron. Dynamite had once been used to loosen a load of frozen ore; the blast hardly improved the car's condition. In its youth the car had carried premium cargoes, and "Rush" cards were always nailed to its doors. But no more; it was now just another has-been, working out its last days.

A more poetic account was written by Strickland W. Gillilan (1869–1954), "The Story of the Freight Car." It begins, "I am a bumped and battered freight car on a side track in the yard; / I am resting—resting gladly, for my life is cruel hard." The car explains its varied service until it is unceremoniously bumped by a yard engine and set off on a new journey. In September 1935 a poem called "Box Cars," by Brian Moore, appeared in *Railroad Stories*, a magazine intended for railroaders. Moore writes about a boy lying on a hillside overlooking track where the trains pass like caravans. The cars pass slowly enough that he can read the names; the western roads make him think about the Cheyenne, General Custer, and cowboys. Anywhere that is far away seems greener to the boy. He comes to love the melodious names painted on the cars, especially the Monongahela, the Lackawanna, and the Piedmont & Northern. The train is a messenger of far-off places, of memories, and of magic—at least to this young boy atop a grassy hillside in 1935.

Almost a century earlier, Henry David Thoreau had viewed the passage of the daily freight train on the Fitchburg Railroad in a more realistic way. This busy piece of track did much to disturb the tranquility around Walden Pond: "All day the fire-steed flies over the country." What followed the fire-steed was a parade of goods-laden cars, and the writer, no friend to railways or things modern, correctly recounted the purpose of these trains: "Here come your groceries, country, your rations, countrymen! . . . timber like long battering rams . . . and chairs enough to seat all the weary and heavy laden . . . All the Indian huckleberry hills are stripped, all the cranberry meadows are raked into the city. Up comes the cotton, down goes the woven cloth; up comes the silk, down goes the woolen." Thoreau captured much in these few lines about the actual function of the railway as a mass carrier of goods. But if he was reluctant to praise the newfangled steam cars, his contemporary Walt Whitman was not. His poem "To a Locomotive in Winter" includes these lines: "Thy train of cars behind, obedient merrily following, / Through gale or calm, now swift, now slack yet steadily careering; / Type of the modern —emblem of motion and power, pulse of the continent."

If a boxcar could be considered romantic, could it be considered beautiful? The answer seems simple enough: No, a boxcar by definition is as ugly as a mud fence. Surely this is true if the classical model of beauty is applied. Nothing so rude as a freight car could be considered pleasing, nor does it conform in any way to the elements of beauty as outlined by the art critic Bernard Berenson. Victorians, who loved embellishment, must have found boxcars a singular disappointment. It is true that a few, like the 1857-era Illinois Central boxcar illustrated in Chapter 2 (Fig. 2.21), showed a fine crown molding around the top of the body, and some of the private freight cars, especially the reefers, had fancy paint schemes. But most were plain and Spartan in the extreme. The fresh paint was soon roughed up by the weather and by the major and minor dents and bangs freight cars are subject to in normal service. All became seedy-looking fast. Passenger cars were washed and revarnished on a regular basis, but not so freight equipment.

Beauty has, however, another side. Rustic good looks were well understood even in Victorian times. A rugged chasm was considered quite handsome. In modern times we have come to love weathered barns, primitive furniture, and country baskets. Things as well as persons not conventionally handsome, like Abraham Lincoln, have come to be admired. Opinions on aesthetics can go topsy-turvy so that what was once considered ugly is now, if not beautiful, at least very fashionable. The stark, earthy, industrial look is now far more interesting than the classical elements once favored. The lean, clean, uncluttered, functional look is very much in, and while all this will change, as fashion is always in flux, boxcars are also now very much in. They are beautiful in their simplicity. Thomas Wolfe "knew" that freight cars were beautiful, though he could not find the exact words to explain why. He just knew it was true and expressed this conviction in one of his lengthy novels, *The Web and the Rock*.

The poetic and romantic appeal of the boxcar was not obvious to the many observers who saw freight cars more in terms of money boxes. They were one of the greatest money makers of the age, but only if carefully managed and driven hard. If treated like pack mules, they would surely earn their oats. Marshall M. Kirkman compared boxcars to trouser pockets in his 1908 work, *The Science of Railways*. If the pockets are full, the owner wears fine raiments, but if empty, the garments are shabby, and their owners have about them something that suggests the undertaker and the scrap heap. Kirkman went on to say that empty cars are like empty stomachs—they offer no nourishment. Full cars, like full stomachs, confer vigor, energy, and health, while empty cars lead to idleness, decay, and premature death.

Traffic and Tonnage

The provisioning of America was a major assignment that involved the cartage of the products of farms, mines, forests, and factories on a continental scale. This was no Mom-and-Pop operation. It was a big business that employed hundreds of thousands of workers, millions of dollars in equipment and trackage, and a bureaucracy comparable only to that of the federal government. The railroad networks operated ceaselessly, seven days a week, twenty-four hours a day, to move the nation's freight in a timely and economical fashion.

A railroad could move a mountain, were it required, one trainload at a time. And each year the rail system moved rich mountains of grain from the prairie states to the portals of the East Coast. They back-marched to the hinterland in a steady parade, laden with manufactured goods from eastern workshops. Tea, spices, and silk poured in from the West Coast, bound for distribution marts scattered along a 3,000-mile trade route. Riches from the Pacific Islands and Asia rattled over the land in steam-powered caravans rather than by the legendary camel trains. Most of the goods carried were far more prosaic, of course, consisting of everyday items like coal, salt, and lumber. The wealth represented by these commonplace bulk items was far greater than all of the silk or tea in the world, and railroads were far better suited for handling them than for the carriage of light, high-value commodities. Young America's special need was in the cheap carriage of bulk goods—a job perfectly suited to railroads.

RAILROADS AND U.S. ECONOMIC HISTORY

The railroads and the nation grew and prospered together. When the railway age opened in 1830, the United States was a raw, unexplored, and underdeveloped land. Its basic industry was agriculture. Its trade was conducted by seaport merchants. Its commerce was largely water-borne on river, lake, and coastal vessels.[7] Its small population, 12 million, was concentrated along the Atlantic coast. Everyday life, frankly, differed little from that of an ancient Mediterranean country. Things began to change radically over the next generation as industrialization transformed the nation and its people. Farm workers became factory workers. Merchants became industrialists. The population shifted from the land to the cities. Mineral riches, untouched by native peoples, were exploited, as were the forests and the rich virgin soil of the prairies. The population not only mushroomed; it began to migrate westward at a rapid pace.

As the pace of industrialization quickened, the national wealth soared. In just one decade, 1870 to 1880, it grew by 87 percent—and this was real growth, not a product of inflation. By 1890 the United States was a formidable economic power, outproducing the other great industrial powers of the world, including such giants as England, France, and Germany. Such rapid economic growth was the prime mover in creating the vast traffic for American railroads. It also created the basic need for an efficient, coordinated national rail system. A big-time economy could be sustained only by a class one transportation system, and the fragmented, short lines of the antebellum period were simply not up to the job. The dramatic changes in the national economy were thus directly reflected in railroads, their equipment, and their operating procedures. It was all a matter of necessity.

Railroad history was also shaped by other profound shifts in the national makeup. Provincial economies began to consolidate into a truly national market soon after the Civil War.[8] Standard products distributed by national chains began to replace locally made goods. The full effects of industrialization were felt by the 1880s, when the American railway system was filling in the final missing links. A maturing of the national economy led to global markets and an increasing flow of goods across the broad continent of North America, all of which swelled the sides of the ever-ready boxcars. Wheat grown in the Dakotas was swept into the cavities of cars greedy for more grain.[9] They then clanked overland to New York or Boston for transhipment to Liverpool and, eventually, more distant distribution in world markets. The humble boxcar was thus a major player in the international grain trade.

The process just explained actually started before the railroad era, and it might well have developed unimpeded through canal and highway carriage. At least this is the contention of the economic historian Robert W. Fogel.[10] Fogel's thesis has been challenged by several other authorities, and I do not intend to contribute to that debate beyond a few comments on the practical aspects of the nature of early road and canal freight transport. The highway system at the beginning of the rail-

FIGURE 1.6

The mighty flow of goods from the nation's interior to the seacoast created a huge traffic for American railroads. Busy dockside yards like this one at Locust Point (Baltimore) help sort out the cargoes for transhipment. (Smithsonian Neg. 49766)

road era, 1830, was primitive even by the standards of the day. After considerable debate, the federal government had refused to fund more than a few major roads. The so-called National Road, linking western Maryland to eastern Ohio, was open by this time, but it was a narrow, winding trail unsuited for the economical movement of heavy goods. The states and private toll roads helped fill in the highway map, but the system remained incomplete and best suited for slow-speed local carriage. More will be said about the deficiencies of the wagon roads in the rates section of this chapter.

Canals, on the other hand, were a time-tested and efficient large-scale mover of bulk goods. This fact had been demonstrated centuries earlier in China and France, although Americans tended to accept the idea only after it had been accepted in the mother country, England. In 1792, when railroads were little more than rudimentary mine tramways, there was talk of building a waterway across Upstate New York to connect the Hudson River and Lake Erie. New York was blessed with the only natural break in the Appalachian barrier. While others talked and studied and planned, New

York went to work on the dream of establishing a great avenue to the western interior of the nation. The God-given pathway through the Mohawk Valley allowed a nearly water-level route for the 364-mile-long canal.

Construction began in 1817–1818. The great ditch was finished in October 1825, much to the consternation of New York's rivals, particularly Philadelphia, Baltimore, and Boston. The Erie Canal was, unlike most pioneer projects, a remarkable commercial success. Bulk goods, notably grain and flour, came over the Great Lakes to Buffalo, then across the canal to Albany and south on the mighty Hudson River to New York. The traffic was so robust that it was necessary to enlarge the canal only ten years after the opening. Success always prompts imitation, and a canal-building mania ensued just at the time railroads were being introduced. None of the canals, except for a few short coal haulers, matched the Erie's success, and many were in fact dismal failures. Indiana's canal system proved particularly ill advised.

Pennsylvania's combined canal-railroad, opened between Philadelphia and Pittsburgh in 1834, might be classed as a partial success. It was hastily

PARADE OF THE CARS

FIGURE 1.7

built to challenge the Erie. Pennsylvania had no Mohawk Valley and so was obliged to overcome its peaked topography with a series of incline planes linked by short sections of track. At the eastern end was a more conventional stretch of railroad connecting Philadelphia and Columbia. The longest part of the journey was made by canal, a water route that required 174 locks—far too many; the more level Erie needed only 84.[11] By 1844 the Pennsylvania State Works, as the freakish canal-railroad was called, was hauling only about 75,000 tons annually, or barely 20 percent of the volume moving over the Erie.

Regional rivalries, really more like wars of the city-states, spawned transport alternatives to the Erie Canal. Massachusetts built the Western Railroad, not in the expectation of stopping the great ditch's flow of tonnage, but to draw some of it eastward to Boston rather than allowing it all to go on to New York. The port city of Baltimore launched the Baltimore & Ohio Railroad, a project more difficult to complete than its projectors imagined in 1828, for the road took a quarter-century to finish. Tiny Georgetown began work on the Chesapeake and Ohio Canal at the same time.

All of these efforts had little negative effect on the Erie Canal. Its traffic continued to swell, although it began to lose passenger traffic and better grades of freight when parallel rail lines opened, but as a bulk carrier it remained very strong until around 1880. In the long run the canal's greatest achievement may have been a shift in traffic flow from north and south to east and west. Before 1825 almost all bulk goods went to New Orleans through the Ohio-Missouri-Mississippi river net-

work. This was an indirect and costly route, for in truth the goods really wanted to go directly east to market in Atlantic seaports. But there was no way to accomplish a direct routing, and so goods went south by river to New Orleans, where they were transhipped through the gulf around Florida and north to Baltimore and other port cities along the East Coast. After the canal opened, bulk goods could go directly east, eliminating thousands of miles of needless transit. New Orleans lost, New York gained. It was that simple. When the railroads came into the transportation field strongly after, say, 1850, this strong east-west traffic flow became more pronounced. By the late 1850s cotton, once a market item exclusive to New Orleans, began to flow in sizable quantities over the Pennsylvania Railroad. In 1859 the Pennsylvania was carrying 17.9 million pounds of the fluffy white fiber.[12] At just about the same time, Chicago replaced New Orleans as the emporium of the West because of this turnaround in traffic flow.

TRAFFIC PATTERNS

By 1865 the railroads were developed to the point that they were beginning to dominate the nation's freight business. A fairly dense network was in place in the more settled parts of the nation, and a thin iron line was working its way across the far western regions. A clear traffic flow had emerged as well—raw material, lumber, grain, produce, and cattle from the West and manufactured goods, machinery, textiles, and furniture from the East. In terms of tonnage, the greatest flow was from the West. In terms of dollar value, the greatest flow was from the East. Over the next generation, as the iron network was completed, another pattern emerged, and that was ten major traffic corridors.[13] They are listed below in order of their importance:

1. Trunk lines—roughly everything bound by Chicago, the Ohio River, southern New England, and the Atlantic Ocean.
2. Atlantic coast route—New York to Atlanta.
3. Chicago to Atlanta—mostly the Louisville & Nashville and its connections.
4. Mississippi Valley—the Illinois Central and the Mobile & Ohio.
5. Western grain—Northern Iowa to Michigan-Wisconsin.
6. Southwestern gulf—mostly Texas.
7. Northern transcontinental—the Great Northern and the Northern Pacific.
8. Central transcontinental—the Union Pacific.
9. Southern transcontinental—the Southern Pacific and the Santa Fe.
10. Pacific coast route—the Southern Pacific and the Union Pacific.

The traffic was not at all evenly divided on these corridors, and this was truer in the last century

FIGURE 1.8

Cheap transportation also created industry in the smallest hamlets. This little steam-powered grain or flour mill was one of tens of thousands of small shippers dependent on rail transport. What appears to be a flour barrel is visible through the open car door.

FIGURE 1.9

For most of the nineteenth century the big traffic was on the eastern lines. Western roads like the Central Pacific moved relatively meager tonnages during this period. Shown here is a westbound train on the CP leaving Cisco, Calif., in 1868. The flatcars are carrying hay bales. (Railway and Locomotive Historical Society Collection, California State Railroad Museum, Sacramento)

than it is now, for population and industry were far more concentrated then in the northeastern section of the country. The Far West was thinly populated, Florida was largely unsettled, the South had not yet industrialized, and California was one of the least, rather than the most, populous of states. The 1880 U.S. Census recorded that 80 percent of the traffic was carried by forty-four of the trunk lines. The top ten were all eastern roads, and Pennsylvania led by far, with a traffic of 3.1 million tons, or about 10 percent of the national total. The New York Central was a close second, with 2.5 million tons. The only line that might truly be described as western to register over 1 million tons was the Burlington. Far western lines like the Central Pacific and the Union Pacific were well down the list, with annual tonnages of just over 0.4 million tons each. These western giants, large in terms of mileage at least, were outranked by pipsqueak eastern carriers like the Lehigh Valley, the Lackawanna, and even the Central of New Jersey. New England was barely represented in the top forty-four, and of those listed, only the Boston & Albany made much of a showing, at 0.37 million tons. The inclusion of the little Fitchburg Railroad underscores the position of New England as a commercial center in 1880. Not a single road from the Old South made it to the top forty-four, which illustrates the weakened condition of this region fifteen years after the war.

Before leaving this interesting ranking of American railroads, it must be noted that principal subsidiaries were listed separately. If the Fort Wayne, the Pan Handle, the Northern Central, and other lines controlled by the Pennsylvania were totaled, the parent company's portion would be not 3.1 million but more like 4.7 million tons, or roughly one-eighth of the national total. It can also be shown that a big road like the Pennsylvania carried as much tonnage in six days as the South Carolina Railroad carried in a year.

In 1888 *Poor's Manual of Railroads* made a comparison of some major eastern and western trunk lines that demonstrates the continued imbalance in traffic between rail lines in these regions. It also shows that the western roads' traffic was growing at a faster rate than that of the eastern roads as the century progressed. The results of this comparison are given in Table 1.1. The concentration of traffic seemed only to intensify as the century drew to a close. In 1895 L. F. Loree stated that fully half of the traffic was carried on only 10 percent of the route mileage.[14]

The concentration of traffic must be explained in at least one more way: the exact nature of the traffic. Railroads serving coal fields, for example, came to be known as coal roads. Some lines like the Reading were, for all practical purposes, conduits for coal, for they carried almost nothing else. More diversified trunk lines like the Pennsylvania

TABLE 1.1 Traffic on Major Eastern and Western Lines (billions of ton-miles)

Year	Eastern[a]	Western[b]
1865	1.6	0.5
1870	3.6	1.2
1875	5.9	1.9
1880	10.3	4.1
1885	11.3	6.5

Source: Poor's Manual of Railroads, 1888.
a. Pennsylvania; Pittsburgh, Fort Wayne & Chicago; New York Central; Lake Shore & Michigan Southern; Michigan Central; Bangor & Aroostook; and Erie.
b. Chicago & Alton; Chicago, Burlington & Quincy; Milwaukee & St. Paul; Chicago & North Western; Chicago & Rock Island; and Illinois Central.

might report 60 percent of its traffic as coal. In more modern times, as the anthracite fields played out, more southerly lines like the Chesapeake & Ohio and the Norfolk & Western became the nation's coal roads. Railroads in central agricultural states became known as granger lines because of the prodigious quantities of foodstuffs carried. The Bangor & Aroostook was called the Potato Line, just as the St. Louis Southwestern was called the Cotton Belt. Railroads in Texas became petroleum carriers after the big oil fields came in. Some railroads that, like the Philadelphia, Wilmington & Baltimore, traversed well-populated areas became big passenger roads. The PW&B received roughly half of its revenues from passengers during most of its corporate life (1838–1881).[15] Other roads that traveled the lonely back country devoid of major cities hauled few passengers. The Norfolk & Western received only 15 percent of its revenues from passengers even when such traffic was at its peak.[16]

Even nature played a role in the concentration and timing of rail freight traffic. The branches of many granger lines stood unused for eleven months out of the year. They were starved for business and witnessed the passage of only a few carloads. But when harvesttime came all hell broke loose, and for thirty days the rails were made shiny by the passage of cars. Traffic volume equaled or exceeded that of the busiest eastern trunk line, at least for this brief period. Rails, ties, bridges, men —all were overtaxed by this floodtide of wheat. But once the fields were cleared, the branch fell back into sleepy quiescence. Busy times and slow times were not limited to branch lines. The main stem could be similarly affected. The harvesttime effect on traffic volume might be less, however, than the effect of the economic climate. In prosperous times the long trains rolled at peak cadence to keep up with factories working three shifts. Times were fat, traffic was up, men and machines worked at maximum level. And then the economy would slow and go flat, and lean times were upon the land. The cars, once so busy, stood idle in yards. Engines, cold and silent, rusted to rails. The men were on furlough. The railroad, usually so noisy and active, appeared like a sleeping giant awaiting morning reveille. Nothing affected traffic volume more directly than the national economy, and the most extreme tonnage peaks and valleys were the result of a business upsurge or slowdown.

Feast days naturally affect passenger traffic, but they can also influence freight traffic. Christmastime involved not only a booming express traffic in presents but an upswing of specialized foodstuff shipments to city wholesale markets. Sweet potatoes from North Carolina, fine white potatoes from Maine, oranges, grapes, and lettuce from California—all these treasures of the good earth came in carload lots. The demand for peppers and turnips and turkeys brought in more cars. Christmas trees, bundled and stacked on flatcars, rolled in to brighten homes and apartments. Big cities like Boston might receive five hundred carloads of Christmas trees during yuletide. An even heavier traffic had been under way for months in anticipation of the winter season. Most railroads stockpiled mountains of coal near major metropolitan areas throughout the summer and fall, so that a goodly supply would be on hand for the heating system. There must also be coal for the Christmas stockings of bad little boys and girls, a custom of the time that has all but disappeared.

There were other seasonal traffic patterns less obvious than these. The end of navigation on the Great Lakes meant an end to ore traffic on the Chicago & North Western.[17] Once the boats were frozen in or felt the threat of a serious freeze, the busy ore trains ceased running. Like the boats, the cars were stored for the season, but the locomotives and train crews were sent south and west to help with the grain traffic. The wheat harvest generally coincided with the end of navigation, and so the railroad was able to reemploy both men and equipment. Unfortunately, the ore cars were too specialized to carry anything but a single specific cargo. Maintaining plant and equipment for use during only a portion of the year was costly and wasteful, but it was a fact dictated by the most powerful forces known to man.

NATIONAL AND REGIONAL TONNAGES

The actual size of the traffic was measured simply in the number of tons moved or in ton-miles (1 ton moved 1 mile). Ton-miles became the industry's preferred benchmark, because it indicated both the volume and the distance of carriage. Many early railroads reported tonnage figures in their published annual reports to the stockholders. A few would offer comparisons on the past tonnage performance of their burden trains, as freight trains were called in their earliest years. But no agency, private or governmental, gathered such statistics on a national basis during the pioneer period of American railroading. A few state railroad commissions performed this function, but only on a per-state level. *Poor's Manual of Railroads* began offering totals for the entire nation based on the reports it received, but because not all roads bothered to respond to the publisher's pleas for data, the totals are, of course, incomplete. Nothing very solid appears to have been offered until the 1880 U.S. Census special report on transportation, and nothing that can be considered definitive was available until the Interstate Commerce Commission began to publish its annual statistical volume several years later. And so American railroads had been moving freight for half a century before reliable national tonnage figures were available.

A few students of railroad history have endeavored to fill in this statistical gap. In his 1870–1871 volume, Henry V. Poor estimated current U.S. rail freight tonnage at 92.5 million tons. This figure was based on a per-mile formula Poor concocted from actual figures reported by several eastern state railroad commissions. Poor decided that his readers needed a comparison with the past to appreciate the progress made in the last twenty years, and so he offered a guestimate for 1851: around 5 million tons. This blue-sky number was conjured up by following the tonnage figures reported by the New York State Railroad Commission multiplied by the national mileage. It is at best a very crude estimation, but it at least has a factual basis. When Poor prepared the 1870–1871 numbers using a similar formula, he decided to deduct one-third of the total as duplicate tonnage, because about one-third of the traffic moved over more than one railroad. Poor reasoned that if each railroad reported this interline tonnage, an inflated total would result. It is probable that the one-third reduction was not applied to the 1851 figure, because so little actual interchange took place.

Many years after Poor tried to measure the size of the nation's rail goods traffic, Slason Thompson published a small book about railway statistics of the United States. He concluded, without offering any reasons, that the national railways moved not 5 million but 10 million tons in 1851.[18] Thompson then claimed that 63 million tons were carried in 1867, while Poor had estimated only 47 million tons for 1866. Modern business historians have been as perplexed or perplexing as their predecessors when it comes to synthetic statistics. In his 1965 book on railroads as a big business, Alfred D. Chandler offers some tonnage statistics that vary considerably from what Poor had reported nearly a century before.[19] Poor maintained that overall tonnage was 30 million in 1861, while Chandler reports it as 55 million. For 1870 Poor said that our rail lines moved 92.5 million tons, but this time Chandler contends that Poor overestimated by 20 million tons. All of this means that the experts disagree to a major degree, and that in fact no exact figures are available—which should come as no surprise to anyone who has studied American railroading before the Interstate Commerce Commission began to collect and codify the numbers. Table 1.2, then, is presented as only a crude guide rather than a precise reckoning.

If tonnage can be only crudely measured, at least for the pre-1880 years, what can be said about growth? Once again, conjecture and conflicting numbers seem to rule in the early period. The New York State Railroad Commission reports show remarkable growth for the period 1851 to 1871. During the first ten years the annual growth rate was given as 50 percent. The pace slowed down after 1861 to 23 percent, but even though reduced, the growth rate was surely a healthy one. In tonnage terms, the Empire State railways carried 3.9 million tons in 1860; 7.2 million in 1865; and 13.3 million in 1870. Fishlow's estimates for the national system for the same general period show somewhat more erratic growth.[20] His estimates, given in ton-miles, show the following upward swing: 1.1 billion in 1852, 1.7 billion in 1855, and 3.2 billion in 1860.

Yet another growth pattern can be discerned when we examine the big three eastern trunk lines, the Pennsylvania, the New York Central, and the Erie. In a twenty-year period (1860–1880) their combined traffic jumped by 1,217 percent![21] The numbers in ton-miles are as follows: 0.6 billion in 1860, 2.6 billion in 1870, and 7.4 billion in 1880. As the wood-car era came to a close, ton-mileage growth continued to spiral, but when reported for the national system, the increases are much more moderate than the heroic jump of the big three. A growth rate of just under 7 percent per year sounds perfectly manageable, but the railroad industry found it to be a staggering burden and struggled mightily to satisfy the demand for service.[22] Delays, blockades, and car shortages became the order of the day, but more will be said on these matters later in this chapter.

The scope of freight operations in the United States can best be explained by a worldwide comparison. By 1900 we were far and away the largest railroad freight carrier.[23] Britain carried less than one-half our tonnage, Germany only about one-third, Russia about one-tenth. When it came to cars, half of the world's fleet resided within our borders. In another comparison, made a decade later, it was noted that the United States moved five to six times the goods handled in France.[24] It would seem no exaggeration, then, to claim that no nation made greater commercial use of freight cars than the United States.

That the growth or decline of freight traffic depended on the national economy seems obvious enough, and experts in the field such as Emory Johnson contended that goods traffic was more sensitive to the business cycle than was the passenger side of railroading.[25] When times were good, car loadings climbed, and when times were dull, they fell. The soundness of these very rational-sounding premises might be questioned, however, by data generated during one of the nation's most serious depressions, the Panic of 1873. In that year the New York Central moved 5.5 million tons, and it actually showed a steady or moderately improved tonnage level through the long depression of the 1870s.[26] Late in the decade traffic rebounded, and a substantial improvement was recorded by 1879, when some 9 million tons of freight were moved. The net revenues from freight operations remained about the same during this period.

The Pennsylvania Railroad's performance mirrored that of its rivals except for 1876, when the big trunk line registered a measurable dip in both tonnage and net income. To the west, the Chicago & North Western also registered a flat traffic pattern during most of the 1870s, with a sharp rebound at the end of the decade.[27] Net operating revenues seem to have stayed rather steady for the whole industry at around $185 million between 1873 and 1878 except for 1877, and so it would seem that from a traffic and revenue standpoint, traffic was not dramatically reduced. But the growth of traffic was definitely affected in the negative; new construction was discouraged, and the weaker lines were forced into receivership, but the existing stronger lines seemed to have survived in a steady

TABLE 1.2   Tonnage of U.S. Railroads

Year	Tons (millions)	Ton-Miles (billions)
1851	5[a]	1.1[a]
1860	NA	3.2[a]
1861	30	NA
1866	47	NA
1870	93	8.0–10.0[b]
1880	290[c]	32.3[c]
1882	350	39.3
1885	437	49.1
1888	590	65.4
1890	691	79.1
1891	704	81.2
1894	674	82.2
1897	788	97.8
1900	1,000	141.1

Source: Based on figures appearing in *Poor's Manual of Railroads*, unless otherwise noted.

Note: NA = not available.

a. Estimate from Albert Fishlow, *American Railroads and the Transformation of the Ante-Bellum Economy* (Cambridge, Mass., 1965).

b. Author's estimate.

c. U.S. Census.

if not expanding fashion during the dark days of the Grant administration.

In 1880 *Poor's Manual of Railroads* contended that traffic was up by over 47 percent on most trunk lines compared with 1875, but net earnings on freight operations were up only a paltry 3.8 percent. It was well understood that operating costs were inelastic, largely because of the large debt service common to the industry, but this would not explain such a poor profit margin. The culprit was lower rates, accountable to greater competition, especially between parallel trunk lines. Between 1875 and 1880, *Poor's* noted, rates had gone down by 42 percent. A bigger traffic did not guarantee improved profits. The congested eastern lines faced higher operating costs and greater competition from neighboring lines, while the lanky western roads enjoyed the luxury of long hauls and few neighbors. Of the ten most profitable lines in the nation, seven were located west of Chicago in the most sparsely settled and least industrialized section of the nation.[28] The Pennsylvania towered above all other roads with gross receipts of $96 million in 1888, but the relatively minor St. Paul, Minneapolis & Manitoba (later the Great Northern) realized a net profit 13 percentage points greater on only a fraction of the receipts of the eastern giants. Mileage was an equally fake measure of a railroad's importance, traffic volume, or earnings. The New York Central, for example, listed only 1,400 miles of line, which was small compared with the Union Pacific's 6,200 miles, but in 1888 the Central grossed $63 million against the UP's $19 million.

So much for the development of the national rail traffic; what of the individual trunk lines? To begin at the beginning, we might look at a pioneer trunk line, the Baltimore & Ohio. Before its tracks had progressed much beyond the environs of Baltimore, the following receipts from the West were reported for a single week in April 1832: 1,721 barrels of flour, 8 hogsheads of tobacco, 127 tons of granite, 31 tons of iron, 16 tons of lime, 36 tons of paving stone, 3 tons of planks, 3.5 tons of leather, 5 tons of tan bark, and 2.75 tons of horse feed.[29] In 1832 this was a mountain of goods representing 320 carloads. By modern standards, the week's traffic was a pittance and the cars that conveyed it were only overgrown handcars. But to the fledgling rail industry, this record was an enviable one that proved the iron road's utility and future prospects. The heavy cartage of flour would prove to be the road's major eastbound traffic for many years. In 1832 over 12,000 tons were carried, and by 1848 that number had increased almost fourfold.[30] Coal had displaced flour as the leading tonnage cargo two years earlier, but flour remained a major item for many years. In 1856, for example, it amounted to 91,000 tons and maintained its number two position.

During this same period Pennsylvania's Allegheny Portage Railroad offered a less promising demonstration of freight movements by rail. Its capacity was estimated at 288 cars in a twelve-hour day, but in reality the number handled was more like 100 or even fewer per day.[31] Progress over the short line was clumsy and awkward, because of the repeated hitching and unhitching of the cars at each incline. The transfer over the so-called levels by locomotive and horse power added to handling delays. It required seven hours to make the 36-mile journey. The line was open only eight or nine months of the year because of the freezing of the canal, and so the figures for the busiest six months, April to October 1836, show that 14,300 cars or just over 37,000 tons were moved. The Allegheny Portage was a freak, and so even at the time it was not seen as a fair example of what a normal steam railway could handle. The truth of this notion was borne out by the eastern rail branch of the Pennsylvania State Works, which carried 871,854 tons for the period 1835 through 1841.[32]

Elsewhere, pioneer railroads traversed mountainous territory with good results. The Western Railroad built to connect Boston and Albany showed gross earnings of $420,000 on its freight operations just four years after opening in 1841. A healthy flour trade developed, and as with the Baltimore & Ohio, this cargo would be the Western's chief freight item for many years. In 1847 flour shipments jumped to 51,000 tons, more than double the usual figure, because of the Irish potato famine. The Western profited more from the Irish need for foodstuffs, because much of the flour was shipped during the winter season when rates were higher. The road's 1865 annual report noted an important shift in traffic after the new Albany bridge encouraged bulk grain shipments from the West directly to Boston. The golden flow of wheat was so brisk that it was necessary to borrow cars from the New York Central.

With a gauge of 6 feet, the Erie surely seemed prepared to grapple with massive quantities of goods, and even before the line reached its Lake Erie terminal, it was handling respectable quantities of goods. In 1850, with a fleet of just 784 cars, it moved 131,000 tons of goods.[33] Within a decade that figure had multiplied nearly tenfold. Year after year the Erie's traffic seemed to rocket. Measured in decades, it progressed thus: in 1870, 4.8 million tons; in 1880, 8.7 million; and in 1890, 16.2 million. Despite its shaky financial history, the Erie was destined to become a major east-west carrier.

Other roads that aimed at a Lake Erie connection were clearly destined for far more modest traffic. Cincinnati's premier steam railroad, the Little Miami, would form the southern end of a rail line running from Sandusky to the Ohio River. This overland portage had become a reality by 1846.[34] In that year so much produce piled up at the stations that it could not all be hauled away. More cars and locomotives were immediately ordered. Two years later a drought depressed production of major staples such as flour and whiskey, and so traffic dropped noticeably. By 1854 the Little Miami's freight business had matured measurably. Merchandise traffic had more than doubled since 1851, from approximately 25,000 tons to 54,000 tons. During the same period lumber shipments jumped from 540,000 board-feet to 2.4 million board-feet. In 1851 only 431 cattle were carried, while in 1854 the number was 17,139.

The Lake Shore & Michigan Southern was more in the mainstream of North American rail freight traffic than the Little Miami. It ran along the southern shore of Lake Erie between Chicago and Buffalo, and so it offered a direct pathway for the most concentrated flow of east-west traffic on the continent. This choice property, consolidated in 1869 from several smaller midwestern lines, was soon taken over by the Vanderbilt interests. This vital western link was a free-flowing conduit between Chicago and New York that ensured not only a healthy traffic but fat dividends as well. The Vanderbilts seem to have understood something unrecognized by more modern railroad consolidations: it takes a combination of healthy and not sick properties to make a successful merger. The strength of the Lake Shore is evident from the figures shown in Table 1.3.[35]

Development in the New World was measured in decades rather than in centuries. The degree of development and the size of the population were preeminent factors in traffic, and so in nineteenth-century America most of the traffic was found in the eastern half of the continent. Car loadings seemed to become scarcer the farther west the iron trail extended. As late as 1882, some thirteen years after the Central Pacific's completion, the *Railroad Gazette* considered its traffic to be trifling for a major trunk line.[36] The daily traffic amounted to 567 tons westbound and 422 eastbound. It was all handled in two trains, each composed of thirty small 10-ton-capacity cars. Through traffic was very light, with the great preponderance of movements involving local haulage. The Central Pacific was essentially an agrarian road with no mineral or merchandise cargoes. Its main commodities were salmon, tea, wool, and beans. All of this was very disappointing, considering the lease to the Southern Pacific, the net result of which meant only additional track mileage with little added traffic. The CP's situation might be better understood by comparing it with that busy eastern trunk line just discussed, the Lake Shore & Michigan Southern (Tab. 1.4). The traffic-starved Central Pacific managed to keep its revenues up by charging a very high tariff, 2.1 cents per ton-mile, or about three times that charged by the

TABLE 1.3 Performance of Lake Shore & Michigan Southern

Year	Ton-Miles (billions)	Earnings (millions of dollars)	Rate per Mile (cents)
1870	0.6	8.6	1.5
1875	0.9	9.6	1.0
1880	1.8	13.8	0.7
1885	1.6	8.8	0.5
1890	2.1	13.5	0.6

Source: *Railroad Gazette,* May 4, 1895, p. 248.

TABLE 1.4 Central Pacific versus Lake Shore & Michigan Southern (1881)

Railroad	Mileage	Freight Cars	Tons Carried
Central Pacific	1,215	4,600	2.7 million
Lake Shore	540	15,600	5.1 million

Source: *Poor's Manual of Railroads.*

Lake Shore. Shippers and ultimately the public benefited from a proximity to well-traveled roads like the Lake Shore.

It is necessary to return to the Northeast to study the biggest rail freight carriers. Of these, the very largest were the archrivals, the New York Central and the Pennsylvania Railroad. Each vied to be king of the traffic world. The Central had the most favorable grade of any railroad in the world, considering its length.[37] It connected major populations within a heavy industrial belt, and it linked the agricultural middle of the continent with the greatest Atlantic seaports. Unfortunately, it paralleled the Great Lakes and the Erie Canal, so its through traffic, especially in bulk goods, was not as great as it should have been. Its operating ratio was high (often 67 percent), but its revenue history was salvaged by a large and profitable local traffic. The Central concentrated on marketing services and so hoped to boost traffic and revenue by finding new customers.

The Pennsylvania, meanwhile, had a shorter, more direct route to Chicago than any other line. This was a great advantage for obtaining through traffic, and through traffic was the prize every trunk line lusted after. The Pennsylvania was aggressive in reaching out to make connections with major cities like Baltimore, New York, and St. Louis. It quickly leased independent roads for this purpose, while most of its rivals only talked or hesitated. The Pennsylvania worried over ways to cut operating costs and rationalized all phases of its operations. Conversely, it worried less about public or shipper goodwill and sought to gain more traffic by building more cars, and so to grow rich on mileage payments from its rivals. These policies paid off handsomely, for the Pennsylvania dominated eastern traffic, and it became the larg-

TABLE 1.5 New York Central versus Pennsylvania Railroad

Year	Railroad	Tons Moved (millions)	Gross Freight Earnings (millions of dollars)
1860	New York Central	1.3	4.9
	Pennsylvania	1.3	4.1
1870	New York Central	4.1	14.3
	Pennsylvania	7.0	15.2
1880	New York Central	10.5	22.1
	Pennsylvania	26.0	29.7
1890	New York Central	16.2	NA
	Pennsylvania	60.6	45.7
1900	New York Central	37.5	34.2
	Pennsylvania	108.8	64.3

Source: Poor's Manual of Railroads.
Note: NA = not available.

est freight carrier not only in the nation but in the world. The traffic battle of the titans might be best summarized by a statistical comparison. The figures given in Table 1.5 are for the parent companies only and not their western subsidiaries.

Traffic Promotion

As a service business, railroads realized that it was necessary to seek business actively. Some shippers were captive, on-the-line customers, but much of the market was more competitive, not only because more than one railroad might serve the area, but also because shippers might choose alternatives, particularly water transport. And so the railroad must make itself known directly to all prospective shippers. This could best be done by direct contact in the form of a salesman or, as he was known within the industry, a freight agent. A friendly, helpful, gregarious person was best suited for this job. He was expected to make friends for the line he represented, to build up the tonnage, and to take traffic away from the competition, by whatever means necessary. The campaign for traffic was a fundamental game because of the railroad's almost total dependence on it for revenues. It was a battle for survival, and so every employee from the common trainman to the chairman of the board understood its importance.

The role of the highest executive in traffic promotion will be explored later in this discussion when we come to deal with rates and pooling, but for now, let us review the work of the freight agent. His job was similar to that of any salesman of the day. He made every effort to serve his customers, to make them happy, to be their friend, and to solve their shipping problems. He would offer small gifts—a calendar at the New Year, a bottle of

whiskey at Christmas, a cigar when a new baby came into the world. A larger customer might require more attention. One could lose a dozen small shippers, but the loss of a big one could not be allowed, and so when Mr. Big complained about a lack of empty cars, inefficiencies in switching, or problems with demurrage charges, the freight agent moved quickly. Mr. Big might also receive a new siding at no charge; he might well be given a free pass or the temporary use of a private car. He would be pampered and courted in all ways. Yet best of all, from his viewpoint, Mr. Big would be offered a special low rate and so could compete more easily with rivals who were forced to pay standard shipping fees, unless they too could work the rebate game. The freight agent surely witnessed both the honest and the not-so-honest side of railroading.

In addition to the on-line freight agents, many larger railroads operated offices in cities far away from their tracks. The Atchison, Topeka & Santa Fe, for example, might have a representative in Atlanta, with the expectation that he could find shippers whose westbound traffic was routed over the AT&SF rather than the Southern Pacific. Advertising appeared in trade papers, more as a matter of goodwill than for the direct impact it might have on shipments. As in all things promotional, just keeping the company's name before the public was thought worthwhile. Low-cost ads in the *Cotton News* or the *Timber Tribune* could surely do no harm. An even cheaper form of advertising was available on the great flat sides of the freight cars themselves. These rolling billboards were covered with the railroad name, a distinctive emblem, and, often, a catchy slogan such as "Southern Serves the South," "The Bee Line," or some other succinct phrase that indicated speed and efficiency.

Gathering a larger share of the traffic market was a selling job handled by the freight department in every major railroad. In its formative years the Pennsylvania Railroad did not feel that such a bureau was needed, and it was content to use contract agents to solicit and handle freight consignments.[38] By early 1851 a freight department was established with very broad powers and responsibilities. Its duties were not limited at all to the promotional side of the business; rather, it was to receive and forward shipments and manage the freight stations. It was to establish rates, handle damage claims, and deal with all aspects of emigrant travel. This range of responsibilities was considerably narrowed in 1863 because of growth in the volume and complexity of the Pennsylvania's freight traffic.

Seeking a larger share of an existing market was only one approach used by railroad managers. Creating new on-line traffic was far more imaginative. Western lines were especially drawn to this method, because they so often traversed unsettled country. As late as 1872 the Chicago, Burlington & Quincy explained the problems in a public statement: "No road proves a good investment unless its local trade and passenger traffic is heavy. No such traffic can exist except in a well-tilled and well-settled region."[39] At this time the Burlington was attempting to peddle some six hundred thousand acres of land-grant land. Western Europe was being blanketed with circulars, printed in German, French, Bohemian, and English, extolling the benefits of resettling in the New World, precisely along the tracks of the Burlington. Generous credits and long-term mortgages were offered, as were free tickets to visit the farm sites. Temporary housing was available, as well as special reduced rates for the transit of furnishings to the new lands. The revenue from land sales was welcome, short-term revenue, but the growing traffic generated by the farms, towns, and industries that followed the settlers meant long-term traffic and income. The Illinois Central had been active in colonization work since the early 1850s and sold about 2 million acres of land between 1854 and 1870.[40] Orchards and vineyards followed the colonists, and soon the Illinois Central was running special fruit trains to Chicago, much to the profit of the farmers and the railroad.

Most railroad land agents were happy to accept any emigrants ready to settle along their lines. But at least a few were far more particular. Some in fact sought out very specific settlers, and not without a reason. Both the Santa Fe and the Burlington & Missouri River encouraged the migration of Russian Mennonites during the 1870s.[41] These folks were actually German Protestants who had settled in Russia along the Turkish border during the reign of Catherine the Great (herself a German). The area they settled in was a high, windy tableland, suitable only for dry farming. The Mennonites became experts in the cultivation of a hard, short-stemmed wheat called Turkey Red. The American agents who came upon these people rightly saw Turkey Red as just the right crop for our arid Great Plains. The Mennonites were ready to move, because the current czar had rescinded the privileges granted in Catherine's time. The U.S. agents promised religious freedom and good land, albeit not much rain. And so the Mennonites packed up, bag and baggage, plus plenty of Turkey Red seed, and sailed to the New World.

Red wheat was perfectly suited to the climate of the Great Plains and soon became a standard crop on new farms that for centuries had known only prairie grass and buffaloes. Other varieties of hard wheat were found suitable for lands to the north that were thought unsuitable for any form of commercial agriculture. The bounty that came forth to fill hundreds of thousands of boxcars during each harvest must be credited, at least in part, to railroad land agents. The work of the agents went on even after most of the land-grant land was sold. The Great Northern, which always boasted that it was a purely private capital road not dependent on government subsidies, had a very active land department. Again, the real purpose was to promote traffic, not just to sell land. The Great Northern's chief colonizer was an Austrian named Max Bass, who managed to sell millions of acres of land.[42] In just two years (1896–1898) he boasted of settling ninety-five thousand people along the Great Northern tracks.

As the American frontier closed, the railroad land agent was replaced by the agricultural agent, whose job was to stimulate farm production and so raise the tide of traffic. Once the farms were established, the railroads wanted them to flourish. Most of this effort was educational in nature. In addition to publications, the favored method was exhibit trains, where locals might learn about the latest improvements in tillage, fertilizers, and animal husbandry as the train rolled from station to station.[43] Exhibits were mounted inside baggage cars to show off better seeds for alfalfa. Lectures were given in coaches, made into classrooms, on how to fight insects, hog cholera, or soil erosion. Experts from state agricultural colleges, the Farm Bureau, or special groups like the Holstein Association were hired to speak aboard the farmers' specials. In addition to the trains, some lines planted plots to show new plant varieties. The Long Island Railroad established its own model farm to illustrate ways to grow good crops in sandy soil. In 1911 the Great Northern set up an exhibit in Madison Square Garden, of all places, to show off the latest ideas on wheat farming. The aid to farmers went beyond advice in at least a few cases. When boll weevils devastated Arkansas cotton fields, railroad agricultural agents brought in tomato plants suited to the soil and climate of the area. Dynamite was given to residents in the upper peninsula of Michigan to blast away tree stumps so that potatoes could be planted. When wheat failed in eastern Colorado, a local railroad helped farmers establish dairy farms by supplying cream-rich cows.

Traffic promoters did not ignore industrial shippers, though they don't seem to have received much attention until around 1891, when the Milwaukee Road opened a special office to encourage the establishment of factories along its lines.[44] Good sites were documented and presented to potential developers. A brick manufacturer would be told about an unexploited clay deposit, or a cement maker would be urged to take over an untouched limestone bed. Manufacturers could find maps and data on towns, markets, labor pools, and laws that would help them decide on a new factory site. Most other larger railroads added an industrial division to their freight departments, and most of these worked with local on-line communities that wanted new manufacturing plants. Inducements that came to be expected were cheap land prices, low taxes, and reduced freight rates.

Types of Traffic

By 1890 railroad freight traffic was classified in six broad categories. This system assigned shipments by product designation, so that lumber, naturally, was listed under forest products and hogs under animals. The breakdown is given here in order of its importance as reflected in an Interstate Commerce Commission report for 1900:

Minerals included coal, coke, stone, metallic ores, sand, and the like. This class—sometimes designated as "mines"—represented the largest single group of shipments by far, at 52.5 percent.

Manufactures included machinery, wagons, furniture, cement, bricks, iron and steel rail and sheets, and products such as naval stores, petroleum, and sugar that might just as well have been classified elsewhere. This class represented 13.4 percent of the tonnage.

Forest products included lumber, wooden shingles, and the like. This class represented 11.6 percent of the tonnage.

Agricultural products included all common foods, grain, and fruits, but also things like hay, cotton, and tobacco. This class represented 10.3 percent of the tonnage.

Merchandise and unclassified was not explained but presumably included clothing and other consumer goods. This class represented 9.4 percent of the tonnage.

Animal products included all domestic animals and fowl plus wool, hides, leather, and dressed beef and fish. It was the smallest of the big categories, at 2.8 percent.

TABLE 1.6   Selected Major Products Carried by U.S. Railroads (millions of tons)

	1880	1890	1900
Grain	42.0	49.4	64.9
Flour	7.4	11.2	15.3
Cotton	3.9	4.9	5.9
Produce	7.0	NA	13.1
Livestock	10.8	15.4	15.6
Dressed meat & packing house products	NA	6.1	10.4
Coal	89.6	184.3	310.9
Lumber	25.4	52.8	76.0
Petroleum	7.7	8.7	8.3
Iron & steel	11.6	21.0	49.3
Merchandise	58.2	NA	42.5

Source: U.S. Census (1880, 1890) and Interstate Commerce Commission statistics (1900).

Notes: Does not represent total traffic moved. NA = not available.

The industry's overall average would not be an accurate model for individual railroads or even groups of individual regional lines. The big trunk lines running east out of Chicago were not really mineral haulers; their major cargoes were food-related.[45] In 1894 77.5 percent of their traffic was food products, with about half of it represented by livestock and dressed beef. Grain and flour represented 32 percent, salt meat 5.7 percent, and lard 2.8 percent of the seven roads' total freight traffic. Conversely, one of the coal roads might ascribe 80 percent of its overall traffic to minerals, while a line in Georgia might report half of its tonnage in forest products. The variety of goods carried was never even and might fluctuate markedly on any one line when a new mine opened or an established industry moved away or closed.

Some idea of the traffic growth of certain products during the final twenty years of the nineteenth century can be gained from Table 1.6. Some items, like petroleum, show a rather flat level of growth, while iron and steel exhibit healthy expansion. Petroleum had rather few uses during this period other than as lamp oil. During the same time, however, the iron industry was experiencing unprecedented growth. Just why merchandise tonnage declined during a time of general expansion in all areas of material life is more difficult to explain; however, it is possible that those compiling the data for 1880 and 1900 used differing yardsticks when defining what should be included or excluded from this category. Such variable judgments are in fact what makes statistics such an uncertain measure of human activity, including railroad traffic.

Table 1.6 offers a bird's-eye view of railroad traffic with special attention to some of the more important products transported. In the next few pages I shall elaborate on what some of these commodities meant in terms of railroad traffic. The groups selected were chosen to illustrate some of the traffic groups related to the basic needs of civilized life—heating, food, shelter, and clothing.

COAL

By far the single most important item carried by rail was coal. It was the very foundation of the railroad freight business and fundamental to its financial success. Coal indeed gave birth to the industry, whose origins can be found in horse-powered tramways established in northern England during the eighteenth century. Rail transport was introduced into this country about a century later on the gravity railways at Mauch Chunk and Honesdale, Pennsylvania, which were created to service coal mines. The practical success of these two small lines in moving coal from the mountaintop to nearby canal landings demonstrated the general utility of railways for other transport purposes. The most literal lesson of the Pennsylvania grav-

FIGURE 1.10

Coal was the principal traffic source and a major revenue producer for American railroads. In some areas it was necessary to use incline planes to move coal cars up or down steep grades. This scene on the Reading at Mahanoy Plane dates from about 1880. (Mechanical and Civil Engineering Collection, Smithsonian)

ity lines was to expand the idea for the long-distance carriage of coal by rail.

And so the Philadelphia & Reading Railway was chartered in 1833 to construct a railway to connect the coal mines around Pottsville with the tidewater part of Philadelphia.[46] The projectors of the P&R built their 90-mile railroad carefully on gentle slopes from the coal fields to the Delaware River. It was not the quick or cheap construction job so typical of most early U.S. railroads. This was a line intended, from the start, for big tonnages and the efficient movement of coal. Gentle grades, sturdy track, and strong bridges drove construction costs to $8 million, or roughly half again more than most other U.S. lines of the period were spending. The money was well spent, making the P&R an economical conveyor of coal. During the first year of full-scale operations, 1842, the line was moving 6,000 tons a week. Within seven years coal shipments exceeded 1 million tons a year; by 1855 annual tonnage was over 2 million tons. Fifty trains a day were needed to move the loaded cars

and to return the empties. Proof of the railroad's efficiency in moving this big traffic was the virtual ruin of the Little Schuylkill Canal, which had formerly moved the bulk of this traffic. In 1859 the little Reading was declared the largest rail freight carrier in the nation, with a traffic comparable to the Erie Canal's.[47] It was moving coal at just 40 cents per ton, yet its earnings were larger than those of big trunk lines like the New York Central.

The fine traffic and profits possible in the coal business were not lost on other railroads in the area. Other lines began to build into the rich anthracite fields of East Central Pennsylvania with the intent of delivering coal to New York City rather than to the smaller port and market of Philadelphia. The distance was 50 miles longer, but potential market sales were far greater. This scheme for carrying coal to New York was slow in coming about and was implemented in a rather casual fashion. Not much happened in fact until around 1855, when two small railroads opened within the coal fields. Both the Delaware, Lack-

FIGURE 1.11

The anthracite fields of north-eastern Pennsylvania were the first to be commonly exploited, and many regional railroads grew rich from the traffic they produced. Shown here is the Palmer Vein Colliery in 1885. (The Hagley Museum and Library, Wilmington, Del.)

FIGURE 1.12

Western coal was not a major factor in the national fuel market during the wood-car era, but a few lines like the Union Pacific serviced mines like the one at Rock Springs, Wyo. The scene dates from about 1890. (Union Pacific Photo)

awanna & Western and the Lehigh Valley soon developed expansive ideas. The Lackawanna patched together a direct route to the lower end of New York Harbor via trackage rights over the Central Railroad of New Jersey.[48] The tonnage was over 3 million tons in 1875 when the Lackawanna/Central of New Jersey contract expired. Fortunately, the Lackawanna had leased the little Morris & Essex and so had its own line directly into Hoboken. The M&E, a pristine commuter line, was suddenly confronted with one hundred thousand carloads of coal a year that had to be sandwiched into an already busy passenger schedule. The number of trains increased as the Lackawanna attempted to channel ever more coal to its Hoboken terminal; by the early 1880s it was approaching 5 million tons a year.

Meanwhile, the Lehigh Valley Railroad tried to find a new way to reach Manhattan.[49] Until 1869 its coal jimmies were handled by the Morris & Essex, but the Lackawanna lease had abruptly ended that avenue to the big city. In desperation the Lehigh took over the Morris Canal, which at least had a valuable terminal in Jersey City. The canal was a weak link, but it provided a temporary solution to the Lehigh's landlocked status until a new railroad could be built to Perth Amboy. The rivalry of the anthracite roads boiled up again in 1871 when the Central of New Jersey, once a friendly and willing bridge line, decided to push directly into the coal fields. After the Lackawanna shifted its traffic to the M&E, the Central of New Jersey was deprived of its once-profitable coal business, a business that amounted to no less than 80 percent of its freight traffic.[50] The line's puny passenger business could not sustain the company. Tracks upgraded for heavy coal trains were little used, and heavy locomotives maintained for this purpose were idle. And so the Central struck out at its former business partners, friends no more, and leased the Lehigh & Susquehanna, which took it directly into the coal fields. Black diamonds once again rumbled over the tracks to the Central's Elizabethport coal docks.

During the time when the anthracite roads were fretting over access to Atlantic ports, they tried to find western outlets so as not to be trapped in the very competitive eastern markets. The Delaware & Hudson began piecing together a western main line in the late 1860s in the hope of reaching Ohio and other more western states.[51] The soundness of this move was demonstrated in a few years, when the line's east- and westbound coal traffic began to even out. The Lehigh Valley also understood the good sense of a western outlet and opened a line to North Fair Haven, New York, in 1872.[52] A big pier jutting out into Lake Ontario transferred coal to lake vessels servicing towns bordering the Great Lakes region. The Lackawanna was a trifle slower in reaching westward, but it too established a lakeside coal distribution facility near East Buffalo around 1890.

North America was rich in coal deposits, and the anthracite fields of Pennsylvania represented but a small parcel of what was available. The Pennsylvania fields contained some of the best coal found anywhere in the world and were the first to be commercially exploited in this country. Great beds of lower-quality soft coal, however, stretched from eastern Pennsylvania to southern Illinois. More coal was found in the West, notably Colorado and Wyoming (Fig. 1.12). Serious efforts to dig out bituminous coal began around Cumberland, Maryland, in 1842.[53] Within six years coal was a major item on the Baltimore & Ohio, representing 42 percent of its eastbound tonnage. In just two years, by 1850, coal represented 57.5 percent of eastbound traffic. On the eve of the Civil War one-third of the railroad's freight equipment consisted of coal cars, and one-third of its freight revenue was from bituminous traffic. The B&O could move a ton of coal 178 miles from Cumberland to Baltimore for just $2.07—a fee considerably higher than that charged by the Reading, as previously noted, but still cheap enough to encourage ever-larger shipments. And the traffic did swell. In 1878 it exceeded 1 million tons, and just four years later it topped the 2-million-ton mark. Such a rich traffic naturally attracted predators, and the B&O soon had an unwelcome neighbor, the Chesapeake & Ohio, pushing into its territory from the south.[54] The C&O began soft-coal train service in 1873 and was soon trundling hundreds of thousands of tons to the Newport News coal docks. In 1888 the C&O reached Cincinnati and soon found that the midwestern market was stronger than the traditional eastern one.

The Illinois Central never considered itself a coal road during its early years, but the big coal fields in southern Illinois argued otherwise.[55] One of the major towns in the area in fact called itself Carbondale. The rising tide of coal traffic on this supposed granger line could not be ignored. Starting in 1860 it was 61,000 tons; in 1870, 200,000 tons; in 1880, 532,000 tons; in 1890, 1.5 million tons; and in 1900, 5.7 million tons. During this forty-year period coal and coke had risen to one-third of the road's overall freight traffic. And so the dependence of the railroad industry on coal for a goodly portion of its livelihood can be seen in example after example. Here were two industries dependent on one another. It was a long-lasting, beneficial, and, in the main, contented relationship.

Moving the coal from minemouth to marketplace was only one phase of the railroad's job. What to do with the stuff upon arrival became a major problem. When times were slow the coal simply sat in cars until a buyer could be found, but under normal business conditions the pressure was on to unload quickly and return the cars to the mines for

FIGURE 1.13

This photograph shows the top of a B&O coal pier at Locust Point (Baltimore) in about 1900. The two outside tracks receive loaded cars, which discharge their coal to ships lashed alongside the pier. Some of the coal chutes are visible in their upright position. The center track is for the return of empty cars. (B&O Museum, Baltimore)

reloading. Idle cars earn no revenue. Much of the coal was transhipped. Some went into ocean vessels for the European market. Some went into smaller coastal vessels bound for New England or the cotton states' ports. The remainder was piled up in huge outdoor mounds to wait out the warm weather and the return of the home heating season.

The Reading's great coal dock near Philadelphia at Port Richmond was probably the first large facility of its kind in the United States. It was opened in 1842. A description of Port Richmond published ten years later portrays a very substantial facility covering forty-nine acres.[56] Most of the open space was a sprawling yard needed for the waiting loaded and empty cars. Along the shore were twenty piers, ranging in length from 342 feet to 1,132 feet. The coal jimmies rolled along the top of these elevated piers and dropped their coal into ships stationed alongside via chutes. Once the cars were empty, they were pulled back and away from the pier while a new set of loaded cars took their place.

As many as one hundred ships could be handled at one time. A ship could be loaded with 150 tons in just under three hours. Port Richmond's piers have been repaired, replaced, and enlarged countless times in their century and more of service. Nothing remains of the original docks save the site, but the rumble, roar, and dust of offloading coal continue, albeit at a slower pace, as they have since four-wheel wooden cars first appeared at that place on the Delaware River nearly 150 years ago.

The Baltimore & Ohio was not many years behind the Reading in establishing a coal pier. The site was Locust Point near Baltimore, but away from the busy inner harbor.[57] It was staked out in 1845 and opened four years later. A photograph dating from around 1900 shows the Locust Point pier as it appeared late in life (Fig. 1.13). The picture shows the top of the pier; the loading chutes can be seen on either side of the docks. The center track was used to retrieve the empty cars. Some years before this picture was taken, the B&O decided that a larger coal dock was needed, and so the

PARADE OF THE CARS

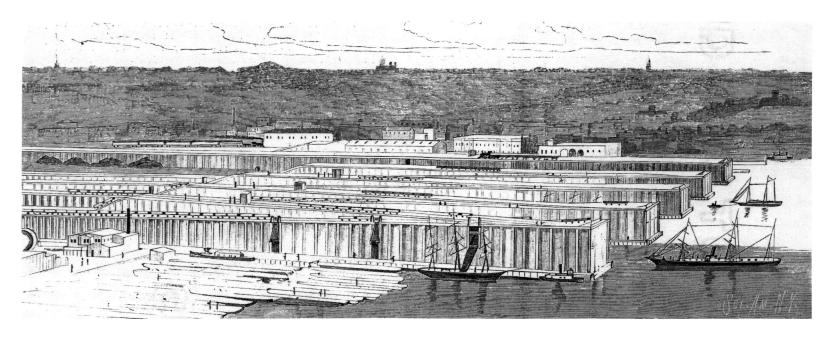

FIGURE 1.14

Loaded coal cars came rumbling
down from the mines to tidewater. The Lackawanna maintained
a semiautomated high-level coal
pier at Hoboken, N.J. A steam
hoist pulled the loaded cars to the
top of the pier, high above the
waiting ships. The empties
returned to the yard by gravity.
(*Scientific American*, April 15,
1882)

Curtis Bay piers were opened in 1884. The need to
remove such a dirty and noisy facility away from
the city placed the new pier several miles south
and east of Baltimore. Curtis Bay was an inlet
more accessible to ocean shipping than the old
Locust Point piers.

Around 1879 the Lackawanna opened an innovative coal pier in Hoboken designed to handle the
cars more expeditiously and so to speed the unloading process (Figs. 1.14, 1.15).[58] Most of the operation was done by gravity. Loaded cars coasted
down grades from the receiving yard to a weighing
shed immediately in front of the coal pier. Brakemen rode the cars and worked hand brakes to keep
them under control. After being weighed, the cars
were propelled in pairs by a cable to the top of the
pier about 60 feet above the water; the brakeman
stayed with his car. When on top of the pier, the
hopper doors were opened and the coal was discharged via chutes into ships waiting alongside
the pier. The empty cars rolled along slowly to an
inclined bumper stop that reversed their direction;

from there they rolled backwards onto a second,
inclined, return track that sent them back some
distance from the pier to the yard reserved for
empty cars. The brakeman rode the empties back
to this last-mentioned yard, so that at no time was
any car running wild or unattended. The facility
had five piers, each 1,000 feet long, that could accommodate forty to fifty vessels. A very mixed
fleet of barges, canal boats, schooners, and tugs
was assembled each day to take away the coal delivered by two thousand coal jimmies.

Major cities consumed prodigious quantities of
coal, but that consumption was seasonal in nature,
especially when home heating is considered. The
railroads and coal dealers found it necessary to
stockpile a large reserve of fuel so that the cities'
coal cellars might always be full (Fig. 1.16). Even in
summer, when every hearth in town was cold,
long caravans of coal drags rolled slowly toward
the storage terminals. New York City was a special
case with its appetite for 10.5 million tons of coal
each year.[59] The city's elevated railways burned up

FIGURE 1.15

The Lackawanna's coal pier, viewed from shoreside, shows a yard full of empty and loaded cars with the pier itself rising above in the background. Locomotives backed the trains into place and then departed. The loaded cars rolled down sloped tracks to the foot of the incline. (*Scientific American*, April 15, 1882)

FIGURE 1.16

The inability of railroads to deliver from producer to consumer is well illustrated by this view of a coal yard in Hedford, N.J. Four men were required to transfer the coal from a car to the wagon. Another labor-intensive transfer was necessary once the wagon reached the customer. (William B. Cooper Photo, Collection of Everett F. Mickle)

FIGURE 1.17

200,000 tons of coal annually in the 1890s. Storage piles of 500,000 tons each were stationed in Jersey City. These were drawn upon as the demand grew.

GRAIN

If bread is the staff of life, then grain was just as surely a major pillar in the structure of railroad traffic, for it was the biggest single commodity in the food group, and hence a bulk item, to move by rail. Grain was even more important when reviewed on a regional basis, for it was the principal traffic on many western railroads. In the first few decades of railroad operations, grain was mostly a local consumable. Corn, barley, and oats were largely used for cattle feed. If grain was sent any distance, it was economical to do so only if it was concentrated into high-value goods like flour or whiskey, but the Erie Canal changed all of this and opened a cheap way to send grains long distance to the seacoast where they might be exported overseas. North America became Europe's breadbasket, particularly in times of crop failure.

A strong export market together with cheap transit encouraged the establishment of grain fields in Ohio, Indiana, and other midwestern states. By 1850 40 percent of our wheat was being grown west of the Alleghenies, and within a decade half of it was being grown in this region.[60] Wheat farms were flourishing in Iowa and Wisconsin by midcentury. As the area under cultivation expanded, production climbed at an impressive rate; between 1840 and 1860 it in fact doubled.[61] The introduction of hard red wheat soon expanded wheat growing on a vast scale into the semiarid Great Plains. Before the railroad age, these grasslands seemed forever safe as havens for the buffalo and Native Americans, but that preserve was swept away by the ever-expanding wheat farmers. The temperate climate of North America plus the moderate rainfalls of the central and northern Mississippi Valley proved ideal for wheat growing, and so in time North America was supplying two-thirds of the world's wheat. This house of plenty ended fears of famine, so long as an economical supply line was available, and that supply line was very largely the railroad and its ever-ready handmaiden, the boxcar. The west-to-east parade of wheat-laden cars was already substantial in 1860, for in that year the Wabash was transporting 336,000 bushels.[62] The bulk shipping of grain had begun in earnest by the end of the same decade.[63] There was no need to use bags; half-doors called grain doors were fitted at the normal side-door opening of the boxcars, and loose grain was simply dumped in.

The western granger lines naturally perfected the bulk shipping of grain, but the eastern trunk lines were also much involved in this traffic, and none more so than the New York Central, which was the largest grain carrier between Chicago and the Atlantic seaboard. By the 1880s the Central had captured almost half of this traffic and was hauling around 40 million bushels a year.[64] By the first years of the new century the New York Central was routinely operating seventy-car grain trains, each carrying 72,000 bushels.[65] During the 1890s the Erie Canal at last began to lose its once-massive grain traffic to parallel railroads, which came to dominate this field and would have monopolized it save for the Great Lake steamers. Rail shipping was faster than water transit, but this advantage was not seen as the deciding factor. Water transit required transfers, and each transfer resulted in added cost and possible loss, confusion, and inaccurate billing.[66] By 1870, once the railroads had gotten their act together vis-à-vis the interchange of freight cars on a national scale, through shipments had become common. Paperwork was simplified, and several middlemen, notably freight forwarders, and the tedious physical transfer of the grain itself were eliminated. Low through rates on bulk shipments also encouraged grain dealers to ship by rail.

Moving the sea of golden grain evolved into a predictable, seasonal pattern, beginning in May with the gathering of empty cars. Any railroad that could spare any boxcars sent them to marshaling yards near the middle of the country. The car service departments sent crews out into the yards to perform light repairs.[67] Floors and interior sheathing were patched to eliminate cracks and holes and so prevent leakage of the cargo. Roofs were checked for leaks, and, of course, the journals, wheels, and brakes were checked over. Extra locomotives, fuel, and operating crews were placed on standby, for all must be ready by June when the winter wheat ripened in Texas, Kansas, and Oklahoma. No matter how well the railroads were prepared, too little or too much rain could ruin the crop. Hail-

FIGURE 1.18

A large portion of western grain traveled halfway across the continent to Atlantic seaport terminals for transhipment to European consumers. The yard and elevators shown here are at Locust Point (Baltimore) in the 1890s. (Smithsonian Neg. 49766)

storms, rust, and cutworms took their toll as well, and so all railroad operating officials learned to monitor these matters, for there was little point in mustering more men and equipment than the size of the harvest warranted. A poor crop called for a much-scaled-down operation, but a bumper crop meant that the harvest specials would gallop day and night. Once the cars were unloaded, they returned to granger states to reload for Fort Worth, Chicago, Duluth, and other major grain centers. The winter wheat rush was hardly over when operation swung north to Nebraska, Montana, and the Dakotas for the onslaught of the spring wheat harvest in July and August. The grain season generally ran its course in three months, and when it was over, both men and machines were ready for a rest.

Not all grain went directly from the fields to eastern seaports; some was consumed in the heartland itself, but far more was stored in trackside elevators. Most towns in the Wheat Belt had three or four elevators, each with a capacity of from 10,000 to 50,000 bushels. Farmers would bag their grain and take it just a few miles by wagon to the nearest elevator. The grain moved as a bulk commodity after reaching the elevator. It might be shipped immediately or held back for a better price later, all depending on the judgment of the dealer. Grain was commonly sold, shipped, stored, resold,

and reshipped several times before actually being consumed. During each of these moves, the railroads would earn revenue.

Once a dealer decided to ship his cargo eastward, his grain would likely end up in one of the big elevators in Boston, New York, or Baltimore. The Baltimore & Ohio built a 500,000-bushel elevator in 1869 at its Locust Point terminal.[68] This huge structure was erected just at the beginning of mass bulk grain shipments by rail (Fig. 1.18). As the prairies disgorged their plenitude, the B&O was required to build two more elevators in 1871–1874. A behemoth joined its smaller sisters in 1880. Elevator "C" with its 1.4-million-bushel capacity was proclaimed the largest such structure in the world.[69] New York, however, was the grain capital of the nation, and as early as 1872, just three years after large-scale movements by rail had begun, the port of New York was handling 75 million bushels of grain a year.[70] The New York Central erected a substantial elevator on the Hudson River that held over 1 million bushels (Fig. 1.19). This great barn, always surrounded by a herd of arriving and departing boxcars, measured 380 feet long and 150 feet high. In 1887, a second elevator boosted the capacity of the facility to 2.3 million bushels.[71] Five hundred cars were handled each day, and separate storage facilities were provided for other grains and flour.

FIGURE 1.19

Grain was the largest cargo of the foodstuffs group, and the New York Central was a major rail carrier of grain. Thousands of carloads arrived each year at the 61st Street elevators. (*Leslie's Illustrated Newspaper*, Nov. 19, 1877, Library of Congress Photo Copy)

FRESH PRODUCE

Civilized man is a finicky eater who craves novelty at the table, but a highly varied menu was possible only for the rich until around 1880. Before that time the mass of the population was forced to accept a bland, repetitious diet of easily stored food such as salt pork, dried fish, turnips, potatoes, and cabbage. Fresh vegetables and fruits were available only during short local harvests. For most common folk, a really tasty meal came only on feast days. Long-distance rail transit, however, radically changed the way ordinary people ate. A wonderful variety of viands now came to market, at prices all but a very few could afford, aboard ventilated and refrigerated cars. Some traveled across the continent to deliver lettuce, oranges, or dressed beef during all seasons of the year. It was now possible to enjoy a healthy and varied diet in all regions of North America because of economical rail transit.

What might be called a wholesaler's meals on wheels began in earnest around 1880. There was a hustle and quick-drill character to the business, for unlike coal and grain, fresh produce demanded speed in transit. Fruits, meats, and vegetables must be rushed to market on preference trains to avoid spoilage. No one would buy wilted lettuce or rancid beef. Here was one segment of the railroad freight business where fast schedules and a hurry-up atmosphere were clearly understood by all employees.

It was not so in the beginning, for in their early years railroads had been short, local carriers. Foods had been carried almost since the first day of operations, but before 1850 few railroads exceeded 100 miles in length. There was comparatively little bustle about the transit of food over the road, and it tended to bump along at the same leisurely pace common to all trains of the time. This is not to say that the railroad industry was indifferent to the carriage of food to market. It represented a fine source of revenue for most of our pioneer lines, but because the distances involved were so modest, there was nothing very special about it.

No exceptional provisions need be made for a load of potatoes rumbling some 50 miles down the line. Fresh fruit received a little more attention, however. In August 1845 the Camden & Amboy advertised in local newspapers that special accommodations would be provided for those engaged in the peach harvest in the Hightstown area.[72] Farmers were charged 10 cents per basket and could expect to be dropped off at the Barclay Street pier near Manhattan's chief produce center, the Washington Market. Those traveling with twenty or more baskets paid freight rates and received a free round-trip ticket. Presumably, baggage cars were provided to handle the big producers. The in-season carriage of fresh fruit also swelled the coffers of other New Jersey railroads serving New York Harbor. In June 1851 the Paterson & Ramapo carried nearly 1 million baskets of strawberries to the New York market.[73]

These local movements helped the regional farmers and satisfied nearby consumers, but only in season. Such service did nothing to vary diets year-round. By midcentury the long-distance carriage of foodstuffs began to emerge. The earliest re-

FIGURE 1.20

Taking berries to the big-city market was the beginning of the produce business on many U.S. railroads. This rare photograph, dating from May 1868, shows the loading of strawberries at Vineland, N.J. (Collection of Edward T. Francis)

corded example, dating from 1851, involved the shipment of butter from northern New York to Boston in insulated, iced cars.[74] A few years later, meat and produce were being shipped from Rochester to New York. At the same time (1855), the Illinois Central began accepting shipments of peaches and apples in southern Illinois for Chicago.[75] Rates ranged from $1.50 per hundred pounds to $150.00 for a carload. The cars were attached to passenger trains to speed delivery. By 1857 the line was carrying 6,000 tons of fruit northward. A decade later, the Central began running berry trains between the same terminals.[76]

The carriage of fresh fruit was at last expanded beyond the provincial level when the Illinois Central transported berries from Louisiana to Chicago, so that the residents up north might have fruit out of season. And so it was not until the late 1880s that the real potential of rail food transit was realized on the Central. Once that had happened, the line decided to institute very fast strawberry specials. Hence the Thunderbolt Express came into being in 1892. The Central was not particularly pioneering in organizing its berry business, but it proved very innovative when it came to bananas. This tropical fruit was popular in a few southern ports like New Orleans, where it arrived

on ships bound from Central America. A minor official of the railroad thought that the strange yellow fruit might sell well in Chicago. Shipments began in 1880, and a new market was established. During the next year 331 carloads of bananas went north to Chicago via a sea and rail route easily 2,000 miles long. Long-distance, out-of-season food transit was clearly under way. An exotic fruit was made commonplace and popular by cheap long-distance transport, aided by some imaginative marketing.

Railroads in the Old South worked to exploit the Yankee craving for fresh produce out of season.[77] In May 1884 officials of railroads in Georgia and Florida convened in Atlanta to plan for the bumper crop of watermelons expected for that season. Their job was to assemble the necessary supply of ventilated cars to move the melons to the nearest seaport where they would be transhipped to Philadelphia, Boston, and Baltimore (Fig. 1.21). Within just a few years the southern roads had arranged for direct, all-rail shipments to northern markets.[78] The first melon train of seventeen cars left Valdosta, Georgia for Boston in 1887. The 1,300 mile journey was routed over ten railroads.

On the opposite side of the continent, California growers were planning a long-distance invasion of

FIGURE 1.21

Watermelons are being loaded aboard ventilated cars at these Atlanta team tracks. The long-distance transit of produce improved the American diet and generated added revenue for the nation's rail networks.

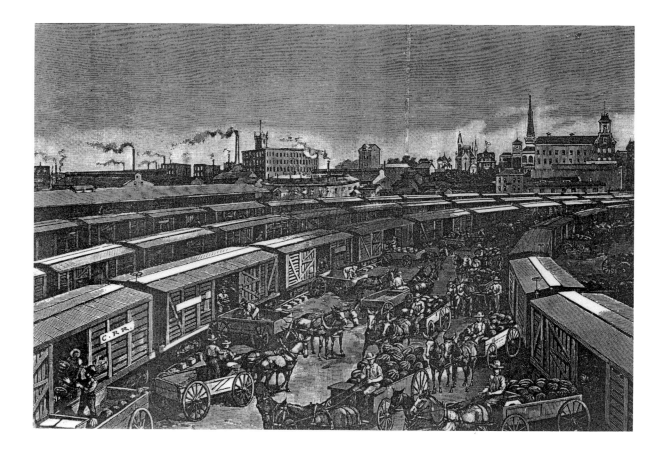

the eastern produce market. Blessed with a sunny climate and an unusually long growing season, southern California was ideally suited to supply out-of-season produce to the large northeastern market. The Union Pacific began to acquire ventilated fruit cars even before the transcontinental line was open. As early as 1871 growers in Riverside, California, were attempting to introduce citrus groves. Production was very poor until Brazilian varieties of trees were adopted, and then the harvest became bountiful.[79] Within a few years twenty thousand acres of citrus trees were in place in the Riverside area. By the end of the century eighteen thousand carloads of fruit were being shipped annually from the Los Angeles area alone. During this same period 100 million pounds of raisins, 160 million pounds of prunes, and some twenty-three hundred carloads of lima beans were shipped eastward from the Golden State each year. Such a cornucopia of produce added significantly to the revenues of the western railroads and their eastern connections.

LIVESTOCK

The cattle trade, much like produce farming, grew from small-scale local production into large-scale, long-distance agribusiness. In the beginning of the railroad industry, beef cattle, sheep, and hogs moved a short way to market. The extension of cattle raising to the West provided the rail network with long hauls, and it opened western lands, too arid or rocky even for red wheat, for grazing.

Cattlemen had long recognized the folly of long overland drives that wasted the herd's fat and bulk and so greatly lessened its value when the market town stockyard was reached after so many weeks on the trail. Canals were deemed too slow and riverboats reached too few cattle areas to satisfy this need. But the railroad could thread through valleys, climb hills, and skirt along the lake shores to reach all but the most remote cattle rancher. Even before the western lands were opened for this purpose, a large traffic had developed back east. In 1860 the New York Central figured that about a quarter of its traffic was accountable to the transportation of domestic animals.[80] In that year it moved 215,000 head of cattle and 380,000 hogs. The Central's connections to the West funneled animals from Ohio, Indiana, and Michigan. The Erie enjoyed a similar trade.

The Baltimore & Ohio yearned to share in this fine traffic, and so in the late 1850s it began to cultivate the western cattle traffic with new cars, feedlots, watering stations, and improved facilities for drovers.[81] Before these improvements were offered, most of the B&O's cattle trade had come from nearby places like Winchester (Va.) and Cumberland (Md.); but now the railroad hoped to attract steers and hogs from as far away as Kentucky. Rates were reduced by $10 per carload, so that a load of fat cattle, ready for the slaughterhouse, went from Wheeling to Baltimore for $80. Regular stock cattle went for $20 less. Faster schedules were offered to cattle shippers; regular freight re-

FIGURE 1.22

Cattle had been an item of rail traffic since the 1830s, but this segment grew in scope and importance as the grazing moved to the Far West. Cattle are shown here being loaded in Montana for the Chicago market. The year is 1904. (Smithsonian Neg. 85-6891)

quired forty-eight hours between Wheeling and Baltimore, but the live-beef specials made it in thirty-six hours.[82] This was some twenty hours slower than the line's best passenger train, but it was nonetheless rapid traveling for a freight train in 1858, and every hour saved meant fatter profits for the shipper. These policies paid off, for traffic on the line almost doubled between May 1859 and May 1860.

Large-scale cattle production shifted ever westward to take advantage of cheap land beyond the Mississippi, where it was possible to convert a $4 range steer into a $40 beef steer. The key was a cheap and sure way to get that fatted steer to market. In 1867 western roads like the Kansas Pacific were ready to cooperate and offered shippers cheap rates on large-scale shipments.[83] Some settlements like Abilene became cow towns where range cattle were gathered at the railhead for a one-way trip to Chicago's packing houses. One million head were shipped out of Abilene between 1867 and 1870. Wichita, Ellsworth, and Dodge City followed Abilene, because more stock-gathering places were needed to handle the growing herds encouraged by economical rail rates and cheap barb-wire fencing. Omaha prospered in its cattle trade because of its terminal position on the Union Pacific. Between 1876 and 1878 receipts at Omaha on the UP doubled, from 48,000 head to 96,000 head.[84] The town expected the total to rise to half a million head very soon, and a Union Stock Yard was organized to handle the anticipated traffic.

Omaha and the other would-be cattle centers were nothing compared with the greatest cow town in the world, Chicago. Its position as a rail center guaranteed its success in this important segment of the food market. Chicago was so central that it proved the perfect way station between the beef-producing West and the beef-consuming East. The number of cattle arriving grew in a Malthusian progression. In 1855, roughly 10,000 head arrived; in 1860, 155,000 arrived; in 1870, 532,000 arrived; and at the end of the century, the number was more than 3 million. In 1901 over 291,000 carloads of cattle, hogs, and sheep arrived at Chicago's gigantic Union Stock Yard.

The cattle trade was important to the Texas railroads. In 1893 it accounted for 9 percent of all tonnage, or 142,000 carloads, and was the third largest revenue producer.[85] On the national level, cattle and other domestic animals represented a much smaller segment of the total traffic. But even if the national tonnage was only about 2 or 3 percent of the total, many railroads depended on the carriage of cows to help pay their expenses. Live animals surely proved troublesome to manage. In terms of damage claims, they were among the most costly commodities handled. A certain number were sure to sicken or die en route. Some of the earliest state and federal legislation regarding railroad cargo management involved the cattle trade. Animal lovers pressured railroad managers to take a greater interest in the humane treatment of domestic creatures aboard the cars (this subject is discussed further in Chapter 4). The railroads were thus forced to handle cattle trains with extra care.

The cost of this handling was returned to shippers in the form of very high rates—generally 2.5 to 3.0 cents per ton-mile. In addition, railroads sought to limit their liability by requiring shippers to agree to a fixed schedule of values—a sheep was set at $5, a cow at $75, and so forth. If an entire carload was lost, the claim value was limited to $1,200.[86] Shippers were also made responsible for the safe loading and unloading of their animals. Station agents made sure that shippers did not overload the cars, for this would often result in death or injuries. Double decks were permitted to increase capacity when small creatures were being shipped, but again this was done at the shipper's risk and expense unless the car had been built as a double-decker. High rates and restricted claim payments were the only way to turn a profit in the cattle trade.

Wild animals are almost never shipped by rail except in circus trains, but there was one great creature native to the American plains that at least briefly contributed to the volume of animal products moving by rail. Most settlers viewed the bison as an impediment to civilizing the frontier. The sooner it became extinct, the better. Sportsmen and professional hunters were encouraged to slaughter the huge, hairy beasts, and during the 1860s and 1870s around 1 million were killed each year.[87] Opinions on this subject have changed greatly since settlement times, but in that era about all the buffalo had to offer was its hide. Millions of hides were shipped east, as were large

FIGURE 1.23

Lumber was essential to just about every phase of American life. Rough-cut lumber was shipped by flat or gondola cars, as shown by this early-twentieth-century scene on the Denver & Rio Grande Railway near Lobato, N. Mex. (Fred Jukes Photo, R&LHS Collection)

quantities of buffalo tongues, but most of the meat was left to rot where the creature fell.

By 1880 most of the herds had been killed. It seemed that the buffalo traffic was at an end, but then someone discovered the value of salvaging the skeletons, bleached white and dry on the prairie. The bones fetched $8 to $14 per ton for fanciful things like hatracks, buttons, and combs. Most, however, were ground up for fertilizer, glue, and bone china. Bone pickers loaded the remains into wagons and headed for the nearest railhead, where their glistening cargoes were dumped into railroad cars. One old fellow eventually filled three hundred cars at Dodge City with bison bones.

LUMBER

Wood was the basic building material of the nineteenth century. It was used for every article of daily life, from furniture to homes to industrial buildings.[88] The transportation of lumber was, accordingly, essential to the public well-being and the earning capacity of the nation's railway system. It generally represented about 10 percent of all railroad traffic. Its importance was, of course, far greater on southern and Pacific northwestern lines, which originated considerable on-line timber. On some western lines, lumber was the largest single cargo in terms of car loadings.[89] It was not always such a great revenue producer,

however, for it generally traveled at low rates; this was especially true for the rough-cut boards that made up a large part of the early railroad lumber traffic. Lines in the Far Northwest were ready to grant especially favorable rates to eastbound lumber to reduce the number of empty-backhaul cars.

The timber resources of the nation were truly magnificent, for a heavy forest covered most of the land. As the woods of the East Coast were consumed, lumbering operations moved into the Midwest and Southeast. The southern yellow pine belt outproduced all other lumber regions, and strong timbers from this region provided the framing for New England's textile mills as well as most of the wooden freight cars described later in this volume. The final major lumber area to be exploited was the Oregon-Washington Douglas fir forests. The railroads enjoyed many long hauls, if only at low rates, from the Pacific Northwest.

COTTON

Cotton was not a major commodity in railroad traffic terms, amounting to only about 0.5 percent of overall tonnage, but it was important as a source of revenue, and it was most assuredly important in terms of clothing and the textile trade. Because cotton was raised everywhere from Virginia to Missouri, just about every southern line benefited from the cotton traffic. Some lines, like the South

FIGURE 1.24

Cotton generally traveled long
distances between plantation
fields and spinning mills. In this
case, cotton is being transferred
from ocean ship to Boston &
Providence boxcars at the
Lonsdale Wharf in Providence
around 1875. (R&LHS Collection)

Carolina Railroad, were built primarily to haul the white fluffy fiber. The merchants of Charleston, chief backers of the South Carolina, hoped that the new railroad would divert the cotton traffic away from their rivals in Savannah.[90] In 1840, after a decade of operations, the South Carolina moved nearly 60,000 bales. Traffic jumped notably later in the 1840s after branches opened to Columbia and Camden. In 1850 around 285,000 bales (each bale weighed 320 pounds) were carried. This traffic was satisfying to the shareholders of the line, but it did not spell the end of Savannah, for that port continued to receive a plentiful allotment of cotton via river steamers.

Texas entered the railroad age many years after South Carolina, but when it did so in 1855, the first line built near Houston was inspired by King Cotton.[91] The little Buffalo, Bayou, Brazos & Colorado succeeded in linking cotton fields with Houston, and in the first years of operations it moved 8,000 bales at one-third the cost charged by the oxcart operators. The mature and much-expanded Texas railroad networks of 1893 carried over a million tons of cotton and listed it as the number one revenue producer.

Elsewhere in the nation, railroads were taking cotton shipments away from river steamers at a steady rate. In 1873 New Orleans continued to receive 68.8 percent of its cotton by steamer, but this proportion declined to 19.8 percent in 1890 and to 9.5 percent in 1904.[92] Railroads never monopolized the cotton traffic, because a certain amount of it continued to go via coastal steamers. This was particularly true for bales bound for the New Eng-

FIGURE 1.25

Less-than-carload shipments were a troublesome but necessary part of the railroad freight business. The two cars parked at the freight house end of this Boulder depot around 1890 are undoubtedly standing by to unload or accept LCL shipments or to exchange cargoes with the narrow-gauge cars parked nearby. (Photo by Joseph B. Sturtevant, Western History Department, Denver Public Library)

land mills. The Merchants & Miners Transportation Company began operations in 1854 and specialized in moving cotton bales between Savannah and Providence. A fine photograph dating from around 1875 depicts this operation (Fig. 1.24).[93] One of the M&M vessels peeks out from one end of the Lonsdale Wharf in Providence. Wagons are used to move the bales out from the receiving shed after being offloaded from the ship. They are then stacked on loading platforms where stevedores maneuver them into waiting boxcars with handcarts. In the foreground, the *Useful*, the Boston & Providence Railroad's plucky little switcher, pulls a cut of loaded cars away from the wharf to a classification yard. From there they will travel north to the great mills at Manchester, Lowell, or Lawrence. In this one scene we can see the various modes of transportation available to nineteenth-century shippers.

LESS-THAN-CARLOAD SHIPMENTS

There are a great many specific classes of cargo that will not be treated here, such as bricks, steel plate, and glassware. All these items, and many others not named, were important traffic. All were potentially profitable and, hence, sought after by railroad freight agents. Before closing this discussion of specific cargoes, we should discuss one element of the freight business that was not particularly sought after, because it was so rarely profitable. It was named LCL, for less-than-carload lots.

These were smallish shipments weighing 50 to 100 pounds each. They were too big for express or the U.S. mails, but they were too many and too small to work well in the normal boxcar mode of operation that was the basis of the railroad business. It was a complex, labor-intensive, and costly business. The sorting and handling of so many

small crates, boxes, casks, parcels, bags, and carboys, all going to different places and consignees, could be nothing but trouble. Yet these small shipments were vital to industry and home alike. A hot-water heater would likely go LCL, as would repair parts or replacement mining gear.

The railroads were under a great deal of pressure to handle these shipments as part of their public service mission, and so LCL business became a necessary, if unwanted, part of the freight business. The industry was not, however, willing to undertake this segment of the trade as charity; high rates were charged. The Interstate Commerce Commission had hardly begun operations before it was hearing complaints from small shippers, early in 1888, about disproportionately high LCL rates.[94] The railroads' representatives countered that if anything the rates were too low. LCL made very poor use of a car's capacity; partially filled cars lost money. The industry figured that the lowest average carload was 14 tons, an admittedly poor showing at a time when most cars were in the 25- to 30-ton range; but LCL was far worse, at 4.3 tons per car. The cars might as well run empty. In addition, each LCL car would carry cargoes for eight to twenty-five consignees and be routed to a minimum of three and as many as seven destinations. Through cars were all but impossible, and LCL cars were routinely unloaded and re-sorted several times en route. With so many destinations, a final terminal or sensible route was impossible. A box going from Baltimore to Willard, Ohio, must be unloaded at both Toledo and Tipton before it could reach its destination.

Most railroads figured the cost of handling LCL at $2 per mile, or exactly double the normal carload (CL) cost. The Pennsylvania Railroad contended that this figure was far too conservative and

THE AMERICAN RAILROAD FREIGHT CAR

that LCL costs were twelve times greater than CL costs. Industry spokesman Albert Fink stated that the LCL rate of 38 cents per hundredweight per 1,000 miles was too low, not too high. Fink went on to point out that railways in Germany charged the equivalent of $2.25 for the same service. Claims and counterclaims as to actual costs and fair rates were never really settled, but LCL traffic just would not go away—that would not happen until the advent of the motor truck—and so the railroads resolved to make the best of a bad thing.

Special yards and transfer houses were established to unscramble and rationalize the LCL traffic. Long, low sheds with eight to ten tracks sorted boxes and bags from two hundred to five hundred cars a day. Some of these sheds were 900 feet long. The Baltimore & Ohio started what might be the first instance of a hub sorting operation in 1888.[95] All LCL shipments were sent to Baltimore for resorting before going forth to their destinations. Placing the hub terminal at one end of the line rather than near the center would result in considerable backhauls—hardly a model for others trying to find cost-cutting methods. The trap, peddler, or ferry car was a far better idea. These cars traveled from station to station, picking up LCL shipments and taking them to a regional freight terminal, where they would be sorted and placed in cars aimed at the final destinations. Peddler cars reduced the number of boxcars needed for LCL movements and helped relieve congestion at main freight houses.

The organization of freight-forwarding companies after 1900 helped relieve railroads of direct involvement with small shippers and the hassles of LCL shipments. Nothing could have made the industry happier than to retreat completely into the wholesale transport business. The freight forwarders would solicit shipments from small businesses and consolidate them into carload lots. They charged their customers LCL rates but paid the railroads' regular CL rates and so realized a profit.

Rates, Rebates, and Regulation

Rates were probably the single most contentious subject associated with the railroad industry. Shippers, naturally enough, felt that rates were too high. Left-leaning politicians echoed this belief to their followers, and it is likely that a sizable portion of the public accepted these allegations as true. Periodic revelations about special rates, rebates, drawbacks, underbilling, and other shady practices increased the suspicions of the public. Railroad leaders tried to explain away such activity as nothing more than normal business practice. Big shippers expected preferential treatment. Everyone did it. A little grease was necessary to keep the wheels of industry turning. According to the railroads, rates were actually too low to maintain the plant and equipment properly and yet pay a fair return on investment. A badly maintained railroad could offer only poor service, and one that could not attract new investment could never improve itself.

And so the debate raged on and off for over a century starting around 1870. It became a staple for the public forums that led to the Granger Movement, followed by state regulation and finally to federal oversight. The reform spirit of America was caught up not so much in passenger or employee safety as in freight rates. The main force in this often emotional segment of our domestic history was over just how much to charge for sending a boxcar loaded with dried apricots from Kennebec to Keokuk. What was fair, just, and reasonable? A few cents per ton could stir passions over the righteous doctrine of common carriage, whereby service was open to all on the same terms with no rebates or discrimination over matters such as the distance of travel or the prominence of the shipper. The ideal of fairness and evenhanded treatment is among the noblest of doctrines. But these aspirations are difficult to maintain, in a very pure sense, in the business world, where the maximization of profits is omnipotent. The railroad tried to sell its service for as much as it would command. The shipper would buy only at the lowest possible charge, and if he could, he would seek a better price from another railroad or an alternative means of transport. And so the negotiations seesawed back and forth until a compromise was effected, one that essentially amounted to the old chestnut about whatever the traffic will bear.

HOW RATES WERE SET

Rate making is not a simple procedure, because the establishment of charges, tolls, or a price for the carriage of goods involves over a dozen factors.[96] These range from obvious considerations such as the value of the shipments to less obvious ones such as the exact nature of the packaging. The worth or value of the goods relative to their weight was the first consideration in the rate-making process. It was among the most simple and obvious: furs traveled at a high charge, sand and gravel at a far cheaper toll. It might be noted that very costly goods, jewelry, currency, and precious metals were not considered proper for freight transit, and shippers of these precious items were referred on to the express department. At first blush it might seem perfectly reasonable to charge substantially higher rates for furs than for gravel, but on reflection, why should one shipper pay significantly more to send his furs than another shipper sending an equal amount of gravel the same distance? It costs the railroad little more to carry a carload of fur than one of gravel. The claim value should either shipment be lost or damaged would, of course, be quite different, but the overall costs of moving the cars across the countryside are equal. And so why should not both shippers receive equal billing?

A basic inequity in rate making now comes to the fore. Furs and other high-value goods can absorb higher transit costs because they are very small in proportion to their market value. The transport costs are then absorbed by the shipper and/or the buyer. The low intrinsic value of gravel, however, will allow for only very low transport tolls. If high rates were charged, gravel could travel only a few miles before it would double in price and so become unsalable. In 1870 Henry V. Poor used the example of sending a bushel of wheat valued at 75 cents to market by wagon at a cost of 20 cents per ton-mile.[97] At such high rates, wheat could not be shipped economically for much over 125 miles. The railroads, conversely, were able to carry wheat at 1.25 cents per ton-mile, and so the farmer could hope to reach markets 1,600 miles distant and still be competitive with other grain producers. This was possible only because railroads agreed to grant special rates and so surely discriminated against shippers of more valuable commodities. These rates were not simply low; they were artificially low, according to a study published in 1885 by a Yale faculty member, Arthur T. Hadley.[98] Hadley claimed that railroads moved bulk commodities like coal and grain at rates below the average costs of doing business, and so bulk ladings were thus subsidized, in part at least, by the higher-rated shippers, a practice now called a cross-subsidy. The railroads claimed that such low-rated tonnage was welcome because of its large fixed plant, normally plagued by overcapacity, that earned nothing while idle. Moving cars below the cost of service sounds stupid, but it brought in cash for immediate expenses, and so as a short-term revenue producer the practice of low rates made sense. If the loss was marginal, the traffic was worth having.

The second factor on the list of rate-making judgments was the matter of fragility. Here the question involves the degree of risk during shipment. How well do the goods withstand the bumping and banging attendant upon railroad shipping? Every time the train starts and stops, the load shifts inside the cars. If the engineer is skillful and steady, these movements are minimal and the cargoes arrive intact. But the best of train crews can do little to mitigate rough track, leaky car roofs, or inadequate dunnage. The more the cargoes were handled, the higher was the chance of damage. Before the era of interchange, goods traveling any distance were subject to repeated handling as they were transferred from one railroad to another. Each transfer was a fresh opportunity for breakage, loss, or damage. Railroad claim agents and rate makers thus became extremely anxious when

shippers came forward with glassware, pottery, furniture, pianos, and other items easily injured in transit. The greater the risk of damage, the higher the rate. There were items, such as carboys of acid, that some railroads would carry only at the owner's risk. Conversely, rate makers were not at all troubled by coal, gravel, pig iron, or other hardy products. These might go at very low rates because the chance of damage claims was so remote. If a carload of coal rolled over, the railroad would recover most of it and send it on its way with little loss. If it was accidentally splattered with paint, the loss was trivial. If it sat out in the rain or the sun for weeks on end, no one would care. Yet should the same happen to a carload of some more fragile material, big losses would result.

And so exactly what was shipped became another important element in rate making. Apples would seem simple enough—one rate should do for all. Oh, but not so, for there were ripe apples, green apples, dried apples, and canned apples. Each traveled at a different rate. Just how the apples were packed added to the complexity of rate making, because it affected both their ability to survive the journey and the ease of handling at the beginning and end of the trip. All but canned apples might be shipped in bags, boxes, or barrels. If they were going to a cider mill, they might be shipped bulk in a boxcar. The packaging and even the projected market were factors for consideration. Apples intended for juice making might arrive a little battered and bruised, but those destined for a fancy fruit stand must arrive pristine. The care in shipping and packing would vary considerably in this case, as would the liability and likelihood for a damage claim. A single juice apple was nearly worthless, but a choice eating apple might be worth 5 cents.

Next on the list of considerations were the matters of bulk and weight. How dense were the goods presented for shipment? If weight was the sole measurement, a shipper of light, fluffy goods like cotton or hay might benefit greatly, and so cubic measure became an important yardstick for determining rates. Cotton, and to a lesser degree hay, was compressed for shipment to overcome its bulkiness. Bailing also made it easier to handle in transit.

Weight and bulk were relatively noncontroversial aspects of rate making, but distance traveled was a matter of considerable contention. Measuring the toll by distance sounds simple enough, but railroad managers contended that their short-haul costs were actually higher than their long-haul expenses. They argued that their costs were not directly proportional to distance of travel; a 200-mile trip did not cost twice as much to perform as a 100-mile trip, because the loading and terminal costs were the same for both trips, and so the short haul was more expensive when calculated on a per-mile basis.[99] There was a degree of truth in this proposition because terminal costs were a significant part of overall operating costs, as were loading and unloading charges.

However, the real differential in long- and short-haul business was actually the disparities between a captive and a free-market situation. Short-haul traffic tended to be local in nature. Shippers were often unable to ship any other way, and the single railroad that served them held them as captive customers and might charge almost any rate it wished.[100] Long-distance shipments were far more competitive. A load of grain might go via any one of six rail routes from Chicago to New York; hence the competition for the long-haul traffic was competitive, and the individual railroads were driven by market pressures to offer low rates. With its high capital costs and massive debts in the form of mortgage bonds, the railroad industry found its debt service representing two-thirds of all costs. Fuel, labor, and repair costs were minor by comparison, but unlike the debt service, these expenses diminished when the traffic was down. Not so the interest charges; they came due whether the railroad was moving one thousand or ten thousand carloads a week. The railroads wanted to keep traffic volume up, even if the profit per car was small, to maintain a cash flow to satisfy the insatiable debt service. This is why long-haul and large-volume shippers could expect low rates. Their business was essential to keep the encumbered rail network solvent. Big shippers felt that it was a natural right to be accorded more favorable charges, because wholesale buyers were customarily accorded a bargain. Buy one axe handle and it costs a dollar; buy a thousand and they are 30 cents each. Big volume begets discounts.

In a similar way, shippers sending out goods in carload lots received far more reasonable rates than those shipping in less-than-carload lots. We have already addressed the conundrum of the LCL traffic earlier in this chapter. Yet even the carload rates were not so simple, because as the rate-making process was refined late in the nineteenth century, rules were established for each commodity moved.[101] The normal standard was 15 tons, despite the fact that by 1885 most cars were designated in the 20- to 30-ton range. Some goods received rather peculiar carload limits; church furniture, for example, was figured at only 5 tons per car, presumably because it was bulky, while school furniture might weigh 12 tons. The carload rate limit was 18 tons for copper bars but 20 tons for wheat.

Rate makers also considered requirements for special equipment and the need for speedy delivery. Shippers always wanted special cars to assure easier, safer handling of their goods. The railroad's reaction was, usually, Why can't it go in a boxcar? The precedent had been established with the cat-tle cars. What would shippers want next? Refrigerator cars, then tank cars, then special oversize flatcars—why, this requirement for exceptional equipment was getting out of hand! It was also getting very expensive, and those costs were quickly passed on to the shipper. Some of the goods sent on single-purpose cars were not particularly urgent. A load of crude oil might bumble along and arrive at the refinery a few hours or even days late with not much material inconvenience to its owner, but this was not true for a load of beef steers, celery, or Christmas trees. These last-mentioned items must get there on time. In the case of Christmas trees it was because of the market—the market was time-urgent, all must be sold before December 25. The other items also required prompt delivery, but more for preservation reasons. Cattle out on the line too long lost weight and sickened. Celery might wilt and so lose its resale value. Foodstuffs, except for bulk grain, were inherently more expensive to ship than lumber because they required attention while in transit. Attention meant men and facilities—ice stations, feedlots, salt, and water—all of which cost money. The rates for special handling and special cars that were suitable only for a limited traffic were legitimately higher than those for less troublesome goods.

Where the goods came from, the direction they were traveling, and the season of the year could also materially affect rates. Before the era of state or federal regulation, railroads might charge whatever they wished, and so in this happytime of laissez-faire many railroads, like the Lake Shore & Michigan Southern, charged more for westbound (2.02 cents per ton-mile) than for eastbound (1.56 cents per ton-mile) traffic.[102] Why was this so? The answer is simple enough. When these rates were posted in 1868 the majority of the road's traffic was eastbound; the cost of doing westbound service was greater, largely because of all the empty cars, and so the shippers were expected to pick up the difference. By 1872 the Lake Shore rates had dropped and evened out somewhat, but westbound shippers were still penalized by 0.36 cents per ton-mile. Many years later, in a curious reversal of such conservative strategies, some railroads would offer shippers cheaper rates for cars moving in the direction of the weakest traffic flow so as to fill up the non-revenue-producing empties. Railroads began to figure costs not as averages but as marginal costs. They concluded that the tide of traffic with the most empty cars would benefit most from lower rather than higher rates.

In the early days of railroading the season of the year could also affect rates. The difference was often substantial. According to a bill of lading to be discussed later (Fig. 1.27), the Pennsylvania Railroad jumped its winter rates 25 percent over its summer levels. Figure 1.26 reproduces an Albany & Buffalo winter schedule of tariffs dating

FIGURE 1.26

Published tariffs appeared at the beginning of the railroad era. The breakdown of rates became even finer as time went on. (New York Central Railroad)

from 1844–1845; in this case no comparison is made with summer rates. The fact that different charges were in effect was justified by rationales. Operating costs were greater in the winter because of snow removal, frost damage to rails, wheels, and axles, and the somewhat greater fuel consumption. Once again these were actually relatively minor costs compared with the railroad's interest payments, but it sounded reasonable to the public. One suspects that the reason for the stepped-up winter rates was to maintain revenue when traffic slumped after the fall harvest rush was over. In addition, many water carriers such as the Erie Canal were shut down in winter, and so the railroads might charge as much as shippers were willing to pay.

Rates varied not just with the transit of the earth but with the location of the shipper within the United States. Costs tended to be higher where traffic was less abundant, and so southern and western lines tended to charge far more than the northeastern roads. In 1870, for example, the northeastern lines were averaging 1.6 cents per ton-mile and the western roads 2.4 cents per ton-mile.[103] The Denver & Rio Grande was in fact notable for its high rates, which came to three to eight times those charged by its eastern counterparts. The costly little narrow gauge got away with its fancy prices because most residents living in the shadow of the Rocky Mountains were accustomed to high transport charges, and because most of the traffic was local, they had little choice but to pay whatever the Rio Grande asked.[104]

Some measure of uniformity came about in April 1887 when the Trunk Line Traffic Association adopted the Official Freight Rate Classifica-

FIGURE 1.27

The traditional four-class tariff system is concisely outlined on this bill of lading dated August 13, 1858. (Pennsylvania Railroad)

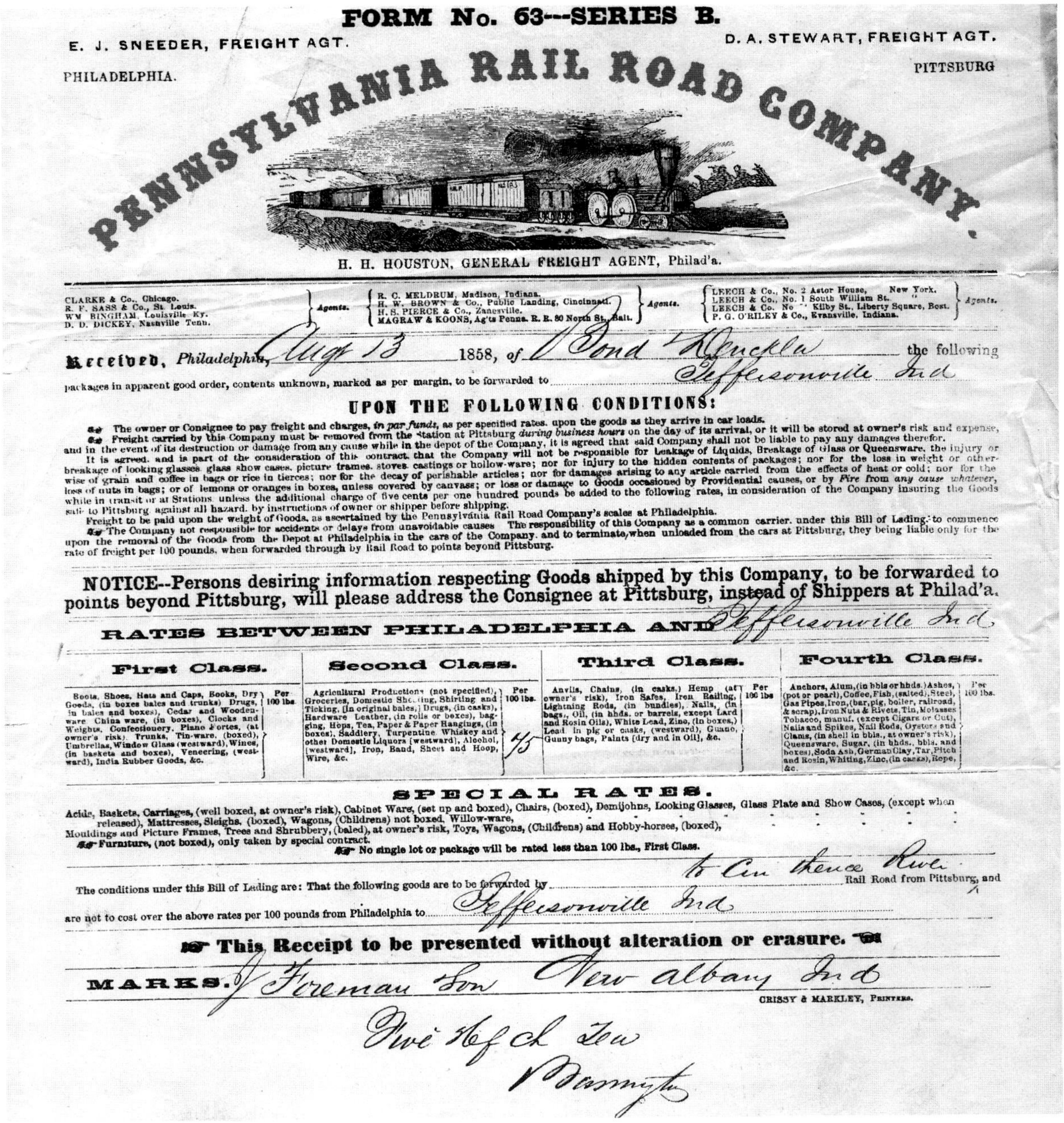

FORM No. 63---SERIES B.

E. J. SNEEDER, FREIGHT AGT. PHILADELPHIA. D. A. STEWART, FREIGHT AGT. PITTSBURG

PENNSYLVANIA RAIL ROAD COMPANY.

H. H. HOUSTON, GENERAL FREIGHT AGENT, Philad'a.

CLARKE & Co., Chicago. R. F. SASS & Co., St. Louis. WM. BINGHAM, Louisville Ky. D. D. DICKEY, Nashville Tenn.	Agents.	R. C. MELDRUM, Madison, Indiana. H. W. BROWN & Co., Public Landing, Cincinnati. H. S. PIERCE & Co., Zanesville. MAGRAW & KOONS, Ag'ts Penna. R. R. 80 North St., Balt.	Agents.	LEECH & Co., No. 2 Astor House, New York. LEECH & Co., No. 1 South William St. " LEECH & Co. No. " Kilby St., Liberty Square, Bost. P. G. O'RILEY & Co., Evansville, Indiana.	Agents.

Received, *Philadelphia,* Aug 13 1858, *of* ... the following packages in apparent good order, contents unknown, marked as per margin, to be forwarded to Jeffersonville Ind

UPON THE FOLLOWING CONDITIONS:

The owner or Consignee to pay freight and charges, *in par funds,* as per specified rates. upon the goods as they arrive in car loads.

Freight carried by this Company must be removed from the Station at Pittsburg *during business hours* on the day of its arrival, or it will be stored at owner's risk and expense, and in the event of its destruction or damage from any cause while in the depot of the Company, it is agreed that said Company shall not be liable to pay any damages therefor.

It is agreed. and is part of the consideration of this contract. that the Company will not be responsible for Leakage of Liquids, Breakage of Glass or Queensware, the injury or breakage of looking glasses. glass show cases. picture frames. stoves. castings or hollow-ware; nor for injury to the hidden contents of packages; nor for the loss in weight or otherwise of grain and coffee in bags or rice in tierces; nor for the decay of perishable articles; nor for damages arising to any article carried from the effects of heat or cold; nor for the loss of nuts in bags; or of lemons or oranges in boxes, unless covered by canvass; or loss or damage to Goods occasioned by Providential causes, or by *Fire from any cause whatever,* while in transit or at Stations. unless the additional charge of five cents per one hundred pounds be added to the following rates, in consideration of the Company insuring the Goods safe to Pittsburg against all hazard. by instructions of owner or shipper before shipping.

Freight to be paid upon the weight of Goods, as ascertained by the Pennsylvania Rail Road Company's scales at Philadelphia.

The Company not responsible for accidents or delays from unavoidable causes. The responsibility of this Company as a common carrier. under this Bill of Lading, to commence upon the removal of the Goods from the Depot at Philadelphia in the cars of the Company. and to terminate when unloaded from the cars at Pittsburg, they being liable only for the rate of freight per 100 pounds, when forwarded through by Rail Road to points beyond Pittsburg.

NOTICE--Persons desiring information respecting Goods shipped by this Company, to be forwarded to points beyond Pittsburg, will please address the Consignee at Pittsburg, instead of Shippers at Philad'a.

RATES BETWEEN PHILADELPHIA AND Jeffersonville Ind

First Class.		Second Class.		Third Class.		Fourth Class.	
Boots, Shoes, Hats and Caps, Books, Dry Goods, (in boxes bales and trunks) Drugs, in bales and boxes), Cedar and Wooden-ware. China ware, (in boxes), Clocks and Weights. Confectionery. Piano Fortes, (at owner's risk). Trunks, Tin-ware. (boxed), Umbrellas, Window Glass (westward), Wines, (in baskets and boxes), Veneering, (westward), India Rubber Goods, &c.	Per 100 lbs	Agricultural Productions (not specified), Groceries, Domestic Sheeting, Shirting and Ticking, [in original bales,] Drugs, (in casks), Hardware Leather, (in rolls or boxes), bagging, Hops, Tea, Paper & Paper Hangings, (in boxes), Saddlery, Turpentine. Whiskey and other Domestic Liquors [westward], Alcohol, [westward], Iron, Band, Sheet and Hoop, Wire, &c.	Per 100 lbs. 45	Anvils, Chains, (in casks,) Hemp (at owner's risk), Iron Safes, Iron Railing, Lightning Rods, (in bundles), Nails, (in bags,, Oil, (in hhds. or barrels, except Lard and Rosin Oils), White Lead, Zinc, (in boxes,) Lead in pig or casks, (westward), Guano, Gunny bags, Paints [dry and in Oil], &c.	Per 100 lbs	Anchors, Alum, (in bbls or hhds.) Ashes, (pot or pearl), Coffee, Fish, (salted), Steel, Gas Pipes, Iron, (bar, pig, boiler, railroad, & scrap), Iron Nuts & Rivets, Tin, Molasses Tobacco, manuf. (except Cigars or Cut), Nails and Spikes, Nail Rods, Oysters and Clams, (in shell in bbls., at owner's risk), Queensware, Sugar, (in hhds., bbls. and boxes), Soda Ash, German Clay, Tar, Pitch and Rosin, Whiting, Zinc, (in casks), Rope, &c.	Per 100 lbs.

SPECIAL RATES.

Acids, Baskets, Carriages, (well boxed, at owner's risk), Cabinet Ware, (set up and boxed), Chairs, (boxed), Demijohns, Looking Glasses, Glass Plate and Show Cases, (except when released), Mattresses, Sleighs, (boxed), Wagons, (Childrens) not boxed, Willow-ware, Mouldings and Picture Frames, Trees and Shrubbery, (baled), at owner's risk, Toys, Wagons, (Childrens) and Hobby-horses, (boxed),

Furniture, (not boxed), only taken by special contract.

No single lot or package will be rated less than 100 lbs., First Class.

The conditions under this Bill of Lading are: That the following goods are to be forwarded by to Cin thence River Rail Road from Pittsburg, and are not to cost over the above rates per 100 pounds from Philadelphia to Jeffersonville Ind

This Receipt to be presented without alteration or erasure.

MARKS. J Foreman Son New albany Ind

CRISSY & MARKLEY, PRINTERS.

tion.[105] Within about two years most southern and western lines had adopted the Official Code, but other lines and traffic groups held out, notably the Transcontinental Freight Bureau. A uniform national rate system was never really effected. Even after the establishment of the Interstate Commerce Commission and passage of major federal legislation, such as the Hepburn Act, rate making continued to be handled on a regional basis. With the passage of the Staggers Act in 1980 the federal government largely withdrew from the railroad rate-making process in a dramatic policy turnabout.

Regionalism sometimes worked for the benefit of local shippers. Those located in the low-tariff Northeast benefited as a whole. Less developed regions could benefit as well. Special cheap rates were not pious acts of charity, to be sure, for the railroads were hoping to profit in the long run by temporarily aiding a struggling industry along its line. It was simply smart business. For example, Washington State was a natural place to raise apples, but the big market was on the other side of the continent. Local orchards could never hope to compete with growers in Virginia or New York unless railroads like the Northern Pacific were willing to grant especially low rates.

The story of the domestic raisin industry is a more striking example. Production began in California in 1876.[106] The local market was minuscule because of the state's small population, and long-distance shipping costs put the East Coast market out of reach. The Spanish exporters monopolized the trade until western railroads dropped to the absurdly low level of 0.0175 cent per 100 pounds. The eastbound raisin traffic began to swell as California grape producers geared up for larger produc-

GREATLY REDUCED. . . . . **ADOPTED NOVEMBER 1st, 1849.**

NAMES OF THE SEVERAL Depots, AND FREIGHT STATIONS, ON THE RAILROAD.	RATES in Cents per 100 lbs. for ARTICLES IN CLASSES 1. 2. 3. 4.	CLASS No. 1.	CLASS No. 2.	CLASS No. 3.	CLASS No. 4.	Special Rates.	NAMES OF THE SEVERAL Depots, AND FREIGHT STATIONS, ON THE RAILROAD.
Pier	8 6 4 3	ACIDS, in glass,	ALE, Porter in bottles	ASHES, pot & pearl	ASHES, leached		Pier
Piermont	10 8 5 5	BASKETS, loose	ALCOHOL, in wood	Do. dry house,	ANCHORS,		Piermont
Blauveltville	11 9 6 5	BLINDS, loose	BAGGING, cotton or tow,	ALE, BEER, in casks	APPLES, in bags,		Blauveltville
Clarkstown	12 9 6 6	BOOKS, b'x'd at owners' risk of chafing,	BASKETS, in nests	ANVILS, Axletrees,	BEANS, PEAS dry,		Clarkstown
Spring Valley	13 10 7 6	BOOTS and Shoes,	BATTING, cotton,	BACON, HAMS, in	BEANS, PEAS dry,		Spring Valley
Monsey	13 10 7 6	BROOMS, brushes,	BEEF, Mutton, fresh	BEANS, PEAS green	BEETS & sim'lr roots		Monsey
Suffern	11 11 7 7	BUFFALO Robes,	BERRIES, Bells,	BEEF, salted in bbls.	BOARDS & Plank		Suffern
Ramapo	15 12 8 7	CANDY, in boxes,	BOILERS for engines	BLINDS, in shooks,	BRAN and FEED,		Ramapo
Sloatsburgh	15 12 8 8	CARDS, for cott. n or Wool,	BRANDY, RUM, in casks,	BONES, Horns Hoofs	BRICK, TILES,		Sloatsburgh
Monroe Works	16 13 9 9	CARRIAGES, new	BROOM CORN, in bales,	BOXES Barrels en'ty	BUILDING STONE		Monroe Wks.
Wilkes	17 11 9 9	CARPETS,	BURLAPS or Duck,	CANDLES, Cheese,	CABBAGES and sim.		Wilkes
Turner's	18 11 10 9	CHINA ware,	BUTTER, Lard,	CHARCOAL,	CIDER, VINEGAR,		Turner's
Monroe	19 15 10 9	CLOCKS, boxed,		CHAIR STUFF, COPPER Sheets	CLAY, MARL, COALS, mineral		Monroe

*NOTE—The character † refers to the estimated weights or some other particular relative to the Article under the head of Conditions. The * refers to the Special Rates.*

FIGURE 1.28

Only the upper portion of this 1849 Erie tariff sheet is reproduced here, but it shows the variety of rates and classifications between various stations along the line. (New York Historical Society)

tion. By 1891 the Spaniards had been driven from the U.S. market. Once the situation was under control, the railroads felt free to raise their rates and did so with such a free hand that the toll rose from 0.0175 cent to $1.10 per hundredweight. Small shippers sending raisins by LCL were required to pony up $2.20 per hundredweight.

The idea of helping infant industries was an appealing one, particularly when the sponsor was paid off so well. The problem was that everyone wanted to be classified as an infant industry in danger of collapse unless special, cheap rates were offered. Standard Oil felt entitled to reduced tariffs via an elaborate scheme of rebates. Perhaps less extreme was a nail manufacturer in Pittsburgh who claimed that he could not possibly compete with a competitor in Williamsport who, being some 150 miles closer to New York City, had an unfair advantage because of lower shipping fees.[107] The Pittsburgher paid 50 cents per hundredweight, while the Williamsport maker paid just 38 cents. The railroad agreed to drop its rate to 38 cents to even out the competition. Such juggling of rates that amounted to special favors was a form of rebate that could make or break a manufacturer. By helping the Pittsburgh firm, the conspiring railroad may well have doomed the Williamsport business.

THE EVOLUTION OF A COMPLEX RATE SYSTEM

In the beginning, rate making was a simple business. Railroads at first were ready to copy the teamsters' elementary rules of just two classes of freight. One was heavy, the other was light. Heavy goods were charged so much per hundredweight, and light stuff was figured at so much per cubic foot. The prices quoted were normally not on a per-mile but on a per-destination basis; that is, a hundredweight of hardware might travel between Pittsburgh and Philadelphia for 60 cents. Simplicity soon gave way to complexity because of the pressure from shippers for better, or at least special, rates. At the same time, railroad managers must have seen this demand as an opening to juggle rates. A complex rate structure offers more opportunities to obscure, confuse, and so subtly inflate charges. The canals already saw this method of doing business as a good thing.

By the 1840s many railroads had settled on a rate schedule of four basic classes. First-class freight included high-price goods such as clothing, carpets, furs, and shoes and moved accordingly at the highest rates. The other categories were a more or less arbitrary division of lesser-value goods into second, third, and fourth classes. Just why anvils were listed in third class and anchors in fourth class seems inexplicable. The rate maker's designations were surely a matter of his best judgment. His sagacity and fairness were often tested for items not on the schedule. Baby buggies, for example, might be variously classified as vehicles, toys, or sporting goods.[108] The cost of transporting the buggy could be materially affected by which category was assigned.

Rates might be considered in more concrete terms by examining the published schedules issued by our pioneer rail carriers. Fortunately, some of these items dating from the 1840s through the 1880s have been preserved and a number are reproduced here. An even earlier notice, published in the 1835 *Planters and Merchants Almanac*, outlined charges on the South Carolina Railroad.[109] Back in an age of rate innocence there was absolutely no breakdown as to type or class of material being moved. All were equal. Prices were given from Charleston to each major station on the line up to the end terminal at Hamburg, South Carolina. Only the teamsters' old rule about light and heavy goods established any differentiation over what was shipped. The price between Charleston and Hamburg for light goods was 14 cents per cubic foot. The rate for heavy goods was figured at 50 cents per hundredweight. These fees included fire insurance and one week's storage. Rate structures rapidly became more complex in the Old South, as evidenced by a tariff issued in July 1847 by the Georgia Railroad.[110] First-class goods moved at cubic-foot rates, while most other material, even drugs and confectionary, were priced at a 100-pound rate. Special rates were given for molasses by the barrel or hogshead. Salt was priced by the bushel or the Liverpool sack. Plows, wheelbarrows, and other items of small farm machinery were carried at the same rate.

The system of four major classes was in evidence in the northern states by the 1840s as well. A small broadside is reproduced here as Figure 1.26. Its title—the Albany & Buffalo Railroad—is rather misleading, for the line's actual title was the Utica & Schenectady, which is given in smaller type below the engraving. This document is worthy of study not just for the specific prices it reveals but for the exceptions and exclusions it records. A willingness to compromise with shippers is evident from the large number of first-class items that are accepted at second-class tolls. Note that prices are given as per 1,000 rather than the customary 100-pound units. One wonders how successful the U&S was in disdaining liability save for locomotive sparks and negligent employees.

The Pennsylvania Railroad provided shippers with a condensed version of its four-level tariff schedule printed directly on each bill of lading (Fig. 1.27). Thus the common shipping document performed double duty. In this example, dated August 13, 1858, we can see that five chests of China tea were shipped from Philadelphia to Jeffersonville, Indiana. The rate was 75 cents per hundredweight, and it should be noted that part of the journey was made by riverboat. The roundabout nature of nineteenth-century transportation is evident from this shipment in that after traveling eastward from China it had to backtrack some 600 miles by rail and river. Six years before this bill of lading was issued, the Pennsylvania issued a summary notice of its freight rates, reproduced in Table 1.7. Note that winter and summer rates are given.

The New York & Erie tended to be more elaborate in its listing of specific cargoes, as is evident from just the upper portion of its omnibus tariff manifest of November 1, 1849 (Fig. 1.28). The basic four-class system was considerably supplemented by special rates for domesticated animals, flour, iron bars, and other products to which Erie rate masters gave separate consideration. A few years later the Erie seems to have broken away from the four-class system to a straight-out individual listing of goods with a price per ton or per piece.[111] The October 1852 printed broadside listed about two hundred items, from acid to well buckets. It covered only the branch line, the Union Railroad in New Jersey. In this document the Erie offered to move a cask of beer for 30 cents, a thousand bricks for 93 cents, a barrel of molasses for $1.87, and a well bucket for 5 cents. All prices are between New York City and Paterson, New Jersey, and they include cartage and ferriage.

The Erie may have established no immediate trend with its individual commodities listing, for there is evidence that many other major lines such as the Illinois Central and the Lake Shore continued to honor the old-fashioned quartet plan of rate

TABLE 1.7   Freight Rates of Pennsylvania Railroad

	FREIGHT Between Philadelphia & Pittsburgh	Winter Rates	Summer Rates
1st class	Dry goods, books & stationery, boots, shoes, hats & carpeting, furs & peltries, feathers, saddlery &c. }	$1—	75d p100 lbs
2nd class	Brown sheetings & shirtings in bales, queensware, glassware, groceries, except coffee, hardware, hollow ware, machinery, oilcloth, wool &c. }	85d	60d " "
3rd class	Butter in firkins & kegs, candles, cotton (in winter), tallow, tobacco in leaf or manufactured (Eastw.d) &c. }	75	50 " "
4th class	Bacon, cotton (in summer), coffee, lard & lard oil, through. Pork fresh in full carloads, at owners risk. }	65	40 " "

Office Penns.a R.R. Co. }
Philad.a Dec. 15, 1852 }

H. J. Lombaert
Superintendent
Altoona, Penn.
H. H. Houston
Gen. Freight Agent
Philadelphia

Source: Author's photocopy of an original in a private collection, current whereabouts unknown.

Fitchburg, Rutland & Saratoga Line's Classification.

Subject to ***Difference in Classifications*** adopted by Western Roads, and to the Government Tax.

Camphene, Burning Fluid and Varnish will only be taken at owner's risk by rail, at first class rates. Gunpowder, Friction Matches, and other combustible articles will not be received or transported over this road. Molasses will be received for transportation at the owner's risk of leakage. Coal Oil, Kerosene Oil, or Oil made in part from these oils, will not be shipped by Propellers on the Lakes.

FIRST CLASS.

Agricult'l Implements, (special contract)
Barrels of all kds. empty
Baskets, twice first class.
Bath Tubs
Batts
Bed Springs, twice 1st class
Billiard Tables, boxed, owner's risk
Bellows
Bird Cages, box'd twice 1st cl.
Boots and Shoes. Blinds
Brooms and Broom Brushes,
Broom Corn pres'd
Books, Buffalo Robes, once and half first class rates
Burning Fluid, owner's risk of leakage Bread
Cabinet Ware, set up, boxed, double 1st class rates. Same, knocked down & well box'd
Camphene, owner's risk of leakage
Caps, Cards, Carpeting
Carriages, well box'd, twice 1st class rates, owner's risk
Cassia, in mats
Castor Oil, in cases or cans
Chairs, matted or boxed, 3 times first class, owner's risk, released
China Ware, in boxes
Cigars, twice 1st class
Cigars, packed in boxes weighing not less than 100 lbs.
Clocks and Weights
Confectionery, Corks, Coffins
Cotton, in square bales
Cotton Waste, Cotton Yarn
Covers and Sieves
Deer Skins, in bales
Demijohns, at owner's risk, double first class
Domestic Sheeting, Shirting, Tickings, Denims, in bales or boxes
Drums, 4 times first class
DRUGS AND MEDICINES.
Dry Goods, in boxes and bales
Duck; Farm Wagons, in pieces
Feathers, twice 1st class
Figs, in drums, Fire Crackers
Fish, fresh, prepaid
Furniture, second hand, well bxd, accomp'd by passengers
Furniture, knock'd down, well boxed, first class rates
Furniture, matted, 2½ times first class rates, released
Furniture, set up and boxed, twice first class rates
Furs, in bales, twice 1st class
Garden Seeds, Glass Ware
Grapes, in kegs
Guns, Rifles and other Firearms Hats, Hemp Carpet
Hides, loose dry; Hair in sacks
Household Goods, (not Furniture), well boxed
India Rubber Goods
Indigo: Ink, in glass
Iron Castings, loose, under 100 lbs., owner's risk
Joiners' Work
Lead, Bar and Sheet
Lead Pipe, on reels or in roll
Leather, loose
Lemons, owners risk
Liquors in glass, at own's risk
Liquor, in wood
Looking Glasses, boxed, not over 3 ft in length, released
Looking Glasses, boxed, over 3 feet in length, twice 1st cl.
Machinery, unboxed
Marble, wroug't (owner's risk)
Mattrasses, twice first class
Mats and Rugs, Measures
Metallic Coffins
Mineral Water, in glass
Moss, in sacks, Mouldings
Musical Instruments
Nuts, in single sacks
Oils, in glass, owner's risk
Oranges, owner's risk
Oil Cl'th, bxs, 12 ft l'ng or over
Oysters, in kegs or cans, fresh
Oysters, in kegs and cans, pickled, not strapped.
Pickles and Preserves, in glass owner's risk,
Paper in boxes; Peltries
Piano-Fortes (owner's risk)
Paintings & Pictures, well bx'd contents of each pkg not to exceed $200 in value, twice
Palm Leaves [1st class
Paper Hanging, not boxed
Picture Frames [1st class
Pill Boxes, in csks or bxs twice
Plate and Looking Glass, box'd at owner's risk of breakage
Porcelain Ware, in bbls or bxs
Porter and Ale, in glass
Printing Presses
Printed Matter, in sheets, bx'd
Quicksilver, in iron flasks
Rattan, Refrigerators
Russia Bristles,
Retorts, clay, (owner's risk)
Scales and Scale Beams, not boxed; Scythe Snaths
Scythes, in bales
Sewing Machines, boxed
Sheep and other Skins, in bales
Shingle Machines, Sizing
Show Cases, 3 times 1st class
Sleighs, well bx'd, 3 times 1st cl
Snuff, in jars, Stationery
Steam Boilers, 30 feet and under; over 30 feet, 1½ 1st class
Stoves, at owner's risk
Stove Plate, at owner's risk
Stove Pipe, Sweet Potatoes
TEA. [bales
Tin Ware, box'd, Tobacco, in
Tobacco, cut, in boxes or bbls
Tools, mechanics'
Toys, boxed, 1½ first class
Trees and Shrubbery, boxed, at owner' risk
Trees and Shrubbery, baled, at owner's risk, 1½ first class
Traveling Bags, Trunks
Tubs, Twine, Umbrellas
Veneering not boxed [age
Varnish, at own's risk of leak-
Wadding
Wagons, children's, not box'd double first class
Wagons and Hobby Horses, boxed, 1½ first class
Wagons and Hobby Horses, knocked down, in boxes and
Wagon Felloes & Bows [crates
Wax, Whalebone, Whips,
Willow Ware, double 1st class
Wheelbarrows, [risk.
Wine, in bxs. or baskets, own.
Wood, in shape, manufactured
Wire Cloth, Wooden Ware
Wool, domestic, sacked, once and a half 1st class
Woolen Yarn
Yarn, Carpet, pressed in bales

SECOND CLASS.

Ammonical Water
Antimony, crude; Apples, dried
Baking Powd's. Berries. [gunny
Beesw'x; Bells; Bags, except
Bagging; Burlaps;
Blue Vitriol; Binders' Boards
Boiler Felting in rolls or bales
Boiler Flues, Copper and Brass
Borax in bbls. or bxs. Bed Cords
Bottles, Brass & Pewter Faucets
Brimstone in bxs or kgs; Butter
Caloric Engines: Candles
Capstans: Carpet Lining
Carriage Springs, Axles & Boxes
Cassia in bags or boxes
Cast Iron Grain Mills
Chair Stuff, in rough
Chain—cotton, woolen and hempen; Chains, loose
Cheese, in boxes and casks
China Ware, in casks; Chocolate
Clover and Grass Seed
Clove Stems, in sacks; Cocoa
Cocoa Matting; Coffee Mills
Coffee, ground, in boxes or bbls
Coffee Extract
Congress and Bedford Water, in bxs or bbls; Copper Bottoms
Copper & Brass Vessels, in bxs or casks; Copper, in bxs & casks
Copper, in plates, sheets, bolts, pigs, wire, nails and rods
Copying Presses
Cotton Waste, pressed in bales
Cream Tartar, in boxes or kegs
Crockery, boxes or barrels
Currants, dried; Cutlery; Dates
Dessicated Meats & Vegetables
Dye Woods, in bags or bbls
Emery; Extract Logwood
Figs, in casks or boxes
Flax, ~~boxed; Flax Seed~~
Forks, hay and manure
Gas Fixtures, boxed; Ginger
Glue; Grass and Clover Seed
Grass and Hempen Mats
Groceries, (assorted, not otherwise specified); Gum Copal
Gums, (not otherwise specified)
HARDWARE.
Hair, pressed; Hemp Machines
Hemp; Herring, in boxes
Hides, dry, in bale; Hinges
Hoes; Hollow Ware; Honey
Hooks; Hops Ink, in casks
Iron Castings, boxes or casks
Iron, hoop and sheet Isinglass
Iron Railings; Iron Safes
Lamp Black, in casks or bbls
Lead Pipe, in casks
Leather, in rolls or bxs; Linseed
Liquorice, in mass, in bxs or mats
Liquorice, in stick or root [risk]
Lithographic Stones, at owner's
Machinery, boxed
Madder, in bbls, boxes or kegs
Marbles, in casks and boxes
Marble slabs or blocks, under 4 in. thick, owner's risk
Milk, condensed
Moss, pressed, in bales
Mustard Seed, in bags or casks
Nails, in bags
Nuts, in d'ble sacks, bbls. & casks
Oakum, Oil-Cloth, (not otherwise specified) [strapped
Oysters, in kegs or cans, pickled
Paints, in boxes or cans, (not otherwise specified)
Pain Killer, in bbls; Prunes in
Palm Leaf, pressed [casks
Paper-Hangings, in boxes
Pasteboard; Peaches, dried
Pins, in original boxes; Pipes
Porcelain Ware, in casks or hhds
Plumbers' Materials, in boxes or casks; Printing Paper
Printers' Ink~~, in bbls or~~ kegs
Pepper & Spice in boxes;
Rags, in sacks; Raisins, strap'd
Rubber Belting
Rubber Car Springs, loose
Rubber Packing and Hose
Saddlery; Sand Paper
Scales and Scale Beams, boxed
Spirits Turpentine, in bbls. owners risk; Stove Blacking
Scythes, in boxes Sardines
Seeds, (not otherwise specified)
Shoe Pegs, in barrels
Shovels & Spades; Shot, in bags
Snuff, in casks, bbls or boxes
Soap, Castile and Fancy
Syrups (not molasses), and Coloring, in wood
Tin Foil, in boxes; Tow, boxed
Tacks in bxs; Type; Tile, Drain
Veneering, boxed Vices, iron
Weavers Syrup, in bbls.
Willow Reeds, in bundles
Wire Fencing Window Glass
Wire, (not otherwise specified)
Wood Screws, in casks or boxes
Wrapping Paper
Zinc, in rolls and sheets

THIRD CLASS.

Ale and Beer, in wood
Alum, in bbls and hhds
Anvils; Axes; Axle Grease
Barilla; Bark and Cob Mills
Black Lead, in bbls & boxes;
Bones; Barley, pearl;
Chain, in casks; Chickory
Cider, in bbls or hhds
Cream Tarter, in bbls or hhds
Crucibles
Dye Woods, in stick
Gas Pipe, wrought.
Gambia, in bgs or bale
Gum Shellac, original package
Gum Kowrie
Herring, in kegs Hides, green
Horse Nails, in boxes or bags
Honey, in casks or bbls
Hoofs and Horns;
Iron Shutters
Junk, Jute
Lightning Rods [in hhds
Millstones, finished; Madder,
Mahogany, and cedar, in board, plank, and scantling; Manilla
Nails, in boxes
OILS, in hhds and barrels
OYSTERS and CLAMS, shells, in bbls, at owner's risk
Potash, in boxes or kegs
Pepper and Spices, in bags
Pickles, in bbls or casks; Putty
Pickles, Preserves and Meats in cans, boxed
Pumice Stone, in casks;
Rags, pressed in bales [or rolls
Roofing Iron; Roofing, in boxs
Rubber Car Springs, in bxs & csks
Salts, Epsom, in bbls
Scythe Stones;
School Slates, boxed
Shoe Blacking, in bbls. or boxes
Saleratus in boxes
Soda, in boxes; Split Peas;
Starch in boxes; Sumach
Tallow. Terre Japonica
Tobacco, in boxes or kegs
Tobacco, in hhds, unmanufact'd
Vinegar
Volute Car Springs, boxed
Water Pipe, wrought
Wool, for'n, press'd; Wire Rope

FOURTH CLASS.

Antimony, metal; Anchors
Barytes
Brimstone, in bbls and hhds
Bleaching Salts; Burr Blocks
Bath Brick;
Cements; Chalk
Coffee, in double sacks
Coffee, single sack, owner's risk
Car Wheels and Axles; Clay
Crockery, Earthen and Stone Ware, in crates and hhds., owner's risk
Chain Cables
Copperas, in bbls or boxes
Earth Paints
Fence Wire; Floor Tiling
Fish, pickled and dry salted
GUNNY BAGS, in bales
Grindstones; Guano
Horse Shoes, in packages
Iron Nuts, Bolts, Washers & Rivets in bxs or casks
Iron—pig, bar, band and boiler
Iron and Coal Facings
Iron Castings, plain, (not machinery,) over 100 lbs. each piece
Lead, in casks or pig
Lime, in casks; Locomotive Tyres; Mahogany Logs
Marble slabs or blocks, 4 inches thick and upw'd, owner's risk
Millstones, in rough
Molasses, owner's risk leakage
Nails and Spikes, in kegs
Nail Rods
Paints, in barrels or casks
Potash, in barrels and casks
Plaster; Pitch
Railroad Chairs and Spikes
Railroad Iron
Rice, Rigging, Rope, Rosin
Saleratus, in kegs
Saltpetre Salt
Salt Cake, Stone, unwrought
Soda Ash
Soda, in kegs
Soap, common
Spelter, in slabs or casks
Steel, in boxesor bundle
Sugar
Sugar, in double sacks
Sulphur, in bbls.
Shot, in kegs
Telegraph Wire: Tar [casks
Tin, pig or plate, in boxes or
White Lead and Zinc Paints
Whiting;
Water Pipes, cast iron
Zinc, (sheet,) in casks or case

SPECIAL CLASS.—Manufactured Marble.—Carriages, not boxed.

CONDITIONS AND RULES.

The destination, name of the consignee, and weight of all articles of Freight, must be plainly and distinctly marked, or no responsibility will be taken for their miscarriage or loss; and when designed to be forwarded, after transportation on the route, a written order must be given, with the particular line of conveyance marked on the goods, if any such be preferred or desired.

The Companies will not hold themselves liable for the safe carriage or custody of any articles of freight, unless receipted for by an authorized agent; and no agent of the Line is authorized to receive, or agree to transport, any freight, which is not thus receipted for.

No responsibility will be admitted, under any circumstances, to a greater amount upon any single article of freight than $200, unless upon notice given of such amount, and a special agreement therefor. Specie, drafts, bank bills, and other articles of great intrinsic or representative value, will only be taken upon a representation of their value, and by a special agreement assented to by the Superintendent of the receiving road.

The Companies will not hold themselves liable at all for injuries to any articles of freight during the course of transportation, arising from the weather, or accidental delays, or natural tendency to decay. Nor will their guarantee of special despatch cover cases of unavoidable or extraordinary casualties or storms, or delays occasioned by low water and ice; and may be stored at the risk and expense of the owner. Nor will they hold themselves liable, as COMMON CARRIERS, for such articles, after their arrival at their place of destination at the Company's warehouses or depots.

Carriages and Sleighs, Eggs, Furniture, Looking Glasses, Glass and Crockery Ware, Machinery, Mineral Acids, Piano Fortes, Stoves and Castings, Sweet Potatoes, Wrought Marble, all Liquids put up in glass or earthen ware, Fruit, and Live Animals, will only be taken at the owner's risk of fracture or injury during the course of transportation, loading and unloading, unless specially agreed to the contrary.

Gunpowder, Friction Matches, and like combustibles, will not be received on any terms; and all persons procuring the reception of such freight by fraud or concealment, will be held responsible for any damage which may arise from it while in the custody of the Company.

It is further stipulated and agreed, that goods shipped to points west of Chicago or Milwaukee, shall be subject to a change in classification and corresponding change of rates beyond those points, and Government Tax.

Cases or packages of boots and shoes, and of other articles liable to peculation or fraudulent abstraction, must be strapped with iron or wood, or otherwise securely protected, or the companies will not be liable for diminution of the original contents, and the companies will hold the freighter, in all cases, to bear the loss arising from improper packing.

It is also agreed between the parties that the said companies, and the railroads and steamboats with which they connect, shall not be held accountable for any deficiency in packages if receipted for to them in good order.

All articles of freight arriving at their places of destination, must be taken away within twenty-four hours after being unladen from the cars—each Company preserving the right of charging storage on the same, or placing the same in store at the risk and expense of the owner, if they see fit, after lapse of that time.

FIGURE 1.29 (opposite page)

By 1869 the breakdown of tariffs had become more refined and specific. Note that foodstuffs traveled at different rates, depending on packaging. (Collection of Richard F. Dole)

FIGURE 1.30

A sample page from the Lake Shore's 1881 published tariff. (Dibner Library, Smithsonian Institution)

making. But even those remaining loyal to the four-class system became increasingly careful to provide an even longer enumeration of exactly what specific goods fell into what specific class. An example of this more fastidious approach is shown by the Fitchburg's Fast Freight Line listing that appeared on the reverse of a bill of lading dated January 26, 1869 (Fig. 1.29). A perusal of this dense document set in very fine print—a lawyer's delight—will reveal many disclaimers whereby things like stoves, billiard tables, and almost every liquid in a glass container went at the owner's risk. The printed tariffs of the 1860s tended to list more exceptions and special rates than earlier tabulations. A local tariff printed for the Illinois Central in April 1864 shows no fewer than fifteen special

rates.[112] These included a promise to return empty bags to grain and potato shippers, a rate of 35 cents per mile to move new or used locomotives, and a flat rate of $8 per carload of agricultural implements when moved 15 miles or less from Chicago. Otherwise the rate was 20 cents per mile.

Perhaps the policy of charging one and one-half or even double first-class rates for some goods had a greater effect on shippers than the special rates alluded to. The formalities of rate making had advanced some distance beyond the simple handbill tariffs published in the infancy of railroading when the Lake Shore issued a thirty-nine-page book for local rates effective April 4, 1881. This small folio volume admonished shippers to label their goods clearly and not to use chalk. It promised to for-

Lake Shore & Michigan Southern Railway.

LOCAL FREIGHT CLASSIFICATION.

A.

Acetate of Lime, L C L	3
Same, C L	D.
Acids, C R	D 1
Same, O R, L C L	1
Same, O R, C L	B.
Agricultural Implements, owner's risk of chafing and breakage, C L	A.
Any shipment of Agricultural Implements, (except Threshers, Separators, or Clover Hullers,) requiring an entire car to transport it, will be charged the full car load rate.	
Same, less than car loads, owner's risk of chafing or breakage, unless otherwise enumerated, as follows, viz:	
Clover Hullers, Threshers or Separators, when shipped singly, and without power, will be charged half car rate.	
Fanning Mills, Sulky Horse Rakes, Cultivators, (wood,) Corn Planters, Shovel, Gang and Sulky Plows, and similar light and bulky machines, set up	D 1
Fanning Mills, Sulky Horse Rakes, Cultivators, (wood,) Corn Planters, Shovel, Gang and Sulky Plows K D, flat, and tied in bundles	1
Grain Cradles, set up, actual weight	3-1
Grain Cradles, K D, in boxes or bundles	1
Wooden Horse Rakes, Mowers, Reapers, Harvesting Machines, Corn Shellers, Cultivators, (iron,) Plows, Feed Cutters, Harrows, Seed Drills, Grain Drills, and Shearing Machines	1
Plows, K D, boxed or in bundles	3
Harrows, K D, boxed, in bundles or in pieces	2
Harrow Frames, without teeth, K D, in bundles	3
Cradle Fingers	1½
Same, K D, boxed or in bundles	1
Alabaster, manufactured, O R	2
Alabastine, L C L	4
Same, C L	D.
Alcohol, in cans, packed in boxes or kegs	1
Alcohol, Highwines, Spirits and Whiskey, in wood, O R L	2
Alcohol, Highwines, Spirits and Whiskey, in wood, O. R. L. value not to exceed $20 per bbl., in case of loss or damage	4
Ale, Beer and Porter, in glass or stone, well packed, in boxes or bbls., O R of breakage or freezing	3
Ale, Beer and Porter, in wood, O R of freezing, leakage or fermentation, L C L	3
Same, in wood, O R of freezing, leakage or fermentation, C L	4

Allspice	1
Almanacs, in bdls. prepaid or guaranteed	1
Same, in boxes, prepaid or guaranteed	2
Almonds, in sacks	1
Same, in boxes, bbls or casks	2
Alum, in boxes, kegs or bags	3
Same, in hhds. or bbls	4
Ammonia, or Ammoniacal Water, in carboys, O R	1½
Ammonia, Carbonate of, in jars, O R	1
Same, in kegs	2
Anchors	4
Animal Pokes	1
Antimony, crude	3
Anvils	4
Apples, green, in bags, boxes or baskets, O R, prepaid	1
Apples, green, in barrels, prepaid or guaranteed, at owner's risk of freezing or decay, as follows, viz:	
Less than 20 bbls	2
Over 20 bbls. and less than C L	3
Car loads of 140 bbls. or over	4
Apples, cider, in bulk, C L, only taken by special contract ; enquire of General Freight Agent.	
Apples, dried, L C L	3
Same, C L	4
Apples, Canned. Same as Canned Goods.	
Apples and Cider, in bbls., when in full car loads, mixed	4
Apple and Fruit Butter, Sauce or Jelly, in glass or stone, well packed, O R	2
Apple and Fruit Butter, Sauce or Jelly, in wood or in cans, L C L	3
Same, C L	4
Apple Parers	2
Arrow Root, in boxes	1
Argols	4
Arsenic, crude, in kegs or bbls	3
Ash, pot, pearl or soda, in cans	3
Same, in bbls. or casks, L C L	4
Same, in bbls. or casks, C L	C.
Ashes, wood, C L	D.
Ash Boilers and Kettles. Same as Kettles	
Asphaltum, L C L	4
Same, C L	C.
Axes	3
Axle Grease	3
Axles, Car and Locomotive	4
Same, Carriage	4
Same, Wagon	4

B.

Baby Jumpers, boxed	1
Bacon, loose or in bags, L C L	2
Same, C L	3
Same, in boxes, bbls or hhds	4
Bags, except gunny, in bundles	2
Same, except gunny, in bales	3
Bags, gunny, in bales or bundles	4
Bagging, except gunny	3
Same, gunny	4
Bags, empty, used in transporting lime, plaster, grain or potatoes over the road, when returned to point of shipment	4
Baking Powders, less than 50 boxes	2
Same, 50 boxes and over	3
Band Boxes, in bdls., O R	3-1

Same, boxed	1½
Banister and Stair Rails, in bundles, L C L	2½
Same, boxed, L C L	3
Same, finished, C L	B.
Same, in rough, C L	C.
Barilla	3
Bark, Extract, for tanners, in bbls	4
Bark, Tan, C L	D.
Bark, Slippery Elm, in boxes or bales	1
Bark, ground, in bags or casks	4
Bark and Cob Mills, L C L	3
Same, C L	A.
Barley, Pearl	2
Barley, Common. Same as Grain.	
Barrel Covers, wooden, in pkgs	3

PARADE OF THE CARS

ward all goods with reasonable dispatch but would not agree to send goods on any particular train or at any specific time. When it came to listing individual items and rates, few could hope to better the Lake Shore, which got down to such details as babies' jumpers and slippery-elm bark. Two pages from this document are reproduced in Figures 1.30 and 1.31.

The complexity and size of the classification code grew decade by decade. In the 1840s the list might easily have been recorded on a single letter-size sheet. Twenty years later this was still possible, but only if very small type was employed. At the beginning of the twentieth century the list of some eight thousand items required a small book. By 1919 the Consolidated Freight Classification had swollen into a fat five-hundred-page volume listing fifteen thousand commodities, from abrasives to zirconium.[113] In 1956, when tariff regulation was in full flower, the Interstate Commerce Commission expanded the list to some seventy-five thousand items, and the carriers wrangled over the correct rate for such obscure commodities as yak fat.[114]

Just how an item was rated could materially affect the cost to the shipper. In the old four-class system the difference between first and fourth class was nearly 100 percent. Feather dusters, for example, were listed as first-class goods and so traveled at 44 cents per hundredweight, while steel beams were fourth-class items and so could be shipped for only 23 cents a hundred.[115] Because

The Lake Shore & Michigan Southern Railway Co.

CLASSIFICATION.

EXPLANATION OF TERMS USED:

THE NUMBER OF THE CLASS IS GIVEN OPPOSITE EACH ARTICLE, THUS:

1.	Stands for	First Class.
2.	" "	Second Class.
3.	" "	Third Class.
4.	" "	Fourth Class.
1½.	" "	Once and a half First Class.
D 1.	" "	Twice First Class.
3-1.	" "	Three Times First Class.
A.	" "	Class A, Special Car Load Rate.
B.	" "	Class B, " " " "
C.	" "	Class C, " " " "
D.	" "	Class D, " " " "
S. R.	" "	Special Rate column in Tariff.
O. R.	" "	Owners Risk.
O.R.B.	" "	Owners Risk of Breakage.
O.R.L.	" "	Owners Risk of Leakage.
K. D.	" "	Knocked down.

The rate on articles in car loads, classified in Classes A, B, C and D, is given in the Tariff for car loads of not exceeding Twenty Thousand pounds.

When a class rate is given on any article in car loads, the minimum weight to be charged for, will be Twenty Thousand pounds.

The Classification given for Boilers, Lumber, &c., in car loads, provides for such articles at a maximum length of 28 feet, that being the length of an ordinary car. When longer cars can be obtained the same Classification will apply, providing the length of the article to be shipped does not exceed the length of the car. When more than one car is required for the transportation of such articles, you will charge car load rate for each car required.

All excess above Twenty Thousand pounds will be charged at a proportionate rate per hundred pounds, until it reaches the limit that may be fixed for the weight to be loaded in cars. For all weight above the limit, *double the regular rate* will be charged.

☞ Articles not enumerated, will be charged the same as similar or analagous articles.

TABLE 1.8 Tariffs for Selected Commodities
(per hundredweight, New York to Chicago)

Year	Stoves	Coffee	Sugar	Soap	Grain
1867	$1.90	$1.62	$0.83	$1.29	$0.75
1880	0.40	0.40	0.40	0.40	0.40
1893	0.25	0.25	0.25	0.26	0.25

Source: Cassier's Magazine, June 1894, p. 114.

TABLE 1.9 Class Rates (per hundredweight,
New York to Chicago)

Year	1st Class	2d Class	3d Class	4th Class
1877	$1.00	$0.75	$0.60	$0.45
1887	0.75	0.65	0.50	0.35

Source: Slason Thompson, A Short History of American Railways (New York, 1925), p. 243.

every shipper was seeking a bargain, freight roads sought ways to accommodate their customers. They were aided in such maneuvers by the size and complexity of the classification book. Some railroads would attempt to hide cut rates under obscure headings. If a cannery was looking for a good rate, the agent knew to look under "Peas and other canned goods" rather than under the general listing of canned goods. These ruses were intentional and were meant to confuse.[116]

There were many things wrong with the railroad rate structure. It was complex, it was irregular, and it was discriminatory. But there was one thing very good about it: rates were cheap, and they were getting cheaper as time went on. George Rogers Taylor estimates that the cost of shipping was reduced by 95 percent between 1820 and 1860.[117] Some of this decline is accountable to the general decline in prices, but most of it is the result of real reductions in shipping costs due to more efficient methods of overland transportation. Some years before Taylor's study was published, Arthur T. Hadley estimated that freight rates had dropped by around 50 percent between 1850 and 1880 and that they would continue to decline as the railroad network expanded and more productive methods were adopted to move an ever-increasing tonnage.[118] At the time of Hadley's statement (1885) the United States was benefiting from the lowest freight rates in the world—around 1.125 cents per ton-mile average. By 1901–1902 the average had fallen to just 0.75 cent per ton-mile, while the railway systems of Britain, France, and Russia charged an average fee of over 2 cents per ton-mile.[119] Only Germany managed to approach the U.S. level, with rates averaging 1.6 cents per ton-mile.

It was not always so. America had once been a land of very high priced transportation. During the early years of the Republic wagon masters charged 30 to 70 cents per ton-mile.[120] Merchants could afford to ship only the most valuable goods any dis-

tance. The transit of cheap, bulky goods was most circumscribed. Teamster rates had declined by the early 1820s because of better roads and a general deflation that drove all prices down. When railroads were introduced in the 1830s, wagon rates had crept back up to around 20 cents per ton-mile. Shippers were naturally delighted when steam cars were able to offer rates of just 6 cents per ton-mile. Prices varied, as might be expected. The Mohawk & Hudson, which carried very little freight in its earliest years of operation, charged 8 cents per ton-mile, while the Philadelphia & Columbia offered exceptionally low rates during the 1830s that came to an average of only 2 cents per ton-mile.[121] The Austrian engineer F. A. Ritter Von Gerstner reported in 1840 that railroad rates in this country averaged 7.5 cents per ton-mile.[122] By midcentury 4 cents per ton-mile seemed more normal, and by the eve of the Civil War this figure had declined to just over 2 cents. In 1860 bulky goods such as wheat traveled for as little as 1.5 cents per ton-mile.[123] Yet astonishing as this last-named rate may seem for the time, the Erie Canal offered to carry grain for only 1 cent per ton-mile.

Greater bargains were to come in the decades following the Civil War. The sampling of tariffs given in Table 1.8 illustrates the decline of rates between the late 1860s and early 1890s.[124] Another comparison for roughly the same time period, given in Table 1.9, shows the decline of the basic class rates.[125] The reductions reflected here do not take into account the many objects reclassified downward, so that shippers realized greater savings than before. By 1887 ton-mileage charges had slid to 1 cent. The fall in freight rates was more rapid than the general decline in prices because of the hard-currency policies of the federal government.

The benefits to the average citizen were considerable. The rail transport costs for a year's supply of meat and grain equaled only about one day's wages, or $1.25.[126] Reductions in the cost of transporting other necessities of daily life from coal to clothing all helped improve the standard of living for ordinary people. Just how much you saved depended on where you lived. Rates were cheapest in the East, where the average rate was a full 50 percent below the national average.[127] A big volume of traffic and considerable competition explained this difference. The South and West endured the highest rates, while the biggest price gougers of all were the Denver & Rio Grande and the Southern Pacific, which billed their customers three to eight times the rates charged by the eastern trunk lines.[128] The Rio Grande got away with high rates because it had almost no competition.

The decline in rates meant that the profit per unit carried would decline unless operating costs could be reduced in a like manner. Larger cars, longer trains, and faster schedules helped reduce operating costs to a degree, but the real salvation of freight profits lay in increasing volume. If rates fell by 400 percent and operating costs remained relatively constant, then the railroad must boost its tonnage, say by 500 percent over a given time period. From the tonnage figures given earlier in this chapter it can be seen that such a result was easily achieved by most roads. In 1855, even before large-scale freight operations had fully taken hold, roads such as the Erie realized 51 percent of their net profit from freight operations.[129]

Figures available from the Pennsylvania, reproduced in Table 1.10, tell a more complete story of rates and profits. The Pennsylvania managed well for a trunk line with a mixed traffic, many branch lines, considerable perishable car loadings, and too many empty-car miles. The New York Central suffered from the same problems and in the late 1890s reported earnings of only $1.84 per train-mile.[130] Roads with rather light traffic like the Great Northern earned a more respectable $2.73 per train-mile during the same period. Coal roads are usually thought to be prosperous, but the Chesapeake & Ohio brought in only $1.38 per train-mile, one of the lowest earnings figures in the industry, presumably on account of its one-way traffic. The ideal operation for profitable railroading was that of the Bessemer & Lake Erie, which earned $5.38 per train-mile. This road enjoyed a splendid two-way traffic, hauling coal north and iron ore south. Most of its trains ran through from one end of the line to the other with no setouts or pickups. Empty cars, costly yards, and branches were all but unknown. Profits were guaranteed on such a railroad even with low-tariff traffic.

The exact level of rates was established more by competitive factors than by any scientific method. The cost of service or a fair return on investment should have been the basis for establishing rates, but such a rational approach was normally overruled by a more arbitrary judgment based on what the shipper might be willing to pay or what rate was being offered by a competing line. Even when an honest effort was made to perfect realistic rates, experts like Octave Chanute were ready to admit that cost data was so variable and imprecise that finding a scientific base for rate making was almost impossible.[131] Most of the cost figures given in annual reports and texts of the time were often little more than broad, general estimates.

Rate making might be supposed as the exclusive work of a gallery for dreary clerks sitting on high stools hunched over fat volumes of classification books. And so it was on one level, but the very highest officials also took a hand in the fussy details of the rate-making process. Superintendents, general managers, treasurers, presidents—all had their say, but there were even higher authorities who would speak to the subject of just how many

TABLE 1.10 Pennsylvania Railroad, Rates and Profits per Ton-Mile

Year	Rate (cents)	Cost (cents)	Profit (cents)	Ton-Miles (billions)
1855	3.0	1.9	1.0	0.1
1865	2.2	1.4	0.7	0.4
1875	1.3	0.7	0.5	1.4
1884	0.7	0.4	0.3	3.0

Source: Railroad Gazette, Dec. 18, 1885, p. 807.

mills per ton-mile would be added or deleted for carriage of stove bolts or cotton waste. Members of the board would from time to time intrude into such details of the railroad's operations. Surely it was normally done as a matter of broad policy, low rates versus high rates, rather than the actual picking over of the precise rate for one traffic item. It was sometimes done because of a concern for the overall good of the territory served or, conversely, for some direct gain to the director himself.

The board of the Western Railroad (Mass.) was bedeviled for many years over the desirability of high or low rates for through traffic.[132] Part of the board pushed for low rates to rebuild traffic for the port of Boston. Cheap rates would pull traffic away from New York, but the railroad's managers argued that the railroad could not afford lower rates. Income was needed to service the line's great debts and to provide dividends so that shareholders would feel they were receiving a fair return on their investment. The trick was to compromise, with rates just high enough to produce necessary revenues but not so high as to drive customers away. This debate was a longstanding one, and in the end the compromise faction won.

The Baltimore & Ohio's board was involved in a similar squabble during the 1850s.[133] Those directors representing the state of Maryland and the city of Baltimore, both large shareholders in the corporation, pushed for low rates to encourage industry and commerce in the area. Cheap tariffs would also attract wheat, cattle, and coal to the port of Baltimore for export. These public officials found a few supporters, oddly enough, among certain other directors who happened to be major capitalists. The B&O was but one of their investments. Most of their money was tied up in other enterprises that would benefit directly from cheap rates, and because these operations were potentially more profitable than the railroad could ever hope to be, they became champions for low tariffs. The railroad's management, and the other board members loyal to them, argued against giveaway rates because of the cash flow needs of the railroad. All agreed, however, on a concession for the coal trade to encourage the new mines in the region of Cumberland, Maryland. By the time of the Civil War the B&O management had beaten down the call for low rates and instituted a policy of high rates to discourage traffic. The road was overwhelmed by demands for service and could not handle what was being offered. Higher rates both dampened the demand and brought in extra revenue to offset losses from wartime damage and interruptions to service.

Low-level rate policy was occasionally formulated by individuals who were both railroad directors and large-scale shippers. The Milwaukee Road's board was in fact top-heavy with big industrialists during the late nineteenth century.[134] William Rockefeller, a brother and partner of John D., and Henry M. Flagler, another chief executive of Standard Oil, were board members. Standard Oil was hardly bashful when it came to asking for rebates. Another fellow director was the beef baron Philip D. Armour, who was not only a big-time shipper but also the owner of one of the largest private freight car fleets in the nation. His direct interest in favorable rates was obvious. The opportunity for conflict of interest seems clear enough as well.

RATE WARS, POOLING, AND REBATES

Rates were normally established at a lower level of management than the board of directors, and most of the fluctuations in rates were modest adjustments of a few mills per hundredweight. But from time to time, particularly during recessions when traffic was down and competitive fevers were rising, savage battles called rate wars broke out. The guiding principle seemed to be business at any price if it would cover marginal costs, rather than no business at all.[135] Let us suppose that several railroads have reduced their rates to the lowest level sufficient to cover expenses, at 25 cents per ton. One of their rivals drops his rate to 20 cents per ton and so captures most of the tonnage. The competitors must do the same. A few will undercut the 20-cent rate and so on until rates are well below the average cost of doing business. But a failure to do so will mean no traffic, and hence no income whatever to pay interest and other costs. And so this ruinous course of rate cutting would be pursued until all the parties were exhausted.

The wars were generally short, perhaps a few months in duration. They almost never settled the question of who would prevail in any given territory or area of traffic. The peace that followed was temporary and would likely last only so long as times were bountiful and each railroad had all the traffic it could handle. James F. Joy, president of the Michigan Central, remarked in 1872 that the rate war of the previous year had destroyed all hopes for profits and seriously affected the value of railroad property.[136] He felt that these wars were chronic and that there was little hope for wisdom or long-term peace. Suspicion and distrust led to a breakdown of peace, and the cycle started all over again.

This was especially true when railroads entered new markets.

Rate wars are often thought to be a phenomenon starting in the 1870s, but there is evidence that they existed some years earlier. The Madison & Indianapolis, for example, complained about ruined profits because of competition from other lines in the mid-1850s.[137] Profits had slipped from 6.1 percent in 1849 to 3.3 percent in 1855 because its monopoly had been broken by competing lines. Ohio's Little Miami Railroad complained in its annual report for 1856 that the reduced prices of its competitors had pushed income below the point of a just remuneration.

But these early rate competitions were mere skirmishes compared with the battles ahead. Real warfare broke out after the big trunk lines became firmly established between New York and Chicago. Most of these battles were over the mammoth grain trade. In 1869 the New York Central and the Pennsylvania got into a duel over freight rates that pushed first-class charges from $1.88 per hundredweight down to 25 cents for goods traveling between New York and Chicago. Fourth class fell from 82 cents to 25 cents as well. Sanity returned, and rates rose back to between $1.00 and $1.50 for first class and between 60 and 80 cents for fourth class. Stability reigned between 1870 and 1874, but the war flared up again in 1876, pushing first class back down to 25 cents and fourth class to 16 cents. Peace returned in 1877, and then in 1881 a furious rate battle opened that went on for eight months. The pattern of war and peace was now well understood, and to no one's surprise hostile action resumed in 1884. These epic struggles for survival produced no clear victor, and a reengagement could always be expected.

In 1870 the New York Central and the Erie got into a duel over western cattle rates.[138] The normal carload rate between Buffalo and New York was $160. This was dropped to $140, and then as the fight grew more heated it went down to just $40, or well below the actual cost of service. At this point Cornelius Vanderbilt, president of the New York Central, lost his temper and dropped the price to a single dollar. At this point officials of the Erie seized the opportunity to make a fast profit in the cattle business and shipped animals at a furious pace over Vanderbilt lines. The crusty old capitalist was beaten, but in an unexpected manner. The Baltimore & Ohio's extension to Chicago in 1874 only intensified the rivalry among the eastern trunk lines. Grain wars were said to recur like epidemics of smallpox. At the time of the B&O's entry into the Chicago market, the normal fourth class rate was 45 cents per hundredweight.[139] It soon dropped to 30 cents. In another rate battle a few years later it went down to 20 cents. By the middle 1880s it was down to just 10 cents. Shippers and public benefited while the railroads and their shareholders lamented over dividends lost and profits never likely to be regained.

Rate wars could take unexpected turns. Big roads that tried to use them to punish a smaller rival often regretted having started the fight. In 1884 the New York Central started a rate war with the newly opened New York, West Shore & Buffalo, which paralleled its Hudson River line on the opposite bank.[140] The West Shore was built as a blackmail line by enemies of the Vanderbilt system and was leased in 1886 by its intended victim. At the time of the 1884 rate war the West Shore was already in bankruptcy, and the losses it sustained by moving freight below cost meant rather little, since the corporation was no longer accountable for its debts. The Central, on the other hand, was a very solvent company. It also had a far larger volume of tonnage; hence its losses were calculated at four to five times those of the West Shore. Rate wars could backfire even when handled by the most successful of railroad managers.

The antidote to rate wars was pooling. The purpose of each pool was to divide traffic and revenues through cooperation of the competing parties. Slicing up the traffic pie guaranteed everyone a piece and averted unseemly fights to grab the whole pie. The division was not always an equal one, for the stronger roads normally received the larger slices. While pooling seemed to be a sensible cure for the dread rate war malady, it proved a temporary remedy at best. A dip in the economy almost always resulted in a traffic decline, and pool members felt compelled to go after whatever business was left. The pool agreement dissolved in the frenzy of a rate war, and then after all the combatants were exhausted, peace was restored by forming a new pool. And so the cycle kept repeating itself.

Pools were regional associations or federations formed to stabilize and regulate the income among several railroads for through traffic. Local traffic was not the concern of pools. Local traffic was not competitive, and each individual line was free to handle it in its own best interest. Through traffic on the other hand, was often very competitive, and so there was a need to soothe the combative spirit that naturally developed. Cooperation, sharing, and a peaceful division of the spoils were the goals of any good pool. After all, no right-thinking capitalist really wanted free and open competition.

Because the pooling agreements amounted to a restraint of trade and so were illegal, these alliances were largely secretive and informal. No one could go to court and sue any member of the pool for failing to abide by his illicit agreement. The unenforceability of pooling was the root of its shifting and temporary nature. Advocates of pooling like to emphasize its public benefits, such as reliable service because of more predictable income. Pools removed the incentives for secret rate cutting and so lessened discrimination. These claims were true to a degree, of course, but such ends could surely be achieved through nobler means. Thinking along these lines led, rightly or wrongly, to state and then federal regulation of rate making. But more on this topic toward the end of our discussion.

As in any good scam, several types of pools evolved. The most popular was the money or cash pool. Typically, each railroad retained only 50 percent of its nonlocal revenues—the surplus must be paid into the pool. These revenues were kept in a common pot to be split up periodically among the members of the alliance. Assuming an equal division and four members, each railroad would then receive 25 percent of the pool. In some pools the more dominant members would demand a larger percentage, and because might meant right, they generally got their way.

Money pools were popular because they were so easy to manage, especially when it came to dividing up the profits. This proved more difficult in the tonnage pool scheme. These were created on the premise of the existing share of the market. The Trunk Line Association formed in 1877 prorated the westbound tonnage from New York to Chicago among the pool members in this fashion: New York Central, 33 percent; Erie, 33 percent; Pennsylvania, 25 percent; and Baltimore & Ohio, 9 percent.[141] Territorial pools tended to be even more vague when it came to the question of revenue sharing. During the late 1870s the Lake Shore and the Pittsburgh & Lake Erie agreed to keep out of each other's territory by building no parallel or incursive lines across an east-west line running 70 miles north of Pittsburgh.[142] At the very time of this agreement the Burlington & Missouri River and the Union Pacific picked the Platte River as their common border.[143] The B&MR would develop lines to the south of the Platte, while the UP was free to do the same north of this shallow 1,600-mile-long stream.

Some railroads protected their territory by building long skirmish lines of branches to discourage poachers from entering their preserves. Secondary parallel lines served the same purpose, as did the consolidation or buyout of smaller independent lines. These tactics were outside the realm of pooling and were possible only for roads with ample capital. The American Railroad Network grew to its excessive size in part because of pooling incentives to build parallel or duplicative lines. Territorial pools were sometimes concocted by rival capitalists. C. P. Huntington and Jay Gould sat down in November 1881 to divide up Texas.[144] Huntington represented the Southern Pacific and Gould the Texas & Pacific. Their written agreement, kept secret until 1893, was a complex document much too lengthy to summarize here, but let us consider a few of its terms. The income of most lines was divided equally, but the T&P would receive only one-third of the Galveston line's earnings; however, Gould was to receive trackage rights into New Orleans.

A few lines attempted to resolve their competitive urges through the creation of traffic pools wherein specific commodities were divided among the participants. In 1882, for example, the Wabash, the Chicago & Alton, the Vandalia, and the Ohio & Mississippi became involved in a dispute over the exact apportionment of the cattle trade moving out of St. Louis.[145] All the members were given 22 percent of the trade except for the O&M, which being the weakest link received only 12 percent. Nevertheless, the Vandalia demanded and got a bigger chunk of the cattle traffic. Its share rose to 26 percent.

About seven years earlier yet another scheme for traffic pooling had been organized by a major eastern shipper.[146] Yes, curiously, some pools were organized by shippers rather than railroads. The shipper became what was known in the trade as an evener; he agreed to persuade the other big cattle dealers to give a reasonable percentage of their business to each of the major trunk lines. The railroads in return would establish a uniform and fair rate that would prove at least moderately profitable, but, of more importance, they would be guaranteed a steady traffic. Not only did the shippers benefit from uniform rates, but each was entitled to a drawback of $15 per car. A more effective traffic pool was worked out in 1872 by the anthracite coal roads. These lines controlled not only rates and the avenues of traffic but the production of the coal itself.[147] This was restraint of trade on a grand scale, and it lasted until 1876. It was renewed in 1878 and stayed alive on and off for some years afterward. The antitrust suit was not taken up until 1920.

Pooling in all its variations was unnecessary so long as American railroads remained small regional carriers concerned solely with local traffic. But by 1854 some lines had developed into lengthy roads reaching a third of the way across the continent through their western connections. The disruptive competition developing over through traffic became obvious to leaders such as J. E. Thomson.[148] Thomson convened a meeting at the St. Nicholas Hotel in Manhattan with the presidents of the Baltimore & Ohio, the New York Central, the Erie, and the Grand Trunk. Some fifteen traffic agreements were signed between 1857 and 1860. These treaties were tenuous at best, and most broke down within a few months after the St. Nicholas conference. There was talk of a $100,000 fine for violating the accord, but the parties could not agree even on this point. Yet the railroad leaders realized that even short-term pooling agreements were in their best interest. The reaction to this premier pool was rather heated, judging from statements appearing in a contemporary Syracuse newspaper that spoke of the agreement as "swindling and roguery."[149] Fears were raised as well about price fixing and monopoly, and this at a time when long-distance transport by rail was just getting under way.

The fears raised by the irate Syracuse editor were a trifle premature, for nothing much developed in the way of large-scale pooling until 1869, when the infamous Iowa Pool was established.[150] The opening of the transcontinental railroad created another highly competitive through-traffic situation as three railroads—the Rock Island, the Burlington, and the Chicago & North Western—nervously eyed the Council Bluffs–Omaha gateway that would form the eastern terminal of the transcontinental's traffic. It seemed sensible to divide the traffic up as an equal-partnership money pool. The informal agreement worked surprisingly well and lasted longer than most such schemes, a full fourteen years. It fell apart soon after two alternate transcontinentals were patched together in 1882 and the Union Pacific lost its monopoly. The Iowa Pool members made connection with the new transcontinentals, and so the old agreement was rendered obsolete. The Iowa Pool was reborn as part of the larger and more broadly oriented Western Traffic Association.

The formation of pools became something of a trend after 1875, and new ones seemed to crop up everywhere. There was the South Western Railway Rate Association, which concentrated on grain traffic; the Trunk Line Association, which represented the busy northeastern lines; the Transcontinental Pool; and the Central Freight Association. The last-named group maintained offices in St. Louis with a staff of seventy. The most prestigious of all these associations was, ironically, the Southern Railway & Steamship Association. The railroads of the Old South were generally seen as small players in the big game of national freight handling, but the head of the association, Albert Fink, was among the most respected railway officials in the nation, and the SR&SA's reputation was mostly reflected glory from its chief.[151] The Southern was started in 1873 by four lines operating out of Atlanta. It reorganized two years later and named Fink as its commissioner. Under his skillful direction the Southern became one of the most powerful and well-disciplined pooling operations in the nation, with thirty members. Admittedly, the SR&SA's success was also owing to the fact that its territory would be expensive to enter, and hence few rivals were tempted to do so. During good times members shared in healthy revenues, but in lean times Fink, the benevolent Bismarck, the all-wise rate king, persuaded the participants to pay in as much as 80 percent of their incomes to keep the pool afloat. Even after Fink's departure to become commissioner at the

New York–based Trunk Line Association, the Southern rolled along at a fairly steady pace until the Panic of 1893. The big drop in traffic then caused the Southern to fall apart.

The success of the regional pools in ending rate wars and stabilizing the revenues and traffic suggested the good sense of forming a national pool. The work of the Joint Traffic Association began in 1877. Its success was middling at best, and the waste, inefficiency, and instability of the periodic rate wars were particularly bothersome to the great financier J. P. Morgan. Morgan hated waste, inefficiency, and instability. When the western roads fell into another major rate war in 1888, Morgan decided to intervene. The efforts of the nineteen-member Transcontinental Pool had proved ineffective. Books were examined, fines were imposed on rate cheaters, and employees who violated the pool's agreements were recommended for dismissal.[152] But to no avail; rates were cut, special deals were made, and the war raged on. Morgan called a meeting of western railroad leaders at his Madison Avenue home in December 1888. Order was restored. A sixty-day truce was agreed to, but Morgan wanted a long-term peace. A new interstate pool was established during the next year with Morgan's blessing, but like so many of its predecessors, the new association soon foundered and sank after only one year of life. Its chief architect, C. F. Adams, president of the Union Pacific, withdrew from the scheme in February 1890. If the master of Wall Street could not bring rate stability to the industry, who could? Federal regulation would in the end answer this need, but peace came at a considerable price to both the industry and the public.

The goal of the pools was industrial peace at a higher cost to the public than might be necessary, but pools actually encouraged considerable inefficiency. The weak and inefficient lines were in effect subsidized by the strong, well-managed ones. The sharing of revenue between strong and weak roads was a form of blackmail. Poorly built, redundant railroads were thus kept in business. Railroads that should have been abandoned were kept running, a drain on the general economy. Federal regulation did not correct this situation, and so real market forces were never allowed to winnow out the weak members of the iron network. Hence the country created about twice as much mileage as was really required.

On the other hand, the pooling associations did introduce at least one tangible improvement into the world of rate making. This was the Mac-Graham rate system for calculating tolls between major cities.[153] This scheme was particularly useful to pools that were trying to figure rates quickly for several lines between given points when each line would report a different distance. James Mac-Graham, a clerk with the Pennsylvania Railroad, devised such a system around 1870. He used New York to Chicago, 920 miles, as the base or 100 percent charge. All other locations were longer or shorter, unless they too happened to be separated by 920 miles. According to MacGraham, the charge to St. Louis was 117 percent, to Pittsburgh 60 percent, to Peoria 110 percent, and so on. The job of the rate maker was greatly simplified, and all the rates were equal no matter how far the cars traveled. The car might go the long way or the short way, but the rate was the same. Shipments between New York and Indianapolis might go the long way on the New York Central via Albany and Cleveland, or by an even longer route via Baltimore, Parkersburg, and Cincinnati on the Baltimore & Ohio; by short routes on the Erie via Marion-Salamanca, or by shorter routes yet on the Pennsylvania via Pittsburgh and Columbus. But whichever route the pool chose, the rate was figured at 92 percent. MacGraham's plan was adopted by the Trunk Line Association in 1871 and came into general use five years later.

Pooling was seen by the industry and most big shippers as a private affair. Such arrangements were none of the public's business, despite the fact that such discriminatory practices were inherently unfair and illegal. Pooling was specifically outlawed by the Interstate Commerce Act of 1887, yet the traffic-sharing arrangement went on quietly as though Congress had taken no action whatever. The Gould-Huntington agreement was discovered accidentally by a Texas State Railroad Commission auditor some six years after the creation of the ICC.[154] At least one granger line looked back to the good old days of pooling when such arrangements were essential to roads such as the Burlington.[155] Its competitive position was increasingly eroded during the 1880s as new railroads built across its territory. Pooling arrangements were about all that kept it solvent. Once this protection was stripped away, the road's only realistic option was to sell out to a stronger rival, and so in 1901 the once-mighty Burlington became a property of James J. Hill. In 1897 the Supreme Court in effect ruled that collusive pricing violated the 1890 Antitrust Act. Yet at this time the ICC had no authority to regulate rates. Ironically, once it was given such authority it developed what amounted to the largest collusive pricing arrangement ever conceived, imposing rates and traffic regulations on a national scale. Traffic was channeled via the ICC's rate-making power to keep weak, inefficient lines alive, while shippers labored under a huge railroad cartel created not by malefactors of great wealth but by well-meaning guardians of the public trust. Pooling was legalized in 1920, but the rate bureaus were stable, and neither the industry nor the ICC saw any benefit in returning to this obsolete method of traffic management.

The practice of pooling might be viewed as an

innocent parlor game compared with the practice of rebates. All shippers looked for special rates, but only the largest or most influential ever struck much of a bargain. A discount was not such an evil in itself, but when it was given on a selective basis, then one of the finest ideals of English Common Law, equal treatment for all, was violated. These under-the-table kickbacks lack the ethics expected of any legitimate business and offended the puritan sensibilities of the American public. Traffic agents might be likened to bagmen maneuvering around the underworld with plain brown envelopes stuffed with payback money. There was just something repellent about the whole business of rebates, and the secrecy surrounding them suggested that all involved realized what shady business it was.

The most celebrated rebater was most likely John D. Rockefeller and his great Standard Oil Company. In 1867 one of Rockefeller's lieutenants, Henry M. Flagler, engineered a rate reduction plan with the Lake Shore that dropped the shipping price per barrel of oil from 69 cents to a mere 27 cents.[156] The carrier required that shipments of oil by water carriers be suspended and that Standard Oil guarantee to send 60 barrels a day for one year. There was nothing wrong with this deal, a discount for volume trade, except that it helped put Standard Oil's competitors out of business because they were not accorded the same rebate. Other railroads were enticed to cooperate with Standard Oil so that just four lines paid the big Cleveland refiner $10 million in rebates in just six months during 1878 and 1879. Ten million dollars was not just a vast sum for the time; it very nearly equaled the net earnings of the New York Central Railroad for 1879.

Standard Oil put the squeeze on western lines as well. The Central Pacific offered refunds to Standard during the 1870s and 1880s and was ready to refine and expand the arrangement in 1892 when a new agreement dropped the rate per 100 pounds between Ogden and Sacramento from $1.25 to between 82.5 and 90 cents.[157] In his 1908 autobiography a pious J. D. Rockefeller attempted to explain away the nasty rebate talk with bromides about the normality of discounts for large shipments. These were only a natural outcome of unfettered competition between railroads, pipelines, and steamers. Each offered a lower rate to attract volume shipments away from his rival, and Standard Oil was only seeking to achieve the best bargain— it was a natural law of trade. See, there's nothing at all wrong with rebates.

Certainly a great many shippers would have agreed with J.D.R.'s reasoning, particularly if they too could get in on the rebate game. A grain dealer named Peavey received a 1-cent-per-bushel rebate from the Union Pacific.[158] Some shippers demanded more than special rates and rebates. The

owners of a Colorado smelter pressured the Union Pacific to do business with a certain bank that happened to be owned by officials of the smelter. If the shipper did it, so did some railway officials. During the 1870s the superintendent of the Indianapolis, Cincinnati & Lafayette obtained a special rate on coal to supply a yard he operated in Cincinnati as a side business.[159] Many years earlier a much larger coal supplier, the Lehigh Coal & Navigation Company, had forced shipping charges down 21 cents per ton.[160] Private car operations offered another avenue for rebates. The Milwaukee Road was said to have paid the Northern Refrigerator Transit line, a major beer carrier, some $30,000 in rebates.[161]

There were many artful schemes for deception in rebating. It was not always a simple or direct transaction. Payments were hidden or purposely misrepresented. The weight or value of a shipment might be deliberately underestimated so as to lessen the transportation cost. For example, 30 tons of coal might be recorded as only 24 tons, or a load of costly veneer might be sent under the rate for common lumber. Such little tricks were called underbilling. Phony claims were another way to launder rebate moneys. A waybill is made out for, say, 75 barrels of whiskey but only 73 are actually shipped; the shipper claims a loss of 2 barrels. The railroad winks at this little fraud, and the shipper receives his rebate in the form of a lost-goods payment. It all looks perfectly legal on the books.

REFORM AND REGULATION

Few businessmen, railroad officials, or public officials failed to recognize the abuse and unfairness connected with rebating. There was hardly a railroad in existence that had not been created by one or more state charters, and all of these documents contained some provision for equality in rates to all shippers. That these provisions were flagrantly violated was also clear to any knowledgeable person involved in the transportation business. Some states attempted to throttle the rebate practice, but without much effect. Federal law as represented by the Interstate Commerce Commission was equally ineffective. The Elkins Act of 1903, aided through vigorous enforcement by the Roosevelt administration, did much to curb rebates, but it was not until the 1910 Mann-Elkins Rate Act was passed that the old evil of rebates was put to rest.

Reform is not always the product of righteous crusaders. Railroad rate reform came about largely because of the agitation of the Granger Movement, a national amalgamation of small farmers organized in 1868 to improve the lot of the agricultural community. A chief goal of this organization was the creation of low freight rates for foodstuffs. The Grange in effect wanted a rebate for its membership and so was really not so much reform-minded as intent on cheap rates for wheat, corn, and hogs. Rightly or wrongly, the Grange blamed high railroad rates for the plight of the small farmer, when in reality most of his economic problems were more likely accountable to the deflation brought on by the hard-money policies of the federal government, insufficient credit because of too little liquidity at the local banks, and a decline in farm product prices. The Panic of 1873 exacerbated the growth of the Granger Movement and the demand for cheaper freight rates.

The Grange and its populist following soon seized on the scheme of government regulation as the means to lower railroad rates. During the 1870s the farm states of Illinois, Minnesota, Iowa, and Wisconsin enacted laws to regulate freight rates. Wisconsin's law—named for a state senator, R.L.D. Potter—fixed rates and established classification. The lowest rate became the maximum possible, and even this was reduced by 25 percent.[162] The railroads of the region fought the Potter law, but when the final appeal came, the Supreme Court upheld the act and in general sustained the broader notion of rate regulation. In another case the high court ruled that in the absence of federal statutes, state governments might act to protect their citizens even when the matter under dispute could technically be classed as interstate in nature. The nation was moving in the direction of public regulation of rate making.

The notion of state rate regulation had been around for several decades before the Grangers began their St.-George-versus-the-dragon act. The Camden & Amboy's charter of 1830 contained an article setting its maximum freight rate at 8 cents per ton-mile.[163] Three years earlier the South Carolina legislature had stipulated that the South Carolina Railroad might not charge more than 35 cents per hundredweight per 100 miles.[164] In 1836 the railroad appealed for relief from its charter restriction, and the legislature agreed to a raise. A second relief bill in 1839 sent rates so high that traffic was lost to riverboats, and the railroad was obliged to drop them to a more competitive level voluntarily.

Other states such as New York took a direct hand in freight rate charges many years before Oliver Kelley and his friends had their first thoughts about organizing the Patrons of Husbandry and the need for cheap produce rates. The New York lawmakers policed railroad freight charges to protect the state's own Erie Canal operation. Lines anywhere near the canal were forced to add canal tolls to their charges and so were rendered noncompetitive. In a similar vein the Ohio railroads were required to charge the exact rates offered by the state system of canals in 1852. It may well be true that many of these early laws, except for the New York acts, were rather passively enforced, but

the point to be made is that government rate regulation was in existence from the beginning of the railroad industry.

At the very time the Grangers were firing up political enthusiasm for rate regulation, a movement began for the construction of cheap freight railways. A prominent champion of such schemes was Lorenzo Sherwood (1810–1869), a native of New York who migrated to Texas in 1849 and became a railroad promoter and a state legislator.[165] Sherwood may have been only a local politician, but he was thinking on a national scale when he created the National Anti Monopoly Cheap Freight Railway League in 1867. Sherwood envisioned a seven-line system 4,000 miles long that would operate like a toll road. Anyone, especially farmers and others beaten down by the allegedly unfair rates of the established railroads, might run cars over these lines. The fees would be only half those charged by the establishment, and the lines would run directly from the Farm Belt to the seacoast or to some other convenient junction.

The fact that rates could not likely be kept so low and the impracticability of giving anyone and everyone access to the tracks did not seem to discourage enthusiastic audiences from the Cooper Union in New York to gatherings in faraway Houston. Senator Harland of Iowa liked the scheme so much that he introduced a bill in Congress to fund a major line with federal moneys. Harlan's action in fact became a model for other congressmen, for hardly a session sent by without the introduction of a similar bill. By 1873 one of these madcap schemes went so far as a War Department survey for a cheap freight line between Omaha and New York with branches to Chicago and St. Louis.[166] An Illinois congressman wanted to establish the Continental Railway at public expense to connect Iowa with New Jersey. Congressional enthusiasm for such schemes does not seem to have abated until around 1884.

By this time the populists seem to have concentrated their energies once again on federal regulation, for in truth the last thing the nation needed was more redundant railroad mileage. The story of the creation of the Interstate Commerce Commission in 1887 has been told so often and so well that there is little reason to give more than the briefest summary here.[167] It is important to note that the language creating the ICC spoke directly to matters of fairness, rates, rebates, and pooling. That the ICC seemed unable to do very much about these matters led to a series of laws intended to strengthen its power and its control over rate making. The law was amended in 1889 so that fines of up to $5,000 or two years in prison might be imposed. Rebate abuse was targeted in the 1903 Elkins Act when both the railroads and the recipients of such discriminatory favors were made liable to penalties. Departures from published rates

THE AMERICAN RAILROAD FREIGHT CAR

were made a misdemeanor. The 1906 Hepburn Act allowed the ICC to regulate maximum rates, inspect accounts, and control the charges imposed by private freight car lines. More vigorous enforcement of existing laws during the Teddy Roosevelt presidency did much to satisfy the populist demands for controlled capitalism.

But in truth, the rate problem persisted. The goal of just and reasonable rates was never really resolved. Shippers and the public continued to demand lower rates while the railroads complained, with some justification, that their earnings were too low to attract the capital necessary for betterments and improved service. More laws were passed and endless studies were made, but the situation remains unresolved, even to this day.

FREIGHT TRAIN OPERATIONS

The story of railroad freight operations is a continuing tale of problems, breakdowns, and deficiencies answered by little more than laments and partial solutions. Care must be taken, then, to focus on the ultimate efficiency of a system that provided cheap, dependable, and reasonably swift transportation service. Certainly its performance was well beyond the sluggish canal barge or the lumbering highway wagon. Compared with the transportation alternatives available in the nineteenth century, the railroad was a paragon of efficiency. Nothing else could match it for flexible routing, large tonnages, and low rates.

On a good day the railroad ran like a Haydn symphony. There was a reassuring, steady rhythm to its operation. The timing was perfect, the melody graceful. The players followed the conductor obediently, and the trains rolled on in perfect clockwork fashion. But then one of the players would miss a beat, and the performance would fall into disarray, for in truth a railroad is more complex than even the largest symphony orchestra. The concert hall is a protected arena, and the players are all highly trained professionals. The railroad, however, is out in the elements at the mercy of the weather, it is staffed by a more variable group of players, and its plan of operation is subject to change without notice. Sour notes and miscues were in fact the order of the day, and with so many possibilities for error, something always seemed to go wrong.

Running dozens of trains in opposite directions over a single track was simply asking for trouble. The lack of signaling and the dependence on train orders, flags, lanterns, and other primitive means of communication added to the likelihood of confusion and delays. Equipment, track, and bridges were prone to a certain number of breakdowns and failures even when well maintained. But the chief cause of mischief remained the operating personnel. Most were untrained, underpaid, and overworked. Training consisted of on-the-job experience and the expectation of not thinking but just obeying the rule book and keeping your mouth shut. The great majority of breakdowns, accidents, and delays were accountable to human error. If only you could run trains without people, you would have a great railroad. But nineteenth-century railroading was labor-intensive; almost nothing was automatic or even semiautomatic, and so the opportunity for mistakes was very high. The tonnage was moved over the line, but not without difficulty or costs, and these costs went beyond fuel, depreciation, and track repairs. A tally for frustration, anxiety, sore backs, and a general weariness at the day's end is hard to enter into the ledger book of train operations.

The Divisional System

It might be assumed that there is nothing more to railroad train operations than hitching a locomotive to a string of coupled cars. Then off they go, a quarter-mile of loose iron and creaking timber that raises a light cloud of dust as it grinds away toward a far horizon. Missing from this simplistic notion is the larger job of planning, scheduling, and coordinating the movement of dozens of trains and thousands of cars. The management of so much wood and iron in motion called for a system that evolved along military lines.

Just like a Roman military highway, the railroad was divided up into divisions, each about 100 miles long.[168] The Roman road was a long pathway marked by forts or camps separated by about one day's march. These havens offered safe refuge to those moving a long distance and in need of a place to rest and regroup before moving on. Things were a little different on the railroad, for it was not safety or rest that division points offered, but rather a logical beginning or end to the journey of the freight trains. One hundred miles was considered a day's work, or about all that man and machine could hope to perform in a ten-hour period.

And so it was Joliet to Bloomington, then Bloomington to Carlinville, Carlinville to Kirkwood, and so on like a bucket brigade passing on its heavy container from one marathon player to the next, until the final station was reached. At each smoky division town, the engine and crews were always exchanged for fresh replacements (Fig. 1.32). In most cases some cars would be dropped off and new ones added to the consist. Sometimes it was necessary to reorder the entire train and run it through the classification yard, an action that materially slowed things down. More rarely the train was left untouched and only a new engine, caboose, and crew were added, but such special

FIGURE 1.32

Freight trains traveled over railroads in segments called divisions, each about 100 miles long. Engines and crews were changed at the end of each division. This Hartford & Connecticut Western train stands at West Winsted, Conn., sometime in the 1880s. (Thomas Norrell Collection, Smithsonian Institution)

treatment was normally reserved for cattle or refrigerated trains. The basic scheme of operation, then, was one of slow, halting movement from yard to yard. You might expect that once a car in Charleston was loaded for Xenia, it went straight through. Hardly; each car went through a parade of division stops and was likely switched off onto several other railroads, because there was no single or direct railroad from Charleston to Xenia. The wooden cars and coal-burning locomotives have long since been retired, but the division-to-division, yard-to-yard plan of operations prevails to the present day.

The divisional system was beset by many inefficiencies; chief among them was low productivity. The repeated makeup and breakup of trains resulted in lost and idle cars, unpredictable service, and high labor costs. Idle cars meant idle assets, or at least poorly utilized assets. For all its failings, the system did work; it did manage to move goods. The deficiencies of the divisional system were to a large degree unavoidable. Just about all trains were a mixture of through and local cars. Some, by the nature of their destination, had to be dropped off somewhere along the line. Dispatchers tended to add new cars to a train that had just dropped some off, to keep its tonnage up and also to help clear the yard of idle cars. In 1880 56 percent of national railroad freight traffic was classified as through and 44 percent as local.[169] These ratios varied greatly from region to region. In the Far West at the

same time the proportion was reversed: 61.5 percent of traffic was local and 39.5 percent was through. But even in regions where most of the traffic was through, there remained the need to change engines for servicing (especially clearing the fires) and for relief of the operating crews.

The divisional system came into being early in the history of railroading, or about the time any line grew to much more than 100 miles in length. Understandably, as a railroad grew physically, so did the problem of managing it properly, especially at the operating level. A general manager sitting in an office in Jersey City needed lieutenants out along the line. A superintendent for each division became a part of this line-of-command management scheme. They were not to be independent princes, but rather loyal subordinates to the general manager. The division superintendents might have several assistants such as a yardmaster, a roundhouse foreman, and a repair shop supervisor.

Meanwhile, as the railroad expanded and the number of trains increased, the general manager added more subordinates back at headquarters in the nature of a trainmaster or a chief dispatcher, whose main concern was the development of schedules to smooth the flow of traffic. Station agents, telegraphers, and towermen became the field agents of the trainmaster. A master mechanic was needed to oversee the repair and design of locomotives. On smaller lines he might double as

THE AMERICAN RAILROAD FREIGHT CAR

master car builder, but on bigger roads this was a separate position, and on even larger systems several subordinate regional master car builders might be created. There would be a road or track master and a bridge and building superintendent and on and on as the layers of management grew to administer the growing railroad. It was all done in good military fashion, with the policies and commands at the top communicated to ever-lower echelons out in the field. Just how the trains were to be run was thus often decided in an office far removed from the yards and tracks that carried them.

The Interchange of Cars

The interchange of cars among the nation's railways is perhaps the most distinctive and important aspect of freight train operations. This free exchange or lending of valuable capital goods between rival concerns is unprecedented in American business history. The interchange agreements reached just after the Civil War in effect took a fragmented collection of several hundred independent rail lines and turned it into one unified system so far as the movement of through freight was concerned. A single car could go from one end of the system over a dozen railroads and never be unloaded or turned back. If it broke down en route, it was repaired at predetermined rates and returned to service. If the car was destroyed, the owner would be fairly compensated, as he would receive a mileage or rental fee while the car was in service on another railroad. And so bitter rivals, such as the Missouri Pacific and the Rock Island, would happily borrow each other's equipment and tender payments back and forth for the privilege of doing so.

The interchange system worked because everyone benefited. The shipper saw his goods go from terminal to terminal without costly and slow loading and unloading procedures at each junction. The railroads benefited from these savings plus the added boon of a greatly expanded supply of cars. In an informal way, a national freight car pool was created. Though there appears to be no exact model elsewhere, the interchange system might be likened to a Mercantile Library where members might freely borrow all the books they wanted. They agreed too return them in good time and in good condition. If a book was lost or damaged, the member faced only modest fines. A major difference was that the member railroads were really borrowing from each other rather than from a central or public institution, so it was actually a trade association or interindustry lending program.

Such a large-scale operation was fraught with problems despite its many obvious benefits. Like a borrowed lawnmower, interchange cars were often roughly used and slow to be returned. It is, after all, a fundamental human weakness to treat the property of others less carefully than your own. The record keeping required for cars scattered to all points of the compass was prodigious, and these labors kept thousands of clerks busy. Every major railroad was forced to create a new bureaucratic branch called the car service department. Even so, cars were lost despite the most dogged efforts of clerks and the car tracers who literally searched out every likely siding and yard. Even when the cars came home, there were squabbles over the repairs made by foreign lines' shops. These involved workmanship or the cost, and if the two antagonists could not resolve the matter, it went before the arbitration committee of the Master Car Builders Association. Fixed rates had been established for specific repairs and for the loss of an entire car, however, which rendered arbitration necessary in only the most unusual cases.

When cars were exchanged only with neighboring lines or when only very small numbers were lent, the problems of interchange were minor. But once the system went national, around 1866, there was a need for a continental control mechanism. The freight car had suddenly grown up. It was no longer a pampered child living in the shadow of protective parents. The boxcar was pushed out into a rude and competitive world to migrate across a sometimes unfriendly land. Once out on the road, its only protection beyond the doubtful kindness of strangers was the Master Car Builders Association. This was a rough game for wooden cars, and most were lucky to last out their sixteen-year life expectancy. The problems of interchange were compounded by the scale of the operation, for literally millions of exchanges were made every year as cars traveled back and forth from one railroad to another. In 1881 the Pennsylvania Railroad exchanged forty thousand cars a day.[170] It was figured as a general rule that 40 percent of the cars on any given railroad were foreign or borrowed cars.[171] Full and empty, hundreds of thousands of look-alike cars rattled around the countryside or stood idle in yards often thousands of miles from home.

THE ADVANTAGE OF NOT BREAKING BULK

The greatest single reason for interchanging cars was to avoid "breaking bulk." This was a shipping term for the loading and unloading of goods or cargo. There is no way to eliminate the breaking of bulk entirely; it must occur at the beginning and end of every journey. But to do so midway is undesirable because it is expensive and time-consuming. Each time goods are handled, they are also subject to damage or pilfering. The rehandling of shipments at every junction and transfer point became anathema to every practical railroad manager. Railroads were willing to do almost anything to avoid it, even if it meant dealing with their worst competitor. In 1853 the five little railroads linking Albany and Buffalo were paying 28 cents per ton to transfer freight between their cars.[172] The porters were thus adding about 10 percent to the cost of carrying goods between these cities. During the course of handling, some boxes would be punctured by the hooks used to shift barrels and boxes about. With too much rough handling, a box might burst open, but no matter; it would be "recoopered" at the owner's expense. During another transfer, a box might be stealthily opened and part of its contents removed by a night watchman, who would then stuff in rags or blocks of wood to fill in voids and keep the weight up to that shown on the waybill. At another terminal the shipment might be accidentally split up and part of it sent to the wrong destination. With so many clerks, porters, watchmen, and warehousemen involved, it was often difficult or impossible to fix the blame on the guilty party, and so the shipper was left to accept the loss or sue the railroad.

The cost of breaking bulk was seen as a great tax on shippers and ultimately on the public. Estimates on the cost of transhipping varied from 7 cents to 34 cents per ton. But even when the lowest rate for breaking bulk was used, the Boston Board of Trade in 1863 figured on an annual cost of $0.5 million just on shipments between Chicago and Boston.[173] Several years later it was calculated that the breaking of bulk added 20 percent to the cost of shipping grain between Chicago and New York. The need to eliminate the unnecessary transfers of goods was surely obvious to anyone who studied transportation matters.

POLITICAL IMPEDIMENTS TO INTERCHANGE

Not all parties involved were disinterested or public-spirited. The porters, teamsters, and others directly involved in the transfer business clearly wanted it to continue. The influence of these workmen was surely small, but their cause was championed by politicians and business leaders who did not want to see the local economy weakened in any measure, no matter how beneficial it might be to the national economy. And so city and state governments made it their business to maintain the transfer points, because the breaking of bulk was good for the local economy.

Railroads typically stopped on the outskirts of large cities and were not allowed to cross through to connect with major lines on the opposite side. In Cincinnati, for example, the Marietta & Cincinnati ended on the far eastern side of the city. All shipments were transferred by wagon to the other side of town so that they might go north or west by rail. Finally, around 1860, a track was built along the waterfront, but even then the cars were moved by horses one or two at a time. This awkward transfer was repeated in Baltimore, Philadelphia, and elsewhere. Things could become violent, as

they did during the Erie War when some of the citizens of this lake port city actually tore up the track and burned bridges when efforts were made in 1853–1856 to avoid a transfer between the broad-gauge Erie and the Ohio-gauge lines.[174] The Erie Railway was already hampered by its state charter, which forbade it to connect with another railroad leading into Ohio, Pennsylvania, and New Jersey. The mayor of Pittsburgh followed the Erie model by trying to block a physical connection between the Pennsylvania and the Pittsburgh, Fort Wayne & Chicago even after a bridge had been completed over the Allegheny River in 1857.[175] The idea of through service between Philadelphia and Chicago without a transfer in Pittsburgh was apparently unthinkable to the mayor of the Iron City. For nearly a year His Honor managed to keep a 200-foot-long gap in the tracks, much to the delight of the porters, and no doubt much to the despair of travelers and shippers.

Some states were no more helpful, and rather than cooperate with efforts to promote more efficient regional or national transportation, they sided with parochial interests. The Virginia state legislature, for example, refused to give either the Baltimore & Ohio or the Pennsylvania Railroad permission to bridge the Ohio River.[176] Both major east-west lines were thus forced to struggle along with cumbersome ferryboats just to keep the porters content at places like Wheeling. The situation changed radically during the Civil War, and while Virginia was out of the Union, the federal government gave permission to build the bridges. The new state of West Virginia saw no reason to undo what Washington had approved, and so within a few years the teamsters and ferries tied up for the last time as the trains rolled overhead on new iron bridges high above the water.

GAUGE DIFFERENCES

Even when railroads were ready to exchange cars and local governments did not block the way, the American railroad industry was crippled by the self-inflicted obstacle of gauge differences. Just why so many varying track gauges were selected by supposedly rational businessmen is too complex a story to delineate here beyond a very brief summary. It is difficult to understand why our forefathers felt that the continent required something more inventive than George and Robert Stephenson's eminently sensible track width of 4 feet 8½ inches. Personal vanity, misguided intelligence, and regional rivalries appear to be the major explanations. New Jersey and Ohio adopted 4 feet 10 inches, while most southern lines felt that 5 feet was best. Many roads in between compromised on 4 feet 9 inches, but not the Erie and its connections, which felt that nothing less than 6-foot gauge would do. New England trimmed 6 inches off the Erie's dream gauge to make a connection with Canada's Grand Trunk, and for reasons no one can now explain, the Pacific Railroad of Missouri adopted the 66-inch track width.

Through connections were actually rather rare during the antebellum period because so few trunk lines physically joined at their end terminals. Wheeling was looked upon as a dumping-off point, for example, where rail cargoes borne 380 miles by rail from Baltimore were piled up in a levee-side warehouse for transhipment by riverboats. The idea of railroads as a portage between two bodies of water persisted into the 1850s. Even when freight was ferried across the river for other rail lines going west or north, a transfer of loaded cars was not possible because of the slightly broader Ohio gauge (4 feet 10 inches). Yet ironically, a Baltimore & Ohio advertisement boasted of the great through line opened between Baltimore and St. Louis in 1857. Goods and passengers might pass nearly halfway across the continent by rail. Forgotten in all the braggadocio were the two major river crossings by ferryboat and three gauge changes. Each change meant delay and added cost. Mention was just made of the Ohio gauge facing the B&O at Wheeling, but once the cars reached eastern Cincinnati, all goods and passengers were teamed across town to the depot of the 6-foot-gauge Ohio & Mississippi Railway. No interchange of through cars between Baltimore and St. Louis was possible over the Great Through Route until 1871, when the bridge over the Ohio at Wheeling was opened. The Central Ohio Railroad changed to standard gauge, as did the Mississippi & Ohio at the same time. Shippers waited another three years for a rail bridge over the Mississippi to carry freight trains directly into St. Louis.

George R. Taylor and Irene D. Neu worked out this breakdown of American railway gauges as they existed in 1861: 4-foot 8½-inch gauge accounted for 53.3 percent of the track; 5-foot gauge for 21.8 percent; 4-foot 10-inch gauge for 9.9 percent; 5-foot 6-inch gauge for 8.7 percent; and 6-foot gauge for 5.3 percent.[177] Midcentury shippers were, then, resigned to several breaks in bulk for almost every long-distance shipment. However, once the merits of interchanging freight cars became established at the end of the Civil War, the railroads began to put their houses in order. The need for a uniform national gauge became increasingly obvious. A few lines like the Morris & Essex changed to the standard or Stephenson gauge at an early date—1866, in this case—but many others simply did not have the capital or the will to make the traumatic changeover of both track and rolling stock. Most chose instead to limp by with compromise solutions and to face the challenge squarely at some more distant time. Laying a third rail would permit dual-gauge operations, but not at a small price. If a standard-gauge line wanted to accommodate broad-gauge cars, extra-long ties would be required. In addition, all dual-gauge trackage involved not only a third rail, an expensive item, but special switches. The operation of mixed-gauge cars was surely awkward and dangerous, since nothing really matched or coupled easily.

Wisely or not, the Great Western of Canada laid a considerable length of third rail in 1867 just to accommodate standard-gauge freight cars and so open an easy exchange with the Michigan Central. The Cincinnati, Hamilton & Dayton put down a dual track in a scheme to link the broad-gauge Erie with the Ohio & Mississippi in 1864. The CH&D link, it might be noted, was less ambitious in track mileage but more ambitious in an engineering sense, because it used four rather than three rails—a straddle-gauge system. The broad-gauge track stood entirely outside the original 4-foot 10-inch track. The switches must have been wonderful to behold. The Grand Trunk spent a considerable sum in 1874 to make a dual-gauge passage over tracks opened some twenty years earlier to Portland, Maine. At one time, the 66-inch gauge was deliberately selected so that no easy connection with the standard-gauge lines radiating out of Boston could be made. Thus hundreds of thousands of dollars were spent to undo what never should have been done in the first place.

Other railroads sought to overcome the difference in gauge by exchanging trucks. This scheme was cheaper than dual-gauge track, at least in first-cost terms, and it worked fairly well so long as the interchange of cars was not too large. For these reasons it was favored by southern lines, which had little capital and relatively small traffic. All along the borders of the Old South, car hoists and transfer yards were stationed. The Louisville & Nashville had nine hoist stations scattered around its system.[178] The hoists cost $3,000, a costly item in 1870 dollars, and it took five to six minutes to switch a car from one set of trucks to another. It is unclear if this time included rehooking the brake gear as well. Yet even if this optimistic transfer time was correct, how well could the transfer trucks be expected to match the center plate of the car bolster? During this period almost nothing matched from one railroad to the next. The same would be true for side bearings. Consider also the storage of the orphan trucks; acres of these sat around for days or weeks awaiting the return of the loaner body now waddling along on its borrowed set of wheels. If the original body and trucks were to be reunited, it was necessary to return the body to its point of entry rather than to send it on freely to wherever the traffic might carry it. This would surely increase empty-car mileage—just what every railroad manager worked to eliminate. The truck exchange scheme was awkward, expensive, and just not smart.

A third makeshift solution offered to solve the

problem of differing gauges was the broad-tread wheel. Our forebears were in the main wise and prudent men, except perhaps for those advocating the use of these pernicious devices. In fairness it should be added that broad-tread wheels were never promoted as a permanent solution to the gauge problem. They were actually suitable only for transfers between nearly alike gauges and so primarily saw use in shifting cars between the standard and the 4-foot 10-inch lines. So long as very slow speeds were observed, the fat wheels could wobble in their uncertain course across the countryside. They were actually tried several years before the advent of large-scale car interchange, according to an article appearing in the *American Railway Review* in October 1860. Other inventors tried to introduce split- or shifting-axle wheel sets to solve the variable-gauge problem. Such devices could more readily handle the width changes necessary for shifts between standard and broad gauges. They proved less reliable than broad-tread wheels and so were not widely used. More on this subject can be found in Chapter 7.

As freight tonnage built up and trunk lines consolidated, the advantages of car interchange became more obvious. The free exchange of cars had in fact become a necessity by 1870. Big railroads united under one management, such as the Vanderbilt system, could not work together efficiently if the eastern portion of the line was 4 feet 8½ inches and the western end was 4 feet 9½ inches east of Toledo but standard once again between Toledo and Chicago.[179] The Baltimore & Ohio's western subsidiary, the Ohio & Mississippi, was a rather silly ally with tracks over a foot wider than those of its parent. Large-scale conversions started to happen in 1871, when both the O&M and the Maine Central gave up the broad gauge. The Grand Trunk and the Lackawanna converted a few years later. The Erie had narrowed its tracks, except for branch lines, to the national norm by June 1880. By this date most main-line trackage was standard (or at least 4 feet 9 inches).

The notable exception was the railroads of the Old South. It was almost as though the former Confederacy had elected not to agree to any Yankee notions, even on matters of track gauge. Five-foot gauge was traditional, after all, and tradition was honored in the South. The South Carolina Railroad had started the trend in 1830 at the suggestion, ironically, of a northern engineer named Horatio Allen. Most other railroads of the region, except for those in North Carolina, had copied the Allen gauge. It seems that the promoters of North Carolina would do nothing to help the port of Charleston and so built to Stephenson's width to make an exchange of cargoes more unlikely. During the war some of these very lines were rebuilt to Allen's gauge and then were once again shifted back to 56½ inches when all of the

South converted to the national standard in 1886.

Some new southern railroads were in a dilemma on what to do about the gauge business. When Cincinnati built its air line to Chattanooga in the 1870s, it reluctantly picked the southern gauge and built a large transfer yard in Ludlow, Kentucky. Just six years after it opened, 335 miles of very expensive railroad was converted to standard gauge. These developments were, of course, not entirely clear when work on the Cincinnati Southern began, and the safest course appeared to be the one chosen. A more daring decision was made in 1884, when the Illinois Central decided to convert its lines south of Cairo, Illinois, to match the standard-gauge width of its northern lines. The Mobile & Ohio followed the IC's lead during the next year. At last the other roads of the South felt that it was time to go along with the Stephenson gauge. Months of preparation led to June 1, 1886, a Sunday, when the rails were shifted inward 3½ inches at an estimated cost of $2.5 million.[180] Now at last a boxcar might roam freely over all the main-line tracks of the nation save for an insignificant mileage controlled by the narrow-gauge roads.

INTERCHANGE ON A NATIONAL SCALE

The interchange of freight cars has no concrete origins. It began in fuzzy and uncertain ways decades before a complete railroad network had been laid across North America. It started as a local phenomenon; whenever two railroads came together at a common junction, an exchange or loan of equipment was sure to evolve (Fig. 1.33). The exchange of goods and passengers would naturally encourage a joint use of cars to the benefit of all parties. In 1836, for example, the Utica & Schenectady and the Mohawk & Hudson ran baggage cars through to Albany.[181] By early 1843 through passenger trains were running over five independent railroads stretching between Buffalo and Albany. Freight cars were running between these cities by and perhaps before 1849; sadly, the record is not particularly clear on this point.[182] In 1838 the Boston & Worcester Railroad recognized its obligation to move foreign cars at reasonable rates but reserved the right to do so only with its own locomotives and train crews.[183] This pioneering interchange was actually limited to the Western Railroad, but it soon developed into an arrangement so free and open-ended that the two railroads were said to use each other's cars indiscriminately. At about this same time, in 1841, the Baltimore & Ohio and the Winchester & Potomac signed an interchange agreement that permitted the use of each other's cars for through shipments.[184] While American rail managers were working on limited local arrangements, a far grander version of car interchange was being established in Great Britain.[185] The idea for a national interchange of both freight and passenger cars was circulated in 1841.

FIGURE 1.33

Once the interchange of cars
began, freight trains became a
mixture of cars from railroads all
over the country. This Michigan
Central freight paused briefly
sometime in the 1880s for the
making of this photograph.
(Thomas Norrell Collection,
Smithsonian Institution)

The scheme was based on the bank clearinghouse that had so effectively handled the exchange of drafts and notes between financial institutions in all parts of the British Isles. Operations began in January 1842 with only five members; however, by 1846 the British Railway Clearing House had forty-six members. Cars ran without inhibition over the tracks of all the cooperating lines. User fees were collected on a mileage basis, with a monthly settlement of accounts. Car repairs and damaged-goods claims were handled by the clearinghouse as well. A national car pool was thus established that might prove a model for other nations. The *American Railroad Journal*, described the British system in the most flattering terms in its issue for August 8, 1846. The editors spoke in wonderment about the exchange of over 180,000 freight cars on British lines during the previous year. They urged American railroad managers to organize a clearinghouse here as well.

American railway officials, however, remained indifferent to the clearinghouse idea, except perhaps for regional exchanges. A few lines, such as the Ohio & Pennsylvania, pursued the idea with some vigor, but the scope of the operation was largely intra- rather than interstate in nature.[186] In 1854 the O&P was moving freight cars from a dozen different railroads between Pittsburgh and Cincinnati in just thirty-five hours. Obviously, several railroads other than the O&P were required to make a through transit of cars between those cities, and so a sort of mini-clearinghouse operation was under way here that involved the

basic requirements for a larger interchange system. Just a year before the report was published about the O&P operation, Charles Ellet, Jr. (1810–1862), a prominent civil engineer of the time, announced his vision for a transcontinental interchange of freight cars.[187] Though Ellet did not live to see his dream realized, he missed the beginnings of national interchange by only a few years.

The interchange system at last came of age because of a wartime traffic emergency. In 1863 goods piled up at Pittsburgh awaiting cars to take them west.[188] Through cars were not possible because of the 1½-inch gauge difference. In normal times the traffic flowed with reasonable dispatch, but this was wartime and the freight business was booming. A shrewd businessman named William Thaw understood the potential for profit in this sea of stranded barrels, boxes, and bags. Broad-tread wheels, mentioned earlier in this chapter, would allow the passage of through boxcars all the way from Philadelphia to Chicago. Shippers desperate for service would gladly pay a premium for through shipping. Thaw and his associates chartered the Union Transportation Company, better known as the Star Union Line, in January 1864 and had cars rolling within a few months. Others had operated private cars before Thaw, but no one succeeded in the field so well as this veteran freight man from Pittsburgh. The origins and growth of the private or fast freight lines will be explained in more detail in a separate section of this chapter. For the purposes of its promotion of the interchange idea, the private freight car system exercised a profound effect and so must be at least briefly addressed here.

Thaw's success was quickly followed by a multitude of imitators. The Empire Line, then the Red, Blue, and White lines all came in quick succession. They multiplied so fast that the car builders most affected met in West Albany in July 1864 to discuss how to handle the appearance of so many foreign cars. Several more informal meetings were held until September 1867, when the Master Car Builders Association was formed. Among its chief concerns was the formulation of rules and procedures for dealing with interchange cars. The MCB Association in effect was taking up the job that should have been handled by a national clearinghouse, though it tended to dwell more on the repair of damaged cars and the development of design standards. The individual railroads were left to establish a car service division to handle billing and the tracking down of cars.

There is no question that by the late 1860s the private car lines had pushed the industry into the large-scale interchange of cars, but it is also true that the railroads willingly plunged once they fully understood the merits of interchange. In 1874 an official of the Boston & Albany stated that a substantial number of foreign cars were running over his line—cars from nearly three hundred different railroads. Taylor and Neu claimed that virtually all through freight was being carried by interchange cars amounting to at least half of all freight tonnage.[189] By 1874 the Star Union and most of the other private freight car lines had already been taken over by the railroads and were no longer independent operations.

THE EVILS OF THE INTERCHANGE SYSTEM

Once under way, it became clear that the interchange of freight cars would not be easy.

The record keeping necessary to document millions of exchanges each year was a nightmare task in a time when there were so few computing aids and every transaction was a matter of handwritten entries on slips of paper or in ledger books. The circulation of copies was difficult and uncertain; hence lost or missing cars were commonplace. There was also the bickering over repair and replacement charges for interchange cars. These matters became a subject of endless complaints and horror stories. Some of it might be discounted as just the normal grumbling that is part of any workplace operation; but even so, there was a real basis for unhappiness with a system so fundamentally desirable but requiring reform in some of its details.

The grumbling started rather early in the history of interchange. In 1868 the Pittsburgh, Fort Wayne & Chicago found that cars lent to other roads were poorly used and averaged only 20.5 miles per day, which at 1.5 cents per mile produced the very meager revenue of 30.8 cents per day.[190] This was poor return on a vehicle worth $500. Revenues actually declined, because mileage rates eventually dropped to 0.75 cent per mile. The casual misuse of other people's property is too fundamental a human trait to have bypassed the interchange of freight cars. A correspondent to the *Railroad Gazette* complained about it at length in 1873, not many years after the system had been inaugurated.[191] One of the most common complaints was the use of foreign cars for local traffic rather than for through service. Such a practice directly violated the basic point of exchanging cars: expedite long-distance or through traffic. Individual railroads were expected to supply their own cars for short-haul or local shipments. When interchange cars were kept in local traffic, they generated small mileage and hence earned little for their owners.

But lazy yard clerks didn't care much about the earnings of a competing railroad and would use whatever empty happened to be available. The same attitude prevailed for freight agents eager to please a shipper; they too would load up foreign cars for short hauls, ignoring the obligation to route the car back to its owner. These low-level operating officials regularly frustrated the imperative that borrowed cars should be returned home with all due dispatch. It was simply more expedient for them to hold cars from any railroad to satisfy their own immediate needs. This practice was especially common when traffic demands were strong. When traffic was down, the foreign cars were left to languish in yards and sidings despite the pleas of their owners for prompt return.

New England became notorious for abuses in this area. It was called the graveyard of railroad freight cars, home to managers who consistently found it cheaper to borrow than to build or buy a car.[192] Cheap mileage rates prompted thrifty New England managers to use as many foreign cars as possible to move local traffic; in 1901, for example, 105 million out of 171 million car-miles on New England lines were run with borrowed cars for which they paid a measly rental of $687,853.[193] Their earnings were a respectable $17,088 per mile on freight operation—not surprising, considering the all but free use of other people's equipment. At the time, New England railroads owned only 11.4 freight cars per mile. The Yankees had always been stingy about buying new cars. Twenty years earlier they made a poor showing for one of the busier and more prosperous manufacturing regions of the nation. In 1883 the middle states owned 1,714.4 cars per 100 miles of track; New England, 635.9 cars; the western states, 483.4 cars; the southern states, 283.2 cars; and the Pacific states, 191.8 cars.[194]

Railroads with light traffic could not be expected to invest in more cars than they could use, but neither was it right for big-traffic lines to freeload off the others in what had been set up as a cooperative venture. Such sharp trading reveals a serious evil in the system. The quick-to-borrow-and-slow-to-return habit could not be broken, because there were no effective incentives to do so. A central clearinghouse might have resolved this problem, but American railroad managers were too independent to adopt such a bureau, and so it was left to economic conditions to determine just when a car would return. When times were good they tended to stray far from home, but when business fell off they would mysteriously come trundling back like so many lost sheep that had at last discovered the right trail.

If it is easy to misplace your own property, it's even easier to misplace someone else's, and the railroad borrowers were no exception. They seem to have done a superior job of losing one another's cars. It was a giant hide-and-seek game to locate missing and lost cars. Surely the greater number could be located by telegraph: Have you seen RI 26882? The answer by immediate wire: Yes, it's still here at Riverside. In other cases a car might simply be lost without record: Last seen at Lexington's northbound yard so many days ago, but somehow it was mixed into another cut of cars and presumably went north to some unknown destination.

The trail often became very thin, and so railroads employed field agents variously called car tracers, searchers, or detectives. They roamed the tracks looking for vagrants that might be buried in the outermost reaches of a big-city yard or on a windswept prairie siding. The car tracer could spend weeks or even months looking for just one castaway. In 1883 the hunt led one tracer to an abandoned distillery in Texas.[195] The side track into the distillery had been torn up, but there was his car, now occupied by Indian squatters. Once the Indians had been evicted by a local sheriff, a temporary siding was put in place and the car was returned to service. Some years later a flood near Kansas City carried twenty cars away. Off came the roofs, doors, and trucks, but the hulks of all but one car were eventually found well downstream in a field.[196] The owner insisted on payment for the single lost car, and so the search went on. Fourteen months later a car tracer found the trucks of the missing car under the debris of a collapsed barn, but the body was not to be found until the tracer's curiosity was aroused by a rectangular "stable" situated across the field. At first glance it was difficult to identify the stable as a boxcar, because it was so well camouflaged by patent medicine signs painted on the sides. The owner admitted that the stable had appeared on his property after the late flood, and so another missing car was returned to work.

Some cars took a Rip Van Winkle–like leave from home and were gone for as long as seven years.[197] Others, while not absent for more than a few months, traveled on erratic odysseys that involved considerable backtracking. Consider the sixteen-month journey of a boxcar that left Indianapolis on December 4, 1886, bound for Boston with a load of corn.[198] Bad weather delayed its arrival in Boston until early January. There it sat until the corn was resold to a firm in Medfield, whence the still-loaded car was sent on after a few days' delay. Medfield was reached on January 24, 1887, but the consignee delayed unloading until March 17—no wonder car mileage was so abysmally low. Finally the car was loaded with coal—yes, boxcars occasionally carried coal—and was sent west, but it became involved in a minor wreck and was sidetracked. Once back in service, the car was sent south rather than west, and now—after a year of incredibly complicated back-and-forth movements—the owners lost track of its whereabouts. At last the wanderer came to rest in Augusta, Georgia, long enough to be discovered. Once the owners were apprised of its location on February 11, 1888, the car was sent north. But progress was never swift, even for lost property, and the wanderer came home a little the worse for wear on April 17, 1888. Untypical? An extreme case? Not according to accounts of the time nor in the opinion of the author of the story just related, Theodore Voorhees, then assistant general superintendent of the New York Central. There were tales of lost cars that required over two hundred tracers and telegraphs to bring home. There were also tales of bad faith and bad management that kept an Indiana railroad's car away for twenty-seven months.[199]

A lost car was at the worst simple mismanagement, but a stolen car was fraud. Whereas the lost-car problem was forever being prattled about in the trade press, theft was a dark side to interchange that received rather little exposure. Perhaps this is true because theft was so rare, but it did exist and it began, as might be expected, at an early date. Around 1860 or 1870, the date is uncertain, A. V. H. Carpenter was employed by the Central Vermont as a car tracer.[200] Carpenter found one of his missing flock in service on an unnamed New England railroad with new trucks and a fresh coat of paint to disguise its origins. Once the fraud was exposed, payment was made to the Central Vermont. About a century later, car theft on a grander scale was reported involving an Illinois short line and the mighty Penn Central Railroad.[201] Some 277 boxcars were purloined in this interchange caper. New numbers, lettering, and paint were not quite enough to hide the identity of these stolen cars.

Even the most honest railroads sometimes contributed to the sorry story of the missing car, simply by misnumbering a car during a repair. A workman picks up the wrong stencil or forgets to shift to the next digit, and a car under repair becomes a missing person. Such a story was related in 1892

PARADE OF THE CARS

about a misnumbered car and a tracer who traveled thirteen weeks and some 12,000 miles looking for it.[202] This bloodhound of a tracer went from New York to Pittsburgh, then to Cincinnati, Chicago, St. Louis, Kansas City, Galveston, and finally San Francisco, only to go back to Galveston. After more moving around, the car always just ahead, it was pursued back to its home territory of Buffalo. Some practical railroad men felt rightly that such an expenditure of time and money for one car was hardly worth it, for the tracer's salary and travel expenses would approach the cost of a new boxcar. Such frantic effort was not economical, and a car tracer with no better sense should be fired.

Inventory Management

An ideal of American business at this time was to keep procedures informal. We seemed always ready to oppose the more highly disciplined and structured European models. This policy was admirable in many ways, but in the case of car interchange it was informality carried too far. American railroad managers had a big inventory of cars, but it was under poor control. As we have seen, cars drifted from railroad to railroad in disarray. By 1905 we are talking about 2 million cars, a $1 billion investment. These cars earned money only when loaded and moving, but they spent too much of their time idle and empty. Better management of cars was essential to industry profits, and so it was necessary to reform the interchange system. Business as usual was no longer good enough.

FEES AND FINES

Talk about change started at least as early as 1873, when a switch from the mileage system of payment to the per-diem system was advocated.[203] Per diem would add a time factor to the cost of borrowing cars—a fundamental change in rules designed not only to prompt the return of the cars but to force borrower railroads to make better use of them. Under the mileage system the borrower was in no hurry to return an idle car, because it cost him nothing to keep it; fees began only when the car was moving. There was no penalty for keeping a car or reward for returning it. However, under the per-diem scheme an idle car cost just as much as a moving car, so even though the fee was a small one, railroads would be more eager to return all foreign cars before the end of the day to avoid as many interchange fees as possible.

The arguments for the per-diem system might seem persuasive, yet there were powerful vested interests supporting the old mileage system. It worked very well for shippers who liked to use freight cars as temporary warehouses. So long as it cost the local railroad nothing, as was true under the mileage system, little pressure was brought to bear on an offending shipper. Grain jobbers were notorious for holding full cars, sometimes for weeks, until a buyer could be found. Yet these idle cars cost their owners dearly, for during good traffic times a boxcar could earn from $5 to $10 a day, and that was very good money in the 1890s.[204]

Shippers were not the lone friends of the mileage system; some railroads—once again the New England lines—were making a nice profit by underloading foreign cars. This shady technique was a great way to boost car mileage.[205] In one instance it was found possible to place the small cargo from sixty-four cars into just fourteen cars. One car was being sent from Vermont to Canada with a single tub of butter. It is no wonder that car mileage was so high, and that the railroads were paying out $44 million per year (1895) in mileage fees. A car loading inspection service was instituted by some roads to end this knavery, while other lines insisted on a minimum load of 3,000 pounds per car to qualify for mileage payments.

After many years of debate, the per-diem system was finally adopted in July 1902.[206] The daily fee was 20 cents, with an additional penalty of 80 cents per day if kept over thirty days. Advocates of per-diem payments optimistically predicted that there would be a massive return of cars to their rightful owners and that some railroads (but surely not the New England lines) would see their fleets grow by 10 percent in a few days' time.[207] It was hoped that the poor handling that left 150,000 cars out of service each day would also end with the demise of the old mileage system. Such great expectations proved a trifle optimistic. To begin with, the private car lines were exempted and continued to operate under the mileage system. Of more consequence, the per-diem rate of 20 cents was simply too low. The car hogs persisted in profiting from other people's cars and could well afford to do so when the rental was so cheap and the earnings so high. Even when the per diem was raised to $1, in 1920, the potential daily earnings were five or six times greater. The industry never could agree on a realistic rental fee for interchange cars.[208] It did not reflect market value but was an institutional fee agreed to in a cooperative fashion, apparently in the belief that establishing a more realistic price might jeopardize the free and open interchange of cars.

Shippers were routinely blamed by railroad officials for the late return of cars and hence for the inefficiencies of the interchange system. Shippers' lazy habits and their penchant for treating boxcars as impromptu warehouses were the root cause of car shortages, low mileage, and all the other ills blamed on the interchange system. If slothful customers would just unload cars upon delivery, the empties could be quickly turned around for new loads. What could be more simple? Many shippers were slow to unload for good reason. Boxcars offered free warehousing. Produce dealers would sell directly from refrigerator cars, a process that could take quite a few days when dealing with small retail grocers who bought in small quantities. Grain speculators kept loaded cars tied up for weeks waiting for prices to go up or another buyer to come along. Iron furnaces thought nothing of keeping loaded coal or coke cars around for weeks until they were ready for the fuel. Why unload the cars and then have to move the pile again? It was simpler to leave it in the cars and shift them around when needed. Technically, all these offenders were subject to fines called demurrage for failing to release cars promptly. These charges were often substantial. In a rate card of 1864, mentioned earlier in this chapter, the Illinois Central threatened to fine shippers $3 per day if cars were not unloaded within twelve hours. In 1881 the Lake Shore was a trifle more lenient and gave consignees twenty-four hours to unload or face a charge of $2 per day. The railroad reserved the right to clear the car for the customer and bill him for this service.

Around 1910 Emory R. Johnson and Grover G. Huebner noted that the demurrage rules had softened noticeably.[209] Shippers now had forty-eight hours to remove their cargo. The fine was now only $1 per day. There was no threat of forceable unloading plus a fat bill for doing so; rather, the railroads expressed a willingness to extend free time in the event of special circumstances such as floods or blizzards that might hamper the unloading process. This increasingly temperate approach is exactly why cars spent so much time at the loading dock. Shippers realized that the rules and fines, no matter how tough-sounding, would not be enforced.[210] No freight agent wanted to confront one of his good customers about clearing his stuff out or paying off a stack of demurrage charges. Mr. Shipper would likely tell Mr. Agent to go to the devil and that he would thenceforth ship on the Hocking Valley line rather than the damned old DT&I. Keeping customers happy, particularly those having access to service on another railroad, was such a preoccupation that demurrage was largely a paper tiger.

In 1883 it was suggested that the superintendent be made responsible for enforcing the demurrage charges and so relieve the freight department of this unhappy role, but no official was eager to alienate the railroads' most valued customers. Four years later a scheme was developed to take the heat off individual railroads and their officials by making enforcement a cooperative venture.[211] And so a Demurrage Bureau was set up in Omaha to enforce late fines in that area. Individual railroads felt that this pooling arrangement would allow them to keep the goodwill of even the most tardy shippers. The plan worked well enough to be repeated at other large terminals elsewhere in the nation. But for every step forward there were counteractions to move the enforcement of demurrage backwards. In 1902 the Pennsylvania imposed a progressive track storage fee to discourage the use of team tracks as produce warehouses. The dealers appealed to their state representatives, who passed a law reducing the fee to $1 a day and in effect voiding the intent of the charge. Other states showed no interest in freeing up cars and sided with the shippers by passing legislation that increased the free time allowed for unloading from the customary twenty-four hours up to ninety-six hours.[212]

CAR MILEAGE

Railroad managers were made more aware of car mileage than ever before after the introduction of the interchange system. Even at 0.75 cent per mile, big numbers would accumulate when thousands of cars were in motion. Total freight train mileage was surely growing; in 1883 it was recorded as 350 million miles, but within a decade it had grown to 531 million miles.[213] If roughly 40 percent of this was accountable to interchange traffic, it can readily be understood that very large sums were involved even at the low fees charged. The size of the business was not the critical factor, however; what commanded railroad managers' attention was the low productivity recorded in mileage figures, particularly when focused on the daily travels of individual cars. With total fleet mileage slipping ever downward, the largest and best-managed railroads in the nation watched in dismay as the daily tallies continued to flag. The Pittsburgh, Fort Wayne & Chicago, the Pennsylvania Railroad's main line west, reported a 21 percent decline between 1868 and 1881.[214] In 1868, or just about the time interchange began in a serious way, the Fort Wayne recorded a respectable 47.6 miles per day per freight car. Most other railroads would have been delighted with this number, because 40 miles per day was considered good, but the Fort Wayne's managers watched helplessly as the figure fell to 37.7 miles per day by 1881. The parent company actually came in just below its western subsidiary, with 37.5 miles per day. Another PRR property, the Star Union Line, did far better and was able to boast of an average daily car mileage of 44.8 in 1882; however, this figure represented a disturbing decline from 1876, when the daily figure was a very robust 88.8 miles per day.[215]

The industry as a whole could not seem to stem the decline in daily mileage; in fact, the growth in traffic, track mileage, and equipment only seemed to make matters worse. Between 1883 and 1892 daily mileage fell another 22 percent, so that by the early 1890s the average travel was just 24.7 miles per day.[216] As the new century opened, an industry expert lamented that average daily mileage ranged between 18 and 24 miles per day.[217] As the wood-car era receded, daily mileage continued to hover in the low twenties.

The yearly reports only confirmed the daily reports of a distressing and seemingly irreversible decline. Arthur M. Wellington gave figures for 1887 that ranged from 11,000 to 15,000 miles per year.[218] Coal cars and flatcars were the most poorly used of the entire car fleet; their annual mileage was only about half of the figures cited above. On the other hand, line cars—that is, boxcars used for through freight—might register up to 35,000 miles per year, but because they were a minority, the average figures remained depressingly low, and they were getting worse. In 1894 10,000 miles per year was said to be, sadly, the common record for an American freight car.[219]

Dormant freight cars were ill-used freight cars. They should be rolling day and night to make the maximum return for their owner. They should never grow old in long service but be used up quickly from hard but profitable employment. The Western & Atlantic Railroad kept accurate records for 1882 that show just how important it was to get maximum mileage out of a car, especially when it was new.[220] So long as traffic was good, a car could be expected to pay off its first cost and make a 9 percent return during its first five years of service. After this time the likelihood of a profit became far less certain, for mileage tended to go down as repair costs went up. The W&A figured that a wooden car was used up economically after ten years, despite the industry's standard depreciation of sixteen years. The policy was to run them hard, then replace them. Old veterans running over the road were a false economy, and those who kept cars in service for twenty years were wasting money on repairs that would be better spent on replacements. Here are the numbers as reported by the W&A's car department: a car with one year of service would be run 13,149 miles per year; with three years, 10,475 miles; with five years, 9,881 miles; and with ten years, 7,656 miles.

The causes of low car mileage became something of an obsession with railroad managers. Many of the reasons were obvious, and some have already been mentioned—for example, the slowness of shippers to unload and the reluctance of railroads to enforce demurrage penalties. In some instances railroads actually encouraged the practice of free warehousing by a "no-bill" policy.[221] This policy allowed coal wholesalers to hold loaded cars at mine sidings or in out-of-the-way yards waiting for demand or prices to go up. The no-bill policy meant that demurrage was indefinitely waived. The seasonal nature of certain traffics, primarily bulk goods like grain and coal, also created great peaks and valleys in car utilization. Fall and winter brought on the greatest traffic surges as harvesttime poured forth grain and the mines pushed to fill the coal cellars of homes, hospitals, and office buildings. Once again cars were needed desperately for a few months and then would sit idle for the remainder of the year.

In addition, shippers were picky about the cars offered for their use. They wanted the new, bigger cars, to gain the maximum from carload rates. The older, smaller cars sat idle waiting for the next grain rush, all of which supports the Western & Atlantic's contention that only new cars showed a good mileage record and profit. Conversely, railroads did not want to offer a good car for a nasty load of fertilizer, lime, or grease, and so another car that might be going in the right direction would sit and wait for a cleaner load while the local agent scanned the yards for an old, dirty car.

Meanwhile, mileage went down. In fact, most of the delays were accountable to the railroads themselves. Uncooperative shippers, the seasonal nature of traffic, and the weather were factors in low mileage figures, but the industry's way of handling trains played a far larger role. Low average speeds were a major cause of low mileage. Obviously, 10 miles per hour would not generate high daily or annual mileages. Yet speeds between yards were actually the best part of freight train operations, for once a train entered the yard limits, the wheels might as well have fallen off. Cars commonly spent eighteen hours out of twenty-four in a yard. The slow and deliberate work of shifting them around for the makeup and breakup of trains added a maddening delay to all schedules. Because of yard delay, cars might actually average only 2 mph during any trip. Yard time also represented delays because of inspections and repairs. Too many cars were set aside for trivial repairs, and a less liberal use of the inspector's blue flag could do much to get things moving.[222] But even the most kick-'em-in-the-pants railroaders would accept that some repair time was necessary, and that for about 8 percent of the year every car would be an invalid at the repair shops.

Finding reasons for low car mileage rarely touched upon interchange itself, but there is no reason not to include it. The casual habits associated with borrowed tools and clothing could be expected to carry over to railroad cars. The fact that someone else's boxcar was standing idle in a yard for days on end would bother only the most conscientious of managers. The large-scale decline in daily car mileage began with the interchange system, and so logically the system itself must share in the blame. Interchange by necessity materially lengthened the trips of individual cars, and so instead of passing through three or four yards, a car might pass through dozens before reaching a final terminal. More yard time meant more idle hours. The Pittsburgh, Fort Wayne & Chicago reflected that its car mileage record was far superior when its rolling stock remained at home, and that as the length of runs doubled between 1870 and 1882, its cars spent far more time standing than moving.[223]

For all the reasons just given, cars remained stationary for too much of the time. In a report published by the American Society of Civil Engineers in 1883, it was claimed that cars stood for twenty-one hours a day; only one-eighth of the fleet was in motion while the rest were stationary in yards, sidings, and repair depots.[224] In 1894 another estimate recorded that only about seventy-five thousand out of a fleet of over a million cars was in motion at any one time.[225] In 1902 *Railway Age* published an equally pessimistic statement, which said that a typical car was in motion only one and a half to two hours each day, and that some moved no more than 10 miles on the average per day.

EMPTIES

While railroad efficiency experts fretted over low car mileage, the problem of empty cars awaited a solution as well. Even if mileage figures could be improved, the efficiency of rail freight operations would hardly improve if a large number of the cars in motion were empties. Unloaded cars only increased operating expenses while adding nothing to the coffers of the revenue-hungry rail industry. Freight men had been voicing concern about one-way traffic and empty backhauls since the 1850s.[226] Even before long-haul interchange traffic was inaugurated, the empty-car problem was familiar and seen as inevitable. Bulk goods went to market and so the down trains were filled, but only small quantities of finished goods went back on the up train, and so the majority of those cars rattled along high on their springs with doors ajar to reveal empty interiors. Mixed-traffic lines such as the Concord stated that two-thirds of their cars returned empty. Big coal roads such as the Reading could find no suitable goods for their coal cars on the return journey. In its annual report for 1855 the Erie spoke with pride about its efficient eastward parade of trains composed of 26.3 million loaded cars and only 295,450 empties. The westbound traffic presented a less favorable picture—11.5 million full cars and 14 million empties.

The imbalance was to increase rather than diminish as railroads spread west and a more complex national traffic developed. In 1869 the Kansas Pacific, for example, reported a huge imbalance in its traffic.[227] While 141,341 tons went to the east, only 34,177 returned from the same direction. Part of the KP's problem was its newness and the unsettled state of the region it served, but even railroads in more settled areas experienced significant empty-car problems. The Louisville & Nashville showed the following results: in 1867, 3.6 million miles run by loaded cars versus 1.2 million by empties; in 1870, 6.8 million versus 1.9 million; and in 1873, 10.5 million versus 3.1 million.[228] The Lake Shore tried to streamline its operations in the hope of upping efficiency but to little avail, at least insofar as the ratio of empty cars was concerned. In 1879 the numbers ran as 1.63 million full cars and

579,000 empties. In 1894 the numbers were actually a little worse—1.58 million full cars and 698,000 empties.[229] One-third empties to two-thirds loaded cars seems to have been an average mix on most railroads of the time. The two major trunk lines faced an even higher ratio of empty-car miles. The mighty New York Central reported in 1885 that 50 percent of its westbound cars were empties. The Central's statistics would have improved greatly when balanced against the eastbound traffic, but the Pennsylvania, supposedly the most scientifically run railroad property in North America, could do little better than 50 percent when its 1881 figures were published.[230] The total was 2.16 million miles for loaded cars and 1.08 million miles for empties.

Some traffics were particularly given to empty return trips. Coal has already been mentioned. The fact that coal hoppers were essentially single-purpose vehicles did not help matters. In a few instances they were used to carry grain, but the cost of cleaning would largely eat up any profit unless the car were reassigned to that traffic, and this did nothing to solve the return-load question. Hopper cars could not be readily used for anything but coal. Flats too often went home empty; the New York Central said that 70 percent of its platform cars returned without benefit of a backhaul. Tank cars might appear to be about as single-purpose as any rail vehicle in service, but some clever operators in the South found a way to put a payload aboard both ways.[231] The car would go south with kerosene and unload in New Orleans. It then went empty only as far as Mobile, where it picked up a tankful of turpentine; thence north, where it was refilled with kerosene. Nothing, however, could beat the versatile boxcar for return loads. Merchandise, lumber, machinery, grain, and even coal were routinely sent away in this standard workhorse. In 1896 the boxcar was said to have the best record for acquiring a backhaul, and that only about 20 percent of its mileage was recorded in the empty category.[232]

Railroads tried to find ways to reduce the empty-car syndrome. Cheaper rates were offered to induce westbound traffic; some instances were given in the rate section of this chapter. Rate reductions of 25 percent were reported in 1885.[233] But efforts to reduce empty-car miles seem to have made no material impact on the problem. In fact, one change in the interchange rules actually worsened the problem, because rather than wait for return loads other lines would start the car home empty to avoid being fined for breaking the code.[234]

While general managers fussed over the backhaul issue, others pointed to the horrors of partially loaded cars. Why worry about empties when most loaded cars were carrying only about half of their rated capacity? This was hardly making maximum use of equipment, and this indeed was what the whole mileage and backhaul debate revolved about. In 1881 the Pennsylvania Railroad reported an average car loading of only 8.1 tons, yet a typical car of the period had a capacity of 15 to 20 tons. Many years later the average car loading was given as only 55 percent of capacity.[235] Shippers were commonly criticized for overloading cars and so doing structural damage, yet there is evidence that they were guilty of just the opposite. Most shippers apparently reasoned that it was cheaper to take a whole car at the carload rate than to pay the premium for LCL shipment. Yet underutilization was also costly to railroads, because it meant that they needed more cars than were actually necessary to move a given traffic.

The railroads generally had more cars than needed because they were so inefficiently used, but genuine shortages would occasionally develop. These were often short-term and caused only temporary inconvenience and disruption to the trade. Then again, major car famines occurred in 1887 and 1901; millions of bushels of wheat were stranded in the Northwest, cotton moldered on station platforms, and sawmills shut down, all for lack of cars.[236] But while shippers might suffer from time to time because of too few cars, the railroads had to absorb big losses from having too many cars when the economy turned down. In late 1907 some 209,000 cars stood idle because of a business recession; the number had almost doubled by the spring of the following year. The real issue was not too many or too few cars, but poor use of existing cars, and this issue, sadly, has never really been resolved by the American railroad industry. For all the problems related here, it must be remembered that the advantages of the interchange system outweighed its disadvantages. Much about railroading was imperfect, but for all its failings, it functioned in ways useful to society.

Train Size

Long trains moving at very slow speeds were long viewed as the optimum in freight train operations. The greater the tonnage, the greater the efficiency, and so wherever traffic would warrant it, long trains would be found creeping across the American landscape. This theory or mode of operation was in fact the very basis of steam railway operations. The locomotive was seen as both a large unit of power and a major investment that must propel a great quantity of goods to be economical.

England, the motherland of railways, provided an exact model for America. The American edition of Nicholas Wood's popular *Practical Treatise on Rail-Roads* (1832) contained a detailed description of a freight train on the Liverpool & Manchester propelled by the locomotive *Planet* on

December 4, 1830. The eighteen-car train carried American cotton, barrels of flour, and sacks of oatmeal and malt, all of which came to a total weight of just over 51 tons. The open cars and their oilcloth covers weighed a little over 23 tons. The locomotive, tender, and fifteen-man crew added another 11 tons, making a total train weight of 86–87 tons. Despite an adverse headwind and the stiffness of so many new bearings, the train maintained an average speed of 12.5 miles per hour over most of the road. Such feats must have impressed pioneer American railway men, for their first efforts were rather feeble in comparison. In 1831 the Baltimore & Ohio sent out an eight-car train filled with flour barrels.[237] Its gross weight was given as around 30 tons, less the locomotive. The open cars weighed but 1.5 tons each and had a capacity of between 2 and 3 tons. About the only active locomotive on the line at this time was the 3.5-ton *York*; hence the total weight of the train would have been around 34 tons.

It would not be many years, however, before British railway managers were looking to the United States for lessons in the operation of large-tonnage trains. By 1835 the Baltimore & Ohio had increased the size of its grasshopper-style engines, weighing from 7.0 to 8.5 tons and capable of moving 211 tons on the level.[238] Reports of amazing trainloads, considering locomotive size, began to come out of New England in the late 1830s. In March 1837 a Boston newspaper reported in some detail on the performance of a new engine recently produced by the Locks & Canals Machine Shop in Lowell.[239] The little 10-ton Planet-style engine pulled a forty-nine-car train filled with cotton bales, groceries, and coal. The cars and the cargo weighed, respectively, 95 and 159 tons. The engine and its tender weighed a little over 17 tons; thus the total train weight was 271 tons. The train, stretching out for 820 feet, was hauled 10 miles out of Boston to Woburn in just over fifty-one minutes. Long trains were not exceptional in New England for that time, because the same report noted that a smaller L&C-built engine, the *Patrick*, pulled a thirty-five-car 201-ton train from Boston to Lowell, 26 miles, in two hours and fourteen minutes. In 1840 the neighboring Boston & Worcester normally ran trains of twenty-five cars loaded with 75 tons of goods. Some thirty-car trains were operated as well.[240]

The smart performance of the little Lowell engines began to tarnish considerably when railroads with steeper gradients and less tangent tracks became their proving grounds for everyday operations. On the Western Railroad the hills and curves of the Berkshire Mountains reduced cargo loads to 59 tons for a 10-ton locomotive and to 85 tons for a 14-ton locomotive; the number and weight of the cars are unknown, but if we estimate twenty cars at 4 tons each, the total train weight would have been 139 tons, less locomotive and tender. Somewhat later in the 1840s the Baltimore & Susquehanna found that its little Lowell Planets—the line had some very small 8-tonners—could handle only 52 tons and be expected to maintain a 12-mile-per-hour schedule.[241] Here again the B&S was a curving, climbing sort of railroad best described as serpentine, all of which greatly impinged on locomotive performance and so very directly on train length.

The small wood burners of the last century could propel sizable trains of 300 tons and more so long as the track was level, yet even a slight incline could markedly reduce the engine's pulling power, because it was required to move the train vertically as well as horizontally. M. N. Forney calculated that 6 pounds of force was needed to move 1 ton on a straight, level track.[242] Thus 120 pounds was needed to move a 20-ton car on the same track. If a moderate grade of 1 percent is encountered, 20.6 pounds is required to move 1 ton—in other words, a power requirement of 412 pounds for a 20-ton car. These figures are calculated at 5 miles per hour, but if the speed is taken up to 30 mph, the energy requirement almost doubles.

These inescapable physical laws forced even the most unscientific railroad managers to assign shorter and lighter trains to a locomotive of a given size on mountainous or even semimountainous railroads. Actually, just one good hill could undo most 25-ton American types. The Erie, hardly a line noted for a rugged profile, limited the load for its freight engines to thirteen cars between Piermont, on the Hudson River, and Port Jervis, New York.[243] The same class of engine could comfortably move a train of thirty cars on the more level Delaware Division that ran from Port Jervis to Susquehanna. In 1857 the Baltimore & Ohio acquired some large ten-wheelers, large for the time at least. These fast freight engines were rated to move 800 tons on the level at 20 miles per hour.[244] They were ideally suited to cattle trains but could double in passenger service if necessary. Their pulling power began to drop noticeably on grades. When an incline of just 0.4 percent was encountered, train weight capability dropped to 600 tons; and when the 2 percent grades near Piermont came into view, the stout little ten-wheelers were reduced to trains of only 150 tons. The Baldwin Locomotive Works was a little less conservative when it came to rating what an engine could haul on a given grade. In 1870 it rated one of its 33-ton ten-wheelers in this way: 1,230 tons on the level, 570 tons on a 0.4 percent grade, and 155 tons on a 2.0 percent grade.[245] In 1876 the Pennsylvania figured that one of its 2-8-0 freight locomotives could move seventy cars or 1,470 tons on the level, but that its pulling power was cut exactly in half when a fairly moderate grade of 0.8 percent was being climbed.[246]

And so just one good-sized hill on the line could reduce train size materially, or call for larger locomotives and/or helper engines stationed at each major grade to push the train over the top. Engines of a given tractive effort were accordingly rated at so many cars, taking into account the gradients likely to be encountered. In some cases it was necessary to consider the direction of travel, for gradients might be more favorable in one direction than another. Using the Erie's Delaware Division as an example once again, 2-8-0s were rated at sixty-five cars eastbound but only forty-five cars westbound because of the difference in slope.[247] While some allowance was normally made for loaded and empty cars, the grades going west must have been formidable, because most of the cars were empties and so train weight for a given length should have been considerably less.

Most traffic men dwelt on size limitations imposed by the rise and fall of the land. There were, however, other factors involved in the tonnage an engine could be expected to haul over a given piece of trackage. The condition or quality of the track itself added to or diminished the friction of a train. A rough, uneven strap-rail line offered more resistance to train travel than did a smooth, well-ballasted line laid with T-rail. A strong headwind could retard the progress of a train at least somewhat, though this was a factor rarely worked into the rough-and-ready rules followed by the average train dispatcher. The resistance offered by curves was more easily measured, and on a road with enough curves it could make a noticeable difference in train performance and size. Forney offered the empirical formula that each degree of curve added 1 percent more resistance to the movement of the train.[248] Gentle or large curves of just a few degrees added relatively little resistance to the train's progress; a 2-degree curve, for example, has a radius of 2,865 feet and posed no particular restraint to a train of the wood-car era. Yet some main lines, notably the Baltimore & Ohio, had 10-degree curves (573-foot radius) that caused the flanges to squeak and increased the resistance of passage to around 6 pounds per ton, or about double what it would have been on a straight, level track.[249] Curves, particularly sharp ones, had a very measurable effect on train resistance and hence train size. Most early railroads were a succession of curves, snaking back and forth across the landscape to avoid a hill, gully, or creek. Based on the 1880 Census report, Wellington estimated that every mile of railroad had at least one curve. In 1896 a survey of the Erie's curvaceous Delaware Division showed that it curved for 70 percent of its length.[250]

Published comments on train size were common enough, but in the main only the ones of exceptional size were considered newsworthy. Who wanted to read about an ordinary or typical freight train? The larger and more spectacular the train, the more certain it was to be an event staged to generate publicity. The 1837 test of the Locks & Canals engine discussed earlier was very likely an operation cooked up by the locomotive's maker to promote sales. The fact that a very large train was hauled is not questioned, but the feat was perhaps less extraordinary than its promoters would ever admit, considering the flatness of the railroad chosen—at random, no doubt—for the test. In May 1842 an even more spectacular train was run over the Western Railroad.[251] This sixty-eight-car monster carried cattle, pigs, wool, flour, and shingles and registered a total weight (including engine and tender) of 356 tons.

Data on more typical trains for the 1830s and 1840s is difficult to uncover. Henry Tanner's 1840 *Canals and Railroads of the United States*, which is reflective of the state of domestic operations for the previous five years, speaks mostly of track construction and hardly touches on train size. Tanner does mention the largest trains on the Philadelphia & Columbia: thirty-five cars with a cargo amounting to 105 tons, for a total train weight (including engine and tender) of 190 tons. Baldwin confirms these statistics in a catalog published around the same time.[252] This slender document includes a letter dated September 13, 1839, from the line's master mechanic, John Brandt. Brandt stated in the letter that an engine of Baldwin's manufacture hauled a thirty-five-car 187-ton train up a grade of just under 1 percent at between 8 and 12 miles per hour. Elsewhere in the catalog Baldwin mentions freight trains of fifteen cars (5.5 tons each) on the Long Island Railroad and a seventeen-car train of 85 tons, less engine and tender, on the Boston & Providence. On a similar scale, Von Gerstner reported fifteen-car trains on the Petersburg Railroad for this period.[253] He gave the car weight as 2.25 tons and cargo capacity as 3 tons, which made for a train weight of around 75 tons; adding 17 tons for locomotive and tender brings the total weight to 92 tons. In another note on the Petersburg Railroad for the 1830s, it was stated that because so few passengers were carried on this early Virginia line, mixed trains were often run instead. They were composed of eight to ten freight cars plus one baggage and two passenger cars.[254]

By the late 1850s train size appears to have crept up to lengths of around thirty cars and weights of 400 tons, less locomotive and tender, on trunk lines like the Erie. The Buffalo & State Line gave thirty cars as an average length between 1856 and 1858 and reported running trains as long as fifty-four cars.[255] Its zealous crews were known to tie down the safety valves in an effort to boost the tonnage on the road, yet their enthusiasm for ever-higher steam pressures blew up one of the better engines on the line. About this time, the New York State Railroad Commission reports began to comment on freight train size. The 1855 report gave average train size as 18.2 cars with the cargo weighing 71 tons and the cars 180 tons, for a total weight of 251 tons. Just how the mix of large and small roads affected the average figure is hard to determine, but it may be taken as a fair average until better data comes to hand.

Of course, large trains were running elsewhere, and reports of the elephantine variety are not difficult to find. In the winter of 1850 the Boston & Maine operated a sixty-one-car train from Boston to Great Falls, seventy-four miles.[256] The weight of 391 tons seems rather light for the number of cars; perhaps most of them were empty four-wheelers. Despite light snow on the tracks and a 1 percent grade at Bradford, the train completed its run at an average speed of just over 14 miles per hour. Three years later the Michigan Central outdid its Yankee counterpart several times over.[257] A 1,000-ton train, of which 544 tons was cargo, was moved in 110 cars. At the same time, a 118-car train was reported on the Michigan Central. It is not clear if these big trains were the norm or just an occasional exercise to gain a little press attention. There was no pretense on the Philadelphia & Reading, for its very first train, run in December 1839, stretched out a full eighty cars. Just two months later a new engine named the *Gowan and Marx* pulled 101 four-wheel coal cars weighing 423 tons from Reading to Philadelphia. The grade was a favorable downhill slope, but even so this remains a creditable achievement for a stumpy 11-ton locomotive. Of more consequence, by 1845 the Reading was regularly running 500- to 600-ton coal trains, and by the latter part of the next decade its *average* coal train weight had reached 708 tons.[258]

The decade of the 1860s does not show much change from the previous ten years. Car and engine size was nearly static, and if the data of the New York Railroad Commission can be believed, average train weight actually went down somewhat from what it had been in the 1850s. By 1870 things were once again astir, and managers embraced big locomotives, larger cars, and longer trains to move mounting tonnages (Fig. 1.34). Big Consolidation (2-8-0) locomotives were seen as a positive traffic cure-all. They might be double the weight of a standard eight-wheeler, but they upped operating costs by only 14 percent.[259] They could move larger trains faster and so reduce overall labor and operating expenses by 300 percent. These claims might have been just a little too optimistic, but the Erie surely benefited from the switch to Consolidations in 1876.[260] Before that time, average train size had been 106 tons and twenty-two cars; after the switch, train size went up to 228 tons and thirty-eight cars while operating costs per mile dropped from 9 cents to 5 cents. In fairness, it might be noted that double-tracking and

FIGURE 1.34

This Boston, Hartford & Erie local freight—pictured at Franklin, Mass., around 1870—was short at the time when forty-car trains were common. The train stands on a siding awaiting the passage of a superior train. The conductor stands on the ground looking up at the engine crew, while the brakemen remain where they belong, on top of the cars. (Smithsonian Neg. 77-13160)

grade reduction contributed to these savings. The Baltimore & Ohio became a Consolidation convert in 1873, and within a dozen years it owned 180 freight engines of this type.

The Reading, long devoted to ten-wheelers for its coal traffic, had also become a 2-8-0 devotee by 1880. In June of that year the performance of one of these engines was recorded; the report offers an interesting account of train size and operating practices of the time.[261] The engine left Philadelphia with fifty loaded cars. At Jenkintown it picked up thirty more and at Bound Brook yet another twenty, to make a train of one hundred four-wheel coal jimmies each weighing 8.55 tons. The train reached Elizabethport, about 78 miles north of Philadelphia, after eight hours. The train was not under way during all of that time. Ten hours and seventeen minutes was spent on sidings waiting for superior trains to pass. The return trip with 119 empty cars—again, all were four-wheel jimmies—required only six hours and ten minutes.

Moving large numbers of jimmies was hardly a novelty on any of the anthracite lines, but the record must have been established in July 1879 when the Lehigh Valley conveyed 593 empties in a single train.[262] The fact that a train of 547 cars had been moved a few weeks earlier would suggest that trains of this size were not that unusual.

It is difficult to conceive of freight trains more than a mile long in the 1870s, yet there is a record of it. Exaggeration was not unknown, even in what should be the most reliable of sources. The 1875 *Master Car Builders Report* claimed that the average freight train contained thirty-five cars, which sounds reasonable enough; but it is surely fantasy to say that all cars were fully loaded and that a typical train weighed 665 tons. A train having even two-thirds of the cars fully loaded is unlikely; a more reasonable estimate would be one-half, thus accounting for empties and partially loaded cars. The train weight then comes to a more believable 490 tons. Yet even this may be too high, for the

PARADE OF THE CARS

best average cargo loading any of the major trunk lines could muster in 1875 was 166 tons, or 184 tons shy of the MCB's top figure.[263] A more realistic average train weight figure for 1875 might be around 400 tons.

As the 1880s opened more conservative statistics were offered on train size. Thirty-car trains weighing 313 tons (cars and cargo, less locomotive and tender) were said to be typical on most trunk lines.[264] Exceptions to this general rule were the biggest carriers such as the New York Central, which was reported to be running fifty-car freight trains weighing 1,000 tons. In the late 1880s the Pennsylvania was outdoing all of its rivals by operating 2,445-ton freights as the normal order of business on its Middle Division.[265] In 1892 the average train size for trunk lines was far less than that reported for industry leaders such as the Pennsylvania. At this time a dozen major trunk lines reported average freight train weights, again cars and cargo only, of just 553 tons.[266] Forty-seven cars was given as the average length; however, the growth in car size improved the productivity, in theory at least.

The question remained: How many empties and partially loaded cars were in the consist? The answer was the same as always: Too many. Adding more cars to a train involved more than just buying bigger engines to pull them. The efficiency of larger trains came at a price that was hardly cheap. Longer sidings and more double track were one requirement. And better couplers were surely a necessity once train length went beyond fifty or sixty cars. The loose and frail link-and-pin-style couplers were being tested beyond their limits even on trains of this size.[267] Janney-style couplers, improved draft gears, and air brakes encouraged the operation of longer trains as the nineteenth century came to its close. In 1898 the Pennsylvania moved a 130-car coal train weighing 5,212 tons, of which 3,693 tons was the coal itself.[268] This monster train, nearly three-quarters of a mile long, was handled by a single Consolidation locomotive.

The exploits of the largest or heaviest train in the world made good headlines, but some more consistent and meaningful measure of train size was needed for the industry to gauge its own growth and productivity. Around 1880 more attention was given to the trainload, or the actual weight of the cargo carried on each train. This was the true payload for the railroad, and so it became an industry gauge for train size. Indeed, some railroads had recorded trainload statistics since the 1850s, and we can offer these numbers for the Pennsylvania: in 1853 trainloads averaged 75.4 tons; in 1863, 94.3 tons; in 1873, 116.8 tons; and in 1883, 196 tons.[269] Data from this far back is not readily available for other lines, but figures from later years offer an interesting comparison for lines in all regions except for the Far West, as shown in

TABLE 1.11 Trainloads for Selected U.S. Railroads (average tonnage per train)

Railroad	1873	1883
Boston & Albany	75	103
Chicago & Alton	115[a]	188
Chicago & North Western	107	121
Cleveland, Columbus, Cincinnati & Indianapolis	79	218
Illinois Central	83	100
New York Central	129	200

Source: Arthur M. Wellington, *The Economic Theory of the Location of Railways* (New York, 1893).
a. Estimate.

TABLE 1.12 Trainloads for All U.S. Railroads (average tonnage per train)

Year	Tonnage
1880	129
1883	129
1888	176
1890	175
1893	184[a]
1896	199
1900	271

Source: U.S. Census, 1880; Slason Thomson, *Railway Statistics of the United States* (Chicago, 1925); and *Railroad Gazette*, July 12, 1895, p. 463.
a. Only 170.4 tons, according to *RRG*.

Table 1.11 for 1873 and 1883. Topography, traffic patterns, traffic volume, and management skills would explain the growth or lack thereof exhibited in these figures. Another view of trainload development is provided by the national figures given in Table 1.12.[270] The authority of these numbers is difficult to guarantee, except for the 1880 item, which is taken from the U.S. Census report for that year. The other numbers are taken from Slason Thompson's 1925 book on railway statistics, which in one instance (1893) differs rather markedly from figures presented by the *Railroad Gazette*. Resolving such differences is, sadly, beyond my capabilities.

Train Crews

Railroading, especially in the wood-car era, was a labor-intensive business. In round figures, the number of employees equaled the number of cars; in 1860 the estimate for overall railroad employment was eighty thousand, and the figure for freight cars was the same. This does not mean, however, that a forty-car freight train would have a crew of forty, for train crews represented only around 40 percent of all railroad employees; a large number were in maintenance and clerical positions. It is also true that a good portion of train crews were engaged in passenger service, and so

FIGURE 1.35

Brakemen gather in an easy pose on the roof of a ventilated boxcar loaded with cattle, supposedly near Shasta, Calif., around 1890. The top-hatted gent standing to the left—a trifle overdressed for the rough-and-ready occupation of brakeman—is presumably an official, included as a token of dignity. (DeGolyer Library, Southern Methodist University, Dallas, Tex.)

those working the freight trains might represent only about 20 percent of all railroad workers. This would vary greatly from railroad to railroad. Some overall numbers of total railroad employment are available, though they are largely questionable before 1870. Remember that only about 40 percent of the numbers represent train crews. In 1840 there were 5,000 employees; in 1850, 18,000; in 1860, 80,000; in 1870, 163,000; in 1880, 418,000; in 1890, 750,000; and in 1900, 1,018,000.[271]

Freight train crews were made up of engineers, firemen, conductors, and brakemen. The first-named were, of course, on the locomotive, at the head end of the train. The conductor, stationed at the rear in the caboose, was the chief or captain of the train. The engineer might command a higher salary, but he remained subordinate to the conductor. A brakeman was assigned to either end of the train to do triple duty as a flagman, brakeman, and switch tender. Other brakemen were stationed atop the cars. Some railroads had a brakeman for every five cars, which could swell crew sizes considerably in pre-air-brake times, and so in the period before 1890 a forty-car freight train might have a crew as large as eleven men. Some railroads were more sparing in their use of brakemen, but even so, seven men would be needed to move a train of this size unless it was on a level line with rather little traffic. Once the majority of freight

cars had air brakes, the normal train crew consisted of five to six men.[272] We have employment figures for one division on the Erie, the 104-mile-long Delaware Division, which might well represent worker ratios for the industry as a whole.[273] Assigned to freight service were 40 conductors, 160 brakemen, and 160 engineers and firemen. The passenger trains were manned by a total of 145. The tracks were maintained by 450 workers, while 80 men worked as switch tenders, indicating a large number of manually operated switches. Another 375 men labored in the repair shops. No figures were offered for office and clerical personnel.

DUTIES AND HAZARDS OF TRAINMEN

Brakeman was the entry-level job for train service. It required little education or skill and was attractive to many a young lad fresh from the farm or the emigrant boat. Many saw it as a better alternative to the drudgery of farm life or the prison atmosphere of the factory. Realistically, a young person with little education and no developed skills did not have many job opportunities, and to those of a more restless nature who were ready to leave the hometown, the choices were limited to the military, seafaring, and railroading. Rather than go to sea, fresh-faced lads, some hardly more than boys, took to the ladders and brake staffs of a rolling boxcar. It was a fine life for a hearty outdoorsy person content to pursue an overalls sort of career (Fig. 1.35). It was not a job for the sensitive, dreamy, or faint of heart. The work was physically demanding; climbing on and off moving cars during switching maneuvers or clubbing down the brake wheels called not just for physical agility and strength but for attention to the job at hand. A misstep could result in death or injury.

The noise and frustration associated with such rough work tended to make the men involved in it equally rough and heavy-handed. Like teamsters and wagon masters, train crews were not notable for refinement of speech. Indeed, they often gave vent to their anger with coarse sayings and plain old-fashioned profanity. Yet the rudeness and dangers of the work were actually appealing to some, because it relieved the boredom of what was often a routine existence.[274] The daily risks were a way of proving one's manhood. To a common workingman who rarely could play the hero, escaping from a big wreck with only a few bruises was a small price to pay for all the attention that resulted. Danger, excitement, virility—these were the ancient attractions that drew young men into military and seafaring careers long before the first steam locomotive turned a wheel. Railroads offered just one more variation on this fantasy of manhood. Some young brakemen surely used the job to flirt with the pretty country girls who might walk near the tracks looking for a stray goat or lamb. There he would be in the middle of nowhere, a young brakeman standing far back from the train to flag. To an innocent farm girl, he would appear very official and manly with his flares and flags. One can imagine the conversation that evolved as the girl sought the help of this attractive stranger.

Other members of the crew might just enjoy sitting as they waited on the siding for a superior train to pass. It could be a few minutes or even a few hours. Longer breaks were used for napping, reading, or fishing, or maybe just for pitching stones in a nearby creek or pond. Even when the train was under way, the brakeman enjoyed a wonderful view of the countryside, especially on a fair day. It was a fine thing just to stand up and look out over the land as the train rolled along. With an agreeable conductor, there was even a family atmosphere on the job. A big pot of coffee was kept on the caboose stove. Some conductors or their helpers, the rear brakemen, were adept cooks, and so a meal could be found far from the home terminal. The best conductors were more like kindly fathers than hard-driven foremen, and at least some young brakemen were lucky enough to find such a boss. Others, of course, were not so lucky.

The glamor of the trainman's life receded rapidly for most new employees. It seemed so fine at first to stand up on a car roof mastering the hand and lantern signals or just learning how to slip a hickory brake club through the spokes of a hand-

wheel with all the effortless skill of a pro. In no time a clumsy farm boy was made over into a self-assured, cocky brakeman (Fig. 1.36). He would soon learn that this was a grim and dangerous job, and that what seemed only a game was in reality more of a contact sport in which one played catch-me-who-can with death. A coupling mishap meant only a smashed hand or a missing finger, but a fall from the cars could result in a broken back, a severed leg, or death. The brakeman was like a bareback rider in the circus, except that there was less to hold on to. He was expected to spring from car to car, a task not too difficult for a young, sprightly man on a dry day. But it became far more treacherous on wet nights, especially if the track was rough.

Even when just standing, the brakeman was at risk, because a hot cinder down the back could set him to dancing in an effort to shake it out, and that dance could lead to a fall under the wheels. He was not even safe when clinging to a brake wheel. He arcs over the wheel, almost doubled over to gain extra leverage, and twists with all his might. Then the whistle screams, one long release brakes. The brakeman straightens up, but as he does so the wheel lifts up unexpectedly off the stem, for the hold-down nut's threads are stripped, and the hapless brakeman falls over backwards and off the car. Loose grab irons, steps, and ladders were other mechanical hazards awaiting the unsuspecting brakeman. Low bridges and tunnels could sweep him off a car as well; however, most lines strung up "telltales," short lengths of rope made like a giant fringe that would brush over him as the train passed underneath so as to alert him that an obstruction was just ahead and that it was necessary to stay low.

In the event of a wreck the hapless brakeman became a mere projectile. In November 1859, for example, an axle broke on one car of an Erie freight train near Lanesboro, Pennsylvania.[275] Part of the train was passing over a bridge at the time; nine cars derailed because of the broken axle. The brakeman on the ninth car was propelled as if shot from a cannon, landing 100 feet from the bridge. The poor fellow died five days later from his injuries. By the late 1880s it was estimated that one thousand trainmen a year were killed and another four thousand to five thousand injured.[276] Most of the deaths were accountable to men falling off trains, and the majority of these were probably accountable to a simple loss of balance. The lucky ones bit ballast, rolled over a few times, and had nothing worse to show for it than a lot of bruises and a few broken bones. This good shaking up was followed by a painful walk to the next station. Others were not so fortunate. Some fell off at night only to be discovered the next morning as frozen corpses lying alongside the track.

Those members of the crew who managed to

THE AMERICAN RAILROAD FREIGHT CAR

FIGURE 1.36

Brakemen were admonished never to sit on the brake wheel, for a hard bump could send them flying backwards onto the tracks and under the wheels. Like so many of his carefree fellows, this trainman considered himself to be immortal. (Smithsonian Photo 78-16478)

survive had plenty of other troubles to worry over. The weather was hardly kind to men condemned to ride a boxcar roof, for little else was so exposed to the baking sun or freezing wind. Even on a good day the train itself could whip up one hell of a dust storm as the engineer charged downgrade to make up time. The engine would roll and jump as the cars, half-hidden in dust and smoke, zigzagged crazily down the track behind it. The brakemen were half-blinded by the smoke, dust, and cinders generated by the engineer's sprint. Blizzards and freezing rains made the brakeman's exposed position even more miserable. Nor did he care much for a gale wind, which was only intensified by the speed of the train. In a bad storm the men would literally have to crawl along the roof walk boards to do their job. Bad weather could present some unexpected hazards as well. Big storms would drive cattle and buffalo into railroad cuts to seek shelter from the driving wind.[277] Along would come a train, running blind because of poor visibility, only to crash into the herd lodged square on the tracks. This was a problem mainly in the Far West, where the open range permitted the animals to roam freely.

The Wild West offered other dangers not normally encountered on the tamer, eastern side of the continent. Indians, mountain lions, and poisonous snakes were occasionally found in the shadows when the brakeman walked back the prescribed half-mile or so behind the train to flag. One could never be too sure about sounds in the night out in the wilderness; perhaps it was only a jackrabbit, but it sure sounded like a mountain lion. Cowboys were a more likely source of trou-

ble. A young brakeman on the Santa Fe in the 1870s remembered a fight that broke out between two rival cowboy groups riding as drovers on a cattle train.[278] Three cowboys were wounded and one killed; the train crew escaped injury by just staying away from the caboose during the trip. The cowhands proved to be a problem both on and off the train. They liked to shoot out lights; headlights, marker lamps, brakeman's lanterns, they were all fair game, especially when the cowboys were in town celebrating. Crews on the Santa Fe found it prudent to run dark and silent when passing through certain towns on a Saturday night.

Cowboys might be found only on the range, but tramps were on hand everywhere, and the freight train was their favorite way to travel.[279] Most were harmless vagrants who quietly slipped on and off the slow-moving trains. Riding the rods meant literally that—lying out crossways on the four truss rods fastened to the car's underside. Sometimes they would place a board on top of the rods to make a more comfortable perch. Other tramps were more brazen and would board the trains in groups ready to fight or even kill any trainman who threatened to put them off. On the other hand, some trainmen were brutal in their treatment of hobos and used excessive force to evict them. Others were not above taking a bribe for allowing a tramp to complete his ride.

Broken wheels and axles imperiled the trainman from time to time, as did hotboxes (overheated journal boxes) and broken rails. But the mechanical failure he most feared was the break-in-two. This could happen because of a broken pin or link. Sometimes the whole coupler, draft gear and all, would pull loose, especially on an old car with rotten draft timbers. Experienced trainmen listened for a sudden change in the sound of the locomotive's exhaust; this usually signaled a break-in-two. The forward end of the train would run away with the locomotive while the rear end coasted along. The head end could be rammed from behind should it stop before the wild rear section could be brought to a stop as well. However, most break-in-twos happened on a summit, where the strain on the couplings at the center of the train became most severe. The head end would roll forward downgrade while the detached rear end ran backwards down the opposite slope. Sleepy or inattentive crews could be in very deep trouble, because cars with no engine could very quickly become a runaway train.

Air brakes largely eliminated the runaway problem; when a train broke in two, the air hoses would separate and the brakes would automatically clamp on in full emergency. But air brakes did not become common on American freight cars until the very last years of the nineteenth century, and so railroaders in the wood-car era listened with more than half an ear for any sudden change in the stack sounds of the engine up ahead. Janney couplers also helped prevent the dreaded parting of the train, because they were more substantial than the traditional link-and-pin couplers. Yet they too could be held accountable for the trainman's most vexsome problem. Sags or sharp dips in the track would allow Janney-style couplers to disengage, as one car dropped down into the dip. This was not a problem just of undermaintained branch lines; it was found on at least one major trunk line, the Erie, according to a report published in 1896.[280]

The peril to life and limb might be lessened when the train stopped moving, but the duties of the brakeman continued. The rear and front brakemen did double duty as flagmen. When the train made an unscheduled stop on the main line because of a mechanical breakdown or for any other reason, it was essential that the flagman scurry up or down the track to warn approaching trains of their presence. When the train made a regular stop, it was almost always directed into a siding, safely off the main line. Unscheduled stops normally blocked main tracks, and because most lines were single-tracked the danger of a front- or rear-end collision was very real, especially on hilly or curving lines where visibility was not good. The flagman must go back or ahead half a mile and stand guard until he was whistled in by the engineer when the train was ready to proceed. It was usually necessary for the flagman to move quickly and clamber aboard before the train gathered too much speed. The head flag might stand by and hop onto the train as it passed. In the daytime the flagman would handle a red flag to signal approaching trains. At nighttime lanterns replaced flags, but fusees (flares) or torpedoes could be used. The latter devices, of British origin, were small explosive caps strapped to the railhead.[281] They were ignited by the locomotive wheels running over them; the loud report alerted the engine crew to stop, danger was ahead.

After the adoption of air brakes, the head brakeman rode up front on the engine. He not only performed the flagging duties just mentioned but would unlock and open switches when cars were picked up or set off on sidings. He was responsible for relocking the switches and was called on to demonstrate his skill in jumping on and off the footboards of the moving locomotive during these switching operations. Being surefooted was a primary requirement for this part of the job. The head-end brakeman was also expected to help the fireman by pushing coal down from the rear of the tender's bunker and to help manage the waterspout at tank stops. Coupling, uncoupling, and handling the air-hose connections were other responsibilities, but for all these jobs the head brakeman was luckier than his comrades stationed back on top of the train, for he at least could find temporary shelter in the engine cab. The rear brakeman,

ensconced in the caboose for most of the workday, was even more fortunate.

The brakeman job was an entry-level position. It was where you started out in the railway scene and where you might stay if you performed in an unpromising, lackluster way. Better men advanced to become passenger train brakemen or freight train conductors. Some went on to become minor railway officials, and a very few, such as Charles E. Pugh of the Pennsylvania Railroad, rose to general manager.[282] Such grand elevations were rare, and most aging brakemen were quite satisfied to be made freight train conductors. This was surely a better life for a middle-aged man who was no longer up to clambering up and down car steps or racing across icy roof boards. Being sheltered, dry, and warm inside a cozy caboose was far more agreeable, particularly as one's joints stiffened.

The conductor managed the train. He was its chief, and if it was necessary to vary operations, even in the smallest way, it was the conductor who decided what was to be done—so long, of course, that the authority of the chief dispatcher was not being interfered with. The conductor would oversee and even, as the occasion demanded, participate in on-line switching operations. He would watch passing trains for dragging brake beams, hotboxes, or other defects and signal to the conductor of another train whether all was well or not. He would watch for problems on his own train from the cupola of the caboose. This elevated tower was a fine place to sniff for the distinctive odor of a hotbox. He kept an eye on the air-brake pressure gauges mounted in the caboose. A drop in the train line pressure indicated a problem, and if the pressure fell too much the brake would come on automatically. It was far more advisable to stop at the first convenient place and check the system out rather than experience a wheel-crunching emergency stop and risk tying up the main line. The conductor carried extra air hoses and gaskets for just such emergencies. Most of the conductor's time, however, was actually spent on paperwork. He might be the captain of the train, but he was also its clerk. He was the agent responsible for the train and its contents, which depending on the cargo could be worth upwards of $100,000.

The conductor's report, also called the switch list or manifest, was the basic document prepared by the conductor (Fig. 1.37). This sheet listed each car in the train, the initials of the railroad, the car's number (which was very often five digits), where it was picked up or set out, whether it was loaded or empty, its position in the train, and its contents and destination.[283] The latter data was taken from waybills, which accompanied every shipment en route and recorded basic information about the shipper, the cargo, and the consignee. Some switch lists might call for data on the train crew, the when and where of helper service (extra en-

FIGURE 1.37

Conductor's reports or switch lists documented the train's consist and cargo in considerable detail. These reports were maintained by the freight train conductor. Only the upper portion of the form is shown; it should show one hundred cars. (Marshall M. Kirkman, *The Science of Railways*, [Chicago, 1908])

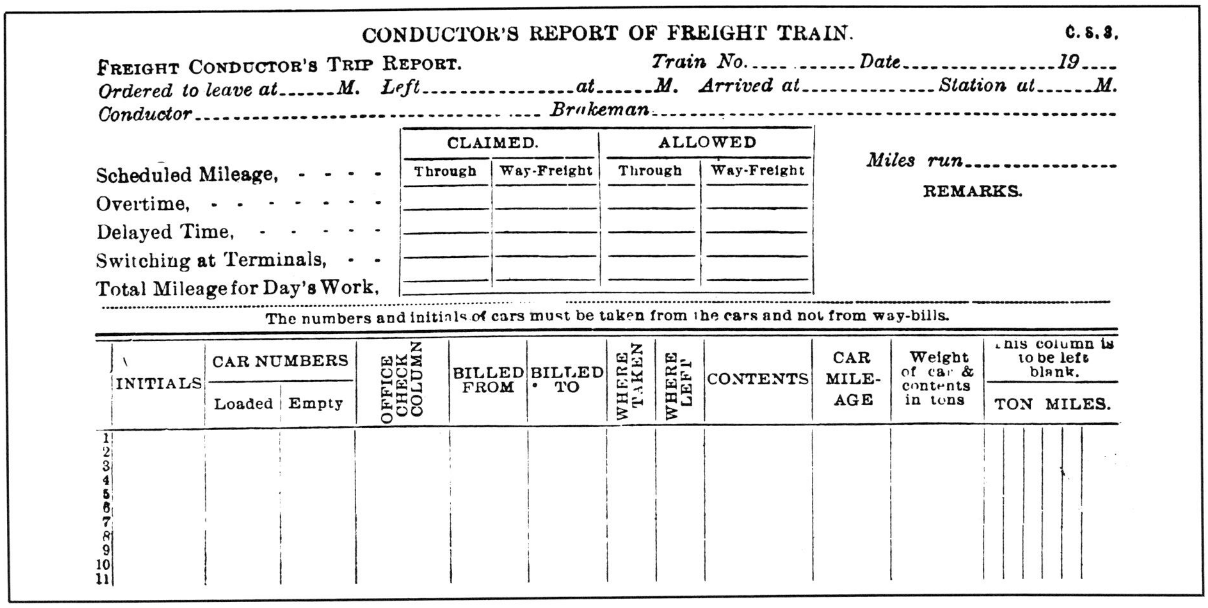

CONDUCTOR'S REPORT OF FREIGHT TRAIN.
C. S. S.

FREIGHT CONDUCTOR'S TRIP REPORT.
Train No._____ Date_____19____

Ordered to leave at_____M. Left_____at_____M. Arrived at_____Station at_____M.

Conductor_____ Brakeman_____

	CLAIMED.		ALLOWED.		
	Through	Way-Freight	Through	Way-Freight	Miles run_____
Scheduled Mileage, - - - -					REMARKS.
Overtime, - - - - - -					
Delayed Time, - - - -					
Switching at Terminals, - -					
Total Mileage for Day's Work,					

The numbers and initials of cars must be taken from the cars and not from way-bills.

	INITIALS	CAR NUMBERS		OFFICE CHECK COLUMN	BILLED FROM	BILLED TO	WHERE TAKEN	WHERE LEFT	CONTENTS	CAR MILE-AGE	Weight of car & contents in tons	This column is to be left blank.
		Loaded	Empty									TON MILES.
1												
2												
3												
4												
5												
6												
7												
8												
9												
10												
11												

FIGURE 1.38

Freight train conductors were kept busy filling out reports to satisfy the paper-happy supervisors at headquarters. Time out for supper was a matter of record on this report. (Bob Richardson, Colorado Railroad Museum, Golden)

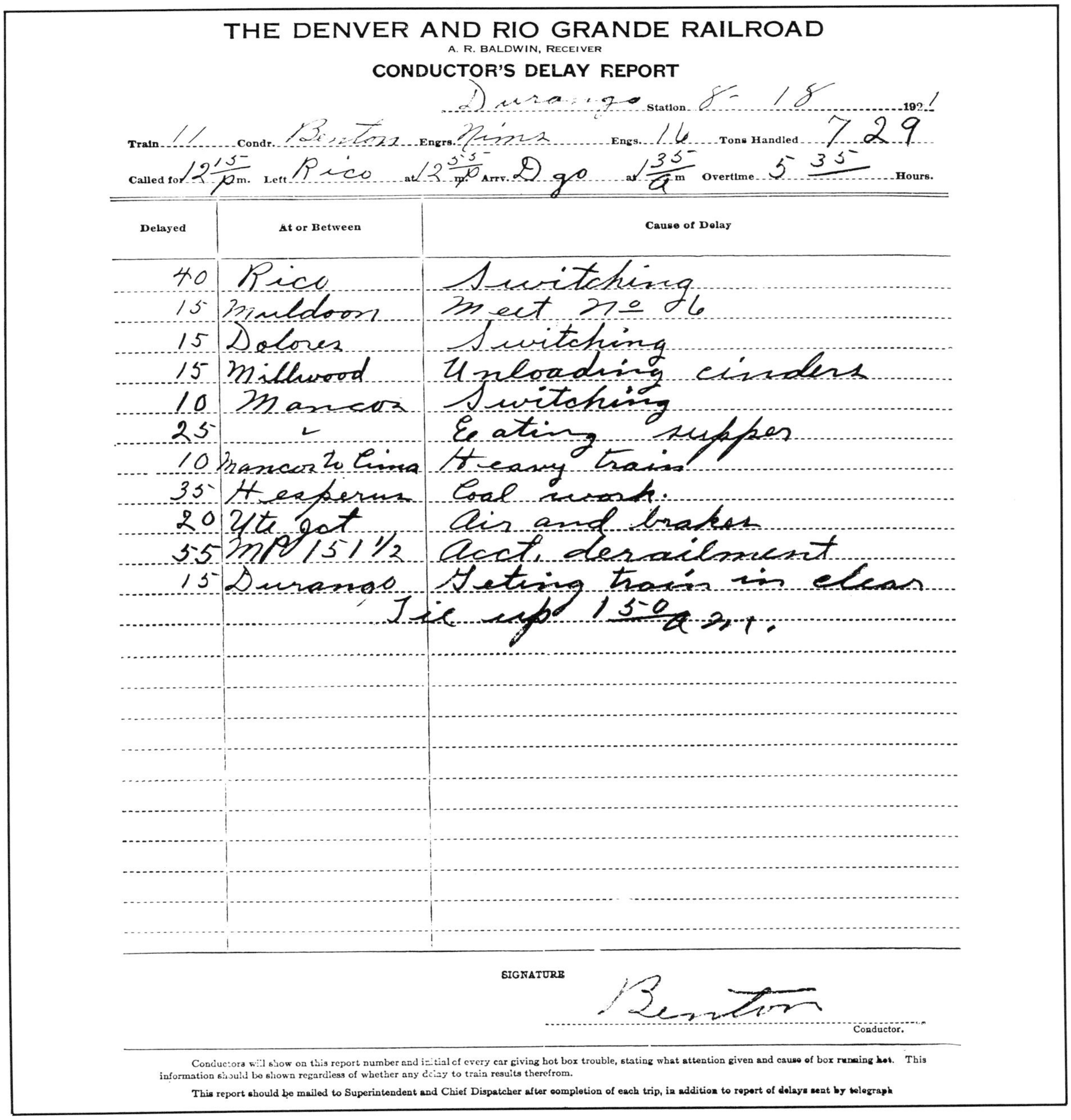

THE DENVER AND RIO GRANDE RAILROAD
A. R. BALDWIN, Receiver
CONDUCTOR'S DELAY REPORT
Durango Station. 8-18 192 1

Train 11 Condr. Benton Engrs. Nims Engs. 16 Tons Handled 729

Called for 12 15 pm. Left Rico at 12 55 m. Arrv. Dgo at 1 35 am Overtime 5 35 Hours.

Delayed	At or Between	Cause of Delay
40	Rico	Switching
15	Muldoon	Meet No. 06
15	Dolores	Switching
15	Millwood	Unloading cinders
10	Mancos	Switching
25	"	Eating supper
10	Mancos to Cima	Heavy train
35	Hesperus	Coal work
20	Yte Jct	Air and brakes
55	MP 151 1/2	Acct. derailment
15	Durango	Getting train in clear
		Tie up 1 50 a.m.

SIGNATURE

Benton

Conductor.

Conductors will show on this report number and initial of every car giving hot box trouble, stating what attention given and cause of box running hot. This information should be shown regardless of whether any delay to train results therefrom.

This report should be mailed to Superintendent and Chief Dispatcher after completion of each trip, in addition to report of delays sent by telegraph.

PARADE OF THE CARS

gines used to push the train up long or steep grades), and the condition of the "car seals." Boxcar doors were not locked or nailed shut. Rather, the door was slid shut and latched. A flimsy tinplated strap was then slipped through the latch and crimped or sealed. Seals were put on both side doors and end lumber doors if the car was so fitted. The conductor had to walk the length of the train on both sides to inspect the seals. His labors at manual record keeping were intensified on those railroads that required him to prepare running slips—individual cards or slips duplicating the information from the switch list.[284] These slips were handed over to the car marker as soon as the train came to rest in the yard so that he could mark in chalk the destination on each car as the train was broken up.

The record-keeping skills, modest as they might appear to a modern audience, were far from simple, because of the need for accuracy. A quick-and-dirty type of record was not acceptable. It must be perfect and readable. All the numbers must be checked and double-checked. Sloppy work would not do, and in a time when figuring was beyond the capacity of the average workingman, the conductor's job was simply not suitable for everyone. Brakemen who were too old for their job often stuck with it anyway simply because they were not up to handling the paperwork that went with being the head man (Fig. 1.38). Detailed record keeping for trains, cargoes, and consists was not a late development, nor even something that appeared only after the beginning of interchange. It went right back to the beginning of the railroad freight business. In his 1928 history of the Baltimore & Ohio, E. H. Hungerford reproduced an original freight train manifest prepared on July 9, 1832, by the conductor of a train running between Baltimore and Frederick, Maryland.[285]

WORKING CONDITIONS

Nineteenth-century workers were accustomed to low wages and dangerous working conditions. Most also accepted the six-day week and the ten-hour day as the price of remaining part of the active work force. A sixty-hour week would be entirely unacceptable to a modern worker, but during the wood-car era most railroaders worked a seventy- or even an eighty-hour week as a routine thing. Railroading, like the steel business, was a twenty-four-hour industry. It was not and could not be a nine-to-five, Monday-to-Friday operation; trains had to move every day at all hours to keep the traffic inching along from terminal to terminal. When factories worked a double shift, railroad crews must be ready to replace empty cars with full ones. When shippers demanded Monday-morning deliveries, railroads were obligated to work Sundays to see that the cars were in place when and where requested.

The normal workday was ten to twelve hours, or enough time to move the train over one division of the railroad.[286] The distance covered was between 100 and 150 miles. The dispatcher was normally much too optimistic, and most train crews were prepared to spend twenty or even twenty-four hours on the road. There was no overtime, and so it cost the railroads nothing in the way of extra wages if the crews were in effect performing double duty. The crews' wages were figured at 15 cents to the mile, or just 15 percent of the train's operating costs.[287] The crew was paid by the day or trip; time by the hour was not really a factor. Most freight train crews ran twenty-six trips a month, and mileage was figured at between 2,600 and 3,500 miles per month. Once the men reached the next division terminal, they were obliged to lay over before taking a train home. Ideally they would turn around after eight to ten hours of rest, but there was little about railroading during this period that was ideal. If times were busy, the men were turned around and ordered to take a train back to the home terminal. There was no time to rest, eat, or clean up; just sign the dispatcher's book and take No. 6 out immediately.

The evils of this system were recorded in an accident report of 1883.[288] A freight train conductor had been at work for eleven days without rest. He brought a train in at 2:00 P.M. and fell into bed without benefit of a meal, nor did he trouble to undress. Only two hours later the call boy came to fetch him; he was ordered back to work, and to disobey would cost him his job. Once out on the road, the exhausted man was too sleepy to observe the normal operations of the train under his care. The train pulled into a siding, and a careless brakeman forgot to close and lock the switch. A following train ran into the standing train, doing some damage to the cars, yet fortunately no one was killed or seriously injured. Even so, the conductor was discharged for negligence of duty. Here was an instance of a sober, faithful employee unfairly discharged because he was simply too tired to hold his head up. The injustice of such cases suggested the need for legal limits on the hours of service, which did not come until 1907, when federal law limited the hours of service for train crews to sixteen hours. This limit was strengthened greatly by the eight-hour day established by the Adamson Act of 1916.

It must be admitted that not all sleepy train crews were products of overwork, nor were they always found out. In the 1850s a freight train pulled into a siding near Smyrna, Tennessee, on the Nashville & Chattanooga Railroad.[289] It was about midnight in this lonely and quiet part of the state. The late hour and the stillness soon lulled the engineer to sleep. He slept so soundly that he was not awakened by the noise of a passing train. In fact, he was not even sure if the train had passed. Nor, indeed, was any of the crew, for they had all dozed off as well. They decided to move out onto the main line but dared not pick up speed for fear that the other train might just be late and could come screaming around the next curve. And so the freight crept along with a flagman walking well in advance until they approached Nashville. They blamed their lateness on a derailment and were believed because derailments were so common at the time. A derailment happened in fact on every other trip, but little damage was normally done because speeds were so slow.

Complaints about long hours and overwork subsided when traffic was down. Now there was time to catch up on one's sleep. There was too much idle time, in fact, for a workingman could not exist very long with no work. The railroad paid only when he was putting in hours on the train. Long, payless furloughs prompted many a railroader to seek work in another field, for in slow times only men high up on the seniority list found steady employment on the railroad. Even then a crew might find itself stranded in the next division town waiting for a return train to take them home. Three- and four-day layovers were not uncommon.[290]

Think of idling around a smoky railroad town far from home and family. Cheap hotels and bars were about all that were open to the poor trainman. A 50-cent-a-night hotel was too expensive for trainmen on the North Pacific during its first years of operation, and many would simply spend the night in a vacant boxcar.[291] Food was another basic problem for a man away from home. Some railroads like the Santa Fe ran eating houses and issued employees meal tickets, often called pie cards.[292] Sometimes the pie cards were honored in local restaurants of modest quality. In other instances no food was available because of the location or hours of work. Twenty-four-hour restaurants were not common, especially in isolated junctions and terminal towns.[293] A wife could be counted on to pack a lunch, but the unmarried trainman was left to his own devices.

One of the most annoying problems of railroad train service was the irregularity of the hours. You could be called at just about any hour to report for duty. A call might come in the middle of the night, during Sunday dinner, or in the middle of a daughter's birthday party. There was no way to plan ahead for household chores, vacations, or any other personal business. Family life suffered as well; because Dad was away so often, he became a stranger to his wife and children. Meals and sleep were taken on the fly at odd hours. Trainmen were expected to live near the roundhouse or yard office, and this normally meant no more than a mile away. He must be easily findable, and so an off-duty man could not wander off unless he had been specially excused from work for the day. A blanket excuse was limited to illness, a death in the family,

or some other major event. The call boys were always on the prowl and combed homes, apartments, boardinghouses, and hotels in their search for the men picked off the board by the dispatcher. Once any one individual was found, he was told to sign the book carried by the call boy. He was now on official notice to report at the yard in time to take out No. 6 bound west for Logansport at 6:38 A.M. Call boys were instructed never to recruit men if they were found in a saloon or dance hall, because intoxicated crews were not wanted; but it was also understood that not many trains would run if this rule were too scrupulously observed.[294]

Never having a Sunday off, unless the national economy was in a deep depression, was another major grievance railroad workers had with their employers. Protecting the sanctity of the Sabbath was one of the great moral issues of the nineteenth century. Church officials, community leaders, and even some businessmen spoke out passionately about preserving the Lord's Day. It was, after all, a divine commandment and so nothing to take lightly. The U.S. Centennial was accordingly closed on Sunday in deference to these feelings. And so when railroad workers urged the suspension of train service, they were not without support. Obviously, many railroad officials agreed with the idea of no Sunday trains. In 1854 New England railroads were said to be scrupulous in the observance of Sunday as a day of rest; even the Iron Horse reposed quietly in his stall.[295]

Matters were handled with a greater laxity elsewhere. The Hudson River Railroad ran only mail trains on Sunday, but it began to reinstate trains of all types because of the service offered by its competitors. In the West the Chicago, Burlington & Quincy stopped all Sunday train operations in 1859, but it too began to resume some service in later years.[296] By 1880 the nation's railroads had all but abandoned the Sunday prohibition and ran trains freely.[297] It was noticed that papers were printed, markets were open, and streetcars ran on Sundays, and the world seemed no more sinful for it. Despite the shift toward godlessness, 450 engineers on the New York Central sent their boss, William H. Vanderbilt, a petition to end Sunday trains.[298] They had petitioned before but hoped that this year their unending labors might be lessened. A seven-day schedule ruined their health, made them prematurely old and worn out. Vanderbilt was generally more responsive to the appeals of labor than some of his counterparts in the railroad industry, but there is no indication that he was ready to end Sunday service.

LABOR RELATIONS

The job might be dangerous, poor-paying, and hampered by long hours, yet the railroads appeared to have no problem hiring employees. Some managers were in fact picky about who they would have on the job. One reforming superintendent discharged 60 percent of the brakemen in order to get rid of the rowdy element or those given to drinking or pilfering the cars.[299] In some cases a new superintendent would order a mass firing so that he could bring in his own followers or lackeys. Such actions went square in the face of the seniority system that had become a sacred cow among railroad workers with union support. Job security and recognition of loyalty and long service were the main virtues of the seniority system. At the same time, advancement or appointment to choice jobs had little to do with brains, skill, ability, or extra effort. Whatever the merits or defects of the tenure system, it did prevail and was a basic element in anything to do with railway labor. Everything, down to who ran what train, was based on seniority.

This is not to say that management had no handle on its employees. The grip of railroad managers on their men was a stern and direct one. It was also a very necessary one, for good discipline was essential to protect life and property. Operating rules might strike the outsider as inconvenient or even silly, but these procedures were neither arbitrary nor casual. They had been developed over long years and reflected countless lessons in how and how not to operate trains. Follow the rules exactly, and an accident will never happen. Ignore or bend the rules, and you might speed up things a bit or just slip by unhurt, but the odds are that you will cause an accident. It was therefore the general policy of all railroads to enforce their rules rigorously. Employees were required to memorize the rule book and were commonly tested on their knowledge of this catechism of safety.

When an accident occurred—and small mishaps were a daily occurrence—a hotheaded supervisor might be tempted to fire those he thought responsible, on the spot. But it occurred to others that a more thoughtful and considered process should follow every accident or serious infraction of the rule book. George R. Brown (1840–1916) devised such a system around 1875.[300] He had started at the bottom of the railroad employment ladder as a water boy, only to rise through the ranks as a telegraph lineman, then dispatcher, and finally superintendent. He could see the discipline problem from both the employee and the managerial viewpoint. He came to understand the need for a fairer, less hurried approach to judgment.

For starters, the accused would keep on working while the investigation was under way. Demerits would be assigned according to the nature of the offense. Demerits served as an early warning to the employee to be more careful—he was being watched. If he continued to earn more demerits, he would be disciplined according to the number acquired. For fairly minor offenses he would be let go temporarily without pay, or reduced in rank.

For major offenses—let's say he willfully broke the rules and demolished a train—he would be fired outright. Most employees made mistakes, and so most earned demerits, but these could be eliminated in part or entirely by long periods of good service when no new black marks were earned. The system was not perfect and was still subject to the inconsistencies of human judgment; however, it was about the best plan developed and so became something of an industry standard. Brown did not publish a book about his code until 1897.

The acceptance of Brown's plan for discipline on a more human level is indicative of the trend toward greater fairness and compassion in employee relations. Paternalism was evident as early as 1869, when the Central Pacific opened a hospital in Sacramento for its workers.[301] Within a few years it had grown to 125 beds. Each employee paid 50 cents a month to help defray costs, but the corporation was accountable for most of the expenses. Curiously, the scheme was credited to C. P. Huntington, as tight-fisted and conservative a capitalist as ever breathed, yet he saw an advantage in keeping his employees fit. After the Central and the Southern Pacific combined, Huntington's idea for hospitals spread, so that in 1883 1,166 patients were treated by company doctors. Huntington extended the medical aid scheme to the Chesapeake & Ohio Railroad in the same year, but rather than build a hospital he contracted with existing facilities in Richmond, Virginia, in Lexington, Kentucky, and elsewhere. Meanwhile, other railroads, especially in the West, built hospitals or at least established medical departments. Most concentrated exclusively on the treatment of railroad employees. The Illinois Central and the Chicago & North Western had medical departments that also attended to accident victims.

The Vanderbilt lines saw no reason to join the faddish notion of establishing railroad employee hospitals, reasoning perhaps that an abundance of doctors and medical facilities already existed in the densely settled Northeast. What was needed, however, was a wholesome home away from home. Trainmen spent at least half of their free time away from home. The first thing to confront a trainman tired and lonely after a hard day on the road was the station bar. A few drinks after work would not hurt. Feeling mellow and cheered, our trainman moves down the street, where he falls under the dangerous allurements of further drink and vice. There was plenty of liquor and bad companionship to be found in nearly every major rail terminal, and without the support and moral uplift of his family, the best of men could fall in with bad company.

The alternative to this easy road to degradation was a sober, Christian setting where a poor workingman could find shelter at an affordable cost.

FIGURE 1.39

The Brotherhood of Railroad Trainmen issued poster-size certificates to its members that illustrate the benefit program available to the family of a trainmen. The story of the trainman's membership, workday, death, and memorial service is depicted in tabloid-style vignettes. (Smithsonian Neg. 73-4619)

Coverage at first was limited to disability, accidents, and death. Payments were relatively modest; during the 1880s an accidental death brought the survivors $1,030, while a natural death yielded only $373. Even so, a poor widow was far better off with these amounts than she would have been before the Relief Association existed, when she could have expected nothing. A pension program started in 1884, and within three years 165 pensioners were drawing an average payment of $125 per year.

The employee insurance idea had actually been pioneered by craft unions some years before the B&O program got under way. The Brotherhood of Locomotive Engineers established a mutual insurance organization in 1867 and was paying out $2.5 million to widows and orphans by 1888.[305] Union membership certificates emphasized the insurance protection aspects of the organization. Large illustrated certificates were given out by all of the craft unions. A typical example from the Brotherhood of Railroad Trainmen is reproduced in Figure 1.39. The motto of the organization—"Benevolence, Sobriety, Industry"—is gracefully supported by a pair of cupids and a guardian angel. The benefits of union membership are described in the vignettes that decorate the certificate. In one scene the member buys a policy for his family. He is next shown leaving home, then standing tall atop a boxcar. The train then plunges into a river; boxcars are shown bobbing around in the water. We now see a memorial service at the lodge hall and, finally, presentation of the insurance check to the widow.

The union movement started late and in a surprisingly passive fashion, considering the privations of the common American worker. The first railroad unions were not in fact concerned with the lower levels of labor; they concentrated on the needs of the more skilled workers. Pay, work rules, hours of service, and perpetuation of the seniority system were the chief concerns of craft unions. The first of these groups, the Brotherhood of Locomotive Engineers, was organized in 1863 more along fraternal than trade union lines. Within a year or so the BLE had sixteen hundred members.[306] Authorized strikes were rare, though some wildcat stoppages did occur under local lodges rather than national sponsorship. In January 1875 the BLE signed an agreement with the New York Central that guaranteed each engineer a minimum wage of $3.50 per day for any trip under one hundred miles.[307] For journeys beyond this limit, the pay would be figured at 3.5 cents per mile; however, no extra pay was authorized for Sunday work. Elsewhere the BLE seemed to do less well and cautiously avoided participating in the great railway strike of 1877. It lost the nearly yearlong battle with the Burlington (1888–1889) over seniority rules, but despite this setback, the ranks of the

The British founders of the Young Men's Christian Association had no idea of creating a chain of modest hotels for U.S. railroad workers when they established the organization in the 1840s.[302] A branch opened in the United States in 1851, and in 1869 a hotel-like operation began in New York City, sponsored by the YMCA. In 1872 some rooms in Cleveland, Ohio, at the Lake Shore station were set aside for railroad men. Here they could find a clean room and Bible-reading sessions. They would have to go down the street for booze and girls. Cornelius Vanderbilt, though hardly a bluenose himself, liked the idea of the railroad-sponsored YMCA because it was good business to keep the boys sober and rested if they were about to run the trains. Vanderbilt and his elder son, William, became loyal patrons of the YMCA. By about 1910 there were 240 railroad YMCAs located in all parts of the nation.[303]

Worker benefits, by modern standards at least, were nonexistent. Paid holidays and vacations were a dream of the future. Retirement and health insurance were things that only Socialists talked about with any hope. Some railroad managers who were hardly extreme liberals began to understand, like Bismarck, that it was simply good politics to promote some of the benefits workers seemed to desire so passionately. A little conciliation rather than confrontation might in the end win labor's cooperation and so benefit the company. This notion gained currency after the bitter and destructive 1877 railway strike. In 1880 the Baltimore & Ohio Railroad established a Relief Association that was essentially a mutual insurance organization.[304] Most costs were borne by employees, but the company did pay a portion of the expenses.

BLE had swelled by then to over twenty-five thousand members. Other unions were established in the following order: conductors, 1868; firemen, 1873; and trainmen, 1883.

As more workers became union members, railroad managers found it increasingly necessary to consider the needs of their employees more carefully. Some lines, such as the Chicago & North Western, adopted a generous pay policy and were blessed with labor peace even during the big strikes of 1877 and 1894.[308] Most other railroads felt less generous, and when business was poor they cut payrolls ruthlessly. Soon after the Panic of 1873, many eastern trunk lines cut remuneration by 10 percent.[309] Labor stood fast and accepted this cut as a cost of the depression. However, a second 10 percent decrease in July 1877 elicited a vigorous protest. The firemen at Martinsburg, West Virginia, walked off the job protesting the cut and the general lateness in awarding payment. Some paydays were two and even four months behind. The strikers at Martinsburg rioted when the state militia was sent in and repulsed the governor's small army. Federal troops were next sent to restore order, if not the cooperation of the firemen.

The strike spread throughout the East and reached as far west as Omaha. Things got especially hot in Pittsburgh. Here the issue was not so much wage cuts as the reduction in the number of trainmen employed. To cut crew size, the Pennsylvania was running long trains powered by two locomotives. A protest strike erupted, and with coupling pins and hickory brake clubs the strikers hammered anyone attempting to take out the doubleheader freight trains. Passenger trains were not subject to the strikers' wrath, because they were operated as before. All freight service was stopped. Tensions and tempers began to mount and then exploded when federal troops were brought in. On the evening of July 21, 1877, the strikers, fueled by a generous supply of whiskey looted from cars standing in the yard, expressed their frustration with the great and powerful Pennsylvania Railroad: they burned the depot, the roundhouse, and most of the nearby rolling stock. The damage was estimated at between $5 million and $10 million—a rather high price to pay for the few thousand dollars saved by cutting the wages of men who earned barely enough for subsistence.

The lesson of 1877 was lost on George M. Pullman, the owner of the largest sleeping car company and railroad car manufacturing facility in the world. Pullman's firm was so profitable that he scarcely knew how to dispose of its cash surplus. The panic of 1893 drove down car orders and profits, but the business was hardly threatened or in danger of failure. Three-quarters of the costs of building a passenger car were in plant and materials. These costs were basically fixed, but the cost of labor, which constituted one-quarter of a new car's price, was more easily manipulated; and so between September 1893 and May 1894 wages at the Pullman Car Works were reduced 25 percent. A number of men had already been laid off, and rents had been raised in company-owned model towns. All of these events were more than the shop men were able to endure, and so a strike began on May 11, 1894. Pullman did not believe in unions, employee rights, or strikes, and he steadfastly refused to negotiate.

The newly organized American Railway Union, the first general railroad workers' union, decided to stop handling Pullman cars as a way to help the workers at the Pullman Car Works. Eugene V. Debs (1855–1926), as president of the 150,000-member ARU, believed in the sympathy boycott, but the federal government sided with Pullman. Debs was arrested and jailed for contempt of a U.S. court. The ARU collapsed, and railway labor returned to the multiple fiefdoms of individual craft unions. The brotherhoods actually proved rather effective in negotiating better working conditions and wages. Federal legislation affecting safety and hours of service was due in part at least to the dogged lobbying efforts of the brotherhood.

Schedules

The trainmen operated the trains, but they did not run the railroad. This was the job of the dispatcher.[310] He in turn was governed by a master plan devised by the general manager or the trainmaster, commonly called the schedule or time card. This elaborate working document outlined the progress of all the trains on the railroad, be they freight, passenger, or work trains. It should not be confused with the far simpler printed timetable intended for travelers or travel agents. The working schedule was an internal document intended exclusively for the staff of the operating department. It delineated the position and time of each train so that dozens of trains might safely overtake or pass one another on a single-track railroad.

For example, if express train No. 1 is to keep on schedule, it is necessary for train No. 6 to get out of the way at Springfield and remain on a siding until the express has passed. If the weather cooperates and traffic is normal, the system written down on the schedule works like a Swiss watch, but should delays occur, the dispatcher must improvise. The schedule may be the foundation of railway operations, but everyone experienced in train service understands that the time card is in reality only an idealized outline of how trains should behave but rarely do. One delay can affect half a dozen trains. In this ballet of heavyweights, the dispatcher must be a skillful choreographer. He must smoothly direct slow-going gravel trains off the main so that a fast-moving produce train can pass. His problems are compounded by second sections, work trains, convention specials, and even handcars.

The orderly, safe, and prompt movement of trains is basically a job of keeping the slow ones out of the way of the fast ones. When playing checkers with 1,000-ton trains, the player must have more than skill and concentration. He must be decisive and sure of his judgments. He must have an exact knowledge of his division. Exactly how long is each siding? Can a class 2100 engine pull twenty loaded and twenty-six empties up Xenia hill and make it to Sherman siding in sixteen minutes? Are the water tanks close enough together to allow a beef special to make it over the division before the commuter rush starts at 5:15? The dispatcher must know these facts before he can successfully start to reshuffle the operating schedule. Tampering with the schedule, especially on a busy section of railroad, works only when you know all the facts. Each little detail is important. Each small change can affect several trains. Each change is a potential problem; any lapse on the dispatcher's part can cause an accident or, in a worst-case scenario, a disaster. The dispatcher's job is clearly a high-pressure and stressful one. It could make one irritable, dyspep-

tic, and old before one's time. Early retirement was not an unnatural choice for many dispatchers.

The time card involved a number of rules or rights that were hardly democratic. Trains were not all equal. Inferior trains got out of the way of superior ones with no questions, arguments, or hesitation. Express passenger trains, as might be expected, were the most superior of trains. Yet certain freight trains, particularly produce and cattle trains, would outrank passenger trains, and certain specials like silk trains, with cargoes of extraordinary value, would outrank even the premiere extra-fare passenger train. Of course, the operation of a silk special was occasional, and so for normal day-to-day purposes, trainmasters lived by an order of four classes or rankings.[311] First-class trains were passenger expresses and, during the season, special freights, as already noted. Second-class trains would cover less prestigious passenger trains and through freights. Third and fourth class covered peddler freights and work trains. Each train had a number that increased in size as its ranking descended; the premiere passenger trains would have very low numbers, while the most humble freight would surely have a double-digit designation. It was common to assign even numbers to east- or northbound trains; the east-

FIGURE 1.40

Two pages from the Atlanta & Charlotte Air Line Railway's Dec. 29, 1878, employee's timetable. Note the train meets listed in the remarks column. (Collection of Richard F. Dole)

WESTERN DIVISION.

NO. 5—LOCAL FREIGHT—Eastward.

Miles bet. Stations.	STATIONS.	TIME OF ARRIVAL.	TIME OF DEP'RTURE	REMARKS.
		A. M.	A. M.	
	Atlanta		5 00	Lower Office.
11	Goodwin's . . .	5 50	5 50	
4	Doraville . . .	6 09	6 22	Let No. 3 Pass.
5	Norcross	6 46	7 00	
6	Duluth	7 24	7 32	
5	Suwannee . . .	8 00	8 10	
6	Buford	8 38	8 47	
7	Flowry Bnch	9 15	9 25	Meet No. 10.
3	Odell's . . .			
6	Gainesville	10 05	10 25	Meet No. 2.
2	New Holland . .			
4	W. Sulph. Springs	10 50	10 50	
7	Lula	11 17	11 30	
1	Bellton . . .	11 35	11 45	Meet No. 6.
		P. M.	P. M.	
7	Longview	12 15	12 20	
6	Mount Airy . . .	12 47	12 55	
7	Ayersville . . .	1 25	1 25	
7	Toccoa	1 52	2 42	
5	Tugalo	3 05	3 05	
4	Fort Madison . .			
4	Harbin's . . .	3 45	3 45	
5	Westminster . .	4 10	4 20	
9	Seneca	5 05	5 22	Meet No. 4.
13	Central	6 20		

Start promptly—Run steadily.

Lose no time at stations and wood-racks that can possibly be avoided.

The Conductor must see that all his hands are prompt in getting the wood on the Tender. The Enginemen will oil their engines themselves and see that firemen help throw on the wood.

Freight Train Conductors will educate their train hands to carry out Rule 46 to the letter.

Observe the difference in meeting and terminal points. Freight Trains leave terminal points on card time, regardless of over due Freight Trains which must be kept out of their way; but if Passenger Trains are due and not arrived, Freight Trains will not leave terminal points without special orders.

WESTERN DIVISION

No. 6.—LOCAL FREIGHT—Westward.

Miles bet. Stations.	STATIONS.	TIME OF ARRIVAL.	TIME OF DEP'RTURE	REMARKS.
		A. M.	A. M.	
	Central		4 35	
13	Seneca	5 30	5 40	
9	Westminster . .	6 20	6 30	
5	Harbin's . . .	6 52	6 52	
4	Fort Madison . .			
4	Tugalo	7 32	7 32	
5	Toccoa . . .	8 00	8 40	Let No. 2 pass.
7	Ayersville · ·	9 15	9 30	Meet No. 3.
6	Mount Airy · ·	10 10	10 17	
6	Longview	10 45	10 50	
7	Bellton . . .	11 21	11 36	Meet No. 5.
1	Lula	11 40	11 55	
		P. M.	P. M.	
7	W. Sulph. Springs	12 22	12 22	
4	New Holland . .			
2	Gainesville . . .	12 50	1 07	
6	Odell's			
3	Flowry Bnch	1 50	2 03	Meet No. 9.
7	Buford	2 30	2 40	
6	Suwannee . . .	3 06	3 16	
5	Duluth	3 39	3 54	Meet No. 1.
6	Norcross	4 17	4 25	
5	Doraville . . .	4 50	4 57	
4	Goodwin's . . .	5 12	5 12	
11	Atlanta	6 00		Lower Office.

Read carefully the Rules for Running Trains by Telegraph, and if not sure that they are fully understood, do not attempt to use them.

Engines of Passenger Trains must not carry a red signal without special orders from this office.

Special attention is called to Rules 14 and 15.

Enginemen will note carefully Rule No. 74.

Observe Bulletin Board daily, and note new orders.

All eastern bound trains must come to a full stop before crossing the Switch at Reedy River.

OPERATORS—Notice change in Telegraph Signals, and study carefully Rules for Running Trains by Telegraph; particularly Rule 102.

Note change in numbering of trains.

Catasauqua and Fogelsville Rail Road Co.

SIGNALS AND RULES FOR TIME TABLE, No. 25.

SIGNALS.

1. **ONE WHISTLE,** "Stop! Down Brakes," **TWO WHISTLES,** "Start! Up Brakes," **THREE WHISTLES,** "Back the Engine or Train," **FOUR WHISTLES,** calls the Signalman. **THREE LONG AND TWO SHORT WHISTLES** for Road Crossings and Curves.

2. Irregular and prolonged whistling signifies that cattle are on the track, and is a signal to apply the brakes.

3. **RED,** always signifies Danger! Stop! **BLUE,** Caution! Move Slowly! **WHITE,** All Right, Pass On.

4. A Lamp, Hat or Hand moved Up and Down means, **"STOP."** A Lamp moved Across the Track means, **"BACK,"** if swung Over the Head, it means **"MOVE AHEAD."**

5. Any signal violently given is to be considered a signal of danger, and means Stop.

6. A Red Flag or Red Lantern upon the front of an Engine, signifies that another Engine is following.

7. Any Engine carrying flags for another Engine running on the same schedule, in passing any Engine on the road will blow Four Short Whistles.

8. All persons working upon the main track must give notice of any obstruction by exhibiting red flags or red lanterns, at least 700 feet from the point of danger in each direction.

9. When running, **TWO TAPS** of the bell signifies, stop at the Next Station. **ONE TAP,** Stop Immediately. **THREE TAPS,** Run Slow. When standing, **TWO TAPS,** Start. **THREE TAPS,** Back the engine or train.

RULES.

1. Passenger trains will have the right of road for 20 minutes beyond schedule time, after which freight trains will proceed with great caution, keeping 20 minutes behind schedule time, until the trains pass, unless otherwise ordered by schedule or special order.

2. Gravel and other irregular trains must keep clear of the regular trains, and be off the track at least 10 minutes in advance of the regular time of any of them, and always, when standing on the main track, have a flag out in each direction.

3. The Conductor of an extra train, following a regular train, must advise the Conductor and Engineer of the same in all cases, and will see that the regular train carries flags for it.

4. Any train for which flags are carried will have the right of road for 20 minutes behind the schedule time of the train carrying the flags.

5. All trains will be run as near schedule time as possible, and under no circumstances leave a station in advance of it—always take the safe side in case of doubt.

6. Five Minutes will be allowed in all cases for variation of watches. If delayed trains should meet between stations, the one shall back out which shall cause the least delay and danger.

7. No. 5 and 6 will keep out of the way of No. 1.

8. No. 3 will approach Trexlertown carefully, and see that No. 7 is out of the way, and will not leave Trexlertown for Breinigsville until 2.20, and will not leave Trexlertown for Alburtis until 2.36, unless No. 8 has arrived.

9. No. 7 will leave cars on the main track at Trexlertown, while they go to Alburtis and return.

10. The Engineer will be with his engine in time to have it in running order and before his train at least ten minutes previous to the time of departure.

11. Conductors and Engineers will both be held responsible for the safety of the train. They will, therefore, in case of doubt or danger, consult or advise together.

12. Brakemen must be at their places when the train starts, and not leave them until it stops; they will be under the charge of the Conductor or Engineer, and shall assist in filling the tender with fuel, cleaning the engine, &c.

13. The Conductors of Freight Trains will always see that the brakes are applied to any cars that are left upon the side-tracks, to prevent their running or being blown off the switch.

14. Run slow and with great care from Rittenhouse Gap to Lock Ridge, and occupy TWO MINUTES in crossing JORDAN BRIDGE.

15. Sound the whistle or ring the bell before approaching and while passing towns, stations, road-crossings, gangs of workmen, and all short curves.

16. See that all switches on the main track are always right for that track and locked, except at the moment of using; after using a switch, the conductor, or if there is none, the Engineer must see that it is secure for the main track, excepting Rule 17.

17. The switch at Alburtis will be left for the branch by all trains. The switch at Trexlertown will be left for the branch by No. 5; by Nos. 1 and 2 while on the branch, by No. 7, by No. 3 while on the branch, and by No. 4.

18. The Conductor of each train must see before its departure that his cars are in good order, safe and oiled ; and cars not in good running order must be put off immediately for repairs.

19. Permit no Engine or Train to stand on a public street or road crossing.

20. Use the greatest care in giving and observing signals, and, in cases of doubt, always take the safe side.

21. Whenever any schedule train is more than 20 minutes behind its own time, it becomes irregular, and will not run within 10 minutes of the schedule time of any opposite train, unless ordered by schedule or special order.

22. When two Trains moving in opposite directions are both late, each will wait 30 minutes beyond the schedule time of the arrival of the other train, and then start, running the curves until the other train is passed or heard from.

23. All persons employed on the road or trains, while upon duty, must abstain wholly and entirely from intoxicating drinks: any person being intoxicated will be immediately dismissed.

C. W. CHAPMAN,
SUPT. AND ENGR.

Catasauqua, Pa., January 1st, 1879.

FIGURE 1.41

Small railroads could compress their rules and signals onto a single sheet, while large railroads required a small book to list them all. The inconsistency of rules and signals from railroad to railroad was a dangerous factor in the operation of trains. (Collection of Richard F. Dole)

bound Broadway would then be train No. 2. West- or southbound trains were given odd numbers; the westbound Broadway would be train No. 1. East- and northbound trains of equal rank had preference over west- and southbound trains. For some unexplained reason the opposite was true west of the Mississippi River. Such timetable rules may sound a trifle arbitrary, but on single-track railroads the establishment of preference between up-bound and downbound trains was a practical necessity. There were no disputes or arguments; every train crew understood its obligation to pull off the main line at meeting points. The inferior train must be on a siding with switches closed and locked five minutes before the superior train was due to pass.

Employees were given printed timetables listing all scheduled trains. Such documents, common by 1870, provided essential information to train crews, station agents, and track repair foremen. Knowledge of traffic on the road was critical for the safety of all concerned. On a major railroad an employee's timetable, even when limited to a single division, would be a very large and cumbersome document. When printed as one sheet, it was as big as a roadmap. If it was presented in bound form, it would be a sizable booklet. Small railroads could compress their timetables into a single sheet or perhaps just a few letter-size pages. All such timetables were clearly dated, and every worker was instructed to discard the old time sheet when a new one was issued. Reference to an obsolete timetable could result in delays and accidents. Some railroads included their operating rules or signal codes on the reverse of the sheet. One railroad, the Atlanta & Charlotte Air Line, made space at the bottom of the page for catchy sayings such as "Start promptly—Run steadily." Particular stress on knowing the operating rules is also evident on this schedule (Figs. 1.40, 1.41).

Time was the master of railway operations. This was especially true before the time of the fixed, trackside signal, which was essentially a twentieth-century development. The governance of train operations came down to the essential of a time-interval system. Train No. 6 was to be at Omaha at exactly 5:06 P.M., train No. 20 was to be at the same place at 8:20 P.M., and so on. If each train was at a specified place at a specified time, the timetable system was a success. Time, then, was critical; but curiously, absolutely precise timekeeping was not really essential. Railroad time was fixed to a central clock at a major terminal; hence the big clock at the Cleveland station became the source for official time on the Big Four. It did not really matter if the master clock was a few minutes fast, so long as every other clock and watch on the railroad agreed with it.[312] Consistent time was the key to safe and efficient operation. It was necessary for headquarters to know if all station clocks were regulated so as to agree with the master clock. It became the duty of conductors to check the station clock at each stop with their pocket watches—which had been set with the master clock at the beginning of the run. After the introduction of the telegraph, clock settings were handled by wire.

Communications

It was a rare day when something did not happen to disrupt or slow down train operations. A small delay to one train tended to compound so that all trains were affected. Rules were developed for this reason. When a train was late, crews did not strike out on their own; they first resorted to the rule book for guidance. Before the telegraph there was no way to contact the dispatcher for instructions. Crews were instructed to wait on the siding for late trains and to proceed out onto the main track only after a reasonable wait (half an hour was normal) for the delayed train. Caution was the underlying philosophy of the rule book. Go ahead, but go slowly, watching always for an unexpected obstacle or problem ahead. The engineman peered ahead for signs of smoke on the horizon, which might be his first indication of an approaching train. A distant whistle or the glint of a headlight far up the line were other danger signals watched for by alert trainmen.

TRAIN ORDERS, TELEGRAPHY, AND
STATION SIGNALS

Written instructions, called train orders, from the dispatcher to a train crew could modify normal timetable instructions and so proved useful in unscrambling schedule problems. The distribution of train orders over any distance was all but impossible before the introduction of the telegraph.

Commercial telegraphy began in 1844, but no one recognized its merits for train dispatching for almost another decade. When a train left a station, it was almost like a ship going to sea. No one knew its whereabouts until it reached the next terminal. An expectant stationmaster might look in vain for the morning freight. It was overdue by an hour. Without a telegraph he could not even be sure if it had left the last station. One could only wait; again, smoke on the horizon was usually the first hopeful sign. If impatience overtook reason, the stationmaster might venture down the track on foot, by handcar, or on horseback. He would find the train down the line somewhere delayed by a derailed tender, a washed-out culvert, or some other problem. Or he might never find his quarry, because the train had been held up at the last station for any number of reasons. There were a great many unknowns in railroading during its pioneering days. Much of the mystery was eliminated when an exasperated general manager on the Erie manned the wire and ordered all trains to stand clear so that his train could move ahead. This historic telegraph of 1851 opened a new era in train dispatching.

Railroad telegraphy was by no means a sudden phenomenon, however, and it was not immediately used for everyday train operations. One railroad in New York reported in 1853 to a railroad safety committee of the state legislature that telegraph dispatching was dangerous and confusing. Even the progressive Pennsylvania Railroad used the telegraph sparingly at first, and then only for emergencies. Around 1855, for example, an accident tied up the Western Division of the PRR.[313] All freight trains were stationary, and passenger trains were moving at a walking pace with a flagman in the lead. The superintendent, Thomas A. Scott, was away from the office at the time, and so his very young and brash secretary, Andrew Carnegie, took it upon himself to man the telegraph and send out orders, as he had done in the past in similar circumstances. The point of the story is that telegraphy was still seen as an extraordinary rather than a routine tool for directing train operations. Just when it was adopted for everyday use is uncertain; there seems to be little evidence on the subject. However, if the massive number of telegraph dispatches from the U.S. Military Railroad (now in the National Archives) is any indicator, telegraphic train dispatching had come into extensive use by the early 1860s.

Some railroads still resisted the use of such radical devices for some years after the Civil War. At least two major New England railroads would have nothing whatever to do with telegraph dispatching into the 1870s.[314] But elsewhere trainmasters, station agents, and especially dispatchers wondered how they had ever managed the traffic fluctuations that so characterized train service. In truth, the telegraph did not put the dispatcher in direct contact with the train or its crew; he could only call towermen and stationmasters along the line. Thus his contact with the train operatives was indirect. The dispatcher would telegraph the station agent at the Madeira depot to pass on an order to train No. 6, which he expected to pass that point in half an hour. The order might tell the crew to stop at the first siding and allow No. 13 to pass. Whatever the particulars of the message, the station agent would take it down and then immediately repeat the message to the dispatcher via telegraph to make sure he had recorded it correctly. Written copies were then made and fastened to train-order hoops. The agent would set his manual station signal to indicate that orders were to be picked up. As the train passed by, at a reduced rate of speed, one hoop was passed up to the engine crew and a second hoop to the conductor on the caboose. And so in this clumsy, laborious way the telegraphic message of the dispatcher reached the train crew.

Signal masts at depots along the line could control train movements through telegraphic messages sent by the chief dispatcher (Fig. 1.25). Once again this was a manual system; the station agent manually set or moved a lever to change the position of the semaphore from *clear* to *caution* or *stop*, depending on his instructions. At night a kerosene lamp illuminated a glass lens fixed to one end of the signal; the white lens indicated *all clear, safe,* or *go ahead*. Red, of course, indicated *stop*, but green, the modern color for *go ahead*, indicated *caution, slow down,* and *be prepared to stop*. These semaphore masts became sentinels along the line that divided railroads into segments called blocks. Space then became a factor in train operations. Dispatchers began to think in terms of space as well as time. Ideally, if only one train were in a block at any given time, no collision could possibly occur, but not all railroads could afford the luxury of one block for one train. "Permissive block" signaling permitted other trains to enter an occupied block so long as they proceeded with caution. This helped move trains and traffic, but it greatly reduced the margins of safety.

FIXED TRACKSIDE SIGNALS

Manual block signaling entered a more sophisticated phase in 1863 when the Camden & Amboy installed illuminated banner signals along twenty-five miles of its road from Trenton to Kensington.[315] It was a stop-or-go system, with a clear glass panel for *go* and a red glass for *stop*. The glass panels were raised or lowered by ropes. It was in no way an automatic system, but it was probably the first instance of a fixed-signal system in this country, because the signals were placed along the tracks as needed, rather than being tied to a station or terminal location. At night kerosene lamps backlighted the glass panels. Trains were governed by the signal rather than train orders, which represented a significant difference from normal operating procedures. The schedule remained the basic agenda for operations, and the dispatcher could still alter train movements through written train orders, but by and large, train crews now kept their eyes on the boxy signals mounted on posts along the tracks. The system worked well enough that the C&A extended it on to New Brunswick and then all the way to Jersey City, so that by 1870 ninety miles of track was controlled by manual block signaling.

The Pennsylvania adopted the C&A plan in 1873 for part of its main line in the vicinity of Pittsburgh. The system was extended along the entire main line, with signal towers placed every mile or two, between 1875 and 1876. The banner signals were worked with ropes by the towerman, who was in telegraphic contact with other signals up and down the line. The chief dispatcher could call any towerman by wire as needed. The Reading concocted a similar system of manual block signal towers starting in 1863.[316] These distinctive octagons looked something like windmills because of the large, slatted mast or vane mounted on their roofs. The towerman could turn the vane so that the white side faced an approaching train to indicate *go*; if red showed, the train was to stop, but if blue showed, the engineer was to slow down. At first the windmill towers were placed only at major junctions, tunnel mouths, or other congested places along the railroad, but in time forty of them were spread along just about the entire length of the Reading's main line. Replacement of the windmills with electric signals did not begin until around 1896.

Mechanical interlocking was introduced in England in 1856 for major junctions and terminals to coordinate the movement of switches and signals. The first such machine to appear in the United States was a British import brought over for service near Trenton by the United Railroads of New Jersey (formerly the Camden & Amboy) in 1870.[317] Some five years later a second and larger imported interlocking machine was installed by the United Railroads at East Newark. At the same time, the New York Central installed some interlocking apparatus of domestic manufacture. The dangers and inconvenience of manual switching persuaded other railroad managers to adopt interlocking machines for especially busy junctions elsewhere in the country. The pace of installation was surely slow, for such expensive apparatus could only be justified by special needs.

For all the wonderful complexity and ingenuity of the interlocking machines, they remained nonautomatic, manual signals. Some forward-thinking men dreamed of an automatic railway signal. They wanted some device or system that was not dependent on an operator, a telegraph message, or a hand signal. Operators were in fact the weak link in the signaling network. They were prone to be inattentive, drowsy, bored, or negligent. Such human failing caused accidents. What was wanted was a mechanical system, independent of human control, that was worked by the train itself. When the train passed, the signal turned red, thus warning following trains to stop.

FIGURE 1.42

A big Manchester ten-wheeler drags a long freight past a semaphore and a disc signal on the Boston & Lowell around 1885. Fixed signals were rare on American railroads at this time and were usually in place only at stations and major junctions. (R&LHS Collection)

When train No. 1 left the block, the signal, now well to its rear, would flash white and the waiting train might pass on.

The concept was obvious and simple. Working out the details and perfecting a workable system was another matter. This is a story too complex to go into fully here, but let us just briefly consider the work of two pioneer investigators in this field, William Robinson and Thomas S. Hall.[318] Both men were Yankees, and both began work on automatic electric railway signals in the late 1860s. Robinson concentrated on track circuitry and devised a closed-circuit scheme with insulated joints that became a basic feature in railway signaling. He obtained a patent in 1872 and spent considerable energy on demonstrations and promotions, but he seems never to have won a sponsor for a large-scale installation. The ultimate soundness of his ideas was recognized by George Westinghouse, who in 1880–1881 bought up Robinson's Union Electric Signal Company and made it part of the newly formed Union Switch & Signal Company.

Thomas S. Hall of Stamford, Connecticut, being a pragmatic New Englander, decided to invent a better railway signal after being involved in an 1866 railway accident.[319] Hall felt that automatic signaling was needed to save labor costs and to eliminate the fallible human element from train control. Hall used wheel-actuated treadles mounted next to the railhead to trigger an electri-

cal switch that caused a solenoid to move cloth-covered wire hoops in front of a circular glass window. One hoop was covered with red silk, the other with white silk. An oil lamp illuminated the window opening at night. The apparatus was housed in a distinctive banjo-shaped enclosure. Batteries supplied the current for the signal. A test installation was made near the inventor's home in 1868, but the first commercial application was not made until 1871, when the Eastern Railroad equipped sixteen miles of its line with Hall's apparatus. The inventor died in 1880, but his firm continued on for another forty-five years. Enclosed-disc signals were installed on many railroads, including the Lehigh Valley, the Reading, and the Illinois Central. The simplicity and reliability of Hall's design did much to promote the cause of automatic electric signaling in this country.

SIGNALS AND TRAFFIC DENSITY

By 1880 the technology for sophisticated railway signaling was available. The will to install it and the cash to pay for it, however, were not. For the most part, railroad managers saw more elaborate signaling apparatus as a luxury that could be justified only for high-traffic lines. The old time card/dispatcher system was perfectly adequate for most train operations, and moneys available would be better spent on steel rail and other improvements. Fancy signal apparatus was a low priority on most railroads. The Cincinnati, New Orleans & Texas Pacific Railway, however, was an exception. It began to install automatic electric signals in 1891.[320] Within nine years 80 percent of the line was equipped with electric signals. Other roads were slow to follow; in 1900 only 2,280 miles of U.S. railroads were controlled by automatic signals.[321] The numbers soon began to improve; by 1902–1903 3,350 miles, by 1908 10,800 miles had been so equipped.[322] The percentage increase during these years was impressive, but in fact only a very small portion of the national system was protected by first-class signaling. In 1908, for example, only a meager 5 percent of American railroads had automatic signaling.

The operating departments might argue convincingly that most traffic densities were just not high enough to justify the cost of automatic signaling. They could no doubt make a good case. The same tight-fisted policy seems less credible when applied to manual signaling, yet it held true there as well. By early 1903 only about 30,000 miles of line were under the manual block system, only 6,500 miles over what it had been a decade earlier. Slow progress, indeed. By 1908 just over 58,000 miles, or roughly 38 percent of the total system, were manually blocked. A radical change in viewpoint seems to have come about by 1910. Automatic block signaling suddenly seems to have come of age. Railroads were now ready to bypass the transition from the time card to manual blocking for a straight-on plunge into electric signaling. By 1920 nearly all major main-line trackage was protected by automatic signals.[323] The rapid conversion from indifference to enthusiasm is difficult to explain, but it may well be accountable to a late realization that more double track, more powerful locomotives, and larger cars were not enough to handle mounting traffic. Signaling offered a cheaper alternative, and so it at last became a welcome part of railway operations.

Elaborate trackside signaling was hardly needed on railroads running only a few trains a day. Wellington estimated in 1887 that even under the most unfavorable conditions, most railroads would operate a minimum of one passenger and one freight train per day.[324] Most railroads exceeded Wellington's pessimistic estimate, at least during good times. The little Sacramento Valley Railroad, California's premiere railway, was an example of good traffic generated by travel to and from the gold fields.[325] In 1856, with only the first 22 miles of the line in place, six trains per day were operated. Four of these were for passengers. A freight train ran up the line but returned as a mixed passenger and freight. Small railroads in less active regions of the nation, such as the St. Paul & Pacific, could be expected to run fewer trains despite a far greater mileage. In 1870 the 76-mile-long St. Paul & Pacific ran two mixed passenger/freight trains per day.[326] During the immigrant season two supplemental freight trains were operated to handle the extra traffic. On branch lines, work trains were expected to do double duty in picking up freight cars dropped off by regular trains.

The St. Paul was a pioneer road serving an unsettled territory. Larger traffics were reported by railroads in more developed regions, as might be expected. The Louisville & Nashville reported an average of 3.9 trains per day in 1867 and 11.5 trains per day in 1873.[327] This total included freight and passenger service. The New York Central, one of the largest trunk lines in the nation, reported surprisingly modest statistics for the same period.[328] In 1868 it ran from twelve to fourteen freight trains per day each way. The number had doubled by 1872, and the figures would be much inflated if passenger service was included as well, as is true for the L&N figures. The 1880 Census claimed that the average number of trains over U.S. railroads was 3.9 per day. Such a low average would confirm Wellington's estimate that a fairly large portion of the national system saw no more than one or two trains per day. Certain segments of the system, however, carried a goodly number of trains and so needed double track, heavy rail, deep ballast, steel bridges, and block signaling. One such road was the Pittsburgh, Fort Wayne & Chicago, which regularly handled 23 freights per day, or 8,452 per year in 1881.[329] The Lake Shore was even busier; in 1879 it ran 58,927 trains (2.2 million cars) over what must have been one of the busiest railroads in the nation.[330]

As in so much to do with the early railroad freight business, data on the number of trains operated is fragmentary. It is also somewhat contradictory. From the sources used in the foregoing discussion, one would assume that if three trains per day were reported, it meant just that. Superficially, these numbers are correct; on the average, let's say, three trains per day were scheduled. *Scheduled* is the key word here, because a railroad would schedule no more trains than was necessary to keep its plan of operation simple. However, it was a common operating practice to run a large number of sections to accommodate the traffic. Hence freight train No. 6 was rarely just one train; it might be followed by one, two, three, or more duplicates.

Why so many small trains operated in series? The reason was small locomotives and cars, light track, and bridges. The operation of a single monster train to handle a given tonnage was not practical or advisable, except on a few roads such as the Reading. So the cargo was divided up among a number of small trains that ran in a closely spaced parade, normally just a few minutes apart, but all under one train number. Because freights moved so slowly, the trains could operate with reasonable safety bunched close together. Technically, train No. 6 would consist of a single locomotive and 40 cars, but when all the sections were added together, it was in reality 6 locomotives and 250 cars. This practice goes back to the 1850s on the Baltimore & Ohio, although earlier examples may exist.[331] In 1855 that line scheduled three freights per day, but each one was made up of as many as eleven sections each. Later in the century the Erie outdid the B&O's passion for extras in a grand fashion.[332] Only a morning and an evening freight were shown on the official schedule, but ninety-eight extra freights were routinely operated.

MANUAL SIGNALS

Since the first days of railroading, train crews have employed a variety of simple, cheap signaling devices such as flags, lanterns, and, of course, their own hands and arms to convey messages to one another. This form of signaling is best described as manual signaling and is distinguished from trackside or fixed signals such as semaphores or other types of mast devices.

Railway hand signals were surely borrowed in large part from existing military and maritime practice, some of which is of ancient origins. Some hand signals seem almost instinctive; the beckoning motion for *come ahead* or *come here* hardly needs to be explained. Or the sign whereby one or both hands are made to paddle away from the person signaling would be readily understood around the world to mean *go away* or *back up*. So too when both hands are raised above the head at a forward angle, palms out and fingers up; this seems always clearly to mean *stop* or *danger ahead*.

Such natural hand motions were commonly used, but some railroads felt the need to develop their own set of motions. Here are some of the refinements used in the mid-1880s.[333] *Go ahead* or *start* was indicated by a sweeping apart of the hands at eye level. *Stop* was indicated by a downward motion of the hand with both arms extended. A sweeping motion of the hands overhead meant *back slowly*. Some railroads published diagrams to indicate their hand signals, and a few of these have been reproduced in recent volumes on railroad history. There appears to be agreement on the *stop* signals (two arms held up overhead) and the *all right* signal (one arm held straight out perpendicular to the body.[334] Two positions were employed to indicate a *slow down* or *caution*. In one case one arm was held downward at a 45-degree angle, but on another railroad one arm was held vertically overhead to convey the same message.

The switching of cars involved many special hand signals. One of the most common was that given to the engineer to give a car or group of cars a shove or kick forward. This was indicated by the brakeman, who would wiggle one or both fists near his shoulder. The faster the wiggle, the harder the engineer was to kick cars. The *cut off* sign was made by holding both hands closed before one's chest and then parting them suddenly by moving the arms outward. At this instant the engineer shut off steam, applied the brakes, and allowed the cars to roll ahead under their own momentum onto a siding or wherever the brakeman wanted them and had lined up the switches ready for the cars' short, free-wheeling journey.[335]

Trainmen and yard crews devised special signals for their own use. On the Erie, for example, a slap on the cheek meant *place cars in track B1*, or a pat on the seat of the pants and then the raising of one hand meant *bring in all but the last car*. Hand signals could also convey very precise and important information affecting a train's safety. Crews were obliged to scrutinize every passing train for dragging brake rigging, loose cargoes, or overheated bearings. A developing hotbox gave off a strong odor, while a fully matured one produced smoke. To alert a crew of this problem, a conductor or brakeman would stand on the rear platform of the caboose and signal to the conductor of the train passing on the opposite track. He would hold his nose to indicate a bad odor, and hence the presence of a hotbox or overheated journal bearing. He would next signal the approximate location of the hotbox by touching his neck if it was at the head of the train, or his waist if it was at the middle, or his lower leg if it was at the end.

During daylight fairly complicated hand signals were possible. Colored flags were sometimes used, especially by a track repair crew watchman or crossing guards, for some hand signaling as well. But once the sun set, hand signaling was basically useless, and train crews resorted to lanterns. Recognizable gestures and body language must now be translated into tiny moving lights just barely visible from one end of the train to the other. Lantern signaling was far more limited than daytime hand signaling, because only about four patterns were possible. Raising and lowering the lantern normally meant *start* or *go ahead*, yet on some railroads it meant *stop*.[336] A lamp swinging back and forth meant *stop*. When it was swung in a small circle, the engineer was being told to back up. If the swing was done in a large circle, the message was that the train had broken in two.

Although the number of distinctive lantern signals was limited, the railroads' individual definitions of what they meant were at considerable variance. In 1878 the railroads entering Chicago showed great independence in interpreting the bobbing of lamps.[337] This lack of a common language must have caused not just misunderstandings but accidents as well. On three railroads, up and down meant *stop*, while on six other roads the back-and-forth or swinging motion indicated this basic message. There was almost no unanimity when it came to the *back up* signal, although three lines selected the up-and-down motion used elsewhere for *stop*.

Confusion was less universal when it came to the selection of colors for flags and lamps. All seemed to agree that red represented *stop* or *immediate danger ahead*. White was also generally accepted to represent *go, safe, all clear*, or *move ahead*. The right color for *caution* proved to be more of a problem. Both green and blue were used long years before yellow became accepted. Green was in fact the common designation for *go slow, look out*, or *be ready to stop* until the advent of electric signaling around 1910.[338] Concern over the possibility of a false *clear* signal should a colored lens fall out prompted some railroads and signal manufacturers to select green as the color for *go*. White or clear indicated a problem, and crews were instructed to stop. Once green became the tone for *go*, a new color was needed for *caution*, and so yellow came to the fore. During the green-for-*caution* period—basically, pre-1910—a few lines preferred blue. The Reading was a major line using blue for *caution*. The Pennsylvania could not make up its mind on the subject, and so it used both for a time in the 1870s.[339] Its block signals used blue banners, but green flags or lanterns were used to denote the *go slow* warning. After yellow became the accepted color for *caution*, blue was used to indicate that a car inspector was around or under the train and that it must not be moved until the blue flag or lamp had been removed.

Flags, lanterns, and hand gestures were essentially passive in nature, and trainmen needed something more assertive for emergencies. Gunpowder had been startling viewers in China for centuries before steam railroads were even envisioned. The flash and bang of fireworks, meant only to amuse, was redirected in the 1840s for the more serious business of railroad safety. Two devices were contrived for this purpose: the torpedo (an exploding cap) and the fusee or flare (a bright-burning stick about the size of a candle). The torpedo, or detonator, was introduced in England in either 1837 or 1841, primarily as a fog signal.[340] Fogs were common in that rainy land, and so visibility was generally poor; hence the need for an audible signal was very real. The torpedo was nothing more than a small watertight canister filled with gunpowder and a few caps. It was strapped on top of the rail and would explode when the wheel of a locomotive or car passed over it. Trainmen generally fastened several down in the event that one proved to be a dud. Torpedoes appear to have become a common trainman's signal by the 1850s.

Fusees or flares were a visual signal but were far more arresting than any other artificial light available to ordinary trainmen because of their bright, intense light. They would burn upside down, on their side, or even underwater; hence they were perfect for the adverse weather conditions when emergency signaling needs were at their peak. An engineer peering through a blowing snowstorm would not see a flag or lantern until he was on top of the flagman and it might be too late to stop, but he could see the sparkle and glare of a fusee from a greater distance. A few hundred feet could make the difference between a safe stop and a collision. Fusees were often dropped off slowly moving trains to warn following trains to slow down, stay back, or be alert for a train ahead.[341]

Any train coming upon a lighted fusee was obliged to stop and wait for it to burn out before proceeding. This provided the time interval needed by the train ahead to get out of the way. If it was mechanically crippled or just overloaded and underpowered, the rear flagman would continue to drop lighted flares onto the track from the rear platform of the caboose. If his train broke down and stalled, he would hop to the ground and walk back the prescribed distance to warn following trains to stop. He carried a flagman's kit of flags, torpedoes, and flares to be used as the occasion dictated.

Most flares were designed to burn for five minutes, but ten- and fifteen-minute models were also available. The more elaborate fifteen-minute flares burn first red, then yellow, and then white so that engine crews could estimate how far ahead the preceding train might be. The typical flare or fusee was a cardboard tube about an inch in diameter filled with a mixture of strontium nitrate, potassium perchlorate, and sulphur; sawdust or sand was used as a filler. The pull-off cap was made like a safety match striker to ignite the flare. Most were made with a spike or sharp nail fixed to one end of the cylinder. A practiced flagman blessed with a good aim and considerable luck could impale the spiked end of the fusee in a crosstie. Yet even if he missed the target, the fusee would burn on no matter how it landed, but it would be more visible if it landed in an upright position.

The engine crew was the recipient of the various signals just described and only rarely resorted themselves to lanterns, flags, fusees, or torpedoes. The engine's bell offered a fairly passive warning to the trainmen and pedestrians that the engine was about to start or indeed was under way. The bell was normally used only in a terminal or road-crossing setting. On the other hand, cab bells were a means of communication between the conductor and the engine crew. The large brass gongs were mounted inside the cab roof. A rope passed back through the cars (on a passenger train) or over the car roofs back to the caboose (on a freight train). A conductor could send simple communications the length of the train just by pulling the rope. One ring of the gong might indicate *go*, two gongs *stop*, and three gongs *back up*. The codes would, of course, vary from railroad to railroad, and the signal bell system was suitable only for the simplest type of messages. Signal bells were commonplace on passenger trains but more rare on freights because of the length of the train. The difficulties of maintaining a workable cord increased with the number of cars. Too much rope slack rendered the system unworkable, and with too little, the rope would break. Reels mounted high in the caboose were used to adjust the length and tension of the gong rope.

The most forceful trainboard signal was the locomotive's whistle, which gave the engineer a mighty voice to warn trespassers off his path or to announce to other members of the crew his intentions about starting or stopping. To the layman the sound of the whistle is just a lot of idle tooting, but in fact it is almost always a studied code that has a clear meaning to the train crew. They are conditioned to listen carefully to the number of notes and the duration of each. The number of notes sounded has meaning. A long or a short blast or a combination of longs and shorts can vary the meaning as well. The whistle code as it finally evolved came down to this for the most elementary signals: one short—*stop* or *apply all brakes*; two longs—*start* or *release brakes*; three shorts—*back up*. Four (or five) shorts recalled the flagman to the train because it was about to get under way. Two longs, one short, and another long blast indicated an approaching highway or road crossing. This signal was generally repeated for very busy crossings, especially if vehicular traffic was evident. The final long shriek was held as the locomotive passed over the crossing. A rapid repeat of shorts indicated *cattle on the track, be prepared to hit a steer or apply the brakes*. The succession of whistle hoots would ideally frighten the cattle off the track, but the crew still had to be alert to what was ahead.

The fact that the pattern of whistling had a definite meaning and was not just random tooting played a clear role in the safe operation of railroad trains, but the effect was diminished because no universal code prevailed. In 1881 a survey of some two hundred American railroads revealed a decided lack of uniformity in the whistle code.[342] There appeared to be agreement on only about one signal—three shorts for *back up*. Otherwise each railroad seemed to follow its own system of whistle signaling. Some would use one short for *stop, brakes on*, while others would use two shorts for the same message. *Brakes off* could be designated by two shorts, or one short, or two longs. Five shorts could mean *recall the flagman, note flag on the engine*, or *train broken in two*. If railroad workers spent a lifetime with one company, a universal code might not have much significance, but railroaders tended to drift from one road to another. There was in fact a class of drifters, known as boomers, who preferred to move on as the mood struck them. They might study the rule book for each new employer but in a time of panic or forgetfulness react all wrong because they were confused by the whistle signal of a new railroad. Their mistake might go unnoticed, or it might wreck a train.

The industry belatedly took notice of this problem in 1884 at the Southern Time Convention.[343] A uniform train signal code was adopted by the railroads in that region. In April 1886 most major trunk lines came to adopt the Standard Signal Code, which governed all forms of railway signaling as practiced by trainmen. Some major lines such as the Union Pacific and the Chesapeake & Ohio were not at first convinced of the merits of this plan and did not adopt it until some time later.

Obstacles to Smooth Running

The operation of trains, especially in the wood-car era, was a wearisome thing. There was so much that could go wrong. The weather was forever acting up with a succession of storms, floods, and blizzards. Once winter's ice and snow abated, the spring rains sent streams over their banks. Bridge abutments were scoured loose, track roadbeds were washed out or made spongy. Even when the weather was more benign, the physical plant would break down. Because of uncertain metallurgy, bad design, or indifferent maintenance, things mechanical kept breaking. Rails and rail joints, axles, wheels, and tires would crack from time to time. Springs and bolts would snap. Wooden parts, such as ties and car frames, failed through dry rot.

Delays and accidents could result when even a small part of the system was out of adjustment. If the track gauge spread just an inch or so, a train could be dropped onto the ties. A slipped eccentric, caused by the loosening of two small set screws, could stall a locomotive and so tie up the line for hours. A side rod key is a small item on a locomotive, but its loss can likewise bring operations to a halt. Bad water can cause an engine boiler to foam or prime and so reduce power or damage the reciprocating parts of the locomotive. A poor grade of coal won't make sufficient steam, and so the train runs behind schedule or stalls altogether. An inexperienced fireman can't always keep up steam even with good coal, and an inept engineer lacks the skill to get the most out of his engine and so can't keep his train on schedule. And so the reasons go on about why trains ran late or not at all, and why the job of moving large tonnages over the immature railroads of the last century was such a terrible ordeal.

THE PHYSICAL PLANT

The industry liked to blame delays on nature, or better yet, acts of God, but in fact most were the result of poor planning and human error. The failure of railroad management to anticipate the rise and fall of traffic levels was chronic and resulted in periodic blockades. Goods would pile up at freight houses while loaded cars jammed the yards. A shortage of locomotives was often the cause of such backups. In 1882 the New York Central experienced a sizable blockade of freight cars at its Syracuse yard.[344] Twenty-five hundred cars a day rolled in from the West, and there were just not enough locomotives to move them all out. Repairing stored engines or buying new ones was a slow business, and shippers were obliged to wait. The Baltimore & Ohio faced the same problem in December 1889 when it moved 2,114 carloads of corn during the first twenty days of the month, or 334 more than it had carried during the same period one year earlier.[345] This fairly modest increase in tonnage caught the road unprepared. New engines on order were not ready. The Mt. Clare shops were full of repair work and could do nothing to help in the short term. Hence the corn shippers were obliged to stand by or seek another avenue for their commerce.

Equipment shortages were common after the period of slow traffic that normally followed a panic. If the economy revived too suddenly, most railroads were badly prepared to respond in a speedy fashion. The Union Pacific's Idaho Division was found particularly ill prepared for a revival of business in 1884.[346] The locomotives were in such poor condition that they could hardly pull themselves. Those capable of running were pressed into snowplow service, and so the trains stood idle for days at a time. Cars and water tanks were falling to pieces. There was so little lamp oil that train crews could not signal properly at night. Traffic was turned away because the railroad was not functional; everything needed to provide service was in short supply.

If equipment shortages could not be rectified in short order, other aspects of the physical plant were even less easily reordered. The great majority of the American railroad network was single track (Fig. 1.43). In 1890, when most of the system was in place, 94.8 percent of it was single track.[347] Eleven years later it remained 93.5 percent single track, mainly because it would have cost too much to install a second track. Many railroad managers were convinced that the cost was not justified except in certain special instances. Single-track lines were cheap in terms of first cost but expensive in terms of convenience and safety. If single-track lines were operated like a one-way street, with trains following one another in an obedient procession, no one would find them objectionable. Faster trains could run around the slower ones parked in sidings along the way. But of course, railroad traffic was a two-way business. It might be heavier in one direction than the other, but it was nonetheless bidirectional, and this is where the delays, frustrations, and dangers came into play.

Two well-known incidents in railroad history illustrate these problems. First, let us consider congestion on a single-track line as witnessed by the abortive effort of James J. Andrews and his group of co-conspirators to destroy the Western & Atlantic Railroad in April 1862.[348] They planned to seize a train near Atlanta and burn major railroad bridges as they progressed northward toward Chattanooga, so disabling a vital Confederate rail link. The morning mixed train was scheduled to meet only two southbound trains; hence Andrews assumed from the official timetable that he would face few obstacles in his flight northward. On the day before the raid, all trains ran on schedule. It was dry that day, but it rained on the day of the raid. The rain was enough to delay all trains. In addition, two extra freights were on the line, and this cost Andrews considerable time—more than an hour on one siding. Normally, such delays were dismissed as routine problems inherent to single-track railroading; a little rain and two extra trains were really not enough to slow down the operation significantly. In this case the price was a heavy one, for Andrews and his men were captured and imprisoned. He and seven others were executed as spies. Had the W&A been double-tracked, Andrews's mission might have been a success and he would have been remembered as a Union hero rather than a martyr.

The death of John Luther "Casey" Jones was a more direct result of single-track railroading.[349] Jones was a fast runner and had been disciplined more than once for ignoring the first rules of engine running: caution, prudence, and safety first. In April 1900 he took a southbound passenger train, No. 1, out of Memphis at 12:50 A.M.. It was running ninety-five minutes late, and Jones was determined to make up most or all of that time on the straight, level track of the Illinois Central. There were no stops, and the only train meets were far away at a little town call Vaughn, Mississippi, where he was to pass a few inferior trains.

While Jones was blasting along toward the south, things were getting a trifle complicated at Vaughn. Two freight trains—No. 83 southbound and No. 72 northbound—were trying to occupy a single siding that was too short by just a few cars, so that one end of this nose-to-nose pair protruded out onto the main line. This meant that a "saw by" would be necessary if another train were to pass on the main line. The two freights would position themselves so that one of the switches was clear and the main-line train could pass that point. Once it had pulled by the clear switch, a flagman would open that switch so that the two freights could back partway out of the siding and thus clear the other switch at the far end of the siding. The way would now be clear for the train on the main line to continue on its way. All this awkward switch throwing and back-and-forth movement were one price of single-track railroading.

FIGURE 1.43

Single track was nearly universal
on American railroads in the
wood-car era. This Erie freight
must hurry along to the next
siding to clear the main line for
preference trains. This scene is at
Bluffton, Ohio, around 1890.
(Gerber's Studio, Bluffton, Ohio)

The crews on freights 83 and 72 were accustomed to saw bys, and they had performed the maneuver successfully with a local passenger train just before Jones's train appeared on the scene. While the two freights were repositioning themselves on the siding, an air hose between two of the cars broke, and the trains were stalled with several cars protruding onto the main line from the north rather than the south end of the siding. Jones's orders advised him of the saw-by meet at Vaughn, so he had been forewarned to slow down. A flagman walked 3,200 feet north of the switch. Torpedoes were fixed to the railhead 500 feet south of that point. Meanwhile, the freight train crew tried to replace the air hose and pump off the brakes so that they could move the train and thus clear the switch and the main line. They too knew that the No. 1 was due soon.

It was dark and difficult to see with only lanterns for illumination. The work went slowly. Jones, however, was moving rapidly toward a scene that called for caution, not speed, yet he sailed by the flagman at 70 miles per hour, too fast to slow down, much less to stop. The freight train crew

scattered when they heard him coming. The accident was not a notable one and would be forgotten today except for the song that was published a few years later. A caboose and one boxcar were destroyed. The locomotive was damaged, but not seriously. No passengers or train crewmen were seriously injured, but the engineer, Casey Jones, was killed, another victim of fast running on a single-track railroad.

Single-track railroads were not very forgiving. They could carry a large traffic if carefully managed and all operating personnel observed the rules scrupulously. But small mistakes proved costly in terms of time, confusion, and even lives. Some American lines began to copy the British enthusiasm for double-tracking at an early age, but the movement was a modest one at best. The Reading was a pioneer in this area, adding a second track to its 90-mile main line in 1843–1844. Even at that early date, this double-tracking constituted little more than a token segment of the national system. If ever there was proof of Carnegie's maxim "Pioneering don't pay," the Reading's decision to double-track was it. The clear space between the tracks was only 4 feet—sufficient for the midget cars of the 1830s, but too narrow for the wider cars that soon came into fashion. The loading gauge was spread in 1862 and once again in 1885.[350] The cost of these reconstructions must have proved discouraging to a railroad that had attempted to do it right from the start but had built on too small a scale.

The New York Central Railroad completed a fairly large-scale double-tracking project—Buffalo to Albany, 300 miles—in the late 1850s.[351] The Hudson River Railroad had rebuilt most of its main line in the same fashion by 1865. The Pennsylvania added a second track on the mountainous western end of its main line, Altoona to Pittsburgh, at the same time.[352] No western road seems to have been smitten with double-track fever, except for the Lake Shore, which set out to double its entire main line from Buffalo to Toledo by early 1873.[353] The Lake Shore had been on a betterment binge since its birth in an 1869 merger. Double-tracking of the eastern end of the road began soon after the merger. Next came a new main track south from Ashtabula, Ohio. Long passing sidings were put in on existing single-track portions. Old yards were enlarged, and big new ones like Collingwood east of Cleveland were opened. The Panic of 1873 ended the improvement program for nearly eight years, but as traffic and revenues recovered, the progressive Lake Shore management got back to work. Not only was the double-tracking program pushed forward, but portions of the main line were rebuilt to reduce maximum grade from 37 feet per mile to 26.4 feet.

The actions of the Lake Shore won few converts. Bits and pieces of the American railroad network were converted to double track. A few sections of main line, notably on the New York Central, saw four tracks, but these were the exceptions, and rare exceptions at that. The arguments for capacity, convenience, and safety were not convincing enough for most cost-conscious railroad managers. The best argument was the idea that safety on a single-track railroad depended too much on human reliability and foresight. It was also noted that cost did not double with double-tracking, because the main cost of construction was in cutting and filling for the right-of-way, and this was only 40 to 50 percent greater for double than for single track.[354] The cost of ties and rails would indeed be double, but this was somewhat offset by reductions in the number of switches required. Single-track lines, if at all busy, required many sidings, and this meant lots of switches, which were costly to install and maintain. Comparatively few sidings were needed on double-track lines.

The deciding factor in whether or not to double-track was traffic density. In general, a railroad with fifty trains or fewer each way per day could operate efficiently if it had a good track, moderate grades, plenty of sidings, good locomotives, skillful dispatching, and disciplined crews.[355] Several more necessities must be added to this long list of qualifiers. More and longer sidings were needed for a busy single-track line; the spacing was figured at one every 4 to 5 miles. Each siding must hold two trains, and in 1902 when this wish list was published, a typical freight train was sixty cars. Water tanks must be placed no more than 20 miles apart. Larger locomotives and better signaling always helped. Rather than spend money on a second track, railroad executives were admonished to put their capital into lower grades, easier curves, heavier rail, and deeper ballast for existing single-track lines. Yet the defenders of the single-track system were ready to admit its inadequacy should the normal mix of traffic include many express passenger trains, fast mails, and perishable and other fast freights. Single-track lines could handle a large and heavy traffic so long as most of them were moving at the same general speed. The introduction of many fast trains then called for a second track, especially if the overall volume was much above fifteen hundred trains per month.

The capacity of a railroad depends very directly on the quality of the track, be it a single or a double line. As a capital-poor nation, the United States tended to favor cheap, expeditious construction methods that were economical in the short term only. The strap-rail track introduced on the pioneer lines proved cheap in first cost, but it was an operational and maintenance disaster. The combination of a wooden structure placed directly on the soil and shod with a thin iron strap produced a weak, unstable, and short-lived track. The mistake of strap-rail construction was discovered too late, and between 2,000 and 3,000 miles of track were built in this fashion, only to be rebuilt at great cost with solid iron rails. The lesson of inferior construction had not been learned, however, for American railroads still tended to buy the cheapest grade of rail, which was filled with cinders and laminated iron. It would break down, crush, and wear out in three years.[356]

The effect of poor rail and track on train speed, safety, and reliability is obvious, yet inferior track was a fact of life on early American railroads, and it was a self-inflicted cause of delays and accidents. Some lines understood the true economy of buying the best-quality Welsh or English wrought-iron rails, which were virtually free of cinders. The Illinois Central was one line that would not risk running its trains over cut-rate iron. Steel began to succeed iron for rail making in the 1860s, with wonderful results. It was not so much stronger than iron, but it was homogeneous and so unlike wrought iron that it did not split and break down. Wellington claimed that it revolutionized track maintenance and offered for the first time a durable and dependable path for the Iron Horse and its train of cars.[357] Hadley was even more enthusiastic about the merits of steel rails.[358] Bessemer rails allowed railroads to run at something like their full capacity with the certainty that the rails would not fail. Larger cars and far larger payloads were now possible at reasonable cost.

The size of the rail had a great deal to do with the traffic it was expected to bear. On an average-duty line with, say, ten trains a day, 60-pound-to-the-yard rail was sufficient, and there was no need for, or saving in, using a heavier rail.[359] On more active lines with twenty-seven or more trains a day, 75- or even 80-pound rail was advisable. Rail was sold by weight, so railroads tended to use the lightest rail possible. As speed became more of a factor toward the end of the nineteenth century, leaders in the railroad field adopted heavier rail. Fifty-six-pound rail, long an industry standard, gave way rapidly to heavier weights after 1880. In 1884 the New York Central adopted 80-pound rail. In 1891 the same line began to buy 95-pound rail, and during the next year it acquired some 100-pound rail. The Pennsylvania and several other eastern trunk lines began to experiment with big section rail as well, but only the Boston & Albany was serious enough to relay its entire main line in 1897 with 95-pound rail.

If other railroads were a trifle slow to jump on the big-rail bandwagon, they did begin to pay more attention to other track improvements. Details such as tie plates and better ballast helped greatly. More ties per mile strengthened the track structure. Deeper ballast and drainage ditches made it more stable. Mud and cinder ballast gave way to gravel or crushed-stone ballast. Good track, however it might be achieved, ensured the safer and more timely operation of trains, all of which increased the railroad's overall efficiency and ability to move tonnage.

CURVES AND GRADES

Early American railroads were built cheap by necessity. A favored means of holding down construction costs was to follow the natural lay of the land and so avoid cuts and fills. This produced a meandering sort of railroad suitable for slow speeds. Such indirect routing added unnecessary mileage between terminals, and it created hidden dangers because visibility around the curves was so poor. Creeks and rivers offered a natural path followed by many pioneer railroaders in search of the easiest, and not necessarily the best, route between any given locations. The back-and-forth wanderings of our premiere steam lines as they sought to dodge a hill here and a dale there are reminiscent of the natural paths trodden by cattle.

The Baltimore & Ohio's original main line out of Baltimore followed the Patapsco River in a corkscrew pattern that must have made the train crews dizzy as it zigged and zagged around its 400-foot-radius curves. Some curves were as short as 318 feet. When it came time to rebuild the line because of periodic floods, the B&O decided to eliminate some of the worst curves.[360] A minor reconstruction in 1838 eliminated some curves, but a major rebuilding was not attempted until 1899–1907, when the 61-mile-long road was shortened by three miles by means of deepening side-hill cuts and building new tunnels. All of this costly work came to $3 million.

Squealing wheel flanges were found on railroads in all regions of North America. Mountainous lines had the most curves, but even fairly level lines were made up largely of wandering tracks. The Erie's Delaware Division was reported to have curves on 70 percent of the line.[361] The Marietta & North Georgia was nicknamed the Hook and Eye because of its many curves, and it retained this name even after it became a division of the Louisville & Nashville. The railroad was completed across some very rugged territory near the junction of the Tennessee, Georgia, and North Carolina state borders in 1890. Engineers twisted the line through the mountains in curve after curve. A double-reverse curve near Tate Mountain became known to train crews as the Hook. The Eye was a spiral loop that crossed over itself elsewhere down the line. In contrast, a few roads found it possible to build absolutely straight or tangent lines wherever the terrain was level and true. The coastal plain in North Carolina allowed the Seaboard Air Line to build a 78-mile-long tangent between Wilmington and Hamlet.

As capital for new railroads became more plentiful and engineering standards improved, most lines attempted to keep curvature to 1 degree (5,730-foot radius) or 2 degrees (2,865-foot radius).[362] On mountainous sections of track, curves as sharp as 10 degrees (573-foot radius) were tolerated. The extra friction in rounding such a curve was comparatively minor. The principal objection to curves was the limited visibility, and hence the danger of accidents and the need to slow down. The cost of building a railroad with easy curves through mountainous territory was more than the wealthiest promoters could afford, and so curves were a given on rail lines among the peaks.

Grades were of more concern to everyday railway operations than curves because inclines, even moderate ones, could so adversely affect train performance. Surveyors were under considerable pressure to find easy grades and, whenever possible, water-level routes. Pulling a train of cars on the level involves little more than overcoming the rolling friction of the wheels and journal bearings. A small engine can propel a fairly sizable train under such conditions, but the moment an incline or a grade is introduced, the locomotive is called upon to lift as well as propel. The power requirements mount steeply as the incline increases. This problem is made worse on a railway by the fact that the adhesion or traction between the smooth rail and wheel is so small. For this reason mainline railroads will take all reasonable steps to keep grades at 1 percent or less.

Such noble intentions were often abandoned, however, once construction crews were confronted by major mountain crossings. In the 1840s and 1850s civil engineers laying out the Baltimore & Ohio main line through what is now northwestern West Virginia admitted defeat. To hold construction costs down to a reasonable level, they were forced to adopt grades as high as 116 feet to the mile (2.2 percent). Unofficially, this became the industry's standard, and it became a rule of thumb when laying out lines far to the west of the Appalachians. In certain instances even steeper grades were adopted, but at the cost of markedly higher operating costs and slower-than-normal schedules.

A worst-case scenario might well be the Denver, North Western & Pacific, often better known as the Moffat Road in honor of its chief promoter, David Moffat of Denver.[363] Construction on this peak-to-peak railroad, from Denver to Salt Lake, began in 1902. It was built across rather than around the Central Rockies and so offered a short route between its major terminals, but the extreme up-and-down nature of the railroad made it expensive to operate. Plans for a major tunnel were abandoned because all capital had been spent before that part of the line could be started, and so a high-level track was pushed over Rollins Pass at 11,600 feet above sea level. Fearsome grades of 4 percent were necessary to accomplish this crossing. The snow removal problem was horrendous at this elevation. This cost plus fuel requirements consumed half of the railroad's revenues. Needless to say, the enterprise failed, and the Rollins Pass High Line was abandoned as soon as the Moffat Tunnel opened in 1928.

Steep grades normally required extra locomotives, called helper engines, to push or pull trains over the incline. If a series of grades were encountered, an extra engine might go with the train. A pair of engines on the front end of the train was called a doubleheader (Fig. 1.44). In the case of a single big hill, the helper engine could cut off after the train passed over the crest of the grade, and it then pulled into a siding while the train passed and went on its way. The helper would then back out onto the main line and return to the bottom of the grade to await the next train. Normally, conventional locomotives were used for helper service, but on some lines massive specialized locomotives were required. The Reading, for example, built a mammoth twelve-wheel tank engine in 1863 to push trains up the Falls grade near Philadelphia.

The maintenance of any pusher service was an expensive and troublesome business, for it was not just the cost of the engines and crews but the cost of the side tracks and water tanks needed for such operations. Pusher engines also interrupted and delayed the normal flow of train service. Operating departments did all they could to avoid pusher service and relied on a system of ruling grades. They tried to calculate what a given class of engine could pull over the worst grade on the division. Once established in terms of either tons or cars, that class of engine would not be asked to move a train of a greater size. Thus trains might be overpowered on long parts of a run because of one bad grade on the line. In theory, however, they could go from terminal to terminal unassisted.

Not included in these crude calculations were the marginal factors that could prevent a train from making the grade. An ice- or snowstorm could hamper a train's ability to overcome a grade it might easily surmount on a dry, clear day. The condition of the locomotive itself had a measurable effect on just how much tonnage could be moved. Badly maintained engines would not perform up to their rated capacity. The skill of the crew could also make the difference between making a grade and not doing so. Being a good fireman required more than just shoveling in the coal. It was an art, and skillful stokers knew how to keep her hot and on the move. The same was true for the engineer. Some men seemed to have a special gift for coaxing extra power out of the worst old teakettle.

Getting trains over the line even with the best equipment and train crews was never an easy task. A report published in September 1855 illustrates the obstacles even small grades could present in the days of wooden cars.[364] A new locomotive, the No. 210, moved out of Port Jervis on the Erie with a thirty-car train weighing 510 tons, less the engine, tender, and caboose. All went well until the first major grade was encountered. It was hardly a formidable incline, at 45 feet to the mile, but it was enough to stall the train. The blame was put on an overfull boiler that caused the engine to prime. The train backed down to the bottom of the grade and dropped off five cars. It was then able to surmount the grade at 10 miles per hour. At the next grade two more cars were cut off. At Chester Junction it was necessary to drop off seven cars. Some of the cars dropped off were picked up by doubling—a maneuver that will be explained below. The train and its weary crew laid over at Turners for the night; appparently the Erie did not operate freight service after dark, at least not at that time. The next morning the 210, coupled onto twenty-four cars and a caboose, moved off downgrade to the terminal on the Hudson River, Piermont. An exact time is not given, but the first day's journey from Port Jervis to Turners covered just 40 miles. The second leg, Turners to Piermont, was only 27

FIGURE 1.44

Steep grades called for doubleheaded engines, and sometimes for a pusher on the rear as well. Such costly and inefficient railroading prompted most lines to reduce grades wherever possible. The scene here is near Chama, N.Mex., on the Denver & Rio Grande in 1908. (Fred Jukes Photo, R&LHS Collection)

miles, yet nearly two working days were consumed as the engine struggled to overcome what at best can be characterized as moderate grades. The apparent need to double on just about every hill would explain why the train's progress was so slow. This was a tedious way of running trains, and it was normally the result of poor planning by the dispatcher.

When a train stalled on a grade in the days before air brakes, the engineer would signal the brakeman to set the brakes on the rear cars. A brakeman would then uncouple the cars at the center of the train. The engine would pull ahead with its lightened load and after surmounting the summit would seek the first siding. The switch is unlocked and opened, the front half of the train backs in, the engine pulls out, and the switch is closed and locked. Meanwhile, the rear end of the train sits down the grade blocking the main line. A flagman has been sent out back along the tracks to protect the rear end of the train. The brakeman at the head of the stranded cars looks up the line for the return of the engine. After a time it comes clank-

ing backwards down the track. If it is dark or foggy, care must be taken that the engineer does not crash into his own train. He must peer backwards into the dark to catch the tiny speck of light given off by an oil lantern. When he returns, the trainmen quickly couple up, and the engineer whistles for the rear flagman to return to the caboose. He must hurry, for all concerned are anxious to clear the main line as trains pile up in front of and behind the cut-in-two freight. After a reasonable time elapses for the flagman to return, the engineer whistles *brakes off*; the conductor has signaled that the flagman is aboard, and the train takes off up the grade that had earlier stalled its progress. The train runs parallel to the siding and cuts off the rear of the train. The engine then moves forward just beyond the switch and stops. Again the switch is unlocked and opened; the engine backs in and couples to the first car. The front half of the train pulls out; the switch is closed and locked. More whistling and calling them in once the train is rejoined.

The train is now ready to go forward on its journey again, but all of this backing and forthing to double up the hill has taken time. It must have consumed an hour, and if this happens repeatedly—as well it might on a hilly railroad with small engines—a freight train's time over the road could be twice that shown on the schedule. A busy main line simply could not afford doubling except in an emergency. It would be required to provide larger locomotives or helper engines, or to run shorter trains. More marginal lines might accept such delays because so few trains were affected.

Cuts and fills could ease sharp curves and steep grades. Earth moving was a costly effort and so was attempted only on as small a scale as was necessary. Tunneling was an even more expensive cure for grades and curves. A railroad manager thought long and hard before agreeing to the construction of a tunnel. Accordingly, our pioneer lines were almost totally devoid of tunnels, and those that were attempted were modest in bore and length. Such an evasive policy became more difficult to honor after the rail network began its westward march. Once the Baltimore & Ohio advanced beyond Cumberland, Maryland, in 1849, tunneling became almost commonplace. There was no reasonable way to forge a path through the Alleghenies without drilling dozens of tunnels.

The largest bore was the 4,100-foot-long Kingwood Tunnel located 88 miles west of Cumberland.[365] It took nearly three years to drill the passage through the mountain. It was closed in 1855 soon after opening because the roof proved unstable. An arch lining of brick and iron plates required another year and a half of work. Meanwhile, trains were run on a temporary shoofly track that ran around and over the mountain. The first shoofly used while the original bore was under way was a miserable affair with a 10 percent grade. A locomotive could boost only one car at a time. A second, easy-grade line was constructed in 1855, but the incline was still a formidable 6 percent. One locomotive could push four freight cars (32 tons) at 6 miles per hour over the 2.12 mile long shoofly. This primitive and costly operation continued until 1857. Seventy-seven miles west of Kingwood was the Boardtree Tunnel. Its completion was delayed by the failure of a contractor, and so the railroad decided to build a 6-mile-long switchback over the mountain. The grades on this zigzag line averaged around 6 percent. A contemporary account recorded that the locomotives acted "as if they were angry with their loads" and pumped out such volumes of smoke that the sky was blackened as if by a volcanic eruption.

Far to the south of the B&O Main Line, work started in 1849 on the Blue Ridge Railroad Tunnel, near Waynesboro, Virginia.[366] This project was on a scale with Kingwood; the bore was 4,273 feet long. Following the B&O's example, a steep temporary track was built over the tunnel site so that service, especially for construction materials, could be introduced before the tunnel's opening. The 4-mile-long line had grades of 5.7 percent and several very sharp curves. Six-wheel Baldwin flexible-beam tank engines managed to huff and puff up this terrible railroad with three loaded or four empty cars. Virginia's worst piece of track was finally abandoned in 1858 when the $0.5 million Blue Ridge Tunnel was opened.

Big projects seem to inspire more and even bigger projects. In 1853 work began on an ambitious 10,000-foot-long tunnel running east from Cincinnati under the Deer Creek Valley.[367] After about 3,000 feet of digging, the project was abandoned. An even more colossal digging project got under way back in New England, but unlike the Cincinnati bore, this one, the Hoosac Tunnel, was eventually finished.[368] The construction site was in far northwestern Massachusetts. The drilling of the 4.5-mile-long tunnel was an on-again-off-again project of the worst sort that went on for over twenty years and drove costs up to between $17 million and $20 million. It finally opened in 1875 and was taken over by the Fitchburg Railroad. In later years it was acquired by the Boston & Maine.

Location engineers could largely ignore the whole subject of tunneling as railways fanned out across the flatland of North America's interior, but it again became very much germane once the western range of mountains was reached. By 1870 tunneling techniques had been greatly advanced, thanks in large part to compressed-air drills and nitroglycerine. Even so, it remained a laborious and costly task. Take the case of the Santa Fe and its conquest of Raton Pass in New Mexico.[369] A 2,000-foot-long tunnel was needed at the crest of the pass to hold the grade to a reasonable angle. Following a now-familiar pattern, the Santa Fe built a temporary track over the tunnel site. In this case it was a 5.5-mile-long switchback, which opened in 1878. A special 2-8-OT locomotive could pass through the six switches with nine cars in just fifty minutes.

WEATHER

A little rain can slow down railway operations— a fact that hardly needs much explanation, because the effects of precipitation on all forms of transit, ancient and modern, should be obvious. The effect is heightened as the intensity and duration of the rain increase. If a little rain can slow things down, a lot can stop trains dead in their tracks. Floods can in fact destroy a railroad. Spring thaws abetted by seasonal rains can soon have a creek or river over its banks. The railroads' propensity for following streams therefore led to considerable damage as culverts collapsed, bridges washed away, and tracks sat idle under widely broadcast waters. Even after the water receded enough for the trains to venture forth again, great care was taken at first because of spongy ballast that had become so soft that a locomotive would sink in or roll over. Experienced crews learned to be wary before crossing a bridge after flooding. The track might well be suspended in midair like a ladder fallen over a chasm, the bridge structure having disappeared with the flood.

Floods did not always wait for spring. Heavy rains in the Ohio River Valley area in February 1884 dumped large quantities of water into regional streams because of the frozen ground. Railroad property from Pittsburgh to Louisville suffered.[370] Service was suspended from two to seven days, depending on the location. Many bridges disappeared, as did a new line of track between Benwood and Parkersburg that was just about ready for traffic. A repair shop was partially carried away by the raging waters. Much traffic was diverted to the north, which represented revenues never to be regained.

Five years later the Pennsylvania Railroad suffered even heavier damages from flooding along 200 miles of the main line.[371] The middle division east of Johnstown lost six bridges and sustained heavy damage to a dozen others. There were twenty-three mudslides or washouts. Mud 10 feet deep covered 50 miles of track. Fifty-six thousand crossties were carried away, and most telegraph lines were down. But the worst was yet to come. On May 31, 1889, an old earthen dam collapsed, sending a 40-foot wall of water down the Conemaugh River. The Johnstown flood was under way. In addition to the terrible human toll, this same flood inflicted major damage to the Pennsylvania's main line—the heaviest and most prosperous freight line in the nation. Three major bridges, including a splendid stone viaduct, were swept away. An engine house, a machine shop, and several water tanks disappeared, as did 561 freight cars and 34 locomotives. It took two weeks of feverish work just to reconstruct a temporary track for passenger service, and no freights reappeared for almost seven weeks. The losses in tonnage and revenues must have been staggering.

If water in its liquid state could prove a mighty impediment to railway operations, it was an even graver threat when frozen. The operators of our fledgling railroads were surely dismayed by the scheduling problems caused by the winter's first snow. Driving wheels would spin with no effect in moving the train forward. Small drifts could stall the light trains of the 1830s. Brooms were fastened to the pilot beams to sweep the white stuff off the rails. Men and shovels helped clear the way. Switch tenders built fires to thaw the switch points. In all, it was a costly lesson in railway operations. A good snow could and often did shut the railroad down. No trains could run, hence no revenues were generated.

The urgent need to reopen prompted such innovations as the snowplow and flanger. The smallest plows were V-shaped contrivances fastened to the front of the engine. Such a device sounds very simple and obvious, but the superintendent of the Boston & Worcester found it necessary to post a directive in November 1835 to all engine crews to make sure that the plow faced the right end forward when attempting to clear the track of snow.[372] Four-wheel and finally eight-wheel plows supplemented the engine-mounted snow shovelers. Dozens of designs were presented for the perfect snowplow, but most were only variations on the wedge or bucker style of plow, which depended on brute force to batter its way through snowdrifts.

Anywhere from one to fourteen locomotives would be hooked behind a plow to clear the track of major drifts.[373] The technique was crude and direct. The plow and its convoy of locomotives would slam into a drift at 60 or 70 miles per hour and keep going until they stalled. Then they would pull free and back up a few miles for another charge. Meanwhile, track crews armed with shovels would clear away all they could of the snow that fell into the newly made canyon. Back would come the snowbuckers, whistles screaming and throttles wide open for another hit at the mountain of snow. Again they would hit, and again snow would fly 10 feet into the air. Foot by foot, inch by inch, the track was cleared as the plow slammed repeatedly into the snowbank. Boards were often fastened over the cab windows to protect the engine crew from broken glass should the snowbank collapse in against the cab. With the normal openings for vision cut off, a trainman would climb up on the boiler backhead and peer out through the cab roof hatch. From here he could see ahead over the tall snowplow, and he would call down to the engineer about what was ahead.

A well-organized general manager would insist that plows and flanger repairs be started late in the summer, for the Storm King could blow in as early as September in the Dakotas or even northern Iowa. A fierce storm might come in with 40-mile-per-hour winds and dump snow for several days. Two or three feet of snow might accumulate, but the drifts could easily swell to 10 and even 15 feet.[374] Sometimes the stuff would crest almost to the tops of the telegraph poles. The winter of 1888 produced a snow blockade the likes of which had not been seen since the record snows of 1866. New England and Middle Atlantic region railroads suspended service for one or two days; even the great New York Central's four-track main line was shut down for twelve hours. Western railroads seemed to suffer even more from shutdowns than their eastern counterparts. Some lines remained closed for ten days at a time and remained a litter of broken-down cars and locomotives bereft of fuel or water.

The Union Pacific was absolutely devastated by a snowstorm that started in mid-December 1871.[375] The railroad closed down for twenty-eight days; not a train ran during this period because of the monster storms that raged on and off. The sidings and yards were packed with stranded freight cars. Some trains were delayed for two months. The severity of the storm had much to do with the poor performance of the Union Pacific, but the newness of the railroad must have been a factor as well. The railroad was not well equipped, some of its track was temporary in nature, and as a long, thin, and immature enterprise it was no match for King Winter.

The Union Pacific's far western connection, the Central Pacific, handled its own snow problem with more aplomb.[376] This is not to say that the CP's challenge was a minor one. Its Sierra Nevada summit near Donner Pass (7,017 feet) is one of the snowiest places in the world, with an average snowfall of 31 feet per year. Even before operations began over this portion of the line in 1866, it was clear that extraordinary methods were necessary to keep the line open. All the snowplows and shovels in North America could not keep this line open; hence the decision was made to roof it over. Within a few years 30 miles of snow sheds had been built of massive timbers to support the tons of snow that came to rest upon them. The trains ran through long, dark wooden tunnels oblivious to the raging storms of the Sierra Nevada. Not all parts of the mountain crossing were protected by sheds, and so plows and shovelers remained very much in evidence on the CP roster. The sheds, for all their merits in keeping the line open, were costly to build and maintain. They were also a fearsome fire hazard. In 1890, when rotary snowplows had proved to be an effective countermeasure to snowdrifts, the Central Pacific adopted them with some enthusiasm. Gradually the snow sheds were cut back in length, and the exposed tracks were kept clear by the rotary plows.

Snow was not the only agent of water capable of closing down railway operations. Ice was equally adept at stopping trains, injuring crew members, and damaging valuable rolling stock. Ice would wrap itself thickly over roof walk boards, steps, handhold irons and brake wheels, brake shoes, and even railheads. It could weld switches closed and fill in frogs, crossovers, and flangeways of road crossings, causing the light engines and cars of the time to derail. The track structure became so brittle that rails tended to break more easily. Ice posed perhaps the greatest menace in the rivers crossed by railroad car ferries. This was a special problem in the pre-1870 period, when most major river railroad crossings were handled by ferries rather than bridges. Cake ice in the river might only slow down the car ferry operation, though in a fast-moving current it could damage or even sink a boat if the cakes were large. If the cold weather was severe enough, the stream would freeze solid and no boats could move. Indeed, some boats might perish in an ice crush. Whatever happened to the ferryboat, the railroad was effectively shut down.

The Philadelphia, Wilmington & Baltimore was a busy north-south main line that was regularly forced to suspend operations because of ice at its ferry crossing over the broad mouth of the Susquehanna River where it empties into the Chesapeake Bay.[377] In 1849 operations were shut down for six weeks. Passengers desperate or determined enough would walk some 3,500 feet across the frozen river to board trains waiting on the other side. Freight had no such option, and so it piled up in cars and warehouses on either side of the river. In 1852, when another prolonged freeze set in, the railroad decided to take a more active role in moving goods. A temporary track was laid over the ice and some 1,378 carloads or 10,000 tons of freight, mail, and baggage were sent across this natural bridge. The cars were pushed down one bank and allowed to roll free as far as they would go. They were then picked up by a team of horses and towed to the other side, where ropes were hooked on and a locomotive hauled them up the riverbank and to the main track.

The principle of the Susquehanna ice bridge was copied elsewhere from time to time. In December 1867 the Union Pacific built a low-level trestle over the frozen Missouri River between Omaha and Council Bluffs.[378] A goodly number of cars rumbled, slowly to be sure, over this ad hoc structure. The scheme worked so well that it was repeated in 1868 and 1869. Frozen rivers could also be most unaccommodating when the ice began to break up and flow. The Erie lost a major bridge over the Delaware River in February 1857. Its replacement was almost ready when a flood took that one out too. In 1875 an ice dam built up on the Delaware River to a height of 50 feet. Erie engineers attempted to blast it away but were unsuccessful, and when the dam broke, the ice once again carried the bridge away.

RIVERS

Large rivers, especially those deep and swift of current, formed natural barriers to railroads. Many early lines were nothing more than lengthy overland portages between rivers, lakes, and oceans. The Erie ran between the Hudson River and Lake Erie; the Baltimore & Ohio between the Chesapeake Bay and the Ohio River; the Illinois Central between Lake Michigan and the confluence of the Ohio and Mississippi rivers. Even the transcontinental was in reality only a go-between for the Missouri and Sacramento rivers. The thought of bridging these streams appears to have been a hesitant one, and in some cases twenty years would pass before stone piers and a superstructure offered a solid path over the waters.

There were some exceptions to this aversion to crossing major rivers. The B&O had little choice but to cross the Potomac at Harper's Ferry in 1837 if it was to proceed westward, and so a relatively cheap covered wooden bridge was erected. The 800-foot-long bridge was made in six spans.[379] In 1835 the Philadelphia & Trenton found it cheaper and quicker to purchase an existing highway bridge over the Delaware River at Trenton, New Jersey, than to build a new structure specially designed for railroad use. But this option was seldom found elsewhere, and new and very expensive bridges were almost always required. The Cumberland Valley Railroad found it necessary to cross the Susquehanna River at Harrisburg, yet even the cheapest wooden bridge cost $122,000, or 14 percent of the total cost of the railway.[380] The 4,300-foot-long bridge was divided into twenty-two spans. The river was wide but shallow at this place, and so footings did not represent much of a problem. The bridge was lightly built to hold down the cost, and so only the smallest locomotives were permitted to cross it. Other big railroad bridges on a similar plan were built at Richmond, Virginia, over the James River in 1838, and at Rockville, Pennsylvania, in 1848, but in the main, railroads tended to stop dead at the muddy banks of any major river.

The getting of goods and people across the water was relegated to steam-powered ferryboats. Get on, then get off and get on again. Another transfer and breaking of bulk, another costly, time-consuming, annoying change of mode. It was the worst feature of transportation in nineteenth-century America, this constant putting on and taking off of goods and people. The Great Through Route was so much advertising puffery. It existed only in the optimistic imagery of the traffic salesman. Each big river crossing meant a ferry transfer because the railroad was too cheap, poor, or cautious to build a bridge. Most railroads simply did not have the means to put up such expensive engineering monuments. Others rightly concluded that traffic levels did not justify such costly solutions. The inconvenience of the ferry transfer would just have to be endured until more tonnage warranted the construction of a bridge.

The first large-scale railroad ferry operation in North America was the Philadelphia, Wilmington & Baltimore's crossing of the Susquehanna at Perryville, Maryland, in 1837. Freight traffic was not significant on the line, and so goods were transferred item by item, at least during the early years. Tracks were fitted to the top deck for baggage and mail cars, but passengers made the transfer to and from the cars and the boat saloon on foot. A new boat replaced the original PW&B ferry in 1854. It had three tracks mounted, once again, high above the main deck, which had a capacity for twenty freight cars. It is not clear if the 1837 boat carried freight cars. In 1845 the whole notion of car ferry service made a quantum step forward when the Camden & Amboy's *Transport* entered service. This big sidewheeler had an enclosed deck with loading access at each end. It transferred freight cars between the Morris & Essex's Hoboken, New Jersey, terminal and the C&A's South Amboy wharf.

The car ferry idea moved west as the railroad network expanded beyond the Appalachians. By 1857 the Baltimore & Ohio had two river crossings on the Ohio that involved more than the shifting of a few barrels, bags, and boxes. But the transfers at Wheeling-Bellaire and Parkersburg-Belpre were handled via the time-honored strong-back method. The transfer of bulk goods must have been especially difficult. Goods that might otherwise have been sent loose in a boxcar would often have to be bagged because of the ferry transfer. So long as traffic levels and the labor cost of porters remained low, there was not much impetus to devise a better system. Then again, the physical transfer of cars would be rather pointless anyway because of the difference in track gauges on the two sides of the river. Broad-tread wheels made it possible to exchange cars east and west between Baltimore and Chicago. In 1860 a ferryboat equipped to transfer cars was put into operation so that through service might belatedly begin on the Parkersburg line of the B&O.[381]

When rail connections were directly opposite the ferry, boat transfer was relatively efficient, but this was not always the case. When the B&O began to interchange passengers and goods with the Marietta & Cincinnati, a ferry ride of some nine miles upriver from Parkersburg slowed the transfer considerably. In 1860 the M&C built a branch to Belpre that cut the distance in half. The Illinois Central endured an even more awkward ferryboat connection at its southern connection from Cairo, Illinois.[382] The river/rail connection involved a 24-mile excursion to Columbus, Kentucky. The largest IC car ferry could handle only a dozen cars; hence it took a day of back-and-forth trips just to handle one train. In 1872–1873 the rail lines were extended north so that a more direct transfer could be made. The final solution came in 1889, when a bridge replaced the ferry.

Poverty was the normal justification for using car ferries rather than bridges, and such a justification was likely a true one in the case of the Scioto Valley Railway. It operated a car ferry from Coal Grove, Ohio (near Ironton), to the Kentucky shores from about 1880 to 1890. A surviving photograph of the obscure operation shows it to be economical in all ways (Fig. 1.45). The equipment is small and noncustom. A standard sternwheel towboat

FIGURE 1.45

Car floats or ferries were used at dozens of water crossings, especially where bridges could not be justified. The scene here is Coal Grove, Ohio, on the Ohio River, the eastern terminal of the Scioto Valley Railway, in about 1890. (Norfolk & Western Railway Neg. 31077)

pushed a simple barge. Rails have been laid on the decked-over barge, and hog bracing has been added to stiffen the vessel for its substantial load. The barge's overhead trussing might be likened to a boxcar truss rod's support turned right side up.

If the Ohio was periodically crossed by railroad car ferries, so were other big rivers across the nation. Some were in service just a few years, such as the Union Pacific's ferry between Council Bluffs and Omaha. This service was abandoned in 1872. By contrast, car ferries ran across the broad waters of the Mississippi between New Orleans and Algiers for nearly eighty years. Another enduring car ferry was located in the Far West at the Carquinez Straits just north of San Francisco.[383] Not long after the opening of the transcontinental railroad, the Central Pacific was seeking a faster route to Oakland and, indirectly, the great Pacific port of San Francisco (Fig. 1.46). This new route involved crossing the Carquinez Straits, a portion of bay water too wide for a bridge within the budget of

the CP. Funds were, however, available to build a first-class ferry boat, and so the 424-foot-long *Solano* was completed in 1879. This substantial sidewheeler could carry forty-eight freight cars and two locomotives. They just did not make them any bigger than the *Solano* or its newer sister, the *Contra Costa*. Both of these vessels continued their labors until 1930, when a $12 million bridge ended car ferry operations between Port Costa and Benecia.

The Great Lakes, spreading out like five giant fingers in the midcontinent, created a water barrier to regional railroads of some magnitude. Long-distance car ferries were about the only practical means available to make a crossing. Because a well-documented study of this subject has already been published, little more than a brief summary will be offered, and interested readers are urged to consult George W. Hilton's book directly.[384] It is enough to say that the first car ferry operation on the Lakes was started in 1858 by the Buffalo &

FIGURE 1.46

The Central Pacific built a 2-mile-long wharf out into San Francisco Bay from Oakland in 1871. Cars were carried back and forth across the bay by the transfer ferry *Thoroughfare*. Some transhipment between freight cars and oceangoing ships was also taking place in this circa-1900 photograph. The long wharf was closed in 1918 or 1919. (Southern Pacific Photo)

Lake Huron Railway between Buffalo and Fort Erie. This service was begun specifically to avoid the need to break bulk and so is an important and early link in the interchange movement. The Lakes were hardly crisscrossed by car ferries, but a fair number came into service over time; nearly ninety boats were built for this purpose.

Moving freight cars around major port terminals was sometimes better done over the water than on the tracks. In the case of the nation's busiest harbor, New York City, there was almost no alternative. Special barges outfitted with tracks and tie-down devices to hold the cars in place were more commonly called car floats. Tugboats were used to push them around the harbor, sometimes singly and sometimes in small clusters. Cars were end-loaded on floats via hinged or adjustable bridges that made a transition between the end of the dock and the vessels. The level of the car float relative to the dock could vary considerably, depending on the tide, and so there was a need for an adjustable connection between the two. The floats were used in two ways. In the first method, the cars remained on the float and were placed alongside a dock or warehouse for loading or unloading by stevedores. In the second method, cars were taken from the New Jersey shore across the Hudson River to Manhattan Island and then rolled off onto a dock where they might remain for loading and unloading. Or they might be pushed by a switcher over street trackage to a team track or an in-town warehouse. The Baltimore & Ohio's inner-city team track at 23d Street is shown later in this chapter in Figure

1.58. These cars were isolated and had no way to get home other than a long ride on a car float to either the B&O's own Staten Island trackage or the Central of New Jersey's Jersey City terminal. Once back on the car float, the tugboat acted as a switch engine to jockey the cars over the waves to their next destination.

The origin of the car float system is, like so much to do with the railroad freight business, obscure and uncertain. The earliest record uncovered so far suggests that the idea was first tried by Herman Haupt during the Civil War as part of the U.S. Military Railroad's operations between Alexandria, Virginia, and the Richmond, Fredericksburg & Potomac railhead at Aquia Creek, some 60 miles to the south on the Potomac River.[385] Haupt, a former general superintendent of the Pennsylvania Railroad, was appalled at the delays and cost of moving goods to Union armies at Falmouth, Virginia. The cause of this waste was the need to load and unload at either end of the steamboat journey that connected the railroads ending, respectively, in Alexandria and Aquia Creek. If only the loaded cars could be ferried between these two points, the breaking of bulk might be entirely avoided. There was no time to build special equipment, and so Haupt acquired some canal barges, lashed them together, and fastened tracks crosswise over their decks. Sideways loading was not ideal, but it worked, and each makeshift car float could handle eight cars. A steam tug began towing two floats at a time in November 1862. It required about an hour to load eight cars and another six hours of travel to reach Aquia Creek. Hundreds of cars were moved in this fashion, at a considerable saving in time and cost. Because the Potomac is a tidal river, Haupt devised hinged or floating bridges to form an adjustable transfer between the dock and the barge.

Scores of soldiers and railroad men must have observed this very public operation during its several years of service, and yet the scheme did not reappear until several years later when it was reinvented, or at least reintroduced, in New York by John H. Starin (1825–1909), a manufacturer of toilet goods who became interested in transport problems in the 1860s.[386] Starin established a lighterage line to act as a link between the various railroads and steamship lines serving New York Harbor. With the patronage of Cornelius Vanderbilt, his firm prospered, and in time he became the largest barge and tug operator in the nation. The transfer of goods between ships, rail cars, and warehouses involved considerable hand labor, and so Starin turned, perhaps unknowingly, to the method used earlier by Haupt. Car floats were introduced around 1866 and soon proved their worth in expediting the movement of goods between all the modes of transit involved. An added benefit of the car float system was the relief it offered to congestion on the ferryboats connecting the Jersey Shore and Manhattan. Goods bound for the great city were formerly loaded onto wagons in Jersey City, Hoboken, and elsewhere. These clumsy teams would then descend onto the already crowded ferries and proceed to push their way onto the equally congested streets of New York City. By sending loaded cars to Manhattan via car floats, all the ferryboat traffic and at least some of the street haulage could be avoided. The car float system was certainly well used. According to figures published in 1912, there were six thousand men and nearly two thousand car floats active in the New York railroad navy, which handled ten thousand freight cars each day.[387]

Ferryboats and car floats were normally stop-gaps—cheap, temporary expedients resorted to only because capital was not available to bridge a river. In some cases, such as New York Harbor and the Great Lakes, marine operations continued because there was no practical alternative, but elsewhere a bridge was put in place as soon as funding could be found. Big bridges ranged in cost from $1 million to $3 million, even in the 1860s. This was a large enough expenditure to make most railroad managers wonder if they and their shippers might just have to make do with the inconvenience of a ferry for a while longer.

If sky-high costs were not enough of a discouragement, there was political opposition to bridge building. We have already seen such a stricture in Pittsburgh, but the citizens and leaders of Troy, New York, and Omaha, Nebraska, were strongly opposed to railroad bridges, primarily because they would end the ferry operations and the employment of local porters and draymen.[388] Forget efficiency and the interest of the shippers; these are matters of no consequence. The Union Pacific agreed to place the transfer yard in Council Bluffs to win support for a bridge over the Missouri. To placate the commercial interests of Omaha, so-called dummy trains ran back and forth over the bridge so that the transfer of cargo could go on as before when the ferries were still running. Cars were loaded in Council Bluffs, then run across the bridge to Omaha where the goods would be transferred into waiting empties before proceeding westward. The eastbound freight was handled in the same fashion. The real value of the $2.87 million eleven-span iron bridge, opened in 1872, was thus greatly reduced. Its value as a traffic expediter was only realized when regular through trains could cross back and forth freely and the foolish dummy train scheme was eliminated.

Structures such as the UP bridge over the Missouri can be traced back to even earlier spans over swift, broad streams. One of the most notable of these pioneers was the Rock Island's 1,582-foot-long wooden bridge over the Mississippi at Davenport, Iowa. Over 1 million board-feet of lumber was used to complete the bridge between 1853 and 1856.[389] Almost another decade would pass before more large river rail crossings were attempted. In 1865 the Pittsburgh & Steubenville implanted an iron bridge over the Ohio at Steubenville.[390] This effort marked the first railroad bridge over the Ohio River. It was followed rather quickly by a succession of railroad bridges at Louisville (1870), Parkersburg and Benwood (both 1871), and Cincinnati (1872). Bridge action picked up on the Mississippi at about the same time, with spans crossing over the Father of Waters at Dubuque and Quincy in 1868. The greatest achievement and the most enduring was James Buchanan Eads's magnificent steel-arch bridge at St. Louis, which served as a highway and rail carrier upon its opening in 1874. Today the rail service on the bridge has been abandoned.

Once the commitment to major bridge construction was manifest, the decision to build on a substantial and nearly permanent plan soon followed. Wooden bridges were a cheap expedient, but they were prone to rot and fire. The 1856 Davenport bridge had hardly been open a month before one of its spans was lost to a fire caused by a steamboat collision. Likewise, the Cumberland Valley's 1838 covered bridge was entirely consumed by flames in 1844. The Baltimore & Ohio's Harper's Ferry bridge was torched by the Confederates during the Civil War. So reads the dismal record of wooden bridges.

Experiments with iron railway bridges go back to the 1840s. The Reading produced a small number of cast- and wrought-truss bridges during this period, one of which has been preserved by the Smithsonian. The Reading was encouraged to adopt metal trusses because armed watchmen and bulldogs were needed to protect the railroad's bridges from pyromaniacal canal men, who succeeded in destroying two major spans. The cost of the watchman, timber bridge replacements, and the maintenance of wooden structures made iron bridges all the more appealing. Wendell Bollman began a more ambitious program of iron bridge building for the B&O in 1851 that led to the replacement of most major wooden bridges along the main stem within approximately fifteen years. Bollman also designed or built iron railroad bridges on western lines in this country and South America. In 1865 Andrew Carnegie, who had about as good a scent as anyone for a business opportunity, organized the Keystone Bridge Company. This firm was created specifically to capitalize on the railroad industry's acceptance of iron, and later steel, bridges.

Yet it is also true that wooden trestles sufficed at thousands of locations on main and secondary lines. Figure 1.47 shows one of these on a Burlington branch in Iowa. The appeal of wooden trestles was their simplicity and low cost. Fire protec-

FIGURE 1.47

**Wooden railroad trestles were
common in all regions of the
United States. Like most other
wooden structures, they were
cheap, temporary, and flammable.
The example shown here, from
about 1900, is on a branch of the
Chicago, Burlington & Quincy
called the Burlington & Western
near Brighton, Iowa. (Collection
of George Krambles)**

FIGURE 1.48

**Pioneer railroads were sometimes
obliged to build massive bridges
to cross rather minor streams—
witness this lattice truss near
Saxton, Pa., on the Huntington &
Broad Top Mountain Railroad.
The picture dates from around
1860. (Smithsonian Chaney
Neg. 22886)**

PARADE OF THE CARS

FIGURE 1.49

The Pennsylvania Railroad originally crossed the Susquehanna River at Rockville on a wooden bridge. It was replaced by an iron truss, shown to the left of the present (1902) stone viaduct. Air brakes were universal at the time this photograph was made, yet two brakemen are still visible on the roofs of cars in the peddler freight pulling off of the bridge. (Smithsonian Chaney Neg. 21646)

tion was offered by water barrels stationed along the top deck, and by periodic daily inspections performed by track walkers. Train crews were also asked to keep an eye out for trestle fires. Trestles were ideal for crossing ravines, especially when only minor streams were encountered. In the Far West the ravine might be a very deep gulch, dry for at least most of the year. The Union Pacific put a high (130 feet) and long (560 feet) trestle over Dale Creek in 1868. Sometimes trestles were combined with truss bridges when a wide valley with a fairly minor stream blocked the path of the Iron Horse. A photograph dating from about 1860 shows such a combination structure (Fig. 1.48). The scene is just east of Saxton, Pennsylvania, where a branch of the Juniata River is crossed by the Huntington & Broad Top Mountain Railroad. The near end of the bridge is a lattice truss stiffened out with bow arch supports. A Baldwin flexible-beam 0-6-0 is pulling a train of four-wheel coal jimmies, some of which appear to have iron bodies. In all, this one small print offers a most informative picture of early-nineteenth-century American railroading.

Rugged landscape and rivers were the chief generators of need to construct sizable bridges. For this reason mountainous railroads like the Baltimore & Ohio or the Denver & Rio Grande Western were a succession of bridges. Even flat prairie roads were not entirely devoid of spans. When the Santa Fe came to build its 450-mile air line between Kansas City and Chicago, it crossed no fewer than fifteen rivers, six of them big ones.[391] Railroads en-

tering New Orleans faced a somewhat different bridging problem, because the most direct route from the north was blocked by a very large, shallow lake. The Cincinnati, New Orleans & Texas Pacific Railway overcame this obstruction by building a low, 22-mile-long wooden trestle across Lake Pontchartrain. The crossing posed no unusual engineering challenges, for the water was only 12 to 14 feet deep; but the length of the structure increased the construction cost materially, and the railroad found itself saddled with a perpetual maintenance project.

Yards

There are really only three facts one needs to know about yards. They are the principal cause of delays in the movement of freight by rail. Their primary function is to make up and break up freight trains. And they do not serve, except in a minor way, as storage facilities for unused cars. There is much more that can be said on the subject of yards and terminals; indeed, it would take a book-length study to cover all aspects of this single topic, but once again only a brief summary will be presented here.

THE MARSHALING OF CARS

A railroad classification yard might be likened to a giant sorting table, 0.5 mile wide by 2 miles long, made up of several dozen parallel tracks. The sorting or separating and combining of cars is done

FIGURE 1.50

The trainman's strike of 1877 shut down the Lackawanna yard at Scranton, Pa., making it exactly what it was not intended to be—a storage place for unused locomotives and cars. Armed guards, visible in the foreground, stand by to prevent the type of destruction that took place in Martinsburg, W.Va., and Pittsburgh. (R&LHS Collection)

as one might spread out a collection of color-coded parts and divide all the green, red, and blue pieces into separate piles. The real work of the yard, then, is the receipt and dispatch of trains. Incoming trains are broken up or disassembled by switch engines, and the cars are individually placed on separate tracks to make up new trains. Cars are thus classified or marshaled, and both of these terms are commonly used to describe a railroad yard.

The drilling of cars was inherently a slow, labor-intensive, inefficient operation. Switch engines moved at a crawl only to stop and couple, proceeding tortoise-like along the tracks in a stop-and-go pattern. Most switching movements were only about 500 feet long. The train would then halt and wait for a switch to be thrown (turned) or a coupling made. All of this lost motion resulted in 23 miles of travel to break up a twenty-car train.[392] This is not to say that yards were inactive places. They were in fact very busy; it's just that the pace was slow. They were noisy, smoky places as well, and anyone who lived near one definitely lived on the wrong side of the tracks. The snorts and whistles of a dozen engines vied with wheezing brakes and the heavy crash and bang of cars coming together. Repeated start-ups tended to make the locomotives smokier than normal.

Inadequate facilities combined with inefficient methods and improper supervision limited the ability of some yards to deal with more than one train at a time. Indeed, there might well be three receiving tracks, but because only one set of car in-

spectors and a single switching crew were available, the other two trains must wait their turn, and that could mean hours of sitting.[393] Not a car could be moved until the inspectors had looked over every journal box, air-brake hose, and draft gear. The main-line train crew could at least cut off their engine and caboose and depart, but the switching crew must wait until the inspector's blue flag was removed. Meanwhile, freights would back up along the main line. They would be required to stand by on a siding, perhaps a mile from the yard, and wait for one of the receiving tracks to clear. Here were men weary after long hours on the road, marooned just short of the final destination, impatient, bored. They could be made very uncomfortable by the weather. If the yard crews worked too slowly, the main line itself would become blocked after all the sidings were filled up with waiting trains. Delays were clearly accountable to more than floods, snowstorms, and broken rails. The chief villain was that necessary evil, the classification yard.

Shippers were greatly annoyed by the delays normal to the classification system in the best of times. In the worst of times they were forced to endure a complete suspension of services. In 1899 a car famine resulted in New York City where one thousand loaded cars awaited sorting and unloading.[394] An even larger number stood idle at Jersey City. The call, as usual, was for bigger yards, more switchers, more men, and—why not?—more cars. The railroads countered that the blockade

FIGURE 1.51

Cars standing idle in yards graphically represent the inefficiencies of railroad yards and explain why car mileage and productivity were so abysmally low. Travel time between terminals was generally good; getting out of the terminal was the problem. (Foote Collection, Manitoba Archives)

was only seasonal and the yards were more than sufficient for eight months of the year; it was only during the grain rush that service came unglued. A few years later a far more serious traffic blockade and car shortage developed in Pittsburgh.[395] Once again the blame was placed squarely on yard facilities. Increased traffic and a scarcity of locomotives played a part in the immediate crisis. New locomotives could be obtained without enormous difficulty, but finding more flat space for yards and sidings amid the hills, valleys, and streams of western Pennsylvania proved more of a challenge. The long-term solutions involved building a new locomotive shop at Altoona and establishing new or at least enlarged yards just outside the Pittsburgh area. In the short run only a huge effort by men and machines could break the blockade. In one weekend the hardworking crews managed to move out twenty thousand cars.

Overall data for the nation's switching service is comparatively rare, but information published in 1912 may well provide a fair yardstick for the late wood-car period.[396] In this statement it was claimed that one-third of railroad trackage and fifty thousand men were devoted to switching operations. The cost per car handled ranged from 30 to 60 cents, depending on the efficiency of the yard. Chicago railroads paid out $6 million per year just for costs accountable to yard delays. In 1889 switching costs in Chicago were figured at 3 cents per hundredweight; a boxcar loaded with 20 tons would thus cost $12 just to move through the classification process.[397] J. A. Droege estimates terminal cost per mile per car at 25 cents, compared with only 0.75 cent per mile on the main line. It would be difficult to paint a picture of the

classification yards' inefficiency any bleaker than the one evident in these cost estimates. The railroad did a fine job in moving cargoes *between* yards and cities, but once they entered the yard limits, any notion of fast service ended. Cars might stand around for hours or even days, thus thwarting the valiant but futile efforts of railroad managers and operating crews to make a timely delivery of goods.

YARD ORGANIZATION

Yards were a domain separate from main-line tracks. These two basic entities of the same railroad operated cooperatively, but they were governed independently. The yardmaster ran every inch of track within the yard limit signs. The rules for train operations were fairly informal; there were no written orders and few signals. With so many short to-and-fro movements, written orders and schedules would be a management nightmare. Switching crews were simply ordered to do their job and stay out of the way of any main-line trains that passed through yard trackage, for scheduled trains had preference even inside the yard limits. Matters were handled in a far more formal manner on the main line, where written orders and a fixed timetable were rigorously enforced by the dispatcher. Separating the yard from the main tracks relieved the dispatcher from the governance of thousands of train movements each day.

To help maintain the separateness of yards and main lines, yards were placed to one side of the main tracks. Having the through tracks pass through the center of a yard was considered very poor design. To the casual observer, yards might appear to be nothing more than a great mass of parallel trackage; but in fact they are almost always divided into multiple subdivisions. Even the simplest yard is divided into an eastbound and a westbound pair. On a few lines the division might be north and south rather than east and west. Most yards were more complex than this basic bidirectional scheme, and they might contain five secondary yards plus segregated tracks for special storage or servicing needs.

The ground plan for a typical run-through division yard is shown in Figure 1.52. The subsections of the yard are apparent at a glance from the drawing, and a little more study reveals that almost everything is done in duplicate. There are a west- and an east-end caboose track, and west and east distribution, storage, receiving, departure, and repair yards. There are even two icehouses, as well as an inbound and an outbound freight house. The stock pens and team tracks are located only at the west end of the yard, and there is only one engine house, near the yard's center. Notice also that the westbound main line runs up over the top of the yard, while the east main runs along the outside of the yard at the bottom of the drawing.

A train entering the yard limits would aim for the receiving yard, which is A in our diagram. After coming to a full stop, the road engine was uncoupled and shunted over to the ash pit—engine house area for servicing. The caboose would at the same time be pulled off and moved to the caboose track by a switch engine. The conductor would walk over to the yard office to drop off his manifest or wheel report. The conductor has already given running slips to the car marker, who has recorded basic data for each car in the train. The car marker or "mud hop" is one of several outdoor clerks who never sit at a desk or enjoy the shelter of an office. He will walk the length of the train marking each car with just a few letters and numbers to tell the switching crews where to place each car. These cabalistic chalk marks would have no meaning to the layman, but 93 BK 7/14 would be easily recognized by any trainman as meaning train No. 93, bound for Youngstown, classified on July 14.

While the train was being marked for distribution, the inspectors checked over each car for obvious mechanical defects. Inspectors looked for a multitude of problems ranging from broken side sills to loose grab irons. Major problems such as a broken sill called for major repair work, and the car would be sent to the car shop. If it was loaded, its

cargo must be reloaded into another car. Naturally, this laborious process was avoided if at all possible. Most minor defects were handled where the car stood by roving mechanics who brought tools and parts directly to the car. Sometimes separate RIP, or repair-in-place, tracks were maintained. Car doppers followed the inspectors. They would flip open the journal-box lids to check the waste and oil level. More oil was added where needed. In some cases it might be necessary to clean out and repack the oil cellars.

After the chalkers had finished their work, a messenger took the running slips to the yard office, where they were compared with the conductor's manifest. The train was now ready for distribution. A switching locomotive coupled onto the rear car and pushed the whole train toward other parts of the yard complex. The cripples would go to their repair track. Less-than-carload cars would be dropped off at the inbound freight house. Empties with no immediate assignment would be shoved over to the storage yard, marked as S in Figure 1.52. A few loaded cars occasionally ended up in the storage yard because their waybills were not ready. While the paperwork was being prepared, these "hold-for-order" cars rested with the empties. Some yards had separate hold tracks for HFO cars and for unconsigned loaded cars whose records were temporarily lost. Long-term storage was normally not handled by division yards. They were too active and land was too valuable for this purpose, at least when traffic and business levels were normal. Railroads would send cars out into the countryside and allow them to molder away on some rustic siding far from the big-city centers. Rather than clog up a busy classification yard with one thousand boxcars standing by for the wheat harvest, the Union Pacific would line them up on remote trackage somewhere in Nebraska. The Canadian Pacific dedicated the big Transcona yard near Winnipeg for the storage of grain cars in the apparent belief that this was more economical than the U.S. system of piecemeal rural warehousing.

The real work of the yard was done in areas devoted to distribution and sorting—B and K in Figure 1.52. This is where the outgoing trains were made up. The tracks were ordinarily numbered or named (Fig. 1.53). The fundamental goal was to assemble cars for a specific destination, in proper order, on each one of these tracks. And because the

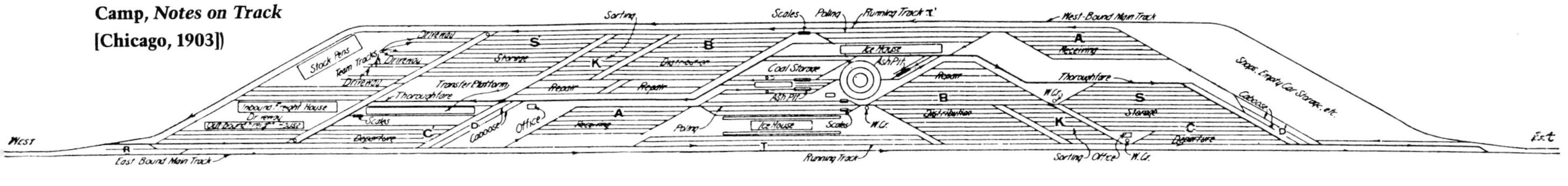

FIGURE 1.52

A modern style of yard evolved by the late 1800s with a basic east-west arrangement plus several subyards intended for specific jobs such as *A*, receiving; *B*, distribution; *C*, departure; *K*, sorting; and *S*, storage. Note that each yard is laid out in the ladder plan of track arrangement. (W. M. Camp, *Notes on Track* [Chicago, 1903])

PARADE OF THE CARS

FIGURE 1.53

Yards worked day and night in a clumsy effort to make up and take apart trains. The brakeman in the foreground signals the track number with his hand lantern to a switching crew, barely visible down the track ahead of him. Note the brakeman on top of the cars, the signal tower to the right, and the signal mast to the left side of this engraving. (*American Railway*, 1889)

cars were received more or less in disorder, the problem of putting them in the right order, on the right track, took considerable shuffling. This is why there was so much back-and-forth running as switching crews tried to line up the cars. If the track number was used, track 1 might be reserved for all cars going beyond Syracuse, while track 2 was for Albany and Troy. Track 3 would be for Fishkill and tracks 4 through 10 for New York City's 69th Street yard. The designations would be a little different in a terminal yard such as Boston or Jersey City. Here certain tracks would be set aside for export cars destined for dockside tracks. Others might be held for local or nearby station destinations, still others for through or long-distance destinations.

The switch engine approached the distribution yard with its train of cars in the lead. If the first or lead car was bound for Fishkill, the engineer aimed for track 3. He would just enter the track and come to an abrupt halt; the car was uncoupled at the same instant and so was kicked into track 3, where it would roll freely under its own momentum. A brakeman rode on the roof ready to wind down the brake and bring the car to a stop just as it coupled onto the cars already in place on the same track. This method was called tail switching, supposedly because the engine held the train of unsorted cars by its tail. Kicking called for some skill and practice. Too light a kick, and the car failed to reach its destination, which meant wasting time while the engine and its head-end train entered the track and gave it another push. Too heavy a kick could result in a hard coupling, with possible injuries to men and draft gears. Cars were in fact more likely to be damaged during switching operations than in any other phase of their normal service life. Headwinds and temperature affected the free-rolling

qualities of cars, and so engineers had to develop a feel for just how much force to apply. Empty and loaded cars acted differently, as did cars with out-of-square trucks or dragging brake shoes. There was considerable art and very little science to tail switching. The brakeman or rider would climb down off the stopped car and scamper over the tracks to the switching train, then climb back up to the top and wait his turn to ride on another kicked car. The switcher would keep drilling back and forth until all of the cars had been put into the proper track.

Simply placing the cars on the right track might seem a big enough job for yard crews to handle, but the need to place them in a more or less precise order made the job even more difficult. The first car in the train would be the first to be dropped off; hence it was placed next to the locomotive, and so the train was arranged. The head-end cars were destined to be dropped off early in the trip, and so on to the last car. New cars picked up en route were added to the rear. Normally, only part of the train would be dropped off during the run. Variations to this basic rule for car orders were made for explosive and inflammable loads. These were placed in the middle of the train for safety reasons; in case of a collision they would be less likely to explode or ignite. An empty flatcar was never put at the head end of the train, because it was more likely to derail. The car-order regimen was further upset during the transition from hand to air brakes—roughly 1888 to 1902. During this time it was necessary to put all air-brake cars at the head end of the train and the hand-brake cars at the rear. Thus cars bound for Fishkill might be at either end of the train. Train crews accustomed to hardship, inconvenience, and stupidity must have seen this situation as just another tedious part of the workday.

After hours of work, the train on track 3 is ready. A crew is summoned by the call boy. The train is moved to the departure yard, where a road engine and a caboose are attached. The crew boards with their orders, manifests, and the blessing of the dispatcher, and the Fishkill local freight pulls out onto the main line. As she rolls slowly out of the yard, a clerk scribbles down the number of each car as it passes on train register sheets. This data will be telegraphed ahead to the next division terminal to let the people down the line know what car and cargoes are on the way.

Flat yards worked by small locomotives were the traditional form of yard during the wood-car era. Some antebellum railroads used horses for yard work. Others adopted obsolete locomotives for the purpose. By the 1850s custom switch engines had begun to appear (Fig. 1.54). These light, four-wheel machines were often called Pony engines, perhaps in honor of the horses they were intended to replace.

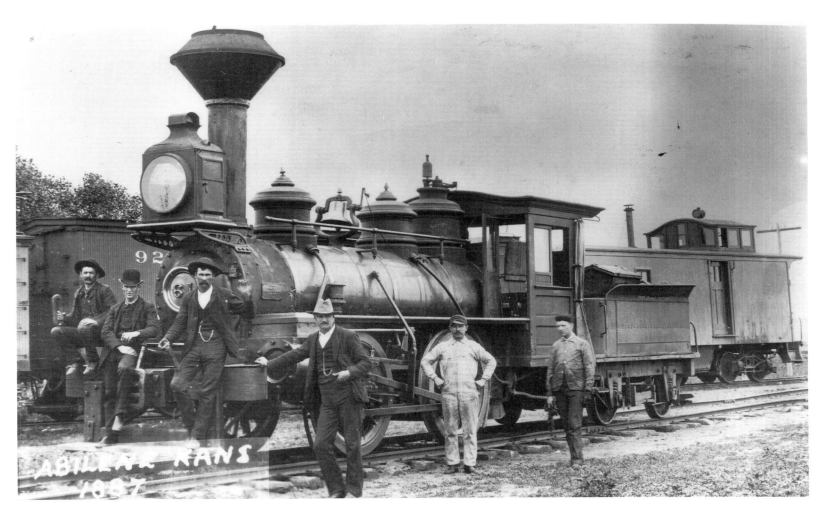

FIGURE 1.54

A Santa Fe yard crew pauses
briefly at the Abilene, Kans., yard
in 1887 for an unknown photographer. The engineer and fireman
stand back toward the cab of this
1877 Hinkley switcher. A supervisor, perhaps the yardmaster,
stands by the front bumper.
Meanwhile, three brakemen
strike a pose on the pilot deck.
The man at the far left holds a
link; the center man leans on a
pin cocked back and ready to fall
into place in the front coupler;
and the man to the right holds a
flattened pin against his leg.
(Union Pacific Historical
Collection)

The ladder or backbone style of track plan was
favored for yards because of its compactness and
the good visibility it afforded the engineer.[398] This
plan allowed him to see from one end of the train
to the other, and to observe the brakeman's signals
in the frequent coupling and switch-throwing operations involved in the switching of cars. The ladder track is on an angle of about 30 degrees to the
run tracks, which are in effect the rungs of the
ladder. This scheme, in a repeat pattern, is shown
in Figure 1.52. Another illustration, Figure 1.55,
shows a photographic image of ladder tracks in the
Reading's Rutherford yard near Harrisburg, Pennsylvania. Notice the main-line tracks to the right
of the yard, the Hall (banjo-style) electric signals
described earlier in this chapter, and the swayback
wooden boxcar in the left foreground. This yard
was a busy interchange junction for the Reading,
the Pennsylvania, and the Western Maryland
railroads.

Rutherford was essentially a junction yard that
happened to be located outside a major city; hence
the land was cheap and available. It could be
planned and laid out in the most efficient and economical manner. But this was not always possible
elsewhere. Land around big cities was expensive,
and most of it was already occupied. Hence yards
around Boston, New York, and other established

metropolises usually had to be sandwiched into
whatever space was available (Fig. 1.56). This
surely hampered the layout and arrangement of
the tracks. In the case of Pittsburgh, as already
noted, flat land was simply hard to find. In other
cases poorly arranged yards reflected poor planning or no planning at all. There was a natural tendency to keep expanding old yards in a haphazard,
jerry-built fashion as the traffic grew. There was no
overall plan and no long-range planning. Soon the
yard looked like a medieval city. It was all very pic-
turesque, but hardly efficient.

METHODS OF SWITCHING: POLING,
TOW ROPES, AND HUMP YARDS

During the 1880s several busy railroads adopted
the pole or stake method of switching to boost
yard productivity. Shallow pockets or dimples
were fastened to the corners of cars and to the
bumper beams of the switching engine. These
pockets could accommodate wooden poles, approximately 6 inches in diameter and 10 feet long.
The pole was held at an angle between the car and
the locomotive, which stood opposite one another
on parallel tracks. The pressure of the locomotive
pushing on the pole kept it in place. The pole was
generally tapered slightly and had iron bands on
each end. When not in use, it was carried in

FIGURE 1.55

The Reading's Rutherford yard near Harrisburg, Pa., gives us a clear picture of a classification yard late in the wood-car era, 1905. A ladder track is visible in the center of the picture. (The Hagley Museum and Library, Wilmington, Del.)

FIGURE 1.56

Boxcars clustered like bees around their hive while grain was exchanged between lake boats and rail cars. The scene is in Detroit at the Michigan Central's yard on the Detroit River in 1885. (Burton Historical Collection, Detroit Public Library)

THE AMERICAN RAILROAD FREIGHT CAR

FIGURE 1.57

A Pennsylvania Railroad class M six-wheel switcher with two hopper-bottom gondolas stands by a coal yard delivery track in the great Altoona yard in about 1895. Note the dimple-like pole pockets cast into the front beam of the locomotive, and the pole itself hanging down from the tender frame. (W. A. Lucas Collection, Pennsylvania Historical and Museum Commission, Strasburg)

brackets fastened to the tender frame. Figure 1.57 shows a Pennsylvania 0-6-0 switcher with its slopeback tender and a pole clearly in place. The shallow, round pole pockets can be seen on both corners of the cast-iron bumper beam of this engine.

The Fitchburg Railroad built a yard at Williamstown, Massachusetts, in 1887 specifically to test the poling method of switching cars.[399] It was a long, thin yard with seven tracks. One engine could handle all the cars of each eastbound train. A very light switcher could be used because it moved only a few cars at a time, whereas in the traditional tail-switching method, the whole train was shoved around, dropping off one or two cars at a time. In the pole method the train stood still while the engine ran back and forth on a parallel track picking off just a few cars at a time. The Fitchburg claimed savings of 40 percent in mileage and 20 percent in overall costs. The operation was so speedy that up to thirty-five thousand cars were handled each month.

The Pennsylvania picked up the poling idea around the same time and put it to work at its Altoona yard, but with a refinement.[400] Here, cars were moved not by the switching engine alone, but by a pole car pushed by the engine. The pole car had poles fastened to the two front corners so that

they could be aimed or adjusted by a lever and held more securely in the pole pocket of the car being moved. The short, flat pole car was ballasted with old car wheels to make sure that it held to the rails, for the push it got from the locomotive was often vigorous. At Altoona, pole cars were worked in pairs. The train being disassembled sat on a center track, and the pole cars and their engines ran on each side. First the right-hand engine and pole car would pick off a car and push it away. Once it was clear, the left-hand engine and pole car would go to work. As a team they could break up a seventy-five-car train in an hour. The Pennsylvania considered pole switching 100 percent more efficient than tail switching.

There were some problems with poling that eventually caused it to be abandoned. The poles could jump out of the pockets, becoming a deadly hazard to any brakeman standing nearby. If the pushing force was great enough, the poles could splinter into a fearsome bunch of wooden lances to skewer the closest trainman. Equipment was also endangered to a degree by poling, because it still involved the kicking approach to moving cars. The engine would give the car a good hard push before slacking off and so allow the car to roll free to its destination. It was always possible to be too energetic in this maneuver, and engineers known for flying starts sent cars sailing down the rails too fast for the rider to stop them before crashing into a cut of cars on the track ahead. Such mishaps were most common at night when visibility was so poor.

The origins of poling may never be discovered, but some indication of just how far back it dates can be gained from locomotive builders' photographs, which in many cases can be dated exactly. One such picture is in the Baldwin negative collection at the Pennsylvania State Railroad Museum in Strasburg. It shows a four-wheel switch engine built in January 1868 for the Detroit & Milwaukee Railroad that is equipped with pole pockets. Evidence of earlier application may well exist, but we can say with assurance at least that the pole method was around just after the Civil War. So far as cars go, about the earliest graphic record is the 1882 New York Central boxcar drawings reproduced in the 1888 *Car Builders Dictionary* (Fig. 95).

Tow ropes for parallel track switching appear to have been used on a few roads around the time poling was introduced. Nothing has been found in existing railroad literature, but several photographs attest to the use of rope as a car-moving aid. Perhaps the earliest of these is a rear-end view of the locomotive *Gov. Stanford* on the Central Pacific, made around 1870. A long, thick hawser is draped back and forth on hooks across the rear of the tender tank.[401] A coupling link is neatly tied to one end of the rope, thus forming a convenient attachment to the common link-and-pin coupler of the day. Back east the Reading and the Baltimore & Ohio took a liking to ropes as a switching auxiliary.[402] The chief advantage of the method was that it saved a certain amount of switch throwing, for the engine did not have to be on the same track as the cars being moved. At least some yard switches involved spotting cars in front of a warehouse door or by the icehouse. The movement might be just a few feet, and if a parallel track was present, the pole or rope method saved both time and effort. In some cases where an industrial siding was sturdy enough to handle a car but not a locomotive, the rope or pole method, where applicable, might prevent a derailment. Ropes were preferred over cable and chains because they could give or stretch a little when under a heavy load that could break a less elastic connection. The B&O at first tried large chains with grappling hooks on one end, but sometime after 1865 it adopted ropes, judging from the record of surviving photographs.

Poling and rope switching were essentially stopgap measures that surely improved yard efficiency, but neither method was very effective, and both were inherently dangerous. A far more promising invention in car sorting, which appeared on European railways in the 1870s, employed gravity rather than a battery of steam locomotives to propel the cars.[403] A single locomotive would push a string up a grade at one end of the yard. Cars were cut off individually or in clusters at the crest of the grade or hump and allowed to roll downhill free of all constraints except for the rider, who would work the hand brake as necessary. The switches were prealigned for each movement so that cars intended for, say, Fitchburg would roll onto track 1. In more sophisticated yards, switches were lined up from a control tower placed above the hump. At smaller or older hump yards, switches were set by hand; hand signaling indicated which track was to be lined up for the next car coming down the incline. The riders or brakemen walked back to the top of the hump ready for the next car, or sometimes the railroad offered a return trip via a flatcar powered by a switcher.

The ease and rapidity of classifying trains was the chief advantage of the hump or gravity system, but it also offered attractive savings in labor and equipment costs. In 1894 the Pennsylvania Railroad's gravity yard in Philadelphia handled twenty-three hundred cars a day with ease and could process nearly double that number if necessary. Two locomotives and sixteen to twenty-four men were employed, as was a switching tower. A flat yard of equal capacity would have required more trackage, five locomotives, and thirty men. The pilot Philadelphia hump yard worked so well that a major gravity yard was planned to replace several small flat yards in the area. Hump yards began to appear in all areas of the nation as the reputation of their efficiency spread. The Illinois Central built a large double-hump yard in Harahan, Louisiana, about 9 miles from New Orleans. This big north-south yard had 48 miles of track and could service up to thirty-six hundred cars a day. An even bigger yard opened near Chicago around 1902 with a daily capacity of five thousand to eight thousand cars. It had 450 switches, of which 120 were electropneumatic and could be operated by one man in a signal tower. To boost productivity and safety, the Chicago Transfer Yard was lighted by arc lighting at night.

For all its advantages, the hump yard did not replace the conventional flat yards or tail switching. Most major yards were converted to or replaced by hump yards, but many operations were too small or seasonal to justify the expense, and so once again poverty preserved the old-fashioned ways of train operations.

CONGESTION

Yards tended to grow around big population centers—a logical enough phenomenon, because cities were natural destinations for goods and products, even those passing on to another place. But from the railroads' viewpoint, big cities were the very worst location for yards and terminals. Congestion and traffic already abounded. The most-

FIGURE 1.58

Team tracks were sandwiched
into any open urban space, such
as the B&O 23d Street yard in
New York City. Grain bags are
being unloaded from a Toledo &
Ohio Central car at the right side
of the picture, taken in about
1890. (Smithsonian Neg. 49769)

needed sites were already occupied, not for sale, or for sale only at a very stiff price. Yet the cars must find a home. They could not always rest on a grassy siding in Indiana, and so it was a matter of making do. Yards and freight sheds were put where space was available. This usually meant the worst part of town. The problem grew with the size of the city and its state of development. Locating a big yard in Jersey City was a vastly different thing from locating one in Omaha, even in 1870. Most choice—and surely all cheap—industrial sites in Jersey City were gone by this time, while one might still move around the environs of Omaha with relative freedom.

With these generalities in mind, it is understandable that the last place in the United States a railroad official would seek for the job as yardmaster would be New York City (Fig. 1.58). It was the busiest, the most complex, the most congested, and the most costly rail terminal in the nation. It was surely also the most inconvenient, because the broad waters of the Hudson River cut off the rail terminals at the New Jersey shore and so made

an orphanage for fifteen railroads all longing for a home on Manhattan Island. The costly car float system has already been described. Only the New York Central had a rail yard with a direct all-rail connection to the north on Manhattan Island. The Baltimore & Ohio, the Pennsylvania, the Lehigh Valley, and all the other railroads faced a tangled web of trackage on the New Jersey shore. The area was a litter of yards, junctions, crossovers, drawbridges, and street crossings. By 1890 there were 378 acres covered by yards, 5 million square feet covered by piers, sheds, and freight stations, all of it manned by forty-seven hundred employees.[404]

Why maintain such expensive facilities in such an inhospitable environment? The answer is simple—traffic. In 1890 rail freight traffic into the New York City port area constituted 28.5 percent of the national total.[405] Within a decade it had risen to a phenomenal 41.6 percent. Class one railroads carried 72.5 million tons into the New York City port in 1870; 437.0 million tons in 1885; 636.5 million tons in 1890; and 732.0 million tons in 1900. One-third of the nation's exports left through New York, and fully 60 percent of all imports came in through the same port.[406] The railroads had no choice but to follow the traffic, and that clearly meant servicing the Big Apple.

The original sheds and team tracks strung out along the waterfront at Hoboken and Jersey City soon proved inadequate. In 1874 the Pennsylvania Railroad retreated from its old Camden & Amboy Jersey City location and developed a yard and piers at nearby Harsimus Cove.[407] It featured a huge freight house measuring 120 by 1,000 feet. Five years later it was necessary to expand the facility materially. By 1900 Harsimus Cove was simply overwhelmed by the growing traffic of the city. Nine hundred acres of flat land was purchased north of Bayonne for the four-thousand-car-capacity Greenville yard. Other railroads were forced to develop yards and terminals to the north of Hoboken. The Erie and the West Shore both built on a grand style in Weehawken. The West Shore terminal, built between 1881 and 1885, was an ambitious and well-planned facility that stretched along the Hudson River shoreline for over a mile. It included nine piers, a stockyard, a slaughterhouse, two meat-packing plants, and an assortment of grain elevators and freight sheds. Car floats made a convenient connection directly across the river to the New York Central's big West Side yard.

Sensibly, every trunk line serving New York City made a concerted effort to perform as much switching as possible outside the immediate area of the nation's busiest port. The New York Central, for example, established major yards well to the north of Manhattan. In 1893 the yards in Buffalo were described as being more than double the size of those in New York City.[408] There was room to spread out and occupy thirty-six hundred acres, an area of property unthinkable in Manhattan because of land prices. Just one of the Central's Buffalo facilities was the cattle yard on the east side of town. This 230-acre yard would load, on a busy day, 375 cars of cattle, 150 of hogs, and 75 of sheep. The Central also had sizable yards at Suspension Bridge, Syracuse, Rochester, and West Albany. The DeWitt yard at Syracuse served primarily as the makeup yard for westbound freights. The West Albany supplemented the work done at DeWitt. By 1912 it boasted a sixty-five-hundred-car capacity and had sixty-five tracks plus four receiving tracks and 73 miles of trackage.[409] The normal work crew was 226 men. More traffic called for more cars, which meant more yards, which after a time resulted in more delays and confusion. The periodic blockades or stoppages already mentioned were the result of the unrestrained growth of the American economy and the inability of the railroads to keep up with that expansion.

The terminal situation in the West was no better than that in the eastern cities. St. Louis was sec-ond only to Chicago as the nation's midcontinent exchange place for freight cars. The flow of traffic was greatly hampered by the Mississippi River, which formed a major barrier to the movement of cars between the several railroads that converged there. The sorting of over 2 million cars a year was handled in forty classification yards and over one hundred interchange points.[410] But so far as the railroad industry was concerned, Chicago was the real hub, for here is where most of the east- and westbound cars met. It claimed to be the greatest rail center in the world. This was no idle claim, for by 1889 6 million cars a year were passing through the Windy City.[411] Three hundred switch engines scrambled around the yards, while another one hundred locomotives performed interyard transfers. Just after the turn of the century, around 650 freight trains a day entered or left Chicago on thirty railroads. Such a concentration of traffic and tonnage is hardly surprising, considering the number of railroads radiating from this great midwestern rail hub.

Bigness did not necessarily breed efficiency, however. Yards worked two twelve-hour shifts, but each switcher and its crew handled only thirty-two cars per shift. For much of the time the engine and its human helpmates stood idle waiting for more cars to arrive or for a signal to turn clear. More time was lost in the slow transfer haulage of interchange cars between yards. Some yards were 12 miles apart, and that meant a tedious trip across town with innumerable street crossings and railroad crossovers. Because transfer trains were inferior to almost all main-line traffic, they spent considerable time on sidings waiting for a clear track. The city insisted on keeping transfer trains short so as not to block wagon and streetcar traffic more than necessary at road crossings, but this meant lots of short trains rather than a few long ones, and so traffic tended to back up on the railroads. The plethora of yard trackage in the metropolitan area hardly helped to streamline the transfer of cars between outer yards, city yards, interchange yards, and storage yards. Costs were accordingly high, and railroads' switching costs came to 3 cents per hundredweight.[412] This works out to 60 cents per ton! Compare this with the cost of moving freight on the main line. A thirty-five-car merchandise train cost only about $1.03 per train-mile to operate, or just a tiny fraction of what it would cost once inside the yard limits.

Chicago-area railroads adopted all of the reforms tried elsewhere in the nation to improve yard efficiency. Poling helped a little and hump yards helped a little more, but in all, the problem was unsolvable. The Chicago & North Western felt that expansion was the answer to yard efficiency and so kept expanding the Proviso yard until it was the largest in the nation. The great ten-thousand-car Clearing Yard was another brave attempt to un-

THE AMERICAN RAILROAD FREIGHT CAR

FIGURE 1.59

Cars for small shippers were
spotted on house or team tracks
for loading and unloading shipments. A carload of lumber is
being taken away by this horse-
drawn dray around 1900. (California State Railroad Museum,
Sacramento)

tie the Gordian traffic knot. The creation of belt
railroads specifically designed to move cars
quickly between the yards of regional railroads
proved to be the most effective reform for big rail
gateways like Chicago. In 1882 the Belt Railway of
Chicago began operations as a clearinghouse for
cars between five railroads.[413] Eventually it would
serve a dozen lines. In 1883 the BRC handled
58,700 cars; by 1900 this number had jumped to
almost 675,000. Other belt lines opened in Chicago as the benefits of a railroad whose main purpose was the transfer of interchange cars became
better understood. The Indiana Harbor Belt and
the Baltimore & Ohio Chicago Terminal are two
well-known examples. In the late 1880s St. Louis
followed Chicago's lead with the Manufacturers
Railway and the larger Terminal Railroad Association of St. Louis.

Many trainmasters came to the belief that the
best solution to the problem of yards was to avoid
them altogether whenever possible. In 1882 the
New York Central was using the solid or block
train system.[414] Cars were grouped together at Buffalo for various destinations—Albany, Boston,
New York, and so on. These trains were not classified en route to each division yard but went
straight through to their destination. Engines,
crews, and caboose would be changed at division
terminals, but the trains remained intact. This
system worked only for through cars. Most trains
were handled in the same old way because most
freights were of the peddler variety that picked up
and dropped off as they traveled over a division,
making classification mandatory at the end of the
run. Block trains often involved single commodities such as fruit or dressed beef. Cattle and
tank car trains were also adaptable to block train
operations and were in many ways the precursors
of the modern unit train. In 1892 the Pennsylvania
began operating special block grain trains between
Chicago and New York.[415] By avoiding yards and
needless classification, these trains made the trip
in only four and a half days.

In some special cases block trains were operated
over several railroads. Each railroad would insist
on its own crews, mostly for safety reasons, but
would refrain from breaking up the train. Operations of this sort were called run-through trains. In
another effort to avoid needless switching, railroads learned to use greater care in loading less-
than-carload cars. Rather than put goods in several
cars, all items bound for Xenia would go into one
car, which could then be dispatched directly to
that Ohio city. This produced a straight car, or one
that went straight to its destination. Sadly, most
LCL cars remained mixed cars or ones with cargoes for several destinations.

UNLOADING

Once the freight car broke free of the final classification yard, it was ready to discharge its cargo
to the consignee. Just how this was done would depend on what was being shipped and who was receiving it. A large business house would have its
own private siding set down beside the factory or
warehouse. Small shippers would unload at team
tracks or the nearest freight house. What was being
shipped could further define just where the loaded
car would be parked. A load of cattle destined for
auction would go to the local stockyard, while a
carload already purchased by Swift & Company
would go directly to that packing house. Grain
would normally go to large central elevators,
though again some of it might be consigned to an
individual flour mill. Coal cars would be shifted
around to scores of small dealers located all over a
city in the days when home heating was dependent on solid fuel. Likewise, lumberyards were
scattered just about everywhere, and all but the
smallest would have a siding ready for a car or two
that showed up periodically to resupply their
stock.

In the wood-car era very few factories lacked rail
sidings. Rail cars brought in the new materials and
hauled away the finished product. Even a moder-
ate-sized city would have three hundred to five

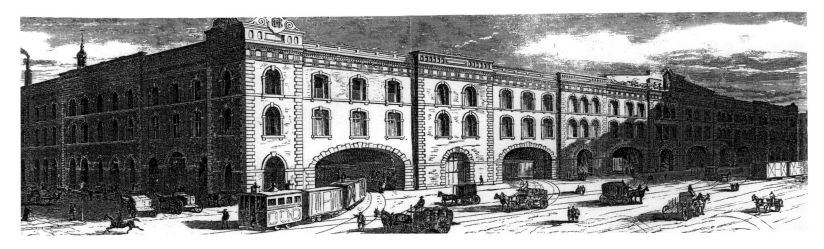

FIGURE 1.60

hundred private industrial sidings.[416] Larger cities like Philadelphia would boast of thirteen hundred private sidings. To accommodate shippers of smaller carload lots, railroads maintained semi-public sidings called team tracks. These were located as close to the central business district/manufacturing part of the city as possible. Most big railroads had several team tracks in various locations as a convenience to their customers. Because more than one railroad served most large communities, team tracks might be found clustered all around the commercial neighborhoods. The term *team track* came from the horse and wagon or teams sent down to haul away the contents of the car. The wagons were usually backed up to the car door to load and unload (Fig. 1.59). In a well-designed team track, the space between tracks was wide and well paved. The pavement was made level with the top of the rail or track to facilitate loading by making the car floor and wagon about even. At small towns or in suburban stations, team tracks were not really needed for the modest local traffic, and so a house track was used. This was little more than a short, designated siding, placed near the depot.

The typical freight house was generally the plain counterpart of its more elegant half-sister, the passenger depot. Small-town freight stations were sometimes little more than sheds. In some communities the passenger and freight depots were built as a combination structure; one such station at Boulder, Colorado, was pictured earlier in Figure 1.25. The notion that small-town life is slow and unhurried was reinforced by the languid pace at the depot. Goods and cars would stand around for up to three days until the consignee came to pick them up.[417] Only about one-third of the goods was picked up on the first day. Combination stations might work well enough in rural areas, despite the casual habits of the local residents, but in big cities they could prove intolerable. In 1851 the Cincinnati, Hamilton & Dayton opened a dual-purpose depot in Cincinnati that soon had passengers fuming as they stumbled

their way over and around boxes, barrels, and bags to and from the trains.[418] The porters were no doubt just as frustrated as the passengers and must have longed for a place of their own—a wish that was not answered until 1864.

Large, well-equipped freight houses had been around for two decades. The Boston & Worcester built a model freight house at South Cove, Boston, in the 1840s that covered 1.25 acres.[419] It had four tracks with platforms level with the car floors to facilitate loading and unloading. A track scale was built in, as was a transfer table. The latter was used to shift cars from track to track. Separate storehouses were erected for cotton and grain. The B&W's connection, the Western Railroad, built an even larger warehouse in 1845, on an island near Greenbush, New York. Steam cranes shifted goods between canal boats and railroad cars so efficiently that five hundred barrels of flour could be transferred in just two hours. It soon became fashionable for trunk lines to erect showplace freight terminals. The Baltimore & Ohio put up a big one 10 miles east of Parkersburg in about 1856, large enough to cover seventy-five cars and house sixteen thousand barrels of flour. About ten years later the Michigan Southern opened a fine new freight terminal in Chicago.[420] It was made on a smaller scale than those just described, measuring 52 feet by 600 feet, but its fire-resistant construction, stone walls, and slate-covered roof were considered noteworthy. It too had an internal transfer table so that cars could be moved easily from track to track.

Never to be outdone, Cornelius Vanderbilt decided to build the grandest railroad freight terminal of them all. It was located on the West Side of Manhattan, well down on the island close to the present-day entrance to the Holland Tunnel.[421] The site had been a beautiful private garden called St. John's Park covering four or five acres in a once-fashionable residential neighborhood. Vanderbilt bought the park in 1867 and quickly uprooted its fine trees, gravel walks, and elegant cast-iron fence. A substantial three-story masonry building was erected in about one year's time (Fig. 1.60). Six broad, flat-arch doorways allowed twelve tracks to enter at the ground level. One hundred cars could be parked on these tracks. Goods were stored on the upper floors. The building was clearly envisioned as more than a utilitarian depot for the sorting and storage of freight. It was to be a monument to commerce, industry, the port of New York, and, not so incidentally, Cornelius Vanderbilt. An imposing bronze entablature, on a heroic scale, ran the length of the building. At its center was a larger-than-life statue of Vanderbilt. The St. John's terminal, surely the grandest of all American freight homes, served on until around 1934.

Railroad freight terminals, be they rustic sheds or magnificent stone edifices, shared several common traits, notably high costs in terms of labor and time. In 1874 Octave Chanute, a consulting engineer for the Erie, conducted a cost study at his employer's Jersey City freight house.[422] This study revealed the cost of loading to range from 4 cents to 57 cents per ton, depending on the commodity. It cost 8 cents to load a barrel of sugar but only 6.4 cents to unload a barrel of flour. It was cheaper to unload than to load, just as it is cheaper to tear a house down than to build one. Obvious things were proved by this survey, such as that loading coal by steam shovel cost half as much as doing it by hand. Overall New York City terminal costs— loading, unloading, switching, lighterage, and so on—came to $3.07 per ton, or 60 percent of the charge rendered to move a ton from Chicago to New York. Around 1890 the overall operating costs of New York rail terminals were figured at $10 million.[423] Not many years later the cost of delays alone in the various Chicago rail terminals was estimated at $6 million per year. Hence the steam railroad was a genius in carrying goods between major terminals but a fool in delivering cargoes to the consignees once they had reached the terminal.

Train Speeds

The hurry-up-and-wait nature of train operations explains the generally slow rate of travel associated with freight service. Surely railway managers searched for ways to expedite the movement of goods between shippers and consignees. Yet curiously, they were consciously devoted to a policy of slow running speeds. Fast running was dangerous, and it cost money. Speed lost money. Rapid transit used up fuel, lubricating oil, and equipment. It damaged tracks and in just about every way possible boosted operating expenses. Railroad managers wanted none of this. They wanted a speed that inflicted the least damage on the train, the track, and the cargoes being carried. The most economical speed was figured to be about 10 miles per hour, or just fast enough to get a train over one

division in a single working day. That was fast enough, heaven knows. After all, God must have intended freight to move slowly, for it had always been so. Over the road, wagons rarely did better than 2 mph, canal boats went a trifle slower, and the best western river steamer did well to match the pace of a railroad freight train.[424] Few men could maintain 10 mph for very long. Thus it was settled; the right and proper pace for boxcars was fixed at 10 mph. Railroads pursued more direct routes than waterways, and so even if they did not move much faster than a packet steamer, they would arrive in less time. The river distance between Pittsburgh and Cincinnati, for example, was 470 miles, while the rail distance was only 311.

Slow running was a deliberate policy rigorously enforced by the trainmaster. In 1859 the Baltimore & Ohio's trainmaster was mightily upset by reports that some of his freight crews would race ahead so that they might stop somewhere along the line for an hour or so to play a game of tenpins.[425] Once the game was finished, off they would tear, pell-mell down the track fast enough to get to the home terminal on time. Did they go double or triple the normal running speed? No one could tell, but this wicked and reckless habit was instantly stopped, or at least so the trainmaster ordered.

Elsewhere it was commonly understood that freight train crews would stop on any pretext whatsoever.[426] It might be to drink from a spring, drop off a package, visit a friend, fish in a stream— any reason would do. Such loafing on the job was not easily detected, especially on more remote portions of the railway system, but every superintendent was certain that it took place. Some officials would drive out into the countryside and hide under a trestle to catch the boys whooping her up to 20 or 30 miles per hour in an effort to reach the next station on time.[427] Managers not given to a personal spy system would install speed recorders in the caboose, which could record each start and stop plus the train speed at any given time. If the engineer poured on the steam and the cars began to sway, the speed recorder would mark that the crew had been walking the dog. Speed recorders, introduced in the 1870s, were understandably much hated by train crews, who generally referred to them as Dutch clocks. Installation and maintenance costs, plus train crew sabotage, prevented Dutch clocks from becoming standard equipment.

To be fair, not all trainmen sped just to defy official rules or to make time for idle and foolish pursuits. Some found the slow pace simply stupefying. Extra-slow running was not necessarily safe, for it tended to induce sleep. Crews dozed at their duty stations as the train lumbered along the track at such a gentle and silent gait.[428] Sleepy firemen might overlook a low boiler water level; the brake-

man might doze through the call for *down brakes*; the engineer might fail to notice a stalled train just ahead. Such problems seem to have done little to shake the management's belief in slow running. Common freight trains continued to be called names like peddler, drag, way, and even dead freights, none of which conveyed much feeling of excitement or speed. Even the term *way freight* has the suggestion of slowness about it, for it indicates the process of picking up and dropping off cars along the way.

The record for actual running times confirms that slow travel was the rule for freight trains in the wood-car era. Part of this was due to the prescribed speed restrictions and part of it, particularly in the early part of the century, to operating conventions of the time. Some railroads had no headlights on freight locomotives, and so goods trains were required to lay over at sunset. This added greatly to running times. The Western & Atlantic in antebellum times took two days to run freights between Atlanta and Chattanooga because of the mandatory layover.[429] By the early 1870s headlights had become standard on all main-line locomotives and layovers were a relic of the past, yet freight trains seemed to grind along as slow as ever. Boxcars rarely traveled between Chicago and Pittsburgh (468 miles) in less than four days, and some required fifteen to twenty days to complete the journey.[430] An official of the Union Pacific lamented in a statement published in 1883 that cars rarely made it between New York and Council Bluffs in nine days, though schedules always showed six days. Cars moving over the same route in two or three weeks were not unusual. Matters were somewhat improved in 1896 when *Railway Age* reported that ordinary freight service between Boston and St. Louis was a dependable five days.[431] Some preference freight covered the same 1,000 miles in just three days. Now *that* took some fast running, to maintain an average speed just under 14 miles per hour.

Safety was the sacred cow used to justify slow train speeds. Even in the late 1880s fretful voices were raised about the wisdom of increasing freight train speeds.[432] The application of air brakes was just beginning, and it would be over a decade before the conversion was finished. Even so, air brakes were not enough. The track and bridges must be brought up to a standard capable of sustaining a 1,000-ton train traveling at 30 miles per hour. Almost no American trunk line was ready to handle such a punishing load. Trucks must be materially improved in design, manufacture, and maintenance before speeds of even 25 mph could be safely attempted. The standard arch-bar truck was a rickety affair that worked fine at 10 mph. It might be out-of-square, with loose brasses, no dust shields, and half a dozen defects and still perform safely at the old steady gait of 10 mph; but double or triple the speed and disaster is sure to follow. One can almost hear the old-guard master car builders chanting their mantra: Slow speed is safe speed, to go fast is wicked and expensive, go slow, go slow, go slow. . . . Such caution was no doubt the safe course to follow, but the dogma that a freight train cannot travel safely at much over 10 miles per hour was occasionally disproved by runaway trains. In 1875, for example, a Lackawanna coal train broke in two.[433] The rear section proceeded to roll ahead freely and soon gained considerable speed. The engine and front group of cars raced ahead to avoid being overtaken by the runaway rear section. The runaway was soon up to 60 mph, and yet the train held to the track with perfect safety. Wheels, axles, bearings, they all performed as if made for fast running. Eventually the front end recaptured its free-wheeling second half, and the brief holiday of the coal cars was brought to an undramatic conclusion. Episodes like this one demonstrated that the primitive wooden cars of the time could run at speeds well beyond what railroad managers considered to be prudent.

It is also true that train crews themselves were not always prudent. Those intent on getting over the division on time tended to break the rules, including the sacred dictum about speeding.[434] They did not bother to honor the five- to ten-minute cushion for meets. They would slide into the siding, sometimes with just seconds to spare in clearing the main track for a superior train. Such crews ignored speed limits. They did not waste time in switching cars either; they would kick them into a siding fast and only rarely bother to uncouple the air hoses—they would pop off anyway. Anything to save a few minutes and then dash off to the next stop. By contrast, the by-the-book conductor would proceed slowly, never exceed the posted speed limit, always allow a good margin of time for meets. This same careful man always ran late. Ironically, he was frequently rebuked by the trainmaster for his tardiness, while his fast and loose comrades were only scolded when finally caught.

A favorite excuse for fast running claimed that it was necessary to do so to boost heavy trains over steep or long grades. A timid engineer or one who

FIGURE 1.62

ILLINOIS CENTRAL RAILROAD CO.

FAST FREIGHT SCHEDULE
FROM
CHICAGO.

	TO		TIME.	NUMBER OF HOURS.
LV.	CHICAGO,	ILLINOIS,	11.00 P.M.	
AR.	PADUCAH,	KENTUCKY,	6.45 A.M.	31.45
"	OWENSBORO,	KENTUCKY,	9.00 A.M.	58.00
"	CAIRO,	ILLINOIS,	6.20 P.M.	19.20
"	LITTLE ROCK,	ARKANSAS,	11.45 A.M.	60.45
"	TEXARKANA,	TEXAS,	1.30 P.M.	62.30
"	WACO,	TEXAS,	7.20 A.M.	80.20
"	FT. WORTH,	TEXAS,	6.00 P.M.	91.00
"	MEMPHIS,	TENNESSEE,	7.00 A.M.	32.00
"	NASHVILLE,	TENNESSEE,	9.55 A.M.	34.55
"	CHATTANOOGA,	TENNESSEE,	10.05 P.M.	47.05
"	ATLANTA,	GEORGIA,	7.40 A.M.	56.40
"	VICKSBURG,	MISSISSIPPI,	2.40 P.M.	63.40
"	BIRMINGHAM,	ALABAMA,	3.00 P.M.	64.00
"	SHREVEPORT,	LOUISIANA,	1.00 P.M.	86.00
"	JACKSON,	TENNESSEE,	3.40 A.M.	28.40
"	JACKSON,	MISSISSIPPI.	7.45 P.M.	44.45
"	NATCHEZ,	MISSISSIPPI.	6.30 P.M.	67.30
"	BATON ROUGE,	LOUISIANA,	1.40 A.M.	74.40
AR.	NEW ORLEANS.	LOUISIANA,	6.00 A.M.	55.00

slavishly followed the speed limit would stall on the hill. You had to kick her in the hind end and lay on the steam to get over the grade. Wasn't it better to go fast for just a little while downgrade than to tie up the railroad? The kinetic energy gained by running faster than the rule book allowed surely helped maintain schedules, and most superintendents were ready to accept some bending of the rules to accomplish that end. At least one railroad felt that increasing freight train speeds from 10 or 12 to 18 or 20 miles per hour actually saved fuel as well. In the middle 1870s the Lake Shore ran a fuel consumption test with a Mogul locomotive pulling 650- to 700-ton trains that showed better fuel consumption at the higher than the lower range of speeds.[435] Some years later, however, the Northern Pacific ran tests that showed a marked increase in operating costs as speed increased.[436] Moving a 1,050-ton freight train at 15 mph cost 60 cents per mile. Increase the speed to 20 mph, and the cost went up to 68.6 cents per mile. At 35 mph the cost soared to $2.60 per mile.

Ironically, while the industry seemed wedded to a policy of slow freight train speeds, it exhibited at least some interest in fast freight as well. The so-called private freight car lines that proliferated just after the Civil War all emphasized the notion of fast service. These claims were no doubt more promotional than real, but they did illustrate the desire of shippers for such service. The getting of cattle to market quickly—or at least the talk of doing so—goes back almost to the beginning of railway service. Speed took on more meaning in railroad circles once the refrigerator car became a commercial reality, around 1880. Shippers tried to convince railroads that products other than lettuce

and berries were time-urgent. In May 1879 the Pennsylvania ran a special four-car tobacco train from New York to Chicago in just twenty-three hours and fifty-five minutes.[437] This record run was made at an average speed of 38 miles per hour, despite one or two stops for hotboxes. The Lake Shore was more interested in raising the speed of its normal freight traffic than in the occasional demonstration train. Its fastest freight ran eastbound over the Sandusky-to-Erie Air Line prior to 1876.[438] This portion of the road, which featured low grades and gentle curves, was opened to westbound freights as well in 1876. The line's fastest freights ran between Chicago and Buffalo in thirty-eight hours and thirty minutes, with an average speed of 13.6 mph. That was considered good time in 1879. By 1894 double-tracking, grade reductions, and improved track alignment had reduced the time to thirty-two hours and twenty-five minutes, or 16 mph.

Western roads were also ready to display their competence in the fast freight business, and so in August 1885 the Northern Pacific ran a trainload of tea from Tacoma to St. Paul (2,017 miles) in just 123 hours and 25 minutes, for an average speed of 16.5 miles per hour.[439] A few years earlier a trainload of canned goods traveled across the continent from Portland, Maine, to Portland, Oregon, in just fourteen days. Bananas, being more time-sensitive than canned goods, seemed to inspire greater speeds. In March 1894 the Santa Fe ran a hot-shot banana special from Galveston to Chicago (1,385 miles) in fifty-three hours, for an average speed of 26 mph.[440] The Illinois Central had its own crack banana train during the same season that raced from New Orleans to Chicago (914 miles) in thirty-five hours and fifty minutes. By the fall of 1896 the Illinois Central was ready to go public with a timetable for fast freight service.[441] The roadway, motive power, and operating departments had cooperated to offer regular and not just occasional fast freight operations. A schedule is reproduced in Figure 1.62. It should be noted that the times shown here are considerably more conservative than those indicated for the more spectacular banana specials. The speed of the Chicago-to-New Orleans train, for example, is more like 16 mph than the 26 mph of the banana special.

The Illinois Central was hardly alone in the fast freight business. It was one of those fads that rose and fell like hemlines. The Santa Fe began Red Ball Service in 1892 to expedite high-value merchandise shipments.[442] The Frisco came forward with both Red Ball and Green Ball fast freight service. Most other railroads from the Southern to the Chicago & North Western introduced fast schedules as well. The industry devised catchy generic names for such trains; they were called manifests, preference, and symbol trains. A symbol train might be called the JP1, which stands for Jersey City to

PARADE OF THE CARS

Pittsburgh westbound. Its eastbound counterpart would be called the JP2.

The notoriety of the fast freights conveys a false idea of their actual importance on American railroads. Fast freights were relatively rare even at the close of the wood-car era, around 1910. They represented only 10 percent of all freight trains. Yet speed was one way to improve car mileage, railroad revenues, and profits. If the railroad is a ton-mile factory and more ton-miles per day are its goal, then it is necessary to increase either train size or the speed of each train. Increases in speed were accomplished in a slow and hesitant manner at best. In 1920, after decades of effort, the average freight train speed on class one railroads had crept up only to 10.3 miles per hour.

THE FLEET

Makeup of the Fleet

It would be satisfying to be able to present a solid and uniform table of data on the early freight car fleet, to enumerate the exact type and quantity of cars in service for any given year. But sadly, the data needed for such a table is not available, and what is available, especially for the first half of the century of American freight cars, is largely a confusion of numbers. This lack of a reliable data base represents a major void in what is normally considered to be one of the best-documented of American industries.

About the first official number to appear is the 1860 Census figure of eighty thousand freight cars, but the roundness of the number suggests that it is nothing more than an estimate. Because no hard numbers existed, railroad journalists attempted to fill the void with manufactured statistics. In April 1861 the editors of *Railroad Record* cooked up some figures based on the number of cars per mile of railroad in New York and Pennsylvania.[443] They estimated that there were, on the average, 4.25 freight cars per mile, a not implausible figure to be sure; but their estimate of national rail mileage at 27,000 miles was off by over 4,000 miles, according to a more modern estimate. And the *Railroad Record*'s total for freight cars came to 114,750, a rather dramatic rate of growth in just four months over the 1860 Census figure. Someone or something was clearly wrong here.

Other unofficial estimates appeared from time to time. In 1869 W. G. Hamilton published a small handbook called *Useful Information for Railway Men*, which contained the figure of 160,000 freight cars for the railroads of the United States. It is difficult to believe that the fleet size had doubled in just nine years; either the 1860 Census figure was remarkably below the actual mark or Hamilton's estimate was too high. How do we validate these numbers? Where do we turn to verify the count? What repository holds the raw data to prepare a new total? These questions appear to generate nothing but negatives. Modern scholars such as Albert Fishlow attempted to compile statistics for the earliest decades of American railroading, but the result was, with all due respect, not much better than the efforts of the railroad trade press as enumerated above, except that Fishlow tended to side with the more conservative estimates of the past.[444]

By the 1870s a trifle more light and truth seems to have entered the realm of railroad car statistics. Curiously, the 1870 Census appears to ignore the whole subject of the railroad industry, except for employment. In 1874 two compilations appeared in print that not only closely agree but are based on an actual count of equipment on a railroad-by-railroad basis. These listings may not be complete, but they are thought to be nearly so, or at least very comprehensive. The first appeared in February 1874 and was drawn from a list of 850 railroads published in December of the previous year.[445] The total given was 338,427 freight cars. In August 1874 the second compilation, based on *Poor's Manual of Railroads*, gives a U.S. freight car total of 345,920.[446] Given the nine months' difference in reporting dates, the increase in fleet size by seven thousand cars or so is plausible. These numbers seem reasonable enough for us to project backwards to 1870 for an estimated freight car total of three hundred thousand.

We seem to be on solid ground at last in 1880, when the U.S. Census conducted a highly detailed survey of every railroad in the nation. The total for that year is given in Table 1.13. Yet even into the last decade of the nineteenth century, there are

TABLE 1.13 Growth of U.S. Railroad Freight Car Fleet

Year	Mileage	Number of Cars
1830	23	50
1840	2,800	6,500
1850	9,000	30,700
1860	30,600	100,000
1870	52,900	300,000
1880	93,300	539,200
1890	163,600	1,109,900
1900	193,300	1,365,500
1910	240,400	2,135,100

Source: Figures before 1880 are largely estimates based on so many cars per mile of railroad. Figures after 1880 are a composite of information from the U.S. Census, *Poor's Manual of Railroads*, and the yearly statistical reports of the Interstate Commerce Commission.

Notes: Main-line railroads only. List does not include private, fast freight, or leased cars.

TABLE 1.14   U.S. Freight Cars by Type

Year	Box	Flat	Cattle	Coal	Tank[a]	Refrig.[a]	Caboose
1880	194,500	70,500	28,600	183,500	3,000	300	NA
1890	408,900	131,600	57,300	354,100	2,000	8,500	NA
1900	657,000	134,500	55,100	454,500	2,800	14,500	17,600
1910	966,600	154,000	78,800	818,700	7,400	31,000	27,200

Source: Interstate Commerce Commission statistics.

Note: NA = not available.

a. Railroad-owned; does not include privately owned cars of this type. Thus tanks and reefers are greatly underrepresented in this table. In 1890 there were an estimated sixty thousand private freight cars; most of them were either tanks or refrigerator cars.

conflicts and doubts about the veracity of the numbers presented. Just as an example, let us examine the numbers for 1890 as reported by four sources: the Interstate Commerce Commission reported 1,160,000; *Poor's*, 1,061,970; *Railroad Car Facts* (1938 ed.), 1,109,952; and Slason Thompson's *Railway Statistics of the United States* (1925), 918,491. This wide span of totals may be due more to what was counted than to poor arithmetic. Unless this was qualified, the unsuspecting reader might assume that each compiler was counting only U.S. revenue freight cars, when in fact some might have included Canadian and/or Mexican cars as well. Others might count or exclude fast freight cars, private cars, work cars, and leased cars.

Take the example of the Northern Pacific. This railroad had 15,434 revenue freight cars in 1890, according to *Poor's*, but also 3,730 leased freight cars and 1,474 work or service cars. This last group included ballast cars, snowplows, track scale, and 1,260 push- and handcars. Some railroads such as the Pennsylvania had yet another separate category for trust cars. These were revenue cars to be sure, but they were counted separately, and they were numerous—in 1880 the Pennsylvania had 21,473 of them. Ownership of these cars was held by a bank or trust company until the railroad paid them off over a five-year period. Because they were, technically at least, not owned by the railroad, they were counted separately and thus may not have been included in any one of the various "totals" made for that year, depending on the judgment of the compiler.

The usual totals for freight cars as reported by the trade press and government agencies tended to exclude sizable numbers of freight cars because they were not owned by the industry itself. Technically, leased or car trust equipment was owned by a bank, not a railroad, and so might be excluded even though they were in revenue service. In 1893 154,000 leased freight cars were in service in this country. Eliminating them from the count could throw the total off by more than 10 percent. In the same year there were over thirty-four thousand cars in railroad-controlled fast freight lines and around seventy thousand in private or shipper lines. Not counting these brings the total error to about 20 percent. Service or work cars, which were nonrevenue in nature, amounted to 39,700 in 1894. Were they counted in the official totals, or excluded? No one can say with any certainty at this late date. It's not so much what was included or excluded as the uncertainty of what was counted or not counted. As in all things statistical, the reader is well advised to be skeptical, wary, and alert to deception, whether intentional or not.

The problem of assembling overall numbers for the U.S. freight car fleet should prepare the reader for the difficulties of establishing numbers for the various types of cars employed on our early lines (Table 1.14). Realistically, it can't be done for the whole system much before 1880. Some notion of how important flatcars were compared with boxcars can be gained from the study of individual railroads. A sample is given in Table 1.15, which is based on data included in *Poor's* or the railroads' own annual reports. Table 1.16 offers another sample of car preference on a major freight carrier, the Pennsylvania.

About the earliest systemwide breakdown was offered in August 1874 by the *National Car Builder* based on data drawn from *Poor's*. It should be noted that Canadian cars are included, but this swells the total by only about 4 percent, and so it is worth reproducing the listing here as a picture of the national car fleet:

Box, 124,009   Stock, 16,722
Coal, 86,887   Oil, 4,000
Flat and gondola, 76,198   Ore, 2,600
Four-wheel (mostly   Caboose, 2,000
   coal?), 47,284   Lumber, 200

Table 1.15 offers comparative figures from 1880 to the end of the wood-car era.

Few readers can fail to see the railroad's undying love of the boxcar. It remained the overwhelming favorite of railroad men until the motor truck and container forced its late retreat. The boxcar was always number one. Why, you could use it to ship anything, from baby bottles to bathtubs; a sturdy foursquare boxcar could take it from here to hell and back. It was the all-purpose, one-size-fits-all

TABLE 1.15   Freight Cars of Representative Railroads

Type	1865	1870	1875	1880	1885	1890	1895	1900
Baltimore & Ohio								
Box	1,151	NA	4,476	6,579	9,632	NA	NA	29,909
Flat	88	NA	NA	NA	NA	NA	NA	1,315
Cattle	250	NA	NA	NA	1,223	NA	NA	346
Coal	1,206	NA	2,868	2,855	4,185	NA	NA	15,575
Gondola	558	NA	2,694	2,879	5,128	NA	NA	12,618
Caboose	7[a]	NA	207	240	443	NA	NA	621
Misc.[b]	138	NA	1,262	1,750	485	NA	NA	1,845
Total	3,398	5,438	11,507	14,303	21,096	25,985	26,635	62,229
Philadelphia & Reading								
Box	601	1,153	1,791	4,236	NA	NA	NA	3,526
Flat	2,140	826	NA	NA	NA	NA	NA	100
Cattle		93	114	149	329	NA	NA	205
Coal	11,499	13,494	14,975	14,976	15,000?[c]	18,559	18,111	21,867
Gondola	NA	NA	NA	2,563	5,873	NA	NA	5,772
Caboose	NA	NA	NA	NA	NA	NA	NA	NA
Misc.	578	155	212	741	2,033	7,101	9,058	354
Total	14,217	15,169	18,515	20,220	27,471?[c]	25,660	27,169	31,824
Chicago & North Western								
Box	2,000	3,674	4,273	7,090	11,668	14,949	19,221	22,573
Flat	611	908	1,025	1,690	2,203	2,197	3,645	4,145
Cattle	109	315	404	942	1,925	1,861	2,881	3,381
Coal	53	84	NA	NA	NA	NA	NA	NA
Gondola	NA	NA	NA	NA	450	1,950	3,250	5,500
Caboose	83	150	140	173	321	451	546	NA
Misc.[d]	214	851	1,957	2,960	3,867	4,820	5,651	5,247
Total	3,070	5,982	7,799	12,855	20,434	26,228	35,194	40,846
Northern Pacific								
Box	0	0	NA	631	4,808	3,180	9,504	13,923
Flat	0	0	NA	660	2,631	3,332	3,694	5,382
Cattle	0	0	NA	NA	789		747	847
Coal	0	0	NA	135	693	1,790	2,468	2,381
Gondola	0	0	NA	NA	NA	NA	NA	NA
Caboose	0	0	NA	37	171	300	313	311
Misc.	0	0	NA	341	75	6,832		1,442
Total	0	0	1,635	1,804	9,167	15,434	16,726	24,286
Central Pacific[e]								
Box	NA	NA	NA	2,551	4,714	2,355	3,408	NA
Flat	NA	NA	NA	2,016	3,191	1,667	1,579	NA
Coal	NA	NA	NA	NA	NA	NA	NA	NA
Caboose	NA	NA	NA	79	138	104	116	NA
Misc.	NA	NA	NA	317	0	194	747	NA
Total	NA	3,200	3,560	4,963	8,043	4,320	5,850	NA

Source: Poor's Manual of Railroads and individual railroads' annual reports.

Note: NA = not available.

a. Listed as drovers' cars.

b. Unspecified freight cars or type unknown. Does not include nonrevenue cars.

c. In an apparent error, the 1885 volume of *Poor's Manual of Railroads* lists 30,679 wooden four-wheel coal cars and gives a total of 42,381 coal cars of all types on the Reading.

d. Primarily ore cars. No "Misc." listing given in *Poor's*.

e. The Central Pacific was leased to the Southern Pacific in 1885 and lost its operating identity around 1899.

THE AMERICAN RAILROAD FREIGHT CAR

TABLE 1.16 Freight Cars of Pennsylvania Railroad

Year	Box	Cattle	Coal	Gondola	Caboose	Car Trust
1853	221	143	28	NA	NA	NA
1860	1,442	300	133	NA	NA	NA
1871	2,713	1,400	1,355	2,762	2	6,211
1880	3,060	1,827	886	5,536	255	21,473

Source: *Poor's Manual of Railroads* and Pennsylvania Railroad's annual reports.
Note: NA = not available.

car. Shippers felt just the opposite, of course. They always wanted some special style of car better suited to their individual shipping needs. They would have a different style of car for every load. And so while the railroads remained fiercely loyal to boxcars, shippers schemed and screamed for exotic rail freight equipment. The railroads would give in on small things like grain doors and end lumber doors. They would even agree to a few high-cube boxcars for barrels or furniture or buggies. Some boxcars were insulated to carry beer, and some were outfitted with big doors to facilitate loading. In general, however, shippers faced a stone wall of resistance when it came to the supply of customized freight equipment, because railroads saw them as single-purpose vehicles able to serve only a limited market. The boxcar, conversely, was a universal car that could service the broadest possible market. And so when it came time to purchase more rolling stock, railroad managers naturally opted for the standard, garden-variety boxcar.

Costs

Freight cars were individually cheap. They were made from common materials: wood, cast iron, and steel. Many parts cost just a few cents each. Collectively, however, they represented a huge investment and were a major consumer of capital in an industry that was forever trying to hold down its capital needs and bonded indebtedness. A good wooden boxcar rarely cost more than $500, a reasonable enough figure for such a useful vehicle. However, when you need several hundred thousand of them to move the nation's freight, assembling the fleet did not come cheap. Locomotives and passenger cars were very expensive compared with freight stock, yet relatively small numbers would meet the national needs. In 1906, for example, $101 million was invested in locomotives, $25 million in passenger cars, but $252 million in freight cars.[447]

The unit price for pioneer-period cars was low because the cars were small, elementary in design, and made from cheap native materials. Four-wheel gondolas and boxcars of the 1830–1840 period ranged in price from $125 to $280. Table 1.17 offers more examples of prices throughout the wood-car era. The introduction of eight-wheel cars

inflated prices markedly, so that around 1840 a boxcar cost $700 to $875, depending on its quality. By the 1850s boxcar prices had fallen to the $500 to $600 level, here again depending on the quality of the car. Refinements like a tin rather than a canvas roof or eight-wheel rather than four-wheel brakes would affect the cost. The method of payment also affected the price. A car builder in Cambridge, Massachusetts, offered the same car at $562 for cash and $625 on credit. The credit price would be adjusted to reflect just how much cash was put down, or how solid the purchaser's credit history might be.

Five hundred dollars is a good general number to remember when discussing cars of the wooden era, for despite an increase in car size, it held true, by and large, for nearly half a century starting around 1850. This was true because of the stability of labor and materials costs. The money was sound, backed by gold, and so was not subject to the inflationary spiral characteristic of modern times. Because of very conservative fiscal policies, the American economy was deflationary much of the time. Hence the price of most capital goods was largely stable. There were, of course, occasional jumps, especially in times of economic upheaval. The largest of these was during the last years of the Civil War, when short-lived but uncontrollable inflation ran wild. The Erie reported paying $1,000 each for freight cars, while Uri Gilbert, a car builder in Troy, New York, recalled selling boxcars for $1,200, only to see the price tumble back to between $400 and $450 at the war's end.[448] Less radical price jumps happened at other times. Between 1898 and 1900 an upswing created demand for more cars to handle a spiral in traffic, which in turn caused car prices to jump 38 percent.[449] In less prosperous times the demand for new cars fell off, and some builders would accept orders at cost just to keep some of their men employed.

If shippers could be satisfied with a diet of boxcars alone, the railroads could avoid the need to buy other more expensive cars. The demand for more specialized—and so in almost all cases more expensive—cars threatened to drive up the industry's rolling stock investment. For the most part, however, the industry neatly sidestepped this issue by passing it on to the private car lines. If the shippers wanted exotic cars, let them assume the financial burden as well. Refrigerator cars were

PARADE OF THE CARS

TABLE 1.17   Cost of U.S. Railroad Freight Cars

Year	Railroad	Type	Source	Cost
1831	Baltimore & Ohio	Gondola, 4-wheel	Annual Report	$126
1831	Baltimore & Ohio	Box, 4-wheel	"	143
1835	Boston & Worcester	Box, 4-wheel	Baker Library Ms. Collection	269
1837	All roads	Box, 8-wheel	Baldwin Letters, Richard Imlay	875
1838	Western Railroad	Box, 4-wheel	a	280
1840	Erie (6-ft. gauge)	Misc., 4-wheel	b	900
1852	Baltimore & Susquehanna	Box	See text, Chap. 2	570
1853	Amherst & Belchertown	Box	Railway & Loco. Hist. Soc. *Bulletin* 47	558
1853	All roads	Flat	*American Railroad Journal*, June 25, 1853	520
1853	All roads	Box	"	500–600
1855	Baltimore & Susquehanna	Coal, 4-wheel	Original specifications	160
1861	Baltimore & Ohio	Box	Annual report	500
1861	Baltimore & Ohio	Gondola	"	400
1871	All roads	Box	*Engineering*, Dec. 11, 1871	735
1871	All roads	Flat	"	575
1871	All roads	Gondola	"	625
1873	Virginia & Truckee	Caboose	c	750
1876	Pennsylvania	Box	d	556
1876	All roads	Milk	*National Car Builder*, Oct. 1879	650
1879	Chicago, Burlington & Quincy	Box	*National Car Builder*, April 1880	456
1879	Chicago, Burlington & Quincy	Stock	"	431
1879	Cincinnati Southern	Box	*National Car Builder*, Sept. 1879	438
1879	Cincinnati Southern	Caboose, 4-wheel	"	500
1879	Cincinnati Southern	Caboose, 8-wheel	"	825
1879	Union Tank Car Line	Tank	e	750–800
1883	Western & Atlantic	Box	f	506
1884	New York, West Shore & Buffalo	Box	*Engineering News*, Jan. 5, 1884	494
1885 (c.)	All roads	Coal, 4-wheel	g	205–250
1888	All roads	Box	*Scribner's*, Aug. 1888	550
1888	All roads	Flat	"	380
1888	All roads	Refrigerator	"	800–1,100
1889	Montgomery, Tuskaloosa & Memphis	Box	*Railroad Gazette*, Nov. 1, 1889	440
1889	Montgomery, Tuskaloosa & Memphis	Flat	"	325
1889	Montgomery, Tuskaloosa & Memphis	Gondola	"	500
1894	All roads	Flat	*National Car Builder*, Jan. 1894	380
1894	All roads	Coke, double-hopper	"	540
1894	All roads	Ore, hopper	"	450
1894	All roads	Fruit, ventilated	"	700
1897	Pittsburgh, Bessemer & Lake Erie	Hopper, steel	*Railroad Car Journal*, April 1897	1,000
1898	All roads	Box	*Railroad Car Journal*, April 1900	483
1900	All roads	Box	"	671
1902	All roads	Refrigerator	*Railway Review*, Oct. 11, 1902	900–1,000
1902	All roads	Palace stock	"	650
1902	All roads	Tank	"	610
1906	All roads	Box	*Railroad Gazette*, Dec. 28, 1906	1,050
1910 (c.)	All roads	Box, steel-frame	h	1,250
1910 (c.)	All roads	Box, wood-frame	h	655
1910 (c.)	All roads	Hopper, wood-frame	h	618
1910 (c.)	All roads	Hopper, steel-frame	h	1,140

a. F. A. Ritter Von Gerstner, *Die innern Communication der Vereinigten Staaten von Nord-America* (Vienna, 1842–1843).

b. E. H. Mott, *Between Ocean and Lake* (New York, 1901).

c. Lucius Beebe, *Steamcars Comstock* (Berkeley, 1957).

d. James Dredge, *The Pennsylvania Railroad* (London, 1879).

e. A. H. Z. Carr, *John D. Rockefeller's Secret Weapon* (New York, 1962).

f. Arthur M. Wellington, *The Economic Theory of the Location of Railways* (New York, 1893).

g. J. Luther Ringwalt, *Development of Transportation Systems in the United States* (Philadelphia, 1888).

h. F. J. Krueger, *Freight Car Equipment* (Detroit, 1910).

TABLE 1.18   Cost of Pennsylvania Railroad Boxcar (1876)

Item	Quantity	Cost
Labor	—	$47.06
Castings at 2 cents	1,733 lbs.	34.66
Malleable castings at 8 cents	122 lbs.	9.76
Wrought iron at 4 cents	2,569 lbs.	102.76
Oak at $21	1,225 ft.	25.73
Carolina pine at $30	1,065 ft.	31.95
Pine at $20	1,253 ft.	25.06
Hickory at $30	53 ft.	1.59
Ash at $40	70 ft.	2.80
Cast-iron chilled wheels, 33-in. diam., at $12.50	8	100.00
Box lids at 26.5 cents	8	2.12
Iron axles at $10.50	4	42.00
Locks at $1.40	2	2.80
Union springs at $8.25	4	33.00
Phosphor-bronze at 32 cents	82 lbs.	26.24
Sole leather at 32 cents	4 lbs.	1.28
Volute springs at $2.14	2	4.28
Follower plates at 80 cents	4	3.20
Square nuts at 7 cents	89 lbs.	6.23
Nails at 3.5 cents	75 lbs.	2.63
Split keys at 11 cents	1 lb.	0.11
Washers at 7 cents	5½ lbs.	0.38
1½-inch screws	7½ gross	3.75
3-inch screws at $1.54	⅓ gross	0.52
1¼-inch screws at 44 cents	½ gross	0.22
2½-inch screws at 88 cents	¼ gross	0.22
Putty at 5 cents	2½ lbs.	0.13
2¼- by ⅝-inch wood screws	2	0.05
5- by ⅜-inch wood screws	5⅔ lbs.	0.81
3½- by ⅝-inch wood screws	2⅓ lbs.	0.16
3- by ½-inch wood screws	1⅔ lbs.	0.14
⁵⁄₁₆-inch brake chain at 10 cents	30½ lbs.	3.05
Paint at 12 cents	34 lbs.	4.08
Black varnish	1 qt.	0.33
Shellac	—	0.28
Paint at $1.58	2 gals.	3.16
Roof	1	32.22
Proportion of fuel and stores	—	1.37
Proportion of superintendent of motive power & chief clerk's time	—	0.69
Total		556.82

Source: Dredge, Pennsylvania Railroad.

about the most expensive style of freight car built in any numbers. The very special sixteen-wheel flats and a few other highly specialized cars might cost more, but their numbers were so small that they did not really affect the overall car investment picture. But reefers were counted in the thousands, and because they cost about twice as much as a standard box, the railroads were pleased to have someone else pick up the bill on this item. This is why in 1890 the railroads owned only eighty-five hundred refrigerator cars, while the private lines held title to around fifteen thousand. In a similar fashion, railroads were content to allow shippers to stock their own tank cars for the transport of petroleum. About the only expensive freight car the railroads could not fob off on the shipper lines was the caboose. Cabin cars, as they were also called, carried the crew but no freight, and so they were nonrevenue vehicles. Costly because of the interior fitting and special trucks, they were equal in price ($800) to a tank car. But they were only a small segment of the overall car fleet and so their cost was seen as necessary and affordable.

Toward the end of the nineteenth century, other factors combined to increase car prices. The push for automatic couplers and air brakes boosted costs by almost $100 per car. Railroad managers shared the general enthusiasm for improved safety so long as it did not cost anything. But when they began to add up the costs for applying safety appliances to nearly a million cars, their resolve began to flag. Cost was in fact the strongest single reason that automatic couplers and air brakes were so slow in coming. Both had been applied to passenger equipment a decade earlier, because the number of locomotives and cars involved was comparatively small. But the refitting of the freight car fleet came about only after federal law required it, and even then it was done in a tardy and reluctant fashion. The industry showed more enthusiasm for steel underframes, starting in the late 1890s. They too boosted costs, in this case by around $200 or $300 per car, but they also increased capacity. And so unlike brakes and couplers, this added expense helped boost a vehicle's earning capacity. There was, then, an income incentive to buy steel underframes rather than brakes and couplers.

Like all things mundane, cars suffer the ravages of time. Hard use and battering of the elements reduce them to a state of dilapidation, and so each year a portion of their original cost is written down. Six percent per annum or sixteen and two-thirds years was the nominal life expectancy of a car, and so that amount was deducted on a yearly basis according to a formula developed by the Master Car Builders Association.[450] The running parts wore out first, as might be expected. Wheels would last around five years; axles were good for ten. The wooden parts of the car were written off in the fifteenth year. In the end only the iron parts were salvageable, with most of the value in the wheels; the scrap value was figured at $128.[451] The trucks might be reused, with appropriate repairs to bolsters and oil boxes, under a new car. The same held true for the air-brake apparatus.

The sixteen-year rule was more of an accounting convention than a doctrine honored by all car builders. One western road retired cars when only eleven years of age.[452] Perhaps only a new roof was needed, but the policy on this line was to push them into the scrap line even so. The Chicago, Burlington & Quincy would scrap rather than repair a car no matter what its age if the repair cost came to two-thirds of the value of the car when new. Attitudes were remarkably different in New England. Railroads there would patch and repair cars thirty, thirty-five, and even forty-five years of age. Some were so old that one wag insisted they had B.C. inscribed after their road numbers. The idea that a car, especially a wooden one, should run for all time was clearly wrong and represented false economy. In 1887 the Fitchburg began to break with the ultra-conservative and started to scrap old, undersize cars when they came due for repairs.

To conclude this discussion of costs, Tables 1.18–1.21 reproduce listings dating from the 1870s through the first decade of the twentieth century.

Private Freight Cars

Privately owned railroad cars operating over American railroads appear to contradict the whole concept of the common carrier system.[453] The phenomenon of private cars was nonetheless a robust one, and during the wood-car era approximately 10 percent of the revenue-producing freight car fleet was made up of cars of this type. Private cars were owned by hundreds of small operators, from the Globe Soap Company with only two cars to giants such as the Armour Packing Company, which controlled several thousand. The tenure of private cars might seem rather precarious because of their tenantlike status while running over the

TABLE 1.19   Cost of Chicago, Burlington & Quincy Boxcar (1879)

Item	Material	Labor	Total
Frame	$52.85	$6.79	$59.64
Floor	10.76	1.12	11.88
Roof	25.49	3.34	28.83
Siding (outside)	22.70	4.04	26.74
Siding (inside)	5.08	0.83	5.91
Trusses	5.89	1.13	7.02
Drawbars	26.08	2.95	29.03
Brakes	7.33	2.16	9.49
Doors	9.00	2.71	11.71
Trimmings	13.29	1.38	14.67
Outside fittings	—	4.85	4.85
Trucks	227.82	12.12	239.94
Painting	5.25	2.16	7.41
Total	410.54	45.58	456.12

Source: National Car Builder, April 1880.

TABLE 1.20   Cost of Typical Boxcar (1898)

Body			Trucks		
Item	Quantity	Cost	Item	Quantity	Cost
Lumber, sills, etc.	1,073 ft.	$19.31	Lumber, white oak	242 ft.	4.60
Lumber, flooring	684 ft.	7.87	Lumber, basswood (or other)	5 ft.	0.08
Lumber, misc. short	63 ft.	0.90	Castings, gray iron	1,392 lbs.	20.18
Lumber, siding	820 ft.	21.32	Castings, malleable	42 lbs.	1.11
Lumber, roofing	760 ft.	13.37	Wrought iron	1,355 lbs.	24.89
Lumber, fascia	60 ft.	0.92	Square nuts	65 lbs.	1.30
Lumber, lining	259 ft.	8.71	Wrought washers	2 lbs.	0.04
Lumber, running boards	72 ft.	1.01	Wheels, 33-in. diam., 600 lbs.	8	48.00
Lumber, door rails	42 ft.	0.63	Axles, MCB, 60,000 lbs.	1,968 lbs.	29.52
Lumber, oak	1,316 ft.	25.00	Journal bearings	100 lbs.	13.00
Castings, gray iron	1,345 lbs.	19.50	Bolster springs	4 nests	8.06
Wrought iron	3,184 lbs.	57.31	Spring cotters	8	0.03
Square nuts	115 lbs.	2.30	Waste	14 lbs.	0.70
Wrought washers	19½	0.39	Car oil	8 gals.	0.60
Nails	108¼	1.84	Journal-box lids	8	0.64
Screws	72	0.73	Paint materials	—	0.25
Draw springs	2	1.68	Labor	—	6.00
Cotters	19	0.07	General expense	—	12.60
Brake chain	3 lbs.	0.08	Total, trucks		171.10
Couplers	2	15.00			
Turnbuckles	4	2.16			
Paint materials	—	1.75			
Labor	—	68.63			
General expense	—	21.19			
Total, body		286.07			
			Total, body and trucks	457.17	

Source: Railroad Car Journal, April 1900.

TABLE 1.21   MCB Price List for 40-Ton 36-Foot-Long Boxcar (c. 1910)

Item	Quantity	Hours	Net Cost
Body bolster (wood)	58 ft.	10	$ 4.43
Body bolster truss block	3 ft.	0.5	0.23
Brake ratchet wheel	5 lbs.	0.5	0.27
Brake shaft	68 lbs.	1	1.77
Brake wheel	14 lbs.	0.5	0.54
Carline	15 ft.	3	1.25
Center plate	66 lbs.	3	1.64
Center sill	140 ft.	35	13.30
Corner post	16 ft.	3	1.28
Dead block	14 ft.	3	1.21
Door handle	2 lbs.	0.5	0.18
Draft gear	—	4	8.96
Draft timber	38 ft.	7	3.01
End door	—	—	1.95
End post	11 ft.	3	1.11
End sill	56 ft.	7	3.64
Flooring	780 ft.	12	30.18
Grab irons	—	—	0.25
Intermediate sill	128 ft.	32	12.16
Journal box	98 lbs.	2	1.85
Journal-box lid	—	0.5	0.32
Queen post	14 lbs.	1	0.66
Roof running board	8 ft.	0.5	0.40
Sheathing per board	4 ft.	—	0.22
Side bearing (body)	16 lbs.	2	0.98
Side sill	145 ft.	25	11.08
Side door	—	—	3.65
Arch bar—top	102 lbs.	3	3.01
Arch bar—bottom	112 lbs.	3	3.24
Arch bar—tie bar	48 lbs.	1	1.32
Journal bearing (bronze)	—	—	2.04
Spring plank	18 ft.	10	3.03
Spring plank hanger	40 lbs.	3	1.62
Truck bolster	88 ft.	10	5.48
Truck bolster truss rod	34 lbs.	2	1.25
Truck center plate	66 lbs.	3	1.64
Truck column	70 lbs.	2	1.46
Truck spring	23 lbs.	2	1.06

Source: Krueger, *Freight Car Equipment.*

Note: Feet = board-feet (1 by 12 by 12 inches).

tracks owned by the operating railroads. They were in fact entirely dependent on the good will and cooperation of the railroads. They were moved on regular trains, powered and manned by the railroads' own locomotives and crews. Yet because some of the private car owners were also major shippers, the railroad industry tended to treat them in a deferential manner. Added to this was the fact that a few private owners were men like Philip D. Armour and J. D. Rockefeller, who were not only business leaders but board members on certain railroads, and it is easy to understand why railroad managers treated the private car lines so kindly.

The origins of the private car movement go back to the beginnings of public railways in this country. The logic that all rolling stock should be owned by the operating carriers did not seem altogether evident in the 1830s. This was particularly true when public money was involved in the construction of a particular line. Why should a taxpayer be deprived of his right to operate a car over a rail line built in part or in whole with tax money? For this reason the Pennsylvania State Works was looked upon as a public highway, and for a time individuals might offer their own cars for carriage. This was done at first by horses, but after a few years the state furnished locomotives and crews. Yet even when state funding was not evident, the privilege of a state charter suggests the idea that railroads were simply another form of public highway, and so the Boston & Providence Railroad was saddled, at least for a time, with the obligation of handling private cars on a toll system. This archaic phase of private car operation died out—happily, at least from the railroads' viewpoint—around 1845. And it remained at rest for almost another twenty years.

Some care must be taken when discussing private cars, because the term is not as simple as it might appear. Not all private cars were in fact privately owned. A large number were actually railroad-owned, under the somewhat misleading banner of the fast freight lines. Major operations like the Red, Blue, and White lines boldly painted up their cars in the best private car tradition. They maintained their own offices and solicited freight as independent agencies, when in fact all three of these lines were wholly owned subsidiaries of the New York Central Railroad and its affiliates. Some of the early fast freight lines, notably Merchants Despatch and the Star Union, had started out as private and independent operators but were taken over by the host railroad. Even so, their identities as private car lines were maintained.

The railroad-owned private cars, which might best be termed the fast freight lines or pseudo-private lines, were operated on two systems. The most favored was the cooperative plan, wherein each railroad contributed a certain number of cars. Earnings were divided proportionally. Cooperatives were normally created only between a parent railroad and its subsidiary lines. The second plan, called the commission plan, involved a mileage fee and a commission of between 2 and 15 percent on the freight charges.[454]

In addition to true private cars and railroad-owned fast freight lines, there was a more specialized category, the shipper lines. These were true private car owners, but rather than owning boxcars in the hope of seizing a piece of the general goods trade, they almost always owned specialized cars meant to service a specific industry. Meat packers, for example, would purchase and operate

PARADE OF THE CARS

their own fleet of refrigerator cars; an oil refiner would acquire tank cars; a pickle maker might see the advantage of owning a few vinegar cars; and so on. A shipper would, then, buy cars to service his own transportation needs. Simple? Yes, too simple, because some shipper lines were not involved in the manufacturing or processing of the products they carried. Some shipper lines were truly middlemen who owned fleets of specialized cars that were made available to any customer needing such equipment. This could be done on a one-time or a long-term basis, depending on customer needs. The renting and leasing of cars only confuses the exact meaning of the term *private freight cars*.

FAST FREIGHT LINES

The useful service rendered by express companies in moving small shipments by rail suggested the possibility of doing so with larger consignments. The express companies, at least in their earliest years, operated out of railroad-owned baggage cars that were part of the train's normal consist. A shipper wanting to expedite carload-scale shipments needed more than a corner of a baggage car. In about 1855 William Kasson of Buffalo began to offer through service from New York to points as far west as Chicago.[455] He had experience in the long-distance moving of locomotives over the various gauges existing between those points and so was already organized to handle less bulky cargoes needing to go a long way. Kasson had no cars of his own for his dispatch freight service, and so he simply used regular railroad-owned cars for the customary charges; but he did manage to effect a quicker exchange of cargoes en route and so hasten deliveries. His scheme worked so well that his business was bought by a syndicate that rechristened it the Merchants Despatch Transportation Company.

At about the same time (1857 or 1858), J. B. Chichester started a fast freight line on the Lehigh Valley with seventeen secondhand cars.[456] Through various connections, service was offered from New York and Philadelphia to the western states. By 1860 Chichester had forty-five cars, some of which were fitted up to carry produce. The line reportedly expired because of the Civil War traffic emergency. However, the Merchants Despatch continued to flourish. In fact, it was expanding its operations under the patronage of the American Express Company, which began what was called a first-class freight service on the Erie.

Undoubtedly there were others flirting with the idea of expediting freight movements by rail during these pre–Civil War years, but no one appears to have seized hold of the notion with a very firm or sure hand until 1863, when William Thaw (1818–1889) of Pittsburgh decided to enter the fast freight business.[457] Thaw was a rich boy who made good; his father had founded the first bank in Pittsburgh. Thaw left the family business to make his own fortune in the canal and steamboat trades. He became an associate of the Pennsylvania Railroad officials through the shipping business. Because of these friendships, he eventually became a director of the railroad. When Thaw approached the Pennsylvania with his freight lines scheme, he came as an insider. The railroad was happy to cooperate with anyone who was on such good terms with J. E. Thomson and Tom Scott. Nor did Thaw encounter many difficulties in attracting partners, or the $1.4 million in capital needed to start up the Western Insurance & Transportation Company. The name was soon changed to the Union Transportation & Insurance Company, but it was better known as the Star Union Line, or more simply the Union Line.

Care must be taken not to belittle Thaw's accomplishments, for he seems to have possessed a clearer understanding than most transportation men about what was needed to reform the railroad freight business. In 1863 that need was simply to provide through service and so end the repetitive loading and unloading of cars at each terminal. This scheme, realized in part by Kasson and perhaps others, would involve the running of loaded cars all the way to their final terminal. What seems obvious and elementary now was not so readily understood at a time when the breaking of bulk was the norm. Thaw and his associates planned to overcome the minor gauge differences between Philadelphia and Chicago by the use of broad-tread wheels. Technically, this scheme had some drawbacks, but it worked well enough that through service would be possible on all but truly broad-gauge lines.

Officials of the Pennsylvania were happy to assist not only their friend Thaw but themselves in the process, because the Erie and its broad-gauge western connections, the Atlantic & Great Western and the Ohio & Mississippi, were ready to offer through service all the way to St. Louis. In addition, freight was piling up in Pittsburgh in such volume that all the freight houses were filled to overflowing, as were the warehouses rented to handle the growing mountain of boxes and bags. This flood of goods was due in part to the wartime traffic, but also to the temporary suspension of navigation on the Ohio River due to low water.

The Union Line was not ready to help untangle the snarl until early 1864, but the one thousand broad-tread-wheel cars did much to relieve the situation. The new line guaranteed delivery from New York to Chicago in just six days. Shippers willingly paid 25 cents per hundredweight over normal freight rates for such astonishing speed. The Pennsylvania was ready to pay a most generous mileage rate as well, which came to 2 cents per mile. Union set up sales offices, freight terminals, and even a car float service. Thaw's enterprise seemed to gain strength each year until the early 1870s, when a group of outspoken shareholders attacked the cozy arrangement between Union and officials of the railroad.[458] The charges of collusion and diverted profits were steadfastly denied by the officials accused, who at the same time told the Union Line that its contract would not be renewed. A plan was then quickly devised for the sale of Union to the Pennsylvania, which occurred on July 1, 1873. Nothing, however, was done to downgrade or curtail Union's operations. If anything, it was expanded as a wholly owned subsidiary of the Pennsylvania system. By 1888 the Union Line boasted a fleet beyond ten thousand cars. This bureau of the Pennsylvania was a valuable solicitor of freight, and the railroad wanted to encourage shippers to believe that they got faster service when they consigned goods to Union Line cars. Savvy freight agents came to understand the promotional value of pseudo-private car lines like Union when attempting to sell service.

The Empire Line, established in 1865, escaped the ire of Pennsylvania Railroad stockholders apparently because it operated primarily on a subsidiary line called the Philadelphia & Erie.[459] Joseph D. Potts, the head of the Empire Line, expanded the service from the typical fast-merchandise style of operations into lake boats, grain elevators, oil pipelines, and railroad tank cars. The tank car operation was called the Green Line, not to be confused with the southern fast freight line of the same name. Potts was an able and energetic manager, so much so that by 1873 the Empire and the Green Line were carrying over 760,000 tons. Traffic had almost doubled by 1876, but Potts's ambition to enter the oil-refining business set off a mild panic at the head offices of Standard Oil. It was made clear that Potts and his expansive scheming must go, if the Pennsylvania Railroad wished to keep the continued patronage of the growing Rockefeller empire. The solution was to terminate Empire as a private operation and to enfold it into the Pennsylvania's corporate structure, where its policies might be more carefully controlled—especially those policies found repugnant to a major shipper. In October 1877 Empire was purchased by the Pennsylvania for $4.6 million, but the pipelines were spun off to Standard Oil.

Mention has already been made of the Merchants Despatch Transportation Company and its pioneering operations over the New York Central and its connections to the West. It began much like the Union Line, as a child nurtured by inside connections with trunk line officials who were ready to pass on large and small favors to friends. This warm relationship lasted through the Corning and the short-lived Fargo administrations on the New York Central.[460] William G. Fargo and his brother James were prominent in the express business, including the American Express Company,

which controlled MDT. All of this was to change just a few years after the Vanderbilt takeover of the Central. If favors were to be passed out, they should by right go to the cronies of Vanderbilt. Accordingly, a joint stock company was formed in June 1871, with capital set at $3 million and the shares divided among the New York Central and its subsidiary roads, including the Lake Shore and the Big Four. Loyal Vanderbilt lieutenants, such as Augustus Schell and J. M. Toucey, were among the incorporators. In short, MDT was no longer a true private car line, but merely a straw fronting for its real owner, the New York Central. As such, it prospered and grew. By 1880 it had shifted its traffic sights away from dry goods and fast freight to refrigerated transit and so evolved into the Central's own refrigerator car line. Its fleet grew steadily from thirty-four hundred cars in 1888 to sixty-six hundred in 1900.

SHIPPER LINES

The private car scheme was for all practical purposes dead by 1875. Most or all of the lines had been taken over by their host railroads. The advantage of fast freight and direct shipment without breaking bulk had been nullified by the universal acceptance of freight car interchange. The private car lines had really nothing to offer shippers or the railroad industry. Or so it would seem, for no sooner had the private lines died out than they were reborn. In their new guise as shipper lines, they offered specialized cars to carry commodities not suitable for normal boxcar transit. Shipper lines were thus especially strong in petroleum and refrigerator cars. The railroads' reluctance to invest in specialized or single-use rolling stock encouraged the creation of this second-generation private car fleet. Initially, meat packers and oil refineries were prompted to start these operations simply to distribute their products, but it soon became apparent that the operation of private cars was profitable in itself, and so a most attractive second reason was found for buying more private cars.

Shipper lines appear to have originated with the western meat packers, especially in the Chicago area. George Hammond was among the first to send dressed beef east by refrigerator car in 1868. Initially he appears to have worked with the Blue Line but was to operate his own car line sometime later. By 1885 Hammond had six hundred refrigerator cars in service; by 1900 this number had doubled. Other western packers soon understood the profits possible by shipping dressed beef to the populous East Coast, and very soon Swift, Dold, Cudahy, and others were active in the shipper line trade.

Philip D. Armour was rather late in sensing the value of private refrigerator lines when he bought his first cars in 1883, but no one proved more enthusiastic once introduced to the potential of the simple yet effective technology of an insulated boxcar with an icebox mounted in each end. Armour not only added more cars; he acquired or created more lines. Fanciful names such as Barbarosa, Tropical, and Continental Field were invented, perhaps in an effort to disguise the growing power of the Beef Trust's monopoly. At the same time, Armour began to supply cars for the carriage of produce, and so his operation was clearly stepping far away from the limited concept of a shipper line intended to serve the immediate needs of its owner. Armour was in the transportation business, and his operation was functioning as a private car line in its original status as a common carrier. The Armour Lines even provided cars to other smaller meat packers who could not afford or could not be bothered to operate their own cars. In 1900 Armour ran no fewer than seventeen car lines and twelve thousand cars. The meat packers combined to control 80 percent of all the refrigerator cars running in the nation.

Slower progress was made in the growth of shipper lines for cattle. The railroad had traditionally been more responsive in supplying cars for this trade. It was not until the humane cattle car craze hit in the 1880s that private cattle lines showed much vigor. Mather, Hicks, Burton, and the others described in Chapter 4 were created during this period to provide self-watering and self-feeding cars, all of which promised to deliver cattle to market fat and healthy. Several of the smaller firms were consolidated with Street's Western Stables Car Line in 1902 to form a very large fleet of approximately eighty-three hundred cars. Street's expired around 1924 for unknown reasons. The cattle car lines never did as well as the other shipper lines, though some, such as Mather, actually leased cars to the railroad industry.

Petroleum offered a richer profit potential for shipper lines. Tank cars proved a valuable weapon in the hands of the larger refineries to control or if necessary to ruin their competitors. It all started out innocently enough in the 1860s with large wooden tubs fastened to flatcars. Within a few years horizontal iron tanks had replaced the wooden models. The railroads were content to allow the oil companies or private operators like the Green Line to provide the cars. In 1873 Rockefeller acquired a small tank car line with the purchase of the Star Oil Company. He came to realize that whoever controlled the transportation of oil also controlled the oil market, and so he became intent on monopolizing the tank car business. Standard Oil's Union Tank Line did not manage to squeeze out all competitors, but its fleet of ten thousand cars (1910) surely dominated this specialized field of transportation.

In more recent years shipper lines have taken a more prominent share of the overall freight car fleet.

While cattle cars and refrigerator cars have largely disappeared, other types have appeared in large numbers under private car ownership. In times past, hopper cars were almost all railroad-owned. At present, however, large numbers of hopper cars are privately owned by electric utilities, which need steam coal in prodigious quantities.

PSEUDO-PRIVATE LINES

Private cars might be divided into yet another category—the railroad-controlled fast freight lines, or what might be called the pseudo-private car lines. These were outright railroad-sponsored corporations, financed, set up, and controlled by the railroad itself yet presented to the public as independent operations. These puppet operations were created to solicit freight that might go elsewhere, and they seemed to serve little function beyond marketing and traffic generation. Most of the pseudo-private car lines worked on the cooperative plan in which each railroad contributed a fixed number of cars to the line as required for the anticipated traffic. The cars were managed by the private car lines, which also controlled mileage payments. Earnings were divided proportionately among the partners; those with the most cars received the largest payments.

The New York Central found the pseudo-private car scheme very appealing and may indeed have originated this plan of freight service by establishing the Red Line in October 1865.[461] The basic scheme was to offer through freight cars between Boston, New York, Chicago, and all points in between. It was the beginning of this service that led to the creation of the Master Car Builders Association, for it marked the true beginning of long-distance interchange service. The Red Line was a cooperative that included the New York Central and all its associated lines to the north and west. Each railroad was required to furnish three new cars per mile of line. They were painted an English vermilion, coated with varnish, and had broad-tread wheels. In less than a year the Central had set up a second pseudo-line, the Central Transit Company, or the White Line.[462] The New York Central supplied 100 cars and the other cooperating roads put in 30 to 40 each, making a total of 270 cars. Each car represented one vote in matters affecting operations. The Blue Line came into being in 1867 and was unique among the three New York Central pseudo-lines in that it had cooperative partners outside the usual family of associated roads, which included the Burlington and the Illinois Central. The first year of operations was surely successful, with 91,500 tons going east and 55,400 going west.[463] Some six hundred Blue Line cars carried grain, cotton, wool, and dressed hogs. Each car averaged 30,000 miles, a figure to gladden the heart of any traffic manager. A third rail on the Great Western Railway (Canada) and the new iron car ferry between Windsor and Detroit helped facilitate the through car service.

The success of the Red, White, and Blue lines naturally spawned many imitators. Even John W. Garrett, an outspoken foe of private car lines, genuine or pseudo, felt obliged to start up a fast line sponsored by the Baltimore & Ohio in 1866.[464] A few years later, when the B&O had at last spanned the Ohio River and its western connection to St. Louis relaid its track to standard gauge, Garrett organized a new company-sponsored line called the Continental Fast Freight Line. One thousand cars proved inadequate for the popular through line, and another eight hundred were required by the spring of 1873.

New England reacted positively to railroad-sponsored private car lines in 1869 when the National Despatch Line was founded by Grand Trunk, Vermont Central, Rutland, and the Boston & Lowell.[465] Its cars had dual-gauge trucks so that through service was possible between New England and South Central Canada. When the Grand Trunk converted to standard gauge a few years later, the dual-gauge trucks were retired, but the National Despatch rolled on profitably until 1914, when the corporation was dissolved. The Vermont Central perpetuated the name for many years as part of its traffic agency, and one branch of the old National Despatch continued on as the Chicago, New York & Boston Refrigerator Company, later a subsidiary of the Canadian National Railway.[466]

In 1868 several of the major southern railroads responded to the growing fad for private car lines by organizing the Green Line.[467] The Louisville & Nashville was joined by the Western & Atlantic, the Macon & Western, and several other roads to form a railroad-owned cooperative fast freight line. Eventually, over twenty railroads were partners in the Green Line, which operated over twenty-two thousand cars. Yet unlike its northern counterparts, the Green Line did not flourish, because its cars were misused for local rather than through service. And so the Green Line was dissolved in the early 1880s.

The western railroads moved into the private car business somewhat later than the eastern roads, as might be expected, because they were constructed so much later. In 1881 the American Refrigerator Transit was organized to provide service on the Wabash and the Missouri Pacific. Its operations were expanded to cover the railroad properties controlled by Jay Gould. This line was a success and had grown to forty-five hundred cars by 1910. Just a few years after ART came into being, the Santa Fe created its own refrigerator line, named, not surprisingly, the Santa Fe Refrigerator Despatch.[468] This operation continued until almost the end of refrigerator car operations on the Santa Fe. Western private car operations were growing apace in 1906 when E. H. Harriman organized the Pacific Fruit Express to provide refrigerator car service in the Union Pacific and the Southern Pacific, both of which were under Harriman's control at the time. PFE went on to become one of the largest railroad-controlled private car operations in the country.

CAR LEASING

Some private car operators found it profitable to rent some of their cars to small shippers who could not afford, or did not want, to take on the responsibility of operating their own line of cars. In other cases trunk lines would lease or rent private cars for their own use. Armour leased insulated cars to the Moerlein Brewing Company on a per-diem basis for between $16 and $17 a day.[469] In cases of a long-term lease, the car would be lettered for the lessee. The Mather Stock Car Company worked out a more elaborate lease plan with the Baltimore & Ohio in 1895 for five hundred cattle cars.[470] The lease ran for five years and called for a rental of $10 per month plus 6 mills per mile for all mileage in excess of 1,667 miles per month. The railroad-owned fast lines were also active in the car rental trade. Both the Union Line and the Red Line leased a large number of cars to Rock Island during the 1880s.[471] In 1888 the Union Line had 3,150 cars in service on nonsubsidiary lines of the Pennsylvania, which included the New Haven, the Naugatuck, and the Milwaukee.

Car leasing went big-time in 1871 with the creation of the United States Rolling Stock Company, which was established by James McHenry (1817–1891) and his associates to supply locomotives and cars to the Atlantic & Great Western Railway.[472] Firms of this nature had already been organized in England and Germany, and so McHenry was able to raise upwards of $5 million through English investors. Approximately four thousand cars were supplied to the A&GW. Profits of 12 percent and more were optimistically predicted. To add a measure of prestige, George B. McClellan, late a Civil War general and a former railroad executive, was persuaded to accept the presidency of the USRS. Rates ran from 75 cents a day for flats and gondolas to $1.50 a day for boxcars. All went well for the first year or so, until the Atlantic & Great Western slipped into increasing financial difficulties following the Panic of 1873. Receivership followed, and the Rolling Stock Company was paid with receivers' certificates rather than cash or gold. McClellan resigned as the fortunes of the USRS declined alarmingly. Because the A&GW was broad-gauge, there was only a limited market for its leased equipment elsewhere.

Somehow the firm survived the rigors of the long panic. It then decided to expand into the car-building trade to boost the profits of its languishing rental business. During the early 1880s car shops were built or purchased in Alabama, supplementing car works at Urbana (Ohio) and Chicago. In 1884–1885 a large plant was built south of Chicago at a new town named Hegewisch in honor of the current USRS president. In 1882 a car trust scheme was adopted to enable railroads to acquire cars on a lease-purchase plan spread out over five years. None of these plans seemed to solve the lingering problems of the Rolling Stock Company, and in 1887 its capitalization was voluntarily reduced to reflect its losses over the years. In 1890 voluntary receivership held off creditors until the business was reorganized in May 1892. About a year later, the property was sold off, ending this interesting effort to find a new way to supply freight cars to American railways.

Just about the time U.S. Rolling Stock failed, another leasing scheme started up under the name of the Southern Iron Car Line.[473] It worked in association with the Huntingdon (Pa.) Car Works, which had been involved in producing cars with gas-pipe frames. The railroad industry had shown marked reluctance to adopt the gas-pipe or tube frame, but the organizers of the Southern Line apparently reasoned that if you can't sell them, maybe you can rent them. Ten thousand were built starting around 1888, and several large rental contracts were made at the rate of $7.50 per month. The lessee was responsible for repairs, and Huntingdon Car made a fine profit selling parts for this purpose. Other contracts were made on the regular mileage formula, but in this case the lessor was responsible for repairs, and tube cars proved to be such high-maintenance vehicles that the Southern's profit was very meager. Two of the Southern's major customers went into receivership and canceled their contracts. Traffic was down in general, and so the leasing company had a large number of cars that no one wanted. Bankruptcy followed, and the Southern bondholders realized 7 to 10 cents on the dollar on the day of settlement.

In the long run, car leasing turned out to be a fine investment for those skillful enough to master the boom-and-bust nature of the railroad freight business. One of the most successful of these firms started operations in Chicago in 1898 with a handful of secondhand refrigerator cars. At first it was called the Atlantic Seaboard Dispatch, but it became better known by its later title, the General American Transportation Company.[474] It steadily expanded into tank and stock cars and eventually became a major freight car builder. Equally successful was the North American Car Company of Chicago, incorporated in 1907. It too grew and expanded, proving that good profits were possible in the car rental business. North American was active in all the normal specialties of shipper lines, including poultry cars.

COSTS, CRITICISMS, AND CONTROLS

Because private freight car lines affected transportation charges, they attracted a fair amount of public attention despite the arcane nature of their operations. Nothing seemed to generate more attention than how much it would cost to ship goods. Shippers always seemed to suspect that they were paying too much, and so if the fault wasn't the railroad's, perhaps it was the private car line's. The private operators hastened to point out that they received exactly the same tariffs as those levied by the railroads. Hence the shippers were not being overcharged. True enough, but what they failed to note was the mileage fees paid by the railroads to the private car lines. This extra cost must somehow, if only indirectly, be charged back to shippers as part of the cost base of railroad operations. Even after the introduction of the per-diem system in 1902, private cars were by necessity paid on the old mileage system. Three-quarters of a cent per mile may sound trivial, but a private car could expect to earn 75 cents to $1 a day empty or full, while a regular interchange car was limited to a 20-cent-per-day fee.[475] In other words, a private car could earn five times as much as a regular railroad car. Just switching a private car around a large rail center, such as St. Louis or Chicago, could add 30 or 40 miles to the distance traveled and hence the revenue earned.

These pennies and dollars added up to big money. In 1896 it was estimated that the railroad industry paid out $7.7 million in mileage fees even when the money paid to railroad-controlled pseudo-private lines was deducted. Large shipper lines such as Armour benefited handsomely from the mileage system and collected about $1 million each year this way. Indeed, Armour and the other refrigerator line operators enjoyed very good profits, especially between 1904 and 1905, when 17 to 20 percent was realized on their investments in refrigerator cars.[476] The Beef Trust added to its revenue by demanding icing fees two or even three times what was normal. No wonder they were such enthusiastic supporters of the private car lines. If Armour won an exclusive contract, as happened on the Illinois Central around 1900, icing fees were jumped from $30 to $84 even though none of the costs had changed.[477]

Fruit cars earned only 12 to 15 percent for their private owners. *Only* 12 to 15 percent! Such a rate of return was looked upon with envy by just about every railroad in the nation. And this was revenue just from the mileage fees. The tank car operators claimed a return of just 3 to 4 percent, yet they were more than compensated by the difference in what they would pay the railroads if oil was transported in nonprivate cars. Back in 1865, when the Empire Line began operations as a purely private operation, the difference in costs was obvious. A shipper was charged $152 to send a car of oil to New York City from Corry, Pennsylvania.[478] Of this amount, the railroad received only $67. After the establishment of Interstate Commerce Commission controls, such rate disparities were not permissible, but the shipper lines would still benefit from the difference between the legal tariff and the actual cost of transportation.

The private lines found other ways, legitimate and illegitimate, to boost their incomes. Some earned commissions by acting as freight solicitors. The Erie, for example, paid the Missouri River Despatch Line 12.5 percent for acting as its agent. The railroads' greed for traffic led them to make foolish deals with some shipper lines. The Lackawanna agreed to give a Chicago meat packer free use of its New York stockyard facilities, plus a fee of 3.5 cents per hundredweight on cattle that passed through the yard, plus a mileage fee of $13.71 per car traveling between New York and Chicago.[479] This overly generous package cost the railroad more than $245,000 between 1888 and 1890. Once again, no wonder meat packers were so keen to have a private car line.

Despite a convincing record of excess profits, at least one private car operation had the temerity to complain about inadequate earnings. A. C. Mather claimed that his cattle cars earned only about $9.00 per month, while repairs cost $8.50.[480] Mather actually had a certain point to make, for the stock car lines seemed to have had far less leverage with the railroads than meat packers or petroleum refineries, and the mileage rates granted were far lower. The cattle cars' mileage rates were pushed down to 6 mills per mile in the early 1890s. Such a large reduction was not long tolerated by the Oil Trust, however, because of the large traffic it controlled compared with the cattle lines. In 1894 some ninety-five railroads belonging to the South Western Traffic Association agreed to end all mileage payments on empty private cars and to reduce the compensation for loaded cars to 0.5 cent per mile. Standard Oil was determined to maintain the existing and more profitable status quo. When resistance was met, Standard Oil sought out what was called a weak sister: the company needed just one railroad in the area that could not resist the golden flow of unlimited tank cars. The Chicago Great Western, which had never carried a single car for Union Tank, became the willing traitor. For a month or two the CGW enjoyed a wonderful oil traffic and prosperity, all of which ended when the other lines broke ranks and agreed to the terms imposed by Standard Oil.

The mileage issue raised even larger questions about the need for and the contributions of the private car fleet. Was it really needed, and exactly what service did it perform for the transportation community? With at least some justification, private car proponents spoke grandly of its contribution in establishing the interchange system as part of the start-up history of the early fast freight lines. Even Arthur T. Hadley, who had no vested interest in the matter, felt that the cooperative private lines benefited the trunk roads by handling through freight shipments and thus effecting a

savings of 6 percent.[481] The most common benefit cited, however, was service performed by providing so much specialized and expensive equipment, especially refrigerator and tank cars, that the railroads would not or could not provide. The mobility of the private car fleet formed a national pool for special shipments and so relieved the railroad industry of its obligation. It also relieved the railroads of making the $100 million investment needed to finance such a fleet and so reduced the railroads' capital needs. The public benefited from the fresh meat and vegetables made possible year-round by the largely private refrigerator car fleet. The cost of kerosene, the common lamp fuel, was held down by the efficient transport offered by the tank car flotilla, which was again largely in private ownership.

These arguments were countered by a host of counterarguments, generated by a widely held suspicion that most private car operations were somehow heavily involved in fraud and abuse. The private car operators were seen as having contrived an elaborate and artful swindle to cheat shippers and the public alike. Railroad stockholders wondered if their profits were being skimmed off by the ever-so-clever and secretive private operators, whose work was facilitated by cooperative railroad officials making unethical, if not illegal, deals with fast freight lines. The larger issue of aiding and abetting monopolies gained more focus as the nineteenth century closed.

The notion of insider deals was very bothersome to railroad investors. Railway officials should be representing the interests of the stockholders first and not scheming on ways to enrich themselves through their positions inside the corporation. Ever since the Credit Mobilier blew up in 1872 as a national scandal, railroad stockholders had worried about what sort of sweetheart contracts their managers might be ready to hatch next. Secret construction companies were a fine way for an unscrupulous railway official to increase his net worth. But once the line was finished, a new insider game had to be found. Private car service, even when it was railroad-controlled, was too good an opportunity to be overlooked. Officials of the Vermont Central took stock in the National Despatch Line, of which VC was a cooperative partner, but these same officials did all they could to direct cargoes to NDL cars rather than to those of their own railroad.[482] New cars were built for the NDL in VC shops at just a little over cost because of the intercession of these same officials. The NDL paid a dividend of between 10 and 12 percent, while the Vermont Central languished in receivership and paid its shareholders nothing. Even where conflict of interest could be avoided, why should the railroads pay out such large mileage fees, especially to shippers who should, in fact, be paying the carriers and not the other way around? A study conducted by the Interstate Commerce Commission in 1889 and published two years later showed that three shipper lines received $100,000 in mileage fees alone.[483] Some private lines earned enough from mileage fees to buy new cars every two years, if they desired. Because most private cars were involved in through service and hence long runs, they racked up high mileages, while the railroads were left with low-value cargoes and short hauls. Why all of this favoritism and preference to the private lines?

The abuses, real and alleged, of private lines brought on some highly partisan attacks. One critic, John R. Wheeler, head of the Illinois Central Railroad Commission, spoke bluntly of the private operators as parasites and vampires who were sucking the lifeblood out of railroad revenues.[484] They were a heavy tax upon the unsuspecting American public. Wheeler seemed particularly upset by the practice of granting officials of the private lines passes for free rides on passenger trains. One railroad superintendent spoke of them in 1885 as a fungus growth, while another critic called them a host of pestiferous solicitors! Clearly, the private lines had raised some ardent critics given to overheated language, but there was enough legitimacy in these arguments to spawn several investigations. Perhaps the earliest was that conducted by the Ohio state senate in 1867.[485] This appears to have been more a fact-gathering hearing than anything else, and at this early point in the history of private freight lines, most of the abuses of the system had yet to be invented. The U.S. Senate made its own investigation six years later, but once again it was largely only an airing of issues.

The Interstate Commerce Commission began looking into the matter not long after it began operations, even though it had no authority to deal with private lines. In 1903, however, it issued a sharply critical report on the evils of the system. At the same time, *Railway Age* published a twelve-part series on the problems of private cars. Congress took its first action against the private lines in 1903 by adding a clause to the Elkins Act that prohibited the payment of rebates to such operators by the railroads. Two years later, a bill was introduced to end all private car operations, but it failed to pass. In 1906 private lines were specifically placed under the jurisdiction of the ICC.

Too late, J. O. Armour came out with his apologetic volume, *The Packers, Private Car Lines, and People*. In the end, Armour need not have worried, because the dragon was not slain by either the U.S. Congress or its bureaucratic handmaiden, the ICC. Private car lines not only survived but flourished over future decades. Good times and bad, they seemed to turn a fine profit. When the railroads were in bankruptcy, the private lines were solvent. Today they are a growing part of the freight car fleet.

Car Design

American freight car design was a conservative, craft-dominated trade. It was never at the leading edge of the engineering world. Structural theory and mathematical calculations were hardly considered in a craft that seemed dedicated to rules of thumb and cut-and-try methodology. Car builders were almost to the man practical individuals, steeped in utilitarian values and faithful to the traditional methods. New designs were based on old designs. When a larger car was needed, the floor sills were simply increased proportionately in size. The bolster timbers were duplicated, but on a slightly larger scale. Wheels and axles also tended to be scaled up or down for different load capacities rather than altered in basic shape or material. Radical solutions to design problems were viewed with considerable suspicion. Car designers simply could not be described as adventurous, imaginative, or especially curious. But at the same time, they had the strength of a cautious and conservative approach to mechanical design. Few costly mistakes came of following an established, time-tested design. Care in examining the works of earlier designers, learning from the experiences of others active in the field, and an element much valued by those loyal to the shop culture of the time, common sense, resulted in a sound if not very flashy design.

This is not to say that no one experimented or tried to find new solutions to old problems. Some designers and builders were forever tinkering with the details of car technology, and some were ready to go well beyond the details. F. E. Canda, an executive in the nation's largest car-building firm, was a notable example of a radical thinker inside the industry; his work is examined in more detail in succeeding chapters. Yet Canda's innovations were apparently never seen as much more than a series of interesting novelties worthy of study but not emulation. And even he came out as a conservative in the end, for he wanted to preserve wooden construction when steel fabrication was at last coming into acceptance.

The prevailing conservatism of car builders can be explained by several factors. First among these was the obvious fact that the railroad's main business was transportation and not technical innovation. The dependable and economical movement of goods and people was its fundamental goal. Dependable equipment was essential to fulfill this mission. Railroad managers were therefore reluctant to change any procedure or design for a new one, no matter how promising, because it might disrupt service. Ideas that promised to save time and money were worthless if they failed out on line and tied up the railroad. This pressure to keep the railroad running discouraged experiment, just as it encouraged a loyalty to tried and true methods.

Because the railroad was dominated by the needs of the operating department, all else was secondary. The mechanical department had no alternative but to provide equipment of a known design. Even if the operating people were ready to accept some new ideas, the car designer was hardly free of all constraints because of the absolute requirement that new equipment be compatible with existing rolling stock. The changeover to Janney-style couplers was an example of this problem. Because the conversion required nearly a decade to complete, the new Janney couplers had to be made so that they could hook up to the old standard links and pins then in service. The same makeshift transition was necessary with air- and hand-braked cars and with the mix of wood- and steel-framed cars. Another constraint to innovative design was cost. Car designers were not free to spend as they liked or felt was necessary. Because so many cars were needed, they must by necessity be cheap. Cheap cars tended to be those produced on conventional plans, and so once again innovation was squelched.

Despite these constraints, American car builders managed to produce a wonderfully practical product that performed with great dependability. Low-cost cars were built from simple native materials. Like the western riverboats of the same period, they were cheap rather than durable and had a short service life. In nearly every way they were a credit to the journeymen who created them. They were strong but not overbuilt, simple but not primitive. They were about as unpretentious as any of man's contrivances, solid and true but not extravagant. In terms of their dead-weight ratio—that is, the weight of the car compared with the load it could safely carry—the American boxcar had no rival. It was strong yet light. If a part failed because it was too light, it was made heavier in the next lot of cars.

By the time interchange came into acceptance, the notion of long-life cars had begun to disappear. They were seen as more a short-term vehicle, almost a throwaway item. Why build a heavy, expensive car when it would be in the scrap line after about sixteen years of service? Lightness and short service life seemed to go together. By the 1880s designers began to accept an increasing role of metal in car construction. Tie and truss rods were by then well accepted. Iron-reinforced body and truck bolsters did much to increase the durability of boxcars that had floor sills the same size as those used in an earlier time for cars of half the capacity. The changes came on slowly as in the finest Darwinian docudrama, where all progress evolved in steady and orderly fashion. John Kirby, the veteran car builder for the Michigan Southern, reflected that the improvement in freight cars was "the result of a gradual unfolding of ideas."[486]

In their effort to balance strength against carrying capacity, car builders adopted the doctrine of just good enough. The vehicle should be made no more durable than was necessary for it to do its job in the short run. This philosophy was indigenous to the American railway as a whole. Everything about it was provisional—wooden tracks, single tracks, timber bridges, and an absence of stations except at major towns and cities. The rationale was not to build a capital-intensive railway when one was not needed. As traffic developed and income was received, the line could be rebuilt and upgraded. And so it was with cars as well, but with a variation. So long as cars remained on the home line, sturdy, well-made cars made sense, when a long service life could be expected. This consideration was reversed by the interchange system, which in effect turned freight cars into common property. The indifferent and careless usage inherent to common property mandated an era of cheap construction. Car builders were never asked to design the most perfect or durable car possible, but rather the cheapest and simplest car possible. The emphasis was quick-and-dirty solutions and the most value for the least expenditure.

The basic boxcar design that evolved was little more than a box on a bridge carried by eight wheels. The wheels were clustered together at each end of the car in four-wheel sets called trucks. The frames of the trucks were fabricated from a large number of small iron straps, bolts, and castings. The heavy bridgelike floor frame was generally made from eight 4- by 8-inch timbers called sills. The sills were bolted on top of heavy cross-timbers called bolsters. The frame not only supported the cargo; it was also required to sustain the draft or pulling stresses of the train.

The framing elements were structurally the most important elements of the car, and yet opinion varied on just how big to make them. Some builders felt that a total cross-section of 30 square inches was sufficient, while others insisted that nothing less than 45 square inches would do.[487] In an effort to hold weight and cost down, most car builders tended to skimp on framing, and many employed undersize bolsters. In a survey of older 30-ton-capacity cars made in 1903, bolster strength varied from 36,000 to 84,000 pounds each.[488] Thirty thousand pounds might just do the job, but good practice called for a 100 percent margin of safety. So a strength of 60,000 pounds was considered just right. But as the survey showed, many car builders erred with either too much or too little material. Such a lack of science was normally covered over by the addition of plates or rods. In fact, metal was increasingly used to strengthen wooden cars as the nineteenth century came to a close and the common freight car became increasingly a composite iron/wood fabrication. The devotion to built-up construction involving hundreds of small wooden and metal parts might be explained by the tradi-

FIGURE 1.63

The major parts for a late-model wooden boxcar are shown in this diagram. (Smithsonian Institution)

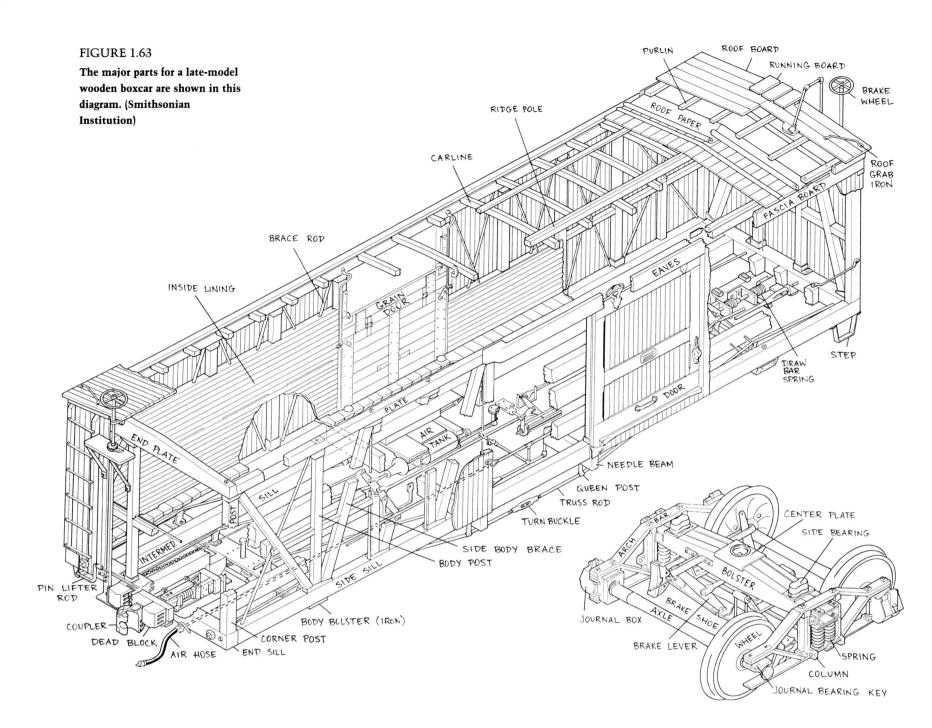

TABLE 1.22   Size of U.S. Boxcars

Year	Railroad	Body Length (feet)	Weight (tons)	Capacity (tons)	Journal Bearing Size (inches)
1871	Hannibal & St. Joseph	28	9.5	10	3¼ by 6⅛
1875	Pennsylvania	28	9.3	14	3¼ by 7
1880	New Haven	30	10.5	14	3½ by 7
1881	Louisville & Nashville	30	10.0	18	3¾ by 7
1885	Pennsylvania	34	15.1	30	NA
1888	Lehigh Valley	35	14.6	25	4 by 7½
1895	New York Central	35	15.0	30	4½ by 8
1898	Pennsylvania	34	15.8	40	NA
1900	Baltimore & Ohio	36	16.3	30	4¼ by 8
1901	Big Four	37	16.5	30	4¼ by 8

Source: Data from drawings of cars pictured and described in Chapter 3.
Notes: Wood-frame cars only. NA = not available.

tional methods devised by millwrights centuries earlier. A better understanding of basic wood-car construction can be gained from a study of Figures 1.63, 1.64, and 1.65.

Perhaps the single most obvious change in freight car design was size. Until about 1870 car size seemed frozen at 10 tons capacity. As tonnage increased during the post–Civil War period, bigger cars were seen as a quick solution to the traffic problem. In rapid succession 15-ton cars were supplanted by 20-tonners, and so on until the 40-ton car came into being in the early years of the twentieth century. A better understanding of this evolution can be gained from Table 1.22. The figures given there make it clear that no exact ratio between car weight and capacity existed. Some of this is accountable to the size of the wheels and axles; bigger wheels and axles alone could increase the capacity of a car. The skill of the designer in arranging the framing was a factor in car capacity, as was the railroad's simple faith that a 15-ton car could carry 40 tons just as readily as 30. Some of the stenciled-on capacity ratings were as much wishful thinking as anything else.

America's enchantment with big freight cars began in 1835, when the eight-wheel or double-truck car was accepted as standard practice. This design lent itself to enlargement, especially in length, as was not possible with the European four-wheel goods wagon. Width and height were limited to the loading gauge or clearances on the railway. As these became less restrictive, width gradually expanded to 10 feet and height to 15 feet on most lines. Length was for all practical purposes unlimited, for the eight-wheel cars could accommodate just about any curve. The practical limits of the frame kept most wooden cars to 40 feet (or less) in length.

Big cars were seen as efficient cars. The bigger the car, the greater its payload and earning power. Shippers loved big cars, because they represented a bargain in terms of carload lots. Traffic agents found them to be handy for sales promotion. Traffic managers championed big cars because they promised to keep the tonnage moving and so clean up clogged yards and overflowing freight depots. The cost of moving a big car was only marginally more than that of hauling a standard car. L. F. Loree figured that it cost only $23.10 more per year to haul a 40-ton car than to haul a 30-tonner.[489] And that cost could be recovered in a simple revenue run of 857 miles. Loree went on to claim that jumbos reduced yard space requirements, investment costs, empty-car movements, the number of trains, and switching and labor costs. If all of this was true, no wonder bigger cars became so popular.

Most railroad managers seemed to agree with Loree's statement, and had it not been so, then even larger cars would not have appeared. The trend in that direction is undeniable. Even so, critics of big cars were not silenced or intimidated by historical trends. In 1855 the superintendent of the Philadelphia & Columbia wondered about the desirability of large eight-wheel cars that were so big that their bolsters bent under the load.[490] He wondered too about the need for 9-ton-capacity cars when the average load was less than 2.5 tons. Several years later, no less an authority than John B. Jervis expressed his doubts about jumbo cars.[491] He preferred the American eight-wheel style of car but thought that cars more the size of those then used in England were really more economical. They would be easier on the tracks, carry more goods for a given weight, and be less likely to be destroyed in an accident. They could also be moved around yards and sidings by hand and so eliminate the need for switch engines.

The soundness of Jervis's reasoning is open to

FIGURE 1.64

The principal members of a wooden car frame and truck are shown in this end elevation. In this design, the car has an iron body bolster and a wooden truck bolster.

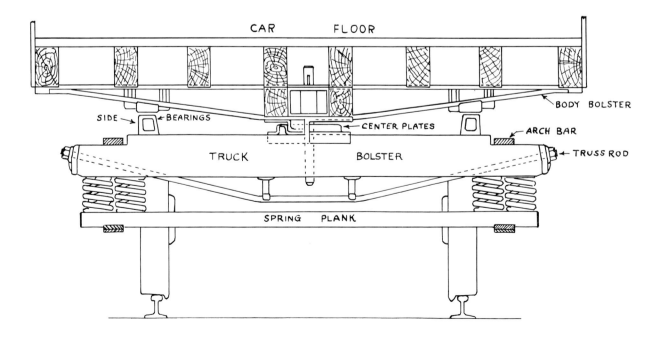

PARADE OF THE CARS

FIGURE 1.65

The underframe of a wooden boxcar is well shown in this engraving. Note the iron body bolster, the wooden draft timbers, and the Potter-style coupler. (*Railroad Gazette*, Sept. 24, 1886)

question on just about every one of these points, except perhaps the matter of track wear. At the time Jervis was writing (1861), weak wrought-iron rail was literally crushed by the weight of heavy trains, but the substitution of steel rail largely eliminated this problem. The homogeneity of steel permitted far larger wheel and axle loading and, hence, far larger cars. Trestles, culverts, ballast, and ties were also upgraded to sustain the heavier loads, but these expenditures were looked upon as a necessary cost if railroad capacity was to be improved. Big cars could not go everywhere, however, since private industrial sidings could not support them, just as many coal trestles could not handle 40-ton hopper cars; and so older or smaller cars were used to service them. In 1889 an official of the Ohio Falls Car Company wondered about the wisdom of bigger and bigger cars.[492] He contended that average car life had been reduced by around 50 percent, and that the dead weight was actually increased by the large wheels and axles required to support them. This added weight accelerated the wear of all running parts; where, then, was the saving?

Wear and weight were actually trivial details compared with the major flaw of the big car as seen by its critics. Oversize cars almost never carried a full load, so why build them in the first place? A survey conducted in 1899 found that the loaded cars weighed in the St. Louis area carried only 57 percent of their rated capacity, or 17 tons each.[493] Even 30-ton cars ran only one trip in ten with a full load. Hence the efficiency of the big car was seen as more a question than a fact.

The Car-Building Industry

An increasing demand for new freight cars was created during the wood-car era by the growth in railroad traffic. This need for new cars was augmented by the necessity to replace worn-out equipment. Most of the new rolling stock was produced by private contract builders. Railroad car shops were basically repair facilities and so generally produced only about one-quarter of the new cars produced each year.[494] The contract builders produced around fifty thousand new freight cars a year, on average, starting around 1875. They were also kept very busy with repair work when traffic was high and the railroads were pressed to find more equipment to handle the tonnage.

ECONOMICS

During good times the railroads, and the supply industry that served them, prospered. At times the demand for new equipment exceeded the capacity of the commercial shops to satisfy it. On the other hand, a dip in economic conditions could depress the new-car market, because railroads tended to withhold or even cancel orders when traffic showed signs of a slowdown. During a deep recession the car-building industry all but closed down. These wild cyclical swings in demand made the car market unstable and unpredictable. Car shops opened and failed with depressing regularity because of the feast-or-famine nature of the business. Only the very lucky and the very strong managed to survive long in this precarious trade. A chart of car production looks like a series of angry, irregular sawteeth. In 1878, for example, some 31,000 freight cars were produced; the number climbed to 109,000 in 1881 and then fell to 7,000 in 1885.[495] In 1900 activity was once again very high, with the commercial shops manufacturing 116,590 freight cars.[496] At the same time, the railroad repair shops were especially productive, turning out 143,134 cars; hence the total production of freight cars in 1900 was over 259,000 units.

The ups and downs of the car business are clearly reflected in the history of the individual firms as reported by the trade press. Jackson & Sharp of Wilmington, Delaware, had been in business barely a decade when the Panic of 1873 struck. By 1875 the company was down to forty employees from six hundred and planned to close completely by summer unless new orders were received. The small Atlantic Car Works in Salem, Massachusetts, remained closed for four years and reopened to make not cars but furniture. The Ohio Falls Car Company of Jeffersonville, Indiana, was closed for nearly two years following the 1873 crash. Times improved and the moribund industry slowly came back to life. By 1880 Jackson & Sharp was so busy that it could not keep up with the new orders and desperately looked for ways to expand shop space and find more workers. Hard times returned, however, following the Panic of 1893. By the following year one-third of the commercial car shops had closed for lack of work. Some proprietors sought orders below cost. Pullman built a group of stock cars at a loss of $12 each just to keep key men on the payroll. Smaller shops did not have the cash reserves to finance such paternalistic policies and so simply shut their doors, as did the Laconia Works in New Hampshire. Its workers were furloughed for sixteen months.

Idle car plants presented opportunities for not a few enterprising speculators. Michael Schall of York, Pennsylvania, had been in the railroad car business for a number of years before taking over the boarded-up plant at Middletown, Pennsylvania, in 1879.[497] The shop had opened a decade earlier but closed in 1874 and was standing idle when Schall purchased it at the very low price of $10,000. The shop was worth $75,000 or more. Schall made one of his best managers, Arthur King (a practical car builder), a partner in the new enterprise. King did well in running the rejuvenated car plant and seemed on his way to a comfortable retirement until 1891, when Schall failed because of overspeculation. It looked like the Middletown Car Works would be swept away in Schall's failure, but King prevailed upon the several creditors involved to allow him to carry on the business independent of his former partner. Short-term notes amounting to $50,000 foredoomed the enterprise. The Panic of 1893 was poorly timed for King's fragile enterprise; receivership followed in 1896 and continued until 1901. A local banker, intent on preserving the town's largest employer, backed King through the 1907 panic. Relief came two years later when the Standard Steel Car Company of Pittsburgh purchased the Middletown Car Works for $265,000. But King's fortune was hardly made, for the sale was in effect a liquidation. The creditors received 91 percent of what was owed them, and King lost all his equity.

The business of car building was made even more risky by the financial instability of the industry it served. Most American railroads were cash-poor and debt-heavy. They could buy equipment only by signing short-term notes or trading in stocks or bonds. This paper was of questionable value and was usually traded at a discount. Yet the selling of cars was based largely on the willingness of car builders to accept paper rather than cash in payment. Just about every car builder in the nation got stuck with some worthless paper from time to time. A weak firm such as the Taunton Car Company might be forced into bankruptcy by worthless paper.[498] Large builders with more ample capital developed easy payment plans for cash-poor customers. In 1884 the West Shore Line bought one thousand boxcars from Haskell & Barker of Michigan City, Indiana, for $494,000.[499] A 10 percent down payment was to be followed by twenty quarterly payments, bringing the total cost to $514,659. The U.S. Rolling Stock Company worked out a lease/purchase plan with the Montgomery, Tuskaloosa & Memphis for $381,500 worth of cars that included a 15 percent down payment and three payments laid out over eighteen months.[500]

Car builders were relieved of much of the risk associated with payments by working through the car trust plan. In such plans, the builder is paid up front for his wares, and the railroad pays the car trust on the installment plan. The first of these financial schemes is said to have been the Railroad Car Trust of Philadelphia, established in 1868 at the suggestion of Edward W. Clark, president of the Lehigh Coal & Navigation Company.[501] Ten years later the Car Trust Company of New York was formed with a capital of $3 million.[502] Clement R. Woodin, a car builder from Berwick, Pennsylvania, was one of the founding partners of this firm, which by 1886 had issued over $34 million in car trusts. The Gilbert Car Trust was formed in 1879 by the head of the Gilbert Car Works in an effort to shift the credit risk from the builder to the banker.

It would be a mistake to suggest that car building was all a matter of risk and large-scale bankruptcy, because excellent profits were also possible in this area of manufacturing. Fortunes were indeed made. Capital growth was sometimes remarkable in firms like the Michigan Car Company, established in 1864 with a capital of just $20,000. By 1892 accumulated earnings, together with a consolidation with the Peninsular Car Company (also of Detroit), had pushed its capitalization to no less than $8 million.[503] The partners of the business prospered as well. One of the founders, James McMillian, started his business career as a clerk but grew so rich in the car business that in 1892 his net worth was estimated at $6 million. He had one of the finest homes in Detroit, a country estate, a large art collection, and a reputation for supporting local charities. In like fashion, Barney & Smith of Dayton, Ohio, developed through good times and bad into one of the largest car builders in the nation—reportedly the very largest until Pullman opened his big plant near Chicago in 1881. Barney & Smith's prosperity was reflected in the growth of its capital from just $10,000 in 1849 to $3.4 million in 1892.[504] Ohio Falls also increased its capitalization to reflect a growing net worth. In the early 1890s it was producing about $3 million worth of cars each year.[505] In 1898 its net earnings reached $220,000, and preferred stockholders received a 14 percent dividend. Good profits were surely possible in this risky trade.

The money earned was, in the main, honestly won, yet as in all business, a shady side can be found. An early instance was the unexpected failure of the Norwich Car Company in 1853.[506] It appears that the owner, one Abner T. Pearce, issued large amounts of spurious paper, some with forged signatures. Pearce absconded to California, then to South America with his ill-gotten gain. Some years later a partner of Michael Schall discovered that Schall had received a payment from the Pennsylvania Railroad for cars delivered but had poured the funds into his own account rather than into the firm.[507] Outraged by this intelligence, Schall's partners sold out and left the firm. The Illinois Central freight car repair scandal was big news in 1910.[508] Padded repair bills, kickbacks to officials of the railroad, and excess profits in the hundreds of thousands of dollars gave the car-building trade a tarnished reputation.

The tendency to build cheap cars from inferior materials was perhaps a more significant criticism of the industry than the individual acts of larceny just recounted.[509] Poor wheels, defective axles, and green lumber were resorted to either to keep prices low or to boost profits. The blame for such practices can not in truth be laid entirely upon the car builders, for railroads were forever looking for cheap cars. They sought the lowest bidder rather than the best maker. Car builders producing inferior cars could defend their actions by explaining they were only responding to market forces. The

TABLE 1.23   U.S. Railroad Car Building Industry (1900)

Establishments	Employees	Capital	Freight Cars Built
65 commercial shops	33,453	$88.3 million	116,590
1,296 railroad repair shops	173,652[a]	NA	143,134

Source: U.S. Census, 1900, pt. 4.
Note: NA = not available.
a. Only some of these employees were employed in new-car construction. Most were involved with repair work.

demand was for cheap cars, not good cars.

The growth of the car-building industry paralleled the growth of the railroad network. Most early cars were manufactured by carriage builders or railroad repair shops. By 1850 forty-one establishments were engaged in this trade.[510] They employed 1,554 persons, produced $2.4 million in products, and represented $896,000 in capital. Within twenty years these numbers had grown greatly; by 1870 there were 170 establishments, 15,931 employees, $31 million in products, and $16.6 million in capital. By 1900 car building had become a big business, as can be seen in Table 1.23.

The car-building business was a growth industry during these years. It was also a widespread one not limited to any one region. Just about every state was home to a car builder at one time in its history. Most of the big plants, however, were located in the industrial northeastern states, and just about every major car builder was found in a belt stretching east and west from New York to Illinois. By the middle 1890s 90 percent of all new railroad cars were being produced in this region of the United States. Pennsylvania alone had fifteen to twenty plants so closely clustered that it was difficult to travel much over 50 miles without seeing another one. The Midwest had a major share of important freight car builders by 1875. Names like Haskell & Barker (Michigan City), Michigan-Peninsular (Detroit), Barney & Smith (Dayton), and Wells, French & Company (Chicago) were recognized by railroad men as the big producers. Pullman helped tilt the production balance toward the Midwest by opening his giant car plant near Chicago in 1881.

New England could account for a dozen commercial car shops, some dating back to the 1830s and 1840s. Yet early beginnings did not guarantee success, and only two, Bradley in Worcester and Wason in Springfield, developed into major car-building plants. The Laconia Works in New Hampshire was a weak third, but most New England shops seemed to die young. The Old South seemed more likely to generate a healthy car industry because of its fine timber, especially its stands of yellow pine, the favorite lumber for floor sills. However, the weak industrial tradition of the cotton states and the limited rail networks and traffic of the region did not encourage the formation of such an industry. A few shops opened in the antebellum period, such as Burr & Ettinger's small facility in Richmond, Virginia. But in the main, there was little activity. The Wason Car Works in Chattanooga, established in 1873 by a brother of T. W. Wason of Springfield, Massachusetts, seemed to open a new era in southern car building, for plants soon began to appear in Lenoir (Tenn.), Raleigh (N.C.), Waycross (Ga.), and elsewhere. Nothing truly major in the way of a car plant located in the South until the steel-car era, when Pullman opened its Birmingham, Alabama, freight car shop in 1929.

The commercial car-building trade had only token representation in the West. St. Louis and nearby St. Charles had important plants, but there was little to be found west of Missouri. The lack of local timber generally discouraged car builders from locating in the prairie states. The Marshall Manufacturing Company of Marshall, Texas, ignored the conventional wisdom of locating only near a major grove and began car building in 1875. A fire ended this fledgling enterprise nine years later. California became home to a few small builders, such as Kimball in San Francisco and Carter Brothers in Newark, but neither was more than a marginal producer, despite an abundant timber supply. The great forests of Oregon likewise did surprisingly little to stimulate a local car-building industry. Carload after carload of excellent Douglas fir was simply shipped to builders in the East.

The competition between the regional car builders appears to have been rather modest. In 1879 Uri Gilbert, a car builder of Troy, New York, admitted his inability to compete with western car builders—meaning, no doubt, midwestern builders—because of their easy access to good lumber.[511] Gilbert went on to say that his competitors did not necessarily build better cars with the excellent wood available to them; they did not season it properly. He also contended that his western competitors used inferior iron.

There were approximately two hundred commercial car builders active during the wood-car period.[512] The origins of these firms are mixed and variable, as might be expected. Certain patterns can be found for just about all the known manufacturers.[513] Most of the earliest builders were originally carriage or coach makers. Osgood Brodley and Davenport & Bridges fit into this mold; but even at a later date, William E. Rutter, founder of the Elmira Car Company in 1851, and Thomas Carter, a partner in the Newark, California, car plant, were both experienced carriage builders. Carpenters and lumbermen were also naturally drawn to car building. Thomas W. Wason was very much a hands-on mechanic. He worked as a carpenter, sawmill operator, and bridge builder before entering the car trade in 1846. His first shop was too small for even one car, and one end of the car stood outside in the weather. E. E. Barney of Barney & Smith started life as an academic but turned to sawmilling in middle life as a more remunerative trade. In 1849 he turned to the even more profitable trade of car building. In a similar manner David E. Small of York, Pennsylvania, made the transition from a family lumber business to a door-and-sash mill to a car plant. The firm of Billmeyer & Small flourished for nearly half a century. Ship building was another established woodworking trade, but one that seems to have produced only a single car builder—James Patten of Bath, Maine, who after many years as a ship builder opened a railroad car shop in 1872. The panic scuttled this effort to shift careers. Patten's plant closed in 1877 and eventually was recycled as a streetcar barn. Ironically, Harlan & Hollingsworth and Jackson & Sharp, both of Wilmington, Delaware, took up ship building in a major way not many years after entering the car-building field.

Manufacturers of agricultural machinery drifted into car building in sizable numbers. The equipment and skills needed to make reapers, threshers, and rakes were equally useful for making railroad cars. John L. Gill of Columbus, Ohio, for example, started out making stoves, then agricultural machinery, and in 1862 he turned to railroad cars. Jackson & Woodin of Berwick, Pennsylvania, began as plow makers, but by the late 1860s they were producing mine cars and then freight cars. By 1880 Berwick had become one of the largest freight car makers in the nation. The LaFayette Car Works of Indiana was a failed reaper plant that reopened in 1880 to produce freight cars. It too prospered in this line of work, though it never achieved the status of its rival in Berwick.

Most car builders depended on their own resources to start up operations. Personal savings and bank loans were the way most car shops were funded. Murray Douglas & Company of Milton, Pennsylvania, started up in this way. The partners bought the tools and a stockpile of materials of a small foundry and moved it by raft downstream to the site of their small operation. A local doctor lent them money to get the car end of the business going. Within a decade they were the largest employer in the area. Big investors were sometimes induced to bankroll car shops, occasionally with disastrous results. Tracey & Fales opened its Grove Works in Hartford, Connecticut, in 1849 with the backing of the town's most prominent citizens. The plant was the best that money could buy, but that proved no guarantee of success. A boiler explosion in 1854 killed fifteen to twenty employees. A terrible flood came a few months later, ruining all the stock and bringing on the failure of a business that had seemed assured of success. Far to the west of Hartford, William C. Ralston, a wealthy banker and civic leader in San Francisco, sponsored the Kimball Car Company in 1868. Like the Grove Works, Kimball's shop was carefully constructed and well equipped. But the reluctance of local railroads to patronize the firm, plus the bankruptcy of Ralston in 1875, meant the end of this pioneer West Coast car builder. Rich backers could mean success as well. In 1872 the Ensign Car Works opened in Huntington, West Virginia, with the backing of two very wealthy men—W. H. Barnum, a car-wheel maker from Connecticut, and the Wall Street baron C. P. Huntington, the source of the town's name. Ensign never wanted for capital, car wheels, or orders from railroads controlled by Huntington.

Railroads tended to favor car builders located on their own line, because the builders were also shippers. Whenever possible, a railroad would order new cars only from on-line builders. But during high-traffic times, when cars were desperately needed, railroads were ready to accept equipment from any supplier that could deliver them, and so the home-purchase rule was abandoned. And as we have just seen, a western railroad like the Southern Pacific could be under pressure to patronize a car builder in faraway West Virginia, just because one of its major investors had an interest in the car plant as well. There is one example of even more direct involvement of a railroad with a commercial car builder. In 1888 Mt. Vernon, Illinois, was destroyed by a tornado.[514] The Louisville & Nashville served the town and did not want to see the community die. It sent in nineteen hundred carloads of building supplies to rebuild houses, stores, and a recently abandoned railroad car repair shop. The car shop was reopened as a commercial enterprise and so provided much-needed employment. It succeeded and, in time, produced forty cars a day and employed fourteen hundred men.

Community support was actually a more common form of subsidy for fledgling railroad equipment manufacturers. Most city fathers looked for ways to attract new factories that would employ workers who would in turn enrich landlords and shopkeepers. St. Charles, Missouri, suffered a loss of employment in 1867 when the local railroad car repair shop was closed. It stood idle for six years until it was reopened with funds provided by local citizens. At first the business did poorly because of the 1873 panic, but it gained strength and prospered for many years as a branch works of American Car & Foundry. In other cases some communities conspired to steal each other's businesses. In 1881 Youngstown, Ohio, induced the Pittsburgh Car Works to move its operations out of the Steel City with a present of free land. This firm proved more fickle than most, because in 1901 it moved once again, to Niles, Ohio. In 1873 the little town of Oxford, Pennsylvania, became involved in a car plant sponsorship that also embraced employee ownership.[515] The town lent the workers $50,000 to start up the Oxford Cooperative Car Company. Each worker promised to donate $10 a month to retire the loan. The Panic of 1873 soon put this noble experiment to sleep. The little shop closed within a few years because of hard times. A gale blew down the paint shop, and in 1880 a fire burned the main building. The workingman's dream was put down cruelly.

CAR SHOP DESIGN AND OPERATION

To most observers the typical car plant would look like a random collection of low buildings on a large, flat lot. There is some truth to this observation, for most car plants were more evolutionary than planned complexes. They might start up in a small way with three or four buildings. As the business developed, the erecting shop was enlarged and a new paint shop was built. After a decade the complex would have grown to seven or eight structures. In other instances the grounds and buildings were systematically laid out and opened as a complete, ready-to-go installation. An early example of this was the Buffalo Car Works, which opened in 1853 with fifteen buildings.[516] A ground plan reproduced in Figure 1.66 shows the arrangement of the facility. The largest building was the freight car erecting shop, which measured 60 by 280 feet, designated as item 7 on the plan. This shop could build twenty-four freight cars a week. The general arrangement here was rectangular, so that long parallel tracks could run through the buildings.

The Harrisburg Car Company, which opened the same year as the Buffalo works, was laid out on

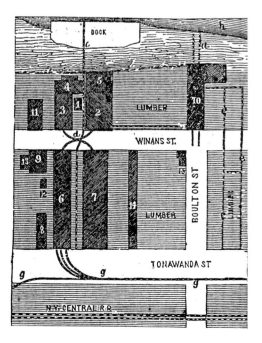

FIGURE 1.66

Ground plan of the Buffalo Car Works. (*Railroad Advocate*, Sept. 20, 1856)

FIGURE 1.67

The Harrisburg Car Co. was laid out in a U shape with a central courtyard for the storage of lumber and other raw materials. (*U.S. Railroad & Mining Journal*, Oct. 1857)

a different scheme (Fig. 1.67). It was arranged in a more or less U shape, with a big courtyard in the center for the storage of lumber. The foundry, planing mill, machine shop, and erecting shops were thus all strung together in the various legs and stems of this continuous building. Some years later, in 1868, Kimball's new shop in San Francisco was built on a similar plan, but it was an L rather than a U shape.[517] The L was a three-story brick structure measuring 50 by 275 feet. The open sides of the L were fenced in by a one-story wooden shed, so that the center open courtyard was enclosed. Both schemes were compact and so put the various departments in close proximity. Yet it was a bad plan, on two counts. First, a fire could easily destroy the whole facility, because it was essentially one structure. In a complex consisting of many structures sufficiently separated, only one might be lost in a fire. Second, long, one-story buildings were far more suitable for the manufacture of large items like railroad cars. The transfer of bulky materials and parts was much harder in multistory buildings.

The multiple-building scheme was adopted by most car builders. The Patten car plant in Bath, Maine, included six buildings when it opened in 1872.[518] Three smaller buildings were occupied by the office, foundry, and machine shop. The paint shop was one of the three larger structures; it measured 75 by 162 feet and had a manually powered transfer table. The larger of the two erecting shops had a single story and measured 75 by 220 feet. It had eight parallel tracks. The smaller erecting shop was two-storied and measured 75 by 150 feet. Patten employed two hundred men and could produce four freight cars per day plus six passenger cars per month. The Middletown Car Works in Pennsylvania was not materially larger than Patten, but it built twice as many cars and employed only 175 men.[519] Middletown had eight buildings situated on four acres of land and held another six acres in reserve for expansion. The buildings were as follows: erecting shop, 70 by 225 feet; repair shop, 30 by 130 feet; planing mill, 45 by 150 feet; carpentry, 50 by 120 feet; blacksmith, 40 by 160 feet; foundry, 60 by 160 feet; machine shop, 50 by 50 feet; and oil and paint shop, 20 by 35 feet.

The investment required for an average or me-dium-size car plant is recorded in a pamphlet issued in 1882 or 1883 by John L. Gill, owner of the Gill Car Company of Columbus, Ohio.[520] The land was valued at $68,850, the building at $82,250, the machinery at $86,700, and the locomotives, tools, and other machinery at $36,000, which added up to a total of $273,800.

Just about every car plant of any size had a transfer table. These rolling platforms could move up and down a broad-gauge track and connect a dozen or more ladder tracks branching off to the right or left. The platform had a standard-gauge track long enough to hold one or two cars. The ability to connect a large number of tracks without any switches was the big advantage of the transfer table. The larger ones were steam-powered, while the smaller ones were pushed along by a few men or worked with a capstan. The Gilbert Car Works had an out-of-doors transfer table at one end of its erecting shops to shift finished cars from any one of five tracks in the building. The new cars were moved by the table to a runaround track on one side of the building and so delivered to a connecting railroad (Fig. 1.68).

Ohio Falls, which had a transfer table between its several erecting shops, set up a parallel row toward the back of the property (Fig. 1.69). The multiple buildings scattered around the twenty-acre lot were connected by a private railway serviced by two small shop locomotives. Twenty-two turntables facilitated the shifting of cars between the buildings via crosstracks. Some plants, however, depended almost entirely on the transfer table for the movement of cars and materials. The Indianapolis Car Works was one such plant. It had two shops with a table in between to service forty-five tracks or stalls.[521] Each stall held two cars each.

The champion of the transfer table scheme for plant layout was surely the Wason Manufacturing Company of Springfield, Massachusetts (Fig. 1.70). Its old downtown plant, dating back to the 1840s, was cramped, dark, and ill-ventilated.[522] The new shop, built in 1872–1873, was located in suburban Brightwood on a sixteen-acre plot with as light, airy, and conveniently arranged a model factory as could be erected. The buildings were arranged in two ranges about 500 feet long on either side of the transfer table track. The iron-framed table—45 feet wide and weighing 9 tons—ran on a track 1,000 feet long that extended beyond the buildings, so that storage tracks and the lumberyard could be serviced as well. The table was the heart of the operation, for it picked up cars loaded with raw materials and took them to the right spot in the factory complex. It moved partially finished cars from place to place when necessary and, of course, deposited completed cars on the out track. The table had a vertical boiler and a 12-horsepower engine that reeled itself along a stationary chain. It could travel the 1,000 feet of track in just two minutes,

FIGURE 1.68

This ground plan of the Gilbert
Car Works shows the arrangement of its shops in 1880. Note
the transfer table at the upper end
of the freight shop. (*Poor's Manual of Railroads*, 1880)

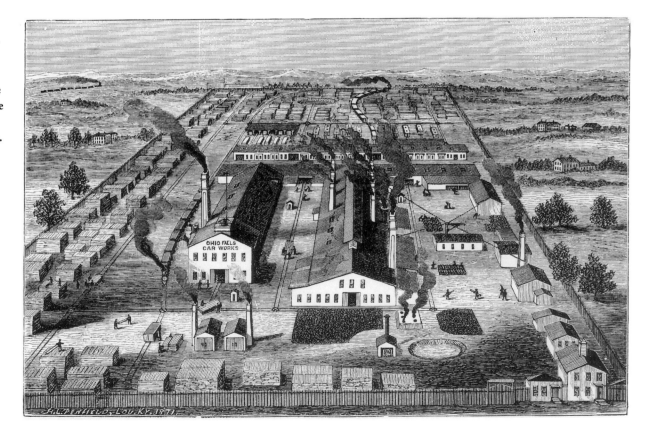

GILBERT & BUSH COMPANY, Troy, N. Y.

ESTABLISHED 1830.

Builders of Railroad Cars of Every Description.

WATER FRONT 875 FT.

BRANCH OF MOHAWK

LUMBER SHEDS LUMBER SHEDS

D&H.C.Co. D&H.C.Co.

EXTRA FACILITIES FOR PACKING AND SHIPPING FOR EXPORT.

COAL HOUSE | BLACKSMITH SHOP | COAL HOUSE | BOILERS | ENG. ROOM

BLACKSMITH SHOP 230 X 40 | MACHINE SHOP

SAWING & PLANING MILL 100 X 180

FREIGHT SHOP 300 X 100

COAL. | COAL. | STORE ROOM | YARD

PAINT SHOP 80 X 60 | WOOD MACHINE SHOP 250 X 40

SHED FOR MOUNTING CARS ON TRUCKS

DRY HOUSE 50 X 90

BRANCH OFFICE,

BOREEL BUILDING,

115 Broadway,

NEW YORK.

TRANSFER TRACKS

FOUNDRY 90 X 135

UPHOLSTERING | PAINT SHOP 430 X 74 | CONSTRUCTING SHOP

URI GILBERT, *President and Treas.*
WILLIAM E. GILBERT, *Secretary.*

EDWARD G. GILBERT, *Vice-Pres't.*
R. E. RICKER, *Gen'l Manager.*

Special attention given to Sleeping, Drawing Room & Passenger Cars.

FIGURE 1.69

The Ohio Falls shop at Jeffersonville, Ind., is shown in this 1872
bird's-eye view. Small turntables
and a transfer table located at the
rear of the property facilitated the
movement of cars around the
plant. (*National Car Builder*, Feb.
1872)

FIGURE 1.70

The Wason work was considered a model for all American car builders. The transfer table was central to the operation of this plant. (*Poor's Manual of Railroads*, 1875)

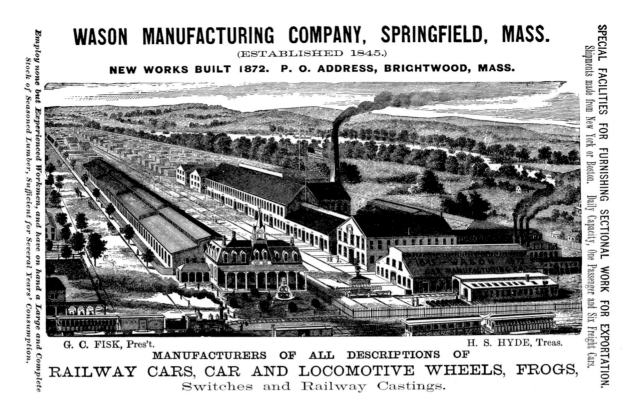

WASON MANUFACTURING COMPANY, SPRINGFIELD, MASS.

(ESTABLISHED 1845.)

NEW WORKS BUILT 1872. P. O. ADDRESS, BRIGHTWOOD, MASS.

Employ none but Experienced Workmen, and have on hand a Large and Complete Stock of Seasoned Lumber, Sufficient for Several Years' Consumption.

SPECIAL FACILITIES FOR FURNISHING SECTIONAL WORK FOR EXPORTATION. Shipments made from New York or Boston. Daily Capacity, One Passenger and Six Freight Cars.

G. C. FISK, Pres't.   H. S. HYDE, Treas.

MANUFACTURERS OF ALL DESCRIPTIONS OF

RAILWAY CARS, CAR AND LOCOMOTIVE WHEELS, FROGS,

Switches and Railway Castings.

and it serviced ninety lateral tracks. When fully opened in the summer of 1873, Wason was thought to be the best-arranged car-building shop in the nation. Perhaps it was, but one wonders what happened when the transfer table broke down, or how convenient and efficient the plan seemed when more than one section of the shop needed something moved. At times the transfer table must have been more a bottleneck than a master of distribution.

Whatever plan the car shop designer might adopt, careful consideration was always given to reducing the danger of fire. Car shops were particularly susceptible to fire because of the stores of lumber, paint, and oil kept on hand. Sawdust and wood shavings added to the fire hazard, as did the forges and iron furnaces. Dividing the complex into many small buildings rather than just one or two large ones helped reduce the chance of loss. That way, if the foundry caught fire it alone would be lost, assuming that there was sufficient space between it and the next building. Lumber piles were generally kept some distance away from any building on the lot, for the same reason. Empty spaces were seen as necessary fire barriers, not wasted land. Buildings made from fire-resistant materials were another obvious precaution. Brick buildings with slate or tin roofs were common, although some plants had at least a few wooden buildings for economic reasons. The Allison Car Works in Philadelphia carried fire-resistant construction to an extreme, for in 1860 it was reported that even the doors and roof framing were of iron.[523] Even so, Allison's supposedly fireproof shops burned to the ground in May 1863, and its

new shop in West Philadelphia went up in smoke nine years later.

Fire controls became part of the daily routine. Smoking on the job was prohibited. Heating by steam rather than stoves was adopted by the more prosperous builders. The Tredegar closed down early on winter days because it feared the hazard posed by artificial lighting, candles, coal, gas, or oil.[524] As early as the 1870s some plants had sprinkler systems, and many had independent water supplies. Ohio Falls had a large cistern under the shops plus two ponds as backup water supplies. The shop also had two steam-powered fire engines and fire hydrants placed around the grounds. The Sacramento shops had five fire brigades and conducted drills every two weeks. Starting in 1869, Altoona maintained a full-time fire brigade of eighty-six men, its own firehouse, a steam pumper, and a chemical engine. Fire alarm boxes were strung up throughout the shop buildings.

An abundant supply of selected dry lumber was the car builder's stock-in-trade. A good wood supply was as essential to the car builder as a well-stocked pantry is to a restaurant. It was a matter almost always mentioned in advertisements or promotional pieces. The potential customer was assured that a good supply of seasoned lumber was on hand to meet their needs. The Michigan Car Company stored 3 million board-feet of lumber for future use and dried 50,000 feet a day in kilns on the property. Ohio Falls kept 6 million board-feet on hand in eleven lumberyards. Large repair shops such as Altoona kept no less than 11 million board-feet on hand for the repair of old cars and the construction of new ones. Impressive as these stockpiles might appear, they were quickly used up. In 1887 the Michigan Car Company used 50 million board-feet to produce ten thousand cars. During the same period, Barney & Smith used two hundred carloads, or 1.5 million board-feet a month. In 1883 *Railway Age* estimated that thirty car plants in this country consumed 200 million board-feet annually.[525]

Many shops ran their own sawmills, and most could find good lumber available locally. The Buffalo Car Company, mentioned earlier, brought log rafts down the Niagara River from Grand Island during the 1850s. A great gang saw sliced up the logs into eight sills at a time. Harrisburg received eight or more log rafts a year on the Susquehanna River, which furnished about one-quarter of its lumber needs. Harrisburg and other northern builders were forced to import yellow pine from the South. Ohio Falls made much of the fine yellow pine it brought from southern Mississippi. White pine came largely from Michigan. The lumber was gone over by inspectors who sorted and graded the wood. Defective pieces were rejected. The boards were marked and graded for easy retrieval, and the best wood was saved for car roofs, for the life of a car depended largely on the durability of its roof.[526] Conservation of lumber was not a large issue in a time when much of the land was still covered by virgin forest. Scrap, chips, and to a degree even the sawdust were used for boiler fuel. The LaFayette Car Works attempted to make better use of end pieces that were too short for car work.[527] These bits of scrap were made into wagon wheel felloes for which LaFayette found a good market.

WORKING CONDITIONS

Being a worker in a car shop was little better than being a trainman, except that the hours tended to be more regular. The ten-hour day and six-day week were common for just about all workingmen of the time. Holidays were rare, benefits were few, and wages were low—usually around $1.50 per day. Paid vacations, health insurance, retirement plans—these were all in the future. If business was slow, men were let go with little notice. A few larger shops would try to find work even at a loss to keep their best men employed, but the smaller, weaker shops could not afford such largess. When times were hard, a car builder could expect to find himself out on the street. Wason's payroll accordingly rose and fell with the health of the economy. In slow times it might be 300 men, and in fat times it would rise to 700. Some shops became very major employers, and villages like Dayton, Ohio, became cities because they were home to such firms as Barney & Smith (Fig. 1.71). In 1859 Barney & Smith employed just 150 men. By 1867 the number had risen to 350; in 1873 it was 800, then 1,200 in 1884 and

TABLE 1.24 Employee Breakdown of Michigan Car Company (c. 1877)

Wood shop	
Woodworkers	50
Laborers	10
Helpers	40
Machine shop	
Engineers	7
Firemen	4
Machinists	61
Blacksmith shop	
Blacksmiths	42
Helpers	50
Erecting shop	
Trackmen[a]	20
Bottom men	37
Carpenters	76
Paint & other shops	
Painters	31
Repair crews	70
Yard laborers	60
Teamsters	8
Clerical & administrative	125

Source: David L. Hay, "Before Ford: Assembly Line Techniques," paper given at 1985 SHOT meeting, Dearborn, Mich.

a. Perhaps truckmen?

1,600 in 1888. In 1908 Barney & Smith had 3,500 on the payroll. Of course, Pullman ran the biggest car shop in the nation; in 1907 it had 5,680 men just in the freight car department. Table 1.24 gives some idea of how a body of around 700 workers might be divided up inside the plant—in this case, the Michigan Car Company in about 1877.[528]

Just what it was like to work in one of these shops is poorly recorded. No car builder, at least at the workingman level, appears to have left a written record of his daily life. One can assume that it was less than delightful. An 1871 account of a visit to the Chicago, Burlington & Quincy's shop in Aurora, Illinois, offers a partial picture.[529] It was described as a very crowded wooden structure with every nook and corner occupied by men, supplies, and machinery. Some men were required to work in lofts directly under the roof that were miserably hot in the summer. No one, not even a lowly car shop worker, should be forced to work under such conditions, especially for a ten-hour day. Even where the factory buildings were airy and light, workers were subjected to rules and a disciplinary code not far removed from that of a military or even a penal institution. Strict attention to hours and the work at hand might be expected. Care of the company's tools and supplies was also a natural enough requirement. But the lists of dos and don'ts went on to include a great many other items, mostly prohibitions. Altoona forbade visitors, including book agents and peddlers. There

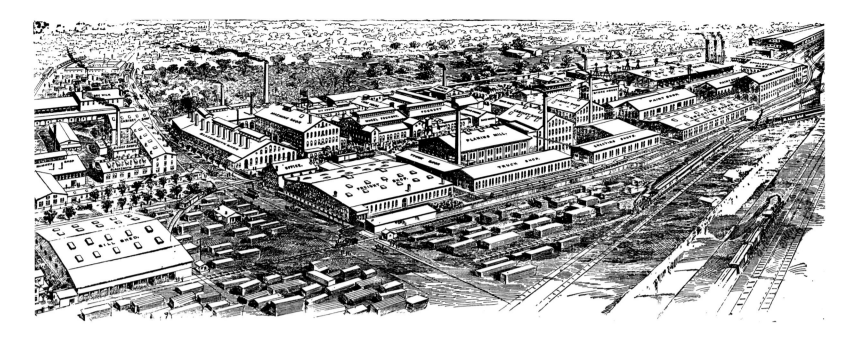

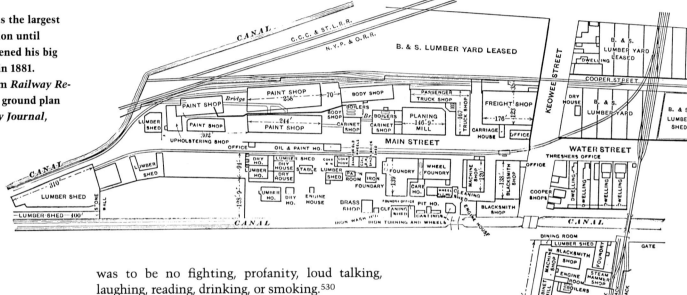

FIGURE 1.71

Barney & Smith was the largest car plant in the nation until George Pullman opened his big shop near Chicago in 1881. (Bird's-eye view from *Railway Review*, Jan. 27, 1894; ground plan from *Street Railway Journal*, Aug. 1894)

was to be no fighting, profanity, loud talking, laughing, reading, drinking, or smoking.[530]

Although unions were not mentioned in the rules, they were discouraged. Harlan & Hollingsworth, of Wilmington, Delaware, wanted no part of independent, talkative workers who might want to organize the men into any kind of union. Such troublemakers were labeled "sea-lawyers." In the 1870s it became the policy of the H&H management to fire any sea-lawyer found in the ranks.[531] Anti-organized-labor sentiments surfaced elsewhere. The works at St. Charles, Missouri, dismissed every man who joined the Knights of Labor, contending that workers could not be subject to rules made outside the car shop.[532] Job H. Jackson was less autocratic. Indeed, his workers might join any group they liked, but if the shop went union and demanded higher wages, he would simply close the plant. A few car builders solved labor problems by using convicts who could not strike or agitate for better working conditions. The South Western Car Company in Jeffersonville,

Indiana, used inmates from the local state prison, as did the North Western Car & Manufacturing Company of Stillwater, Minnesota. The new central prison in Toronto was established by the Ontario government in 1874 specifically to supply labor for the building of railroad cars.[533]

Competition from convict labor was a less serious challenge to car shop workers than the piecework system—a scheme favored by some shops starting in the 1880s. One laborer might produce ten truck bolsters a day, while another would finish only six. In the old wages-by-the-day system, both would be paid the same; but under the piecework scheme, the more productive man would be rewarded with more pay. Most piecework systems involved a quota, so many truck bolsters per day, plus rigorous inspection to assure that fast procedures were not turning out substandard products. A survey conducted by the Master Car Builders Association tentatively concluded that the piecework system reduced labor costs by 35 percent.[534] Studies conducted at the Pennsylvania Railroad's Fort Wayne shops claimed savings of 50 percent. This concern over labor costs is a bit puzzling, considering that labor represented only about 10 percent of the overall cost of building a freight car.[535] In any case, management's enthusiasm for piecework was somehow never shared by labor.

Car builders tended to locate in out-of-the-way places, because housing costs were lower there than in the urban centers. Cheap rents would lessen demands for higher wages. Pullman carried this notion to an extreme, building his own towns far from high-rent districts, well south of Chicago.[536] In the model industrial towns of Pullman, housing and utilities were provided by the employer. The cost of these services was somewhat below the conventional tariffs elsewhere, but many of the workers resented living in a company town where everything was run to suit the puritanical notions of the proprietor, George M. Pullman. A thirsty worker had to travel down the road to find a tavern, because no saloons were permitted within the corporate limits of Pullman. Jackson & Woodin of Berwick, Pennsylvania, had similar views about liquor, but since it did not own the town, it simply bought every saloon and hotel in the community.[537] Berwick became a strict temperance town in 1881.

Other employers took a less dogmatic stand on prohibition. The Erie Car Works, for example, opened a boardinghouse for its bachelor workers around 1870 that prohibited the sale of strong drink anywhere near the building.[538] This three-story boardinghouse provided shelter for one hundred men. Besides sleeping chambers and a dining hall, it provided a reading room with a supply of current magazines and newspapers. The manufacturers of Wilmington, Delaware, which included two major railroad car builders, hoped to foster sobriety through the Fountain Society, which met in a workingman's club called the Holly Tree Inn.[539] The Fountain Society encouraged the consumption of water rather than ardent spirits. Compliant workers who frequented the capitalist-sponsored club might be favored with home loans at low rates, dispensed by local businessmen to the deserving sober. Paternalism of a simpler kind was practiced by car makers as by other manufacturers of the time. The company-sponsored summer picnic might involve a steamboat excursion or trip to the county fair. A ham or turkey at Christmas assured many a poor family a feast rather than just another meal. Michael Schall, owner of the Empire Car Works, gave each employee a dollar at Christmastime.[540] Whether it was spent in saloon or butcher shop was up to the recipient.

The Assembly Process

The actual building of cars was a frenetic and noisy business. It was much like assembling a town of bungalows all at once. The circular saw screamed its way through the sill timbers while the mortising machine crunched out square holes. Nearby, the forge and machine shop hammered on the metal parts as the great central engine labored to keep the belts turning. Above all this racket was the rapid beat of a hundred hammers as carpenters nailed the car together.

The car shop was surely a busy and dusty place, but just how were the cars put together? The data available on the process of manufacture is rather slender, but it is likely that the earliest freight cars were put together much as a country carpenter would build a house. That is, each board and timber was measured, cut, and bolted or nailed together, and the car would go together one piece at a time. This method was surely supplanted at an early time by the assembly using premade parts. The car builder/craftsman who made and assembled his own parts was replaced by a worker who simply put ready-made pieces together. This latter method was in full swing by 1856 when a description of the Buffalo Car Works was published.[541] In this account, which regrettably offers few details, mention is made that five men worked as a team to assemble each car. Four of the men started work at each corner while the fifth helped the others as needed. When the body was finished, trucks were rolled under the trestles supporting the body above the floor. The body was then lowered into place, and the car was ready for painting. Giant sawhorses or trestles supported the body 3 or 4 feet above the floor and acted much like a ship's cradle. This facilitated the assembly process by allowing access to the underside of the frame.

It was a build-in-place or fixed-position system of assembly, traditional for the erection of ships, large machine tools, and steam engines (Figs. 1.72, 1.73). Wood and iron parts fabricated elsewhere in the shop complex were delivered to the setting-up shop for assembly. Thus floor sills, carlines, brake wheels, and hundreds of other parts were placed at some convenient location on the floor so as to be nearby yet not directly underfoot. Workmen would fetch the parts as needed during the putting-together process. The supply of parts would be replenished as needed by the woodworking shop, the forge, and the foundry. The trestle system lent itself to stalls or relatively short tracks

FIGURE 1.72

Building cars in place, in much the way a ship or building is erected, was a time-honored system during the wood-car period. Note the trestles, visible beside the walkway post in the picture's center. The walkway gave workers access to the upper level of the car. This outdoor scene actually shows old cars being rebuilt rather than new-car construction. (Pullman Standard Photo)

that might hold only one or two cars. While this method might appear inefficient or at least limited for the assembly of large numbers of cars, the Indianapolis Car Company, built on this plan in 1881, was praised as being modern and efficient in an article published eight years later.[542] It had forty-five stalls large enough for two cars each, all of which were connected by a transfer table. The delivery of parts to so many locations must have been notably inefficient despite the praise offered at the time regarding the plant's floor plan.

A better arrangement had been in use for some years, and this involved a long rectangular building with parallel tracks. Cars were built in place in long lines on their own wheels rather than on trestles. The trucks in effect took the place of the trestles and so eliminated the tedious process of removing the trestles and installing the trucks. Because the cars were built in lines, materials might be placed in just a few places rather than in many, as was true of the old stall system. In 1873 the Empire Car Works of York, Pennsylvania, had two parallel tracks, each 260 feet long.[543] The north tracks were used for materials handling and the south tracks for the assembly of cars. Preassembled trucks were placed on the south tracks. Bottom men assembled the frame, bolster, and draft gears. Then the top men nailed together the body and roof. The finished car was pushed down the tracks, where the painters took over. Some builders would assemble the trucks in place on the setting-up track and then follow with an in-place assembly of the frame and body.

The long-track system appears to have won over most of the trade and remained in favor into the twentieth century. The Milwaukee Road refined the long-track system at its West Milwaukee shops, according to a description published in 1899.[544] The tracks and stockpiles of unassembled parts were arranged so as to avoid any unnecessary handling of materials. The erecting shop had five tracks. Each one was 624 feet long. Track 1 was devoted to the assembly of trucks. The others were used for the in-place assembly of cars. The frame/draft-gear or bottom men worked in four-man teams, with two men working at either end of the frame. The frame, less draft gear and brakes, took two hours to assemble. The top gang, which consisted of a five-man crew, put the body and roof together. The time required to finish this work was not recorded, but it was noted that the exterior siding was nailed on in one hour.

In 1907 a technical reporter found the old-fashioned build-in-place method alive and well in the Lackawanna's modern freight car shop in Scranton, Pennsylvania.[545] This model shop featured a 31-foot-high ceiling in the center bay, overhead crane, and concrete floor. Elevated platforms ran along the track to facilitate assembly of the body. The trucks, floor frames, and roofs were preassembled and were moved by the electric cranes to the center tracks, where the assembly took place. These parts were given one coat of paint prior to being moved to the assembly area. The finished car was pulled outside by a cable attached to an electric winch. In good weather the floor was

PARADE OF THE CARS

FIGURE 1.73

The traditional build-in-place system is shown in this 1882 interior view of the Ensign Car Works, Huntington, W.Va. Note the wooden brake beams and truck bolsters stacked in the left foreground. A trestle stands in the center foreground. (ACF Industries Photo)

nailed down outside to clear the shop floor so that another car might be started. Final painting was handled in a separate shop.

A few contemporary accounts speak of freight car building as if raw materials entering the back door were magically converted into finished cars rolling out the front. They spoke of the rapid assembly of cars in such enthusiastic terms that at least a few modern readers have assumed that moving assembly lines were involved. This is surely not true, at least in the Henry Ford sense of the moving assembly line, but the railroad car industry did devise a similar, if slower-paced, version of the moving assembly line. It was more of a stop-and-go system, with the assembly operation progressing in stages. The sequence of assembly followed the same logical pattern already established in the track system—trucks, frame, body, roof, and painting—except that the car did not stay in place. Rather, it moved down the tracks on its own wheels until it reached the end of the line as a finished vehicle. This method was called the progressive system, because the car grew progressively more complete as it passed down the track.

The origin of the progressive system, like most matters concerning the car-building trade, are uncertain, because the subject matter was not considered of enough general interest to be well recorded. Even the trade press of the time followed its development with relative indifference. The earliest claims are the recollections of old employees made half a century and more after the event. Such a claim is made for Haskell & Barker in 1889, when the old stall system was supposedly abandoned for the progressive system after a major fire made a new shop necessary.[546] Michigan-Peninsular, the American Car & Foundry Company's Detroit works, began the progressive system in 1902, according to a recollection published forty-seven years later.[547] This later claim, however, is verified by a description published in the November 21, 1908, *Railway Review*. The advantages of the progressive system prompted the parent company, AC&F, to adopt it at its Berwick, Pennsylvania, shop as well. Faster assembly was the chief benefit, but a saving in shop space and a better division of labor were also noted. Competitors of AC&F soon took on the staged-assembly technique. The Pressed Steel Car Company adopted it late in 1908, using ten to thirteen assembly stations, depending on the style of car being manufactured. The frame was flipped in the seventh or eighth position, again depending on the style of car. The brake apparatus was applied more easily with the frame upside down.

The system was finding favor outside the United States at the same time. The Canadian Pacific's large repair shops at Angus, Montreal, built cars in seven stages.[548] The semifinished cars were hooked together in a train by rods 10 feet long. At intervals the train was pulled forward one station so that the progressive work of assembly might proceed. The forward motion was very slow, so that the men could finish up the tag ends of work as the car proceeded to the next station. Angus employed seven hundred men in the freight car setting-up shop. All were directly involved in assembly work save one who packed and mixed up oil and waste for the journal boxes; six who applied the air-brake apparatus; one who looked after the bolt supply; four sweepers to clean out the cars before painting; and thirteen laborers who carried and stocked precut lumber. This crew built twenty-eight boxcars a day.

The assembly sequence was as follows. *Stage 1:* trucks in place, short truss rod posts are bolted to the wooden body bolster, and the truck center pins are put in place. *Stage 2:* the frame is assembled and the draft rig is put on, as are the needle beams and truss rods. *Stage 3:* four men nail down the floor. *Stage 4:* the body and roof frames are assembled. *Stage 5:* twelve men nail on the siding and metal roof. *Stage 6:* the inside lining, roof hardware, and brake pipes are put on. *Stage 7:* the running boards, brake gear, and trimming are put on. The car is then swept out and sent to the paint shop. Painting takes two days. The car is weighed, stenciled, given a final inspection, and sent on, ready for service. The time spent at each station has been reported for Pullman's freight car plant in Chicago.[549] That system used thirteen positions, and the men were given fifteen or twenty minutes to finish the work before a whistle sounded and the train moved forward to the next station. Between 150 and 180 men worked on the line.

There is an urgency about every manufacturing process, because the faster a product can be completed, the cheaper it is to produce. Because freight cars were made in such large numbers, and because they were such a rough-and-ready product, and because unit cost was such an important consideration to both the seller and the buyer, car builders were more than a little concerned about finding ways to produce them quickly. Organization and planning were, then, very important considerations in the car-building business.

Making the assembly of cars easy and convenient became a paramount consideration. Parts must go together with the minimum of fitting and fussing. Because cars were not precision machines, parts might fit together loosely. Tolerances were crude. Holes could be slightly oversize for an easy fit, and nothing was lost in the way of utility. It was equally important to have a good supply of ready-to-go parts near at hand to make the assembler's job as easy as possible. Extra steps took time away from the job, so parts were stacked separately in a logical order, within easy reach. Laborers were on hand to move the heavy parts into place and so save the skilled carpenter's time and strength for the more technical part of the work. Some plants used overhead supply tracks and small pushcars on the main shop floor to distribute materials.[550] Everything was thought through so that the materials need travel the minimum distance from the back of the shop to where the finished car exited the erecting floor. One writer spoke of the assembly process as a continuous one, as parts seemingly flew together and the finished car issued gaily forth, ready for service. Pullman found that two sets of nailers were faster than one. The first team set the nails and a following team drove them home. Trucks were done in a similar fashion. One gang loosely assembled all the parts, then a second gang tightened the bolts and squared the frame.

The concept of premade, easy-to-assemble parts predates the car-building trade. Almost every part of a typical freight car was fully fashioned before the assembly process even began. The axles were forged, turned, and mounted to wheels. The floor sills were cut, mortised, tenoned, and drilled. The siding was planed, joined, and cut to its exact dimensions. And so on with every wooden and iron piece. The assembler needed little more than a hammer and wrench to do his job. Making the individual parts faster was part of the background operation that helped cut the time required to finish a car. In the 1870s the Erie Car Works used a six-spindle drill for making holes in arch bars for truck frames.[551] One man could produce two hundred bars in one ten-hour workday. A wood-boring machine had four heads so that different-size holes could be drilled without constantly changing the bits. The Erie plant was also proud of its nut-and-bolt machine, which could turn out six thousand pieces a day. Other plants used time savers like gang punches to speed up the making of parts.

The actual time required to manufacture a car cannot be calculated from the information available. The time required to mine the ore and fell the trees would have to be taken into account, as would each step in the manufacturing process, and such figures have yet to be found. Source data exists on assembly time, but it must be evident that this represents only a portion of the total time needed to make a car. In the late 1870s, for example, the Michigan Car Company estimated two and a half hours to assemble a freight car.[552] At almost the same time, the Chicago & North Western's Chicago shops produced one car per hour with a three-man team for each car, and it is no exaggeration to say that the work was done in a lively fashion.[553] In 1892 Pullman was producing one new freight car every fifteen minutes.[554] A dozen or more men worked on a single car to maintain this hectic pace of production. Yet surely the actual assembly time must have exceeded fifteen minutes, and the rate of one car for that short interval must be due to multiple assembly teams. In 1884 Pullman had shown what fast work was all about by using twenty-nine gangs of four men each to assemble flatcars for the Vicksburg, Shreveport & Pacific.[555] The layout work was done as usual on a Saturday. Work began in earnest at 7:00 A.M. on Monday. By 9:15 A.M. the first car was finished. The twelfth was completed by 10:40 A.M. and the hundredth by 5:00 P.M. This record was set at a time when Wason, for example, could assemble only twenty cars in a ten-hour day. The speed and agility of Pullman's assemblers seem even more impressive when it is considered that each car contained 855 pieces plus 680 nails.

Fast time on the assembly floor normally brought forth cigars and refreshments for the men. But some car shop superintendents, after reading these reports, must have been convinced that their men were dogging it. The employees at the Central of New Jersey shops in Lambertsville, New Jersey, required about ten hours to assemble a four-wheel coal jimmy.[556] Their languid pace surely brought forth no cigars or refreshments. The local master car builder decided that a car-building contest might prompt faster production, and so he invited two local carpenters in as a test. To the undoubted annoyance of regular shop employees, these inexperienced workmen put a car together in just six hours and twenty-eight minutes.

Repair Shops

The railroad industry found it necessary to operate repair shops for the maintenance of its freight car fleet. Routine repairs were done at yards in outdoor settings called RIP (repair-in-place) tracks. More complex fix-up jobs were done in division shops. Almost every trunk line had a main shop where the most involved repairs were undertaken. Some of these installations were major industrial complexes covering fifty or more acres. In 1889 there were just over seven hundred railroad-operated repair shops in the nation.[557] That number grew to 1,292 in 1899 and then declined slightly to 1,145 in 1909. The number of shops is not a full measure of their productivity; although only 173,595 workers were employed in 1899, by 1909 over 282,000 were on the job.

When it came to mending ailing cars, few places were busier than a large trunk line repair shop. In the early 1880s the New York Central's big West Albany shops repaired sixty thousand to seventy thousand freight cars each year. Other railroads were equally busy with repair work, and not a few undertook the construction of new freight cars as well. Statistics for this activity were poorly re-

ported in the nineteenth century, but a few known examples help explain the productivity of railroad workshops. In 1872–1873 railroad shops were reported as building 21,678 freight cars, or 38 percent of the new construction for that period.[558] In 1900 railroad shops built some 143,000 freight cars, in a rare instance actually outproducing the commercial car builders. The numbers fell off considerably in 1909, when railroad shops finished just 13,900 cars. Rather than attempting a shop-by-shop survey, we shall instead briefly consider three major railroad workshops that were active in car building.

It is natural to start with the largest railroad shop in North America, the Pennsylvania's Altoona shops.[559] This was a company town laid out by the railroad in 1849 at the beginning of its long grade over the Alleghenies in western Pennsylvania. This terminal for helper engines was developed into a systemwide repair facility that grew to over two hundred acres and employed twelve thousand men at its peak. Operations began in 1852 with a modest cluster of brick buildings not much larger than might be found elsewhere in a normal division yard. The demand for more repairs and new cars prompted a separation of car and locomotive work. The engine shops remained in the old facility while an entirely new car shop complex was built in 1869–1870 (Fig. 1.74). A sixty-two-acre plot about a mile east of the old shops was laid out for a dozen buildings, the largest of which was a doughtnut-shaped freight car shop. This forty-sided polygram measured 433 feet in diameter and had a hole in the center 100 feet across. There was one through track and thirty-eight working tracks with a capacity for 114 cars. The turntable in the center of the building measured 65 feet 4 inches. The normal staff for the freight car shop was 250 men. The capacity of the shop was five hundred new cars and two thousand repairs per month, but these numbers naturally fluctuated according to traffic levels. Because of the short tracks, nothing approaching an assembly line was possible, and the trestle or stall operation was used. Hence the productivity of the roundhouse was impressive for what is inherently an inefficient layout.

In addition to the roundhouse, several other parts of the car works complex were directly involved in the fabrication of new freight equipment. The woodworking shop, foundry, and forge turned out the basic parts. A kiln or wood-drying house 30 feet by 65 feet had three tracks for lumber-rack cars. Steam heat kept the interior a toasty 170 degrees. In this atmosphere white pine was dried in four days, poplar in three, and yellow pine in six. The kiln processed 120,000 board-feet of lumber per month. Separate paint shops were maintained for freight and passenger cars, because the process of finishing was so different. Passenger cars were done up in a lengthy, multistage process

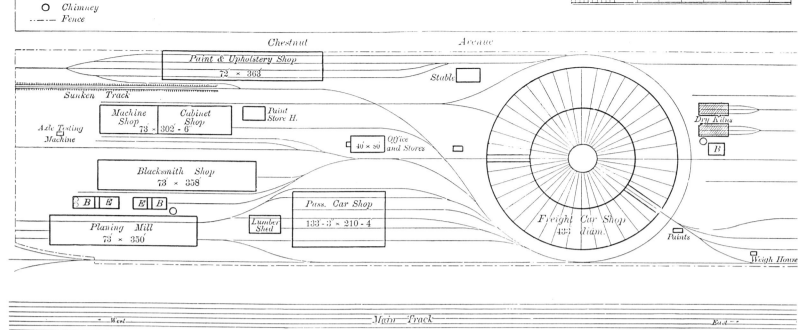

E   Steam Engine
B   Boilers
O   Chimney
----- Fence

Scale of Feet

FIGURE 1.74

Ground plan of the Altoona shops, built by the Pennsylvania Railroad, 1869–1870. A second freight car shop was erected some time after this diagram was published. (*Railway Review*, Aug. 20, 1884)

to produce a fine, pianolike finish, while freight cars were given a quick in-and-out treatment more on the order of common house painting. Even so, the freight car paint shop at Altoona was a sizable structure measuring 108 by 393 feet.

Altoona was largely a planned workshop area that was enlarged from time to time, but always in a thoughtful manner reflecting a long-term plan. Most other railroad shop complexes grew by fits and starts with no thought to the long-term consequences of the latest addition. By 1890 the Baltimore & Ohio's Mt. Clare shops in Baltimore were such a patchwork of small, antiquated buildings that the railroad's management was said to favor a Chinese-style spirit of contentment with past achievements.[560] Another jumbled collection of elderly shop buildings, though less ancient than those standing in Baltimore, could be found in Louisville, Kentucky. Here the Louisville & Nashville had its principal shops that developed in a haphazard manner around the nucleus of the Kentucky Locomotive Works.[561] The KLW opened around 1852 as a private contract firm intent on producing locomotives and cars for railroads of the area. The firm's substantial physical plant did not assure success, however, and it expired within a few years. The shops were leased and then purchased in 1858 by the L&N.

As the L&N system expanded, so too did the need for larger, more commodious repair shops. A major betterment program was undertaken between 1868 and 1872. Toward the end of this renovation scheme, a reporter from the *Railroad Gazette* was moved to say that the shops had lost their careless and dilapidated look. As revised, the shops occupied a plot three city blocks wide and

six blocks long, which was covered by some twenty major and a dozen or so minor structures. A new car shop was built on a twelve-acre plot to the west of the original grounds. During the early 1880s these shops were reported to build two and repair twenty freight cars a day. As the old century came to a close, the L&N management realized that the old patch-and-fix method of expansion had come to an end. The city of Louisville had closed in around the 10th Street shops, eliminating any economical way to enlarge the facility, and so it was abandoned for a new site to the south of the city. When the South Louisville shops opened in 1905, they were among the finest in the nation. They were capable of producing not two but forty-two freight cars a day. Employment there eventually reached forty-two hundred. Merger madness during the 1970s made South Louisville redundant, and so this once-busy repair and construction terminal was closed in 1988.

In most eastern cities a major railroad workshop was just another factory, so many low-rise brick buildings punctuated by a series of smoking chimneys. But in the Far West a railroad workshop was often a major industrial complex. The Central Pacific's Sacramento shops were considered the largest industrial establishment in the state for nearly a generation after their opening in 1867 (Fig. 1.75).[562] The Central Pacific might well have preferred to maintain a more modest facility, but its remoteness from the industrial East made self-sufficiency a necessity. All the basics needed for everyday operations, from bolts and axles to wheels and journal bearings, might take weeks to rumble across the continent, and so those responsible for keeping the trains rolling realized that a

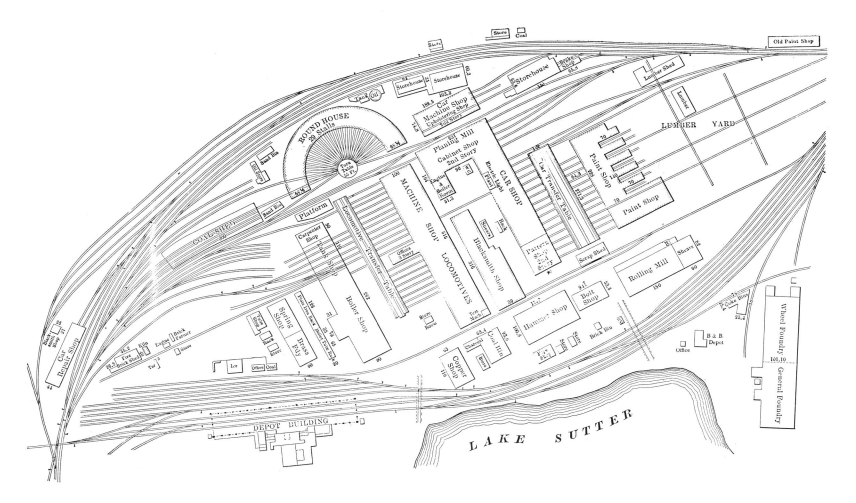

FIGURE 1.75

The Central Pacific's Sacramento repair shop was one of the largest industrial sites in the Far West. It had been much enlarged, as shown in this ground plan. (Copied from *Master Mechanics Magazine*, March 1892)

well-equipped shop was fundamental to continued operations. The main shop building was a sizable two-story brick structure, measuring 99 by 238 feet, that featured a truss roof so that the work floors were clear of posts. By 1872 investment in the facility had grown to $1.2 million plus another $1 million invested in supplies. The production of new freight cars was generally a little over five hundred per year during most of the nineteenth century. Such volume was not particularly impressive, however; the facility was used more as a parts factory than a car works. In 1892 Sacramento was reported to be makng 40 axles and 122 car wheels a day. An unquantified number of bolts, tie rods, and springs were also produced. The foundry melted 60 tons of iron for castings each working day.

By 1880 the Sacramento shops were described as occupying fifteen acres and employing six hundred men. Employment had been as high as fifteen hundred when times were busy, as indeed they were once again in 1888 when twenty-seven hundred were engaged. New buildings were under construction, meaning that several hundred of the new job holders were obliged to work out of doors. Never had the facility been so crowded or rushed. The success of C. P. Huntington in expanding the

Central and Southern Pacific rail empires had only added to the backlog of work at Sacramento. More expansion followed to improve productivity, so that by 1892 a thirty-three-acre area was occupied. Some of this was low, swampy land, a perfect breeding ground for mosquitoes. The borders of the grounds were planted with thirty-three hundred eucalyptus trees, in the belief that these odiferous plants would make the shop area a fever-free zone. It is likely that the railroad-sponsored hospital located a few thousand feet from the main gate of the shops was little more effective than the eucalyptus trees in combating malaria. Even so, men and machines managed to carry on through the years, and freight car building continued at Sacramento until well into the 1950s. These shops were active as a locomotive repair center until 1992.

This summary of the car-building industry ends our general discussion of freight traffic and railroad operations. The remaining chapters of this volume are devoted to the design and construction of individual types of freight cars and their associated components.

2   THE PIONEERS

AMERICAN FREIGHT CARS BEFORE 1870

Few subjects of interest to historians of technology are as poorly documented as the earliest U.S. freight cars. Original papers, drawings, and contemporary documents on this subject are almost nonexistent. Early railroad annual reports rarely offer more than the number of such cars in service. Newspapers of the day would occasionally describe a new passenger coach, particularly if it was luxurious and handsomely outfitted, but never a word or a line of type was offered on the lowly freight car. The occasional drawing published in the antebellum period has come down as almost the only concrete engineering record available on these pioneers. It is true that freight traffic was not remarkably important to American railroads during their very first years, but that traffic grew quickly, especially when measured in economic terms; it soon overshadowed passenger revenues. And yet the compilers, publishers, and archivists of railroad history have devoted their attention to the more glamorous subjects. The locomotive, passenger car, and depot are comparatively well recorded as a result.

Despite a paucity of information, it is possible to assemble at least a partial picture of the first generation of American railway freighting vehicles. Many areas are cloudy and incomplete to be sure, but several trends can be traced. As with locomotives, British models were soon revised for local operating conditions. The modified British four-wheel plan was quickly abandoned, with the exception of the coal jimmy, for the more stable double-truck car. Covered cars for merchandise were found necessary for the North American climate, while the British were long content with canvas-covered open cars because of their comparatively mild weather. American cars grew in size and capacity, and by the 1840s we were operating what Europeans considered jumbo-size rolling stock. U.S. mechanics remained loyal to all-wooden construction throughout the nineteenth century while railways elsewhere in the world were beginning to adopt iron frames. More attention was paid to the special needs of shippers; not

everything could be conveniently or economically shipped in a boxcar, and so at a surprisingly early date, cattle, container, and covered hopper cars were introduced.

It would be difficult for many present-day railroad men to believe that these supposedly modern ideas were not only suggested but actually tested and introduced into everyday service during the 1830s. The story of America's pioneer freight cars ends in the last years of the 1860s, when the interchange system was adopted on a national basis. Freight service lost its parochial innocence; freight cars now rattled across the land, far from home. They were required to develop a new toughness to endure thousands of miles of unchaperoned travel. Individualism gave way to standardization, so that off-the-home-road repairs might be more practical. By 1870 the pioneer period was over.

The Earliest Examples

Freight cars first appeared on remote industrial railways far removed from public notice. During the mid-1820s the average citizen knew nothing about railways in general, and few would have known that such a system of land conveyances was beginning to appear in the eastern coal fields and quarry sites. Yet it was on such short and primitive tramways that the design of the first American freight cars was worked out. The first of these roads to carry enough traffic to qualify as a commercial operation was the Granite Railway of Quincy, Massachusetts, which opened in October 1826.[1] The main line ran from the granite quarries about 2 miles to a nearby river, where the stone was placed aboard ships for transhipment. The 5-foot-gauge railway and its horse-drawn rolling stock were designed by Gridley Bryant (1798–1867), apparently without reference to existing British prototypes. If Bryant sought inspiration, he appears to have drawn upon the ancient Egyptians, for his selection of massive stone rails and ponderous timber-framed cars would surely have pleased Imhotep, architect to the Pharaoh. The

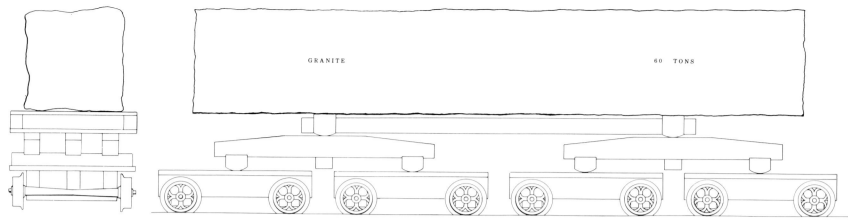

GRANITE   60 TONS

FIGURE 2.1

first series of cars were carried by four high wagon wheels. The stones were carried below the axles on a platform raised or lowered by a hand-powered winch fastened to a wooden truss frame that stood above the wheels.

A replica of one of these stone cars, reportedly based on original drawings, was built at the time of the line's centennial. A lithograph published in 1826, however, shows three four-wheel gondola-like cars loaded with stone blocks being pulled by a single horse. The driver sits in the lead car. A wooden track is shown, which together with the cars casts some doubt on the accuracy of the print despite its contemporary origins.[2] And yet the simple cars shown suggest that Bryant may have reconsidered his original design for the elaborate winch cars and devised a simpler and cheaper plan. In 1828, according to Bryant's own statement, he began building very simple four-wheel cars with small cast-iron wheels about 18 inches in diameter. The frames were heavy oak timbers. The wheels were mounted loose on axles. It was soon found that very large stones weighing around 10 tons could be carried on a simple timber bridge spanning two of the four-wheel carriages. The very small wheels kept the center of gravity low, and because speeds were limited to about 4 miles per hour, very large loads could be carried over the line. In about 1834 it was necessary to ship several very large stones destined to be shaped into columns for the new Boston Court House.[3] These stones weighed up to 64 tons each, but Bryant met the challenge of moving this unprecedented load by bridging four of the stone wagons into a single giant flatcar (Fig. 2.1). The movement of such enormous weight demonstrated the railway's almost unlimited capacity to move heavy loads. Bryant's equipment served well, and most, including the curious winch cars, remained in service until the line was abandoned around 1866.

Some 200 to 300 miles south and west of Quincy lay America's pioneer coal fields, where limited mining had been taking place since the earliest days of the Republic. These operations would remain small until a better way was found to carry coal from the pithead to market. A combination of railway, canal, and river was the initial solution to the Lehigh Coal & Navigation Company's transport problem. Soon coal went to market from an open-pit mine at Mauch Chunk, Pennsylvania, located on the Lehigh River only 75 miles from Philadelphia.[4] At first, wagons carried coal to the river's bank, a system found deficient in both capacity and economy. In the spring of 1827 a 9-mile-long 42-inch-gauge wooden railway was opened. Mules hauled empty cars upgrade over tracks that followed the original wagon road to the mines.

Short trains of fourteen loaded cars, manned by one or two brakemen, rolled downgrade to the loading docks on the Lehigh River. A rope running the length of the train connected the brake levers from car to car. The cars weighed 1,600 pounds each and carried about 3,000 pounds of coal. The down trip took only thirty minutes and must have provided a thrilling, if dusty, descent for the brakemen. The return trip of empty cars, powered by mules, took three hours. The mules quite willingly rode in special open rail cars to the bottom of the railway after completion of the ascending journey. By 1831 the switchback gravity line had 115 mules, 21 mule cars, and 308 coal cars. It delivered 42,743 tons of coal from the mountaintop mine to the riverside loading shed during the same year.

In around 1840 the Austrian engineer F. A. Ritter Von Gerstner visited the property. He noted that the cars were small, lightly built vehicles weighing from 1,600 to 1,700 pounds each. The capacity is given as 2 tons, which appears too great and conflicts with the figure given in Lehigh Coal & Navigation reports as noted above. Either Von Gerstner's estimate was simply in error or the managers of the switchback knew something about car design that escaped succeeding generations. Some twenty years after Von Gerstner's visit, a German named A. Bendel inspected the Mauch Chunk Railway.[5] The line had been remodeled in 1844 with a second track and steam-powered incline planes for the return of empties, but the loaded cars were still sent downhill by

FIGURE 2.2

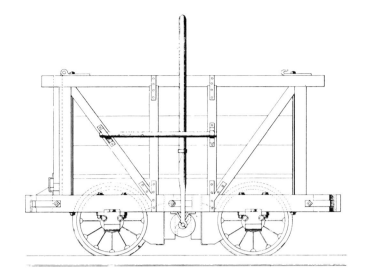
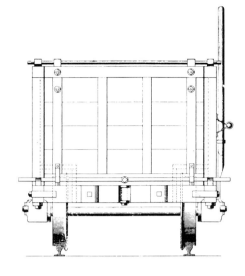

This Mauch Chunk & Summit Hill four-wheel coal jimmy is from a drawing in a German work on American railroads published in 1862. The data was gathered three years earlier and most likely represents a design of about 1850. (A. Bendel, *Aufsatze Eisenbahnwesen in Nord-Amerika* [Berlin, 1862])

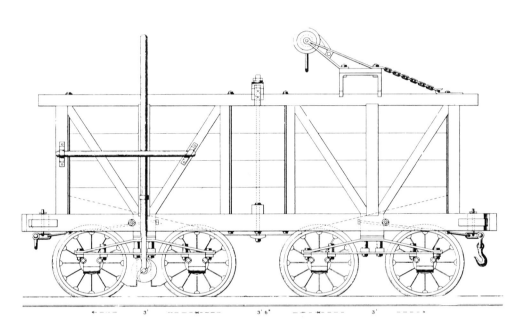
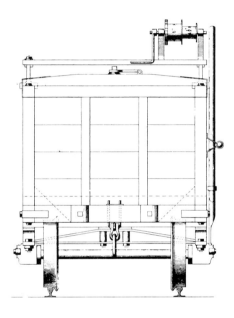

FIGURE 2.3

The Mauch Chunk & Summit Hill eight-wheel hopper car shown here carried 4 long tons of coal. The design, like that of the four-wheel car pictured in Figure 2.2, dates from around 1850. (Bendel)

gravity. Both four- and eight-wheel cars were employed at the time of Bendel's visit (1859). The four-wheel cars then in use may have been an advanced design over those introduced in 1827, but there is no hard evidence to prove or disprove this speculation. The car depicted in Bendel's drawing is surely light and small (Fig. 2.2). It would hold about 90 cubic feet of hard coal. Weight was given as about 1 long ton (2,240 lbs.), capacity as about 2 long tons. One end of the car was hinged so that it could be emptied by tilt dumping. The spoked 24-inch-diameter wheels protruded into the body in boxes. Note also the wooden brake shoes and large brake lever. The body of the car measured 7 feet long by 5 feet 3 inches wide by 40 inches deep. The wheelbase was 4 feet.

The eight-wheel coal hopper was a larger version of the four-wheel car; the wheels and braking system were identical (Fig. 2.3). The body was 11 feet 6 inches long and had the same width and height as its four-wheel counterpart. The wheelbase of

each truck was 3 feet, while the overall wheelbase was 9 feet 6 inches. Weight and capacity were exactly double those of the four-wheel jimmy. The small winch and framework mounted to the top of the car were for the continuous braking system. A light wire rope connected the brake levers on each car. The cable was wound up and released by a brakeman seated on the plank that spanned the open hopper. It would appear that the plank, winch, and light iron framework were a unit that could be mounted or unmounted as required so that the brakemen could always be on the last car of the train. The chain fastening suggests this option. Just why the drum has two diameters is not clear.

The Delaware & Hudson Canal Company's mine railway at Honesdale, Pennsylvania, in the northeastern part of the state, was similar in many respects to the Mauch Chunk line. It was a narrow-gauge (51 inches, in this instance) wooden tramway built as a portage between hillside mines and a

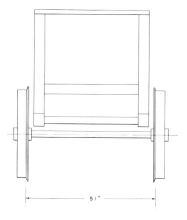

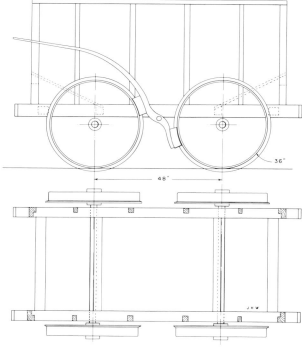

FIGURE 2.4

This Delaware & Hudson coal jimmy of about 1830 is representative of the first hopper cars used in the United States. (Traced from a drawing at the John B. Jervis Library, Rome, N.Y.)

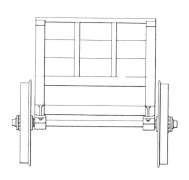

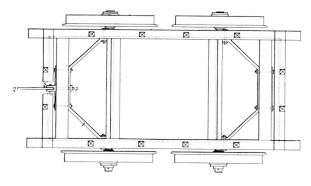

FIGURE 2.5

The Danville & Pottsville in northeastern Pennsylvania acquired cars like this one in the 1830s. It is very similar to the D&H car shown in Figure 2.4. This car was slightly wider because it was standard-gauge. The body was 8 feet long and 50 inches wide. The wheelbase was 44 inches; the wheels, 34 inches in diameter, were loose on the axles on one side. A wooden board connected the brake levers on several cars. (M. Chevalier, _Histoire et Description des Voies de Communication aux Etats Unis_ [Paris, 1840–1841])

canal/riverboat operation intended to carry coal to a big-city market. The D&H line opened in 1829. Some of the first cars were purchased in that year from the West Point Foundry in New York City.[6] A drawing for a D&H car of this period is held in the Jervis Library (Rome, N.Y.); a modern tracing is reproduced here as Figure 2.4. John B. Jervis was chief engineer of the line. William J. McAlpin (1812–1890) executed the original drawing. The wheelbase was 4 feet and the cast-iron spoked wheels measured 3 feet in diameter. One wheel was mounted to the axle, while the other turned

loose so that the car could more easily negotiate curves. This scheme was devised by Horatio Allen, a junior engineer whom Jervis sent to England in 1828 to study the state of technology in the homeland of railways.[7] These same cars were briefly described in the 1832 edition of Nicholas Wood's _Treatise on Rail-Roads_.[8] The car's weight was given as 1 ton with a capacity of 2.5 tons; again, as with the Mauch Chunk cars, the weight-to-capacity ratio seems extremely optimistic. The axle diameter was given as 2½ inches. Von Gerstner visited the line in about 1840 and found 350 cars in service.[9] All were four-wheelers except for five double-truck cars. Drawings for a similar four-wheel coal car used on the Danville & Pottsville Railroad are shown in Figures 2.5 and 2.6. In later years the D&H adopted a 5-ton-capacity wooden eight-wheel car (Fig. 2.7). Photographs of these cars, with their rudimentary wood-beam trucks, show lettering on the frames that indicates a weight of 5,500 pounds.[10]

These very first American cars taught few profound lessons, nor did they establish many lasting design patterns. But they did help establish one important fact—that railways were practical carriers for heavy goods. These slow-moving animal- and gravity-powered tramways demonstrated, in the most concrete way, that railways were more than visionary; they were a practical and economical mode of transportation. The curious and primitive vehicles that rattled over this long-forgotten trio of lines were the ancient kin of the generation of cars that have come down to modern times.

Innovations

Regrettably, the example established by these pioneers went unrecognized by several of the more prominent engineers involved in creating the first public railways in the United States. They belonged to a school that condemned the efforts of Jervis and Bryant as an unscientific backwater, best forgotten and left buried in the trash heap of the industrial past. What was needed were radical, bold, and scientific schemes.

FRICTION WHEELS

Freight cars, for example, could not be just a simple container with plain bearings. Ross Winans (1796–1877), a self-trained engineer associated with the Baltimore & Ohio, began experiments as early as 1826.[11] A small test car built in England about two years later demonstrated his friction-wheel idea, which promised to reduce friction substantially. Winans claimed that a horse could pull a loaded car weighing 8.75 tons on a level track with conventional bearings; however, with his friction-wheel bearing the horse could move 56 tons. This same impressive advantage was demonstrated by the inventor upon his return to Baltimore. The man was a genius; he revolu-

FIGURE 2.6

Details of iron hopper doors and latch for Danville & Pottsville hopper car. (Chevalier)

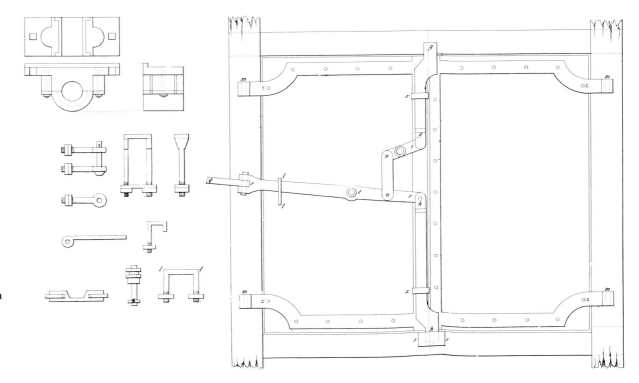

FIGURE 2.7

This Delaware & Hudson wooden hopper car, 5 tons capacity, displaced the four-wheel cars shown in Figure 2.4. The photo dates from 1866, the design likely from a decade or two earlier. (Smithsonian Chaney Neg. 21628)

THE PIONEERS

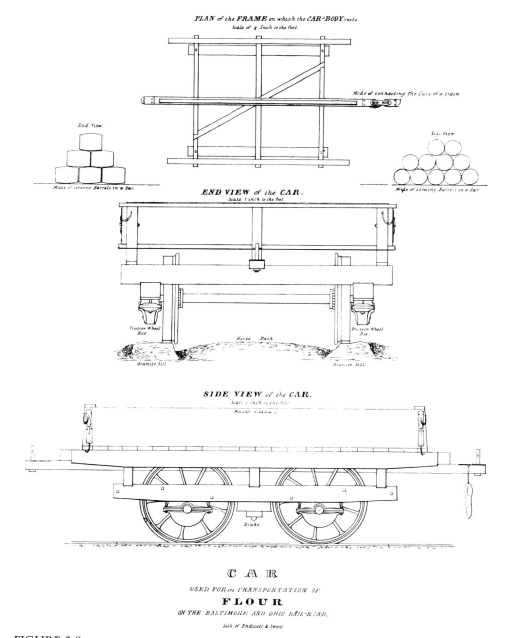

PLAN of the FRAME on which the CAR-BODY rests.
Scale of ¼ inch to the foot.

Mode of connecting the Cars of a train

End View.

Side View.

Mode of stowing Barrels in a Car.

Mode of stowing Barrels in a Car

END VIEW of the CAR.
Scale 1 inch to the foot.

Friction Wheel Box

Friction Wheel Box

Horse Path

Granite Sill

Granite Sill

SIDE VIEW of the CAR.
Scale 1 inch to the foot.

Brake

C A R

USED FOR the TRANSPORTATION OF

FLOUR

ON THE BALTIMORE AND OHIO RAIL-ROAD.

Lith of Endicott & Swett

FIGURE 2.8

Baltimore & Ohio gondola or flour car fitted with Winans's ill-fated friction-wheel bearings. (From a drawing in B&O annual report, 1831)

tionized rail transport beyond the claims of even its most optimistic advocate. Very small locomotives would now be practical for moving large payloads, and because the locomotives would be so light, trestle and track structures could be constructed more economically.

The B&O hastily outfitted several construction cars with Winans's friction wheels to demonstrate the merits of the plan. William Howard (1793–1834), another engineer associated with the B&O, devised his own variation on Winans's idea and was awarded a U.S. patent on November 2, 1828. Winans himself did not obtain a patent until October 11, 1828, although he claims to have submitted a model three years earlier. The first order of B&O freight cars, completed sometime in 1830, had Winans's friction wheels (Fig. 2.8), as did some of the B&O's first passenger cars. The author of one of the first railroad books published in the United States, Thomas Earle, approved of Winans's and

Howard's efforts but claimed that he had perfected an even better design for a frictionless bearing.[12] Earle was optimistic in his belief that a practical and strong freight car could be developed that would weigh only one-quarter of its rated capacity—a hope that in over 150 years has yet to be realized.

The illusions and dreams expounded by these men at first captured the attention of some leading railroad promoters, to say nothing of the daily press. And yet it has been proved, decade by decade, that top railroad officials—who are generally lawyers or financiers—and newspapermen are not the best judges of engineering matters. The misguided friction-wheel adventure came to a sudden end when the bearings on the B&O test cars exhibited a propensity for rapid wear. It was also discovered that because the axle was in effect floating, the wheel sets were free to wander back and forth, making the vehicle unstable even at fairly low speeds. Just how to apply brakes was yet another problem for cars with friction-wheel bearings.

BRITISH INFLUENCES

While the friction-wheel fiasco was under way, less imaginative men were investigating established railway car design. Actually, there was rather little to study. The existing American prototypes were few in number and elementary in form. Textbooks on the subject were just about nonexistent and tended to be general works with few if any specifics on car design. Those that did treat the subject tended to emphasize patents or other untested ideas, as was true with Earle's book. However, the first major public railway had opened in England in 1830, and it was to this source that our more practical car designers turned for inspiration. The Liverpool & Manchester Railway and its engineering staff headed by George and Robert Stephenson were seen by many as the best source of advice on freight car design. They were a father-and-son team active in railway engineering for nearly a decade and were without peer in reputation and accomplishment. Just about every major engineer associated with early U.S. railroad construction, from Horatio Allen to George Washington Whistler, consulted with the Stephensons in England.[13]

Sketches and notes on the smallest details of British practice were prepared by the information seekers from the New World. Some of it was studied and rejected, but much was brought to North America and duplicated. The tracks of the Boston & Lowell, the Allegheny Portage, and a few other lines were faithful replicas of the Liverpool & Manchester. George W. Whistler fabricated several dozen locomotives at Lowell, Massachusetts, that were almost exact copies of Stephenson's Planet-class engines. Whistler, or one of his associates,

THE AMERICAN RAILROAD FREIGHT CAR

obtained a drawing for a goods wagon from the Stephensons while the Boston & Lowell was under construction. The original wash drawing, dated 1832, was beautifully executed and offered a very detailed picture of what is presumed to be a Liverpool & Manchester freight car.[14] Small cars of this general description were used on the B&L to carry cotton.[15] An old resident of Lowell describes them as having had canvas covers tied over the bales.

Another drawing for a nearly identical car, which is specifically identified as being for the Liverpool & Manchester, was found in an ancient scrapbook dealing with the Petersburg Railroad of Virginia.[16] The book, which contains original engineering tracings and newspaper clippings from the 1830s, is believed to have been assembled by Moncure Robinson (1802–1891), the principal engineer involved in the construction of the Petersburg line and one of the many American technicians who visited Britain at the opening of the railway era. It is possible that Robinson made the tracing during a visit to England, for it shows evidence of having been executed in haste; many of the details were freehand without the aid of a compass or straight edge. The original is too crumpled and folded to reproduce well and so was redrawn for inclusion here as Figure 2.9. The elegant proportions of the wooden parts, together with the well-shaped ironwork, suggest a vehicle devised by a carriage or coach maker. The Petersburg line received a sample car and ironwork for fifteen "transportation cars" from Liverpool on August 31, 1832.[17] While there is no absolute evidence, it would seem logical that the sample car was a standard L&M vehicle as pictured in Figure 2.9.

The Petersburg Railroad was not alone in importing cars and car parts from England. At least one other southern line also went overseas for its initial rolling stock. The Pontchartrain Railroad, running out of New Orleans, ordered four passenger and eight freight cars from an undisclosed supplier in Liverpool in November 1830.[18] During the following year, according to the published annual reports of the line, axles, wheels, and other car hardware were secured from England and Baltimore because local suppliers were unequal to the task. The cars ordered in 1830 finally landed in New Orleans in June 1832.

From the few surviving drawings of the earliest American freight cars, the Liverpool & Manchester's low-sided four-wheel gondola set something of a pattern for U.S. car builders. The secondary frame used to elevate the floor above the wheels was adopted in each instance (Figs. 2.8, 2.9, 2.10). The Baltimore & Ohio flour car shown in Figure 2.8 is the oldest example. That drawing, published in the B&O annual report for 1831, shows a design credited to Benjamin Latrobe, the architect/engineer employed by the B&O for much of his career.[19] These cars, apparently first built in 1830, are very similar to their British counterparts. The timbers are heavier, while the ironwork is less elaborate and more sparingly employed. Note that the sides can be removed and that Winans's friction wheels were applied. Just how the brakes worked remains a mystery. The car was very

FIGURE 2.9

This Liverpool & Manchester gondola of 1830 was the inspiration for the first American freight cars. (Traced from a drawing presumably made by Moncure Robinson)

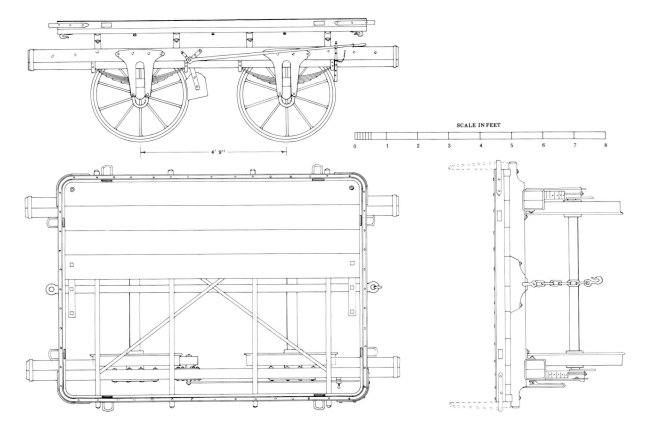

THE PIONEERS

small—really not much larger than a handcar. The body measured 9 feet 6 inches long by 7 feet 8 inches wide. The wheelbase was only 44 inches, and the wheels themselves were 30 inches in diameter. A wooden perch pole or center sill has been substituted for the forged-iron drawbar of the British car. This would be cheaper in first cost, but it was of course far weaker than a metal drawbar. The lack of springs was another ill-advised economy that the B&O soon came to recognize. In its annual report for 1833 the company stated that springs were essential for both passenger and merchandise cars to preserve the equipment as well as the track.

Meanwhile, the Petersburg Railroad made an attempt to Americanize the Liverpool & Manchester goods wagon. The plan was similar to the B&O's effort, resulting in a much coarser and cheaper vehicle (Fig. 2.10). Here again heavier timbers and less ironwork are evident. The wheelbase was cut back to 4 feet. The body measured 10 feet long by 6 feet 3 inches wide. The side sills were made from 3½- by 6-inch timbers. Springs were eliminated with the notion that the track would provide a perfect guideway—a notion soon upset by the realities of American construction standards. Even a well-made and well-maintained track could not avoid the depressions and voids encountered at every rail joint and switch. These imperfections could damage an unsprung car and its contents, even at slow speeds. A separate drawing, also copied from the Petersburg scrapbook, offers excellent detail on the bearings and wheels of this low-sided gondola (Fig. 2.11). The most interesting detail is the half bearing and oil cellar of the journal box, the earliest documentation so far uncovered for what was to become the standard arrangement for railroad car bearings. The journal itself is very small, measuring 1¾ inches in diameter by 4½ inches long. The cast-iron wheels are also of a very light pattern. The spokes are ½ inch thick, the tread 1 inch thick. The hub was split to compensate for the uneven cooling of the casting, which would have caused cracking at some point, until better foundry practices were perfected that permitted the casting of spoked wheels with solid hubs. Wrought-iron rings ¾ inch thick by 1 inch wide were fastened on both sides of the hub to hold it together.

These drawings date from about 1832. One hundred cars were in service when the line's 1835 annual report was published. Several years later Von Gerstner visited the property and found two hundred freight cars.[20] He gave their weight as 4,500 pounds and their capacity as 6,000 pounds. The first cars, according to Von Gerstner, were fitted with wrought-iron tired wheels imported from England, but later cars were made with cheaper cast-iron wheels. Two wheels mounted on an axle cost $60.

The two car designs just described are excellent illustrations of both the need and the propensity of American mechanics to copy and modify British rolling stock designs. The British model was found to be too refined and costly. At the time, Britain was the workshop of the world. It was the major

FIGURE 2.10

The Petersburg gondola of about 1832 shown here was inspired by the L&M car shown in Figure 2.9. This car weighed 4,500 pounds and carried 6,000 pounds. (Traced from a drawing presumably made by Moncure Robinson)

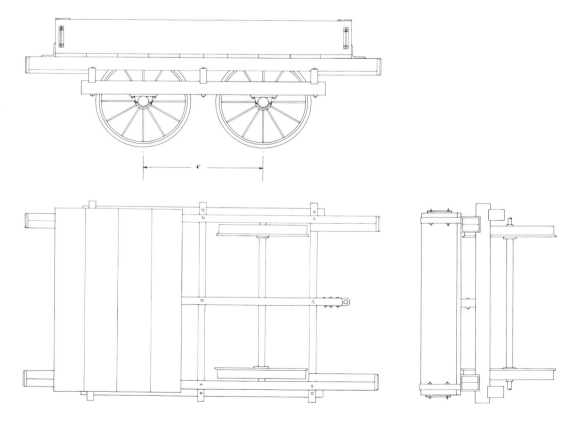

THE AMERICAN RAILROAD FREIGHT CAR

FIGURE 2.11

Journal, wheel, and axle details
for the Petersburg car of about
1832. (Traced from a drawing
presumably made by Moncure
Robinson)

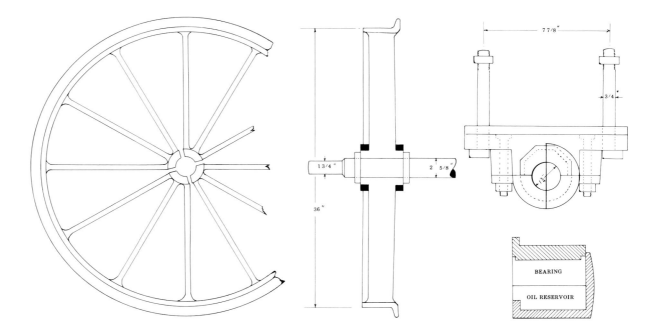

producer of iron and was home to thousands of ar-
tisans skilled in the shaping and forging of iron. Its
timber resources were, conversely, very limited.
The situation in the United States could hardly
have been more different. We produced little iron
and had relatively few metalworkers, but we had
vast forests. Fine timber was cheap and abundant,
while iron was a costly imported material, and so
domestic mechanics tended to substitute wood
whenever possible. The result was a heavier, less
durable, and less refined vehicle, but one quite to
the liking of our pioneer railwaymen, who seemed
to prefer mass to elegance and found rough-and-
ready structures more suitable to the needs of a society just entering the industrial age.

Closed Cars

The open, four-wheel gondola was the standard
freight car during the first years of American railroading. It continued to hold this position in England until recent times. It was the most elementary
and adaptable style of freight vehicle imaginable
and was capable of transporting just about every
class of goods presented for carriage. Bulk materials could simply be dumped in. Whiskey, oil, or
flour packaged in barrels were readily handled.
More fragile goods could be carried so long as a
canvas cover or tarpaulin was employed.

This system worked almost flawlessly in England because of its comparatively mild climate and
its coke-fired locomotives. But conditions were
not the same in North America. Winters were far
harsher, and wood-burning locomotives threw out
showers of glowing embers and sparks. An open
car of merchandise provided numerous nooks and
crannies for sparks to lodge and burn. Tarpaulins
were equally susceptible to incineration. In 1834
the Petersburg Railroad suffered fire losses of

$5,000 because of burning cars.[21] The line blamed
these rolling conflagrations on combustible cloth
covers and stated that a new, cheaper cover had
been adopted to protect cotton shipments from
sparks. Regrettably, the precise nature of the new
covers was not explained. The Allegheny Portage
faced a similar problem and assigned one man
with water buckets and brooms to every two
cars.[22] Such solutions were obviously costly and
short-term.

The Baltimore & Ohio did not comment on the
fire problem but concluded in its published report
for 1832 that valuable merchandise suffered "dep-
redation and injury" when carried in open cars,
even when protected by canvas. The report went
on to say that freight cars must be designed with a
view to the different commodities conveyed, and
that seventy "house" cars or boxcars had been
placed in service. We have no illustrations of these
vehicles, but documentation exists for other first-
generation covered cars. A car of this type is pictured in a lithograph of the Belmont Incline on the
Philadelphia & Columbia Railroad printed by J. C.
Wild in 1838.[23] Other P&C boxcars are shown
in a watercolor rendering of the same scene by
David J. Kennedy, now in the collections of the
Historical Society of Pennsylvania. These four-
wheel cars had highly pitched roofs and outside
body framing.

More complete information exists in the form of
a drawing for a covered car built for the Mohawk &
Hudson Railroad in about 1833 (Fig. 2.12). This
railroad, like all early New York lines paralleling
the state-sponsored Erie Canal, was discouraged
from carrying freight by a high tonnage tax that
was suspended only in winter when the waterway
was closed because of freezing. Thus freight operations were under way in the worst-weather
months, and so from the very beginning the M&H

FIGURE 2.12

This Mohawk & Hudson covered gondola of about 1833 is the ancestor of the modern boxcar. The drawing omits the spoked wheels and leaf springs. (Smithsonian Neg. 83-1628)

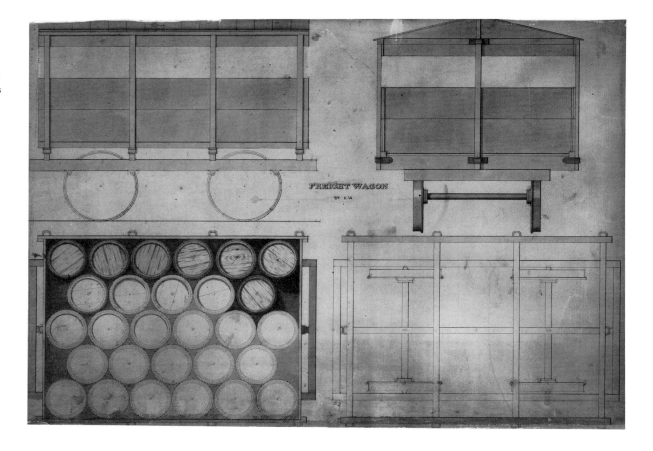

was dissuaded from using open cars. Only a substantial roof and body could offer effective shelter against Snow Belt winters.

The frame is similar in plan to the Liverpool & Manchester gondola pattern except for the notable lack of cross-bracing. Iron pockets fastened below the lower side sill indicate leaf springs. The 36-inch-diameter wheels were most likely spoked. The car measured 12 feet long by 8 feet 6 inches wide and was about 9 feet high from the top of the rail to the peak of the roof. The wheelbase was 6 feet. Just how the car was loaded is a matter of speculation; it is possible that the end boards could be lifted out. It is also obvious that the body could be lifted off, turning the vehicle into a flatcar. The omission of the upper body board suggests that provision was thus made to carry cattle. Asa Whitney (1791–1874), machinist and superintendent of the M&H, designed these cars and stated that 150 were constructed in the railroad's own workshops in 1833 and 1834.[24] Von Gerstner gave their weight as 4,100 pounds and their capacity as 6,000 pounds.[25]

Other railroads were not slow to adopt covered freight cars. The advantages were obvious and the extra costs modest. The Western Railroad in Massachusetts found that a closed car cost $280, or only $30 more than an open car.[26] Its eastern connection, the Boston & Worcester, noted a somewhat greater difference and stated that a closed freight car cost $285, or $55 more than an open car. Considerable detail on a B&W closed car of this period is given in the specifications reproduced in Table 2.1. In June 1837 the Boston & Worcester had forty-six closed and twenty-one open cars operating in the general merchandise trade.[27]

The early acceptance of closed cars in the northeastern states is not surprising, but their adoption in the Deep South illustrates that the idea was embraced even in regions that might have been tempted to remain loyal to the British open gondola. The South Carolina Railroad, for example, noted in its published report for 1834 that 99 of its 149 freight cars were of the "high covered" pattern. In its 1840 annual report the same line noted that only six of its cars were open. It would appear that by 1840 the open British-style car was regarded as obsolete by American railroad managers.

The Introduction of Eight-Wheelers

During the same years that the shift from open to closed bodies was under way, an even more fundamental transformation in the general arrangement of American railroad freight cars was taking place. The double-truck or eight-wheel car displaced the single-truck four-wheel car. Here again we find an early break with British and European railway practice, which remained loyal to four-wheel vehicles well into the twentieth century. Yet by 1840 most U.S. lines had come to adopt the double-truck car because it was so much better suited to our special operating problems. American lines were built cheap and fast. Financial re-

TABLE 2.1 Specifications for Boston & Worcester Covered Car

Our Mdze. car wheels are 33 inches diameter & average 315 pounds each—Baltimore Patent Cast Iron. The axles are of wrought iron, weigh 164 lbs.

Journals are 2 inches diameter—axle in the wheel 3⅜, swell 3½, tapering down to 3⅛—made at Hanover, Mass. by Loring & Sprague, and we have never broken one—

Cost of a Mdze. Covered Car

Wood Work		$75
2 Rough Axles, 328 lbs. @ 7½ cts.	24.60	
Turning & Steeling Journals $9—	18.	
4 Wheels 1260 lbs. ¢5 & expenses	66.40	
Fitting Wheels to Axles	28.	
		137
4 Tallow Boxes 76 lbs. @ ¢5½	4.18	
4 Steel Springs 3½ by ⅜ thick N York Steel 200 lbs.	16.	
Labor making springs & box bolts per car	7.	
8 Spring Boxes 64 lbs. @ ¢5½	3.58	
8 Bolts for Springs 10 lbs.	1.67	
Irons & Chains on hauling perch 56 lbs. @ 9 (¢12½)	7.	
Waste of Materials	2.57	
		42
Zinc to cover roof 80 lbs. @ ¢8	6.40	
Labor & Solder	3.	
		9.40
Painting, 2 coats Slate Color		5.75
Total Cost		$269.15
Weight of such a car is about 4200 lbs.		

Source: Baker Library Manuscript Division, Harvard School of Business Administration.

sources were limited, traffic density was low, and distances between major population centers were great. Meager capital and low earnings forced the construction of inferior track and bridges. Grading was minimal, ballast sparse, and rail light, while ties were widely spaced. Only a flexible vehicle, with its load spread over as long a distance as was practical, could run over the flimsy, uneven trackage of America's backwoods railways.

Four-wheel cars tended to be short and rigid. The weight was concentrated on a short length of track, which would crush and break down a weak structure. If the wheel sets were more widely spaced to spread the load, the car could not traverse short curves or switches, making the car even more unsuitable for the provisional style of railway characteristic of young America. Even if it could force its way through curves and stay on the track, the flange and rail wear would be objection-

able. Double-truck cars could be long and rangy, thus spreading the load over a longer distance, yet they could still navigate short curves because the individual wheel sets of each truck could be placed close together. Because the trucks were connected to the car frame by center pins, they were free to turn or swing, and thus the vehicle possessed a remarkable flexibility. A single eight-wheel car might perform the same service of two or three four-wheelers with a saving in weight because of fewer end bulkheads, draft gears, and couplers. The eight-wheel car also offered greater opportunities for carrying larger overall loads and larger individual items; for example, long beams or pieces of machinery could be more easily accommodated. Car designers had greater freedom and could improvise and expand within the basic eight-wheel car arrangement.

All these advantages made the eight-wheel car attractive for both passenger and freight vehicles. However, it was found especially advantageous for freight service, where the question of axle or wheel loadings was so much more important. Passenger cars were not called upon to carry heavy loads. Even when fully seated they rarely weighed much more than 10 percent above their empty weight. Not so with a freight car. It could weigh double its empty weight when filled with goods, and the practice of overloading, with shippers eager to cram cargo to the roof ribs, was not uncommon. The weight borne by each wheel was well above what was safe for the wheels, axles, rails, or crossties. The results could be disastrous. An overloaded journal bearing would overheat. The train, if stopped in time to treat the inflamed bearing, would tie up the railroad and delay all transit over the road, for single track was the norm. If not detected in time, the journal might burn off and cause a wreck. An overloaded wheel, axle, or rail is liable to break, with the same result, while an overloaded tie will crush, leaving an unsupported rail that might turn over or spring out of gauge. Hence limiting axle loading on freight cars was critical in an age of wrought iron and primitive track construction.

The most effective solution was to add more axles per car, thereby spreading or redistributing the weight of the car and its load. Broken axles were the principal cause of accidents, according to the South Carolina Railroad's report of October 1834. This problem was directly related to the overloading of freight cars. While the South Carolina line despaired over the destructive habits of its shippers, the Boston & Providence sought to regulate the loads borne by its freight vehicles. The frigid New England winters made iron brittle, and so the B&P insisted that cars be loaded one-sixth less than normal during the winter season. In extreme cases, as we shall see later, twelve- and even sixteen-wheel cars were introduced for ex-

FIGURE 2.13

The B&O firewood cars introduced in 1830 were among the
very first eight-wheel cars in the
United States. Note that the
"trucks" are in reality four-wheel
flour cars. (*Scribner's Magazine*,
Aug. 1888)

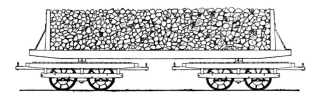

traordinary shipments. But in the main, the eight-wheel arrangement was found sufficient for most general service. Where not required, additional wheel sets only added to the cost, weight, and friction of the vehicle.

The first eight-wheel car was most likely a makeshift affair, just as the first bridge must have been no more than a tree fallen over a stream. One can imagine a group of trainmen looking about for a way to carry some long timbers or rails. And then it came to them: Take two four-wheel cars, separate them a short distance to act like piers, then lay the load across the pair like a bridge. The cars could shift and turn to a degree under their linear cargo and act in concert as if all three were a single vehicle. A perch pole between the single-truckers made it all the more secure and mechanical.

This very combination happened sometime not long after the Pennydarren tramway opened in Wales around 1800.[28] Small four-wheel tramcars were hitched together as widely spaced twins for the carriage of large iron bars. Mention has already been made of a similar marriage of stone trucks on the Granite Railway. In a similar pragmatic exercise, contractors on the Baltimore & Ohio seized two small four-wheel flatcars to haul long bridge timbers to a construction site in April or May 1830.[29] During the following winter a more refined eight-wheel car was assembled to carry cord wood into Baltimore from the outlying woodlands along the railroad. The weather was unusually harsh, and the demand for fireplace wood was great. The experimental horse-drawn car worked so well that more were constructed late in 1830 and early 1831 (Fig. 2.13). These were followed by a set of curious open-side-framed eight-wheel "trussell" cars used to carry light, bulky loads such as empty barrels or hay. The B&O also on occasion used them to haul cattle.[30] A car builder employed by the B&O during this period claimed that twenty to twenty-five cord-wood and "trussell" cars were in operation by 1834.

The good service of the double-truckers encouraged the fabrication of boxcars on a similar plan. A contract was given to J. Rupp and H. Schultz to build 110 24-foot-long boxcars.[31] They were to have cast-iron body bolsters and shingled roofs. Construction began in December 1834, but slow delivery of bolsters, wheels, and axles delayed their completion. The railroad's report for 1835 mentioned that forty-eight of the new "burden cars" were ready for service. The remainder were not completed until the fall of 1836. They were similar

in appearance to the Winans patent drawing shown in Figure 2.14, except for the shingled roof. In 1838 Von Gerstner stated that the B&O was now committed to the eight-wheel car and that many of the four-wheel cars, though only a few years old, were abandoned and going to ruin on side tracks. The line had 175 double-truck boxcars on the property. They weighed 4.5 tons and carried 5 tons.

The appearance of these pioneers is illustrated by a drawing appended on November 19, 1839, to Ross Winans's patent of October 1834. The drawing is believed to date from 1839 or thereabouts and was sent to the Patent Office only after litigation had commenced over Winans's spurious claims regarding the eight-wheel car.[32] Of interest is the arch roof, the outside body frame, the hinged double doors, and the long bar latch and lock. Individual brake levers operated brakes on wood-frame trucks. Notice the absence of truss rods. The body and truck bolsters are cast iron. The car measures 29 feet 6 inches long, while the body itself is 25 feet long by 7 feet 4 inches wide. The truck wheelbase is 36 inches. This is a very small car by today's standards, but it was a giant for its day and helped make the Baltimore line a leader within the industry.

Just a few years after this drawing was delivered to the Patent Office, other measurements were recorded for a B&O boxcar.[33] They gave its weight in long tons (2,240 lbs.) as 4.82 and its capacity, also in long tons, as 5.76. This car, built in 1841 or 1842, had journals 2.75 inches long by 2 inches in diameter. The bearings were chilled cast iron, but some cars on the line used Babbitt (a soft, white-metal alloy) bearings instead. The wheels were only 31 inches in diameter and the wheelbase for each truck was 41 inches.

Karl Von Ghega, an Austrian engineer, offered dimensions on a B&O boxcar that showed only a slight increase in size over the dimensions given in the 1839 Winans drawing, in his book *Die Baltimore-Ohio Eisenbahn*, published in Vienna in 1844. Interest in the car-building ideas of Mt. Clare, the B&O's principal workshop, was also evident in a drawing reproduced in a general work on railway engineering published in Paris in 1857 (Fig. 2.15).[34] The drawing bears a strong resemblance to the 1839 Winans drawing. The text confirms only that it is of U.S. origin. I believe it is a B&O car of the 1840s because it seems to be a link between the Winans car of 1839 (Fig. 2.14) and the Tyson car of 1856 (Fig. 2.27). The French engraving shows several interesting features in addition to the curious live-spring trucks. Notice the sliding side doors, the roller side bearings at the ends of the cast-iron bolsters, and the metal brake levers.

It was not long before railroads and commercial car builders outside Baltimore began to adopt and promote the eight-wheel car. In a letter dated February 17, 1837, M. W. Baldwin, a prominent Phila-

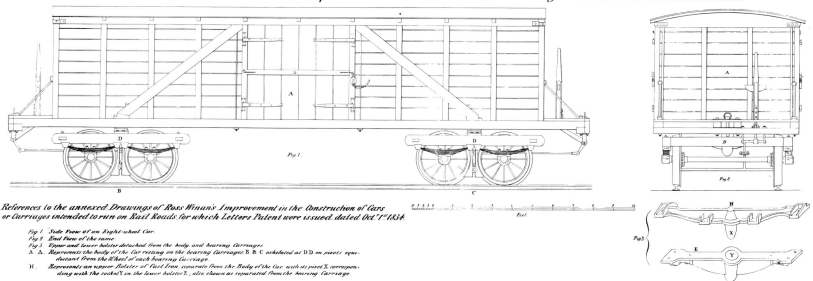

References to the annexed Drawings of Ross Winans Improvement in the Construction of Cars or Carriages intended to run on Rail Roads, for which Letters Patent were issued, dated Oct.r 1.st 1834.

Fig 1   Side View of an Eight-wheel Car.
Fig 2   End View of the same.
Fig 3   Upper and lower bolster detached from the body, and bearing Carriages.
A . A . Represents the body of the Car resting on the bearing Carriages, B & C exhibited at DD on pivots equi-
         distant from the Wheel of each bearing Carriage.
H .   Represents an upper Bolster of Cast Iron, separate from the Body of the Car, with its pivot X corrospon-
         ding with the socket Y in the lower bolster E., also shown as separated from the bearing Carriage.

FIGURE 2.14

B&O boxcar. The drawing was sent to the Patent Office five years after Winans's patent was granted in October 1834. The design presumably dates from the late 1830s. Note the cast-iron body and truck bolsters and center pin at the lower right. (Peale Museum, Baltimore)

delphia locomotive builder, stated that eight-wheel cars were superior and were favored by his neighbor Richard Imlay, who produced a first-class 10-ton double-truck merchandise car for $875.[35] Betts, Pusey & Harlan, car builders of Wilmington, Delaware, also pushed double-truck cars in their prospectus of October 1839. They offered a first-class boxcar with a body 7 feet 6 inches wide, 22 feet long, and 6 feet high at the sides, with a zinc roof and cast-iron wheels, for $800.[36] A cheaper car of the same size with a painted and sanded canvas roof sold for $700. Both styles were offered with either end or side doors. This firm, better known by its subsequent corporate name of Harlan & Hollingsworth, was even by 1839 well known in the car-building field and supplied cars and rolling stock to railroads throughout the United States. The prospectus just mentioned was printed in both French and English and quoted prices in both U.S. dollars and pounds sterling. Hence it was intended for foreign customers as well, which means that the gospel of the eight-wheel car was being broadcast overseas as early as 1839.

At the very time that the Harlan prospectus was being distributed, Von Gerstner was conducting his survey of American railroads. He found several roads well stocked with four-wheel cars that had now realized their mistake and promised to adopt the eight-wheel plan once their existing stock was worn out.[37] The southern roads seemed especially taken with the eight-wheel car. The South Carolina Railroad had twenty-two in service at the time of Von Gerstner's visit. The Georgia railroad had fifty boxcars, which Von Gerstner described in some detail. The bodies were 27 feet 6 inches long, 8 feet 6 inches wide, and 8 feet high. The cast-iron wheels were made with wrought-iron-bar tires, cast integrally. The car weighed only 5 tons and was rated at 6 to 7 tons capacity. It cost $750. The

chief engineer of the Raleigh & Gaston Railroad thought that the eight-wheel freight car was among the greatest of improvements. In his published annual report of May 1840 he claimed that they were coming into general use.

New England showed considerably less enthusiasm for the double-truck car and seemed ready to stand by the pattern established by the British goods wagon. This same conservatism was evident in New England locomotive design, where the inside-cylinder engine remained in favor for a decade or more after having been declared obsolete elsewhere in the United States. This point is underscored by the persistence of the Locks & Canals machine shop at Lowell, Massachusetts, in cranking out virtual duplicates of Stephenson Planet-class locomotives into the mid 1840s. As late as the 1850s the Eastern Railroad was arguing that the old-fashioned English-pattern car was best suited to its needs.[38] The line claimed that such cars offered less friction and wheel wear and hence could be hauled over the line with less power. The Boston & Lowell agreed completely, and both roads were stocked exclusively with four-wheel freight cars long after they had gone out of fashion elsewhere.

The four-wheeler, of course, was the subject of periodic revivals as late as the 1930s. Examples of four-wheel boxcars and gondolas are shown in subsequent chapters of this book. Most railroads found, however, that the four-wheelers were unstable, hard-riding, and more inclined to derail than double-truck cars. By 1870, of course, once the interchange of freight cars had become prevalent, even the most reactionary New England master car builder was forced to yield on this question. But there was a special class of car, the coal jimmy, that remained firmly in the four-wheel camp until almost 1900. Thousands of these diminutive coal

THE PIONEERS

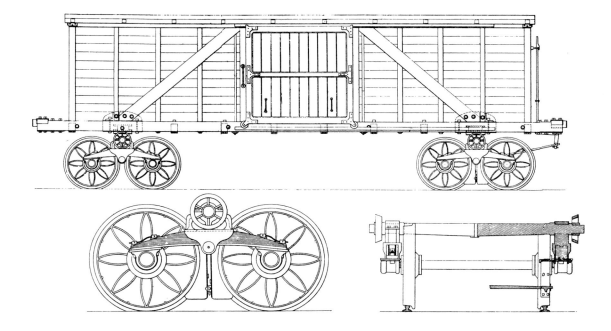

FIGURE 2.15

This drawing of an American boxcar, almost assuredly of B&O origins, was published in a French book on Railroad engineering. (Emile With, *Nouveau Manuel Complet: Construction des Chemins de Fer* [Paris, 1857])

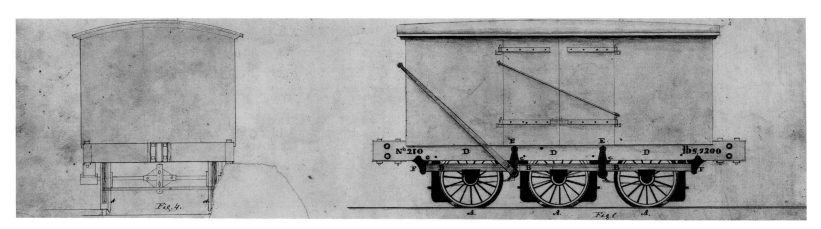

FIGURE 2.16

James Millholland's six-wheel boxcar as illustrated in his patent of September 23, 1843. Cars of this type were used on the Baltimore & Susquehanna. (National Archives)

carriers were found in service on the Lehigh Valley, the Pennsylvania, the Lackawanna, and other eastern lines. More on the history of the jimmy will be told later in this chapter and in Chapter 5.

The shift from four- to eight-wheel cars was early, abrupt, and relatively universal—a pattern that does not agree with the normal evolutionary trend in engineering developments. This most likely occurred because the merits of the double-truck car were established before a very large investment had been made in four-wheel rolling stock. Rather little railroad mileage was completed in this country before 1840, and hence the number of freight cars was comparatively small. Regrettably, no hard-and-fast figures are available, but there were hardly more than five thousand freight cars in existence in the United States in the late 1830s. In addition, single- and double-truck cars could be mixed, and they would work together well enough in the slow-speed service typical of the pioneer lines; hence the introduction of the new-style car did not render the old plan useless. The old cars could be used up and then discarded.

While the great majority of railway mechanics were content with the changeover to eight-wheel cars, a few independent-minded builders sought a more novel plan. Lewis J. Germain of Catskill, New York, patented a design for a six-wheel undercarriage on May 7, 1839 (No. 1,145). The wheel sets worked together on the radial-axle plan to assist in going around curves. The idea was tested but made no progress beyond that elementary level of success. A more determined friend of three-axle cars appeared in the person of James Millholland (1812–1875), a prominent railway master mechanic. Early in his career he was appointed mechanical head of the Baltimore & Susquehanna Railroad. Sometime in the early 1840s Millholland decided that six-wheel cars were a worthy compromise between the old and new styles of freight car. In its published report for 1843 the B&S reported having twenty-nine new six-wheel cars in service. They weighed 8,500 pounds, carried 12,000 to 14,000 pounds, and cost only $450 each.

On September 23, 1843, Millholland obtained a patent (No. 3,278) that contained a description and drawing of a boxcar already in service (Fig. 2.16). The body measured 14 feet long by 7 feet wide. The wheelbase was 8 feet, the wheels were 33 inches in diameter, and the white ash springs, another of Millholland's ideas, measured 12 feet long, 7 inches wide, and 2 inches thick. The wooden leaf spring is perhaps the most novel adaptation of a cheap native material for car-building purposes. In the patent specification, the car's weight is given as a mere 7,200 pounds, which is very light for a car of this size.

More six-wheelers had been built by the time Millholland left the Baltimore & Susquehanna for the Reading; the B&S had 81 six-wheel freight cars in a total fleet of 377 cars.[39] There were also three passenger cars built on this curious plan. Millholland had no success in convincing his new employer of the merits of the hexapod, although the Reading applied thousands of his wooden springs to four-wheel coal cars. Only the Baltimore & Ohio showed much interest in developing the six-wheel freight car. It produced several hundred iron coal cars on six-wheel undercarriages, which are described and illustrated in Chapter 8.

Specialization

It might be thought that the transition from open to closed cars together with the introduction of eight-wheelers would have fully occupied America's car builders during the 1830s. But the first decade of any new technology is usually a busy, unsettled period with many unsolved questions as well as numerous changes on several levels happening all at once. And so it was with railroad cars. Not only were their size and arrangement undergoing radical alterations, but many new types of cars were being introduced. Boxcars, hoppers, and flats might satisfy the majority of shippers, but special situations called for a greater variety of conveyances, fine-tuned to satisfy the exceptions and needs that have always been part of the transportation business. And so at a surprisingly early date, containers, cattle cars, and covered hoppers joined the national rail car fleet.

CONTAINERS

The container, which is commonly regarded as the most modern of innovations, is actually rather ancient. The scheme can be traced back to England around 1790 when boxes, each containing a ton of coke, were transferred among road carts, barges, and rail cars during their journey to market.[40] Other instances of English containers dating from around 1800 involving the coal and limestone traffic are also recorded. In one instance metal boxes were employed, while in another the containers were nothing more than wicker baskets. But the

FIGURE 2.17

The Liverpool & Manchester introduced containers for coal in about 1830. A drawing in Nicholas Wood's *Practical Treatise on Rail-Roads* **(London, 1832) shows rollers rather than slides on the containers, and the wood-bodied rather than the metal version.** (*Dictionnaire Technologique Chemin de Fer* **[Paris, 1835]; metric measurements)**

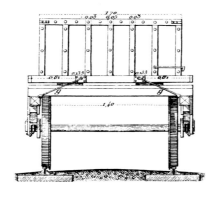
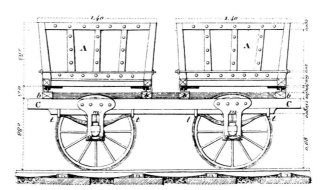
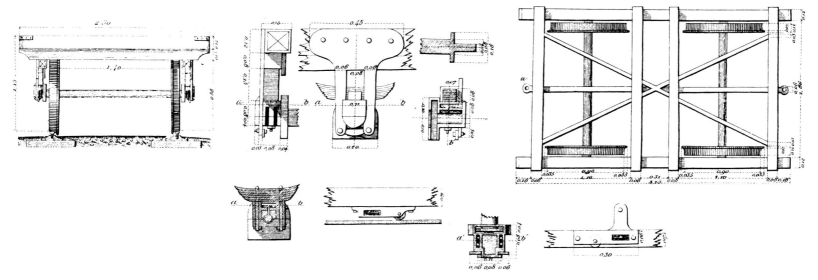

THE PIONEERS

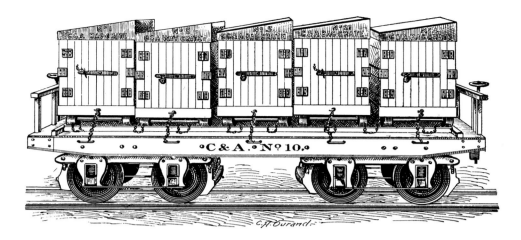

FIGURE 2.18

Baggage container cars were used on several eastern U.S. lines starting in the 1830s. This example was operated by the Camden & Amboy. (Reconstruction prepared by C. H. Ourand, an artist employed by the Smithsonian in the 1880s and 1890s)

particulars of the apparatus are less important than the understanding of the need to devise a technique for avoiding the breaking of bulk.

The process of unloading and reloading a cargo is costly in labor and time. Damage and loss to the materials being transported is another result of transhipment. All these perils and costs can be minimized by keeping the shipment in a package or container that is itself transhipped. This method is useful for connections, be they between differing modes of transit—rail, water, highway—or where rail lines of differing gauges meet. The idea was hardly new when the Liverpool & Manchester Railway opened in 1830 and the Stephensons adopted the method for the coal traffic (Fig. 2.17). Both wooden and metal boxes, two to a car, were used for this purpose. The boxes or containers were transferred between narrow-gauge mine cars and standard-gauge flats. The upkeep of the containers and the weakness of the carrying vehicle caused Stephenson to abandon the idea in about 1833, but a private contractor offered container service for general merchandise for some years afterward.

Meanwhile, in the United States the same general scheme was being adopted for baggage handling. Bags were placed in boxes or luggage crates fitted with small wheels. They were rolled onto flatcars. This system was especially valuable where trains transferred passengers to connecting steamers, and for that reason it was used by the Camden & Amboy, the Old Colony, and the Philadelphia, Wilmington & Baltimore in the 1830s, 1840s, and 1850s (Fig. 2.18).[41]

The container system was most usefully employed for freighting purposes on the combination canal/railroad constructed between Pittsburgh and Philadelphia by the state of Pennsylvania. The so-called Public Works opened in 1834 and offered even more interruptions to transit than the cumbersome combination system might suggest, for there was a series of incline-plane railways in the most mountainous, western portion of the line. Hence a cargo would be loaded and unloaded a

dozen times if it went the length of the State Works. Fragile or valuable goods were in danger of damage or theft. Bulk goods like grain, coal, or vinegar were liable to spillage and contamination. It didn't take long for shippers to understand the monstrous deficiencies of the combination canal/railroad system. And because it was a state-financed system, private shipping companies were allowed to operate over it as public carriers.

Two of these freight forwarders, men who were daily confronted with the defects of the State Works, came forward with competing schemes for sectionalized canal boats or lift-off bodies for freight cars.[42] Sectionalized canal boats were suggested in 1827 by Canvas White, an American civil engineer active some years before the State Works opened. John Elgar (1780–1854), an engineer from York, Pennsylvania, developed White's idea and obtained a U.S. patent on November 7, 1835. The long, narrow boats, roughly 7 by 80 feet, were cut into 20-foot sections and held together by latches once in the water. The separated boat sections would be carried by flatcars over the rail portions of the line, including the incline planes. The cargo would never be touched once it was placed in the boat, and hence according to the White/Elgar system, no breaking of bulk was necessary. Elgar, who had previously built an iron-hull steamer, suggested that the canal boats be built of iron; or if wood sides were preferred, the bottoms could be made of iron.

At the time that Elgar was working on his patent application, John Dougherty (c. 1803–1886) of Hollidaysburg, Pennsylvania, a principal in the freight line called the Reliance Transportation Company, mounted a small canal boat on a flatcar and sent it by rail over the mountains. Very soon he was operating boats much on the White/Elgar model. His boats measured 82 feet long by 8 feet 9 inches wide. They weighed 10 tons, carried 20 tons, and could be broken down into four sections. The boats were loaded and unloaded in an ingenious manner. An inclined track was built into the canal at a terminal basin, and the flatcars were eased down the track below the surface of the water so that the boat sections could float on or off as desired. Thus even a fully loaded section was handled in the gentlest possible manner. Service began around 1837 with two- and four-section boats. The trip between Pittsburgh and Philadelphia took six days. Service was extended to Baltimore as well. Von Gerstner witnessed the operation of these improbable vehicles in the streets of Philadelphia. He said that it was indeed an amazing sight to see the huge hulk of a canal boat, perched high off the ground on a railroad car, as it turned swiftly out of a warehouse drawn by a team of horses.

For all its ingenuity, Dougherty's business did not prosper, and he was forced out in about 1839. A prolonged controversy developed with the state, which had taken over operation of the sectionalized canal boats. Dougherty demanded large payments, but the state contended that traffic was down because of the lingering effects of the 1837 panic, and that revenues were accordingly modest. Amid this controversy, on February 24, 1843, Dougherty obtained a patent (No. 2,973) that included an excellent drawing of one of the eight-wheel flatcars used to transport the canal boats (Fig. 2.19). The drawing is far more detailed than the average patent illustration, which usually doesn't go beyond the crude schematics needed to illustrate concepts. Dougherty's illustrations are more like working plans. The inclusion of so much detail and the presence of a graphic scale make it possible that these drawings are more in the nature of prototype drawings. They are among the best documents we have for an American railroad freight car of this period, and while hardly typical, they do represent the thinking of the time as well as the construction techniques used.

Despite his elegant patent and an intense newspaper campaign, Dougherty never beat the state, and he died poor and embittered. The sectionalized boats, however, remained in service until about 1854 and were abandoned just before the State Works was sold to and abandoned by the Pennsylvania Railroad. Critics of Dougherty complained that the boats were overweight, leaky "butter tubs," but G. E. Sellers, who observed the system in operation, contended that it actually worked well and that its principal defamers were jealous competitors.

About the time Dougherty was starting up his operation, a Pittsburgh freight forwarder, James O'Connor, inaugurated a container service that was similar to modern concepts of such service. Separate bodies were fastened to flatcars that could be disengaged from their undercarriages and then moved by cranes to waiting canal boats. An engraving shows large hook eyes fastened to the car tops for this purpose (Fig. 2.20). Precisely when O'Connor began operations is apparently unre-

FIGURE 2.19

John Dougherty's patent issued on February 24, 1843, shows a flatcar used to carry canal boats over the Portage Railroad between breaks in the Pennsylvania State Canal. The design predates the patent by several years. A complete side elevation of the car has been assembled from elements in the original patent drawing to represent the original vehicle more clearly. (U.S. Patent Office)

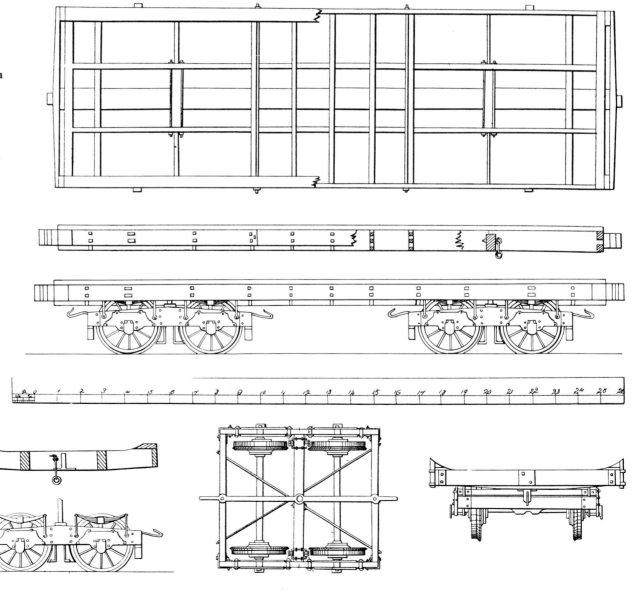

THE PIONEERS

PATENT PORTABLE CAR BODY LINE.

M. BURKE & CO.
SUCCESSORS TO JAMES O'CONNOR & CO.
From J. O'CONNOR'S
Pittsburg, Wheeling, and York County Depot,
NORTH STREET, BALTIMORE.
Goods carried from BALTIMORE to PITTSBURG in the same Car Bodies,
without being re-packed, handled, or separated on the way.
☞ Not accountable for casualties by Fire. ☜

FIGURE 2.20

James O'Connor of Pittsburgh introduced flatcars with removable car bodies in about 1837. His line operated in competition with Dougherty's sectionalized canal boat service pictured in Figure 2.19. (Smithsonian Neg. 82-11222)

corded, but a bill of lading dated August 31, 1838, indicates that service was under way by that date; it is possible that operations had started a year or two earlier.[43] By 1843 O'Connor's successor, M. Burke & Company, was advertising that its portable car bodies were serving Baltimore, Philadelphia, and Pittsburgh.

Advertisements appearing in the Cincinnati newspaper the *Daily Atlas* for May 1, 1845, indicate that some form of container service extended into what was then the western United States, so far as major centers of population were extended. Three firms—the Pittsburgh Transportation Line, the Citizens Line, and the United States Line—all offered to handle shipments via portable boats. James O'Connor was listed as a partner and as Baltimore agent for the Pittsburgh Line, with service to Baltimore rendered by the Baltimore & Susquehanna Railroad. The Citizens Line maintained a Cincinnati office and noted that it transferred "cars through from Port to Port without removing the loading." Because no through rail service was possible between the several cities (or ports) mentioned in the ad, it is assumed that the term *cars* actually referred to containers or portable boats.

In 1849 the Richmond, Fredericksburg & Potomac Railroad fitted two flatcars with a dozen small crates specifically intended for small freight shipments. The Richmond line's arrangement was clearly patterned on the baggage crate scheme already mentioned that was employed by several eastern railroads a decade earlier.[44]

COVERED HOPPERS

If the existence of containers before 1830 seems remarkable, the operation of covered hopper cars is even more impressive. Information on these vehicles is scanty at best, but two lines are known to have used them. Von Gerstner mentioned seeing special grain cars on the Mad River & Lake Erie Railroad during his 1838–1840 tour.[45] He said that they were filled from the grain elevators by spouts and that at the end of their journey the loaded cars were spotted above lake boats, so that their cargo could be emptied onto the vessels through hatches in the floor of the cars. Each car carried from 3 to 5 tons of grain. Similar cars were in service in northern New York on the Tonawanda Railroad. The line began handling a sizable grain traffic when it opened in 1836–1837. Cars with hinged roofs for loading the hopper body and a valve at the bottom were constructed to move the grain to market more efficiently.[46]

CATTLE AND COMBINE CARS

In preindustrial America foodstuffs were the major item of the freighting business. In the first years of operation pioneer railroads gave first consideration to the transport of raw materials and farm products; hence the early creation of special cars for coal, stone, and grain is not surprising. Another abundant agricultural product offered for shipment was livestock. In a time before effective canning, mechanical refrigeration, or refrigerator cars, it was necessary to send animals to city markets. Cattle were abundant in the countryside, but they had little cash value until delivered to the consumer. Long overland cattle drives offer romantic material for novels and films, but they were an economic disaster for ranchers and farmers. Some of the herd would sicken and die in the course of such a long journey, while others simply strayed away. Those that reached their destination lost the very flesh they had acquired at pasture because of the rigorous overland march. Transporting stock by rail got them to market quickly and at full weight, or nearly so if they were fed and watered en route.

The idea was so obvious that the Liverpool & Manchester began building wagons for the purpose even before the line opened.[47] In May 1831 reports were printed about the noisy journey of forty-five Irish pigs over the line. Contemporary prints by Ackerman show a solid livestock train on the Liverpool line dated November 1831. Three of the cars were double-decked, had roofs, and carried pigs. Three were slat-sided open cars that carried cattle, goats, and pigs respectively. A final semiclosed car carried a lone horse. The line is known to have provided special cars for horses with padded interiors and food bins in 1833.

The carriage of cattle started early enough in the United States, but there was considerable resistance to providing cars specially made for that purpose. The first recorded cattle shipment went over the Baltimore & Ohio in 1830 in open-top, slat-sided cars used for a variety of cargoes, including empty whiskey barrels. *Hazard's Register* for April 18, 1835, mentions the shipment of two fine head

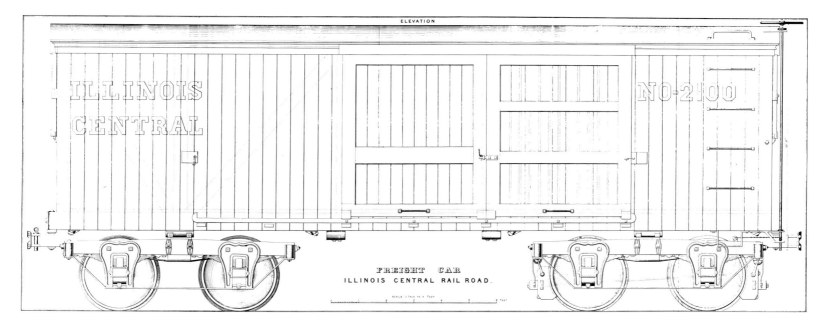

ELEVATION

FREIGHT CAR
ILLINOIS CENTRAL RAIL ROAD.

FIGURE 2.21

This Illinois Central combination merchandise/cattle car dating from around 1855 is shown with the solid door covering the central opening into the car's side. The open-barred door used for cattle is stored at one end of the sliding-door track. When it was in use, the solid door would be locked into place at the opposite end of the track. (Douglas Galton, *Report on the Railways of the U.S.* [London, 1857 and 1858])

of cattle from Lancaster County to Philadelphia by an enterprising farmer named Herr. The newspaper understood it to be the first such shipment by rail—at least in the Philadelphia area—and went on to report that the trip took only a day, whereas on foot the beasts would have traveled six to eight days. And Mr. Herr, who traveled with his tiny herd, would have required the services of one or two men if the journey had been made by road. The animals arrived ready for the butcher and did not require rest or feeding to rebuild their flesh.

Until 1860 there appears to have been a heavy reliance on boxcars fitted with barred doors for transporting cattle. These compromise cars were very much like the ventilated boxcars of later years that were intended for produce rather than cattle. This unwillingness to provide special or at least single-purpose cattle cars was based on two considerations. The traffic itself was so light, at least on some lines, that the expense of special cars wasn't justified. The movement of cattle was also seasonal. Few farmers wanted to feed their herds through the winter season when pastures were barren, and so the big selling season began in the fall.

Specific examples of combination box/cattle cars were found on several midwestern lines in the 1850s. John Kirby, the veteran car builder for the Michigan Southern who started with the line in 1854, recalled the first such cars as having openings in their sides with ¾-inch iron bars placed on 6-inch centers.[48] These grills prevented the cattle from poking their heads outside. Shutters were placed over the opening to convert the cars for merchandise purposes. Kirby went on to say that the trip from Chicago to New York passed over eleven divisions with a break in gauge at the middle of the route, thus making the journey a week long.

During the same period the Illinois Central used combine cars for which we have detailed drawings (Figs. 2.21, 2.22). This car had no side openings as mentioned by Kirby, but rather double sets of doors on the sides and ends. One door was solid, making the car ready for general merchandise, while the other was open, with iron bars, and so was suitable for livestock. The end ventilator openings were also useful for loading lumber and became a relatively common auxiliary on many standard boxcars. The body measured 26 feet over the end bumpers. The truck wheelbase was 52 inches, and the wheels measured 33 inches in diameter. This drawing, from Douglas Galton's report on U.S. railways to the British Board of Trade, is one of the finest available graphic representations of a first-generation American freight car. In addition to the splendid detail exhibited in the engravings, note the absence of truss rods, a design characteristic to be commented on later. A later Illinois Central car of similar design is shown in Figure 2.23.

The Michigan Central introduced its own plan of combine car in the summer of 1859, a design that appears to be very similar to the cars in operation on the Michigan Southern.[49] The road built 120 of these new 31-foot 6-inch-long cars in its own shops. They had side openings 13 by 30 inches with iron bars. The doors and shutters were so contrived that the vehicle could be changed from a boxcar to a cattle car in just two minutes.

The road intended to use them for grain, but it is difficult to imagine how this could be done after the car had been home to occupants given to answering nature's call without the slightest inhibition. Car sweepers can be expected to do only so much, after all.

When the Michigan Central cars were introduced in 1859, the emphasis given to their being

THE PIONEERS

FIGURE 2.22

End view of Illinois Central multipurpose boxcar. Here the barred or ventilator door is positioned over the opening. The brakewheel staff has been shortened much below its actual height to fit into the drawing. (Galton, *Railways of the U.S.*)

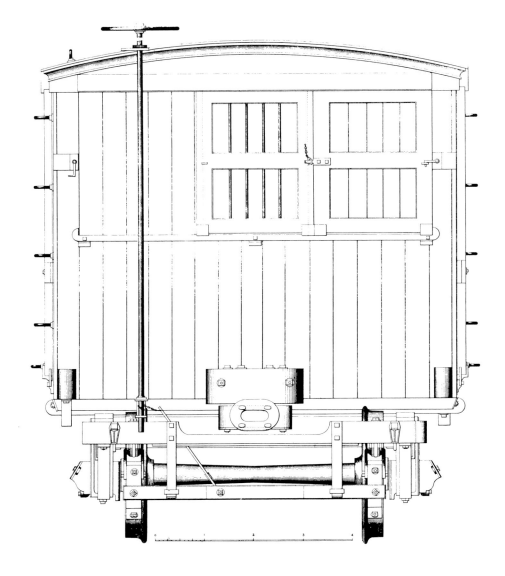

FIGURE 2.23

An Illinois Central car very similar in design to those illustrated in Figures 2.21 and 2.22 is shown at the end of its career in this inventory photo made around 1890. The barred ventilator doors are gone, as are the wood-beam trucks and wrought-iron couplers. (Railway and Locomotive Historical Society Photo)

covered—that is, having roofs—implies that open-top cattle cars were common before this time. Again, this is a difficult question to answer because of the incomplete information available. The Baltimore & Ohio, for example, is known to have used open-top cars in 1830, and its annual report for 1860 lists twenty-one open stock cars. As late as 1858 the Central of New Jersey reported ten open and eighteen covered stock cars.[50] In 1849 a New England brakeman lost his lantern one night while standing on the roof of a freight car. While reaching for the falling light, he fell through an open hatch and landed amid the agitated cattle. He was rescued at the next stop but was in "no condition to appear in public."[51] The story not only confirms that roofed cattle cars were in existence before 1850 but also explains one unexpected hardship faced by freight train crews of that time.

Railroad managers also found the livestock trade troublesome on many other accounts. Cattle, sheep, and pigs were almost as troublesome as human passengers. They required shelter, food, and water, if not heat and light. If these amenities were not attended to, the animals might die en route, presenting the company with a settlement claim.[52] And then to add to the complexity of this business, the herd masters wanted to accompany their flocks during the journey. As early as 1839 the New Jersey Railroad & Transportation Company agreed to add a special car to accommodate those who wished to travel with their animals.[53] It was not explained whether this was the ancient ancestor of the drovers' caboose or merely a coach.

Exactly twenty years later the Erie specifically announced its readiness to furnish caboose cars, as well as faster schedules.[54] Both news items would be welcomed by those accustomed to long weeks on the road with their herds. In June 1851 it was stated that it required seventy-five days to drive a herd from Kentucky to New York City.[55] The hardship endured by both the herd and its masters is difficult to calculate, but the dollar cost was recorded at $20 a head. A combined rail/lake boat route reduced the time of the journey. The cattle

FIGURE 2.24

This Pennsylvania Railroad cattle car of 1863 had a removable double deck. Sheep or hogs were carried when the deck was in place, cattle or horses when it was removed. The wheels at the right side show strengthening ribs on the rear of a solid plate; they are not open-spoked. (J. E. Watkins, *History of the Pennsylvania Railroad* [proof sheets, 1896])

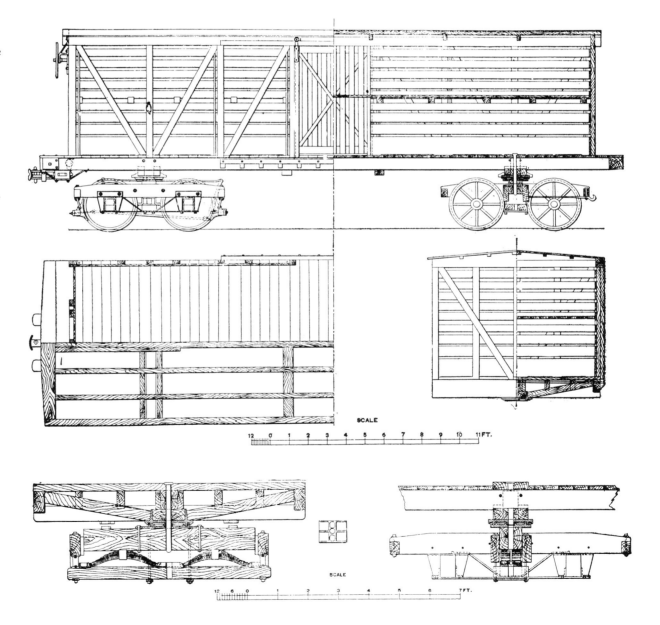

SCALE

12 0 1 2 3 4 5 6 7 8 9 10 11 FT.

SCALE

12 6 0 1 2 3 4 5 6 7 FT.

THE PIONEERS

were driven to Cincinnati, where they were loaded onto cars of the Little Miami and passed over the line and its northern connection to Cleveland and thence by water to Buffalo. There the herd transferred to rail cars for the final portion of the journey.

Just how long the Erie's 1851 trip took is not recorded, but more precise data is available for stock trains on the Pennsylvania Railroad nine years later.[56] The journey from Pittsburgh to Philadelphia took only forty-four hours. Traffic was reported as being very large—so large it would appear that the road was inspired to develop a better cattle car. Drawings for a 31-foot-long car of a rather modern outline were prepared in 1863 (Fig. 2.24). Here we have a fully matured cattle car design that remained popular on the Pennsylvania Railroad for many years except for the trucks and floor frame. The body measured 28 feet long by 9 feet wide. The end platform offered a perch for the brakeman. The pitched rather than arched roof marks a break from old-fashioned designs, but the wood-beam trucks and the absence of truss rods are the main clues that this car dates from the 1860s and not a decade later. Note that the leaf springs are placed below the bolster inside the side framing members. The cutaway drawing of the trucks at the right side shows the back of single-plate wheels, and the strengthening ribs should not be mistaken for open spokes. The main frame is peculiar in having seven rather than six or eight longitudinal members, as was common for most wooden freight cars.

This design was replaced in 1869 by the Pennsylvania's first standard stock car, the class KA. Its general dimensions and body details were nearly identical to those in the 1863 drawing. It also incorporated a removable second deck that could be used for hogs or sheep. The car weighed 19,000 pounds, but its capacity varied widely, depending on what was being transported. A load of horses was rated at 16,000 pounds, cattle at 14,000 to 18,000, and sheep at 9,000 if the single deck was used or 14,000 if both decks were in place.

Construction Details

During the same years that specialized cars like these were being introduced, the more common type of rail freighting vehicles were going through a slow evolution. *Slow evolution* may be something of an exaggeration, because the standard workhorse of the fleet changed little from the mid- and late 1830s, when the first eight-wheel boxcars were introduced, to the end of the Civil War. For thirty years the majority of boxcars remained about the same. Typically, they had an 8- to 10-ton capacity, an arch roof, wood-beam trucks, no truss rods, and a body length of 24 to 28 feet. Only one truck had brakes. About the only noticeable

change during this expanse of years was the displacement of spoked by plate wheels. The obvious explanation for this superficial lack of progress was that the design was adequate for the job. Before 1860 few individual railroads were more than 100 or 200 miles long. Their traffic was essentially local in nature; hauls were accordingly short, and overall tonnage was modest. Hence simple 10-ton cars made perfect sense.

A fine visual record of an 1845 boxcar has survived through an advertisement published for Davenport & Bridges, car builders of Cambridgeport, Massachusetts.[57] The car exhibits an outside truss frame, an arch roof, sliding side doors, and peculiar inside-bearing iron trucks (Fig. 2.25). Spoked wheels were still used. Note that the side sills of the floor frame extended beyond the end sill. An extra set of dead blocks was attached on either side of the coupler, as were safety chains, a rarity for freight equipment. Davenport & Bridges was a major car builder that supplied equipment to many New England railroads. A guidebook published by the Western Railroad (Mass.) in 1847 includes an engraving of a very similar car.[58]

A somewhat more antique car but one that shows similar features was depicted by George Bruce & Company, New York City, in a type specimen book issued in 1848 (Fig. 2.26). The hinged side and end doors and very large end platforms suggest that the engraving is based on a prototype dating some years before 1848. The pitched roof should not go unnoticed and is an early instance of a break with the much-favored arch roof. The interest in outside framing went beyond the northeastern states, for an illustrated advertisement for the firm of Burr & Ettenger, machinists and car builders of Richmond, Virginia, shows a long boxcar with outside frames. It is an unusually long car that also features double sliding doors and arch-bar trucks.[59]

John Kirby (1823–1915), master car builder for the Michigan Southern, added a few fragments to our knowledge of early freight cars in recollections published in 1894.[60] He recounts his experience in boxcar construction as practiced in 1848. Regrettably, the remarks were very brief, but he did emphasize the amount of hard labor required of the men engaged in that trade. Machinery was scarce, and the planks were ripped and planed by hand. This observation was confirmed by another old car builder, H. M. Perry, who said that the larger timbers were rough-cut at a sawmill but the smaller material was sawn and planed by hand—making car building a time-consuming, labor-intensive occupation.[61] Kirby went on to describe the 1848 cars as having spoked wheels and a painted and sanded cotton duck roof. The cloth was, of course, stretched over a thin wooden decking. The trucks used elliptical springs because cheap coil springs were not yet being manufactured. In June 1853 the

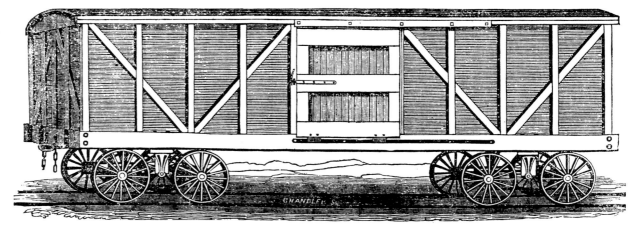

FIGURE 2.25

Davenport & Bridges, car builders of Cambridgeport, Mass., advertised cars of this design in 1845. The inside-bearing iron-frame trucks are thought to be atypical. Wood-beam outside-bearing trucks like those shown in Figure 2.21 were more common. (*American Railroad Journal*, June 12, 1845)

FIGURE 2.26

This engraving is from a printer's type sample book published in 1848, but the prototype design might easily date to a decade earlier. The hinged side doors agree with Figure 2.14, while the end doors correspond with the Harlan & Hollingsworth (Wilmington, Del.) prospectus of 1839. (George Bruce & Co., New York, 1848)

American Railroad Journal described the typical eight-wheel boxcar in the following words: "The size of an eight-wheel box freight car is usually 28 feet by 8½ feet and 6½ to 7 feet deep. The wheels generally of 30 inches diameter. The cost of such a car, with oak frame, is generally $500 to $600. For an open platform car, otherwise the same, $520. The weight of a box car is from 12,000 to 16,000 lbs., according to the strength of its frame, axles, etc."

A year earlier the Baltimore & Susquehanna Railroad ordered ten boxcars from Scott & Bolster of Baltimore. The specifications, which are really more a bill of material, give a good verbal picture of these vehicles. Unfortunately, no drawings accompanied the following list:

Articles of Agreement made and entered into this twenty-fourth day of March AD 1852 between Scott & Bolster of Baltimore, Maryland of the one part and the Baltimore and Susquehanna Rail Road Company of the other part as follows, to wit

The said Scott & Bolster for the consideration hereinafter mentioned doth covenant and agree to build Ten Eight wheeled house cars at Five hundred and Seventy five dollars for each Car, according to the drawing and specifications given by the Master of Machinery of the said Company. The bottom Sills of each car will consist of Yellow Pine, Studding to be of White Ash, Cross floor beams Yellow Pine, Top Sills, Yellow Pine, Bumping front, White Oak, Roof and outside weatherboarding best quality Cypress Timber, tongued and grooved. The drawing will give the dimensions of the iron work, Axles to be of the best charcoal hammered iron, Wheels to be made of cold blast iron, weight of wheels to average 470 lbs. the wheels to have iron feathers and banded according to the drawing. The Brass work to be Babbited, the truck of each car is to be made of the best quality of White Oak free of knots, each car will be required to have three coats of fire proof paint of a dark brown color, and a break [*sic*] at both ends of each car. . . .

All the materials of which the cars are constructed must be approved by the Master of Machinery and their Construction shall be under his inspection and at anytime during the progress of Construction he may condemn or reject materials, The letters B & S RRCo and number and weight of the car must be painted on each car or put on in Metallic letters, The weight of each car when complete not to exceed 16,500 pounds, no modification of this Contract to be made without the consent of the President of said Company and if any such modification is made to be entered on the contract.[62]

Drawings do exist, however, for a boxcar designed by Henry Tyson, master of machinery on the Baltimore & Ohio (Fig. 2.27). This car mea-

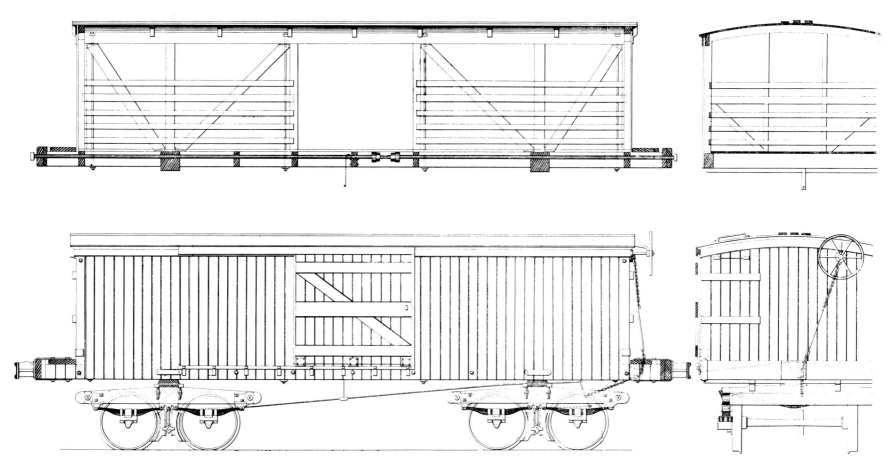

FIGURE 2.27

B&O boxcar of 1856. (Galton, *Railways of the U.S.*)

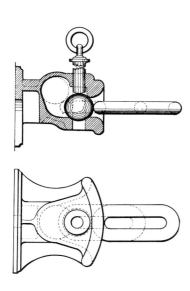

FIGURE 2.28

Detail of Joseph T. England's semi-automatic coupler used on the B&O boxcar depicted in Figure 2.27. (*Railway and Locomotive Engineering*, May 1921)

sured 28 feet 8 inches over the couplers.[63] The body was 24 feet long, 8 feet wide, and 6 feet 2 inches high at the center. It weighed 6 tons and carried 9 tons. It had a continuous iron drawbar with curious semiautomatic couplers patented by Joseph T. England on December 4, 1855 (No. 13,369) (Fig. 2.28). The wood-beam trucks, leaf springs, and single-plate cast-iron wheels are standard fare for the period. But notice the primitive wood-block steps at each end of the car and the very long end platforms. The 30-inch-diameter wheels are somewhat undersize, but the double brakes were a novelty and a necessary one on the B&O because of its mountainous trackage in western Virginia. Very few American freight cars are believed to have been so equipped at this early date. The brake rig is not fully explained by the drawing, but it could hardly be more simple.

FRAMING

The construction and framing of pioneer cars remain something of a mystery because of the scarcity of engineering drawings for the period. And much of what has survived gives only a suggestion of the actual structure and fabrication. John Kirby, when speaking of the freight equipment of 1848, stated that the main frame consisted only of two outside sills with a number of cross-timbers mortised into the sills.[64] The frame was like a giant ladder built of heavy timbers. This style of frame

is illustrated in the 1856 Baltimore & Ohio boxcar drawing (Fig. 2.27). The B&O used wrought-iron drawbars in place of center sills. Even with a drawbar, the ladder-style frame was both primitive and weak.

There is evidence that by the 1860s additional longitudinal sills were being introduced to reinforce the car's foundation. Surviving drawings for B&O boxcars of 1862 and 1866 at the B&O Museum in Baltimore confirm this improvement, as does the Hartford, Providence & Fishkill drawing reproduced in Figure 2.29.[65] This latter drawing exhibits a fundamental weakness despite the employment of four longitudinal sills. Notice that they do not run the full length of the car but are cut in two and mortised into the body bolster. This, of course, greatly weakens the frame. Notice also that the truss "rods"—which incidentally are ½- by 2½-inch flat bars and not rods at all—stop just beyond the bolster and do not connect into the end sills. A very similar floor frame is evident in the 1861 Philadelphia, Wilmington & Baltimore boxcar illustrated in Figure 2.30, indicating that this defective plan was not used by the Hartford line alone. H. M. Perry, a veteran New England car builder, recalled the unhappy result of cars with floor frames of this style. If the engine started with a hard jerk, the car would often break in two at the bolster, and as the front end of the car obediently followed the train, the rear three-quarters would

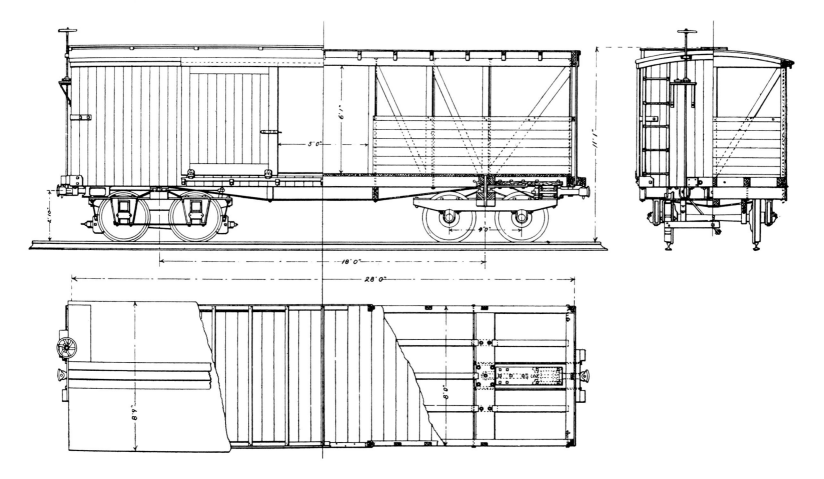

FIGURE 2.29

This Hartford, Providence & Fishkill boxcar was built in 1864. The center sills are broken at the body bolster, which materially weakens the frame. What appear to be truss rods are actually flat iron straps. Note the wooden safety beams placed inside the wheels; the truck's outer frame is iron. (*Railway Master Mechanic*, June 1914)

ignominiously collapse on the track, scattering the lading along the right-of-way.[66] Not all master car builders were so misguided. The Dougherty flatcar patent of 1843 shows a very strong frame with a heavy wooden center sill. Perry came to abandon the split floor sill only after an 1869 visit to the Illinois Central's car shop, where one-piece center sills were employed.

The Camden & Amboy was a line much given to distinctive rolling stock designs. Its curious locomotives and passenger cars with shad-belly wooden frames were at variance with equipment on most American railroads of the time. Its boxcars were atypical as well. Early photographs dating from the 1860s that show these cars were uncovered at the Vineland (N.J.) Historical Society by Mr. William Coxey, who kindly furnished me with prints. The cars are not the true subject of the prints and so they are not reproduced here, but they do show the main design features. They had very deep wooden side sills made with tapering ends. Some of the cars had a horizontal outside brace midway on the body. All have very large door openings, fairly large end platforms, and arch roofs. Some had rudimentary wooden end railings.

More information on these cars is drawn from a description published in 1884.[67] During the 1860s the Camden & Amboy built a very solid boxcar using six sills, in anticipation of what would be-

come the norm after 1870. The side sills measured 4½ by 8 inches, the center sills 4 by 8, and the intermediates 3½ by 8. The body bolsters were 5 by 14 inches in cross-section. The bodies measured 27 feet 6 inches long but—typical of the period—had no truss rods except for two transverse ⅞-inch rods at the bolsters. The outside sheathing was ⅞ inch thick. The oak carlines (roof rafters) measured 2 by 2 inches and were placed on 16-inch centers. The cars weighed 9 tons and were said to be lighter, cheaper, and more serviceable than many modern cars in service in 1884. The bodies were tight and straight, a quality accountable to good craftsmanship, careful design, and the use of only the best timber.

The need for strong frames was underscored each day for every car builder by reports of cripples due to overloading. It was in the shipper's interest to pile as much into a car as it would hold, without regard to its posted capacity. Cars rated at 10 tons were filled with 14 and 15 tons of merchandise, sometimes as much as 17 tons.[68] Experienced car builders therefore understood the necessity of using select and well-seasoned timber for framing material. They warned against using green lumber, some of which found its way into cars after having left the sawmill a scant forty-eight hours before. The debate over what kind of wood to use was not very pronounced during these early years, when virgin forests still covered much of the land.

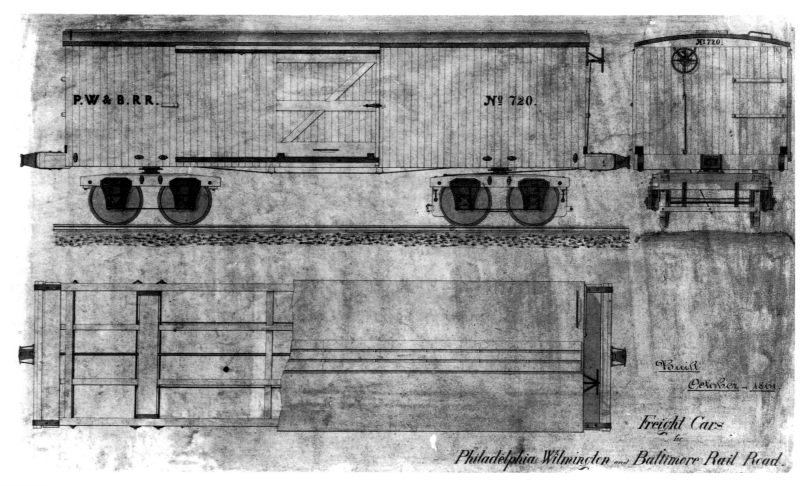

FIGURE 2.30

This car was built for the Philadelphia, Wilmington & Baltimore by Harlan & Hollingsworth in 1861. The body measured 27 feet 10 inches long by 8 feet 5 inches wide. The length over the couplers was 33 feet. The truck wheelbase was 44 inches. Note that the truss rods, two in number, terminated in the body bolsters rather than in the end sills. (Smithsonian Neg. 81-8117)

White oak and yellow pine were preferred, with the latter gaining in popularity as the best stands of eastern oak began to disappear. In an advertisement appearing in the 1868 *Ashcroft Railroad Directory*, the Ohio Falls Car Company of Jeffersonville, Indiana, promoted yellow pine from southern Mississippi as being unequaled for durability and stiffness. As we have seen, this same material was being used in the early 1850s by the Baltimore & Susquehanna Railroad. Oak remained in favor for body bolsters and end sills until nearly the end of the wood-car era in this country.[69]

TRUSS RODS

Iron rods are an efficient and relatively cheap way to stiffen and tie together wooden structures. They were really like very long bolts and were generally threaded on both ends and fitted with washers and large square nuts so as to not cut into the wood once tightened. Tie or truss rods were much used in roof trusses for large buildings, bridges, and railroad cars. One tends to think only of the longitudinal rods under the floor frame, because they protrude outside the car and so are visible. Once fully developed, they ran the length of the car and were fastened at the end sills after humping over the body bolster. If the car became loose, it could literally be pulled together by

tightening the nuts on the ends of the rods; if it sagged, it could be made straight by adjusting the turnbuckle between the queen posts. The strengthening effect of this deep truss under so much tension was significant and was the cheapest way to increase a wood car's capacity without adding very much to its weight. It was, in short, godsent and must have been seen as a special blessing by every practicing car builder.

With all these virtues, why was the idea so slow to gain popularity? The idea was neither new nor untried; a few passenger cars dating from the 1830s had longitudinal truss rods, and yet few freight cars before the 1860s appear to have had them. The freight cars of the Civil War period were well documented by Matthew Brady's camera, and few except for flats appear to have been so equipped. Two drawings reproduced in this chapter dating 1864 and 1861 (Figs. 2.29, 2.30) show scrawny, underdeveloped trusses set on very low wooden crosstimbers instead of queen posts. In the Hartford, Providence & Fishkill car the rods are actually straps fastened without the possibility of adjustment. And in both cases there are only two rather than four or six rods—an unusual feature for boxcars much before the 1870s; after 1880 truss rods in sets of four became a standard feature of floor framing. Deeper queen posts and even heavier rods made floor-frame trusses one of the most distinc-

FIGURE 2.31

Starting around 1870 the mortise-
style end sill gave way to the
stronger butt-end plan shown to
the upper left. (*National Car
Builder*, Aug. 1874)

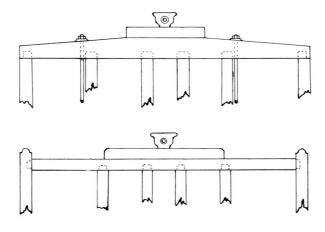

and this was the practice of extending the side sills
a few inches beyond the end sill to form a primitive buffer. This practice had an unexplained appeal to car men, and examples of it will be noticed
in several drawings reproduced in this chapter
(Figs. 2.22, 2.25, 2.31). Despite its popularity, it
produced a weak frame and did not work well once
truss rods were introduced, because the pressure
exerted by the rods was placed on mortised joints
rather than on the butt ends of the sills, which
were much better equipped to withstand the load.
The protruding sill ends were also a positive menace to trainmen walking between the cars to couple and uncouple. The old pattern had largely disappeared by the middle 1870s.[71]

ROOFS

Boxcars must have more than solid framing to
sustain their cargo. They need weatherproof
bodies to protect shipments, and for this a tight
roof is essential. The arch roof, as already noted,
prevailed during the pioneer period (Fig. 2.32). Just
why it persisted for so long is difficult to establish,
for it seemed to offer no special advantage over the
simpler peaked roof that came into general favor
for boxcars around 1870. Perhaps it was because of
the pattern established by passenger cars, where
architectural appearance was more important. Car
builders may simply have championed the arch-
roofed freight car for so long because it was both
familiar and graceful.

In an arch roof, the carlines or roof rafters were
one piece, which called for a long, high-grade piece
of lumber bent to shape. This was a trifle expensive and elaborate for a freight car. Because the carlines ran crosswise, the sheathing or thin wooden
roof boards had to be nailed on lengthwise to the
car body. Should a leak develop in the surface
of the roof covering—a certainty in any wooden
structure subjected to constant twisting and vibration—the water would follow the joints between boards until it passed through a crack into
the car. In a peaked roof, however, purlins—small,
longitudinal wooden strips—were fastened to the
carlines. The roof boards, nailed to the purlins,
were now parallel to the carlines and hence ran
transverse of the car. These boards were very short
compared with those used for arch roofs, and they
sloped down from the peak to the car sides. If water found its way under the tin or canvas covering,
it tended to run down the joints of the tongue-and-
groove boards rather than puddling. Because of the
incline and short passage, the water would tend to
flow out to the roof's edge before leaking inside.
The scheme didn't always work, of course, water
having a perverse way of penetrating any structure
no matter how well arranged, but the odds were
more in favor of this scheme.

Peaked roofs had other advantages. The carlines
were easily cut to shape on a table saw and did not

tive and visually prominent features of the American freight car in the last quarter of the nineteenth
century.

Truss rods had been used in less visible ways at
an earlier date. It was common to run vertical rods
inside the body from the top rail, near the door
opening, straight down and through the outside
sill. A cross-timber was often used at this point to
catch the bottom end of the rod. Several examples
are illustrated elsewhere in this chapter. The body
was further stiffened by placing rods at an angle opposite the wooden truss members—again, inside
the body walls, where they were not visible. An
early example of this practice can be seen in the
Illinois Central car of around 1855 (Fig. 2.21). Note
the ends of the angle rods with their neat little
cast-iron pads or washers on the underside of the
side sills near the door opening. Truss rods were
also used to tie the structure together side to side
by running rods alongside of, through, or up and
over but always parallel to the body bolster (Figs.
2.27, 2.29). Dougherty's patent drawing of 1843
(Fig. 2.19) also shows good use of transverse tie
rods.

A more substantial and expensive method of
cross-bracing was introduced in 1852 on the Illinois Central.[70] This was the all-iron body bolster
fabricated from wrought-iron plate. This scheme
was rather slow in coming into general use, as will
be outlined in subsequent chapters. The cast-iron
bolster used on the earliest Baltimore & Ohio boxcars was abandoned without explanation, and one
can only assume that this was done because of the
brittle nature of the material, which did not withstand the shocks and hard service of its office. It is
also possible that the car department decided that
cast-iron bolsters were simply too expensive because iron was so much dearer than timber. Yet another explanation would be simply the preference
of the presiding master car builder. Thatcher
Perkins favored iron trucks, while his successor
J. C. Davis, in a seemingly retrogressive step,
adopted wood-frame trucks.

There is one more detail of framing that began to
wane during the last years of the pioneer period,

THE PIONEERS

FIGURE 2.32

This scene is on the Pennsylvania Railroad at the Duncannon Bridge around 1870. The arch-roof boxcar in the foreground illustrates several details of early freight car construction. The design, if not the car itself, probably dates from around 1855. Notice the absence of truss rods, the wood-beam trucks, the metal corner braces, and the end steps. (Pennsylvania Railroad Photo)

FIGURE 2.33

This Lake Superior & Mississippi boxcar produced by Harlan & Hollingsworth shows many of the reforms coming into favor by the late 1860s. These include peaked roofs, butt-end sills, truss rods, and one-piece center sills. The body measured 28 feet 1 inch long by 8 feet 3 inches wide. Its length over the couplers was 31 feet. The truck wheelbase was 54 inches. (Smithsonian Neg. 81-8129)

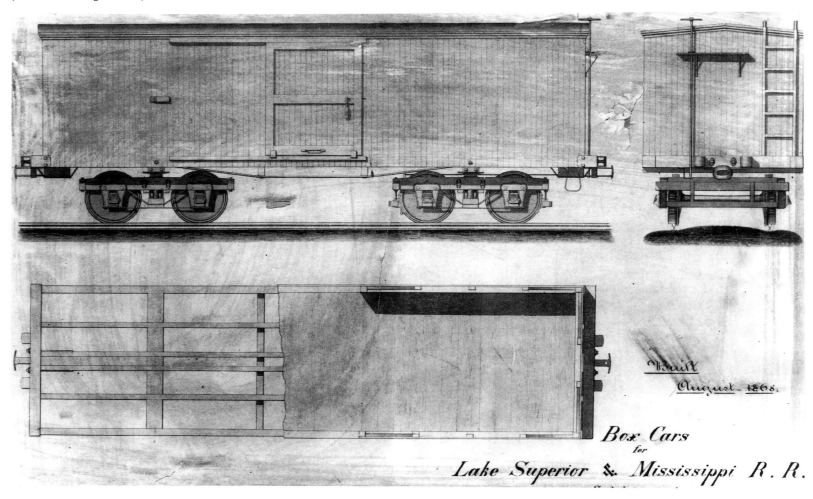

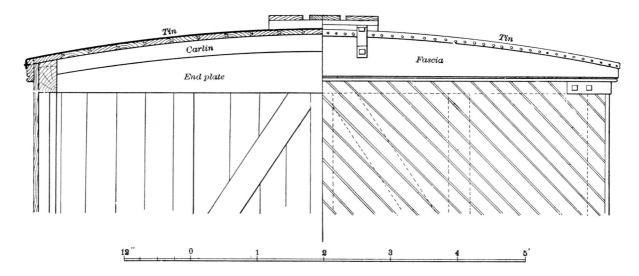

have to be steamed and bent to shape as they did for arch-roofed cars. The peaked roof was in all simpler and more effective than the arch roof, and once its qualities were understood it became the common standard. One of the earliest examples is the Bruce engraving of 1848 reproduced earlier (Fig. 2.26). A Lake Superior & Mississippi car made by Harlan & Hollingsworth in 1868 shows the plan in its final form (Fig. 2.33). Drawings for the old-fashioned arch roof and the more modern peaked roof appeared in William Voss's 1892 *Railway Car Construction,* and while the book appeared late in the wood-car era, the engraving reproduced in Figure 2.34 illustrates the parts and assembly methods generally used for this segment of freight car construction.

Methods for waterproofing board roofs had been developed centuries before railroads were even envisioned, and these traditional coverings were simply adopted by the car-building fraternity without further invention. Zinc rolled into thin sheets and soldered at the joints was an excellent material for this purpose, and one favored since antiquity for its durability and low maintenance. It was, however, comparatively expensive in first cost. The extra expense would be trifling for a cathedral or palace, but it did seem considerable for a short-service-life structure such as a railroad car. Even so, zinc was employed for this purpose. Harlan & Hollingsworth used it for its best-grade freight cars, as mentioned in its 1839 prospectus. The Boston & Worcester required it in the specifications dating from 1835–1840 printed earlier in the chapter (Tab. 2.1).

Tin-plated sheet iron was a far cheaper but very serviceable roof covering. As with zinc, the joints were attached and sealed by solder. Just when tin was first used for railway cars is difficult to determine, but it may not have been so applied until rather late in the pioneer freight car period. There is mention of it in a letter from Perkins to John Garrett, president of the Baltimore & Ohio, dated June 9, 1860.[72] Possibly it was new only to the Baltimore line, because tin roofs for building purposes long predated this application. The Hartford, Providence & Fishkill was using tin in 1864, but how much earlier the line had adopted this material is not specified.[73]

Tin must be periodically painted, unlike copper or zinc, to protect it from the weather, and because the coating is so thin, it cannot withstand much traffic lest it be punctured or broken. A brakeman struggling to maintain his balance could be expected to tread none too lightly off the roof walk boards and onto the fragile tin surface. Ice and snow only made the brakeman's ambulations more unsteady and the chances of roof punctures more real. Heavier galvanized sheet metal offered a solution, but only with a greater investment. Material costs of even a few cents more per square foot could add up rapidly when figured for freight cars, which were purchased in lots by the hundreds or thousands.

Canvas was by far the cheapest covering available. It was painted, and while the paint was wet fine sand was sprinkled on to offer a less slippery surface for trainmen. The sand would also slightly improve fire resistance against sparks from the engine. Painted canvas was low in first cost, yet like all provisional economies it proved costly in the long run, for it required frequent renewal and maintenance. It was easily torn and much more likely to develop leaks than just about any other roof covering commonly available. Damage claims over spoiled merchandise must have done much to promote its retirement in favor of more durable materials. Except for tin, all the coverings just described gave way to double-board roofs around 1870, a development to be described in Chapter 3.

PAINTING

Drawings and photographs document the size and structure of cars, but regrettably, only the most meager information exists on the colors and finishes used in the time before color photography. The assumption that boxcars were always painted the familiar brown-red appears falacious. The Boston & Worcester specifications dating from 1835–1840 (Tab. 2.1) state that the finish was to be a slate color. The Northern Central's specifications for 1852 call for dark brown paint and metal lettering. The rendered drawing for a Philadelphia, Wilmington & Baltimore car of 1861 shows a pale yellow body, tan roof (canvas coated with sand?), black hardware, and red wheels (Fig. 2.30). Coal cars produced during this period by the same builder, Harlan & Hollingsworth, are painted black. A photograph taken in 1858 of a wreck on the Naugatuck Railroad shows a boxcar painted in white or very light yellow or gray.[74] Admittedly, the evidence on this subject is fragmentary, but neither can any specific evidence be found for boxcar red, a shade so universally associated with American railroading in more modern times.

Other Special-Purpose Cars

In an effort to increase the revenue mileage or productivity of the car fleet, builders began modifying ordinary boxcars for a greater variety of cargoes. Combination cars for cattle and general merchandise have already been mentioned. Both the Michigan Central and the Illinois Central carried this idea a step further in 1855 and 1856 by inaugurating combination passenger/freight cars.[75] Seats were installed for emigrants going west and removed for grain and other products going east. A far more basic and sensible idea for making the boxcar suitable for dual carriage emerged just a few years before these emigrant cars entered service. Hugh Gray, master car builder for the Galena & Chicago Union (Chicago & North Western), is said to have introduced the grain door, thus making the boxcar a more efficient carrier of bulk commodities. The idea was admirably simple. It consisted of a close-fitting half-door mounted inside the door opening. Loose grain could be loaded inside the car up to the top level of the grain door. The

regular outside sliding door was then closed to protect the contents. When not needed, the grain door was moved away from the opening. This was probably the most successful and long-lasting scheme for converting the boxcar into a practical dual-purpose vehicle. The cost of the plan was minimal, and it did not interfere with the vehicle's use for merchandise traffic. Examples will be noted throughout Chapter 3.

COAL CARS

Coal became a common fuel after the Civil War and began to displace wood for home heating, locomotive fuel, and iron making. As the eastern Pennsylvania anthracite fields and soft-coal deposits in western Maryland and southern Illinois were exploited, coal became a major source of traffic. Just how best to carry it remained a question, and a variety of cars were developed for this purpose. The industry agreed on very similar designs for boxcars, but there was a diversity of opinion on the best form for coal cars. Most of the hard-coal carriers such as the Reading, the Lehigh Valley, and the Central Railroad of New Jersey came to champion very small wooden four-wheel hoppers commonly called jimmies. Other lines not having such a large traffic, like the New York Central, were content to copy this design. For most of the nineteenth century jimmies prevailed in numbers if not in overall capacity. The Baltimore & Ohio favored iron pot hoppers, while the Reading seriously considered four-wheel iron hoppers for around thirty years and then began a gradual conversion to wooden eight-wheel cars. Other lines found low-sided gondolas satisfactory, and most were willing to ship coal in almost any style of car when traffic was up and equipment was scarce.

The jimmy was obviously an enlargement of the very small hoppers that first appeared on the Delaware & Hudson and Mauch Chunk gravity lines in the late 1820s. Drawings for these pioneers, with their narrow hoppers and wheels outside the floor frames, appear earlier in this chapter (Figs. 2.2, 2.4). Regrettably, no drawings exist for the first

FIGURE 2.35

Philadelphia & Reading jimmy of around 1860, featuring Millholland's wooden plank spring. The journal boxes were bolted to an ash spring located just below the main frame. (Reading Co.)

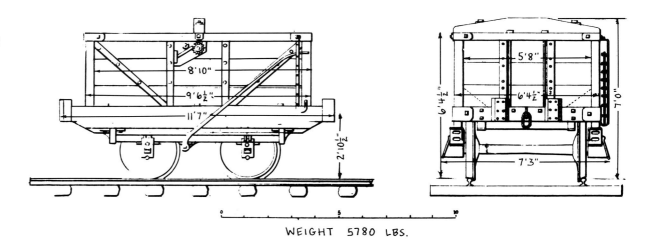

WEIGHT 5780 LBS.

THE AMERICAN RAILROAD FREIGHT CAR

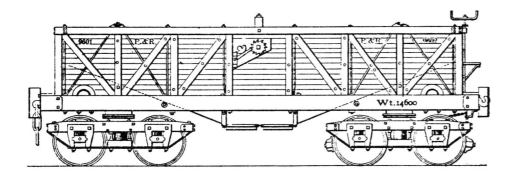

FIGURE 2.36

Eight-wheel hoppers began to succeed jimmies on the Philadelphia & Reading in the early 1860s. This car carried 10 tons of coal. (*Locomotive Engineering*, Aug. 1896)

Reading jimmies, but a published report of 1840 notes that they had 36-inch wheels, a 54-inch wheelbase, and a weight of 1.50 to 1.65 long tons.[76] Capacity was not given, but it must have been similar to that of the cars just mentioned.

By the 1850s the jimmy had doubled in capacity. This was achieved by raising the body over the wheels and making it wider. The overall length was not materially increased. Smaller wheels, usually 30 inches in diameter, made the new body arrangement possible without unduly raising the floor level. Drop-bottom hopper doors were standard. Jimmies weighed 3.0 to 3.5 tons and carried 5 to 6 tons of coal. Specifications for an 1855 Northern Central car of this type, together with a drawing for a Philadelphia & Reading car dating perhaps a few years later (Fig. 2.35), will more clearly illustrate the jimmy in its early maturity:

Specifications for constructing 50 Coal Hopper Cars for $160 ea. Oct. 16, 1855

The wheels to be 30 in. diameter, made by Messrs. Whitney of Phila., to be drawn on the axles tightly, according to the gauge furnished.

Axles to be made of best charcoal iron equal to those of Mr. Lawrence' manufacture, journals & wheel seats cleanly turned off to the exact size of drawing of patterns, wheel boxes to be made like drawing, with brass bearings, lined with Babbit metal truly fit to the journals.

The Springs to be made of two pieces of good ash 2¾ × 6 in. × 7 ft. 6½ in. long, with gum cloth under each end and in the centre as shown on drawing.

Wheel boxes bolted on the springs 51 inches from centre to centre with a piece of gum under each box.

The end Buffers must be made of the best white oak of two pieces 2⅜ in. thick, being 4¾ × 8 and secured by wrought iron straps passing around them, and bolted to the side sills with two ¾ in. bolts as shown by drawing, a cast iron plate to be let in & bolted to buffer head as shown on drawing.

Pulling Straps to be made of 3 × ⅝ and 3 × 1 in. wrought iron and arranged as shown on drawing.

The Side Sills of the body to be of N.C. Pine or best white oak 8 in. deep by 4 in. thick, the intermediate sills, two in number, to be 3 × 14 from end to end of same material.

The Body to be 10 ft. long, 7 ft. 9 in. wide, 37 in. high from top of sill to top of body and 48 in. from bottom to top of Car, Corner posts to be 4 × 5 in., intermediate posts & braces 4 × 2½ in. all of good white oak or N.C. Pine, the Sides and Ends to be Weather boarded with 1 in. White pine boards, the floor to be of 1¾ in. White oak, two braces of 2 × ⅝ in. wrought iron across the bottom and bolted to inside Sills to support the floor, the hinges to be of 2 × ⅝ in. wrought iron, to extend all around each drop door to which the boards are bolted with ½ in. bolts, two vertical tie rods on each side and ends of ⅝ in. round iron as shown on drawing, top rails of best White oak or N.C. Pine 3½ × 3½ in. with a beam across the top of White oak 4 × 8 in. in centre and 4 × 5 in. at ends, a rod through the centre, as shown on drawing to hold up the drop doors, & two vertical tie rods bolted to the inside sill, running through the cross beam as shown on drawing, two coupling bolts to be chained on each Car, one of the bolts of each Car to run through the cross sill with a key in the end as shown on drawing.

The Brakes on one side, so as to break [sic] on two wheels only the blocks to be shod with 3½ by 3½ in. wrought iron, the back of the blocks where the lever operates upon them, is to be lined with ¼ in. wrought iron, with a flanch 1½ in. deep on each side.

The Cross Sills to be of good White oak 8 × 10 in.

The Cars to have two heavy coats of Mineral paint.[77]

These cars were built by George W. Ilgenfritz and John A. Nevin of York, Pennsylvania.

The Reading remained loyal to the general design shown in Figure 2.35 until 1873, when it shifted to eight-wheel wooden hoppers.[78] Meanwhile, the iron-bodied hoppers once so much in favor were being retired or placed in storage. In 1875 the line had 5,434 wooden jimmies in service and was building more as late as 1884. But in the early 1850s attention shifted to developing a 10-ton wooden eight-wheel hopper. The side elevation reproduced in Figure 2.36 was drawn from memory by C. H. Carruthers, who worked around these vehicles as a car checker when a boy. He describes them as having side sills 4 by 8 inches by 19 feet long.[79] The wooden hopper bottoms were covered with sheet iron; in later years self-supporting iron plates formed the hopper bottoms. The trucks were cushioned with cylindrical rubber springs. The cast-iron plate wheels were only 28 inches in diameter. English-style hook couplers with large-link chains were used. The cars weighed 7.3 tons. They were painted black with white lettering. The double numbering system was used because they were considered the equal of two jimmies, and hence each end of the car had its own number. The design proved successful, and by 1870 the Reading had 5,723 cars of this type in service, according to figures printed in *Poor's Manual of Railroads*. Before the century closed they had all but replaced

THE PIONEERS

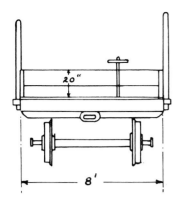

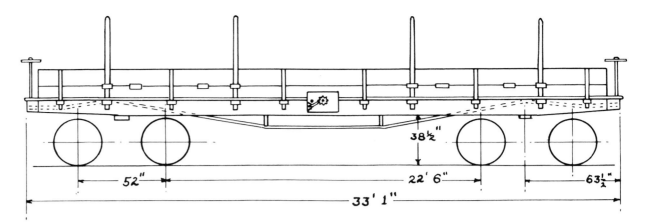

FIGURE 2.37

This Pennsylvania Railroad drop-bottom gondola, designed in 1869, was a makeshift effort to simplify the unloading of coal from non-hopper-bottom cars. Weight was 18,500 pounds, capacity 28,000 pounds. (Traced from an engraving in *Engineering*, Nov. 2, 1877)

what had been one of the largest jimmy fleets in the country. A listing in Chapter 5 (Tab. 5.1) more clearly shows the evolution of the Reading's coal car fleet.

Specialized coal cars were a necessity for lines with a voluminous on-line traffic, but other roads whose coal shipments were more limited were less inclined to invest in equipment suitable only for this traffic. Gondolas, while not a very efficient vehicle for the carriage of bulk materials such as coal, could be used for lumber, rail, pig iron, and any number of other products. Hence when the Michigan Southern acquired its first coal cars in 1865, it purchased low-sided gondolas.[80] These multipurpose vehicles had 16-inch-high wooden sides and a capacity of 10 tons. It could carry as much coal as a hopper, but the labor of unloading it was considerably greater—an expense and inconvenience that were tolerable so long as the tonnage handled remained moderate.

The Pennsylvania developed a drop-bottom gondola in 1869—probably not the first, but the first for which we have documentation—that overcame the unloading problem, at least in part (Fig. 2.37). The floor was flat, and so the coal not directly over the hatches had to be shoveled into the opening. Other big coal haulers facing an equipment shortage pressed low-sided gondolas into coal service, a fact documented by an 1871 photograph made on the Baltimore & Ohio somewhere near Connellsville, Pennsylvania (Fig. 2.38). The cars, dating from a decade or more before—at least in design—weighed 13,600 pounds and carried 20,000 pounds.

FLATCARS

Flats or platform cars were the simplest form of freight car employed for revenue service. They were suitable for the carriage of anything that either did not require protection from the weather or was too bulky to fit inside a boxcar. Lumber, rail, bar iron, and even hay were common cargoes. Rail and timbers were self-supporting so long as they spanned the body bolsters, and in such cases the load-limiting factor was not the car's framing but rather the trucks, springs, wheels, axles, and jour-

nals. Machinery—particularly oversize loads like boilers, machine tools, and farm apparatus—was often shipped by flatcar, not because it was so immune to the elements but simply because it could not be moved in or on any other style of equipment, except perhaps a gondola. Flats having no sides or ends were more easily loaded, but some gondolas were made with stake sides that could be lifted out of staples or pockets fastened to the side sills. Conversely, flatcars generally had stake pockets to receive upright posts or stakes that assisted in confining loose materials.

Historically, flatcars can be traced to the very beginnings of American railroading. The B&O flour car and the Petersburg Railroad gondola pictured earlier (Figs. 2.8, 2.10) are examples of the origins of the flatcar. Dougherty's giant eight-wheel flats for the Portage Railroad (Fig. 2.19) illustrate its progress into the double-truck era. Flatcar drawings typically show the special attention given to heavy framing. Flats were at the same time the least-talked-about freight car, and documentation for these early years is extremely scarce. Fortunately, the arrival of an unusual shipment prompted an occasional photograph of the otherwise ignored platform car. Examples are shown in two early prints reproduced nearby (Figs. 2.39, 2.40).

The first was taken in 1859 on the Milwaukee & Mississippi Railroad. The cargo, a curious river ice boat, so attracts the eye that one is tempted to ignore the vintage platform car shown somewhere on its journey to the Mississippi River. But a careful study of the photograph reveals some interesting facts about flatcars of the 1850s; this veteran shows enough service scars that we can safely assume it had seen several years of operation and so must date from 1855 or earlier. Notice the deep side sills tapered at each end and the absence of truss rods. The end sill is mortised into the side sills and extends beyond them. The neat stake pockets appeared to be cast and are most likely malleable iron. The arch-bar trucks are well developed, and the apparent absence of springs is no doubt explained by their placement inside the wheels, between the truck bolster and spring

FIGURE 2.38

A string of veteran B&O gondolas await coal loading somewhere near Connellsville, Pa., in 1871. (Smithsonian Neg. 49645)

FIGURE 2.39

The curious ice boat *Lady Franklin* en route to the Mississippi River on a flatcar of the Fox River Valley Railroad sometime in 1859. (Milwaukee Road Photo)

THE PIONEERS

FIGURE 2.40

A U.S. Military Railroad flatcar is used to test a portable bridge truss in Alexandria, Va., in about 1863. The truss rods and wooden brake shoes are clearly shown. (National Archives Neg. BA1748)

FIGURE 2.41

Harlan & Hollingsworth built this car in 1860 for a railroad in Cuba. The platform measured 25 feet long by 7 feet 7 inches wide. The truck wheelbase was 4 feet. (Smithsonian Neg. 81-8106)

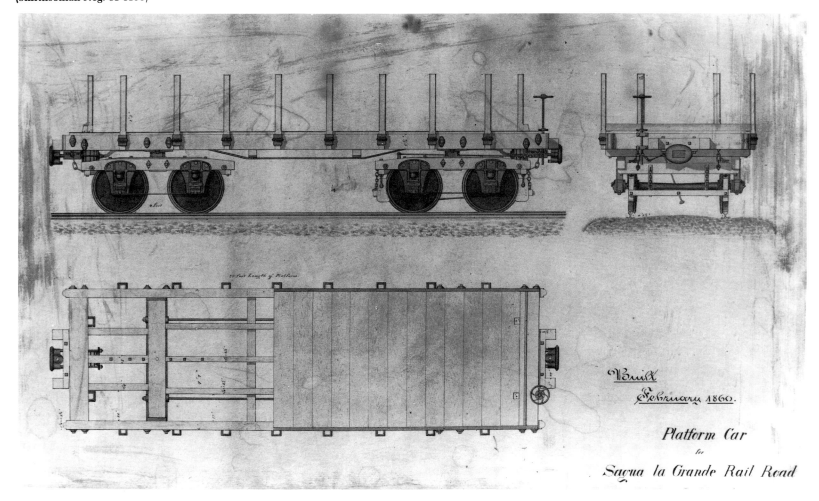

Platform Car
for
Sagua la Grande Rail Road

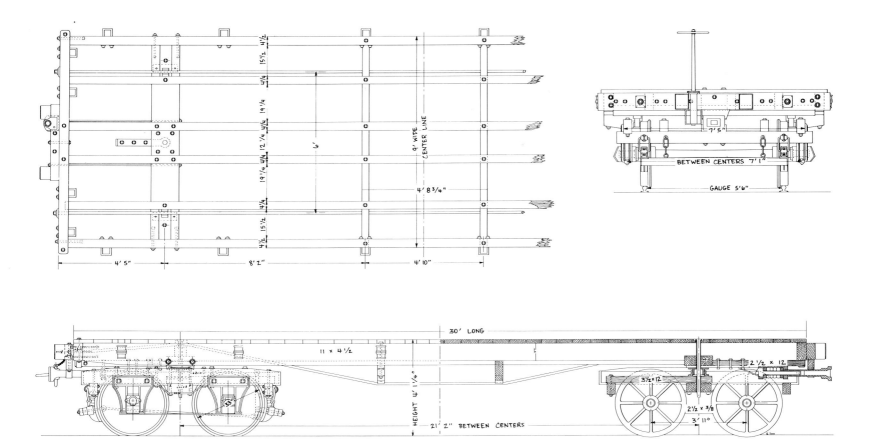

FIGURE 2.42

The Portland Co. built a group of broad-gauge flatcars for the Bangor & Piscataquis Railroad in 1869. The design is somewhat obsolete even for the late 1860s and might well represent a car plan of around 1855. (Traced by E. Tone from a drawing in the Maine Historical Society, Portland)

plank, which would make them invisible in the photograph. Only one end of the adjacent flatcar is shown, but it documents one item of major interest to car construction: the flat-bar truss rod visible on the extreme right-hand side of the picture between the bottom of the side sill and the truck.

The second photograph recording a vintage flatcar was taken during the Civil War by or for Matthew Brady, at the Alexandria, Virginia, depot of the U.S. Military Railroad to document load tests for a portable bridge truss (Fig. 2.40). This car is a revealing record of freight car construction of the 1860s. It shows early use of truss rods; note that only two are used, and these are fastened to the side sills. There are no iron queen posts; the rods bear on wooden beams. The side sills extend beyond the end sills, showing the old style of car framing soon to be considered obsolete. Other items of interest are the wood-beam trucks, the wooden brake shoes, and the two stakes upright in their simple strap-iron pockets.

An even better understanding of platform car design can be gained from a Harlan & Hollingsworth drawing for the Sagua La Grande Railroad of Cuba (Fig. 2.41). Nothing about the car suggests that it was specially created for a foreign purchaser; it is most likely a fair representation of what this builder was producing for its domestic customers. The car measures 25 feet long by 7 feet 7 inches wide. The side sills are heavy and measure 9¾ inches by 5¾ inches. The three lighter intermediates are broken at the bolsters. The side sills ex-

tend beyond the massive end sills. Notice the two longitudinal truss rods, sans queen posts, placed well inside the frame, and the heavy transverse truss rods at the end sills and body bolsters. The stake pockets are iron castings fastened to the side sills with U-bolts or massive staple nails. Typically, the brakes are mounted on only one pair of trucks. The wood-beam trucks have 33-inch wheels placed on 4-foot centers.

Some nine years after the Cuban flats were delivered, a group of platform cars were produced for the Bangor & Piscataquis Railroad by the Portland Company of Portland, Maine. The original assembly drawing is preserved at the Maine Historical Society (Fig. 2.42). The design, rather old-fashioned even for 1869, represents a plan that might well date back to the middle 1850s. Despite the 5-foot 6-inch gauge, the platform measured only 9 feet wide. The frame measured 30 feet long over the end sills. The side sills measured 4½ by 11 inches, while the intermediates were somewhat smaller, at 4¼ by 9 inches. Note the slight taper at the ends of the side sills. Two truss rods were used to stiffen the frame. The wood-beam trucks had a wheelbase of 3 feet 11 inches. Single-plate 33-inch cast-iron wheels were used. Cast-iron link-and-pin couplers were attached to the draft gear by a stem and key. Rubber springs cushioned the coupler. The draft gear was securely fastened by a flat bar with an eye at one end. The long truck king pin passed through the eye. Wooden dead blocks helped protect the coupler from buffing damage.

The Photographic Record

In the foregoing discussion, photographs have been mentioned as one source of information on American freight cars. They are, of course, a wonderful record because they show an exact likeness and not the reconstructed image of an engraver or artist. But it must be understood that very few photographs of freight equipment exist much before the 1860s because the art of photography, except for portraiture, was still in its infancy. Even so, readers will find surviving examples, particularly post-1860, available in published sources. One of the earliest views, not already mentioned, is an ambrotype of yards and grain elevators at the mouth of the Chicago River made in 1858. In the foreground are several boxcars, one of which bears a strong resemblance to the drawing of the Illinois Central car shown in Figure 2.21.[81]

Possibly the largest number of early U.S. freight car photographs are among those made during the Civil War. Cameramen working for Matthew Brady and others, with no particular interest in the subject of railroad rolling stock, inadvertently became the most faithful boxcar photographers in the nation. The views made in and around the U.S. Military Railroad shops in Alexandria, Virginia, between 1862 and 1865 are filled with images of box-, flat-, and cattle cars. Arch roofs, wooden trucks, and other features are well recorded. One view made just outside Alexandria on March 28, 1863, shows a splendid view of a boxcar with a broken axle. Further afield, photographers were on hand for the fall of Chattanooga, Nashville, and Atlanta as the war moved south. In one yard scene at Chattanooga, three freight cars stand in the foreground. The first is a boxcar with a peaked roof, a roof-mounted brake handwheel, and horizontal siding. Next stands a ventilated box with an arch roof and small end doors. Behind it is a great rarity, or what at least appears to be: an open-top stock car.

Good images of freight cars, often overlooked, are visible in other widely reproduced Civil War pictures. One scene at Hanover Junction, Pennsylvania, provides an excellent view of a peak-roofed boxcar, belonging to either the Pennsylvania Railroad or the Northern Central, parked on a side track. It shows the large handwheel and iron grab irons. The left side of the picture is often cropped to emphasize the passenger train and depot, because layout artists assume, no doubt, that no one is interested in so mundane a subject as an 1860 boxcar. The familiar view of the ruins of the Atlanta roundhouse after Sherman's devastating visit includes a Georgia Railroad boxcar with an arch roof and horizontal rather than the more common vertical siding. Happily, all the pictures mentioned here have been reproduced in numerous published works on both the Civil War and rail-

FIGURE 2.43

This fine end view of a boxcar at Promontory, Utah, was taken by A. J. Russell in the summer of 1869. (Oakland Museum, Oakland, Calif.)

THE AMERICAN RAILROAD FREIGHT CAR

road history, but the most complete and convenient collection is offered in George Abdill's picture history, *Civil War Railroads*, published in 1961.

A. J. Russell, once an associate of Brady, was commissioned to record the construction of the Union Pacific Railroad during the 1860s. Many of his negatives, now at the Oakland Museum in Oakland, California, include freight cars. One of the best is so clear and sharp that it is hard to believe the negative was exposed well over a century ago (Fig. 2.43). In another view two flatcars are pictured at a construction site on the Green River. Those doubting that flats during this period used two rather than four truss rods are urged to study this photograph. Like the Civil War pictures described earlier, Russell's images have been widely published.[82] Early published views of eastern equipment are more widely scattered, but one convenient source is the volumes covering the Pennsylvania Railroad and its numerous subsidiaries by E. P. Alexander.[83]

The first ten years of the American freight car witnessed many fundamental design changes. The pace and degree of innovation were so intense as to be revolutionary. Open cars gave way to closed bodies at the same time eight-wheel undercarriages began to succeed the four-wheel variety. Radical transformations are to be expected with a new technology when nearly every mechanical detail is experimental, untested, and open to question. By the early or middle 1840s matters had settled down. A consensus on what constituted good car design was reached, and uniformity, at least on general arrangement, was achieved. Details of framing, roof coverings, and trucks varied, of course, from road to road. Car size and capacity gradually increased during the pioneer period so that by the 1860s vehicles with a 10-ton capacity were common. Spoked wheels had been largely abandoned by the 1850s in favor of cheaper, stronger plate wheels. Some roads were using arch-bar trucks, but the wood-beam variety remained very popular.

So long as freight traffic remained small in volume and confined to short, on-line movements, radical improvements in car construction were unnecessary. Why develop more rugged, large-capacity equipment to carry a wagonload of beets 40 miles? The cars never left the home road and were ever under the watchful eye of their owners. Light, small cars were perfect in their place. But new conditions evident in the early 1860s called for a searching reappraisal of freight car design. The little arch-roof 10-ton boxcar that had served so well for nearly twenty years would soon become an antique and would disappear as completely as the Conestoga wagons they had made obsolete a generation earlier.

3 CARS FOR GENERAL MERCHANDISE, 1870 TO 1899

The basic, general-service car to emerge from the pioneer period of American railroading, discussed in the last chapter, was the familiar boxcar. It was seen as the universal goods wagon that carried lumber, grain, furniture, machinery, or barreled whiskey equally well. Just about every product of man and nature found shelter and carriage within its wooden walls. Its essential characteristics were well formed before 1870. The eight-wheel undercarriage, rectangular body, and sliding side doors were already a national fixture by this date.

The development of the boxcar over the next thirty years should not be dismissed as an expanded orthodoxy. True, its capacity grew from 10 to 15, 20, 30, and finally 50 tons while its cubic volume doubled, from roughly 1,000 cubic feet to 2,000, by 1900. Its dead weight (weight-to-capacity ratio) plummeted nearly 40 percent during the same time period. Malleable iron and steel began to replace cast and wrought iron. Far more metal than ever before was introduced to strengthen the structure. Air brakes and automatic couplers were widely adopted after 1890. In 1870 only the most visionary car builders would have predicted so many changes. Even fewer could have foretold the astonishing growth in traffic that created the demand for larger and better rolling stock.

Veteran car builders could look back over the maturing years of the wooden boxcar with warrantable pride. They had accomplished much by working within the borders of a traditional design and conventional materials. Wooden beams and iron rods were artfully combined for ever-larger vehicles while tons were literally shaved off. The lower vehicle weight saved fuel and reduced track maintenance. More tonnage was moved for less money. Shippers, railroad investors, and, presumably, the consumer benefited. The car builders could not help but indulge in a little self-congratulation during their annual meetings at Atlantic City, Saratoga Springs, and such other watering places of the day. Yet in the background there was an undercurrent of criticism, some of which occa-

sionally emanated from those inside the profession. If there was reason for pride, so was there cause for criticism.

David L. Barnes, a consulting engineer much involved in car and locomotive design, said to a meeting of the Western Railroad Club that American freight cars were structurally unsound.[1] They were, in his opinion, too weak to withstand normal use without regular failures. Ideally, a car should run year after year without constant frame, bolster, and draft-gear repairs. Just a few years after Barnes's remarks George L. Fowler, editor of the *Railroad Gazette*, took the car builders to task by saying that the American boxcar staggered down the track in such an unsteady fashion that one expected it to collapse under its cargo. Thousands stood idle each day awaiting repairs, while most of the operating fleet was in a state of "chronic decrepitude."[2] Too much dependence on rule-of-thumb formulas result in "eyeball designing," whereby a draftsman would add an inch to the sill size and a quarter-inch to a truss rod and then figure that the vehicle's capacity had been augmented by 25 percent. Fowler admitted that calculating stress was far more difficult for moving vehicles than for static structures, but he called for more science and less dependence on experience—advice no doubt much resented by old-line car builders.

Many years before, Fowler's predecessor at the *Railroad Gazette*, M. N. Forney, longtime secretary and honorary member of the Master Car Builders Association, had been even less kind.[3] He declared that freight car design not only lacked science but was not a precise mechanism, as evidenced by the rudeness of design and the crudeness of individual parts. Not too many months after Forney's editorial, an anonymous writer, who would only identify himself as a civil engineer, allowed that the nation's car builders were for the most part practical mechanics and not engineers.[4] As craftsmen they had a good feel for materials and their application, but their knowledge was limited to existing designs and proportions. They did not

FIGURE 3.1

Some railroads like the Portland & Rochester were still buying small, old-fashioned boxcars like this Portland Co. product of 1871. Note the arch roof, wood-beam trucks, and lack of truss rods. (Thomas Norrell Collection, Smithsonian Neg. 85-28723)

FIGURE 3.2

The fully developed wooden box-car is represented by this Baltimore & Ohio car built in 1896 by Mt. Vernon. It could carry 20 tons more than the car shown in Figure 3.1, yet it weighed only about 6 tons more. Note the peaked roof, truss rods, arch-bar trucks, air brakes, and Janney couplers. (Baltimore & Ohio Railroad Photo)

compute. Nor did they understand the theory of structures and the function of their individual parts. A skilled designer, such as an engineer, made the best use of materials and did not waste labor and capital. The chronic breakdowns of freight cars were not a matter of age or hard use but the result of ignorance and quackery.

While these heavy and not entirely accurate broadsides went unanswered, car builders surely understood that the wooden freight car was an imperfect vehicle in many of its details. That it was underbuilt and overused there was little argument. To hold down weight, structural members were kept to minimum dimensions, with little material being allowed for a margin of safety. To hold down cost, cheap and easily available materials were employed. And everyone connected with shipping knew that cars were overloaded. In earlier years the *Railroad Gazette* actually commiserated with car builders over this problem, putting the blame squarely on errant shippers.[5]

The Philadelphia, Wilmington & Baltimore, finding its cars overloaded by 5.5 to 8.0 tons, billed its customers 25 cents per hundredweight for overloading.[6] Others contended that it wasn't always the shippers' fault. James J. Hill found as many as four weights painted on the sides of cars and discovered that track scales were generally unreliable.[7] Another western line fraudulently increased the capacity of some of its cars with nothing more than a stencil and a paint brush, only to see that portion of its fleet go to pieces.[8] Even shippers acting in good faith might load to excess in such circumstances.

The overloading of certain components was one area where most car builders tended to agree with their critics. Journal or wheel-bearing loadings were very high. A British correspondent was disturbed to find journal loadings in the United States at 153 pounds per square inch, or about 40 psi greater than in England.[9] Yet the Master Car Builders Association had even more disturbing fig-

CARS FOR GENERAL MERCHANDISE

FIGURE 3.3

Progressive railroads like the Pennsylvania began to adopt larger, stronger boxcars in the middle 1870s. Truss rods—only two at first—and arch-bar trucks are features of the improved postinterchange freight car. Hand brakes and link-and-pin couplers, however, remained a standard until after 1890. (Pennsylvania Railroad Photo)

ures to offer in a report published in 1882.[10] The builders figured that in the old days of toy equipment, a 10-ton boxcar would register journal pressures of 167 psi. They worried about the new 30-ton car just coming into favor. Pressures were expected to reach 379 psi, more than what was carried by most locomotive journals. The problem would only be aggravated by greater freight train speeds, for the current "creeping along" rate of 12 miles per hour would not be tolerated much longer. It was also acknowledged that the old 10-ton limit was dictated not by weak floor frames but by undersize journals and axles. There was admittedly much reform needed in the area of wheels and draft gears—a matter treated in Chapter 7.

The Debate over Size

Most car builders were disinclined to contribute to the debate over the merits and faults of American rolling stock. Their energies were directed to its fabrication and maintenance, and if they had a common goal, it would appear to have been the creation of ever-larger cars.

THE BIG-CAR MOVEMENT

The argument for larger cars has been explained elsewhere in this volume as the economics of scale. The savings resulting from fewer, larger units moving the same tonnage meant shorter trains and lower labor costs. Car builders found the big-car bargain especially attractive because when just 1 square foot of floor space was added, cubic space inflated geometrically. As all architects had realized ages ago, a 1-foot square equals 1 cubic foot, but a 2-foot square offers 8 cubic feet. The Michigan Central capitalized on this law of

space and gained 324 cubic feet by adding only 1 foot in length and 11 inches in height to a new class of boxcars.[11] Of more consequence, weight and costs did not increase proportionately. The Pennsylvania Railroad calculated that a 40-ton car cost only $50 more and weighed only 3,000 pounds more than a 30-ton car.[12]

The Pennsylvania's mechanical chief, Theodore N. Ely, became an advocate of big cars in the wake of the 1873 panic. Like all railway officials of the time, he was seeking ways to reduce costs. Ely concluded that an increase in train size was the most direct way to accomplish this goal. This meant larger locomotives and bigger cars. In 1876 he began building 15-ton-capacity boxcars. These units offered a 50 percent jump in weight capacity, which required stronger trucks and axles. Ely adopted double-plate wheels made from cast iron toughened by a 10 percent mix of steel. The cubic area was not much larger than that of the old 10-ton standard, because Ely figured that longer cars would require more yard space and longer sidings. He figured that adding even 2 feet to the length of a car would require a 33.3 percent enlargement of yards and sidings. The cost would more than cancel the estimated productivity gains of the longer cars. Longer cars did gradually win acceptance, but the examples given in this volume show that growth in length was far more conservative than the growth in weight capacity. Cars of 20- and 30-ton capacities might both measure only 30 feet in length. The industry as a whole tended to side with Ely on the question of car length.

THE SMALL-CAR COUNTERMOVEMENT

While Ely and his advocates pressed on with the big-car movement, a minority opinion was voiced by other railroad men who were critical of the in-

FIGURE 3.4

This end view of a Pennsylvania boxcar shows the peaked roof, brake handwheel and shaft, steps and cast-iron dead blocks, and link-and-pin coupler. (Pennsylvania Railroad Photo)

FIGURE 3.5 (*middle*)

The B&O was one of the first trunk lines to build 20-ton-capacity cars, as shown by this diagram drawing. The car is just 4 feet longer than the line's old 10-ton car. (Baltimore & Ohio Railroad)

FIGURE 3.6 (*bottom*)

The growth in size and capacity over twenty-six years can be seen by comparing this class M-8 B&O car with the earlier M class shown in Figure 3.5. (Baltimore & Ohio Railroad)

discriminate championing of big for the sake of being big.[13] Large hopper cars made wonderful sense because they were always filled. The same applied to boxcars used for grain traffic, but boxcars used in general merchandise trade were rarely loaded to capacity, and in the mid-1880s few individual shippers needed more than a 20-ton car. The big-car men were wasting time and money by sending half-loaded cars over the line. C. A. Seley, an experienced engineer serving both the Norfolk & Western and the Rock Island, felt that the railroads' own traffic departments were the source of the big-car demand. L. F. Loree agreed with Seley and contended that big cars were unwieldy, dangerous, and unprofitable. During the early part of his career he served as general manager for the Pennsylvania Railroad's lines west, which served America's densest industrial and agricultural belt where traffic levels were very high. Even so, average car loadings ran only 15.56 tons. In a letter published by the *Railroad Gazette* on July 6, 1900,

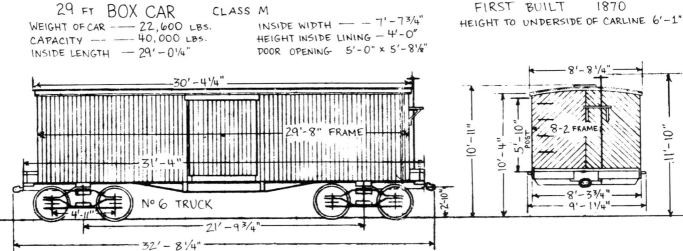

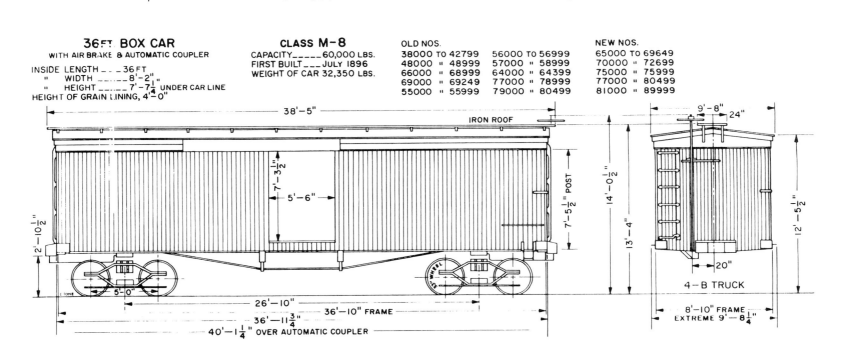

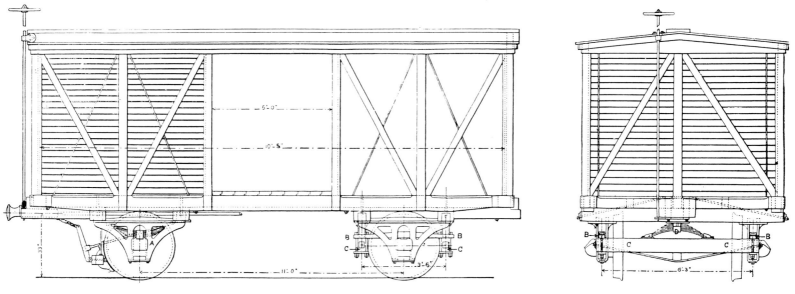

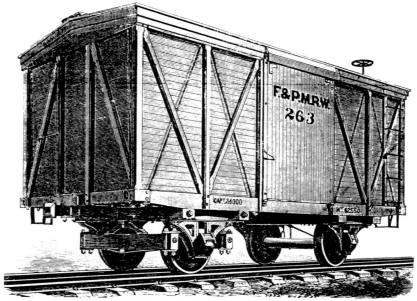

FIGURE 3.7

A vain effort to reintroduce four-wheel boxcars was made around 1880. The superintendent of the Flint & Pere Marquette Railway devised this little car, which featured a three-point suspension. (*Railroad Gazette*, Oct. 8, 1880)

Loree wondered about the need for 30- or 40-ton cars in the light of his experience.

There was a modest counterdemand that flourished briefly during the traffic-starved years after the 1873 panic. The small-car champions attempted to revive the obsolete four-wheel car to produce an economical 10-ton freight vehicle. The New York Central built several hundred boxcars of this description at its London, Ontario, shops in 1879.[14] They measured 20 feet 2 inches long by 9 feet wide and had a wheelbase of 10 feet. Weight was given as 11,000 pounds. They were said to be very useful for small shipments of local goods between way stations. Another New York Central property, the Michigan Central, was, ironically, using similar cars for grain—precisely the type of commodity they were least suited to carry. Yet the Michigan line insisted that they were cheap and not as prone to derailment as their detractors claimed. No drawings for these cars have been uncovered, but an unidentified engraving for a car of

this description appears in the 1879 *Car Builders Dictionary* (Fig. 13).

A year after the New York Central's experiment another midwestern road, the Flint & Pere Marquette, began building a somewhat larger version according to the plan of its superintendent, Sanford Keeler (Fig. 3.7). This car measured 19 feet 5 inches long inside and had an 11-foot wheelbase. It weighed 12,550 pounds and carried 24,000 pounds. One pair of wheels was mounted to transverse equalizers and leaf springs to effect a three-point suspension. Keeler's design appears to have made no progress beyond the home road, but less exotic forms of single-truck boxcars enjoyed a limited popularity in New England. At least one line, the Eastern, appears never to have made the transition completely to double-truck equipment. *Poor's Manual of Railroads* for 1872–1873 lists the majority of its freight cars as four-wheelers. Among these were ninety-three boxcars. In the same volume, the neighboring Boston & Maine

THE AMERICAN RAILROAD FREIGHT CAR

TABLE 3.1 Car Fleet of the Lake Shore & Michigan Southern

1879		1894	
Number of Cars	Capacity	Number of Cars	Capacity
100	11 tons	2,368	12 tons
2,683	15 tons	3,587	15 tons
7,737	12 tons	3,596	20 tons
		3,397	22.5 tons
		3,332	25 tons
		4,390	30 tons
		49	40 tons
10,520 total cars		20,719 total cars	

Source: *Railroad Gazette*, Feb. 15, 1895, p. 97.

FIGURE 3.8

John Kirby, a leading car builder of his time, was building cars of this style in 1878. It is a mixture of new and old ideas. What appears to be truss rods is really more of a draft gear, for it does not support the floor but rather connects the draft timbers. The metal bolsters, arch-bar trucks, double brakes, and iron corner braces represent advanced features. (*Railroad Gazette*, Dec. 20, 1878)

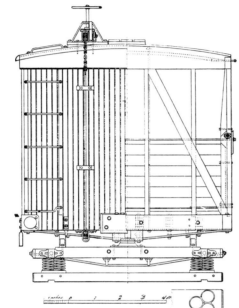

listed eighty-four four-wheel boxcars. Photographs of some of these cars numbered in the 7000 and 8000 series have been preserved, which indicates that these vehicles were still in service as late as 1900.[15]

The nearby Boston & Lowell tested a four-wheel boxcar built in 1880 to the design of Charles Barrett.[16] It weighed 7 tons and carried 12 tons. Special journals and axles with plain ends allowed enough free lateral movement for Barrett's car to navigate the sharp curves found in the Lowell factory district. To the south of New England, the Central Railroad of New Jersey had a number of four-wheel boxcars for the cement traffic. Several of these were sold to the Singer Manufacturing Company and subsequently became part of the Branford Trolley Museum's collection. Hidden away on a seldom-seen storage track, they represent an obscure and little-recorded episode in American freight car history.

The four-wheel boxcar was an interesting curiosity, yet it formed such a small fragment of the overall American car fleet that it only underscores the mainstream trend toward big cars. Industry-wide statistics for the makeup of the car fleet are not available, but some idea can be gained from the example of several trunk lines. The Lake Shore & Michigan Southern, a New York Central subsidiary, offers an illuminating pattern of change, as illustrated in Table 3.1.[17] Several points are obvious from this table other than the tremendous growth in the number of cars available for service in just fifteen years. Even the variety of sizes is overshadowed by the growth in size. In 1879 the LS&MS had no freight cars with a capacity greater than 15 tons, but in 1894 the great majority of its cars ranked 20 tons and over.

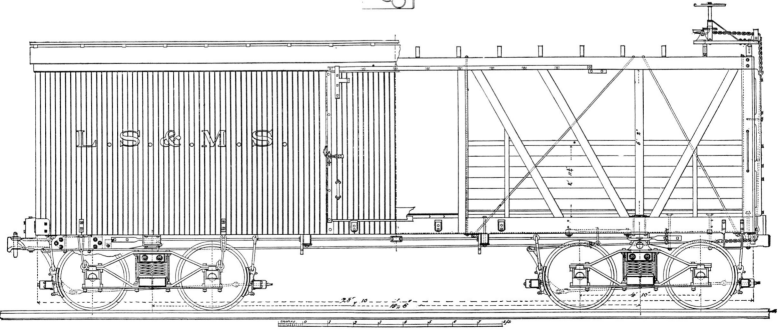

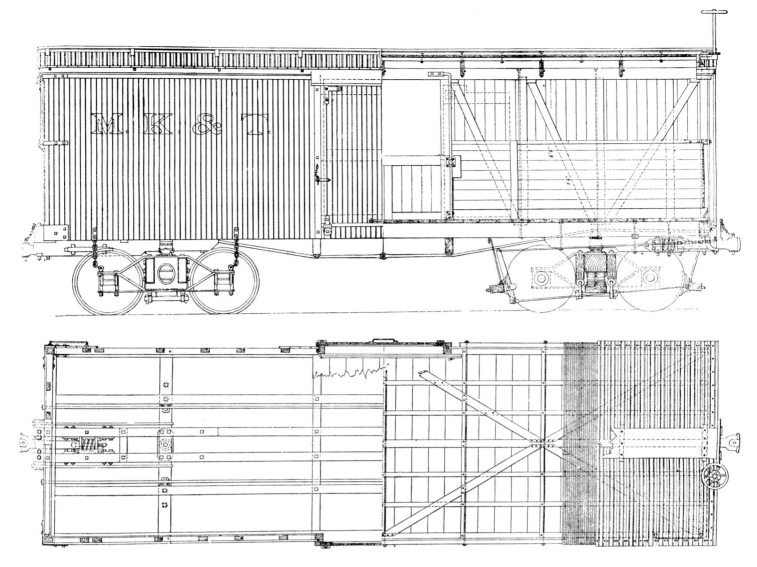

FIGURE 3.9

George W. Cushing produced a more advanced design in 1879 that featured divided truss rods, metal bolsters, grain doors, end doors, and a double-board roof. The body measured 28 feet 2 inches long by 8 feet 6 inches wide. (*National Car Builder*, Sept. 1879)

By 1899 many other major lines had adopted 30-ton-capacity cars.[18] The Baltimore & Ohio was among the leaders, with twenty-five thousand of its forty thousand cars rated at this size. The New York Central had twelve thousand 30-tonners, while both the Illinois Central and the Union Pacific reported that fully half of their freight cars were of this capacity. Elsewhere, less progress was evident. The Chicago & North Western estimated that fewer than one-third of its cars were rated at 30 tons. Major roads like the Reading, the Great Northern, and the Northern Pacific registered only about one-quarter. The Santa Fe was designated as the big railroad having many small cars. Of its twenty-five-thousand-car fleet, only two thousand were even in the 20- or 25-ton class. Medium-size lines like the Erie had almost no 30-ton cars. It was estimated that on the whole the average capacity of U.S. freight cars could hardly be more than 22 to 25 tons. The average size was surely held down by the older cars in service and by established traffic patterns that did not require larger vehicles. But it does seem peculiar that western roads like the Great Northern were not leaders in this move-

ment, because big cars were so suitable for grain.

Twenty-ton boxcars were hardly known before the late 1870s. The who, when, and where of the big-car movement, in common with so much of freight car history, appear to be unrecorded. Twenty-five-ton cars appeared by the early 1880s and 30-ton cars by the middle of that decade. The Pennsylvania adopted a standard design for a 30-ton box in 1888. The Master Car Builders Association was somewhat tardy in following these developments and did not adopt a standard axle for cars of this size until 1890.

HIGH-CUBE CARS

The 30-ton boxcar was probably about the practical limit for a wooden car of normal construction intended for normal interchange service. Shippers of light, bulky goods like empty barrels, furniture, and carriages required cars of greater volume for the most economical transportation. The construction of these high-cube cars goes back to the 1860s, when the Baltimore & Ohio built a special car for the carriage of barrels.[19] Despite these early origins, not many others offered large-volume box-

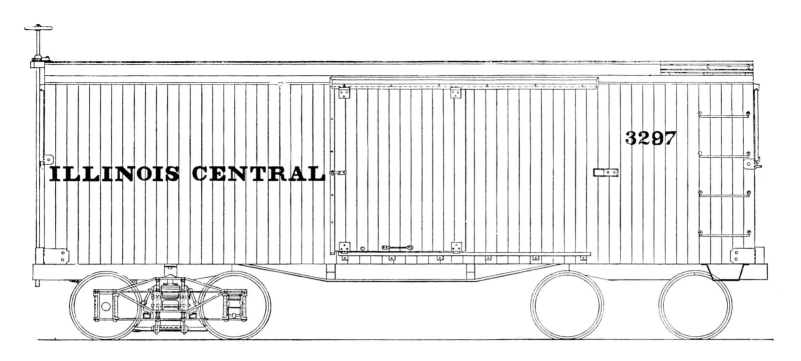

FIGURE 3.10

The Illinois Central car shown
here is post-1870, modern in all
respects except for the arch roof.
The body measured 28 feet 6
inches long by 8 feet 6 inches
wide. As with most published
descriptions for this time, weight
and capacity were not listed.
(*National Car Builder*, Dec. 1881)

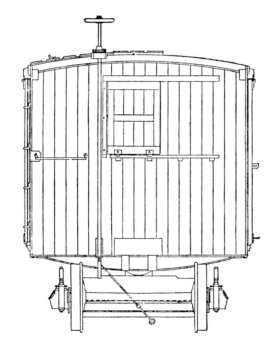

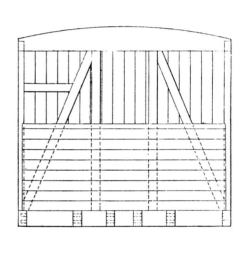

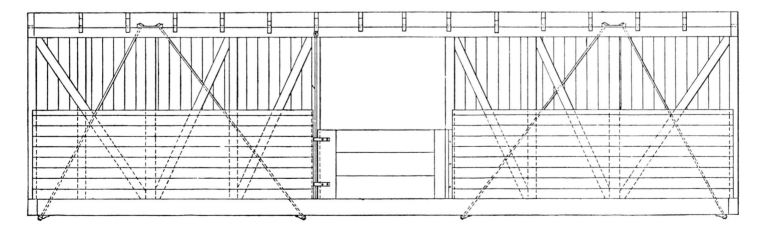

FIGURE 3.11

Very large boxcars were developed for light but bulky cargoes such as furniture and carriages. The 46-foot-long car shown here and in Figure 3.12 was designed in 1895 by J. N. Barr for the Milwaukee Road. Notice how low the car is carried on its wheels; this was done to maintain overhead clearances. (*American Railroad Journal*, Sept. 1896)

FIGURE 3.12

More details of the Milwaukee furniture car. Note the double iron body bolster and mortise details for the end and side sills. The four inside sills are clustered very close to the centerline to allow a clear space for the wheels. The car had a capacity of 30 tons. (*American Railroad Journal*, Sept. 1896)

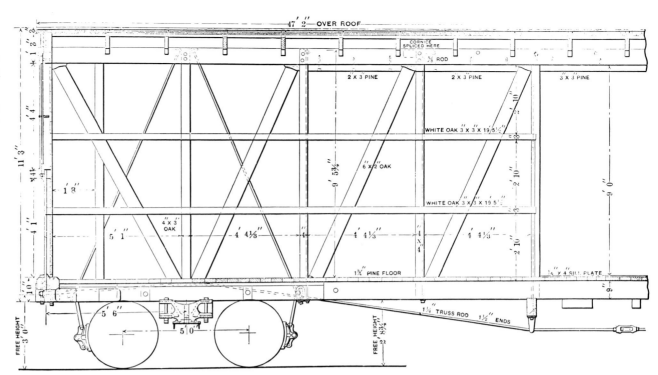

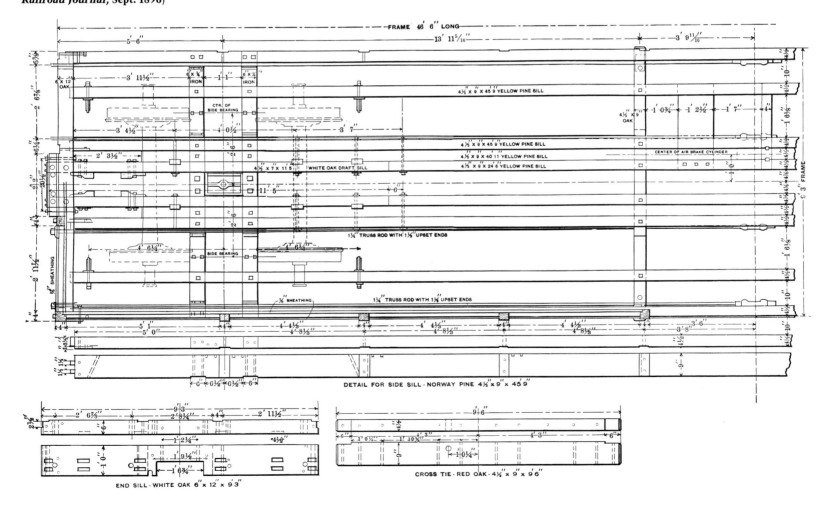

THE AMERICAN RAILROAD FREIGHT CAR

cars much before 1885. In that year the Chicago & North Western produced a 20-ton car of this type that measured 37 feet 6 inches long by 8 feet 5 inches wide inside the body.[20] The door opening was about 6 feet wide.

The Burlington produced a 25-ton variation that had a 39-foot-long body.[21] The frame and trucks were specially constructed to keep the overall height of the vehicle within acceptable limits. The bottom of the sills was only 30 inches from the top of the rails. These were clustered toward the centerline to create an open space for the wheels. The iron body bolster was built up into framing members so that its top plate was just below the flooring. The body posts were oak rather than pine, and the body was braced with ¾-inch-diameter vertical rods. The interior was fully sheathed to stiffen the body further and make it suitable for grain traffic. Because of the car's greater bulk, the permissible levels for wheat, corn, and oats were lower than in a normal 25-ton boxcar. Cars of this design were in successful use from 1891 on.

Several years later the Milwaukee and the Illinois Central began producing 30-ton-capacity high-cube cars. The Milwaukee's design had eight 4½- by 9-inch sills (Figs. 3.11, 3.12).[22] A great mass of timber was created at the frame's center because the draft timbers were bolted on parallel to rather than beneath the center sills. Double iron bolsters were used to spread the load on the frame sills. The body measured 46 feet long by 11 feet 3 inches high. The door opening was equally gargantuan for the time, measuring 5 feet 6 inches by 9 feet. The body was set very low on the trucks. The Illinois Central car was equal in size and registered an area of about 3,400 cubic feet. Sills 9 inches deep were supported by six rather than the normal four truss rods. Despite an inverted body bolster, and low trucks, the car measured 14 feet 3¾ inches over the top of the brake-wheel staff.

The buggy car was an outgrowth of the high-cube furniture car that was created to satisfy a major adjustment in the carriage business. Carriages were produced locally by small shops until late in the nineteenth century, but after that time large mass-production shops began to take over the market for cheaper vehicles. Cincinnati became a center for this trade and was soon a shipping center for thousands of buggies, station wagons, and surreys destined for journey to all points. The vehicles were bulky but light and so were ideal for high-cube cars. But they needed very long doors,

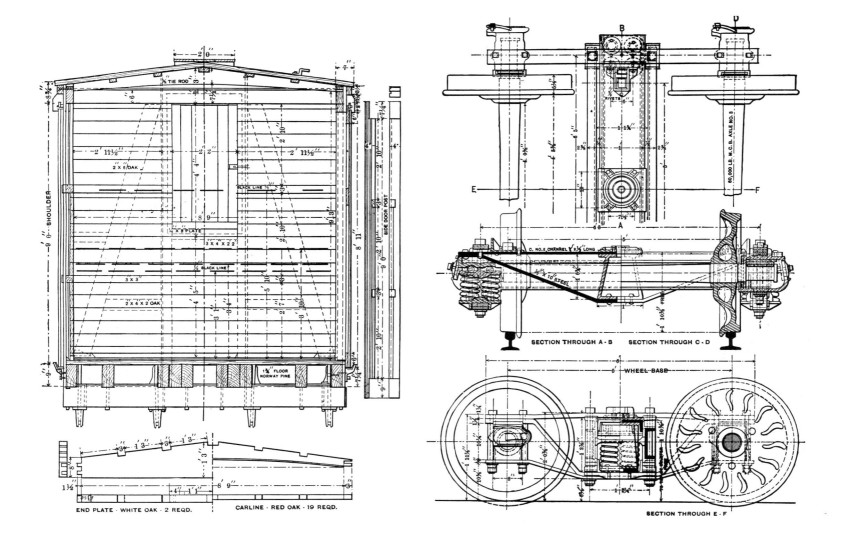

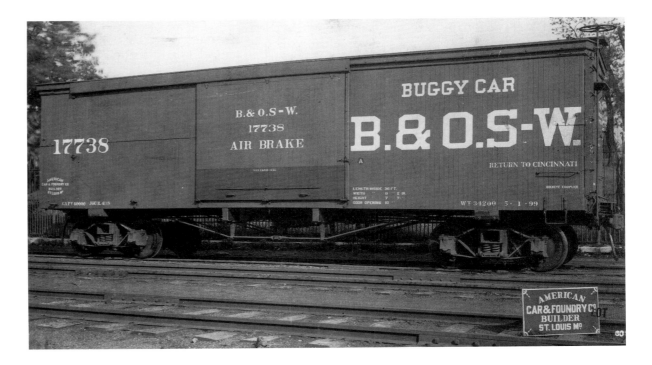

and so a new class of car was born. A 30-ton Baltimore & Ohio South Western buggy car is shown in Figure 3.13. Note that it is marked for return to Cincinnati. This car is not unusually large. It is in fact not much longer than a conventional boxcar of the period, but it does have unusually large doors.

A far larger car intended for the same traffic was produced for a St. Louis firm, Mansur & Tebbetts, an implement and carriage agency, by the Missouri Car & Foundry Company in 1893.[23] It and its sister were surely the largest wood-framed boxcars ever produced. These mammoths measured 60 feet long and 9 feet 2 inches high inside. The same firm had fourteen smaller cars measuring 52 feet inside, still giants for their day. These cars were too large to travel in the East, where clearance restrictions were more critical, and so their journeys were restricted to western travels. Mansur & Tebbetts took advantage of these wooden giants by painting their vast sides as rolling billboards.

Furniture and buggy cars proved that very large wooden vehicles were possible to construct. At least one car builder, Ferdinand E. Canda, elected to build a large boxcar that was big in capacity as well as volume. Canda, who once built cars in Chicago in the 1870s, later worked for the Ensign car plant in Huntington, West Virginia. He then moved to New York, where he became acquainted with financier and railroad magnate C. P. Huntington. Steel cars had become a commercial reality by the late 1890s, yet old-line car men like Canda felt that the wood car's potential had yet to be realized. Huntington agreed to back the project in a big way. One might expect a cautious project for a few sample cars, but at Huntington's order the Southern Pacific agreed to purchase two thousand Canda cars from Ensign.

The result was a 40-foot-long car with a 50-ton capacity—a truly astonishing achievement, considering that Canda had followed conventional designs to create this extraordinary series of vehicles (Figs. 3.14, 3.15). The floor timbers were only slightly enlarged over common 4- by 8-inch sills. The center sills measured 5 by 9 inches, the side sills 4½ by 9, and the intermediates 4 by 9. Two extra truss rods were used, and the center pair was made rather stouter than usual, being 1¼ inches in diameter. The queen posts were very deep, at 33 inches. The body had more than the usual number of tie rods, and it was necessary to use extra-large 5- by 9-inch journals for the trucks. Canda agreed with certain other car builders that mortises and tenons created weak joints, and this he avoided by the use of malleable-iron pockets. Some wooden structural members were reinforced with channels or plates. Yet none of these features was new or unusual. Canda had simply borrowed the best of established wood-car construction techniques to create a vehicle of unusual size and capacity. Its long-term significance was marginal, however, because Canda's creation could not turn aside the steel car. The limitations of composite construction were already understood, and so the Canda car was the last stand of an old knight against a younger challenger.

The Push for Lightweight Construction

If size was a preoccupation of Victorian car builders, then so was weight. And the goal of increased size was counterbalanced by the desire for reduced weight. Since the 1830s hopper cars had achieved a fine weight-to-capacity ratio, while boxcars rarely registered less than 1 ton of weight

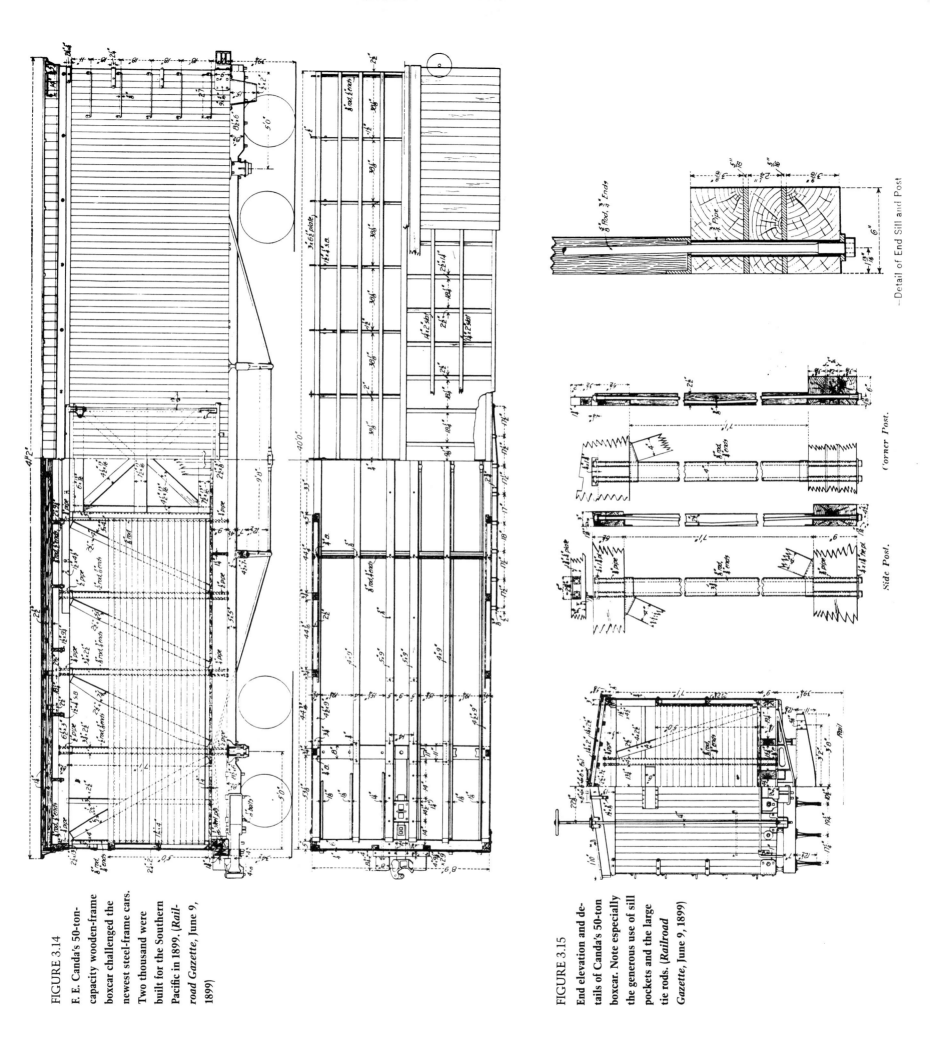

—Detail of End Sill and Post

Corner Post.

Side Post.

FIGURE 3.14

F. E. Canda's 50-ton-capacity wooden-frame boxcar challenged the newest steel-frame cars. Two thousand were built for the Southern Pacific in 1899. (*Railroad Gazette*, June 9, 1899)

FIGURE 3.15

End elevation and details of Canda's 50-ton boxcar. Note especially the generous use of sill pockets and the large tie rods. (*Railroad Gazette*, June 9, 1899)

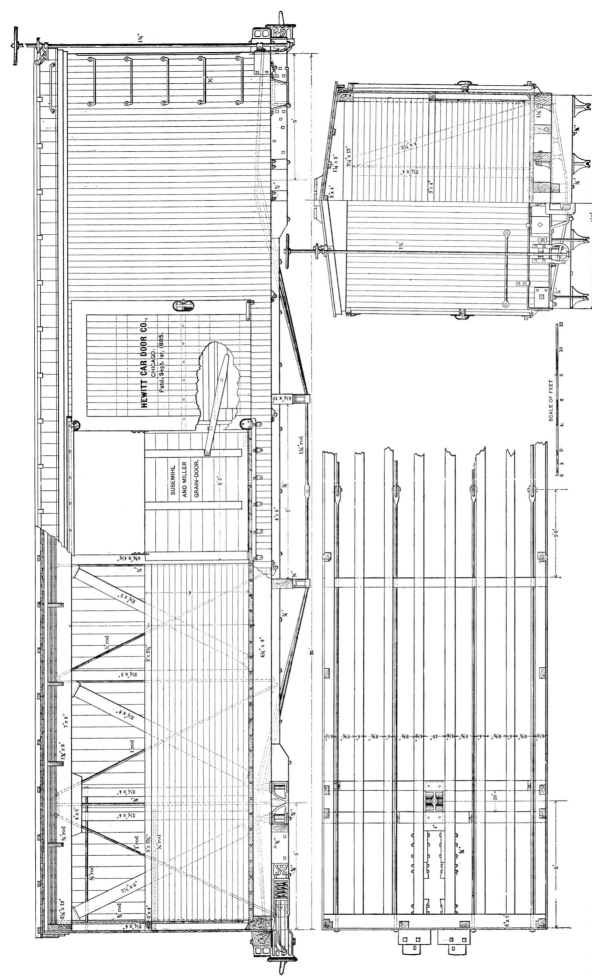

FIGURE 3.16

This Michigan Central boxcar of 1886 has an eight- rather than the more common six-sill frame. Note the double iron bolster and the Ames-style coupler made by the Cleveland Malleable Iron Co. The car weighed 27,850 pounds and carried 60,000 pounds. (*Railroad Gazette*, March 25, 1887)

HEWITT CAR DOOR CO.,
CHICAGO.
Patd. Sept. 1st, 1885

SUSEMIHL
AND MILLER
GRAIN-DOOR.

SCALE OF FEET

for each ton of capacity. And then around 1880 boxcars also crossed over the line and joined hoppers in the one-for-two category. One might assume that some radical alteration in basic design had taken place. But such was not the case. Floor timbers were about the same size. At the very most an extra pair of truss rods was added. What had happened since 1870 was the phenomenon of the big car. A modest enlargement in floor space had resulted in a geometrical expansion in capacity. A British engineering journalist was astonished to find that the American boxcar had grown from a 10-ton rating to a 20-ton capacity with the addition of only 1,500 pounds in new material.[24]

Feats of this nature were being attempted as early as 1870. The Baltimore & Ohio introduced its class M boxcar during that year. To the casual observer it looked like another old-fashioned arch-roof house car. But it registered a mere 22,600 pounds on the scales and offered a capacity of 40,000 pounds. The Erie constructed a broad-gauge boxcar during 1873 that weighed only 17,700 pounds, less wheels and axles.[25] It measured 29 feet 2 inches long by 8 feet 8 inches wide and had a volume of 1,390 cubic feet. The trucks would add about 6,500 pounds to the dead weight, bringing total weight to around 24,200 pounds. Its known capacity was 36,140 pounds; hence the weight-to-capacity ratio was almost 1.5 to 1. After four years the car was reported to be still in service and showing no signs of deterioration because of its light construction.

The Erie's achievement was an experiment, but within a decade it had become commonplace. Other lines tried to improve on the now-standard 2-to-1 ratio. In 1886 the Michigan Central produced a 30-ton boxcar that weighed only 27,850 pounds (Fig. 3.16).[26] A car of the same capacity built by the Big Four (Cleveland, Cincinnati, Chicago & St. Louis) registered a dead weight of 33,500 pounds.[27] The extra 5,650 pounds was justified as being necessary for greater durability and dependability. A lighter car was possible, as demonstrated by the Michigan Central and other roads, but the Big Four believed that this was a false economy, because their cars were regarded as among the strongest and most durable. The cost of moving the extra 5,650 pounds was offset by a car that remained in revenue service, and hence earned a greater income than one standing in a repair yard. No direct proof was offered for this opinion, but it was an argument used by car builders favoring heavy construction.

William Voss, a leading car builder of the day and author of the only text on the subject published during the nineteenth century, refused to take sides on this crucial issue. Voss saw valid arguments on both sides of the question and felt that both viewpoints were worthy of consideration. And so the business of car building continued with strong convictions on the question of light

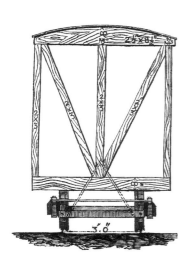

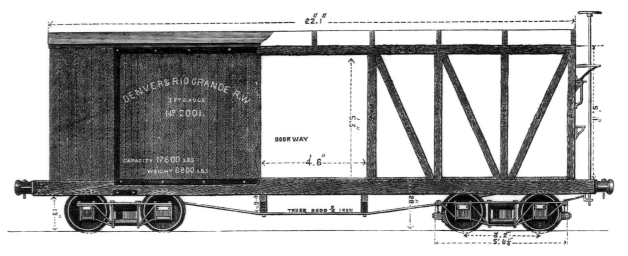

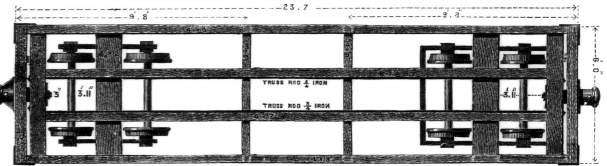

FIGURE 3.17
This Denver & Rio Grande narrow-gauge boxcar, made by Billmeyer & Small of York, Pa., was essentially a scaled-down standard-gauge car. (*Engineering*, Dec. 11, 1871)

CARS FOR GENERAL MERCHANDISE

and heavy cars. In the mid-1890s the Chesapeake & Ohio attempted to better the record by producing a 34-foot 30-ton boxcar that weighed only 25,000 pounds.[28] This series of cars remained a lightweight standard until Canda came forward in 1899 with his 50-ton box with a dead weight of 33,100 pounds.

The lightweight question was for the most part discussed rationally by the two opposing factions, both of which could offer logical reasons for their viewpoints. But a considerable degree of fantasy was introduced into the debate during the 1870s by the narrow-gauge proponents.[29] In their zeal to promote the supposed advantages of tracks having a width less than 56½ inches, they laid claim to results that occasionally countered the natural laws of physics and mechanics. In attempting to prove the superiority of narrow-gauge railways, it was said that small-gauge cars were proportionately lighter than standard-gauge cars. The first eight-wheel boxcars produced for the Denver & Rio Grande were offered as proof (Fig. 3.17).[30] Built by Billmeyer & Small of York, Pennsylvania, these cars measured 23 feet 7 inches long by 6 feet wide. Drawings show them to follow conventional plans for a standard American freight car of the time. These drawings show nothing that would make them different from, much less superior to, any other car then in service. Side sills measured a hefty 4 by 8 inches, or about the same as in a full-size car. The wheels were definitely undersize, however, measuring only 20 inches in diameter. The cost was reported at $450, or again about the cost of a standard-gauge car. Weight was a mere 8,800 pounds, less than half that of a standard-gauge car of that date. Capacity was lettered on the car as 16,000 pounds, but the accompanying text claimed that it was actually 17,600 pounds.

Whichever weight figure is correct, the narrow-gauge car did indeed offer an unmatched weight-to-capacity ratio. Standard-gauge cars were to match the 2-to-1 ratio within a few years, as already noted, but in 1871 the friends of the narrow gauge had won the contest handily. Or so it would seem on the surface. The carrying capacity of the car is measured by more than weight; it also involves volume. In no way could the contents of a car measuring 28 feet long by 8 feet 6 inches wide be put into a narrow-gauge vehicle with a cubic area of 792 feet. Part of a second car would be needed for the extra 200 to 300 cubic feet available in the standard-gauge car. On reflection, the dead-weight advantage of the narrow-gauge car is fallacious. In addition, items of extra width or height that might fit into an ordinary standard-gauge car could not be housed by its small-gauge counterpart. Yet it was not the merit or lack of merit inherent in the rolling stock of narrow-gauge lines that doomed them to extinction; it was their inability to interchange cars on a national basis.

Construction Details

The term *house car*, preferred by the Baltimore & Ohio helps us to think of the boxcar as a building, which it resembles in many ways. The main elements are present. The trucks, bolsters, and floor sills are the foundation, while the body and roof make up the superstructure. But unlike a house and its foundation, the body and underframe of a boxcar must work together as a unit.

The body frame was essentially a skeleton for the protective sheathing or outer walls, but it was also expected to act as a truss for the floor sills. The large truss rods below the floor stiffened and literally held the structure together. In a well-designed car the framing members were made rigid in proportion to their strength, and all elements worked in concert to carry the loads imposed on the structure. Because of the central door opening, a continuous side-panel truss like those found in passenger cars of the period was not possible, and so the floor sills and truss rods carried the load. Passenger cars with their long span are thus more bridgelike than the short-span freight car, which might be likened to a culvert.

INNOVATIONS IN FRAMING

The most notable change in floor framing after 1870 was the placement of the body bolster below the longitudinal sills. This elemental change totally avoided cutting and splicing the floor timbers at the bolster, a process that greatly weakened the car. It also simplified construction because no mortise-and-tenon joints had to be prepared, which saved time, labor, and costs. The single disadvantage was the difficulty of maintaining floor levels at existing heights, because part of the working space between the bottom of the sills and the truck bolster was now occupied by the body bolsters. As weight capacities increased, so too would the size of the bolster, and in time an all-wooden bolster might simply have grown too bulky to fit the available space. This is one explanation for the growing popularity of metal bolsters after 1870.

Butting the floor sills against the end beam was another change that became nearly universal after 1870. Like the one-piece floor sill, it was both simpler and more rational than the old scheme, whereby the end sill was mortised into the side sills. The Rock Island's veteran master car builder, B. V. Verbryck, persisted in following the old plan for attaching end sills into the mid-1880s, despite its obvious defects (Fig. 3.19).[31] The use of only two truss rods was also obsolete by this time, and yet Verbryck's cars were said to withstand the most rigorous interchange service, including regular passage through the Chicago freight yards. Their durability was credited to the long and well-fastened draft timbers, which are clearly shown in the drawing.

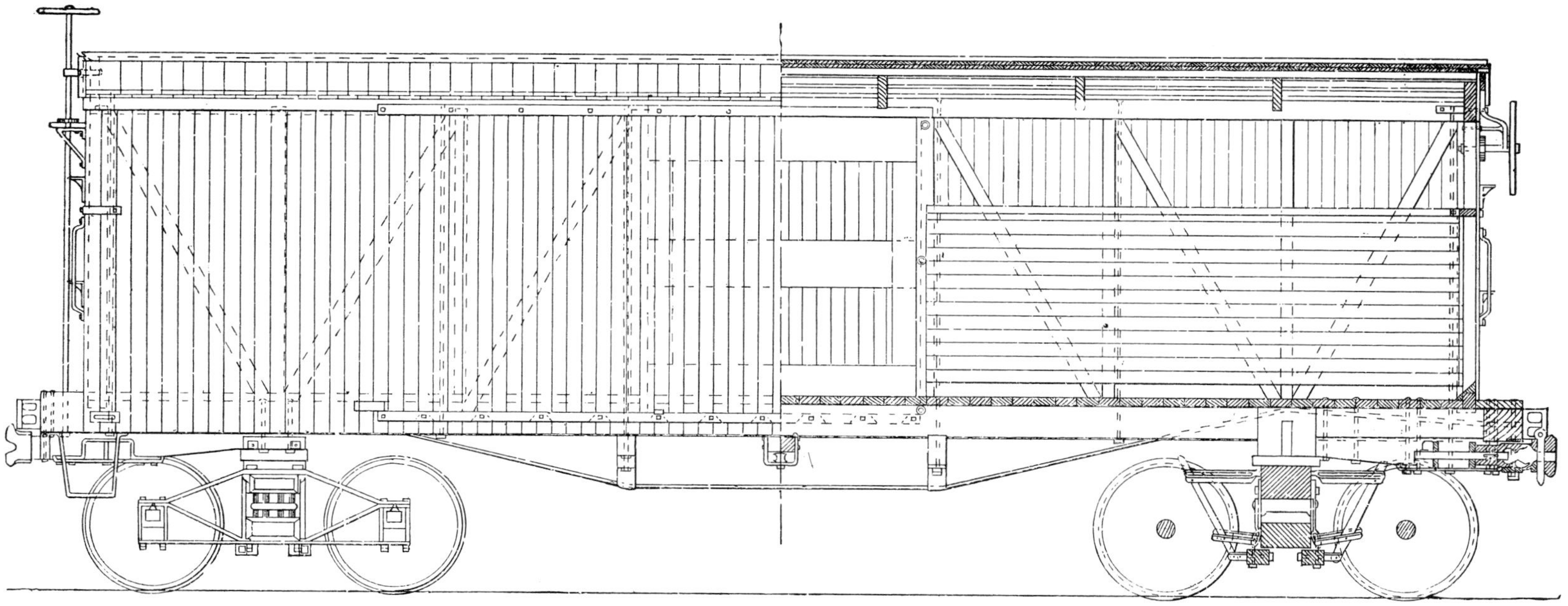

FIGURE 3.18

The Pennsylvania Railroad became a champion of large-capacity cars after 1875. This 20-ton-capacity car, produced in 1880, continued to use wooden bolsters. The body measured 28 feet long by 8 feet 7¾ inches wide. (*National Car Builder*, Aug. 1884)

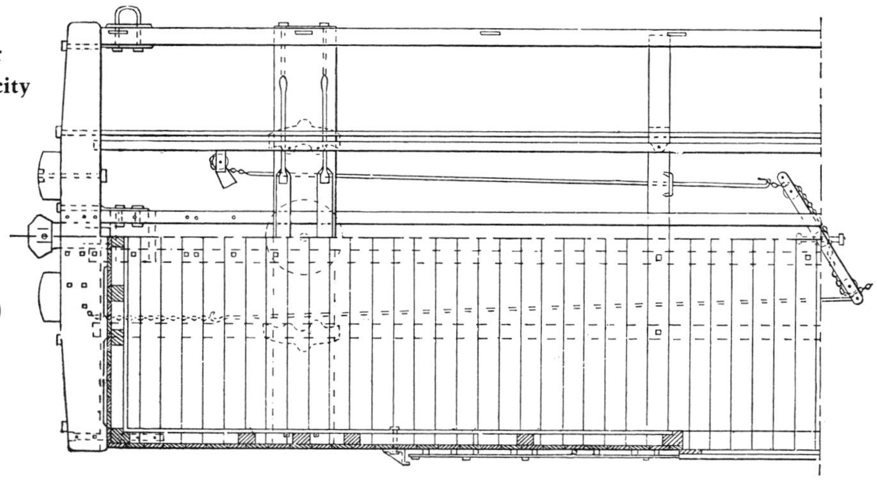

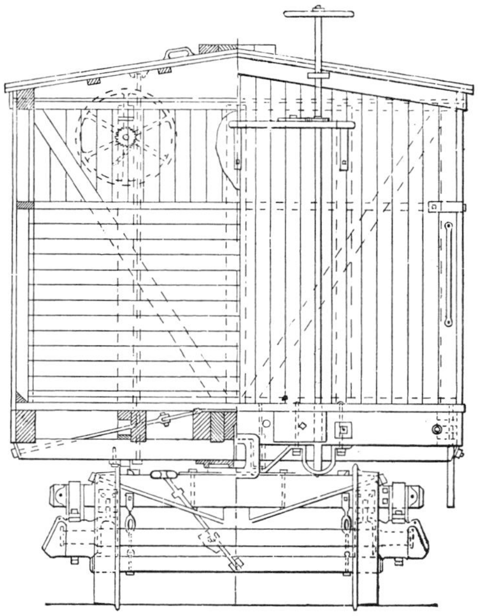

GENERAL DIMENSIONS.

Total length of frame	29 ft.	10¾ in.
Total length of body	28	0
Total width of body	8	7¾
Length of running board	28	4½
Width of running board	—	18
Width of door opening	5	4
Width of door	5	6

BODY TIMBERS.

Side sills	4¼x 8½ in.	x 28 ft.	9 in.	Yel. pine.
End sills	8½x 9¼	x 8	10	Wh. oak.
Center stringers	3¼x 8	x 28	9	Yel. pine.
Intermediate stringers	3¼x 8	x 28	9	"
Cross bearers	4¾x 7½	x 8	8	"
Bolsters	5½x14½	x 8	8	Wh. oak.
Draught timbers	3¼x 7	x 4	4	"
Block for center lever	4¼x 7¾	x 1	1	"
Blocks over center plates	7¼x 6¾	x 1	3	"
Blocks for saddles	4¼x 7¼	x 1	3	"
Pieces for foot of end posts	2¾x 3¼	x 8	2	Yel. pine.
Corner posts	3¾x 4¾	x 6	5	"
Door posts	3¾x 4¾	x 6	5	"
Intermediate side posts	3¾x 2¾	x 6	5	"
Intermediate end posts	2¾x 3¾	x 6	3	"
Center end posts	2¾x 4¾	x 6	3	"
Side braces	3¾x 2¾	x 7	long.	"

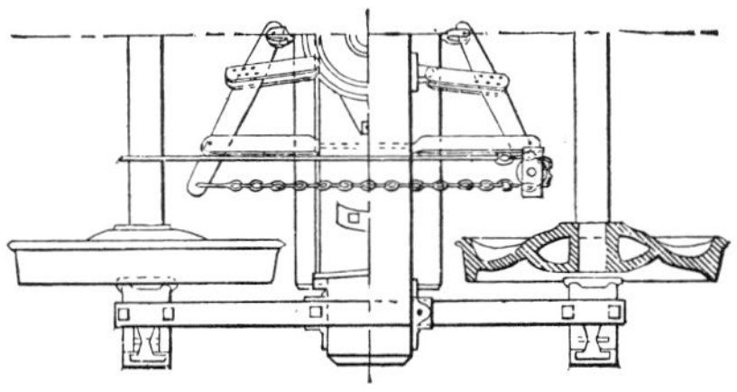

Half Plan of Truck.

End braces	2¾x 3¾	x 7	2 in.	"
Side plates	3¾x 5½	x 28	3	"
End plates	2¾x13½	x 8	8	"
Carlins	2¼x11	x 8	8	"
Center purlin	2¼x 4¼	x 28	3	"
Intermediate purlin	2 x 3¼	x 28	3	"
Door strip (stops)	2 x 3¼	x 6	6	"
Step for end brake	1½x 8¼	x 2	6	Wh. oak.
Fascia boards (side)	1 x 4	x 28	6	Wh. pine.
Fascia boards (ends)	1 x 4	x 4	9	"
Running boards	1½x 6¼	x 28	6	"
" "	2½x 6¼	x 28	6	"
Guide Rail (top)	2¾x 3½	x 11	5	Yel. pine.
Guide rail (bottom blocks)	1 x 2	x 10 long.		"
Cappings for sides	1¾x 3¾	x 11	1 in.	"
Cappings for ends	1¾x 3¾	x 8	8	"
Siding	1 x 4½	x 7	2	Wh. pine.
Platform planks	2 x11¼	x 9 long.		Yel. pine.
Door stiles	1 x 6¼	x 4	4 in.	Wh. pine.
Top rails for doors	1 x12	x 5	8	"
Bottom rails for doors	1 x12	x 5	8	"
Middle rails for doors	1 x12	x 5	8	"
Siding for doors	1 x 4½	x 6	3	"
Inside lining for sides	1 x 4½	x 11	1	"
Inside lining for ends	1 x 4½	x 8	2	"
Cant boards for sides	2¾x 2¾	x 11	1	"
Cant boards for ends	2¾x 2¾	x 8	2	"
Roofing (well seasoned)	1 x 5¼	x 4	9	"
Flooring (planed side up)	2 in. thick x 8 ft. 8 in. long.			Yel. pine
Pieces for covering nuts	1¾ x 2¾ in. x 12 ft. long.			Wh. pine.

TIMBER IN TRUCKS.

Top bolsters	8¼x12½ in.	x 7 ft.	5 in.	Wh. oak.
Bottom bolsters	6¼x12½	x 7	5	"
Brake bars	4 x 6½	x 5	10	"
Truss blocks	3 x 6½	x 1	3½	"

The axles have 3½ x 7 inch journals and 4¾ inch wheel seats. Maximum carrying capacity of car, 40,000 pounds.

FIGURE 3.19

The Rock Island's car builder followed an obsolete design by mortising the end beams into the side sills and employing only two truss rods. (*National Car Builder*, Aug. 1884)

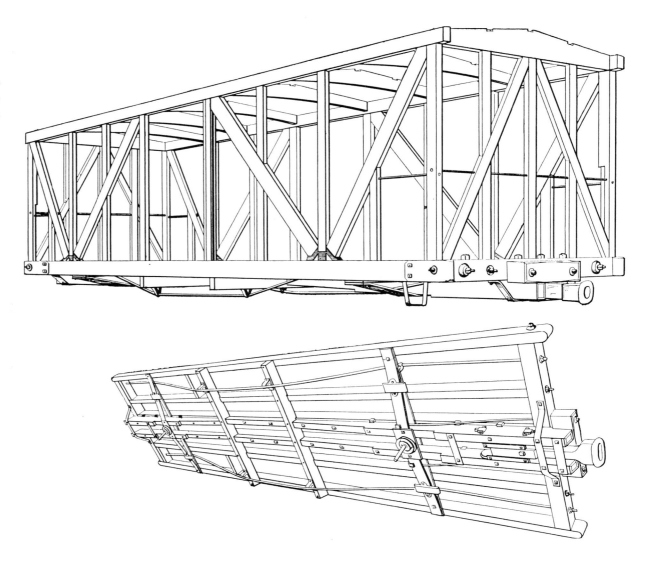

FIGURE 3.20

The arrangement of a late-model wood-frame boxcar is nicely shown by this car rolled over onto its side. The position of the truss rods from the end sills, under the body bolster and over the queen posts, is particularly clear. Notice also the metal body bolsters and king pins. (California State Railroad Museum, Sacramento)

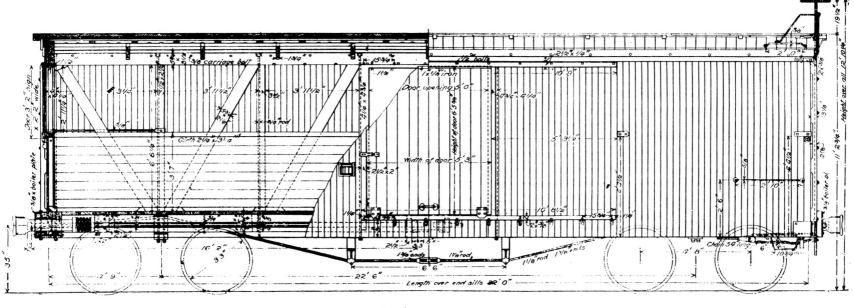

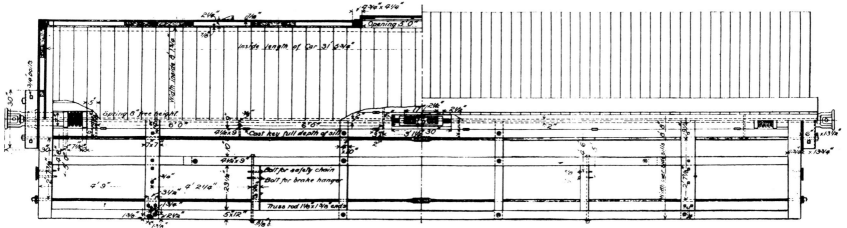

FIGURE 3.21

This standard-gauge Denver &
Rio Grande car of 1891 featured
an unusual floor frame, a continuous drawbar, and a body set low
on the wheels. Air brakes were
not common for freight cars,
except on a few western lines
like the Rio Grande. Traditional
link-and-pin couplers prevailed,
however. (*Railroad Gazette*,
Oct. 9, 1891)

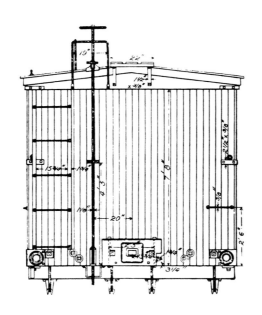

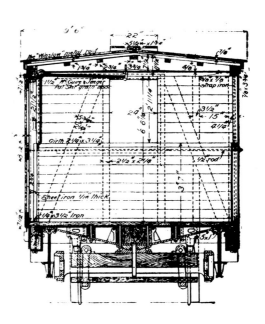

Another pocket of reaction where old ideas prevailed was found in the Central Railroad of New Jersey. Ironically, its boxcar was featured in the first *Car Builders Dictionary* (1879), which was intended to reflect the latest thinking in American railroad car architecture. This volume had been in preparation for several years before its publication, but one still wonders why such an antique design was put forward. What is shown in this instance is a car whose origins appear to be more in the 1850s than the 1870s. It has an arch roof, end beams so large that they form small platforms, two shallow truss rods devoid of queen posts, and a body bolster that is weak because it is cut into the floor timbers. The trucks are of a very old-fashioned wood-beam pattern and have a short wheelbase of 42 inches. The modern car builder of 1879 could profit little from this model and was well advised to look elsewhere for inspiration.

But there seems little danger that car builders would be misled by bad textbook examples, since the profession was not remarkably bookish. Other than the several editions of the *Car Builders Dictionary*, no single volume appeared on the subject until Voss's 1892 *Railway Car Construction*. Inspiration seems to have come from field observation and the work of others. In the area of floor framing, a dull uniformity prevailed for nearly thirty years. Timbers were always six or eight in number, with the former being favored. They were commonly 4 by 8 inches in size, although a few designers specified slightly smaller intermediate sills, shaving off as much as an inch in either dimension. Spacing was made as even as possible, but the center sills were placed fairly close together to accommodate the draft gear. In cars with low-set bodies, like furniture cars, the intermediate sills might be closer to the centerline than was usual, to allow space for the wheels.

To provide a level floor when loaded, an old carpenter's trick was borrowed. The sills were made with a camber, or slight upward bow, of as much as 2 inches greater than a true, straight line. The camber did not add strength to the floor; it only meant that the floor would bend down to a level rather than sag when loaded. The practice, so common for barns and covered bridges, was cited as an evil when applied to car building, for it strained every part of the structure and added an extra load of 8 to 9 tons.[32] The flexing of the body and roof was also injurious, because it opened up joints and seams. Water gained access, damaging cargo and encouraging rot within the car's own fabric.

A few alternative plans for wooden floor framing were offered that showed some promise, if not widespread acceptance. One of the best was the continuous draft timber, which was in effect a doubling of the center sills. Rather than short, bolt-on timbers at each end of the center sills, single full-length beams were fastened directly beneath the center sills. They were generally of oak. In combination, the pair could produce a massive piece of lumber. Such timbers on the Milwaukee Road measured 4½ inches wide by 14⅞ inches deep.[33] Strength of fastening and a stiffer backbone were the chief merits of this scheme, but it was encumbered by a failing more serious than the extra weight and costs involved. If the draft gear were torn loose with enough violence to damage the draft timber badly, then the entire piece would have to be replaced. This involved considerable labor because the trucks had to be removed and the floor pulled up before the long oak beam could be unbolted. Broken draft timbers were the most common failing in wooden cars, which explains why this idea was not more widely accepted. Spliced draft timbers made repairs easier but reduced the strength offered by double continuous beams.

A more radical plan was offered by the Denver & Rio Grande in 1891 to eliminate the draft-timber problem.[34] Both the center sills and the draft timbers were dispensed with and a heavy iron bar 2 inches in diameter was substituted (Fig. 3.21). The Rocky Mountain mechanics had unknowingly reinvented the old continuous-drawbar scheme championed by the Baltimore & Ohio since the 1850s. The plan was mechanically sensible, but it surely cost more than the wooden parts it replaced and so remained unpopular. The D&RG frame was unusual in other respects as well. The side sills were unusually large, measuring 5 by 12 inches. The two intermediate sills were of a more conventional size, 4½ by 9 inches.

The end sills were subject to constant and heavy battering. During switching moves, the end of the

FIGURE 3.22

This Milwaukee Road end view captures the look and feel of a late-century wooden boxcar. We are confronted with the wooden ladder and a pair of deadly cast-iron blocks, together with an equally dangerous link-and-pin coupler. (*Trains* Magazine Collection)

FIGURE 3.23

Wooden bolsters, though obsolete on many railroads by 1875, were used until the end of the wood-car era. The example shown here has a three-piece truss rod. This elaborate arrangement was justified because it could be disassembled without removing the floor. (William Voss, *Railway Car Construction* [New York, 1892])

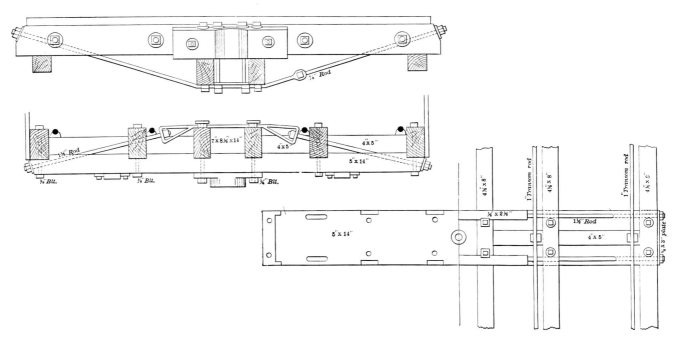

FIGURE 3.24

Seven body-bolster designs are shown in this series of engravings. Figures 1 and 2 show a complex and weak plan that required a secondary rod to hold up the bolster ends. Figures 3 and 4 show other, less complex, plans for bolsters made flush with the bottom of the sills. Figures 5 and 6 show cast-iron brackets and washers where the bolster can be placed below the center and intermediate sills. Figure 7 is an iron bolster bereft of its posts. Figures 8 and 9 show a below-the-sills bolster with a split or three-piece truss rod. This was considered a good, strong design. (*Railroad Gazette*, April 27, 1888)

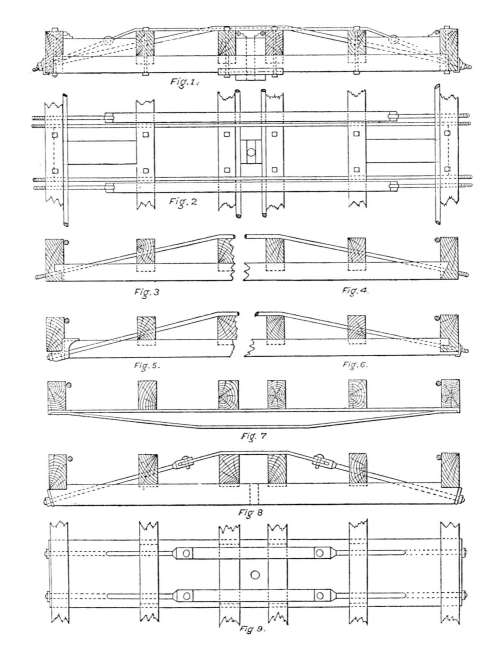

CARS FOR GENERAL MERCHANDISE

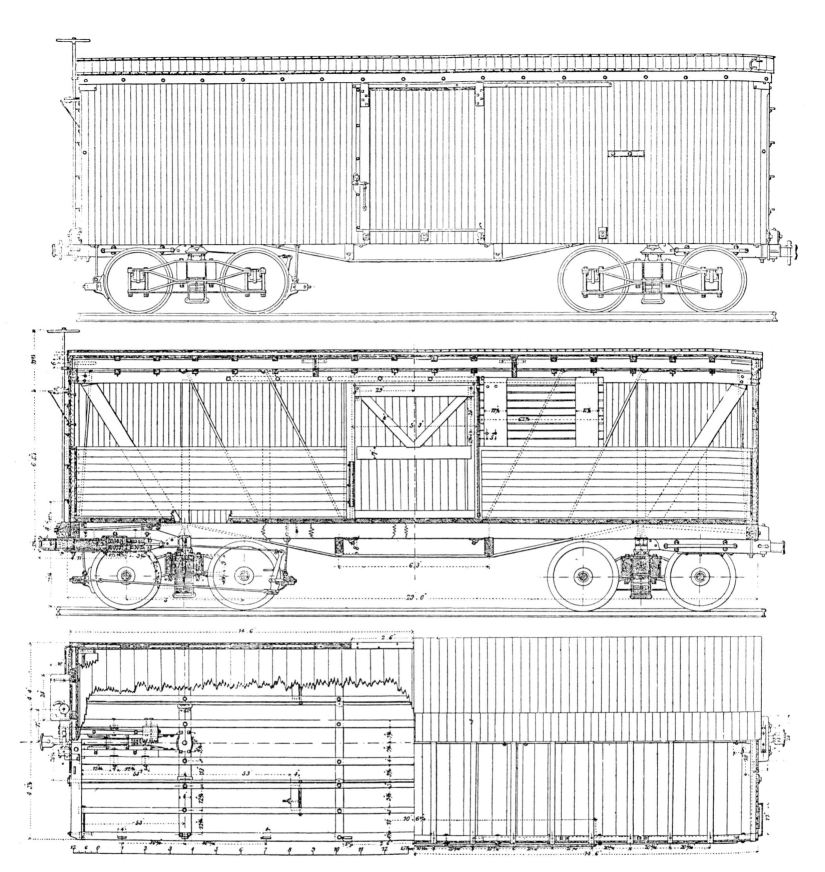

FIGURE 3.25
Leander Garey, the New York Central's superintendent of cars, designed this car in 1875–1876. It featured a Winslow interior iron roof and iron body bolsters. The design was adopted as standard on January 1, 1876. (*Engineering*, May 26, 1876)

THE AMERICAN RAILROAD FREIGHT CAR

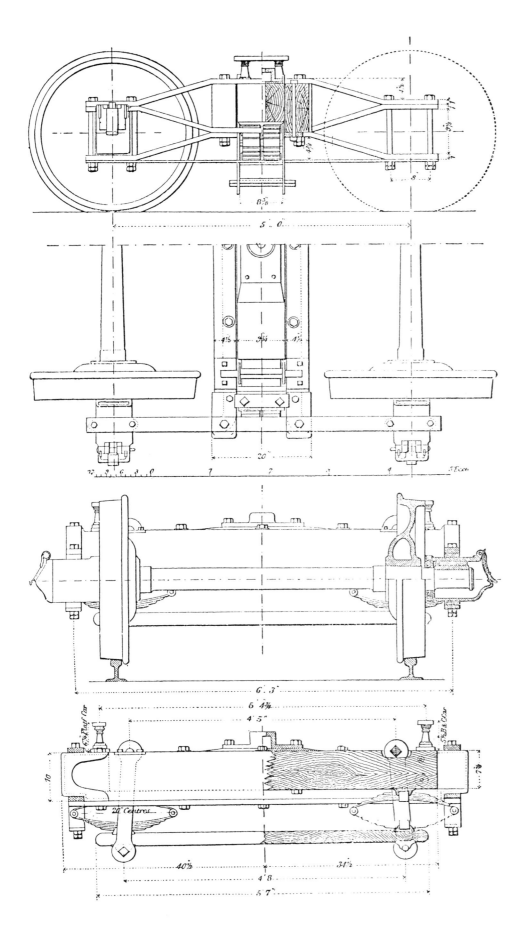

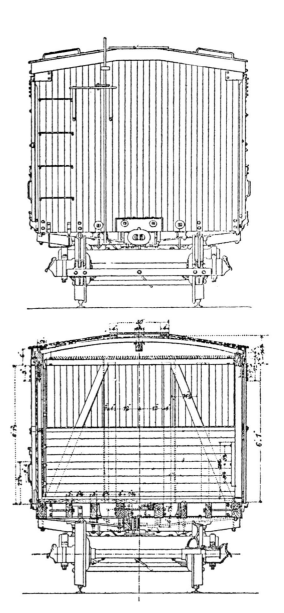

frame was almost always subject to a heavy bump or blow. The sills also served as an anchorage for the truss rods. Mass and strength were required. Soft wood would never do, for washers at the ends of the truss rods would sink into the timber, allowing the rods to go slack, and hence the car would sag. For this reason white oak was always used, and timbers 8 inches square were common. Mass was needed here also because part of the wood was cut away for mortises to receive the longitudinal sills. Wrought-iron corner brackets, generally made from 5/16-inch plate, were often used for added strength.

WOOD VERSUS METAL BOLSTERS

The body bolster was the cross-framing member subject to even greater stresses than the end timbers. All of the weight of the car was transferred to the trucks through the body bolster. Like a bridge's piers, it must be strong and dependable. Its office was considered so demanding that it was the first major structural freight car member to evolve from wood to metal. Wooden bolsters, however,

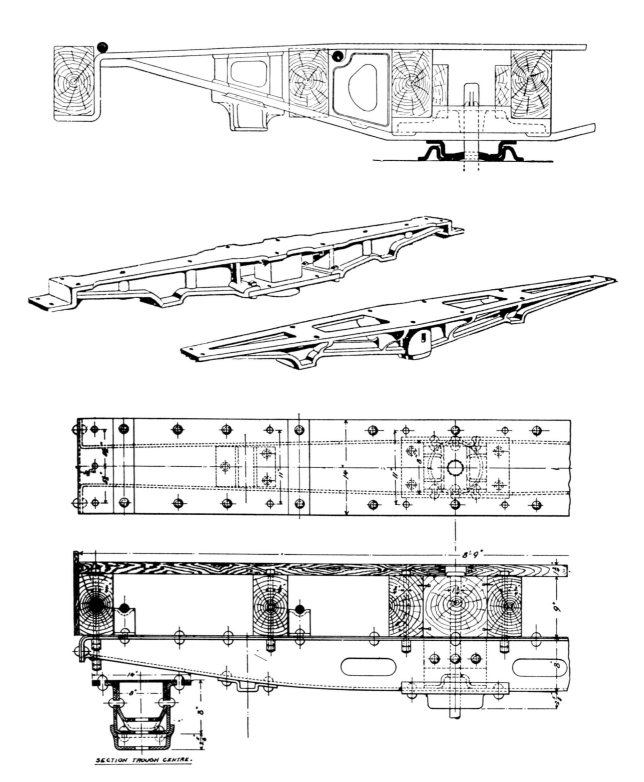

FIGURE 3.26

Iron bolsters were placed on the same level as the floor sills by this arrangement. A more secure mounting of the side sills was effected by bending the ends of the top plate into an L shape. The large black "dots" on the left side of the drawing represent two of the side rods. A pressed-steel center plate is shown at the bottom center of the drawing. (*National Car Builder*, March 1892)

FIGURE 3.27

Cast-steel body bolsters for wooden cars were introduced around 1895. Two variations are shown here. (Marshall M. Kirkman, *Science of Railways*, volume on cars [Chicago, 1908])

FIGURE 3.28

Pressed-steel body bolsters were promoted by C. T. Schoen of Pittsburgh as early as 1890. They were lighter and stronger than the fabricated iron bolsters they were intended to succeed. (O. M. Stimson, *Modern Freight Car Estimating* [Anniston, Ala., 1897])

SECTION TROUGH CENTRE.

continued in at least limited use until the end of the wood-car era because they were cheap—cheap at least in first cost. Timber measuring 5 by 14 inches was common, but even pieces of this size required truss rods for greater stiffness. The rods varied in diameter from about 7/8 to 1 1/4 inches. They were often made in three pieces so that they might be removed without pulling up the floorboards (Figs. 3.23, 3.24). Though more complex and costly than plain rods, the three-piece units were justified by lower repair expenses. Ease of removal was recognized by experienced car men as

being as essential as strength.[35] Composite bolsters built up from wood and iron plates sandwiched together produced a cheap, strong bolster, but like most such assemblies they became loose, weak, and noisy as the wood dried and shrank and the bolts lost their holding power.

The limited space available for bolsters was not enough to accommodate a wooden bolster of sufficient strength. Ideally, the bolster would be stiff and not sag. The weight would rest squarely on the center plate and touch the side bearings only when the car was leaning into a curve. Wooden bolsters

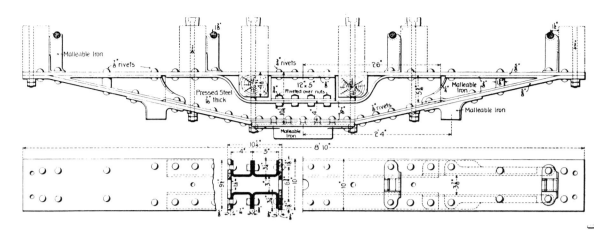

FIGURE 3.29

The Northern Pacific had one thousand 35-ton-capacity cars built on this plan by the Michigan-Peninsular Car Co. and the Illinois Car & Equipment Co. in 1898. An unusually heavy metal bolster was used, as shown in the detail accompanying the drawing. E. M. Herr designed the car. (*Railroad Gazette*, Sept. 30, 1898)

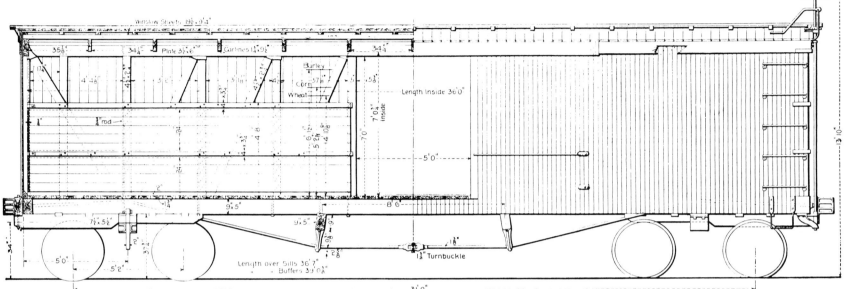

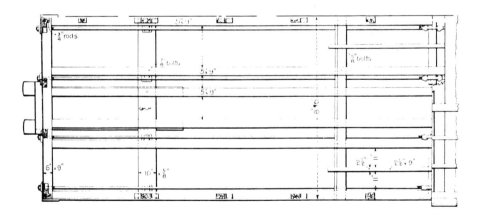

were not capable of such performance, and even when cars carried only 10 tons, they would bend and settle down on the side bearings. This inhibited the free turning of the truck and thus increased flange wear. This fact was obvious even in the pioneer period, as noted in Chapter 2, and all-iron body bolsters had become standard on a few lines by 1870. A simple truss fabricated from 1- by 6-inch wrought iron formed the common metal bolster. While this was hardly an overpowering assemblage of metal, an opinion was offered in the 1880 *Master Car Builders Report* that iron bolsters

were far too massive and as built could support 100 tons.[36] They were seen as a useless expense—in the opinion of one member. Others found that a stiff enough bolster had yet to be built, and that too many bent down and developed a permanent sag so that the car rode on its side bearing.[37] But the iron bolster was the best cross-framing member available at the time, and it became a fixture with big lines like the New York Central (Fig. 3.25). At first the top plate (tension member) was only 3/4 inch thick, while the bottom plate (compression member) was 7/8 inch thick. Seven years later thickness of the plates was reduced to 5/8 and 1/2 inch respectively, but the width was increased from 6 to 8 inches. The greater bearing surface helped prevent the bolster from digging into the floor sills. By the late 1890s some lines, such as the Milwaukee, had expanded bolster width to 10 inches.

Iron bolsters were ordinarily bolted to the underside of the sills, but in an effort to gain depth and hence strength, some lines began to build them into the frame so that the bolster surrounded the sills (Fig. 3.26). The top plate was placed on top of the floor sills, while the bottom plate passed below. The ends were bent down to form an L-shaped bracket for the outside sills. Bolsters of this design were particularly useful for furniture cars, since they helped lower height. Their drawback was, of course, the fact that they were buried within the structure and could not readily be removed.

Another variation was the double iron bolster, a device used on passenger cars since about 1880. It was simplified for freight car application about a decade later. The purpose of the double bolster was to spread the load and relieve the sills of too great a concentration of weight by making a larger bearing surface. This was to become more important as larger-capacity cars came into service. Failures occurred not only because of crushing due to the small area of the bearing, but also because the bolt holes, so necessary to hold the frame components together, also weakened the sills at this stress-burdened portion of the car. The Michigan Central began using double bolsters in 1886 with such good results that they were applied to subsequent cars (Fig. 3.16). Two 5-inch bolsters spaced 10 inches apart did much to end the broken-sill problem.[38]

By the mid-1890s car builders wanted something stronger than the traditional wrought-iron body bolster. As they inched reluctantly toward all-steel construction, wooden cars were fitted with more massive metal bolsters. In 1893 the American Steel Foundry of St. Louis began to market trucks with cast-steel bolsters. Not long afterward a companion body bolster was developed. Made very much like the standard wrought-iron bolster, it could be bolted to the underside of a wooden car (Fig. 3.26). In 1896 the Rock Island began tests on a more elaborate malleable-iron bolster developed by two employees, G. A. Akerlind and J. J. Carroll.[39] The results were satisfactory, and two hundred new cars were similarly equipped during the summer of 1897. More were required a year later, but the foundry was unable to deliver, and so the Rock Island turned to another supplier that was able to furnish only steel castings. They served well, and so by early 1899 the Rock Island had fifteen hundred cars fitted with cast bolsters. Railway suppliers like Simplex and

FIGURE 3.30

This New Haven drawing, published in 1880, is one of the earliest examples of truss rods fitted with turnbuckles. Other notable features of the car are the metal trucks and their wooden safety beams. (*National Car Builder*, Oct. 1880)

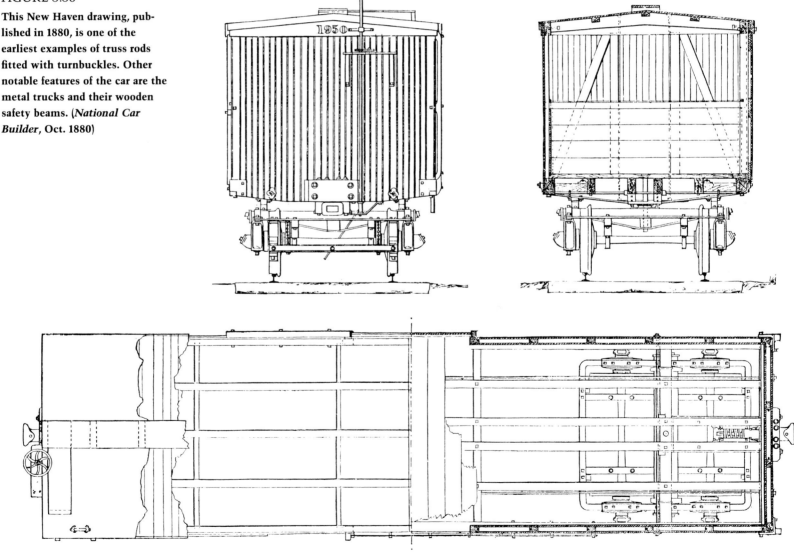

THE AMERICAN RAILROAD FREIGHT CAR

Commonwealth began to produce cast-steel bolsters commercially, and they were soon a standard freight car fixture.[40] At the same time, pressed-steel bolsters, notably those designed and manufactured by Charles Schoen, began to gain in popularity. Schoen's first design appeared in 1890 and consisted of pressed-steel pieces riveted and bolted together. Within five years the design had been simplified and strengthened. Even though ½-inch steel plate was used, Schoen claimed that his bolsters saved 300 to 500 pounds per car. Examples of early cast- and pressed-steel bolsters are shown in Figures 3.27, 3.28, and 3.29.

TRUSS RODS

Truss rods are the final element of freight car framing to be treated here. Their emergence and adoption were discussed in Chapter 2. Two rods per car, bearing against the needle beams or, at best, having very truncated queen posts, was common practice until the middle 1880s. After this time four rods and queen posts became the standard arrangement. Rods were often joined in the center by a curious arrangement that, while reducing strength, did allow for easier removal should repair or strengthening be necessary. Adjustment was made with the end nuts. Turnbuckles served as a point both for adjusting the rods and for taking them apart for repair purposes. The earliest dated example uncovered is for a New Haven drawing published in 1880 (Fig. 3.30).

Truss rods were rarely less than 1 inch in diameter and hardly ever greater than 1⅜ inches. The ends were upset and forged ⅛ inch larger in diameter so as not to be weakened by the metal cutaway for the threads. In a 34-foot boxcar it was estimated that the rods carried a load of 25,000 pounds.[41] Assuming that the four rods had a load limit of 7,000 pounds per square inch each, they were clearly being worked to their maximum. Larger-diameter rods or deeper queen posts would help introduce a greater margin of safety. Loose rods, which place a strain on the floor sills, were a familiar maintenance problem. The alternate

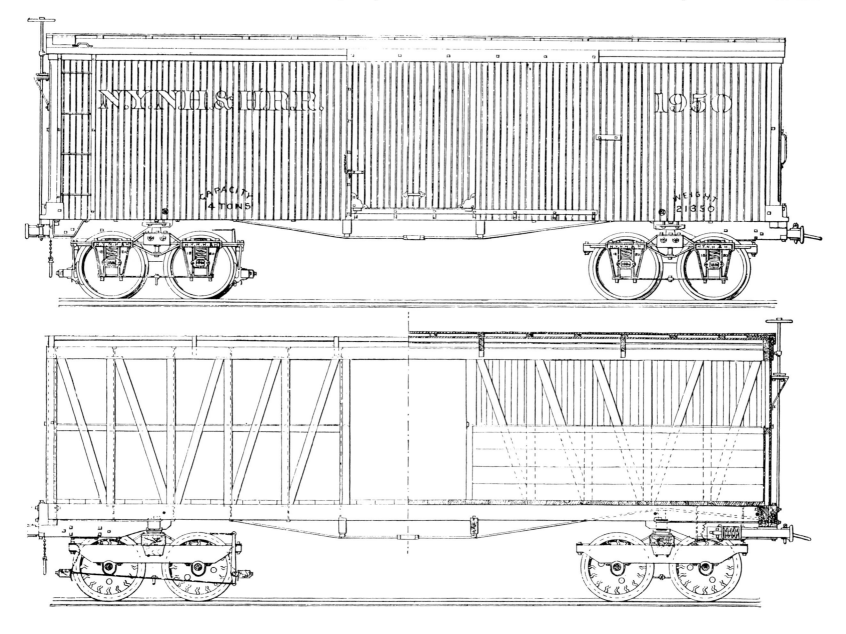

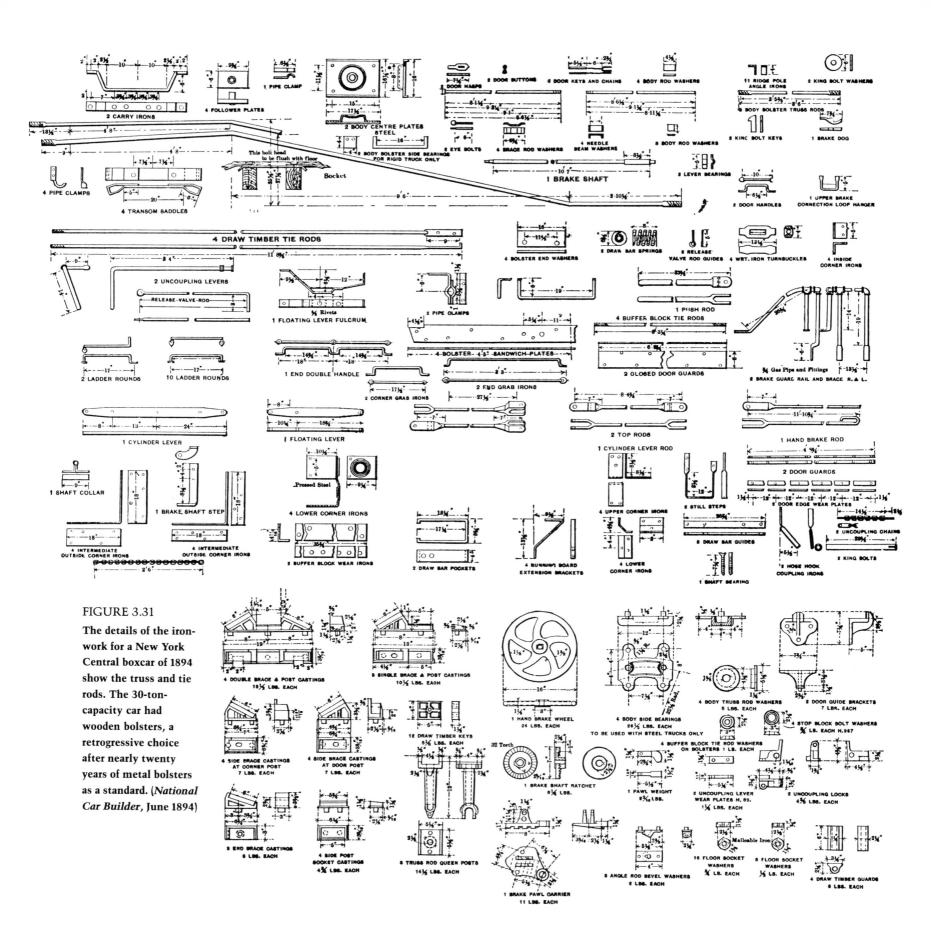

FIGURE 3.31

The details of the iron-work for a New York Central boxcar of 1894 show the truss and tie rods. The 30-ton-capacity car had wooden bolsters, a retrogressive choice after nearly twenty years of metal bolsters as a standard. (*National Car Builder*, June 1894)

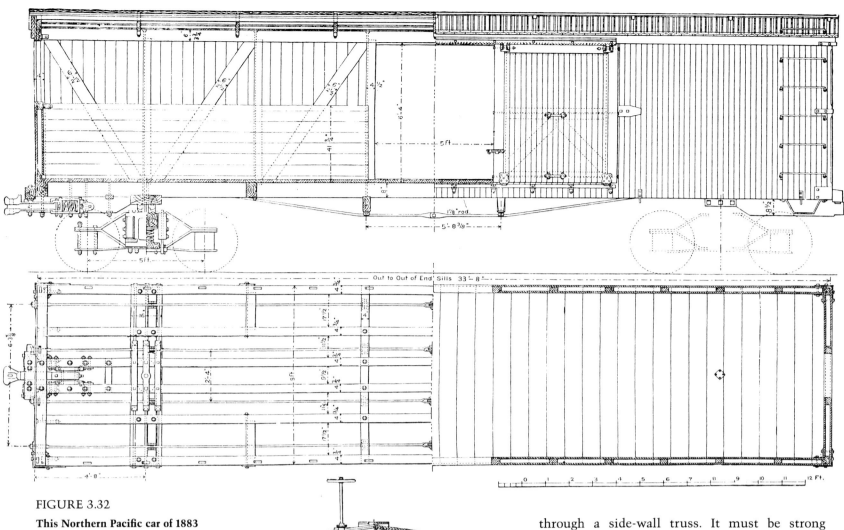

FIGURE 3.32

This Northern Pacific car of 1883 featured four truss rods. They are joined at their centers but without benefit of turnbuckles. The car was built by the Peninsular Car Company of Detroit. (*Railroad Gazette*, Aug. 31, 1883)

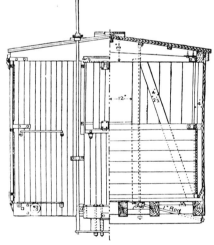

swelling and contraction of the wood tended to loosen the truss rods, as did the vibration caused by running over the track. If overtightened, the queen posts might pull loose and lean inward, thus allowing the rods to go limp. Extreme heat, as was encountered in the southwestern deserts, made them elongate. The defects of composite wood and metal fabrication were amply illustrated by the truss rods' erratic behavior.

STRENGTHENING THE BODY

A boxcar body was expected to do more than shelter its cargo. It was to help support the load through a side-wall truss. It must be strong enough to withstand the battering of loose cargo such as flour barrels that might be sent flying by sudden starts and stops. It must also be capable of withstanding the outward pressure of a bulk load like grain. Despite the demands for strength, the countervailing need for light weight resulted in body framing little greater in mass than one could expect to find supporting a common front porch. Most members were not larger than 2½ by 5 inches. The door and end posts were generally 4-by-4s, and the only piece of any mass was the top plate, which measured 4 by 6 inches.

Car builders copied freely from bridge engineers, and the familiar Pratt and Howe trusses were used, often in a modified form. The so-called bastard Howe became the most commonly used style of truss for boxcar bodies after around 1875 (Fig. 3.33). Before this time, and notably for cars bereft of truss rods, the Pratt truss was favored.[42] The Pennsylvania Railroad decided to employ the full Howe truss in 1885 in an effort to strengthen side walls so that they would not bulge out under the pressure of bulk loads like grain.[43] A stronger body resulted, but 1,000 pounds was added to the cars. The Pratt truss remained popular on a few lines, such as the Lehigh Valley, well into the 1880s.

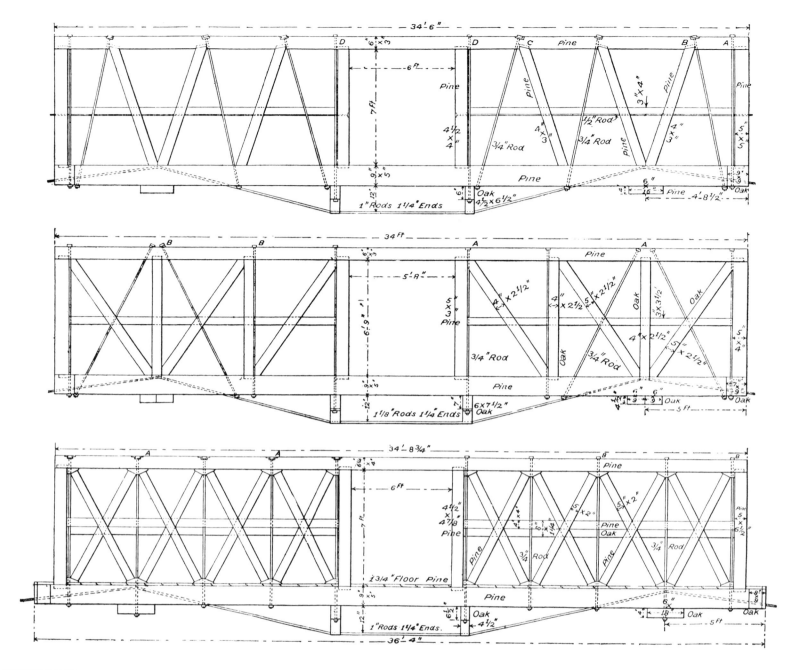

FIGURE 3.33

Three styles of boxcar body side frames. The center plan is the popular bastard Howe truss. The bottom plan is a true Howe as employed by the Pennsylvania Railroad. (*Railroad Gazette*, Nov. 4, 1887)

Whatever style of truss was employed, iron tie rods were used to hold the wooden members together. In the bastard Howe they were simple vertical rods running between the top plate and the outside floor sill. They were placed hard against (or sometimes cut into) the vertical posts, and were generally ⅝ to ¾ inch in diameter. Twenty was the usual number for a 30- to 34-foot-long car. Some roads like the Michigan Central used bigger, 1-inch-diameter rods, starting in 1887, for greater strength. The tie rods, like the main truss rods, were worked nearly to the breaking point, some being subjected to 10,000 pounds per square inch, or a load factor large enough to fracture wrought iron.[44] Other tie rods, particularly those directly over the body bolsters, carried almost no load and so were eliminated by some car builders. Lest tie rods come loose and so default on the job of hold-

ing the structure together, it became customary to hammer over the ends so that the nuts could not become loose.

End framing was handled very much like the side frames. The posts or compression members were square timbers (at the corners and doors) and thick planks for angle or tension members. This style of construction was thought too weak, and plans were offered in 1894 calling for heavier end timbers and tie rods so that shifting loads like lumber would not knock out the ends of the cars. This scheme was adopted as evidenced by a photograph of an old Baltimore & Ohio box taken in 1922 at the American Car & Foundry (Jackson & Sharp) plant in Wilmington, Delaware (Fig. 3.36). The end framing is made up entirely of square or solid timbers. The siding has been stripped away, presumably in preparation for rebuilding, a process

THE AMERICAN RAILROAD FREIGHT CAR

very much in order considering the dry rot evident in the side and end sills. Many years before this photograph was taken the B&O devised another plan for stiffening car bodies. The end boards were fastened on in a herringbone pattern. This idea went back to an 1862 design prepared by Thatcher Perkins. The idea was carried west to the Louisville & Nashville by Perkins, who became master mechanic of that line in 1869. It appears in a drawing of an L&N car published in 1881 (Fig. 3.37).[45] The B&O and its subsidiaries used herringbone ends into the 1890s, as shown by a photograph taken at the Mt. Clare repair shops, Baltimore (Fig. 3.38).

The body siding was usually clear white pine, measuring 3 to 4 inches wide by ⅞ inch thick. It normally had tongue-and-groove joints with decorative V-beading. Yellow pine was sometimes used but did not hold the paint so well because of its sap content. Vertical siding was standard throughout the wood-car era, but as might be expected at least one car builder, Fitch Adams of the Boston & Albany, advocated horizontal siding.[46] Adams generally had independent ideas and frequently expressed them at Master Car Builders Association meetings. His idea for horizontal siding, though against the conventional tide, did much to stiffen the body, while ordinary vertical siding did rather little. Vertical siding surely shed water better than Adams's plan for horizontal boards, but one suspects that few car builders could accept the odd appearance offered by the B&A henhouse on wheels.

Adams pushed one idea in car building that did gain popularity, and that was the use of pocket castings to hold body framing members in place of conventional mortise-and-tenon joints. Mortise joints not only required the cutting away of material and hence the weakening of both pieces being joined; they also tended to work loose in a structure subject to constant vibration such as a railway car. As the tenon worked in and out of the mortise, it wore down in size while the hole grew large. A cavity was thus formed for the collection of water, and the ensuing rot helped shorten the vehicle's useful life. The pocket or shoe castings were used as early as 1876 by the New York Central, although their inventor and first use remain unrecorded. They not only eliminated mortise joints but also prevented the end crushing of timbers. Adams felt that they simplified and speeded car construction, but Voss contended that they were regarded only as an extra expense and were not used for that reason by some railroads. Around 1900 the Western Railway Equipment Company of St. Louis began marketing cast-iron pockets to secure frame sills to end sills. They were bolted to the end sill and had removable bottom plates or keys so made that the longitudinal sills could be removed for repair or replacement without detaching the shoe (Fig 3.40).

Internal sheathing was at first used only for cars in grain service, but as the interchange system grew it became harder to predict what commodities might be present for carriage, and so inside planking became more common. It was usually not carried much more than halfway up the body because of the load limit of the car and the size of the grain doors. Most cars were marked for the major grains, with the denser ones like wheat having the lowest mark and light ones like oats the highest. Inside boards or linings were always fastened horizontally. A poor grade of lumber—poor by nineteenth-century standards, that is—was used for lining. Better wood was needed for floor-boards, which were subject to heavy use and some abuse. Planks 1¾ to 2¼ inches thick were used. Both tongue-and-groove and shiplap joints were used, the latter being preferred because they made it easier to disassemble the floor for relaying in case it had to be pulled up for repair.

ROOFS

Devising a good roof was a challenge car builders never seemed to conquer. The requirements were a collection of contradictions. It must be watertight, light, strong, durable, safe to walk on, and cheap. A great variety of materials and techniques were tried, but none was entirely satisfactory. A survey conducted by the New York & Harlem in 1878 found that the interchange cars passing over its line had roofs in varying degrees of bad order.[47] The data was gathered railroad by railroad—the worst recorded 65 percent with leaking roofs and even the best was 7.4 percent. Fourteen years later J. C. Barber, master car builder for the Northern Pacific, found the situation little improved after inspecting cars in yards located in New York, Chicago, and St. Paul.[48] He found that fully 25 percent of the cars had leaking roofs. The cost in damaged merchandise claims to the industry must have been enormous, and yet Barber suspected that even worse conditions would be found by a more systematic survey. William Voss, writing at the time of Barber's study, could only lament that the perfect car roof was a problem still to be solved and was likely to remain one.

Tight, inexpensive roofing was available at the time for homes, barns, and other structures, but it was not subject to the flexing and vibration endured by the typical railroad car, which resulted in open joints and ruptured surfaces. In addition, brakemen were forever walking over the roofs as part of their duty. This was especially true before the time of air brakes. Drovers attending to their cattle would also clomp over the roofs. Their great, heavy boots and pointed cattle pikes inflicted no end of damage, particularly to tin or canvas roofs. To make matters worse, it was necessary to keep the pitch of the roof to a minimum for the crews' safety, yet the flatter the roof, the more likely it

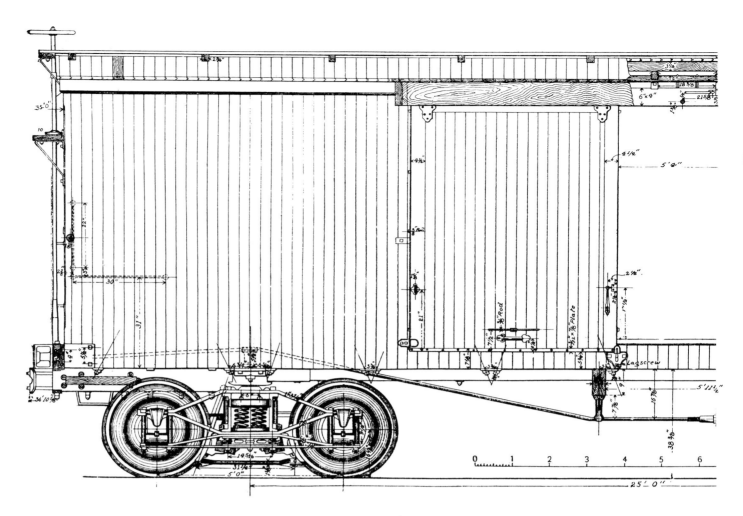

FIGURE 3.34

This Lehigh Valley 25-ton boxcar, designed by J. S. Lentz, used a Platt truss for the body side frame. Note the use of X tie rods to stiffen the roof and body. The car weighed 29,300 pounds. (*Railroad Gazette*, June 8, 1888)

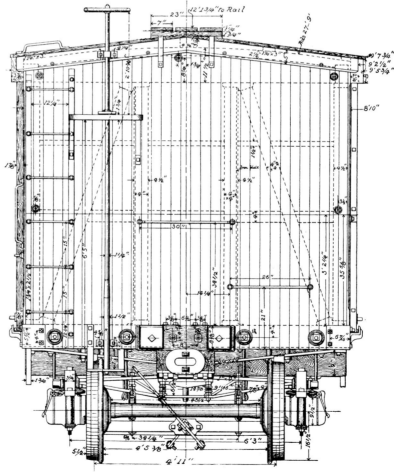

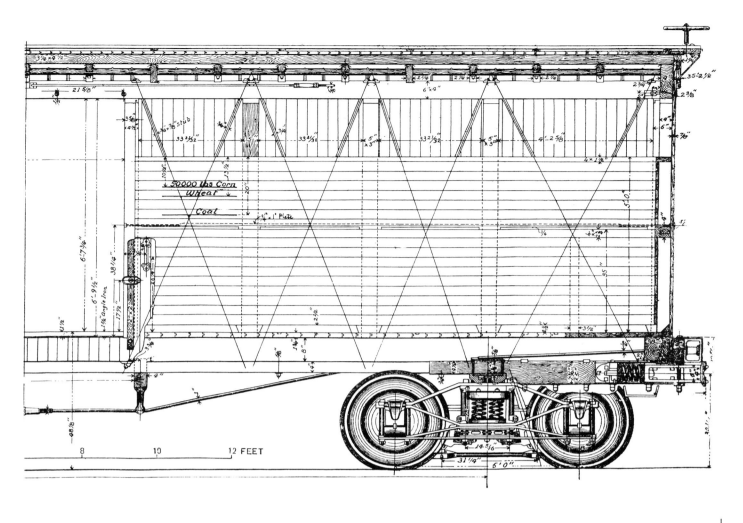

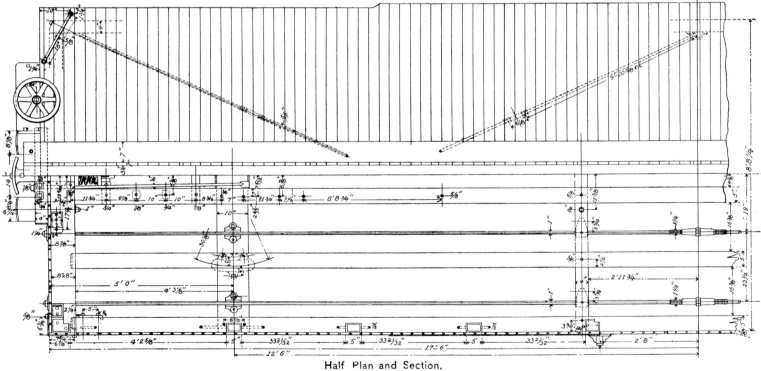

Half Plan and Section.

CARS FOR GENERAL MERCHANDISE

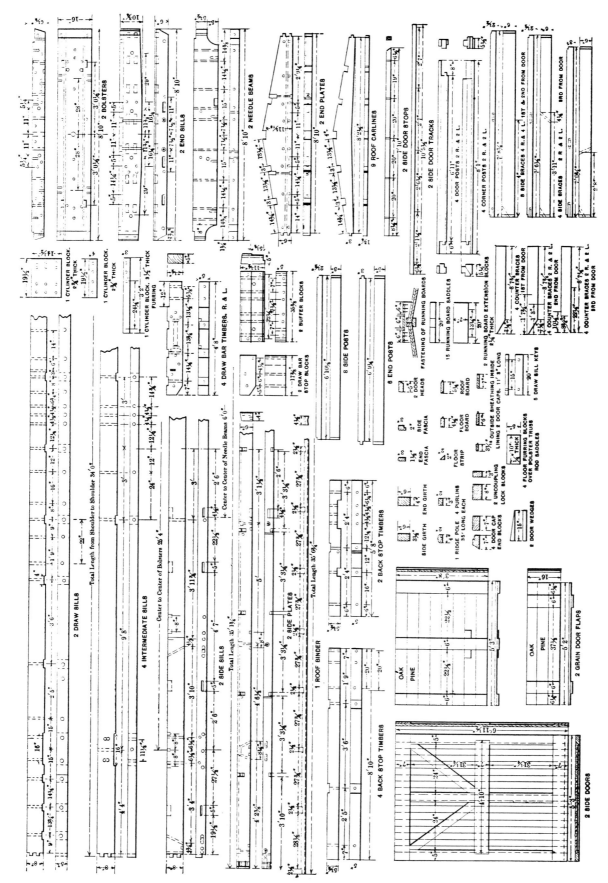

FIGURE 3.35

Wooden parts for a 30-ton New York Central boxcar of 1894. Notice the many mortises, tenons, and bolt holes in each timber. (*National Car Builder*, May 1894)

FIGURE 3.36

A late-model wooden-framed B&O car with its sheathing removed shows both the body framing and the effects of many years of service. The scene is the yard of American Car & Foundry, Wilmington, Del. (Jackson & Sharp), 1922. (Hall of Records, Dover, Del.)

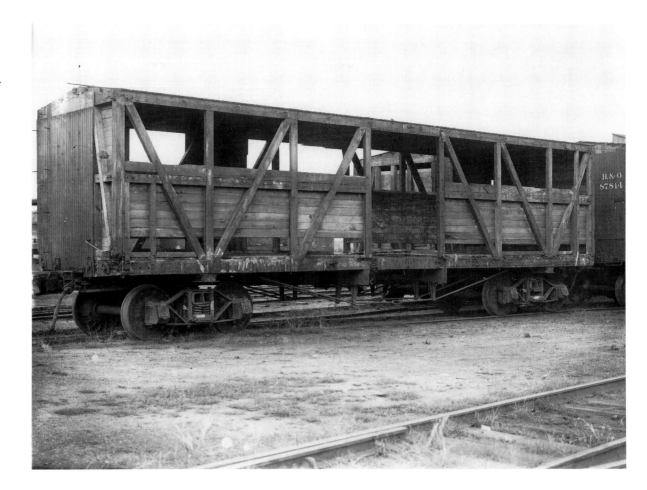

FIGURE 3.37

Mechanical superintendents such as Thatcher Perkins carried ideas from one railroad to another as they changed jobs. The herringbone end pattern was used on the B&O by Perkins and came to be adopted by the Louisville & Nashville after he moved south in later years. (*National Car Builder*, April 1881)

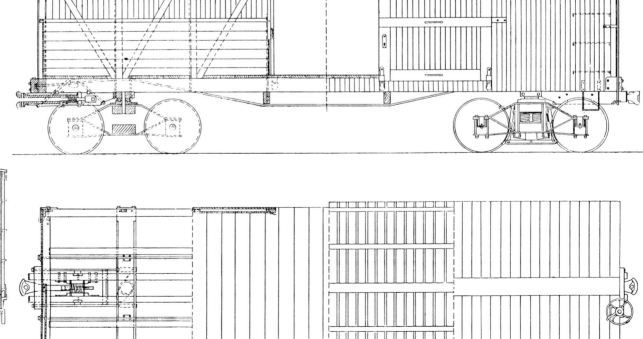

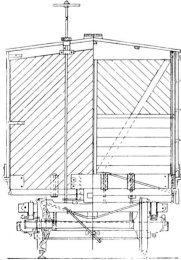

225   **CARS FOR GENERAL MERCHANDISE**

FIGURE 3.38

A veteran Baltimore & Ohio & Chicago boxcar undergoes repairs at the Mt. Clare shops in about 1900. Both the car and its trucks have wooden bolsters. Car 27333 has herringbone wooden ends. (Smithsonian Neg. 49864)

FIGURE 3.39

The beaded vertical siding favored by American car builders is clearly shown in this Big Four boxcar produced by Pullman in 1901. (Pullman Neg. 5701)

FIGURE 3.40

Pocket castings had been used in place of mortise joints in freight car bodies since at least the 1870s and possibly earlier. The idea was adopted for connecting the side and end sills of the main frame. Note that the bottom keys can be removed to disassemble the frame. (F. J. Krueger, _Freight Car Equipment_ [Detroit, 1910])

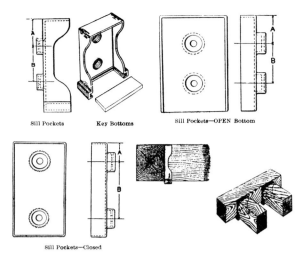

Sill Pockets   Key Bottoms   Sill Pockets—OPEN Bottom

Sill Pockets—Closed

was to puddle water and hence leak. Is it any wonder that leaks were epidemic?

Vain hope for improvement was aroused when the old arch roof gave way to pitched or hip roofs around 1870. The arch roof tended to break down with age and traffic near the center, resulting in pools and leaks. Tin was seen as the answer to the need for a durable covering for the new hip-style roof. Being fireproof was seen as a decided advantage in the spark-strewn path that followed every steam locomotive. It was successful when applied to the flat-roof townhouses then being built by the block in nearly every Victorian city. But these homes were not traversed daily by a small army of heavy-footed trainmen. The rocking of the cars opened their soft-solder joints to the sky. Tin was considered a twenty-year roof, but some railroads were realizing not much more than three years. By 1884 the Master Car Builders reported that it was out of favor, having been largely replaced with double-board roofs.

The double-board roof had the appeal of simplicity and economy (Fig. 3.41). It was practical only for short spans, because no horizontal joints were possible. Two layers of narrow boards were nailed together with their longitudinal joints overlapping. Any water that crept through the upper joint would run down the underlying board to the edge before finding the inside joint. Because water runs were so short—never more than, say, 54 inches—the plan was a reasonable one. As an extra precaution, tiny gutters were made on the centers of the underlying boards exactly beneath the joint of the surface boards. The gutters were nothing more than small grooves milled or cut into the boards' surface.

Experience with double-board roofs demonstrated that they were not necessarily so cheap or simple.[49] Only the best lumber could be used. Common grades of pine would rot out after two years, while the best clear pine would give six years of service. Green lumber would shrink and warp, while seasoned boards would swell when wet and pull up the nails. Water worked its way under the boards, and if the drain grooves became clogged with cinders, rot began out of sight between the two layers of wood. A better-than-average carpenter was needed to fit and assemble a board roof; hence labor costs were relatively high. About 250 boards were used in each roof. In the _National Car Builder_ of April 1892 Barber noted that the cost of clear pine had nearly doubled in recent years. But despite all these shortcomings, the application of board roofs continued until the end of the wood-car era. Painting or asphalt coating between the two layers was tried in an effort to extend service life. The top surface was also painted, but asphalt coating was not attempted because of the fire hazard.

The combination wood/sheet-metal roof was the most ingenious and perhaps the most successful effort to find a truly practical weatherproof top for the American boxcar. It was a double-layered arrangement that might be called a roof within a roof. The upper surface was a single layer of boards, giving it the appearance of a double-board roof, but a relatively poor grade of lumber could be employed because the exposed layer was meant only as a walking or protective surface for the metal subsurface suspended just a few inches below. It worked something like a giant drip pan, but one that covered the entire surface rather than just one portion. The force of the falling rain was broken and partially deflected by the upper wooden surface. The water that penetrated ran down the metal surface and escaped through an opening running the length of the car just behind the fascia board.

A. P. Winslow of Cleveland, Ohio, who may have been the inventor of the combination car roof, received a patent on August 9, 1859 (No. 25,071), that featured an arched inner roof of corrugated iron (Fig. 3.42). Winslow made little progress until Julius E. French joined the firm around 1870.[50] Sales began to mount, and by 1874 twenty-seven thousand roof kits had been sold. An advertisement appearing in the 1877 _National Car Builder_ claimed that forty thousand Winslow roofs were in service. Imitators were bound to flourish following such a success, and soon car builders had a host of combination roofs from which to choose. National, Anchor, Empire, and others offered their special variety of galvanized-metal inner roofs. Costs for the kits, less the cost of installation, were reckoned at $27.20 to $30.00 each.[51] To no one's surprise, outer roofs of pressed galvanized metal were also offered. A waterproof joint to connect the sections remained the problem with all metal roofs. Various caps were tried, as well as rolled and crimped joints. Aluminum was suggested in 1892 as perhaps the first mention of this material for freight car construction.[52] It was light and would not rust, yet no one appears to have

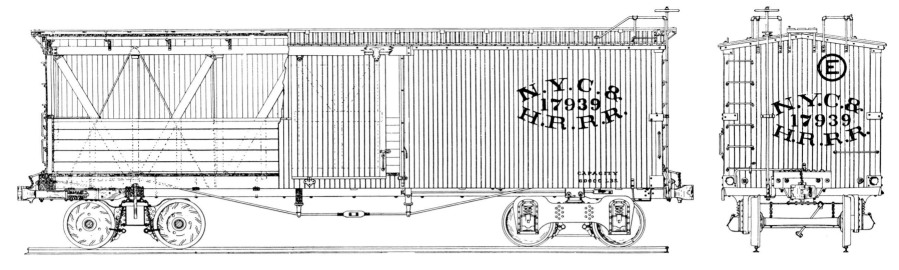

FIGURE 3.41

This 30-ton New York Central & Hudson River boxcar of 1892 had a double-board roof. The body measured 35 feet long by 8 feet 6 inches wide. Pressed-steel trucks made by Sampson Fox were tried on fifty cars. (*Locomotive Engineering*, Jan. 1892)

FIGURE 3.42

A. P. Winslow's double roof, patented in 1859, consisted of an arched metal roof covered and protected by an outer layer of boards. (*Car Builders Dictionary*, 1884)

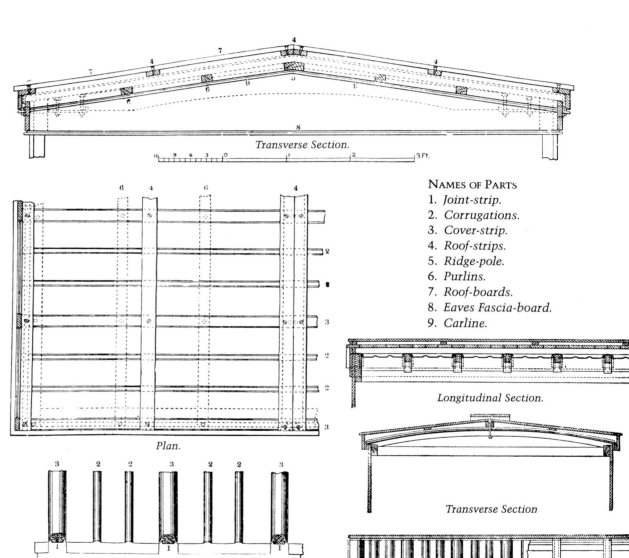

Transverse Section.

NAMES OF PARTS
1. *Joint-strip.*
2. *Corrugations.*
3. *Cover-strip.*
4. *Roof-strips.*
5. *Ridge-pole.*
6. *Purlins.*
7. *Roof-boards.*
8. *Eaves Fascia-board.*
9. *Carline.*

Plan.

Longitudinal Section.

Longitudinal Section.

Transverse Section

WINSLOW CAR ROOF.
(With flat roof-sheets; the most recent and common standard.)

Plan.

WINSLOW CAR ROOF. *(With curved roof-sheets.)*

SECTION OF JOINT STRIP.

tested aluminum until many years later. Other inventors perfected a cheaper variety of combination roof. In about 1880 C. B. Hutchins began marketing a double roof with a "plastic" undersurface made from tarpaper and asphalt. John M. Ayer, better known for his work with refrigerator cars, tried to sell sheet-rubber coverings for boxcar subroofs.

Roofs of heavy galvanized sheet iron were introduced in the 1890s to eliminate the need for the elaborate double roofs just described. Because the metal was on the surface of the wooden roof sheeting, they were called outer roofs. The metal was heavy-gauge to withstand the traffic so injurious to the old-fashioned tin roofing. Solder was replaced by rolled or crimped joints that produced a long ridge where each sheet abutted its neighbor. By flanging the metal and bending it up and over, a watertight union was formed that was strong enough to stand considerably more "working" from expansion or contraction than could be tolerated by soft-solder connections. A number of outer roof systems were developed such as the Hutchins and Excelsior shown by the drawings reproduced in Figures 3.43 and 3.44. Most of these ideas were adapted from ideas used in domestic and industrial roofing.

The problems with roofs explain the need for roof walk boards. They were introduced almost at the beginning of covered freight cars to offer a convenient and sturdy path for the brakeman. Ideally, he would confine himself to this narrow walkway and thus help preserve the fragile waterproof surface on each side. The running board was self-defeating in a way because of the blocks, nails, or screws needed to attach it. These fastenings punctured the roof skin and became yet another source of leaks.

LUMBER

The unflinching loyalty to wooden cars may appear misplaced or even foolish from today's perspective, but from the nineteenth-century viewpoint nothing could have been more natural. North America's virgin forests offered an abundance of superb building material practically for the taking. The virgin forest has long since disappeared and the building lumber available today is a pulpy, tree-farm hybrid that no respectable Victorian carpenter would have allowed in his workshop. In-

FIGURE 3.43

The Hutchins board and painted canvas roof was introduced in the late 1870s. Hutchins was also active in the design and operation of refrigerator cars. (*Poor's Manual of Railroads*, 1892)

FIGURE 3.44

This pressed-metal exterior roof with rolled joints was made by the Excelsior Car Roof Co. of Saint Louis. (Stimson, *Modern Freight Car Estimating*)

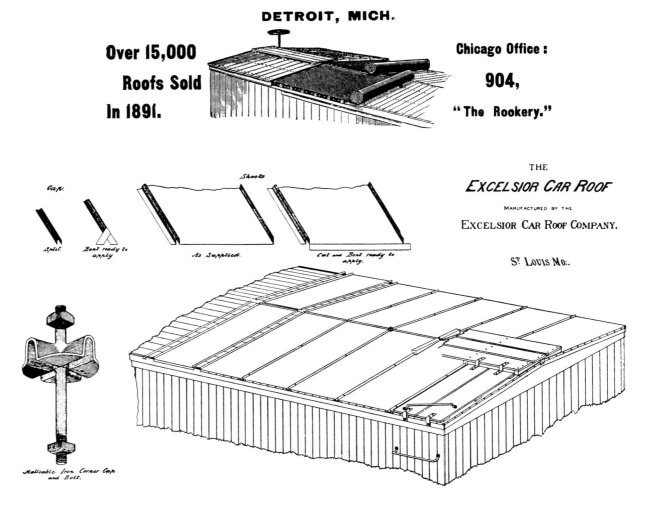

CARS FOR GENERAL MERCHANDISE

deed, much of the wood offered at present-day retail lumberyards would have been sent off to a crate factory, even in my childhood. But car builders just a century ago could demand and purchase prime lumber at modest cost for such mundane products as freight cars. Ordinary construction lumber at this time was virgin, long-leaf yellow pine, free from large knots or sap and wind shakes.[53] The big stands of white oak were gone by around 1880, but there was still an abundance of Norway, southern, and Georgia yellow pine. The harvest of big firs in the northwestern states was just beginning. Yellow pine then became the standard framing material, with oak being used only for end sills, posts, draft timbers, and carlines. Hickory, elm, or oak was used for brake beams, while a few lines would specify lesser woods like hemlock for nonstructural parts like roof walks. White pine was preferred for siding and roofs.

Green or unseasoned lumber was about the worst problem car builders had to fear from their basic building material. Raw timber shrinks, cracks, and is susceptible to rot. But the fault was not always with the supplier. In the rush to complete cars, builders would push sawmills for timber on short notice.[54] Trees were thus often turned into finished cars within a matter of days. If the logs were cut in winter, the sap would remain frozen inside the cut lumber until the first warm days of spring. The defects and horrors of green wood were a common topic among car builders, as reflected in the trade journals of the period. But experienced car builders like F. E. Canda admitted that even fairly well seasoned lumber might lose as much as 1,100 pounds of water in a large car.[55] J. S. Lentz, who practiced as a car builder longer than just about any other American, was not particularly concerned about green lumber. In 1888 he mentioned a boxcar built four years earlier with green wood that exhibited no particular ill effects.[56] It had even maintained its original camber.

The quantity of wood used in a typical boxcar was roughly 3,300 to 4,300 board-feet. Translated into hundreds of thousands of cars, this suggests entire forests vanishing. While freight cars were hardly the principal user of the nation's timber production, they did consume a respectable percentage of the better building lumber, and car builders grew ever more nervous about available supplies during the late 1890s. This fear was inflamed between 1897 and 1907 when lumber prices rose an unprecedented 78 percent.[57] Prices then fell to more normal levels, but during those critical years the industry decided to adopt steel cars. It would be interesting to know how much the short-lived lumber crisis and the attendant rumors about future supplies had to do with the conversion to steel construction.

THE INCREASING USE OF METAL

By 1890 American car builders must have sensed that the wooden age was over, yet they sought ways to postpone the coming of the steel car. Elaborate compromises were devised to prolong the ancient ways of car construction. Superficially, wood remained the predominant material, but gradually more and more metal was being worked into the structure. Most of it was hidden from view, as in the body tie rods, but the trend toward more metal was unmistakable to anyone familiar with the details of freight or passenger car construction. Iron body bolsters and four truss rods were common by 1885, as has already been explained. In reality, the car builders had come to accept composite construction without ever admitting it. When compared in terms of material weights, freight cars were roughly 40 percent iron and 60 percent wood. The better portion of the metal was in the trucks, notably in the wheels, but the body was nearly 25 percent metal.

A 20,000-pound Louisville & Nashville car dating from 1881 had the following breakdown of metals in its body: 1,075.5 pounds of cast iron, 956.5 pounds of wrought iron, and 546 pounds of bolts, for a total of 2,578 pounds.[58] Assuming that the trucks weighed 10,000 pounds, the wood used weighed 7,422 pounds. Similar data is available for a 27,300-pound Chicago, Burlington & Quincy boxcar built several years later.[59] In this case the trucks weighed 10,946 pounds, the metal in the body 4,850 pounds, and the wood 11,504 pounds. In terms of volume only did wood exceed metal. At least a few car builders felt that the ratio would have to be increased more in favor of iron if more durable cars of greater capacity were ever to be introduced.

An early effort in this direction was mounted by the Pittsburgh Car Works in about 1875. The firm's manager, George W. Bitner, advocated floor sills reinforced with light but deep channel iron.[60] They were clad directly to somewhat undersize wooden timbers. The body bolsters were also of composite construction. Bitner's idea never progressed beyond the experimental stage, and nearly twenty years were to elapse before it was again considered. In 1895 the Baltimore & Ohio experienced such good results from reinforcing some of its locomotive tenders with channels that the line's mechanical department decided to adopt the idea for wooden boxcar frames.[61] Early in 1896 a 30-ton boxcar was produced with two 10-inch channels in place of the normal wooden center sills. The Chicago, Burlington & Quincy liked the idea of a metal backbone and removed the massive 5- by 15-inch double wooden sills from one of its larger-capacity boxcars. Two 10-inch channels were substituted. Neither of these efforts was actually so

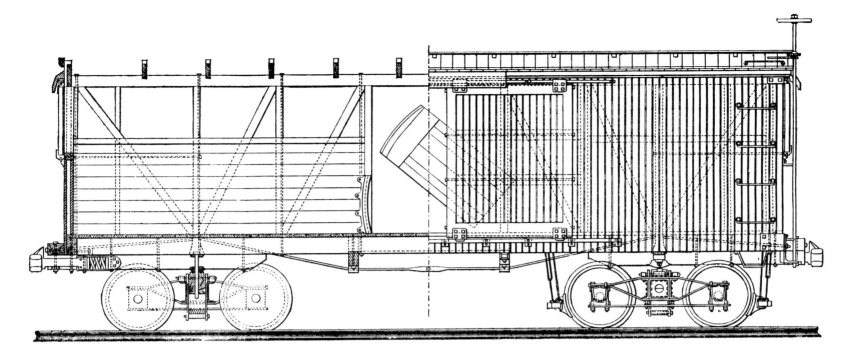

FIGURE 3.45

A Chicago, Burlington & Quincy car designed in 1879. The body measure 28 feet long by 8 feet 9 inches wide. Weight and capacity are not recorded, but the body is known to have contained about 3,200 pounds of metal. The car cost $456.12. Compare this drawing with the Burlington boxcar of 1886 shown in Figure 3.46. (*National Car Builder*, April 1880)

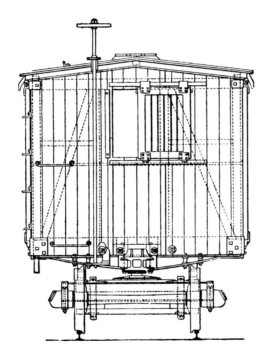

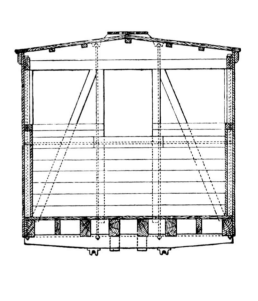

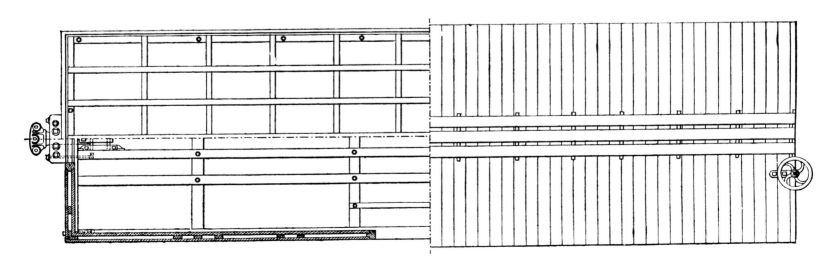

CARS FOR GENERAL MERCHANDISE

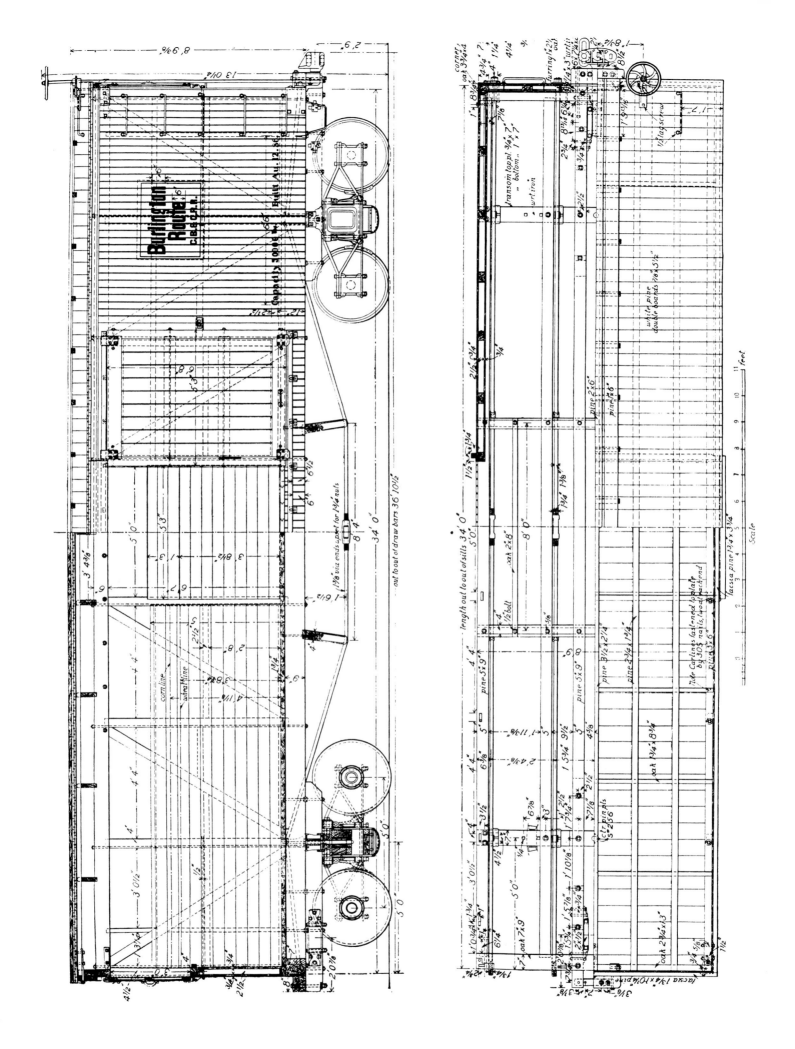

THE AMERICAN RAILROAD FREIGHT CAR

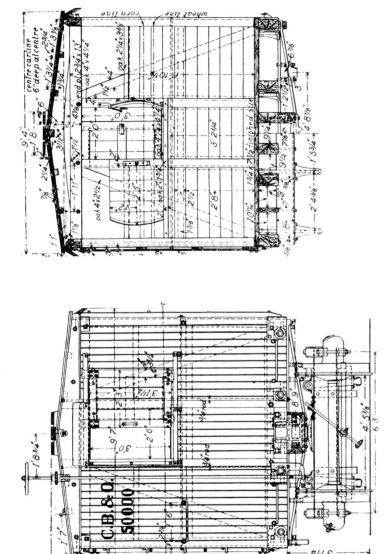

FIGURE 3.46

In just seven years the Burlington's standard boxcar showed measurable growth. The body was 6 feet longer and 7 inches wider. The car weighed 27,300 pounds and carried 50,000 pounds. (*Railroad Gazette*, June 21, 1889)

FIGURE 3.47

The neat finish and glossy paint of a new Illinois Central Boxcar are evident in this photograph made at the Pullman Chicago car plant in 1893. (Pullman Neg. 2442)

233   CARS FOR GENERAL MERCHANDISE

FIGURE 3.48

A decorative shield helps dress up this Rock Island car built in 1886 by the Peninsular Car Co. Note the offset end door, single pair of truss rods, and link-and-pin couplers. (ACF Industries Photo)

very pioneering when the history of the iron car is considered (Chap. 8); however, the willingness of two main-line mechanical departments to dally openly with metal underframing was viewed as something novel. Within a few years pressed-steel and channel-iron shapes were replacing old-fashioned wooden carlines. The boxcar was becoming ever more metal. The steel freight car was being slipped into general practice, unofficially, through the back door.

As the wooden era closed, even experienced car builders like Eugene Chamberlain of the New York Central were seeking ways to save weight and lessen maintenance costs.[62] He felt that cast iron was too weak and heavy for the many small parts regularly made from common gray iron. These parts included couplers, draft-gear blocks, queen posts, washers, and post pockets. He favored malleable iron and contended that 1,300 to 1,500 pounds could be saved by using it in place of cast-iron fittings. It was easy to see the dead-weight savings in even a short train, for in fifteen cars 10 tons would be eliminated. The extra cost of malleable castings would be offset by lessened fuel and maintenance expenses. Malleable iron was calculated as being 80 percent stronger than cast iron. It was almost as strong as a forging but was, of course, less expensive. In June 1898 the *Railroad Gazette* contended that except for wheels, gray-iron castings were no longer used on the better class of freight cars.

Two railway supply manufacturers named Sampson Fox and Charles Schoen agreed with the need for stronger car parts, but they argued for pressed steel rather than malleable iron. By 1890 they were making some progress as more lines began to specify pressed-steel components.[63] Fox had sold center plates to upwards of fifteen railroads. Schoen plates were used on seven thousand cars. His new Pittsburgh plant was also stamping out drawbar attachments, stake pockets, and corner plates. Both manufacturers offered pressed-steel trucks, and both were eager to produce entire cars of steel, an ambition that would be realized within a few years. Meanwhile, Fox and Schoen noted with some satisfaction that their parts weighed only one-third as much as cast iron and were three to four times stronger.

PAINTING

The utilitarian character of freight cars would suggest that they are painted only to preserve the structure. One might as well say the same for a barn, but even a modest countryside tour will turn up a surprising number of handsomely painted barns. And so it is with boxcars, whose great flat sides called for paint and lettering. To traffic managers and public relations men, this broad surface was free advertising space. And many of the illustrations in this book show the symbols, logos, and fanciful examples of the sign painter's art that decorated boxcars, particularly after 1880 (Figs. 3.48, 3.49, 3.50). Intertwined letters, maps, shields, badges, even stylized sunsets were painted on the vertical boards of America's great wooden fleet. Just how much goodwill or freight traffic these elaborate creations generated is questionable. One official insisted that there was only one real justification for the expense of painting a car, and that was to provide a foundation for the numbers and lettering so essential to the yard clerk's record keeping.[64] Cars were painted for appearance only, because they were used up in service or destroyed in a wreck before they grew old enough to decay.

This extreme opinion surely held true for a portion of the fleet, but car builders agreed that sixteen years was a fair average life, and so paint for preservation purposes did seem reasonable. Painting the exterior was obvious, but many roads called for interior applications in areas prone to rot. Painting between double-board roofs was one such precaution. The tops of the sills and mortise-and-tenon joints were occasionally painted. A few roads like the Chicago & North Western and Nickel Plate used creosote or brand-name preservatives—Fernoline, for example—for the same purpose. Just how effective these treatments were is again very questionable. It has been demonstrated over the history of wood preservation that surface coatings have little long-term effect and that only pressure-treated timber provides any real gain in life expectancy. The best application methods are of little avail unless the preservative itself is an effective agent. In 1895 the Pennsylvania Railroad specified a liquid distilled from Georgia pine that contained 5 percent tarry matter, 45 percent tar acids, and 50 percent neutral oils.[65] A directive published by the same railroad in 1893 ordered all mortise-and-tenon joints to be brush-coated with wood preservative instead of paint as formerly specified.[66] It was not to be done if the joints were glued, as was the case for passenger cars. Preservative was to be applied wherever two pieces of wood came into contact. The floors and the tops of all sills of stock cars, gondolas, flats, and any cars used in the manure trade were to be coated.

The question of color was mentioned in Chapter 2, and again even after 1870 there appears to be little specific documentation for boxcars being painted boxcar red. Surviving specifications—essentially what was published in the trade journals—almost never mention color but usually state only that three coats of the best mineral paint were to be applied. The color, if ever mentioned, was to follow a sample provided to the contractor, but these samples are, of course, now lost. Of the few specific instances where color was described, not one includes red.[67] In 1872 the Boston & Albany was reported to be painting new grain cars a flame yellow or orange color. Four years later the Missouri, Kansas & Texas specified brown for its boxcars. The Virginia & Truckee Railroad, a short line serving the gold fields of Nevada, appropriately painted its freight equipment yellow during the 1870s and adopted more somber shades only in later decades. According to the one other report published late in 1897, the Boston & Maine had adopted brown with white lettering. Previously its freight equipment had been slate, a lead gray, with black ironwork and yellow lettering. Was boxcar red a phenomenon of the twentieth century, or was it used previous to this time and escaped attention simply because it was so familiar? Artists never painted freight cars, and photography was all black and white, as were engravings, and so the visual record is of no assistance.

There is a simple counterargument to the foregoing discussion of the absence of boxcar red. Natural pigments were the only coloring agents available to the nineteenth-century painter. One of the most abundant of these was red oxide, which was derived from iron ore. The best red was made from the rich Lake Superior ores.[68] Poor ores like those

FIGURE 3.49

Bold lettering and emblems helped advertise the owning railroads. The admonition to post no advertisements stenciled on the side door was often ignored. This car was built by Pullman in 1897. (Pullman Neg. 3850)

CARS FOR GENERAL MERCHANDISE

FIGURE 3.50

The Southern Pacific used its
Sunset Route logo on equipment
of subsidiary lines such as the
New York, Texas & Mexican
Railway. This car was completed
by Pullman in 1895. (Pullman
Neg. 3061)

found in the South produced a dirty brown color called Tennessee Mud, and so boxcar red could be almost any brown-red tone, depending on the quality of the oxide-producing ore. Because it was cheap and in plentiful supply, red oxide was a common painter's ingredient and was for this reason, if no other, employed for rough work like freight car painting.

The method employed to apply the paint is not in doubt. The familiar brush and bucket served well until almost the end of the nineteenth century, when spray-painting came into limited favor. Compressed-air tools became common around this time and offered a powerful and reasonably flexible medium to propel rivet hammers, chisels, and wrenches before electricity came into more general use. Compressed-air painting was a natural offshoot of this minor revolution in machine and repair shop practice. The Southern Pacific reportedly began spray-painting freight cars in the 1880s.[69] Techniques were perfected over the years, so that by the middle 1890s a body could be coated in just twenty-five minutes. Spray-painting was satisfactory for rough work, and while no one during this period would have considered it for passenger cars, it made rapid progress in the freight equipment area. By 1896 major railroads like the Illinois Central were completely won over to spray-painting. The saving in labor costs and time was entirely in its favor. A 35-foot boxcar and its trucks were painted by brush for 163.75 cents in ten hours and fifty-five minutes or by spray for 57.75 cents in two hours and fifty-one minutes.[70] Only the stenciling was performed by hand. Spray-painting consumed 7.5 pounds less paint per car than brush-painting. If this was a fad, as some critics maintained, then railroads should busy themselves in seeking out more fads.

Critics of the squirt gun could, however, raise two serious objections to painting by "wind." It

was dangerous to painters, particularly if lead paint was used. One master car painter admonished his comrades to consider the risks of killing "our boys in the springtime of life." Some supervisors required their men to fasten sponges over their mouth and nose, but these crude masks were so uncomfortable and ineffective that most men tore them off once the foreman left the job. Overspray was a problem too, because almost all early spray-painting was done out of doors. It would drift over the yards and shops, depositing an unwanted layer of paint on every surface; this may be one source of the old chestnut about painting the town red. Spray booths and better face masks eventually answered the most obvious hazards.

A final criticism of spray-painting was that the mixture had to be thinned significantly before it would pass through the nozzle. Such thin, watery paint didn't offer much protection against the weather. Old-line car painters could not reconcile themselves to the commercial ready-mix paints that dried in a matter of hours rather than days.[71] Real paint, according to veteran car painters, was custom-made according to the individual formula of the paint shop boss. It contained no artificial driers and was heavily laced with linseed oil. Several days of drying time was needed between coats. But the rush to get cars out of the shop and back into revenue service ended these leisurely procedures around 1875. If the transition from handmade to factory-made paint was seen as a lamentable decline in standards, then the more rapid conversion from brush- to spray-painting must have been viewed as unspeakable.

Flatcars

Flatcars, like boxcars, were used for the carriage of a wide range of goods, as explained in Chapter 2. Essentially, they were best adapted to carry mer-

FIGURE 3.51

Six truss rods could not prevent the noticeable sag in this Pittsburgh, Fort Wayne & Chicago flatcar. The scene is probably the grounds of the Columbian Exposition, Chicago, 1893. (Mechanical and Civil Engineering Division, Smithsonian Institution)

chandise or nonbulk shipments, and so they most logically belong in this chapter. Very large, high-capacity flats will, however, be treated in the chapter on special cars (Chap. 6). Whatever could withstand the weather or would not fit inside a boxcar would generally go by flatcar. Cargo would range from the very light, like hay, to the very heavy, like machinery. Lumber, particularly large timbers, rail, pipe, and other heavy durable goods were common flatcar ladings.

Shippers had an even easier time overloading flats than boxes because there were no walls or roof to limit the quantity of material piled on the car. Big pieces of machinery were also dropped onto the platform car, with damaging results, as evidenced in the photograph of the Pittsburgh, Fort Wayne & Chicago car shown in Figure 3.51. Six truss rods were not enough to prevent the noticeable sag in this 30-ton-capacity flat's middle. Extra-heavy side sills, often 4 by 12 inches, were used to strengthen the platform car. The flat was probably the first of all American freight cars to be regularly fitted with four truss rods. The large-capacity units received six rods. And some cars were equipped with a second upper set, forming a double set of rods. They ran between the body bolsters and were hidden from view. But none of these measures prevented the making of cripples, and flats were seen as the most short-lived freight equipment.[72] They seemed the least able to resist the bumps and rough handling of railway freight service. Cars broke in two at their ends just over the bolsters. Some broke apart at the middle. Since the cars had no body or roof, water gained access to framing members and rot helped reduce life expectancy. And so the simplest and cheapest of freight cars tended to render the fewest years of service.

A record of the last wooden-frame flats built for the Baltimore & Ohio in 1901 confirms this notion to a degree.[73] The 490 cars in the P-9 class survived like members of any community (Fig. 3.52). Some died young, while others ripened into old age. A small number saw only a few years of service and met their early demise in accidents, sustaining damages so major that rebuilding was impractical. Fewer lasted on into the mid- and late 1920s—a remarkable achievement, considering that heavier and stronger steel cars were steadily growing in number after 1900. Wood-frame cars were vulnerable to crushing when placed between two steel cars, like a chestnut on an anvil. Yet most of the P-9s were retired at around fifteen years of age, which corresponds almost exactly with the Master Car Builders Association depreciation formula of 6 percent a year.

Under the right circumstances even a flatcar could become a centenarian. What may be the oldest eight-wheel American flatcar is now owned by the state of Nevada. It is one of several relics purchased by the state in the early 1970s for a projected rail museum (Fig. 3.53). The car, No. 308, was built in 1876 by the Detroit Car Works for the Virginia & Truckee Railroad at a cost of $477.[74] It was one of seventy identical cars built for the V&T in 1875–1876. Flatcars loomed large on the Nevada short line, in a proportion greater than was normal for most other U.S. lines, where flatcars were not a sizable part of the fleet. Yet the V&T served the gold and silver industry; mine props and oversize machinery, like boilers and stamping mills, were regular cargoes best handled by flatcars. The 308 measured 29 feet 8 inches long. It weighed 15,900 pounds and carried 24,000 pounds. Like all other early V&T freight equipment, it was painted yellow.

The No. 308 ran on the home road, which may explain its long life, for surely the rigors of interchange service would have reduced it to splinters long before its retirement in 1936. Actually, it had seen little service after 1923, but because of the

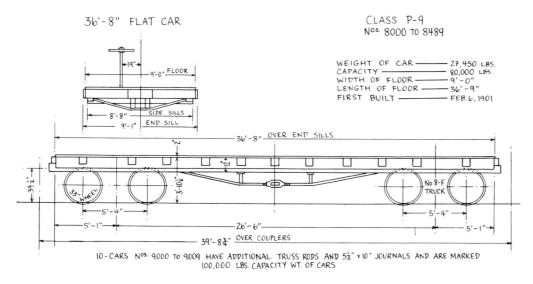

36'-8" FLAT CAR

CLASS P-9
Nos. 8000 TO 8489

WEIGHT OF CAR ——— 27,450 LBS.
CAPACITY ——————— 80,000 LBS.
WIDTH OF FLOOR ——— 9'-0"
LENGTH OF FLOOR ——— 36'-9"
FIRST BUILT ————— FEB. 6, 1901

10 CARS Nos. 9000 TO 9009 HAVE ADDITIONAL TRUSS RODS AND 5½" x 10" JOURNALS AND ARE MARKED 100,000 LBS. CAPACITY WT. OF CARS

FIGURE 3.52

The wood-car era was at an end when the B&O purchased its class P-9 flatcars in 1901. (Collection of the late L. W. Sagle)

FIGURE 3.53

The Virginia & Truckee's platform car No. 338 was built at the Carson City shops in 1891. This car is very similar to the V&T's No. 308 described in the text. The car is now in the Nevada State Museum collection. It was renumbered first as No. 161 and later as No. 140. (Drawing © Michael A. Collins, 1982)

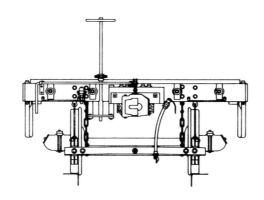

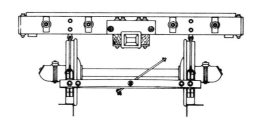

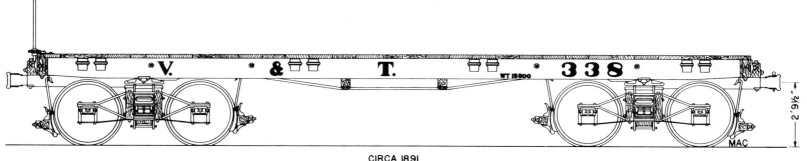

CIRCA 1891

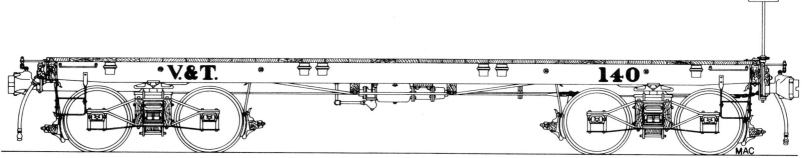

CIRCA 1933

dry desert climate, it suffered little while in service or in out-of-doors storage. Air brakes and Janney couplers were installed in 1897–1898, and so there was no reason not to use the 308 for light loads despite its advanced years. Not long after retirement, it was sold to Paramount Studios. Its movie career added only a few hundred miles to its travel history, but the wet Los Angeles–area climate slowly reduced the well-preserved relic to a rotten wreck. Little remains of the car today except for its hardware. Restoration will result in more of a replica than an original because, sadly, none of the remaining wooden members can be salvaged. Had the 308 remained in Nevada, it would, like the artifacts of the Pharoahs, have survived the ages.

Flatcars underwent a steady, if moderate, growth in length and capacity during the last two decades of the nineteenth century. The Virginia & Truckee car just described, as a representative of the 1870s, exhibits a conservative weight-to-capacity ratio: roughly 8 tons to 12 tons. A ratio of 2 to 1 became common during the 1880s and 1890s. Some lines like the Tennessee Midland received platform cars from Pullman in 1888 that weighed 17,400 pounds and carried 50,000 pounds (Fig. 3.54). Thirty-ton capacities were common by 1890, as were lengths of 34 to 36 feet, a noticeable gain over the 29-foot standard of the 1870s.

Cost Questions

The wooden box- and flatcars described in the foregoing discussion evolved from small vehicles suitable for modest local traffic into large, rugged carriages capable of moving a vast interstate tonnage. The post-1870 American freight car traveled

FIGURE 3.54

This Tennessee Midland platform car is a garden-variety specimen common on all American railroads of the period. It weighed 17,400 pounds and carried 50,000 pounds. The No. 5039 was completed by Pullman in 1888. (Pullman Neg. 119)

far from its home road in interchange service. That wooden-frame boxcars of 50-ton capacity could evolve from 10-ton cars in twenty-five years represents a major achievement in straight-line development, for no fundamentally different construction techniques were devised. The advances in size were achieved by following the most conservative possible course, and by simply doing more of the same on a slightly larger scale. The offspring of this unadventurous policy surely moved the traffic during the period of major industrial growth that followed the Civil War.

Just how effectively or economically the job was handled by wooden cars remains a basic question that is very difficult to answer. Critics of the wooden car claimed that they were inefficient and unequal to the job. But these critics tended to be advocates of iron rolling stock; their arguments were self-serving, for they sought to displace the existing wooden fleet with vehicles built after their own patented designs. Most car builders of the time must have understood the defects of the wooden freight. It was not the ideal vehicle any more than a frame house is the ideal structure. It was flammable, subject to rot, and structurally weak. However, these disadvantages were offset by the advantages of low first cost and ease of fabrication.

Any mechanic, even the greenest novice, understands the relative ease and speed of building in wood as opposed to masonry or metal. The level of skill required is less for wood than for any other common building material. Fairly simple and inexpensive tools are adequate for carpentry, which is surely not true for major metal- and stoneworking. The labor pool in nineteenth-century America was largely unskilled and hence better adapted to carpentry. Workers drawn from the farm and frontier understood the techniques of cutting timber because it was so much a part of their daily life. The abundant supply of prime timber was another

obvious argument for wooden cars. Long-leaf yellow or Georgia pine is one of the strongest natural building materials available. This wonderful tree, now commercially extinct, was once plentiful. New England's textile mills depended on it for wide-span floors. It was also a longtime favorite with car builders.

First cost was another important consideration for railroad managers of the time. America was a capital-poor, agricultural nation. Local funding for major construction projects like railroads was difficult to arrange in a nation with such small cash reserves. Investment costs had to be kept low, particularly in the early years of the enterprise, and so short-term economies were in favor. Wooden trestles, station buildings, and cars, though more ephemeral and costly in the long run, looked very good to American railroads whose capital needs were like the Sahara's thirst for water.

If wooden cars seemed to be the most rational choice from the standpoint of materials, labor, and available capital, how did they measure up in terms of maintenance and repairs? Were their repair costs reasonable? This is a critical question, for if it could be shown that repair costs were inordinately high, then the wooden car was no bargain, no matter how cheap or easy it was to build. Just how to determine reasonable cost, then, became the problem. Information on the subject is neither abundant nor consistent.[75] The Boston & Worcester gave one of the earliest cost figures for car maintenance in its annual report for 1837. Repairs came to just a little under 5 cents per mile, or about $500 per year. In 1847 the Baltimore & Ohio reported spending $58,469 to maintain 983 freight cars.[76] On a ton-mile basis, this works out to 2.1 cents. On a per-car basis, it comes to almost $60 per year. Other estimates made late in the century range from $20.00 to $40.50 per year.[77] The latter figure included oil and waste, which can be construed as more an operating than a repair expense.

FIGURE 3.55

A standard-gauge platform car carries two narrow-gauge flats ready to leave Pullman's car plant yard in 1892. (Pullman Neg. 1466)

In some other examples freight and passenger repair costs were presented as a single total.[78] In the late 1860s the Massachusetts Railroad Commission reported 6.5 cents per mile as a repair cost base for all railroad cars operated within the state. *Poor's Manual of Railroads* reported maintenance figures for many major railroads during the wood-car period, but again in the form of both passenger and freight cars. Passenger cars—complex vehicles with seating, lighting, toilet, and heating systems—naturally cost far more to maintain, but then again their numbers were significantly less. Even so, exactly what portion of the total car expense account did they consume?

Tables 3.2 and 3.3 offer a few samples. The figures given here are hardly conclusive and may not even be representative. They are surely open to interpretation. Why was the Missouri, Kansas & Texas spending only about $600 per year per car (freight and passenger) on maintenance while the Erie was spending nearly double that? Why were there so many ups and downs? Most railroads figured that about 10 percent of their operating costs were accountable to freight car repairs; however, according to the *Railroad Gazette* of December 18, 1885, the Baltimore & Ohio was spending 25 percent. Why? Was it due to the construction of the cars, their age, the mileage they ran, or the level of their maintenance? All these critical factors are unexplained. The information to document these matters might be even more elusive than raw cost data itself. And if car maintenance costs are subject to scrutiny, what about maintenance costs for locomotives, buildings, and track? Could those costs have been markedly changed by engineering design changes? Again, we have more questions than answers.

Accounting procedures may explain why maintenance costs varied so greatly from railroad to railroad. Exactly what was charged to the repair account? Was it done in the narrowest manner—that is, only for costs involving the repair and replacement of broken and worn parts and structural members of in-service cars? Such information might be ascertainable if the account books of a car department were available for inspection. Such documents are virtually unavailable, however, and the figures offered in more public documents, such as published railroad annual reports, do not give detailed breakdowns. In fact, many such reports don't bother to separate the costs of locomotives, passenger cars, and freight cars. Even so, certain clues make one skeptical about the veracity of the repair cost figures offered to the public. In 1878 the Atlantic & Great Western reported freight car repair costs for its four thousand–plus fleet at just over $203,000.[79] It was noted that renewals were included in this figure. Renewals would include rebuildings, major and minor, and so could inflate the repair cost total considerably. Interestingly enough, the cost for freight car journal oil was listed separately as $7,044.12—a cost that was normally lumped into the general operating cost of the car department. During the same period the Baltimore & Ohio admitted to an accounting convention that had an even greater inflationary effect on the repair accounts.[80] The cost of new cars was charged to the repair account. This procedure helps explain why repair costs zoomed up and down so dramatically, and why they might be so much greater on the B&O than on other railroads of equal size.

Car costs should tower above all other expenditures if this single area of railroad equipment were

TABLE 3.2   Railroad Maintenance Costs, by Category

Railroad	Maintenance Costs (millions of dollars)				Miles of Railroad	Number of Cars
	Cars	Locomotives	Right-of-Way	Total		
Erie	1.80	4.70	3.20	17.39	1,600	14,000 freight
						450 passenger
Pennsylvania	3.00	6.00	4.80	22.32	1,300	25,000 freight
						1,300 passenger
Missouri, Kansas & Texas	0.33	1.60	1.60	3.80	1,500	5,200 freight
						100 passenger

Source: Poor's Manual of Railroads, 1888.

TABLE 3.3   Railroad Maintenance Costs per Train-Mile (cents)

Railroad	1860	1870	1880	1884
N.Y. Central	8.9	15.4	14.4	10.3
Erie	11.7	12.0	8.0	8.9
B&O	8.7	5.5	20.6	NA
Reading	8.9	13.5	12.1	14.1

Source: Arthur M. Wellington, The Economic Theory of the Location of Railroads (New York, 1893).

Notes: In 1870 the Erie spent $88.02 per car; by 1881 the cost was down to $34.19. NA = not available.

as defective as described by its harshest critics. They should stand out in crimson ink on the balance sheets to alarm not just the bookkeepers but railroad managers and shareholders alike. But this does not appear to have been the case. Car repairs were significantly less than locomotive repairs, as shown in Table 3.2. Steam engines were far more complex than cars, but there were far more cars than locomotives, and so the comparison is more balanced than one might expect. There appears to have been no inordinate hue and cry over locomotive repair costs, and yet that figure was two or three times the dollar total for car repairs. The comparison is meaningless unless reasonable cost is determined. Is $200 fair, or $500, or $1,200?

And can a fair and reasonable sum even be named, considering the huge variables in mileage, age, and levels of maintenance? In good times, equipment was operated at its maximum, and there was little idle time for preventive maintenance. This resulted in very high rebuilding costs at a later time, costs that would not appear until several years after the season of hard use. Conversely, in dull times, repair accounts might be artificially low simply because low use resulted in minimal refurbishing. Idle cars required no servicing, and there was no money to pay for it anyway.

Old cars would require more attention than new ones in normal service. A line having nothing but veterans would surely spend more on maintenance than a neighbor blessed with a more modern fleet. The Erie was able to cut its repair costs by nearly 70 percent because of more uniform car

design and the replacement of obsolete cars in the 1870s and 1880s.

Prosperous roads would tend to spend more money on repairs than impecunious lines. Fussy master car builders would lavish more time, attention, and funds on their flock than their slovenly brethren. Car builders given to overhauls and betterments would spend more than someone content with minor repairs limited to a temporary quick fix. Deferred maintenance could always make the cash accounts prosper, at least temporarily. And so a line spending only $100 per car per year might continue paying dividends at the expense of maintaining its equipment. But needed repairs could not be postponed forever, and the line had either to suspend service or to begin a more costly crash repair program.

Conversely, the industry was not encouraged to make elaborate repairs, because freight cars were so abused in interchange service. It was more economical to patch up and make minimum repairs than to discard worn cars and replace them with new ones.

Railroads with poor track and a history of derailments and wrecks could expect to spend more on repairs. And what about a line that included the cost of new cars built in its repair shops as part of maintenance figures? In 1883 the Baltimore & Ohio was spending around $299,000 on repairing approximately seventeen thousand freight cars, but the sum—which works out to only about $17 per car per year—also includes the expense of new cars built at Mt. Clare and other shops elsewhere on the line.

Repair depots represented a major investment on the part of the railroad industry, as already noted near the end of Chapter 1. In 1895 L. F. Loree estimated that American railroads spent $50 million per year on freight car repairs, or about $40 per unit.[81] Of this total about 60 percent was spent on labor. The 1880 Census estimated that 20 percent of the railroad labor force found employment in the repair shops. Of this number, one-quarter were said to be carpenters, and it can be safely assumed that most of these woodworkers were engaged in car work. Even so, a greater number were engaged in running-gear repairs, for Arthur M. Wellington contends that three-quarters of the repair work involved the trucks—especially the wheels, journals, and axles.[82] Each year repair crews replaced 1.5 million car wheels and 175,000 axles. Drawbars demanded 10 percent of car toad's attention, while body repairs were trifling and rarely accounted for more than 5 percent of all repairs.

Were freight car repair costs fair and reasonable? This fundamental question remains unanswered. Could they have been lowered? Unquestionably, but by how much it is impossible to say. Could a repairless car have been perfected? Not likely, so long as railroads remained imperfect systems, devised and managed by such fallable creatures as men.

The wooden car was viewed as the best compromise available until almost the end of the nineteenth century. There were trade-offs, to be sure. Low cost and ease of construction compensated for less strength and a shorter service life. Railroad managers were aware of the alternatives of iron and steel, but they consistently remained loyal to the wooden ark of railroading until all doubt had been removed that its usefulness had at last come to an end. The real measure of the wooden car's success was the billions of ton-miles of freight carried during its eighty years of prominence.

4   CARS FOR FOOD

Sustenance is one of the most basic human needs. Life ends, armies fail, nations fall for want of food. In ancient times chronic food shortages were caused as much by a want of transportation as by crop failure.[1] The rapid, dependable, and economical means of transporting foods has been a fundamental need since the beginning of civilization. Most edibles are bulky and relatively low in value; if they must move long distances, therefore, cheap transit is necessary.

Cheap and certain transit could add to the value of foodstuffs, particularly those raised far distant from the markets. This was especially true in nineteenth-century America, where rich farmland was separated by 2,000 miles or more from the population centers clustered in the eastern cities. The agricultural frontier was constantly advancing westward because of improved river and rail connections. Railroads united the truck farm, the cattle range, the citrus grove, and the feedlot with the marketplace. They literally became food lifelines for the nation and the world. The fruits and vegetables that grew so abundantly in the South and Far West were of almost no value locally, for that market was small and the supply was abundant beyond imagination. But transported to Boston and Philadelphia, these same viands were eminently saleable—a true cash crop.

One can hardly imagine how delightful fresh produce must have appeared to our ancestors, accustomed as they were to a dreary diet of salt pork, cabbage, and perhaps a few dried apples for most of the winter. But it was summer, or nearly so, in California and Florida. The seasonal nature of eating was ended by more efficient transportation. Farming became a year-round endeavor within the boundaries of a single nation, perhaps for the first time in history. Before 1880 only the very rich might taste an orange or fresh berries in midwinter; now all but the very poor could afford such luxuries. The ancient monotony of dried fruits, salt meat, and smoked fish gave way to an abundance of fresh meats and vegetables.

Once-isolated but fertile lands, like the Salinas Valley, flourished like new Edens. Enormous orchards in North Carolina, Virginia, and faraway Washington State became economical. Poor or arid lands in the West became productive pastures. The agricultural work force need no longer cluster in the East. Rapid settlement of unoccupied territory was encouraged, for good or ill, by the union of agriculture and railroads. Eastern city populations might double or triple every decade, yet there was no fear of starvation because of the abundance deliverable by rail. This market expanded as millions of immigrants poured into the United States. And it was a nation of big eaters. Many workers labored out of doors or in factories where physical exertion encouraged hearty appetites. Few people bothered with diets except for religious reasons, and no one but scientists thought about calories. To be plump and rosy-cheeked was to be pleasing. Ten-course meals were seen as a feast, not an overindulgence. In a time when Grover Cleveland was president and Victoria was queen, being portly was positively fashionable. Thin people were considered sickly, consumptive, or underfed.

Moving food to the market was an obvious job for the railroads. It was also an unavoidable task for any common carrier, and while the traffic was large, it was not necessarily profitable. All provisions are perishable. Many can be moved short distances in favorable weather without damage, but a head of lettuce or a steer sent 3,000 miles by freight car is bound to be on the road for days or even weeks. It can be exposed to freezing mountain temperatures and desert heat, all in the same journey. Unless very careful attention is paid to the load in transit, the lettuce will rot and the steer will sicken and die.

The movement of perishables was an exacting business. It called for attentive management and well-trained crews. It was labor-intensive. Checkers and ice handlers were needed at every main terminal. They had to be on duty day and night. Extra shifts called for more men and overtime pay. Failure to open a ventilator, fill an ice bunker, check the salt, or open a trap could result in the loss of an entire carload. Small details were important, and doing it tomorrow was not soon enough. It must be done on time, in an exacting manner. If the weather changed, the staff must be alert and ready to act. A sudden warm day in the spring called for more ice; an unexpected freeze required the installation of heaters. Administrative and loss claims were greater than for any other phase of the transportation business, outside of passenger traffic. Railroad managers accordingly preferred the carriage of lumber and coal, which never got hot, cold, wet, or hungry.

Advancing technologies helped all food handlers, and some major improvements took place not long after the railroad became part of the transportation scene. Canning had become a practical success by around 1830. Natural ice refrigeration was being used to a limited degree at the same time and was adopted for rail shipments after 1870. The big shift to dressed-beef as opposed to live-cattle shipments after 1870 was perhaps the most profound change ever to take place in the meat trade. Insulated cars cooled by blocks of ice allowed the safe shipment of dressed beef between Chicago and eastern cities. Ventilated boxcars were effective in moving fresh vegetables. Less successful were efforts to perfect self-feeding and self-watering cattle cars, at least for very long-distance travel. A variety of highly specialized cars were developed to move vegetable oil, vinegar, and live chickens late in the nineteenth century.

All this equipment improved service and lowered claims, but at a cost. Investment was considerable, and most of the cars were complex and expensive. They were generally good for only one product and couldn't be used for general traffic. Thus, private freight companies or shippers themselves provided cars that the railroads were reluctant to own or maintain. However, not all food traveled in exotic special cars. Wheat, corn, and oats were sent in bulk by boxcar. Flour, sugar, and other common commodities were bagged or barreled for shipment by boxcar. Sugar beets and cane went to the mill in open cars. And not all food tonnage was direct from the field to the consumer. There was a great deal of transhipping or double shipping. Cattle went from barnyard to pastureland that was often a great distance away. Grain traveled from granaries to feedlots, distilleries, or flour mills. Unprocessed foods traveled again after being cooked and packaged. And the by-products of the food industry, especially those from animals, were considerable. Fats, hides, and hair are the most obvious examples.

Livestock Cars

The carriage of live animals to market was already a large trade by 1870. Railroads also realized additional revenues from related products such as lard, bone meal, glue, and hides. The supplies needed by meat and leather processors—salt, ice, barrels, tanbark, dyes, and chemicals—all added to rail traffic and earnings. After 1875 dressed and canned meat promised yet another source of income. In the beginning, the cattle trade was a relatively simple and profitable business. For short trips moving live animals to market worked very well. The stock arrived fresh and ready for consumption. But as grazing operations moved farther west because of cheaper pastureland, the problems of stock transport grew more complex and worrisome. Animals confined inside a rocking cattle car for 1,000 miles or more, bereft of food or water, sickened and sometimes died. Those that survived their odyssey were wasted, and few could be classed as prime beef. Damage and loss claims mounted in direct proportion to the journey's length, and the cattle trade became something of a nuisance, if not a source of negative cash flow.

Feed and water stops helped alleviate the problem, but a more radical change in the meat trade did far more to solve the problem. Midwestern meat packers began to prepare beef, pork, and lamb for the eastern market starting around 1870. The animals traveled a comparatively short distance to slaughterhouses in Kansas City, Chicago, and similar points where they were killed and dressed. The carcasses were loaded into refrigerator cars for the big consumer markets of New York, Philadelphia, and Boston. Within a decade of its introduction, the refrigerator car was established as an important food-carrying vehicle. After this time cattle movements were once again generally limited to essentially local carriage. Railroads tended to champion the cattle car over the refrigerator because it was simple and cheap. But no vehicle or method will long be tolerated by shippers if it damages or destroys the goods in transit.

THE SCOPE OF THE CATTLE TRAFFIC

Some idea of the vastness of the cattle traffic can be gained from shipments to New York City. In 1881 thirteen thousand head of cattle were consumed in one week.[2] One hundred fifty cars a day were unloaded in the Jersey City stockyards. The New York Central's big hogyard at 40th Street received ten thousand swine each week. Table 4.1 details the quantities of cattle and dressed beef delivered to New York over the New York Central, Erie, Pennsylvania, and Lackawanna railroads.[3] These figures show not only the volume of business but the decline of the cattle trade and the growth of the dressed-beef trade. The tilt toward dressed beef was even more pronounced than might at first appear because so much of the cattle's weight was not edible, and hence so far as actual consumption was concerned, the percentage of dressed beef was larger than the gross tonnage shown. New York was in fact slower to accept

TABLE 4.1 Cattle and Dressed-Beef Deliveries to New York (tons)

Year	Cattle (Stock Cars)	Dressed Beef (Refrigerator Cars)
1882	366,487	2,633
1883	392,095	16,365
1884	328,220	34,956
1885	337,820	53,344
1886	280,184	69,769

Source: Railway Review, Jan. 29, 1887, p. 62.
Note: Delivering railroads = New York Central, Erie, Pennsylvania, and Lackawanna.

dressed beef than, say, Boston, which by 1886 was receiving 52,431 tons of cattle and 153,544 tons of dressed beef annually.

New York and Boston were on the end of the western livestock pipeline, while other cities, notably Chicago, were midway transit points. Before the Civil War Cincinnati was the hog capital of the nation. It did a lively cattle-shipping trade as well, but the rich midwestern farmland became too valuable for grazing, and cattlemen moved to Missouri, Iowa, and Kansas. Cheaper land in the arid Great Plains to the west of these states was next settled for cattle raising. By 1880 most beef stock was raised west of the Mississippi River. Hogs and some sheep were still raised in the Midwest. Stockmen sent their frisky products to Chicago rather than Cincinnati because it was closer and because a stockyard, serviced by brokers, offered a convenient sales outlet.[4] While the agents looked for buyers, the cattle rested and fattened in pens at the yard.

Chicago Union Stockyard opened in 1865. Kansas City and Cincinnati created identical facilities in 1871, which were in turn copied by St. Louis,

Omaha, and a dozen other Corn Belt cities during the next decade. But none could equal Chicago's mammoth yard for size or smell. The stench, which could spread for miles on a breezy day, may have offended the ordinary citizen, but it was the sweet smell of money to cattlemen. Just 14 years after it opened, the Chicago yard occupied 350 acres of land. It had space for 10,000 cattle, 120,000 hogs and 5,000 sheep. Its seven hundred employees unloaded as many as fifteen hundred cars in a single day. By 1881 it was handling thirty thousand head of cattle and one hundred thousand hogs per week. Over the next twenty years the yard nearly doubled in size, and its annual traffic in cattle, hogs, and sheep exceeded 16 million. Chicago railroads delivered over 290,000 carloads of cattle alone. On a busy day sixty cars would be unloaded. Meat packers naturally located nearby. Men like Jacob Dold, Philip D. Armour, and Michael Cudahy at first specialized in salt pork but then turned to dressed beef once ice refrigeration made preservation and transhipment practical. By 1882 the Chicago packers were sending out fifteen hundred carcasses to the east each day.[5] Nineteen years later they were shipping over seventy-five thousand carloads per year.

The source of Chicago's fortune in beef was the great grasslands to the west. Plains too dry for corn or even wheat could support grazing cattle. Men who ventured west seeking gold and silver soon discovered that there was more bullion to be found in T-bones than in mines or ores. Long overland drives from Texas to eastern cities occupied four months. The labor costs and the weight loss suffered by the animals reduced the profits for such journeys. After the Civil War rail lines began to penetrate the unsettled western lands, and stock shipments by rail became more convenient. A new industry was born, as were new towns, at the rail-

FIGURE 4.1

Pennsylvania Railroad stock car built at the Altoona shops in April 1876. Dimensions are shown in the diagram drawing reproduced in Figure 4.3. (Pennsylvania Railroad Photo)

CARS FOR FOOD

FIGURE 4.2

head. One of the best known of these was Abilene, an obscure settlement before the tracks of the Kansas Pacific reached it in 1867. Near the stockyards were riotous dance halls and saloons where the ranchers and cowpokes might refresh themselves after their dusty weeks on the trail.

Some thirty-five thousand longhorns were sent away during the first year of rail service. Only four years later the total reached six hundred thousand steers per year. New cattle lands were opening far to the north as the Northern Pacific Railroad pushed inland. The buffalo and the Indians made way for ranchers who coveted the open space, water, and grass of the northern ranges. The railroad hauled yearlings and two-year-old steers for grazing in Montana and Dakota territories. Later in the year they carried the fattened cattle to Chicago. In 1886 the NP moved 4,718 carloads (90,200 head) through St. Paul, bound for Chicago.[6] In 1891 the figure reached 9,508 carloads (208,460 head). Transhipping increased cattle car mileage and railroad revenue. Mention has already been made of the movement of yearlings to far western pasture and

thence to market at a later date. Cattle were also taken from Texas and other western grazing states to Kansas, Iowa, and other "corn-surplus" states for feeding. Once fattened, they were again loaded aboard cars for the packers in Chicago or Kansas City. And so a steer might experience several rail journeys before reaching his final destination as Sunday's roast. Some, indeed, might travel halfway around the world. A longhorn in Texas might find himself sent by rail to Philadelphia and thence by ocean steamer to Amsterdam or London. Some 139,000 head, for example, were exported from eastern U.S. seaports between 1878 and 1882.[7]

The volume of traffic just outlined explains the need for a sizable number of livestock cars. The numbers are impressive, but one should not assume that cattle cars constituted a majority of the total fleet. In general, they rarely accounted for more than about 5 percent. The earliest national estimate appears to have been published in 1874, when 16,722 cattle cars were reported to be in service.[8] The 1890 Census reported 57,351, but this figure does not include those privately owned and operated, which would surely account for another ten thousand cars. The growth of the fleet is explained in part by the example of the Chicago & North Western Railway. *Poor's Manual of Railroads* reported 315 cattle cars on that road in 1870; 942 in 1880; 1,861 in 1890; and 3,381 in 1900. The number of stock cars owned by individual lines varied considerably, as might be expected. Big roads with a heavy livestock trade naturally had the most cattle cars. Here are a few examples from the April 1897 *Official Railway Equipment Register*: Central Railroad of New Jersey, 27; Boston & Maine, 61; Chesapeake & Ohio, 442; Louisville & Nashville, 729; Milwaukee Road, 2,441; Rock Island, 2,399; Burlington, 3,000; and Santa Fe, over 3,200.

The number of cars and the volume of traffic are evidence for a prodigious income. In terms of gross income this is certainly true. But just how profitable was the cattle-hauling business? Not at all, according to the industry. J. A. Droege, in his book-length study of the rail freight business, concluded that no class of freight was more subject to losses than livestock.[9] The cargo was truly perishable. Feed- and waterlots had to be maintained at intervals along the line. Crews were needed to maintain these facilities as well as to assist with loading and unloading. The transportation of drovers, one to every two stock cars, was a costly nuisance, and they required free return transportation. To avoid delays and possible sickness or more rest stops, cattle trains were run on extra-fast schedules. They could mean annoying delays to passenger trains. It surely meant greater maintenance and fuel costs. Only the best engine crews could be assigned to cattle trains. Rough handlers, men prone to make fast starts and stops, would knock animals down, making extra work for claim agents. Finding crews who were at once fast runners and smooth train handlers was not always easy. Air-brake equipment required special maintenance to eliminate "kickers" that might result in rough or sudden stops.

All these requirements resulted in extra work, time, worry, and costs for the operating and mechanical departments. At the same time, shippers wanted the cheapest rates, and where parallel lines existed they sought and received rebates. In 1878, for example, the carload rate from Chicago to New York was $115. The customary rebate was $15. Even where no parallel railroad existed, returns were found unsatisfactory.

The Northern Pacific figured its gross revenue on a typical sixteen-car stock train traveling 1,160 miles at only $1,672.[10] This worked out to 3.76 cents per car-mile after deducting the cost of first moving the empty cars to the loading station. The NP received 10 cents per mile for moving local cars, a rate that rendered a fair return on investment. The NP, furthermore, complained that the number of stock cars returned empty stood at 95 percent.

The problem of revenueless, empty return trips was, of course, not unique to the cattle trade. But because these cars were designed especially for live animals for whom ventilation was a primary need, the semiopen side bodies were suitable only for low-grade commodities like pig iron, bark, and rough lumber. Only rarely did a railroad enjoy the lucky coincidence of a particular shipment of this nature wanting transportation to the destination of the empty cattle cars. The Missouri, Kansas & Texas claimed that it used its cattle cars for grain haulage by sealing up the sides and ends of the body part-way with temporary siding.[11] This may have been practical out of season, but it is hard to imagine readying for the grain trade a car fresh (or not so fresh) from delivering a load of heifers. The car cleaners would have to be unusually energetic to complete the task in any reasonable time. There was also a seasonal rhythm to the cattle trade. In the spring yearlings and the like were moved from breeding farms to distant pasture. In the fall mature animals were taken to market, to avoid the cost of winter feeding. During the intervening months most stock cars would stand idle. After reviewing the account ledger many railroad managers were probably ready to leave the whole beastly business to private cattle car operators.

REPRESENTATIVE STOCK CARS

The shape, arrangement, and details of the common stock car were established before 1870, as noted in Chapter 2. The horizontal-slat sides with large openings between the boards provided air, light, and a degree of vision to the occupants of the vehicle. The semiopen sides might appear no more oppressive or frightening than a pen or high fence. Horizontal siding was universal, and it was almost always combined with the same bastard Howe truss so much in favor for boxcar bodies. The body-frame members were nearly always fastened top and bottom with iron pockets. Only the Reading appears to have deviated from this scheme, producing cars with vertical slats in the 1870s or 1880s. This deviation from common practice produced a vehicle suggestive of a circus car. Floor frames, trucks, roofs—in general, all followed the same design, techniques, and materials employed for boxcars.

The growth in size and capacity followed a similar pattern. During the 1870s few cars exceeded 28 feet in length or 10 tons in capacity. By the early and middle 1880s 34 feet was a common length. Capacity doubled to 20 tons during this same period. The new standard for the 1890s was 30-ton cars; however, length rarely exceeded 36 feet. Fifteen to twenty full-grown cattle was the normal load per car during the final thirty years of the nineteenth century. Another half-dozen animals might be fitted in if the car was loaded with smaller, leaner Texas steers. If sheep or hogs were carried, the number carried per car could be greatly multiplied. These creatures were short enough for a temporary second floor or deck to be installed. Capacity was doubled, and as many as 240 animals could be placed in a single car. The idea of double-decking goes back to the 1830s, when the idea was apparently first applied by the Liverpool & Manchester Railway. Double-deckers were fairly common but never universal. In 1909 it was estimated that of the seventy thousand railroad-owned stock cars, about 20 percent had twin decks.[12] Most or all of these decks contrived to be collapsible or removable so that cattle or other cargoes could be carried. Because swine were commonly carried in two-deck cars, it seems correct to suggest that this was the earliest form of piggybacking.

The standard Pennsylvania Railroad stock car of the 1870s is one of the best documented and may surely be accepted as fairly typical of the time. The 28-foot-long body was carried by a frame with eight sills and two truss rods (Figs. 4.1, 4.2, 4.3). The large end timbers formed a small platform that gave the car a slightly old-fashioned look. The side-view photograph reinforces the information in the drawing, but the end-view photograph is worthy of some extra study because it is so rare and because it illustrates several interesting details. At the roofline note that the purlins (longitudinal support timbers) are set into the end carline and stick out beyond it. The plain, open-slat end wall, bereft of a door, means that lumber and rail were not intended alternative cargoes, as was common with many cattle cars on other lines. Numerous examples of end doors are shown elsewhere in this chapter. Notice the wooden end ladder and the two horizontal tie rods. The brake wheel mounted on the end bulkhead was so placed for easy access if the car was under a low overhang, such as a car shed or loading platform canopy; the

Pennsylvania called them tunnel wheels. The upright wheel at the opposite end was more convenient to the brakeman under normal conditions. Notice the large square nuts marking the end of the truss rods just above the wheels. The ends of the riveted-over wrought-iron coupler are clearly evident. But perhaps the most interesting detail of all is the wheels. They appear to be of the wide-tread variety, which made the car suitable for interchange on the slight variations of standard gauges that separated the Pennsylvania and its several subsidiary lines.

The Pennsylvania expanded the same general design in 1884. The body was lengthened to 35 feet, capacity reached 15 tons, and weight was listed as 20,200 pounds.[13] The following year brought a variation on the standard stock car that featured a true Howe-truss side frame (Fig. 4.4). Weight was just 200 pounds above the 1884 standard, but its capacity was 5 tons greater. Several years earlier the Pennsylvania's great rival, the New York Central, had completed plans for a 34-foot-long stock car.[14] The design, published in 1882, reveals several interesting peculiarities (Fig. 4.5). The frame consisted of eight sills rather than the usual six. They were placed very close together, while the intermediates were very undersize. No truss rods were employed. The drawbars were long and reached in toward the center of the car, where they were attached to two separate cross-timbers. The cross-timbers, in turn, were connected by 1-inch-diameter rods, thus effecting a weak, continuous drawbar. Small end doors, placed at floor level, made the loading of long objects convenient. The corners were strengthened by pressed-iron corner plates. They were embossed with poling pockets as an aid to switching crews. The roof was of the common double-board variety. The purpose of the roof hatch is uncertain. It may have been for inspection or feeding purposes. The pipe handrail around the brake wheel was a common New York Central safety feature.

The Baltimore & Ohio introduced its class L-1

FIGURE 4.3

Diagram drawing of a standard Pennsylvania stock car. (*Engineering*, **Nov. 2, 1877**)

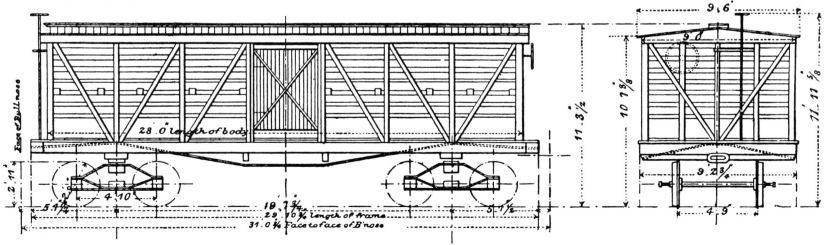

THE AMERICAN RAILROAD FREIGHT CAR

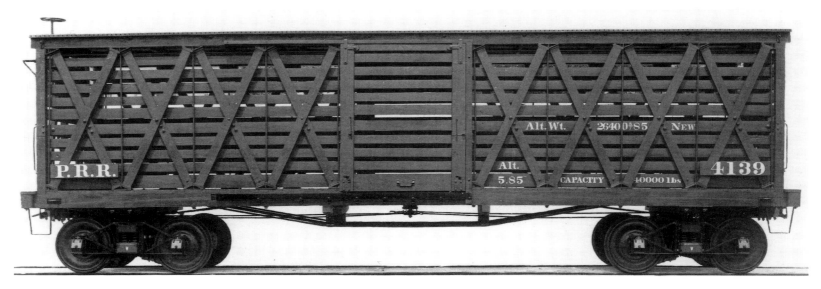

FIGURE 4.4

stock car during the same year that the NYC car just described made its appearance. The essentials of the L-1 are shown in the diagram drawing (Fig. 4.6). Successive designs showed no increase in size or capacity but a noticeable jump in dead weight that might be explained by heavier framing, trucks, and wheels, or by the addition of air brakes and Janney-style couplers. The L-3 (1886), for example, weighed 5,250 pounds more than the L-1. A photograph of a late-model wooden-frame B&O stock car is also included here (Fig. 4.7). The end of the car, showing many construction features, illustrates the final development of the wood-frame cattle car that lapped over into the twentieth century. The 30-ton-capacity vehicle is listed as a single-deck Mather car, yet the picture reveals no patented or special features such as feed or water pans. It appears to be a simple garden-variety cattle car.

The importance of cattle cars to the American West has already been mentioned, and fortunately drawings for cars actually used in this epic trade were published in 1880 and 1881 (Figs. 4.8, 4.9, 4.10). The Missouri, Kansas & Texas was at the very heart of the western livestock industry, and its superintendent of machinery, George W. Cushing, understood the need for serviceable equipment to handle such traffic. The design, like the two other western examples shown here, appears to be perfectly standard, which indeed adds very much to its value as a historical reference (Fig. 4.8). The 28-foot-long car had wooden body bolsters, end doors, a double-board roof, and cast-iron couplers.[15] It had swing-motion trucks for an easier ride. The wheelbase was 4 feet 9 inches. No weight or capacity was registered. The Santa Fe car also dates from 1880 and is very similar to the MKT car, except that it was 30 feet long and had no end doors (Fig. 4.9).[16] Weight and capacity were not given. George Hackney was master car builder at the time and is the presumed designer. The side

ladder is a detail that should not be overlooked. The last example in our western trio is from the Chicago, Burlington & Quincy Railroad, a major mid- and far western line very much involved in cattle transportation (Fig. 4.10).[17] The 20-ton-capacity car measured 30 feet 6 inches long by 8 feet 9 inches wide. Thielsen iron-bolster trucks were used. Total cost was figured at $511.24, of which $56.86 was charged to labor.

The slow progress of the standard stock car is shown by the accompanying illustrations of three more easterly cars (Figs. 4.11, 4.12, 4.13). The first of these is a fine builder's picture of a Michigan Central car dating from 1886. The 20-ton-capacity No. 6815, measuring 34 feet long and weighing 12.5 tons, was produced by the Michigan Car Company of Detroit. It offers a lesson in car construction, for many elements of boxcar body construction are illustrated in its skeletal outline. The trussing, both timber and iron rod, is nicely shown. Notice the very prominent cast-iron post pockets and how neatly the ends are finished solid in boxcar fashion. The doors, made with a wooden frame and iron bars, were an alternative to the common pattern of all-wooden ventilated closures. This style of door was often found on ventilated boxcars. Only two truss rods are used, and this at a time when four rods were common. There are no turnbuckles, but the rods are split and bolted together at the center for easier installation and removal. The peculiar semiautomatic link-and-pin coupler is made on the Ames patent. The cast-iron dead blocks above the couplers and the outside-hung wooden brake beam are also worthy of attention.

The Lehigh Valley was a coal road, yet enough stock moved over the line to warrant drawings for a 20-ton-capacity car in 1889 (Fig. 4.12).[18] This is an unusually fine drawing, and the road's master car builder, J. S. Lentz, appears to have expended no less care than if he had been designing an

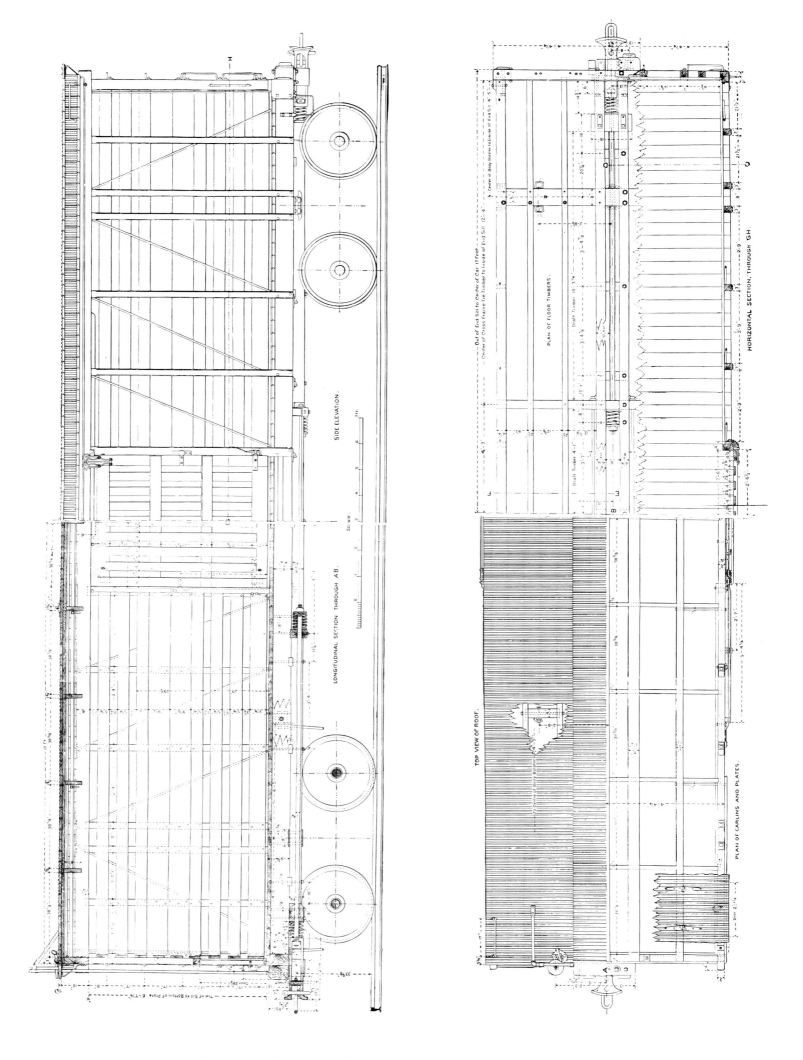

SIDE ELEVATION.

LONGITUDINAL SECTION THROUGH AB.

Scale.

PLAN OF FLOOR TIMBERS.

HORIZONTAL SECTION, THROUGH GH.

TOP VIEW OF ROOF.

PLAN OF CARLINS AND PLATES.

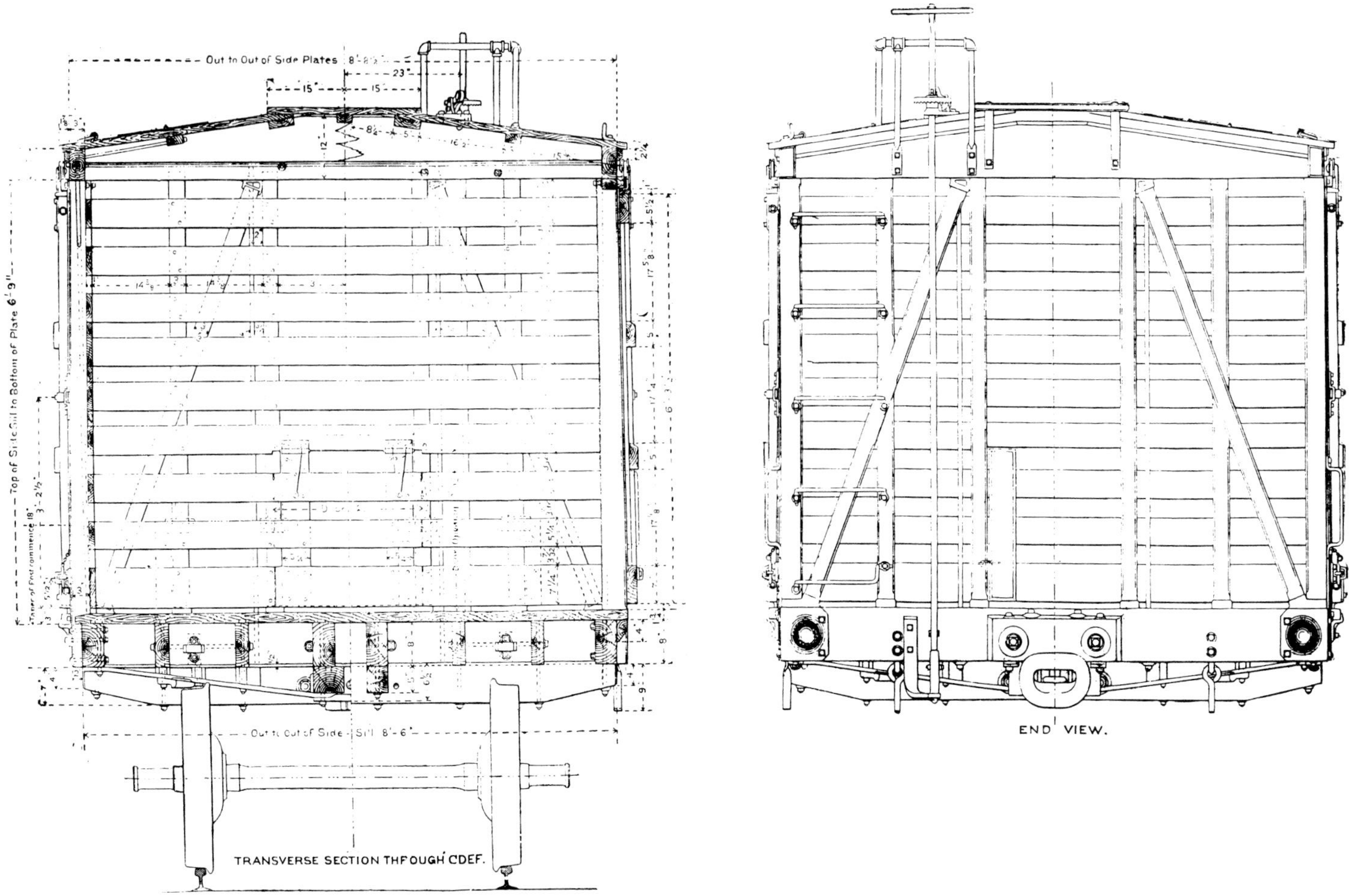

FIGURE 4.5

This New York Central stock car, designed by Leander Garey, incorporated several novel framing variations. (*Railroad Gazette*, May 26, 1882)

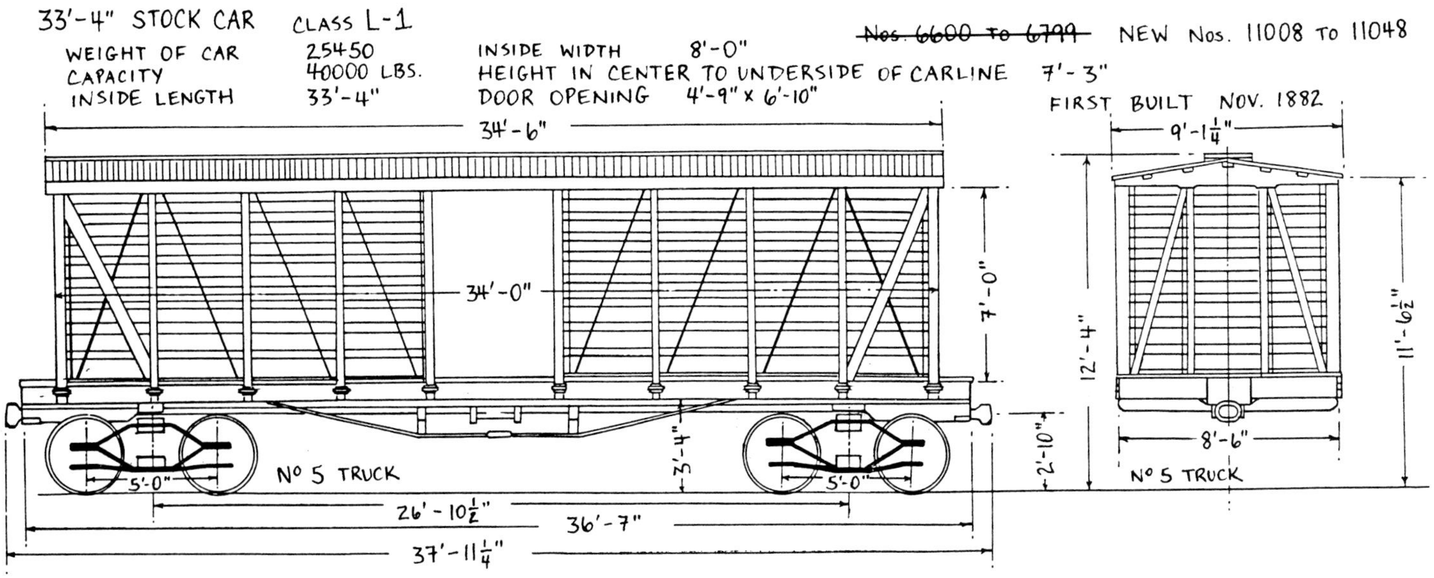

FIGURE 4.6

Diagram drawing of a Baltimore & Ohio class L-1 introduced in 1882. (Baltimore & Ohio Railroad)

FIGURE 4.7

This 1907 wood-frame B&O stock car is classified as a Mather car, yet it exhibits no obvious features associated with Mather's patents. The car was rated at 30 tons capacity. (Smithsonian Neg. 49863)

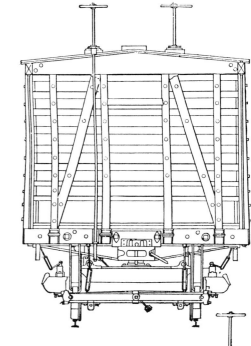

FIGURE 4.8

The body of this Santa Fe stock car measured 30 feet long by 8 feet 6 inches wide. (*National Car Builder*, Nov. 1880)

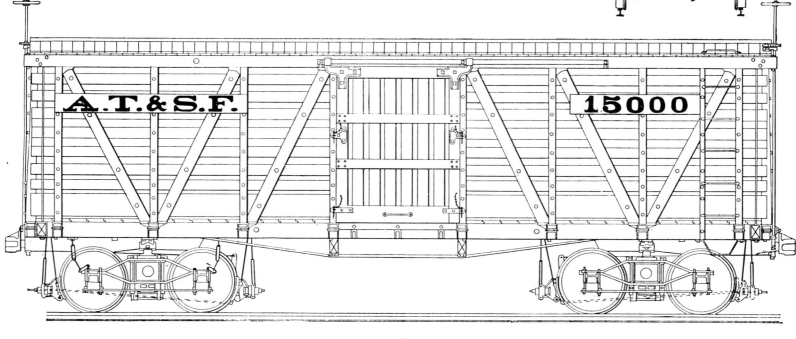

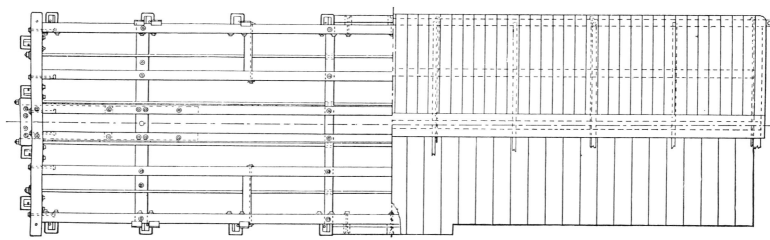

Floor Frame.   Floor and Roof.

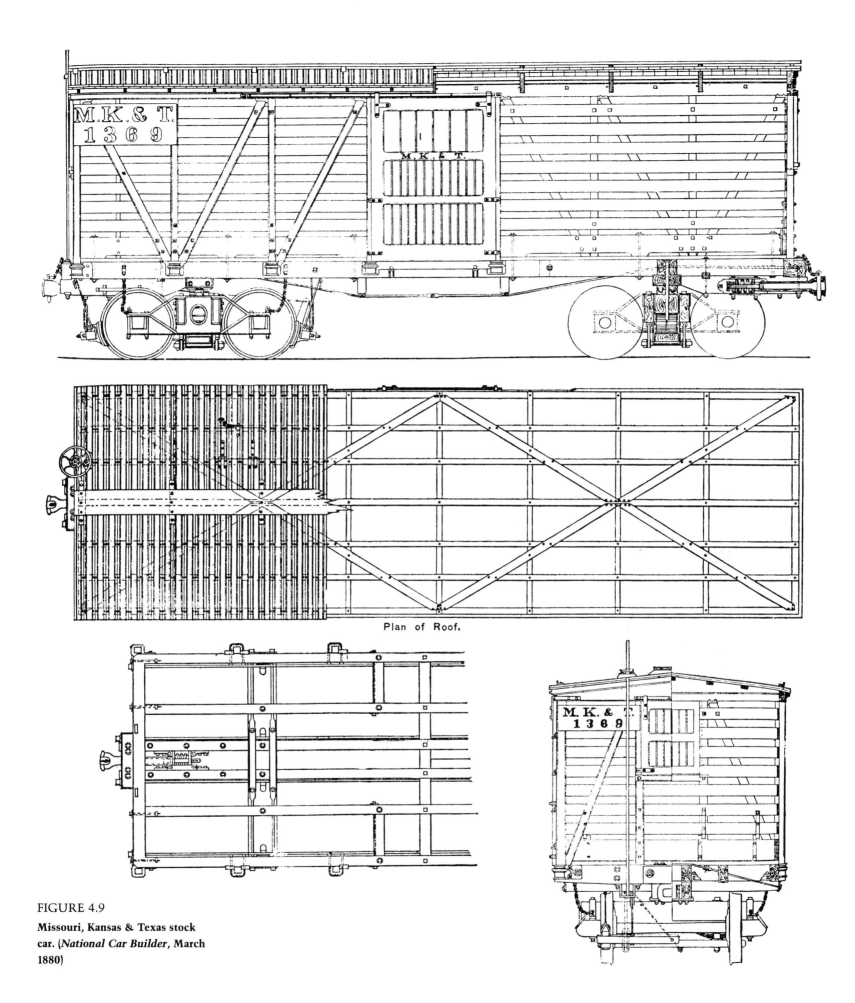

FIGURE 4.9

Missouri, Kansas & Texas stock car. (*National Car Builder*, March 1880)

CARS FOR FOOD

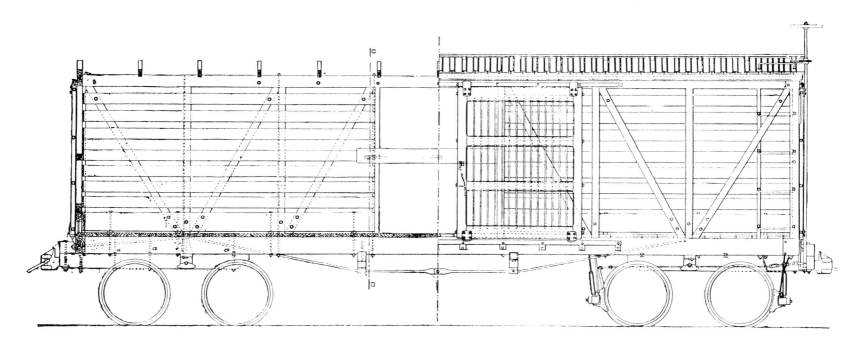

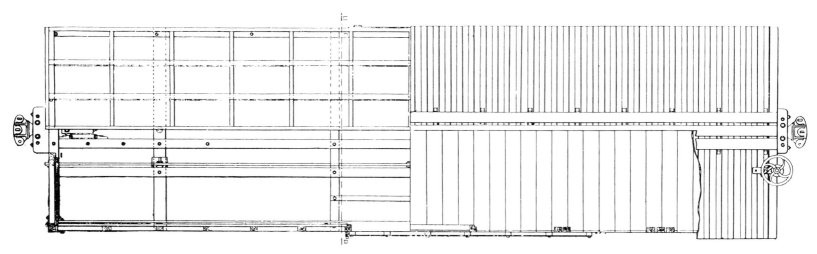

FIGURE 4.10

Burlington stock car. (*National Car Builder*, Sept. 1881)

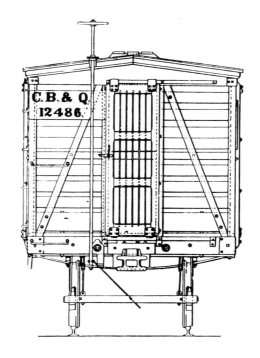

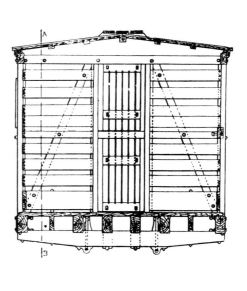

FIGURE 4.11

This Michigan Central stock car was built in 1886 by the Michigan Car Co. of Detroit. (Collection of Arthur D. Dubin)

elaborate palace passenger car. Lentz's design featured double end doors for ventilation and loading and an iron bolster. The side boards were oak, while the floor frame was yellow pine.

W. S. Morris, superintendent of motive power for another coal road, the Chesapeake & Ohio, set out to design a very lightweight cattle car in around 1895 (Fig. 4.13).[19] The car measured 36 feet inside. It was rated at 30 tons capacity yet weighed only 27,300 pounds, which was 5,000 to 6,000 pounds less than a conventional car of this size. Just how Morris achieved his goal is not clear from the surviving drawings. The sills were conventional in size and number. However, Morris apparently made good use of pressed-steel and malleable-iron parts to reduce overall weight. One hundred cars were built on this plan.

HUMANE CATTLE CARS

Few subjects seem to bore the public more than freight service. The type of equipment used, the volume of traffic, its importance to the economy and the smooth flow of everyday life—all this is of no consequence or interest. It becomes a topic only if the service is interrupted by a strike or a natural catastrophe.

There is at least one exception to this general indifference, and that is the controversy over the humane shipment of animals by rail, a debate that

peaked during the 1880s. It is common to think of farm animals as stupid brutes able to withstand just about any treatment. "Strong as a bull" and "tough as an ox" are commonplace expressions that underscore this attitude. Compassion and sentiment are usually reserved for household pets, who are pampered, fed, and often treated better than the human members of the same household. But the inhabitants of the field and barnyard are living creatures who, though strong and dumb, require a certain level of feeding and care just to survive.

At least a few individuals felt strongly enough about the subject to carry the work of humane societies beyond household pets and draft animals. They took pity on the plight of livestock en route to the market. Though they were destined for the restaurant or family supper table, their last days and hours should not be ones of unnecessary torment. The suffering of animals because of thirst, hunger, and bodily damage was seen as a scandal.[20] They were crowded in barren cars in the most brutal fashion. Injuries from pikes and clubs were common. The weak or injured fell to the floor, where they were trampled or gored by their stronger fellows. There was no space to lie down, no water or food.

Cattle might be left standing for days if there was a delay on the railroad, and even when things

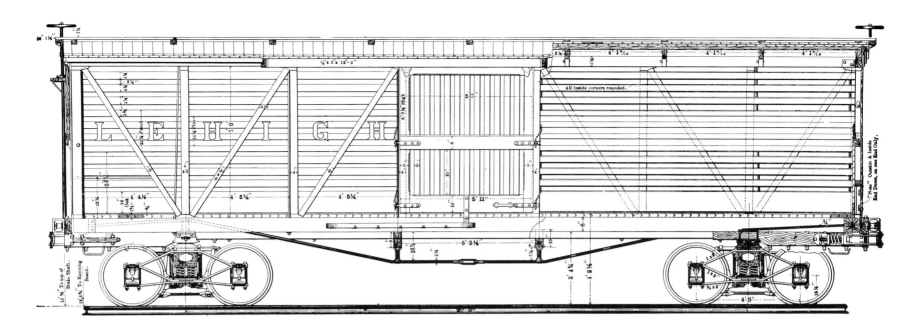

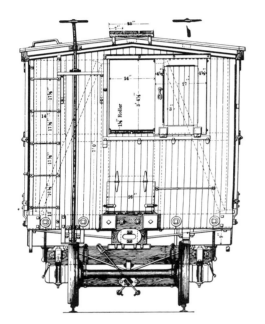

FIGURE 4.12

This Lehigh Valley stock car was designed by John S. Lentz. (*Master Mechanics Magazine*, Oct. 1889)

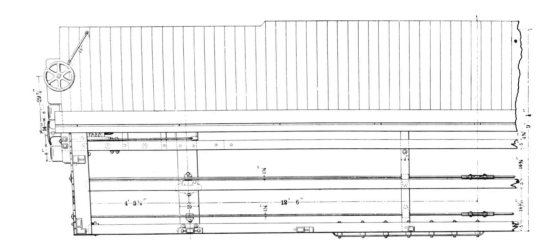

FIGURE 4.13

A lightweight cattle car designed by W. S. Morris around 1895 for the Chesapeake & Ohio. (*American Railroad Journal*, April 1899)

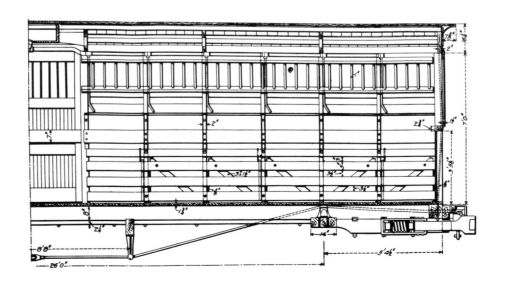

went well, loading and reloading at feed stops were sometimes more damaging than keeping the herd on the cars; poking and beating happened only at a stop. Cattle half-crazed from their frightening journey were not usually disposed to eat and drink, particularly if the stop was of a very short duration. They were confused and shaken by the rough-riding cars, a situation only aggravated by efforts to run fast to get the stock to market before more died during the trip. Rough handling of trains caused everything not nailed down to fly into a heap at the far end of the car. Broken legs and unintentional gorings were the result. Some ranchers, in an effort to get even more of a payload, would load hogs with cattle. The results were often deadly, because the hogs would bite and tear at the cows' legs and udders.

There was much talk and some real effort to improve the speeds of cattle trains in order to lessen both the suffering and the losses just described. Solid trains were made up of stock cars, with trackage rights over lesser trains, but schedules rarely worked precisely on busy single-track lines. On August 29, 1879, the *Pittsburgh Commercial* spoke with some wonder about the fast flight of a stock train from Chicago to Jersey City that made the journey in only about forty five hours. Just a few years earlier twice that amount of time was thought good, and a more normal schedule for Chicago-to-New-York stock trains was almost seven days.[21] By 1882 the same trip was down to about five days with private car lines claiming to make the run in only fifty-four hours—assuming that all went as scheduled.[22] Five years later the Burlington boasted of improvements to stock train speeds due to the installation of double track on its line between Aurora and Chicago (163 miles).[23] When still single-tracked the trip took sixteen to eighteen hours, but on the double line the time was ten hours. The improvement was noticeable, but it was still rather slow traveling for creatures on a thousand-mile journey. And the pace hardly quickened over the next several decades. In 1909 stock trains were said to average no better than 17 to 19 miles per hour.[24]

Critics of the livestock trade tried to gain greater public support with claims about unhealthy food. It was argued that abused animals produced meat unfit for human consumption.[25] Animals neglected became fatigued and feverish. They infected healthy animals in the close confines of the cars. It is true that cattle staggering out of stock cars after several days on the road presented a dismal sight with their sunken flanks, bruises, sores, and ulcers. They might be tired, thin, and beat-up, but they were not necessarily unhealthy. The quality and quantity of their flesh were more seriously affected than its purity. A more realistic argument here was the poorer quality of the beef delivered by stringy, dried-out steers. The weight loss was often no less than 10 percent, which means that the fleshy and best-tasting beef was wasted away en route. In 1873 it was estimated that 120 million pounds of edible meat was lost this way each year, resulting in a loss of $8 million.[26] This loss was, of course, augmented by the death of cattle in the cars. A congressional inquiry reported that 6 percent of the cattle and 9 percent of the sheep shipped by rail died on their way to market.[27]

The weight-and-death-loss argument appeared to be the best way for the humane groups to win support from the industry, but this was not the case. Reducing transport costs was of greater consequence than saving a few head of cattle. Carload rates were the lowest available, and the more a shipper could push into the car, the lower his shipping costs. Cattle raising had become a big business by 1880, and ranchers were accustomed to losing a certain percentage of their herd each year to disease, storms, floods, wolves, and rustlers, so why not a few more to the railway? Dead cows were thrown out along the tracks to be eaten by hogs, dogs, and other scavengers. Some carcasses went off to illicit butchers or were sent straightaway to the glue factory. Ranchers thought it was enough to pay high rates to railroad-operated feed yards, and anyway, didn't they ensure the public health by quick-liming the cars between shipments?

Since the railroads and cattle industry wouldn't act, the humane groups next tried state and federal governments. In 1869 Illinois passed a law requiring that livestock be unloaded every twenty-eight hours and given a five-hour rest.[28] Similar legislation was proposed in the Ohio legislature, and the matter was considered by the Massachusetts Railroad Commission during the next year. Federal legislation was passed in 1872 and became law in March of the following year. It copied the Illinois law and specified fines of from $100 to $500. But there was apparently no personnel or machinery to enforce the law, and so however decent and reasonable its intent, the law had no effect on reforming the shipment of cattle by rail. Effective federal legislation was not enacted until 1906. Meanwhile, those concerned about animal rights turned to the American public.

The American Humane Association was formed in Boston in 1877 with the backing of such prominent and wealthy citizens as Nathan Appleton.[29] The group pressed for reform, with no great result or public attention until October 1879. At that time a contest was announced for the creation of a humane cattle car wherein animals could rest, feed, and drink. The AHA hoped that the contest would appeal to practical railroad men, not just to cranks and amateur inventors, because a provision of the 1873 federal law would relieve the railroads of the need to unload cars if the stock was adequately provided for aboard the train. It was

CARS FOR FOOD

optimistically supposed that the common stock car would, like other evils, be forced into a sullen march to oblivion by the forces of reform. A few railroad officials such as E. T. Jeffery, general superintendent of the Illinois Central, agreed to serve on the judging committee. Wealthy supporters of the association contributed $5,000 for the award—a very substantial sum in 1879. The contest generated considerable excitement and succeeded in focusing attention on the plight of animals in transit. Individuals who had never before considered the subject became champions of the cause.

The committee of judges was soon overwhelmed by inquiries, models, drawings, and descriptions. News of the prize spread to Canada, England, and Switzerland. By mid-1881 480 models and 243 plans had been received. Some had merit, but most repeated old ideas or were simply notable for their impracticability. Comparing these plans with the 111 existing patents—the first issued in May 29, 1860—the judges could find little that was truly new, nor anything that would raise the comfort and safety of cattle without a sacrifice of capacity. The greed of shippers would never allow reduced capacity in a sentimental solicitude for animal comfort. And in the end, the AHA decided not to award the cash prize. A swindle was suggested by the contestants, but there appears to be no reason to question the integrity of the judges. It is possible that the best talent refrained from entering the contest because they did not want to surrender their patent rights to the AHA. To mollify the contestants, a few prizes were given out in 1883. A. C. Mather received a gold medal in that year.

The long-term effects of the AHA contest are questionable. It certainly elicited a broad public response to a subject of minor general interest. It also stimulated a flurry of patents and the formation of several private stock car companies that emphasized safety and comfort for animal travelers. Efforts to introduce extra-fare cattle cars came just a few years after Pullman began to use the term *palace car* to designate luxury passenger travel. The comfort and well-being of passengers both human and otherwise could be improved, but for a price. These private firms invested a great deal of energy and money in creating cars and service, but the effectiveness of their efforts was seriously questioned by Angus Sinclair, editor of *Locomotive Engineering*.[30] In 1892 Sinclair made a 5,000-mile tour of western railroads and claimed not once to have found a palace stock car being serviced. The drinking troughs were never watered, nor were the hay racks ever filled. He thought them to be a terrible fraud.

There were upwards of ten thousand palace stock cars in service when Sinclair penned his attack in 1892. Most were owned by private car lines operated by Burton, Hicks, Mather, and Canda. Nearly all of these firms acquired and operated a thousand and more cars each, and at least a few remained in business into the twentieth century. There were others, to be sure, like Lincoln and Tallman, that failed after just a few years. Each of these firms offered cars with onboard feeding and watering devices. Cattle were given more space than that offered in conventional cars, and some provision was usually made to separate them into pens or stalls within the cars. The apparatus contrived to feed, water, and separate the livestock was not so very remarkable or inventive, and yet the stock car lines individually possessed several patents each covering the details of the particular system. In most cases it is impossible to distinguish any notable degree of superiority among the plans worked out by Mather, Hicks, or Burton, but each attempted to safeguard his idea, no matter how conventional, from a possible competitor. A fee beyond the normal freight cost was charged. It was similar to the first-class or sleeping car surcharge that gave the Pullman Company its revenue. Operators of palace stock cars argued that the increased transportation costs were recovered because of less weight loss and a better mortality rate. Faster service and improved trainboard conditions resulted in fatter, healthier animals, or so it was claimed.

The first patent for an improved stock car was issued on May 29, 1860, to Lee Swearingen. More patents followed, but it was not until very late in 1870 that an actual test was made of one of these schemes. Zadok Street of Salem, Ohio, was granted two patents on August 30, 1870 (Nos. 106,887 and 106,888). Street's first car journeyed from Chicago to New York later the same year.[31] The trip was made in just ninety hours, which might seem painfully slow even by the standards of the time, yet Street claimed that cattle shipments normally required an additional three days. His car had water receptacles in each stall supplied from a tank below the floor. Feed boxes were replenished via tubes, from rooftop hoppers. Only six cows shared the car, a luxury of space normally reserved for officials of the railroad. The car was big for its time. It measured 36 feet long and was rated at 19,600 pounds capacity. With such a big car and so few passengers, Street's rates must have been high for the journey to pay. Yet nine years later the same Zadok Street was still very active in the cattle transport business.[32] He traveled 18,000 miles in six months to visit over thirteen hundred shipping stations, and he reported poor facilities, overcrowding, and inhumane treatment both in and out of the cars.

Street's proselytizing apparently persuaded another resident of tiny Salem, Thomas J. McCarty, to join the cause for better stock cars. McCarty received a patent on October 7, 1873 (No. 143,414).

FIGURE 4.14

George D. Burton's original 1882–1883 plan for a humane stock car is shown in these two advertising engravings. (*Master Mechanics Magazine*, June 1887)

Several months before the patent was granted, McCarty ran two cars with sixteen cattle in each over the Pennsylvania Railroad from Chicago to Jersey City in around forty-five hours.[33] This was fast time indeed, being one-half of Street's record and very similar to passenger train speeds. Despite this impressive run, nothing more is known about McCarty's plans, yet during the next year James Montgomery of Chicago organized a Palace Stock Car line in his name and announced that he controlled the Street and Lee patents.[34] The operation of Montgomery cars was reported upon favorably in 1882, indicating that the business was of some duration and substance. The internal partitions were made up from folding wooden gates. They could be folded out of the way so that the car could receive ordinary cargoes. Feed and water were provided. The cars had special trucks, well sprung, so that the cars rode like a Pullman even at 40 miles per hour. The manager of the Jersey City stockyard, after unloading one of Montgomery's trains, congratulated him for its being the only one in his experience to arrive with no dead or injured animals.

Yet by the early 1880s Montgomery was compelled to share the deluxe cattle trade with several other operators who began to offer identical service. George D. Burton (1855–1918) of Boston began service between Chicago and New England in 1883. His first patent and sample car had been created the year before.[35] The first car measured 33 feet long and had two compartments and four side doors (Fig. 4.14). Each compartment held six animals. The space was so generous that they could lie down in a comfortable bedding of straw. The bins held 2 bushels of feed. Water troughs were also provided.

Burton also planned a 48-foot-long three-compartment car for eighteen cattle. This scheme was not actively pursued, however, for he devised a new single-compartment plan in 1887 wherein a 36-foot-long car would safely carry eighteen head.[36] Combination water/feed pans ran halfway down the length of the car on one side, continuing the rest of the way down on the opposite side (Fig. 4.15). Water was supplied from a standpipe or tank crane normally used to fill locomotive tenders. The water was directed into a funnel-like opening in the car roof and sent to the troughs by supply pipes. The idea was commonly used by most palace stock cars and was not original or unique to Burton. Feed was manually shoveled into the troughs through small outside doors. The grain box was below the floor. The car weighed 26,800 pounds and was rated at 20 tons. It cost $720, and like all palace stock cars, it had removable or foldaway partitions and feeding apparatus so that it might be cleared for regular boxcar loadings. To improve it for normal cargoes a new design was tried in 1889.[37] The upper sides of the body were extended 9 inches to leave space for the hay racks and water pans and thus remove them entirely from the body of the car when in their folded position (Figs. 4.16, 4.17). A tilt lever, manipulated from the roof, worked the position of the racks and pans. This 20-ton-capacity car carried twenty animals within its 38-foot-long body.

While these design changes were under way, Burton was expanding his fleet. In 1883 there were only twenty-three cars. Within four years the

FIGURE 4.15

Burton's single-compartment
stock car as revised in 1887. (*Journal of Railway Appliances*, 1888)

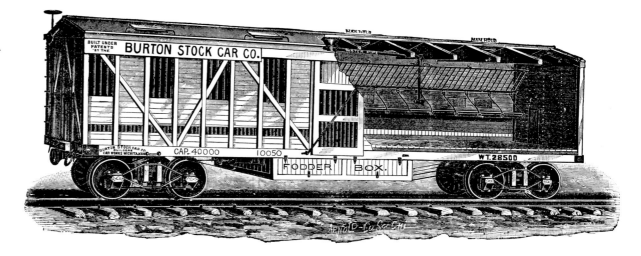

FIGURE 4.16

The final Burton stock car plan,
introduced in 1889, had water
pans and hay racks contrived to
fold back into the upper body of
the car. (*Railroad Gazette*,
March 22, 1889)

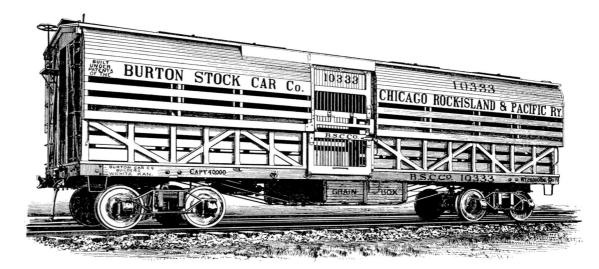

FIGURE 4.17

Cross-section detail of Burton's
1889 design. (*Railroad Gazette*,
March 22, 1889)

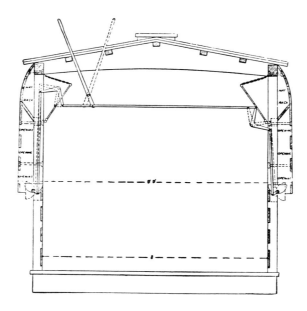

number had grown to seventy, and by June 1887
there were one hundred Burton cars. A car shop
was opened in Wichita about this time, and the
number of Burton cars grew quickly; by the end of
1888 there were twelve hundred.[38] About half of
these were leased to the Rock Island. The Burton
company continued in business until around 1900.
Some idea of its operations can be gained from the
rates listed in the April 1897 *Railway Equipment
Register*. All fees are in addition to the regular railroad freight charges. A flat fee of $10 was charged
for trips under 100 miles. Sixteen dollars was collected for trips of 100 to 400 miles, and 4 cents per
mile was charged for trips over 400 miles.

Henry C. Hicks of Minneapolis obtained three
patents in 1881 (Nos. 240,250, 279,162, and
281,067) and a single patent in 1883 (No. 288,535)
for a convertible stock/boxcar. Partitions were
raised to or lowered from the ceiling by chains and
bars worked by handwheels mounted on the sides
of the car. Water pans and mangers or feed boxes
folded up into the sides of the car, and like the partition, their position could be adjusted from outside the car (Fig. 4.18).[39] By 1890 the Hicks car was
improved with a removable double deck, which
was stored above the hay racks. The deck, made in

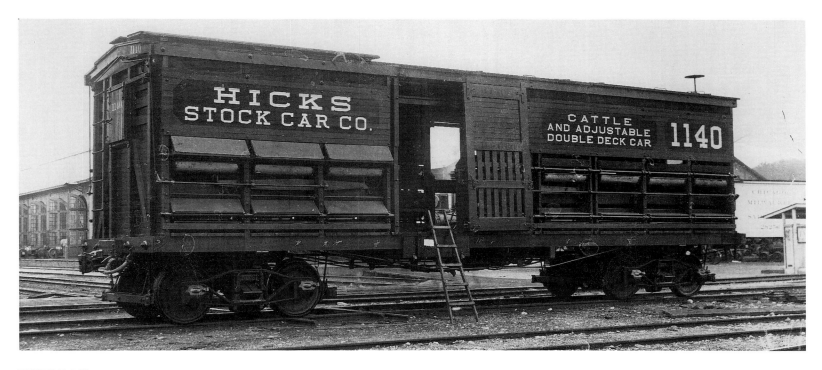

FIGURE 4.18

The feed and water pans built into the lower sides of this Hicks car are clearly evident. Notice the vertical handwheel in the roof just above the letter *S* in the word *Hicks*. (Kalmbach Publishing Co. Collection, Milwaukee)

sections, could be assembled in as little as two minutes, according to the Hicks company. All partitions, mangers, and drinking pans were smooth, with rounded edges, to protect the cattle. The extra equipment added only 1,000 pounds to the car's total weight. The racks held enough hay for five days and could be replenished through roof hatches. A water tank in the roof supplied the drinking pans. The Hicks Humane Live Stock Car Company was organized in Chicago in 1884.[40] Hicks was no longer an officer by 1890, and the Hicks company became part of the Consolidated Cattle Car Company before 1897. The exact size of Hicks's operation has not been determined, but a report published in the *Railway Review* of May 31, 1890, noted that the Santa Fe had ordered four hundred Hicks cars and was considering the acquisition of more. Hicks was taken over by J. W. Street's Western Stable Car Line in 1902 (Fig. 4.19).

F. E. Canda was mentioned in Chapter 3 for his work in designing a 50-ton-capacity wooden boxcar. Canda was a man of obvious talent and daring, and we would expect his plan for a palace stock car to be exceptionally neat and mechanically sound. The drawings for his designs support that judgment. Canda was not a pioneer like Street and did not even enter the field until several years after Hicks and Burton. His first car, which appeared in 1888, possessed no remarkable new features (Fig. 4.20).[41] Partitions could be raised on tracks for under-the-roof storage. The folding feed and watering apparatus was not new. That the eight compartments could be set up in three or four minutes was not unusual. The car was rather big for its time, measuring over 39 feet long. It is the attention to detail and the sure hand of an experienced car builder that make Canda's convertible palace

stock car somewhat better than those of his competitors. The clerestory-like roof certainly gave it a distinctive appearance.

In 1893 Canda revised the basic design, replacing the eight cross-fences with a single central partition (Fig. 4.21).[42] He continued to use the easy-riding Chisholm truck, which carried the weight on the bolster ends. The truck was an even greater curiosity because it employed no center plate. The 1893 car weighed 32,900 pounds and carried 40,000 pounds. In 1890 the Ensign Car Company built one thousand cars for the Canda Cattle Car Company.[43] Within two years Canda had twenty-four hundred cars, and considering his close ties with Ensign, it is likely that they were all built by that West Virginia firm of car builders. In 1902 Canda's cattle car line became part of Street's Western Stable Car Line.

Just about everything involving humane affairs seems to begin by happenstance, a casual introduction or some minor incident. And so it was that Alonzo C. Mather (1848–1941), a wholesale clothing merchant from Chicago, became interested in cattle cars.[44] He was educated to be a physician but preferred to follow a business career. He knew nothing about railways except that they were necessary for business travel. While on an eastern trip in 1881 his train was delayed for twelve hours because of an accident. A cattle train stood on an adjacent track. During the night Mather was kept awake by the rustle and mooing of the animals just feet away from his sleeping compartment. When day broke, he could see the reason for their distress. A large, powerful steer was pacing furiously from end to end in one of the cars. In his desperate search for food or water the beast had injured or killed some smaller occupants of the car.

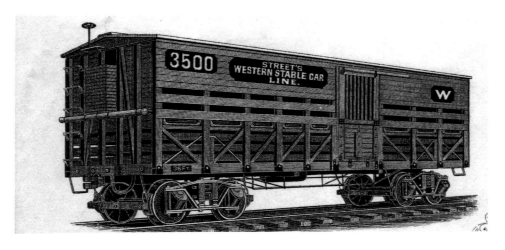

FIGURE 4.19

J. W. Street's stable car featured hinged sides that folded partway out to facilitate the replenishing of the feed racks. This idea was covered in U.S. Patent No. 336,872 (Feb. 23, 1886). (From a letterhead in the Warshaw Business Americana Collection, Archives, Museum of American History, Smithsonian Institution)

Mather counted five dead and wondered about the fate of the others, for their journey was far from ended.

Saddened by that early-morning sight, Mather began investigating the problem of shipping live cattle by rail. The day before he hadn't even realized that the problem existed. He probably also didn't know that at least a dozen other men had been wrestling with the same problem for nearly two decades. Mather wasn't a car builder or even a mechanic, and it seemed unlikely that he would succeed in this specialized area. Compared with Hicks, Burton, and the others already in the business, he was a child. His first effort was amateurish and primitive.[45] He used loose boards placed in slots to form pens or compartments. Such stalls would be labor-intensive to handle, and not very secure. He used wooden barrels for watering, and a farmer rigged up some hay racks for feeding. Mather's first efforts were no doubt received with some ridicule, but he pressed on. He either developed the skill needed to produce a practical car or, more likely, hired an experienced car builder to help him. He claims to have spent $10,000 before perfecting a vehicle able to withstand the hard usage and abuse that befall the typical stock car. Mather contented that individual stalls or tethers were not needed, for so long as the creatures had something to eat, they would remain peaceable.

By the late 1880s the Mather car was well advanced over its first model and was as perfect as any of the other patented cars then in service, which is not to say that any of them was a total success.[46] Their ability to deliver cattle safe and healthy is not documented beyond the claims and advertisements of the palace stock car lines themselves. If one chooses to believe Angus Sinclair, they were a humbug. But Mather, after a struggle, had managed to produce a workable car, as shown in Figures 4.22 and 4.23. What is of more importance is Mather's business ability. He managed to form a large and long-lasting private cattle car line. By 1893 he was operating 1,260 cars and within

two years had 3,000. The business grew so steadily that by 1930 the Mather company was shipping 5 million cattle a year. It seems evident from existing photographs—an example is shown in Figure 4.7—that the self-feeding and self-watering features were abandoned in later years and Mather operated conventional cars.

Street's Western Stable Car Line, of Chicago, was a relative latecomer to the palace cattle car business. Its manager, J. W. Street, is of an uncertain relationship to the Zadok Street mentioned earlier in this chapter. It is certain, however, that the Street firm was well established by 1886 and was offering service from Chicago to both Boston and New York.[47] The cars were much on the order of those already described for Hicks, Burton, and Canda. The 36-foot-long cars carried from eighteen to twenty-four head, depending on weight. Street charged $10 per carload above regular freight charges, which were around $100 per carload between Chicago and New York on railroads that would freely accept private palace stock cars. On railroads that imposed extra charges, the Street company required an additional $15 fee from shippers. A carload of horses cost $5 more. Street began service on the Northern Pacific in 1886. Within five years the company was carrying, over several lines, 96 percent of the cattle traffic between St. Paul and Chicago.[48] The Street Company grew rapidly over the next few years. It acquired both the Canda and Hicks lines and by 1902 controlled and operated about nine thousand cars.[49]

The several inventor-operators described here showed that there were only so many ways to outfit a stock car with feed racks and watering pans. The Hicks, Burton, and Mather cars differed largely in details and showed few remarkable differences one to the other. But there was at least one would-be inventor in this field who would handle things very differently. A. D. Tingley of New York City understood the essentials of humane cattle transport to be the feeding and watering of stock.[50] Yet he felt that the various schemes for special apparatus aboard the cars were not the right way to solve the problem. Such apparatus simply added weight, cost, and complexity. Stock cars should be as simple and light as possible. And so Tingley planned for feeding and water apparatus mounted at the trackside.

A few boards midway up the side of the car would be lowered or removed so that the cattle could stick their heads out. After the train pulled into a feeding station, Tingley's feeding containers, mounted on extendable pipe frames, would be moved into a convenient position for the cattle to feed and drink. Only a few men were needed to manage the feeding equipment, fewer than it took to unload and load the herd at a conventional stockyard. Keeping the cattle aboard the train also eliminated the wear and tear of the offloading pro-

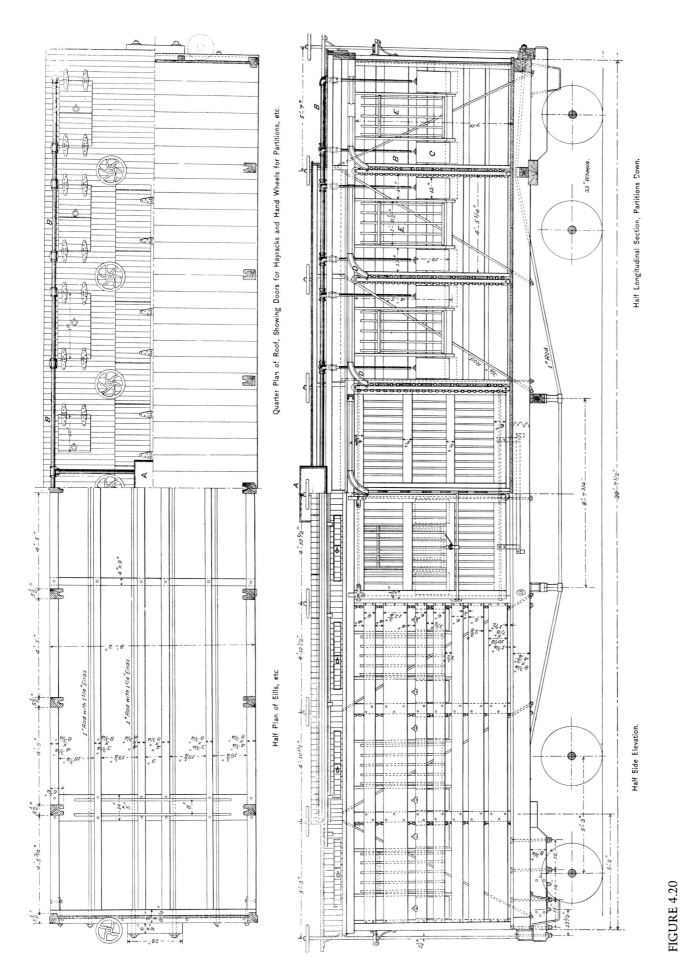

Quarter Plan of Roof, Showing Doors for Hayracks and Hand Wheels for Partitions, etc.

Half Plan of Sills, etc.

Half Longitudinal Section, Partitions Down.

Half Side Elevation.

FIGURE 4.20

F. E. Canda's first style of stock car was introduced in 1888. (*Railroad Gazette*, March 2, 1888)

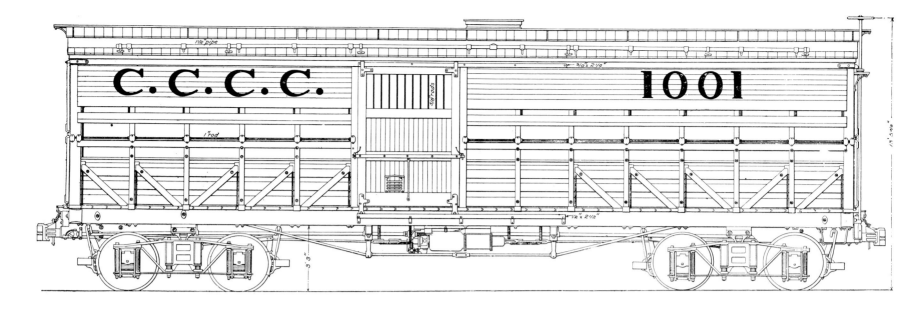

FIGURE 4.21

Canda's second plan was far more complicated than the 1888 design. It was the most ingenious of the several palace-style stock cars used in the United States. (*Railroad Gazette*, July 7, 1893)

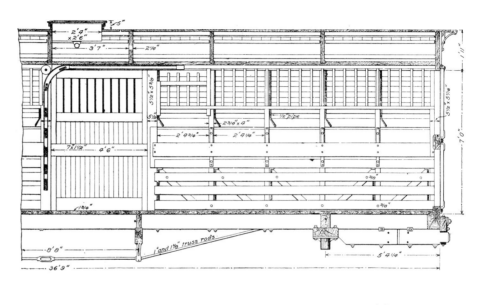

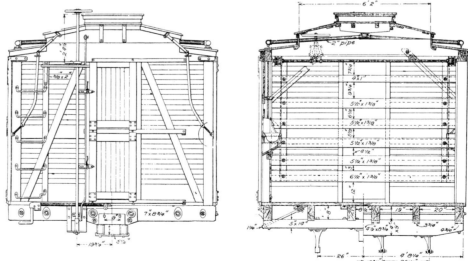

FIGURE 4.22

A Mather stock car of 1889 shows end levers used to work the folding hay racks. The boxlike structure across the end of the car is a water tank. (*Journal of Railway Appliances*, March 1889)

FIGURE 4.23

This cross-section drawing of a Mather car shows the method for replenishing the mangers. (*Journal of Railway Appliances*, March 1889)

cess. The Union Line Stock Feeding Company was organized in 1880 to exploit Tingley's idea. Construction of feeding stations between St. Louis and New York apparently began during the following year, but there is no evidence that the system was ever put into operation.

The reaction of the railroad industry to the rise of the palace stock car was predictable. At first, suggestions for such service were ignored, or it was said that there was no need or demand for it. But once premier cattle car transport was established and gaining favor, then the railroads joined the race to capture this business by providing their own palace stock cars. Some lines took direct measures to discourage the operation of privately owned cattle cars over their tracks. The Burlington refused to pay Burton the customary 0.75-cent-per-mile empty backhaul fee. The dispute was aired before the Interstate Commerce Commission in 1887.[51] The railroad contended that the palace stock cars were too encumbered with interior apparatus to handle return loads conveniently. Burton argued that this was only an excuse used to discourage private car shipments. Two years later a similar dispute arose between the American Live Stock Transportation Company and the Lackawanna.[52] The railroad was accused of refusing to move American's cars and did everything possible to encourage patronage of the railroad-owned Lackawanna Live Stock Express Company.

A few railroads anticipated the competitive attraction of the palace stock car before many private operators entered the field. In 1877 the Pennsylvania Railroad's Altoona shops produced a car that could feed and water in transit.[53] A charge of $5 above the regular tariff was imposed. The Pennsylvania continued offering cars of this type for many years, as evidenced by the photograph reproduced here as Figure 4.24. During the 1880s and 1890s just about every railroad with cattle traffic pur-

chased some palace-style cars. Surviving Pullman negatives show cars of this type being produced for the Chicago Great Western, Chicago & North Western, Omaha, Kansas City & Eastern, and Atlantic & Pacific railroads. A print of a car belonging to the last-named railroad is shown in Figure 4.25. Notice the roof feed hatches and water pipes and pans. The pans are turned 90 degrees to the empty position. An interesting example of a railroad-built-and-designed palace car is shown in Figures 4.26 and 4.27. William Garstang, superintendent of motive power for the Big Four (the Cleveland, Cincinnati, Chicago & St. Louis Railway), is credited with the design of this 36-foot vehicle.[54] Most of the features of the car are obvious from the drawings and are again more or less identical to those used by the private car operators, differing only in the details of construction. The attention to an easy-riding and stable truck is evident from the cross-section detail shown in Figure 4.27. The swing motion and clusters of coil springs are clearly shown in this drawing.

HORSE CARS

The term *horse car* normally designates a street railway vehicle, but there is yet another class of main-line railway conveyance properly designated by the same name. Such vehicles were entirely different from the tramcar variety, where the animal propelled the car. In the steam railroad context the horse was a passenger. The cars employed for the purpose were very similar to the palace-style cattle cars described in the preceding pages. Before 1880 it is almost certain that horses were shipped in conventional stock or ventilated boxcars, and so it seems logical to include horse cars in this chapter, even though horse flesh is not common in the North American diet. Were this book broad enough to cover French railway practice, it would be even more appropriate to include equines in this chapter on food-carrying cars, because horse meat is considered a delicacy in that land so well known for its gourmets.

In nineteenth-century America horses were kept as draft or sporting animals. As such, they were individually more valuable than if raised as a source of food, and their security and well-being in transit were of more concern. As early as 1833 special padded cars with feed and water apparatus were provided for this service in England. It remains in question whether such precautions were taken this early in the United States, for there is no apparent record of its being done much before the middle 1880s. Burton, Street, and most of the other private stock operators solicited horse passengers as well. The accommodations provided were surely adequate for a workaday steed, but there was an elite market requiring even greater care and attention. Racehorses or prize breeding stock could be extremely valuable. That they were

FIGURE 4.24

A Pennsylvania Railroad stock car
with roof feed hatches. It was built
in May 1891 by the Peninsular
Car Co. of Detroit. (Smithsonian
Chaney Neg. 26317)

FIGURE 4.25

This elaborate palace stock car
was built by Pullman in 1889.
Note the shallow oblong pressed-
steel water pans turned sideways.
They are parallel to the number
1330 on the door. (Pullman
Neg. 364)

also often high-spirited and temperamental underscored the need for special treatment. Only the
most careless owner would entrust a costly thoroughbred to the confines of an ordinary cattle car.
There were also wealthy families who showed exceptional concern for their "pets"; though they
might not be superior animals, it would never do
to have Dolly or Hank mixed in with a lot of common horses. And so when moving the household
during a shift in season, let us say from Great Neck
to Bar Harbor, Dolly and Hank went first class, just
like the rest of the family.

As the science and hobby of cattle breeding grew
in popularity, a new market grew for the carriage
of prize bulls. Cattle bloodlines became a matter of
grave discussion and conspicuous consumption.
After investing thousands in a champion Angus,
only the safest and fastest means of transit could
be considered. The business possibilities created
by this specialized transportation need were belatedly recognized by Harrison Arms (d. 1917), a
livery stable operator from Toledo, Ohio.[55] In 1885
he organized the Arms Palace Horse Car Company.
The cars, built with clerestory roofs, end platforms, and passenger-style trucks, looked like baggage cars and were clearly intended for passenger
train service. They were divided into two compartments and had a total of sixteen individual

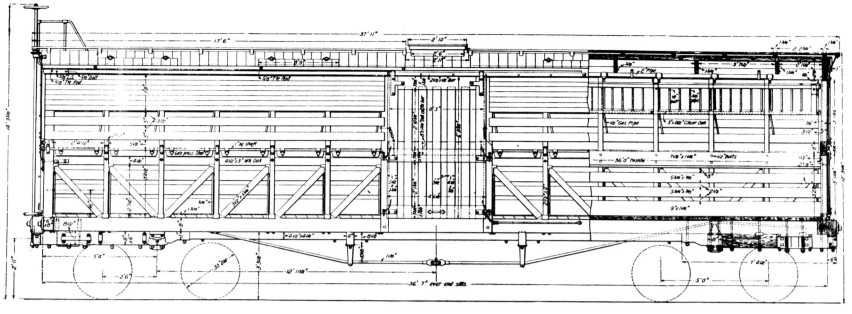

FIGURE 4.26

This Big Four palace stock car was designed in 1895 by William Garstang. The water-tank access hatch is located at the center of the roof. (*Railroad Gazette*, Dec. 27, 1895)

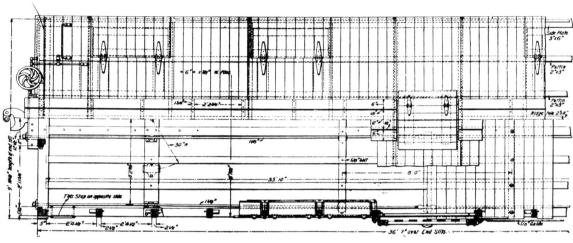

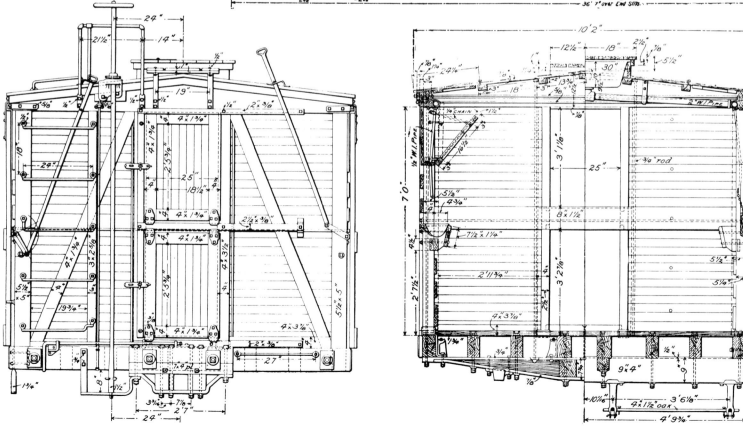

CARS FOR FOOD

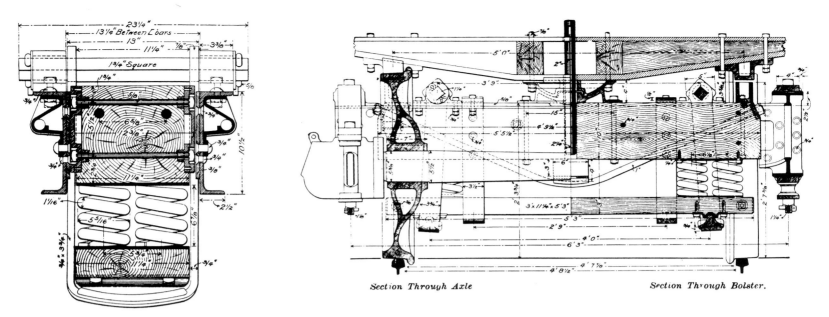

Section Through Axle   Section Through Bolster.

FIGURE 4.27

Cross-section detail of easy-riding swing-motion truck for Big Four stock car. (*Railroad Gazette*, Dec. 27, 1895)

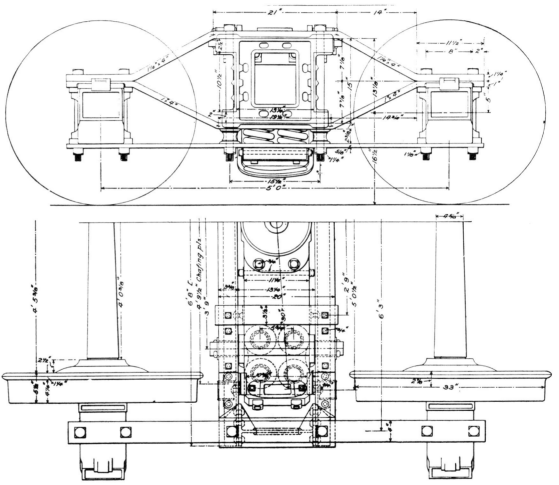

stalls (Fig. 4.28). By the late 1890s Arms had acquired the express horse delivery business of Burton and Keystone. The Burton cars were much like those already operated by Arms, while the Keystone cars were long boxcar-like vehicles (Fig. 4.29). The Keystone cars were considerably more utilitarian in appearance than the Burton or Arms cars and were undoubtedly intended for animals of lesser value that might travel as part of a freight train. The Keystone portion of the Arms fleet grew to one thousand cars.[56]

POULTRY CARS

The most specialized rail vehicles engaged in the transport of animals were those used to carry live poultry. In terms of odor, noise, and the number and skittishness of the occupants, these henhouses on wheels surely surpassed all other cars.

FIGURE 4.28

Express-style cars operated by George D. Burton and Harrison Arms starting around 1885 for the transportation of thoroughbred horses. (*Railway Review*, Jan. 18, 1890)

FIGURE 4.29

The Keystone Co. operated a plainer style of car for the carriage of horses. This 48-foot-long vehicle carried eighteen horses in cross-stalls. It was built by Pullman in 1889. (Pullman Neg. 634)

FIGURE 4.30

Chickens and other live poultry were carried in this curious vehicle with wire-mesh sides. This car was built by Pullman in 1890. (Pullman Neg. 1280)

CARS FOR FOOD

FIGURE 4.31

Continental Live Poultry's car held seven thousand live chickens. This giant wooden vehicle was built by the Youngstown Car Works in 1890. (*Railway Review*, July 19, 1890)

One pities the poor attendant who rode in the central compartment for days at a time.

The idea of special cars for sending live birds to the market appears to have originated with William P. Jenkins, an Erie freight agent.[57] He worked with James L. Streeter, a poultry dealer from Muncie, Indiana. Jenkins received his first poultry car patent on August 26, 1884 (No. 304,005). Three more patents were issued in 1885 and 1888. The Live Poultry Transportation Company was organized at about the time the first patent was granted. Cars were built by Ohio Falls, Terre Haute, and Pullman. By 1897 the firm had two hundred cars in operation. One of these had been shown a few years earlier at the Columbian Exposition.[58] It measured 34 feet 6 inches long and incorporated ninety-six wire-mesh compartments or coops. Small gutter-shaped drinking pans were placed in each coop. They were supplied by a roof water tank that was placed at the center above the attendant's compartment. The attendant regulated the water and parceled out feed. He also removed dead or ill birds. In winter a coal stove warmed the custodian and those birds near the center of the car. Cloth was fastened over the car sides in winter to protect the chickens. Most Jenkins cars built after 1890 were set low on the wheels, much like a furniture car (Fig. 4.30). This allowed a high body and hence extra room for more chickens. An earlier and slightly smaller Jenkins car was rated at 8 tons capacity and could carry five thousand chickens and two thousand geese or fourteen hundred turkeys.

A rival firm was started in 1890 to exploit James Nolan's patent.[59] Several very large cars were built for the Continental Live Poultry Car Company. They held seven thousand chickens in 120 coops (Fig. 4.31). Drinking water was supplied from barrels mounted within the roof. The Nolan cars were impressive giants, but in this instance bigger didn't prove better, and the Continental Line disappeared after only a few years of operation. The basic idea was eminently sound, however, and

poultry cars remained in commercial service until around 1960.

Refrigerator Cars

All food is subject to decay. There is no way to stop or reverse this natural process, but it can be retarded. Lowering the temperature is one of the most effective methods. Root cellars and springhouses had been used for food preservation by farmers since ancient times. The benefits of reduced temperatures combined with an insulated environment were thus well known long before the introduction of railways. Natural ice, cut from ponds, began to play a role in food preservation at the beginning of the railway age. It was then a natural evolution that led to the construction of insulated boxcars cooled by blocks of ice. All varieties of food—berries, fruits, vegetables, fish, and meats—were sent to market in refrigerator cars.

Despite a vast new market backed by an intense demand for service, railroads were reluctant to adopt refrigerator cars. This was particularly true for the meat trade.[60] The reasons were simple enough. The railroad industry had invested lavishly in cattle cars, stockyards, and feedlots. The lines running into Chicago, for example, were major shareholders in that city's Union Stockyard. They had thousands of stock cars that were not adaptable for other shipments. Just when everything was nicely organized and a decent return on investment might be realized, a group of midwestern meat packers—notably Hammond, Swift, and Armour—wanted to ruin everything by shipping dressed beef to the big eastern cities. Instead of shipping the whole cow, they sent only the edible carcass, which equaled only about half of the animal's original weight. It reduced tonnage and threatened to reduce profits. The only sensible thing to do was to raise rates on dressed beef to nearly double those charged for livestock, and to refuse to buy or supply refrigerator cars. But the meat packers were not to be discouraged. They purchased their own cars with such enthusiasm that most refrigerator cars were privately owned (Tab. 4.2, 4.3).

TABLE 4.2   American Refrigerator Cars

Year	Private Lines	Railroads	Total
1880	1,000 (E)	310	1,310 (E)
1885	5,010 (E)	990	6,000 (E)
1890	15,000 (E)	8,570	23,570 (E)
1895	21,000 (E)	7,040	28,040 (E)
1900	54,000 (E)	14,500	68,500 (E)

Source: Poor's Manual of Railroads and Interstate Commerce Commission and Census reports.
Note: (E) = estimate.

THE AMERICAN RAILROAD FREIGHT CAR

FIGURE 4.32

The Pittsburgh & Western was one of many private car lines created to handle perishable shipments. It was established around 1872. (From a letterhead dated Nov. 16, 1874)

TABLE 4.3   American Private Refrigerator Car Lines, 1886

Line	Location	Founding Date
American Refrigerator Transit	St. Louis	
Anderson Refrigerator Line	Chicago	
Anheuser-Busch Refrigerator Car Co.	St. Louis	1877
Armour Refrigerator Line	Chicago	c. 1883
Austell Refrigerator Car Co.	Atlanta	
California Fast Freight Line	Chicago	1889
Goodell Refrigerator Cars	?	
Kansas City Dressed Beef Line (Armour)	Kansas City, Mo.	
Kansas City Refrigerator Car Co.	Kansas City, Mo.	
Merchants Despatch Transportation Co. (NYC RR)	New York	1875?
National Despatch	New York	1869
Post Refrigerator Car Co.	Kansas City, Mo.	
St. Louis Refrigerator Car Co. (Anheuser-Busch)	St. Louis	1877
Swift Refrigerator Transportation Co.	Chicago	c. 1875
Tiffany Refrigerator Car Co.	Chicago	1877
Union Line (PRR)	Pittsburgh	1864
Windisch, Muhlhauser & Bro. Refrig. Car Co.	Cincinnati	

Source: Poor's Directory of Railway Officials, 1886.

Note: This list contains most of the big lines but neglects a few such as Hammond (1868) and no doubt many other smaller operations. Founding dates—added by the author where known—indicate the establishment of the business, and not necessarily the beginning of refrigerator car operations.

Operators like Swift found a friend in the Grand Trunk. It had little through traffic between Chicago and Boston because of its indirect northern route through Canada and upper New England. It was not part of the Chicago traffic pool and so gladly agreed to haul refrigerator cars at reasonable rates. The success of Swift and his competitors was remarkable. In 1880 livestock tonnage into New York City exceeded dressed-beef tonnage by fourteen times, yet in three years it had declined to three times the amount. The packers proved that shipping beef 1,000 miles was practical and profitable. They showed the logic of raising cattle on cheap western rangeland, slaughtering them in Chicago, and sending the beef via rolling iceboxes to the great eastern cities. But refrigerator cars were costly and complex. Double walls, insulation, tight-fitting doors and roofs, hatches, ice bunkers, and drains combined to drive the price of a single vehicle to as much as $1,200 in 1883. Even the cheapest refrigerator was half again as costly as an ordinary boxcar. For this reason private car lines owned and operated most refrigerator cars before around 1910. After that time railroads began to enter the field more aggressively because they had more capital available, realized that the demand for this type of car was not a temporary fad, and saw that a fair return on capital was possible.

EARLY EXPERIMENTS

The refrigerator car, like the sleeping car, has a long history, and like the sleeper it was a child claimed by many fathers. A series of experiments can be recorded between 1842 and 1865. Each appears to have been an independent effort that did not lead to any large-scale operation. The idea for refrigerated transit was hardly an inspired one, considering the existence of the ice trade and home refrigerators. Frederic Tudor of Boston began shipping ice in 1805 and had developed a sizable trade within fifteen years. In the process he developed insulated containers and storage methods that could easily have been copied by an intelligent mechanic. The size, profits, and technology of the ice trade were hardly a secret in business-minded America. Even so, many years were to pass before any effort was made to apply refrigeration to railroad carriage. One obvious reason was that there was so little in the way of a rail network much before that time. And until a more complete system was finished, long-distance shipment of foodstuffs was not possible. Yet even once a fairly dense network was in operation, refrigerator cars were slow to come into commercial use.

In June 1842 the *American Railroad Journal* printed an enthusiastic report on a new type of insulated freight car being placed in service by the Western Railroad of Massachusetts, one that promised to convert "the whole country into a garden for our great cities."[61] Ice was to be used in the summer, while in winter 4 inches of powdered charcoal between the inner and outer walls would protect the car's contents from freezing. A wonderful trade was predicted:

> These refrigerator cars will be used, for the like advantageous purpose, to carry eggs, butter, lard, fresh fish, oysters, lobsters, vegetables, cheese, lemons, oranges, strawberries and all berries and fruits and roots:—being a mode of transportation of great value, for nice delicacies, which bear a good price.
>
> We also learn that it is contemplated that these refrigerator cars shall go with the passenger trains, in twelve hours, through from Albany to Boston; and shall be placed between the tender and the passenger cars, giving additional security to the passengers, in case of accident.
>
> If our Michigan and Ohio friends will put, in refrigerator cars, the fresh meat and the wild game they intend for this market, they can send their cars to Buffalo on the lakes; and from Buffalo to Greenbush, partly by railroad and partly by canal, or wholly by the Erie canal. Then from Greenbush, it can come to Boston quickly and in perfect order, the moment the system now proposed is perfected. In like way a chowder of fresh Massachusetts codfish will readily be obtained at Chicago.

These plans were not visionary even for the 1840s, yet there is almost no subsequent evidence on this pioneering effort. Refrigeration historian Oscar Edward Anderson mentions fish and oyster shipments by rail during the same years in New England, confirming that the experiment lasted at least for a few years.[62] Late in the same decade an unnamed inventor fitted up a boxcar on the Concord & Nashua Railroad with an ice trough at one end and attempted to ship fresh beef. His plans were not well laid, and the project was abandoned after a two-season test.[63]

Another thrifty Yankee named Jonas Wilder (1813–1906) tried to introduce refrigerator cars while in the employ of the Ogdensburg & Lake Champlain Railroad.[64] The line traversed fine grazing land populated by Scots who understood the art of butter making. Wilder said that this "gilt-edge product" became a sorry mess upon reaching the Boston market after several days of unrefrigerated transit. Producer and consumer alike suffered. In June 1851 he persuaded the line's master car builder to refit four boxcars with charcoal-insulated walls. The cars were loaded with ice and butter, with more ice added en route. News of the superior golden table spread resulted in large sales and more shipments by the ice cars. Consumers were willing to pay double for the excellent butter, proving the good economics of refrigerated transit. Not long afterward, Wilder took a job with the Rutland & Washington Railroad. He soon had twenty refrigerator cars on this line transporting butter, cheese, and meat. Heaters were installed in wintertime for potato shipments.

Not long after Wilder began his service, promoters in other parts of the nation introduced similar cars. A Mr. Lyman began the regular operation of cars fitted with refrigerators between New York City and Rochester in 1853.[65] Lyman formed a company to operate cars over the principal railroads of the state, two or three times a month. He successfully shipped corn, cucumbers, oysters, and clams. A lamb dressed a week before was found upon arrival to be in better condition than any available in Rochester markets. A report published in 1857 spoke of large-scale shipments to New York City from the West of fresh meat and poultry by refrigerator car.[66] Recent cargoes included 1,500 turkeys, geese, and chickens and 180 mutton carcasses; all arrived in good condition after a nineteen-day journey.

Reports presented in 1860, mentioning that Lyman was still active in this field, suggest that the experiments started seven years before had resulted in a steady business. The latest of Lyman's cars were zinc-lined with the double walls filled with cork. Ice bunkers 2 feet deep and 4 feet long were placed at either end, and an inclined drip floor and pipes carried away the water from the melting ice. Roof hatches were provided above the ice bunkers. This brief description is enough to confirm that the general arrangement of the standard refrigerator car was already established in 1860. On a trip to and from Indianapolis, one car carried 5 tons of fish and fruit. It returned with a

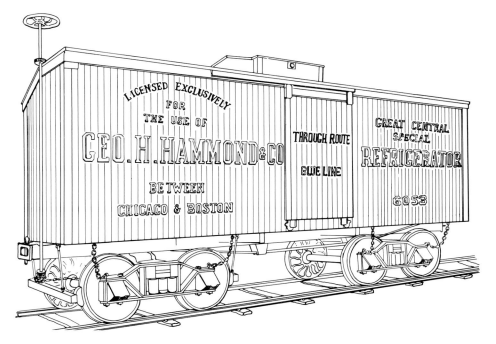

FIGURE 4.33

The Blue Line began refrigerator car service between Chicago and the eastern states in 1867. By early 1868 the line had twenty refrigerator cars in operation. This car was assigned to the pioneer pork packer George H. Hammond. (Rudolf A. Clemen, "George H. Hammond," Newcomen Society in North America pamphlet, 1946)

full load of beef. Going out from New York, the car was re-iced at Troy with 1,500 pounds. Additional supplies of 2,000 pounds and 600 pounds of ice were taken on at two other unnamed stations. The car arrived at Indianapolis with 1,200 pounds unmelted.

A reference to Lyman cars dating from 1867 has also been found. Yet knowledge of Lyman's nearly fifteen years' experience appears to have been limited. Writing many years after the fact, W. W. Chandler, an official of Star Union Line, recalled that he had never seen or heard of a refrigerator car when he remodeled thirty boxcars for the purpose in March 1865.[67] The Star Union was a fast freight line controlled by the Pennsylvania Railroad. Both were major firms whose officials were supposedly well acquainted with developments in the railroad industry, yet none was aware of all the efforts already made by others in this field. Chandler, writing in 1889, naively felt that he was the founder of what was by then a vast industry.

Another would-be inventor of refrigerated transit was Parker Earle of Cobden, Illinois.[68] Earle found southern Illinois an excellent area for growing strawberries, but it was also over 300 miles away from Chicago, the nearest large market. In 1866 he devised a large ice chest with 3-inch walls and a nearly airtight lid. The bottom chamber held 100 pounds of ice; the upper levels were packed with 200 quarts of berries. The berries, considered a luxury, fetched a dazzling price of $2 per quart in Chicago. The trade flourished, and the Illinois Central began operating special strawberry express trains in 1867. Within two years 800 bushels of berries, aboard five cars, moved to Chicago daily during harvesttime. Precooling was practiced, and Earle's ice chests gave way to full-size refrigerator

cars at about the same time. Some strawberries were also going to New Orleans, Pittsburgh, and New York City by refrigerator.

A PROLIFERATION OF PATENTS

By the late 1860s the refrigerator car was a commercial reality, even if it was not widely used. Its utility had been proved time and again. Yet each principal involved in its introduction seemed to approach the subject as though it were an entirely new challenge requiring a novel approach. Most of these endeavors were ill-advised, and the resulting experiments and patents were generally of modest value. It's likely that the simple cars produced by Wilder, Lyman, and others were as effective as the more costly patented cars that began to appear after the first patent for a refrigerator car was granted to the Michigan Central's master car builder, J. B. Sutherland, on November 26, 1867.

George H. Hammond (1838–1886) is on occasion credited with introducing refrigerated rail transport—a claim easily disproved, but he was surely instrumental in enlarging the scale of such operations.[69] Hammond, a Yankee, came to Detroit in 1854 and became involved in the meat trade. He soon came to understand the inefficiency of shipping live cattle to eastern cities. Because 60 percent of the animal was inedible, it would be cheaper to slaughter the cattle near their midwestern source and ship the dressed beef to the big eastern markets. Hammond saw volume as the key to profits, and he wanted to proceed on a massive scale. These beliefs led him to investigate refrigerator cars around 1865. Not being a mechanic, he turned to a neighboring fish dealer, William Davis, who had similar interests and the same engineering pretensions. Unfortunately, Davis did not benefit from the experience of the past but was of the school that only some new scheme would produce a satisfactory car. He obtained two patents in 1868 that described ideas of doubtful utility. At least no features advocated by Davis were used in standard reefer design.

Davis placed narrow, wedge-shaped galvanized-iron tanks at the ends of the car, or in some cases around its interior walls.[70] The tanks were about 5 inches wide at the top and 2½ inches wide at the bottom. They extended from the roofline to within 6 to 8 inches from the floor. Water was discharged through airtight traps. Because of the narrowness of the tanks, the ice had to be crushed into pieces no larger than an egg. Ice and salt were introduced into the top of the tanks—a troublesome task, particularly compared with the simplicity of re-icing a conventional bunker-style car that could accept full-size block ice. About 3 tons of ice filled the tanks; 10 pounds of salt was added for every 100 pounds of ice. The side walls were 9 inches thick with a 3-inch layer of hair felt, the remainder being dead-air space. The body measured

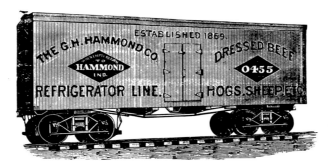

FIGURE 4.34

Hammond began to operate his own refrigerator cars in 1868–1869. This engraving was published in the 1890s but appears to depict a car of the 1870s. (*Railway Equipment Register*, April 1897)

30 feet long and 7 feet wide. Capacity was between 8 and 10 tons, or 120 beef quarters.

The first Davis-Hammond car was built, presumably in 1868, by the Michigan Car Company of Detroit. A Boston newspaper noted the arrival of one in April 1869. The *Railroad Gazette* in July of the same year mentioned the use of Davis cars again "this season," confirming that they had been tried the previous summer. Parker Earle tried a Davis car for strawberry shipments with poor results; the cooling was so uneven that the berries near the brine tanks froze and were spoiled for the market. Another Davis car was said to have carried a load of fruit from California late in 1870 with better results. In September 1869 yet another trade paper noted that forty-four were under construction in Detroit at a cost of $2,000 each. If that reported price is correct, then the Davis car was as expensive as it was impractical. The "V" ice tanks were reproduced in later Davis-Hammond cars at least as late as 1876, when they were described in a British report on rolling stock shown at the U.S. Centennial.[71] The Davis car as originally conceived was never really successful as a year-round meat carrier. In warm weather dressed beef would

occasionally arrive spoiled. The design was modified and thus greatly improved with end bunkers sometime in the 1880s, as noted in the 1884 *Car Builders Dictionary*. The Hammond firm remained loyal to the Davis car and had eight hundred in operation by 1885.

Davis was not alone in believing that novelty was the key to good reefer design. The patented schemes of Tiffany, Ayer, Wickes, Zimmerman, and others were enthusiastically presented, but many, after a few years' service, were quietly abandoned. Most plans amounted to a needless complication and were more ingenious than practical. Railroads yearned for simple cars, capable of being operated by inexperienced and not always intellectually gifted trainmen. Meanwhile, the trade press seemed ready to yield space only to the latest patent scheme for refrigerator car improvements. Reports on more conventional designs were rare much before 1900.

Best known among the early patent promoters was Joel Tiffany (1811–1893), a lawyer who spent much of his life in the Chicago area. His descendants were unable to explain what attracted Judge Tiffany to explore the subject of refrigeration, but the Patent Office records six patents issued to him between 1875 and 1882. The only one devoted to railroad cars was No. 193,357 issued July 24, 1877. The design featured a full-length overhead ice bunker (Fig. 4.35). A clerestory-like structure had hatches for replenishing ice en route. Air ducts built inside the walls were intended to stimulate air circulation. The inventor mentioned the superiority of fans for this purpose but felt that no dependable means of power was available. An engraving appearing in the *Railroad Gazette* just a few days before the patent was issued, reproduced

FIGURE 4.35

Joel Tiffany's overhead-ice-bunker car was patented in July 1877. The clerestory design was abandoned about a year later in favor of an ice bunker built inside the body. (*Railroad Gazette*, July 13, 1877)

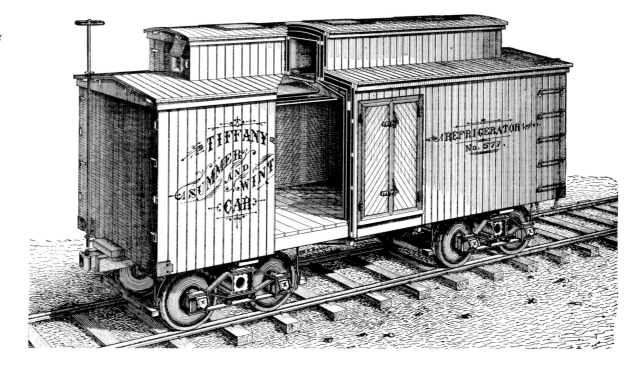

FIGURE 4.36

The improved Tiffany car featured flush roof hatches and an internal overhead ice bunker running full-length just below the roof. (*Railroad Gazette*, April 4, 1879)

here as Figure 4.35, did much to stimulate interest in Tiffany's design. Because it is the earliest illustration of a complete refrigerator car, it has been used to depict all first-generation reefers, when in fact it accurately represents only one distinct plan of car being offered at that time; it should not be accepted as portraying a typical car of the period.

Tiffany's careful attention to insulation may account for the interest in his cars. A series of confined air spaces was formed by thin boards and felt paper. Eighteen months after the patent ninety-five Tiffany cars were in service in this country, with seven more running on European lines.[72] Of more interest, the Atlantic Coast Line had 8-foot-long containers built on Tiffany's designs. Placed on flatcars, they carried vegetables to Charleston where the containers were offloaded onto ships bound for New York. Plans were announced to build more containers and extend service to Texas and Florida. At about the same time (1878), the elevated or clerestory roof was abandoned. The top-ice-bunker scheme remained in favor, and the ice hatches were mounted directly to the car roof (Fig. 4.36).

In May 1879 the Tiffany firm had ten cars built with iron-tube underframes by the National Tube Works of Boston.[73] The wooden car bodies were made according to Tiffany's plan, but the iron underframes were based on designs perfected by B. J. LaMothe some years before. The iron-tube frame apparently did not become a general feature with Tiffany, but new customers, because of or despite this decision, came forward to enlarge his fleet. In 1880 four hundred were reported to be in service.[74] The National Despatch Line was a loyal patron. In February 1880 it had one hundred Tiffany cars in service and an equal number on order. In the fall of the next year the line had three hundred carrying dressed beef between Chicago and Boston in just six days.[75] Within a few years National Despatch was carrying citrus fruit between the same cities in only four days and three hours (Fig. 4.37). Dur-

ing the next few years Tiffany's operation expanded rapidly, in part because of a general expansion of refrigerator car service during this same period. By the summer of 1883 the number of Tiffany reefers had reached one thousand; in less than four years the total was three thousand.[76] They were running on the Grand Trunk, the Union Pacific, the Chicago & North Western, the Milwaukee Road, and the Erie. The firm was said to be well organized and "full of vim and push."[77] Much of its success was due to its energetic manager, Charles F. Pierce.

During these years of growth the full-length overhead ice bunker remained a basic Tiffany feature.[78] The V-shaped bunkers were actually very shallow tanks that sloped gently toward both ends of the car (Figs. 4.38, 4.39). End tanks collected the drip water. A drip pan, located under the bunkers, carried condensate to end tanks to prevent water from falling onto the cargo or the floor. The merits of overhead bunkers were evident from the comparatively small quantities of ice used by these cars. It was claimed that on a trip between Chicago and Boston in the hottest part of the summer, Tiffany cars used only half the amount of ice required by conventional end-bunker cars. Tiffany cars were operated by the National Despatch and its associated lines into the late 1890s, but the design appears to have become obsolete a few years later (Fig. 4.40).[79]

James H. Wickes (d. 1892) was a contemporary of Tiffany whose designs, at least for a few details, remained in favor for many years. His first refrigerator car patent was issued on March 27, 1877 (No. 189,001). It covered internal air ducts and under-the-floor axle-powered fans, details that appear to have had no measurable influence on car builders of the day. But within a few years Wickes had devised a sheet-metal ice bunker that became popular (Fig. 4.41). The details of the design, as it evolved, were described in the 1895 *Car Builders Dictionary*:

There are two ice tanks in each cooling compartment. These tanks are constructed on an oak frame work to which are nailed in vertical and horizontal rows, galvanized iron strips 2 inches wide interwoven in the manner of basket-work. Projecting outward from these strips 2 inches are galvanized iron leaves which largely increase the cooling surface. These tanks are separated from one another, from the jacket and from the walls at the sides at the end of the car by air spaces of about 4 inches. They are supported by 2″ × 4″ oak grate bars 2 feet from the floor. Beneath the bars are many rows of galvanized iron wire, crossing and recrossing from side to side of the car. A sloping bottom or apron of galvanized iron at the bottom of the jacket leads the drip water to the wires. There is another apron of galvanized iron in front of the wires extending to within 12 inches of the floor. On the floor, directly

FIGURE 4.37

The National Despatch Line was a large operator of Tiffany cars. This train carried lemons and oranges between Boston and Chicago in four days and three hours in early February 1884. Each car measured 29 feet long inside and had a capacity of 20 tons. (Smithsonian Chaney Neg. 13904)

FIGURE 4.38

End cross-section of a Tiffany car with a V-shaped overhead ice bunker. Note the dead-air insulation spaces under the floor. Hair and sawdust were common insulation materials of the day. (*Railway Review*, March 31, 1888)

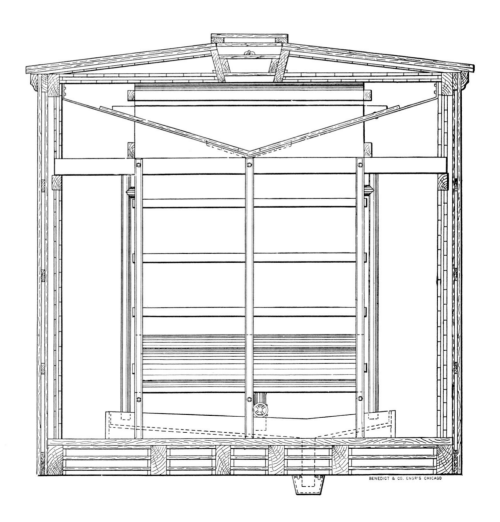

FIGURE 4.39

End detail of longitudinal cross-section of a Tiffany car, showing details of drainpipe, tanks, and plug-type ice-bunker roof hatch. (*Railway Review*, March 31, 1888)

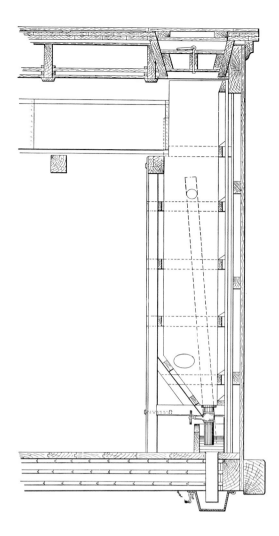

FIGURE 4.40

This 15-ton Milwaukee Road Tiffany car carries a date of 1894 on its exterior, but the prefix letter *A* suggests that this is an alteration date and not the year of construction, which is more likely around 1880. The small size, shallow truss rods, and hand brakes on only one pair of trucks indicate a pre-1890 design. (Kalmbach Publishing Company Collection, Milwaukee)

under the wires, is the drip-pan with a properly trapped drain at each end. The drip water falls from the ice through the grate bars on to the wires and down into the drip-pan. The warm air enters the cooling compartment through the opening at the top of the jacket and descending as it cools comes in contact with the ice, the metal surface of the tanks, the wires, and the spray of drip water about the wires, and re-enters the car through the opening below the apron in front of the wires, having become cooled, dried and purified. Each tank is iced through an opening in the roof, provided with an inner and outer door, each properly insulated.[80]

The basket straps were generally 2 inches wide with 1 inch of open space between them.

Merchants Despatch Line favored Wickes's design (Fig. 4.42). Drawings for one of the line's cars are given in Figures 4.43 and 4.44. This car was built in 1895, a few years after the unit illustrated in Figure 4.42.[81] The New York Central's standard plans for a 35-foot-long 30-ton boxcar were followed, except that the body was made 8 inches higher and 1 foot longer. The drawings are unusually complete and thus record not only Wickes's peculiar features but those more typical of ordinary wooden reefers of the time. Note the use of pressed-steel Fox trucks. While the Merchants Despatch cars were used in freight service, the Erie built large 50-foot-long cars with clerestory roofs in 1886 at its Susquehanna shops for express train service.[82] They look like baggage cars and were cut

FIGURE 4.41

James H. Wickes devised a metal ice bunker that remained popular for many years. This engraving represents a Wickes bunker used by the Erie for the series of express reefers represented by Figures 4.45 and 4.46. (*Railroad Gazette*, Feb. 4, 1887)

up into three compartments (Figs. 4.45, 4.46). In the center was a room for the expressman. At each end were insulated refrigerated compartments fitted with Wickes bunkers.

In 1887 the *Railroad Gazette* stated that nearly two thousand Wickes cars were in operation. The firm built cars for lease or licensed others to build cars using its patents. By 1898 around ten thousand Wickes cars were in service.[83] It is possible that Wickes-type bunkers were being produced in the 1920s, for the 1928 *Car Builders Cyclopedia* contains drawings showing the old basket-weave bunker without specifically mentioning whether they were of Wickes's pattern.

A year before the Wickes and Tiffany patents were granted, John M. Ayer of Chicago suggested a simple but practical improvement for refrigerator cars. He proposed a thin sheet of rubber to form an air seal and thus block the flow of outside air that

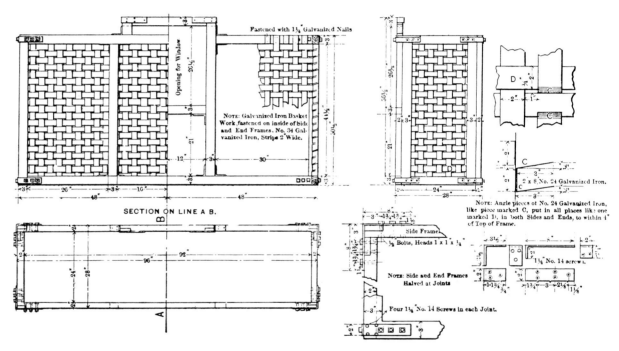

FIGURE 4.42

Merchants Despatch was one of many refrigerator car operators to adopt Wickes's designs. This white car was built by Pullman in 1889. The application of air brakes to freight cars at this date is noteworthy. (Pullman Neg. 664)

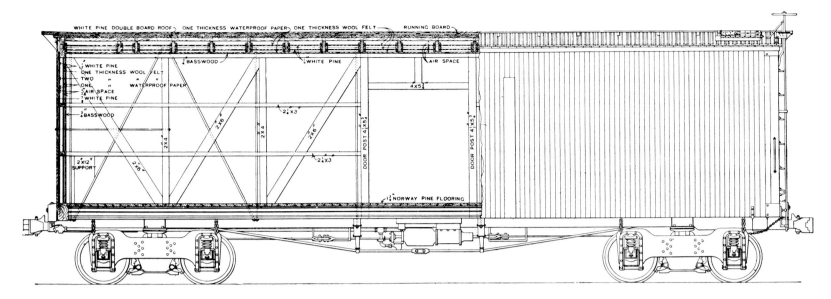

FIGURE 4.43

A standard New York Central boxcar design was modified by Merchants Despatch for this 35-foot-long 30-ton-capacity Wickes refrigerator car. (*Railroad Car Journal*, Nov. 1895)

FIGURE 4.44

The end cross-section of a Merchants Despatch 30-ton refrigerator car shows the continuing dependence on layers of paper, wood, and air spaces for insulation. (*Railroad Car Journal*, Nov. 1895)

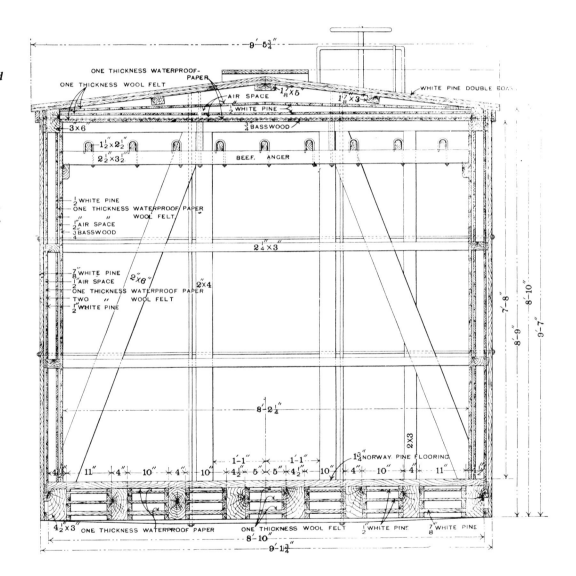

CARS FOR FOOD

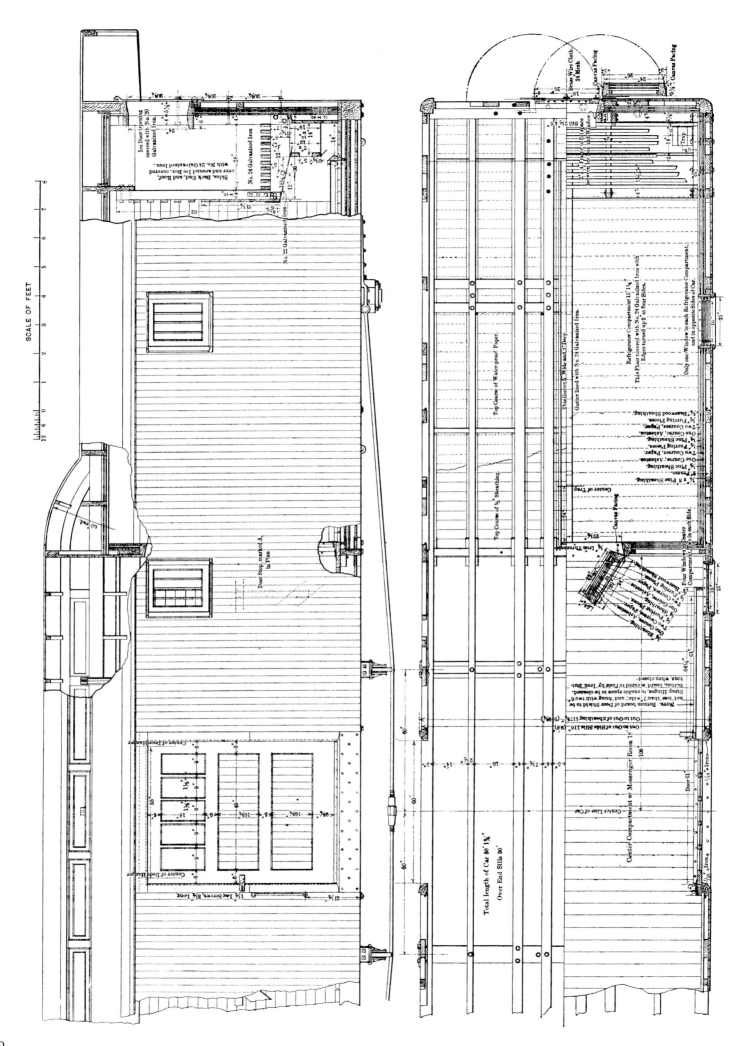

SCALE OF FEET

280

FIGURE 4.45 (*opposite page*)

This Erie express refrigerator car of 1886 had Wickes ice bunkers. The line built around fifteen of these cars at its Susquehanna shops. (*Railroad Gazette*, Feb. 4, 1887)

FIGURE 4.46

End-view details of the Erie's express refrigerator car illustrate the modest insulation applied to even the best reefers of this period. The dependence on layers of paper, wood, and dead-air spaces was considered wholly inadequate by 1900. (*Railroad Gazette*, Feb. 4, 1887)

so easily penetrated the numerous cracks found in any wooden car body. No matter how thick the insulation, the introduction of warm outside air could ruin its effectiveness. In the fall of 1876 the Illinois Central shops produced two test cars on Ayer's plan.[84] A very detailed description of this type of construction was given in a report on the U.S. Centennial prepared by the celebrated British engineer Douglas Galton:

> The Ayer car is about 28 feet long; the outer appearance is that of an ordinary box car, but with a deeper roof. The non-conducting material used is indiarubber sheeting 1/32 inch thick, and paper board 1/4 inch thick, which has been compressed under a pressure of about 45 tons to the square inch. The general construction is briefly as follows: The floor consists of planks laid upon the transoms grooved and tongued together; upon these boards are laid, first indiarubber sheeting. The joints of the indiarubber sheeting have at least 3/4 inch lap, and every lap is thoroughly cemented with rubber. The sheets of paper board are squared at the edges and closely butted together. When the rubber has been laid and fastened, the flooring boards are put down over it. These are sound seasoned Norway pine 1 1/8 to 1 1/4 inch thick, grooved and tongued. In order to form the sides, when the framing is complete, sheet indiarubber is tacked on the inside of the posts, the joints having the same lap as already mentioned, and being made air-tight with rubber cement. The lower edges of the rubber lap 1 1/2 inch over the rubber of the floor, and are secured in the same way. After the sides and ends have been covered with the indiarubber sheeting a layer of paper board, squared and butted is tacked to the

posts over the rubber, so that each joint of the paper boarding shall have a timber backing; over the paper board a second coat of indiarubber sheeting is laid as before; and, after this second coating of rubber has been fixed, the insides and ends are ceiled with seasoned ash timber from 1/2 to 3/4 inch thick. The roof has a double ceiling under it, each ceiling being made of non-conducting material. The first or inner ceiling is made by nailing paper board to the underside of the rafters, indiarubber being laid under the paper board with lap joints, and secured to the sides of the car under the indiarubber; the lining for the inside of the car is made of seasoned ash match-boarding 1/2 to 3/4 inch thick. The second ceiling is formed of match-boarding nailed to the upper side of the rafters. The boards are covered first with indiarubber, then with paper board, next with indiarubber again, and finally with the roof boards. The sides and ends of the car are covered on the outer side with seasoned pine 3/4 inch thick, and not more than 5 1/2 inches wide. The doors resemble the walls of the car and are lined with indiarubber and paper board. The door frames are bevelled outwards, and double rabbeted all round to receive a strip of thick rubber to form the joint. The doors are bevelled the same as the frame, and the joint where the two doors come together is bevelled, rabbeted, and provided with thick indiarubber to form the joint as in the door frames. The door fastenings have a bolt with a wedge arrangement which presses the doors more tightly against the indiarubber joint the harder the bolt is pressed home. Ventilation between the outer and inner covering of the car is rendered continuous by means of grooves cut in the posts. At each end of the car inside is an ice-box, made with as little metal about

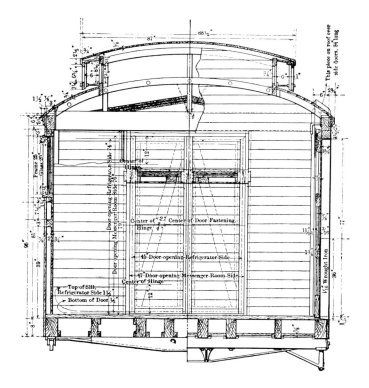

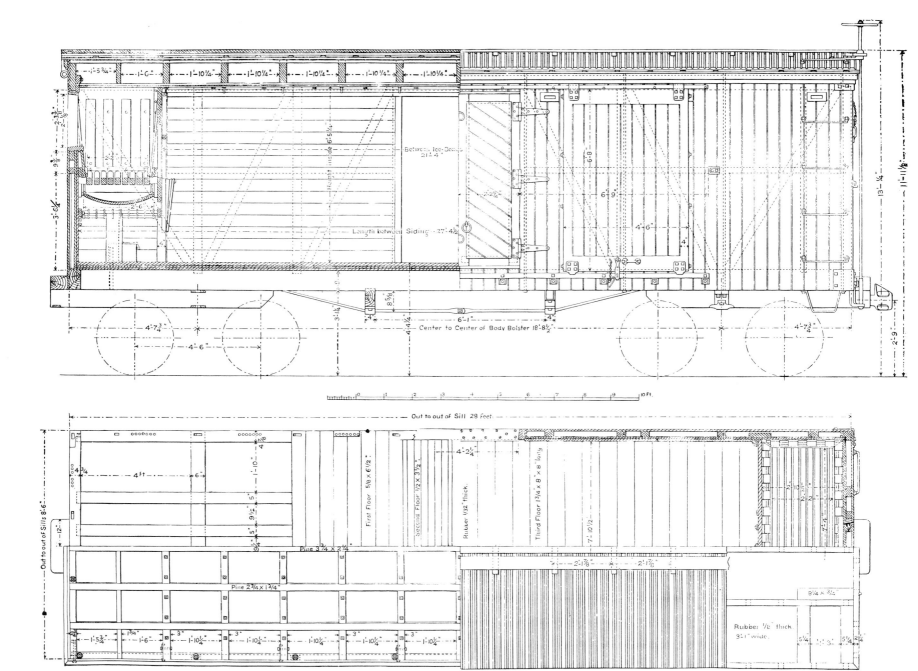

FIGURE 4.47

John M. Ayer attempted to improve car insulation by using layers of sheet rubber in place of paper. However, his dependence on air spaces must have largely defeated his efforts. These drawings, published in 1884 for a Burlington line car, are the most complete available for a reefer of this era. (*Railroad Gazette*, Aug. 15, 1884)

it as possible. The space between the floor of the car and of the ice-box is 3 feet 2 inches, sufficient to permit an ordinary barrel to stand under the ice-box. The bottom of the box containing the ice is grooved, to allow the water from the ice flowing to gutters leading to indiarubber pipes fixed at the side of the car. These pass through the bottom of the car and carry the water to a closed trap, arranged to remove the water from the car and to prevent ingress of air. Protected doors at the end of the car allow of ice being put into the boxes without opening the body of the car.[85]

The efficiency of the rubber-insulated cars was shown by the number of roads that were purchasing them. The Erie, the Baltimore & Ohio, the Chicago, Burlington & Quincy, the Chicago &

North Western, and the Illinois Central were all acquiring Ayer's cars.[86] The claims for good insulation and dry interiors were said not to be exaggerated, and unlike cars equipped only with conventional absorbent insulation, the elastic yet waterproof rubber liner did not become soggy, musty, or sour. The contents of the car were likewise not subject to tainting. A fine drawing of an Ayer car, and one of the most complete available for an early-generation reefer, was published by the *Railroad Gazette* in 1884.[87] It shows a 23,000-pound car built by the Burlington shops (Figs. 4.47, 4.48). The description accompanying the drawing noted that double-lined cars of this type cost from $750 to $800, single-lined cars from $650 to $700.

There were other patentees hoping to rival the success of Tiffany, Wickes, and Ayer, and while a

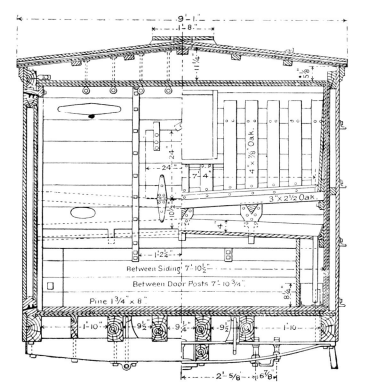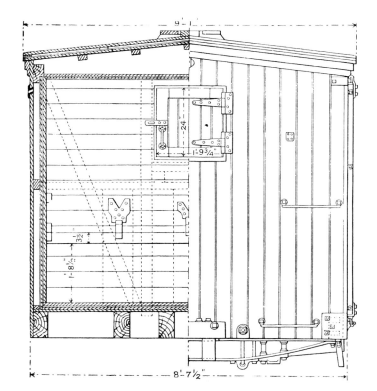

FIGURE 4.48

Ayer's rubber-insulated car featured a particularly awkward and impractical end ice-bunker door, which must have presented problems for the hapless icing crews. (*Railroad Gazette*, Aug. 15, 1884)

complete summary of their efforts would fill more pages than warranted by the results of their labors, a brief mention of a few of the minor contenders is justified. Two of these figures were actually rather well known car builders of their day—Zimmerman and Canda—although today they are obscure names, indeed even to students of railroad history. Arnold W. Zimmerman cannot be discounted as an impractical inventor. He was general manager of the Swissvale Car Company, founded in 1873 specifically to build refrigerator cars of his design. Advertisements for his first plan of car began to appear in 1878 (Fig. 4.49). His claims for superiority were positive, and as he added more improvements, Zimmerman became absolute in the opinion that his cars were the "most perfect" in use.[88] By means of tubes and circulating chambers he hoped to promote air movement within the car and so remove dampness and odors. In 1882 he claimed that four hundred cars of his design had been produced.[89] They were used by the Anglo-American Packing & Provision Company and the Chicago & Alton Railroad. Advertisements and notices appearing as late as 1888 showed that Zimmerman continued to promote his ideas for at least a decade.[90] His plan received wider application through the efforts of Gustavus Swift, who grew increasingly discouraged by the performance of the Davis-style refrigerators. In 1878, the same year of Zimmerman's patent (No. 199,343, Jan. 15), Swift hired an engineer named Andrew Chase to develop a more reliable carriage for fresh meat. Chase borrowed some of Zimmerman's ideas for natural circulation inside the car and added a few notions of his own. The improved car was so good

that Swift concentrated on dressed beef, and by 1881 he was sending three thousand carcasses a week to the Boston area. His Chicago-based car fleet numbered nearly two hundred.

F. E. Canda owned or operated several car plants and was once associated with the Ensign Car Company of Huntington, West Virginia. His car-building experience dated back to at least 1872. A daily association with car construction naturally suggested small improvements possible in all types of cars, including reefers. The changes offered by Canda were not remarkable, and about the only notable ones were the use of four doors and the unusual sliding doors held in place by toggle-lever locks. Cars designed especially for the California fruit trade in 1894 were produced on Canda's plan for the Southern Pacific, but no further reports have been found of a widespread clamoring on the part of other lines for more equipment of this design.[91]

In the summer of 1890 an inventor named J. F. Hanrahan introduced a curious three-compartment car that contained both good and bad features. He placed the ice bunker in the middle of the car. This eliminated the separate bunkers and, in the process, two roof hatches. The separate cold rooms on either side of the bunker were ideal for less-than-carload lots. The central ice bunker was said to offer a better distribution of cold air, and the truss panel at the car's center in place of a door did much to strengthen the structure. However, the four doors were a mischievous source of air leaks, and because of the extra furniture involved, the design was undoubtedly more expensive and costly than the simpler single-compartment style

FIGURE 4.49

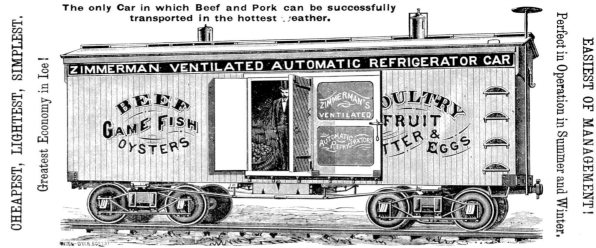

of car. Even so, the idea was intriguing enough that the Michigan Central, American Express, Hammond Packing, and several other firms bought cars on this plan.[92] Pullman built a number for the Colorado Midland in 1897 that were handsomely decorated (Fig. 4.50).[93]

Milk, Dairy, and Heater Cars

The three food-handling rail vehicles discussed here are all related to the refrigerator car. Some were in fact identical to standard reefers. These cars were never large in numbers, but they were often important in terms of revenue.

During the pioneer period of American railroading, most communities received their milk directly from local farms.[94] Because distances were so short, rail transit was not required. An exception to this general rule was Manhattan. The concentration of population and spiraling land values tended to push dairy farms beyond the practical limits of wagon delivery. The city's milk was produced by stabled cows fed refuse from distilleries and breweries. The quality of the milk was poor compared with that produced by grass-fed herds, and adulteration made it worse.

The Erie recognized the traffic potential offered by these special circumstances and began to transport fresh milk to Manhattan in the spring of 1842 from Orange County, New York, the center of a rich dairy region famous for its butter. During the first year about 150,000 gallons were brought to Manhattan. The market for wholesome country milk flourished, and during the second year 1 million gallons made the 80-mile trip from Orange County. By 1845 the Erie was receiving 40 percent of its gross freight revenue from the carriage of

milk. The traffic and revenue continued to expand so that by 1900, twenty-five cars were in daily milk service transporting around 18 million gallons a year. Some milk was transported from as far as 350 miles away. When this service began in 1842, no special cars or equipment were employed. Milk was carried in wooden churns in ordinary freight cars. The trip was fairly short, and all went well during cold weather, but when the warm season began the milk tended to sour before reaching the market. Precooling solved the problem. In July 1843 it was reported that tin tubes filled with ice were being inserted into the containers.

Once the pattern was established, other neighboring railroads sought to copy the Erie's success. Soon every railroad in the area was hauling milk. In 1862 the New York & Harlem, determined to gain a larger share of the milk market, opened a new dairy depot in Manhattan. The Harlem line grossed $270,000 per year from the milk trade, or approximately half of what was received by all milk-carrying railroads entering the city. Between June 1860 and June 1861 nearly 13.8 million gallons were delivered to New York City. The Harlem carried it from as far as 128 miles away. Because transit time varied between forty-eight and sixty hours, the milk was precooled to 40 degrees and carried to the station in large tin cans draped with wet cloths. Trains ran at night to avoid the sun's heat. The New York, Ontario, & Western showed remarkable leadership for such an impecunious line, surrounded by such financial giants as the Erie and the New York Central, when it began to operate solid milk trains in February 1871. It built up this traffic to a very respectable level, so that in 1902 it was the largest single carrier of milk for the New York port area. In August of that year

FIGURE 4.50

J. F. Hanrahan received eight refrigerator car patents between 1884 and 1891. This is one of several central-ice-bunker cars built for the Colorado Midland by Pullman in 1897. (Pullman Neg. 3760)

it delivered 1.5 million gallons out of a total of 8.7 million. The other major carriers involved in that trade were the New York Central, the Erie, the Lackawanna, and the Lehigh Valley, in that order.

The traffic growth was most encouraging, for milk consumption was growing faster than the city's population. In 1886 over 5 million cans (10 gallons each) were delivered. But in 1907 15 million cans were carried, even though the population had increased by only 98 percent. Such growth, together with the prospect for even more expansion, encouraged railroads to buy more and better cars.

It also encouraged faster and more frequent trains. Yet the very term *milk train* is understood to mean exactly the opposite. One thinks immediately of the legendary slow train through Arkansas whose progress was so leisurely that its occupants were not sure if it was going forward or backwards. The milk train stopped at every station, platform, crossing, junction, and siding. It held precedence over no other train, not even the slowest freight. Its creeping pace and ancient equipment made it the subject of popular humor. It is a fact that because creameries were located just a few miles apart, frequent stops were a necessity. But once beyond the dairy country there was no need to stop, and the pace would quicken remarkably. On the Delaware & Hudson, for example, milk trains stopped frequently between Binghamton and Albany (143 miles), but once the cars were exchanged with the New York Central, the milk cars would jingle and dance as the train sped down the Hudson River Valley toward Grand Central. For this reason milk cars were generally fitted with passenger-style trucks. If fast running was not in their normal schedule, the added weight

and cost of high-speed trucks could be saved, and so a degree of fast running was a certainty.

Most milk cars were essentially baggage or express cars. The earlier cars do not appear to have been insulated or fitted with ice bunkers. Only the shelving for milk cans and the exterior lettering made them different from sister cars used for express or baggage. One of the earliest descriptions of a milk car, which is all too brief, was published in 1877.[95] We are told only that several 35-foot-long cars were being built at the Erie's Jersey City shops. Each would carry two hundred cans, and adequate ventilation was provided. An engraving appeared in the 1879 *Car Builders Dictionary* that shows a car with six ventilators on each side. That it depicts one of the Erie cars is only a speculation, but it is most likely similar. It looks like a cross between a boxcar and an express car. Its square body, peaked roof, and sliding side doors give it the look of a freight car, while the overhanging roof, end platforms, and trucks suggest passenger car origins. The plain rectangular windows are inspired by a caboose.

Some fifteen years later a similar car was produced for the New York, Susquehanna & Western by Jackson & Sharp (Fig. 4.51). It is somewhat longer and has no side-wall vents or windows. Two large roof ventilators are visible, and the windows are placed at the ends rather than on the side walls. The doors are recessed, revealing thick body walls that suggest heavy insulation. A similar car built for the Lehigh Valley, also by Jackson & Sharp at about the same time, is recorded in a photograph preserved by the Delaware State Archives, Dover. It differs slightly from the Susquehanna car. The roof has a very high arch, which gives it a distinctive appearance. The roof hatches are of more con-

CARS FOR FOOD

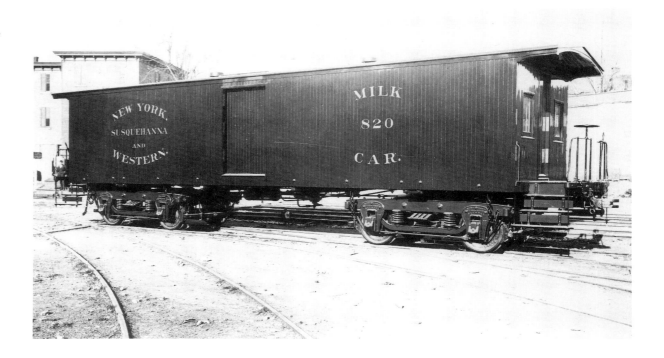

FIGURE 4.51

Jackson & Sharp of Wilmington, Del., manufactured this milk car in 1892 for the New York, Susquehanna & Western. (Hall of Records, Dover, Del.)

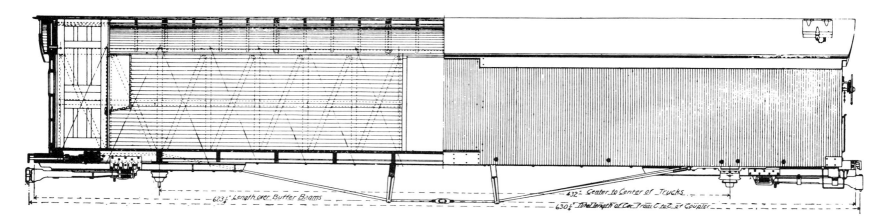

FIGURE 4.52

This arch-roof milk car for the Lehigh Valley Railroad dates from 1900. (*Railway Review*, Feb. 17, 1900)

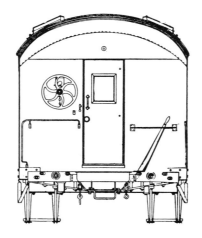
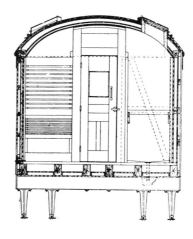
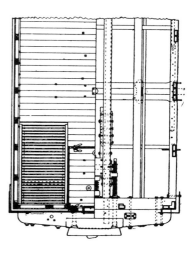

FIGURE 4.53

An express milk car produced by Pullman for the Rutland Railroad while under the New York Central's control in September 1908. (Pullman Neg. 10984)

sequence, because they denote the presence of ice bunkers. This car is lettered Farmer's Dairy Despatch and so may have been intended more for cheese than for milk.

Drawings for a larger Lehigh Valley car dating from 1900 are reproduced here as Figure 4.52. This car, which is specifically identified as a milk car, has the same high arch roof.[96] The car had no vestibules or platforms, a practice that had become common for many head-end cars by the late 1890s. This saved weight and cost and helped reduce train length, which could be a critical factor at stations with short platforms. End doors were installed so that train crews could pass through if required, but loading was done at the side doors. Ice bunkers were serviced from roof hatches. One interesting feature not shown by the drawings was the Fox pressed-steel passenger-style trucks used on these cars. Fox freight-style trucks were relatively common, but the passenger variety was far more rare. The New York Central purchased some platformless milk cars for its then-subsidiary, the Rutland, in 1908 (Figs. 4.53, 4.54). They were about the same size as the Lehigh Valley car just described but had a less pronounced arch roof, four side doors, two side windows, and no ice bunkers. Floor-level ventilators can be seen in the exterior view, but they are not evident in the interior photograph. The interior is open and unfurnished, but it was not uncommon for milk cars to have shelves to increase the can capacity. Some shelves were made with shallow pans to hold ice. Examples can be found in the 1895 and 1898 *Car Builders Dictionary* (see Figs. 271–273c).

Milk cars could also be full-scale passenger cars complete with clerestory roofs, six-wheel trucks, and vestibule ends. The Baltimore & Ohio built around eighteen such cars starting in 1887. Drawings for a 60-foot-long car show what appears, externally, to be a conventional baggage-express car.[97] However, about one-third of the body is segregated as an insulated room with overhead ice bunkers. A photograph of one of these cars, as remodeled, is shown in Figure 4.55. The side windows have been covered over, the vestibules enclosed, and the decorative arch door squared off. This picture was probably made around 1915. The last cars in this series were retired by 1933.

There was yet another variety of milk car that stood in contrast to the B&O cars just described. They looked like nothing more than an elongated boxcar. Drawings for an Erie car of this style show a peaked-roof car with a plain boxcar-like body and reefer doors.[98] Only its length and ends give it away as something special. Short platforms, windows, and doors show it to be a hybrid. This car was designed to carry 16 tons, or 288 10-gallon cans. Drawings for an even more extreme design were published in 1902.[99] Although constructed for the Lehigh & Hudson, it was designed by W. H. Davis of the New York, Ontario & Western (Fig. 4.56). It had insulated double walls and passenger car trucks. The purpose of the small end doors is uncertain. They may be for ventilation. They are too small and set too high for easy crew access. The loading of ice and milk cans could be done far more easily through the side doors. The NYO&W had some milk cars like the one shown here, which were photographed in service during the 1930s. By this time Andrews freight car trucks had

Interior of the Rutland milk car
shown in Figure 4.53. (Pullman
Neg. 10985)

FIGURE 4.55

The B&O express milk car No.
820 is shown late in its service
life. The car was built in 1887.
(Smithsonian Neg. 50727)

THE AMERICAN RAILROAD FREIGHT CAR

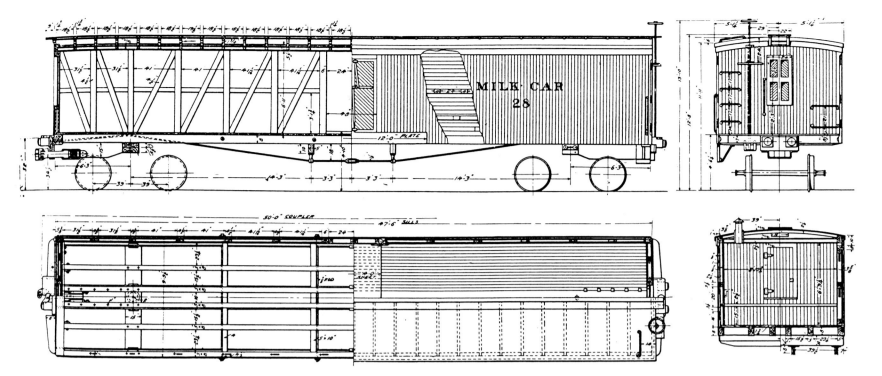

FIGURE 4.56

Very plain boxcar-like milk cars were operated by the Lehigh & Hudson. (*Railway Age*, Oct. 3, 1902)

replaced the original passenger-style wheel sets.

The mass movement of milk in 10-gallon cans remained a standard during the nineteenth century. Such containers were subject to rough handling on the farm, on the trains, and at the dairy. They were, accordingly, made of thick-gauge metal. When filled they were about as much as a single man could handle. But clearly this was not an effective way to handle liquids being transported in such vast quantities. It was labor-intensive. And there was the cost of individually cleaning so many small containers. The cans were also subject to loss, damage, and theft. It seems surprising that no one thought of bulk shipping. Tank cars were a common sight at every terminal and siding. Yet it wasn't until 1903 that a Boston milk dealer began bulk milk shipping. A porcelain-enamel-lined steel tank holding around 3,200 gallons was mounted inside a Boston & Maine refrigerator car.[100] The car was a success, but the idea was not widely accepted for another 20 years.

Milk was not the sole product dairies sent to market. Cheese, butter, eggs, and even dressed poultry were all classed and shipped under the general heading of dairy products. Butter, as already noted, was among the earliest prepared foods to go by rail. It moved over the Erie as early as 1842. Jonas Wilder was inspired to develop a refrigerator car just a few years later for the sake of this same commodity. In August 1873 the Blue Line, a private care operator, was sending five to seven carloads of butter per day to New York and Boston from Illinois and Wisconsin.[101] Refrigerator cars were able to maintain their golden load at 40 degrees. Transportation costs were figured at $250 per carload, which was considered cheap for a ship-

ment valued at $5,000. Good profits encouraged this trade to grow. Chicago became a shipping center for eastbound dairy products, and in 1903 that traffic had reached these impressive annual levels: butter 100,000 tons, eggs 60,000 tons, dressed poultry 30,000 tons, and cheese 10,000 tons.[102] A single refrigerator car carried 144,000 eggs. Refrigerator cars were assigned to this traffic and were boldly lettered as dairy cars. A good example is found in the Union Line's car No. 6699, a Pennsylvania 25-ton-capacity vehicle manufactured at the line's Fort Wayne shops in March 1890 (Fig. 4.57).

Shipping food seems to center on devising ways to lower temperatures, but in actuality *controlling* temperatures was more often the goal. Under favorable conditions produce could be shipped unrefrigerated. During spells of mild weather, fruits and vegetables could travel quite safely in a refrigerator car with the ice hatches open for ventilation. Certain commodities like canned goods, bottled beer, or breakfast cereals required protection from temperature extremes and were thus safely transported in insulated boxcars that had no ice bunkers. Yet other foodstuffs, particularly fresh ones, were far more fragile and required special care. Low temperatures could be as destructive as high ones, and railroads serving northern or mountainous regions saw the need for heated cars during the winter season. Sustained subzero temperatures were common enough to ruin carloads of tomatoes, melons, and even potatoes. Small heaters, fueled by alcohol, kerosene, or charcoal, were used to raise the temperature inside an insulated car to above freezing. The operating department of every railroad engaged in this traffic had

FIGURE 4.57

Dairy cars such as this 1895
Union Line vehicle carried milk,
cheese, and eggs. (Smithsonian
Chaney Neg. 26319)

FIGURE 4.58

An Eastman heater car protected
perishables from freezing during
extreme winter conditions.
Fifteen hundred were in service
when this engraving was
published in 1890. (Walter D.
Crossman, ed., *The Official
Railway List* [Chicago, 1890])

Automatically Heated and Ventilated
FREIGHT CARS.

The only cars that will protect fruit and vegetables, from damage by frost
in the coldest weather, and the

Best Summer Ventilated Cars in Service.

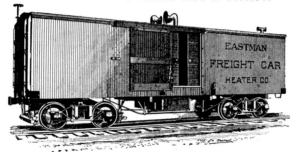

to be prepared to act quickly in the event of sudden
weather changes. In just a few hours they might
have to shift from ice to heaters or face the loss of
hundreds of tons of valuable produce.

The introduction of heaters appears to be yet another unrecorded event in American railroad history. It was so obvious a need that it must have
been adopted soon after the first sending of
produce by rail. Jonas Wilder, mentioned earlier in
this chapter, employed freight car heaters in the
1850s. Whether he introduced the idea remains a
question.

In the early 1880s William E. Eastman of Boston
introduced a special car aimed at protecting fruit
and vegetables from the winter's frost.[103] His special concern was the Maine and eastern Canadian
potato crop, much of which was shipped in the
winter. Four kerosene heaters fired a hot-air furnace mounted in a box below the floor; air ducts
distributed the heat inside the body (Figs. 4.58,
4.59). The idea was not to raise the temperature to

toasty warm but simply to hold it to around 35 degrees, or just enough to prevent freezing. A 35-gal-
lon tank provided fuel for ten days. The body was
insulated, and large roof ventilators were installed
to ensure a proper airflow. By early 1884 three hundred Eastman cars were serving New England.
Some cars ran as far south as New York City. Plans
for a route west to Chicago were under study. The
Eastman car was clearly a success, for its use
spread to the major eastern trunk road. In 1887
over one thousand were in service, and at least a
few were in the transcontinental fruit service.
Within a decade Eastman was operating sixteen
hundred cars. Many railroads contrived their own
heater cars by temporarily installing small heaters, usually charcoal-fired, in the ice bunkers of
conventional refrigerator cars. Eastman's semiautomatic system, though more costly, was superior to such primitive manual operations, particularly when long journeys were involved.

Ventilated Cars

Shippers found that certain produce cargoes required neither heat nor cold. Nothing more than a
good air supply was required for many fruits and
vegetables to reach their destination in market-
able condition. Most fruit is picked before it is
fully mature, and so it gives off heat and gas as the
ripening process continues. If it is housed in a
poorly ventilated enclosure the self-generated heat
and gas will cause it to spoil. Good air circulation
can be induced in a common boxcar by equipping
it with open-barred side doors and end or side ventilators. Cars of this type were introduced in the
pioneer period, as noted in Chapter 2. They were
fairly popular because they could be used for a va-

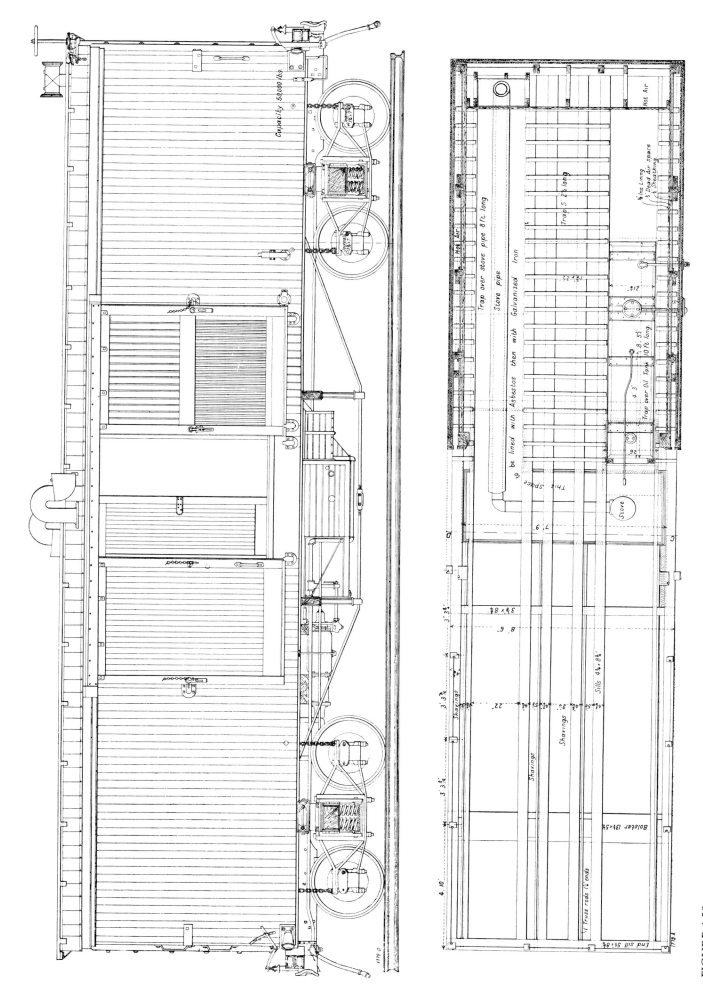

FIGURE 4.59

Details of Eastman's kerosene-fired heater apparatus. (James Dredge, *Record of the Transportation Exhibits at the World's Columbian Exposition of 1893* [London, 1894])

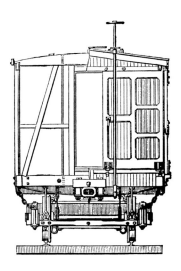
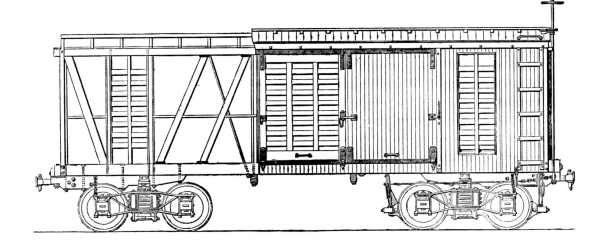

FIGURE 4.60

Hannibal & St. Joseph ventilated boxcar dating from 1871. (*National Car Builder*, April 1871)

riety of shipments. They were a makeshift but adequate cattle car, and when the second set of solid doors was closed, they became an instant boxcar ready to handle merchandise. The special equipment consisted of nothing more than a second set of doors that were not particularly expensive to purchase or maintain. The ventilated box became the most common and successful convertible car in American railway practice. Refrigerator cars could also be turned to ventilator service by opening the roof hatches, and a great deal of produce was shipped in this manner without benefit of ice.

The traffic potential for ventilated cars was realized early in a nation blessed with abundant orchards. Local needs were easily satisfied, and there was often so much fruit that it was left to rot. Transporting out-of-season produce from the South and West was a challenge the railroads were ready to accept. The transcontinental railroad began through shipments from Sacramento to Omaha in August 1869, just months after the union of the Central and the Union Pacific at Promontory, Utah.[104] Through shipments to eastern cities apparently did not begin until June 1887, when a Central Pacific fruit car arrived at Jersey City loaded with plums, apricots, and peaches. A year or two later, Carleton B. Hutchins organized the California Fruit Transportation Company and began large-scale shipments to East Coast cities. Some of Hutchins's produce went to England. By 1895 around forty-five hundred carloads were leaving California each year, and within ten years the number had more than doubled.

In the East the Philadelphia, Wilmington & Baltimore was hauling 76,550 tons of fruit annually by 1875. Peaches were the main crop carried, and they made up 9,077 carloads. Each car had four levels or decks for fruit baskets. The trains were run at night, if possible, to avoid the sun's heat. The ventilated cars had double insulated roofs as an added precaution. In 1870 the now-famous Georgia peach was rarely seen in the North, and as late as 1889 only 150 carloads per year were being shipped

out. Nine years later exports reached 1,650 carloads, and by 1905 a full five thousand carloads were leaving Georgia. Yet this was only a small portion of the southern produce going to market by rail, for as early as 1888 some three hundred thousand carloads were being moved. Ventilated cars were almost never found idle in a southern freight yard between February and August.

The Hannibal & St. Joseph Railroad offered plans early in 1871 for a combination box/cattle/produce car that featured very large side and end ventilator openings.[105] In place of the usual ironbar wooden-frame grill, this series of cars had shutterlike movable wooden louvers (Fig. 4.60). The cars measured 28 feet 4 inches long by 8 feet 8 inches wide and weighed 9.5 tons each. The absence of truss rods and the single-end brakes were normal for the period. Iron bolsters and the Winslow iron underroof were progressive features. Swing-motion trucks made for an easy ride. The truck safety chains were unusual for a freight car. A more detailed assembly drawing was published in 1878 for a Burlington ventilated car designed by the road's equipment chief, George Chalender.[106] It was the same size as the Hannibal car but differed in several details. More conventional ironbar openings were used, which were supplemented by small floor-level cast-iron grills (Fig. 4.61). Chalender's car had truss rods, iron bolsters, and swing-motion trucks. The roof appears to be the common double-board variety. The massive three-hole Potter-style drawbar is a notable feature of this car.

The Central Pacific became something of a patron of ventilated cars because it had so many online shippers requiring equipment of this type. The number of fruit cars developed by the Central Pacific represents a fascinating and varied record. What is surely the oldest surviving ventilated car was built in the line's Sacramento shops in January 1873 for the Virginia & Truckee, a Nevada short line that connected with the Central Pacific.[107] This 28-foot-long vehicle had the usual features

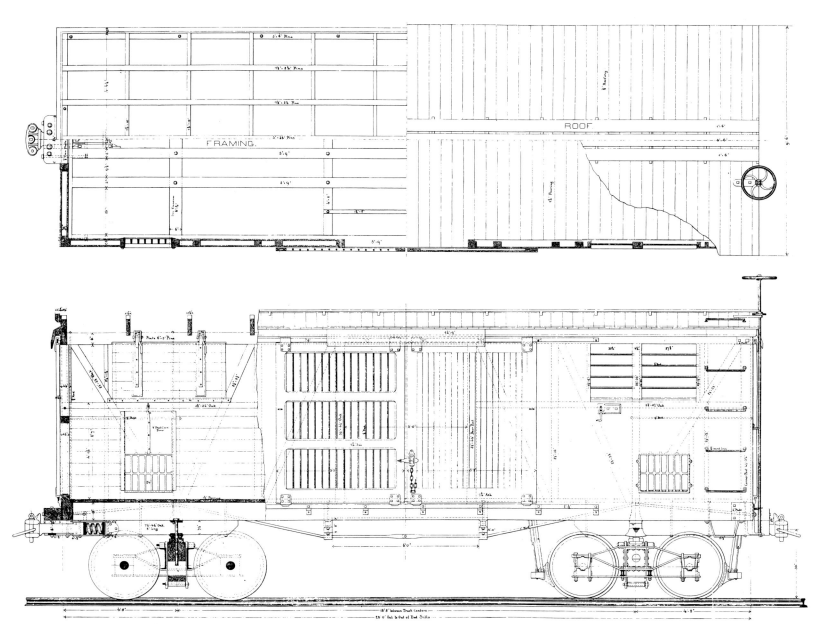

FIGURE 4.61

A Burlington ventilated boxcar designed by the line's equipment chief, George Chalender, in 1878. (*Railroad Gazette*, May 24, 1878)

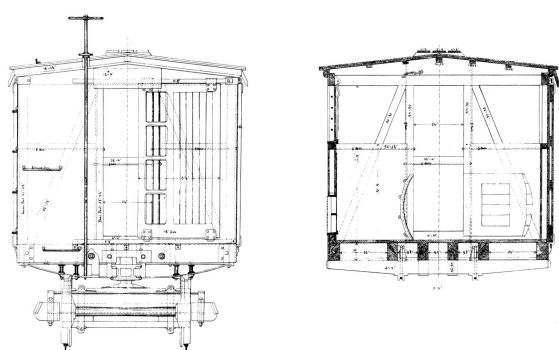

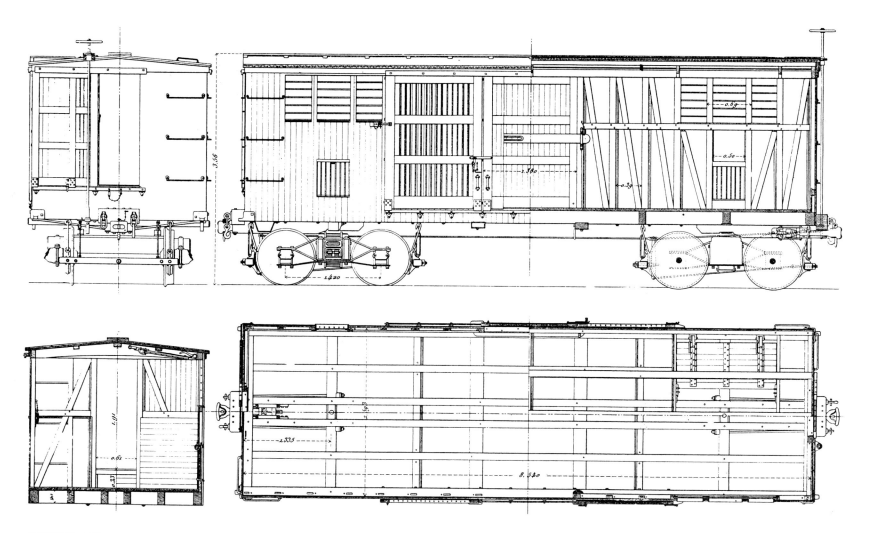

FIGURE 4.62

**A Central Pacific fruit car,
depicted in a French report
published in 1882 but based on
data gathered in 1876. (E. Lavoinne
and E. Pontzen, *Les Chemins de
Fer en Amérique* [Paris, 1882];
metric measurements)**

associated with most fruit cars, except for the absence of external sheathing. In this particular design the body frame was outside, or exposed, rather than being covered and hidden by vertical boards. It weighed 9 tons and carried 15 tons. Like other V&T equipment, the No. 1005 was painted yellow, and like most rolling stock of this silver carrier, it saw an extended service life. Retirement did not come until 1916. It stood as a wheelless toolhouse until 1938, when it was sold to Paramount Pictures for a second career in the movies. In 1971 the relic, badly ravaged by termites and dry rot, was acquired by the Nevada State Museum. The 1005 has since been restored.

The Central Pacific built a great many fruit cars that have not been preserved, and little is known about them save for a few random drawings and photographs. Engravings for a CP ventilated car dating from around 1876 show a rather ordinary-looking vehicle with normal exterior sheathing. It possessed no unusual features for a car of this period except for the safety coupling chains, normally reserved for passenger equipment. This detail suggests that the car was intended for passenger service. The lack of underbody truss rods and the presence of double brakes are also worthy of notice. The French engineers who gathered data

from the Central Pacific gave the plan mixed reviews.[108] They praised the multiple uses possible—cattle, fruit, grain, and ordinary freight. They also liked the roof hatches, which could be used for ventilation, feeding cattle, or loading grain. But they complained that such a versatile vehicle was also heavy, complex, expensive, and maintenance-greedy. In all fairness the draftsmen at Sacramento should not be singled out for special criticism, because the same could be said about most every combine or ventilator car in operation. And since there appeared to be no magical way to create a multiple-use car without some additional contrivance, car builders continued to produce almost identical vehicles over the next several decades.[109]

There was a variant on the boxlike ventilator car that should surprise no one after reading about the hybrid freight/passenger cars created for express refrigerator, milk, and dairy shipments. There was a small but distinct class of ventilator cars with end platforms. Some had canopy roofs, passenger car trucks, and Miller couplers. Cars of this type first appear in two photographs taken by the Union Pacific's official photographer at Promontory, Utah, in 1869.[110] They show several express fruit cars at the loading platform. The lettering on the sides, the barred doors, and the large roof ventilators identify them clearly. Fifteen years later, engravings for a similar Central Pacific car were published in the *National Car Builder*.[111] The presence of Miller couplers is a clear indication that these cars were meant for passenger train operations. That they were also intended for long-distance travel is reinforced by the side lettering: "San Francisco, Chicago, St. Louis & New York Through Line" (Figs. 4.62, 4.63, 4.64). One detail of interest is the padlocks on the end ventilators. These were no doubt necessary to keep tramps or other unauthorized persons from tampering with the setting of the ventilator.

Almost identical cars were found on other western lines. A Union Pacific diagram dated June 1886 in the DeGolyer Library of Southern Methodist University, Dallas, Texas, depicts a fruit express with a 20-ton capacity and a weight of 36,700 pounds. It had passenger car trucks with a 78-inch wheelbase. A photograph of a Northern Pacific fruit express built by the Peninsular Car Company in 1885 is reproduced in Figure 4.65. Air brakes in addition to Miller couplers safely place it in the passenger department's operating plan. A very late version is shown by a Denver & Rio Grande photograph printed in the 1898 *Car Builders Dictionary* as Figure 6.

The great majority of fruit cars were of the common box freight type, and most roads were content to refine old designs as new equipment was re-

quired. A good example of this thinking is the Illinois Central car dating from 1891 shown here in both a photograph and a drawing (Figs. 4.66, 4.67).[112] The arch roof is clearly a relic left over from the 1860s, and the many ventilator doors reveal a penchant for repeating past practice. The small side openings were fitted with solid doors, which allowed more control over airflow and temperature and so must be seen as a small improvement. The length of 37 feet 6 inches shows growth, but the 25-ton capacity seems modest for a car of this size. Light framing may well have been justified if the car was used exclusively for fruit.

J. M. Elliott, president of the Elliott Car Company, Gadsden, Georgia, was inspired to produce something more adventurous than the established design so venerated by industry stalwarts like the Illinois Central.[113] Fruit cars were surely a topic of interest among railroad men and car builders of the South. Elliott's pondering resulted in a patent issued on November 20, 1888. The principal innovation was a double or false floor that was intended to supply air below the load. The secondary floor could be removed when regular freight was the cargo. A sample car was shown at the 1893 Exposition. The Cincinnati, New Orleans & Texas Pacific Railway was impressed enough with Elliott's plan to purchase a number of cars. The 34-foot-long vehicle weighed about 20 tons and carried 30 tons (Fig. 4.68). Elliott's patent caused no discernible revolution in fruit car design, but his idea was

FIGURE 4.63

Details of the Central Pacific express fruit car shown in Figure 4.64. (*National Car Builder*, June 1884)

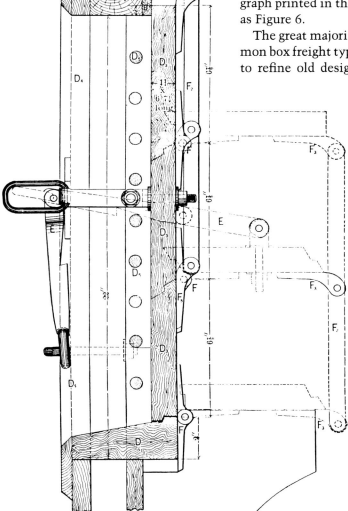

Enlarged Section through Lower Ventilator.

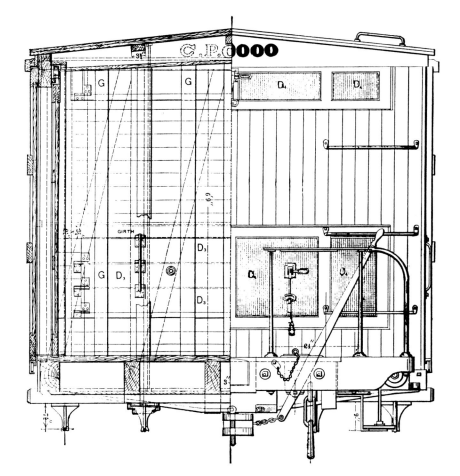

FIGURE 4.64

**Central Pacific express
fruit car. (*National
Car Builder*, June 1884)**

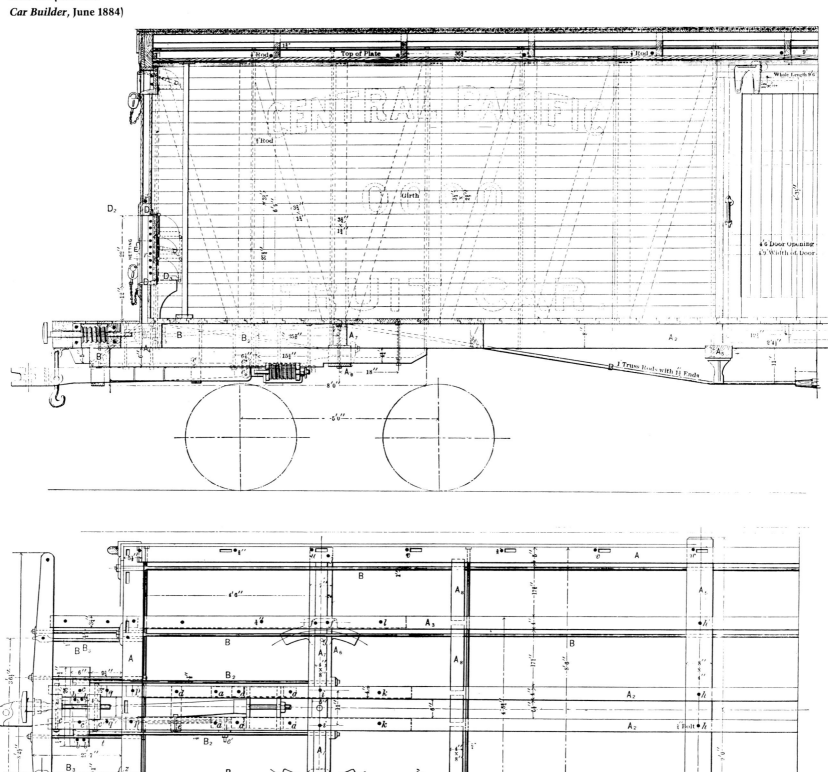

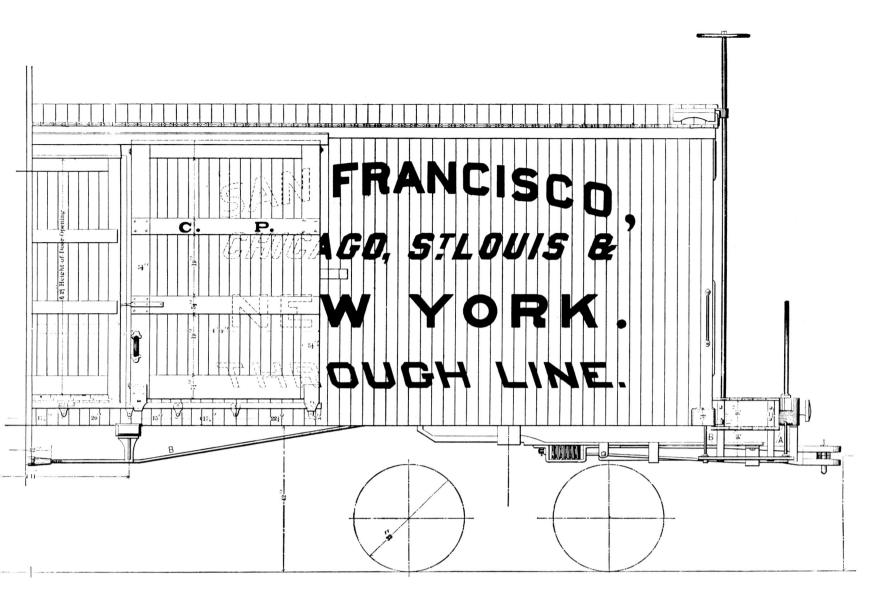

FIGURE 4.65

Northern Pacific fruit express car manufactured in 1885 by the Peninsular Car Co. of Detroit. (ACF Industries Photo)

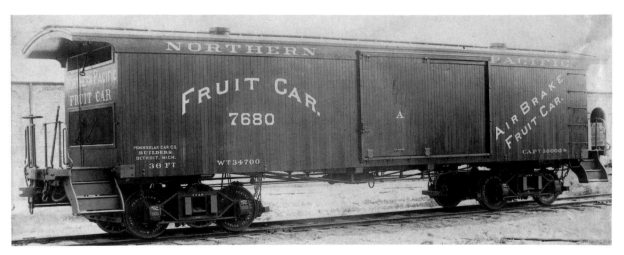

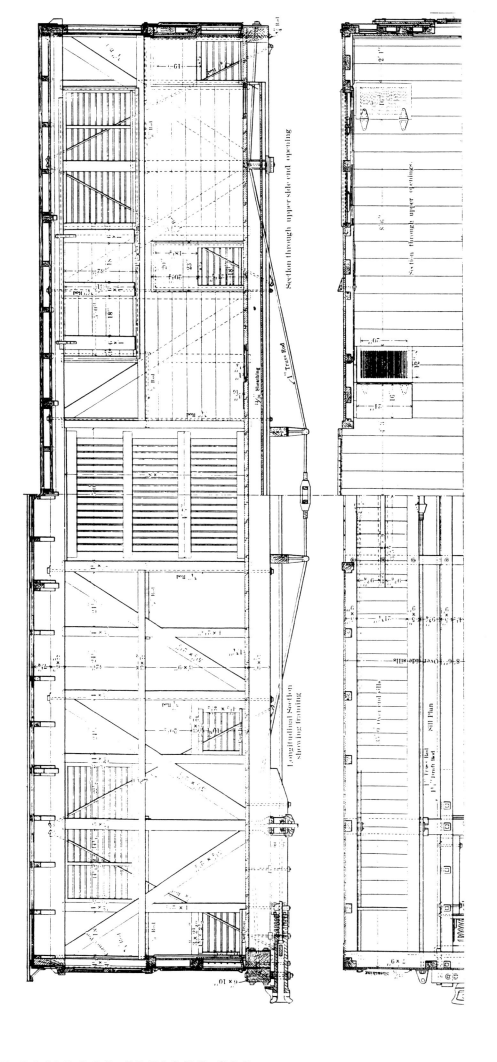

FIGURE 4.66
Illinois Central fruit car dating
from 1891. (Railroad and Locomotive Historical Society
Collection)

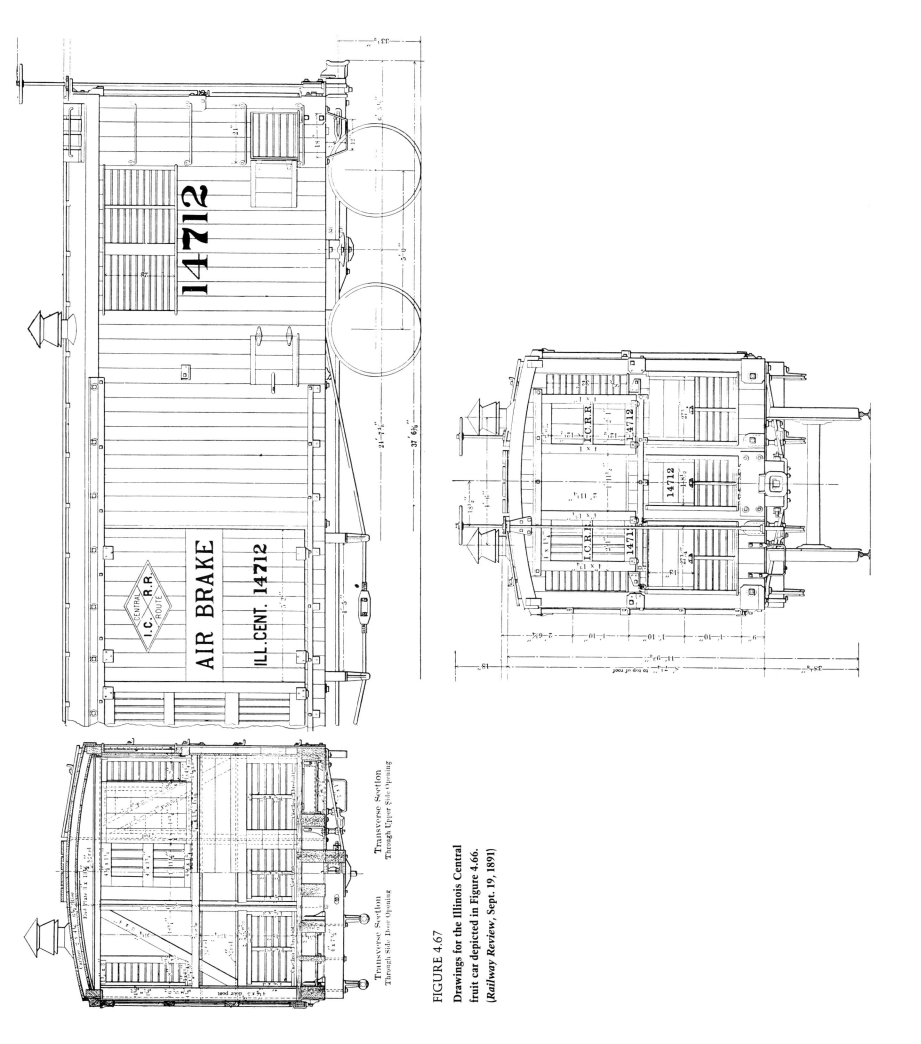

FIGURE 4.67

Drawings for the Illinois Central fruit car depicted in Figure 4.66.
(*Railway Review*, Sept. 19, 1891)

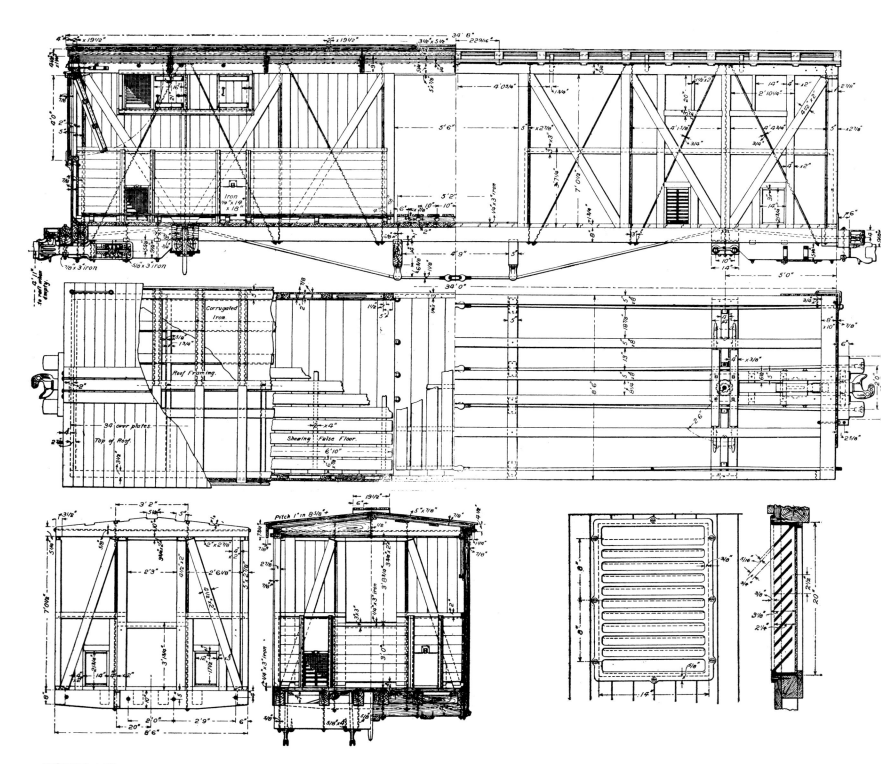

FIGURE 4.68

J. M. Elliott's fruit car was patented in 1888. The example shown here was produced for the Cincinnati, New Orleans, & Texas Pacific in 1893. (*Railroad Gazette*, May 25, 1894)

essentially correct, though not necessarily original. Floor racks did become a standard, at least for refrigerator cars, after about 1900 because of the improved airflow within the car.

The traditional fruit car served its intended purpose well enough, although there were complaints about spoilage because of the dirt and sunlight that naturally penetrated the many openings. Screens helped block out some of the cinders, road dust, and insects but provided little protection from sunlight. The small side vents were abandoned around 1910 because they were found to deflect the flow of air so much that part of the cargo

received no circulation. Side- and end-door ventilators were found to work well together, and so the little low-to-the-floor openings, so favored by old-line car designers, fell from favor. This change was modest compared with the plans of a few progressive food forwarders who would abandon ventilated shipments altogether.[114] The advocates of controlled atmosphere would retard the respiration rate of fruit by lowering oxygen levels. Instead of encouraging the flow of air, they would stop it altogether and increase the carbon dioxide level. The scheme that sounds so advanced was, in reality, older than public railways and had been tried

FIGURE 4.69

Arthur E. Stilwell's curious oyster car was built by Pullman in 1897. The car was fitted with Cloud pressed-steel trucks. (Pullman Neg. 3936)

in France in 1819. In 1894 a Wickes refrigerator car was converted into a controlled-atmosphere unit. Two compartments were fed by five carbon dioxide tanks during the eleven-day journey from San Jose to Chicago. The semiripened fruit arrived in good condition. Efforts to send fully ripened fruits were less successful, and only pears survived the trip in edible condition. Controlled-atmosphere storage proved successful for storage purposes but apparently did not succeed in railroad transit because of the difficulty of maintaining an airtight compartment.

Tank Cars for Special Foodstuffs

As the nineteenth century drew to a close, the food industry began to realize new economies by bulk-shipping in ever more specialized cars. Many products that always traveled in relatively small individual containers were now riding across the land in curious vehicles built especially for vinegar, wine, vegetable oil, and even oysters.

C. W. Goyer, a Memphis sugar refiner, was one of the first in the field.[115] In 1894 he had several 5,000-gallon steel tank cars built by the Memphis Car & Foundry Company for the carriage of molasses and glucose. Formerly, crude molasses had been shipped from Louisiana in wooden barrels. The staves soaked up around 45 pounds of the liquid. Goyer found that he could eliminate this loss, plus the cost and care of the barrels, by substituting tank cars. A steam line inside the bottom of the 6-foot 6-inch-diameter tank heated and thinned the thick contents so that it might be pumped dry in just twelve to fifteen minutes.

At the time that reports on Goyer's cars were being circulated, it was noted that the French were already shipping wine by wooden tank cars. The same idea was adopted in this country. Vinegar

distillers found wooden tank cars well suited to their needs. H. J. Heinz adopted the idea for shipping pickles, and by 1897 he had ten such cars in service.

Wooden tank cars were also found suitable for mineral water, and the Glen Summit Spring Water Company of Wilkes-Barre, Pennsylvania, purchased a tank car from Murray, Dougal & Company of Milton, Pennsylvania, in about 1895.[116] Rudimentary tank cars had been used much earlier for the carriage of water on the Central Pacific Railway, as portrayed in construction-period photographs made by Alfred A. Hart. One picture shows three vertical wooden tanks mounted on a flatcar. In another view we see a rectangular box tank running the full length of a car on a special train being run for Leland Stanford. Whether these cars were carrying extra water for locomotive or potable purposes is unknown, but one suspects that in the rough-and-ready construction camps of the 1860s, no particular distinction was made.

The tank car was employed for at least one more bulk food cargo that is even more exceptional than those just mentioned. Live oysters, especially those from the Chesapeake Bay, had long been recognized as a great delicacy. Transporting them on ice had been attempted almost since the inception of long-distance train travel in this country. In 1866 the Baltimore & Ohio constructed some special express cars to speed oysters to Wheeling, Cincinnati, and Chicago.[117] The fastest train took fifty hours to reach Chicago, and oysters sent on ice for such an extended period could be claimed as fresh or alive by only the most optimistic restaurateur. Alive or dead, Chesapeake Bay oysters were in such widespread demand that a shipment of bluepoints reached San Francisco in November 1869.

Many years later the railroad builder Arthur E.

CARS FOR FOOD

Stilwell decided to find a better way.[118] He would have nothing less than live, fresh-from-the-sea oysters. In 1897 he had Pullman construct a four-compartment 30-ton-capacity wooden tank car for the purpose (Fig. 4.69). Gulf oysters and salt water from Port Arthur, Texas, were loaded into round hatches on top of the rectangular tank. They were unloaded through side hatches upon arrival in Kansas City. Ice was added during hot weather to hold temperatures to a level agreeable to the mollusks as they rattled through midwestern communities whose corn- and beef-eating inhabitants could not be expected to understand the taste for so strange a dish as raw oysters.

This chapter explains the railroads' role as a common carrier in terms of the variety or commodities carried, especially those items that have meaning to the ordinary citizen. The mass movement of coal, iron ore, or machine tools is rather difficult for the average person to understand, but the haulage of eggs, sheep, cheese, and pork makes the once-essential involvement of railroads in everyday life far easier to grasp.

CARS FOR BULK CARGOES

The union between bulk cargoes and railroads has been a long and happy one. Durable materials like coal, gravel, and ore are what nineteenth-century railroaders liked to handle. They were rough, tough, uncomplaining payloads. They never resented wind, rain, sun, slow schedules, or bumpy track. Most bulk cargoes could even withstand a major accident, for it was often possible to send a fresh set of cars to recover the freight dumped around a demolished train. The leakage losses along the road accountable to loose side boards or ill-fitting hopper doors were accepted by most shippers, so long as they didn't exceed 5 percent.

Coal Cars: Jimmies and Other Four-Wheelers

It was a given that coal would become the king of all classes of railroad freight. Vast coal deposits extending from eastern Pennsylvania to southern Illinois made possible this great traffic. Railroads offered a cheap, reliable way to bring coal to the market. Production, accordingly, swelled decade by decade. In 1850 U.S. coal production was 8.3 million tons; by 1870 it had risen to 40.4 million tons and within another decade had nearly doubled. At the end of the century coal production exceeded 269 million tons. At first only the anthracite fields in eastern Pennsylvania were worked, but by 1843 the Baltimore & Ohio had begun to tap the soft-coal deposits around Cumberland, Maryland. The great bituminous fields in West Virginia and Kentucky were not exploited until the Chesapeake & Ohio and the Norfolk & Western entered these areas in the 1870s and 1880s. Even railroads outside the coal fields enjoyed healthy bridge-line coal traffic. The New York Central, for example, was reported by *Poor's Manual of Railroads* to have carried 1.6 million tons of coal in 1880. Thousands of coal cars, in a surprising variety of sizes and shapes, were developed to move mountains of black diamonds from mine to market. Most anthracite lines remained loyal to small four-wheel hoppers, while the B&O favored iron pot hoppers and the remaining coal

movers explored options that ranged from low-sided gondolas to high-sided hopper cars. The following discussion will focus on the various types of coal cars used by American railroads, beginning with the four-wheel jimmies. Readers are reminded that pot hoppers and other iron-bodied cars are treated in Chapter 8.

JIMMIES

The longevity of the four-wheel coal car, commonly called the jimmy, remains one of the puzzles of American freight car history. The continued dependence on very small cars to move increasing quantities of coal seems to contradict the basic Darwinian law about the survival of the fittest—a law that, in the case of coal traffic, should have favored the largest of the species. The jimmies were short on size and capacity, and they should have vanished from the American railroad scene around 1870. With few exceptions the four-wheel freight car had become obsolete a generation earlier.

The persistence of this pigmy race might be explained by reasons other than a stubborn allegiance to an ancient order. Because thousands were already in existence, it may have been decided to perpetuate the design because larger cars could not be easily mixed into jimmy trains. The little cars worked well enough in long trains, and most anthracite lines were, in effect, operating unit trains long before the notion was supposedly invented in the 1960s. The Lehigh Valley, for example, was noted for the operation of very long jimmy trains. In 1891 a train of 225 loaded jimmies was reported to have passed over the line without difficulty.[1] Some years earlier a train of 593 jimmies was operated on the same railroad. One always thinks of trains in nineteenth-century America as being short, but here is a train in 1879 measuring more than 1.5 miles long, composed of an endless brigade of ignoble jimmies.

The jimmies' apparent success when operated in long trains does not mean that they were efficient carriers, however. Their resistance to evolution can be explained by the need for small cars at mine sid-

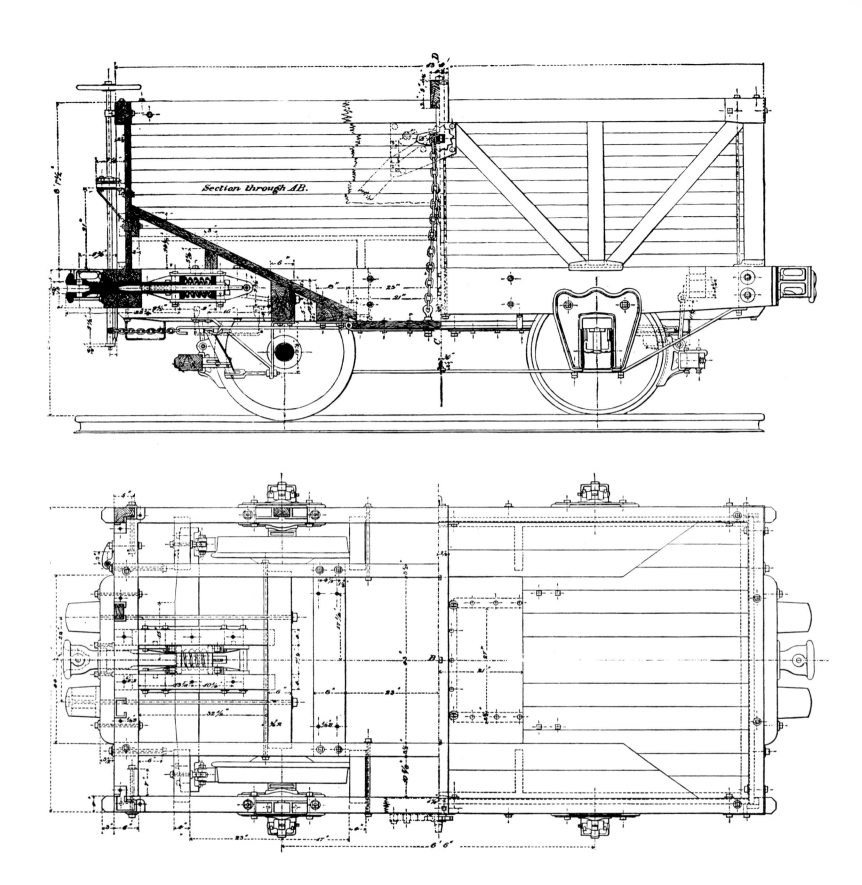

Section through AB.

THE AMERICAN RAILROAD FREIGHT CAR

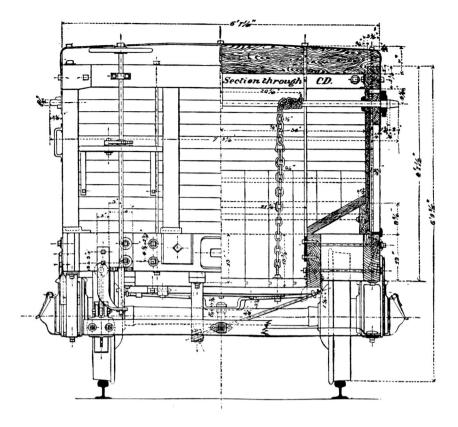

FIGURE 5.1

The New York Central & Hudson River Railroad produced a giant among jimmies, starting in 1876, that carried 8 tons of coal. (*Railroad Gazette*, Sept. 8, 1876)

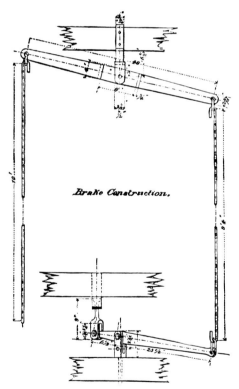

ings with limited clearances, or at old-fashioned loading and unloading docks where the coal chutes were arranged for small cars. The small cars were easier to shift manually than big ones, and a switcher was not always available at coal yards or small mine sidings. Hence reasons having nothing to do with the efficient movement of coal may have kept the jimmy in favor. These are speculations only, and nowhere in the literature can I find a convincing reason for perpetuating the jimmy throughout the nineteenth century.

The number of jimmies in service was prodigious and makes them one of the largest single classes of freight cars for their time. At their peak of popularity, around 1875, we can account for up to fifty-five thousand jimmies.[2] The Central Railroad of New Jersey, the Lehigh Valley, the Delaware, Lackawanna & Western, and the Philadelphia & Reading were the top owners, but the Pennsylvania and the New York Central had moderate to large fleets as well. At least one western line, the Northern Pacific, offered an iron path to over four hundred coal jimmies, in a place far removed from their native surroundings in the hills of Pennsylvania.

The ancestors of the jimmy were the tiny cars operated by the Mauch Chunk and the Delaware & Hudson gravity railways mentioned in Chapter 2. The size and capacity of these pioneers were nearly doubled in subsequent years by an overall enlargement. Wheels were no longer outborne but were now under the body and inside the frame. A wheel well was created by sloping the side of the hopper floor inward. By the mid-1870s the jimmy was no longer a handcar outfitted with a coal box. Some had bodies measuring 13 feet long by 3 feet 6 inches high by 6 feet 7 inches wide. Capacity was as large as 8 tons, although the more typical jimmy was more in the 5- to 6-ton range. Even in its biggest form, the jimmy was never more than half a freight car, and its smallness became more disproportionate as all other freight vehicles progressed steadily from capacities of 10 to 20 to 30 to 40 tons.

The origins of the jimmy are surely to be found in the British chaldron wagons used on primitive railways long before steam locomotion was introduced. Just who devised the bottom trapdoor for unloading is unknown. Nor is it likely that the individual who first thought of the sloping side walls or inclined floors will ever be named. It's likely that no one individual was responsible for any of these basic improvements. They were in use before 1830 and may well predate railways altogether; a scheme so fundamental and obvious may well have been tried by wagon builders before the time of George Stephenson.[3] It appears that the generic American-pattern jimmy was developed by the Philadelphia & Reading. It is characterized by a massive wooden frame and a rectangular hopper-shaped body with outside framing. The floor sloped downward toward a pair of drop doors not only from the

FIGURE 5.2

The Pennsylvania Railroad
produced a more conventionally
sized jimmy that was a virtual
duplicate of the Reading's
standard design. These five views
are reproduced from a French
study based on an 1876 survey
of American railroading.
(E. Lavoinne and E. Pontzen, *Les
Chemins de Fer en Amérique*
[Paris, 1882]; metric
measurements)

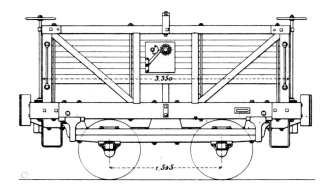
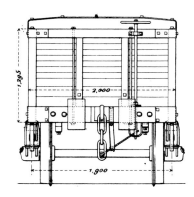
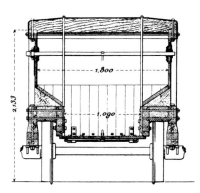
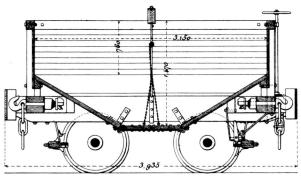
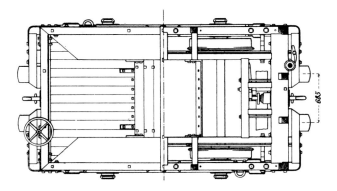

ends but from the side walls as well. The major incline, however, was from the ends. The Reading tended to favor wooden springs, made from long, thin strips of flexible ash, with the axle boxes bolted directly to them. This elemental scheme was introduced by the road's master mechanic, James Millholland. It was widely copied and was a common, though not standard, feature of the jimmy.

Despite the Reading's central role in jimmy development, the best surviving drawings and information are for cars on other lines. Ironically, the most detailed jimmy drawings are for the New York Central—a relatively small-time operater in this specialized area of coal cars.[4] Even so, the Central's master car builder, Leander Garey, produced plans for the largest jimmies built in this country (Fig. 5.1). The car measured 16 feet long over the cast-iron dead blocks and had a 6-foot 6-inch wheelbase. The windup-chain hopper doors were a standard feature, but the double-acting brakes and elaborate draft gear were rather special for a jimmy. Wood springs are definitely not in evidence, and it is likely that the axles were cushioned by coil or rubber block springs not visible in the drawings. The car weighed 9,500 pounds and carried 16,000 pounds. The Central had around fifteen-hundred cars built on Garey's plan.

A more faithful copy of the Reading's prototype was rendered by the Pennsylvania Railroad. It included all the design elements, even down to the wooden springs, undersize journals, hook-and-chain-link couplers, and individual brakes for each

wheel set (Fig. 5.2). Very heavy end timbers and volute draft springs are clearly visible in the drawings. Notice also that the draft-gear shank is fastened by a flat key rather than by a nut. The heavy wooden/strap-iron cross-brace across the middle top of the body is designed to prevent bulging of the car sides. The Pennsylvania design was briefly described and illustrated in James Dredge's 1879 book.[5] His drawing is less clear than the French drawings reproduced in Figure 5.2, but Dredge does offer useful dimensions. The inside body dimensions are given as 10 feet 4 inches long by 4 feet 10 inches deep by 5 feet 10¾ inches wide. The side body planks were 1 inch thick and the floor planks 1¾ thick. The 33-inch-diameter wheels were placed on a 5-foot wheelbase. The car weighed about 4 tons. Its capacity was not given.

The Central Railroad of New Jersey had at least two styles of jimmies. One followed the traditional Reading plan, as shown in surviving photographs (Figs. 5.3, 5.4). In the first view—taken at Elizabethport, New Jersey, in September 1885—the photographer was more intent on engine No. 153, but he managed to include most of a wooden-sprung jimmy in the right-hand side of the picture. The men in the scene help to establish the scale of the car. In the second view—taken at Hampton, New Jersey, around 1890—the two designs are shown standing side by side. The bodies appear identical, but the running gears are different because of the displacement of the wooden springs by conventional pedestals on the 22000-series car. The round symbol painted on all these cars is the

FIGURE 5.3

The scale of this Central Railroad of New Jersey jimmy is well defined by the train crew and switcher No. 153. The scene is Elizabethport, N.J., Sept. 1, 1885. (Smith Collection, Stevens Institute of Technology, Hoboken, N.J.)

FIGURE 5.4

A yard full of jimmies at Hampton, N.J., around 1890. (Collection of Edward T. Francis)

CARS FOR BULK CARGOES

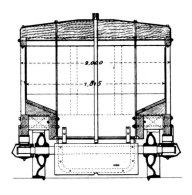

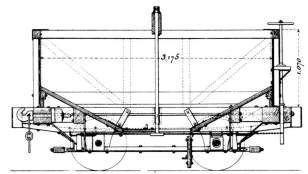

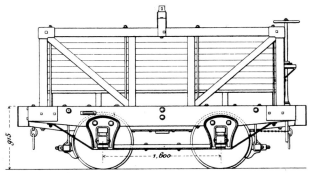

FIGURE 5.5

An 1876-vintage Central Railroad of New Jersey coal jimmy. (Lavoinne and Pontzen; metric measurements)

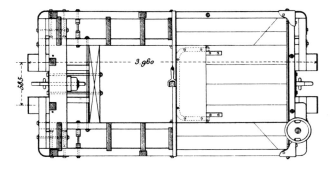

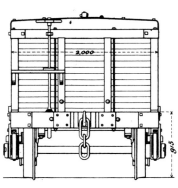

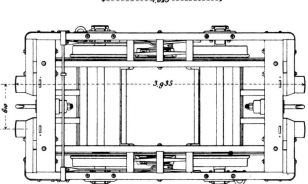

FIGURE 5.6

The Lehigh Valley's jimmy was a faithful copy of the plan shown in Figure 5.5, except for the peculiar rail shoe brake, a device tested but not widely adopted by any coal carrier. (Lavoinne and Pontzen; metric measurements)

so-called fried egg of the Lehigh Coal & Navigation Company, a major shipper and miner of anthracite. The outer band is white, and the large central dot is red. The CRRNJ leased the LC&N's railroads and hence acquired rights to use its symbol. Published drawings provided an even better understanding of the pedestal-style cars (Fig. 5.5). Most features are self-explanatory and follow the familiar Reading design, which includes the brutish hook-and-chain-link coupler. The absence of the windup-chain hopper-door rig is a notable exception. The primitive latch arrangement may not have proved to be much of an improvement. It surely required two and perhaps three men to close the doors. It also could not have been a simple device to open, particularly under freezing conditions. A long-handled wrench was surely a favored tool for dealing with these hopper-door latches.

The Lehigh Valley also used latch closers for its jimmy hopper doors, according to surviving photographs and drawings. The best of these illustrations is reproduced here as Figure 5.6. It follows the CRRNJ plan almost detail-for-detail, with the obvious exception of the brake. The rail shoe-brake scheme was devised early in 1876 by William L. Hoeffecker (1842–1902), son of the road's master mechanic. The design was intended to reduce wheel wear by eliminating the conventional wheel shoe-brake arrangement. Friction-rail brakes were rarely used in the United States. The scheme depicted here is even more exceptional because of the eccentric in place of a crank or lever. The idea was apparently only an experiment, for it doesn't appear in any other illustration known to me. Drawings for an 1883 6-ton-capacity Lehigh Valley jimmy show ordinary shoe brakes.[6]

Yet another variant on conventional braking was offered by the Lackawanna, another major employer of jimmies. Conventional brake wheels were displaced by large side levers very much like the old-fashioned British scheme for hand-braking.[7] It was estimated that the Lackawanna operated a fleet of no fewer than twenty-two thousand four-wheel coal cars. Jimmies were constructed at the Lackawanna's Scranton shops. Twelve hundred were built in 1875.[8] When times were slow and traffic was down, only two were built each day, but during prosperous times, the shop worked to its capacity of twenty cars per day.

The Lehigh Valley, like the Lackawanna, maintained a big shop for the construction and repair of jimmies. According to the *National Car Builder* for June 1878, such work was concentrated at the Packerton shop. At the time, the Lehigh line had 26,635 jimmies in service and normally produced 36 new ones per day. During busy times, three hundred were repaired in a single day and one thousand men were engaged in the repair and construction of the four-wheel fleet.

The jimmy was scaled down for narrow-gauge

FIGURE 5.7

The Billmeyer & Small car works
of York, Pa., produced a scaled-
down miniature of the basic
anthracite jimmy for an
unspecified 30-inch-gauge line. It
is faithful to its standard-gauge
big sisters, even to the wooden
springs. (Thomas Norrell Collection, Smithsonian Neg. 85-21826)

service by Billmeyer & Small, car builders of York,
Pennsylvania (Fig. 5.7). The result was a midget
3,500-pound four-wheeler for 30-inch-gauge lines
with a capacity of 7,000 pounds. Other than its diminutive size, there is little to differentiate this
car from one of its full-size Reading prototypes.
The body and the wooden-spring plan are identical
to those of a standard-gauge car. The latch hopper-
door closer is evident from the squared-off shaft
protruding from the top crossbeam.

OTHER FOUR-WHEELERS

The four-wheel coal car prevailed in forms other
than the ubiquitous jimmy. A number of lines employed fairly long, low-sided gondolas on the four-
wheel plan. They were almost twice the length of
the stumpy jimmies but were more notable for
their long wheelbases. Cars of this type were apparently first introduced on the St. Louis & South
Eastern in February 1875.[9] The line's general manager had been impressed by a four-wheel coal
wagon seen during a European tour. His reaction
was, curiously, the reverse of that of most U.S. railroad men, who viewed the bobbing goods wagons
as an abomination and a reminder of the bad old
days when four-wheel freight cars were the only
rolling stock available. Yet this misguided individual returned to the United States convinced that
his colleagues were entirely mistaken. He had a
sample car constructed. The operating department was skeptical at first and would allow the car
to carry only 8 tons. It was cautiously placed at the
end of the train. After a trial period the load was
raised to 10 tons. After even more trials the general
manager finally seemed vindicated, and in 1876–
1877 another 130 of the cars entered service.

Joseph W. Sprague, president of the Ohio Falls
Car Company of Jeffersonville, Indiana, decided to

improve on the St. Louis line's design.[10] By late
1877 he had a number of sample cars ready for tests
on several lines, including the Louisville & Nashville. The 12-ton-capacity cars measured 20 feet
long and weighed from 5.0 to 5.5 tons, depending
on the type of wood used in their construction.
The rigidity of the 11-foot wheelbase was relieved
by a flexible subframe attached to the main frame
by coil springs (Fig. 5.8). The L&N liked the design and acquired fifty more cars but built them
with more conventional running gears. An 11-foot
wheelbase was again used, but the cars were said
to run well and never developed hotboxes. It appears, however, that the four-wheel gondolas were
not reproduced in great numbers after the initial
test lots were completed.

Several years later, the Delaware & Hudson
Railway reintroduced the idea with a design prepared by R. C. Blackall, superindendent of machinery.[11] The plan was for cars very much like those
just described except for the suspension and the
inclusion of drop doors and windup chains. The
wheelbase of 9 feet 11½ inches was rigid, and despite the apparent complexity of the framing and
the use of both coil and semielliptical springs, the
journals could neither turn nor adjust themselves
to curves (Fig. 5.9). The wheelbase was, however,
about the same as that for a six-wheel passenger
car truck, so Blackall's gondolas could pass wherever normal cars traveled.

The most novel series of four-wheel coal cars operated by a main-line common carrier were the
twelve hundred side-dumpers owned by the Providence & Worcester.[12] In general plan they were
identical to common side-dump cars used for construction and track repairs since the 1850s. In
place of spoils, stream gravel, or ballast, this particular set of side-dump or tip cars carried 7.5 tons
of coal, a load greater than the typical jimmy. Even
so, its body measured inside only 10 feet long by 7
feet 6 inches wide by 25.5 inches deep. The crescent-shaped rocker bearings were massive iron
castings. The general scheme appears to have been
introduced in Michael Berney's patent of September 11, 1849 (No. 6,712). Old T-rail, inverted, served
as a track for the rockers. Previously, the P&W had
used regular gondolas for coal service, but the cars
tended to stand for long periods at coal yards waiting for crews to unload them. With tip cars, coal
was unloaded immediately, and the car was returned for another consignment. The large number used by the P&W indicates that the scheme
worked well. The design of the car was universal
and tested by generations of rough service, and yet
only the P&W appears to have adopted it for coal
service (Fig. 5.10).

THE DEATH OF THE JIMMY

In the spring of 1890 *Engineering News* said that
time was at last running out for the old-fashioned

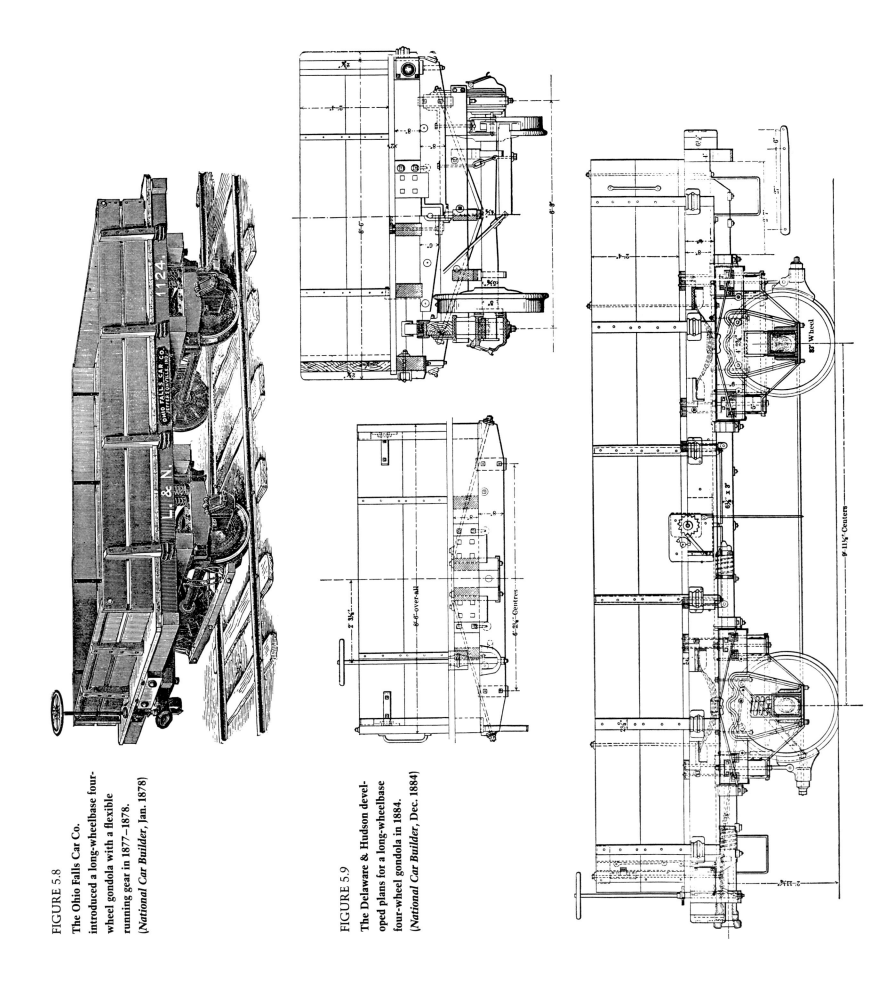

FIGURE 5.8

The Ohio Falls Car Co. introduced a long-wheelbase four-wheel gondola with a flexible running gear in 1877–1878. (*National Car Builder*, Jan. 1878)

FIGURE 5.9

The Delaware & Hudson developed plans for a long-wheelbase four-wheel gondola in 1884. (*National Car Builder*, Dec. 1884)

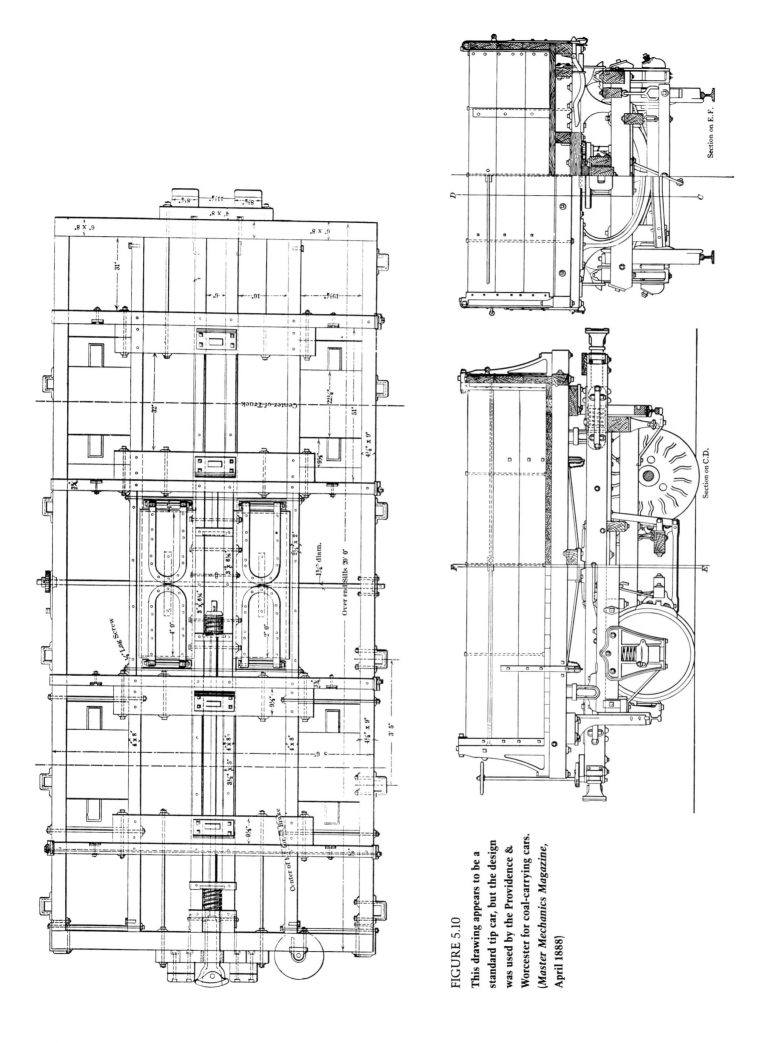

FIGURE 5.10
This drawing appears to be a
standard tip car, but the design
was used by the Providence &
Worcester for coal-carrying cars.
(*Master Mechanics Magazine*,
April 1888)

jimmies, "which for so many years rattled cheerfully over road and were endured as a necessary evil."[13] A month earlier another trade paper had said that the Lackawanna was ready to banish jimmies from main-line service and had built no new ones for ten years, yet the cars held on as if they would run through the millennium.[14] The trainmen hated them, and they were blamed for countless accidents and injuries. They derailed easily, especially when they encountered a depression in the track while empty, or when the brakes went on for a fast stop. Yet despite the contempt heaped upon them, the little four-wheelers continued to bounce over the anthracite lines for almost another decade. They were said to be of low origins—mine-shaft tramcars were given as the parents. No better design came forward, for it is always easier to imitate than to originate. And when the end came, it came swiftly and without mercy. The Lehigh Valley sent thirteen thousand of its dinky coal cars to American Car & Foundry's Detroit plant for scrapping.[15] A special inclined track was built to send fifty-car trains hurtling down a steep grade to a crashing finale. The metal bits were salvaged, but the splintered timbers were given away to the poor for firewood. In early 1899 the Central Railroad of New Jersey sold sixteen thousand jimmies for scrap and replaced them with four thousand eight-wheel hopper cars.[16] The once-familiar little jimmy disappeared entirely from the American railroad scene in just a few months' time. The extinction was complete; not a single one has been preserved.

Coal Cars: Eight-Wheel Gondolas and Hoppers

While the anthracite lines played out the drama of the four-wheel jimmy, other coal-hauling roads evolved an increasing variety of eight-wheel cars. All were of the open-top style and so were roughly related, but only as a casual mix of cousins that were reckoned up by dozens. A wide variety of coal-carrying cars might be found on any one main-line railroad: plain gondolas, drop- or slide-bottom gondolas, single- or twin-hopper-bottom gondolas, single- or double-hopper-bottom cars, and high-sided hopper cars. Any of the above, except for the first two, might have pyramidal hoppers rather than the straight-sided variety. In smaller numbers one could expect to find an occasional side-dump or forward-tilt-dump car in coal service. The Baltimore & Ohio had a sizable fleet of iron pot hoppers that also served on a few associated lines as well (see Chap. 8).

Few railroads were content with a single style of coal car—a decision based on more than whimsy. Gondolas were substitute coal cars when tonnage was temporarily high, but they could be used for other cargoes when mine production was down.

The variety of cars on a single road is also surely accountable to shifts in the mechanical department's staff. One master car builder might be a strong jimmy advocate and boost the line's rolling stock in this direction, while his successor might favor drop-bottom gondolas. Yet one is rarely free to discard the purchases of a predecessor immediately, and the existing stock is retained as long as it is serviceable. Then again, experience or a simple rethinking of needs can convince a master car

TABLE 5.1 Philadelphia & Reading Railroad Coal Cars

Year	4-Wheel Iron	8-Wheel Iron	4-Wheel Wood	8-Wheel Wood
1843	0	0	1,592	0
1844	856	1	1,600	0
1845	1,497	1	1,606	0
1846	3,019	1	1,539	0
1847	3,019[a]	1[b]	1,586[c]	0
1850	2,990	1	1,576	0
1852	2,980	1	1,596	2
1854	2,980	1	1,693	215
1857	2,948	1	1,357[d]	525
1860	2,948	2	1,238[e]	490[f]
1861	2,944	3	1,824	837
1864	2,964	3	2,187	3,134
1865	2,958	3	2,177	3,179
1866	2,897	3	2,132	4,079
1868	2,834	3	2,114	?
1869	2,713	3	2,101	4,234
1870	2,389	3	5,379	5,723
1871	2,284	3	5,373	5,963
1872	2,284	3	5,373	5,963
1875	1,776	3	5,434	7,762
1876	1,728	3	5,434	7,762
1880	1,429	3	5,729	7,815
1885	805	3	?	10,894
1887	403[g]	3	5,715	11,022
1890	?	?	2,976	15,583
1891	?	?	1,664	19,147
1892	?	?	26,248[h]	29,220[h]
1893	?	?	107	18,961
1894	?	?	88	18,438
1895	?	?	23	18,088
1896	?	?	8	18,742
1897	?	?	8	19,309
1898	?	?	?	18,228

Source: P&R published annual reports and *Poor's Manual of Railroads.*
  a. Weight 2.4 tons, capacity 5 tons.
  b. Weight 4.7 tons, capacity 11 tons.
  c. Weight 2.2 tons, capacity 4.6 tons.
  d. Plus 609 private.
  e. Plus 610 private.
  f. Plus 100 private.
  g. Plus 1,442 in storage.
  h. The high figure for 1892 is not explained; perhaps the count included cars from leased lines not reported in other years.

FIGURE 5.11

The New Haven had relatively
little coal traffic and so employed
low-sided gondolas that were
also suitable for other cargoes.
(*National Car Builder*, Feb. 1881)

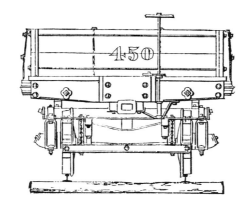

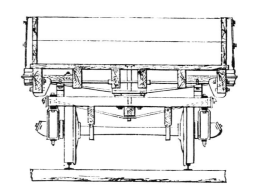

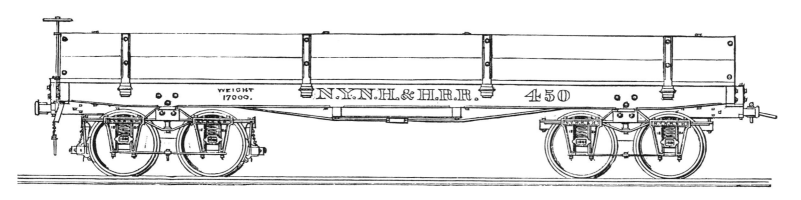

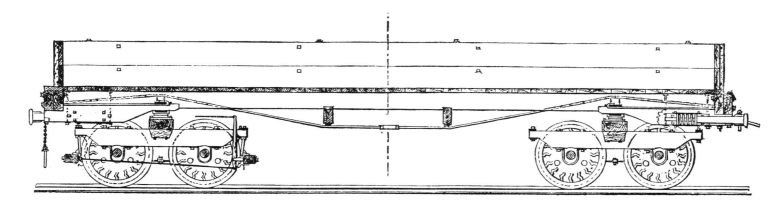

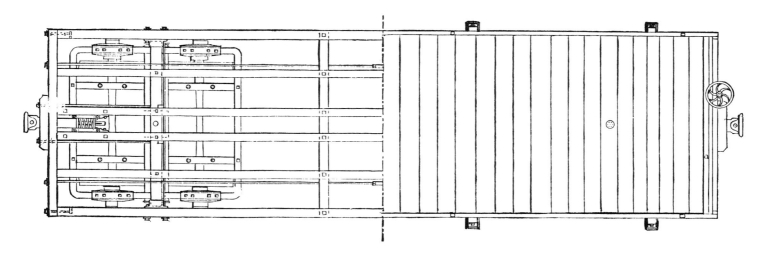

CARS FOR BULK CARGOES

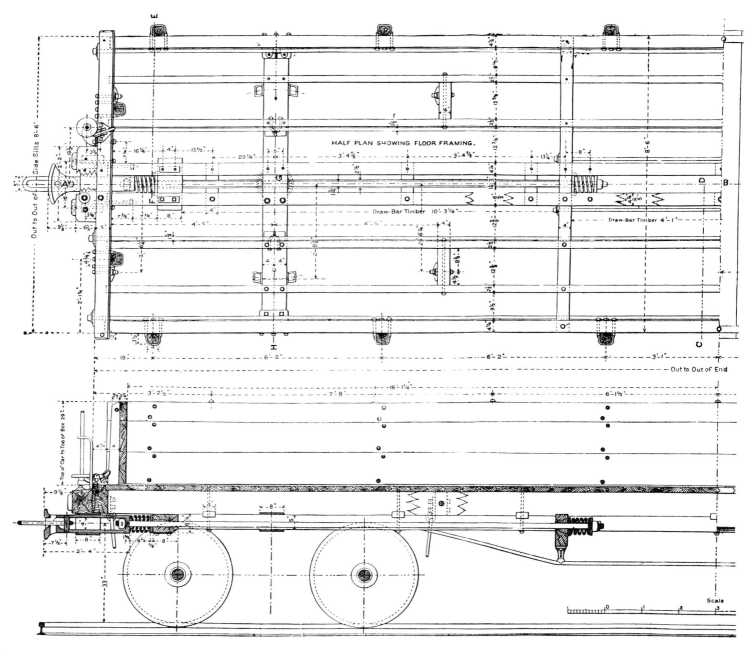

HALF PLAN SHOWING FLOOR FRAMING.

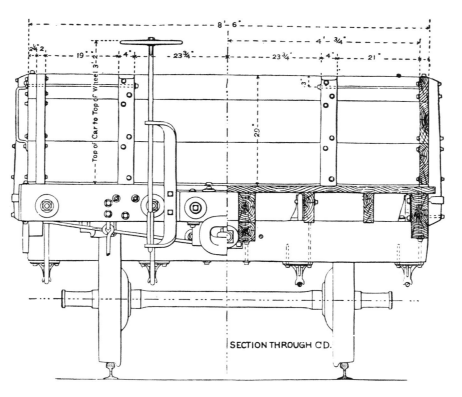

SECTION THROUGH CD.

FIGURE 5.12

Leander Garey prepared this design, which featured a peculiar draft gear, for the New York Central in the early 1880s. (*Railroad Gazette*, June 9, 1882)

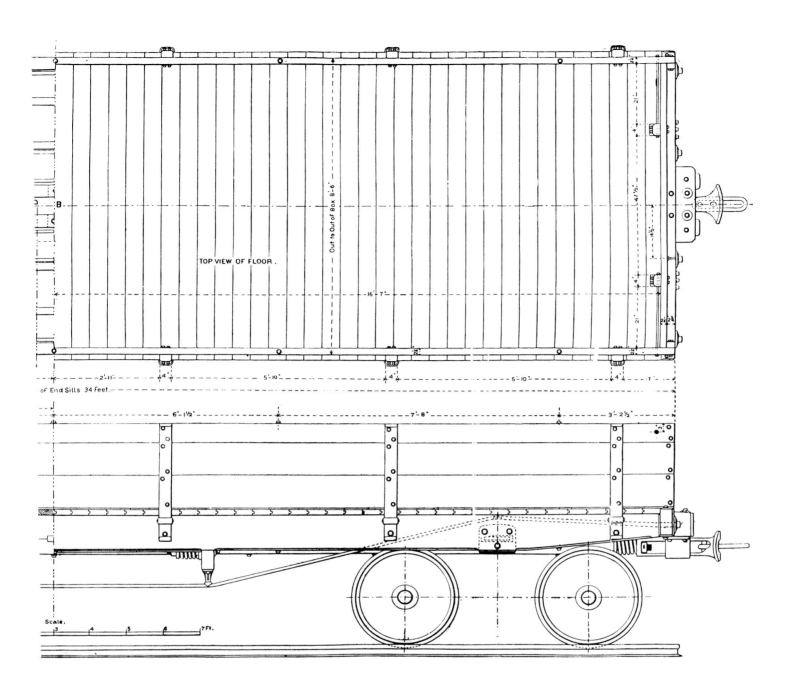

TOP VIEW OF FLOOR.

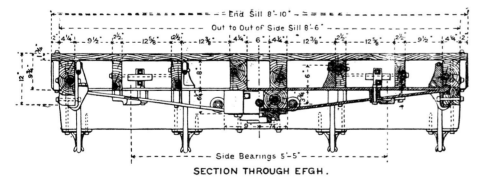

SECTION THROUGH EFGH.

builder to adopt an entirely new plan, after years of loyalty to a given design. Long-term traffic growth might also dictate a radical shift in car size or design.

And so a variety of reasons that are difficult to document can be offered to explain changes in car selection on any one railroad. The Reading, for example, drifted away from wooden jimmies in the mid-1840s to four-wheel cars with iron bodies. By the mid-1850s the line showed a beginning interest in eight-wheel hopper cars. This interest intensified even while more and more wooden jimmies were being acquired. The little iron cars slowly declined in numbers until around 1875, when they fell rapidly from favor. Meanwhile, eight-wheel gondolas were on the rise, to become dominant by 1885. These fluctuations, shown in Table 5.1, are not explained in the available record, and so we can only speculate on the reasons for change.

GONDOLAS

Plain gondolas were the simplest type of coal car. They were also the most inefficient because of their low sides and inability to self-unload via bottom doors or hoppers. Yet they could be used for general freight traffic and served as a tolerable makeshift coal car when the need arose. We can find examples of this familiar general carrier from New England to the Far West that differed only in the smallest details. Drawings for a New Haven gondola were published in the February 1881 *National Car Builder* (Fig. 5.11). They show a plain little car of rather antique appearance, even for the early 1880s, that featured moderately high sides (24 inches) and short-wheelbase iron-frame trucks. The car measured 28 feet 9 inches long and 8 feet wide over the frame. It weighed 17,000 pounds and had a capacity of 38,000 pounds. The frame was, typically, like that of a flatcar with very deep (5- by 12-inch) side sills and smaller inner

sills. The body bolsters were iron, but the truck bolsters were wooden. Only two truss rods, set in toward the car's center, were used. Turnbucklers are in evidence, but no iron queen posts were used. Brakes were attached to only one pair of trucks. These cars, on a plan by Loren Packard, the line's master car builder, were built at the road's Hartford shops.

To the south of Hartford, Leander Garey, master car builder for the New York Central, was at work on a drawing for a larger gondola of more modern outlines.[17] This series of cars measured 34 feet long over the end sills. The body sides measured 29 inches high. The frame was of the flatcar variety, but it incorporated a few unusual features. Most notable was the lightness of the intermediate sills—a mere 2½ by 8 inches. These undersize members were, however, supported by four truss rods. Shallow iron queen posts improved the framing. Iron body bolsters were another good feature of the car. The curious draft gear, with its long cou-

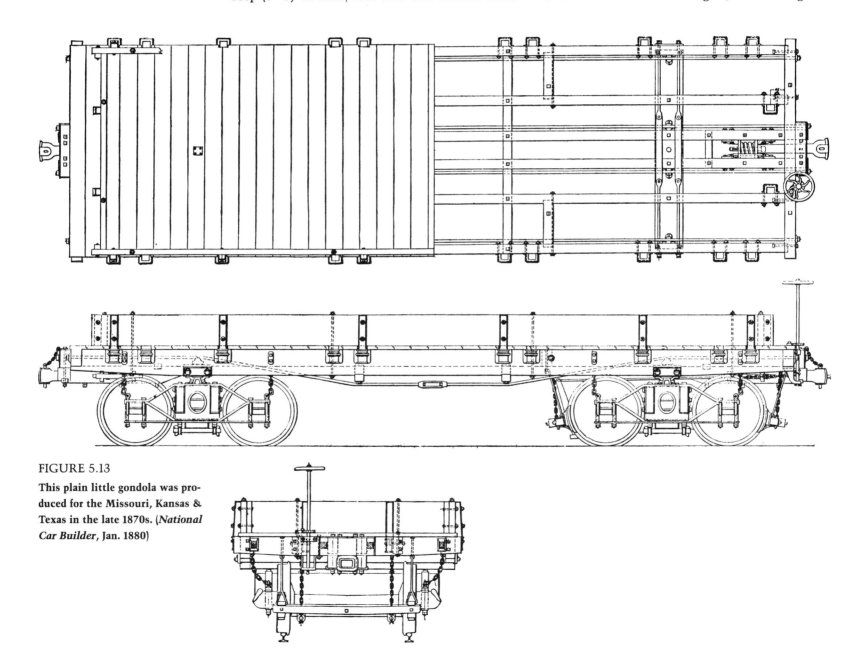

FIGURE 5.13

This plain little gondola was produced for the Missouri, Kansas & Texas in the late 1870s. (*National Car Builder*, Jan. 1880)

pler shank and buffer springs placed near the car's center, was original with David Hoit and was mentioned in Chapter 3. While its details are clearly shown in the drawings, the advantages of this curious arrangement are difficult to surmise (Fig. 5.12).

During the 1880s western lines tended to accept more straightforward plans for gondolas. The drawing for a Missouri, Kansas & Texas platform and coal car with low sides exhibits no design peculiarities (Fig. 5.13).[18] In scale and arrangement it was like the New Haven car. It measured 28 feet ¼ inch long. The truss rods were without benefit of queen posts, but in this case there were four rather than two. The wooden body bolsters would lessen the car's capacity and durability. Brakes were applied to only one of the swing-motion trucks. The truck safety chains are unusual for a freight car of any period, as are the chain hangers for the brake beams. Drawings for a similar car manufactured by the Missouri Car & Foundry

Company of St. Louis were published in the *Transactions* of the Scottish Institute of Engineers and Shipbuilders for 1884–1885 (vol. 28). The author of an article describing American freight car practice, Alexander Findlay, had worked for the St. Louis car builder and used this drawing to illustrate his paper (Fig. 5.14). Cars of this design were undoubtedly supplied to railroads in the St. Louis area. The drawing documents the history of American freight car design even though the engraving itself comes from such a remote and unlikely source.

Later in the 1880s larger and heavier gondolas came into favor, following the general trend toward equipment of greater capacity. An example of this development was the 25-ton-capacity high-sided gondola designed by T. L. Chapman for the Chesapeake & Ohio Railway in 1888.[19] This car stretched out to 36 feet 4 inches over the end sills. The sides stood 42 inches above the floor, a height that required body trusses to prevent side bulging

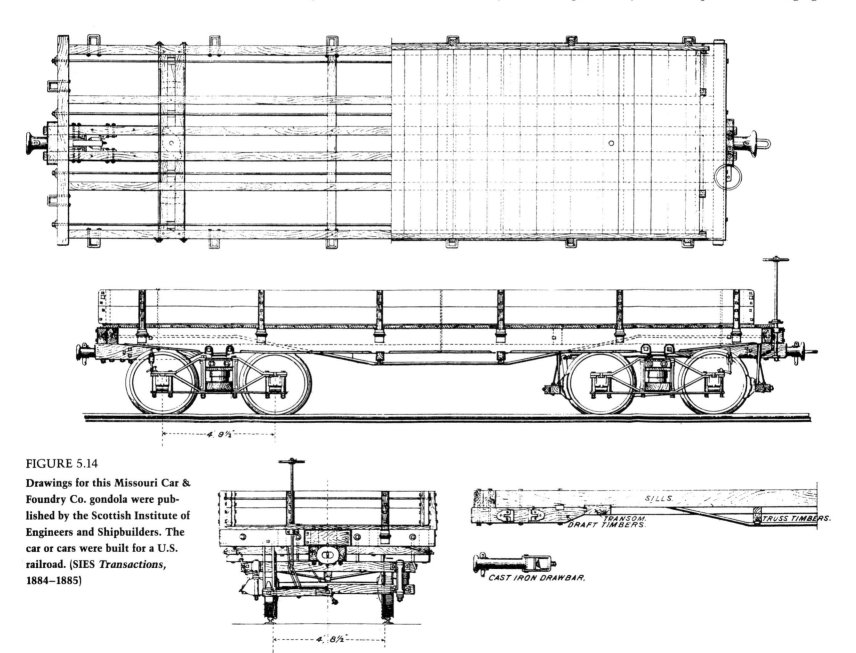

FIGURE 5.14

Drawings for this Missouri Car & Foundry Co. gondola were published by the Scottish Institute of Engineers and Shipbuilders. The car or cars were built for a U.S. railroad. (SIES *Transactions*, 1884–1885)

CARS FOR BULK CARGOES

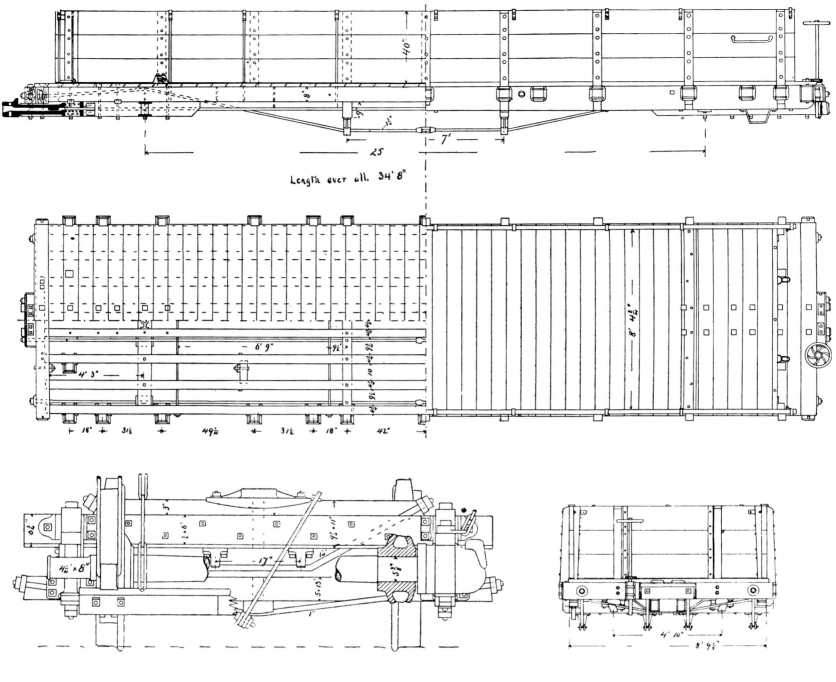

Length over all, 34' 8"

FIGURE 5.15

By 1890 30-ton-capacity gondolas were the standard. The car shown in this drawing was manufactured for the Texas & Pacific. (*American Railroad Journal*, Nov. 1891)

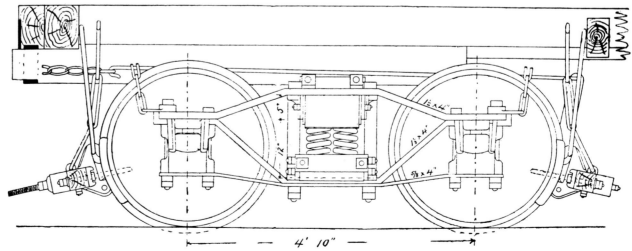

THE AMERICAN RAILROAD FREIGHT CAR

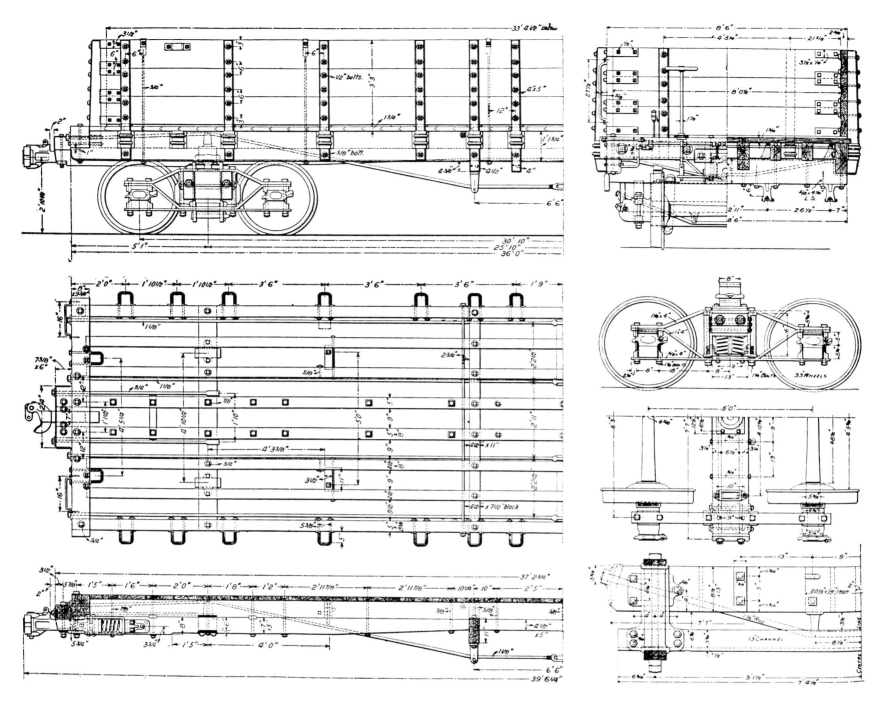

FIGURE 5.16

The Rock Island's belief in heavy framing is well represented in this 1896 drawing for a 30-ton gondola car. (*Railroad Gazette*, March 6, 1896)

when loaded. Both trucks had brakes, but they were manually powered. The link-and-pin couplers and wooden body bolsters would in a very few years be considered obsolete. Unloading such a big car by hand must also have proved a burden to the unlucky crew responsible for servicing the car at journey's end. A few years after the C&O gondolas entered service, the Texas & Pacific Railway received three hundred Spartan 30-ton gondolas from the St. Charles Car Company.[20] They measured 34 feet 8 inches over the end sills and were fitted with iron bolsters. The sides were 40 inches high and 2½ inches thick. The pine floorboards were 1¾ by 8 inches and were made with shiplap joints (Fig. 5.15).

Plain gondolas had matured and grown by the

middle 1890s when the Rock Island adopted its 36-foot-long standard coal car.[21] Capacity was 30 tons, which required deep side sills (5 by 13¾ inches) and four heavy truss rods (Fig. 5.16). The body sides rose 39 inches above the floor. Air brakes, Janney couplers, and iron body bolsters contributed to the car's safety, but as with all plain gondolas, immense manual labor was required to unload it.

DROP-BOTTOM GONDOLAS

Placing hatches or drop doors in the floor was a simple way to improve gondolas for coal, gravel, or ore service. It was a partial solution, of course, because considerable work was still required to move the bulk material toward the center of the car

FIGURE 5.17

Drop doors or hatches were effective aids in unloading bulk loads from gondolas. Note the position of the windup shaft and chains. (William Voss, *Railway Car Construction* [New York, 1892])

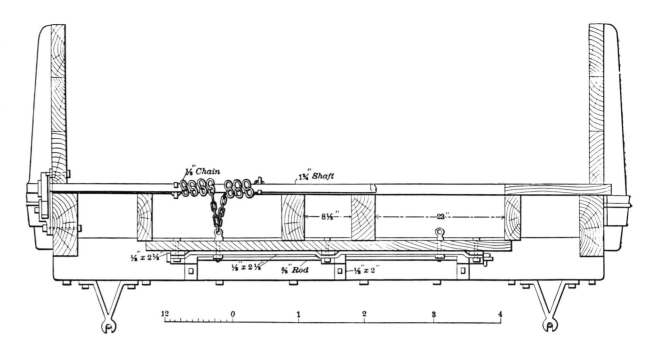

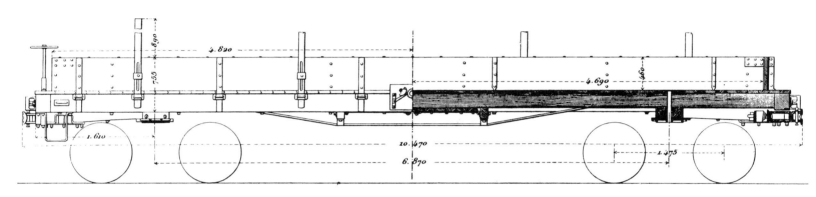

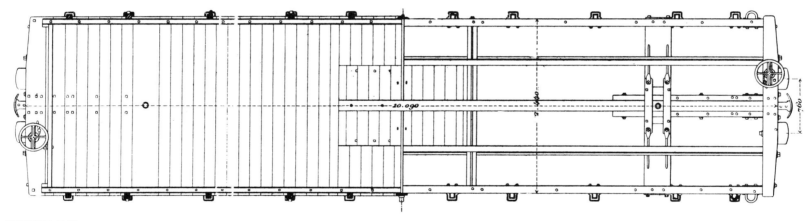

FIGURE 5.18

The Pennsylvania, like many other U.S. railroads, found the drop-bottom gondola to be a grand multicargo vehicle. This drawing dates from around 1876. (Lavoinne and Pontzen; metric measurements)

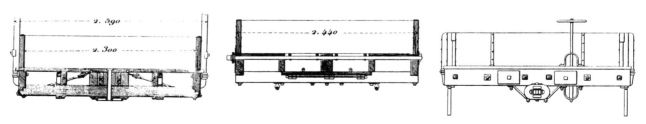

FIGURE 5.19

A comparison of this photograph with Figure 5.18 shows that both represent the same Pennsylvania design. (Pennsylvania Railroad Photo)

where the openings were located, but it was less labor-intensive than attempting to shovel the cargo over the sides. The compromise scheme possessed one additional drawback. The hatches were fitted below the sills so that the windup shaft and chain could be placed level or just below the floor. This was done to make the floor as clear as possible for loads of lumber, machinery, and the like. The arrangement achieved the major goal of producing a multi-use car, but it also created a hazard to any trainman required to walk across the floor. Shallow pits or voids spanned the floor opening, and the brakemen would likely trip or stumble, particularly at night, while making their way along the train. During the age of hand brakes, brakemen frequently had to scurry across and over several cars to perform their duty.

Despite their several failings, drop-door gondolas were popular with most U.S. railroads. The idea dates back to the pioneer period, as noted in Chapter 2. A better understanding of the scheme can be gained from a study of the drawing reproduced from William Voss's 1892 *Railway Car Construction* (Fig. 5.17).

Drawings for an 1869 Pennsylvania drop-hatch gondola were given in Chapter 2 (Fig. 2.37). A very similar car, also of Pennsylvania origins but dating from around 1876, is illustrated in the Lavoinne and Pontzen book on American railways (Fig. 5.18). This fine engraving provides many details not apparent in the crude diagram of the earlier car. The relatively long bed, measuring around 33 feet, was ideal for lumber. The fold-down stakes are shown upright in the drawing and were used to hold lumber when stacked above the low car sides. Photographs of a shorter Pennsylvania plain gondola from the same period show the stakes in their down and secured position (Fig. 5.19, 5.20). Many details pertinent to early car construction are shown in the French engravings. The position and arrangement of the drop doors between the needle

beams and below the floor sills are clearly shown. Notice how carefully the windup-chain shaft is placed just below the floor level. The truss rods, located well under the car, are supported by the shortest possible iron queen posts, which are so truncated that they might better be described as pads. The body bolster is trussed with two rods that are attached to flat bars at the car's center so as not to cut into the center or draft-gear sills any more than is necessary. The cast-iron center plates on the body bolster, the iron dead blocks, and the draft gear and coupler are all carefully delineated in this remarkably clear drawing.

It is easy to understand the presence of drop-bottom gondolas on lines with comparatively small coal traffic, but they were used by major coal roads as well. In evidence are two photographs showing cars on the Louisville & Nashville and the Norfolk & Western (Figs. 5.21, 5.22). The first of these views shows a high-sided gondola built in 1887 by the Ensign Manufacturing Company. The 30-foot-long car with its two slender truss rods boasted a capacity of 30 tons. It was in fact one of 2,698 drop-bottom gondolas that were the most numerous single class of coal car on the L&N.[22] The N&W had a smaller number of drop bottoms, 1,490, which represented about one-third of this big coal road's large coal car fleet.[23] The No. 40910's inside body measurements were 33 feet long, 8 feet 4 inches wide, and 2 feet 6 inches high. The car weighed 27,300 pounds. The windup-chain shaft is just barely visible in the 1893 Pullman builder's view. Within seven years the N&W was building 40-ton-capacity steel-framed drop-door gondolas.[24] These high-sided gondolas had wooden bodies and floors. Four doors helped speed the unloading

The Lake Shore & Michigan Southern, like the N&W, made the transition from 30- to 40-ton drop-bottom gondolas during the 1890s. An example of one of the wooden 30-tonners is shown in an

CARS FOR BULK CARGOES

FIGURE 5.20

An end-view photograph of a
Pennsylvania Railroad drop-
bottom gondola of 1876. (Pennsylvania Railroad Photo)

1893 photograph (Fig. 5.23). Because of its deeper side sills, the car has a more solid look than its N&W counterpart. The air brakes and Janney-style couplers anticipate federal pressure for safer freight equipment. Note how the truss rods are clustered in pairs below the side sills so as to clear the center of the floor for the drop doors. In 1899 the Lake Shore acquired five hundred 40-ton-capacity drop-door gondolas from the Michigan-Peninsular Car Company.[25] Unlike the N&W, the Lake Shore decided to stay with wooden framing and so resorted to 12-inch-deep side sills and six very deep truss rods arranged in pairs. The rods were a hefty 1¼ inches in diameter. Four drop doors were installed behind the needle beams rather than at the car's center. The windup mechanism was entirely below the floor, but of more importance, the doors were made flush with the floor, thus eliminating the dreadful mantraps typical of the traditional design for such cars. The car had Simplex steel body bolsters, air brakes, and Master Car Builders Association standard couplers as specified by their designer, A. M. Waitt. Another five hundred cars on the same plan were built for the Pittsburgh & Lake Erie.

A variation on the drop-bottom scheme was devised by F. E. Canda of the Ensign Manufacturing Company.[26] Canda placed two parallel sliding sheet-metal doors at the car's center. They were moved by ratchet levers and light chains. The levers were placed on opposite sides of the car, as

shown in Figure 5.24. A sample car was built for the 1893 Columbian Exposition in Chicago. No remarkable advantage is apparent, but at least Canda again demonstrated his ability to produce an alternative solution to an old design problem.

HOPPER-BOTTOM GONDOLAS

Another major class of coal car was the hopper-bottom gondola. It was a compromise vehicle designed to carry timber, rail, or pipe in addition to coal, but it was less suitable for mixed cargo than gondolas or drop-door gondolas. It was but one step away from being a full-fledged hopper car. About one-third of the floor space was given over to the pyramidal hopper, while the ends of the floor were flat, like a platform car's. These large flat areas served as broad piers for long objects to rest upon and thus span the open pit formed by the hopper cavity. Like most compromises, this one served neither of two distinct needs very well. The hopper-bottom gondola was a poor flatcar because of the hole in its middle. It was a halfway satisfactory self-unloading coal car because of the large level floor areas at both ends.

To the best of my knowledge, the particulars concerning the origin of this type of coal car are lost. The oldest illustration of a hopper-bottom gondola is a perspective engraving reproduced here as Figure 5.25. It represents a New York Central & Hudson River Railroad car that appeared as No. 8181 in a similar but less detailed engraving of

FIGURE 5.21

This Louisville & Nashville drop-bottom gondola was built in 1887 by the Ensign Manufacturing Co., Huntington, W.Va. (ACF Industries Photo)

FIGURE 5.22

The Norfolk & Western was a major coal road that employed a large number of drop-bottom gondolas. This car was built by Pullman in 1893. (Pullman Neg. 2210)

FIGURE 5.23

The Lake Shore & Michigan Southern purchased five hundred drop-bottom gondolas from Pullman in 1893. Note the pairs of truss rods under the outside sills. (Pullman Neg. 2317)

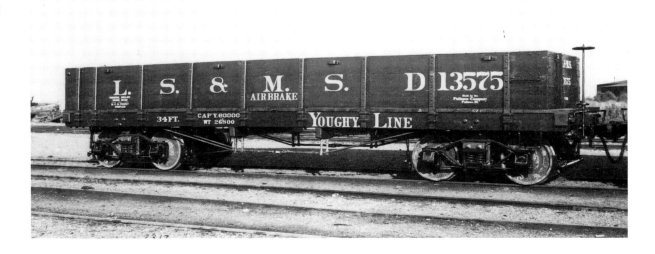

CARS FOR BULK CARGOES

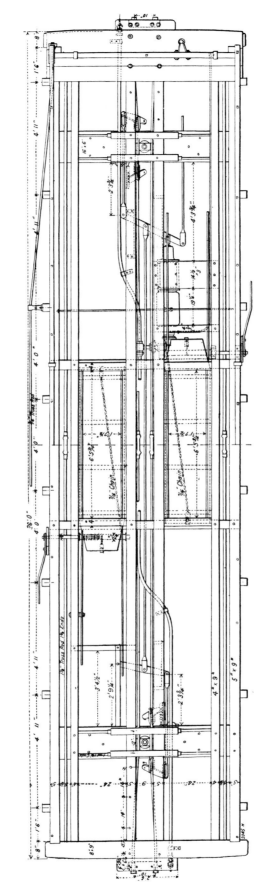

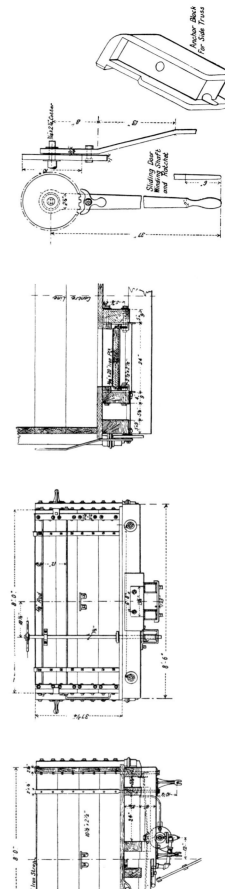

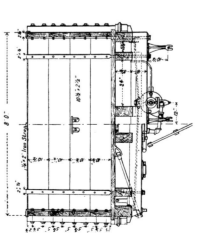

F. E. Canda devised a sliding bottom door that was used by the Ensign Manufacturing Co. Notice the ratchet levers used to open and close the doors that are fixed to the car's sides. (James Dredge, *Record of Transportation Exhibits at the World's Columbian Exposition of 1893* [London, 1894])

FIGURE 5.25

Hopper gondolas, like this New York Central car of the 1870s, were more of a coal car and less of a flatcar than the several examples of gondolas previously reproduced in this volume. (Lavoinne and Pontzen)

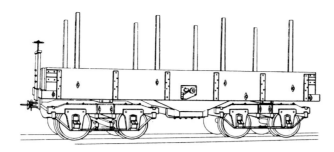

FIGURE 5.26

The Pennsylvania Railroad class GB hopper-bottom gondola was rated at 25 tons capacity. (Lavoinne and Pontzen; metric measurements)

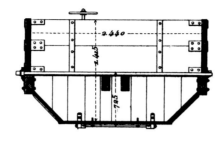

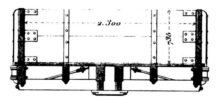

the same style of car in *Asher and Adams's Railroad Atlas* (1879). The car's small size and ancient wooden trucks suggest that the design—though not necessarily this particular car—dated before 1870. The windup-chain apparatus, so plainly shown, is by now familiar to readers of this book. The tall inside-mounted lumber stakes are a less familiar feature, however, and are not easily explained.

The Pennsylvania Railroad class GB coal car is more easily dated, having been introduced in 1874. The car was comparatively short, even for the mid-1870s. The frame over the end sills measured only 26 feet, while the body was a scant 24 feet on the outside. James Dredge gives the weight as 17,350 pounds, while an 1876 photograph of GB No. 8397 shows a weight of 17,700 pounds.[27] The slight variable is undoubtedly accountable to the type of wood used or some alteration in the hardware applied. Dredge gives the capacity as 37,000 pounds, but the car was generally rated as carrying 20 tons. Altoona produced this style of car for $480.69. Technically, the class GB cars are rather ordinary (Figs. 5.26, 5.27). The outside sills were of moderate size, 4 by 10 inches. The hopper was all wooden without the benefit of a sheet-metal lining. The wooden body bolster and draft gear were of an old-fashioned plan. Notice how the center

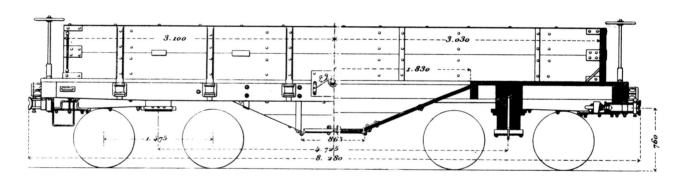

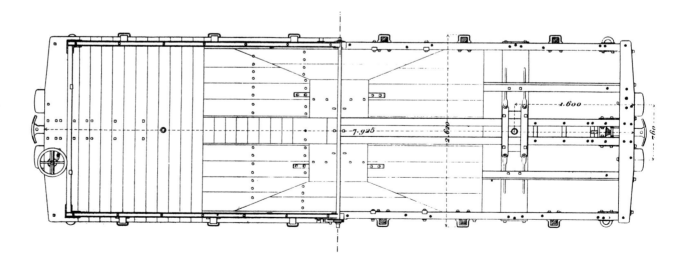

FIGURE 5.27

An Altoona shops photograph of a new hopper-bottom gondola delivered in May 1876. It is very similar to the design shown in Figure 5.26. (Pennsylvania Railroad Photo)

FIGURE 5.28

A larger Pennsylvania hopper-bottom gondola was produced by Jackson & Woodin in 1886 or 1887.

FIGURE 5.29

This Buffalo, Rochester & Pittsburgh hopper-bottom gondola was built by the Peninsular Car Co. in 1886. Note the open wooden hopper doors and windup chains. (ACF Industries Photo)

FIGURE 5.30

A long hopper-bottom gondola that was built for the Delaware & Hudson by Jackson & Woodin in 1893. (ACF Industries Photo)

FIGURE 5.31

The Philadelphia & Reading purchased five hundred long hopper-bottom gondolas from Pullman in 1892. (Pullman Neg. 1987)

sills interrupt the hopper space. They are planked over to provide a narrow walkway for the brakemen. The sides of the car were so low—29 inches—that trainmen could easily step over the ends as they passed from car to car. The design was so well liked by the assistant superintendent of the Erie that it was faithfully reproduced, with only very minor modifications, for service on that road.[28] In the published specifications for that series of cars, it was noted that the 2¾-inch-thick side body planks were of pine, while the 1¾-inch-thick floorboards were made of white oak.

The class GB grew in capacity with the passing years by the mere expedient of raising the car sides from 29 to 47 inches. Weight grew to 20,700 pounds while capacity was enlarged to 50,000 pounds. In 1886 or 1887 the designers at Altoona brought forward an expanded hopper-bottom gondola known as the class GD (Fig. 5.28). It closely followed the lines of its predecessors, except that the frame and body were lengthened by 2 feet. The result was a 30-ton-capacity car that weighed

23,200 pounds. The sides were held in line by truss rods placed high up on the body. Air brakes and Janney couplers boosted the weight of the GD by 2,200 pounds. In 1886 the Buffalo, Rochester & Pittsburgh purchased a group of cars very similar in design to the Pennsylvania's class GD, from the Peninsular Car Company. A photograph of one of these units is reproduced here because it so clearly shows the pyramidal hopper with its doors in the open position. One of the windup chains is clearly visible on the right-hand door (Fig. 5.29).

The Pennsylvania avoided truss rods by keeping its cars short, but other lines liked the idea of low-sided hopper-bottom gondolas and wanted longer frames to make them more adaptable for return loads. We can examine several examples of longer cars of this class outfitted with truss rods. A Delaware & Hudson car of this type, built by Jackson & Woodin, is shown in Figure 5.30. The pair of queen posts in no way interferes with the hopper or its doors. A much longer variation on the same theme was produced for the Philadelphia & Reading

CARS FOR BULK CARGOES

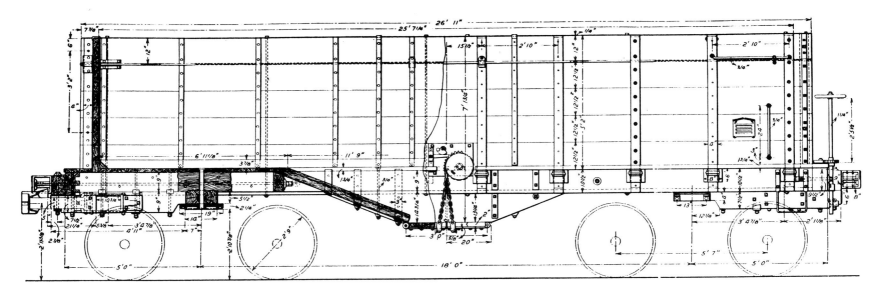

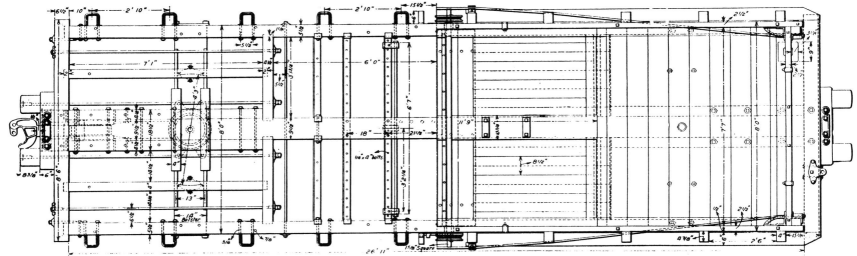

FIGURE 5.32

The Lenoir Car Co. produced 30-ton hopper-bottom gondolas for the Southern Railway in 1896. (*Railroad Gazette*, April 17, 1896)

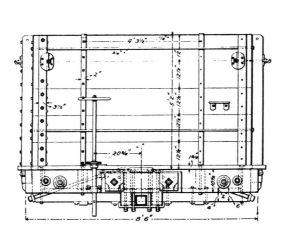

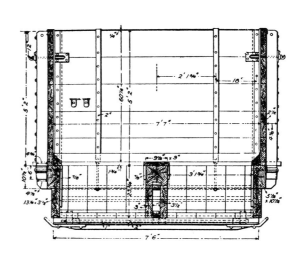

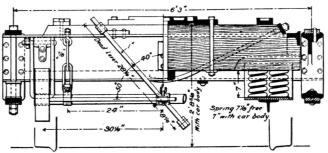

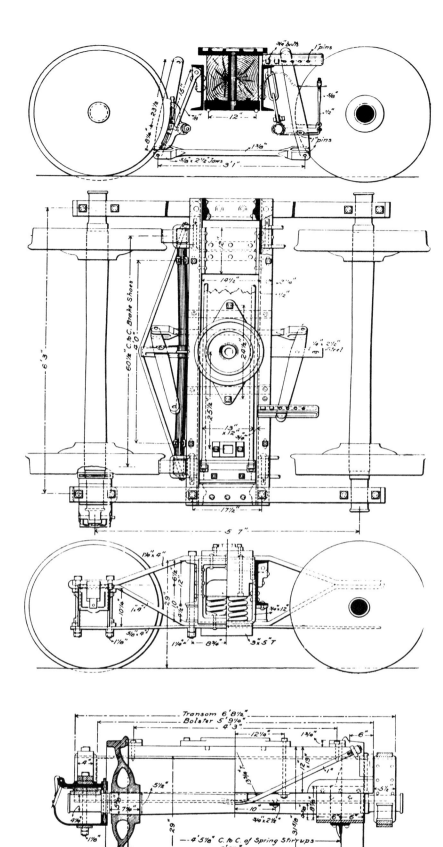

around the same time by Pullman (Fig. 5.31). Dimensions taken from a diagram drawing of a sister car show this series of hopper-bottom gondolas to measure 38 feet 5 inches over the bumper blocks. Cars of this length could carry rail, long timbers, and poles as well as coal, and so were more adaptable to mixed traffic than the short hopper-bottom gondolas favored by other lines.

Some lines ordered hopper-bottom gondolas with straight sides rather than the more traditional pyramidal shape. Only the floor of the hopper itself sloped. Drawings of a late-model 30-tonner built in 1896 for the Southern Railway by the Lenoir Car Company of Lenoir, Tennessee, are reproduced in Figure 5.32.[29] The design had given satisfaction for some years and so should not be taken as the latest thinking for 1896. The short body, wooden bolsters, and dead blocks were old-fashioned features, but Janney-style couplers and air brakes helped make it more acceptable in the beginning of the safety appliance era.

TWIN-HOPPER-BOTTOM GONDOLAS

The twin-hopper-bottom gondola is the final type of coal-carrying gondola in this series. Its origins are uncertain, but it does not appear to have been much used before the 1880s. It is less a dual-purpose vehicle and more a single-purpose coal car than any of the varieties previously discussed. As a general rule, more than half of its floor area was given over to the hopper. The Lehigh Valley showed a preference for this style of car, and we have good engravings for two designs worked out by the road's longtime master car builder, John S. Lentz (Figs. 5.33, 5.34). The first of these was for a 30-ton-capacity car measuring 38 feet 6 inches over the dead blocks.[30] The body was long enough for two stacks of standard 16-foot lumber. The center sills, the windup-chain shafts, and a large central cross-timber formed an open floor for lumber in the hopper areas. Just how well lumber rode on this skeleton bed is unrecorded, but the arrangement was surely not a perfect one. Loading and unloading must have been awkward. Large wooden dead blocks were required so that these gondolas could be interchanged with trains of jimmies. As can be seen from the drawings, iron body and truck bolsters were used. The body and hoppers provided 910 cubic feet of space for coal. The Packerton shops consumed 2,900 board-feet of lumber for each car produced on this plan.

In 1896 drawings were published for a somewhat shorter version of this 30-ton twin-hopper coal car (Fig. 5.34). This series was equipped with King doors in an effort to remedy the evils of the old windup-chain arrangement. It also had very deep cast-iron queen posts of a most elaborate pattern. The two rods were calculated to carry about 27,500 pounds, or roughly half of the car's rated loading capacity. The iron body bolster was at-

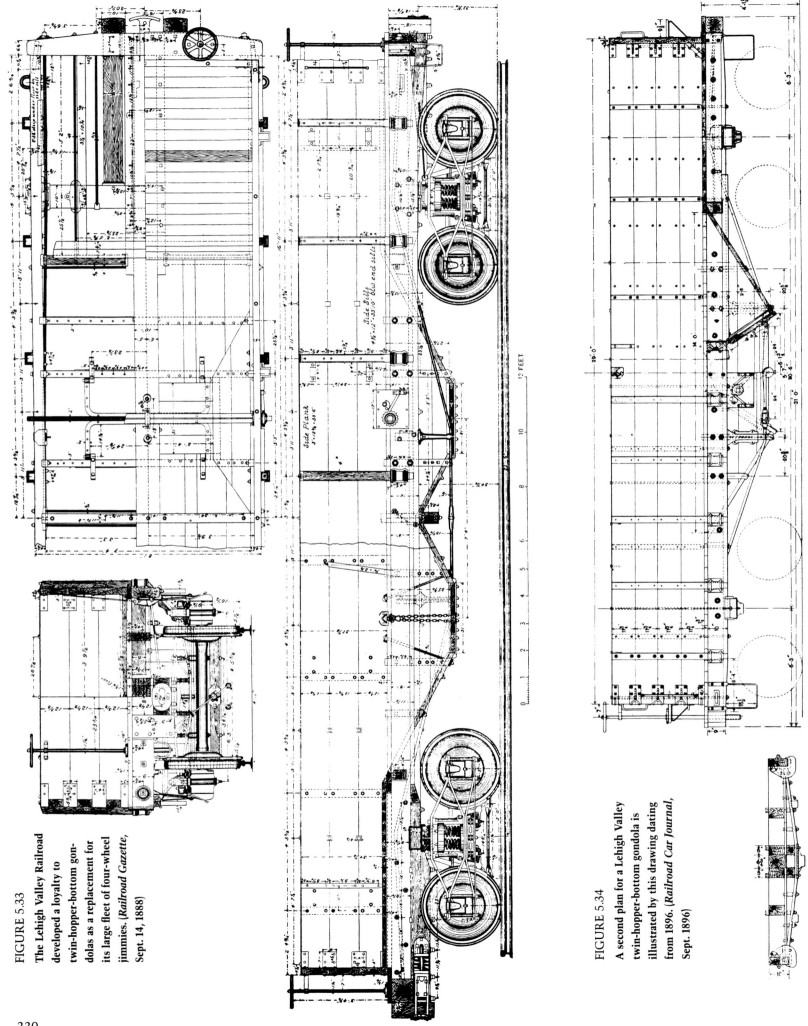

FIGURE 5.33

The Lehigh Valley Railroad developed a loyalty to twin-hopper-bottom gondolas as a replacement for its large fleet of four-wheel jimmies. (*Railroad Gazette*, Sept. 14, 1888)

FIGURE 5.34

A second plan for a Lehigh Valley twin-hopper-bottom gondola is illustrated by this drawing dating from 1896. (*Railroad Car Journal*, Sept. 1896)

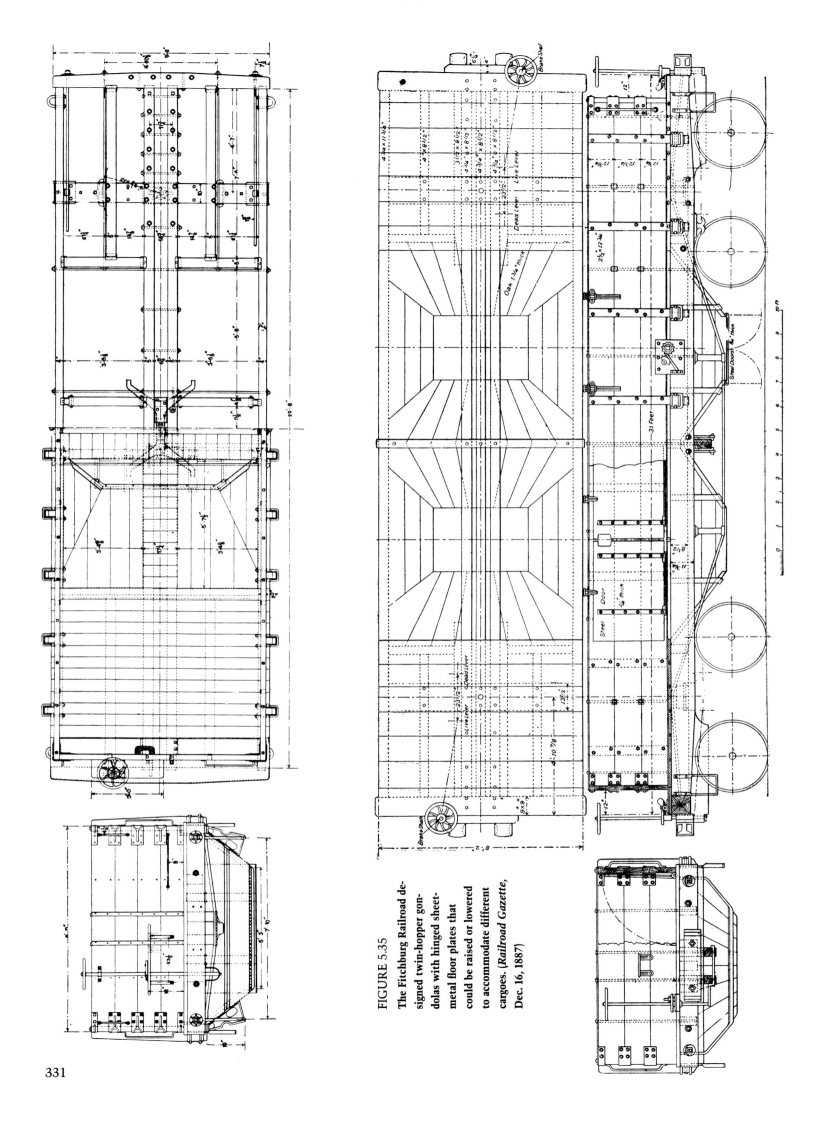

FIGURE 5.35

The Fitchburg Railroad designed twin-hopper gondolas with hinged sheetmetal floor plates that could be raised or lowered to accommodate different cargoes. (Railroad Gazette, Dec. 16, 1887)

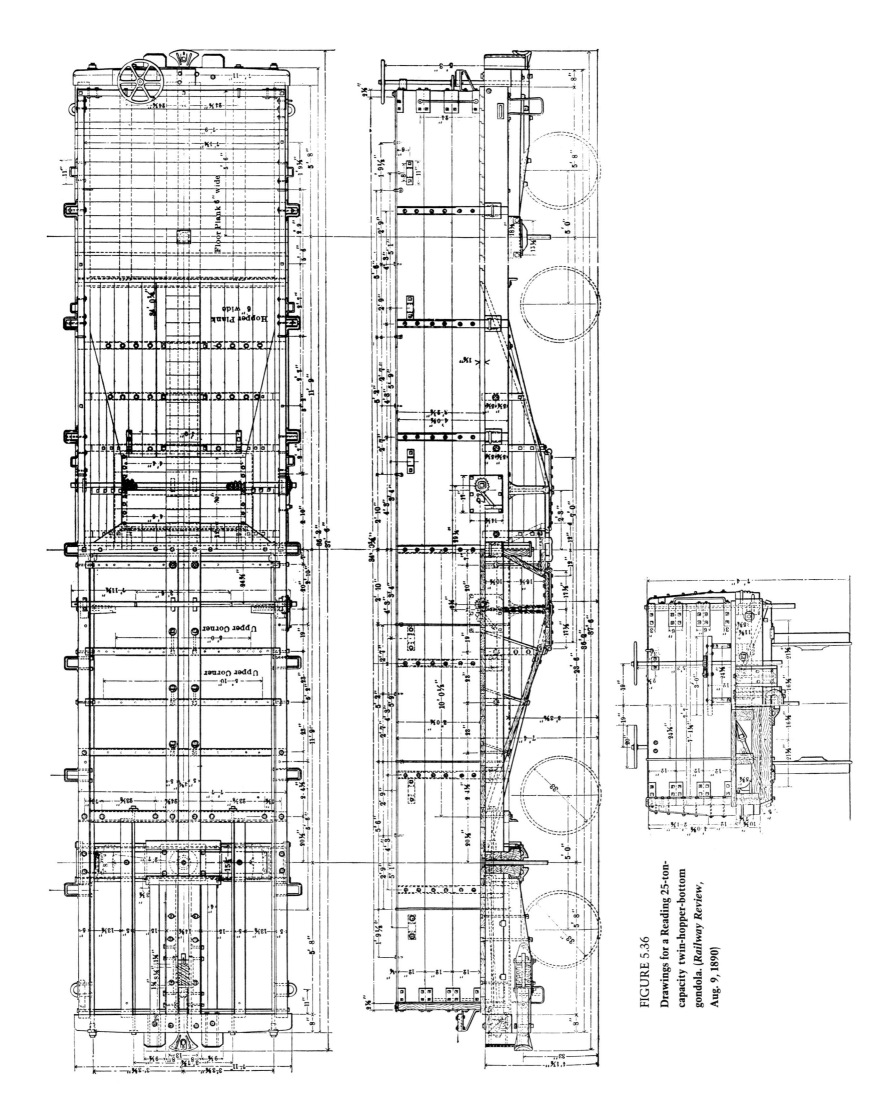

FIGURE 5.36

Drawings for a Reading 25-ton-capacity twin-hopper-bottom gondola. (*Railway Review,* Aug, 9, 1890)

FIGURE 5.37

The Middletown Car Co. produced this 30-ton-capacity gondola in April 1900 for the Reading just as the wood-car era was nearing its end. (Arthur D. Dubin Collection)

tached to the side sills by curiously shaped end castings. Janney-style couplers, air brakes, and Fox pressed-steel trucks completed the equipment.

At the very time Lentz was completing his drawings, J. W. Marden of the Fitchburg Railroad was finishing a group of 25-ton-capacity cars at that road's Charlestown shops.[31] They weighed 24,720 pounds and yielded 940 cubic feet of space. The Fitchburg, like most New England railroads, carried relatively little coal, except for local delivery, and so Marden took special measures to make the car suitable for other cargoes. Eight large plates of ¼-inch-thick steel were hinged to the body's inner sides (Fig. 5.35). They could be folded down to form a continuous floor or raised up and latched to clear the hoppers for coal. The idea was a sensible one, but, of course, it added cost and weight and so was not widely copied.

Even the roads specializing in coal traffic ordered twin-hopper-bottom gondolas in large numbers. The Philadelphia & Reading, looking for a more modern replacement for its fleet of aging jimmies and small 10-ton hoppers, presented Pullman with a series of orders for eighty-five hundred single- and double-hopper-bottom gondolas.[32] The quantity built in its own shops or by other outside builders is unknown, but it can be assumed that Pullman was not the Reading's sole source for new gondola cars. Such a massive buildup in just a few years indicates a major shift in car selection by a big eastern coal carrier. Drawings for a long, low-sided, 25-ton-capacity twin-hopper gondola are reproduced as Figure 5.36. The inside of the body offered a clear space of 34 feet, which again allowed the carriage of two parcels of 16-foot lumber. The center sills are set apart by a 4¾-inch gap that is bridged over with short lengths of floorboards to provide a walkway for trainmen. These cars had two very heavy truss rods measuring 1½ inches in

diameter. The wooden body bolsters were reinforced with two 5-inch channel irons. Schoen pressed-steel center plates were used. The cars did not have air brakes, but they were outfitted with Van Dorsten couplers, a variation on Janney's basic design. Between 1888 and 1894 the Reading ordered or built cars on this exact pattern that were numbered in the 42001–49999 and 57000–59950 series.[33] The exact number constructed is not known, however.

Beginning in 1895 the Reading started acquiring twin-hopper gondolas of 30 tons capacity. Numbered in the 30249–39999 series, most were built by the Middletown Car Company of Middletown, Pennsylvania. These cars were about 4 feet shorter than the series previously described, with an inside body measurement of 29 feet 6¼ inches. This would have limited their utility as lumber carriers. A photograph taken at the Middletown works in April 1900 shows one of these cars as outfitted with air brakes, Janney-style couplers, and Fox pressed-steel trucks (Fig. 5.37). Notice also the pressed-steel corner plates and the cast bracket-style queen posts.

The New York Central produced plans for a solidly built twin-hopper gondola in 1892.[34] The 35-foot 6-inch-long car provided a coal space of 1,091 cubic feet with a normal rating of 30 tons, a figure that could be boosted by 5 tons if the coal was heaped high in the center. Large, one-piece corner irons were considered a good design feature. The hoppers were of normal pyramidal shape with windup chains. While old-fashioned wooden bolsters were used, the cars had air brakes and Janney-style couplers. The air-brake cylinder and tank were ingeniously placed at one end of the car between one of the hoppers and the truck. The more normal placement was between the hoppers under the car or outside the frame at the car's center. Written

CARS FOR BULK CARGOES

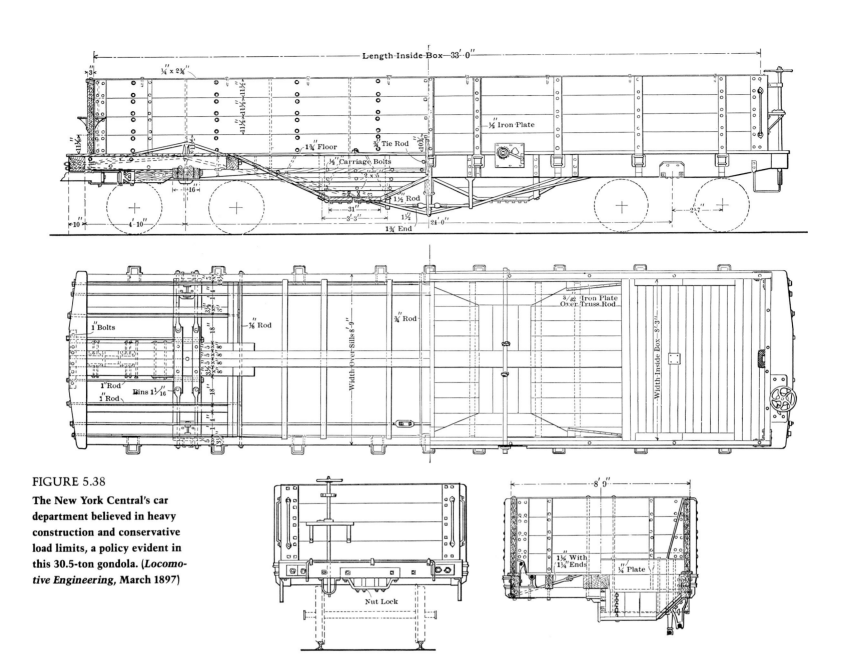

FIGURE 5.38

The New York Central's car department believed in heavy construction and conservative load limits, a policy evident in this 30.5-ton gondola. (*Locomotive Engineering*, March 1897)

descriptions of freight cars almost never go beyond a listing of dimensions, but in this case the color of the car was given as brown with white lettering.

The 1892 general design was improved a few years later by strengthening the frame and covering the hoppers with ¼-inch steel plates.[35] The outside sills grew from 4½ by 12 inches to 5 by 13½ inches. Four, rather than two, truss rods helped support the frame. The 1½-inch-diameter rods dropped a full 30 inches. Coal space increased to 1,220 cubic feet, but the cars were conservatively rated at only 30.5 tons. Inside iron knee braces and full corner plates helped strengthen the body (Fig. 5.38). The car was surely a heavy one, but it was felt that the extra mass of its parts would be favorably reflected in the repair accounts. One small detail that might be overlooked is the small iron steps placed on either side of the body ends to assist the trainmen in their movement over the train. The sides were too high to step over.

Forty-ton-capacity wooden-frame twin hoppers followed the same growth pattern of the general freight car fleet, and a goodly number entered service even after the acceptance of steel cars in the late 1890s. An example is shown in one of the better Pullman freight photographs, reproduced here as Figure 5.39. It shows a sturdy, high-sided car completed for the Rutland in 1901. The outside air-brake apparatus and the pair of massive truss rods are clearly shown. Notice how the body stakes are clustered near the center to prevent side bulging.

HOPPERS

The ultimate coal car was the self-unloading hopper car. Its entire floor area was sloped, and as such it was suitable only for loose, weather-resistant cargoes such as coal, ore, or gravel. It was remarkably efficient for bulk cargoes of this nature, and like all specialized vehicles it did one job very well. Unlike the other cars previously discussed in

FIGURE 5.39

Forty-ton wooden gondolas began to appear around 1900. This twin-hopper-bottom car was built by Pullman in 1901. (Pullman Neg. 5740)

this chapter, it was not readily adaptable for other uses. The idea of an inclined floor can be traced back to the beginning of American railroading. It appears to be of American origin, and the earliest known drawing for such a car, a Delaware & Hudson vehicle, is reproduced in Chapter 2 as Figure 2.4. Earlier British coal wagons had inclined side walls and drop-bottom doors but appear to have had flat floors. The slope-floor idea was perpetuated by the jimmy and in time was applied to eight-wheel cars.

The first large fleet of eight-wheel hopper cars with sloped floors was introduced by the Philadelphia & Reading Railroad. After a slow start in the middle 1850s, more 10-ton-capacity cars of this design entered service, until by 1870 the Reading had 5,723 in service. In 1880 the Reading began to replace its Gunboat 4-6-0 freight locomotives with more powerful 2-8-0 consolidations. The movement toward greater tonnage in a single train was now possible with the new locomotives. Bigger locomotives led to bigger cars, heavier rail, and stronger bridges. As part of a general upgrading of its coal-handling operation, the line's engineering department saw the need for more and larger eight-wheel hopper cars.

In 1884 the old plan was revised and enlarged to double the hopper's capacity (Fig. 5.40).[36] Officially, the capacity was conservatively rated at 35,840 pounds, but 40,000 pounds was a more realistic figure, and the vehicle could safely handle as much as 50,000 pounds. It weighed 18,480 pounds. The body measured inside 21 feet 6 inches by 7 feet by 3 feet 8 inches and yielded 540 cubic feet of space. Two unique design features should be spe-

cially noted. The first of these is the lack of a center sill. The great majority of wooden hopper-bottoms were encumbered by center sills that intersected the otherwise open coal space. The side sills were sturdy but, at 4 by 10 inches were hardly oversize. Large body bolsters and end sills explain in part the structure's ability to withstand the strains and pressures of buffs and drafts. Yet the integrity of the body and frame is better explained by the car's second unique feature, an all-metal floor made from heavy sheet iron. The exact thickness does not appear to have been recorded, but ⁵⁄₁₆ inch is the most likely dimension. The floor sheets were stiffened by angle-iron crosspieces, as shown in Figure 5.40.

The hopper doors were also made of sheet iron, as was the lower part of the hopper sides. Note that the hopper is straight-sided rather than pyramidal. The incline of the floor was not very steep, and because of this it took three minutes to unload after the doors were dropped. The body side parts and angle braces are mounted in cast-iron pockets to avoid mortising, which would have weakened the side sills. The very large corner plates on the frame help stiffen the entire car. The curious U-bolts fastened to the side sills near each end of the car were used as an attachment for ropes or chains in switching movements. They were not used for lifting the car off its trucks, as might be supposed. The curious hook-and-chain-link couplers are clearly shown in drawings reproduced in Figure 5.40. The windup-chain shaft is placed high on the body; because the cars were used only for coal, there was no reason to place the shaft at floor level. The very large center plates associated with the

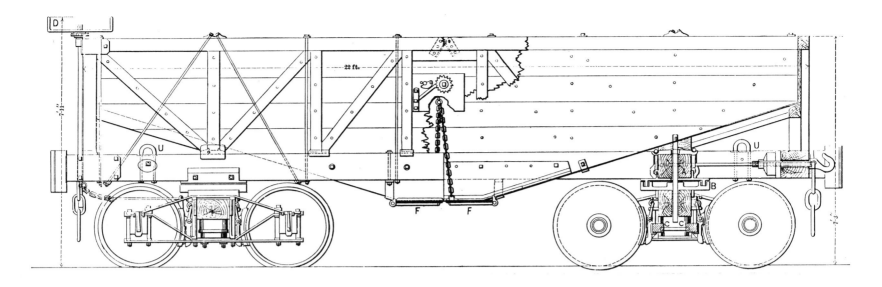

FIGURE 5.40

This Reading car of 1884 was a full hopper intended exclusively for the carriage of coal. The design shown here was considered a model for the industry. (*National Car Builder*, March 1884)

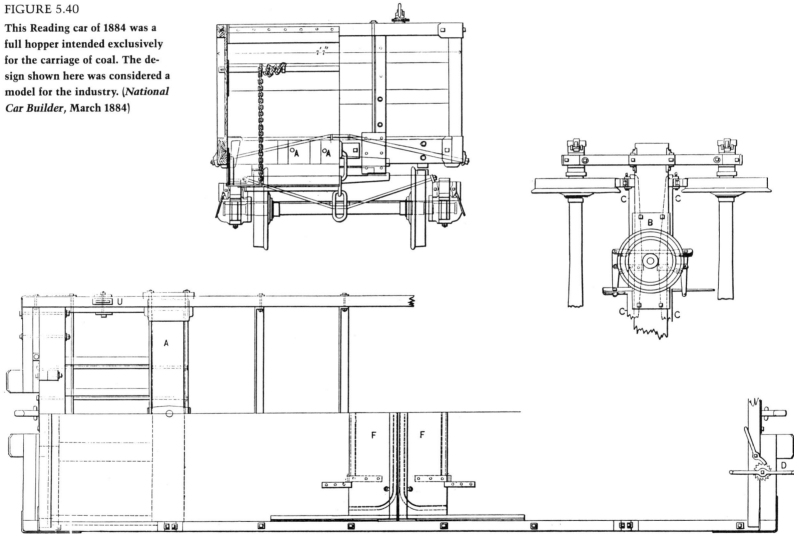

Plan of Floor, Drop-Doors and Truck.

THE AMERICAN RAILROAD FREIGHT CAR

FIGURE 5.41

A string of eight-wheel 20-ton hopper cars stand here at Pottsville Junction on the Reading system, around 1890. (Thomas Norrell Collection, Smithsonian Neg. 85-28477)

trucks are worthy of notice, as is the long king pin dropped down through the hopper sheets. What might appear to be an air-spring cylinder on the trucks is in reality a round, hollow cast-iron casing that houses a cluster of small coil springs.

In 1884 the Reading had 3,134 cars of this design in service. Just how many more were added over the ensuing years is difficult to say, but the number was surely sizable (Fig. 5.41). In the late 1880s the Reading's car department appears to have abandoned the 20-ton hopper in favor of more flexible hopper-bottom gondolas. Even so, the line returned to wooden hopper cars even after the steel-car era was under way. In 1902 the Reading purchased two hundred high-sided, 40-ton-capacity twin hoppers from the Middletown Car Company.[37] These giant, all-wooden arks were hardly a reintroduction of the old standard design just described. They were 6 feet longer, at 32 feet 2 inches, and 2 feet higher. Center sills, truss rods, King hopper doors, Fox trucks, air brakes, and Janney-style couplers constituted other major differences that cannot be dismissed as mere variations on the old plan.

By 1877 the Lehigh Valley Railroad was casting about for something better than the diminutive jimmy.[38] Like many other coal roads, the Lehigh found a model in the Reading's coal car design. In 1885 published drawings appeared that give us a detailed picture of the Lehigh's plan.[39] Superficially, the car is the mirror image of a Reading car of the period. The size and general appearance are surely very similar (Fig. 5.42). A more careful examination reveals many differences in the details, however. The list is actually rather extensive; notice the center sills, truss rods, iron body and truck bolsters, wooden hopper floor and doors, pyramidal hopper, stake body sides, link-and-pin couplers, latch door closers, and top body cross-timbers. All of these features differ from the Reading's design. The shape of the hopper and the presence of four rather than only two doors allowed an unloading time of one and a half minutes, or twice as fast as the Reading's car of the same size.

Far to the west of the anthracite fields, coal cars very much on the Reading pattern were being produced by the Ohio Falls Car Company.[40] How directly they were inspired by the Reading is impos-

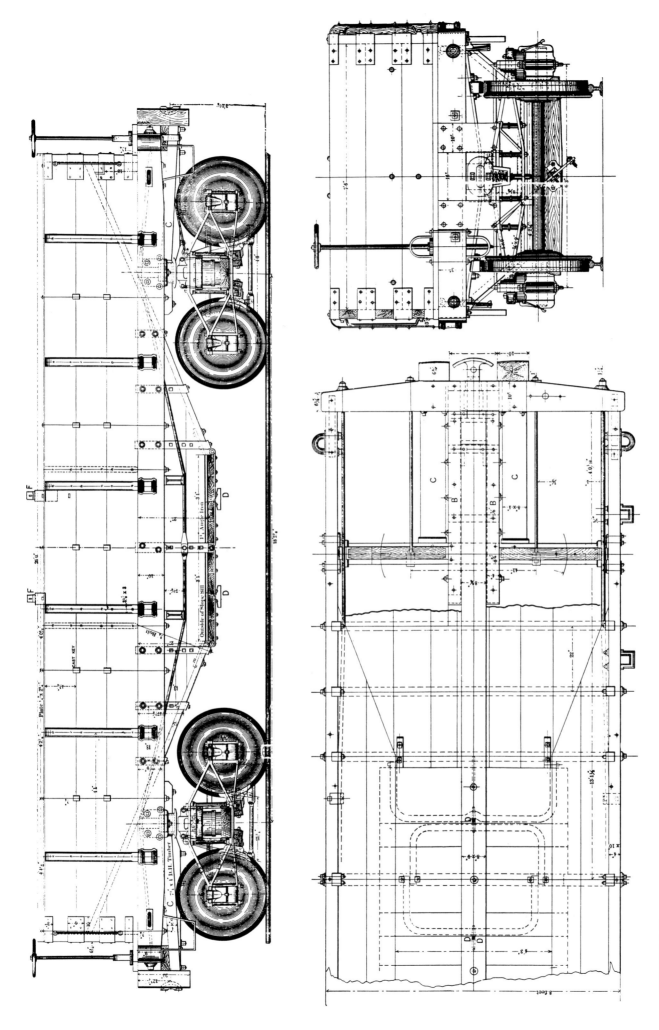

FIGURE 5.42

The Lehigh Valley modified the Reading's basic plan for coal hopper cars, as shown in this drawing. (*National Car Builder*, Jan. 1885)

THE AMERICAN RAILROAD FREIGHT CAR

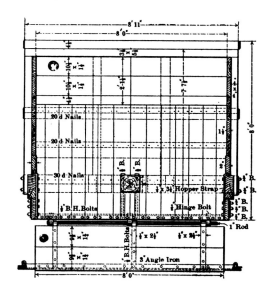

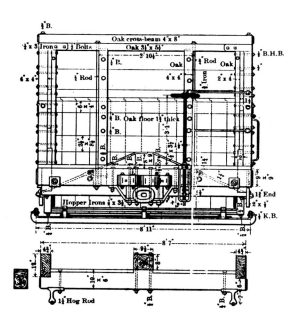

FIGURE 5.43

The Tredegar Iron Works built this 30-ton hopper for the Georgia Pacific in 1888. (*National Car Builder*, June 1888)

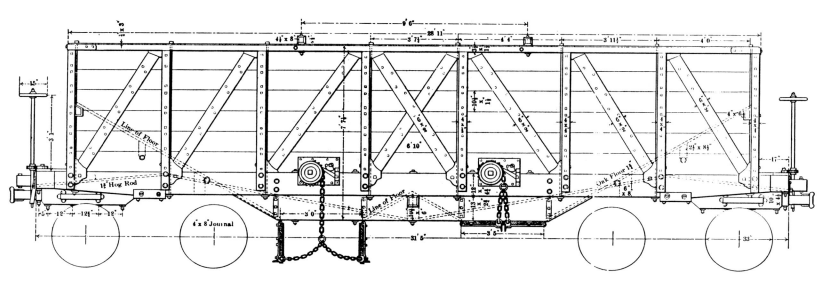

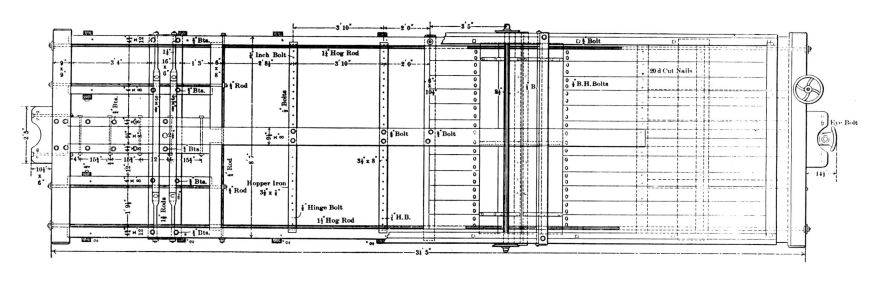

CARS FOR BULK CARGOES

sible to say, but the design was a much more faithful copy than the Lehigh Valley cars just described. Except for the couplers, and the presence of a single top body cross-timber brace, they were the twin of the Reading's standard coal hopper. Ohio Falls delivered a series to the Evansville, Henderson & Nashville and presumably other coal-hauling lines in the Midwest.

HIGH-SIDED HOPPERS

In Chapter 3 we reviewed the steady escalation in freight car capacities, in the post-1870 period, from 10 to 15 to 20 tons and so on until vehicles of 40 and 50 tons were achieved. For general merchandise cars this push for bigness was rightfully criticized as a grandiose gesture not always justified by practical needs. It was contended that most shippers sent goods in small lots that only rarely filled the bigger boxcars.[41] A coal car, by contrast, was almost never sent away partially loaded. Mines wanted to move their stockpiles directly to the market. Here the economies of scale could be di-

rectly realized, and the trend for bigger coal cars was steadily pursued by both shippers and carriers. The push for ever-bigger cars was inhibited more by bridge, track, and wheel loading limits than by a resistance to high-capacity rolling stock. Hopper cars became not only longer and heavier but higher. By 1890 some cars were beginning to take on a top-heavy look that made them appear almost as tall as they were long.

An early example of a 30-ton hopper was a group of cars built in 1888 by the Tredegar Iron Works of Richmond, Virginia, for the Georgia Pacific (Fig. 5.43).[42] They measured 31 feet 5 inches long and had a coal space of 1,290 cubic feet, or more than twice the space of the standard Reading car built just four years earlier. The side sills measured 4¾ by 12 inches, while the single center sill was an 8- by 9½-inch timber. King post truss rods helped strengthen the frame. The oak floor planks were 1¾ inches thick.

About six months after the Georgia Pacific cars were delivered, the Chesapeake & Ohio Railway

FIGURE 5.44

The Chesapeake & Ohio had enlarged its wooden hopper cars to 40-ton capacities by the late 1890s. (Pullman Neg. 4165)

FIGURE 5.45

The Baltimore & Ohio was another advocate of high-sided 40-ton wooden hoppers. This car was built by Pullman in 1897. (Pullman Neg. 3544)

THE AMERICAN RAILROAD FREIGHT CAR

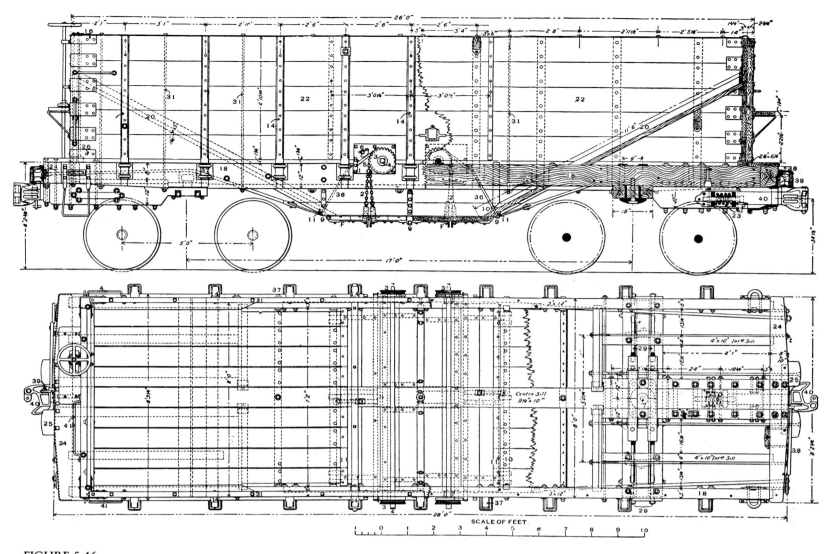

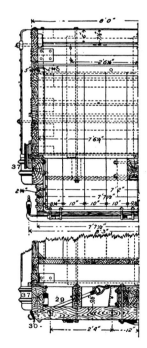

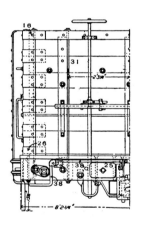

SCALE OF FEET

0   1   2   3   4   5   6   7   8   9   10

FIGURE 5.46

**These drawings for a Norfolk &
Western high-sided wooden hop-
per depict a design similar to the
C&O and B&O cars shown in
Figures 5.44 and 5.45 (*Railroad
Gazette*, June 8, 1894)**

acquired some high-sided hoppers of the same capacity that measured only 26 feet 8 inches long.[43]
The windup chains were mounted outside the
bodies in an effort to prevent the mechanism from
icing up during the winter months. By the late
1890s the C&O was purchasing 40-ton wood-
frame hoppers.[44] Length was just 16 inches longer
than the 1888–1889 model, but the sides were a
full 7 feet high. The side sills measured 4½ by 10½
inches, while the center sill was 9 by 10½ inches.
Truss rods were made flat at the midsections to
slip under the hopper sides. The outside windup
chains were retained for the hopper doors. A photograph of one of these 40-ton hoppers is shown in
Figure 5.44.

Other soft-coal carriers were adopting high-sided
hopper cars during the same period. The Baltimore
& Ohio started in 1890 with designs for 25- and 30-
ton hoppers of this type. The larger version measured 27 feet 3 inches long over the end sills. Originally its sides were 61 inches high, but after 1896
they were raised by 6 inches. A photograph of one
of these 30,400-pound hoppers with its king post
truss rods and outside windup chains can be seen
in Figure 5.45. In 1898 the B&O built an experi-

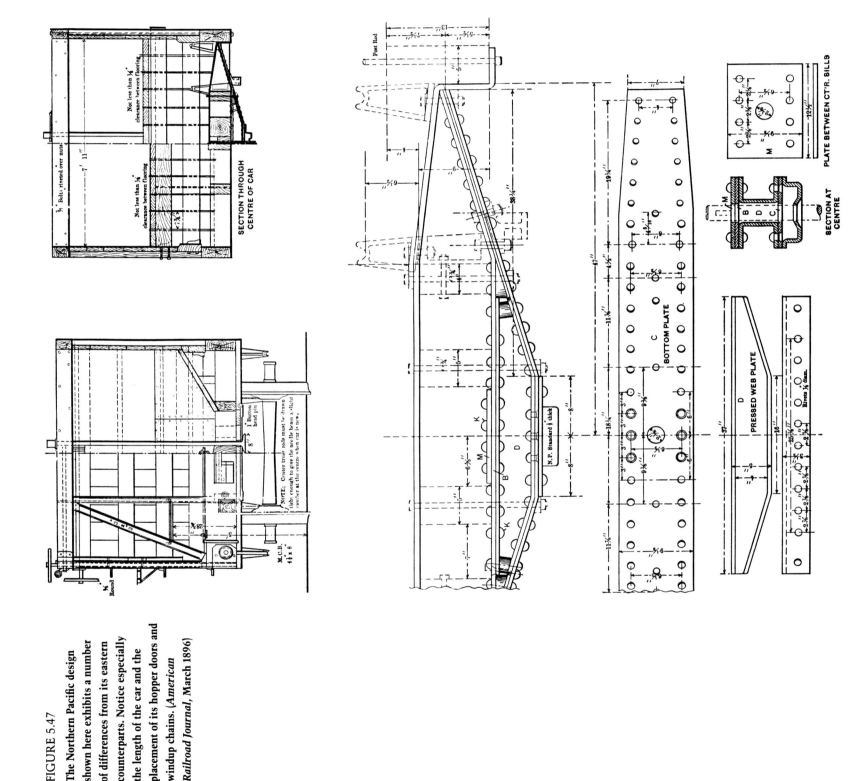

FIGURE 5.47

The Northern Pacific design shown here exhibits a number of differences from its eastern counterparts. Notice especially the length of the car and the placement of its hopper doors and windup chains. (*American Railroad Journal*, March 1896)

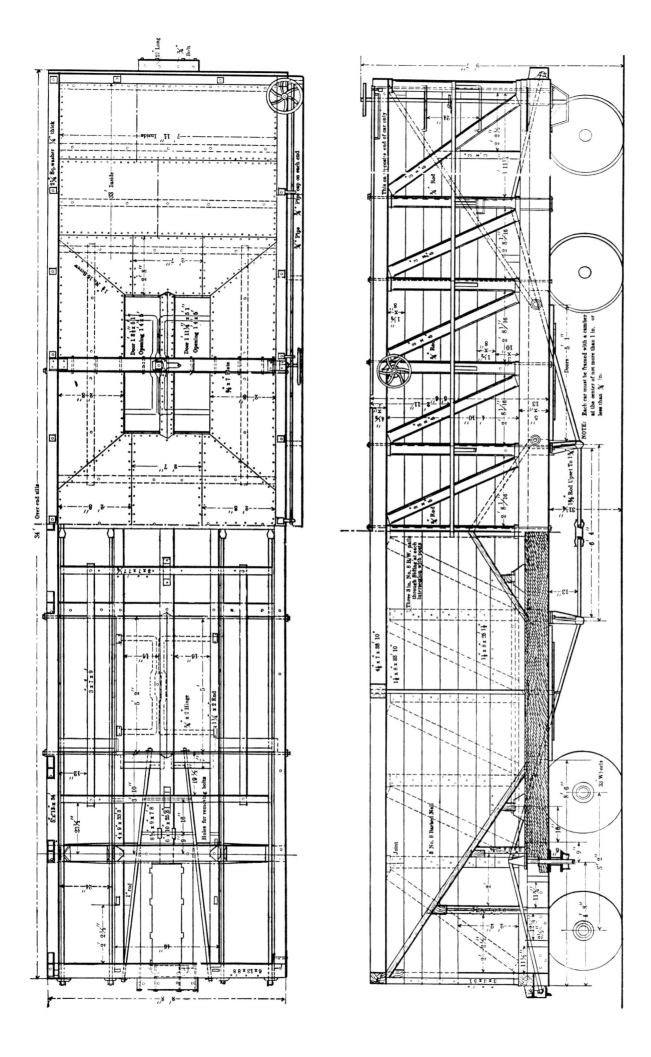

343   CARS FOR BULK CARGOES

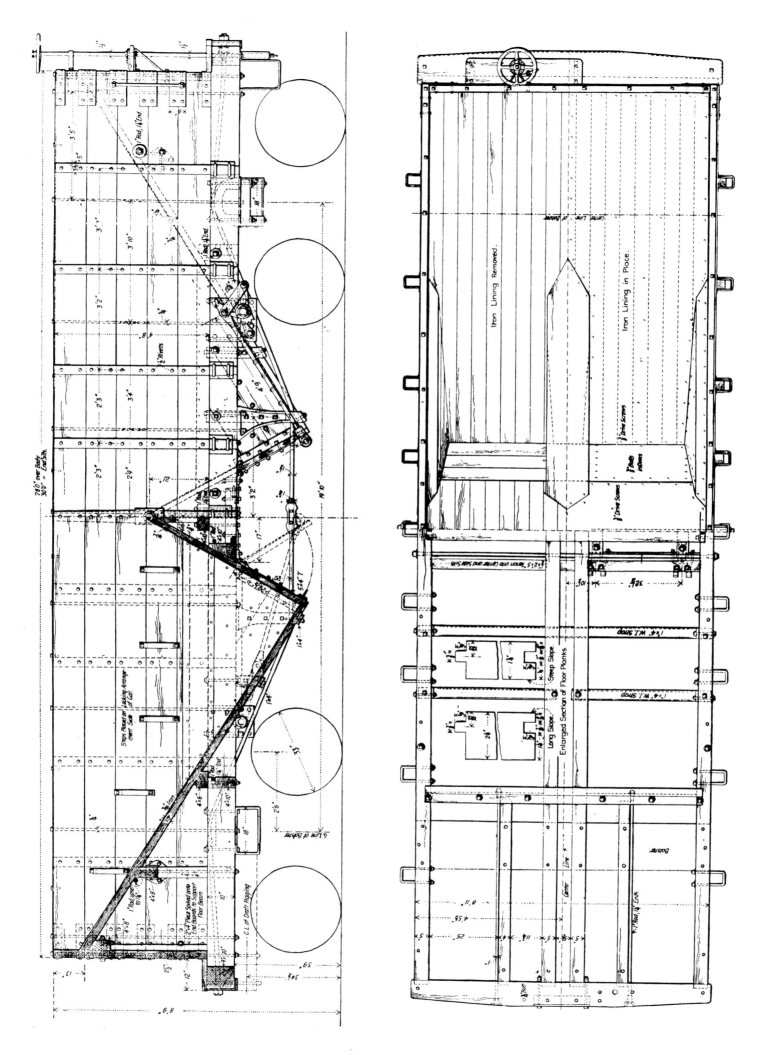

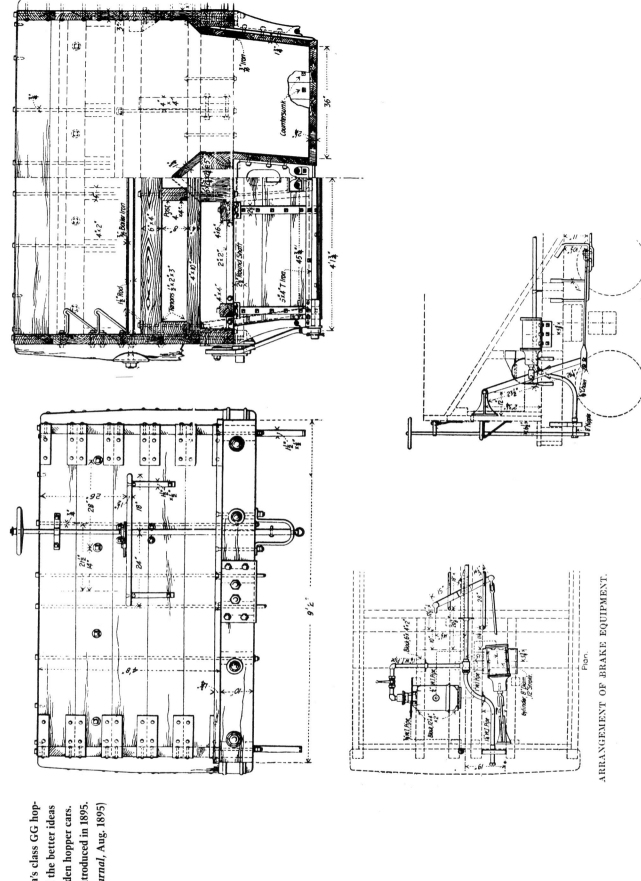

ARRANGEMENT OF BRAKE EQUIPMENT.

Plan.

FIGURE 5.48

The Pennsylvania's class GG hopper car combined the better ideas available for wooden hopper cars. The design was introduced in 1895. (*Railroad Car Journal*, Aug. 1895)

mental 40-ton version of this general style of car, but it was not adopted because of the advancing fortunes of the steel hopper.

The Norfolk & Western, realizing that gondolas were really not the ideal vehicle for massive coal shipments, began to acquire cars very similar to the big hoppers on the C&O.[45] The body sides were relatively short and low, measuring 26 feet by 4 feet 10 inches. Cross-timbers and rods placed near the top center of the body prevented side bulging. A heavy band of strap iron fastened to the top body boards of the sides and ends also helped to stiffen the structure (Fig. 5.46).

Far west of the big Appalachian bituminous fields, the Northern Pacific was at work on its own version of the full-scale wooden hopper car.[46] It was longer than its eastern sisters. Pyramidal hoppers and doors placed longitudinally, rather than transverse, permitted five instead of the usual three frame sills (Fig. 5.47). A center sill was employed, as were four truss rods. The other major peculiarity of these 35-ton-capacity cars was the placement of the windup shafts at the top rather than at the bottom of the hopper space. These were so high off the ground that a catwalk was attached to the car sides about midway up on the body. Handwheels, rather than square ends on the shaft for removable cranks, were another odd feature. The large side sills, 5 by 13 inches, are worthy of notice, but they are not the most massive ever used for this purpose. The Union Pacific was building big hoppers at the same time with 14-inch-deep frame timbers. About two years after the first lot of Northern Pacific cars, just described, entered service, the road's Tacoma shops built another one hundred sisters using Washington State fir instead of yellow pine or oak.

Officials of the Pennsylvania Railroad watched the growing interest in large hopper cars and at first seemed content to pursue nothing more adventurous than an upgrading of their traditional designs. While Altoona declined to act, George L.

Potter, superintendent of motive power for Lines West at Fort Wayne, Indiana, began to study the large foreign hoppers passing through the yards near his office.[47] Potter selected the ideas he liked best and combined them into what became the Pennsylvania's class GG coal car. It was a 35-ton-capacity vehicle that weighed 35,200 pounds. It was claimed that the car could safely carry 7,000 pounds more than its rated limit. It was a fairly short, low-sided piece of rolling stock that at first glance might be mistaken for a gondola (Figs. 5.48, 5.49). But its sloping floors and twin hoppers made it a full-blown hopper car. It measured 30 feet over the end sills. The body alone measured 28 feet long by 8 feet 11 inches wide by 4 feet 8 inches high. The 5- by 10-inch side sills were supported by two heavy (1½-inch-diameter) truss rods. The wooden center sills were placed close together. Schoen pressed-steel body bolsters added to the frame's rigidity. The body was built up from heavy 3- by 11-inch planks. The 2½- by 5½-inch floor boards were covered by ³⁄₁₆-inch-thick steel plates fastened down by wood screws.

The shape and arrangement of the hopper doors are worthy of special attention (Fig. 5.48). Notice that the doors are mounted at an incline rather than being positioned flat against the hopper opening; hence the doors swing out, rather than fall down, to open. They were hinged at the top and held closed by levers and rods rather than by the ancient windup-chain system. It is not certain if this idea was original with Potter, but this plan for doors later became an industry standard. The car was not only fully self-unloading; it could do so quickly and could discharge its load in one-half to three-quarters of a minute. Potter's design was finished in March 1895. An early order for one thousand GG hoppers was divided among Barney & Smith, Union, and the Altoona car shops. Cost was figured at $550 each. Hundreds more were built over the next few years, and before the advent of steel, they were considered one of the finest

FIGURE 5.49

A Pullman photograph shows a class GG hopper produced for one of the Pennsylvania's lines west in March 1896. (Pullman Neg. 3346)

THE AMERICAN RAILROAD FREIGHT CAR

FIGURE 5.50

F. E. Canda designed the largest wooden-frame hoppers for the Central Pacific in 1899. The cars were produced by American Car & Foundry at Huntington, W.Va. (ACF Industries Photo)

American hopper cars in service. Fortunately, a lone example has been preserved at the Pennsylvania State Railroad Museum in Strasburg.

Potter's attention to coal car design was an effort to consolidate the best ideas available for a wooden car of this type. It was left to another car builder, F. E. Canda, to design the largest wooden car of this type.[48] Readers will remember Canda's giant wood-frame boxcar prepared for the Southern Pacific in 1899, described in Chapter 3. He prepared plans at the same time for a 36-foot-long 50-ton-capacity hopper. Except for its size, there was nothing radical about the vehicle. Its floor timbers were of a conventional dimension. The four truss rods were only 1⅛ inches in diameter and were peculiar only because of the flat under-the-hopper sections, connected by links to the round portions of the rods (Fig. 5.50). The twin drop-door hoppers with their windup chains were, at best, old fashioned. The body posts were set in malleable-iron pockets. It would seem that the cast-steel body bolsters were about the only deviation from conventional wood-car construction that Canda was willing to employ. Three hundred cars were built on this plan by the former Ensign Car Company, which had just months before been taken over by American Car & Foundry. The longevity and service records of the 90000-series cars appear to have disappeared, but even assuming that they performed to the expectations of their designer, Canda was waging a losing fight against the rising fortunes of the steel hopper car.

OTHER HOPPER ENHANCEMENTS

While car builders labored to build larger coal cars, some pondered over ways to improve details like hopper doors and trussing. The *Railroad Gazette* wondered why the windup-chain hopper doors remained so popular despite their several defects.[49] Wet loads of coal would freeze around the chain and shaft, making the apparatus inoperable during much of the winter season. Yet no matter the season, the doors had the unhappy tendency to bounce open in transit, spilling the load along the tracks. C. H. Carruthers remembered a train ride on a hot day in the 1860s. The coach windows were wide open to catch the breeze. As they passed a slow-moving freight train, a dense cloud of white dust filled the car that set the passengers to sneezing like the inmates of an allergy ward. The culprit was a hopper car of lime in the adjacent freight train. The doors had dropped open, and it proceeded to dump its load in a billowing trail.

Such incidents could be prevented with latch- or lever-style door closers, but they were never popular until after 1900. Some roads would briefly adopt them and then go back to windup chains. Just why is never explained. An example of such a turnaround is the Lehigh Valley, which used latch closers on its 1885 hoppers but reinstated windup chains in its 1888 hopper-bottom gondolas. In about 1890 George I. King of the Middletown Car Company devised a toggle-lever door latch that was tried by the Philadelphia & Reading and a few other lines (Fig. 5.51). The efforts of G. L. Potter to introduce lever-type door closers on the Pennsylvania system has been mentioned in our discussion of the class GG hoppers. The plans of King, Potter, and numerous other car designers appear to have made little progress until after the first decade of the twentieth century.

Stiffening the frame and body of a heavily loaded hopper car presented yet another set of problems for the car builder. What went on under the floor was of no particular consequence for a flat- or boxcar, but this was not true for a coal car because of the bottom-unloading requirement. The center of the car had to be unencumbered by frame members or brake apparatus. A center sill was generally tolerated, but it was an obstacle eliminated in the better designs. Clustering a pair of truss rods under each side sill was one common solution, and

347

CARS FOR BULK CARGOES

FIGURE 5.51

George I. King devised a toggle-lever hopper-door latch that became moderately popular during the last years of the wooden hopper car. (*Car Builders Dictionary*, 1895)

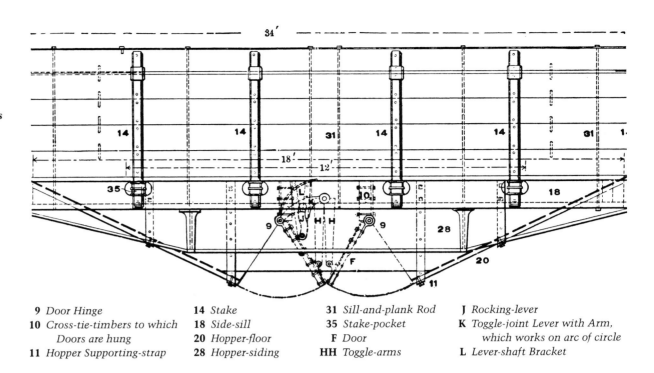

9	*Door Hinge*	14 *Stake*	31 *Sill-and-plank Rod*	J *Rocking-lever*
10	*Cross-tie-timbers to which Doors are hung*	18 *Side-sill*	35 *Stake-pocket*	K *Toggle-joint Lever with Arm, which works on arc of circle*
11	*Hopper Supporting-strap*	20 *Hopper-floor*	F *Door*	L *Lever-shaft Bracket*
		28 *Hopper-siding*	HH *Toggle-arms*	

one especially used for drop-bottom gondolas. In some instances a third pair of rods was run under the twin center sills. These rods were normally placed parallel or all at one level, but some designers would place one rod lower than its mate.[50] An even more peculiar plan is documented by a photograph of a New York Central twin hopper built around 1895 by the Buffalo Car Company. An ordinary queen post truss rod setup is reinforced by a deeper king post truss rod.

Just how to prevent the side of an open-top car from bulging out when loaded was another problem. Short, low sides could resist the outward pressures, but as sides grew longer and taller, additional support was required. Heavy side boards, sometimes as thick as 3 inches, could help. One or more cross-timbers, usually stiffened by tie rods, were one answer, but they obstructed the top opening, which ideally would be kept clear for loading purposes. The lengthwise truss rod blocked out along the car sides was reasonably effective as an antibulging device. Examples appeared earlier in Figures 5.44 and 5.45. Continuous cap irons, screw-fastened to the top of the body, worked well so long as the fastenings held tight. They also served to protect the top edge of the body from the normal batterings of loading operations.

Side-Dump Coal Cars

This discussion has so far emphasized the conventional type of coal cars, but some mechanics envisioned a different plan of cars to handle this traffic. Most sought to adopt schemes used for a generation or more for work cars. Use of the side-dumpers by the Providence & Worcester was mentioned earlier (Fig. 5.10). In 1880 the New England

Car Company of Boston (later renamed the United States Car Company) began promoting the so-called screw-lever dump car, patented on December 18, 1880 (No. 236,121), by Matthew Van Wormer of Dayton, Ohio (Fig. 5.52).[51] A hand-wheel activated a screw and chain-drive mechanism that caused the body of the 20-ton-capacity car to tip over. Van Wormer's creations were used by the Union Pacific, the Lehigh Valley, and the Boston & Albany, as well as by some private lines. One of these operators used them for ore traffic, although most were likely used for ballast service. Another side-dump car was devised by George E. Blaine of Chattanooga, Tennessee, a few years after Van Wormer's mechanism was introduced.[52] A patent was issued on April 14, 1885 (No. 315,892). Blaine used a roller-gear chain-drive arrangement to operate the side-dump car. He too depended on a manually powered handwheel (Fig. 5.53). Two hundred Blaine cars were produced for the Union Pacific. Another lot was also produced for the Utah Northern.

In 1892 A. M. Waitt, master car builder for the Lake Shore & Michigan Southern, began designs for a side-dump car intended for either ore or coal.[53] The first produced were 25-tonners carried on two ordinary four-wheel trucks. The body was stationary, yet the car was self-unloading because the floor was inclined to either side. It was divided into five compartments with hinged side doors. In 1894 the Michigan-Peninsular Car Company produced fifty 40-ton-capacity cars on Waitt's peculiar plan. They measured 32 feet long and weighed 36,500 pounds. In body style they were nearly identical to the 25-tonners, but twelve wheels, or three trucks, were used to carry the load. The third truck was placed at the car's center. The Erie liked

FIGURE 5.52

Matthew Van Wormer patented a side-dump car for bulk cargoes in 1880. (*Poor's Manual of Railroads*, 1881)

FIGURE 5.53

George E. Blaine patented a side-dump car manually activated by a roller-gear chain drive in 1885. (*Railroad Gazette*, July 29, 1887)

Waitt's twelve-wheelers and built a number for ore service.[54]

Some years before any of these side-dump bulk carriers were produced, N. Kirkwood and W. Merrington of McKeesport, Pennsylvania, had invented an ingenious center-dump car. They received a patent on November 1, 1870 (No. 108,810), and proceeded to have a sample car made by the Pittsburgh & McKeesport Car Company.[55] The body consisted of two large open-topped iron boxes, made to tip downward and thus discharge through their open central ends when a pin was released. The inventors claimed that their flat-bottomed car held one-third more cargo than a conventional coal car and eliminated the need for hoppers. However, these advantages were achieved at the cost of the higher price and complexity of the arrangement, and so any supposed benefits were canceled.

Ore and Coke Cars

Specialized hopper cars were developed to move ore, coke, and charcoal. They followed the general plan of the common coal cars already described, yet they varied in size and details. Garden-variety coal cars were used to handle these specialized cargoes, but they could never do so as efficiently as cars designed specifically for the purpose. Ore is far denser and weighs more than three times as much as coal; hence a common gondola or hopper could be only partially filled before its weight capacity was reached. Ore cars, then, were made shorter, stronger, and more compact than the average coal car. They were also made with four rather than two sloping sides, because ore didn't slide as easily as coal. Coke, conversely, was about 20 percent lighter than coal, and so relatively light, high-sided cars were most suitable for this traffic. Most coke cars were in fact converted hopper or gondola cars equipped with slat-sided extensions added to the existing car sides.

CARS FOR BULK CARGOES

FIGURE 5.54

A four-wheel ore car made for the
Cameron Furnace, Middletown,
Pa., by the Harrisburg Car Co. in
about 1875. (Lavoinne and
Pontzen)

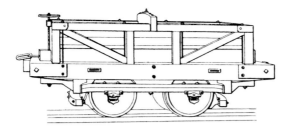

ORE CARS

The origins of the railroad ore car appear to be lost; the who, what, and where do not appear to have been recorded. It is certain that railroads began hauling iron, copper, and other ores at a very early date, but drawings and descriptions of these pioneers have vanished, with a few exceptions. The remains of five Canadian ore cars dating from about 1860 were recovered from the Trent River in 1981.[56] The frames and wheel sets of these four-wheelers have survived in reasonably good condition, but the accident that dumped them into the river, together with the ravages of time, has destroyed the bodies. The pioneer iron-ore steam railroad in the Lake Superior iron district opened in 1857 near Marquette, Michigan.[57] A history of the line's successor, the Duluth, South Shore & Atlantic, relates details of the corporation and its locomotives but says nothing of its ore-carrying stock.

One also finds few specifics in print when it comes to one of the nation's most spectacular gold- and silver-ore carriers, the Virginia & Truckee Railroad of Nevada. This short line has been the focus of several rail historians, including the late Lucius Beebe. The emphasis is on legend, locomotives, and lore of the gold fields with only incidental mention of the ore traffic itself. Gilbert Kneiss in his 1938 history of the line, relates that the V&T carried 112,000 tons of ore during the first six months of 1873.[58] Stephen E. Drew, a knowledgeable historian with the California State Railroad Museum in Sacramento, has provided the following information from his files. The V&T had 120 metal-body, wood-frame, four-wheel ore cars plus a small number of side-dump and combination ore/flatcars. The first sixty iron-body four-wheelers were constructed in 1869 by the V&T shops and the Risdon Iron Works. The remainder were fabricated in later years by the railroad and the Union Iron Works. The ¼-inch-thick metal bodies of the railroad-built units were made by Risdon. They looked like squared-off versions of the four-wheel pot hoppers used on the Pittsburg Railroad. An engraving of one of these is reproduced in Chapter 8 (Fig. 8.19). The V&T cars were rated at 8 tons capacity or 6 yards of ore. They were nearly square in plan view, measuring 8 feet 8 inches long by 8 feet wide over the sills. The undersize appearance was accentuated by 26-inch

wheels. As ore production declined in the Comstock region, the iron cars became redundant. A dozen were rebuilt for narrow-gauge service in 1900. The remaining cars of this type were scrapped seventeen years later.

In addition to the metal-body cars just described, the Virginia & Truckee built around nineteen wooden side-dump cars of the conventional ballast car design in 1869 and 1870. Fifteen of these cars were later regauged for service on the Carson & Colorado Railroad. The railroad's master mechanic, J. W. Bowker, determined to devise an ore car capable of carrying a return payload, designed a platform car in 1872 with a collapsible metal hopper set down into the car's floor. The hopper plates could be folded up to form a solid floor. The hoppers held 10 tons of ore. Eight cords of firewood were carried on the return trip. Bowker built a sample car in the road's repair shop and had five duplicates produced by the Kimball Car Company of San Francisco.

Iron was found in just about every part of the nation, and some of the earliest deposits to be worked were in the mid-Atlantic states. It is from this region that we find the oldest ore car illustrations. An engraving for a small four-wheeler with wooden springs, dating from the 1870s, was published in a French book on American railroad engineering and is reproduced here as Figure 5.54.[59] No specifics are offered on the car except that it was used by the Cameron Furnace, an iron company started in 1856 at Middletown, Pennsylvania. The car is nothing less than a cut-down jimmy. Dating from the same period is a photograph of an ore train on the Crown Point Iron Company's 3-foot-gauge railway near Crown Point, New York. It appears in Jim Shaughnessy's history of the Delaware & Hudson.[60] The cars are shown clearly in this broadside view taken around 1875 to 1880. The wooden four-wheel ore cars have pyramidal-shaped hopper bodies where all sides incline toward the bottom hopper doors. Externally they look like a jimmy, except for the long wheelbase made possible by placing the journal-box pedestals near the ends of the frame.

By 1880 U.S. iron production had so expanded that the country was no longer dependent on imports from Great Britain. The mining of native ore expanded as well, and the Lake Superior ore beds were finally being worked in a major way. Bigger lake boats, ore docks, and mechanized unloading facilities made the transportation of ore from remote Duluth and Marquette to eastern furnaces an economic reality. In August 1881 the Chicago & North Western deposited 270,000 tons of iron ore at the Escanaba, Michigan, docks.[61] Ninety to one hundred carloads a day were normally delivered. The road had 4,000 ore cars in service and had just completed 196 new ones at the Escanaba shops. The Michigan Car Company of Detroit produced

FIGURE 5.55

A Chicago & North Western 20-ton-capacity ore car dating from 1886. (*National Car Builder*, Nov. 1886)

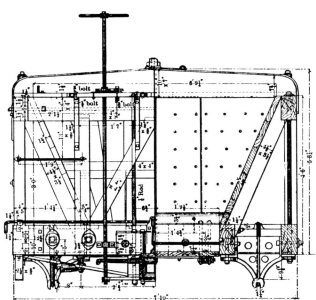

another three hundred for the North Western.

By 1890 the rich Great Lakes ores accounted for half of all the iron mined in the United States.[62] An important link in this long chain between Duluth and Pittsburgh was the railroad ore car. The railroads were short lines running from the mines to the Lake Shore docks. The ore was, of course, once again transhipped when the ore boats reached their Lake Erie port destination. Four-wheel jimmy-style cars were no longer adequate for the tonnage moving west to east. Eight-wheel cars, something on the order of the Philadelphia & Reading's 20-ton hopper cars, were introduced.

The earliest available drawing of this general design is for a Chicago & North Western car of 1886.[63] The 20-ton-capacity car measured 21 feet 7 inches long over the end sills (Fig. 5.55). The frame

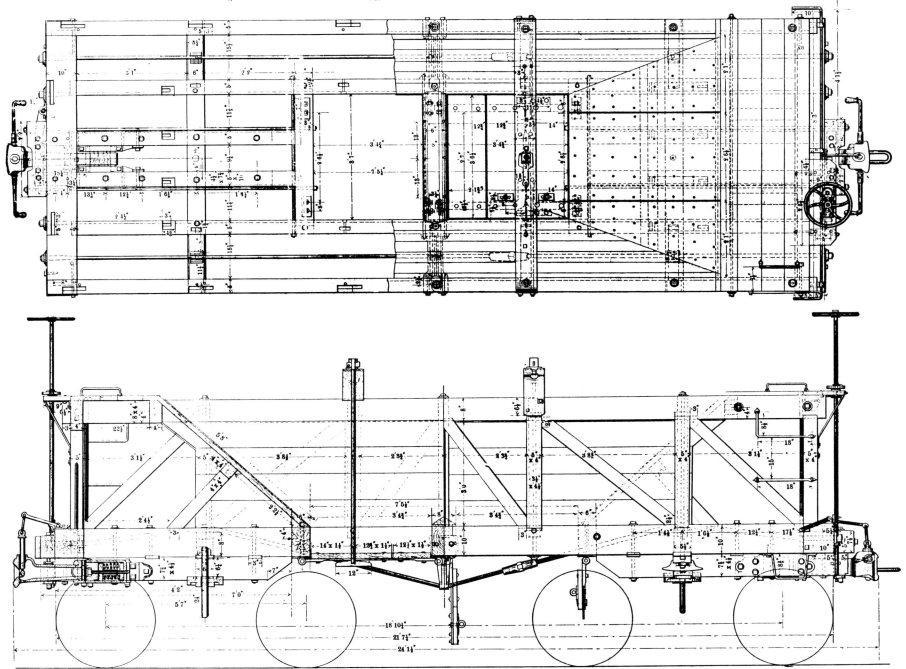

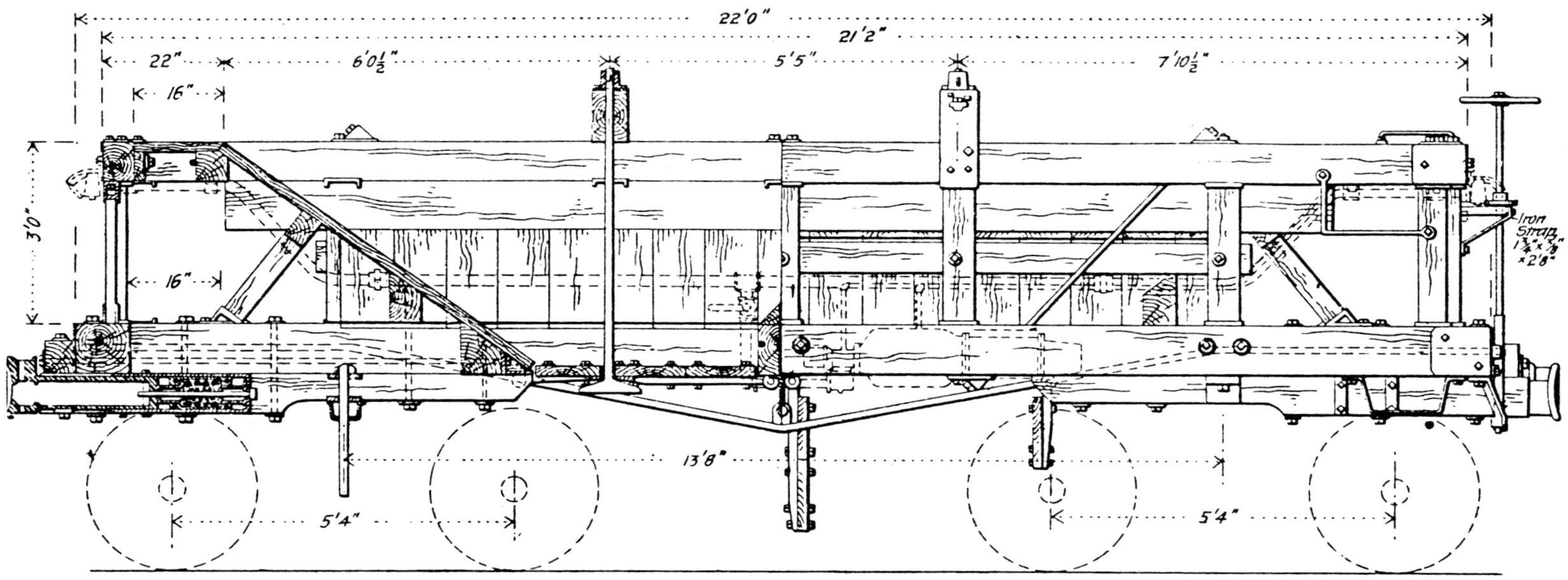

22'0"
21'2"
22"
6'0½"
5'5"
7'10½"
16"
3'0"
Iron Strap
13'8"
5'4"
5'4"

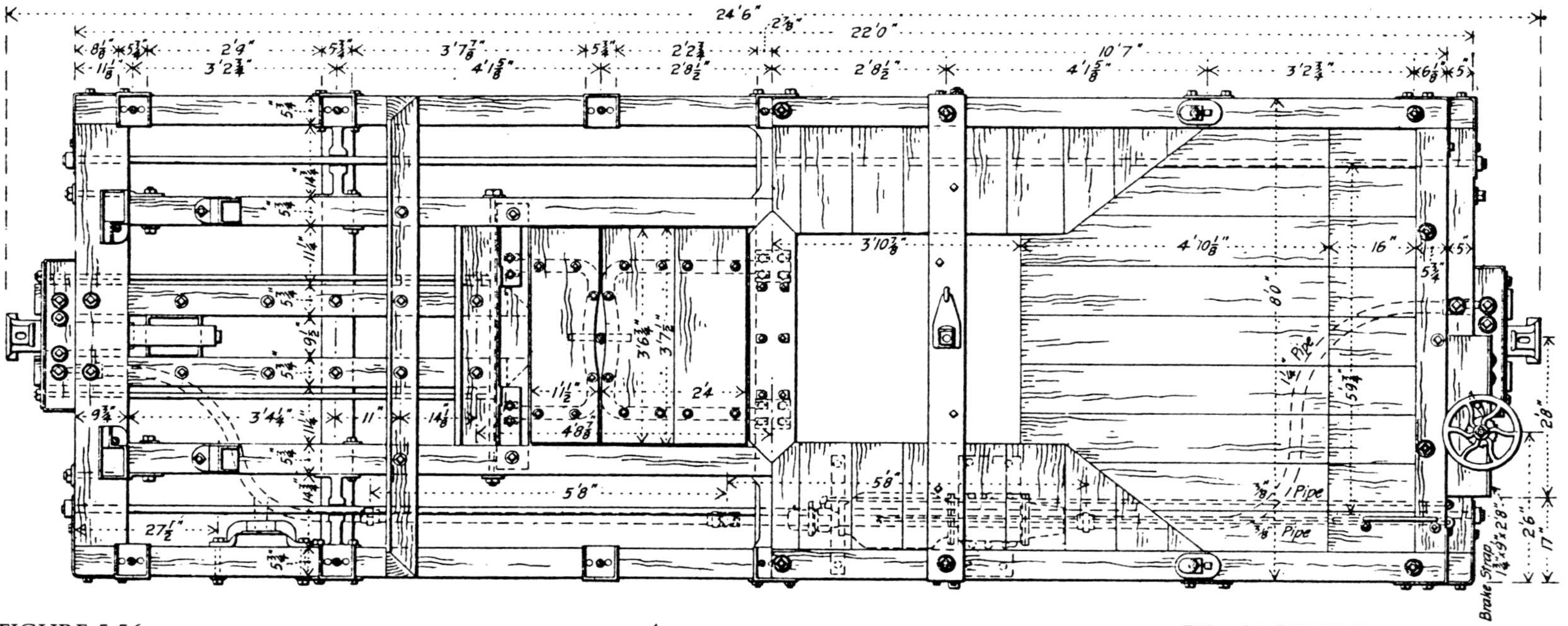

24'6"
22'0"
10'7"
Pipe
Brake Strap

FIGURE 5.56

A Duluth & Iron Range ore car.
(*Engineering News*, Oct. 26, 1893)

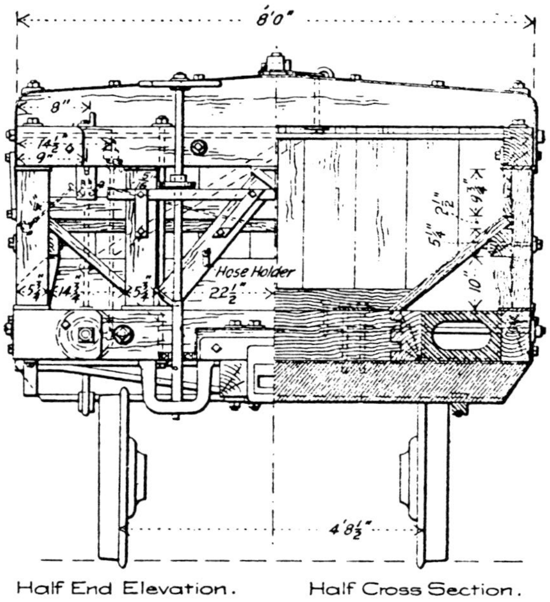

8'0"
Hose Holder
4'8½"
Half End Elevation.
Half Cross Section.

BILL OF LUMBER.

Pieces.	No. of pieces.	Thickness, ins.	Width, ins.	Length, ft.	Length, ins.
Intermediate Sills,	2	5¾	9¾	20	8½
Draft "	4	5¾	9¾	5	5⅝
End "	2	9¾	9¾	7	4½
Center Cross Beam,	1	9	9¾	3	11½
" " " Cap,	1	4½	9	4	6
Cross Timb., Hopper Ends,	2	7¾	9¾	3	11½
Draw Timbers,	4	5¾	7¾	7	9
Buffer Blocks,	2	5¾	7	2	6
Cross Beam on Sills,	2	5¾	9¾	8	0
Upper Cross Beam,	2	7	10	8	0
End Posts,	4	3¾	5¾	2	3
Corner "	4	3¾	5¾	2	3
Center "	2	3¾	5¾	2	3
Side "	8	3¾	5¾	2	3
End Braces,	4	1¾	4¾	2	6¼
Side Bearing Timbers,	2	4	5½	14	3
Supports for Same,	10	4	5¾	0	9⅝
Side Hopper Plank,	2	1¾	9¾	17	8
Bench Caps,	2	4	5¾	7	0½
" " Posts,	4	4	4	0	21
Side Hopper Plank,	34	1¾	9¾	2	4
End " "	10	1¾	9¾	6	0
" " "	4	1¾	9¾	4	10
" " "	4	1¾	8	3	6
Deck Planks,	14	1¾	9¾	1	4
" "	4	1¾	8	1	4
Drop Door Planks,	4	1¾	14	3	6¾
" " "	2	1¾	13½	3	6¾
Brake Step,	1	1¾	9	2	8
Side Sills,*	2	5¾	9¾	22	0
" Rail,*	2	5¾	7¾	21	2
End " *	2	5¾	7¾	7	4
Deck Beams,*	2	5¾	6	7	4½
" "	2	1¾	4	7	½
Addition to Corner Posts,	4	1¾	5¾	2	3

Sizes given are of finished lumber. Lengths are exact with allowance for tenons. All pieces of white oak, except those marked *, which are of Norway pine.

consisted of four 5- by 10-inch yellow pine sills. The sills were clustered toward the car sides so that the full pyramidal hopper could project down to the hopper doors. No center sills were used, and the hopper space was interrupted only by a heavy cross-timber that doubled as the needle beam. This frame plan appears to have been exclusive to ore cars. Two heavy (1¼-inch-diameter) truss rods supported by king posts helped strengthen the frame. The hopper doors were held closed by T-latches fixed on the bottom of upright shafts that passed through the center of large cross-timbers fastened to the top of the body. Notice the large platform at each end of the body, which provided a spacious landing for the brakeman. The trucks had an unusually long wheelbase of 67 inches—presumably to spread the wheel load over a greater track space. The Blocker couplers illustrate one effort to improve the safety of the link-and-pin coupler by provision of a lifting lever for the pin.

The plan of this car can be taken as standard design for just about every Great Lakes ore carrier. That it was original to the C&NW remains a question, but the drawing just described is the earliest known example. Drawings for an identical car on the Wisconsin Central were published five years later.[64] There is hardly a dimension or detail that differs from the C&NW plan, and the sills were identical in number, size, and shape, yet the Wisconsin Central rated its cars as having a 30-ton capacity.

The Duluth & Iron Range Railroad, one of the bigger ore carriers, followed the general C&NW plan but not so slavishly as the Wisconsin Central.[65] The car was slightly longer and lower (Fig. 5.56). The frame plan, latch door closers, and most other details are familiar. But notice that the king post is shorter and that the rods have no turnbuckles. The hopper sides are broken, rather than following a single straight-sided incline, and the support framing of the body is quite different. The timbers are set in cast-iron pockets, and angle rods are used in place of timbers. The brakeman's platforms are smaller. The D&IR used white oak for almost every part of the car, except for the end sills, at a time when most car builders had switched to yellow pine. The car had air brakes; the cylinder was near the center, about level with the outside sill but inside it. Link-and-pin couplers remained in favor, since the cars were not

FIGURE 5.57

Pullman built three hundred ore cars for the D&IR in 1895 that agree with the engraving reproduced in Figure 5.56. (Pullman Neg. 2971)

FIGURE 5.58

Pullman's great Chicago works was celebrated for its palace cars, but the plant was not above fabricating utilitarian ore cars such as the No. 6209, dating from 1897. (Pullman Neg. 4045)

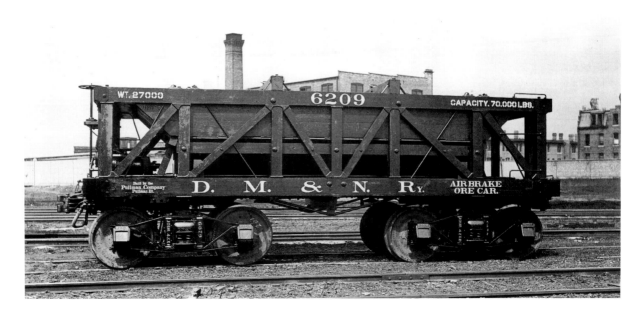

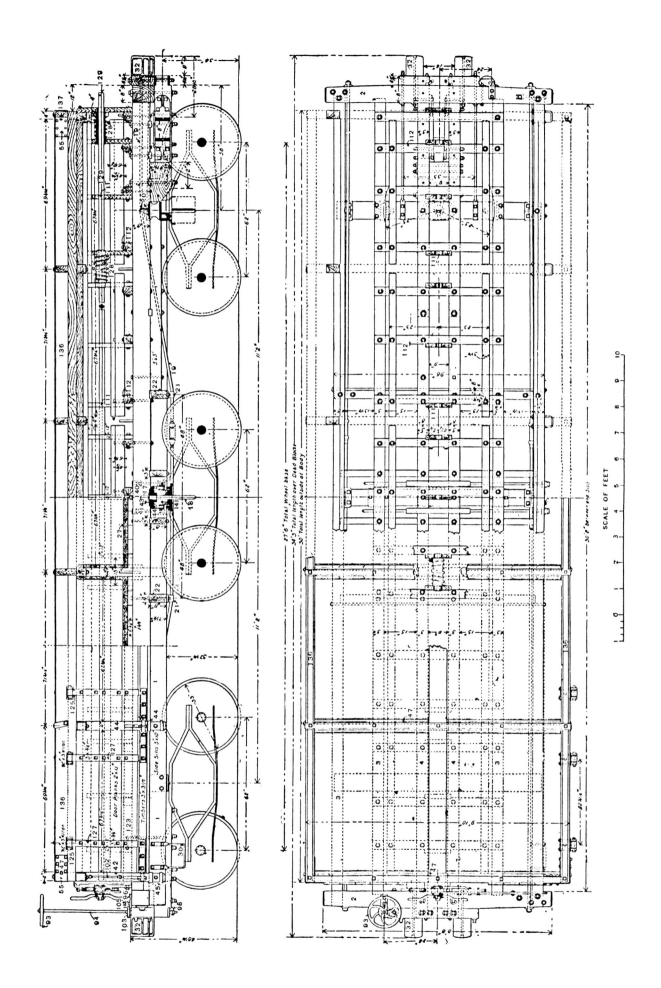

SCALE OF FEET

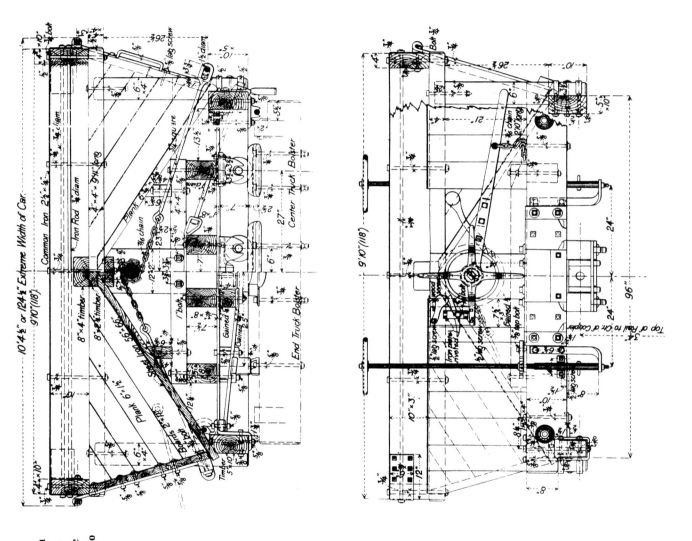

FIGURE 5.59

The Erie Railway devised a plan for twelve-wheel side-dump ore cars with stationary bodies. The inclined floors allowed the ore to slide out after the doors were opened. (*Engineering News*, April 19, 1890)

CARS FOR BULK CARGOES

very likely interchanged or employed in interstate traffic. In 1893 the road had five-hundred of the 24-ton-capacity cars in service. A photograph of a standard D&IR car built by Pullman in 1895 is shown in Figure 5.57. It follows the drawings very closely and shows the odd position of the air-brake hoses. The location of the air-brake cylinder and reservoir tank can be deduced from the T-connection in the air pipe near the car's center.

By the late 1890s the neighboring Duluth, Misabe & Northern was acquiring 30- and 35-ton-capacity ore cars.[66] They were more on the Chicago & North Western pattern (Fig. 5.58). The short length of 22 feet was observed, but the car sides were higher than those on the D&IR. The air-brake apparatus was placed at one end in the open space below the hopper body. The traditional design for a wooden ore car was pushed to its upper limit of 50 tons in 1900 by the Milwaukee Road. The standard length of 22 feet, fixed by ore-dock hopper openings, was faithfully followed, as was the four-sill frame plan. The king post, however, was so enlarged that it almost touched the rails, and four rather than two truss rods were used. The sides reached about 5 feet above the top of the side sills. The massive bridge-truss body frame gave way to stake-pocket-style posts like those found on most gondolas, except that the stakes were, of course, longer. Only two hopper doors were used, and they opened parallel to the rails and were operated by windup chains rather than the latch-type closers. Weight was calculated at only 30,700 pounds. The Milwaukee continued to produce wood-frame ore cars as late as 1913. The road's latest model carried an impressive load of 61.5 tons.

It was natural that attempts to introduce a different plan for ore cars would be made. Two examples have been uncovered. The first of these was put forward in 1890 by the Erie (Fig. 5.59). It was a 40-ton-capacity car carried on twelve wheels—three arch-bar trucks.[67] Cars of this type have already been mentioned in the hopper car section of this chapter. The Erie's car department wanted to build these vehicles as conventional double-truck cars, but the roadway engineers protested, saying that the wheel loadings would be too great for the track and bridges. In 1894 A. M. Waitt built some duplicates for the Lake Shore & Michigan Southern.[68] The 34-foot-long side-dumpers weighed 36,500 pounds. The center truck had a floating center plate that could move 5 inches laterally and was attached to the frame by channel irons and large rollers.

A more conventional-appearing ore car was devised by the Duluth, South Shore & Atlantic in 1895.[69] What appeared to be an ordinary 22-foot-long ore car was actually a convertible ore/gondola car (Fig. 5.60). The sloping hopper floor was attached to the frame by long lever hinges that allowed the floor to fold forward and down so as to cover the hopper opening and make a continuous flat floor. Unlike most ore cars, this one had body

FIGURE 5.60

The Duluth, South Shore & Atlantic introduced an ore car in 1895 with hinged metal floor plates devised so that they could fold down to form a level floor. (*Railway Review*, March 14, 1896)

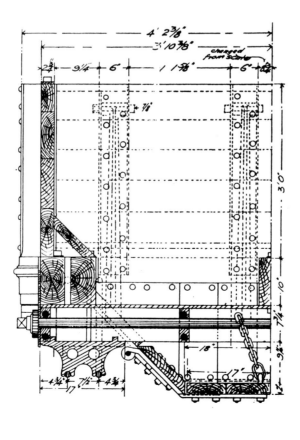

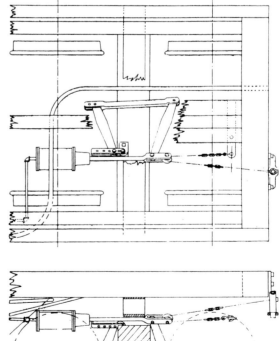

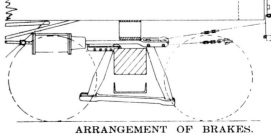

ARRANGEMENT OF BRAKES.

sides that were straight, except for a small angular shelf. The hopper was pyramidal and made more in the fashion of a single-hopper-bottom gondola. Windup chains were used to close the doors. The body was also different from the average ore car in that it lacked the usual massive outside truss framing. Stake pockets and posts together with iron blocks and tie rods held the body planks together. The ends of the body folded down to make the loading of lumber and logs more practical. For this reason the expected corner irons were not applied. The frame consisted of twin outside sills, plus a center sill. Four truss rods helped to stiffen the frame. They were braced by a double king post casting. The car's nominal capacity was given as 25 tons, but it was capable of carrying 33 tons. The trucks had a normal short wheelbase of 4 feet 10 inches rather than the longer spread found on most other ore cars of the period, which might indicate stronger track and bridge structure on the DSS&A. The car had Janney-style couplers and air brakes. The brake cylinder was placed below the frame just in front of the hopper and above one of the axles. The prototype, built in 1895, was so successful that the DSS&A ordered six hundred more from the Michigan-Peninsular Car Company early in the following year.

COKE CARS

The special cars assigned to convey coke were very different from those designed to carry ore.

Coke cars were really not very special. Most were simply drop-bottom gondolas fitted with temporary side extensions. These extensions, generally made with open slats as in a cattle car, rose 3 to 3½ feet above the normal body sides. Coke traffic didn't develop on a very large scale until after 1860, when the Pittsburgh iron makers began to use it in place of charcoal. The coal from around Connellsville, Pennsylvania, was well suited to coke production, and so the Baltimore & Ohio and the Pennsylvania Railroad moved large quantities of this fuel north to the furnaces at Pittsburgh.

Private shippers had their own cars, as illustrated by a 20-ton-capacity car produced in 1889 for W. J. Rainey (Fig. 5.61). The high-sided extension is made solid rather than slatted, like most coke racks, and is not easily removable. Note also the light truss rods and windup shaft for the drop doors. A diagram drawing for an older coke car contrived from the B&O standard class N hopper-bottom gondola of 1879 is reproduced here as Figure 5.62. Slatted panels were fitted between the racks, shown in the drawing, when the car was filled. The racks could be removed by lifting them out of the side pockets if the car was needed for coal shipments. Ten years later the Pennsylvania was employing 30-ton-capacity coke cars based on its class GE drop-bottom gondolas (Fig. 5.63). This was a long car for the time, measuring 35 feet 7 inches over the end sills. The inside dimensions of the body were 33 feet 7½ inches by 8 feet.

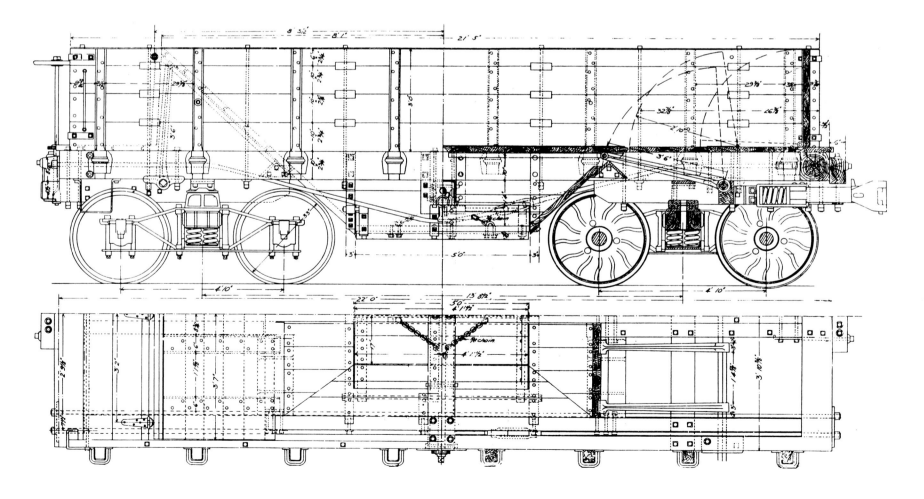

FIGURE 5.61

Pullman produced this high-sided gondola for coke traffic in 1889. The car did not have the conventional open-slatted racks normally fitted on coke carriers. (Pullman Neg. 584)

FIGURE 5.62

The B&O moved quantities of coke from around Connellsville, Pa., in rack cars like the one represented in this 1879 diagram drawing. (Baltimore & Ohio Railroad)

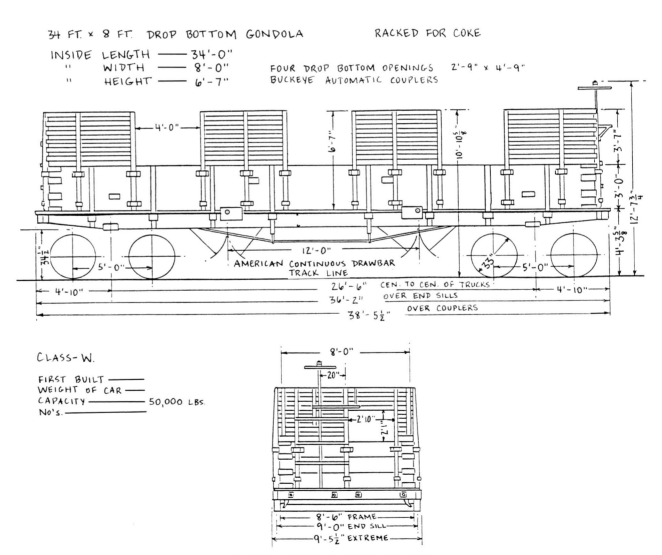

THE AMERICAN RAILROAD FREIGHT CAR

FIGURE 5.63
The Pennsylvania Railroad outfitted drop-bottom gondolas with
rack sides for coke service. This
car was built at Altoona in 1889.
(Smithsonian Chaney Neg. 26326)

The Pennsylvania used a very different type of car for charcoal that was more on the style of a cattle car than a coal hopper. Called the class LA, it was adopted in 1883. A roof, with loading hatches, was deemed necessary to protect the charcoal from weather, a precaution not taken for coke. A few years after the introduction of class LA, the Louisville & Nashville developed a similar car for the coke traffic in the Birmingham, Alabama, coal and iron region.[70] Again, the designs appear inspired more by a cattle car than by a coal car. The open-top, slat-sided vehicle had six small side doors measuring 4 feet by 4 feet 9 inches. There were no drop doors. The coke was loaded through the top, then dumped and shoveled out the side doors. The 32-foot-long car was supported by large 5- by 13½-inch side sills and four truss rods (Fig. 5.64). Two planks served as running boards for the brakeman and ran the length of the body. Designs for these 30-ton units were the creation of the road's chief mechanic, Pulaski Leeds, who supervised their construction at the Louisville shops.

Tank Cars

The other major class of railroad bulk carriers were those designed for liquids. While hopper cars appeared before the public railroad existed in the United States, tank cars were a comparatively late development, and very few were in service before 1870. They came into being to serve the petroleum industry and were specifically created to haul crude- and refined-oil products. By 1890 the once-specialized oil car had been adapted to carry a variety of products. Once the idea was explored, more

and more uses were devised, and very soon the tank car was proclaimed the "Bucket of American Industry."

At the conclusion of Chapter 4 we noted the food industry's exploitation of tank cars to move molasses, wine, vinegar, and oysters. Early in the same chapter, glass-lined milk tank cars were also alluded to, but no mention was made that common cooking oils such as corn, peanut, and olive were carried by rail cars. To this list might be added fish oil and fruit juice. Wooden tanks were used for wine, vinegar, and pickles, but tin-plated, copper-lined iron or steel tanks were used for many other products that might have been contaminated by contact with a ferrous-metal container.

Wood-bodied tanks were used for water service. The first example is apparently unrecorded, but the Central Pacific had some in use during the 1860s construction period of the transcontinental railroad. Pictures exist for both three-barrel vertical and long, rectangular tank cars (Fig. 5.65). Little railroads like the Rio Grande of West Texas and the Virginia & Truckee of Nevada were forced to maintain such cars for many years because of water problems in arid sections of the nation.[71] The Rio Grande's car had two vertical wooden tanks, while the V&T's was outfitted with a long, horizontal wooden tank. One of the V&T cars stood in the Carson City yards until around 1941. Larger cars were introduced in the 1890s to handle potable water sold as a health and status liquid by such firms as the Glen Summit Spring Water Company of Wilkes-Barre, Pennsylvania.[72]

The paint industry became a patron of tank cars

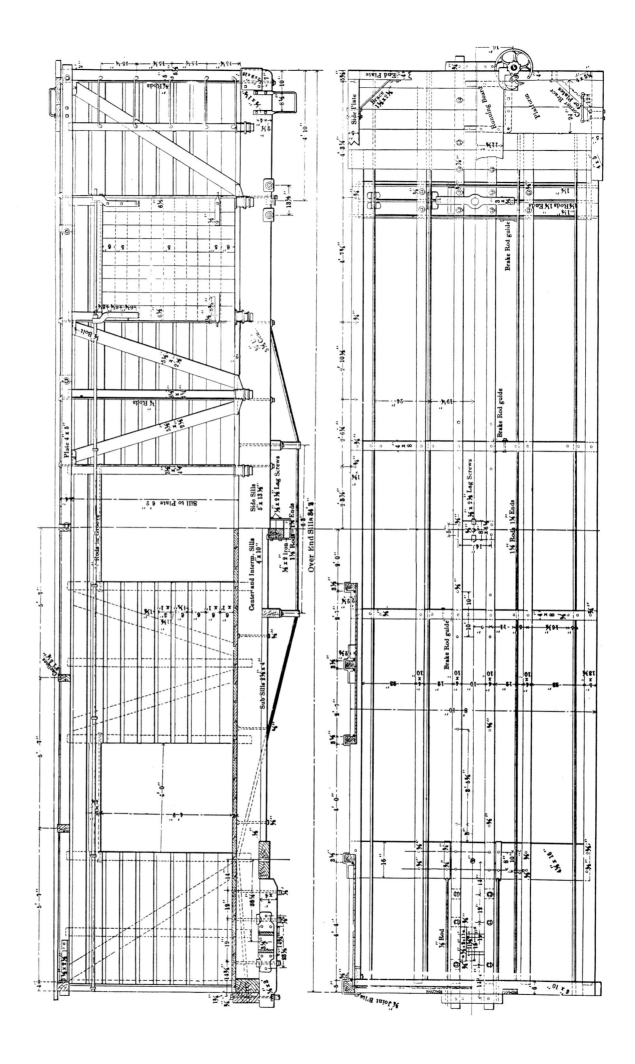

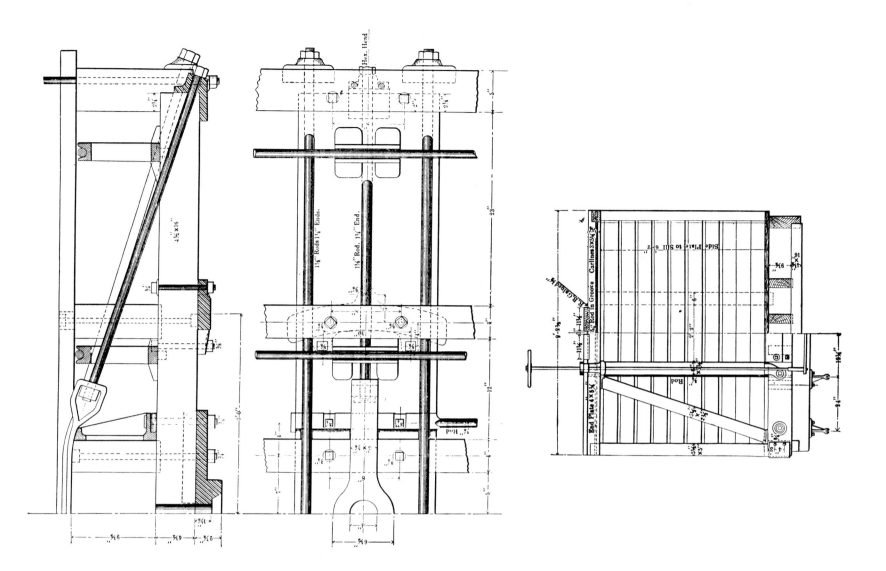

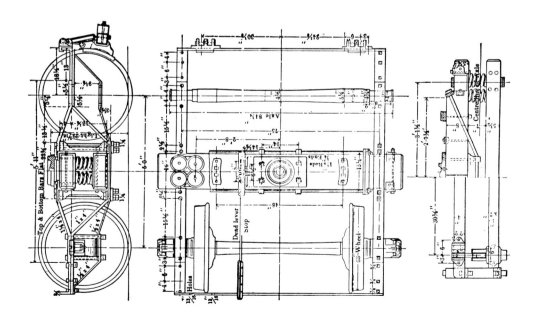

FIGURE 5.64

In about 1890 the Louisville &
Nashville introduced open-top
cars with side doors for the coke
traffic around Birmingham, Ala.
(*Railway Review*, Feb. 7, 1891)

CARS FOR BULK CARGOES

FIGURE 5.65

Special cars to transport water established a pattern for the first petroleum tank cars. The example shown here was in use on the Central Pacific in Nevada during 1868. (Library of Congress)

and used them for turpentine, dyes, alcohol, linseed oil, dryers, and naphtha. Stronger and more dangerous liquids were required by other industries, and tank cars were soon carrying chlorine, caustic soda, coal-tar distillate, insecticides, ammonia, and a host of other vile but necessary products of our technological society. Occasional spills or explosions were the inevitable result of handling such lethal commodities.

In recent times horror stories of small towns endangered by leaking gas or awash in fearsome liquids are reported with ghoulish delight by the national press. This fascination with the deadly acrobatics of a derailed tank car seemed to fill less space in newspapers during the nineteenth century, but we can offer a few examples. Best known is the accident near Poughkeepsie, New York, in February 1871, when the Pacific Express plowed into a derailed tank car train. The oil-soaked wreckage burst into flames, killing twenty-two passengers. A more modest incident happened many years later outside Chicago.[73] A hotbox set fire to a tank car that in time generated enough heat to explode. A crowd had gathered to watch the blaze, which they erroneously took to be a harmless spectacle. Their curiosity was repaid in a most cruel fashion. Four were killed and several others injured by the explosion. Most tank car accidents were less spectacular or deadly than these two, and unless innocent bystanders or passengers were annihilated—train crew members didn't count, for some reason—newsmen didn't seem to evince much interest in the spillage of a ruptured tank.

Tank cars before the oil boom of the 1860s were so rare as to be almost nonexistent. It was the rapid exploitation of petroleum that prompted inventors and car builders to devise special rail cars for the transportation of this valuable liquid. And so before exploring tank car design, it is necessary to consider very briefly the commercialization of petroleum.

PETROLEUM PROMPTS DEVELOPMENT

The successful drilling for petroleum in northwestern Pennsylvania set off speculative excitement on a national scale not seen since the California Gold Rush a decade earlier.[74] Colonel Edwin L. Drake's modest well at Titusville created an oil fever that attracted fortune seekers, sharp traders, and adventurers from all regions of the nation. They began to assemble not long after Drake's well first began delivering black gold on August 27, 1859. Just as John A. Sutter had seen his remote ranch near the American River transformed into a bedlam of gold seekers, so too did residents along Oil Creek witness a dramatic transformation of their sleepy farmlands in Venango County. Pastoral villages became boom towns. Instant new communities, such as Pithole, sprang up overnight in places inhabited months before only by the hare and the fox. Farmland fetching no more than $4 per acre began to command $500, $1,000, and, for choice locations, $7,000 per acre. The two-hundred-acre Holmden farm near Pithole sold for an unprecedented $1.3 million in 1865. It was soon pierced by one hundred wells and numerous oil storage tanks. Most of the virgin forest had already disappeared in earlier timber operations, but the remaining trees and vegetation disappeared as greasy, malodorous, black crude coated the countryside. Even a modest well could produce $150 per day, which was big money in the time of the dollar-a-day wage. A real gusher might produce 4,000 barrels per day and so create a sudden millionaire. Yet there was many a dry hole, and fortunes were lost as well as made, for Mother Nature was both fickle and secretive in her placement of these sandy black pools below the earth's crust.

Petroleum had been known since ancient times and was used for elixirs, for curatives, and occasionally for building purposes. The supply was limited to what could be scooped up from streams or ponds as tiny quantities oozed up to the surface. Only after Drake showed that it could be extracted in large quantities was petroleum seen as something more than a novelty. Scientists quickly pointed out that it could be distilled to produce prodigious volumes of lubricating and illuminating oils. Industrial America had outgrown natural oils such as lard, tallow, rape seed oil, and oil rendered from whales. The potential supply of coal oil was nearly limitless, but production costs were substantial, and the market of 1860 could hardly afford $1 per gallon. Cheap lamp oil, kerosene, was to become the new everyday standard and basis for many a fortune, including that of one John D. Rockefeller. During the first year—really only about five months—Drake produced 2,000 barrels (1 barrel equals 42 gallons). In 1860 more wells were drilled and production reached 500,000 barrels. In the next year 2.1 million barrels were produced, and by 1880 the wells in western Pennsylvania were pumping 26 million. After this time, large oil fields elsewhere in the nation and the world began to challenge the northwestern Pennsylvania wells. Gradually they lost their standing as the world's number one producer, but even today limited petroleum production continues in these pioneer fields.

SOLVING THE TRANSPORT PROBLEM

The production of crude oil grew from a trickle to a flood in just a few years; oilmen faced terrible logistical problems. What to do with this sea of oil suddenly brought to the surface? How to store, refine, and move it to the market was the basic dilemma. Very soon every tank and barrel was filled to overflowing. Open ponds were resorted to, and some oil was dumped into creeks whenever these rude storage reservoirs were full. Tank farms soon dotted the former hay- and cornfields of Venango County. Shipping crude to refineries in Pittsburgh, Cleveland, and New Jersey was the major problem. Before the oil boom, the Titusville area had been a backward agricultural area and as such had little in the way of a transportation system. It possessed no canals or railroads but depended on rural roads and shallow streams.

Teamsters at first enjoyed fat times, hauling wagonloads of barreled oil to the nearest railheads. Three thousand teamsters were at work by 1865. Horses were brutally treated, and many died after a few months of strenuous service. Soon every available horse within a 30-mile radius had been purchased, and new animals had to be imported from a greater distance. The roads became a litter of cast-off barrels, broken-down wagons, and dead horses. Wanting to become rich, teamsters also pushed up their rates and were soon demanding $4 per barrel. Some of these men were earning $20 to $30 per week—then considered a fabulous wage for a workingman. Flatboats offered an alternative for oil going to Pittsburgh, but they were no more economical than the teamsters, and their sailings were inhibited by low water. Collisions, floods, and unexplained sinkings made them an unreliable means of transit.

Yet by these costly and uncertain means, oil was reaching railroads passing nearby at Corry and other points in the area. The Pennsylvania & Ohio Railroad reported moving 27,546 barrels of oil between December 1860 and February 1861.[75] The Philadelphia & Erie, a new line very much in need of revenue, realized $200,000 during 1861 from the oil trade. The obvious need for a direct rail line into the oil fields prompted construction of the Oil Creek Railroad, which opened to Titusville in October 1862. This broad-gauge line, through its connection, reached Cleveland and New York in November 1863. It extended to Oil City early in the next year.

The direct rail connection did not solve the transportation problem, however. The teamsters' monopoly was broken, and cheaper, faster service between the tank farms and distilleries was assured, but all shipments continued via the established barrel system. Barrels could be loaded into ordinary railroad cars. Flats, boxcars, gondolas, and even cattle cars could handle barrels. Railroads wanted shippers to use existing, standard cars because they were available and could also be used for return loads. Specialized cars, made specifically for a given product such as oil, were costly and could be used only for that traffic. They returned empty and hence produced no backhaul revenue. But barrels could be carried in existing cars, and the empty barrels going back to oil shippers didn't travel free.

The railroads found the status quo very much to their liking. Oilmen did not, for many reasons. Barrels were expensive and sometimes scarce. They rarely cost less than $3 each. They were liable to damage, loss, or theft. They generally leaked a tenth of their content and sometimes as much as a third. Efforts to seal their interiors with glue, molasses, and glycerin were generally ineffective. Wooden barrels depend on swelling of the staves for tight joints, but petroleum contains too little water to effect this normal function of a liquid, and so the joints remained slightly open. The barrels also tended to explode because of the gaseous nature of crude oil. There was no practical way to vent each barrel.

James and Amos Densmore—also known for their financial role in the development of a practical typewriter—were among the many oilmen who pondered the problem of oil transport.[76] They came to Meadville in 1861 to join the oil rush. They seem to fit a description of most newcomers to the area as rough-looking, eager men bent on making money by boldness and luck. James was born in Moscow, New York, in 1820 and spent most of his early years as a newspaperman in Wisconsin. Amos, a younger brother, was born in Rochester five years later. Neither had been particularly successful before entering the oil trade, but they flourished in this new business and organized the Densmore Oil Company on July 18, 1863.

James appears to have been the dominant partner. He was described in the following terms by a former business associate:

> A great ponderous, beefy-looking man of nearly three hundred pounds weight, with a florid complexion, a great shock of red hair, shaggy beard, the eye of a hypnotist and the heavy jaw and animal force of a great Hyrcanian bull in *Quo Vadis* . . . restless as a tiger, a born bully with a fierce military spirit . . . he had an unfortunate personality that repelled many people instead of attracting them.[77]

This sketch might be dismissed as a burlesque, but I am assured by Donald Hoke, curator at the Milwaukee Public Museum, that Densmore was a hard-driving, determined person given to eccentricity. Mr. Hoke is the keeper, and a student, of the Densmore Papers, which are housed in the Milwaukee museum. A flattering biographical article in the *National Cyclopedia of American Biography* specifically notes James's "bull-dog pertinacity" and "grim courage." Of Amos we have almost nothing in the way of personal information, but one suspects that he was comparatively mild-mannered, being overwhelmed by his strong-willed brother.

The Densmores analyzed the oil industry's need for bulk carriage of oil. Continued shipment by barrels meant continued losses. Oil had already been bulk-shipped in open barges. The absence of a good waterway limited the usefulness of this method. Pipeline experiments were under way, but even a minimal pipeline network would not exist for many years. What was needed was a special form of rail car for the bulk carriage of petroleum. Correctly isolating the problem, however, was the easy part of the question. Correctly solving it was the hard part. At least half a dozen would-be inventors had, like the Densmores, already recognized the problem. Late in 1862 an anonymous Canadian inventor made a test run of an iron tank car over the Grand Trunk Railway to Portland, Maine, whence the oil was transhipped to England.[78] Regrettably, no description of this pioneer vehicle is known to exist.

A few months later, John Scott of Lawrenceville, Pennsylvania, received the first U.S. patent for a tank car (No. 37,461, Jan. 20, 1863). Scott proposed a boxlike wooden tank lined with soldered tinplate sheets. Partitions were installed to limit surging of the liquids in transit. There is no record that Scott's plan was tested. Later that year John Clark of Canandaigua, New York, obtained a patent showing a variation on Scott's idea (No. 40,458, Nov. 3, 1863). A low, rectangular tank, looking much like a covered gondola car, was divided into eight compartments. A far superior plan was patented on June 2, 1863 (No. 38,765), by Samuel J. Seely of Brooklyn.[79] Seely was active in the design of iron railroad cars and already held several patents on this subject. His idea for a tank car was advanced; it called for a horizontal iron tank and so was on the path that would eventually lead to the modern tank car. His patent drawing shows a relatively short, large-diameter tank fabricated from corrugated sheet iron. No provision for an expansion dome was made.

There was a lull among the tank car inventors for over a year. On January 10, 1865, Joel F. Keeler of Pittsburgh was granted U.S. Patent No. 45,834 (Fig. 5.66). Like Seely, Keeler was on the right track, though not exactly the same track as his Brooklyn-based competitor. Keeler wanted an iron tank car, but he wanted it to be a combination tank/boxcar. He attempted to resolve the longstanding problem of single-purpose bulk carriers by devising a car capable of doing more than one job. Very simply, he would mount a boxcar body on top of a half-round or U-shaped tank. The tank, made of ⅜-inch-thick iron, doubled as the car's frame. The combination of frame and cargo vessel may be the earliest instance of unitized construction in the railroad freight car field. Even more ingenious, Keeler made the troughlike tank underframe in three parts. The large central tank hung low to the rails, increasing capacity and at the same time

FIGURE 5.66

Joel F. Keeler's 1865 patent for a
combination tank/boxcar: *a* is the
end tank, *b* the central tank, *f* the
filler hold, *g* the discharge opening, and *k* the corrugated floor
separating the tank and the main
car body, *K*. (Traced from a Patent
Office engraving)

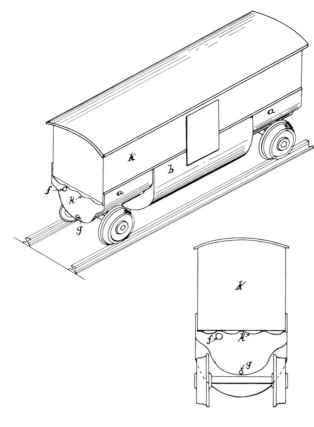

lowering the center of gravity. The two end tanks were made smaller by necessity to clear the wheels. A test car was fabricated at the Wallace Boiler Works in Pittsburgh in October 1865. It measured 28 feet long by 8 feet wide.[80] The possumbelly tanks held 80 barrels of oil. The box body would allow a return payload. For all its merits, Keeler's idea was allowed to die.

At about the same time that Keeler's patent was issued, another scheme was afloat in the oil regions for a new style of tank car. The Howard Oil Tank Car was a combination vehicle like Keeler's, but there all similarities ceased.[81] Howard called for two giant drums, the ends of which had flanged tires so that the drums were both the cargo compartment and the wheels of the car. A frame and overhead platform held the drum wheels together and offered an open space for additional cargo. The scheme was actually an old idea patented on June 24, 1851, by Lawrence Myers of Philadelphia. Myers had envisioned cylinder cars for grain or coal, while Howard sought to apply them to the oil trade. It is difficult to imagine a more impractical or foolish plan. The cargo, be it oil, grain, or coal, would be incessantly tumbled and agitated while in transit. Assuming that the cargo survived its rolling journey, the next wonder would involve the condition of the track after the passage of so much unsprung weight. At least one test car was completed, and the normally astute Pennsylvania Railroad agreed to acquire twenty-five cars from Howard in September 1865. Perhaps this mis-

guided contract was later canceled, for there was little hope of success. It does appear, however, that more than one sample car was tested by Howard, for a surviving photograph reproduced here as Figure 5.67 shows it carrying the numeral 5.

While Howard's folly was being pursued, a more practical idea was being patented by H. J. Lombaert of Philadelphia on September 12, 1865 (No. 49,901). Lombaert offered a simple container arrangement whereby tanks measuring 29 inches by 7 feet, or about the size of an ordinary wagon bed, would carry 208 gallons (5 barrels). The containers were taken by horse-drawn wagon from the wellhead to the nearest rail siding. The container tanks were then crane-loaded onto waiting rail cars and traveled thence to a refinery. The empty barrels could be returned to the oil fields in the same manner as sent. Lombaert thus required no special road or rail vehicles.

The point of describing all these schemes, none of which was adopted, is to show that many inventors were already grappling with the problem of bulk oil shipping before the Densmores became involved with it. Whether they were aware of their predecessors and contemporaries is uncertain, but considering the Densmores' ongoing involvements with the typewriter, other inventors, and Patent Office procedures, it is likely that they had some knowledge of the patents issued on the subject. They were also active in the oil trade and must have known that others were interested in answering the need for a practical tank car.

This supposed knowledge might explain why the Densmores chose vertical tanks mounted on a flatcar. The idea was an obvious one, yet it remained unprotected by a patent. Perhaps it imputes too much logic to the process of invention to suggest that inventors actually research the efforts of others before beginning work on a given problem. It is just as likely that the invention of the Densmore car was nothing more scientific than a chance inspiration. Victorian streets abounded with sprinkler trucks meant to wet down the dust kicked up by wagon traffic. Large wooden tanks had been used for centuries to store wine, vinegar, fats, and other common liquids. Older-style fire engines used wooden tubs for water reservoirs. The Army watered troops and horses from tank wagons. Yet the most likely inspiration for the Densmores was the tank farms they walked through each day. How much creativity did it take to imagine placing one or two of the smaller tanks on a flatcar? Obvious, of course, but apparently the Densmores were the first to seize upon the idea.

Assembling a car on the Densmores' plan couldn't have posed much of a problem, since it was a mere combination of existing parts. One is tempted to call it off-the-shelf inventing. Their first two cars were ready for test on or about September 1, 1865 (Fig. 5.68). Two tanks, made of clear

CARS FOR BULK CARGOES

FIGURE 5.67

A rolling-drum-style car was
tested in the Pennsylvania oil
fields in 1865 by the Howard Oil
Tank Car Co. (Smithsonian
Neg. 30738-E)

pine with close-fitting lids, held about 3,400 gallons of oil between them. The tanks were set at the
ends of a common flatcar. The center of the car
was vacant. Iron rods, bolted through wooden
beams crossed over the top of the vats, held the big
tubs to the car's platform. The first trip over the
broad-gauge Erie was monitored carefully, and the
cars were periodically checked for leaks on their
long journey east. They performed handsomely,
and the Densmores were inspired to fabricate
more vat cars. They were also inspired to protect
their simple yet effective idea by letters patent. A
patent was issued to James and Amos Densmore of
Meadville on April 10, 1866 (No. 53,794). The
specification was the usual rambling but cautious
text that claimed only as much novelty as the patent examiners would allow. It explained little, for
in truth there wasn't much to be said, yet the text
did specify wooden or iron tanks. The model submitted with the application is now in the Smithsonian's patent model collection. The model and
drawings show that oil was loaded through a
small, square manhole on top of the tank and was
emptied through a valve fastened to the low side of
the same tank.

Success breeds imitation, and others began to
copy the Densmores. An official of the biggest fast
freight line in the oil fields, Charles P. Hatch of the
Empire Line, immediately understood the potential of the Densmore idea. He sought to copy it, yet
not so closely that a patent fee might be required.
Early in 1866 he installed three square wooden
tanks inside a boxcar. The workmanship was defective, and the tanks leaked so badly that Hatch's
car was pronounced a failure. He then openly cop-

ied the Densmore model, and soon there were
hundreds of vat cars in operation. Bulk shipping,
even in small vat cars, reduced transportation
costs by 5 cents per gallon, or $170 per carload.
This is exclusive of the savings achieved by eliminating barrels. Within a year after the Densmores
began tank car shipments, the greedy and inefficient teamsters were out of business.

James Densmore, seeing that the tank car was
becoming a good thing, sought ways to protect his
invention. The original patent of April 1866 had
rather little legal clout because of the pedestrian
nature of the invention. It was merely a combination of well-known elements—wooden tanks and
a common flatcar—and the patent claim, correctly, was not allowed to claim anything beyond
this simplistic combination. Densmore understood this but decided that one way to intimidate
imitators would be to create a series of patents. He
undoubtedly hoped that by saying the vat cars
were protected by no fewer than four patents, he
could make his competitors think that the design
was considered novel and hence that some substantial license fees were due the patentees.

Densmore worked with a new business associate, George W. N. Yost of Corry, Pennsylvania.
Yost was one of many mechanic/inventor types
Densmore both supported and dominated during
the long years that the typewriter was under development. In their early association Yost and Densmore concentrated on making money in the oil
business. Their second, third, and fourth tank car
patents were all issued on June 22, 1866. The first
two, Nos. 55,830 and 55,831, covered tank cars
with three vats in place of two. Cars were actually

FIGURE 5.68

In 1865 the Densmore brothers of Meadville, Pa., began bulk oil shipments in wooden tank cars like this one parked at Gregg's Switch near Titusville. (Drake Well Museum, Titusville, Pa.)

constructed on this plan, as evidenced by a photo showing several such vehicles in the service of the Empire Line (Mather Neg. 1922). The fourth patent, No. 55,832, is a departure from the other three. It shows a variety of four- and eight-wheel flatcars fitted with one, two, and three rectangular wooden tanks. No cars are believed to have been constructed on this plan.

The success of this battery of repetitious patents is actually beside the point, because the defects of the Densmore design quickly became apparent. Yes, they were surely an improvement over what existed before, but they were really only a partial solution. The vertical tanks had a limited capacity because they could be made only as wide as the car—roughly 8 feet—and only so high, so as not to become top-heavy. Even as built, they were considered top-heavy and unstable. The absence of baffles allowed the oil to surge about, adding to the sway and roll of the vehicle, especially when going around curves. The vat cars were inherently dangerous because of the fragility of the wooden tanks—a loose collection of boards tied together by bands—and the flammable cargoes they transported. In a major accident the Densmore cars were likely to go to pieces, erupting in a deluge of thick crude over the broken timbers of the wreck. The likelihood of a serious fire was high. And even the twists, bumps, and shaking of normal operations were likely to cause leaks in the wooden tanks.

IRON TANKS

Because of all these obvious failings, it is not surprising that the Densmores' creation was made

obsolete in just over a year by a better tank car. Horizontal iron tanks answered every defect of the suddenly obsolete vat car. They offered greater volume, a lower center of gravity, few leaks, and far greater security in wrecks. Placing a large expansion dome on top of the tank made it possible for the oil to seek a level and to find room to grow in volume if the sun's rays were sufficient to warm it. Gases could collect here and were safely vented through an escape valve.

It was, in all, a rational scheme possessed of such sound engineering principles that it has survived into modern times as the basic tank car arrangement. Just who devised it is uncertain, but surely the germ of the idea was in Seely's 1863 patent already mentioned. As early as March 1867 a Titusville newspaper mentioned iron tank cars as the superior form, even though wooden cars were still predominant. Iron cars were involved in an accident on the Allegheny Valley Railroad in February 1868, which again confirms that such cars were in regular service by that time. Surely some Densmore-style cars remained in operation for many years until being used up in service, but it is unlikely that many were built much after 1868.

The Densmores didn't stay around to witness the demise of their brainchild. They had made their fortunes and left the oil fields in 1867 to pursue their typewriter mania. After many years, and the employment of many mechanics, including Yost, they achieved their goal. Many technical histories speak of their achievements in this area, but none even mentions their adventure into tank car design, which can best be described as a near miss.[82] In 1880 the Supreme Court ruled that the

FIGURE 5.69

MICHAEL SCHALL,
EMPIRE CAR WORKS, YORK, PA.
Manufacturer of ILGENFRITZ & SCHALL'S
PATENT IRON TANK OIL CAR.

Also builder of all kinds of RAILROAD CARS. Has a large stock of Car Materials constantly on hand. Orders solicited and promptly filled.

Densmore patent of May 1866 was "destitute of utility and novelty."

The general shape of the iron tank car was forecast by Seely's patent of 1863, but exactly how the cars first looked or evolved remains a question. Most likely a shipper, like the Densmores, commandeered a flatcar that was then united with a tank fabricated by a local boilermaker. More finished cars were assembled by nearby machine shops, such as the Titusville Manufacturing Company, which made all types of oil-well machinery and was ready to produce railroad cars as well to satisfy its customers. But professional car builders were more particular and yearned to develop sophisticated designs to set their products apart from the ad hoc assemblies of a country mechanic. That a better car would result, based on experience and talent of a professional in the trade, cannot be overlooked, nor can the desire for beating out the competition and collecting a patent design fee be ignored.

George W. Ilgenfritz and Michael Schall, proprietors of the Empire Car Works of York, Pennsylvania, obtained Patent No. 79,573 on July 7, 1868, with these goals in mind (Fig. 5.69). The design reveals nothing unusual in its general arrangement, but it does exhibit some interesting details. The most obvious is the drop or fish-belly shape of the tank's cylinder. It drops slightly below the frame to ensure an absolute discharge outlet and to lower the center of gravity. The tank itself is set very low on the frame to assist in achieving this last goal. The frame could be of metal or wood, according to the patent description, but the drawing and model —the latter is in the Smithsonian's collection— have very heavy section I-beam sills. Note the very long end deck, which in effect creates a collision barrier in the event of an accident and thus protects the tank's ends. The expansion dome was for some reason offset and is not central in the model and the patent drawing, but this placement is not so apparent in the advertising engraving of 1870 reproduced here as Figure 5.70. A manual valve or cock was installed on the dome to vent gases—a

most unwise replacement for an automatic breather valve and a design fault no doubt necessitating one of the earliest recalls in the history of car building. Two filler pipes on top of the car are shown in the model and the patent drawing, but only one is visible in the engraving. The Empire Car Works was a major freight car builder at the time this patent was issued, and so it is likely that the ideas embodied in the patent were actually applied to cars produced by this firm.

The fish-belly idea was subtly improved by a Philadelphia inventor named William G. Warden about three years after the Ilgenfritz and Schall patent appeared.[83] Warden employed the old boilermaker's telescoping-ring technique, but rather than aiming for a nearly uniform diameter by making the rings of only two sizes, he made the center ring or wrapper of iron plate the largest, while making each ring toward the ends smaller. Hence the cylinder tapers out to its ends. The British journal *Engineering* published an elegant engraving of Warden's car, which is reproduced here as Figure 5.70. Mention was made of a patent, but none has been uncovered.

In 1864 Murray, Dougal & Company opened a car works in Milton, Pennsylvania. The new shop was located near the oil fields. The proprietors eventually opened a metal-fabricating plant beside the car shop and so were drawn into the iron tank car business. In 1869 or 1870 the principal partner and a practical car builder, Samuel W. Murray, decided to compete more directly with Ilgenfritz and Schall by developing a patented tank car design. He worked with an employee, Benjamin P. Lamason, later general manager of the Union Tank Line. A patent was issued on February 22, 1870 (No. 100,058), which featured a straight or parallel tank, a large central dome, and a composite iron and wood frame (Fig. 5.71). The tank was placed low inside the two massive wooden floor sills. The outside portion of the frame consisted of large cast-iron brackets bolted to the main wooden sills. A large-diameter pipe passed through the extremities of the brackets and formed the outside sills and, with the brackets, supported the walkway and handrail. A very long rod with nuts at each end passed through the pipe and thus helped tie the frame together. Ample space was allowed at each end of the frame to protect the tank's ends.

In the patent drawing a separate filler plug was placed at one end of the tank, but in the published drawing first reproduced in the 1879 *Car Builders Dictionary*, this feature was replaced by a hatch fixed on top of the central expansion dome. A small handwheel with a very long valve stem was placed inside the dome to one side, just below the hinged top cover. The valve itself was directly below the dome at the tank's central bottom. The tank rested on concaved cross-timbers and the wooden body bolsters. Four heavy straps, bolted to

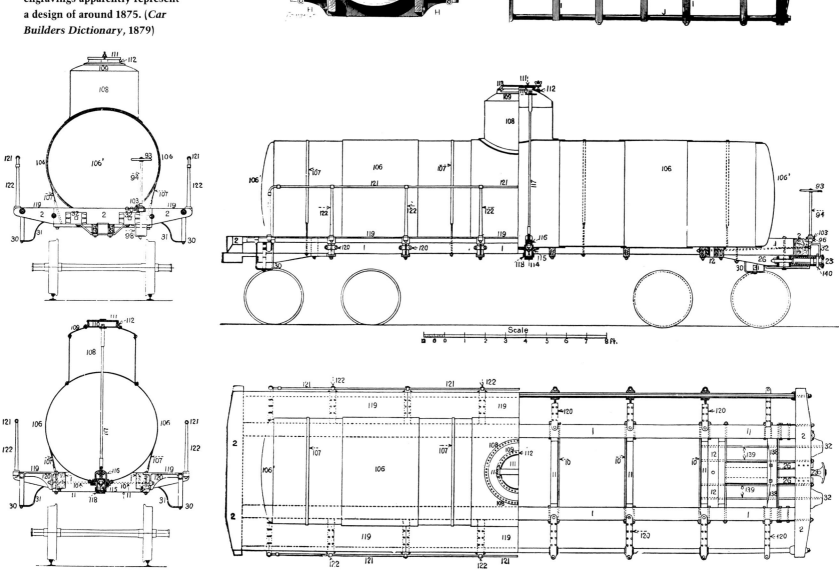

FIGURE 5.70

FIGURE 5.71

Engravings for a tank car that agrees very closely with S. W. Murray's 1870 patent drawing were published in 1879. These engravings apparently represent a design of around 1875. (*Car Builders Dictionary*, 1879)

FIGURE 5.72
This perspective engraving represents a tank car design produced by the Harrisburg car works in about 1875 for the Pennsylvania Railroad, the Union Tank Line, and other buyers of such equipment. (Lavoinne and Pontzen)

FIGURE 5.73
Iron tank cars proved far superior to the wood-tub variety and had rendered the Densmore design obsolete by the late 1860s. This car was built for the Oil Creek & Allegheny River Railway around 1875 by the Titusville Manufacturing Co. (Drake Well Museum, Titusville, Pa.)

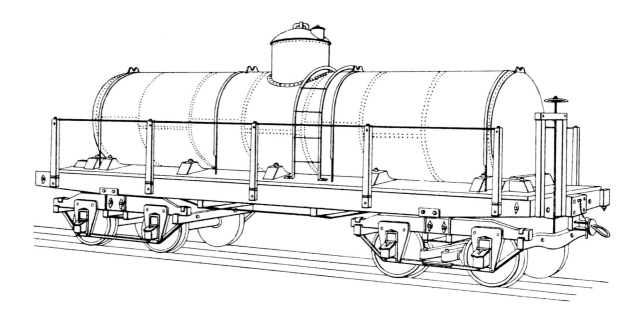

FIGURE 5.74

The Allison Car Works produced this car in the 1870s for the Allegheny Valley Railroad. Notice the offset dome, separate valve, manhole, and spindly handrails. (Collection of A. Andrew Merrilees)

the main frame sills, wrapped around the tank to hold it in place. Notice how the tank's ends were made convexed, for added strength and interior space. The patent drawings—which are unusually detailed and prototypical in appearance, without the toy- or model-like appearance of so many patent illustrations—duplicate the 1879 and 1888 *Car Builders Dictionary* tank car drawings in every detail except for the dome detail just described. There can be little doubt that those drawings are copied from Murray and Lamason's plans. Warden's frame design (Fig. 5.70) was also clearly inspired by the Milton car builders.

The Harrisburg car works started operations in 1853, some years before petroleum had become a common consumer item, and when the oil boom began the company seemed only too ready to produce cars on a very pedestrian plan, devoid of the novel features so beloved of nearby rivals. A perspective drawing dating from the 1870s agrees exactly with an engraving of a car built for the Union Tank Line by Harrisburg.[84] This car appears to be nothing more than a conventional flatcar with wood blocks cut away in a concave shape to receive the tank (Fig. 5.72). The handrail stanchion and brake-wheel bracket are simple wooden posts. A very light iron bar forms the handrail. The gently chamfered ends of the deep sides and the shallow truss rods without benefit of iron queen posts are unmistakable signs of an early flatcar. The wooden trucks contribute to the antique appearance of the car, as does the single-end brake arrangement. Harrisburg improved and modernized its designs over the following decades, but the basic, simple wood-frame plan was retained until quite late in the nineteenth century, as is evident from Figure 5.78.

Photographs of other pioneer tank cars show a few interesting variations on the general scheme already described and illustrated. The Titusville-built car is very similar to the older Harrisburg scheme and shows few novelties other than the raised-metal letters and numerals (Fig. 5.73). This car stands partially incomplete outside its maker's shop awaiting handrails and other finishing details. The light shade of the tank is reminiscent of the bright green favored by the Green Line. Notice the expansion tank on display at one end of the No. 150. Against the shop wall stands an unrolled tank sheet, a convex end, and a valve and stem rod standing upside down. Notice too how the end of the tank is protected by heavy timbers mounted on top of the deck.

The Allison Car Works of Philadelphia built a more distinctive tank for the Allegheny Valley Railroad in about 1870, shown in Figure 5.74. The offset expansion dome and separate filler cap are the most obvious peculiarities of the car. Yet it should not be viewed as a freakish exception, for other similar cars are shown in a series of petroleum field stereo cards issued by an Oil City photographer in the 1870s.[85] The Allegheny Valley's No. 725 has other features worthy of notice. The lack of truss rods for what is a fairly long car with such a light wooden frame is peculiar. The large cross-timbers seen under the side sills are cradle beams for the tank. The flimsy handrail is an insult to trainman safety standards, but the apparent absence of a ladder or steps to the dome or filler cap may be accountable to their being on the other side of the car.

It must have become apparent to most car designers that very heavy framing for oil cars wasn't really necessary beyond what was needed to with-

FIGURE 5.75
The Lima Car Works of Lima, Ohio, built this wood-frame car around 1885 for the Shaunee Oil Co. (Allen County Historical Society, Lima, Ohio)

FIGURE 5.76
Pullman built this car using a tank supplied by an outside contractor in 1889. (Pullman Neg. 356)

THE AMERICAN RAILROAD FREIGHT CAR

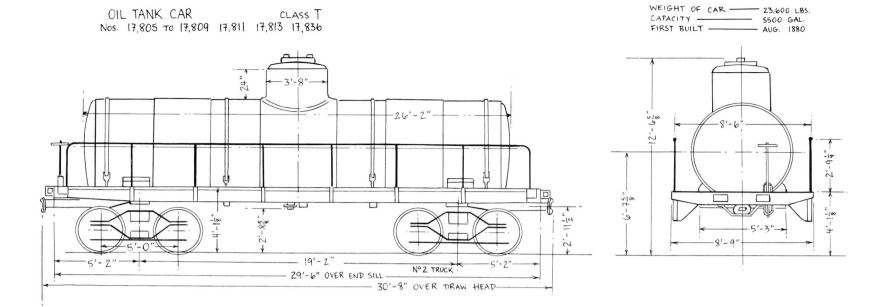

OIL TANK CAR   CLASS T
Nos. 17,805 TO 17,809   17,811   17,813   17,836

WEIGHT OF CAR ——— 23,600 LBS.
CAPACITY ——— 5500 GAL.
FIRST BUILT ——— AUG. 1880

FIGURE 5.77

This diagram drawing depicts a B&O tank car of 1886. (Baltimore & Ohio Railroad)

stand normal buff and draft stress, because the tank itself, being made of boiler iron, was self-supporting. It was most likely for this reason that king post truss rods were sometimes used. Just when this practice came into favor is apparently unrecorded, but a photograph dating from around 1885 shows a car of the Shaunee Oil Company of Lima, Ohio, with this style of floor framing (Fig. 5.75). The car was built by the Lima Car Works, a small concern then controlled by a car works in Lafayette, Indiana. In 1889 Pullman produced some similar cars for a packing house to ship lard or tallow. Like the Lima car, these wood-frame vehicles had king posts for a pair of truss rods. Figure 5.76 reproduces a photograph that shows many features typical of tank cars of the period, including composite wood and iron-plate body bolsters, square wooden railing posts, brakes on only one truck, and link-and-pin couplers. The single-post rod arrangement had also been adopted by Murray, Dougal & Company by the 1890s.[86] The firm had abandoned Murray and Lamason's wood and pipe frame by this time in favor of a simple, all-wood frame.

Most of the early iron tank cars had a capacity of 80 barrels, or around 3,400 gallons. They were thus in scale with other freight cars of the time and fit nicely into the 15- to 20-ton-capacity range. The Empire Line had some 4,500-gallon cars in operation at the time of the Rockefeller takeover in 1877 and was reported to have some 5,040-gallon (120-barrel) cars two years later.[87] The tenth U.S. Census report for 1880 briefly commented on tank cars, saying that they ranged in size from 3,856 to 5,000 gallons in capacity. The report noted that a typical tank measured 24 feet 6 inches long by 66 inches in diameter. The top plates of the cylinder

were made of ³⁄₁₆-inch-thick metal, while the bottom section was rolled from ¼-inch-thick plate. The dished ends were ⁵⁄₁₆ inch thick. The tank weighed 4,500 pounds. By the middle 1880s capacity was 5,500 gallons for new cars, again in step with the general increase in car size.

A diagram drawing for a Baltimore & Ohio class T tank car gives exact particulars for a car of this period (Fig. 5.77). The design was introduced in August 1886. The arrangement and size of the car are given in the drawing. It weighed 23,600 pounds and had a weight capacity of around 20 tons. A more elaborate drawing from around the same period has been uncovered for a Union Tank Line car produced by the Harrisburg Car Company (Figs. 5.78, 5.79). It was far larger than the B&O car. It measured 37 feet long over the end sills. The tank was 32 feet long by 76 inches in diameter and had an estimated capacity of 8,000 gallons, or roughly 30 tons. This magnificently detailed document is surely one of the most complete to survive for any American freight car of the nineteenth century. Because no tank cars from this period have been preserved, it is also the best record available.

An examination of the document reveals some intriguing details. Note that the valve crank or handle is outside the dome, making it more accessible than in the Murray, Dougal & Company scheme, which placed it inside the dome. There is, however, a large pipe cap screwed onto the bottom of the discharge valve so that the contents of the tank could not be emptied should the valve vibrate open or be maliciously tampered with by tramps, children, or other trespassers. The square wooden handrail posts are inclined out at a slight angle to provide a wider passageway for the often portly trainmen of the Victorian era. The neat transition

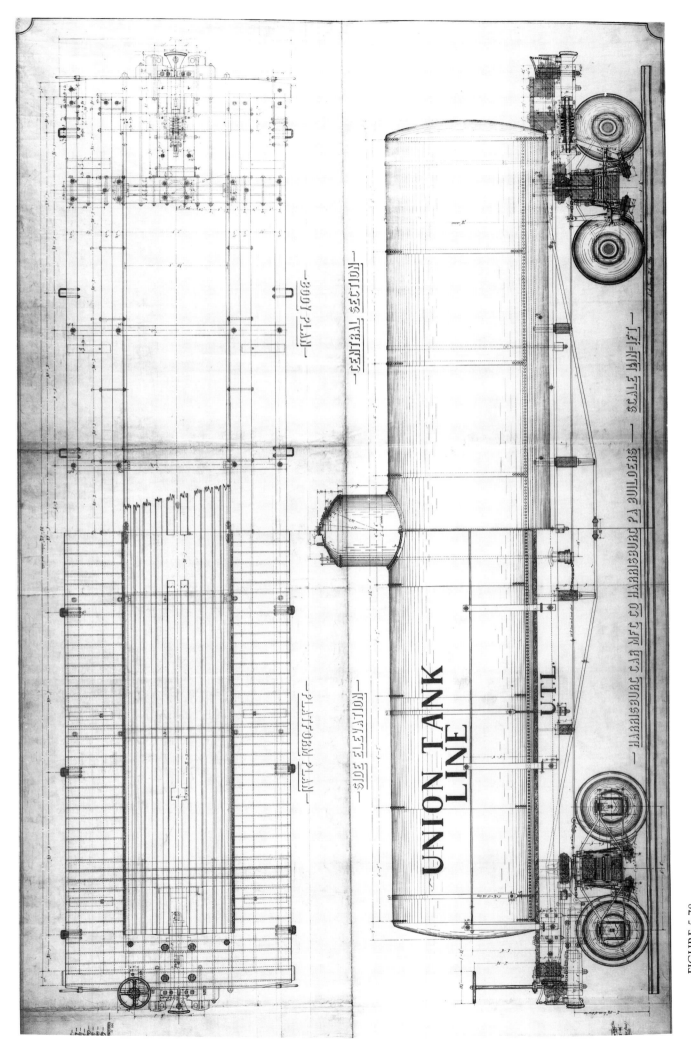

FIGURE 5.78

The Harrisburg car works produced cars on this plan around 1885–1890. (Division of Transportation, Smithsonian Institution)

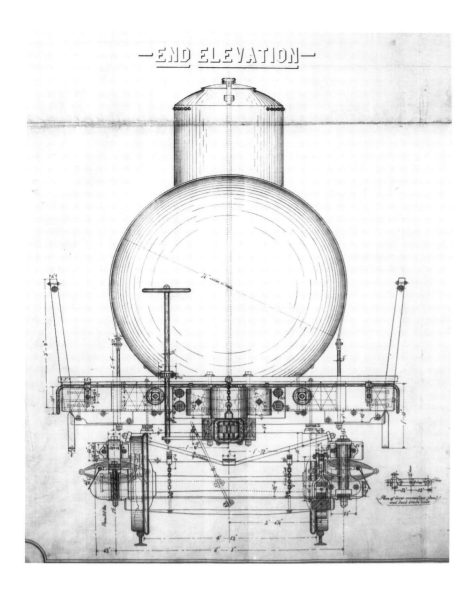

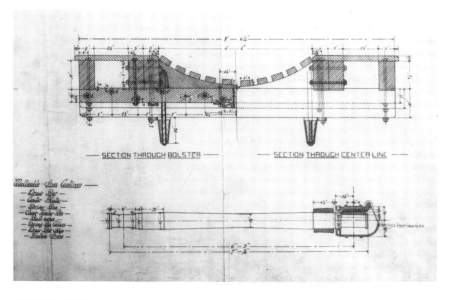

SECTION THROUGH BOLSTER   SECTION THROUGH CENTER LINE

FIGURE 5.79

**End and cradle details from the
1885–1890 Harrisburg drawing.
(Division of Transportation,
Smithsonian Institution)**

of the flat tank straps into a clevis on the tie-down bolts is shown in the end elevation. The cast-iron dead blocks are common for the period, but the semiautomatic couplers contrived by the lift-pin lever and chain are worthy of study. The double-acting hand brakes are a decided improvement in stopping power over the early tank cars shown in this chapter. The trucks appear to be of the common arch-bar variety, and so they are, but the fully elliptical springs are a decided novelty for a post-1870 freight car. A much simplified drawing for a similar UTL car is given in the 1895 *Car Builders Dictionary* (Figs. 48, 375). It was rated at 8,000 gallons with a weight of 24,000 pounds, and it can therefore be safely assumed that the Harrisburg car's weight was about the same.

For all its draftsmanship and completeness, the Harrisburg drawing is not dated. Most features of the car would suggest a date of 1885 to 1890. The firm failed in 1890, which assures that the document was not prepared after that time. The failure, it might be observed, was unexpected, for Harrisburg was a leading car builder and as late as 1888 ran full-page advertisements in the *Car Builders Dictionary* and *Poor's Manual of Railroads* emphasizing that oil tank cars were a specialty.

STEEL-FRAME EXPERIMENTS

Car designers, for the most part, were content to enlarge and refine standard designs. The tank car of 1890 was not noticeably different from its ancestor of twenty years before, except in size. During this age of wooden construction, radicals dreamed of all-metal rolling stock, and a few of them had the opportunity to test their ideas. Ilgenfritz and Schall only proposed the scheme in their 1868 patent, and it was a proposal that lay dormant, at least regarding tank cars, for a full quarter-century.

George L. Harvey of Chicago opened a car shop in 1891 with the sole purpose of introducing steel freight cars, but his efforts were largely ignored by American railroads. He turned to repair work in an effort to keep his business afloat. One of his customers was the Standard Oil Company and its subsidiary, the Union Tank Line, the largest tank car operator in the nation. Harvey and his staff, after a firsthand study of UTL standard design, decided to produce a steel-frame 8,000-gallon tank car.[88] Six months was devoted to the design, which resulted in a 37-foot-long car with a tank measuring 29 feet 3 inches long by 78 inches in diameter (Fig. 5.80). The frame was made from light 6-inch I-beam (13 pounds to the foot), reinforced by 1¼-inch truss rods. The only wood in the car was the running boards, the end deck, and nine long hardwood strips under the tank to cushion it against the channel-iron cradles. The test car was well received, and the *Railroad Gazette* stated that a "considerable number" were being built for UTL by Harvey. The truth of this claim cannot be ver-

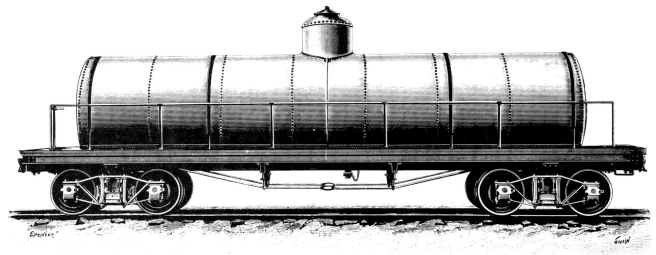

FIGURE 5.80

The Harvey car works attempted to introduce steel-frame tank cars in the early 1890s. This test car was produced for the Union Tank Line. (*Engineer*, April 20, 1894)

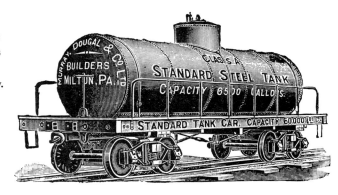

FIGURE 5.81

Murray, Dougal & Co. used this engraving dating from the 1890s in advertisements for tank cars into the early twentieth century. (New York Railroad Club *Proceedings*, 1900)

ified, but it is certain that very few steel-framed tank cars were employed on U.S. railroads until after 1900.

MULTIPURPOSE TANK CARS

Standard single-tank cars served the needs of most big oil dealers, since they tended to ship in bulk, carload lots. But smaller dealers or customers requiring limited amounts of a more specialized liquid, like valve oil or ink dye, required a smaller container. Individual barrels could be used to satisfy most such needs, and a great deal of less-than-carload liquid goods was shipped in this manner. The volume of the traffic is explained in part by the barrel-rack cars maintained by UTL and some other carriers involved in the oil and industrial-liquids trade. The filled barrels were shipped in ordinary boxcars or gondolas but the empties were gathered at central depots and consolidated for reshipment to the refinery or chemical plant. Special high-cube, open-sided barrel cars were used for the return trip. They were very long and high for a freight car of this era (Fig. 5.82). Lengths up to 51 feet were reached by the early 1890s. Such cars could carry 430 barrels, or 20 tons, and were related to furniture or buggy cars of the period. Some had roofs, while others were made with open tops.

Divided or two-compartment tank cars were seen as another solution for less-than-carload liq-uid shipments. Such cars required a central interior dividing head or sheet, two discharge valves, and two expansion domes and filler caps. A small oil dealer might then be able to order half a tank of kerosene and half a load of machine oil, or the same car could service two independent customers or distribution points. Eventually three-compartment oil tank cars came into existence, but this appears to have been a twentieth-century development. The earliest known twin-tank car is illustrated by a very small and rather crude engraving—reproduced in *Asher and Adams' Railroad Atlas*—of an Allegheny Valley Railroad car built by Harrisburg in the 1870s.[89] Again, like so much in freight car history, the origin and date of the twin-tank cars appear not to have been recorded. A later example is illustrated in an 1896 Murray, Dougal & Company catalog referred to earlier in this chapter.[90] It shows a 30-ton-capacity car with two 3,500-gallon compartments. The car weighed 27,600 pounds.

No railroad car seemed less adaptable than the tank car for multipurpose freighting. It was destined for one-way revenue trips and an assured empty backhaul. Considerable ingenuity was expended by inventors to solve the problem. As already noted, some plans were offered by individuals such as Keeler, just before the tank car era began. By the 1870s tank cars had become common enough to refocus attention on the need for a more flexible car. In 1873 the New York Central bought 150 combination oil/boxcars from the Cleveland Bridge & Car Works.[91] Sadly, no description of them was included in the brief note announcing their construction, but it is conceivable that they might have been made after Keeler's plan.

Just two years later, on May 18, 1875, Albert P. Odell was issued U.S. Patent No. 163,515 for a perfectly goofy-looking combination oil/gondola car that was sure to alienate every practicing car builder in North America. The car had an iron-truss side frame and a semicylindrical tank hang-

FIGURE 5.82

ing between the trucks. The half-cylinder tank connected to a rectangular receptacle that reached back over the trucks. A low-sided gondola box sat on top of the tank's roof plates. Ill-conceived as this plan may appear, the patent was purchased by the Union Tank Line in 1883, and it is claimed that about five hundred were built in the 1890s for transcontinental service.[92] No plans for these cars are known to survive, but it is impossible to believe that Odell's design would have been closely followed by any sane car builder of the period. Photographs of a UTL combination oil/boxcar built in 1898 at the Atlas Works, Buffalo, are in the Drake Well Museum's photographic collection. Possibly it was one of the vehicles built for cross-country service. The exterior views reveal nothing about the interior arrangement of the No. 830, which appears to be an ordinary boxcar, except for small ventilator openings high on the sides and king post truss rods. It is clear that no semicylindrical tank is appended to the car's underside.

More practical minds than Odell were at work on the problem of combination oil/merchandise cars. Chief among these was Horatio G. Brooks, founder of the locomotive works bearing his name in Dunkirk, New York. Brooks's knowledge of railroad machinery came from long years at the throttle and in the repair and construction shop. He was no amateur or dreamer but a fully credited member of the shop culture that dominated railroad engineering at the time. One would expect the ideas of such an insider and experienced practitioner to win acceptance within the industry.

Brooks's arrangement was not new and followed Keeler's general plan for a combination frame and tank surmounted by a wooden boxcar body (Fig. 5.83). It was in the details that Brooks hoped to win the support of his fellows. The tank was made rectangular and relatively boxy compared with Keeler's scheme and thus presented fewer fabrication problems and a trifle more space. The tank, a shallow rectangle, overreached the trucks and then dropped down low to the rails between the wheels. It was calculated to hold the then-standard tank car volume of 3,600 gallons. The inlet and discharge valve were combined, and so it was necessary to pump or gravity-feed oil into the tank. Vent pipes ran up through the roof and were placed so as not to interfere with the interior space of the merchandise portion of the upper body. A covered manhole was fitted into the upper metal deck of the tank for inspection and repair purposes.

In September 1876 a sample car was reported abuilding at the Brooks Locomotive Works.[93] It was to have a boxcar body, but Brooks contended that the superstructure might just as easily be constructed as a flat- or cattle car. In October 1876 the sample unit was sent from a Cleveland refinery to the West Coast. It got as far as Council Bluffs, where the Union Pacific refused to accept it for interchange. The car was ignominiously returned to Dunkirk, where it went into storage. After a few years its body was stripped away, and it became a shop flatcar, reported to be still in service in 1902. The Patent Office issued Brooks a patent on May 18, 1877 (No. 190,542), but apparently forgot its original grant, for two Great Northern engineers, Messrs. Toltz and Hill, received a patent twenty-five years later on March 25, 1902 (No. 696,150), for a virtually identical invention.

Brooks's defeat did not discourage others from endeavoring to succeed where he had failed. The plan of attack was different from the possum-belly tank previously tried, though not really any better. The idea now was to install vertical iron tanks at each end of a box. The central space was for merchandise, but the defects of the scheme are obvious, for the car was the worst sort of compromise; it was both a poor tank and a poor merchandise car. Merchandise could be carried in both directions, it is true, but the space was limited by the bulky tanks at each end. And the tanks added considerable dead weight to the vehicle, full or empty. These failings did not discourage M. Campbell Brown of Cleveland from obtaining Patent No. 216,506 on June 17, 1879. Cleveland was

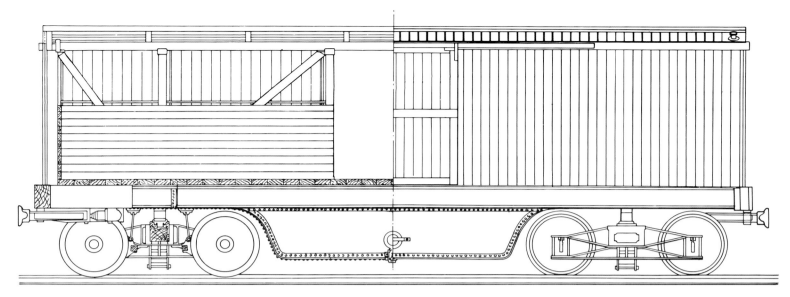

FIGURE 5.83

Horatio Brooks designed this combination box/oil tank car in 1876. A patent was issued in May of the following year. (Traced from a Patent Office engraving)

then home to Standard Oil, and Brown most likely hoped to sell his idea to this major petroleum shipper. He inclined the bottom sheet of each tank toward the drain plug. Roof hatches allowed access to the inlet cap on top of the tank inside the car.

Brown's scheme was revived by Charles L. Rogers of Murray, Dougal & Company in 1897. He received Patent No. 583,660 on June 1 of that year. Rogers's design called for vertical round tanks at each end of the car. An expansion dome, inlet plug, and valve handle were placed in a compartment raised above roof level. The Milton Car Works offered to build cars to Rogers's design with tanks of 3,500- to 6,600-gallon capacities. Just how many were constructed is not known, but at least one sample car was constructed and appeared in a halftone illustration in the firm's 1896 catalog.[94]

The Rogers or Brown car may not seem inspired, but it surely made more sense than the combination grain/oil car patented by John R. Gathright on March 17, 1885 (No. 313,932). Triangular tanks were placed at the ends of a boxcar body, forming a hopperlike compartment inside the car for grain. The tanks were by necessity small, so small that one wonders if they were worth the extra expense and complexity of construction. The oil tanks were connected by narrow rectangular sections that also formed the bottom of the grain space. Side doors opened into the grain areas, as they would in an ordinary boxcar. Externally, Gathright's creation would in fact look like a boxcar, except for the filler-vent pipe at the roof lever and oil discharge valve under the car's center. The inventor called for a raised wooden floor over the tanks to protect the grain from oil spillages or leaks. But this in no way protected it from the odor of the liquid cargo. The pungent smell of crude or kerosene could ruin the resale value of most goods, whether grain, furniture, or clothing, and this was an inherent defect not only of Gathright's plan but of all combination oil/merchandise cars.

THE OWNERSHIP ISSUE

Tank cars can easily be identified even by non-railroaders because of their distinctive appearance. The fact of their largely iron construction also separates them from the common herd of nineteenth-century freight cars, but there was another basic circumstance that made them special, one having nothing to do with their shape or structure. It was the matter of ownership. Very few were railroad-owned. The great majority were privately owned by oil refiners and shippers. One obvious reason, which must by now be familiar to readers of this volume, was the reluctance of operating railroads to invest in specialized rolling stock. The tank car was perhaps the most specialized freight car of all and the least able to refashion itself for any nonliquid backhaul cargo. A refrigerator or stock car could at least function as a poor sort of boxcar. Tanks were not only one-way revenue producers, with guaranteed revenueless return trips; they were also pricey in terms of first cost. Why invest $700 to $800 in a vehicle with such limited earning prospects? The same money would almost buy two boxcars.

The understandably negative view held by railroad men was countered by a positively radiant vision on the part of oilmen. They saw the tank car as both a convenience and a cost cutter. It moved quantities of oil with minimal spillage losses for pennies a mile. Eighty barrels of oil could be sent to New York from northwestern Pennsylvania for $170 less by tank car than if shipped in individual barrels.[95] This transportation saving resulted in a 5-cent-per-gallon drop in oil prices. Because millions of gallons were involved, the savings were substantial enough for oilmen to see real beauty in the prosaic tank car.

Joseph D. Potts, a civil engineer and minor official of the Pennsylvania Railroad, was one of the first to understand the glorious profits made possible by the ugly tank car. With the backing of Thomas Scott, vice president of the mighty Pennsylvania Railroad, he helped organize the Empire Transportation Company in March 1865.[96] This private car line was tied to the Pennsylvania's recently acquired Philadelphia & Erie, and through that association it was able to reach hundreds of miles east and west of Corry, a principal terminal for the Empire Line. Within a few months Potts was operating Densmore-style cars, a design that gave way quickly to iron tank cars. Potts was ambitious and energetic. He purchased thirty-four hundred boxcars for grain and a fleet of four hundred rack cars for barrels. Grain elevators were assembled, as was a line of lake steamers. Pipelines and a great oil terminal at Communipaw, New Jersey, were constructed. By 1875 Empire was operating just over thirteen hundred bright green tank cars.

And then Potts became too ambitious. He wanted to refine as well as transport oil, an objective that frightened the greatest of all oil processors, John D. Rockefeller. Rockefeller was unquestionably the most gifted business mind in the petroleum trade. He most clearly saw its potential and yearned to make it neat and tidy, for he abhorred waste and competition. Free enterprise might be a noble principle, but it didn't really suit Rockefeller's plan for the oil industry, nor did Potts's presumptuous scheme to compete openly with the monopoly that the pious oil titan from Cleveland had begun to fashion.

Rockefeller placed an embargo on all shipments via the Empire Line in the spring of 1877. He diverted everything to his own Union Tank Line, a small property he had acquired in 1873 along with the purchase of the Star Tank & Pipeline Company. The effect on Potts and his backers—principally the Pennsylvania Railroad—was devastating. The Pennsylvania was cash-poor because traffic was down following the 1873 panic. In October of 1877 the railroad sued for peace; it arranged to sell the Empire Line to Rockefeller for $3.4 million. It was a good buy, for the property was well maintained and its traffic was good, despite the depression.[97] In 1876 it moved almost 3 million barrels of oil by rail and only slightly less by pipeline. With the Empire Line in hand, Rockefeller was now well on his way toward dominating the American petroleum industry.

Rockefeller saw more clearly than ever that tank cars were the key to controlling the oil industry. Whoever held title to the nation's tank cars could make or break any oil refiner or dealer in America. It was rightly said to be his secret weapon, and so began the amazing growth of the Union Tank Line. In 1878 Standard Oil established the Tank Car Trust, a holding company organized to buy cars at below-market prices. During the next year Rockefeller leased tank cars from the New York Central as a means to complete his monopoly, which lacked only two hundred cars. More red-painted cars bearing UTL lettering appeared, so that by 1888 Standard Oil owned outright or directly controlled through various subsidiaries fifty-six hundred of the nation's sixty-one hundred tank cars. In August 1891 the Union Tank Line was incorporated, with a capital of $3.5 million. It owned around five thousand cars at the time, and within nine years its fleet had increased by nine hundred cars. Hence it was argued that Rockefeller's Standard Oil Company owned less than half of the nation's thirteen thousand tank cars. In truth, Standard's control was more realistically established at up to 90 percent because of equipment titles held by subsidiary firms.

Public interest in tank cars was about on a par with most matters technical, yet this obscure subject became a center of national attention during the first years of the twentieth century because of a series of articles in *McClure's Magazine*. The internal history of Standard Oil and its motives in corralling every tank car in sight were discussed at considerable length. Arcane firms like the South Improvement Company became instant bogeymen. A public outcry eventually led to a breakup of the giant Standard Oil empire. It was declared illegal in May 1911, and efforts to stimulate competition in the tank car field were soon under way. UTL was separated from the mother company, and in a symbolic gesture, in 1912 it began to paint its red cars black. General American Transportation, a pigmy private car operator established in 1898, began to push into the tank car business, but it did not become competitive in a major way until around 1920. By this time tank cars had retreated from the headlines as the reform spirit died with the advent of the Harding administration. The Bucket of Industry was allowed to continue its pedestrian duties untroubled by the attentions of reformers, the press, or the public.

6 CARS FOR SPECIAL SHIPMENTS

So far we have been discussing the more common varieties of American freight cars. Boxcars, gondolas, and flatcars can hardly be classified as special or curious. Cattle and refrigerator cars might also strike many students of railroad technology as conventional as well. Yet within these ordinary ranks a great variety of nonstandard designs can be found. The prosaic boxcar came in many sizes and shapes, from pigmy four-wheelers to the hulking high-cube merchandise carriers. Similar novelties are noted in the chapters on bulk and food carriers. And so this chapter should not be seen as the exclusive haven for the odd and unusual. Rather, it was created to discuss the several lesser varieties of freight cars, from containers to cabooses, that don't fit conveniently into the previous five chapters.

Jumbo Flatcars

American railroaders could draw on the experience of the earliest North American lines for models of heavy-duty cars. Mention was made in Chapter 2 of the great sixteen-wheel car used on the Granite Railway, as well as the fleet of substantial eight-wheel cars employed on the Allegheny Portage to ferry canal boats. Yet another example dating from the pioneer period that is also worthy of mention is the eight-wheel rail cars used on the Morris Canal inclines.[1] Opened in 1836, the Morris Canal ran across New Jersey from Jersey City to Phillipsburg. In place of locks, incline railways were built to overcome the different levels. The 60- to 80-foot-long boats were carried by massive wooden flatcars on 10-foot-gauge tracks. A heavy wooden bridgelike structure stiffened the cars for their 25- to 30-ton loads. The car frame measured about 41 feet long and was carried by two cast-iron side-framed four-wheel trucks. The fact that railways could handle big loads was well demonstrated by these ponderous vehicles.

Further progress in the area of oversize flatcars is poorly recorded, and no mention of these vehicles is made until about 1850. Railroads were expanding into the western states, but there was still no direct linkage to the eastern lines. Locomotives were consequently transhipped by lake vessels from Buffalo to Cleveland, Detroit, and other lake ports. Others went south by ocean ships to New Orleans, where they were again reloaded onto riverboats. Such shipments were costly, hazardous, and slow because of the indirect routing. The loss of locomotives at sea or in the Great Lakes prompted William M. Kasson of Buffalo to organize the Kasson Locomotive Express Company in December 1852.[2] Kasson had previously been in the express business. He had special "trucks" and loading apparatus built in Boston. Unfortunately, the description of the equipment is so limited that it is not clear whether Kasson used flatcars or independent trucks. Apparently it was the latter, for it was noted in one reference that the driving wheels, trucks, and smokestacks were placed separately in ordinary freight cars. In November 1853 mention was made of special cars being constructed in Cleveland for Kasson's Locomotive Express. No description was offered, but it may be that the truck system was not altogether satisfactory and that heavy flatcars were substituted. As the midwestern railroad boom continued, Kasson's business flourished. In less than a year he moved 150 engines and had contracts for transporting another 200. Kasson continued until about 1866. The business was apparently taken over by the Merchants Despatch Transit, a private freight car line associated with the New York Central Railroad.

At about the time of Kasson's closing, William A. Lovell started up a locomotive express business in Boston to serve the area's three locomotive manufacturers, Hinkley, Mason, and Taunton.[3] Lovell's equipment is documented by several fine photographs, one of which is reproduced here as Figure 6.1. Deep wooden frames set low on the trucks characterize all of Lovell's cars. Some had what appear to be toolboxes mounted near their centers. The numbers in the photos range from 7 to 55, indicating a rather sizable fleet. By the time Lovell began operations in the mid-1860s, most standard-gauge engines could travel just about

FIGURE 6.1

Lovell's Locomotive Express used heavy-duty wooden flatcars to transport broad- and narrow-gauge locomotives. The 5-foot-gauge engine pictured here is ready to leave Taunton, Mass., in May 1871, bound for the Atlantic, Mississippi & Ohio Railroad. (Smithsonian Chaney Neg. 3443)

anywhere on their own wheels. Yet most of the southern lines remained 5-foot-gauge, and so their engines required special handling by flatcars. By 1871 the ill-advised narrow-gauge craze began sweeping the country and so created a new market for locomotive express operators. The narrow-gauge fad at last expired around 1885, and so did Lovell's operation. At this same time, the southern lines were preparing for conversion to standard gauge and ceased purchasing broad-gauge engines. Indeed, most of the Boston-area locomotive builders were closed or about to close, and so Lovell's business had reached its end.

Lovell found competition in the several fast freight lines that were established shortly after the Civil War. In 1879 the Blue, Red, National Despatch, and Merchants Despatch lines placed bids with the Taunton Locomotive Manufacturing Company for the carriage of locomotives.[4] An engraving of a heavy Red Line flatcar, published in an article about the Harrisburg Car Company during the same year, further establishes that the private car lines were equipped to handle oversize shipments.

The idea that very heavy flatcars were always needed to ship locomotives proved fallacious. It depended, of course, very much on the size and weight of the engine. Very light engines could be safely transported in ordinary cars. Photographs again provide the documentation for this generalization.[5] In 1872 Baldwin shipped a narrow-gauge tank engine, weighing around 13 tons, to Texas in a standard Philadelphia & Reading gondola. The car, without even the benefit of truss rods, managed to convey its iron burden to Brownsville, Texas. In another photograph dating only one year later, an-

other more substantial narrow-gauge engine and tender straddle two Lake Shore flats. The engine in question was a 20-ton Brooks Mogul destined for Utah. In yet another published photograph, the Pittsburgh Locomotive Works exhibits its own flatcar, No. 2, used for shipping light locomotives. In this example a very light (12.5 ton) American type with a diminutive four-wheel tender is on board No. 2 bound for Georgia. In at least one instance, a railroad had a special low-level car built specifically for the movement of narrow-gauge locomotives. In 1889 the Virginia & Truckee Railroad contracted with W. C. Allison & Company, Philadelphia car builders, for a 25-ton-capacity flatcar.[6] The 30-foot car was ready by March 1881 to carry a new engine from the Baldwin works, also in Philadelphia, west for service on the V&T's subsidiary, the Carson & Colorado Railroad. Once in Nevada, the car was used to transfer engines from Mound House to the parent road's shops at Carson City. It is also likely that the car was used to convey machinery for the mines in the area. In 1938 the now-obsolete and little-used car was sold to Paramount Pictures. In more recent times it has come into the possession of Short Line Enterprises and is presently stored on the Sierra Railroad.

Not all oversize flatcar loads were necessarily heavy machinery or locomotives. Some cargoes were bulky and light. Some were able to withstand the weather and could not be conveniently stowed inside a high-cube boxcar, and so a singular class of long but lightly constructed flatcars emerged. They were especially useful for moving vehicles like street railway cars, circus wagons, and odd-gauge steam railroad cars. An early example is a

CARS FOR SPECIAL SHIPMENTS

FIGURE 6.2

Jackson & Sharp, car builders of Wilmington, Del., shipped two horse cars, bound for a street railway in Easton, Pa., on an extra-long platform car made especially for such shipments.

photograph taken in 1866 of a Jackson & Sharp vehicle carrying two horse cars consigned to the Easton passenger railway (Fig. 6.2). Each horse car measured about 20 feet long and weighed around 3 tons. The flat accordingly must have measured between 42 and 44 feet in length. The light construction of the car and its trucks is evident from the photograph. In another Jackson & Sharp photograph, dating from the middle 1870s, a narrow-gauge coach rests on another long wooden flatcar lettered "Narrow Gauge Car Express; A Carey, Wilmington, Del."[7] The car has deep fish-belly side sills reminiscent of the ancient shad-belly cars of the Camden & Amboy. The Narrow Gauge Express car probably measured around 50 feet in length. It rode on a substantial pair of diamond-arch-bar trucks.

Jackson & Sharp by no means had a monopoly on long flatcars, for many other car builders owned or leased such vehicles. The 1897 *Equipment Register* listed a number belonging to Brill, St. Louis, J. M. Jones, Brownell, Barney & Smith, and Laclede. The shortest, at 50 feet, were owned by Jones, while the longest, a 65-foot-long giant, was owned by Brill. The smaller cars were rated at 20 tons, the larger ones at 30 tons. The Laclede Car Company

of St. Louis leased long flats from Venice Transportation, a private car operator located in the same city.

Venice started up in the early 1890s to lease and operate flat-, box-, and stock cars.[8] The company was notable for a fleet of 143 flats 60 feet in length used for the carriage of long timbers, streetcars, and wagons. A number were leased for circus trains. Part of the Venice fleet was carried on 28-inch wheels to facilitate low-level loadings. These flats stood only 35 inches above the rails and were nearly 18 inches lower than a standard car floor.

The designer of these stretch flats was F. E. Canda, of the Ensign Manufacturing Company, a builder already mentioned for his novel and imaginative contributions to wood-car construction.[9] Canda observed that long flats tended to buckle up under the extreme buff stresses set up by heavy trains. Loaded cars seemed to handle the same compressions without ill effects (Fig. 6.3). But it was not always possible to provide loads, and so Canda attempted to solve the problem with a counter-truss, which would retard an upward bending of the car's frame. Inverted king and queen posts supported a very light (¾-inch) rod in a cross or diamond pattern that was partially hidden by the very deep side sills. The main truss rods, four in number, were conventionally mounted, but they did have extra-deep (14-inch) queen posts. The first cars on this design appeared in about 1890. They were used to carry long framing timbers from Georgia to car plants in the North. The 41-foot-long vehicle weighed only 23,700 pounds and was rated for 30 tons. Even when overloaded, it retained its camber and could be counted on to run without the breakdowns encountered with conventional cars. Canda added one more detail to assist in loading long timbers, and that was a fold-down brake-wheel shaft. This feature also made the cars useful for circus train service.

The jumbo flats dating from the 1830s are not known to have been used for the general carriage of freight, and they were surely never in interchange service. Yet as industrial America matured, so grew the need for extra-heavy flats of ample capacity. The Pennsylvania was among the first to respond to this need, constructing a low-level sixteen-wheel giant that it labeled the Gun Car.[10] The original purpose of this car, built in 1869, was to carry large cannons for the Army and Navy between government armories, shipyards, and military bases. It is also likely that it was employed for any oversize cargoes requiring transportation over the Pennsylvania Railroad and its connections.

The general arrangement of the gun car is shown in Figure 6.4. It was not remarkably long—31 feet 10 inches—but its width of 9 feet, the massiveness of the frame, and the cluster of wheels below the floor surely set it apart from any other railroad freight car of the period. About the only other rail vehicle of equivalent bulk to roam the rails in recent memory was a fourteen-wheel flat built to carry a naval gun reportedly used by Union forces against Petersburg, Virginia, late in the Civil War.[11] This very special-purpose car was intended for a specific, if not one-time, use and so offered no pattern for the car builders at Altoona. However, experience with sixteen-wheel passenger cars suggested a way to spread the weight over a multitude of wheels, axles, and journals and so enable the great gun car to ease its bulk over the light trackage of the period without crushing rails, ties, bridges, and culverts.

The side sills of the gun car measured 6 inches thick by 17 inches deep. This appears to be an all-time record for single timber size used in car construction. The intermediates were, by contrast, undersize sticks (Fig. 6.5). The side sills were stiffened by 6-by-6 timbers bolted through the 3-inch-thick floor planks to the main side sills. Another pair of even larger longitudinal beams were mounted above the intermediates, both to support the frame and to form a crude cradle for the cannon's barrel. There was no center sill. The coupler thus depended on the end sill for an anchor—a feature that may have proved troublesome. Sur-

prisingly, only two truss rods were used. They were placed toward the car's center, nestled between the pair of double beams inside the wheels. These 1¼-inch rods bent up and over the body bolsters, then sloped down to cup under a wooden king post at the car's center. They were split at both ends at about one-third of their length. Note the eyelet over the innermost wheel.

The running gear rivaled the frame for novelty and size. A pair of arch-bar freight trucks were connected by two heavy wrought-iron equalizing levers. These in turn were fastened together by three large cast-iron bolsters. The center bolster war far deeper than the end ones, and it connected the body bolster to the running gear. The end bolsters over each truck were made very wide to offer lateral stability, but they also had to be made very shallow so as to clear the frame. The iron body bolster itself is worthy of study; notice the continuous top strap that extends from under each side sill like a giant channel iron, behind the sill and under the full width of the car just under the floorboards. The truck's bolsters are composite wood-beam and flat iron plates. The journals appear to be 3½ by 7 inches and are about the only normally proportioned parts on the car. The 26-inch wheels are definitely undersize and were selected so that the floor level could be held to 38 inches above the rails. Why single-plate rather than stronger double-plate wheels were used remains a puzzle, considering the loads anticipated. No springs are visible in the drawing because the nests of steel-coil springs were hidden inside a cylindrical casing between the truck bolster and spring plank. Notice how the truck bolster is cut away at its ends to receive the spring case.

Some time after the gun car entered service the Pennsylvania Railroad issued a diagram drawing that gave the weight as 36,100 pounds and the capacity as 35 tons. One might assume that this was its official weight; however, in his aborted 1896 history of the railroad J. E. Watkins, the Smithsonian's first curator of technology, claimed a capacity of 40 tons. The matter is further unsettled by the December 1885 *Equipment Register*, which rated cars 5102 and 5104—both sixteen-wheelers—at 45 tons capacity each. It is possible that the diagram was drawn late in the service life of the wooden gun cars, when their load limit had been reduced because of their advanced years. In 1888 Altoona decided to render a new sixteen-wheel gun car in iron. This 60-ton-capacity giant is briefly discussed in Chapter 8.

While the Pennsylvania abandoned wood for iron, the Wabash Railroad decided to construct a sixteen-wheel flatcar for the transportation of cable reels for street railway service.[12] In 1890 or 1891 the Wabash planned a 60-ton wooden car so similar to the old Pennsylvania gun car that it is hard to believe that direct reference was not made to Al-

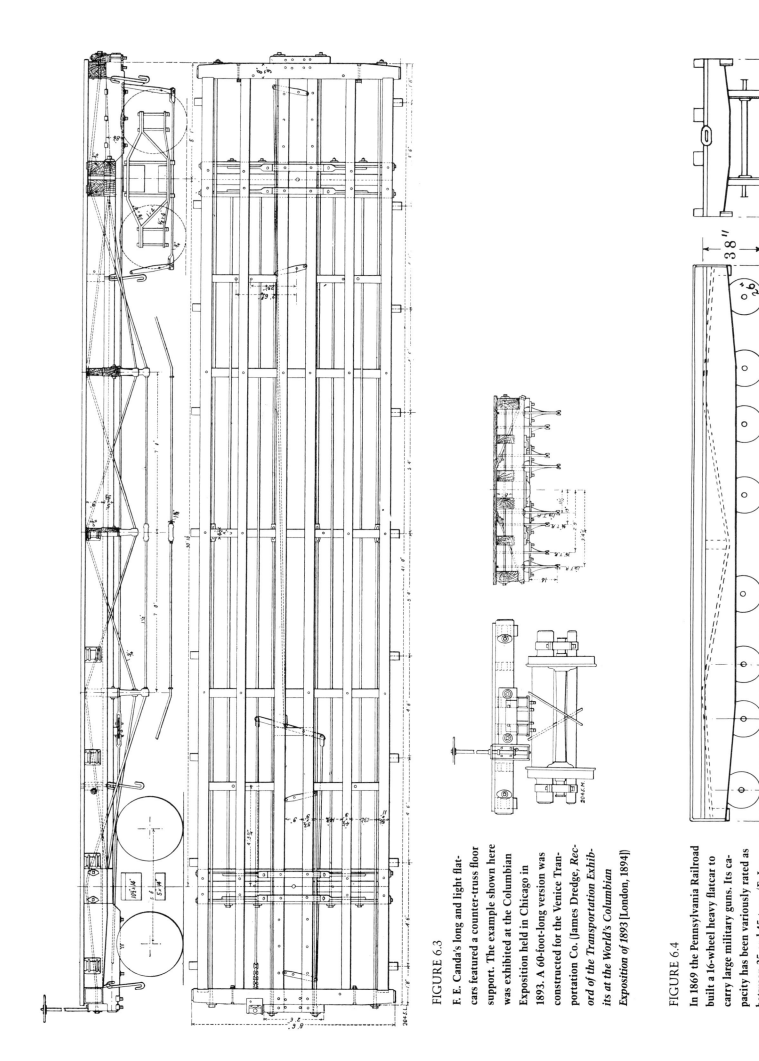

FIGURE 6.3

F. E. Canda's long and light flat-cars featured a counter-truss floor support. The example shown here was exhibited at the Columbian Exposition held in Chicago in 1893. A 60-foot-long version was constructed for the Venice Transportation Co. (James Dredge, *Record of the Transportation Exhibits at the World's Columbian Exposition of 1893* [London, 1894])

FIGURE 6.4

In 1869 the Pennsylvania Railroad built a 16-wheel heavy flatcar to carry large military guns. Its capacity has been variously rated as between 35 and 45 tons. (E. Lavoinne and E. Pontzen, *Les Chemins de Fer en Amérique* [Paris, 1882])

THE AMERICAN RAILROAD FREIGHT CAR

FIGURE 6.5

Details of the 1869 Pennsylvania
gun car's frame and running gear.
Note the very small wheels
employed to lower the floor level.
(J. E. Watkins, *History of the
Pennsylvania Railroad [proof
sheets, 1896])*

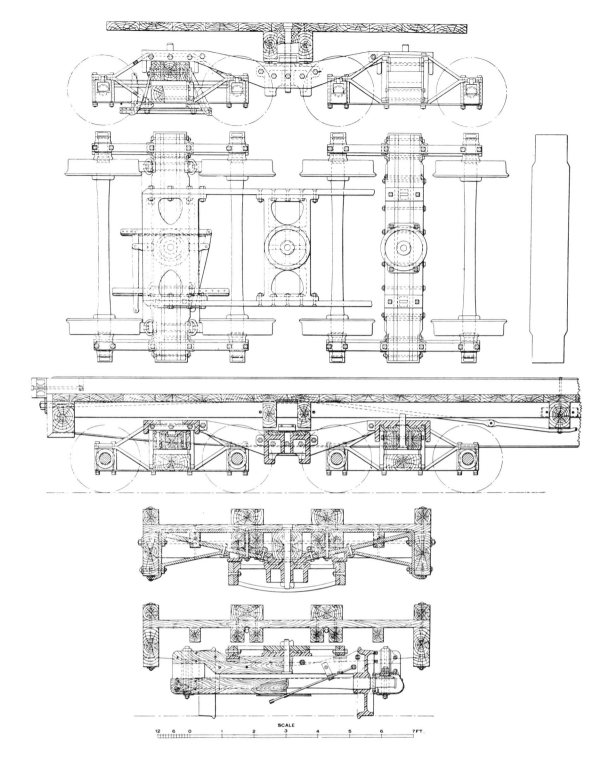

toona's blueprints. The size and general arrangement are nearly identical, while the details of the bolsters and equalizing levers appear to be verbatim copies (Fig. 6.6). There were some important differences, however. Standard 33-inch wheels were used, which raised the floor level to 54 inches above the rail tops when the car was loaded. Raising the level of the car had both advantages and disadvantages. Top-heaviness and clearances for high loads worked against the use of bigger wheels. However, these drawbacks were offset by the greater freedom afforded in laying out the undercarriage. Conventional trucks could be used, and

there was clearance for four heavy truss rods placed near the side sills. The frame could also have center sills, an absolute must for a durable and solidly mounted draft gear and coupler. The brake rigging could be better arranged, and space was also available for the air-brake apparatus.

The side sills of the Wabash car were made from 5½- by 16-inch yellow pine. Parallel cap beams were fastened on top of the 2½-inch floorboards by long bolts passing through cross-timbers and iron brackets. The equalizing levers were fashioned from 2-inch-thick wrought iron and were most likely forgings. The trucks had built-up iron bol-

FIGURE 6.6

The Wabash Railroad built a sixteen-wheel platform car rated at 60 tons capacity in 1890–1891. The car is shown outside the Allis-Chalmers plant in Milwaukee around 1901. (*Trains* Magazine Collection)

sters and 4¼- by 8-inch journals. The 60-ton-capacity car, originally numbered 30002, had a sister, No. 30001, that measured 2 feet longer but carried only 55 tons. One of these cars (or yet another sister) is shown in a photograph made around 1901 (Fig. 6.6). It was renumbered 20006 and bore lettering that is hard to decipher. The weight appears to be 52,553 pounds.

Not many railroads seemed to want heavy-duty flats that were all wheels. A number did respond to the idea of twelve-wheelers that compared in load capacity very well with their sixteen-wheel counterparts. The Reading fabricated a 34-foot-long twelve-wheeler in 1884 with 12-inch-deep side sills. More information is available for a 55-ton-capacity flat built at the Lehigh Valley's Packerton shops to the design of John S. Lentz (Fig. 6.7).[13] Lentz departed from the giant-timber approach and elected instead to adopt composite iron and wood construction. Iron channels 12 inches deep and 5/16 inch thick were clad to 4½- by 12-inch pine sills. Six very heavy truss rods, each 2 inches in diameter, combined with the frame to produce a structure with a strength more than two and a half times its rated capacity. The car's mettle was tested on one occasion by a load of heavy machinery bound for the Bethlehem Iron Company. The shipment weighed 122,724 pounds, or 12,724 pounds over the vehicle's rated limit, yet the floor did not sag perceptibly, and the bearings ran cool for the entire trip.

Lentz's creation was not unduly heavy. It registered 45,700 pounds, with more than half the weight in the trucks. The trucks were designed by one of Lentz's associates, Alexander Mitchell. As a locomotive man, Mitchell offered drawings for six-wheel trucks—all in iron—that employed many elements found in locomotive frames. Actually, Mitchell had produced a similar if not identical design in 1884 for large locomotive tenders. The trucks were also used on a few passenger cars as well. The equalizing levers and forged pedestals were an obvious transfer from the locomotive to the car shop. The top rail, or wheel piece, of the truck was made from two 6-inch channel irons. The double body and truck bolsters used were very much like those on a passenger car. Large side bearings were employed to stabilize the car. Other special features included steel-tire wheels and oversize 5- by 9-inch journals. The link-and-pin couplers and the absence of air brakes were the only obsolete characteristics of this otherwise elegantly conceived platform car. The Lehigh Valley acquired three more cars on this plan, all presumably built at the Packerton shops.

Long before the first Lehigh Valley platform car entered service, John Kirby, master car builder for the Lake Shore & Michigan Southern, built a number of composite-frame eight-wheel flatcars. By 1895 Kirby had retired, but he remained active in car design as a consultant and produced plans for a twelve-wheel 50-ton-capacity platform car.[14] Kirby's car was lighter and longer than the Lehigh Valley car just described. It measured 40 feet long and weighed only 40,113 pounds. Kirby used steel rather than iron channels; however, he too was not prepared to go entirely to all-metal construction, and so the 9-inch-deep channels were backed with heavy wooden sills. Six truss rods also helped disguise its alien metal framing. The double body and truck bolsters were worked out as in typical passenger car practice. The brake-wheel shaft could

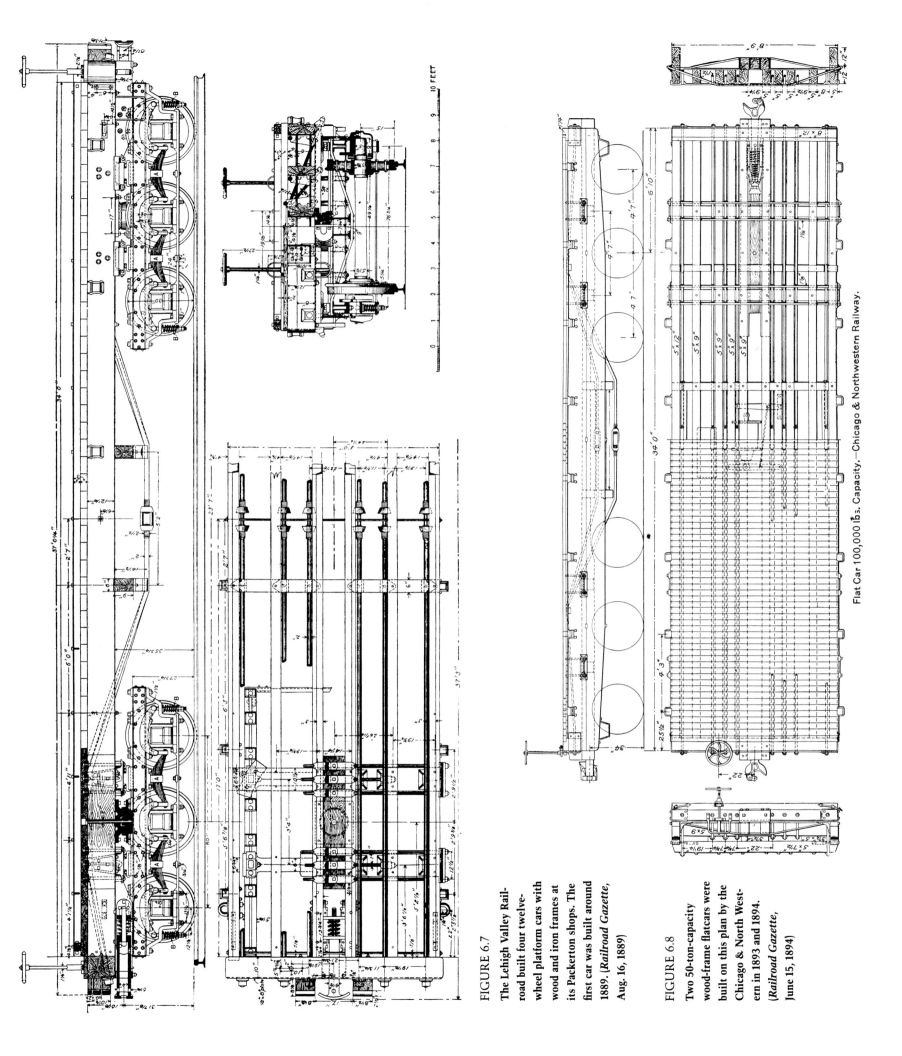

Flat Car 100,000 lbs. Capacity.—Chicago & Northwestern Railway.

FIGURE 6.7

The Lehigh Valley Railroad built four twelve-wheel platform cars with wood and iron frames at its Packerton shops. The first car was built around 1889. (*Railroad Gazette*, Aug. 16, 1889)

FIGURE 6.8

Two 50-ton-capacity wood-frame flatcars were built on this plan by the Chicago & North Western in 1893 and 1894. (*Railroad Gazette*, June 15, 1894)

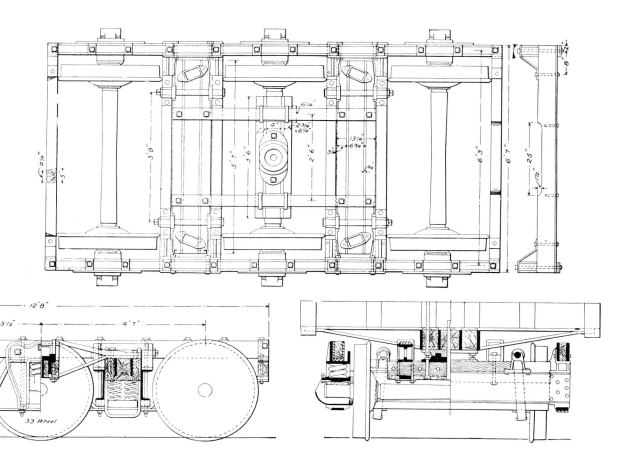

be tilted down below the level of the floor for easier loading. The trucks were built up of light bar stock and looked very much like the passenger car trucks developed in the 1860s by C. F. Allen, master car builder for the Burlington. Kirby figured the cost of his car at $731.88. It was fitted with Janney-style couplers and air brakes.

Not everyone was so experimental. The Chicago & North Western liked the idea of a 50-ton twelve-wheel flatcar but decided to stand by all-wood construction. Two 34-foot-long cars were built by the railroad's own shops in 1893 and 1894.[15] Most of the construction features were conventional, but an inspection of the drawings reveals that the floor frame was almost solid timber; it contained no fewer than ten longitudinal sills (Figs. 6.8, 6.9). The side sills were made up from two 5- by 12-inch beams, stacked one on top of the other. Bolts and metal block keys held them together and in line. Eight 1½-inch truss rods mounted in two levels helped support the weighty wooden frame. Double iron body and truck bolsters were used. The body was set low on the trucks to hold down the floor's height, but the decision to use standard 33-inch wheels resulted in a floor level just over 4 feet above the rails. While this figure was within a few inches of the norm, it was higher than desirable for a heavy-load car. The trucks appear rather conventional at first glance, but they differ from normal passenger car trucks of the period in their lack of equalizers and leaf springs. Swing-motion hangers were provided to help stabilize the car on curves

and rough track. Air brakes and Janney-style couplers were applied. The weight of the car was not mentioned in the brief published description. Perhaps this was only an oversight; then again, it may be that the designers were embarrassed by the heftiness of their creation.

Yet another believer in big-capacity wooden platform cars were the Barre Granite Transit Company of Barre, Vermont. This firm saw no reason to use twelve wheels if eight would do the job. In 1898 it purchased a 36-foot-long open-well car from the Michigan-Peninsular Car Company.[16] For a car rated for 50 tons, it appears to have been lightly framed. The six sills measured only 5 by 10 inches. No continuous center sills were used, because of the open well. Six very heavy 2¼-inch-diameter truss rods did their part in carrying the load. Very heavy body bolsters were assembled from four 9-inch I-beams. Two four-wheel arch-bar trucks carried the frame, but they were beefed up with large journals, 5¾ by 9 inches, and extra-heavy 700-pound cast-iron wheels.

Related to all the jumbo flats just described are a series of very large, low-sided gondolas built in 1899 by the Elliott Car Company of Gadsden, Alabama, for the Lake Terminal Railroad.[17] The little line was controlled by the Lorain Steel Company, which needed special cars to ship lengths of street railway girder rail too long to go in ordinary cars. Lorain also planned to use the cars for double stacks of conventional rails that could be placed end to end. The car required would be 66 feet long,

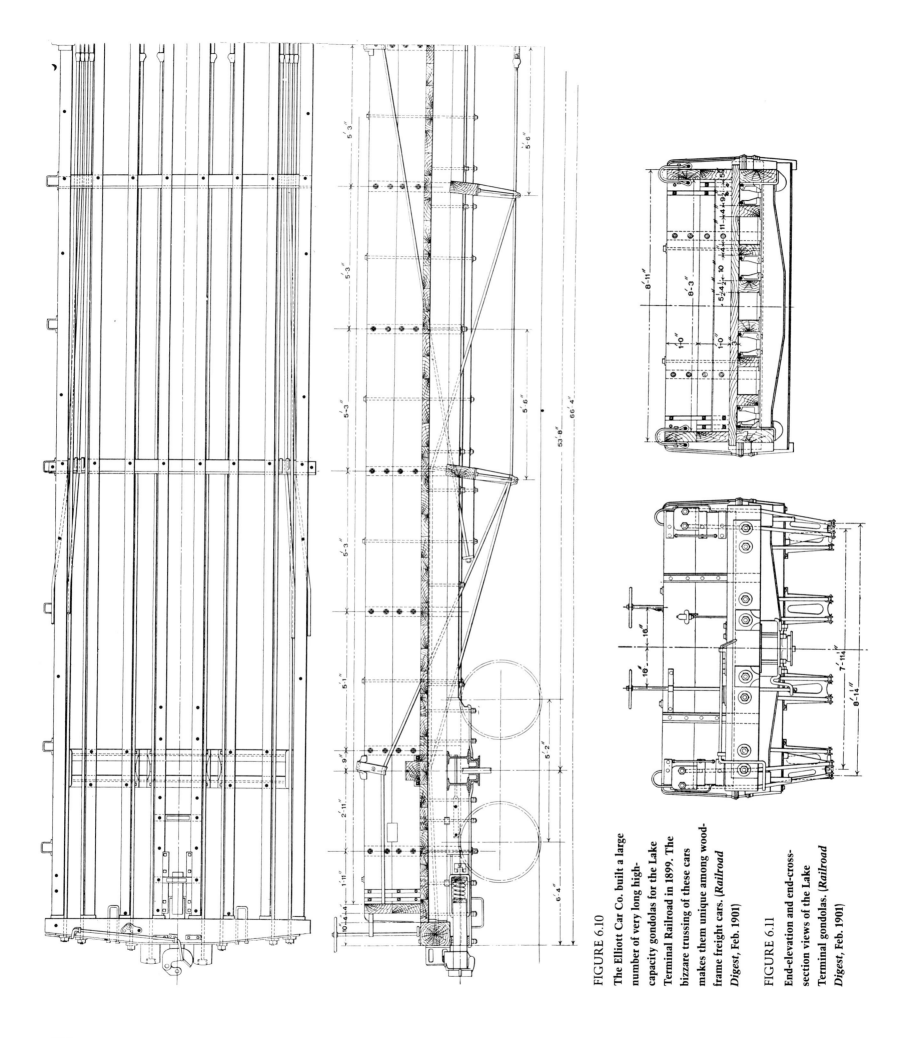

FIGURE 6.10

The Elliott Car Co. built a large number of very long high-capacity gondolas for the Lake Terminal Railroad in 1899. The bizzare trussing of these cars makes them unique among wood-frame freight cars. (*Railroad Digest*, Feb. 1901)

FIGURE 6.11

End-elevation and end-cross-section views of the Lake Terminal gondolas. (*Railroad Digest*, Feb. 1901)

FIGURE 6.12

or about 20 feet longer than had been considered
practical by most wood-car builders. F. E. Canda
had built some 60-foot platform cars in the early
1890s, but they were not intended for loads much
over 25 tons. Lorain wanted an even longer car
with a 40-ton capacity. The challenge was taken
up by a local master car builder, Frank H. Stark of
the Cleveland, Lorain & Wheeling Railroad.

Steel-frame freight cars had already proved their
worth in commercial service. It would also seem
natural for a steel company to favor this form of
construction, but Lorain demurred, and Stark proceeded to design a wood-frame car on a most incredible plan. His combination of timber and iron
might be called a tour de force in truss rods (Figs.
6.10, 6.11, 6.12). Georgia pine sills were used. The
side sills were 5 by 14 inches, while the others
were 4 or 4½ by 8 inches. The floor was laid in ample 3-inch-thick planks. These timber sizes were
not especially notable for a car of this length, but
look again at the drawings and the photograph to
behold the cluster of sixteen truss rods used to
support and tie together Stark's super-gondola.
The rods were grouped in double and triple clusters. There were four sets of posts. An outer pair of
rods rose up above the frame to terminate in a
washerlike steel plate fastened to the body rather
than the end sills. A counter-truss hugged the car's
side as it traveled up and over a king post pointing
up, rather than down, at the car's center. In all,
Stark's plan was most unusual, most remarkable,
and decidedly bizarre. Yet his ideas worked, and
102 cars were built on this plan. Weight worked
out to a little over 22 tons, which is respectable

considering the vehicle's length. The photograph
reproduced in Figure 6.12 probably dates from
around 1920 and shows one of Stark's creations in
service. Though battered from long and hard service, the frame smartly arches up with a good
camber. She appears ready for plenty of tonnage
and many more miles on the road.

Piggybacking and Containerization

In the previous review of extraordinary flatcars,
no mention was made of their more average-size
cousins, which also were assigned some unusual
cargo. The carriage of road vehicles by rail cars is
commonly called piggybacking or, in the more
modern parlance, trailer-on-flatcar. The concept
seems obvious enough and must have occurred to
any number of early railroaders who had witnessed the transport of wagons and carts on riverboat ferries. Ferrying had, after all, been practiced
by just about every advanced society for centuries.

It is natural to assume that piggybacking would
have first been tried in England, the homeland of
railways. The earliest recorded instance of a piggyback car found so far, however, is German.[18] Its
inventor was a Bavarian physician named Josef Ritter Von Baader (1764–1835). Baader somehow
became enamored of mechanics and made three
extended visits to England to study his new fascination. In 1822 he published a two-volume
treatise to expound upon his ideas for transport,
which included everything from horse-propelled
elevated freight railways to compressed-air locomotives. Among his many notions was a scheme

for the rail carriage of farm wagons (Fig. 6.13). Two four-wheel trucks, fitted with small wheels and tied together by a too-slender rope, served as mounts for each set of wagon wheels. The flangeless wheels sat upon the fish-belly rails of a 42-inch-gauge track. A center rail and rollers guided the trucks. An inclined loading ramp was outfitted with a block and tackle to pull the wagons on board the trucks. Once in place, the nest of trucks were adjusted to the wheelbase of the wagon. The horse or horses used to propel a train of Baader wagon cars walked along one side of the track, exactly as they would on a canal towpath. The king of Bavaria liked his subject's ideas and in 1825 permitted Baader to build a small demonstration railway in the palace gardens. But the inventor's difficult personality soon soured the royal patronage, and his plans were not developed for commercial use.

Meanwhile, as the public railway began to flower in Great Britain, the transport of road carriages was almost immediately tried.[19] For the most part it appears to have concentrated on the movement of very fancy vehicles for upper-class patrons who wanted the use of their own vehicles but were equally intent on removing themselves from the more ordinary folk traveling in the coaches, where the titled and untitled were expected to sit together in a distressingly democratic mix. The segregation scheme seems to have worked rather well, except for one fatal flaw. The upholstery and fancy hammer cloths and soft tops of these elegant carriages were a perfect firetrap for locomotive sparks. In one instance the countess of Zetland was forced to flee from her blazing vehicle while en route to London on a fast Midland Railway train.[20] The countess and her party were unable to communicate with the crew, and as the train raced along, the wind fanned the flames, making short work of the expensive coach. There were seven coaches and two vans riding on flatcars in the same train, which indicates that such traffic had become rather sizable by the late 1840s. But no matter how sizable the trade, there could have been little profit in the traffic if the countess of Zetland's experience was at all typical.

There is evidence that the idea was extended to freight-carrying highway vehicles as well. A guidebook produced for the Great Northern Rail-

FIGURE 6.13

Josef Ritter Von Baader, a German inventor, proposed this scheme for piggybacking farm wagons in 1822. (Deutsches Museum, Munich, Germany)

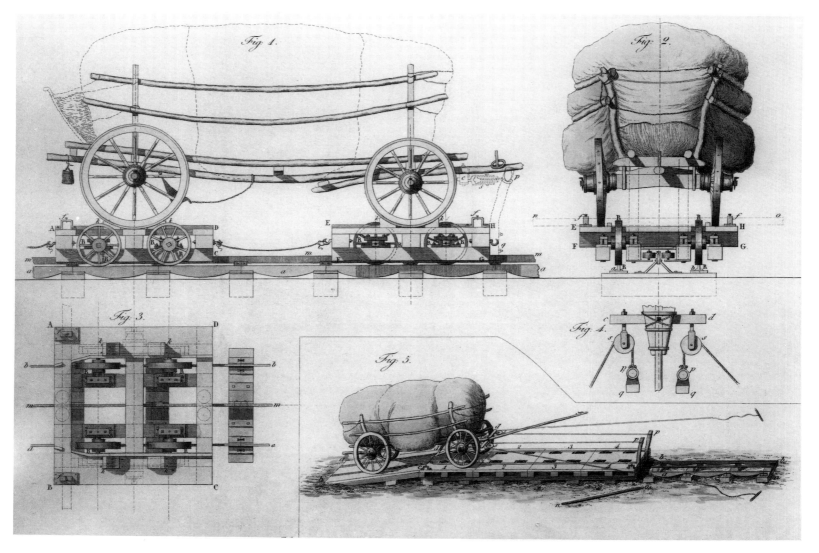

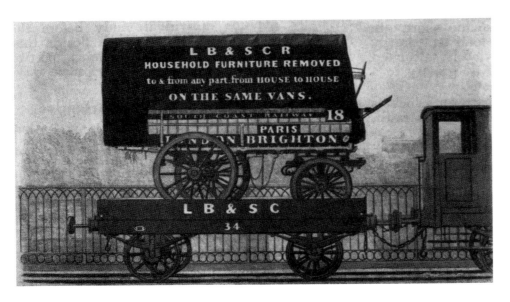

FIGURE 6.14

Lift vans intended mainly for household furnishings were used in Britain before 1860. Some were engaged in international shipments. (***The Locomotive***, Dec. 15, 1936)

way in 1861 includes an engraving of a farm-type wagon with a cloth top seated upon a four-wheel flatcar.[21] During the same period the London, Brighton & South Coast Railway was active in transporting house movers' vans (Fig. 6.14). A business card dated 1860 shows a furniture van atop a gondola car used to carry household goods as far as Paris, with a transhipment of the van from rail car to channel steamer.[22] It is not certain if the rail car was interchanged with French railways, but whatever arrangement was followed, the contents of the van went from door to door without breaking bulk. The scheme was not a short-lived phenomenon but one that grew and was extended so that British "lift vans"—as they came to be called—were later used in transatlantic service as well.[23]

The French also revealed a certain enthusiasm for piggybacking, especially before their rail system was fully developed, for a road vehicle might go part of its journey by rail and then proceed from the end of the line to a more distant destination. In 1843 the Orléans Railway began carrying stagecoaches from Paris to Rouen and Orléans.[24] The bodies were lifted off their running gear and transferred by a traveling overhead crane to an adjacent flatcar. At the end of the run, the body was remounted onto highway wheels. Some carts were carried as well, but the main traffic appears to have been stage wagons.

The effect of these European activities on American practice is impossible to calculate, but whether directly inspired or not, American railroaders began loading carriages on rail cars at an early date. Before regular steam service began, the Baltimore & Ohio began carrying carriages and wagons in long, open eight-wheel freight cars.[25] A Baltimore paper dated December 17, 1830, mentioned the trip of former president John Quincy Adams and several friends over the line from Baltimore to Relay, where the horses, the carriage, and its passengers were offloaded to continue on by

highway to Washington. Adams and his party remained seated in the carriage during the entire trip. The paper also mentioned that the ends of the car dropped down to permit loading and unloading. An engraving from this period used to embellish the B&O's stock certificate shows a carriage being transported on a four-wheel flatcar rather than the eight-wheel car alluded to in the newspaper account (Fig. 6.15). The artist showed a Braithwaite-style locomotive and so may also have depended on a British source for his piggyback depiction. The passenger car is, however, a faithful reproduction of an Imlay coach.

The piggyback idea persisted in America. The Albany & Buffalo published a tariff on December 16, 1844, that included a second-class rate for carriages. Presumably they were carried in open cars, for the railroad would assume no liability for damages due to fire and weather. Many years later, carriages, ammunition wagons, ambulances, and other vehicles were regularly being hoisted aboard flatcars as armies in the North and South moved across the land during the Civil War. Veterans of the war would not have been surprised by the trains of wagons moving over the Union Pacific. One of these movements was captured on a glass-plate negative by A. J. Russell in the late 1860s.[26] There is also the much-published photograph made in April 1868 of thirty Abbott-Downing stagecoaches moving by rail cars. Far to the west of New Hampshire, a wagon ferry service was offered to emigrants needing to cross the Missouri River at Omaha.[27] A long covered car received horses and wagons for a 3-mile journey over the great railway bridge at this location. Horses would not have made the trip so easily in an open car, but they would enter the closed vehicle as if it were a covered bridge or barn.

The average citizen was not likely to witness the piggyback movements just outlined, but just about everyone, even those from fairly small towns, would at some time see the circus train unload.[28] It was a fairly ordinary event between 1880 and 1950. The wagons, loaded with equipment, tents, and animals, were rolled off of the cars and onto the ground by an inclined ramp. They would roll along in a single line from car to car, over ramps that formed short bridges between the cars. A rope, powered by a train of horses, connected the wagons and propelled them along the train to the unloading ramp. When the circus was finished, the wagons were packed up and reloaded onto the cars in the same manner. The roll on/roll off method greatly facilitated the traveling show's movement between towns, and it became widely used after its apparent introduction in 1872.

The omnipresent and efficient circus train obviously suggested piggyback handling for other types of traffic. Late in 1884 Isaac D. Barton, superintendent of the Long Island Rail Road, decided to

FIGURE 6.15

An engraving used on Baltimore & Ohio stock certificates indicates an early awareness of the piggyback concept. Horse-drawn carriages were being moved over the line as early as 1830. (Smithsonian Neg. 85-7061)

try the idea for produce farmers living on the east end of the island.[29] He observed that they were eight hours on the road in reaching Manhattan, while his train covered the same distance in one and a half hours. The produce raisers could arrive in good spirits with fresh teams and merchandise. Farmers, being of a conservative turn of mind, thought the scheme foolish. Saloonkeepers along the old route naturally opposed Barton's plan too, for it threatened to dry up the trade they received from thirsty farmers en route to the big city. But Barton persisted and offered to run a free train to test the idea.

Silly or not, a free ride wasn't to be ignored, and so ten wagons boarded the first farmers' special on January 5, 1885 (Fig. 6.16). The horses were stabled in boxcars, the wagons went on eight flats, and the teamsters rode in a coach. The pioneers pronounced last month's crazy idea a fine success. More than twenty wagons showed up for the second trip, and within weeks the railroad was running three trains a week. By January 1886 special long flatcars, capable of handling four wagons each, and improved stable cars had been introduced. Most trains carried forty-five to fifty wagons, but the total reached seventy-five on heavy days. The round-trip fare for wagons, teams, and farmers was only $4. The service was not very profitable, but the railroad felt that it would help build up the outlying regions of the island and so in time build up traffic as the area developed. For reasons never explained, the service appears to have been terminated around 1893. One logical explanation may be Barton's resignation from the Long Island management in early 1892. It's possible that his successor had little faith in the traffic-building potential of the produce trains and reasoned that the marginal profits were not worth the time and equipment involved.

Just two years after the end of the Long Island service, a small electric interurban line in the Oakland, California, area began carrying express wagons on flatcars.[30] A sturdy box motor pulled two four-wheel platform cars. This service continued until 1906. Elsewhere in the country automobiles were becoming a source of traffic. New motor vehicles were at first shipped in boxcars. As the traffic grew, so-called automobile cars were de-

veloped with extra-wide double doors for more expeditious loading and unloading. In 1923 the Grand Trunk Western was inspired to try the old piggyback method for autos, but with an improvement.[31] The Buick Motor Company was flooded with orders and couldn't ship automobiles fast enough. Buicks went to dealers by road and Great Lake steamers. Every available boxcar was requisitioned, but cars still languished at the factory for want of transport. The Grand Trunk agreed to help the shipper by double-decking some 61-foot-long log cars (Fig. 6.17). The old wood-frame flats were fitted with jerry-rigged collapsible frames made from wood, angle iron, and chains. That this plan was not pursued and perfected at the time represents another missed opportunity for North American railroads, for it was not until the early 1960s that auto-rack cars were commercially introduced.

Piggybacking might seem to be the most natural transition between rail and highway carriage, and yet containerization appears to have been tried at an even earlier date. It has been contended, though not proved, that the ancient Romans used containerlike cages to transport wild animals from Africa to the arenas of the Eternal City.[32] Barrels have been known since ancient times as well, and surely they match the definition for a reusable package that can travel via all modes of transit for door-to-door delivery. Barrels came in a variety of sizes, from tiny kegs to casks large enough to require a team of oxen for movement. They could be rolled into and out of warehouses, wagons, and ships with considerable facility. They safely carried everything from flour, wine, vinegar, and molasses to gunpowder. Some were packed with nails or even dishes. There may be other overlooked examples of reusable units employed in commerce centuries before the beginning of the modern containerization age. Yet whatever the details, the concept clearly preceded the steam railway era.

The Horsehay tramway opened in 1792 near Coalbrookdale, England.[33] It used iron crates, each of 2 tons capacity, to carry coal to a nearby canal. At this junction the crates were lifted by a crane into the boats. In Chapter 8 mention is made of a similar operation on the Peak Forest Tramway around 1797 for the transport of lime and limestone. Other British tramroads using containers at the dawn of the railroad age include the Little Eaton Gangway and the Denby Canal Tramroad. In 1803 the coal mines at Waldenburg, Silesia (then part of Germany), adopted small wooden containers 30 inches deep by 24 inches wide and 40 inches long for the carriage of coal by rail and water.

The advantages of containerization were eloquently stated by the Scottish economist James Anderson (1739–1808) in a volume of essays published in 1801:

CARS FOR SPECIAL SHIPMENTS

FIGURE 6.16
The Long Island Rail Road began carrying farm wagons to market in January 1885. Horses and farmers occupied separate cars. (*Harper's Weekly*, Jan. 31, 1885)

FIGURE 6.17
The Grand Trunk rigged up a makeshift bi-level rack car to piggyback automobiles. Nine Buicks were loaded onto this long wooden flatcar in 1923.

THE AMERICAN RAILROAD FREIGHT CAR

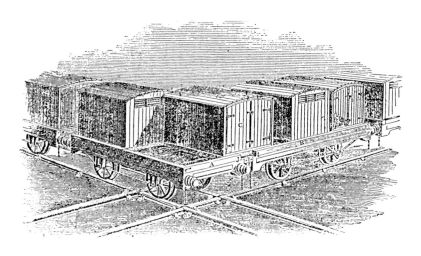

FIGURE 6.18

Henry B. Powell obtained a British patent in 1845 for a container road-rail system that anticipated what would become a common form of shipping a century later. (*Illustrated London News*, June 29, 1846)

Suppose a railway were brought from the wharves to Bishops-Gate Street all the wagons to be made of one form and size, each capable of containing one ton of sugar or other goods of similar gravity. Let the body of these wagons be put on a frame that rests upon the two axles of the four wheels calculated to move only upon the railroad and let each of these wagons be loaded with goods which are to go to the same warehouse or its vicinity. The whole of the wagons being loaded they are moved forward until they come to the end of the road at which place they should be made to pass under a crane. The crane would lift the wagons upon another truck formed for street use and when emptied at the end of the day returned to the railroad track which returns to its point of departure.[34]

Anderson seems to have envisioned containers, or at least lift bodies, for the general carriage of goods. The Liverpool & Manchester tried coal containers briefly during the early 1830s but found larger units for general merchandise more practical. These were said to have been managed by Pickford's for many years.[35] An engraving published sometime in the 1830s shows a two-level goods station on the L&M with seven iron cranes transferring merchandise containers between rail cars

and wagons. During the same period the Canterbury & Whitstable Railway was using containers for rail and oceangoing cargoes.[36]

The debate engendered by the British Gauge Commission report in 1846 rekindled the containerization idea. At least there was considerable talk about the subject. The Gauge Commission dismissed containers as so many "loose boxes for goods"—a system repeatedly tried and one that repeatedly failed.[37] At least one inventor was not dissuaded by the commission's condemnation. Captain Henry B. Powell of the Grenadier Guards obtained a British patent (No. 10,957) on November 18, 1845, to document his idea for a transferable road-to-rail container.[38] The box had an arch roof, end doors, and a vent opening at one end. It was not unlike the lift vans used later in Britain and the United States. Captain Powell employed roller guideways for transferring the bodies between rail cars and wagons. The same system could suffice for transfers between cars of a different gauge (Fig. 6.18).

It is unclear how containerization originated in the United States. It may have been independently conceived, or perhaps it was yet another example of European technical transfer. We can never be sure of how or why the scheme was tried here, but we do know that it was tested at a very early date. In the 1820s the Lehigh Coal & Navigation Company needed a way to get its coal to the market from Mauch Chunk, Pennsylvania. River barges were first tried, but the cost and difficulty of returning the empties upstream in a shallow river prompted the mine owners to consider a unitized boat built up of many small sheet-iron sections.[39] The watertight sections, nicknamed pillboxes, were tied together to form a boat of convenient size. They were then loaded and floated downriver, carried along at least in part by the current. Once unloaded, the "vessel" was disassembled and the empty pillboxes were loaded into wagons for the return trip to Mauch Chunk.

A far bolder adaptation of the sectionalized-boat scheme was embraced by the Pennsylvania Public Works during the 1830s, an operation already described in Chapter 2. That chapter also discussed the baggage crate system used on the Old Colony and other eastern lines. H. J. Lombaert's 1865 patent for oil tank containers was described in Chapter 5, and the ice chests of Joel Tiffany and Parker Earle were mentioned in Chapter 4. These men were by no means the only ones who understood the potential of this simple idea. John Lippitt of Wilmington, North Carolina, patented a portable icehouse in 1877 that was in effect a container fitted with its own wheels for roll on/roll off transit between rail and water transport.[40] Of all the pioneer schemes, the one that looks most like a modern container is the patent granted on August 31, 1869 (No. 94,461), to Joseph P. Woodbury of Boston.

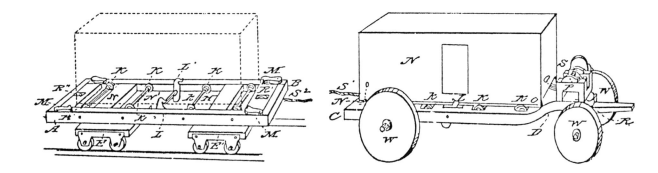

FIGURE 6.19
Joseph P. Woodbury of Boston
envisioned the supposedly
modern idea of trailer-on-flatcar
in his U.S. patent of August 31,
1869. (U.S. Patent Office)

Woodbury had been active in steam rail car design and construction before being drawn into the container arena. It was actually more of a sideshow, for clearly the railroad industry was not very interested in the ideas of Woodbury or anyone else active in container developments. As Woodbury's drawing shows, he would employ rollers to move the container, which he called a "freight-cab," onto and off the car (Fig. 6.19). The street dray would also have rollers and a windlass at the forward end to pull the box off the car. Latches held the "freight-cab" steady when the train was in motion. Woodbury was ready to apply sideways rollers if desired.

The final American Victorian experiment for which we can find any record started very late in the nineteenth century in Washington, D.C. Lawrence Meinberg operated a bakery not far from the Baltimore & Ohio station. He decided to expand into the suburban markets by shipping bread in bulk to outlying towns. Part of his plan involved the use of wood-crate breadboxes about the size of a large laundry hamper (20 inches square by 27 inches long). They were light but sturdy, with metal-reinforced corners and heavy wire running around the middle of the brown-painted box. The lid was hinged. Rope handles provided a good purchase for the expressman. Bread was loaded into the boxes at the bakery, whence they were taken by Adams Express wagon to the depot. At the station the boxes were reloaded into baggage cars for shipment as far away as Purcellville, Virginia. The boxes were dropped off at an outlying station to be picked up again by wagon and taken to the store. The boxes functioned like full-size containers; after making factory-to-store delivery, they were returned to the Meinberg bakery. The box in the Smithsonian collection is lettered "Return to Meinbergs Bread Bakery Wash D.C." It is numbered 578, indicating that the operation was a rather sizable one.

Photographic evidence survives to indicate that the crate containers used by Meinberg were not an isolated East Coast phenomenon. A picture reproduced in *Narrow Gauge in Ohio*, John Hauck's 1986 history of the Cincinnati, Lebanon & Northern Railroad, shows a horse-drawn stake wagon piled high with crate containers. The returnable boxes were used to haul bread, dry goods, groceries, and other small items of merchandise from downtown suppliers to suburban stores. Identical crates were used on the Rio Grande narrow-gauge lines into the 1950s for carriage of express. One of them, perched on a railway express handcart, is pictured in George Drury's *Historical Guide to North American Railroads*.

And yet for all these individual tests, patents, and promotions, the container, and for that matter piggyback service, made no real progress in taking over a significant portion of intercity freight traffic. Nothing of much consequence happened in this area until the early 1920s, when the New York Central attempted to promote containerization for less-than-carload traffic. The Pennsylvania reluctantly followed its rival's example a few years later, but it wasn't until the 1960s that containerization and piggybacking became a factor worth considering in American railroad freight traffic.

Combination Freight Cars

Most railroad managers understood the need to increase car revenue mileage. Everyone connected with the railway freight business recognized that far too many empty freight cars moved across the land, and that most hauled profit-producing cargo in one direction only. It was claimed that 30 to 40 percent of the cars in an average train were empties.[41] This persistent evil prompted an equally persistent effort to devise a multipurpose freight vehicle that could economically carry a variety of shipments and be more adaptable in securing a return payload. Master car builders, commercial freight equipment manufacturers, and inventors outside the industry all tried to develop a dual-purpose freight car. The basic concept was appealing, for it involved the creation of a less specialized, general-purpose car able to handle several classes of freight. Considerable effort was expended in attempting to devise cars adaptable to seasonal traffic, be it berries in the spring or lumber in the winter. The quest for an all-purpose car went on year after year, up to and including the present time.

The idea of flexibility proved to be like so many laudatory goals—better in theory than in practice.

The cruel truth about compromises was embodied in the patchwork mongrels that came forth—fairly good for one class of cargo, ill-suited for others, and actually not well tailored for any single one. The essential problem might best be explained by George Stephenson's statement about design solutions that suffer from too much ingenuity. Multiple hatches, hinged floors, and interior panels all added to the car's weight and cost. The best freight cars were simple. The more parts and gadgets, the greater the maintenance and repair shop time. It took time and labor to set up or take down the special apparatus of the combine car when it was converted, say, from a bulk grain carrier to a self-feeding cattle car.

Not all the multi-use cars were failures. Some were indeed successful, but in just about every case these were the simpler and unpatented varieties. The plain box might be classed as a multipurpose carrier, for the industry used it for a great variety of shipments. Boxcars were, for example, the basic American grain car until around 1960, when covered hoppers began to take over this trade. And of course, throughout their history boxcars routinely carried everything from kitchen stoves to lumber. The multiple-service use of other styles of cars has already been mentioned in previous chapters. The fact that cattle cars carried lumber, pig iron, and oil barrels is not surprising. Less successful were early efforts to carry grain in open-top hopper cars, or the costly and surely inefficient combination oil/merchandise cars. Perhaps the best and most widely used combine was the now-obsolete ventilated boxcar.

The combination freight car has a long, if not necessarily honorable, history. Very early examples, some dating back to the 1850s, were discussed in Chapter 2. The concept remained in favor into the interchange era, and that is where we shall begin our story. A resident of Cornwall, New York, named Stephen W. Wood, offered the railroad industry his ideas for a combination grain/merchandise car in Patent No. 134,181, issued on December 24, 1872. A boxcar was to be fitted with hinged iron plates that might be either set up to form three hoppers or folded down as a level floor. Wood's scheme seems neither original nor imaginative, but at least it was more practical than the plan patented the following year by William Worsley of Little Falls, New Jersey. Worsley would have car builders mount a shallow hopper on the floor of a boxcar. A discharge pipe would empty the hopper, which was filled from roof hatches. But the hopper appears to have been fixed and not easily removable; hence only short items could be placed above the grain space.

No record has been found to suggest that either of these patents was tested or adopted, but it is likely that a scheme patented in Canada by Richard Eaton did actually appear in service, if only as an experiment.[42] Eaton happened to be an English mechanic of an innovative bent, much given to devising locomotive-feed water heaters, car wheels, and other railway-related hardware. He also happened to be master mechanic for the Grand Trunk Railway between 1863 and 1873 and so was in a position to indulge his inventive inspirations. His omnibus patent of June 9, 1877, covered several designs for a boxcar outfitted with fold-down grain hoppers. A secondary hopper was permanently fixed below the floor (Fig. 6.20). Folding back a portion of the floor would greatly in-

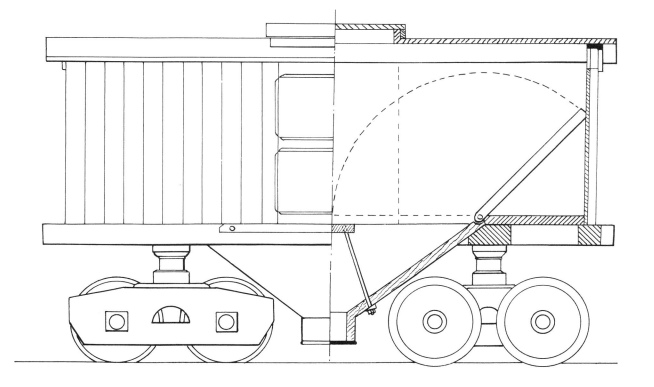

FIGURE 6.20

Richard Eaton's combination box/hopper car incorporated the basic concepts followed by inventors before and after Eaton's Canadian patent was granted in June 1877. Eaton offered several plans, of which this is the most conventional. (Traced from a copy of a Patent Office illustration)

CARS FOR SPECIAL SHIPMENTS

crease the hopper's size. Loading was done through a central roof hatch. In another design Eaton proposed a six-wheel boxcar outfitted with a permanent central hopper built up to the roof. It too extended below the floor. The car was divided into three longitudinal compartments. The center, as already stated, was a grain hopper, while the two outer corridors on either side offered open space for other cargoes. These spaces were accessible through ordinary side doors. Whatever the merits of Eaton's designs, there is some question as to whether the Grand Trunk actually built many of his dual-purpose boxcars, because a minor purchasing scandal had forced his retirement some four years before the patent was granted.

The decade of the 1870s did witness the production of some combination cars for the Iowa Central by the Mowry Car Company of Cincinnati.[43] Boxcars were fitted with very heavy plank sides, as in a gondola car, in the lower portion of the body next to the floor. Coal and other rough bulk cargoes could be safely carried without damage to the normally light interior sheathing typical of an ordinary boxcar. These cars were manufactured in 1873 and 1874. Just how many were made or how well they performed is not recorded, but that more cars on this plan were produced for the Big Four during the 1880s suggests that they were successful. The idea of the combination gondola/boxcar was reinvented in later years and will be discussed shortly.

Perhaps inspired by Eaton's patent, another Canadian, Thomas L. Wilson of Port Hope, Ontario, introduced a twin-hopper-bottom boxcar in 1882 for the carriage of grain, coal, and merchandise.[44] His car used the by-now familiar hinged wooden floor sections that Wilson labeled "flaps." A 34-foot-long 30-ton-capacity car built on Wilson's plan was put in service. After running 16,000 miles it was declared a success, at least by the inventor. Wilson contended that his improvements added only $100 to the cost of a standard boxcar. But the sum is not so trifling as it might appear, for $100 in 1882 money represented about 20 percent of a new car's cost.

A few years after Wilson's trials began, the hopper-bottom boxcar found a new champion in F. L. Joy of Chicago.[45] Joy felt that his car offered a triple threat to the common boxcar because his would carry grain, freight, and fruit. Top and end ventilators added a new range of possibilities to earlier designs (Fig. 6.21). Joy provided a single below-the-floor hopper, which occupied roughly half of the floor and could discharge two-thirds of the 1,000-bushel load. The remaining grain required hand-shoveling. A test car was constructed and was again said to perform well, yet there is no evidence that the plan was tried beyond the test car.

Robert B. Campbell was no amateur or outsider to the railroad industry. As general manager of the Baltimore & Ohio Railroad, he held a position of authority and power. Campbell understood the dilemma of empty-car mileage. It was a daily story. Whether he was aware of past efforts to produce a satisfactory combination car is unknown and perhaps irrelevant, but we do know that he, in partnership with Howard Carlton, secured a U.S.

FIGURE 6.21

F. L. Joy of Chicago built this test car in 1889 to demonstrate his not very original ideas for combination box/hopper cars. (*Railway Review*, Nov. 9, 1889)

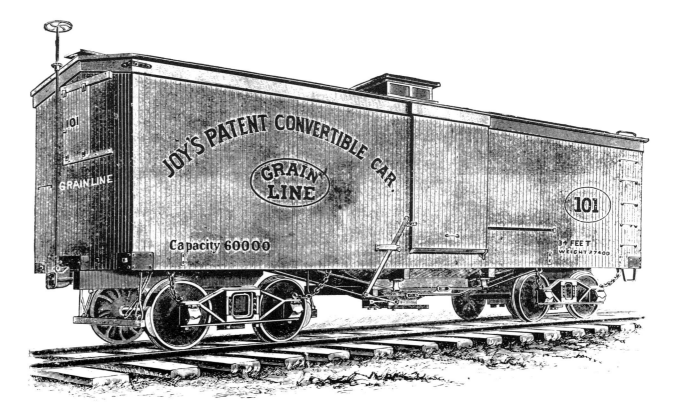

THE AMERICAN RAILROAD FREIGHT CAR

patent on February 1, 1898 (No. 598,136). Their creation was envisioned as suitable for cattle, merchandise, coal, and coke, but in place of the usual hopper bottom, they adopted the simpler drop-door plan.[46] A sample car was built early in 1895 by the South Baltimore Car Company, even before the patent was issued. A series of long, narrow side doors, cut into the sides of the 30-ton-capacity body, served two purposes: ventilation for cattle and loading access for coal and coke. Campbell and his associates adding folding hay racks, roof hatches, and tilting water pans. Small end doors were fitted to accommodate long loads such as pipe or lumber. For a vehicle 38 feet 3 inches long, it had about every gadget known to freight car design. None of the ideas was new, but in combination the inventor claimed to have created something novel. In all, only seven Campbell cars appear to have been constructed.

Another Baltimore & Ohio official, William T. Manning, tried to improve on the designs of his associate.[47] Manning had recently retired as chief engineer and so had the opportunity to study Campbell's design. He decided to elaborate rather than simplify. The old hopper-bottom scheme was adopted, while the cattle car apparatus was abandoned (Fig. 6.22). Manning added oysters to the list of potential cargoes and sent a load of bucket oysters to Chicago to make good his claim. Basically, the car was intended to carry grain, coal, and merchandise. A 1,500-bushel load of corn shipped to Baltimore from Ohio was unloaded directly into an elevator in just twenty seconds. The inventor also claimed that the entire body might be removed to produce a flatcar, but it is uncertain that this feature was ever tested. It appears that only a single oyster/coal/grain/merchandise car was built on Manning's plan. Was it another good idea that neither really failed or succeeded but simply faded away? We may never know the complete answer, but we do know that Manning was a retired consultant at the time his car was unveiled, and so he may have lost some of the influence associated with his former office as chief engineer of the B&O system.

Just at the time of Manning's tryouts, a Chicago attorney, Joseph S. Ralston, was asked to examine some freight car patents.[48] Ralston was not a mechanic, but he somehow became intrigued by the challenge of freight car design. He bought the patents and had Pullman build a few cars. The result was a stock car with drop doors adaptable for grain, coal, or cattle. Interior fold-away panels could be adjusted to form covered interior walls so that merchandise might be carried as well (Fig. 6.23). Note how the upper panels folded up into the underside of the roof, while the lower panels were hinged so that they could fold down to form a secondary floor. While most combine inventors placed the drop doors parallel to the frame sills,

Ralston placed them crosswise. The intermediate sills were eliminated, while the truss rods were clustered toward the car's center. This arrangement would by necessity weaken the frame and hence reduce either the car's capacity or its service life. This defect was corrected within a few years by steel underframing.[49]

It is likely that the Ralston car would have faded rapidly into oblivion, as had its many predecessors, had it not been for the fact that its patron was no ordinary promoter. Other car builders might scoff and ignore his scheme; Ralston decided to start up his own car plant. In 1905 he bought an old factory near Columbus, Ohio, and converted it for freight car manufacturing. While Ralston continued to push his combination freight car in railroad trade press advertisements, his firm produced quantities of a more conventional sort and while so doing evolved into a substantial car-building business that lasted until 1953.

In 1905, just after Ralston's first cars appeared, Spencer Otis of Chicago introduced the National Combination Stock and Drop Bottom Dump car. The American Car & Foundry Company built twenty-five for the Rock Island Lines.[50] The design bore a similarity to Ralston's design, except for a few details. Otis went directly into steel framing, although four 6-by-10s were also used for secondary frame members. Otis raised the floor level some 13 inches above the main frame so that the trap doors might open with less interference from the intermediate sills. He also applied solid sheathing about one-third of the way down the side to eliminate the need for interior closure panels. The drop doors were attached to massive hinges that employed 3-inch-diameter pipe for a center pin. The doors were worked by rooftop handwheels. The design received some praise. Otis claimed that coal could be unloaded for only 0.5 cent per ton. Yet for all its supposed merits, the Otis car rapidly faded into the oblivion that seemed to overtake all purveyors of combination freight cars.

While the railroad trade press devoted some space to the recurring schemes for combination freight cars, they were occasionally given to ridicule. Reporting on a new combination grain/emigrant sleeping car, the December 1881 *Railway Age* speculated in jest on the possibilities for a combination cattle/sleeping car. So many plans had been tried by so many inventors with so little profit that perhaps a marriage of new settlers, most of whom devoted their days to animal husbandry, with cattle might just work. Jokes on such a serious subject tended to fall flat, because railroad men were too conscious of the real cost involved. Added to the losses rung up by empty-car miles were the handling expenses. In 1901 estimates were offered on loading costs at 5 cents per ton and unloading at 1 to 3 cents per ton.[51] Total

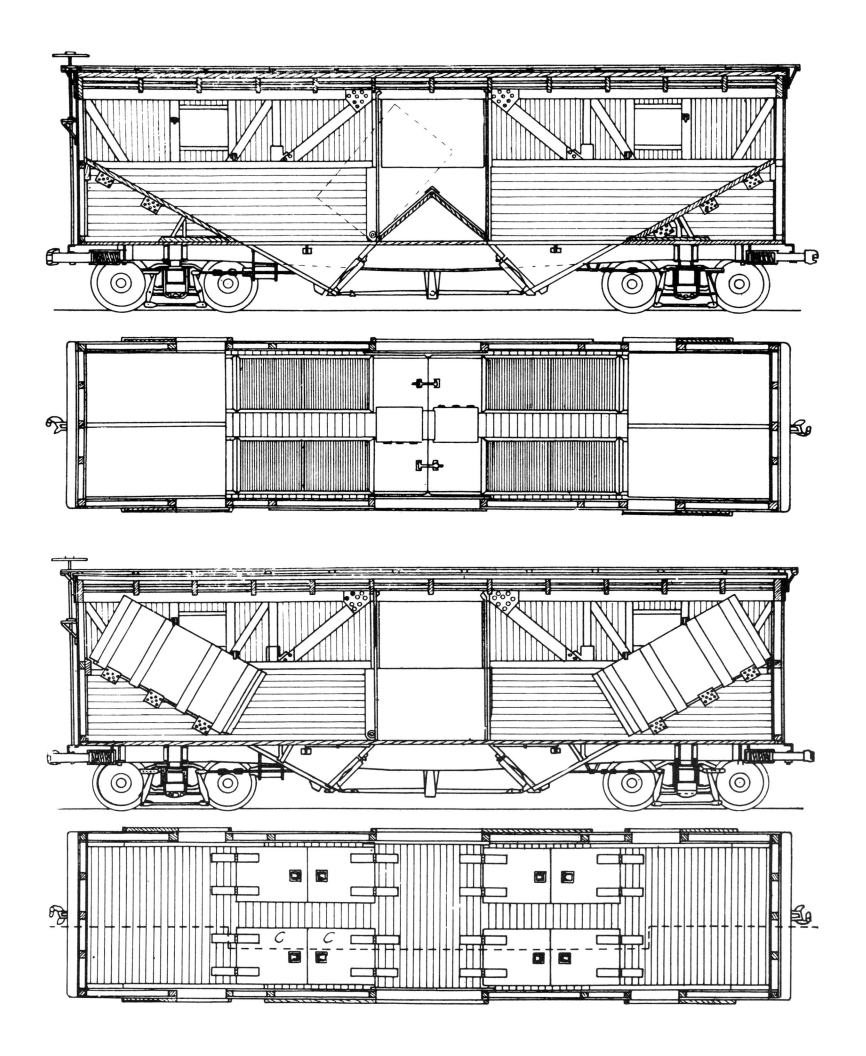

THE AMERICAN RAILROAD FREIGHT CAR

W. T. Manning, former chief engineer for the B&O, devised this combination box/hopper car in which he intended to carry merchandise, grain, coal, or oysters. The two diagrams at the top show the traps down and the floor folded up to form hoppers for grain or coal. Notice the doors high on the sides near the ends, which are cut in to load coal and grain. The two diagrams at the bottom show the car with the traps folded up to make a merchandise or boxcar. (*Railway Review*, May 5, 1900)

handling costs just about equaled the profit of hauling a ton of coal. And so the succession of combination cars continued their parade.

The Handy car was introduced in the late 1890s as a no-nonsense combination car. Its promoter, C. L. Sullivan of Chicago, said that it was free of the gadgets, folding whatnots, and other bothersome novelties of construction that plagued the ordinary run of combination cars.[52] Yet it was quickly noted that the Handy car was not so new or fresh as projected but an unwitting reinvention of the old Mowry combine car dating back to the 1870s. Being a combination gondola/boxcar, the Handy car also followed the general plan of the 1890 Borner car. William Borner was a western freight agent for the Pennsylvania Railroad stationed in Chicago. His February 18, 1890, patent (No. 421,379) can hardly be called gadget-free, for Borner wanted a detachable body so that the gondola/boxcar could be transformed into a flatcar. A neat trick, but not one easily accomplished. The body and roof were made in fold-down sections, resulting in numerous joints and connections. Such a structure would be shaky, costly, and prone to leaks. Sullivan wisely avoided these defects. He also avoided the usual wooden post-and-lintel style of body frame, using instead steel T-angles. The bottom body planks were heavy 2½-inch-thick horizontal boards. Thinner vertical boards enclosed the upper section of the side and end walls. This method of construction increased the interior cubic area by 8 to 10 percent. Large side windows or portals were cut into the side for loading coal and other bulk cargoes. Loading could be done at a tipple or by conveyor machines, but

the lack of drop doors or hoppers meant that unloading had to be done manually. The extra cost of removing cargoes was thought to be offset by the savings resulting from a simpler car, in terms of both first cost and maintenance.

Others in the car-building fraternity agreed, as witnessed by the one thousand coal/grain boxcars produced in 1889 by Pullman for the Georgia Pacific.[53] These 30-ton-capacity cars were fitted with small side doors near each end for bulk loading. Men and shovels were used for unloading through the ordinary center doors (Fig. 6.24). During the same year the Chesapeake & Ohio acquired some very similar cars.[54] The four loading portals measuring 26 by 45 inches were placed near the ends of the car body. The cars were swept clean of coal dust and lined with paper to handle merchandise. In 1896 the Madison Car Works produced a number of 30-ton-capacity boxcars, 34 feet 8 inches long, for the Choctaw, Oklahoma & Gulf Railway.[55] Externally the cars looked like a regular stock car, except for four 39- by 66-inch doors high on the body sides. There were end doors in addition to the side portals and regular side doors for the loading of lumber. The Choctaw line had 250 coal/cattle/lumber combine cars by 1897.

The combination car was tested, reintroduced, and reinvented over and over during the nineteenth century, a pattern that continued into the twentieth. A recent reincarnation of this ancient scheme appeared on the Milwaukee Road in 1977.[56] A hopper-bottom steel boxcar was placed in service for the carriage of grain, or palletized freight or lumber, all in an effort to reduce empty-car moves. Sounds familiar, doesn't it?

FIGURE 6.23

Joseph S. Ralston produced a cattle car with fold-up interior panels and drop doors so that the car might be converted for grain or coal. The engraving on the left shows the car with its panels down to form a solid wall and the drop doors down for unloading. The right-hand side shows the car arranged for cattle. (*Railway Master Mechanic*, July 1904)

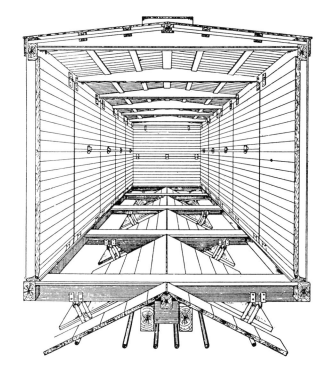
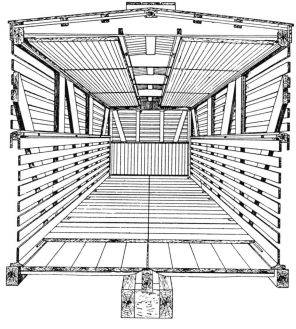

CARS FOR SPECIAL SHIPMENTS

FIGURE 6.24

Pullman built one thousand combination grain/coal/merchandise cars for the Georgia Pacific in 1889. Bulk materials were loaded through the small high-level doors at each end of the car. (Pullman Neg. 543)

The Elusive "Standard" Car

A standard freight car—why, what could be more obvious or sensible? What could benefit railroaders, shippers, and the public more than a cheap, uniform, serviceable freight car? After all, boxcars looked alike, at least to the layman. They were all about the same size and shape, and they were all called upon to perform the same job, so why not start making them exactly alike? A small committee of the nation's top car builders should be able to put together plans for such a car easily, since there was already nearly universal agreement about the basic features and proportions. And yet the simple logic of the standard car was never adopted any more than the idea of a standard radio, automobile, or suburban house. The standard boxcar was like world peace; no sane person would speak against it, yet agreement on important details was never worked out, and the question remained unresolved.

So much seemed to be in favor of a standard car. Construction and repair costs could be cut dramatically. Parts inventories could be streamlined. During a business slump car builders could produce cars ahead of time, knowing that they would sell once traffic levels recovered. Workers would benefit from the steady employment, and railroads would receive better cars that had not been slapped together at the last moment. Even during peak demand periods a better, cheaper car would result, because workers would need no special training to fabricate the familiar standard car. Think of the savings possible if all rolling stock, for example, used a single style of axle. Steelmakers would produce quantities of a uniform size. Axle makers could produce them year-round rather than in a boom-and-bust fashion as demand

rose or fell, or to the custom patterns devised by individual master car builders. Life in the repair shops would also be easier. Again, parts inventories could be simplified and reduced in size. Repair crews could go about their work without puzzling over how to disassemble and repair the specialized creations of the individual railroads belonging to the interchange pool. Broken-down cars would no longer languish for weeks on the RIP (repair-in-place) track awaiting a special floor sill or journal bearing coming by slow freight from halfway across the continent. As matters now stood, 10 to 20 percent of the nation's idle cars were clogging the yards awaiting repair.[57] Accountants and operating officials could offer even more reasons for adopting a uniform car design.

Executing this reform was left to the engineering community. Car builders dutifully picked up on the talk of standardization. In public they spoke of the need to banish confusion, diversity, and waste. The foolish choices of individual design must make way for the harmony and good sense of standard design. Yet it became clear that great counterforces were blocking the path to uniformity. Innovation, change, and experiment are fundamental to mechanical progress. Fixing a design through rigid standardization frustrates creative energy. Conformity is assured only if designs are frozen. Engineers thus faced a classic dilemma. Standardization meant giving up the freedom to innovate.

There were other, more practical factors outside the car builders' control that also inhibited standard design. Basics such as standard wheel gauge could not be adopted because of the lack of uniformity in track gauge, switch frogs, and guide rails.[58] The Pennsylvania Railroad, self-proclaimed as the Standard Railroad of the World, maintained a nonstandard track gauge of 4 feet 9 inches until around

1900. Moreover, car size was hardly stable during the first thirty years of the interchange era. In 1870 a 10-ton boxcar was the industry's standard, but traffic increased every five years or so, and the standard capacity would jump 5 tons. By the end of the nineteenth century 35- and 40-ton boxcars were becoming common. Larger axles and journals and heavier wheels were required each time capacity went up. Automatic couplers and air brakes added to the problems of maintaining old standards. Car design was simply in too much flux to be harnessed to a simple one-off standard. In the end it was this basic truth that confounded the dream of standardization. A standard car in the literal sense was never achieved.

There was, of course, agreement on standard design in the broad sense. Cars were very much alike from railroad to railroad during any given time period. They were all about the same size and shape. Floor frames, body sheathing, and many other details were quite similar. All used cast-iron wheels. Most used arch-bar trucks. Only an expert would see much difference between a Central Pacific and a New York Central boxcar of 1880. The parts were not, however, interchangeable, and so in the narrow sense standardization did not exist. The compromise that was eventually worked out was a *degree* of standardization. Major components like axles and journal bearings were made uniform, though never as a single unit. That is, no one axle was found satisfactory, but a range of standard axles from light to heavy was devised. In other areas basic dimensions or profiles were established. Thus a car builder might build a car as he pleased, but its draft gear must be at a certain height and its width must not exceed so many inches. And so as to not inhibit coupler development, only the contour of the knuckle actually became a Master Car Builders Association standard. Standardization was also effected in the nonabsolute sense by individual railroads that created their own internal systems of standard car, truck, and component design. On larger systems this could affect hundreds of thousands of units.

In a letter published by the *Railroad Gazette* on October 4, 1878, John Kirby, veteran car builder of the Lake Shore, lamented that diversity seemed to be growing in car design despite the efforts of the Master Car Builders Association to spread the gospel of uniformity. Kirby said that the standard-car movement continued to fail because of the vanity of individual car men who believed themselves to be more skilled than their fellows. Curiously, at the very time of his letter, Kirby himself was exhibiting little cooperation with his New York Central compatriots, all of whom seemed intent on pursuing their own ideas in the area of car building. Despite Kirby's apparent duplicity, he did touch upon a very potent force working against standardization. Car builders were sovereigns in their departments.[59] They were accustomed to having things go their way. Subordinates deferred to their judgments. Such authority disposed them to reject the ideas of others. One can imagine a crusty old car builder alone in his office pondering the prospect of an industrywide standard car advocated by the MCB Association. Chairman of the committee and chief architect of design was, let us say, the respected George W. Cushing. Our solitary car builder ruminates: "Now George is a fine fellow and a good mechanic, but I know as much about car building as he does. Why, I can build a better and cheaper car than George can, any day of the week! Just look at the end framing he always uses—the posts are too small, and the tie rods just won't do. I don't care what the other members do, I'll keep on building cars to my own plans." There were loftier motives involved than mere vanity or jealousy. Some car builders found creative satisfaction in the act of design. It offered a positive outlet to men with imagination and energy.

EFFORTS BY THE MCB ASSOCIATION AND OTHERS

The Master Car Builders Association might seem the logical body to push for the creation of a standard car. The organization was established in 1867 specifically to deal with the problems created by interchanging freight cars. Its first actions involved formulating rules for interchange service and, wherever possible, standards for those fittings most subject to renewal—wheels, axles, and bearings—so that repairs could be completed expeditiously even for a car that had strayed 3,000 miles from its owner. The published reports of the organization spoke of little else. Committees met year after year to devise standards for basic components that might please the membership. Yet even when adopted, the standards were as often as not ignored. The organization was a voluntary group. At best, it was a trade association—at worst, a social club. Its standards were really recommendations that might be ignored or honored depending on the disposition of the membership. Where there was no uniformity of opinion, there could be no uniformity of design. During its early years the organization represented only a minority of the nation's car managers, for many individual car builders did not join. By the middle 1880s its position had begun to improve measurably. The railroad industry itself began to support the group, and some roads paid dues for their car masters, feeling that all should belong. In 1886 the interchange rules became binding. Railroads or officials refusing to cooperate could expect an embargo and an end to interchange cars. Voting was now based on car ownership; a trunk line with one hundred thousand cars had more say than a short line with only a few dozen.

The MCB Association started out with the fundamentals of standard car design in mind. In 1869 it attempted to establish a standard level or height for drawbars so that cars from all lines might be coupled. Yet even this basic step was frustrated. The standard did nothing more than specify the level of 33 inches from the top of the rail to the center of the drawbar or coupler. Members were not required to conform to any other specifics. Nothing about a special type of coupler, spring, or shank—the only requirement was a centerline 33 inches above the rail. A decade after the rule was published, it was found that one-quarter of all cars in local service did not conform and hence could not be coupled without special linkage.[60]

Next the organization attempted to grapple with standard screw sizes. There were dozens of standards, and nothing was more fundamental to efficient machinery maintenance than a uniform supply of nuts and bolts. Bolts ⅝ inch in diameter were commonly used in car construction, but minor differences in the number of threads per inch or in thread depth and pitch required the stocking of fifty varieties of nuts. Shop men repairing a foreign car experienced considerable frustration in seeking out the right-size nuts. Octave Chanute, mechanical engineer with the Erie during the 1870s, recalled that the time wasted in picking through the "old nut barrel" resulted in considerable profanity.[61] In 1871 the MCB Association tried to end this madness by adopting the Franklin Institute screw-thread system worked out by William Sellers. Unfortunately, in its haste to establish a standard the association failed to specify the pitch and depth of thread, and so the standard was largely useless. As late as 1882 the organization's annual report noted that the screw-thread problem remained unresolved. The Erie Railway's engineering department put some effort into attempting to resolve the problem and adopted the Sellers system in 1873.[62] Four years later it was discovered that the nuts made in some shops would not fit bolts made in other shops. Taps and dies were made at the various shops along the railroad, but careless workmanship and a failure to reproduce exactly the original thread shapes and sizes resulted in a general nonconformity. Just to see how the competition was doing, twenty-two nuts from cars belonging to sixteen different railroads were removed for inspection. All were supposedly of the ¾-inch size, yet all were found to be oversize and hence nonstandard even in such an elementary dimension as diameter.

With the screw-thread issue still unresolved, another MCB committee attempted to rationalize the journal bearing. For so large a "machine," a railroad car has very few wearing parts. In pre-air-brake days there was little to wear down but the axle bearings. By the early 1870s most car builders had come to adopt axle journals 3½ inches in diameter. There had been no edict to mandate this size; it had simply evolved in the same unofficial way as the conventions of 10-ton capacities and eight wheels. The MCB Association felt that since there was already de facto accord on journal diameter, it shouldn't be too difficult to agree on a uniform length and hence create a standard journal size. If repair shops across the nation could reduce their stock of bearings, axles, and axle boxes to one size, the savings would be remarkable. And so in 1873, with high hopes, the association adopted a standard 3½- by 7-inch journal. A standard journal box was approved the next year. Detailed drawings were published in the MCB's annual reports and in its handbook, the 1888 *Car Builders Dictionary*. The virtues of the system were trumpeted in the pages of the *Railroad Gazette* and other trade papers. The obvious good sense of the plan and the publicity for it were cause for optimism. The Master Mechanics Association voted overwhelmingly to adopt the MCB standard axle for locomotive tenders in 1879.[63] At the same time, the New York Central announced that it had just placed contracts for twenty-two hundred cars, all to have the standard MCB axle and journal.

For all the good news, there was reason for gloom as well. The conversion to standard journals was in truth superficial and halfhearted. Most railroads continued to adhere to their old standard brass sizes, a fact demonstrated by a listing, published in 1879, that is reproduced in Table 6.1. It was, of course, a difficult time for imposing uniformity, because car size was expanding so radically. With capacities jumping from 10 to 15 to 20 and finally to 40 and even 50 tons, it was impossible to stay with the old 3½- by 7-inch journal. Periodically the MCB Association was required to devise larger journal sizes for the bigger cars. By 1899 there were four styles of journals, ranging in size from 3¾ by 7 inches to 5½ by 10 inches. But even allowing for this authorized variety, the industry was doing a poor job of putting its own recommendations into practice. Tough talk about cutting off the interchange-car supply was apparently only that—tough talk. Most roads were too much in need of cars for a dogmatic application of MCB standards. In 1900, when many of the MCB rulings were supposedly mandatory for interchange cars, scant attention was actually given to complying with established journal sizes. It was necessary to stock 108 journal sizes for interchange-car repairs.[64] One car maker manufactured journal bearings and wedges marked with the MCB standard, and yet the parts were undersize and would not work in a standard boxcar. The situation was properly seen as a mockery of the MCB Association and its attempt to bring order out of chaos.

A mood of exasperation began to overlay the published reports of the association. Almost every committee in some way confronted the standardization issue. Most members ignored existing

TABLE 6.1 Axle and Journal Standards, 1879

Railroad	Kind of Axle	Diam. of Journal (inches)	Length of Journal (inches)	Diam. of Wheel-Seat (inches)	Length from Center to Center of Journal (feet)	(inches)	Weight Finished (pounds)
Boston & Albany, Eastern Div.	T	3½	5⅝	4½	5	11⅝	307
Boston & Albany, Western Div.	T	3¼	5½	4⅜	6	¼	345
Boston & Albany	P, F	MCB	MCB	MCB	MCB		340
New York Central & Hudson River	T, P, F	MCB	MCB	MCB	MCB		335
Lake Shore & Michigan Southern	T	3½	7	4⅝	6	3	325
	P	3½	7	4½	6	3	320
	F	3¼	5⅝	4½	6	1⅜	315
Canada Southern	T	3⅜	6	4⅜	6	2	284
	P, F	3½	7	4½	6	3	321
Wabash	T, P	MCB	MCB	MCB	MCB		360
Michigan Central	T	4	7½	5	6	2½	402½
	P	3¾	7	4¾	6	5	350
	F	MCB	MCB	MCB	MCB		394
Chicago & North Western	T	3½	5½	4⅛	5	11	NA
	P, F	3½	6	4½	6	2½	311
Chicago, Rock Island & Pacific	T	3½	6½	4	6	2¾	275
	P	MCB	MCB	MCB	MCB		NA
	F	3½	6	4⁵⁄₁₆	6	3	NA
Chicago & Alton	T	3¾	5⅝	4⅞	6	1⅞	360
	P	3¾	7	4⅝	6	2	320
	F	3¾	6	4⅝	6	2	320
Chicago, Burlington & Quincy	T, P, F	3½	6	4½	6	1½	350
Illinois Central	T, P, F	3⅜	6	4½	6	2	NA
Chicago, Milwaukee & St. Paul	T	3½	7	4½	6	2¼	313½
Union Pacific	T	4	7¾	4½	6	3⅜	330
	P	3¾	7¼	4½	6	4¾	325
	F	3½	6	4¼	5	11⅜	290
MCB		3¾	7	4⅞	6	3	335

Source: Railroad Gazette, May 23, 1879.

Notes: MCB = Master Car Builders Association standard. T = tender. P = passenger. F = freight. NA = not available.

standards and failed to answer questionnaires on basic items such as wheels and brakes. In 1888 one committee chairman complained that only 17 members out of 230 had bothered to respond. The frustration of the long-suffering standardization committee was best expressed by John Kirby, who noted that "uniformity under existing circumstances is an impossibility."

Several years later the circumstances had clearly not changed for the better. The car fleet numbered about 1.25 million in 1897, and in that multitude there were several thousand differing patterns.[65] No individual railroad had even come close to a single standard car. At best they might have half a dozen differing styles of box-, hopper, or stock cars. Their "standard" car tended to be nothing more than the most recent lot purchased.

Standardization zealots sought other ways to achieve their goals. The MCB Association was looked on as a failure. It had failed to motivate its membership. The gospel of uniformity had been preached to death. If the car-building professionals wouldn't listen and act, perhaps higher levels in railroad management could be reached. As early as 1878 John Kirby had already given up on his colleagues, and he looked to the general managers to bring about reform.[66] Car builders reported to their superintendents or general managers. These were the men who directed the railroads' day-to-day operations. These were also men who understood operating costs down to the finest detail, and who knew that car shortages were often aggravated by repair slowdowns. If the car builders wouldn't move on the issue, maybe the general managers could make them do so. George Cushing wrote a public letter of support for Kirby's notion, saying that the MCB Association had indeed failed to harmonize the various design opinions and that the general managers could end the debate on their own authority. This tactic seems to have foundered as well. Years passed, and the general managers failed to stop the car builders' hobby of individual car design.

In 1895 the general managers, acting through the American Railway Association, finally sent a resolution to the MCB Association on the desirability of a standard freight car.[67] Such a weak gesture could hardly force the MCB to do anything beyond reopening the debate. In 1901 the American Railway Association sent a more strongly worded message to the MCB. It was also more specific. The managers now wanted car builders to fix car size as the first step in drafting specifications for the long-dreamed-of uniform car.[68] Dimensions were laid down for a 36-foot-long boxcar. The body would be 8 feet 6 inches wide and 8 feet high, with 6-foot door openings. There was nothing startling about these figures, and the resolution was passed with only one dissenting vote. Subsequent study revealed an unforeseen difficulty: the body was 6 inches too high for passage over sections of the Pennsylvania, New York Central, and New Haven railroads. This error was no doubt gleefully seized upon by opponents of the standard car. Of course, not every car in interchange service could pass over every railroad in the United States, because of variable clearances. Most common boxcars in the interchange fleet could, however, go everywhere.

While the general managers hesitated, friends of the standard car hoped to force its ascendancy through another power play. This more focused attack pressured only the big railroads to agree on a standard design.[69] If, let us say, the Baltimore & Ohio, the New York Central, the Pennsylvania, the Southern Pacific, and perhaps two or three other giants, together with their associated lines, were to get together, a meeting of the minds might be possible. Six or seven men can come to an agreement more easily than two hundred can. If the big boys could agree on a uniform car, the small fry would be forced to conform. The big roads could in time drive up the repair charges for nonstandard cars or refuse to accept them for interchange. The invisible hand of the marketplace would push down the price of standard cars, because so many would be ordered by the big roads that the small roads would find it in their own interest to purchase the cheaper standard car. Trust the marketplace. Trust natural selection. Trust mass production and the notion of the greatest good for the greatest number. These were all cherished nineteenth-century beliefs, but they had little effect on the big railroads. These corporations saw themselves as rivals, not allies. They cooperated rarely and only when necessary. Even traffic-pooling agreements were at best temporary. So much for the notion of using big-road cooperation to bring on the age of standardization.

EFFORTS BY THE TRUNK LINES AND COMMERCIAL CAR BUILDERS

Standardization was achieved to some degree within the major trunk lines. As traffic and car fleets swelled after 1870, the need for uniformity and system forced greater attention to be given to these matters. Cars were being purchased in groups of hundreds and occasionally thousands, and each lot normally followed a single design. In the 1870s some big roads also started to establish standard design classes that often prevailed for decades and were not easily upset by changes in car department personnel.

The Baltimore & Ohio Railroad, for example, established its first standard boxcar design in 1870 as the class M. The design did not actually differ greatly from the earlier B&O house cars illustrated in Chapter 2. The class M had a 20-ton capacity and a 28-foot 9-inch body. It was followed a few years later by the M-2, which carried the same tonnage rating but was 5 feet longer. In 1883 came the M-3, and so on, until in 1890 the M-6 appeared. The length of all these classes varied by only a few inches. The M-5 and M-6 were both rated at 25 tons. The cars were very similar but surely not standard in the literal sense. Basic hardware, like the trucks, was interchangeable to a degree between certain classes. The fact that the B&O boxcar fleet varied even within number series is indicated by the December 1885 *Equipment Register*'s listing for the 8000 through 11273 series. All are listed as having an inside length of 28 feet 2 inches, but the capacities varied between 13 and 20 tons.

Perhaps the mighty Pennsylvania did a better job. It called itself the Standard Railroad of the World, a title no doubt deserved after 1900, but somewhat less secure in early times, judging from a review of its boxcar fleet. Like the B&O, the Pennsylvania began a standard-class car system in 1870. Every few years, as car size grew, a new class would appear. With a few exceptions body lengths stayed nearly uniform until around 1890. The Pennsylvania favored short cars, and most of its units stayed at around 28 feet. Pressures for longer, higher-capacity carriers eventually pushed car length to 34 feet. And so by 1885, according to the *Equipment Register*, every number series of the main-line boxcars was listed as either 28 or 34 feet in length.

The old Erie was perhaps the worst example of a dogs-and-cats car fleet for an American trunk line.[70] It had broad-gauge and standard-gauge cars. In 1874 it counted 11,744 cars representing 230 designs. The mongrel fleet incorporated twenty-seven styles of drawbars, nineteen forms of journals, fifty-three types of journal boxes, and fifty-two designs for brake shoes. In 1876 standard designs were adopted. The road had by then changed over to standard gauge, except for a few branches. Retiring the old broad-gauge cars eliminated some of the freaks. So far as possible, old cars were rebuilt to conform to the new standards when heavy repairs were required. Within a few years car repair costs fell, from $88.02 per car in 1870 to $32.55 in 1881. Some of this saving is accountable to the newer car fleet required by the gauge change, but at least some of it is the result of standardization.

Cornelius Vanderbilt's consolidation of eastern railroads during the 1870s had an unexpected effect on standard-car practice that went beyond the corporate amalgamations of the New York Central, the Hudson River, the New York & Harlem, the Lake Shore, and the various other properties assembled by the commodore through stock purchase and lease. In December 1873 Leander Garey, car superintendent for the New York & Harlem, was appointed master car builder for the entire Vanderbilt system. Overall authority for car repairs, design, and purchasing was centralized. A certain autonomy prevailed for a few years; as late as 1878 John Kirby prepared a new "standard car" for the Lake Shore that differed in just about every detail from the parent road's standard boxcar introduced in early 1876.[71] But as Vanderbilt's control of the leased lines was strengthened, the power of the regional car builders waned.

In February 1880 the regional master car builders from the parent and leased lines were assembled in New York to discuss standard plans for freight cars.[72] Until recent times most of these men had reigned supreme as heads of the car departments of lines like the Lake Shore, the Michigan Central, and the Canada Southern. Now their authority was much reduced, and they were only members of a committee. The talks continued for five days, with each man trying to introduce his most cherished ideas for good car construction. It was agreed at last that each shop should build a sample car incorporating the best features and thinking available on the subject. The samples would be assembled in Buffalo the following April, when an inspection team would rate the offerings of eastern and western shops.

What transpired at Buffalo and afterward is unclear, but in April 1882 the new standard boxcar design was unveiled. It bore a remarkable resemblance to the 1876 New York Central & Hudson River Railroad car designed by Leander Garey. The new car was longer and was rated at 20 tons capacity, but there can be no doubt that whatever the recommendations of the Buffalo committee, Garey's ideas had prevailed. It was announced that the 1882 design was the standard for all the Vanderbilt lines. Ten years later this design was modified for a 30-ton-capacity boxcar. By 1903 Garey was dead and long out of office. The leased lines were apparently again going their own way insofar as car design was concerned.[73] By this time the Vanderbilt Lines had some 150,000 freight cars. So much diversity existed that a systemwide committee was formed to reestablish design standards. It appears that Garey's mission had failed after all, and that the process was to start all over again.

A struggle for standardization had developed in the West as well. The Southern Pacific, like most other big trunk lines, was a consolidation of several smaller companies. None of the equipment matched, and as new roads joined the system, the diversity grew worse. By 1895 the SP controlled 8,500 miles of railroad. It had repair shops scattered all over the western states.[74] Some were so distant that letters spent four days in transit between headquarters and outpost facilities. Since the early 1890s the SP's central mechanical office had tried to impose standards on all regions and subsidiaries, but local master car builders confounded these efforts. They defended their deviations from standard practice as being necessary to satisfy special local needs. Sometimes the changes were very minor, but every alteration, no matter how small, destroyed uniformity. Thirteen hundred standard cars were placed in service in about 1895. Orders were issued that their repair costs and general performance were to be carefully monitored. They were not to be altered in any way by creative mechanics on the frontiers of the Sunset Line's empire. Reports were to be sent to the second assistant to the president, who resided in the New York offices of the Southern Pacific. This requirement suggested that Collis P. Huntington himself might oversee these accounts, and the errors of the most obscure car man would be apparent to the highest authority of the railroad's management.

E. H. Harriman succeeded Huntington as president of the Southern Pacific after 1900. The SP then became part of the Harriman Lines, which soon involved railroad properties all over the nation. It was not simply western lines but more easterly roads like the Chicago & Alton. The mix of equipment was now even more of a potpourri. Yet Harriman felt that standardization, so far as was practical, should be put into effect. In 1902 an ambitious standardization program was announced that would affect thousands of locomotives, passenger cars, and freight cars.[75] Within six years some progress had been made. New equipment was made to the standard series of plans. The Harriman Lines boasted 459 passenger cars, 1,140 locomotives, and 31,845 freight cars, all built to standard design. These numbers look somewhat less impressive when compared with the actual size of the fleets. The Harriman Lines' equipment roster included roughly 2,500 passenger cars, 3,000 locomotives, and 150,000 freight cars. While good progress had been made, much remained to be done. The untimely death of Harriman in 1909 broke up his rail empire and put an end to this promising experiment in large-scale standardization.

A more modest standardization program was under way at the time of Harriman's big push. The Rock Island and Frisco lines attempted to rationalize their joint car fleets in 1905.[76] All cars with less than a 30-ton capacity were recommended for retirement, because they were too small and obsolete to be of further service. They were considered unsafe as well, for most of the small cars had wooden underframes. Only steel-frame cars would be purchased from this time forward. The Rock Island/Frisco fleet had an average age of between twelve and eighteen years. Because the MCB Association had decided that a car was fully depreciated at sixteen years of age, there was a need to restock. A series of standard designs was developed that included interchangeable parts wherever possible. Just how far the Rock Island/Frisco plan was carried out cannot be determined from the existing record.

There was another faction outside the railroad industry that had an interest in advancing the standardization cause. Commercial car builders had a great deal to gain if a standard car could be launched. They stood to gain from simplified production, smaller inventories, lower engineering costs, and the ability to build for stock with a reasonably secure chance for sales when the market revived. The logic of self-interest was lost on some car builders, such as Barney & Smith of Dayton, Ohio. This was the second largest car maker in the nation, but in 1881 it was found to offer eight or nine different styles of journal boxes—none of them compatible with the MCB standard. This failure to cooperate on so basic a standard indicates a profound lack of interest in the concept of uniformity.

However, other builders showed greater sensitivity. The Ohio Falls Car Company in Jeffersonville, Indiana, was headed by Joseph W. Sprague, a believer in the standard car. Even before professional car builders had given much thought to the matter, Sprague's firm was offering standard boxcars, flats, and hoppers. An advertisement in the 1868 *Ashcroft Railway Directory* mentioned that cars were on hand ready for immediate delivery. Ten of each type were ready for delivery in twenty-four hours, and this delay was only to allow time for lettering. Another 120 cars were in frame and could be delivered at a rate of from 12 to 20 per week. Ready-made cars went beyond simple freight cars. Streetcars were standing ready for delivery after lettering. Passenger car bodies were also on hand, ready for trucks and interior trimming of the buyer's choice. Fast delivery and high quality were guaranteed, but it was also understood that the customer must be ready to accept Ohio Falls' design. Just how successful this method of selling cars was cannot be determined. Sprague undoubtedly built cars to the specifications of the buying railroad, and this may well have constituted the majority of his production. The records that might explain the exact nature of Ohio Falls' production have not been preserved so far as can be determined.

Whatever may have transpired, it is known that Ohio Falls never forgot its early commitment. In 1896 the firm circulated a letter to practicing car builders asking them for precise measurements for a 30-ton boxcar.[77] Several railroad clubs were contacted and asked for thoughts on the subject. Most ignored this solicitation as yet another useless exercise. But Ohio Falls persisted and early in 1897 was able to compile a table of dimensions for 30-ton boxcars, reproduced in Table 6.2. A certain thread of similarity can be found in the numbers gathered. Whatever might have come of this fact-finding mission was most likely swept aside by the merger of Ohio Falls as part of the American Car & Foundry Company early in 1899.

All car builders participated in standard design in at least one particular. They did build hundreds and thousands of cars all exactly alike to the order of individual railroads. The pages of the railroad trade press are filled with reports of cars ordered and delivered. In June 1887 the *Railway Review* noted an order by the Reading for two thousand coal cars to the Harrisburg Car Works. Here is a specific instance of a big group of identical cars, representing standardization at one level. Nor is this an isolated example. In December 1900 the following large orders were reported: the Chicago & North Western, three thousand boxcars; the Great Northern, three thousand; and the Missouri Pacific, twenty-five hundred.[78] Most orders were for one hundred to five hundred cars, but each order did not represent a new design.

The practice of letting out orders for identical cars is of more consequence than might be apparent at first glance. The Cleveland, Lorain & Wheeling ordered a big group of 30-ton gondolas in the following fashion: 650 from the Michigan-Peninsular Car Company of Detroit; 150 from Wells & French of Chicago; and 150 from the Mt. Vernon Car Company of Mt. Vernon, Illinois.[79] Car builders were accustomed to manufacturing to the drawings and specifications of their cus-

TABLE 6.2 Dimensions of 30-Ton Boxcars, 1897

	Barney & Smith Car Co.	Locomotive Engineering Journal	Terre Haute Car & Mfg. Co.	Elliott Car Co.	Ohio Falls Car Mfg. Co.	United States Car Co.
Clear inside length	33 ft. 0½ in.	36 ft.	33 ft. 6 in.	34 ft.	34 ft.	33 ft. 4¼ in.
Clear inside width	8 ft. 3½ in.	8 ft.	8 ft. 4 in.	8 ft.	8 ft. 2¼ in.	8 ft. 1¼ in.
Clear inside height	7 ft. 2 in.	7 ft.	6 ft. 9 in.	6 ft. 11 in.	6 ft. 8¼ in.	7 ft. 0¼ in.
Door opening	5 ft.	5 ft. 6 in.	5 ft.	6 ft.	5 ft. 6 in.	5 ft. × 6 ft. 6½ in.
Center to center of center ties	6 ft. 10½ in.	8 ft. 10 in.	4 ft. 6 in.	6 ft. 9 in.	6 ft. 3 in.	8 ft.
Section of side sills	5 in. × 8½ in.	5 in. × 9 in.	5 in. × 9 in.	5 in. × 10 in.	5 in. × 9 in.	5 in. × 9 in.
Section of center sills	5 in. × 8½ in.	5 in. × 9 in.	5 in. × 9 in.	4 in. × 10 in.	5 in. × 8½ in.	5 in. × 9 in.
Section of intermediate sill	5 in. × 8½ in.	5 in. × 9 in.	4 in. × 9 in.	4 in. × 10 in.	5 in. × 8½ in.	5 in. × 9 in.
section of side plate	3½ in. × 7 in.	NA	4 in. × 6 in.	NA	3 in. × 6 in.	3 in. × 8 in.
Section of end plate	3½ in. × 12 in.	NA	NA	NA	3½ in. × 12 in.	3 in. × 13 in.
Height of lining	3 ft. 6 in.	7 ft.	3 ft. 6 in.	4 ft.	4 ft.	2 ft. 6 in.
Truss rod's diameter	1¼ in.	1⅜ in.	1⅛ in.	1⅛ in.	1⅛ in.	1¼ in.
Truss rod's end	1⅜ in.	1¾ in.	1⅜ in.	1⅜ in.	1⅜ in.	1½ in.
Wheel spread	4 ft. 10 in.	5 ft.	5 ft.	5 ft. 6 in.	5 ft.	5 ft.
Upper arch bar	4 in. × 1¼ in.	1 in. × 4 in.	1¼ in. × 4 in.	1¼ in. × 4 in.	1⅛ in. × 4 in.	1¼ in. × 4 in.
Lower arch bar	4 in. × 1⅛ in.	1 in. × 4 in.	⅞ in. × 4 in.	1⅛ in. × 4 in.	1 in. × 4 in.	1 in. × 4 in.
Tie bar	4 in. × ⅝ in.	⅝ in. × 4 in.	⅝ in. × 4 in.	⅝ in. × 4 in.	½ in. × 4 in.	⅝ in. × 4 in.
Set of upper arch bar	NA	6 in.	NA	NA	5 in.	5 in.
Set of lower arch bar	NA	14 in.	NA	NA	12 in.	13 in.
Set of tie bar	NA	NA	NA	NA	1 in.	2¾ in.
Diameter of column bolts	1⅜ in.	1¼ in.	1⅜ in.	1¼ in.	1¼ in.	1¼ in.
Diameter of oil box bolts	1⅛ in.	1⅛ in.	1⅛ in.	1 in.	1 in.	1⅛ in.

Source: Railroad Gazette, Jan. 22, 1897.
Note: NA = not available.

tomers rather than the other way around. The practice goes back to the beginnings of the American freight car. Specific examples are given in Chapter 2. And so limited or segregated standardization existed, but a universal standard car never came into being during the wood-car era. It didn't even come close to being born. One skeptic was inspired to say, "The millennium is just about as near as the standard car."[80]

The standard car seems to fit well into a chapter devoted to special cars, for it was very special indeed: it was the car that never was. Standard design became more of a reality in the steel-car era. Standards in general began to be more closely followed as the Association of American Railroads' Mechanical Division got tougher about enforcing the interchange rules. Components conformed exactly with standard practice, or else the car was embargoed. The variety of standard components like axles and wheels was greatly reduced. In 1924 the American Railway Association's Mechanical Engineering Division voted to adopt a standard box plan. It was fairly well received, and by 1930 twenty-five thousand had been built. The standard car was not mandatory, however, and most railroads continued as before, but plans increasingly tended to follow the basic outlines of the standard car. The standard design was updated and re-

formed over the years. Some of the industry's best talent, like Paul W. Keifer, devoted considerable time to perfecting these designs. Following World War II both Pullman and American Car & Foundry based their mass-production freight car series on these designs. Thousands were built, and by the 1950s it appeared that perhaps the day of the standard car had at last arrived. All was to change radically over the next few decades. Railroad traffic became more specialized, shippers more particular, and the competition tougher. The boxcar became obsolete as highway trucks ran away with merchandise traffic. Shippers would stay with rail transit only if railroads would provide special cars suited to a specific traffic. And so the standard-car notion died once more, never, it would appear, to rise again.

The Caboose

The caboose was both home and office to train crews.[81] It offered a haven from rain, sun, and wind. At a desk in this last car of the train, the conductor kept records of the train's consists, noting cars set out and picked up. Tools and supplies were stored here for on-the-road repairs and emergencies. Chain, rope, jacks, rerailing frogs, air hoses, and coupler jaws were stowed in lockers and cabinets inside and even under the car. Trainmen could nap on benchlike bunks usually fitted with long, solid cushions. The men could drink from a watercooler or the ever-present coffeepot. They could warm themselves by the stove, have a simple meal, or simply visit with other crew members while waiting on a siding or at some other quiet time during the trip. There would be talk of family, friends, and the job. It is this homey, human factor that makes the caboose so appealing. Its warm, smoky interior with its glowing lamp or the occasional face at the window made it the most alive and friendly of all railroad freight cars. Technically, of course, it isn't a freight car—at least not a revenue-producing freight car—for it carries no cargo. But it is so intimately associated with freight train operations that it must be included in any book devoted to that subject.

Yet all the matters just presented are secondary to the main purpose of the caboose, which is to protect the rear of the train. As a marker, an island of light with glowing red lamps or fluttering flags, it signaled all who came upon it that a train was present. If the train stopped on the main line or could not clear the main track because of a short siding, one of the trainmen would gather lanterns, flags, fusees, or torpedoes and walk a suitable distance behind the train to warn approaching trains of their existence. This was basic to safe railway operations, before and even after automatic signaling became common.

The caboose also served the secondary safety function of being an outpost from which to observe overheated journal bearings—the dread hotbox that caused so many accidents. The smoke could be observed or smelled by trainmen watching the cars ahead. Dragging brake gear, a derailed car, or a loose load could also be seen from the caboose. And as trains grew longer, this rear observation point became a help in surveying the whole train. After air brakes were introduced, very late in the wood-car era, the caboose was outfitted with an air gauge so that the train line pressure could be read. The engineer depended on the conductor to signal when pressure was fully pumped up so that all brakes were released and the train could proceed. The conductor also had the option, in an emergency, of applying the brake by opening the caboose brake valve. In short, the caboose was more than a bunk or galley car; it was a field headquarters and an observation tower.

During the nineteenth century a caboose was assigned to the exclusive use of a single crew. Work assignments were scheduled on a first in/last out basis so that a caboose and its crew would remain at a terminal until all others arriving before them had been sent out. The caboose became a lodging room for crews awaiting the return trip. On a long division (150 miles), waits might extend for three or even four days. The men saved lodging expenses, and the railroad benefited because the crew was on hand and immediately available when needed. This rude little house on wheels softened the hardships of travel over the road as well. Freight trains were normally scheduled for ten to twelve hours per division, but because of delays, the actual time was often twice that. The men's strength and health were much bolstered by the minimum creature comforts offered by the caboose. In addition to shelter and warmth, there was the opportunity for a meal. Most crews would stock coffee, oatmeal, and potatoes. Sometimes a single crewman would assume the duties of cook; in other instances, all the trainmen would share in the chore of meal preparation.

ORIGINS

The caboose is familiar to and even loved by many people who otherwise don't have the slightest interest in freight trains, simply because of its unique name. There are few English words in common usage more peculiar than *caboose*. The *Oxford English Dictionary* suggests that its origins are Dutch or French but goes on to admit that its "history and etymology are altogether obscure." There is agreement, however, that the term can be traced back to the eighteenth century and that it meant a small cookhouse on the deck of a ship. Not all railroads adopted the word *caboose*; some preferred terms such as *conductor's car*, *way car*, or *cabin car*. There were also dozens of nicknames devised by train crews such as *crummy*, *shack*, *hack*, and the like. Only the gondola rivals the caboose for novelty in a field notable for the prosaic nature of its terminology.

For all its glamour and popularity, the caboose is essentially of unknown origins. Its earliest history is a matter of speculation and undocumented oral tradition, but it appears that very few cabooses were in operation much before 1850. This conclusion is supported not only by the lack of any contemporary evidence but by the facts of freight operations. The amount of traffic was relatively small before 1850; trains were consequently short, and few railroads measured more than 100 miles in length. Because no cars were interchanged, there was little need for on-train record keeping. Hence the need for such equipment was slight. As traffic, train length, and the duration of travel increased, so too did the need for cabooses. The decision to provide them was likely based on concern for both the well-being of the crews and the safety of the trains. The story of how it happened likely varied from road to road, but I suspect that most followed one of these patterns.

In scenario one the trainmen do it for themselves. It's a cold, nasty day, and the men know they will be out in the wind and rain for the remainder of the day and part of the night. The train always runs late, and fourteen hours without shelter or haven is a grim prospect. It was possible to hunker down in the engine cab for a while, but there really wasn't much extra room on a hand-fired engine, and the cabs tended to be tiny. Just before the train departs one of the men spies a derelict boxcar standing nearby. He persuades the engine crew to add it to the rear of the train. The men climb aboard the old wreck. The roof leaks and wind blows through the cracks. It's a poor but welcome shelter. Along the line a few empty crates and kegs are picked up for furniture. Someone finds a little hay to make a primitive bed. The men volunteer over the next few days to repair the car. The roof is fixed, windows and a stove are added, and in a matter of days the caboose is born. According to the recollections of several trainmen, this is exactly how the locomotive cab was introduced.[82]

In scenario two, a young assistant trainmaster, who had risen through the ranks after starting as a brakeman, pleads with his boss to do something for his men. "It's cold and miserable out there, Boss. Believe me, I rode the cars for years in the heat and cold. We could improve morale and efficiency. I know it would reduce accidents and save the company's property. Surely it would lessen the complaining—just give the men a warm place to sit. It might also persuade them to give up all this union talk. And it wouldn't cost much. I'm thinking of something very plain and simple."

The superintendent isn't sure. "It's not good to spoil the men—might make them soft and lazy. Give a little and they'll want more. Bunks, you say, and a stove? Why, they'll slip away to sleep rather than attend to their duty. Being outside keeps them tough and alert, fresh air never hurt any workingman. Now when I was a young man. . . ."

"But Boss, have a heart, we won't be coddling any one of them. Not one, I promise. I will be responsible. We have those old 100-series four-wheel boxcars. We never use them; they just sit there. We could refit them for a trifle—a few windows, steps, a table or two would do the job."

The superintendent at last agrees with the trainmaster, but the men must do all the work on their own time and at no expense to the company. He adds another condition: If any train is stalled or delayed because of the extra weight of these "concession cars," he will order their instant removal from service. The trainmaster has achieved a yes, if only a qualified one, and again the caboose is born.

In scenario three, the cabin car is introduced from outside rather than inside the railroad. There was a certain wealthy man who was also much given to good causes and local charities. He happened to be a large shareholder in the railroad. During his travels over the line, he observed trainmen clinging to the cars in all extremes of weather. He was disturbed by what he saw and made a point of speaking to some trainmen at the end of his trip.

He was impressed and saddened by their suffering and appealed to the president of the railroad to do something to relieve this wrong. We have a model for such a man of conscience in Lorenzo Coffin, who lobbied so long for railway safety appliances. The pleas of so dedicated and respectable a person could not be easily ignored, and so again the caboose is born.

None of these speculations may be correct, yet they are all supported, in part, by oral tradition. The story of Nathaniel Williams, a conductor on the Auburn & Syracuse during the 1840s, has often been presented as the origins of the caboose. Unfortunately its true source is unknown, but it was apparently first reported long after the fact. Williams, according to the tale, used a boxcar for his office at the end of a mixed train. A box was his seat, while a barrel served as the desk. He carried tools, flags, and chains with him. It is not clear whether his adopted caboose was a particular car or just any boxcar that happened to be at the end of the train. A similar but less debatable story was reported in 1901 about events on the Baltimore & Ohio in 1867.[83] J. Clark Mercer was a freight train conductor with a regular run from Wheeling to Grafton. He and his men would steal hay or corn fodder from farms near the track and throw it into empty boxcars for a temporary bed, for there were no cabin cars. One day most of Mercer's men were sick, yet he was asked to take a train over his regular route. He agreed but asked that a stove be placed in one of the cars for the comfort of his ailing crew. The superintendent not only complied with Mercer's request but ordered a number of boxcars to be outfitted with stoves and end doors. Mercer believed that these converted boxcars were the origin of the caboose when in fact cabooses were by then already in use on other railroads. The date does agree with the time when the B&O adopted cabin cars. It was, however, a latecomer rather than a pioneer.

Turning from legend and recollections, we find the oldest mention of a "caboose" appearing in the *Railroad Advocate* in 1855. It is a very brief notice stating only that such a car was operating on the Buffalo, Corning & New York Railroad, and that the car was supported by an odd style of truck. The BC&NY later became part of the Erie, and that line already had cabooses in 1854, according to an 1897 report.[84] There is an 1859 report concerning drover-style cars or cabooses on the Erie. In the same year the Michigan Central had similar cars.[85] Here again, these are contemporary reports and not recalled history printed half a century later. Earlier historians have often cited an 1859 legal suit as the first concrete mention of a "caboose car," but none has offered a source. This case involved a car on the New York & Harlem Railroad that broke in half after the rear truck derailed near Chatham, New York, in February 1859. Regular

FIGURE 6.25

Atlantic & North Carolina Railroad conductor's car at the Little River Bridge in 1864. (Library of Congress Photo)

fare-paying passengers were being carried, and their injuries were the subject of the suit. The train was moving at only 5 or 6 miles per hour, but the car was so old and weak that it fell apart during the derailment.[86] There is one more very early printed notice regarding cabooses, this one involving a car on the Detroit & Milwaukee Railroad in October 1861.[87]

A search of the patent records produces some interesting curiosities, but nothing that explains much about the origins of the cabin car. A patent was issued on April 1, 1862 (No. 34,829), to Henry C. Glasgow of Chicago for a brakeman's cab. It was not, however, for a distinctive railroad car, but for a small, portable, tentlike shelter that a trainman could set up on the walk boards of a boxcar. The same idea was the subject of a patent granted to Charles L. Heywood of Belmont, Massachusetts, on June 10, 1879. The specification was more detailed than Glasgow's, in that it called for tent bows of rattan, whalebone, or metal and a covering of sailcloth, rubber, or oilcloth. William A. Goodwin of Newton, Massachusetts, pursued a different solution. While the tents proposed by Glasgow and Heywood were light, cheap, and individual, Goodwin wanted American railroads to build a shelter at each end of every car. The compartments would be built low into the body of the car, projecting just above the roofline. Any manager would have seen that the common caboose was more practical and economical than the zany solutions offered by patrons of the Patent Office.

A wider acceptance of the caboose is documented by railroad photographs made during the Civil War.[88] Most of these cars appear to have been converted boxcars, and all had roof lamps and side doors. In one scene on the Orange & Alexandria Railroad, car No. 1258 is pictured with roof vents and a window in the door. Another print shows a former ventilated boxcar with roof lamps, side steps, and a ladder to the roof at one side of the car. It was in service on the Sullivan Branch of the Nashville & North Western Railroad. Four similar cars are pictured in a Nashville yard scene made around 1864. That these vehicles were peculiar to military operations is countered by a photograph of a car on the Atlanta & North Carolina Railroad. The car is lettered "A&NC RR Conductors Car" (Fig. 6.25). Could this be the oldest dated caboose photograph? It is possible that many railroad men from all parts of the nation were first exposed to cabooses during their service on the U.S. Military Railroad, and that they carried the idea home at the war's end. Surely the times were ripe for such ideas as the rail network expanded and interchange of cars began. The work of freight train crews mounted along with traffic and the need for greater safety precautions.

The theoretical espousals of the caboose by Civil War veterans notwithstanding, the next dated images of cabin cars are found among the construction photographs taken along the Union Pacific Railroad in the late 1860s by A. J. Russell. As if the function of these cars was not obvious from their appearance, the word *caboose* was lettered boldly on their sides. These were fairly long, arch-roof cars without cupolas. They did, however, have end platforms, canopies, side doors, and passenger-style trucks. The need for such cars is obvious for trains traversing unsettled territory. Trainmen, construction supervisors, engineers, and laborers needed both shelter and safe living

TABLE 6.3 Cabooses in Service

Railroad	1870	1880	1890	1900
Baltimore & Ohio	140 (E)	240	350 (E)	428
Chicago & North Western	150	157	451	546
Union Pacific	45 (E)	96	210 (E)	182
Pennsylvania	21	397	628	890
All U.S. railroads	2,500 (E)	5,500 (E)	11,000 (E)	17,605

Source: Unless otherwise indicated, *Poor's Manual of Railroads* and published reports of the railroads concerned. The 1900 figures are from Interstate Commerce Commission, *Statistics of Railways in the United States* (Washington, D.C., 1900), p. 31.

Note: (E) = estimate.

and office space. A great many Civil War veterans worked on the transcontinental project, and it is possible that their recent memory of the good service of conductor's cars on the U.S. Military Railroad brought about an early introduction of the caboose on the Union Pacific. There are two Russell photographs that show this series of cars especially well.[89] One shows two cabin cars headed by three locomotives posed on a Weber Canyon bridge. In another scene, caboose No. 8 is shown at Taylor's Mills Station on May 8, 1869. A party accompanying the Leland Stanford train to the Golden Spike ceremony stands nearby. Perhaps this party of VIPs took a few minutes during their journey to see how lesser folk traveled over the about-to-be-opened Transcontinental Railway.

The reasons for the Union Pacific's early espousal of the cabin car may be obvious, but why were some roads so early or so late to accept it? Climate conditions are the easy explanation; in that case Canada and New England should have been the pioneers in the introduction of cabin cars, but in fact a stronger case can be made for southern and midwestern lines, based on the slim evidence available. In most cases we don't know just when cabooses were adopted. *Poor's Manual of Railroads* did report on equipment; but while some roads furnished very detailed accounts, others were more cryptic and a few failed to report anything at all on the subject. Hence it is possible that some of the later roads had cabin cars but simply did not bother to account for them. Even so, the data presented by the earliest *Poor's* helps frame the beginnings of the caboose. The Chicago & North Western was especially dutiful in reporting car data to *Poor's*. Starting in 1864—though this was not the start-up date for the line's caboose operations—the C&NW listed 74 cabooses.[90] In only four years the number had climbed to 137, and it peaked in 1870 at 164. Five years later the number had fallen back to 140. Perhaps declining traffic following the 1873 financial panic explains this retrogression in the size of the caboose fleet. By 1880 the number was back up to 157 cabin cars.

Southern lines blessed with moderate climates reported cabooses as listed in *Poor's* volume for 1869 and 1870. The Atlantic & West Point, the

Georgia Railroad, and the Central of Georgia all listed cabin cars. It seems curious that lines in the American South were willing to provide cabooses for their crews while roads in Canada, where bitter weather was a given, were so reluctant to do likewise. A comprehensive survey of Canadian railways conducted in 1858 by Samuel Keefer included detailed equipment listings. Not one caboose was recorded.[91] Keefer's 1859 report showed that the Great Western Railway had taken mercy on its crews by providing thirty-three eight-wheel conductor's cars. In 1860 Keefer found that the Northern Railway protected its men from frostbite by putting seven caboose cars in service. Big roads like the Grand Trunk held off until the early 1860s, when they began to convert boxcars into cabin cars. According to Canadian Pacific historian Omer Lavallee of Montreal, cabooses were not commonly used in Canada until the 1870s.[92]

Midwestern roads were well represented as caboose operators; the Cleveland & Toledo, the Cleveland, Columbus, Cincinnati & Indianapolis, and the Michigan Southern were listed as owners of such equipment. Yet the pattern of use was not consistent by region, for while the Northern Central claimed sixty cabooses in 1868, the nearby Philadelphia, Wilmington & Baltimore had none. Another neighboring line, the Baltimore & Ohio, was rather late to adopt cabin cars, listing its first in its 1866 annual report. The initial four cars apparently proved their worth, for in 1872 the B&O listed no fewer than 143 eight-wheel cabooses. Within four years the number had jumped to 234. By the end of the century the B&O had over four hundred cabin cars on its roster.

And so it seems that the adoption of cabin cars was governed less by climate or regional considerations than by individual prejudices and predilections. How else can we explain the Union Pacific's early enthusiasm and the Central Pacific's general indifference? As late as 1870 the Central Pacific had only two cabin cars, while the Union Pacific had been using such vehicles since its construction period. In all, the adoption of the caboose was late and gradual, and the now-familiar conductor's car—assumed always to have been a necessary part of American freight trains—was not very common much before 1880. Table 6.3 helps illustrate the delayed acceptance of the caboose.

Though the precise origins of the caboose are likely to remain unknown, its appearance is far from hazy, for no other railroad freight car is better documented. Its image appears frequently in railroad-related publications. From these numerous illustrations the variety of sizes, shapes, and arrangements is apparent. There were four-wheel and eight-wheel cars. There were cabin cars with and without cupolas. Those with lookout towers might have them at one end of the roof, at the center of the car, or just off center. Some had bay windows rather than cupolas. Many early cabooses had side doors, but this feature had largely disappeared by about 1900. Nearly all had end platforms and end doors, but some true bobtails existed with side doors only. There were transfer cabooses used in yard work or interyard service. There were long drover-style cabooses with benches and bunks to carry the attendants who traveled with cattle trains. Most cabooses were built new, but some were converted from boxcars, from troop sleepers, and, in at least one case, from steel ore cars. Short lines and lumber roads were given to the assembly of homemade makeshifts, in which construction values and aesthetics reached a new low. To say that there was a great variety of designs is an understatement. No attempt will be made here to cover every variant—only the major known styles.

FOUR-WHEELERS

Although four-wheel cars were considered obsolete for almost every other purpose, the single-truck caboose found a surprising number of advocates in this country before 1900. They were compact, light, and cheaper than their larger eight-wheel counterparts. Anything that added unnecessary weight to the train was a negative in terms of speed, fuel, and track maintenance. A few extra tons could mean the difference between overcom-

ing a grade and stalling on it. Economy-minded managers liked the four-wheelers for these reasons. Others were said to have liked the midget cabin cars for their rough ride, which guaranteed an alert crew. Should a single-trucker acquire a flat wheel, the car developed a bone-shaking ride that prevented sleep for even the weariest crew.[93]

Critics of the four-wheelers said that they were too small and unstable for regular service. At best they were suitable for the short runs typical of eastern lines, but they would never do on western lines, where trips were long and drovers were regularly carried. This rule was generally followed; four-wheelers were rare west of the Mississippi on trunk-line railroads. There were exceptions, of course, and the Burlington offers but one example of the diversity of opinion on the subject. In January 1885 Charles E. Perkins, president of the line, wrote to his general manager, T. J. Potter, on the Pennsylvania Railroad's good experience with four-wheelers.[94] He felt that real economies could be achieved; it was never possible to be too conservative when it came to operating expenses. Perkins's enthusiasm carried over into the early years of the twentieth century, when the Burlington shops reported the construction of twenty new four-wheelers.[95] Each could sleep five men. Three suspension systems were tested, and crews were allowed to pick their favorite.

The bouncy four-wheel bobber is said to have

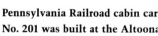

FIGURE 6.26

Pennsylvania Railroad cabin car No. 201 was built at the Altoona shops in July 1876. (Smithsonian Chaney Neg. 25532)

FIGURE 6.27

This late-model four-wheel cabin car was completed at the Altoona shops in April 1907. (Smithsonian Neg. 83-12553)

FIGURE 6.28

The Boston & Albany's No. 32 poses in the Boston yards with its crew in 1884. (Railway and Locomotive Historical Society Photo)

CARS FOR SPECIAL SHIPMENTS

FIGURE 6.29

The New York Central published drawings for this well-equipped four-wheeler in 1883. (*National Car Builder*, Aug. 1883)

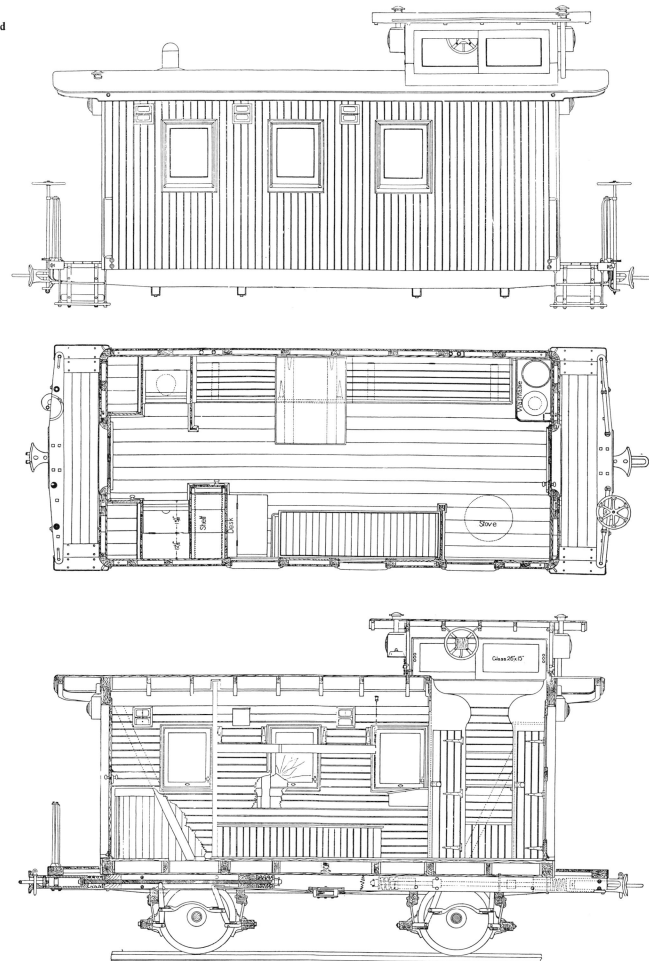

THE AMERICAN RAILROAD FREIGHT CAR

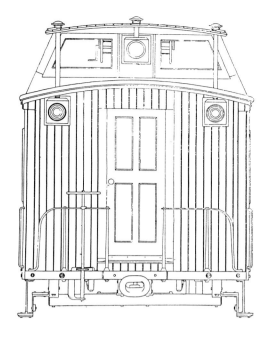

In 1872 the Pennsylvania produced a new design designated as the class NB cabin car. It was somewhat less stingy and sported a few extra amenities, such as platform canopies and brake wheels at both ends. An extra foot was added to the width, and overall weight crept up to 14,900 pounds. Caboose design on the Pennsylvania remained static for another eleven years, until the class NC was introduced. The designers took a less cheeseparing view and produced a car that was less toylike than its predecessors. The body measured 18 feet 5¾ inches by 9 feet 1¾ inches. Weight went up to 20,950 pounds, or almost twice that of the old class NA. Massive cast-iron steps offered the crew safer egress than the skimpy iron-bar step formerly used. The suspension was simplified by eliminating the long equalizing lever. Leaf springs were placed over each journal box.

Ten years after the class NC came into being, the Pennsylvania introduced an almost man-sized version of its traditional four-wheel dinky. The ND had a body 21 feet 1¾ inches long. The width was the same as that of the NC. The new cabin cars weighed in at 14 tons, accountable mainly to the steel frame and the 36-inch wheels. A photograph of an ND built at Altoona in April 1907 shows a straight-sided cupola, a large toolbox under the floor, and conventional marker lamps in place of the built-in sheet-metal lamps formerly used (Fig. 6.27).

The Pennsylvania's influence went beyond corporate boundaries, for as already mentioned, Charles Perkins decided to adopt four-wheelers based on the good reports emanating from the Keystone line. A photograph of the Shenandoah Valley Railroad's cabin car No. 4007 dating from around 1887 illustrates one car built from Altoona blueprints.[98] Curiously, the Pennsylvania's Lines West tended to be more independent and built cars to their own designs that were often only slight variants on the plans of Altoona. Why they could not follow a standard plan is beyond explanation.

The shops at Fort Wayne and Columbus are known to have gone their own way.[99] The Columbus shops came out with a new wood-frame four-wheel cabin car in 1902. It was very similar to the class ND, but it had a highly arched cupola. The sides of the cupola were inclined for tunnel clearances. These cars were carried on second-class 36-inch-diameter passenger car wheels.

Cabin cars are the habitat of ordinary men, and as such they are normally not associated with famous men or national events. The Peary Arctic expedition car mentioned earlier is one possible exception to this rule. Another notable, if not famous, caboose was the Albany & Susquehanna Railroad's No. 10. In September 1883 eight trainmen met aboard the car to organize a union that later became the Brotherhood of Railroad Train-

originated on the Erie Railway in 1854. Cavin A. Smith, master car builder of the line, cut a wrecked boxcar in two and turned it into two broad-gauge four-wheel cabin cars. Short cabooses of this type remained popular on the Erie for another half-century.[96] Admiral Robert E. Peary borrowed an Erie four-wheeler (No. 4259) in 1899 for use as a portable house during one of his Arctic expeditions. The wheels and undercarriage were removed. After four years in the frozen North, the body was returned to the Erie for exhibit at its Susquehanna station.

There appears to have been no greater champion of the four-wheel caboose than the Pennsylvania Railroad.[97] The parent road and its numerous subsidiaries doted on four-wheel cabin cars during the wood-car era. The origins of the caboose on the Pennsylvania are cloudy, but around 1869 a standard four-wheel design called the class NA came into being. It was an extremely small and Spartan vehicle. The body measured a mere 15 feet 1¾ inches by 8 feet 1¾ inches. Tiny end platforms, bereft of even the most meager canopies, brought the total length to just over 19 feet (Fig. 6.26). Weight was a scant 6 tons. Sleeping for six was provided by three double berths. Tall men would have found the sleeping cramped, and the 6-foot 4-inch ceilings more than a trifle low. The culinary equipment was apparently limited to a stove, which also doubled as a heater. The several built-in tin lamps betray a preoccupation with the security function of the cabin car as a beacon for the end of the train. Even more attention was given to the truck, which featured long leaf springs over the journals and a lengthy equalizer with rubber spring snubbers. The primitive bar-iron step and single brake wheel

FIGURE 6.30

The Chesapeake & Ohio remained loyal to the four-wheel caboose concept even after the coming of the steel car. Cars of this design could sleep three men. (*Railway Age*, Aug. 9, 1901)

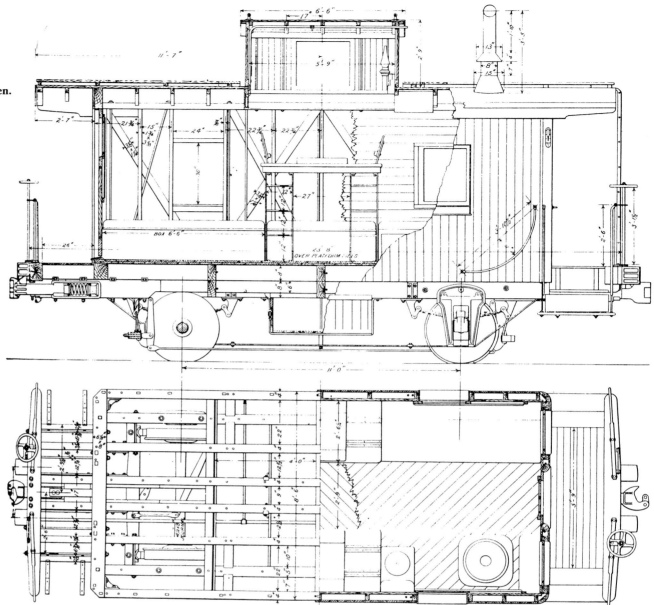

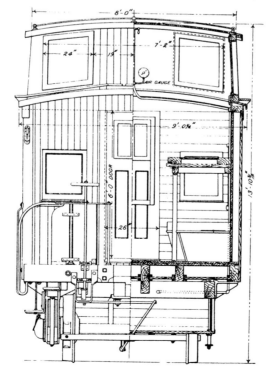

men. In 1924 the successor railroad to the A&S, the Delaware & Hudson, agreed to restore the car and place it in an open shed located in a park in Oneonta, New York. Only the body of the car had survived. Wheels, pedestals, and other parts were taken from another similar car, but the accuracy of the restoration remains in question, for no in-service views of the No. 10 or its sisters are known to exist. The No. 10 is small even for a single-trucker, for it measures only 17 feet long over the end platforms. The body is a scant 12 feet by 8 feet. The car has no cupola, bunks, or cabinets. The interior has long side benches and a stove. A reel hanging from the ceiling once held a rope that could be extended over the train to a signal bell in the locomotive cab. A small metal seat with three legs is attached to the roof. From this insecure perch trainmen could view the train. The car is said to be the oldest preserved caboose in North America. This may or may not be true, for neither its date nor its builder is recorded.

FIGURE 6.31

Pullman finished this sturdy four-wheeler for service on the Tonopah & Goldfield in Nevada in August 1907. (Pullman Neg. 9939)

Illustrations of other, less celebrated cabin cars help document the design and construction of the American four-wheel caboose. One of the most engaging is a photograph of the Boston & Albany's No. 32 taken in the Boston yards during 1884 (Fig. 6.28). Five crew members pose beside their mobile home in a fine display of American workers on the job. The lack of a cupola was made up for by twin roof lamps equipped with fierce-looking bull's-eye lenses. The wooden passenger-style steps were more generous than the iron-bar steps used by the Pennsylvania. The short platform canopies are the only sign of New England parsimony. Horizontal side boards, used instead of the usual vertical sheathing, give the car a distinctive appearance, as does its peculiar truck suspension. Independent equalizing levers fitted with coil springs reach back to the end of the frame, where they hang on a long bolt fitted with a rubber spring snubber.

About a year before this B&A car was photographed, the New York Central completed drawings for a very neat and well-equipped single-truck caboose.[100] The six-window car had its cupola placed at the extreme end of the body. The cupola had inclined sides for tunnel clearances. Lamps were built into the cupola and the end walls of the main body. A bell rope reel was mounted high in the cupola ceiling. The interior was neatly outfitted with folding bunk beds, a washbasin, stove, woodbox, desk, and even a dry-closet toilet. The use of clasp brake shoes is notable and may be one of the earliest known instances of double brake shoes on an American railroad car. The drawings for this series of cars are unusually complete, and except for the truck and suspension, few details appear to have been omitted by the diligent drafts-

man (Fig. 6.29). The car measured 17 feet 4½ inches over the end sills and 8 feet 6 inches over the side sills. The wheelbase was 10 feet.

More than twenty years after this design was introduced, the New York Central's midwestern subsidiary, the Big Four, introduced a new single-truck caboose.[101] It was 6 feet 1½ inches longer than the 1883 plan and boasted an 11-foot wheelbase. The cupola had straight sides and was placed centrally on the body. A strong floor frame was built with eight 4- by 8-inch timbers. While large for a single-truck car, it was too small and too fragile for 1905 main-line operations. Forty-ton steel freight cars had become common by this time. It seems ill-advised that the Big Four's management decided to build more bobbers so far into the steel-car era. But when these cars first appeared, the *Railroad Gazette* said that they were not obsolete, and that four-wheel cabooses were not likely to disappear soon from the American railroad scene. There is evidence that other railroads concurred.

The Chesapeake & Ohio introduced a new style of four-wheeler in 1901 that was very similar to the Big Four car.[102] Though not intended as a "living car," it was nonetheless comfortable and roomy (Fig. 6.30). Space was provided for three sleeping bunks. The space between the sills was covered with boards top and bottom, and the interior was packed with wood shavings as a sound deadener and insulator. A sample car was sent over the railroad for testing and criticism by conductors and crewmen. Other big eastern lines like the Baltimore & Ohio saw no reason to abandon single-truckers just because steel cars were coming into common usage. The B&O's class K-1 four-wheeler introduced in 1886 continued in production until

FIGURE 6.32

The interior of the T&P car
shown in Figure 6.31 had a
varnished natural wood interior.
The pocket steps for entry to the
cupola are visible at the picture's
center. (Pullman Neg. 9941)

1910. More than eight hundred were once in service. By 1924 the number was down to 650, and by 1945 only a dozen were listed in the B&O's annual Summary of Equipment. One class K-1 enjoyed an unusual longevity. It was sold in 1927 to the B&O's Chicago Terminal line and there continued in service until 1954.[103] It is now preserved in a railroad museum in Monticello, Illinois.

Other eastern trunk lines such as the Lackawanna continued to depend on four-wheel cabin cars until the early 1920s, when they were replaced rather rapidly by steel-frame eight-wheelers. The Pennsylvania updated the single-truckers in 1903 by introducing a steel-frame car. Four-wheelers, as might be expected, tended to survive longer off of main-line trains. The B&O Chicago Terminal found them satisfactory for yard work, as did the Belt Railway of Chicago. Short lines offered an even more secure haven for the tiny single-truckers. Examples could be seen until fairly recent times on the Unadilla Valley in New York State to the Tonopah & Goldfield in faraway Nevada. Photographs of an attractive single-trucker built for the latter road are reproduced as Figures 6.31 and 6.32.

EIGHT-WHEELERS

The survival of the single-truck caboose so long into the twentieth century is a perplexing episode in American railroading, which exhibits such a decided preference for ever more massive and heavy rolling stock. No class of four-wheel car lasted longer. Yet not all railroad managers were so enthralled with the compact caboose. The merits of larger eight-wheel cars were understood by nearly all railroads, even those favoring compact models. The earliest reports on cabin cars, dating from the 1850s, refer to double-truck cars, and even during

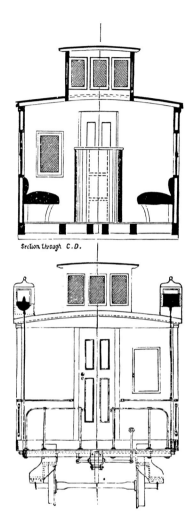

Section through C.D.

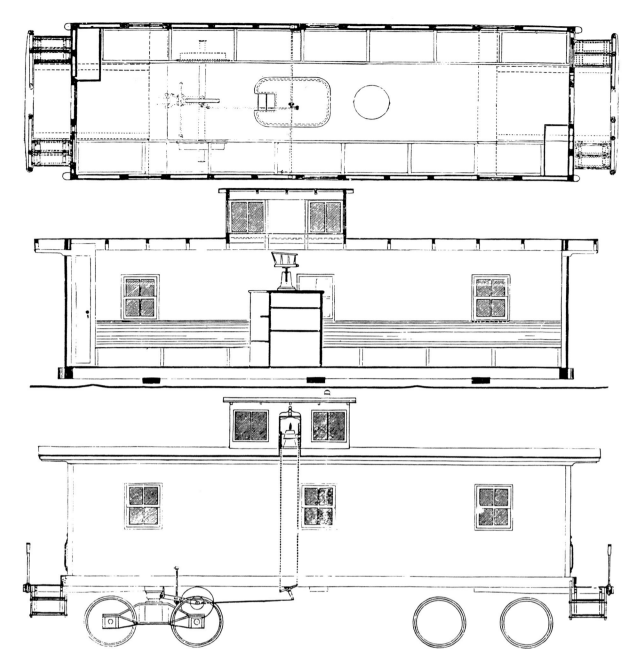

the heyday of the four-wheelers, the larger, more stable eight-wheel caboose was very much present on the American railroad scene. There is considerable pictorial evidence to support this notion. A Baltimore & Ohio train was photographed around 1875 while passing over the Tray Run viaduct. An eight-wheel caboose with a high and narrow cupola is coupled at the head of the train just behind the locomotive's tender.[104] Though small, the image shows features such as three windows per side and end platforms and canopies. During the same period, the New York Central issued a lithograph to publicize its four-track main line. On one track a boxcar-like caboose, with six side windows, no cupola, and end platforms without benefit of roof canopies, brings up the rear of one train.

A caboose belonging to the Big Four is among the best-documented eight-wheel cabin cars of the

1870s.[105] Drawings for it are reproduced in Figure 6.33. It is very similar in appearance to the B&O car just described. The body measured 28 feet long by 9 feet wide and 8 feet high. The narrow cupola, referred to as the "dome-lights" in the *National Car Builder*, was fitted with a revolving chair so that the observer could easily look in all directions. To either side of the cupola were fitted flashing warning lights patented by William L. Needham of Cleveland. The lamp was revolved by axle power and was said to be a very effective warning device. The drawing shows long upholstered side benches and cabinets at the ends of the car, but there is no evidence of a toilet or washbasin. Perhaps these details were simply omitted from the drawing.

Some twenty years later drawings for another Big Four caboose were published in the *Railroad Gazette*.[106] This car was not much larger than its

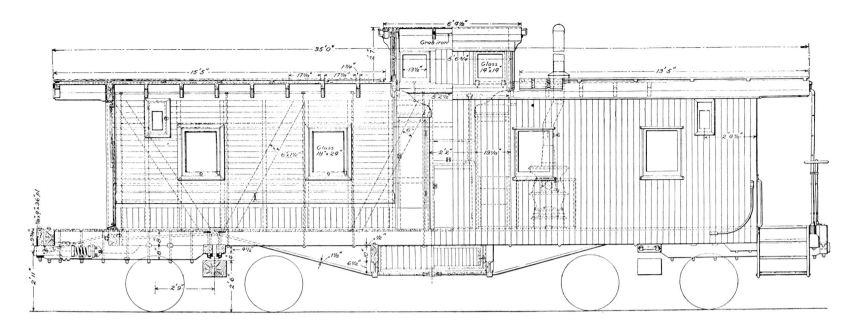

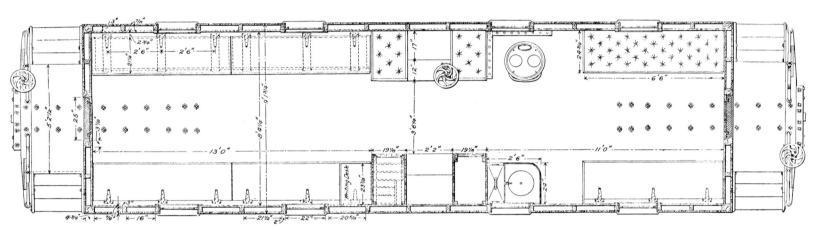

FIGURE 6.34

These drawings for a Big Four caboose are more complete than those shown in Figure 6.33 and were published twenty years later. (*Railroad Gazette*, Sept. 7, 1894)

predecessor, but many of the details had changed (Figs. 6.34, 6.35, 6.36). The narrow cupola gave way to one made to the full width of the roof. The island platform surmounted by a single revolving chair was replaced by more conventional bench seats built into the cupola. Needham's revolving lamps were abandoned in favor of simpler built-in marker lights. This series of cars had ample locker space for oil, coal, and other supplies. Tools and repair parts were stowed in a box under the car. The tin roof was painted brown, while the body was red. The design was credited to William Garstang, master car builder for the Cleveland, Cincinnati, Chicago & St. Louis Railroad. The cars were built by the Terre Haute Car & Manufacturing Company.

Economy-minded managers liked cabooses made without end platforms. These bobtail cars looked like remodeled boxcars and were just that in many cases, but some were built new on this plan. They were shorter, lighter, and somewhat cheaper than the more standard design that featured platforms, steps, and roof canopies. Surely they must have proved less popular with the train crews, for egress through a side door would be

more difficult and dangerous. Trainmen were constantly having to climb on and off the car to switch cars or to flag at stops along the line. One of the oldest known photographs of a bobtail caboose is a print of the Portland & Rochester Railroad's No. 101, built by the Portland Company around 1871.[107] Photographs also exist for similar cars on the Old Colony and Fitchburg railroads. Lest the plan appear to be an isolated New England phenomenon, photographs exist for such cars on far western railroads as well. Reproduced here is a fine 1903 builder's view of the San Pedro, Los Angeles & Salt Lake's No. 215 (Fig. 6.37). Only able-bodied men would find the access through the side door, lacking either steps or handrails, an easy one. Disabled, overweight, or aged crewmen must have cursed every official associated with ordering the 215. The car does appear to have been an easy rider, because of its passenger-style trucks.

The western trunk lines were for the most part strong advocates of full-size eight-wheel cabin cars. In August 1872 the Central Pacific completed some eight-wheelers big enough for ten men. They had end platforms and side doors. They rode on

FIGURE 6.35

Details for the Big Four 1894 caboose. (*Railroad Gazette,* Sept. 7, 1894)

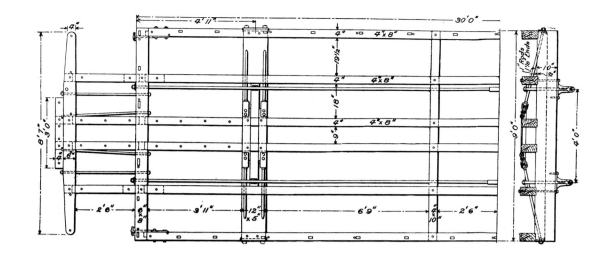

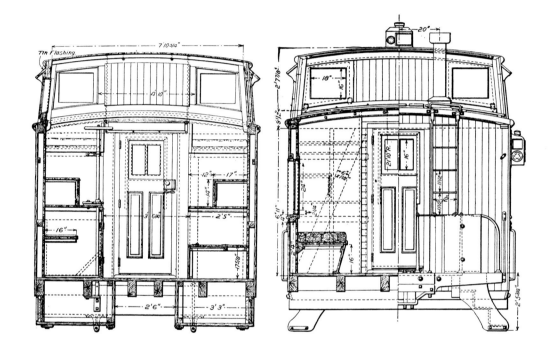

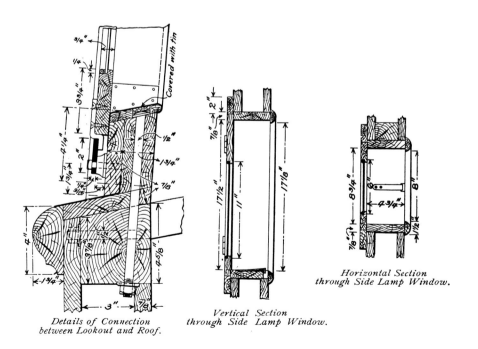

Details of Connection between Lookout and Roof.

Vertical Section through Side Lamp Window.

Horizontal Section through Side Lamp Window.

FIGURE 6.36

This interior of the Santa Fe, Prescott & Phoenix Railroad's caboose No. 1 shows the car as a pristine specimen, a condition not likely maintained by the train crews. The elevated platform and revolving chair are old-fashioned features for a car produced in 1894. (Pullman Neg. 2872)

FIGURE 6.37

This bobtail caboose was built by Pullman in 1903. (Pullman Neg. 7097)

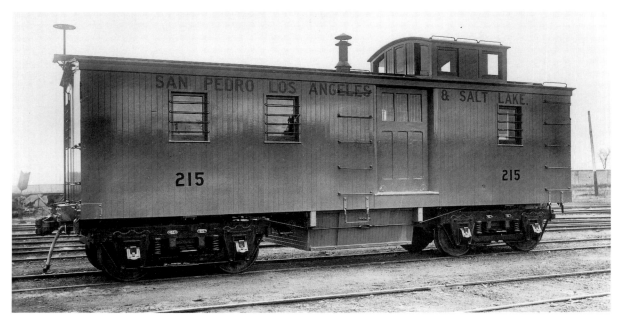

THE AMERICAN RAILROAD FREIGHT CAR

passenger-style trucks and boasted a big tool-box under the body. A picture of one of these Sacramento-built units has been reproduced in several railroad books.[108] Less publicized is a group of forty cabooses completed in 1881 for the Santa Fe by Wells & French of Chicago.[109] These cars were well outfitted for long-distance travel. They had comfortable bunks, a stove, a washbasin, a toilet, and a water tank (Figs. 6.38, 6.39). In pre-air-brake days, a hand-brake wheel was placed near the conductor's chair in the cupola. There was a signal lamp on top of the Bombay-roofed cupola. The car had a Wythe speed recorder. The arch-bar Thielsen-style trucks had metal bolsters and elliptic springs. The body measured 30 feet long by 9 feet 1½ inches wide. The car was painted a buff

color with a black roof. The lettering was Tuscan red with drab shading. The interior had a pea-green ceiling and a vermilion floor.

These fancy western cars may have inspired some of the eastbound roads running out of Chicago to purchase what then, in only partial jest, were called Palace Caboose Cars.[110] In 1883 the following description of one of these luxurious land cruisers was published:

The finest freight cabooses on any road in this country are in use on the Chicago & Atlantic road, the new feeder of the Erie. They are built alike, and are 33 ft. long. On entering you will find a neat conductor's office fitted up with a desk and pigeon-holes over it for way bills, etc. A revolving chair is

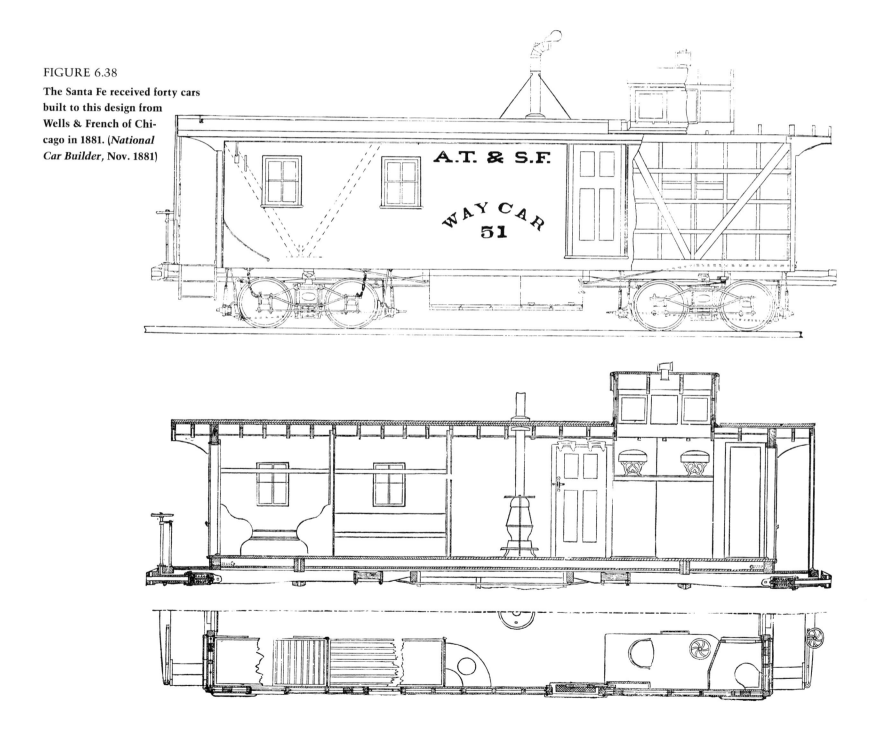

FIGURE 6.38

The Santa Fe received forty cars built to this design from Wells & French of Chicago in 1881. (*National Car Builder*, Nov. 1881)

CARS FOR SPECIAL SHIPMENTS

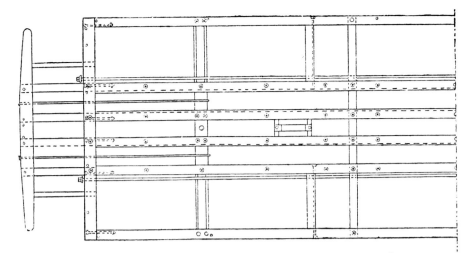

FIGURE 6.39

The frame and end elevation of the Santa Fe car shown in Figure 6.38 are reproduced here. Note the stylish Bombay roof of the cupola. (*National Car Builder*, Nov. 1881)

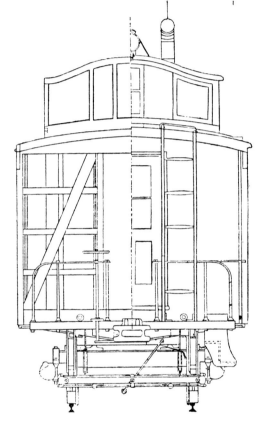

On the ceiling are six beds, or bunks, for the men, made in the same style as in the conductor's room. Long cushioned seats run along both sides, which are used for bedding, clothing, etc., each man having one of his own, with lock and key. The chains, links, pins, drawheads, wrecking frogs, etc., are all carried under the caboose in a box which opens on both sides. Two neat black walnut tables are made so as to let down at meal time from the side of the car. At one end of the car is the oilman's room. He has a zinc sink in which to fill lamps, and a waste box, lantern rack, oil cans, etc., all have a separate place in this little room. Opposite this room is a pantry. In here we find a large box, lined with zinc, for an ice box, and racks to place meat, etc., on during the trip, and a neat cupboard with racks for holding plates, cups, etc. There is also a place under the cupboard for the ironware used in cooking. The sink for dish washing is connected by a pipe with the water cooler in the caboose, and a neat earthen wash basin is set in the corner at the end of the sink, with a large looking-glass over it. A double hanging lamp lights up the centre of the car, together with a side lamp in each room. A time-table framed is in the conductor's room in the caboose, with a fine marine clock over it.

The Union Pacific may not have been ready to indulge its men with palace cabin cars, but it did provide large, comfortable-looking cars during the latter decades of the nineteenth century. Horizontal rather than the more common vertical siding and arch windows gave these cars a pleasing exterior appearance. In one photograph, way car No. 1112 poses at Gibbon, Nebraska, in January 1884.[111] This particular car sports an octagonal cupola.

CUPOLAS

The elevated lookout or cupola is the most distinctive feature of the American caboose. Its origins have been credited over and over again to a claim made in 1898 by an elderly Chicago & North Western conductor named T. B. Watson.[112] Watson relates that in the summer of 1863 he was required to turn over his caboose to another train and was ordered to make do temporarily with a boxcar. His substitute cabin car was not in the best of repair; it had a hole in the roof about 2 feet square. The day was fair, and so Watson decided to view the scenery by perching atop a box placed under the hole. With his head and shoulders above the roof, Watson could observe the train ahead clearly. He suggested that a "pilot house" be added to the railroad's cabin cars. The master mechanic at Clinton, Iowa, liked the suggestion and added cupolas to two cabooses then under construction. And so, according to legend, the cupola was born. Watson himself was not sure that this was so and closed his brief account by asking if any cupolas had been in use prior to 1863. His question was ignored by

placed in front of the desk, and a window on the side and end gives plenty of light. A long box is used for clothing, etc., with a cushion on top, making it a very easy seat. Over this is a bed, which lets down, the same style being used in the Pullman sleepers. Opposite this office is a closet for the use of the men. In the centre of the car is a regular cooking stove with four holes and hot water reservoir, and a railing around the top of the stove so nothing can slip off when going around a curve. By having this kind of a stove the trainmen can cook their meals while on the road, and live in good style and save money, and the company saves what time they would lose by waiting at stations for the boys to get meals while on the road. The coal box is made under the floor on the left-hand side of the stove, and by raising up a cover you find the box.

FIGURE 6.40

popular historians, who simply repeated Watson's recollection as if it were an indisputable claim for invention rather than the more modest account of an old railroader who only wondered if he had been the originator of the cupola.

Watson was a pioneer, but he was surely not the first, for a report published in October 1861 mentions a dozen cabooses on the Detroit & Milwaukee Railroad that were outfitted with "observatories."[113] The projections stood 2 feet above the roof and had windows for fore and aft viewing. Railroad cars with observatories appear to have predated the caboose itself. An engraving published in 1838 shows a passenger car on the Tonawanda Railroad with a fully developed cupola.[114] It was said to have been executed to the design of the railroad's engineer, Elisha Johnson.

The early introduction of the cupola did not mean a universal adoption, for the 1884 *Car Builders Dictionary* contended that the majority of cabooses did not have them. Nor was the word *cupola* commonly used at the time. *Lookout* was then in common usage, with *clerestory*, *dome*, or *observatory* being occasionally used. *Cupola* did not become the normal designation until around 1910.

Most eastern lines built cupolas with slant or inclined sides for better clearances, particularly in tunnels. Western and midwestern lines with better clearances preferred straight sides, while some roads like the Chicago Great Western had cupolas that overhung the car sides very much like saddlebags. The overhanging cupola, though ungainly in appearance, improved visibility and combined the advantages of the bay-window cabooses.

Cupola roofs were normally made in the form of a flattened arch, but some lines like the Pennsylvania indulged in very high arches after about

1905. Far more elegant was the Bombay roof, adapted from the practice with horse cars. The drawings of the Santa Fe caboose of 1881 reproduced earlier (Figs. 6.38, 6.39) illustrate this bizarre variant in cupola architecture. Other lines like the Port Arthur Route, the Kansas City Southern, and the Wabash also favored this design (Fig. 6.40). A 1904 Wabash car with a Bombay cupola is preserved by the National Museum of Transport, St. Louis, Missouri.

Cabooses with cupolas tended to develop sagging roofs. Car builders attempted to overcome this defect by mounting the cupola on the body's side frame, thus placing less weight on the carlines. Other car builders sought to dispense with the cupola altogether by installing bay windows. Trainmen could view the sides of the train from these shallow windowboxes, which projected out slightly from the car's side. The side-bay caboose was introduced on the Akron, Canton & Youngstown Railroad in 1923, but the idea of bay windows for railroad cars was far older. A German engineer observed them in use on the New York & Harlem Railroad during an 1859 visit.[115] They were used on passenger cars so that conductors might better anticipate the train's arrival at stations en route. These rather pronounced lanternlike metal boxes tended to get knocked off in service. The same idea, handled more subtly, was revived and patented in 1877 by George S. Roberts of Meredith, New Hampshire. Other inventors tinkered with the scheme during the last decades of the nineteenth century, but there is no evidence that it was applied to cabooses until the 1920s.

SAFETY FEATURES

The safety function of the caboose was to protect the rear of the train—to serve as a beacon of

FIGURE 6.41

Advertisement for a revolving car
lamp. (*Poor's Manual of Railroads*,
1884)

light and as a lookout station for men alert to misplaced switches, being overtaken by a following
train, a break-in-two, and the other hazards that
trains are liable to. But despite its intended safety
function, the caboose was ironically the most dangerous place on the train. Even the heaviest
wooden frame offered scant protection against a
speeding locomotive. The weekly accident reports
published by the *Railroad Gazette* and *Railway
Age* recount many accidents involving cabooses.
Some were minor, others disastrous.

One such incident that occurred in October
1882 resulted in twenty-one injuries and possibly
eight deaths.[116] Normally, no more than three or
four men would ride in a caboose, but in this instance thirty laborers were riding in a cabin car on
the Hoosac Tunnel line. The caboose was being
pushed from behind when it rammed into another
locomotive, in part because of poor visibility.
Pushing cabooses was a dangerous practice prohibited on many lines. It was particularly hazardous when pusher engines were engaged in helping a heavy train up a grade. But switching the
caboose to the rear of the pusher engines and then
recoupling it to the train was a troublesome operation that was occasionally ignored by lazy crewmen. The time and trouble saved often resulted in
death and destruction. Four-wheel bobbers had a
tendency to derail, and the frames of even the
largest eight-wheelers were known to collapse under the strain of pusher service. Surviving photographs of splintered cabooses are a graphic reminder of the hazards of train service during the
wood-car age.[117]

It was not until the final years of the nineteenth
century that much energy was devoted to solving
this dismal problem. In that year H. H. Sessions, a
leading car designer and engineering head of the
Pullman car works, offered a steel end platform
aimed at eliminating this paramount weakness of

caboose construction.[118] Sessions's patented design was an adaptation of his earlier steel passenger car end platform. The device was adopted by
the Illinois Central, the Rock Island, and other
lines. Yet steel platforms, steel underframes, and
even all-steel bodies never made the cabin car a totally secure haven. So long as human error remains
a part of railroading, the end of the train will continue to be a vulnerable position.

Since the caboose could never be made crush-
proof, a few inventors believed that its security
could be improved by more prominent lighting devices. A smoky marker lamp is not particularly
visible through fog or starlight. The preference for
large or multiple signal lamps is obvious from illustrations in this volume and those reproduced in
William Knapke's 1968 book, *The Railroad Caboose*. Railroad tinsmiths and commercial producers like Adams & Westlake or M. M. Buck devised a variety of lamps meant to be seen and
noticed. Rotating lights, forerunners of modern
Mars lamps, were tried as early as the 1870s, as
mentioned earlier in this chapter. A later variety
was advertised by the Revolving Car Signal &
Lamp Company in *Poor's Manual of Railroads* for
1884. An axle drive unit furnished the power to rotate the lamp (Fig. 6.41).

Another safety device applied to cabin cars during the 1870s engendered hostile crew reactions.
Speed recorders—derisively nicknamed Dutch
Clocks by trainmen—were favored by at least a
few railroad managers as a way to check on freight
train operations. Rotating marker lamps were tolerated because aside from the noise of the drive
mechanism, they were inoffensive. But having a
speed recorder was like having a spy on board who
was ready to tattle about every minor rule infraction its little mechanical inscriber could scratch
down.[119] The Dutch Clocks developed consistent
mechanical problems, which the crews blamed on
poor design but which the maker ascribed to sabotage. One of the most widely used of these devices
was patented on July 28, 1874 (No. 153,470), by
W. W. Wythe of Meadville, Pennsylvania. Wythe's
invention was being tested with good results during the following year on the Lake Shore, the Jeffersonville, Madison & Indianapolis, and the Terre
Haute & Indianapolis railroads.[120] The Santa Fe's
use of Wythe recorders in 1881 indicates that the
device was fairly widely employed.[121] The Railway
Speed Recorder Company of Cleveland was another maker of Dutch Clocks, supplying apparatus
to the Michigan Central, the Union Pacific, and
the Rock Island.[122]

COACH-CABOOSES, DROVERS' CARS, AND WRECKING CARS

Some cabooses were designed as multiple-use
vehicles. This special class was often a combined
cabin, coach, baggage, and express car. Some were

FIGURE 6.42

Jackson & Sharp produced this coach-caboose for the Mahoning Valley in 1891. (Delaware State Archives, Dover)

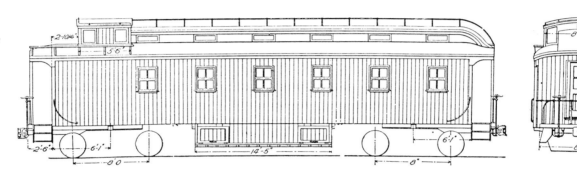

FIGURE 6.43

A Burlington boarding or drovers' car dating from about 1890. (*Car Builders Dictionary*, 1895)

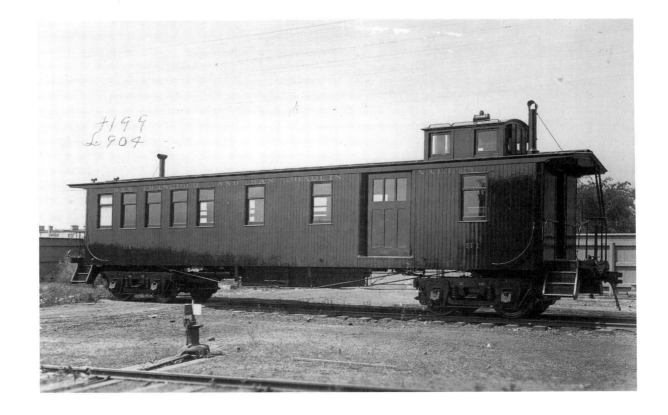

FIGURE 6.44

Pullman built four drovers' cabooses in 1898 for the San Francisco & San Joaquin Valley Railway. The exteriors were painted vermilion. (Pullman Neg. 4199)

CARS FOR SPECIAL SHIPMENTS

center seat was installed so that as many as sixty passengers could be carried within the 30-foot-long body. The coach-cabooses had been retired by the early 1930s and were sold to Paramount Pictures in 1938. One of them is now preserved by the state of Nevada at Carson City. The original specification, reproduced below, is preserved by the Bancroft Library of the University of California, Berkeley:

Dimensions: Car to be 30 feet long, 9 feet wide, 6′4″ post.

Material: Oregon Pine Frame, sheathed with matched redwood inside and out. Oregon Pine floor 1½″ thick, planed and matched.

Roof: of matched redwood covered with good roofing tin and to have two coats paint.

Interior: of car to be finished as per drawing, stove set up. Covers on seats and all complete.

Platform: to have hand rails cast iron drawbars—with rubber springs, timber of good oak.

Trucks: of the kind used in the flat cars with double brakes, hand wheels at each end of car.

Paint: well painted inside and out, to have three coats and varnish.

Less well documented is a coach-caboose built in 1891 by Jackson & Sharp for the Mahoning Valley Railroad. It was more highly finished than the average caboose, but it was surely not a fancy car (Fig. 6.42). The arch-bar trucks had leaf springs for an adequate if not elegant ride. The absence of a clerestory and the presence of a cupola marked this car as a service vehicle. Yet it would hardly have been lettered at one end for baggage if it had not also been intended for revenue passenger service. A very similar car is pictured in a catalog issued by the American Car & Foundry Company in about 1900. It shows a long coach-caboose, No. 117, for the Bangor & Aroostook. It too has arch-bar trucks and a low cupola, but the roof is made in the clerestory pattern. It is possible that the No. 117 was used to carry potato car stove tenders on return journeys from the potato fields.

The Chicago, Burlington & Quincy exhibited a partiality for coach-cabooses, acquiring a rather sizable fleet of what it called boarding cars before 1900. Most or all of these vehicles had been rebuilt from old coaches, some of which dated back to the 1860s.[124] Most had clerestory roofs, cupolas, and bodies 40 or more feet in length. They were very sizable for way cars, which tended in the main to be rather stubby affairs. They rode on long-wheelbase passenger-style trucks and were fitted with high handrails along the rooftop and large toolboxes under the floor. In all, they were a most curious-looking combination of freight and passenger car design elements. A simplified drawing

FIGURE 6.45

This interior view of the drovers' car illustrated in Figure 6.44 shows a section made up for sleeping. The berths nearest the camera remain tight against the ceiling. (Pullman Neg. 4200)

custom-built for this service, while others were old coaches or passenger combines remodeled with a cupola and other features thought necessary for cabin car use. In a few instances, such as the East Broad Top Railroad, an old but unaltered combine car was simply assigned to caboose service. Any passenger, mail, or express that happened along was loaded aboard. Coach-cabooses normally operated on branch lines; however, some hermaphrodites could be found on the main line, though passengers were normally limited to drovers riding along with cattle trains.

The oldest surviving coach-caboose is a car completed in February 1873 by the Kimball Car Company for the Virginia & Truckee.[123] Two cars were produced for $1,900 each. They had arch roofs, side doors for baggage, and longitudinal bench seating for twenty-two passengers. A few years later, the side doors were removed and seating was increased to thirty-five. Around 1890 a

FIGURE 6.46

One end of the San Francisco & San Joaquin Valley drovers' car was set up as a coach. (Pullman Neg. 4202)

Drovers' cars appeared rather early and were known to be in existence before the western railroad movement was under way. While such cars are normally associated exclusively with far western lines, the earliest dated references to such cars are for eastern roads. In October 1859 reports of three drovers' cars on the Michigan Central Railroad speak of a stoutly built car with plush "couches," pillows, and other bed gear for sixteen men.[125] A mirror and washstand were also thoughtfully provided by the car's designer, S. C. Case, master car builder for the railroad and a sleeping car pioneer. Later, in 1859, it was announced that the Erie was operating drovers' cars, but regrettably no other details were offered.[126]

While the eastern origins of the drovers' caboose are certain, it was a style of car widely favored in the West. An early example was built in 1872 for the Central Pacific at the road's Sacramento shops.[127] The car measured 52 feet 2 inches over the couplers and had a 46-foot-long body with twelve windows. It was fitted with side doors, end platforms, a cupola, and passenger-style trucks. It was painted orange-red with white lettering.

Very similar cars were produced by Pullman in 1898 for the San Francisco & San Joaquin Valley Railroad (Figs. 6.44, 6.45, 6.46). One section of the car was fitted with reversible coach seats; however, no plush or fancy upholstery was expended on a lot of dung-laden cowpokes. Pieced wood-veneer seats were thought adequate for the clientele of these cars. In a second compartment normal side bench seats provided more daytime seating. Overhead bunks could be dropped down at night. The lower benches with deeply upholstered cushions provided yet another bed.

Possibly the largest drovers' cars, at least in terms of sleeping capacity, were constructed in 1908 by the Santa Fe's Topeka shops.[128] These cars were only 40 feet long, yet they contained ten sleeping sections and two toilets. Forty beds were provided, specifically for stockmen traveling to the Kansas City markets. These cars were not, incidentally, combination cabooses but were intended exclusively for drovers and were coupled behind the cabin car.

The final example of a drovers' caboose to be considered here is a design developed in 1898 by the Illinois Central.[129] These cars provided seating for twenty-four and sleeping berths for half that number. The seats and berths were Spartan versions of Pullman's standard open-section sleeping car accommodations. While the finish was plain, each car was provided with blankets, linens, and toilet articles. Heating stoves were placed at each end of the car.

One other variety of cabin car must be mentioned before closing our discussion of cabooses. Tool or wrecking train cars became a necessary

of a boarding car is reproduced here as Figure 6.43.

Drovers' cars were closely related to coach-cabooses in size and appearance, although their intended use was less eclectic. These elongated way cars were meant to offer temporary housing for cattlemen who traveled with their stock to market. Drovers were needed to load and unload cattle at feed and watering stations along the line as well as at the origin and termination of the trip. Because such trains might be on the road for days at a time, it was necessary to provide sleeping, eating, and resting areas for the cowhands. Train crews were expected to share the same cars, although the interior space was often segregated at the passageway to the cupola. The train crew normally occupied the rear section of the car. The kitchen and toilet areas were usually shared. The finish of drovers' cars was uniformly plain and utilitarian, a fact obscured by the occasional use of the term *drovers' sleeping car.*

FIGURE 6.47

piece of equipment on almost every railroad property, because accidents were such a certainty. The oldest view of a wrecking car uncovered to date is a tintype of a work train on the Rome, Watertown & Ogdensburg Railroad dating from around 1870.[130] The car appears to be a converted boxcar and is lettered "wrecking car No. 12." It has side doors, a wide step, and two small windows on each side. A stovepipe indicates the presence of heating apparatus. There is a large box on the roof that may be a primitive cupola or a signal box, but the image is too small and indistinct to make this out clearly. A later custom-built wrecking or work train caboose made by the Pennsylvania Railroad in 1888 is currently on display at the Indianapolis Children's Museum. The No. 60 was built to class TA plans introduced in 1877. The 41-foot-long eight-wheel car looks like a conventional cabin car, except for a few minor details. There are, for example, no rail-

ings on the end platforms, presumably to facilitate the removal of large tools. The 33,000-pound car rides on passenger-style trucks. It was retired and restored at the Altoona shops in 1939 for exhibit at the New York World's Fair (Fig. 6.47).

Today, the caboose appears to be doomed to extinction. The railroad industry contends that cabooses are expensive and unnecessary. Electronic black boxes—called end-of-train devices—mounted on the rear coupler of the last car are said to fulfill the safety functions of the old-fashioned caboose. A flashing white light marks the end of the train, while a radio reads the air pressure. The train crew is obliged to ride ahead in the engine cab. And so the once-ubiquitous caboose appears destined to join the steam locomotive and the roundhouse as a relic of American railroading.

FREIGHT CAR TECHNOLOGY

The largest single element of a freight car is its body. Chapters 2 through 6 described the variation in car bodies in some detail. More on the same subject will follow in Chapter 8. This chapter therefore is largely devoted to the second largest component of a freight car: the trucks.[1] Other important auxiliaries such as couplers and brakes will also be treated here.

Trucks

Trucks can be described as a small four-wheel car—an assembly of wheel sets, frame, bearings, and a center plate—that transfers the load of the car body to the rails. Trucks, one at each end of the car, carry and lead the car along the track. A car truck might be described as a weight transference and guidance device. During the nineteenth century freight car trucks ranged in design from elementary unsprung wooden primitives to complex, well-sprung creations made up of an intricate cluster of iron bars, rods, and castings. The subject of truck design attracted the best and worst talent of the railroad engineering world. The ideas and plans proposed and practiced exhibited moments of inspiration, longer mundane interludes, and, as in all human affairs, considerable personal prejudice and perversity.

It is tempting to summarize truck development as a simple evolution from wood to all-metal construction, with each decade witnessing the introduction of bigger and better trucks. On closer examination, we find the story more complex. The definition of good truck design seems always to have been in flux. A railroad would adopt metal-frame trucks as its standard only to revert to wood beams a few years later. A new master car builder could upset years of standardization. Even well-established car builders were tempted to experiment, and so consistency, logic, and predictable patterns are difficult to uncover. This is particularly true when it comes to details like bolsters, center plates, and journal boxes, but then again it is this very diversity that makes the subject of car trucks so fascinating.

There were broad areas of agreement, to be sure. Most railroads remained loyal to wood-frame trucks until around 1870. Only then did the arch-bar truck come to prevail. Short-wheelbase trucks were universally favored for freight car service, from the beginning until modern times. There was a gradual elongation of the wheelbase; in the pioneer period 42 inches was common; by the time of the Civil War, the square truck had been adopted and the wheelbase equaled the track gauge. This measurement expanded, until by the middle 1880s 5 feet was the recognized wheelbase. Most roads had added another 6 inches by 1900. A gradual growth in overall size was also evident with the passing decades. This might best be indicated by weight increases. In 1870 most individual trucks weighed around 2 tons each. By 1880 an extra 500 pounds was common, and so on to the end of the century, when most trucks weighed just under 3 tons each. Much of this extra poundage was accountable to heavier wheels and larger-diameter axles, but all-metal bolsters and thicker/wider side-frame members added to the gross weight.

WOOD-FRAME PIONEERS

Historically, the earliest freight car trucks were a classic product of the ligneotechnic culture, where nearly all useful things were made from timber. Wood-frame trucks were perfectly serviceable for small cars operated at slow speeds. The same economic arguments that made wooden trestles, houses, and railway carriages so practical and cheap held true for trucks. Excellent virgin lumber was readily available at giveaway prices, whereas iron was both scarce and dear, and so any sensible mechanic would make as much of his mechanism as possible from wooden elements.

Another important truism for the first few decades of American railroading is that both freight and passenger car trucks were made on an almost identical pattern. Simple, square wooden frames and short wheelbases prevailed. Even when all-metal trucks were being promoted in this period, little or no distinction was made between freight and passenger applications. Advertisements in the

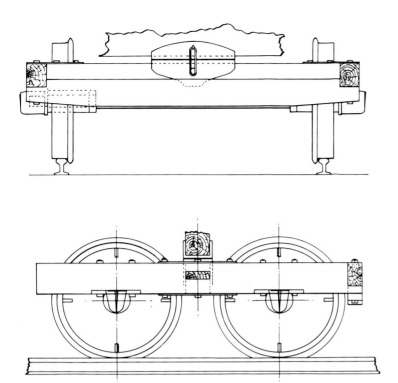

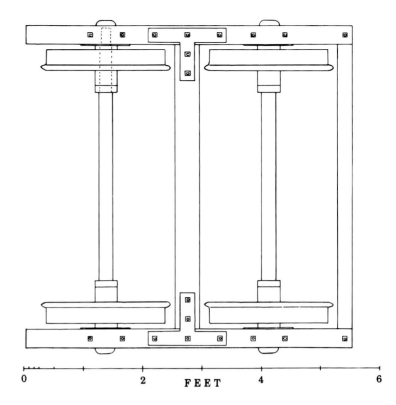

FIGURE 7.1

The American freight car truck in its most elementary form is represented by this Delaware & Hudson Canal Co. drawing. The drawing dates from around 1850. (Copied from an original drawing at the John B. Jervis Library, Rome, N.Y.)

American Railroad Journal during 1845 for the Davenport & Bridges patented car truck show it applied to both types of car. By about 1850 a noticeable schism had developed. Passenger cars had elongated, and bigger trucks were needed to support them. Longer wheelbases, swing-motion bolsters, equalizing levers, and more elaborate springing were applied. No such attention was lavished on the freight truck, for no one cared how well it rode so long as the cars did not derail or destroy their cargoes. And so the freight car truck designs that remained in force on some lines into the 1860s looked very much like the universal trucks used for both passenger and goods cars twenty or more years before. Examples are illustrated throughout Chapter 2, and more will be given in the pages of this chapter.

The Delaware & Hudson Canal Company supplies an example of the elemental four-wheel truck in its most primitive form (Fig. 7.1). The drawing shown here is traced from an original in the Jervis Library in Rome, New York, and while it is not dated, we can surmise that it is not from the years when John B. Jervis was employed by the D&H—1825–1830. The D&H railroad was a narrow-gauge gravity line intended mainly for coal traffic. The original cars were small four-wheelers as described in Chapter 2. They were not replaced until after 1847, when the railroad was extensively remodeled. Although the original track was wooden strap rail, T-rail is shown in the drawing. The plate wheels, rather than the older spoked variety, indicate a design more of the 1850s than the 1830s. Note the long wheel hubs and collars on the axles, which indicate loose wheels—a practice at

variance with the normal method of fixing the wheels so that they turned with the axle. This was done in an effort to minimize flange wear on the sharp curves of the gravity road. A long link, used to connect the truck to the car in place of the conventional king pin, is but another peculiarity of this truck. Because speeds were moderate, the riding quality of the car was unimportant, and so these 51-inch-gauge trucks were made without springs. The axle boxes were bolted directly to the wheel pieces (side beams). The connection between the bolster (crossbeam) and the wheel pieces was mortised; added strength was achieved at these junctions with heavy T-irons. An end piece is used only at one end, another singular feature of this odd little truck. This was presumably done to clear the hopper at the open end. The wheelbase was a minimal 33 inches, and the wheels were very small, measuring only 24 inches in diameter. This truck might be said to represent the first thinking for an American freight car truck.

One of the oldest datable drawings for an American car truck is shown in Figure 7.2. The original tracing, in rather crumpled condition, was found among the Winans papers at the Maryland Historical Society, Baltimore. It was redrawn to show the details of construction more clearly. The original sketch was prepared in 1833 as part of a proposal to provide trucks for the newly organized Hartford & New Haven Railroad. The truck is remarkably similar to a design worked out at about the same time for the Baltimore & Ohio Railroad's Washington Branch cars. Compared with the later D&H trucks, the Washington Branch bogie was a model

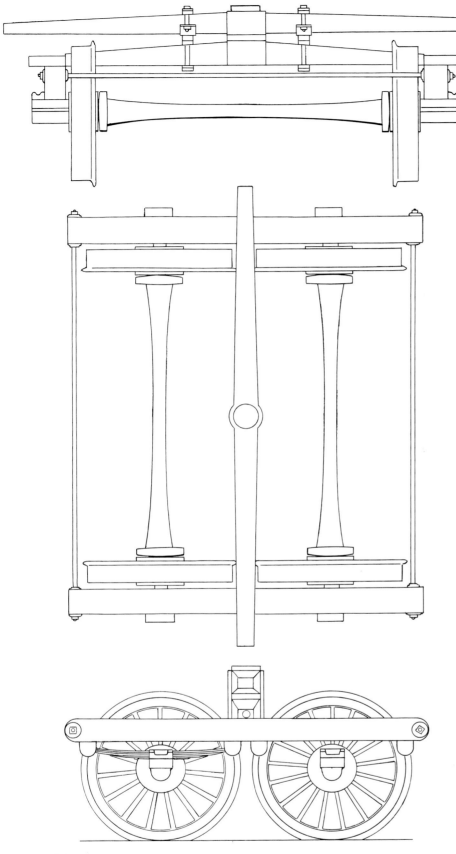

FIGURE 7.2

A truck design proposed in 1833 by George Gillingham and Ross Winans for the Hartford & New Haven Railroad. (Redrawn from an original tracing at the Maryland Historical Society, Baltimore)

of sophistication. It featured double springs, roller side bearings, and wrought-iron bolsters. Regrettably, the first two features are not shown in the truck drawing, but they do appear in the full assembly drawings of the cars.[2] The side pieces are the only major wooden elements in this truck. The rods held the side pieces parallel. The general design is straightforward, but there is one puzzling detail that is difficult to explain: the convex-shaped collars, mounted on the axles against the wheel hubs. Why were they made in this shape rather than being flat? One possible explanation is that the wheels were loose on the axles and that the convex collars created less friction because less surface came into contact with the hubs. The wheels were 31 inches in diameter. The wheelbase was 40 inches.

This general design as revised for freight car service in 1834 or 1835 is depicted in Chapter 2 (Fig. 2.14). Cast-iron bolsters replaced the more costly wrought-iron variety, and wooden end beams, mortised into the side pieces, replaced the wrought-iron tie bars. The roller side bearings were eliminated, as was the double springing. This process of simplification is a common accompaniment to any new technology; elements originally thought necessary prove to be superfluous after a few years of experience.

Another truck from the pioneer period is recorded in the patent of J. Dougherty (or Daugherty, as it is sometimes spelled) of 1843 for a heavy flatcar to transport canal boats over the Portage Railroad. The car, with its trucks, is illustrated in Chapter 2 (Fig. 2.19). Because of the anticipated loads, the truck frame was heavier than was typical for that time; however, it appears to have followed a basic plan of the 1840s. A fixed or rigid bolster with leaf springs above the journal boxes was common for the period. Coil springs were not used until many years later, and car builders seeking a better method for attaching leaf springs to the journal boxes devised a way to place them below the wheel piece. The Baltimore & Ohio worked out such a scheme in the 1830s. In this case the springs were bolted directly to the oil boxes. The plan persisted until at least the late 1850s, when drawings of a B&O box were published in England—see Chapter 2 (Fig. 2.27). The design was simple but not remarkably good in a mechanical sense, because unequal flexing of the long leaf springs would allow the wheel sets to swing slightly askew so that the flanges were no longer perpendicular to the rails. A better, though more costly, scheme involved mounting cast-iron pedestals on the truck frame to serve as guides for the oil boxes. The Illinois Central boxcar of around 1855 illustrated in Chapter 2 (Fig. 2.21) shows this plan.

By the late 1840s the New England Car Spring Company was offering master car builders a new

FREIGHT CAR TECHNOLOGY

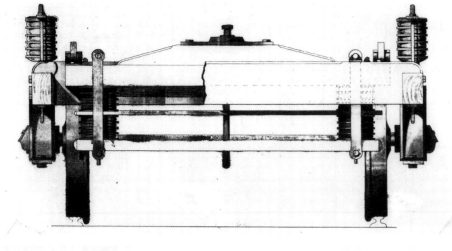

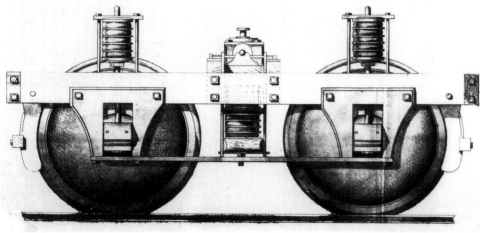

FIGURE 7.3

This wood-frame car truck of 1848 features maximum usage of India rubber car springs. The swing-motion bolster and roller-style side bearings are worthy of notice. The drawing was published for the New England Car Spring Co. of Boston. (*American Railroad Journal*, Sept. 16, 1848)

solution to truck suspension, one that was claimed to be cheaper than tempered leaf springs.[3] India rubber, vulcanized for long life and greater stability, had entered the market and was being adopted for every conceivable application, including that of railway car springs. The design offered by New England Car seemed intent on using the maximum number of rubber springs (Fig. 7.3). Notice that the springs were built up of alternate rubber and iron discs with a hole at the center for an iron rod to pass through. The springs over the axle boxes stood high and could work only if placed between the outside and intermediary sills of the floor frames. Another set of springs was placed between the bolster and the spring plank, well inside and below the truck frame. Notice the very long swing hangers in front of the bolster springs. With a 4-foot wheelbase, this truck could have been used for both freight and passenger cars. A very simple truck suspension could be made by placing a rubber spring on top of the oil box in the open space in the pedestals. An example of such an arrangement is shown by drawings for a New York & Harlem Railroad truck published in the 1879 *Car Builders Dictionary* (Figs. 88–90). Surely the design does not represent current practice for the late 1870s. The truck exhibits features more relevant to the 1850s or 1860s, and it agrees very closely with

trucks on the 1869 Portland Company flatcar depicted in Chapter 2 (Fig. 2.42).

If simplicity and plainness were the mark of good truck design, nothing could compete with the wood-beam trucks produced in 1865 by Harlan & Hollingsworth for the Philadelphia, Wilmington & Baltimore Railroad (Fig. 7.4). The 44-inch wheelbase brought the flanges within inches of the rigid bolster. The small axles, 3¼ inches at the centers, indicate that these trucks were intended for a light car. The outside-hung brake beams were supported by both chains and hangers as a safety measure to ensure that the beams would not fall down on the track if one of the hangers failed. One modern feature was the employment of coil springs; they were encased in a cylindrical canister above each oil box.

Not all railroads were content with so elementary a truck for freight service. The Pennsylvania Railroad, for example, used a truck in the 1860s that incorporated several novel and ingenious features (Fig. 7.5). Just how old the general arrangement may be is uncertain, but it may have predated our drawing, which is dated 1867. The wooden frame is conventional enough, but notice that the bolster fits loose between the two parallel crosspieces called transoms. The loose or secondary bolster rests on inverted, half-elliptic springs and is thus free to float up and down. Lateral motion is also possible because of the rollers under the ends of the springs, and so we encounter an early and very clever plan for a swing-motion truck. The cup-shaped pads under the rollers act as a centering device. The spring plank is stationary and is bolted to the wheel pieces by three bolts on each side of the car. An open iron casting is bolted between these two wooden beams. The suspension is supplemented by rubber springs on either side of the journal boxes. A short equalizing lever, bent in the shape of a flattened U, spans the springs and journal box. This part of the suspension appears to have been given up as an unnecessary refinement, but the half-elliptic spring and floating bolster remained popular on the Pennsylvania Railroad for some time. The roller swing motion was also abandoned, and for heavier cars and tenders the spring plank was strengthened with a king post truss rod. As late as 1879 drawings for this style of truck spring and bolster were published as standard on the Pennsylvania.[4] By this time the wooden frame had been replaced by arch bars; however, the bolster and spring plank remained wooden.

Other railroads adopted wood-frame trucks very similar in appearance to the Pennsylvania truck just described but on a somewhat simpler plan. The U.S. Military Railroad, which happened to be staffed by a goodly number of Pennsylvania Railroad men such as Herman Haupt and Tom Scott, adopted a suspiciously similar truck that is evi-

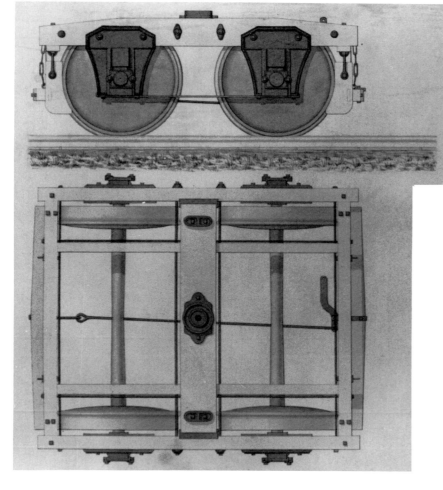

FIGURE 7.4

Wood-frame trucks with rigid bolsters and short wheelbases were the mainstay of American railroading during its first decades. This drawing was produced in April 1865; the design is far older. The wheelbase was 44 inches. Davis spiral springs were placed over the axle boxes. (Smithsonian Neg. 81-8111)

dent in often-reproduced photographs of Civil War railroad scenes. A drawing exists as well, drawn in March 1865 by Harlan & Hollingsworth for the USMRR.[5] The wheelbase is 4 feet 6 inches—just a trifle less than the Pennsylvania truck. The floating bolster is used, but in place of the half leaf springs, canister or rubber springs are used. The side frame is stiffened by an inverted arch bar that runs under the bolster and on top of the tie bar that connects the pedestal bottoms. Drawings for a nearly identical Central Railroad of New Jersey truck appeared in the 1879 *Car Builders Dictionary* (Figs. 92–94). Again, the elder design and short wheelbase—only 42 inches—suggest that the CRRNJ had not submitted its most current design when material was being gathered for the Master Car Builders Association's basic reference work. This truck was somewhat updated by having canister coil springs.

Photographs of early car trucks are almost nonexistent, but such a rare image was recently uncovered in the collection of the late Andrew Merrilees.[6] It corresponds very closely with the USMRR truck of 1865 just described. The bolster and wheel piece are separated by two short pedestals made in the shape of a classical column (Fig. 7.6). The floating bolster is evident, but the springs cannot be seen because they are hidden away in-

side the wooden transoms. The truck was made by the Allison Car Works of Philadelphia. The purchaser and date are unrecorded.

While most other major railroads were preparing to abandon the wood-frame freight car truck for the diamond arch bar, the Baltimore & Ohio was making plans for a clumsy-looking wood-beam truck of massive proportions. The railroad had tried arch bars, but sometime in the 1860s while Thatcher Perkins remained in charge of the mechanical department, the wood beams began to appear. The earliest datable photographs for them are a series of locomotive views made for the B&O in 1864 by a young mechanical department employee named R. K. McMurray.[7] The tenders of just about every engine in the series had the wood-beam trucks.

The first evidence of their application to a freight vehicle is a boxcar drawing credited to Perkins's successor, J. C. Davis, made in September 1866 (Fig. 7.7). At first glance the design seems to be a step backwards into the Dark Ages of railroad technology when cars with no springs rumbled over the tracks of eastern America. Don't be deceived, however, by the rude appearance of the four-stick subcarriage with its boxes bolted directly to the wheel pieces. In photographs the design appears simpleminded, but study the drawing, where it is plain that as in so many other early-nineteenth-century truck designs, the springs are hidden. The bolster is actually made in two pieces. The center is cut out and is free to move up and down. Nests of coil springs placed inside cast-iron cylinders or canisters are located just behind the wheel pieces. The weight of the car pushes down while the springs push up, and so the loose center part of the bolster floats up and down.

The design has some other interesting features worthy of study, such as the bolted connection between the spring plank, the bolster, and the wheel pieces. Notice that the bolts are outside the wheel pieces; if they were drilled to receive the bolts, the side beams of the truck would be greatly weakened. Mortising would make a neater connection between the major wooden pieces; an example of this construction is the Delaware & Hudson truck

FIGURE 7.5

This wood-frame swing-bolster truck is dated 1867, but the general design is believed to be earlier. (J. E. Watkins, *History of the Pennsylvania Railroad* [proof sheets, 1896])

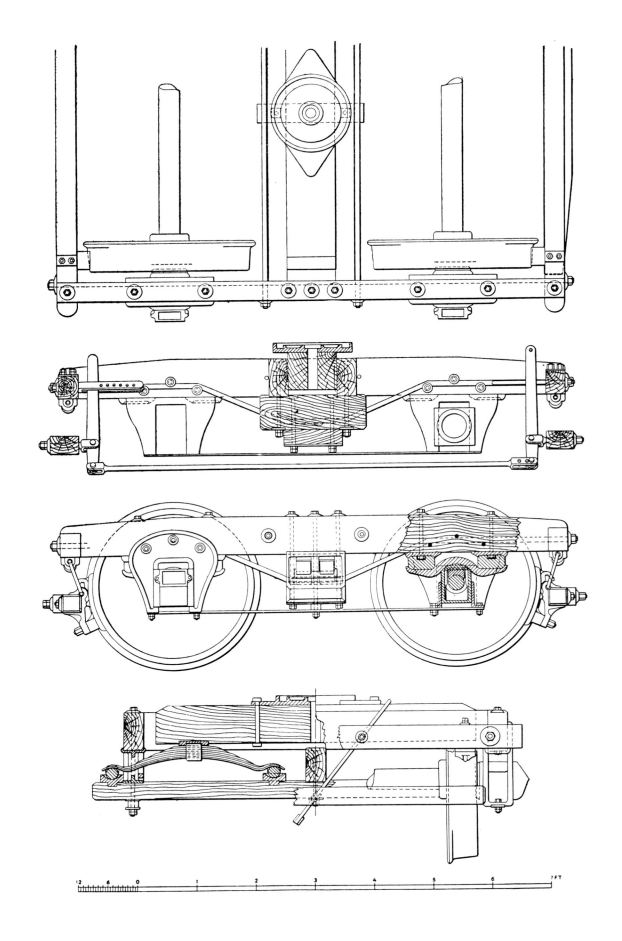

438   THE AMERICAN RAILROAD FREIGHT CAR

FIGURE 7.6

This wood-frame truck was built for an unknown railroad by the Allison Car Works of Philadelphia around 1860 or 1870. Note the safety straps for the brake beams and the ribbed single-plate wheels. (Collection of A. Andrew Merrilees)

FIGURE 7.7

Sometime in the 1860s the Baltimore & Ohio adopted this heavy-timber style of truck. Note the floating bolster and case-style springs. (Copied from a drawing prepared for J. C. Davis, Sept. 1866)

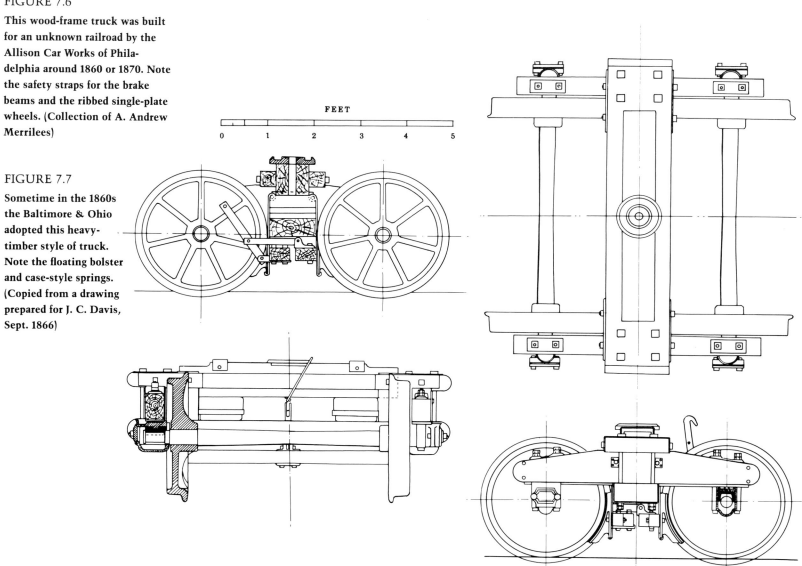

FEET

0   1   2   3   4   5

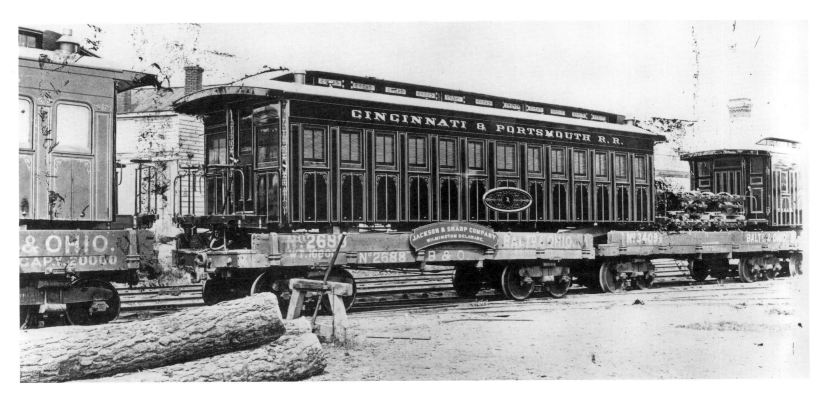

FIGURE 7.8

The wood-beam B&O-style trucks are shown at the Jackson & Sharp Car Works of Wilmington, Del., in 1877. The truck design was abandoned at about this date for new construction. (Hall of Records, Dover, Del.)

shown in Figure 7.1. But it weakens the structure by cutting away a large portion of the timber just where the connection is made. Placing the cross-pieces on top of the side pieces means that nothing need be cut away—though the wheel pieces have a small notch cut into them to receive the bolster. The assembly looks bad, but it is much stronger than a mortised assembly. The wheelbase is a scant 3 feet 11 inches. The wheels are somewhat undersize as well—only 31 inches in diameter; but do not mistake them for spoked wheels, for they are not. The apparent spokes shown in the drawing are actually strengthening ribs for the single plate that connects the hub and rim of the wheel.

Why did Perkins revert to a wood-frame truck? Actually, it hardly seemed retrogressive at a time when passenger cars were all fitted with wood-frame trucks. This practice continued until the end of the century. What was correct for passenger equipment was surely just as proper for merchandise cars. The actual motivation was probably one of economy. An equally serviceable truck could be fabricated from a cheaper material, so it was logical to select wood over iron. Even if the unit saving was modest, the total would grow to a considerable sum if thousands of trucks were involved, which was true over the fifteen or so years that the Perkins/Davis truck was a standard. The economics apparently was not overwhelming, however, for the B&O had few imitators. Perhaps the same independence that made the B&O favor pot hoppers and Camel locomotives is what inspired Perkins and Davis to stand by this distinctive wooden truck.

Davis proved to be an especially strong advocate of the design; a notice published in 1873 said that it remained the road's standard freight car truck.[8] The canister springs had given way to full-elliptic springs that were said to offer the necessary elasticity. The truck was praised as cheap and durable, but in less than three years the wooden truck was reported as abandoned and all new cars would have a more modern form—presumably meaning arch bars.[9] These lumbering reminders of the wooden age were long in evidence, however, on the B&O and its subsidiary lines. A photograph taken in 1877 shows several of these cars very much in service (Fig. 7.8). The fact that they had broad-tread wheels—a fact denoted by stenciling directly on the side frames—shows that they traveled far and wide. Tender trucks on older power carried the Perkins/Davis truck into the final years of the nineteenth century. Regrettably, none appears to have been preserved. The tender trucks on several early locomotives now in the B&O Museum in Baltimore are believed to be replicas dating from around 1927. These roughly made reproductions have one-piece bolsters and no springs. The bolts connecting the nominal spring plank and the bolster go through the wheel pieces rather than beside them, indicating that the replicators either misunderstood the original design or, in their haste to get the job done, did not bother to follow the correct plans.

Before moving on to iron trucks, it should be mentioned that Perkins carried his affinity for wooden-beam freight trucks to his new employer, the Louisville & Nashville. French engineers visiting the Louisville shops of the L&N after the U.S.

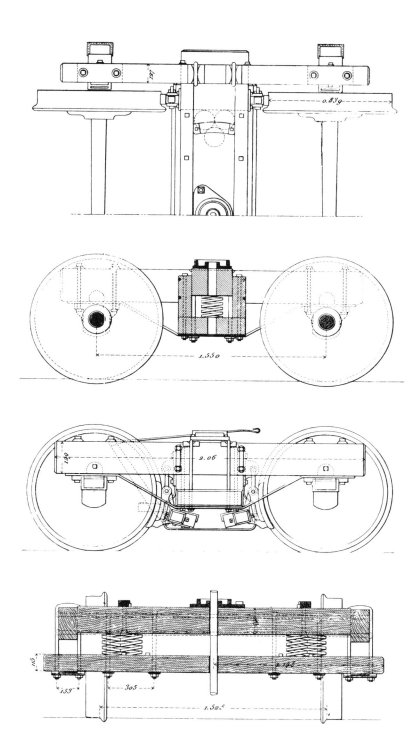

FIGURE 7.9

In the 1860s and 1870s the Louisville & Nashville developed a design very similar to the B&O truck shown in Figures 7.7 and 7.8. It too featured floating bolsters and hidden springs. The track gauge was 5 feet. (E. Lavoinne and E. Pontzen, *Les Chemins de Fer en Amérique* [Paris, 1882]; metric measurements)

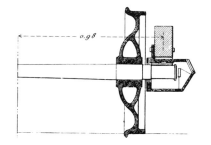

Centennial found wooden-frame trucks, very similar in design to those of the B&O, still being manufactured.[10] The wheel piece in this case was simply a square, wooden timber and had none of the niceties of shape-tapered and rounded ends that characterized the B&O version. The wheelbase was about equal to the 5-foot track gauge. Mechanically they were nearly identical, but by 1876 coil springs had become more common, and so they were adopted for the Blue Grass version. The Frenchmen were so taken with the novelty of the L&N truck that they published drawings of it in their report on American railway practice of the time, published in 1880 (Fig. 7.9). Hence an idea thought by many to be obsolete at the time of the Civil War, was presented as worthy of study by graduates of the Ecole Polytechnique.

EARLY METAL FRAMES

The strongest and most durable truck was one of all-metal construction. Few mechanics would argue over this point, but all would agree that the cost overruled such an approach for undercarriages. Most iron was imported, and the days of cheap steel would await Henry Bessemer. Meanwhile, an abundance of inexpensive white oak kept most master car builders loyal to wooden and semiwooden construction. A few mavericks attempted to push the cause of metal construction. The Baltimore & Ohio used a curious metal truck for both freight and passenger cars starting around 1835 called the live-spring truck. A large leaf spring served as both wheel piece and suspension. The axle boxes were bolted to each end, and the bolster was attached at the center by means of a cast-iron bracket. One of these strange contrivances is shown in Chapter 2 (Fig. 2.15). Loose and unstable as they might appear, they were steady enough for tenders, which, considering the weight involved, was no small job.[11] The Georgia Railroad had about two hundred freight cars in service as late as 1890 with live-spring trucks, though they appear to have had wooden rather than iron bolsters, according to an engraving published in 1891.[12]

An improved version of the same plan was laid down on paper by an anonymous Harlan & Hollingsworth draftsman sometime in the 1860s.[13] No specific customer is indicated on the drawing prepared by this Wilmington, Delaware, car builder —the only label specifies nothing more than steel springs—but the fact that the gauge scales out to 4 feet 10 inches suggests that an Ohio or New Jersey railroad was the intended customer. Metal brackets fasten to the end of the spring, and the axle boxes reach just beyond the wheels and are bolted to wooden end pieces. This would help hold the frame square and so stabilize the jumpy, live-spring truck. The wooden truck bolster is mounted in a massive cast-iron fixture at the spring's center.

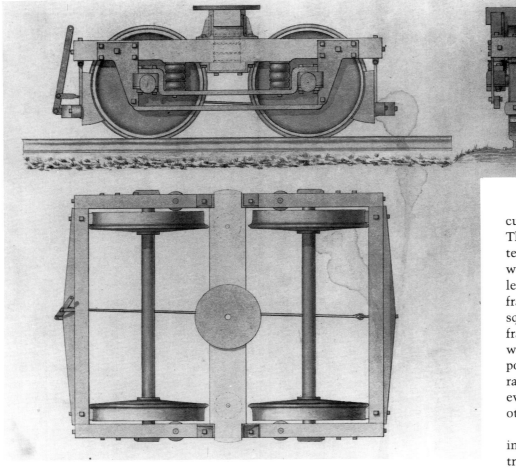

FIGURE 7.10

This peculiar iron and wood truck was produced in 1866 for the Philadelphia, Germantown & Norristown Railroad by Harlan & Hollingsworth. The wheelbase was 3 feet 11 inches. (Smithsonian Neg. 81-8138)

The creative designers at the Harlan plant devised another imaginative composite wood/iron truck in 1866 that is illustrated by Figure 7.10. Large wrought-iron pedestals and an equalizing lever are used in place of the conventional cast-iron pedestals. A heavy iron bar connects the lower ends of the pedestals. The external gum or rubber springs are plainly visible, but another similar spring is shown by phantom lines behind the wheel piece at the truck center. Here we have yet another example of the floating bolster. The short wheelbase—3 feet 11 inches—is a good indication that the truck was intended for freight service.

Charles Davenport, the pioneer car builder of Cambridgeport, Massachusetts, was a leader in introducing metal car trucks. His thinking on the subject was outlined in tangible form in U.S. Patent No. 3,697 (August 10, 1844). The side frame was composed of wrought-iron bars measuring $7/8$ by 3 inches. Davenport came very close to introducing the diamond-arch-bar truck, but in place of the critical inverted arch bar he supported the upper arch bar with two parallel flat bars and so lost the added strength offered by the opposing trusses that worked together in carrying the load imposed upon them.[14] Davenport's design included a large rocker side bearing and a pair of full-elliptic springs attached to the iron bolster. The tie bar was

curved down at its center to clear the lower spring. The bolster was made up of heavy flat bars fastened together to form an X. The center was made with a hole for the king pin, while the ends of the X legs were fastened to the side frames. The cross-frame bolster naturally did much to hold the truck square and so obviated a major criticism of the bar-frame truck. Davenport could boast of lightness as well as strength, for his design weighed 1,200 pounds less per pair. Placing the bearings inside rather than outside the wheels made the truck even more distinctive in appearance than it might otherwise have been.

A New Yorker named Fowler M. Ray, long active in promoting rubber car springs, tried his hand at truck design as well. We examined one of his wood-frame designs earlier in this chapter (Fig. 7.3), but he, like Davenport, wanted to introduce more iron into truck construction. In a patent dated March 21, 1845, Ray employed a large leaf spring with a small bar frame to separate the axle boxes. Roller side bearings and a long pedestal to connect the spring and frame completed the wheel piece. The bolster was built up of two wooden beams separated by spacer blocks. Very light, spoked, cast-iron wheels were employed.

The design was modified considerably, according to an engraving published a year following the patent (Fig. 7.11). The side frame was more like Davenport's semi-arch bar, except that the top rails were missing and the wheel piece and boxes were placed outside the wheels. A full-elliptic spring replaced the massive leaf top spring shown in the patent. Ray had also adopted disc or plate-style wheels in place of the old-fashioned spoked wheels. This is in fact the oldest or one of the oldest known illustrations of an American plate-style car wheel. Ray boasted that his trucks weighed only 1,000 pounds each, which is very light indeed, but the dependence on two bars for the wheel piece seems rather skimpy even for the light cars then in service. One can imagine it failing structurally or simply twisting out of shape. In addition, Ray must have used very light wheels and axles to keep the total weight so low.

Not long after the Davenport and Ray trucks

FIGURE 7.11

Fowler M. Ray of New York was
an early advocate of iron trucks.
The design shown here was
patented in 1845 and lacked only
the inverted strap iron to be a
full-blown arch-bar truck.
(*American Railroad Journal*,
Nov. 7, 1846)

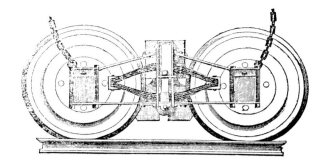

came onto the market another bar-frame truck appeared in New England that proved exceptionally long-lived and serviceable. Its originator and date are unrecorded, but it likely came into being during the 1850s. Men in the industry called it the continuous-frame truck because the side- and end-frame members were made up as a single bar of iron, and so the main truck frame was a single unit (Fig. 7.12). The top bar frame was made of 1- by 3-inch iron bar welded at the corners. It was in effect a copy of the wooden-frame truck side and end pieces made into a more slender but stronger metal model. The bolster was wooden, generally a 5- by 12-inch oak beam truss, with one or two rods. It was made as a plain timber without any mortises, all of which added to the simplicity and economy of the design—characteristics that made it very appealing to the New Englanders who became its champions. Notice that the bolster sits on top of the continuous frame and is attached by bolts and small arch-bar brackets. These connecting brackets also help stiffen the top rail between the pedestals. The inverted bracket has a king post, while the upper bracket rests on top of the bolster. Wooden safety beams, parallel to and just behind the wheels, were bolted to the bolster and ends of the continuous frame. They added a little stiffness to the frame, but their main office was to serve as a mounting for the axle safety straps. These straps were found necessary in the days of wrought-iron axles. The straps would support a broken axle and thus avert derailments.

It seems that the New England truck was bound for the graveyard when the diamond arch bar came into favor during the 1860s. Fitch Adams was determined to retire them when he became master car builder of the Boston & Albany in 1870, but after a short experience, he became convinced that the continuous frame was a very good truck despite its short wheelbase and old-fashioned appearance.[15] These trucks proved strong and square, and it took a tremendous blow to knock one out of alignment. They carried big loads and rarely developed a hotbox. Adams was certain that they could safely handle a 50-ton car, although nothing this large was running when he spoke on the subject in 1888.

William Voss praised the New England–style truck as well in his 1892 book on car building, say-

ing that the strong framing not only remained square but also kept the boxes from spreading.[16] He went on to say that the bolts stayed tight because of a good suspension—an important consideration in frost-prone regions, where it was common for track to freeze rigid and unyielding. On the other hand, the short spring space available above the oil boxes and the lack of any bolster springing would appear to have made them a hard-riding truck. V. D. Perry, a veteran car man for the New Haven, said nothing about their riding qualities, but he noted that they were popular with his shop men because they were so easy to repair.[17] Perry kept the wheelbase to 46 inches, a practice that appears to have continued until the 1890s. By this time most arch bars were stretched out to 60 or even 66 inches. Loren Packard, another master car builder for the New Haven system, found 44 inches a satisfactory wheelbase for such trucks as late as 1880.[18] The *Car Builders Dictionary* published a year earlier included engravings for a Boston & Albany continuous-frame truck with a 62-inch wheelbase. Several years later the B&A sought ways to perpetuate the New England truck by replacing the wooden bolster with twin steel I-beams.[19] The new design tested well and was found good for a 40-ton car.

Non-Yankee car builders might agree that the continuous-frame truck was a superior design in many ways, but they would grumble about the cost, $300 for a pair of freight car trucks—why, that was far too costly! Voss contended that it was never more than 16 to 20 percent more than an ordinary set of arch bars and was worth every penny of it because of the long service life and lower maintenance costs. For all of their virtues, the New England–style trucks remained a minority. All seem to have disappeared except for one lone example preserved by the Seashore Electric Railway at Kennebunkport, Maine.

DIAMOND ARCH BARS

The several flat-bar trucks just described form a link between the wood-beam and the diamond-arch-bar truck that by 1870 had come to dominate the American railroad industry. The diamond truck, so-called because of the gem shape created by the arched top and inverted middle bar of the three-bar side frame, was the single most important freight truck of the wood-car era. It remained popular for nearly half a century and continued to be applied into the steel-car period. Its popularity was based on its simplicity and strength. It could be made by relatively unskilled mechanics from common, cheap materials. The design of the side frame was perfectly rational. The two arches formed a strong truss well able to handle heavy loads. Because the top bar was in compression—not good for wrought iron—it was often made from slightly heavier stock than the lower or in-

FIGURE 7.12

The New England continuous-frame car truck was one of the earliest largely metal bogies developed for freight service. It was introduced around 1850 and remained popular for nearly fifty years. (William Voss, *Railway Car Construction* **[New York, 1892])**

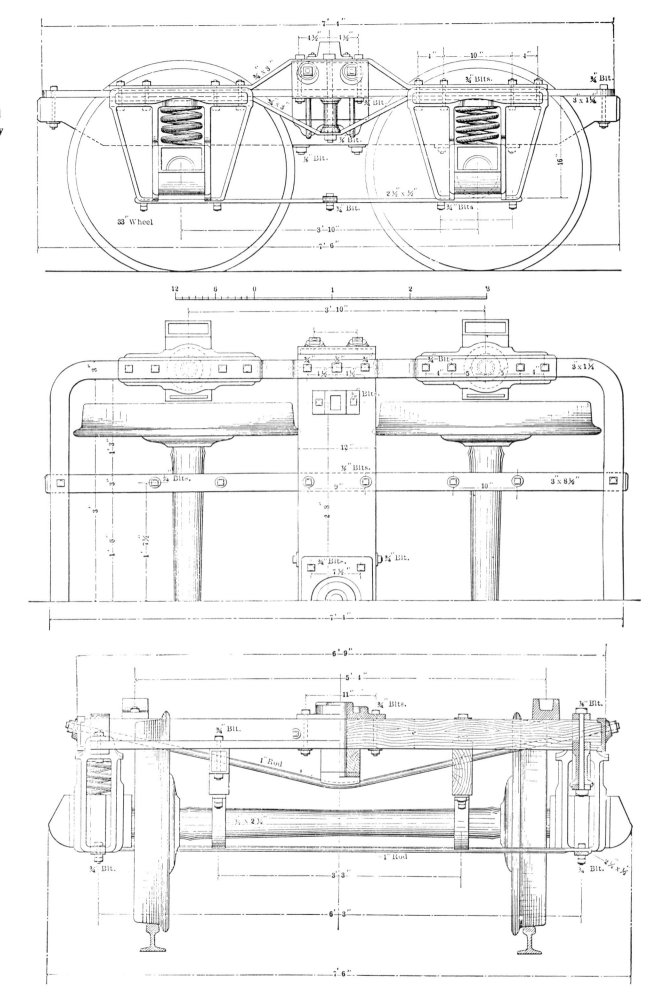

THE AMERICAN RAILROAD FREIGHT CAR

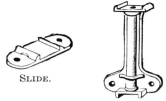

SLIDE.

STANDARD, OR COLUMN.

FIGURE 7.13

Two cast-iron columns, such as the one shown here, offered central support for arch-bar side frames. They also provided a guideway for the truck bolster. (*Car Builders Dictionary*, 1888)

verted bar, which being in tension held up well under loads. The bottom tie bar was not critical to the structure and served mainly to hold the boxes in line, and so it was made of light bars, generally ½ to ⅝ inch thick.

For all its good points, the arch-bar truck was defective in many ways that were well understood at the time. Possibly the main defect was the penchant for composite construction. The early arch bars were only a step toward metal construction. The insistence on using wood for the bolster and spring plank well into the 1890s greatly weakened the structure, making it unsafe and short-lived when subjected to very heavy loads. The wooden elements also contributed greatly to the notorious looseness of the structure, which caused it to run out of square and so led to dragging, extra friction, and excessive flange wear. And a flexible structure with so many bolted joints that was in almost constant motion was bound to loosen up. Keeping the nuts tight on a million and more car trucks was a goal never realized.

While the general plan of the truck was simple, some explanation of how it was put together seems necessary. A typical truck was made in this way. Flat iron bar stock, 1 inch thick and 3 inches wide, was cut to length by a shear. Holes were drilled or punched. The bar was then bent into shape by a bulldozer. The truck wheelbase was kept short, generally 5 feet from the centers of the axles, so that the bars might be kept relatively light. A longer wheelbase would require heavier stock, hence more weight and expense. Long, 1-inch-diameter bolts held the axle boxes in place and tied the ends of the side frame together. The boxes were cast with hollow lugs or columns on each side to receive the bolts. Because the center of the bars, near the bolster, was under greater stress, slightly larger bolts, 1⅛ inches in diameter, were used. These passed through cast-iron or malleable-iron castings called columns (Fig. 7.13). The columns were on either side of the bolster and so formed an opening for the springs and spring plank. The bottoms of the columns were bolted solidly to the spring plank, which was stationary. The bolster was free to move up and down, so far as the springs would allow. Grooved guides or wear plates were bolted to the sides of the bolster, which ran in the track formed by the flat, rear side of the column. The bolster was a heavy white oak timber, generally around 8 by 12 inches, that was strengthened by one or two truss rods made of 1-inch-diameter rods. Wooden bolsters worked fairly well under 20- and 25-ton cars, but greater weights generally proved too much for them. This weak link will be treated more fully later in this chapter.

The lateral stiffness of the truck depended on the connection between the side frames and two crosspieces—the bolster and spring plank. The lack of end pieces was a major flaw in the design. Two light members would have done much to stiffen the frame, but only a very few roads would adopt them because of the added cost, and so the spring plank—which was the only solidly connected crosspiece—was depended upon to hold the truck square. In the case of swing motion, the transoms—crosspieces parallel to the bolster—helped to hold the truck square.

Because large bolts were needed to hold the side-frame members together, a large hole was necessary to receive them. This meant cutting away one-third of the bar and so weakening it at critical load-bearing points. Breaks were not rare, and when they occurred the entire structure failed, though it was more likely that the bolster would break before the side frame failed. Lock nuts were used to keep the structure tight, but they rarely answered this need for long. Keys were also sometimes used, but they served only to keep the nuts from working off and failed to keep them tight.

Side bearings were bolted to body and truck bolsters to steady the car, especially on curves or inclined trackage. Ideally, the weight would be carried by the center plate, but often a bent or broken bolster allowed the car to settle down instead on the side bearings. Most were simple, flat rub plates. Some railroads tried a variety of roller side bearings, which worked well enough when new. By necessity they were small in diameter and so tended to wear egg-shaped because of the limited travel; a full revolution could not be made. Because of cost they were made of relatively soft material, such as cast iron, mild steel, or wrought iron, and so again quickly wore out.

Two exploded drawings of an early- and a late-model arch-bar truck show the parts required for this portion of the undercarriage (Figs. 7.14, 7.15). Comparing the two drawings offers a quick review of the changes in truck design as they evolved between approximately 1875 and 1920. Not only is the more modern truck far heavier; wood has been entirely eliminated as a construction material. Strength and capacity were greatly increased by the massive riveted bolster and the heavy channel spring plank.

Before exploring the origins of this most American of bogies, brief mention should be made of the variety of arch-bar trucks used. The major division is between rigid and swing bolsters. In the former, no provision is made for side motion other than the looseness of the gauge. In the swing bolster, hangers or rollers allow for a certain degree of lateral movement. Opinions on the need and benefits of swing motion were sharply divided, as will soon be explained. Subdivisions of lesser consequence include soft- and low-ride and changeable-gauge trucks.

The importance of the diamond-arch-bar truck is abundantly documented. Its preeminence is un-

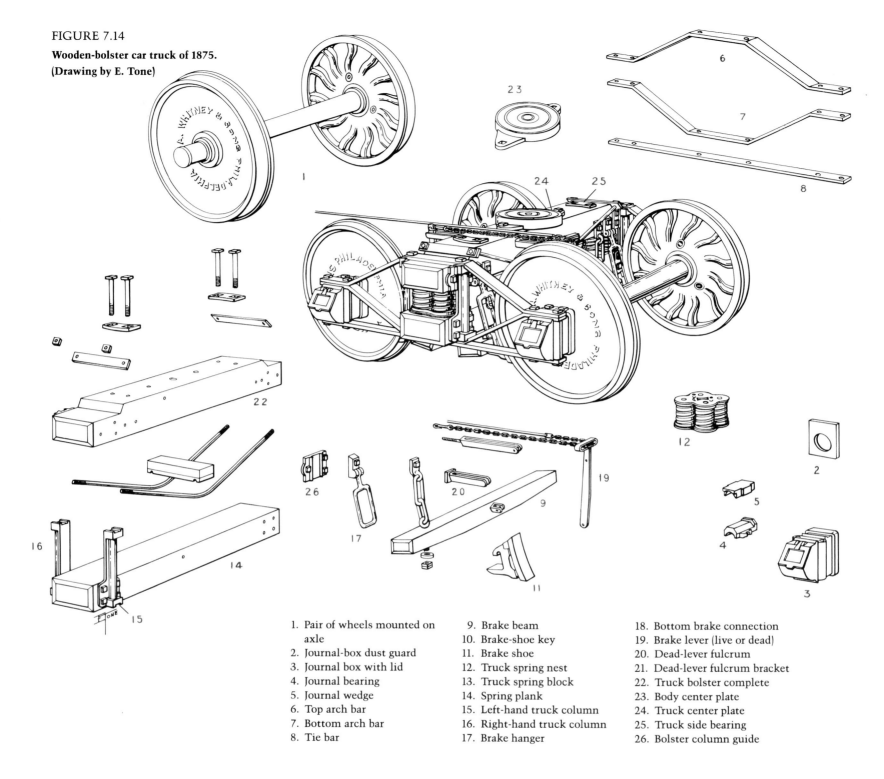

FIGURE 7.14

Wooden-bolster car truck of 1875.
(Drawing by E. Tone)

1. Pair of wheels mounted on axle
2. Journal-box dust guard
3. Journal box with lid
4. Journal bearing
5. Journal wedge
6. Top arch bar
7. Bottom arch bar
8. Tie bar
9. Brake beam
10. Brake-shoe key
11. Brake shoe
12. Truck spring nest
13. Truck spring block
14. Spring plank
15. Left-hand truck column
16. Right-hand truck column
17. Brake hanger
18. Bottom brake connection
19. Brake lever (live or dead)
20. Dead-lever fulcrum
21. Dead-lever fulcrum bracket
22. Truck bolster complete
23. Body center plate
24. Truck center plate
25. Truck side bearing
26. Bolster column guide

questioned, for just about every American freight car after 1870 rode on diamond arch bars. Yet for all its importance, the details of its origins are poorly recorded. Isaac Dripps, former master mechanic for the Camden & Amboy Railroad and superintendent of the Trenton Locomotive & Machine Company from 1853 to 1857, is said, in an obituary published in 1893, to have produced the first truck while at Trenton for the Belvidere & Delaware Railroad.[20] Experienced historians learn to distrust claims appearing in obituaries without further proof. Angus Sinclair repeated the claim for Dripps in his 1907 locomotive history but spec-

ified the year as 1857 and the customer as the Lehigh Valley.[21] Both claims were made long after the fact and appear to be unsupported by more contemporary evidence. No patent for the period supports the Dripps claim; he did, however, receive a truck patent in January 1870 that covers details rather than the basic design.

If Dripps's claim is uncertain, hard evidence exists for arch-bar trucks dating from midcentury. Two drawings, dated June 1857, were prepared by Harlan & Hollingsworth for the Mobile & Ohio Railroad. Both show an arch-bar truck where the top bar is dead flat but the inverted arch is fully

FIGURE 7.15

Steel-bolster arch-bar truck. (Magor Car Corp. [Passaic, N.J.] catalog, 1912)

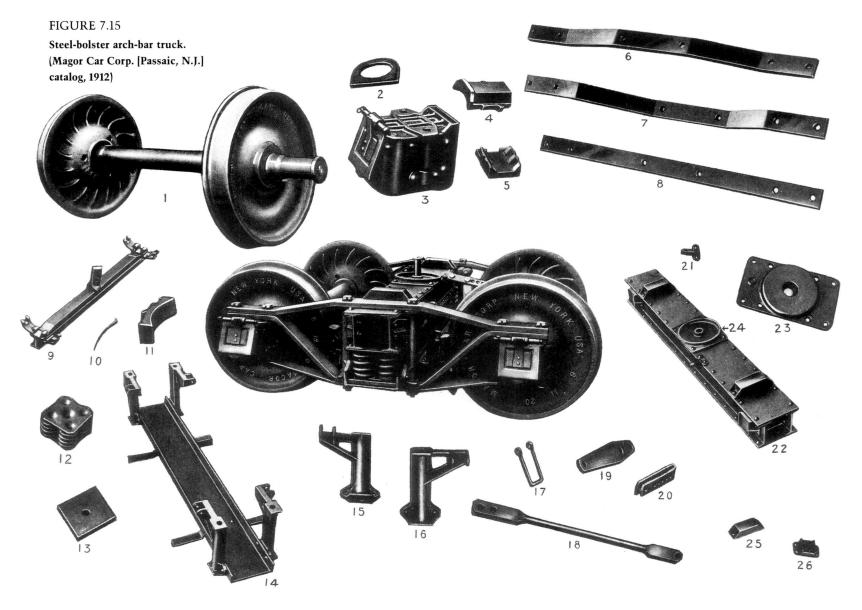

1. Pair of wheels mounted on axle	9. Brake beam	18. Bottom brake connection
2. Journal-box dust guard	10. Brake-shoe key	19. Brake lever (live or dead)
3. Journal box with lid	11. Brake shoe	20. Dead-lever fulcrum
4. Journal bearing	12. Truck spring nest	21. Dead-lever fulcrum bracket
5. Journal wedge	13. Truck spring block	22. Truck bolster complete
6. Top arch bar	14. Spring plank	23. Body center plate
7. Bottom arch bar	15. Left-hand truck column	24. Truck center plate
8. Tie bar	16. Right-hand truck column	25. Truck side bearing
	17. Brake hanger	26. Bolster column guide

developed. The drawing for the more elaborate of the duo is reproduced here as Figure 7.16. Aside from the arch bar, the design shows other interesting features for a truck of this early date. The continuous-top-rail frame is but one notable item. The swing-motion bolster, the metal body bolster, and the volute steel springs are also worthy of study. The wooden transoms are supported top and bottom by a pair of short columns, while the bottom tie bar angles up at both ends for an attachment to the top rail.

As already mentioned, the arch-top bar was used at least as early as 1844 by Davenport, and so it was only a matter of time before a mechanic would combine all these well-known elements into the classic diamond arch bar. Perhaps it was Dripps as claimed in 1857, but the oldest dated drawing that shows all the elements correctly combined is credited to John P. Laird, 1863, master mechanic of the Pennsylvania Railroad.[22] Contemporaries of Laird could not seem to get it quite right. Some betrayed their confusion regarding the theory and workings of the diamond side frame by adding unnecessary bars or, in a worst-case situation, reversing the position of the lower arch bar so that it too was in compression rather than tension. Very elaborate

FIGURE 7.16

The continuous-frame car truck of 1857 illustrates an early form of arch-bar truck. Swing bolsters, wooden transoms, and volute steel springs were part of this Harlan & Hollingsworth design. The metal body bolster is also shown for this 5-foot-gauge truck. (Smithsonian Neg. 81-8101)

engravings of a Grand Trunk Railway truck were published in 1869 not so much to illustrate its wrongheaded side-frame design as to portray its adjustable-gauge axles (Fig. 7.20). Some thirty years later the Grand Trunk had managed to position the arch bars properly, but the road would not give up the redundant bars and was steadfast in its allegiance to an old-fashioned design introduced many years earlier by Leander Garey of the New York Central.[23] The oldest known drawings for Garey's truck were published in 1876, but the design may be older, for Garey had been an active car builder since 1849.[24]

Just what Garey hoped to achieve by his double-arch-bar side frame is uncertain, but whatever the benefits, the deficits included extra pieces, weight, and cost (Fig. 7.17). Two wooden transoms were made from 4½- by 10-inch white oak. Very long swing hangers (25½ inches) were hung from the transoms. Garey was content with the general design and maintained the double-arch-bar side frame for many years. In 1882 the truck was strengthened with channel-iron transoms and a twisted bottom tie bar.[25] The leaf springs were replaced by cluster coil springs encased in cast-iron

cylinders. The 60-inch wheelbase was maintained. Within months of Garey's retirement in 1884 his successor, William Buchanan, was busy with a new plan of truck. The redundant four-bar side frame was abandoned for the simple three-bar design.[26] Buchanan also replaced the complex swing-motion bolster with a rigid one. Coil springs were adopted at the same time.

The Baltimore & Ohio had employed arch bars before adopting the wood-beam trucks described earlier in this chapter. The designers at Mt. Clare seemed to feel that the arch-bar side frame required an extra bar (Fig. 7.18). The function or need for this heavy, flat center bar is uncertain, yet surely it was more rational than the 1868–1869 Grand Trunk design shown earlier. The drawing shown here is copied from an original prepared in 1862 by Thatcher Perkins. The drawing is one of the finest graphic records of a first-generation arch-bar truck. The by-now familiar floating bolster is in evidence. The coil spring canisters placed inside the side frame can be seen in the end view. The general plan shown in Perkins's drawing was used by the Illinois Central, as illustrated by a photograph reproduced in Chapter 2 (Fig. 2.23). Samuel Hayes was mechanical chief of the Illinois Central at the time. Until 1856 he held the same position on the B&O, and so there is some reason to speculate that the design was original with Hayes. If so, it might have been copied by Perkins and so may represent a very early form of the arch-bar truck.

The penchant for the four-bar side frame went beyond even the examples shown here. That it was favored by prominent car builders like Garey, Perkins, and Hayes may have prompted others to copy and experiment. John Lentz of the Lehigh Valley was a late but ardent advocate. He produced cars into the late 1880s with trucks of this type.[27] Some five years earlier the Pennsylvania had tried to give the four-bar side frame a real purpose. A king post truss was inserted between the spring plank and the bottom tie rod to form a strengthening truss.[28] This double-diamond truck had an elaborate metal bolster and a 63-inch wheelbase. It was altogether a most curious plan and is assumed

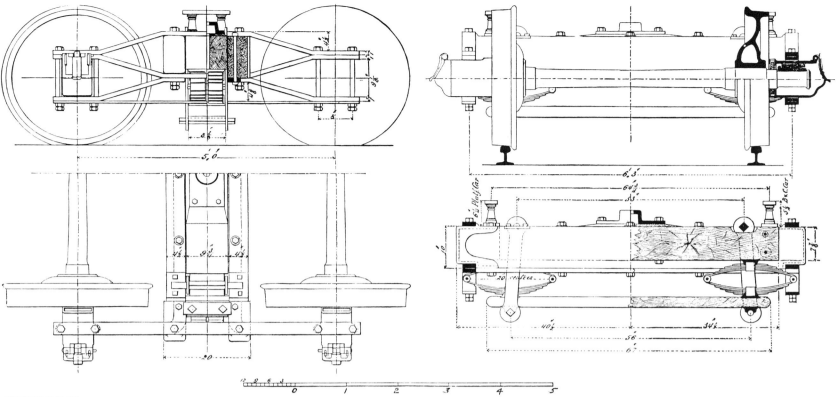

FIGURE 7.17

The New York Central adopted a swing-motion truck in 1876 that featured leaf springs and wooden transoms and bolsters. Notice also the redundant lower arch bar. (*Railroad Gazette*, March 17, 1876)

FIGURE 7.18

The oldest preserved image of a fully developed diamond-arch-bar truck was part of a general design drawing prepared for a B&O iron box in 1862. This illustration was traced by the author from the assembly drawing reproduced as Figure 8.15.

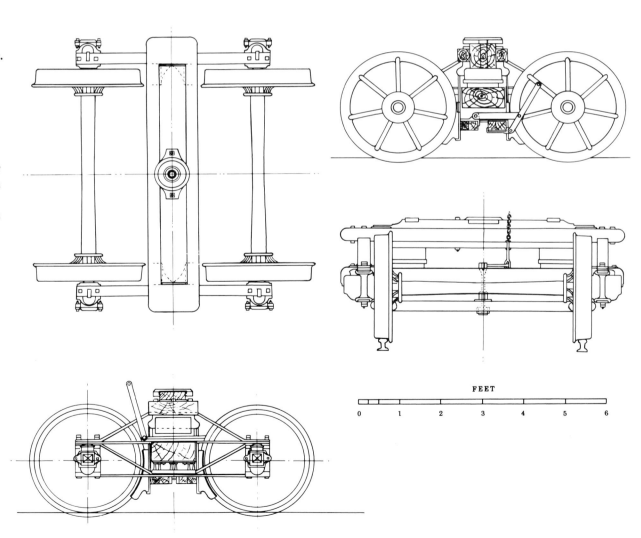

FREIGHT CAR TECHNOLOGY

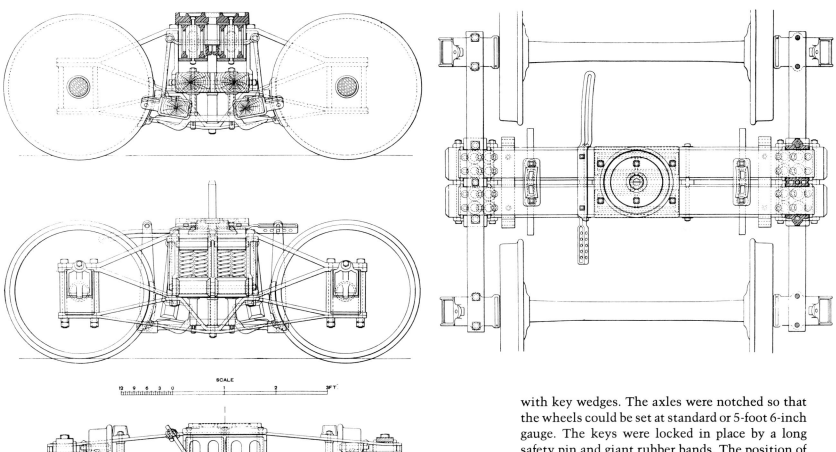

SCALE
12 9 6 3 0   1   2   3FT

FIGURE 7.19

This Pennsylvania Railroad design of 1883 includes many peculiar features. The redundant lower arch bar is bent down to form a king post truss for the wooden spring plank. The metal bolster was relatively rare for the early 1880s. (Watkins, *History of the Pennsylvania Railroad*)

to have been largely experimental in nature (Fig. 7.19).

ADJUSTABLE-GAUGE TRUCKS

Before exploring other aspects of arch-bar trucks, we should study with greater care the 1868–1869 Grand Trunk truck mentioned earlier (Fig. 7.20). Because of its 5-foot 6-inch gauge, the Grand Trunk could not exchange cars with its connecting lines and was thus cut out from the economies of the interchange service just then developing. Just as the Union Line had experimented with broad-tread wheels to solve a similar problem, so too did the National Despatch Line, another fast freight line, develop a way to overcome the gauge difference.[29] National Despatch adopted telescoping axles so that wheels could be reset for a 9½-inch difference in track gauge.

The scheme selected was patented by C. D. Tisdale of East Boston, Massachusetts, with the first patent having been issued in March 1863.[30] Special wheels with extra-large hubs were fitted

with key wedges. The axles were notched so that the wheels could be set at standard or 5-foot 6-inch gauge. The keys were locked in place by a long safety pin and giant rubber bands. The position of the wheel was shifted by a gradually diverging or converging track. In the shift from broad to standard, the keys would be loosened and removed at one end of the tapering track. Workmen in a 4-foot-deep pit removed the keys from below the train. A long shed was built over the pits to protect the workmen. With the keys out, the train was slowly pushed down the track, and the wheels would be forced inward as the train moved along the converging rails. Once at the end, the workers would reinsert and lock the wedges and the train could go on its way. The change could be done in five to ten minutes. When shifting to broad gauge, a third rail set inside the tapering track pushed the wheel out to the wider gauge. Shifting stations were located at Point St. Charles, Montreal, and Sarnia, Ontario. The plan was first tried in November 1863, yet no serious consideration was given to it until early 1868. The tests proved so promising that by late in the following year two hundred adjustable-gauge cars were running between Chicago and Boston via the Michigan Central, the Grand Trunk, the Vermont Central, and several connecting lines in New England. The problems of the northern east-west route seemed to have been resolved, and three hundred more cars were ordered by National Despatch.

Just months later, however, the Grand Trunk announced plans to rebuild its entire line to standard gauge. Major conversions were completed in 1872 and 1873, with all parts of the system having been remade to the Stephenson gauge by September

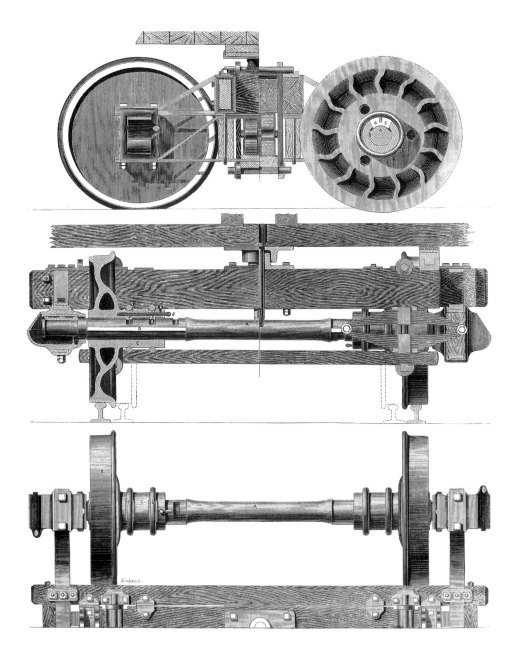

FIGURE 7.20

The Grand Trunk Railway adopted its peculiar style of diamond-arch-bar trucks for changeable-gauge service in 1868. Wheels with very large hubs could slide on the axles for this purpose. Round rubber bands helped hold the keys in place. (*Engineering*, Nov. 19, 1869)

1874. This disruptive and costly conversion might have been avoided had the changeable-gauge trucks worked as well as advertised. Problems obviously had developed. The keyway grooves were said to weaken the axles.[31] Misgivings over the safety of the telescoping axles were voiced as early as 1846, long before the Grand Trunk test.[32] Considerable skepticism was expressed as to the reliability of the workmen charged with loosening and tightening so many wheels day in and day out. Even on the short freight trains of that time, could the men be trusted to pursue their jobs with care? Crouching in a dank pit for ten hours with a rumbling train overhead could be tiresome and lead to boredom and negligence. It seemed like a scenario for disaster. Even if the axle crews proved true to their duty, the normal wear of the shifting wheels would beget loose fits, and even a slight wobble could cause a derailment.

The shifting-wheel scheme seemed inherently

defective. There was apparently only one other attempt to use such a system, and it was much more limited than the Grand Trunk effort. The Erie was also excluded from the free interchange of cars and so was tempted to test William B. Snow's patented scheme, which used a spring bolt in place of wedge keys.[33] A broad-gauge boxcar fitted with Snow's apparatus was sent on a cross-country tour in 1870 to demonstrate the plan. The trip from New York to California required fourteen and a half days, which was considered good time for a freight car of the period. There was talk of building one thousand cars on Snow's plan, but it is unlikely that the program was ever completed.

CLEVELAND TRUCKS

While so many car builders contrived ways to improve freight car trucks by making them more complex, others concentrated on simplifying the design to its most basic elements. One of the most popular arch bars of the purebred variety was the so-called Cleveland truck (Fig. 7.21). The origin of the name is now lost, but it may have been so called for a railroad in the Cleveland area, like the Cleveland, Columbus & Cincinnati, or for the Cleveland Car & Bridge Works. Whoever perfected the plans for the Cleveland truck surely had a talent for eliminating all unnecessary elements and parts. The simple side frame is free of redundant bars. The bolster is a plain-vanilla variety with wooden cross-members strengthened by two truss rods laid on outside the beam with a metal plate, set into a notch cut out of the bolsters' ends. This neat plan avoids drilling, which would weaken the bolster. The truck is rigid and has no provision for swing motion. Notice also that the spring plank is where it belongs—directly on top of the bottom tie bar. There are none of the fussy columns and posts typical of so many other early arch bars. The coil springs are placed in canisters just inside the side frames. The common sense and purity of the design guaranteed a long life, and we find specific references to the Cleveland truck in the 1870s and 1880s.[34] The design became so generic that in time what had once been called the Cleveland truck blended into the diamond-arch-bar truck and so lost its identity by becoming so commonplace.

SWING MOTION

The diversity of opinion common to all human activity held true for freight car trucks as well. The classic simplicity of the Cleveland truck was lost on many car builders, who seemed to find satisfaction in complex answers to simple questions. The swing-motion truck was a product of this school of thinking. Superficially the proponents of swing motion had a logical reason for their intricate and expensive form of truck. On rough track, the wheels were worked back and forth because of uneven or twisted rails. This lateral motion was,

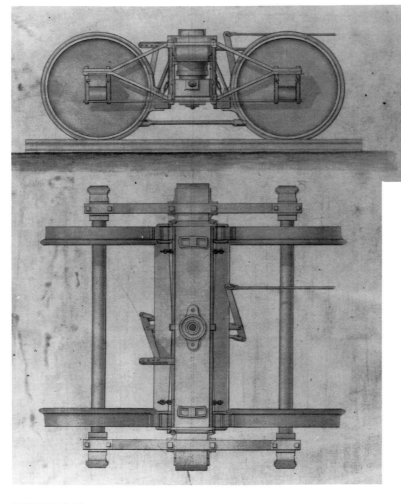

FIGURE 7.21

The Cleveland truck was the simplest and most rational form of the arch-bar design. Note the massive bolster and spring plank. The drawing was prepared by Harlan & Hollingsworth in 1866. The wheelbase was 4 feet 9 inches. (Smithsonian Neg. 81-8113)

of course, destructive to the flanges of the truck frame and the car itself. The car springs might absorb most of the vertical vibrations, but almost none of the sideways motion was cushioned. This movement could be handled only by swing hangers or rollers.

Present-day railroad mechanical engineers find it impossible to believe that swing-motion trucks were once relatively common for ordinary freight cars. Yet it was true, particularly between about 1860 and 1890, that almost half of the nation's car builders were loyal advocates of swing-motion trucks for every variety of railroad freight car.[35] In 1884 the Master Car Builders Association voted thirty-two in favor and thirty-one against swing-motion trucks. This might be construed as a rather small sampling of the entire car-building fraternity, but because most of those attending the meeting represented major rather than minor lines, the vote indicates a greater majority than might at first appear.

The advocates had a plausible checklist in favor of the vibratory truck, as it was sometimes called.[36] The list ran something like this:

Reduces shocks to the car body and trucks.
Reduces shocks to track and switches.
Decreases frictional resistance on curves.
Reduces flange wear and breakage of same.
Reduces likelihood of cattle being thrown down in cars.
Reduces damage to goods being carried.
Reduces wear of journal collars and journal bearing ends.
Assists in coupling cars—because the car can be shifted sideways as much as 3 inches.

A final argument in favor of lateral motion was its universal acceptance for passenger cars. No railroad in the country ran rigid-bolster trucks under its passenger cars, and many lines had adopted swing motion for locomotive leading trucks and tender trucks. Why, then, shortchange freight cars? The extra cost of lateral-motion devices was more than compensated by the long-term savings in wheel and track wear alone, and a smoother, safer ride was surely worth some extra capital investment.

The rigid-truck men would dismiss all these arguments as pure nonsense. Indeed, lateral motion could be justified for passenger cars and locomotives traveling at relatively high speeds, but because freights lumbered along at 10 miles per hour, the same rules did not apply. The lurching of cars at so low a speed was not comparable to that of an express train. The detractors went on to label swing motion as costly, unnecessarily complex, and dangerous. The cost and extra maintenance might be justified for a few hundred passenger cars and locomotives, but freight cars that numbered in the hundreds of thousands multiplied a small unit cost into big money. Swing motion cost 16 to 17 percent more than rigid trucks and 25 to 30 percent more to maintain.[37]

It was all those extra parts that made the lateral motion such a luxury—nine extra parts to fix! The swing hangers, pins, washers, bolsters, and transoms were all too fussy for freight cars. Maintenance was casual and erratic because there were so many freight cars constantly on the move. They tended to receive attention only after they broke down. Because the lateral-motion apparatus was inside the wheels and out of view, it was difficult to inspect and repair. This made the swing-motion truck treacherous, for it tended to break down

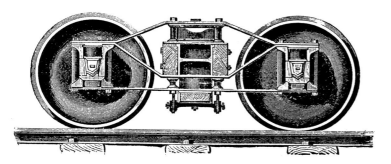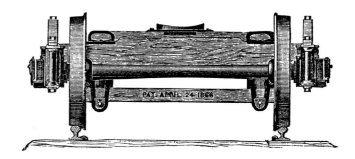

FIGURE 7.22

C. F. Allen's 1866 swing-motion truck became known as the California truck because of its popularity in far western states. The links were made from rods fitted with cast-iron bearing blocks at each end. (*National Car Builder*, Jan. 1874)

when least expected, and no railroad could afford the constant vigilance necessary to maintain it safely. If a pin or hanger failed because of wear or overloading, the spring plank and all associated gear would drop to the track. A derailment was likely, and even a minor wreck could cost thousands of dollars as well as tie up the line for hours.

As one anonymous mechanic who would sign himself only as "One of the old Fogies" said, "It was a truck that rode a little easier but cost a great deal more."[38] And some wondered even about this benefit. During an actual test it was found that a train with all rigid trucks went around curves with less friction than a comparable train with all swing-motion trucks. The test was an incidental part of the famous Burlington brake trials (1886 and 1887). The real hangup on going around curves was the failure of the side bearings; they tended to crush down into the wooden truck bolster and were soon so out of line that they created enough friction to prevent the truck from swiveling freely. This was a real cause of excessive flange wear, and swing motion could not correct this problem any more than it could solve the other major source of wheel and journal wear, an out-of-square truck.

In a final rejoinder to supporters of swing motion, John Player, master car builder for the Santa Fe, contended that ordinary rigid trucks soon loosened up after a few months' service and so offered ½ to ¾ inch of sideways elasticity at no extra expense. In addition, the wheel flanges were fixed loosely inside the track gauge to allow for lateral freedom. And the axles could wander back and forth, for they were not fitted snugly to the bearing ends. Even the rigid truck had some built-in lateral motion, which was augmented slightly by the slope of the wheel tread and flange. The cone of the tread and curved connection between the tread and the flange allowed the wheel, in effect, to climb up and down the railhead and so dissipate some of the lateral motion. In addition, the massive construction of the wheels, axles, and truck frame was meant to absorb the forces imposed on them by the lurching of the cars as they moved along a less-than-perfect track.

The controversial swing-motion truck was born when the steam railroad was still an emerging industry. Its technology was in the development stage, when little seemed fixed or certain. Tracks were rough and uneven, mainly because funds were unavailable to provide a better path for the Iron Horse and its brigade of cars. Inventive Americans, such as John B. Jervis, had devised ways to make the equipment fit the roadbed, with devices like the locomotive leading truck. And so Charles Davenport, a car builder in suburban Boston, came up with the lateral-motion truck, an invention he patented in May 1841.[39] By means of leaf springs and swing hangers, Davenport's truck would absorb part of the lateral motion of the wheels and center itself. A more fully developed version of Davenport's design is shown by the 1848 drawing of Ray's truck (Fig. 7.11). Other refinements can be seen in the 1857 Mobile & Ohio drawing and the 1869 Grand Trunk drawing (Figs. 7.16, 7.20).

A more refined version of the swing-motion truck was patented on April 24, 1866, by Charles F. Allen, one-time mechanical chief for Pullman.[40] Allen's truck had wooden transoms supported top and bottom by short column spacers (Fig. 7.22). The swing links were made from round iron bars welded into the shape of a long chain link rather than the more typical flat rod with eyes forged on each end. Cast-iron blocks were fitted into the ends of the link to form the pin bearing. The design was so popular in the West that it became known as the California truck, and it was apparently much used by the Central Pacific. It was claimed that the easy-traveling California truck saved 20 to 25 percent of the power needed to move a train. The claim that it took the curves "like a boat in the water" was probably no less extravagant. The cup-shaped center plate allowed the car body to slide up or down on elevated curves, while on the level it tended to help center the body. The California truck remained in favor in the Far West until at least the middle 1880s. Drawings were published at the time for a modified Allen truck intended for a low-profile flatcar.[41] The arch bar was flattened, and 28-inch wheels were used to achieve the lower level. A single, heavy casting replaced the usual columns at the side frame's center, and double leaf springs were used in place of coil springs. The top bar of the side frame was made from 1¼- by 3-inch stock, indicating that the truck was intended for heavy service.

George W. Cushing, a veteran car man who

FIGURE 7.23

The Thielsen truck, introduced in 1867, was the best known of all swing-motion trucks. Iron transoms made it a solid, long-lived truck. (*Railroad Gazette*, Dec. 13, 1878)

served as master car builder for several major railroads, was another lifelong admirer of swing-motion trucks. Cushing devised a simple and very cheap lateral-motion hookup that used very long eye bolts running through deep wooden transoms. The swing link itself was rather short. It was attached at the upper end to the eye bolt and at its lower extremity to the cross-pin that supported the spring plank. Exactly when the scheme was introduced is unknown, but an 1879 engraving of it is reproduced in Chapter 3 (Fig. 3.9). Simplicity and cheapness were not enough to promote the fortunes of the short-link swing motion, for it produced a quick, jerky motion that made its vital parts liable to fracture.[42] Cushing himself must have come to abandon the scheme in time, but he remained an advocate of the swing-motion truck to the end of his career and was ready to speak in its defense even at a time when the majority of his peers had turned against it.[43]

The swing-motion truck came into prominence with the introduction of the Thielsen truck.[44] Here was a rugged car carrier whose chief merit was its solid construction. That it included a swing bolster was almost incidental, but for better or worse, this put it in the lateral-motion camp. In design it came very close to being an all-metal truck, except that its creators were too conservative to abandon wood for the bolster and spring plank. The transoms, however, were metal, and this allowed a secure connection between the side frame and cross-frame, thus making the truck a firm unit. The best wood-to-metal connections always tended to work loose as the wood shrank, swelled, or rotted. Large bolts and lock washers were no avail. But in the Thielsen truck the transom and the arch-bar side frames were riveted together using a heavy cast-iron bracket as the spacer where the union was made (Figs. 7.23, 7.24).

The Thielsen truck was actually a joint venture and was the brainchild of two experienced railroad men. Hans Thielsen, a Dane, was chief engineer for the Burlington & Missouri River Railroad, and his partner, George Chalender, was master mechanic for the same road. In 1867 and 1868 they designed and built some test units. A patent was obtained on June 1, 1869 (No. 90,795), with the two men sharing equal rights. The transoms were at first fabricated from ¼-inch plate with angles

riveted top and bottom, but by 1877 one-piece channel iron was used. At first, 10½-inch-deep channels were employed, but 12-inch channels were later adopted. Normally, trucks were junked when the car body wore out, but the Thielsen truck was not ready for the scrapper and would outlast two sets of wooden-bolster trucks. Such good service gave the swing-motion truck a reputation it really did not deserve, but justified or not, it became popular on many lines. The Burlington system, which absorbed the B&MR, naturally became a patron, but other unrelated lines adopted it as well. The Michigan Central, the Santa Fe, and the Central Pacific all became users. In May 1877 the *Railroad Gazette* stated that 3,225 cars rode on Thielsen trucks. About half of these were on the Burlington. Sixteen months later, twelve thousand sets were in service on eleven different railroads.

Around 1880 Leander Garey, head of the New York Central system's car department and already identified as a swing-motion loyalist, came to adopt the Thielsen-style truck. Garey directed the selection and design of car equipment not just on the parent New York Central & Hudson River Railroad but on its several subsidiaries, such as the Michigan Central and the Lake Shore. These connecting lines were major railroads in themselves, and so the total car fleet directed by Garey was vast. It would total around sixty thousand cars in 1885. When Garey completed designs for a new standard series of freight cars in 1882, they were all to be equipped with the Thielsen-style truck.[45] Decisions such as this one are what made the Thielsen truck so important. It was a good design despite the swing-motion feature and undoubtedly served the New York Central system well during its years of ascendancy. But fortune is a fickle thing, even for car trucks, and no sooner had Garey retired than it was abandoned.

Swing-motion loyalists tried to find ways to answer the criticisms of their detractors. Mounting safety straps below the spring planks was projected as a cheap way to end the dangers resulting from a broken swing motion. Two straps made from ¾- by 3-inch iron were bolted to the transoms just to one side of each swing hanger. If a pin or hanger failed, the plank would drop down to the straps and not to the track where it would likely cause a derailment. The scheme was copied from the ancient scheme for axle safety straps. Railroads known to have used spring-plank safety straps were the Chicago, Burlington & Quincy and the Big Four.[46]

John C. Barber, one-time master car builder for the Northern Pacific, devoted much of his career to improving freight car truck design. In theory he liked the notion of swing motion, but he recognized its defects. It was too complex, too fragile, and too hidden away from view to be practical for

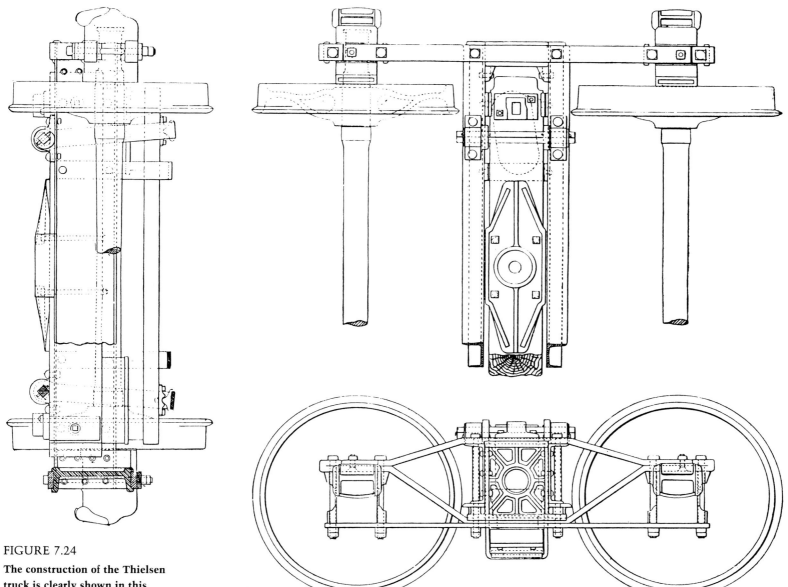

FIGURE 7.24
The construction of the Thielsen truck is clearly shown in this 1881 Chicago, Burlington & Quincy drawing. (*National Car Builder*, Sept. 1881)

so rough a trade as railroad freight service. To simplify and improve, Barber made radical design changes. The swing hangers and spring plank were eliminated, thus exiling the chief offenders so far as safety and maintenance cost were concerned. The transoms and free-floating bolster were retained; the transoms were kept not only because they were needed so that the bolster could rock back and forth but also because they added so much strength to the truck. The bolster was extended slightly so that it could form a double incline pad for two steel rollers about 2½ inches in diameter (Fig. 7.25). An opposing inclined pad was mounted on top of the springs; the rollers were placed inside the pads. The lateral motion was taken up and dissipated by the rollers, which would always recenter themselves to the middle of the pads. The springs were moved outside the wheels and were supported by a bracket bolted to the inverted arch bar and tie rod. The entire mechanism was outside the wheels, readily accessible

for inspection and repair. There was no hidden time bomb sandwiched behind the wheels and transoms waiting to go off without notice.

Barber's design was a miniaturized version of locomotive lead truck centering devices dating back to the 1860s. Whether Barber was inspired by the ideas of Levi Bissell and Alba F. Smith will probably never be known. He tried the scheme in 1888 on a test group of one hundred refrigerator cars.[47] The roller motion worked so well that it was next used on locomotive tender trucks. During the 1890s the Milwaukee Road adopted Barber's design for some furniture cars, and the Drexel Supply Company of Chicago offered the Barber swing-motion truck to an even larger market. Barber entered the supply trade on his own, establishing the Standard Car Truck Company in 1898. By 1903 he claimed to have sold roller-motion trucks for sixty thousand cars and tenders. The roller seat had been moved from under the bolster ends to under the lower ends of the springs. The roller lateral-

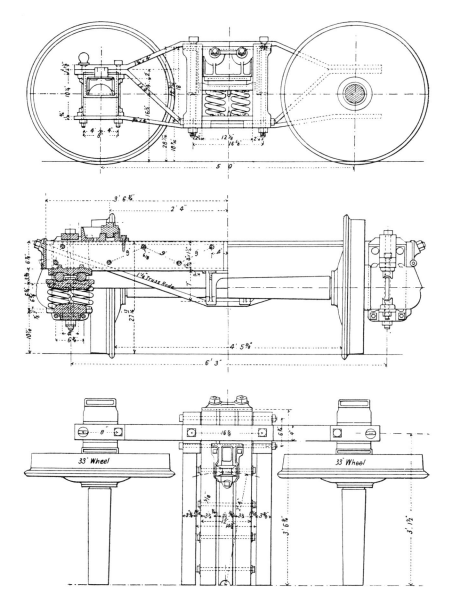

FIGURE 7.25

In the 1880s John C. Barber introduced a curious but effective lateral-motion truck that used small rollers rather than swing links. The rollers are just above the springs. (James Dredge, *Record of the Transportation Exhibits at the World's Columbian Exposition of 1893* [London, 1894])

motion scheme continued to be offered for at least another quarter-century. To make it suitable for 50-ton cars, three 2- by 10-inch rollers were used.

The efforts of Barber and other friends of swing motion failed to save the lateral truck, at least for mass application to freight cars. It was retained for special equipment like cattle cars, cabooses, and tenders. It remained universal, of course, on passenger equipment, but elsewhere it largely disappeared. John Player began a major program early in 1891 to rebuild all swing-motion cars on the Santa Fe to the rigid plan.[48] John Hickey, master mechanic for the Milwaukee, Lake Shore & Western, felt even more strongly than Player on the subject. He would remove lateral-motion bolsters from locomotive and tender trucks as well.[49] By 1890 even the Burlington, original home to the celebrated Thielsen truck, was preparing to banish swing motion from its freight car fleet.[50] The major reason put forward at the time was the inability of swing motion to handle the extra stresses created by the heavier 30- and 35-ton cars then coming into use.

Faster train speeds added to the stresses of freight car trucks. In 1896 the *American Railroad Journal* reported that rigid trucks were almost universal for heavy cars.[51]

CENTRAL-SUPPORT CARS

Schemes more radical than the ill-fated swing-motion idea were proposed to improve railroad undercarriages. We shall explore a few of these before turning to the more important issues of truck history. One of the more bizarre ideas tried was the central-support or three-truck car. By placing a third truck at the car's center, the capacity of the vehicle could be greatly increased. It was like putting an extra pier at the center of a long bridge span. The frame could be made much lighter. Truss rods could be eliminated. In fact, they had to be eliminated, to make way for the center truck. Increasing the number of wheels from eight to twelve meant lighter axle and wheel loads, which in turn meant that smaller and cheaper wheels and axles could be used. It also meant that the track would suffer less wear and tear because the load was being spread.

These were the advantages as seen by the proponents of the twelve-wheel car. Like most mechanical panaceas, it also involved many negatives and a few trade-offs. Was it only a scheme to save the wood-frame car by propping up its middle? Was it just another band-aid cure for an inherently limited and defective structure? And how could it deal with a uniform distribution of weight on track that was generally uneven? The twelve-wheel car could perform well only if the weight was borne evenly by all three trucks, but in a dip or sag, the center truck would carry little or no weight. The full load would fall on the end trucks. When humps or rises were encountered, most of the weight would fall on the center truck. Such overloads would surely lead to breakdowns such as spring, axle, bolster, or wheel failures, any one of which could result in a derailment. And how well would such a queer centipede car track? Would it negotiate tight curves and switches as readily as an ordinary double-truck car?

The small number of center-support cars built would indicate that most of these questions were answered in the negative. Actually, rather little was written about them. The first appear to have been produced in 1880 to the design of I. C. Terry of St. Louis (Fig. 7.26).[52] The inventor received a patent on June 8, 1880 (No. 228,694). Some test cars were produced for the St. Louis, Iron Mountain & Southern Railway about the time the patent was issued. The capacity of the cars was increased from 11 tons to 24 tons—an astonishing improvement, if true. The center truck could swivel but was also equipped with a traveling center plate that could shift from side to side about 10 inches in either direction. The movable center

FIGURE 7.26

An extra truck greatly increased a wooden car's capacity. But the scheme, introduced in 1880 by I. C. Terry, involved too many drawbacks to be widely used. (*Railroad Gazette*, Sept. 7, 1883)

plate was carried by a heavy bolster and eight very small flanged wheels, which in turn ran on T-rail fastened upside down to the underside of the car frame. After a two-year test on an Iron Mountain line, the Burlington announced plans to convert ten cars on Terry's plan. The cost was estimated at $150 per car, which was regarded as cheap considering the increase in capacity that would result. Reports on Terry's car end here, and so it can be assumed that problems developed during subsequent tests that discouraged a wider use of the center-support car.

Not long after Terry's brief encounter with truck design, Leonard Finlay, master mechanic for a narrow-gauge short line, the Hot Springs Railroad, tried to revive the idea.[53] Finlay had Barney & Smith build some 42-inch-gauge cars that measured 36 feet long, or 13 feet longer than a normal narrow-gauge car. They had supplemental trucks attached at the center on a plan patented by Finlay on August 7, 1883 (No. 282,510). It differed little from Terry's plan, except that the rollers were fastened to the truck bolster. The cars weighed 21,000 pounds, which did not tax their capacity. As expected, the line's superintendent was "loud in his praise of the car," designed as it had been by his colleague, the residing master mechanic.

Standard-gauge lines were impressed enough to try Finlay's patented design. In 1885 the Pennsylvania's Fort Wayne shops put center-support trucks under several cars. They were said to be successful. During the next year the Pennsylvania's Allegheny shops produced a number of 40-ton ore cars on Finlay's plan. In 1888 both the Lake Shore and the Erie built lots of fifty and forty ore cars with Finlay trucks. Like the Pennsylvania models, they were of 40 tons capacity. After five years' service no major problems had been encountered. William Voss was cautiously optimistic about the three-truck scheme and published a drawing of the rolling center plate connection in his 1892 book, *Railway Car Construction*. He raised doubts about how well the plan would work for boxcars, because even a slight rocking of the body could open cracks in the roof or side walls. The center-support car was never widely adopted during the wood-car era, though it appears to have received a fair test.

The idea came alive again in the steel-car age, as is evident by a report about a giant steel-framed central-support boxcar patented in February 1892

by Henry C. Hodges, president of the Detroit Lubricator Company. Hodges's 70-foot-long dream never took shape, but the twelve-wheel idea became a reality in Germany a few years later. In 1901 German railways were suffering a car shortage.[54] A quick solution was found by adding center trucks to existing 15-ton-capacity coal cars, which could then carry 25 tons. In 1985 the Canadian National sought a large-capacity grain car suitable for light branch-line service.[55] The result was a variation on the center-support idea, for two articulated bodies were used with the middle truck carrying one end of the twin bodies. Because separate frames were used, the worst defects of the old Terry and Finlay plans would be avoided.

Using more wheels to carry a greater load was hardly a new idea, nor did it necessarily require such a radical approach as the central-support truck. By adding another wheel set to each truck, cars of greater capacity could be produced without sacrificing any of the advantages of the profoundly successful double-truck car. This idea was rather widely applied to the oversize flatcars described at the beginning of Chapter 6. A few car builders like Richard Eaton of the Grand Trunk Railway thought that they could be applied to ordinary merchandise cars.[56] Eaton envisioned a high-capacity boxcar carried on two six-wheel trucks. The extra wheels would spread the load so that heavily burdened cars could operate safely over the lightest track. The trucks were a free adaptation of the standard arch-bar side frame, except that they were double so as to accommodate six rather than four axle boxes. Several sample cars were produced for testing purposes. The inventor had by this time lost his position with the Grand Trunk over a contract scandal, and his influence in the mechanical affairs of the company was therefore greatly diminished. No more was heard about the twelve-wheel boxcar experiment, and Eaton's idea appears to have been abandoned.

Nine years later a remarkably similar design appeared on the Central Pacific.[57] The Sacramento car shops adopted the swing motion from the California-style truck. There is some reason to believe that the Central Pacific's draftsman may have been inspired by C. F. Allen's patent of May 12, 1868, which covered designs for a similar elongated arch-bar six-wheeler. Allen had designed the California-style truck adopted by the CP, so his work was hardly unknown to that line. Allen incorporated equalizers and a blind center wheel set in his design. These two features were not copied in the CP design. The top bar of the side frame was flattened and stretched out for an 8-foot 2-inch wheelbase (Fig. 7.27). Leaf springs were applied. The design was apparently experimental or intended for a very limited number of special-purpose cars, for there is no evidence indicating widespread use. Twelve-wheel freight cars were a

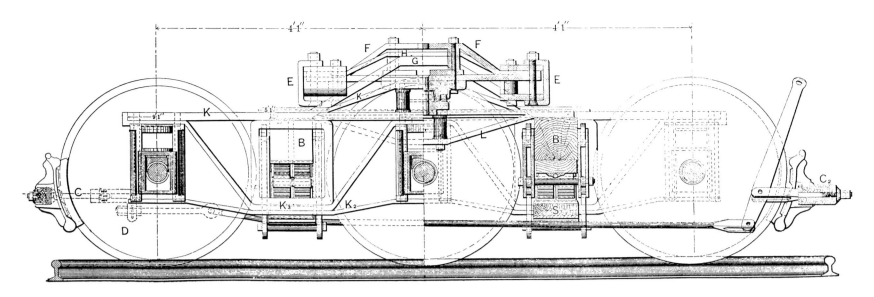

FIGURE 7.27

Freight car capacity could also be increased with six-wheel trucks. The example shown here was designed by the Central Pacific Railroad in about 1883. Swing motion was part of this plan intended for a 40-ton car. (*National Car Builder*, Feb. 1884)

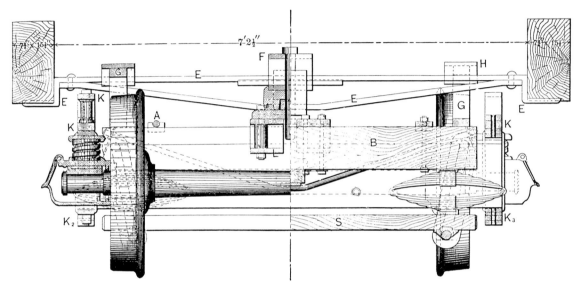

novelty in modern times, with the only major user being the Virginian Railway, which developed a fairly sizable fleet of gondola coal cars in the 1920s.

SOFT-RIDING TRUCKS

Hard-riding trucks for freight cars were considered acceptable by most car builders, who seemed to feel that such equipment was meant to be rough and tough. So long as the cars were not jolted to pieces or did not ride so poorly that the bands flew off the pickle barrels inside, who cared how much they bounced around? An exception was made in certain cases, however, especially when damage claims were likely to be excessive. One obvious special case was the cattle trade. A car lurching from side to side was more than uncomfortable for its occupants; it was dangerous. An animal thrown down could break a leg and become a costly claim. The humanitarian movement that welled up in the 1880s also created pressures for smoother-riding cattle cars. Soft-riding trucks became something of a standard for such equipment.

Swing motion and leaf springs were the common solution, though little else was done to make life aboard stock cars any less jolting for their occupants. The palace stock car companies all seemed to adhere to a standard arch-bar truck modified in this fashion.

A representative design is given in drawings for a Street's Western Stable Car Line truck of 1893 (Fig. 7.28).[58] There is nothing remarkable about the truck except for the rather short wheelbase of 4 feet 8 inches. Besides cattle cars, other users of easy-ride trucks included poultry cars and cabooses. Curiously, the Union Tank Line used easy-ride trucks, but surely not because its cargoes were so fragile; it is more likely that fear of derailments prompted petroleum car operators to accept the extra cost for a steadier-riding car. Drawings for a Union Line tank car are reproduced in Chapter 5 (Figs. 5.78–5.80).

STANDARDS

Like the standard freight car, a standard truck was a subject of much discussion. Considerable

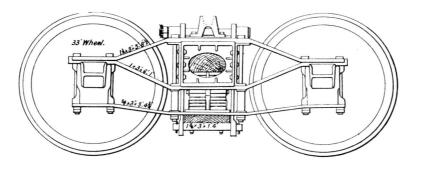

FIGURE 7.28

Soft-riding swing-motion trucks were popular for cattle cars. The example shown here was used by Street's Western Stable Car Line during the 1890s. (Dredge, *World's Columbian Exposition*)

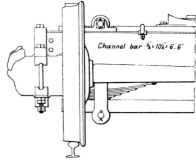

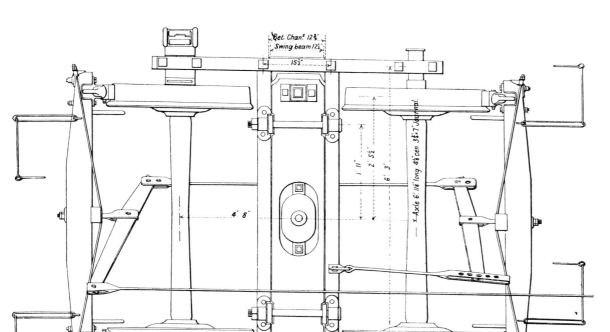

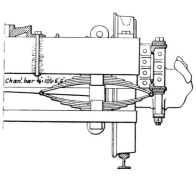

progress had been made in standardizing journals, axles, and wheels by the 1880s. At least several standard sizes had been established with comparative ease by the Master Car Builders Association since its founding in 1867. How well its members adhered to these standards is another matter, but there was a feeling of guarded optimism that what could be done with elements of the running gear might be applied to the entire truck. A standard freight car truck may sound simple and achievable. Yet when it came down to specifics, no universal agreement was possible. The swing-motion debate flared up, with each camp wanting all or nothing. In the end it was decided to have both. After three years of debate, the issue of the standard truck was put to a vote at the 1885 Master Car Builders Association annual meeting.[59] A decision was postponed with the agreement that sample trucks should be produced for inspection at the next year's meeting.

The sample designs offered were so conventional that surely no one could object, or so it was hoped. When the 1886 meeting convened, only the

Michigan Central and the Chicago, Burlington & Quincy had bothered to produce samples (Figs. 7.29, 7.30). In the discussion that followed an inspection, most members agreed that the samples were beyond criticism, models of orthodox practice; no one objected to their general design. But then it began—a picking away at the details. Just where to place the side bearings became a matter for debate. Many railroads adopted a 60-inch center, but others varied greatly. Spacing on the Illinois Central was 42½ inches, while the Boston & Albany placed them on 64-inch centers. The center plate issue was just as confused. A composite drawing reproduced here shows but a sampling of the various designs adopted by individual railroads. Many were similar, but none would fit exactly with another (Fig. 7.30). The standard truck was rejected by letter ballot at the 1886 MCB meeting.[60] Advocates of the standard truck pushed doggedly over the next several years, but to little avail. They met a second defeat in 1894 when it was announced that no fewer than thirty-six alternative plans had been recommended for the MCB stan-

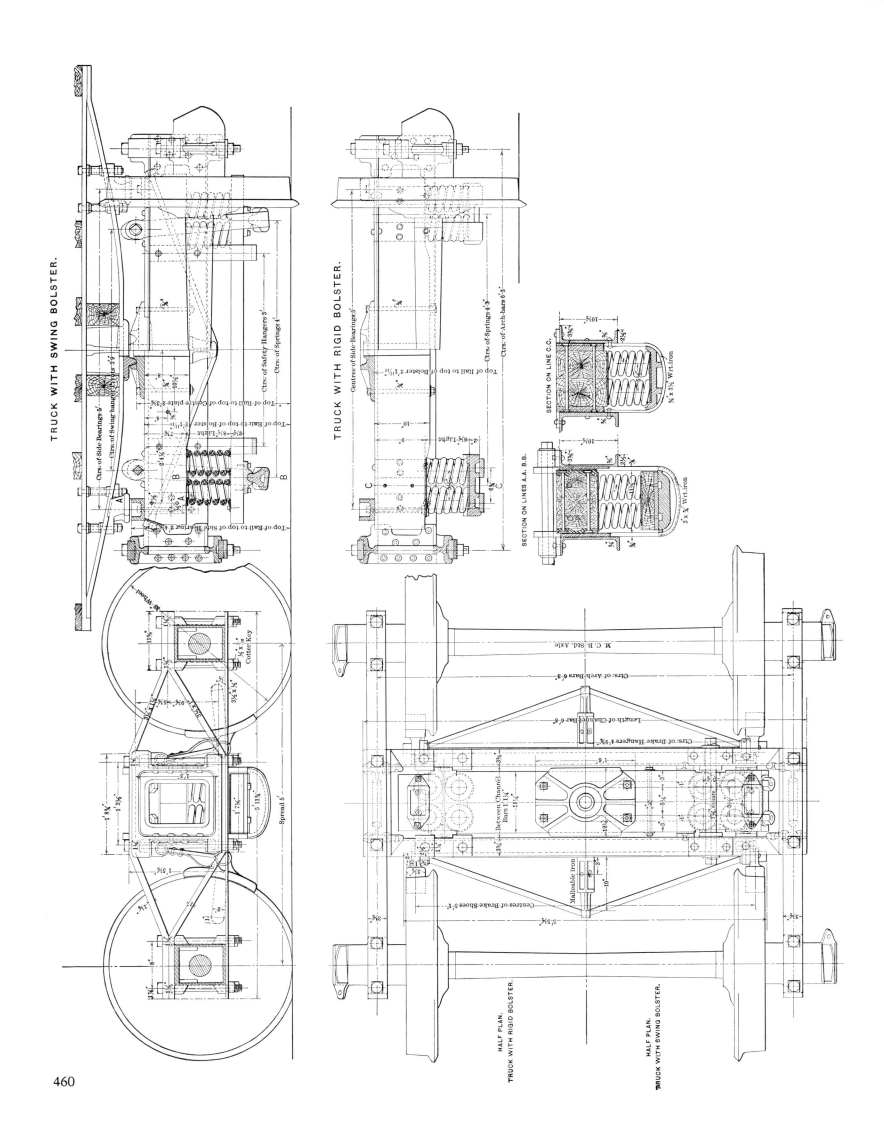

TRUCK WITH SWING BOLSTER.

TRUCK WITH RIGID BOLSTER.

SECTION ON LINE C.C.

SECTION ON LINES A.A. B.B.

HALF PLAN.
TRUCK WITH RIGID BOLSTER.

HALF PLAN.
TRUCK WITH SWING BOLSTER.

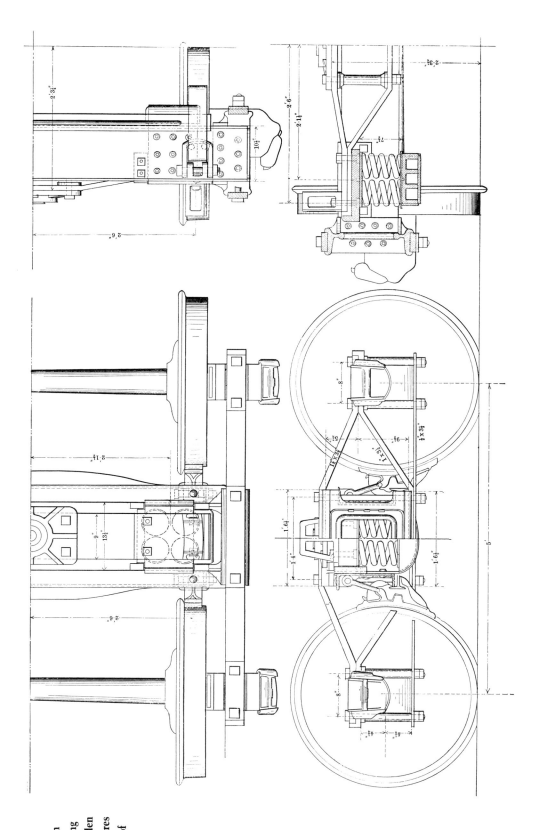

FIGURE 7.29 (opposite page)
In 1886 sample trucks based on
this design were shown at the
Master Car Builders Association
annual meeting in an effort to
win support for a standard swing-
motion design. (MCB Report,
1886)

FIGURE 7.30
The proposed rigid truck shown
at the 1886 MCB annual meeting
eliminated the traditional wooden
bolster and spring plank—features
not likely to win the approval of
old-line car builders. (MCB
Report, 1886)

FREIGHT CAR TECHNOLOGY

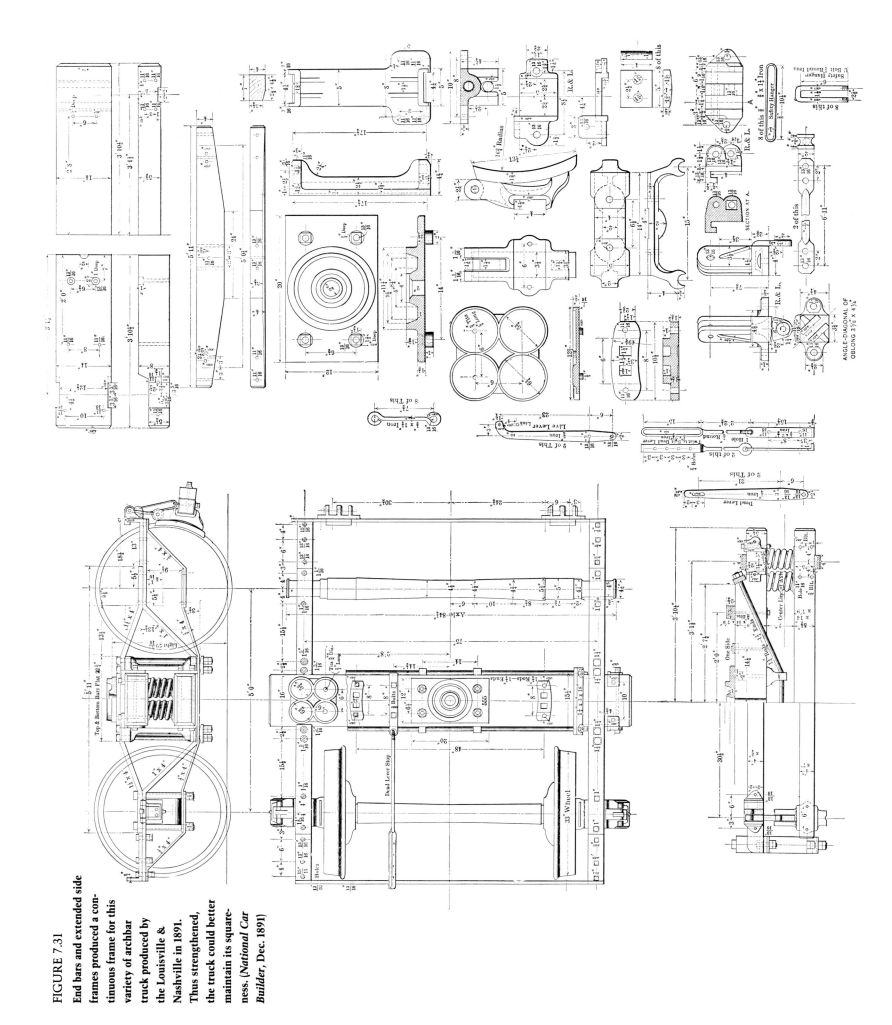

FIGURE 7.31

End bars and extended side frames produced a continuous frame for this variety of archbar truck produced by the Louisville & Nashville in 1891. Thus strengthened, the truck could better maintain its squareness. (*National Car Builder*, Dec. 1891)

dard truck.[61] All hope was then abandoned of reconciling so many differences, and the old order of custom-design trucks was allowed to continue.

STRENGTHENING THE DIAMOND ARCH

As the nineteenth century drew to its close, few car builders would have claimed that the diamond truck was perfect. It did the job and was cheap, but ideal it was not. In 1889 the *National Car Builder* said that although it held together "wonderfully" considering the big loads it carried, the arch-bar truck was hard on the track and hard on itself.[62] In his little book published for railroaders in 1895, Charles Paine offered this candid assessment of the diamond truck:

> There is nothing which so nearly approaches a fortuitous concourse of atoms as an ordinary freight car truck. How it holds together, how it does duty so long, are questions which every mechanic must reflect upon with amazement. It is not strange that every terminal and division yard is full of cripples, and that every truck in every car, except those fresh from the shop, is lacking something.[63]

Both statements combine praise with criticism, but they do not explain what was wrong with an industry standard that represented an investment of so many millions of dollars. Precision was neglected in order to hold down costs. It was crude, the product of unskilled labor, a device put together with a large hammer and a monkey wrench. Tolerances were liberal, and axle centers varied by ⅛ inch and sometimes more.[64] Bolt holes were routinely drilled 1/16 inch oversize.[65] Because of sloppy workmanship, king pins were placed off-center, which inhibited a true and free-running truck. And the minority that came out of the shop true and square were then misaligned by the stress and strain of everyday service. The continual motion of a truck rolling and jolting over the tracks caused bolts to work up and down, making the holes larger and the bolts themselves smaller. Loose nuts—an epidemic problem for all arch-bar trucks—added to the general limberness of the mechanism. All of this looseness meant that the truck was never square and, hence, that the axles were not parallel, which caused a series of problems such as excessive flange and journal wear. It also added to the drag or friction of the train, meaning that extra power—more fuel—was needed to move the train. In an extreme case a badly out-of-square truck could cause a derailment.

How serious was the problem? In one sampling it was found that even a new, tight truck was ¾ inch out of square.[66] After just one year's service the misalignment grew to 1¼ inches and after eight years to 2 inches. In 1883 W. R. Davenport of the Erie Car Works made an inspection of cars near his plant and found them all to have out-of-square trucks, some out by 2⅜ inches.[67] On a normal

curve (5.5 degrees), it was figured, a badly out-of-square truck produces 150 percent more friction than a true truck. In these extreme cases the wheel sometimes slipped rather than rolled because of the angle of the flange against the rail, which resulted in extra drag and the opportunity for a flat wheel.

A number of solutions were offered to reform the ailing diamond truck. Greater precision could be effected by drilling the parts in jigs. Holes could be drilled and reamed to size for a precise fit of bolts, but this was too fussy for the unsophisticated equipment and unskilled farm laborers employed for freight car work. Axel Vogt, locomotive and car designer for the Pennsylvania Railroad, would eliminate the loose-nut problem by using rivets.[68] His scheme was marketed by the Pressed Steel Car Company, though the Vogt truck was never that widely used. Car builders objected that repairs and disassembly would be inhibited by the riveted construction. A surer way to hold the truck frame square was to introduce some form of cross-bracing. Charles Davenport employed X-bracing in 1844, as already mentioned, and the New England continuous-frame truck—also of ancient vintage—addressed the problem. The idea was not new, but it was resisted because of the extra labor, material, and cost.

Trivial as these costs may seem, most car builders would have rather lived with out-of-square trucks than spend a few extra dollars per unit, though not all were so parsimonious. The Lehigh Valley reported the use of adjustable diagonal rods in 1885 to hold the diamond frame square.[69] Six years later the Louisville & Nashville adopted an old plan that proved effective.[70] The ends of the top and bottom bars were extended just beyond the wheels so that cross-braces might be attached at both ends to form a box frame (Fig. 7.31). This plan goes back at least to the 1870s and was much used by Billmeyer & Small for narrow-gauge cars at the time.[71] The American Steel Foundry of St. Louis offered a good plan around 1895 to make the truck square and not a trapezoid.[72] It consisted of an X-brace fastened to the lower side frames in place of the spring plank.

The wooden bolster was actually the weakest link in an assembly fabricated from so many individual pieces. It represents an ancient loyalty to wood as the proper material for car construction. The shift to metal fabrication was slow and grudging. There always seemed to be an effort to forestall and compromise. An agreement to add just a little more iron was made in the hope that some vestige of wood might be maintained. Just why this is true is difficult to explain, but it seems to have prevailed in all areas of car construction from trucks to frames to bodies.

The fondness for wood construction was more than a stubborn or uninformed loyalty to a traditional material. Car builders may not have been notably progressive as engineering professionals, but they were surely practical men. Their judgments were guided by economics and not sentiment. They stayed with wood because it was cheap in the short run. Most understood that metal parts would last longer, break down less frequently, and in the long run pay for themselves many times over. But just as the homeowner understands that slate and copper sheet will make a far superior roof than tarpaper and tin plate, he tends to select the latter because of cost. Car builders worked with limited budgets, and the choice of materials and methods was often expeditious—good enough to do the immediate job, but more costly in the long run. And so they stayed with wood bolsters because they cost $3.08, compared with $23.00 for a metal bolster.[73] It was a choice of having one very good car or seven not-so-good cars available for revenue service.

It seems ironic that there was a willingness to convert wooden wheel pieces to iron-bar side frames at such an early date, for the side frames carried far less weight than the bolster and were less likely to fail under a heavy load. Assuming that the car was in reasonable balance and evenly loaded, each bolster would carry half of the weight, while each side frame carried only one-quarter of the load. Ironically, in compression tests the arch-bar side frames proved far stronger than the wooden bolsters. A test was conducted by the Santa Fe in 1891 with an average diamond-arch-bar side frame.[74] The top bar measured 1⅛ inches thick, the inverted arch bar 1 inch, and the tie bar ⅝ inch. The frame buckled at 128,200 pounds, while the plain wooden bolster broke under a load of only 72,800 pounds. A wood bolster reinforced with two truss rods—a common style of construction—broke at 109,550 pounds. In another test wooden bolsters performed no better.[75] In this case deflection and permanent set or bending were recorded under a pressure of 25,000 pounds:

- 8 × 11½-inch oak with two truss rods deflected 1⅜ inches and set at 1¼ inches.
- 9 × 12-inch oak with two truss rods deflected 5/16 inch and set at 1/16 inch.
- 8 × 12-inch wood and iron composite deflected 1/32 inch and set at 0 inches.

By contrast, a pressed-steel bolster showed no deflection until under a pressure of 100,000 pounds, when ¼ inch was recorded. There was no permanent set until the steel pressing was subjected to 162,500 pounds. A set of 1¾ inches was recorded.

In 1893 a Master Car Builders Association committee made a random sampling of freight car truck bolsters and found none free from defects. The wood had shrunk, split, cracked, splintered, or warped. Metal parts such as the side frames, side bearings, and center plates were loose. Why put up with such a weak, defective element in the truck structure? No wonder nothing remained square or tight. The major excuse was always cost, yet who could afford such an ineffective bargain? In a strange twist of logic, the Union Pacific defended the wooden bolster because it was easier to repair than a metal one![76] The Northern Pacific agreed and contended that it wanted nothing but wooden bolsters and transoms on its property. Orders to car builders confirmed the prejudice in favor of wooden bolsters. As late as 1889 Wason reported that 100 percent of its orders were for wood; Pullman reported 80 percent; and the U.S. Rolling Stock Company reported 50 percent. Only the Harrisburg Car Company stated a 100 percent record in favor of metal.

Late examples could still be found at Pullman as late as 1894. We offer excellent photographs taken in that year at the Pullman Car Works, Chicago, which show a classic wood-bolster truck in all its naked primitiveness (Figs. 7.32, 7.33). These Cleveland-style trucks were complete except, curiously, for the center plates, which were not in place at the time the photographs were made. The Southern Railway felt comfortable using wooden bolsters for a group of 30-ton-capacity coal hoppers it acquired in 1896. The 12- by 13-inch oak beam was materially aided, however, by two channel-iron transoms, and so for even a middle-weight car, total dependence on wood was no longer considered advisable (Fig. 7.34). In another progressive step, the spring plank was replaced by a steel-plate step support.

Metal truck bolsters were surely not a new idea when they began to be adopted into standard practice in the 1890s. Examples going back half a century were given earlier in this chapter. The benefits inherent in metal cross-framing were also clearly demonstrated by the good service rendered by the Thielsen truck. Iron body bolsters were popular and much used by 1890, as has been documented in Chapters 3 through 6. The cheapness of the wooden truck bolster apparently persuaded car builders to hold to this form of construction until engineering considerations overwhelmed economics. Once common car capacities exceeded 30 tons, the wood bolster was doomed. It simply could not safely or dependably carry loads much above this weight, and so when 35- and 40-ton cars began to squeeze out the old, smaller cars, there was no practical alternative other than to adopt the metal truck bolster.

Because railroad repair shops were not equipped to handle large metal components, there was a tendency to develop bolsters fabricated from many small parts. In 1888 the Lehigh Valley had a bowstring arch bolster made up of several dozen thin bars, bolts, plates, and thimbles. An example of this design is illustrated in Chapter 5 (Fig. 5.33). An engraving of a similar bolster for an 1883 Penn-

FIGURE 7.32

Despite all the ingenious variants in truck design developed since the first diamond arch bars appeared, the simple Cleveland style remained much in favor. This example was produced in 1894 at the Pullman Car Works for a refrigerator car. (Pullman Neg. 2524)

FIGURE 7.33

The rugged simplicity of the diamond-arch-bar truck is well illustrated by this 1894 photograph. Note the truss-supported wooden bolster. (Pullman Neg. 2525)

FREIGHT CAR TECHNOLOGY

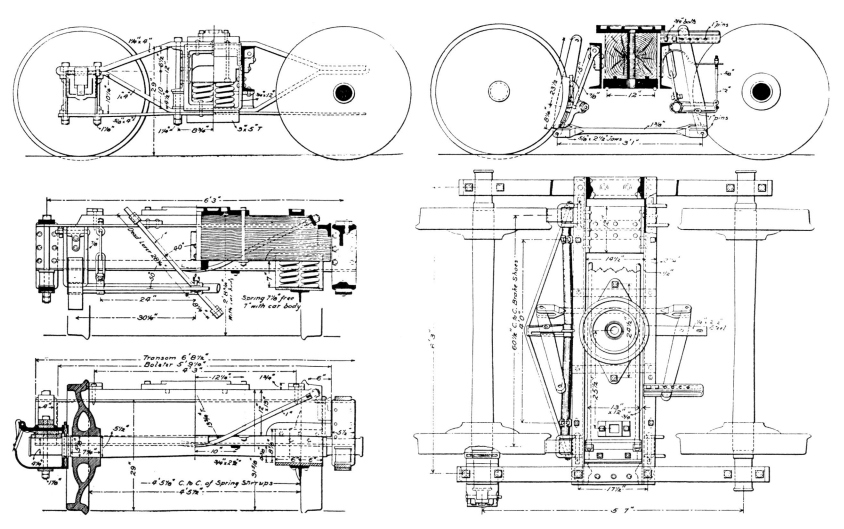

FIGURE 7.34

Car designers were testing the upper limits of the wooden bolster by the late 1890s and would soon be forced to adopt all-metal construction. This Southern Railway design had a floating bolster and no spring plank. (*Railroad Gazette*, April 17, 1896)

sylvania design was reproduced earlier in this chapter (Fig. 7.19). A tendency to break down once the bolts loosened made these needlessly complex fabrications something less than a success. A more direct solution was to use I-beams, which were easily obtainable from any sizable rolling mill or metal supply company. Two I-beams placed side by side with a center plate riveted on made a simple, strong bolster. The Boston & Albany tested the plan in 1886.[77] The Norfolk & Western adopted the same idea in the 1890s, and as cars grew larger the line simply increased the size of the I-beam; when 40-ton cars came into service around 1902, the N&W was using twin 10-inch I-beams.[78] The N&W found them cheaper than the more complex patented varieties offered by commercial makers, who seemed wedded to exotic pressings or castings.

By 1888 big railroads like the Michigan Central were committed to eliminating wood-bolster trucks from all of their freight equipment.[79] Wood bolsters were wracked to pieces and unfit for service after ten years' use. They should be made extinct. Wood-bolster arch-bar trucks cost $200 per pair but another $50 would buy a pair of metal bolsters that would double the life of the undercarriage. The Michigan line was determined to purchase no more new cars with wood-bolster trucks.

All existing cars would be rebuilt with metal bolsters when scheduled for major rebuilding. The Santa Fe found metal truck bolsters and heavier axles a cheap way to boost car capacity.[80] In 1897 it had a betterment program under way that converted 20-ton cars into 25-tonners at a relatively modest cost.

Steel-car advocates like Charles T. Schoen saw a way to draw traditional car builders into the realm of metallic rolling stock by offering a steel truck bolster. Schoen had already established a market for pressed-steel car parts, like center plates and stake pockets, when he unveiled his diamond truck bolster in 1895.[81] The name, if nothing else, was intended to reassure old-line car builders that the merits of the diamond truck were to be carried on. The bolster was composed of two heavy ½-inch-thick steel stampings riveted together by several small connecting brackets and the center plate. The bolster measured 12 inches deep at the center and tapered down to 6 or 7 inches at its end. Schoen claimed that 300 to 500 pounds of dead weight could be saved by using his hollow-metal stampings. Several years later, the Bettendorf Company of Davenport, Iowa, introduced a design that looked remarkably similar to the Schoen diamond bolster, except that no stampings or rivets were employed (Fig. 7.35). Twin I-beams were fastened together with malleable-iron brackets and tubular tenons. The I-beams were cut longitudinally and bent down so as to be smaller at their ends. The bolster weighed 440 pounds and could support loads of more than 100,000 pounds with no permanent set.[82] At 180,000 pounds a permanent set of ½ inch was recorded, but for normal loads bolsters of this type could be depended upon for long years of service.

Another fabricated commercial truck bolster to appear in the late 1890s was the Simplex bolster.[83] The design was patented by Waldo H. Marshall (1864–1923) on August 11, 1896. At the time, Marshall was an obscure mechanical superintendent, but ten years later he rose to prominence as the president of the American Locomotive Company. This bolster was built as a simple king post truss with the arched upper cord made from a 12-inch channel iron. The bottom cord was a flat bar of steel measuring ⅝ by 10 inches. The ends of the bottom cord were rolled over and riveted to the ends of the top cord, thus forming a simple, strong truss (Fig. 7.36). The Simplex bolster was still a current design into the late 1920s.[84]

Around 1891 the ultimate in metal truck bolsters was introduced by the American Steel Foundry Company of St. Louis.[85] A cast-steel truck bolster combined strength and simplicity. Being a single unit, it had no fastenings to loosen.

The idea originated with M. C. Schaffer, master car builder for the Missouri Pacific. The bolster included the center piece and side bearings in a one-piece casting. Its weight was given as between 400 and 500 pounds, and it cost only $10 more than a standard bolster. The cost claim is undoubtedly too optimistic, but no one could argue about the steel bolster's durability. In April 1893 the Missouri Pacific was installing eight hundred Schaffer bolsters per month. By 1895 the design had been refined to include the bolster column and spring seat as a one-piece casting, which was far more solid than the old plan (Fig. 7.37). Longer-than-usual springs could be used because of the special casting and the elimination of the usual thick, wooden spring plank. Longer springs gave the car a better ride. The spring plank was replaced by a rugged X-brace made from two 3- by ¾-inch iron bars. The X-brace held the truck square. The bolster itself was cast as an open girder roughly T-shaped in cross-section. This was later modified to a U shape and then still later to a box shape for greater strength.

The Northern Pacific is also known to have used the American Steel Foundry bolster.[86] The idea was so superior that others surely adopted it as well. Imitators quickly seized upon the plan; the firm of Shickle, Harrison & Howard, also of St. Louis, put a very similar bolster on the market in around 1896.[87] It was tested by the Louisville &

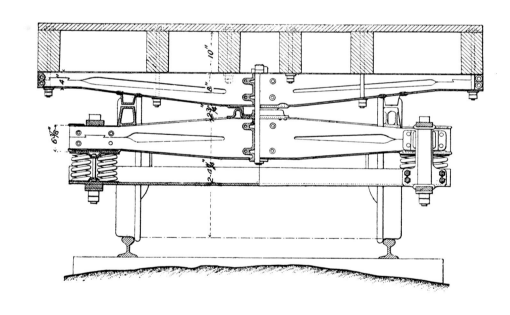

FIGURE 7.35

Steel bolsters greatly strengthened car trucks and became increasingly popular after 1895. The example shown here was patented in 1899 by William P. Bettendorf. It was adapted for an arch-bar side frame. Note the channel-iron spring plank. (*Railroad Car Journal*, June 1899)

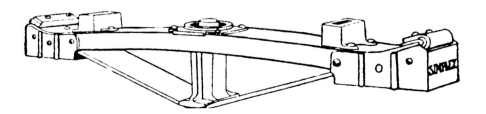

FIGURE 7.36

The Simplex truck bolster was patented in 1896 and remained popular for about twenty-five years. The arched top cord was made from a channel iron. (Marshall M. Kirkman, *The Science of Railways* [Chicago, 1894])

FREIGHT CAR TECHNOLOGY

FIGURE 7.37

One-piece steel castings offered unitized construction to car builders accustomed to elaborate fabrications that were too dependent on the security of their fastenings. This cast-steel truck and body bolster set was offered by American Steel Foundry. The side frames were held square by an X-frame used in place of the normal spring plank. Note that the columns and spring seat were cast as a single unit. (*National Car Builder*, Nov. 1895)

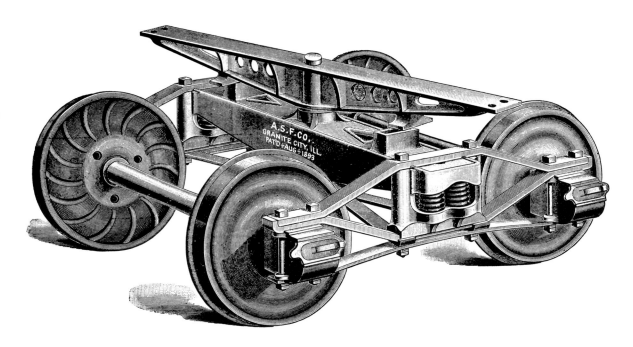

FIGURE 7.38

Another one-piece cast-steel truck and body bolster set was introduced around 1895 by Shickle, Harrison & Howard. (*Locomotive Engineering*, April 1896)

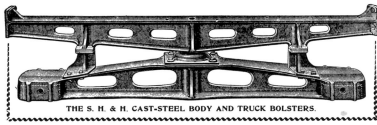

Nashville Railroad on one hundred furniture cars (Fig. 7.38). Within a few years cast-steel bolsters were recognized as the best form of freight car truck bolster and were essentially the industry's standard by 1930.

THE SLOW DEMISE OF THE ARCH BAR

With the end of the wooden era it seemed logical that the arch-bar truck would also find a resting place in the museum of American railroading. Steel-car advocates had won the day, and they pushed pressed- and cast-steel trucks as the only fitting carrier for modern freight cars. The arch-bar side frame was an ancient design and even had the look of a rustic relic. Yet car builders were not altogether ready to abandon their old friend. Misguided sentiment was hardly the reason for this loyalty. Car builders were a practical crowd, if nothing else, and in their worldly way they only championed survivors. For all their obvious failings, arch-bar trucks had proved themselves year after year in some of the hardest service endured by any machinery anywhere in North America. Neglect and abuse could not stop them from running. They were called upon to carry the heaviest

cars in service. Most of the big flats mentioned in Chapter 6 were carried by arch bars. In 1906, when Allis-Chalmers wanted a new 100-ton-capacity flatcar, it had the sixteen-wheeler outfitted with arch-bar trucks—even though steel trucks had been available for over a decade and the company was neither impoverished nor untutored in things mechanical.[88] With a slight increase in size of the bar stock—a few extra-heavy arch bars grew in size to 1¾ by 6 inches—larger axles, and heavier wheels, the capacity of the truck could be greatly increased. The adoption of metal bolsters eliminated the weakest link in the basic design. No case could be made for weakness—the reformed arch bar was a tough customer.

The arch bars' detractors said that they were unfit for high-speed service, and that as speeds increased—as they began to do in the 1890s—the safety factor would render them unfit for mainline use. But again the critics were confounded by actual example. Standard refrigerator cars fitted with ordinary arch-bar trucks were regularly attached to passenger trains for express passage between California and New York.[89] During the Christmas rush season with express cars in short supply, the Pennsylvania Railroad, hardly a slow-speed line, used ordinary boxcars on passenger trains to fill in. Special stock trains sped eastward out of Chicago as a regular matter. Only the stock cars were likely to have swing-motion trucks; the others would likely have been rigid. If the arch bars could be run safely at passenger train speeds, then surely they could handle any schedules developed by the freight department.

Another factor in the extended life span of the arch bar was cost. Exact cost figures have not been found for comparisons among cast, pressed, and arch-bar side frames; the Master Car Builders Association based replacement cost on the capacity of the trucks and the style of bolster (wooden or metal) irrespective of the style of side frame. But consider the mere cost of materials: bar steel was slightly cheaper per pound even than cast iron, and about five times cheaper than cast steel. In 1898 the cost per pound was 1.12 cents for rolled shapes (bars, angles, etc.), 1.5 cents for cast iron, 2.75 cents for malleable iron, 3 to 4 cents for pressed steel, and 4 to 6 cents for cast steel.[90] Just on a cost-of-materials basis, then, arch bars would have been much cheaper than their rivals. The cost of patent license fees must also be considered in the case of almost all pressed- and cast-steel trucks.

A census of freight car trucks does not seem to have been recorded until the arch bar was going out of favor. Photographic evidence as recorded in surviving builders' photographs of new freight cars—the car dictionaries are but one obvious source—indicates that the arch bar prevailed into the early years of the twentieth century. The 1919 edition of the *Car Builders Dictionary* stated that the arch bar was the standard for many years and that only recently had cast-steel trucks come into common use.[91] Nine years later the same source recorded that some railroads continued to favor the arch-bar truck. The Methuselah of car trucks was surprisingly lively in its old age. It had indeed outlived its early critics—Charles Schoen and Samson Fox had both passed out of this life long before 1928. Indeed, the pressed-steel truck itself had passed out of favor after a brief interlude of popularity. But as all veterans understand, there is a difference between a long Indian summer and perpetual life, and time was running out for the ancient order of arch bars.

In July 1929 the American Railway Association proposed prohibiting arch bars for new or rebuilt cars after January 1, 1930.[92] This ruling would apply, as do all such rules mentioned in this case, to cars intended for interchange service. Industrial, work, and home-road cars would not be affected. In June 1930 a more sweeping motion was adopted to prohibit cars equipped with arch bars from interchange service after January 1, 1936. At the time, nearly half of all interchange cars still rode on arch-bar trucks. The breakdown in 1930 was

43.6 percent for railroad-owned cars and 49.5 percent for private fleets.[93] Despite the retirement of much old stock during the 1930s, the economic depression of the time hampered efforts to comply with the changeover ruling. As of January 1, 1934, 34.7 percent of railroad-owned cars had arch bars, while 41.3 percent of private cars were so equipped.[94] It wasn't all a matter of impoverished short lines; big roads were involved as well. The Pennsylvania—the "Standard Railroad of the World"—which boasted the finest car fleet in the land, had 185,000 cars with arch bars at the end of 1936. The chances of converting so much stock in one year's time were hopeless. An extension was granted for two years.[95] More extensions were granted, so it was not until December 1939 that the archaic arch bar was banished from interchange service.

But the arch bar would not soon leave the stage. Thousands of cars in maintenance-of-way or industrial service fitted with arch bars could be seen on American railroads well into the 1960s. A few holdouts may yet be found. Elsewhere in the world, especially in less developed nations, hundreds of arch-bar trucks are still in daily service. Not all are necessarily veteran or vintage. In 1985 I spoke by telephone with a descendant of the founder of the Gregg Company Limited, in Hackensack, New Jersey. The firm had been manufacturing rail rolling stock for many years in Portugal rather than in the United States and was ready to supply new cars of all gauges and types. Arch-bar trucks were featured in the company's most recent catalog, along with the latest cast-steel types. The arch-bar truck may well live on into the twenty-first century.

PRESSED- AND CAST-STEEL CONSTRUCTION

The effort to unseat the long-entrenched arch-bar truck started in 1890 and continued for another decade or so. Designers were determined to develop something more substantial than the wood and strap-iron contraptions that had for so long dominated the truck world. Everyone engaged in this attack on the established order hoped to introduce a stronger, lighter, and more durable truck frame. They tried structural steel, heavy steel stampings, or castings. A few inventors advocated a combination of all three. The most radical designers abandoned the spring bolster for the pedestal truck, which featured a rigid bolster fastened to the side frame. Springs were placed above the axle boxes.

The strongest and earliest advocate of the pedestal truck was a self-assured engineer from outside the American railroad industry. Samson Fox hailed from Yorkshire, England, and gained a special facility in handling sheet steel from long years in boiler manufacturing.[96] In 1887 he applied his knowledge of steel fabricating to railway car un-

FIGURE 7.39

The Fox pressed-steel truck was a remarkable innovation in freight car trucks when it was introduced in 1889, but it never achieved the universal acceptance envisioned by its British promoters. (*MCB Report*, 1891)

dercarriages and trucks. He believed that special, large steel stampings could solve the ills of railway rolling stock by providing a very solid but light supporting structure. The parts were made from cherry-red steel sheet pressed to shape at 850 pounds per square inch in a hydraulic press equipped with suitable dies to shape and flange the part as desired.[97] The stampings, often ½ inch thick, were fastened together by ¾-inch rivets. Traditional hand-riveting was not adequate for the structures Fox envisioned, and so the fastenings were driven home by hydraulic hammers. Fox claimed that his technique of riveting was so perfect that in no instance had a loose rivet been detected.[98] His trucks could support 120 tons without failing. They saved anywhere from 1.0 to 2.5 tons of dead weight per car. They were guaranteed for five years but would surely last fifteen to twenty years.

These claims were probably overdrawn, but there is no question that Fox had created a distinctive and very original style of car truck that warranted careful testing. Fox's design surely started with a clean sheet of paper, for it looks like no other truck of its time. Rather than the usual jumble of small parts, it was composed of just three large stampings (Figs. 7.39, 7.40). All previous trucks had been an open filigree, but Fox's plan was as plain and solid as a toolbox. Its remarkable appearance and the fact that it was so very different must have worked both for and against its adoption. Those seeking an alternative to the arch-bar truck would have been enchanted by its novelty, while those content with the status quo would have been put off by its odd, battleship appearance. The fact that Fox was British would not have added to the truck's appeal, since there was a prevailing opposition to British railway rolling-stock design in the United States.[99]

Yet Fox was keen to promote pressed-steel trucks in North America. He felt that they were in fact ideal for colonial railways, and he pushed

them in India, Japan, and elsewhere. The market within the British Isles was actually rather limited because of the continued preference for four-wheel goods wagons. Fox came to the United States in 1888 to promote sales and establish a factory.[100] A plant was set up in Joliet, Illinois, but rather few orders were tendered. Fox was a shrewd businessman as well as an original engineer, and he sensed that a proper salesman was needed to handle the job of promotion. He found his man in James B. Brady (1856–1917), perhaps the greatest personality in the railway supply industry. Brady's congenial manner made him the star salesman for the mighty railway supply house Manning, Maxwell & Moore. In early days his employment as a messenger for the New York Central offered an opportunity for friendships inside the railroad industry. His talents and connections were turned to promoting the Fox truck.

By 1891 thirty railroads were using Fox trucks.[101] The plant's capacity was one hundred trucks per month, which the Fox management hoped to boost to one hundred per day. This ambition was apparently predicated on the notion that U.S. railroads were going to adopt the pressed-steel units and abandon the arch bar in a wholesale fashion. A year later, some forty railroads were using Fox trucks.[102] The New York Central was an enthusiastic convert, but most other roads were less committed and seemed ready only to sample, test, and think about this new marvel in railway undercarriages. Lines from the New Haven to the Northern Pacific all became at least limited patrons of the Joliet shops. In 1896, just eight years after their introduction, around sixty thousand Fox trucks were in service.[103] Business was so good that Fox was contemplating a new plant for his American branch works.

This thinking was just a trifle optimistic. Reception of the new truck was good but not overwhelming. Most cars continued to ride on arch bars, and the sixty thousand Fox trucks in service seem less significant in light of the fact that the total freight car fleet registered over 1 million cars at the time. Fox remained bullish, however, noting that the good service of his pressed-steel creations on Mexican railroads was the best possible testament to their durability. A new design was created to satisfy the special requirements of tenders. It featured a floating bolster supported by leaf springs; coil springs were also inserted over the axle boxes, as in the original design.

Car builders were actually finding much to criticize in the pressed-steel truck. It cost more than the common diamond truck; a pair of Fox frames cost $110, and by the time wheels, journals, and other fixtures were added on, the cost far exceeded that of the arch-bar truck. The fact that the pressed-steel truck ran square and would outlast a diamond truck somewhat countered the cost

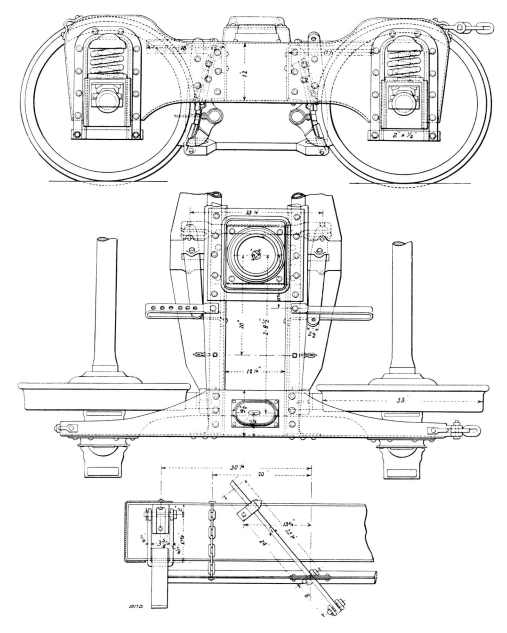

FIGURE 7.40

The simplicity of the Fox truck is obvious when this drawing is compared with the views of arch-bar trucks reproduced elsewhere in this chapter. Note that even the brake beams were formed from heavy-gauge sheet steel. (Dredge, *World's Columbian Exposition*)

question, but it did not silence the major objection that car builders were raising: difficulty with repairs.[104] The most frequent repair involved removing the journal boxes and wheel sets for servicing, and this proved troublesome with the pedestal-style truck because it was necessary to raise the truck very high to remove the wheels. The same maneuver on a diamond truck required only the removal of several large bolts. Fox trucks also developed another failing after prolonged service, one that posed a serious safety problem: the riveted connections between the transoms and the side frames failed because of the shearing strain imposed upon them by the truck's constant up-and-down movement.

One railway mechanical officer expressed the frustration of his colleagues at the 1896 meeting of the Master Car Builders Association.[105] Richard P. C. Sanderson of the Norfolk & Western said that he was eager to try the pressed-steel truck because

he believed something must be found to succeed the old-fashioned arch-bar variety. He was hopeful at first and repeatedly tried to measure the good points of pedestal-style trucks, but after a long trial he found that they exhibited no marked superiority over the common diamond truck. A native of England himself, Sanderson was unlikely to have been motivated by Yankee prejudice against Samson Fox. The decline of the Fox truck was apparently as rapid as its ascent. There was little or no discussion, but pictorial evidence, particularly as noted in the *Car Builders Dictionary*, indicates that by around 1910 the Fox truck was no longer in fashion. Railroads wanting a cheap truck stayed with arch bars, and those wanting a first-class truck purchased cast-steel units.

With the benefit of hindsight we can see that the Fox truck was a failure, but during the 1890s this was not so apparent, and a number of imitators were soon in the field hoping to capture part of the enthusiasm for the new style. From the perspective of history most of these efforts were unimportant, but they generated considerable interest at the time and so are worthy of a brief review.

Charles T. Schoen, the pioneer steel-car maker, whose career is examined in greater length in Chapter 8, attempted to counter his rival, Samson Fox, with his own plan of freight car truck. Schoen's entry into the field was both late and unsuccessful. In 1895 he produced prototypes of a modified arch-bar truck that featured pressed-steel bolsters and spring planks.[106] About the only novel feature was shallow U-shaped arch bars made from steel pressings. This uninspired plan seems to have excited little interest, and so Schoen set about copying the Fox truck. The result was again not very original and in many ways defective.

Superficially the Schoen truck looked like the Fox, except that the humpback ends were eliminated; while this disguised the look of the side frames, it structurally weakened them because metal was taken away where it was needed to support the springs. Strength was added by a channel stamping made with a wide top flange for attachment to the transoms (Figs. 7.41, 7.42). Tie bars, gusset plates, and a main center-panel plate made the Schoen truck a fussy, overly complicated assembly compared with the Fox truck side frame, which was a simple one-piece stamping. In laboring not to imitate his competitor, Schoen succeeded only in producing a truck frame as complicated and full of parts as the very truck he had hoped to replace—the diamond arch bar. Even so, as the pioneer manufacturer of steel freight cars, Schoen was in a position to promote his truck, and a small number—still not very many—were used by the Pittsburgh, Bessemer & Lake Erie for its 1897 steel hopper fleet. Other lines were even less enthusiastic than the PB&LE.

Schoen, no doubt annoyed that so much busi-

FIGURE 7.41

Charles T. Schoen attempted to produce a pressed-steel truck on a plan independent of the Fox patents. The design was overly complex and attracted few customers. (*Railroad Car Journal*, June 1898)

ness was going to Fox, conspired with Axel S. Vogt, mechanical engineer with the Pennsylvania Railroad's Altoona design office, to develop a better all-steel truck. Vogt returned to the orthodoxy of the time-tested arch bar. Blending too much of the old with too little of the new, Vogt created a slightly upscale diamond arch bar likely neither to inspire nor to offend any practicing car builder of the time. The scheme was patented on August 10, 1897 (No. 587,886). The inventor was described as a subject of the king of Sweden and Norway, even though Vogt had resided in the United States since 1870. The top arch bar was a one-piece steel stamping, U-shaped at its ends, with a wide center flange for attachment to the transoms. The inverted arch bar was angle iron bent to shape, while the bottom tie bar was plain flat bar. A floating bolster was employed, but no spring plank was provided. The springs were supported by two small platforms suspended from the transoms. The general plan of Vogt's design is shown in Figure 7.43. Unlike Schoen's own creation, the Vogt truck had a wider application. It was used for some of the PB&LE 1897 hoppers. A few years later, the Rio Grande Western used them on some 50-ton hoppers. In a Pressed Steel catalog issued in 1920 Vogt trucks are shown under cars furnished to the Reading, Tonopah & Goldfield, the El Paso & South Western, and a few other roads. Most of these cars were built around 1915.

John W. Cloud, another mechanical engineer employed by the Pennsylvania Railroad, may have been inspired by the efforts of Axel Vogt to try his hand at truck design. He obtained a patent on May 19, 1896 (No. 560,258), for a pressed-steel design that featured cast-malleable-iron pedestals riveted onto ⅝-inch-thick pressed-steel side panels (Fig. 7.44). The transoms were two 13-inch-deep steel channels. The big advantage of Cloud's design was the space provided for springs by the cast ends. The Fox truck could accommodate only one spring over each axle box.[107] According to Cloud, a single large spring was too stiff and unyielding. On rough

track the front axle was so unresponsive that the rear set of wheels tended to lift up from the track. If the bump was intense enough, a derailment could follow. Cloud's design could handle a cluster of four springs over each axle box, thus guaranteeing a more responsive suspension. It could also be equipped with fully elliptical springs, as shown in Figure 7.44.

Much was claimed for the Cloud truck. The head of the Chicago-based firm marketing the truck, Willard A. Smith, was so extravagant in his praise as to cause skepticism on the part of experienced car builders, who expected only a serviceable product and not the final solution to every mechanical ill of freight car trucks. Smith, speaking before the New York Railroad Club in April 1900, assured his audience that the Cloud truck "obviates all danger from hot boxes, . . . greatly decreases derailments, . . . [and] almost absolutely prevents, or at least reduces to a minimum, the flange wear of wheels." Smith's hyperbole did not completely discourage sales, for thirty sets were provided to Pressed Steel for the PB&LE 1897 hopper. By June 1898 eighteen hundred cars were carried by Cloud trucks, and in 1906 around ten thousand cars were so equipped.[108]

Cloud's design had by this time been taken over by the Kindl Company of Chicago. The firm had earlier roots in a pedestal truck introduced in 1896 by Frederick H. Kindl of Pittsburgh. Kindl boldly copied the Fox truck, but it was fabricated not from pressed steel but from angle irons, channels, and gusset plates (Fig. 7.45). The transoms were formed from two 15-inch channels. Kindl's plan was heavy and complex. Each truck frame involved no fewer than 50 pieces and 252 rivets.[109] Rather than simplifying, Kindl seemed only to add parts. It is not surprising that Kindl dropped his own plan for that of John Cloud.

G. R. Joughins, superintendent of motive power for the Norfolk & Southern Railroad, was active in promoting steel freight cars. In 1896 he developed a truck design suitable for the forthcoming steel fleet.[110] It offered a special advantage over the Fox, Schoen, and Cloud trucks, because the journal box and wheel sets could be removed without more than the normal difficulty. This was accomplished by a hinged leg or pedestal at each corner of the truck frame (Fig. 7.46). Like Fox, Joughins understood the merits of a simple side frame, and so he worked with a 15-inch I-beam. It was split and bent open at each end to form a jaw. The hinged legs of the pedestals were castings, most likely made from malleable iron, but it is possible that steel was used instead. Joughins used both rivets and bolts as fasteners, as is shown in the drawing. Around 1900 the Sterlingworth Railway Supply Company of Easton, Pennsylvania, was producing the Joughins truck.[111] After Sterlingworth was swallowed up by American Car & Foundry, the

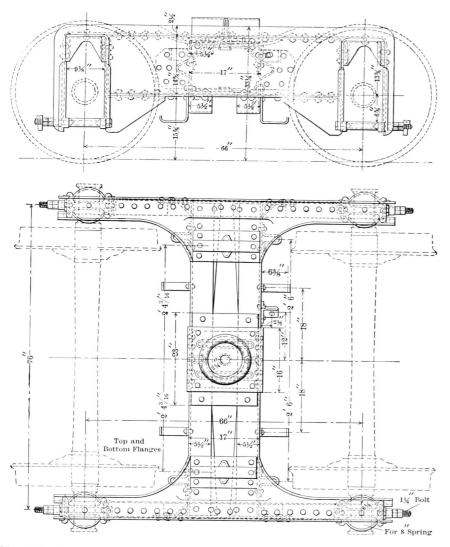

Joughins truck was perpetuated for at least a few years into the twentieth century.[112]

Within months of the introduction of the Joughins truck, Herbert H. Hewitt of Buffalo had introduced a very similar truck.[113] The design was covered by four patents, all issued on November 17, 1896. Hewitt specified a special rolled shape that resembled a common 12-inch I-beam, except that it had a small vertical flange top and bottom that added 5 inches to its depth. The beam was split and spread open to form a space for the springs and axle box. The end leg or pedestal was loose so that the wheel sets could be removed for servicing. Hewitt used a long bolt through the leg to hold it in place.

Beyond Hewitt there are easily another half-dozen imitators of Fox, Schoen, Cloud, and company. Most were not very original or prepossessing. The Buckeye truck, a plain-Jane slab of an I-beam with a spring bolster, was used mainly for tenders rather than cars. The Wright truck was a poor imitation of Cloud's design and appears to have received little attention. Of all these sub-varieties, perhaps the most interesting was the Black Diamond truck produced by the Bloomsburg Car Company to the design of R. W. Oswald.[114] Oswald bent the heavy angle iron roughly into the shape of a Fox side frame. He added an arched angle iron to form a diamond arch bar. The assembly was ungainly in appearance. We know less of its practical virtues, but it appears to have been little used despite the promotions of an established car builder.

FIGURE 7.42

The ideal of strength and simplicity was lost in the Schoen pressed-steel truck, as it was in most of the steel trucks introduced in the 1890s to compete with the Fox bogie. (*Locomotive Engineering*, Aug. 1897)

FIGURE 7.43

The Vogt truck was the old arch bar rendered in pressed steel. The deep top flange helped to ensure squareness. The suspension was dependent on a floating bolster. (O. M. Stimson, *Modern Freight Car Estimating* [Anniston, Ala., 1897])

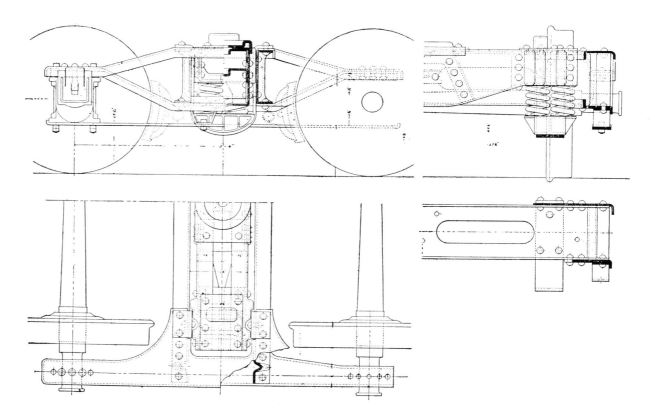

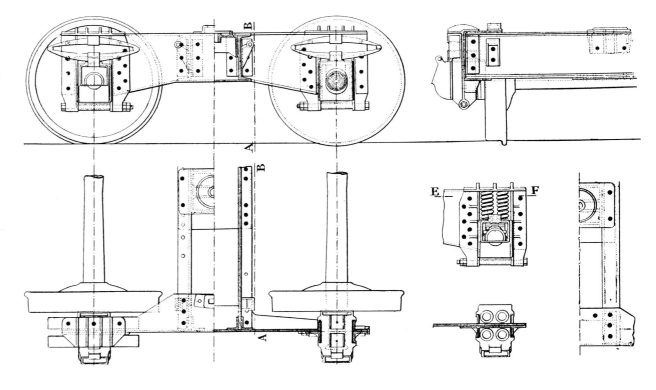

FIGURE 7.44

The Cloud truck, introduced in 1896, was touted as an improvement over the Fox truck because of a superior suspension. It proved moderately popular for about a decade. (*Railroad Car Journal*, June 1899)

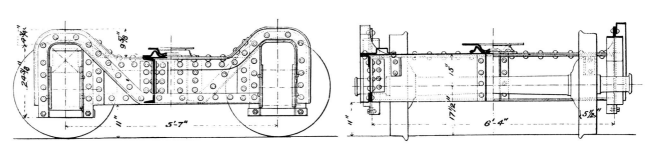

FIGURE 7.45

The Kindl truck dates from 1896 and was a clear imitation of the Fox design. It incorporated 302 pieces and rivets, far too many for it to succeed the diamond arch bar. (*Railroad Car Journal*, Sept. 1899)

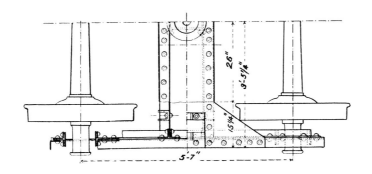

Car builders wearied by the claims of the pressed-steel truck's advocates were equally skeptical of the bright talk about cast-steel trucks that began in earnest in the late 1890s. This time their wariness was misplaced, for cast steel proved to be the ultimate solution to the truck problem. It would end the multiplicity of parts and eliminate loose bolts, broken straps, and sheared rivets. The components had been reduced to the smallest possible number—the main elements were down to just three, a bolster and two side frames. Each was a single, one-piece unit. The axle boxes were cast integral with the side frames. The bolster fitted into an opening in the center of the side frame. It was held in line by a broad grooveway, cast near its ends. The bolster was free to move up and down, but its travel was limited by the springs. The design of the cast-steel truck as worked out by 1902–1903 was simple, massive, and comparatively trouble-free.

Designs were being offered for cast-steel freight car trucks as early as 1884.[115] The designs were worked out by the members of the Master Car Builders Association. The first design strongly resembled the Fox side frame, rendered in cast rather than pressed steel. The second design was very much like a locomotive lead truck of the time, except that the side frame was outside rather than inside the wheels. Despite the radical nature of both plans, there was no discussion as to their merits or lack of same. The car builders just did not seem ready even to consider the matter. The idea was picked up and developed in the 1890s by streetcar manufacturers who were seeking a sturdier undercarriage for the electric-traction cars then coming into regular use. Steam railroads exhibited less interest in the idea. But cast steel was nonetheless being accepted into the main-line car builders' trade by way of body and truck bolsters. Indeed, it

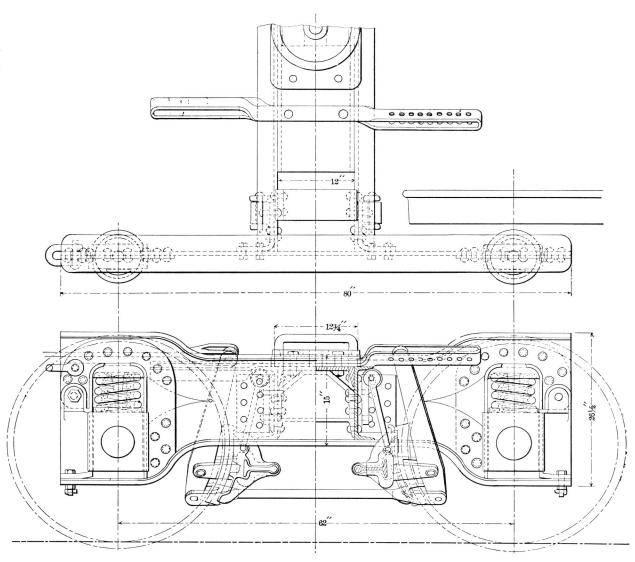

was the provisioners of the cast bolster who attempted to introduce the nation's railways to the
revolutionary idea of car trucks fabricated from a
few strong parts rather than a scrap bin of weak little pieces.

In about 1897 the Shickle, Harrison & Howard
Iron Company of St. Louis marketed a cast-steel
truck under the trade name of Ajax.[116] The firm
started out in 1867 and grew into a major producer
of nails, cast-iron pipe, and iron architectural
building elements. By 1890 it had entered the new
field of steel casting intent on becoming a railway
supplier. The designs were worked out and patented by Thomas M. Gallagher (1840–1912), foreman at SH&H since 1870. The side frame was a
rather flat, shapeless casting made from low-
carbon steel (Fig. 7.47). The springs were over the
axle boxes, making the truck a pedestal style like
the then-popular Fox design. Low-carbon steel was
selected because damaged parts could be heated
and straightened for further service. The bolster
was bolted directly to the side frames.

In 1899 Gallagher modified the design and offered two alternative side frames. Both had open-
jaw pedestals with a bolt/spacer bar at one end to

facilitate the removal of wheel sets. The pedestal
plan was eliminated and springs were placed under
the bolsters at the center of the side frames. The
new plan was to provide low-ride trucks for special
needs like furniture cars. By 1903 the Ajax truck—
now being produced by American Steel Foundries,
an American Car & Foundry subsidiary—had un-
dergone another radical transformation.[117] The
side frame was made with exaggerated humpback
ends and a depressed center. The bolster was attached on a top flange of the side frame with fourteen large rivets. The *Railroad Gazette* for January
13, 1899, had claimed that the Ajax truck was
widely used and gave satisfactory results in service. Yet beyond this optimistic statement there is
no evidence that the Ajax was ever more than a
novelty that helped introduce the notion of cast
steel into the lexicon of car builders.

In 1900 W. E. Symons, superintendent of motive
power for the Plant System, offered drawings for a
truck that looked identical to the Ajax low-ride design of 1899.[118] However, on closer examination it
becomes clear that Symons had advanced the
cause of cast-steel trucks an important step by
eliminating the bolt connection between the bol-

475   **FREIGHT CAR TECHNOLOGY**

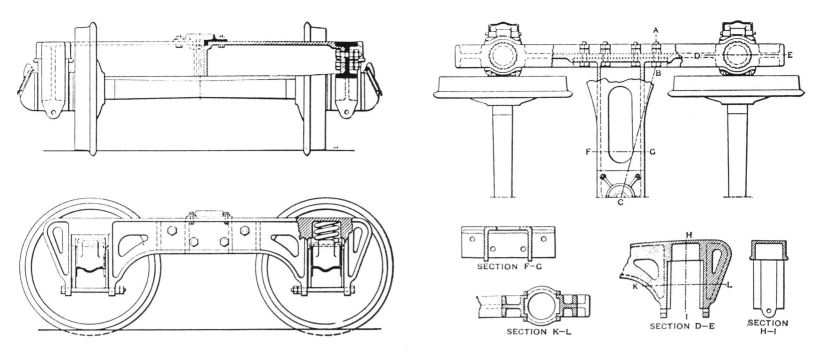

FIGURE 7.47

Cast steel provided the best solution to the truck problem. The Ajax truck, introduced in 1897, was one of the first efforts to exploit cast steel for this purpose. (*Railroad Gazette*, July 30, 1897)

ster (or transoms) (Fig. 7.48). He did not, however, travel the final mile; he inserted guide wedges rather than simply fitting the bolster to the guideways formed by the opening of the side frame. Symons did copy the open end jaw for the axle boxes, an idea that can be traced back to the Joughins truck described previously. The Symons truck was tested on the Plant System with rather mixed results. Even the inventor admitted that it was too rigid for normal track conditions, which were not very good on most southern lines of the period. In 1906 Symons revised the design to correct this defect, but there is no evidence that any railroad outside the Plant System used this form of cast-steel truck.

The cast-steel-truck concept had been around railway engineering circles for nearly twenty years, yet rather little had been done to develop its potential. Car builders who studied every detail of car construction and labored on a daily basis to perfect rolling-stock technology had done little to realize the merits of the cast-steel truck. Perhaps they focused too much on the details of car building to recognize the broader benefits of new materials. Maybe they were too tradition-bound to experiment. Many years later, when the transit industry was looking for a radical new streetcar, the leaders of industry intentionally selected a project manager from outside the traction field in the hope that he would review the problem from a fresh perspective.[119] The wisdom of this approach was vindicated by the very successful PCC car. In a similar vein, it took an agricultural machinery manufacturer from Davenport, Iowa, to grasp the value of cast steel for freight car trucks.

William P. Bettendorf took up truck design in about 1902 because he was convinced that the diamond arch bar composed of 164 separate pieces

was not only complex but weak. In 1902 and 1903 he placed several hundred trucks in service that were composed of only three major components.[120] The side frames were shaped like the old diamond truck except that there was no bottom tie rod. The journal boxes were cast integral with the side frame. Not only did this innovation increase strength and reduce weight; Bettendorf also noted that in a derailment the smooth (boltless) bottom of the side frame would slide over the tops of the ties and not tear up the track. The railroad industry soon came to recognize the genius of this obscure midwestern manufacturer who so clearly saw how to rationalize truck design. Bettendorf established a car plant outside Davenport and was soon rewarded by large orders for cars equipped with his remarkable trucks. In 1913 the Union Pacific gave Bettendorf a contract for $15 million, the largest car order placed to that time. The moral of this story is that sometimes the insight of an outsider is properly rewarded during his lifetime.

Running-Gear Subassemblies

The foregoing discussion covered the general design and construction of freight car trucks. It is now possible to examine some of the subassemblies and details of freight car running gears.

CENTER PLATES

Center plates are an important part of the structure because they performed the fundamental duties of weight bearing and providing the point of rotation on the truck. Center plates also functioned as a bearing or holder for the loose-fitting center pin. We must always speak of center plates in the plural, for two are needed—one for the truck and one for the body. Normally, they were

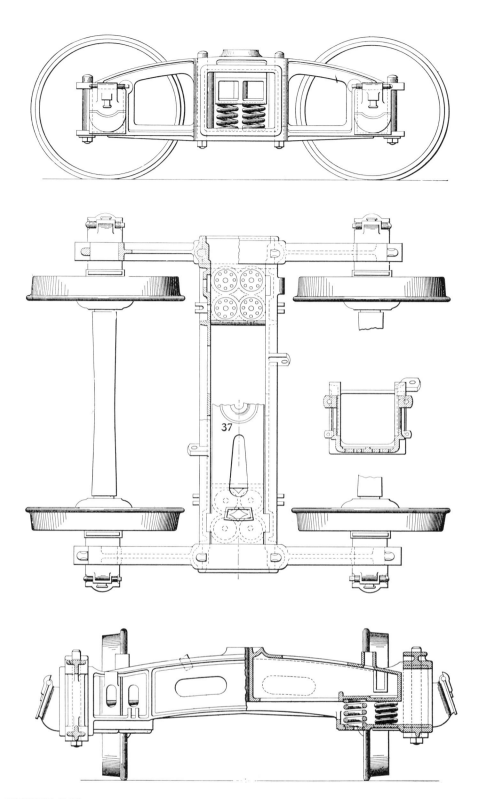

FIGURE 7.48

The Symons cast-steel truck of 1900 moved toward the ideal of unitized construction for car trucks. The efforts of Symons were soon rendered obsolete by William P. Bettendorf. (*Locomotive Engineering*, March 1900)

other lines were ready to give up cast iron for stronger plates made from malleable iron or pressed or cast steel.[121] For smaller cars, 30 tons and under, cast iron was a perfectly satisfactory material because the working loads and stresses were basically all in compression. Because space was not critically limited, fairly bulky—hence strong—castings were possible, and so center plates did not present a very great design problem. There were relatively few failures until car capacities reached 40 tons. The weights carried did vary considerably, and there was no apparent concern about what was considered a safe load. In 1899 one road stated that its center plates were subjected to 288 pounds per square inch, while another felt safe with a load of 3,136 psi.[122]

The diversity apparent in safe load limits was reflected in the variety of center-plate designs adopted. Each railroad had its own standard. Perhaps it would be more accurate to say that each master car builder had his own standard. George L. Fowler, the noted railroad technical journalist, said that no master car builder felt that his life's mission was complete until he had designed his own form of center plate.[123] None of this insular fraternity really seemed to know or care what style of plate was being used on the nearest connecting line. A small sampling of center-plate designs was published by the Master Car Builders Association in 1886. The plate reproduced here, as Figure 7.49, shows twenty-one designs that vary in size from 12 to 27 inches in diameter. Three general designs prevail. The round-bottom or spherical plate, shown in the upper two rows, was the most popular form at the time. It was used as early as the 1860s and became popular because the body could tilt front to rear as well as side to side and so settle the load more squarely on the center plate. This assumed that the body bolster had not already broken down to the point that most of the load fell on the side bearings. The spherical form declined in popularity in the late 1890s, when car builders returned to a simpler, flat style of plate.

The large-diameter plates shown in the central tier of drawings shows the oldest style of center plate with flat edge rings. Being large in diameter, it was supposed or hoped that all of the weight might be borne centrally without resorting to side bearings. The Pennsylvania Railroad was a stalwart advocate of this old form, though it resorted to side bearings as a backup safeguard. In the mid-1880s the designers at Altoona were briefly persuaded to try the fashionable, spherical center plate.[124] In the same manner that the Pennsylvania temporarily adopted iron truck bolsters and then returned to wood, the big railroad was ready to reverse itself as the occasion might dictate. Other major roads like the Grand Trunk and the New York Central favored small-diameter plates that necessitated a clear dependence on side bearings.

made in a male/female configuration so that the ring of the male plate penetrated the ring of the female plate. Both were made with deep grooves or rings.

Cast iron was the material commonly used for center plates during the nineteenth century, although pressed steel came to be rather extensively used late in the century. Some roads, like the Chicago & Alton, were early advocates of pressed steel and had adopted this form as a standard by the middle 1880s. It was another decade before most

FREIGHT CAR TECHNOLOGY

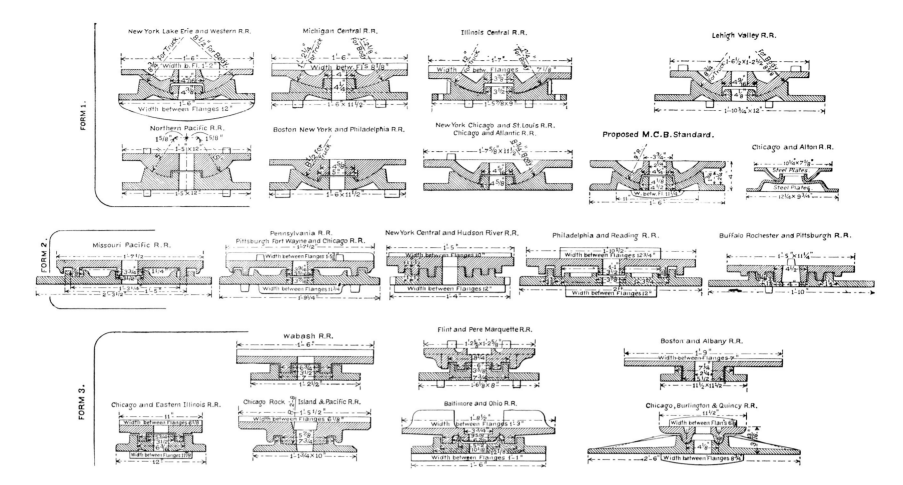

FIGURE 7.49

Some idea of the variety of center plates used on American railroads can be gained from this selection of designs. (*MCB Report*, 1886)

WHEELS

The flanged wheel is perhaps the most distinctive feature of any railway vehicle. A safe, dependable wheel was, of course, essential to railway operations. It would seem to follow that only the very finest wheel would be considered for this service. Yet when it came to freight cars, none but the cheapest variety of wheel could be considered because of the numbers involved. With eight wheels per car, the numbers grew as in a Malthusian nightmare as the fleet multiplied. By the end of the wood-car era—1910—it was recorded that there were 18 million wheels in freight service.[125] Even at only $10 each, freight car wheels represented a staggering $180 million. A cheap wheel was no bargain if it could not handle the rigorous duties it was called upon to perform. It led a stressful and hard life. It was expected to carry the weight of the car and its cargo and to steer the car around curves and along unwavering tangents. The wheel must absorb the jolts and bumps of countless frogs, crossovers, and road crossings.

The search for a safe and cheap car wheel was actually not a prolonged or circuitous one. A few early lines, like the Petersburg, imported spoked wrought-iron wheels from England, while others fabricated wooden-spoked wagon-style wheels shod with iron tires.[126] But the British imports were too costly, and the domestic wooden specimens too weak. Imitating much earlier British practice, American railroad men adopted cast-iron wheels with chilled rims. Cast iron was a cheap, time-tested material that machinists had used for a variety of products. It was reasonably strong and could be easily cast in simple sand molds into almost any shape. It could be machined easily. Foundry men had discovered that cast iron developed an extremely hard crust or chill if poured against another iron surface. The chill, often as deep as 1 inch, could not be machined and so was avoided in normal foundry practice; but in the case of car wheels it was found to produce a tough, long-wearing surface, and so the tread and flange were intentionally formed against an iron ring. While the running part of the wheel was as hard as glass, the plate and hub remained ductile and resilient, being cast in sand. They were relatively soft and could be machined as necessary to receive the axle. The chilled car wheel, a single-wear wheel, became a basic for railway service. Manufacturers were so confident of their products that 40,000- to 60,000-mile guarantees were common during the wooden-car era. In actual service, chilled wheels as finally perfected were considered good for around 500,000 miles.[127]

If there was early agreement on the right material for car wheels, there was a diversity of opinion on several other fundamental matters. Was the wheel to be mounted solid or fixed on the axles, or

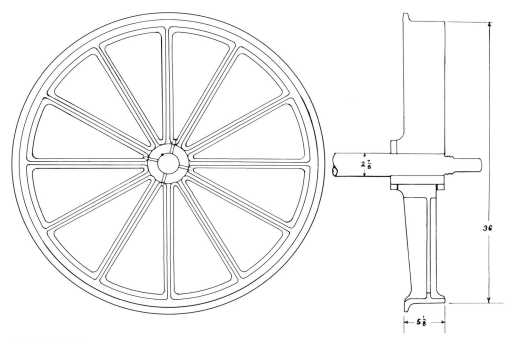

FIGURE 7.50

An early cast-iron T-spoke car wheel—possibly used by the Mohawk & Hudson Railroad in the early 1830s—is depicted in this drawing. (Traced from an original at the John B. Jervis Library, Rome, N.Y.)

was it best to make a loose fit? Before the public railway emerged, pioneers like the Mauch Chunk Railway opted for loose wheels so that the axles and wheels revolved independently. This had a decided advantage when it came to negotiating very tight curves and reducing flange wear. Yet as the hub opening became enlarged, the loose wheel set tended to become unstable. Following the lead of the Stephensons, American railways adopted fixed wheel and axle sets as the simplest and most stable way to answer this fundamental engineering question. The wheel tread was slightly tapered to help alleviate the stress caused by unequal speeds of the inner and outer wheels when rounding a curve. These matters were settled by 1830.

Less settled was the proper diameter for wheels. Larger wheels tended to provide a smoother ride, but their greater mass added to the car's dead weight. Big wheels were also more costly. Even so, there was a prejudice in favor of big wheels even for freight cars during the first decade of American railroading. This preference may well be accountable to the British model and to American inexperience and hesitation to experiment—a reluctance, in effect, to reinvent the wheel. And so most of the first-generation freight cars for which we have specific information were equipped with 36-inch-diameter wheels. In the minority were such mavericks as the Baltimore & Ohio, which started out with 30-inch wheels for its flour cars—introduced in 1830—and gradually worked up to a 31-inch diameter. This size remained in effect on the B&O into the 1860s, although a compromise of 33 inches had become nearly universal in this country by about 1840.

A uniform opinion also prevailed when it came to the proper width for the wheel. In the 1830s overall width was held to 4½ inches, but this di-

mension gradually grew by another inch, as can be seen from the hundreds of car wheels represented in drawings reproduced in this volume. Alas, it became impossible to maintain a uniform width after the interchange of cars began in the 1860s.[128] The slight variations between the Ohio and New Jersey gauges (4 feet 10 inches) and the Stephenson standard (1½ inches less) prompted the introduction of makeshift broad-tread wheels. An additional inch of tread made it possible for cars to travel from east to west, but at the expense of a sure or steady ride, because the wheel set now did not exactly fit either track gauge.

The compromise wheels, as they were called, were like most compromises—unsatisfactory to all parties. The scheme was born during the traffic crisis of the Civil War, when a car shortage developed. William Thaw, a freight forwarder from Pittsburgh and later a director of the Pennsylvania Railroad, is credited with originating broad-tread cars in 1864. What started out as a wartime expedient took firm hold, and soon thousands of freight cars were so equipped. Railroads that could not or would not adopt a uniform gauge chose to go with broad-tread wheels. All who did so came to realize what a costly item these compromise wheels were. They weighed 30 to 60 pounds more than standard wheels, and this added 240 to 480 pounds of dead weight to each car. They cost 10 percent more than a narrow-tread wheel and required special axles, also available only at extra cost. Because they fit neither gauge, they wore out quickly and were, in all, a costly and dangerous expedient. They stayed on the track only because of the extremely slow freight schedules prevailing at the time. Any thought of express service would be ludicrous should compromise-gauge cars be included in the train's consist.

And so it was with some relief that the Ohio-gauge lines began to convert in the late 1870s, for it meant an end to the pernicious broad treads. The Lake Shore reduced its gauge from 4 feet 9¼ inches to standard in 1877 to end its dependence on compromise cars. Later in the same year the Cleveland, Columbus, Cincinnati & Indianapolis converted from 4 feet 9½ inches to the Stephenson gauge. Other roads in the area began to follow, but even during this time of transition the March 1879 *National Car Builder* estimated that one-third of the cars owned by the standard-gauge roads of the nation remained on broad-tread wheels. Their demise was complete, however, by the end of the next decade; the 1888 *Car Builders Dictionary* stated that they were now rarely, if ever, used.

So much for wheel width. What about the centers? The surviving illustrations of the earliest American freight cars are consistent in showing spoked wheels. This evidence supports the notion that new technologies are at first modeled to duplicate the appearance of their predecessors: the new

FIGURE 7.51

Three views of a cast-iron car wheel of the Washburn pattern. To the left is a view of the face of the wheel. To the right is a back view showing the curved strengthening ribs; the three holes are for the internal cores. In the foreground is a sectionalized wheel lying on its face.
(*Scribner's Magazine*, Aug. 1888)

iron wheel looked like the traditional wagon wheel. Considerations other than sentiment may have induced mechanics to stay with the spoked pattern. Weight or production considerations may also have been involved. Some of the earlier spoked wheels were of an extremely light pattern, as is evident from a drawing found in the John B. Jervis Collection (Fig. 7.50). The 36-inch-diameter wheel may well be from the Mohawk & Hudson Railroad and, if so, would date around 1832. Note the fine cross-section of the spokes and the very thin rim and flange cross-section. The four gaps in the hub were necessary to prevent the hub from cracking during the cooling. The openings were filled with lead, and iron bands were shrunk on to hold the hub together. Just why the band was omitted from the elevation drawing cannot be explained, but they are clearly indicated in the end drawing.

The mass of spoked wheels had greatly increased by 1847, when heavy rectangular spokes and thicker rims pushed the weight of individual wheels up to 500 pounds.[129] Hubs remained split, however, thereby reducing their strength considerably. By this time three rather than four gaps were employed. By 1850 most car builders had lost patience with the spoked wheel. Split hubs were bad enough, but a uniform chill was never achieved, and soft spots appeared opposite the spokes so that the tread wore unevenly and flat areas formed.[130] The spokes also acted as fans; the dirt they blew up attached itself to the underside of the train and more grit entered the journal boxes.

A more rational style of wheel center, the disc or plate, had been available for many years. The locomotive builder M. W. Baldwin produced some as early as 1836, and George Lobdell of Wilmington, Delaware, received a patent just two years later for a strong but light double-plate car wheel. So many patterns were devised over the ensuing years that before long over one hundred patents had been issued just to cover the style and shape of the plate for disc wheels. The cheapest variety involved the single-plate wheel. These were plain and straight-sided with ribs on the rear side for added strength. Of the many patterns for stronger and more costly double-plate wheels, none succeeded so well as Nathan Washburn's patent of 1850 (Fig. 7.51). By

1890 it was an industry standard. The double-curving upper plates supported the hub against lateral stresses; the lower single plate curve helped prevent cracking due to unequal cooling. The curved strengthening ribs on the back also lessened the effects of contraction cracks.

Just a few years before the Washburn wheel appeared, another front-ranking wheel producer solved the problem of split hubs. Asa Whitney found, as did others, that plate wheels, particularly those with curving plates, cooled more evenly than spoked wheels, and so splitting the hubs was no longer necessary. Whitney, however, markedly improved wheel quality by the process of controlled cooling. Wheels were no longer dumped out onto the molding floor to cool as quickly as the ambient temperature might carry the heat away. Rather, they were sent immediately to an annealing furnace where the temperature was maintained over a three-day period. Whitney's method, though more costly, was adopted by his competitors once its benefits had been demonstrated.

For all the improvements in wheel making, the cast-iron plate wheel was surely not trouble-free. According to one estimate, over 5 percent failed each year. That might sound like a small percentage, but in 1883 it meant that 267,000 would break.[131] A major cause of this problem can be traced to the casual methods used by nearly every wheel maker. Each manufacturer attempted to create an impression of mystery or alchemical skill. None employed a chemist, and managers of the foundry understood neither the chemical nor the physical properties of the iron they employed. Instead, they worked by intuition, combining known brands of pig iron with a certain percentage of scrap wheels. What saved them and produced a reasonably durable wheel was the high-quality pig iron generally available. Fine ores like those found at the Richmond (Berkshire County, Mass.) or Salisbury (Litchfield County, Conn.) mines registered tensile strengths as high as 41,000 pounds.[132] No matter how little science was applied by old-line wheel makers, it was not difficult to make a good product from such superior iron.

When it came time to produce a stronger wheel, the wheel maker tended to rely on empirical methods. A stronger wheel was made by simply using more iron. In this case mass equaled strength; it was very much like dam building. In the late 1840s a 360-pound wheel might serve very well for a light-service car.[133] There was no reason to increase weight and cost beyond the minimum that would safely do the job. Even so, most car wheels by this time were more likely in the 500-pound range. Wheels of this mass remained more or less typical for the remainder of the century, though those in the 550- to 600-pound class became more common toward the end of the wood-car era. The introduction of steel cars with 40- and 50-ton capacities resulted in weights that seemed

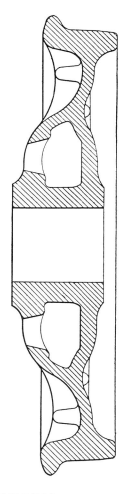

FIGURE 7.52

Greater strength in cast-iron wheels was achieved by increasing the mass. The thick profile shown here illustrates how the Washburn wheel was enlarged to handle 40- and 50-ton cars. (*Cassier's Magazine*, March 1910)

beyond what the old-fashioned cast-iron wheel could carry. A boxcar loaded with 100,000 pounds, plus an allowable 10 percent overload, plus the weight of the car itself resulted in a total weight of very nearly 150,000 pounds.

Some members of the engineering profession felt that steel wheels were now the only way to go, but the wheel makers were not so ready to agree. They simply beefed up the old Washburn design. The rim plate and hub were made thicker, while the brackets on the rear of the wheel were reshaped (Fig. 7.52). The revised double-plate wheel now registered 700 or even 750 pounds and soon proved itself quite capable of handling 75-ton loads. Like the arch-bar truck, it was not ready to join the dinosaurs just because a few critics were saying that its time had come. Of course, no one really pretended that cast iron was superior to steel. The reason it prevailed was that it was so much cheaper. It continued also because a simpler form of cast wheel was developed in 1917 to replace the Washburn pattern: a single-plate design known as the arch wheel, which became standard in 1928. Several million of these wheels were produced each year until cast-iron wheels were outlawed by the Association of American Railroads in 1958.[134] Some were still being produced as late as 1963, but all were gone within five years.

AXLES

In America's wood-car age, timber predominated because it was plentiful and cheap, whereas iron was scarce and expensive.[135] At the beginning of the railway era it was necessary to import iron —even more reason to use it sparingly. Very much attuned to the dearness of iron, car builders were accordingly conservative in using it. When it came to axles, they were made in the smallest size possible. Slender axles saved not only cash but also weight, an important consideration that dominated the thinking of all car builders. For the greater part of the nineteenth century, freight car axles rarely had a nominal diameter much over 4 inches or a weight much beyond 350 pounds. As a further economy, a number of lines used the same axle design for both freight and passenger cars.[136] The idea of thrift was carried to an extreme by car builders intent on purchasing the greatest numbers for the least possible cost. By specifying the cheapest axles and thereby saving $240 per unit, so many extra cars could be obtained.[137]

This expedient would temporarily answer an immediate traffic need, but in the long run someone was bound to pay. It might be the railroad itself, forced to replace the axles within a few years, or it might be some hapless trainman who paid with his life when one snapped under a moving car. The 5-cent cigar of axles was made from scrap wrought iron. Old bits of iron were piled, heated, and hammered into a shaft while red-hot. If good scrap was selected and the smithwork was prop-

erly handled, a reasonably solid axle would result, but too often the work was poorly done because price, not quality, was the first consideration. Cinders or bits of steel confounded the welds. Superior craftsmanship is hardly the norm for any cheap product, and so the scrap axle was in general a rather sorry piece of work. Far better iron axles were produced directly from muck bars or puddle iron, which was new wrought iron. The muck bars were stacked red-hot and hammer-welded together into a rough shaft. These forgings were then machined. But even the new iron was hardly free from such defects as cinders or imperfect welds. A higher price alone did not guarantee a safe axle, because the shaft might appear sound and perfect to the eye yet contain hidden interior flaws that became evident only after a failure.

Railroad men found that a common point of breakage was near the point where the axle entered the rear of the wheel hub. It was thought that the wheel acted as a lever, particularly as it slammed through a rough crossover or switch frog and so broke the axle off at this pivot point. A photograph dating from 1868 shows several wheel sets with axle breaks of this type (Fig. 7.53).[138] It was thought that some of the deadly force of the lever effect might be dissipated by allowing the axle to flex or spring slightly. This was achieved by making the axle slightly smaller in diameter at its center. The taper also saved a little weight and iron and so became a common feature for most railway axles.

The load-bearing capacity of a car was determined just as much by what its axles weighed as by what its floor frame could sustain. And so car builders had to proportion axle size to the car's intended capacity. The diminutive freight carriers of the 1830s might ride safely on axles only 2 inches in diameter, but as 10-ton-capacity boxcars came into use, axle diameters began to swell. Northern roads, with their frostbitten roadbeds, had adopted axles 3 to 3½ inches in diameter by the late 1830s. The Master Car Builders Association approved a standard axle in 1879 that, if not readily adopted by its membership, at least reflects the standard practice of the time (Fig. 7.54). It measured 5 inches at its largest diameter—the weak point just outside the inner end of the hub—and 3⅞ inches at its center. Steadily increasing capacity made for a steady growth in axle size, so that by 1900 axles with diameters of 6⅜ inches at the hubs and 5⅜ inches at the centers were common.

The introduction of cheap steel following Henry Bessemer's discovery seemed to answer the requirements for an economical but safe car axle. The material was homogenous, dense, and strong. But curiously, car builders reacted negatively to the new material. Not just a few cranks or reactionary old-timers, but the whole car-building community seemed united in opposition to steel axles. They were not cheaper than iron; steel axles

FREIGHT CAR TECHNOLOGY

FIGURE 7.53

Wrought-iron car axles typically failed at the inner hub of the wheel, as illustrated by several examples shown here. The wheels all appear to be of the popular Washburn pattern. (Joseph Anthony, *Forms of Railroad Car Axles* [Greenbush, N.Y., 1868])

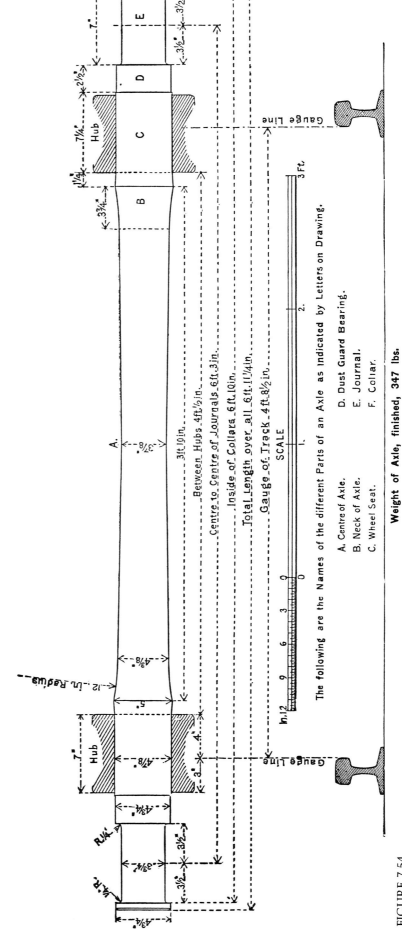

FIGURE 7.54

The MCB standard axle adopted in 1879, with journals made to the standard 3¾ by 7 inches. (*MCB Report*, 1883)

FIGURE 7.55

iron remained in business. Thus the car builders were unwilling converts to the age of steel.

JOURNAL BOXES

In the common parlance of the car-building trade, bearing boxes were always journal, axle, or oil boxes. Axles passed through the wheels so that the point of bearing would be outside the wheels. This provided for a wider base of support and placed the journal box in a more accessible position for inspection and servicing. The journal was actually only a half bearing (Fig. 7.55). The exposed underside of the journal was wiped by oil-soaked waste; the bottom of the journal box was an oil cellar. The bearing itself, often called a brass, was generally made of bronze. It often had a liner of a soft white-metal alloy, such as Babbitt metal. This simple bearing, box, and lubrication system prevailed for most of the history of the American freight car; not until 1960 was it rendered obsolete by roller bearings, which are now the norm for interchange cars.

The exact origins of this durable arrangement are uncertain, but one of the earliest examples was John Elgar's patent of October 1, 1830. Elgar's design embodied an oil reservoir and a dust seal.

were in fact twice as expensive as late as the 1870s, when Bessemer steel production was hardly experimental. Cost, however, was not the prevailing objection. Car builders simply distrusted steel axles. They were notoriously uneven in quality. Even those from the same lot showed marked differences—some being soft while others were brittle. Others would develop rough journal surfaces that cut away the bearings. Even while steel was taking over bridge and boiler work, car builders remained hostile to the idea of steel axles. William Voss reflected the opinion of his fellows in his 1892 *Railway Car Construction* when he stated that iron axles were favored by the majority because they were cheaper and more predictable. A few years later more cars were equipped with steel axles only because so few producers of wrought

FIGURE 7.56

The most detailed drawings for an early journal bearing and box are this series taken from an original B&O tender drawing dating from around 1845. Note the very small journal size of 2 by 3½ inches. (White, *American Locomotives, 1830–1880*, Fig. 130A)

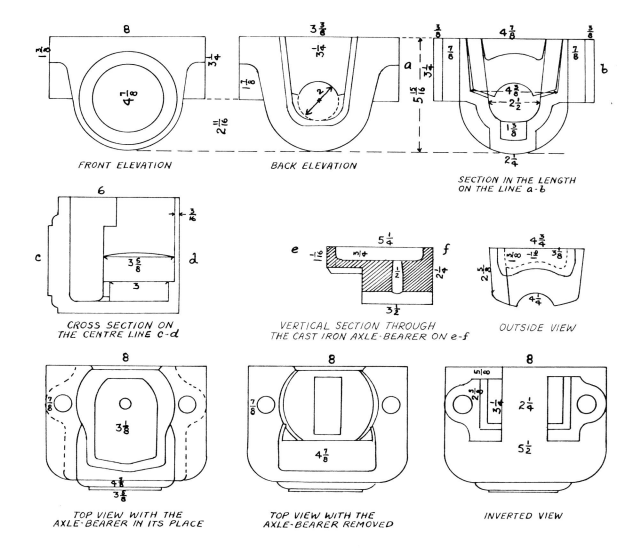

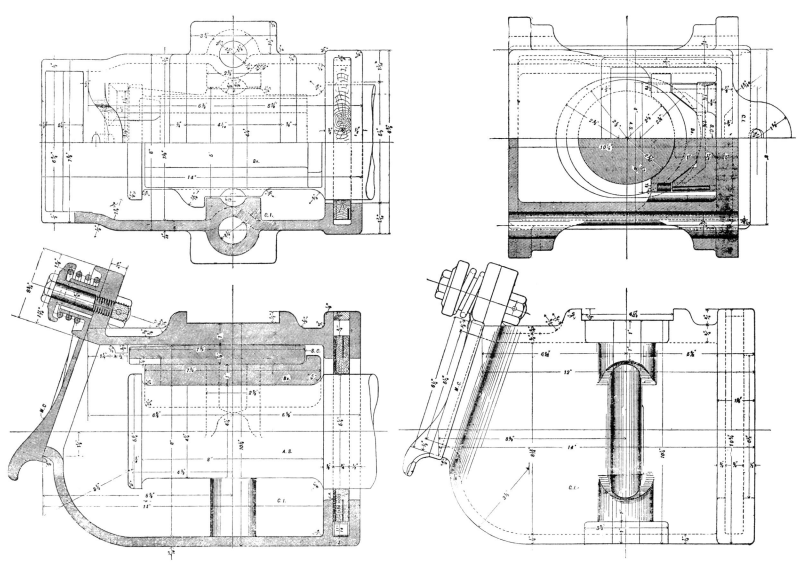

FIGURE 7.57

The MCB standard 4¼- by 8-inch journal box suitable for 30-ton cars. Note details for the Fletcher-style lid. (*Engineering News*, June 14, 1890)

However, rather than employing a soft material for the bearing, Elgar used chilled iron. The axle ends were sheathed with steel bands. The economy or good sense of the scheme might be questioned, for it is surely contrary to conventional practice, but even so it was long used by the Baltimore & Ohio Railroad. A very detailed graphic depiction of the Elgar bearing as it had evolved by 1845 is shown in Figure 7.56.

In 1848 John Lightner, master car builder for the Boston & Providence, devised a simple addition to journal-box design that greatly facilitated the work of car men across the nation. Lightner's improvement consisted of a wedge or pad inserted between the top inside of the box and the bearing itself. The weight of the car held the edge in place, but when the weight was transferred to a jack, it was possible to pull the wedge out of the box; this left a gap so that the bearing could be lifted up and over the collar on the end of the axle and so be removed from the box. The inspection and replacement of car bearings were thus greatly simplified, and Lightner's device became an industry standard.

As has been noted before in this text, the indus-

try was almost always amenable to standard designs in a general sense, but almost never in a specific sense. Car bearings are but another example of this frustrating pattern. The general arrangement was universal for just about every railroad in the land, but the exact dimensions and certain small details might vary just enough that no two railroads could interchange such a fundamental part. The Master Car Builders Association bravely proposed a standard 3¾- by 7-inch bearing in 1869. Various larger sizes were adopted in succeeding years to meet the needs of larger cars. The 4¼- by 8-inch journal came in 1889 and the 5- by 9-inch size in 1896, but the acceptance by a majority of the membership meant nothing, for even those who voted for these standards did not abide by them (Fig. 7.57). And so in 1900 the *American Railroad Journal* gloomily reported that no fewer than 108 patterns of car brasses must be stocked by each car shop in the country because of the continued individualism of the car-building profession.

Some car builders were more worried by such details as how to extend axle life. The constant lateral motion inherent in all cars tended to wear away the lip or collar on the front end of axles. Once they were gone, there was little to check the sidewards lurching of the car, and an otherwise perfect axle would be condemned to the scrap pile. This problem could be countered by providing an end stop inside the box. One of the oldest known schemes was that devised by Thatcher Perkins. A replaceable pluglike pin was set inside the rear side of a very heavy cover plate. The details of the scheme are illustrated in a drawing prepared in May and June of 1862 (Fig. 7.58). How much earlier it may have been used is uncertain. A slightly more elaborate scheme was offered some twenty-three years later by the Raoul Journal Box Company.[139] William Voss advocated end stops in his *Railway Car Construction*, yet the idea never seems to have come into general practice.

Lubrication has been basic to car service since the beginning of steam railroading, for it reduces friction and thereby the power required to move trains. It also lessens the wear on all associated machinery. In the beginning, only animal or vegetable oils were available for this purpose. The natural oils that worked best, such as whale oil, tended to be expensive. Petroleum was not seriously considered for car lubrication until around 1870, and even then it was viewed with some skepticism. After this time it became the basic railway lubricant, and old standards like tallow for valve oil gave way to distillate of petroleum.

The common journal box was condemned by many unknowing persons as a failure because of the persistent hotbox problem. Surely it was not perfect, but for all its inherent defects, the journal

box worked well if properly maintained. Yet if there was ever a neglected child in the realm of simple machinery, it was the freight car journal box. The cause of its woes were many. Overloaded cars, out-of-square trucks, rough journal surfaces, insufficient or dirty oil, and finally waste grabs were among the leading causes of hotboxes. Waste grabs were considered to be a major cause of overheated bearings. Waste is a clump of loose thread stuffed into the underside of the box to act as a wick to lift the oil and brush it against the underside of the journal. However, in rough handling the brass will jump momentarily off the journal; if strands of waste slide up at the same time, the threads will be trapped under the brass when it comes down to reseat itself. And soon the friction between the threads and the bearing sets the waste and oil on fire. Thus deprived of its lubrication, the bearing and journal can overheat to the point of melting. The waste-grab problem was a perennial subject of discussion, and nothing really effective was done about it until long after the wood-car era. The solution was absorbent pads devoid of loose strands, but these did not become widely used until around 1955, and they were actually adopted too late because roller bearings were already being accepted as the standard car bearing. Roller bearings were grease-lubricated, and so the entire oil-cellar scheme was unnecessary.

Much blame for journal-box failures was placed on the lowly car "doper." Poorly paid and unskilled, the oilers were considered lazy, indifferent workers who tended to do more harm than good. They passed over cars with tightly bolted-on

FIGURE 7.58

This B&O journal box dating from 1862 had an end-stop plug fitted into the rear of the bolt-on lid. (Traced from assembly drawing of a B&O iron boxcar reproduced as Figure 8.15)

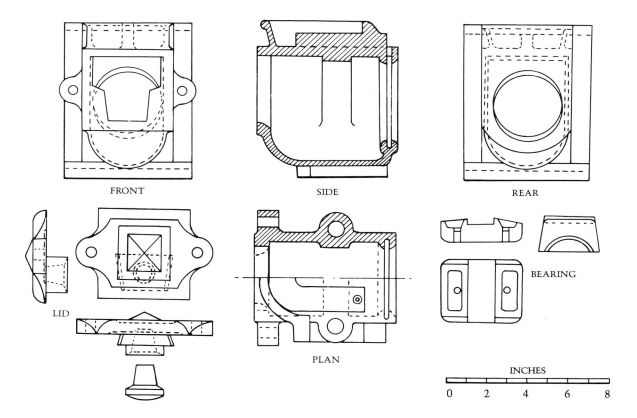

FRONT

SIDE

REAR

LID

PLAN

BEARING

INCHES

0 2 4 6 8

FREIGHT CAR TECHNOLOGY

FIGURE 7.59

The Harrison dust guard was one of many patented schemes perfected to replace the notorious plain wooden-board dust shield. (*Railway Age*, Jan. 3, 1902)

covers. They put in too little or too much oil. The dopers' greatest sin was to overpack the boxes with waste. Not only did this encourage waste grabs; an overpacked box would wipe the oil off the journal rather than wicking it up.

If the humble dopers were at fault, so were the lordly managers of the car department. On a carefully managed road like the Boston & Albany, only two hotboxes a month were encountered; for the same time period—1893–1894—the Michigan Central reported thirty-five a month and the Lake Shore shamefully admitted to eleven hundred. Indifference and sloppy management were the explanation. On a well-run line, freight cars were inspected monthly and repacked as needed, while on other roads cars might run three or even four years between inspections. This was carrying the "Don't fix it if it ain't broke" doctrine to an extreme.

There was a respected school of car men who contended that the true scourge of car lubrication was not waste grabs or fallible car dopers and managers. The real enemy was dust. The cloud of fine grit that swirled, low and sinister, along the path of every moving train was the great destroyer of brasses and smooth journal surfaces, and until it was controlled or eliminated, railroaders must be ready to accept hotboxes as inevitable. From time to time some railroads would sprinkle oil over the right-of-way, but most of these programs were short-lived efforts aimed at enhancing the comfort of passengers. Crushed stone and ballast cleaners would in time help alleviate the dust problem, but these advances came too late to help those active during the wood-car age. Dust was a given, and the job at hand was to find a way to keep it out of the oil boxes.

The common journal box was quite vulnerable to the invasion of dust because it was essentially open front and back. The rear hole proved the most troublesome to secure, for several reasons. It was made oversize so that the axle could be inserted easily. Clearance of 1 inch must also be provided so that the axle could move upward by at least 1 inch as the bearing wore away. This large hole was filled in by a wooden board cut out to surround the axle. Even when new, the rear dust seal was a crude barrier, and with time it became less effective as the

hole wore even bigger. The board tended to split and fall out, leaving the rear opening entirely exposed to whatever might enter.

Efforts were made to improve the wooden dust guard with tension springs and other contrivances. One example is the Harrison dust guard shown in Figure 7.59. Other vendors produced rubber, vulcanized-fiber, and leather guards that were guaranteed not to rot, fall out, or be adversely affected by oil or grease. Despite the claims and good press, the 1889 *Master Car Builders Report* concluded that most guards were defective, and the claims made for them were largely rubbish. Even if the mechanical problems could be solved for the rear seal, there was the problem of inspection. The guard was hidden from view and not easily accessible—out of sight, out of mind. The Pennsylvania Railroad's mechanical department questioned the tendency to push for larger bearings when the real cause of hotboxes was the want of better dust seals.[140]

Closing up the front entrance proved easier. Nothing need move up or down; the opening was simply for inspection and servicing. A plain bolt-on style of cover or lid would answer nicely. Bosses cast on either side of the box were drilled and tapped for hold-on bolts. This basic lid goes back almost to the beginnings of American railroading and remained popular until around 1890. Examples can be found in this and the preceding five chapters. Some of the bolt-on covers were identified as to road by initials cast in, and occasionally they were dated in the same fashion. An early Central Pacific journal-box cover was found buried along the now-abandoned main line near Promontory, Utah. A drawing of the plate is shown in Figure 7.60. The bolt-on plates were cheap and efficient, but they were inconvenient to remove and replace. In pre-interchange days they made good sense for cars in home-road service. Distances tended to be modest and maintenance relatively high, but once the same cars entered interchange service—after 1870—all of this changed. It took too much time to unbolt and rebolt eight covers per car when a train pulled into a junction and was scheduled to leave within minutes. Sometimes covers were removed but there was not time to put them back securely; sometimes they were not put back at all. Even when the lids were properly replaced, the vibration of the trucks would work them loose to the point that they would fall off. According to the 1899 *Railroad Car Journal*, it was common to see cars running for months after losing their covers.[141]

The failings of the bolt-on type prompted other mechanics to come up with a fast-opening, boltless cover. These cast-iron lids slid in a groove cast into either side of the front opening. It was normally V-shaped. C. F. Allen developed a wedge-shaped sliding cover that was attached to a small chain so that it could not be lost. Allen's lid was

FIGURE 7.60

Some railroads cast their initials and the date of manufacture into their journal-box lids. This example was made by the Central Pacific Railroad in the 1860s. The last numeral is indistinct but appears to read as 5. (After an original in the Smithsonian Collection, Gift of John Eldredge, Salt Lake City, Utah)

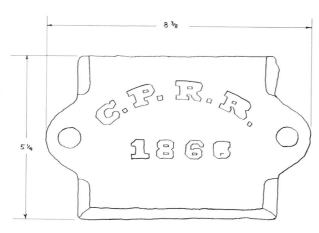

FIGURE 7.61

The cast-iron Hewitt lid patented in 1877 slides up to open. It was held in place by grooved guides. The lid on the bottom is shown in the open position. The top lid is in the closed position. (*National Car Builder*, June 1883)

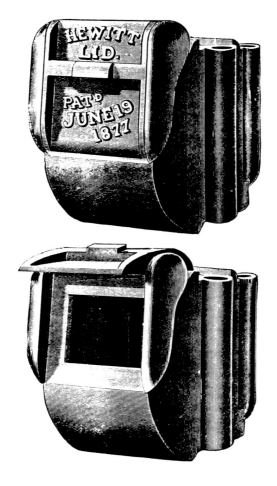

used by Pullman for a time, as is evident from engravings featured in the 1879 *Car Builders Dictionary*. It was also used for his swing-motion freight car truck shown earlier (Fig. 7.22). Another variety of the gravity-wedge lid was patented in June 1877 by H. H. Hewitt of Detroit (Fig. 7.61). It was contrived to slide up and back in a grooved track so as to clear the opening, but it could not be pulled out of the track and so could not be lost. The Hewitt cover is shown in the perspective engraving of the Thielsen truck reproduced in Figure 7.23.

Another form of quick-opening lid was introduced in 1867 by Richard Eaton, master mechanic

for the Grand Trunk Railway.[142] It was actually a variation on the old bolt-on lid except that only one bolt was used. A small coil spring under the bolt pressed down on one end of the lid, firmly enough to hold it in place but not so snugly that it could not be shifted manually. A doper might pull the lid up to open and if all looked well snap it shut. The opening and closing could be accomplished in a few seconds. The lid was fixed and so could not easily be lost. Eaton's lid appealed to J. B. Fletcher, superintendent of the National Despatch fast freight line, and was adopted for all of the line's cars. Fletcher went on to become such a vigorous advocate of the spring lid that railroaders came to identify it with its promoter rather than its inventor, and so it became the Fletcher lid. Its renown spread beyond New England, so that by 1889 the Master Car Builders Association had adopted it as its standard. An example of a Fletcher lid may be seen in the Illinois Central boxcar shown in Figure 3.47. A detailed view of the end and spring is shown in the drawing of the 1890 MCB journal box reproduced in Figure 7.57.

There was, of course, one other obvious way to construct a tight yet quick-opening journal-box lid, and that was a simple hinged cover. The hinge on top of the box was equipped with a flat spring to hold it tightly shut. The who, how, and when of the hinged lid cannot be precisely detailed, but there is pictorial evidence that it was known in the late 1830s. Lids of this general type are evident in patent drawings for an Allegheny Portage flatcar shown in Chapter 2 as Figure 2.19. They appear again in drawings for an antiquated freight car truck, credited to the New York & Harlem Railroad but surely not reflecting the most current design, in the 1879 *Car Builders Dictionary* (Fig. 88). The design of these trucks was presumably that of Leander Garey, longtime car chief for the Harlem line, who carried the idea over to the New York Central and its associated lines. By the 1870s the Pennsylvania had also come to adopt this hinge, and by the turn of the century it had become the industry's standard.

SPRINGS

Selecting just the right springs for a car was a major challenge faced by freight car designers. Springs that worked well for a loaded car proved far too stiff when the car ran empty. The car bounced and jumped as if the trucks were outfitted with iron blocks rather than springs; yet if softer cushions were put in, they could not carry the load of a full car and might crush down or even break under the strain. And so the car builders must find springs that were soft but not too soft, stiff but not too stiff.

In the beginning, pioneer railroads believed that no springs were needed because of the smooth and perfect path afforded by their new iron highways. This notion was quickly disproved as the little four-wheelers began to pound themselves, their contents, and the track structure into very small pieces. Every car meant for train service must have a suspension to absorb and not magnify the imperfections inherent in the railroad track. Most mechanics involved in car construction were content to apply some form of springage to the trucks, but a few felt that springs should also be applied between floor and underframe.[143] A sample car built in 1873 by the Erie Car Works after the patent of one T. R. Timley was tested on the Pennsylvania Railroad. This cushion-floor curiosity prompted another even zanier patent two years later that featured butter churns spring-mounted to the underside of boxcars so that the golden spread might be manufactured as the car bounced along.

The earliest railway springs were of the leaf or elliptic variety. Spring making was an art during this period, and only the most skillful blacksmith would undertake the arcane business of tempering steel into flexible bands. It took a fine eye and some experience to know just when the heat was right. Too much could make the metal brittle, while too little resulted in a soft, limp material unsuitable for springs. Quenching and heating were not work for the unskilled mechanic. Steel itself was a rare and expensive metal before the time of the Bessemer process. Almost all steel was imported, which only added to the cost, even though the spring itself might be locally made.

The traditional leaf spring had one characteristic that endeared itself to all vehicle makers and users. If one or even two leaves failed, the spring continued to function. It might sag and give more than was desired, but unless unduly overloaded, it would usually continue to function, at least after a fashion. The main complaint, other than its cost, was the space it required. This was especially true for large vehicles such as railroad cars. Leaf springs tended to be long and deep when sizable weights were involved. This hampered construction and design freedom when it came to relatively compact structures like car trucks. For this reason car builders tended to adopt more compact coil springs once practical, cheap models became available around 1860. Elliptics remained the standard form for locomotives and passenger cars after this time, but freight cars were largely free of the old-fashioned leaf springs. Illustrations of older-pattern car trucks with leaf springs are given earlier in this chapter and in Chapter 2. A few car users remained loyal to leaf springs for freight car trucks—perhaps because of conservative convictions or special needs that are no longer entirely explainable. Examples can be found in the 1876 New York Central and 1890 Union Tank Line's arch-bar trucks illustrated earlier in this volume (Figs. 3.25, 5.78).

Before the transition from leaf to coil springs took place, American railroads flirted briefly with rubber springs. The soft, springy material was referred to as gutta-percha, India rubber, or gum during most of the nineteenth century. It was a novel material that fascinated mechanics and inventors and was adopted for many purposes. It was suggested as a suitable material for car springs as early as 1825. However, sticky natural rubber proved anything but ideal for car springs or most other applications. Vulcanization greatly stabilized the material, and so after 1839 it did become something of a handmaiden to industry. It was surely cheaper and lighter than steel. One manufacturer of India rubber car springs contended in 1850 that a set of steel springs added 444 pounds in weight and $59.06 in cost compared with his product.[144] Yet the disadvantages were actually overwhelming, and it is surprising that rubber springs remained popular for so long. When new, they were altogether too lively and tended to make the cars dance down the track. After a few years' service, they would crush down into a solid, unyielding block. In the winter they were certain to solidify as the temperature dropped. Few railroads used them much after 1870, and by that time rubber springs were limited to secondary work as cushions or snubbers.

Coil or spiral springs had been used for centuries in a great variety of light-machinery applications, but they were not widely applied to railway vehicles until about 1860. Their primary virtue was compactness. Their ability to fit into a small space made them especially attractive to truck designers. However, they were difficult and expensive to make. The variable quality of the steel available during most of the nineteenth century made the spring maker's job especially difficult. Temperature control for heat treating and quenching was unpredictable because of this lack of uniformity in the material being tempered. When a coil spring failed, the means of suspension was gone; hence early car builders tended to stay with bulkier but more dependable leaf springs.

Rising to the challenge, spring makers sought ways to make the leaf spring obsolete. They rolled springs from round, square, and flat stock. The latter shape produced the volute spring that appears to have gained some measure of success as early as 1855.[145] Technically, volutes are really not a true coil spring but more of a combination leaf/coil spring. Springs of this type manufactured by P. G. Gardiner of New York City were used on the Hudson River Railroad and some twenty other lines. A rival producer claimed that the New York & New Haven Railroad was replacing India rubber springs with volute steel springs (Fig. 7.62). In 1860 the Metallic Car Spring Company of New York took out a double-page advertisement in the *Lowe and Burgess Railway Directory*. Drawings for an arch-bar freight car truck equipped with eight volute springs illustrated the advertisement. Despite

Volute Spring, (Concave Bar.)
Perfect temper, without friction.

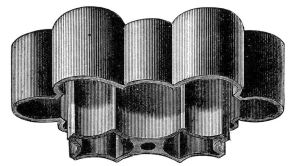

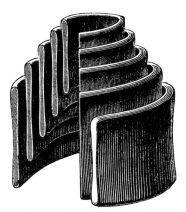

Volute Spring, showing Concave Bar
and spaces between the Coils.

"Dinsmore Spring," Fluted Bar, with
Round Edges, Right and Left Spiral.
This form avoids friction.

Group Rubber Centre Spiral Spring, open, Oil-tempered
Steel Spirals, with open Columns. Any number of
Spirals can be grouped.

Group Rubber Centre Spiral Spring, closed.

FIGURE 7.62

A variety of car springs were available to car builders by 1870. (*National Car Builder*, Nov. 1870)

such heavy promotion, railroad mechanics never seemed entirely satisfied with volute springs for car trucks. The springs did, however, remain long in favor for draft gears.

Around the end of the Civil War, a new style of car spring appeared on the market. On first consideration it seemed unnecessarily complex, but it proved to be a very good idea at a time when coil-spring quality remained so uncertain and cluster or group springs were becoming so popular. As many as ten small coil springs—though six to seven was a more normal number—were placed inside a cast-iron case. The case or holder was sometimes cylindrical but was more often made as a multicylinder unit cast in two telescoping halves. The bottom of the two-piece case fit loosely inside the top. This arrangement can be better understood from Figure 7.62, which shows the top of the case separated from the bottom, revealing the springs. When assembled for service, the springs were entirely hidden from view, so that anyone not familiar with their construction might suppose that the car had no springs at all, or that the massive spring cases were actually rubber blocks.

The basic thinking behind the cluster spring was to combine a group of weak elements to make

a single strong one. If one or even two failed, the remaining springs in the nest could carry the load. Each spring, coiled from ⅝-inch-diameter steel rod, could sustain a load of 3,500 pounds. Thus all the nests in combination offered a comfortable margin of safety in a maximum sustaining load of 42 tons, or twice the load expected for a normal boxcar of the period. The center of the springs was packed with wool or rubber, presumably to act as a damper and hence retard bouncing. A large number of cars illustrated in this volume are equipped with canister springs; this is particularly true for cars dating from the 1870s and 1880s. This style of spring went into an eclipse around 1885 and was replaced by fewer and larger coil springs.

The style of springage that became common in the 1890s remained in place until the present time. Exposed coil springs in groups of four replaced canister springs. In 1896 it was reported that some cars had clusters of three or six but that four was preferred on most roads. Each spring—made from a 1-inch-diameter rod—was 5 or 6 inches in diameter and 7 inches high.[146] A double or compound spring was used to compensate for the weight difference between loaded and empty cars (Fig. 7.62, *lower left*). The smaller and shorter coil served as a cushion for heavier loads.

Couplers and Draft Gears

Nothing is more fundamental to railroad operations than the makeup and breakup of trains.[147] All rolling stock must be free units, easily attached or detached from each other. Only the engine and tender are slave units semipermanently attached to one another. The fastening devices for cars must be simple, rugged, and dependable. They must work under all weather conditions. They must be manageable by the most untutored members of the labor force. They must be capable of thousands of couplings during a long service life and yet be cheap to buy and maintain. The search for such a remarkable mechanism was, as might be expected, a long and arduous one.

LINK-AND-PIN COUPLERS

Early railroads hoped to solve the problem by adopting the simplest form of coupling. Simplicity is often the best choice in things mechanical, but it can sometimes be carried too far. American mechanics chose to combine the coupling and draft mechanisms into one unit and so broke away from the British or European system in which the two were separate and distinct. In the latter plan the coupling consisted of links and hooks at the centerline of the car. The buffers, however, were attached at the corners of the body or frame. The American plan, apparently introduced in the early 1830s, eliminated the corner buffers and threw the whole load of the train's undulating motion onto a single spring normally fastened to the far end of the coupler itself. The load thrown onto the draft spring often proved excessive, and so stops or dead blocks were often resorted to. They transferred the most severe shocks to the end of the car frame. The dead blocks aroused considerable concern for the safety of the trainmen who had to step between the cars to make couplings. (Of course, the European buffers still in use today are no less man-killers than our long-banished dead blocks. Trainmen must still crawl between or under the buffers when making couplings with the ancient links and hooks in service on the majority of European cars.)

Our first-generation railroaders sought and found the simplest possible form of coupler. It was not unlike what a farmer might contrive to hook up his tractor and hay wagon. The tubelike body of the coupler received an oblong link. A hole through the tube, a few inches from one end, received a pin that held the link in place. It was simple, no doubt. It was wonderfully elementary—stark simplicity and bare bones. Cheap, yes, it was very cheap. But as suggested earlier, some things can be too simple. A badly made tool might do the work, but if it does it poorly and inefficiently, it is no bargain. If the tool is inconvenient and dangerous, there is every reason to junk it.

The link-and-pin coupler was very much a poor tool. It made a loose, sloppy connection between the cars with too much give and play. There was so much slack or loose play between the cars that they tended to crash into each other when the train stopped or slowed down. Even at the slow speeds of the time, so much mass lumbering together could cause damage to the cars and their lading. A trainman manning the brakes on the roof could be knocked down by the jolt, or even thrown off the train. A hard jerk could break a pin or link. The links were particularly vulnerable because they were fabricated from wrought iron. Only a blacksmith's weld held them together. A broken link or pin could mean a runaway train. Should the failure happen on a hill, the runaway could end as a serious accident. The loss of life and property made the cheap link-and-pin coupler a bad bargain. Indeed, there was much to criticize and little to praise.

The list of defects goes on. If being sloppy, weak, and unsafe was not enough, the link and pin was also inconvenient. Because there was no standard coupler, every railroad had its own design. Each drawhead, pin, and link was different. Sometimes the variations were slight, but in general no link or pin was interchangeable. When trains were made up or broken apart, train crews might spend hours matching up pins and links in a frustrating game of square pegs and round holes.[148] Inconvenience on this level was not simply irritating; it cost time, and that cost was borne by the railroads and their shippers. Train crews perversely aggravated the situation by stripping idle cars of their coupling hardware. When cars were set off, the trainmen would systematically take all the links and pins for their own stockpile. Whoever came to pick up the cars would then sort his stockpile in hope of matching pin size to hole size. This process required the extra labor of carrying links and pins from the caboose or tender, where the extras were normally stored, back to the cars to be serviced. A small selection could easily weigh 50 pounds. If none of them fit, then it was back to the other end of the train for a second or third set. If nothing worked, that car or group of cars must be left to another crew. Again, more time was lost for both the railroad and the shippers.

Like loose change, links and pins tend to get lost or stolen rather easily. At the 1875 Master Car Builders meeting it was reported that millions of new pins were made each year but no one knew what became of them. At the same time, the Jeffersonville, Madison & Indianapolis Railroad lost over forty-three thousand links and pins each year, an impressive loss for such a small railroad. Big roads like the New York Central figured monetary losses of $45,000 to $50,000 due to truant coupler hardware.[149] Tramps and juvenile trespassers stole links and pins as a ready source of cash at the

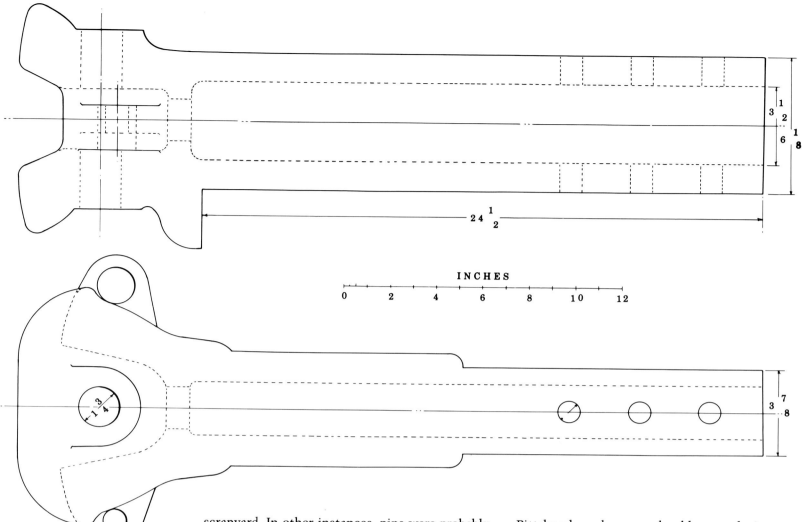

FIGURE 7.63
This late-model Potter three-hole drawhead was made in 1888 by the Portland Co. for the Maine Central. The wrought-iron stirrup for the draft spring is not shown; however, the three rivet holes to secure it are indicated. The draw-head is shown upside down in the side elevation. The dead stop should face up rather than down. (Traced from an original drawing at the Maine Historical Society, Portland)

scrapyard. In other instances, pins were probably taken simply because they were available and easily transportable. The losses, whether from inside or outside the railroad industry, were substantial enough to suggest to any competent manager that some better form of coupling was needed. Yet no one seemed ready to abandon the loose-link coupler, at least not much before 1880.

Several stopgap measures were devised. Chains fitted at the top end of the pin were offered as one way to keep the pin with its coupler. But too heavy a chain interfered with the trainman's work. A light chain was easily broken, and away went the pin.[150] Fixed links seemed to offer a better means of security. A second pinhole just behind the normal hole had a long rivet to hold the link in place. The link could not wander, but it did interfere somewhat with the coupling operation. Yet another approach was taken by Ransom S. Potter of Chicago. Potter proposed a massive three-hole coupler with a fixed link on one side, a chained pin on the other, and an open hole in the center for conventional couplings (Fig. 7.63). The plan was patented on August 30, 1860 (No. 29,815) and was adopted by a number of major railroads, including the Burlington, the Milwaukee, and the Maine Central. The inventor preferred a cast-iron body for cheapness, but Wilson, Walker & Company of

Pittsburgh made a more durable wrought-iron version for the Central Pacific, the Illinois Central, and several other lines.[151] In addition to preserving the company's hardware, Potter claimed that his coupler offered a spare pin or link should either break. He also claimed that it was possible to make a stronger connection between cars by a double link-and-pin coupling.

There does not appear to be any statistical data on the scale of use for Potter's invention. It may not have been used in very large numbers because of its weight and extra cost. Even so, there is pictorial evidence that the Potter coupler was used widely, at least in geographical terms. For example, drawings published in 1878 for a Burlington combination box/ventilated/cattle/car—reproduced in Chapter 4 (Fig. 4.61)—show Potter's coupler. A Milwaukee version of a few years later is shown here as Figure 7.67 and a New England version as Figure 7.63. The latter illustration depicts a cast-iron Potter drawhead used on a group of gondola cars built in February 1888 for the Maine Central. The wrought-iron draft-gear bracket, spring, and bar were omitted from the drawing. The bumper or safety stop should face up rather than down. Note that each hole is of a different size so as to accommodate various-size pins.

A simpler pin-saving scheme was devised by the

FIGURE 7.64

This link-and-pin coupler with a lift chain was made by the Harrisburg Car Co. between 1885 and 1890 for the Union Tank Line. (Traced from an original drawing at the Division of Transportation, Smithsonian Institution)

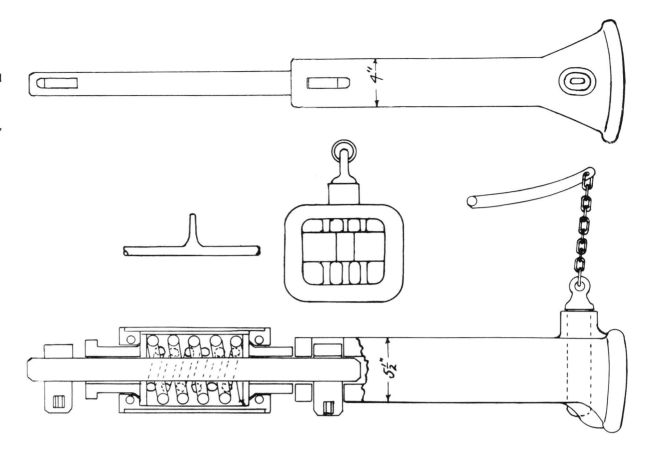

Union Tank Line around 1885 or 1890, as illustrated in a tank car drawing reproduced in Chapter 5 (Fig. 5.78). The details of the coupler and draft gear have been copied separately for greater clarity and are reproduced here as Figure 7.64. The plan is unremarkable in all respects except for the pin lifting lever and chain. At first glance one might be tempted to take this scheme for a safety device, but it actually averts only a little of the danger of the coupling process. Lifting or dropping the pin involves relatively little hazard compared with guiding the link into the coupler pocket cavity. To do the latter, the trainman must stand between the cars. When uncoupling, he can stand beside the track and lift the pin lever; there is little chance of injury even with ordinary links and pins, and so the safety benefits of the UTL plan seem very modest. The lifting arm would, however, offer some measure of safety if an uncoupling were necessary, as in the case of switching cars in a classification yard. The main goal of the design, then, seems to be to save pins by means of the chain attachment.

Tracing the exact origins and early evolution of the American car coupler is an impossible task because of the paucity of information available. The few surviving drawings showing general car arrangement in the earliest period often omit details such as couplings altogether. Specific clues are offered in a few instances, however. One of these is a drawing dating from around 1831 of a Baltimore & Ohio gondola car that shows a simple fixed-pin coupler (Fig. 2.8). Upright pins at each end of the car are attached to a center sill that projects out well beyond the end of the main car frame. A link with two oblong holes is used to connect the pins. A slight chain prevented loss of the link by attaching it to the extended center sill or perch pole. This same general plan was employed by Maine's pioneer railroad, the Bangor, Oldtown & Milford, which opened in August 1837. The link or shackle in this instance was made from leather rather than iron. Three thicknesses of heavy sole leather were held together by copper rivets. Holes at each end of the leather band made it possible to hook it over upright pins fastened to the car ends and so unite the train. How typical this arrangement was will never be known, but it apparently proved unsatisfactory, for by the late 1830s the B&O had adopted a scheme closer to the conventional link-and-pin system. A drawing from that time of an eight-wheel boxcar, also reproduced in Chapter 2 (Fig. 2.14), shows a pin fitted with a lifting ring at its top. A pocket for a link is made between the end sill and the floorboards of the end platform. Presumably a metal reinforcing plate was attached to the end floorboard. This arrangement would be weak and was surely replaced by a more substantial drawhead. By the 1850s the B&O was using a stubby cast-iron drawhead bolted directly to the end sills. There were no draft springs, but a heavy iron rod connected the two drawheads and so formed a continuous drawbar that was sturdy but inflexible.

FIGURE 7.65

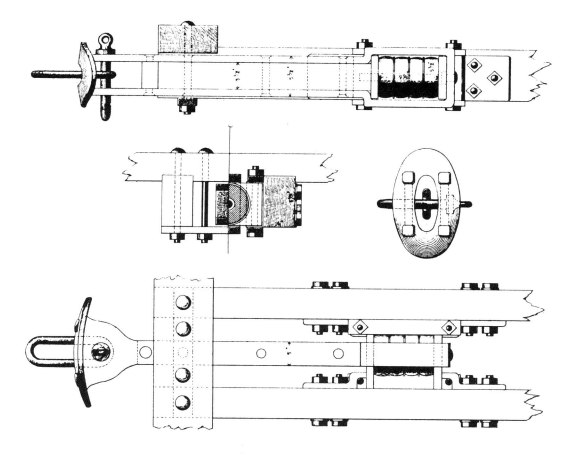

Other railroads adopted the British chain-and-hook system. This simple, loose method of connecting cars provided ample slack between the cars, which made starting heavy trains with light locomotives less of a challenge. It also made smooth stopping an impossibility, and the bumping together of cars in a long train, even at slow speeds, must have created a considerable clamor. The link-chain coupling continued in use on four-wheel coal jimmies long after it had been abandoned for other types of freight cars. Some of these antiquated vehicles were in revenue service until the 1890s. Examples are shown in Chapters 2 and 5. A large number of eight-wheel coal hoppers on the Reading, dating from the 1880s, had link-and-hook-style couplers. Three link-chains were sometimes used to attach drawheads when a considerable difference in height existed. This crude arrangement was reported in the recollections of an Old Colony Railroad official, James H. French, who began railroading in 1854.[152] French went on to note that some connecting roads were still using large leather-covered buffers mounted on the corners of the cars. This added to the coupling woes of the train crew, for the Old Colony cars had dead-woods like most other U.S. lines, fastened near the center of the car. And so the crews were beset with the aggravation not only of matching links, pins, and chains but of dodging man-mashers at both the centers and the corners of the cars.

By the mid-nineteenth century the common link-and-pin coupling was in widespread use for both passenger and freight cars. Wrought iron was the favored material, and pieces were by necessity riveted together. The front plate was often fastened onto the top and body plates by projections peened over square holes cut into the face plate (Figs. 7.65, 7.66). The longitudinal plates measured 1 inch thick by 3 inches wide; they spread out to about 9 inches at the head end where the pinhole was drilled or punched. A wooden or cast-iron spacer piece was riveted or bolted between the top and bottom plates. The entire assembly, including the draft-spring pocket, measured about 46 inches long.

The example shown in Figure 7.65 reflects a generic design observed by a German engineer named Bendel who published a study of American railroads based on data collected in 1859. The general plan had been in use for at least a decade and perhaps longer, for it agrees closely with a drawing published in the September 16, 1848, edition of *American Railroad Journal*.[153] Wrought-iron construction remained popular into the 1870s, and some railroads, such as the Pennsylvania, remained loyal to it as the best form of construction. Cost-conscious car builders were always seeking ways to substitute cheaper components, particularly when it came to freight cars; in a fleet so large, sizable economies could be effected even if the cost of a part were reduced only by a few cents. Cheaper cast-iron drawheads gained in popularity starting in the 1860s. Cast iron was the wonder material of the Victorian engineer and, like the

FIGURE 7.66

This wrought-iron drawbar, like the one shown in Figure 7.65, has a pocket end for the draft springs, but in this design a volute spring is used. Note the column spacers. Wood was commonly used for this purpose. (*American Railway Review*, Feb. 28, 1861)

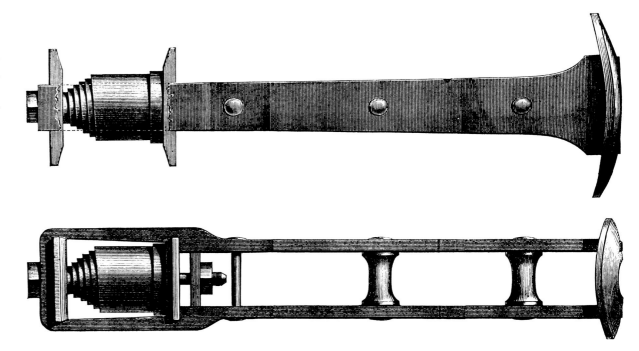

FIGURE 7.67 *(below and opposite page)*

The profusion of individual coupler designs used by American railroads is illustrated by the forty-two cast-iron drawheads kept in stock by the Erie Railway for the repair of interchange cars. Note that none is shown with the draft springs or draft-gear apparatus. (*Railroad Gazette*, Sept. 19, 1884)

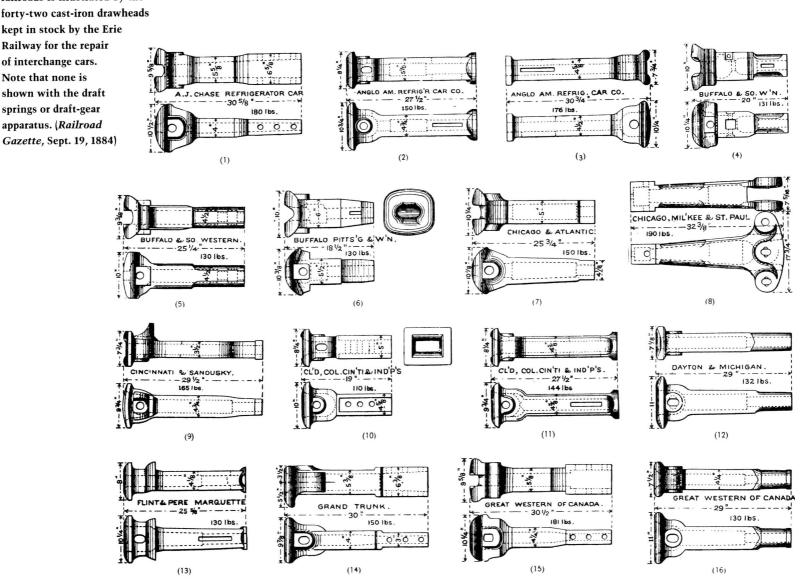

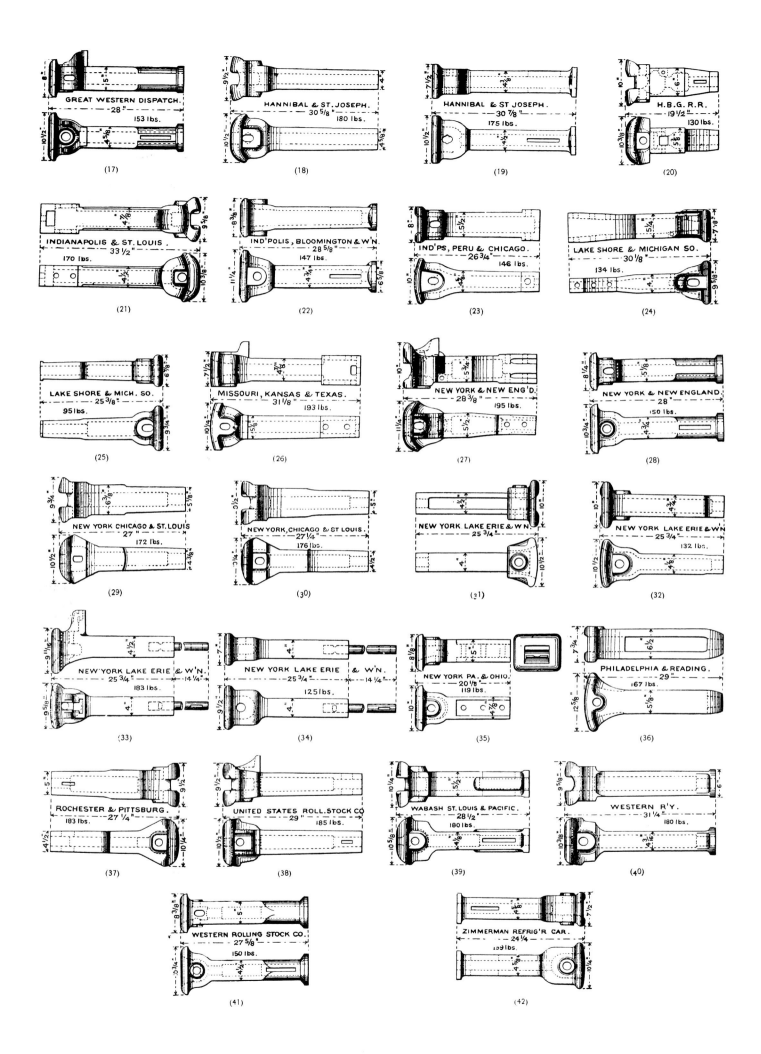

FREIGHT CAR TECHNOLOGY

plastics of today, was called upon to solve every fabricating problem. It was a very good material in many ways: cheap, reasonably strong, easily molded and machined into shape. It did not, however, withstand impact or stress shock very well, and so for something that would measure up to coupler service, it was necessary to make rather massive castings. Weight became one of the trade-offs of the cast-iron drawheads. Discounting the draft-gear parts, a wrought-iron drawhead weighed 100 to 125 pounds, while the cast-iron variety weighed between 130 and 170 pounds.

In a railroad car this weight difference was trivial, but the breakage record of the two types was decidedly against the cheaper cast-iron head.[154] In 1884 the Erie Railway, for example, had to keep no fewer than forty-two different patterns of cast-iron drawheads in stock because they failed so frequently (Fig. 7.67). This large stock was maintained so that the repair of interchange cars could be handled more speedily. Weeks might pass before a special-order casting could be obtained, and the Erie would find its yards choked with foreign cripples. Couplers for cars infrequently interchanged were not stocked; otherwise the Erie might have had hundreds more patterns on hand. It is hardly necessary to comment on the lack of standardization. The Erie found no need to stock wrought-iron couplers because they broke so infrequently.

That drawheads came in many shapes and sizes is amply documented by the forty-two examples shown in Figure 7.67. Links, however, tended to be somewhat more uniform. They were generally made from 1¼-inch-diameter wrought-iron bar stock that was slightly flattened in the forging and welding process. The oblong loops measured 13 inches long by 5 inches wide. Each link weighed 6 to 8 pounds. Common links were made flat, but a certain number of bent links were produced to aid in coupling drawheads mounted at varying heights above the track. Despite the longstanding Master Car Builders Association standard level for drawbars, many railroads ignored this sensible call for uniformity even in its most basic form. Those who cooperated would send out new cars with couplers at the standard height; however, the ravages of interchange service soon resulted in loose and sagging drawheads. Sometimes in-service damage would result in drawheads being jammed permanently to one side, in which case a bent or offset link would be needed to make a coupling.

The fastening pin offered a greater opportunity for car builders to exhibit their independence and creativity. Unlike the cargo of Noah's ark, no two were alike. Some were short, some long, some fat, some thin. Some were round in cross-section, some oblong. Some were 1 inch in diameter, some 1⅛ inches, some 1¼ inches, and some 1½ inches. Head shapes showed the same wonderful spirit of inventiveness. The simplest head was made by bending over the top of the bar to form a right angle. Most others employed a more elegant shoulder. The top was forged as a ball, a stumpy handle, or an open eye. Occasionally the upper part of the shaft above the shoulder was flattened out into a rough disc. Some railroads would stamp their initials on the shaft's side, thinking naively that this precaution would ensure its safe return home should the pin become separated from the car.

THE SEARCH FOR SOMETHING BETTER: SELF-ACTING COUPLERS

Modern historians tend to treat the railroad coupler strictly in terms of employee safety. The link-and-pin coupler is seen as a prime example of industrial malfeasance. The safety issue did not receive much emphasis, however, until rather late in the nineteenth century, and even then it was regarded by most railway officials as more a nuisance or crank's cause than a major problem requiring immediate action. Railroading was a dangerous occupation. Coupling cars was only one of many hazards to be faced each day when out on the line. The railroads hired rough men to deal with such work. Those wanting a safer and more secure occupation should seek employment elsewhere. This attitude may seem hardhearted or callous, but it is supported by the longstanding reluctance of the railroad mechanical leadership to adopt something better than the link-and-pin coupler. They did not perceive trainman safety as a primary problem; if they had, then surely a better coupler would have been adopted decades earlier. This tardiness cannot be explained away by the lack of a practical alternative. Workable automatic or at least semiautomatic couplers had been available since the 1840s. Miller's hook-style coupler, introduced in 1863, became an accepted standard on passenger cars within ten years.

The early promoters of self-acting couplers in fact seemed to sense the railroads' indifference to the safety issue. Early inventors, such as Hopkins and England, emphasized the practical and economic advantages of their couplers. Most of the text of their advertisements and handbills dwelt upon the convenience, security, and cheapness of their devices. Only near the end of the text was the safety matter mentioned. In at least one case the issue of accidents to trainmen was not even raised—in a leaflet issued by car coupler patentee S. D. Locke of Hoosick Falls, New York. Was this an oversight, or an intentional omission? Could it be that the coupler promoters sensed a wariness on the part of railroad managers regarding coupler safety? Or was it that managers were embarrassed by this subject and so became defensive or irritated by assertions concerning the plight of freight train crews? Coupler promoters were suppliers who hoped to sell their wares, and so it is likely that the more astute of them chose to downplay the safety issue in their literature.

As editor of the *Railroad Gazette*, M. N. Forney had every reason to soft-pedal the safety issue as well, because his career and livelihood surely depended on the goodwill of the railroad management. Forney was not a man to compromise his beliefs, however, and he took a strong position on the issue at an early date. He spoke on the coupler safety issue frequently in the pages of his magazine and at the meetings of the Master Car Builders Association.[155] As the secretary of the group, his words were always recorded in the published annual reports of the national professional body of the car-building fraternity. Car coupler manufacturers might tread delicately around this sore issue, but not Forney. He marched right into it. He poured on the guilt and shame. How could respected citizens like the railroad managers so ignore the welfare of the men in their employ? The suffering borne by the injured was sickening to contemplate, but worse, in Forney's opinion, was the "perfectly brutal disregard for the safety of those who are employed in coupling cars."[156] To drive the guilt home, Forney stated that the situation was tolerated because the victims were poor and obscure individuals without prestige or influence. In addition, most of the accidents happened in out-of-the-way places and so went unnoticed by the press or the public. These factors only raised Forney's compassion and made him a lifelong champion of the trainman's cause.

Just how many trainmen were killed or injured during the link-and-pin era will never be known, for no national statistics on the matter were gathered until late in the century. Before that time estimates were based on the spotty data gathered by various state railroad commissions. Between 1877 and 1887 eight states, including New York and New Jersey, recorded 25,929 railroad employee accidents.[157] Of these, 9,851 or 38 percent were coupler-related. In 1888 the Massachusetts Railroad Commission noted that almost half of all employee deaths and injuries were coupler-related. Let us assume, then, that 40 percent of all employee accidents were coupler-related. On a national basis it is estimated that twenty-five thousand railroad men were injured a year between 1870 and 1900.[158] This would mean that something around ten thousand men were injured by couplers during this thirty-year period.

Not many men seemed to die in coupler accidents. Deaths were most frequently caused by falling off the train, particularly from on top of the cars. Every now and again some luckless trainman would place himself between the deadwoods or the couplers themselves and have the life crushed out of him. In other instances he might be knocked down while attempting to manage the coupler and roll under the wheels as the cars were jostled back and forth by an impatient engineer. Maybe he would only lose an arm or a leg, perhaps two legs; or he could be cut in two. Running between cars when making a flying switch was another sure route to self-destruction. One railroader recalls his dance with death after a few months in the yard force:

It was four or five months before I "got it." I was making a coupling one afternoon. I had balanced the pin in the drawhead of the stationary car and was running along ahead of the other car, holding up the link. Just before the two cars were to come together, the one behind me left the track, having jumped a frog. Hearing the racket, I sprang to one side, but my toe caught the top of the rail. I was pinned between the corners of the cars as they came together. I heard my ribs cave in like an old box smashed with an ax.

The car stopped and held me like a vise. I nearly fainted with pain, quite unable to breathe. Fortunately, Mr. Simmons was watching. With the presence of mind that comes of long service, he called at once for the switch rope, and he would not allow the engine to come back and couple the car again, as that would surely have crushed out my life.[159]

Most coupler accidents involved hand injuries. Fingers or the whole hand were crushed or sheared off. Trainmen of this era rarely needed an introduction, for they could readily recognize a brother by looking to see if he had any missing digits. These injuries occurred because the trainman had to hold and guide the link into the coupler pocket while one of the cars was in motion. He could release the link only seconds before the two drawheads banged together. This maneuver was repeated thousands of times each day. The slightest miscalculation, a momentary lack of attention or a misjudgment, and a finger was gone in a flash. Poor lighting, bad weather, and inexperienced engine crews added to the likelihood of losing a digit.

The coupler safety problem primarily concerned freight trains. Passenger cars had been fitted with self-acting Miller couplers by 1875. The number of passenger cars was small compared with the freight car fleet; hence most coupling operations involved the latter. Passenger equipment tended to be switched less frequently anyway. It was rarely interchanged. Many train sets stayed together for weeks or months at a time. Except for diners and sleepers, passenger cars were rarely picked up or dropped off en route. Because they were valuable and carried vocal and high-grade goods—passengers—palace cars and even coaches were by and large treated carefully. The exact opposite was true for freight cars. They were constantly coupled and uncoupled as freight trains were made up and broken apart. Cars were routinely picked up and dropped off en route. The national interchange of cars ensured numerous coupling operations. Nothing on wheels received rougher handling than freight cars. This quick-and-dirty attitude resulted in sudden and unexpected movements that often ended tragically for the trainman, be he a veteran or a greenhorn. If he lost a finger, the odds were overwhelming that it was because of a freight train switching accident.

Care should be taken not to exaggerate the hazards of early American railroading. For all its inherent perils, it was not the most dangerous occupation. Sawmill workers faced being crushed by falling timbers or sliced to bits by huge saws. Steelmill workers were vaporized by molten metal. Chemical workers were dissolved by overflowing tanks of acid. Farm workers were regularly chewed up by threshers, binders, corn grinders, and a host of other machines. Miners unquestionably faced the greatest prospect of death and injury of any American worker.[160] Between 1890 and 1900 twenty-five thousand to thirty-five thousand industrial workers were killed each year in the United States. One million or more were injured annually. The railroads' share of these grim numbers would be approximately 10 percent of the reported deaths and 4 percent of the injuries. Considering the low safety standards characteristic of American industry, railroad leaders probably felt that they were being unfairly singled out for criticism. That both workers and managers had smaller expectations as to job safety is probably also true. The railroad business was, however, more public than many other areas of American industry, and as such it was the subject of considerable public scrutiny and comment. Railroads were also regulated, at least on the state level, and it was in this local arena that the coupler question was addressed starting around 1880.

The Massachusetts Railroad Commission began hearings on the matter in 1880.[161] These investigations led to a state law requiring automatic couplers for new freight cars and the application of the same to all older cars undergoing major repairs effective March 1885. In September 1884, some months before the law took effect, the railroad commission invited the public to submit ideas for the proper style of coupling. Over 170 inventors crowded into the hearing room. Among these contestants showing drawings and models was a Mrs. Moulton of Lynn whose coupler was considered dangerous but workable.

A yard test of full-size couplers was conducted in Boston a few weeks later, by which time the number of contestants had been greatly reduced. The commissioners picked out five finalists—the Cowell, Janney, Ames, Hilliard, and United States brands; however, not one was compatible with any of the others. The Hilliard was a hook-style device that coupled in the horizontal plane, whereas the United States was a loose-link drop-pin style. Both the Cowell and the Janney were swing-jaw or

vertical-plane designs, but the heads were so different that they could never form a union. Despite such problems, major railroads such as the Boston & Maine agreed to re-equip their freight cars with self-acting couplers approved by the commission. In actual practice, conformance with the law was slow and indifferent; by 1888 only five thousand cars had been refitted. Ineffective inspection and enforcement were blamed for the poor showing.

The Massachusetts law inspired other state commissions to act. Connecticut began public study of the problem in 1881 and followed with legislation about the time the Massachusetts lawmakers acted. New York and Michigan followed suit in 1885 and 1886.[162] Like their New England role models, they had difficulty selecting a single coupler. Michigan picked out seven different styles of coupler as its standard. In New York the drop-pin design was favored. Neither state considered the Janney coupler worthy of selection.

Car builders watched the well-intentioned fumblings of the state commissions with growing concern. What was being visited upon them? The selection of couplers was being made by a bunch of amateurs. They were political appointees, do-gooders, or someone's brother-in-law, and not a one had the slightest mechanical sense or training. Who was less well qualified to make judgments on engineering questions than a state railroad commissioner? Members lamented about the state of affairs during the annual meetings of the Master Car Builders Association.[163] Chaos and confusion seemed all around them, but much of the blame was on their shoulders. Who had delayed and procrastinated year after year? Who would not take the time to study this complex issue? And who tended to reject both the sound and the foolish designs? Had the car builders faced up to this issue years before and worked resolutely to develop a practical self-acting coupler, they would not be in this absurd race to find a last-minute solution. The politicians were really only doing their job by forcing the technocrats to act.

In defense of the car builders, there are two plausible reasons for their reluctance to grapple with the coupler problem. It was a complex matter. The number of proposed designs was staggering. It was confusing, annoying, and frustrating to be confronted with such an array of plans. Who had the time and patience to sort through such a volume of material? There were not dozens or hundreds but thousands of car coupler patents. By 1875, before the safety mania began, there were already nine hundred patents devoted to this one branch of railroad engineering. In 1887 the number had grown to over four thousand, and a decade later it reached sixty-five hundred, or 1 percent of all U.S. patents ever issued. Never were there so many candidates for one job, and never were there so many losers and also-rans. Since there could be only one cou-

FIGURE 7.68

James B. Safford's draft gear, introduced in 1874, offered a safety cove or opening to protect car men's fingers during coupling operations. The design gained some popularity over the next twenty years and helped prolong the use of link-and-pin couplers. (*Railroad Gazette*, April 14, 1876)

pler, the selectee would presumably become a wealthy and celebrated individual. The intense lobbying of the coupler inventors became an irritant to just about every master car builder. Receptionists were instructed to shield the boss from any coupler crank who might appear at the door. The railroad trade press ridiculed the silly ideas proposed by the lunatic fringe.[164] It was fun to make light of such a dark subject, but it did little to solve the problem or persuade the car builders to pay attention to a problem put off for too long.

It is possible that the second reason for postponing a decision on couplers was the fear of failure.[165] What if the wrong selection were made? After careful study and testing, the individual master car builder makes his recommendation to his management. The management agrees: "He's the resident expert—if he wants to buy four thousand McBird couplers and is convinced that they are the best available, that's fine with us." Major funding is authorized to purchase and install the McBird apparatus. A few months later, the new couplers begin to fail on a daily basis. Idle cars produce no revenue. The cost of refitting is huge. The master is disgraced. His mechanical judgment is deficient. He is a failure. He is discharged. It would have been safer to wait and make do with the old links and pins. Maybe the Master Car Builders Association would make a selection, and if that did not work then the individuals could blame the mistake on group action. The judgment of the selection committee would be on the line. This line of thinking may explain why no individual car builder wanted to take on the responsibility for reforming his own railroad. There was greater safety behind the shield of group action.

This does not mean that car builders were unwilling to deal with the coupler problem in small ways. A few halfway measures were at least attempted. Sticks or wooden paddles were issued to all workers required to connect cars.[166] The stick was to lift and guide the link into the coupler

pocket and so eliminate the dangerous manual operation. Some railroads mandated use of the safety paddles, and whenever the superintendent was present, the men faithfully obeyed orders. The instant he retired, the men placed the paddles back inside their neat black leather sheathes and grabbed hold of the links as God intended them to do. Forney could not understand the trainmen's "violent opposition" to the use of coupling sticks. It was for their own good, but the men felt that it reflected badly on their professional competence and manhood. Only a sissy or unseasoned hand would use the paddles; the veteran trainman was too brave and dextrous to need them. And so it would appear that the concern of reformers like Forney were wasted on the very men they hoped to protect. Ironically, the indifference to danger was not accountable to negligent managers alone. Worker resistance to safety gear is just as evident today; witness the reluctance to use eye protection and hard hats. The usual reason given is that the safety measure is inconvenient and gets in the way of doing the job. The same reason was no doubt voiced for not using the coupler sticks.

Because there was no way to make the men use sticks, James B. Safford of Albany devised a special drawhead with a side gap or safety cove (Fig. 7.68). The opening provided finger space even after the drawheads had butted together. The safety factor went down considerably if the drawheads were not on the same level, which was so often the case. But even if only a few fingers were saved, Safford's idea was worth adopting because it was easily implemented and cost very little. A patent was issued on January 20, 1874 (No. 146,714). Within two years about fourteen thousand were in service.[167] By 1879 over ninety thousand were in use on over 160 railroads. The number of Safford couplers continued to grow, with fairly major lines like the Chicago & Atlantic and the Hannibal & St. Joseph adopting them. The Interstate Commerce Commission stated that they ranked ninth among the

FIGURE 7.69

An even safer version of the Safford coupler was manufactured by the Kansas Rolling Mill Co. It was stronger than Safford's cast-iron coupler, being made from wrought iron. (*Railway Age*, July 17, 1879)

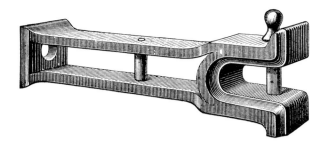

categories: automated links and pins; hooks or catches; spears or harpoons; and the vertical-plane type generally having a hinged or swing-open jaw. The Janney coupler was a member of this last class.

AUTOMATED LINKS AND PINS

The automated link and pin was the scheme most favored by nineteenth-century mechanics. It was the obvious and direct approach of converting the existing standard into something new, very much like trying to make a carriage into a motor car. Efforts along these lines appear to have started with an Erie Railway employee named David A. Hopkins of Elmira, New York. In his patent of August 1, 1854 (No. 11,428), Hopkins placed a spring-loaded block inside the coupler cavity (Fig. 7.70). The block supported the pin, and when it was hit by a link inserted into the drawhead, the block was pushed and the pin fell into place. This became a basic scheme copied, revised, and toyed with by hundreds of other coupler patentees. It was simple, it worked, and it preserved the traditional link-and-pin design. Hopkins's device was applied to a number of cars on the Erie and some of its connecting lines. In all, perhaps several hundred were outfitted. But was it really a safety coupler? It would work only if the drawheads were reasonably level so that the link could enter the coupler opening unassisted. Hopkins used a scored block to hold and in effect aim the link. One wonders about the effectiveness of the grip and the patience of the trainman who had to keep adjusting the angle of the link until it entered the opposing drawhead successfully.

Just a year after the Hopkins patent, another railroad employee, J. T. England of Baltimore, devised another drop-pin style of coupler. England used a ball rather than a spring and block. The link pushed the ball back, allowing the pin to fall into place. To reset the coupler the pin was manually lifted, allowing the ball to roll under it via an inclined pathway cast inside the coupler drawhead. England made no effort to find a way to hold or guide the link into the opposing drawhead, and unless conditions were nearly perfect, the link would need some guidance. Differing coupler levels and the bouncing of the car, which would jostle the link to one side or the other, almost guaranteed a mismatch. England made no provision for a lifting bar and chain, so uncoupling required a manual lifting of the pin. For all its obvious defects, England's coupler was applied to a number of Baltimore & Ohio cars. An example can be seen in Chapter 8 (Fig. 8.8).

Of Hopkins's many imitators, George W. Smillie copied the pin-support device of his mentor the most faithfully. Smillie's borrowing stopped when he devised a shallow dish face that would guide the link to the center hole whence it would plunge in-

forty-one coupler types listed in the commission's 1889 statistical report. In 1879 a fully open-side jaw coupler was put on the market by the Kansas Rolling Mill Company of Kansas City (Fig. 7.69). The larger space offered more finger room and hence greater safety, but the true effectiveness of either the Safford or the Kansas device will never really be known. In any case they did not meet the need for a self-acting coupler that would not require men to stand between cars.

It was the ambition of a great many inventors to perfect and market a safe and practical automatic car coupler. There were all the humanitarian satisfactions one could hope for, plus a very real cash reward to the individual who created a plan acceptable to the entire industry. Royalties on the hundreds of thousands of couplers needed would not be trifling. Some of the inventors were gifted, imaginative men, but the majority were lesser talents who poured forth a torrent of unoriginal, uninspired, and often unworkable devices. We have spoken already about the number of patents granted in this area. For all this activity, no one actually developed a truly automatic coupler, nor did any one inventor reap the full benefits envisioned for the creator of the new industry standard. The winner of the contest, Eli Janney, developed a very good semiautomatic coupler that still, even in its most modern guise, occasionally requires manual maneuvering to open. Janney received some compensation for his specific design, but he was never able to claim royalties for the generic vertical-plane coupler that replaced the entrenched link and pin.

The solutions offered to solve the coupler problem included just about every imaginable mechanical contrivance that clamped, grappled, snapped, speared, hooked, latched, pinned, clasped, or linked. To describe all these devices would result in a very long book that would serve no particular purpose. At the same time, the mental labors of so many mechanics should not be too casually dismissed. The sheer absurdity of some of these ingenious designs is itself worthy of study. The contorted shapes and bizarre mechanisms illustrate just how far some inventors were willing to go for success and at the same time how wrongheaded mechanics can be. Certain couplers pictured in these pages could be mistaken for medieval armaments. Our selection can be broken down into four basic

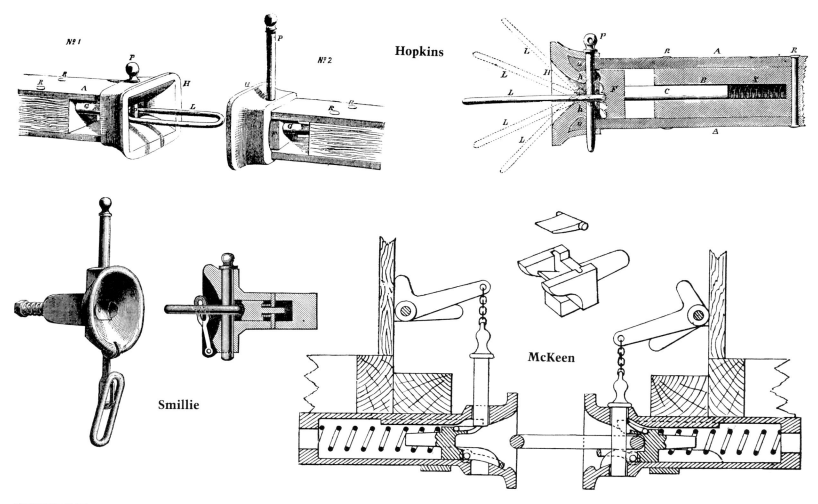

FIGURE 7.70

Drop-pin couplers were offered in many designs as a way to automate the old link-and-pin coupler. Shown here are David A. Hopkins's patents of 1854, George W. Smillie's design, and Thomas L. McKeen's 1884 coupler. (*Railroad Gazette*, Oct. 2, 1885, and June 11, 1886)

ward, pushing back the spring block shaft and so allowing the pin to drop. The inventor hoped thus to overcome a major failing of the drop-pin coupler—the inability to self-direct the link into the coupler pocket. To prevent the loss or theft of the link, Smillie attached it at the bottom of the dish with a secondary link. This distinctive if not comely coupler is shown in Figure 7.70. The Smillie couplers worked well enough that twenty-five hundred cars on the Lackawanna were fitted with them by 1886.

During this same time Thomas L. McKeen of Easton, Pennsylvania, was also active in developing a pin-supported automated coupler. His release mechanism was similar to Hopkins's, but he refined the block into a hinged grip that would hold and guide the link into the opposing coupler pocket (Fig. 7.70). A lifting arm and lever allowed uncoupling from the outside of the cars. A short chain attached the pin to the lever and also prevented loss or theft. The link was loose, however, and so remained fair game to those who would prey upon loose pieces of iron. By 1886 McKeen claimed that three thousand cars on the Lehigh Valley, the Lackawanna, the Erie and the Union Tank Line were equipped with his automated links and pins. McKeen's coupler worked well enough that it was one of a dozen selected by the

Master Car Builders Association for further testing at the 1885 Buffalo trials.

Not all mechanics devoted to preserving the link-and-pin coupler chose to do so by imitating Hopkins. There was another small school that preferred the lever system of pin release. While the general theme was the same, there were hundreds of variations of which we shall explore only a few. S. D. Locke of Hoosick Falls, New York, tested his double-lever drop-pin idea on the Troy & Boston late in 1886 (Fig. 7.71). A link entering the coupler pocket bumped the hook-shaped lever and so removed the pin's support. The pin could be lifted clear for uncoupling; at the same time, the levers would swing back into place by gravity and were ready once again to support the pin. Just how well Locke's apparatus performed is uncertain, but it looks far too fragile to withstand the rough-and-tumble life typical of freight service.

A few years before Locke's endeavors were tested, two residents of Fort Wayne, a major railroad town and home to the Pennsylvania Lines West, combined to produce the Fort Wayne coupler. It featured a single curved lever that was actuated from below (Fig. 7.71). Unlike Locke's design, the Nidlinger and Heath coupler could use a common pin. The drop lever could also grasp and hold the link, the better to guide it into the coupler

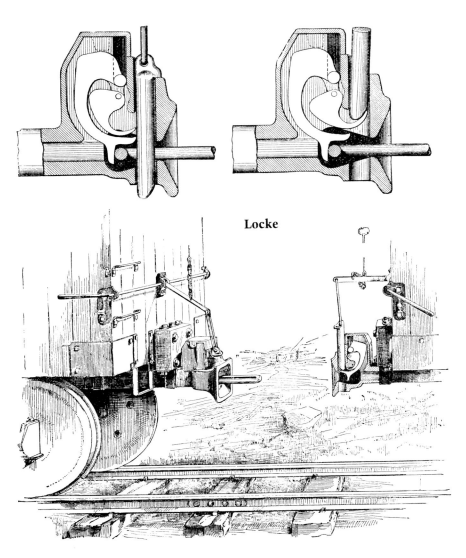

Locke

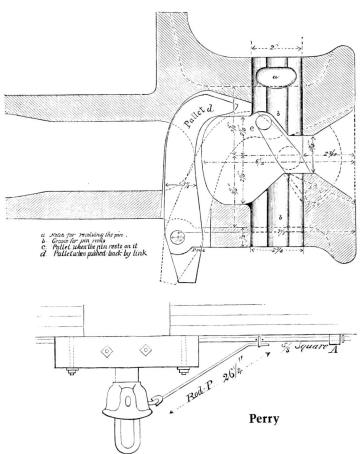

a. *Notch for revolving the pin.*
b. *Groove for pin revio*
c. *Pallet when the pin rests on it*
d. *Pallet when pushed back by link*

Perry

pocket of the opposing car. W. V. Perry of Chicago believed in the lever pin-support design but decided that the mechanism should be worked through a side rod (Fig. 7.71). Twisting the rod brought the lever forward and down slightly to form a support for the pin. When the link entered the pocket it hit the lever, knocking it backwards and so allowing the pin to fall and complete the coupling. In advertisements appearing in the 1883 *National Car Builder*, Perry claimed that several thousand of his units were in use on a dozen or so railroads. The coupler worked well enough that it was selected by the MCB Association for further study after the Buffalo coupler trials.

There was another basic idea on how to reform the traditional style of freight car coupling. This scheme involved a push-up arrangement in place of the more obvious drop-down or gravity-style automatic link and pin. Ezra N. Gifford of Cleveland began tinkering with this idea in the early 1870s. The mature design as it was worked out by the middle 1880s employed a special pin, oval in cross-section and featuring a crescent-shaped end (Fig. 7.72). The bottom point of the pin was balanced on the shelf. Upon entering the pocket the link struck the curved surface of the pin and caused it to jump upward. Gravity would send it back down a moment later, and it would fall through the slot of the link, making the coupling. The fat little pin could be hung up by a lip so as not to couple. The

FIGURE 7.71

The lever-release idea was a variation on the drop-pin coupler. Shown here are S. D. Locke's 1886 design, W. V. Perry's plan, introduced in about 1885, and the Fort Wayne coupler patented in 1882. (*National Car Builder*, July 1883; *Railroad Gazette*, Oct. 2, 1885; composite copied from trade publications of the period)

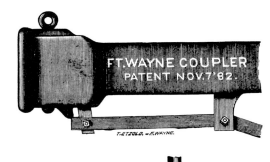

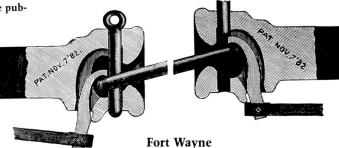

Fort Wayne

FIGURE 7.72

The push-up pin idea was used in the two couplers shown here, the Gifford and the United States designs. Both became relatively popular in the 1880s. (*Railroad Gazette*, Sept. 26, 1884)

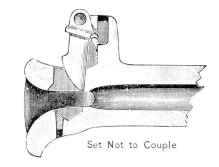

Set Not to Couple

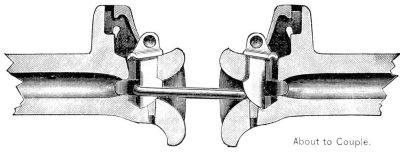

About to Couple.

Gifford

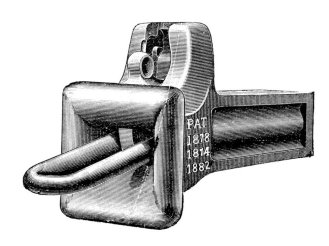

United States

Erie outfitted two thousand cars with Gifford's jump-pin connector, and the MCB Association was impressed enough to recommend it for more study at the conclusion of the Buffalo coupler trials.

An even more successful version of the bump-up pin coupler was marketed by the United States Automatic Car Coupler Company of Boston. Introduced around 1883, some twelve thousand were in use by the spring of 1887. A special pin was made with a deep flange on the front side so that the pin would not rotate (Fig. 7.72). The pin was a steel forging guaranteed not to bend or break for eighteen months. A curved dog was attached to the coupling pin flange by a small secondary pin. The dog served two purposes. Its curving surface helped guide the intruding link under the pin. The dog also made it difficult to remove the coupling pin and so again thwarted theft and loss. A set of couplers, including malleable-iron drawbars and royalty fees, was offered for $13. A goodly number of northeastern roads purchased the United States coupler for testing purposes. It was also approved by both the Massachusetts and Connecticut railroad commissions and the 1885 MCB Association coupler committee.

HOOKS

The internal-hook coupler offered another appealing way at once to preserve and to reform the established link and pin. The pin was altogether eliminated, but the old-fashioned link was retained. Better yet, the new form looked, externally at least, exactly like a bona fide link and pin, which was a likely source of comfort to veteran car men. It was reassuring to see a coupler that looked like a coupler should look. Perhaps the most successful of the internal-hook jobs was introduced in 1874 by Gillman H. Ames of Fairfield, Maine. Ames combined the link and the hook as a one-piece casting (Fig. 7.73). The hook-link was capable of movement for coupling and releasing. It could travel via an open slot up and back to uncouple, then down and forward to couple. The motion was imparted by a manually operated lever-and-chain arrangement. A very large front opening or mouth was needed to accommodate the link-hook and its travels. A steel version of the Ames coupler sold for $20 a set; the malleable-iron model cost $18. The New York Central and several of its subsidiary lines showed at least a passing interest in adopting the Ames coupler. The Massachusetts Railroad Commission put it on its approved list, and the MCB Association considered it worthy of more testing.

The Mitchell and Stanford couplers, introduced some years after the Ames, offered variations on the combined link-hook idea. But most other inventors pursued an arrangement employing a separate hook and link. P. A. Aikman, for example, of-

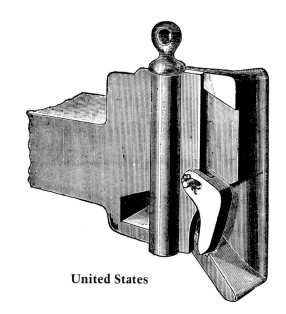

FIGURE 7.73

Three views of the Ames fixed-link coupler. The basic design, introduced in 1872, was adopted by major railroads like the Michigan Central. (*Railroad Gazette*, Oct. 2 and Sept. 25, 1885)

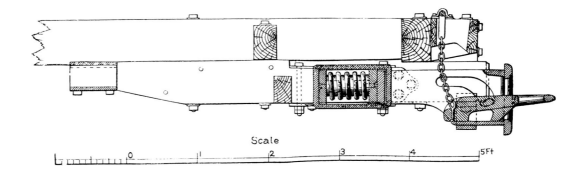

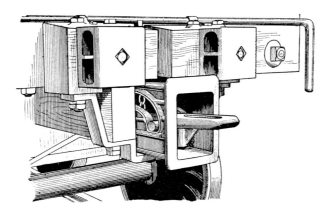

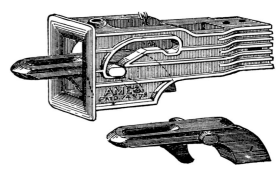

fered a very simple internal-hook coupler that avoided the use of springs, locks, or even lifting chains (Fig. 7.74). A thick, longish hook was moved by a top-mounted treadle that was itself manually worked by a shaft and lever. Aikman's design, patented in December 1885, was tested by the Grand Trunk and was recommended by the Michigan Railroad Commission. At $9.98 per car it was considerably cheaper than the Ames coupler.

C. E. Marks of Flint, Michigan, brought out a coupler at about the same time that was very similar to that of his neighbor, Aikman. Marks used a longer hook and placed a disengagement cam under the hook (Fig. 7.74). He also completely enclosed the hook so as to protect the mechanism from the weather. Marks's plan was also approved by the Michigan Railroad Commission, and three thousand were applied by one railroad within the state. The device passed preliminary tests at the MCB Association Buffalo trials, and so the Marks was one of twelve selected for more testing. S. B. Archer of Saratoga liked the long hook favored by Marks but felt that it must be secured by a spring to overcome the danger that the hook, engaged only by its own weight, was not a sufficient safeguard against an inadvertent uncoupling (Fig. 7.74). Uncoupling was achieved by lifting the hook manually through a light chain fastened near the forward top of the hook. Archer's design was tried on the Delaware & Hudson, but the hopes of the inventor were dashed when his contrivance failed to couple on curves at the 1885 Buffalo trials.

Curtis & Wood of Conshohocken, Pennsyl-

vania, approved of springs to hold the hook secure but altered the normal scheme to achieve this end. The inventors actually employed more of a notched block than a proper hook (Fig. 7.74). The end of the block was, however, made curved, like most hooks, so that the link could bump under the catch. When engaged, a spring and block kept the stumpy hook locked into position. A lifting chain effected an uncoupling. Note that the hook moved on a steep inclined angle. One of the partners, John Wood, Jr., was a practicing commercial car builder; however, this association did not prevent the failure of the Curtis & Wood coupler. It failed to couple on curves with conventional links and would do so only with special linkage; hence it was rejected during the 1885 Buffalo trials.

The final example offered here for internal-hook car connectors is the Union coupler of Boston. It was among the simplest devised and consisted of little more than a hook barren of springs and latches (Fig. 7.74). Considerable ingenuity is shown in the way the hook and lifting link are made as a single unit. To provide clearance for a lift sufficient to uncouple, a very large cavity was necessary, and so the coupler exhibited a large open mouth. The Union was tried by the Eastern and a few other New England lines. The test results were not inspiring, and fears were expressed about the possibility of the hook jumping loose when operating over rough track. The Union coupler's performance at the Buffalo trials condemned it, along with many another ingenious but deficient coupler, to the limbo of those multitudes who tried but failed.

FIGURE 7.74

The internal-hook scheme engen-
dered a number of variations.
Shown here are the Aikman,
Union, Archer, Marks, and Curtis
& Wood designs. All date from
the 1880s. (*Railroad Gazette*,
Sept. 24, 1884, and Oct. 2, 1885;
National Car Builder, Aug. 1886;
and other trade papers of the
period)

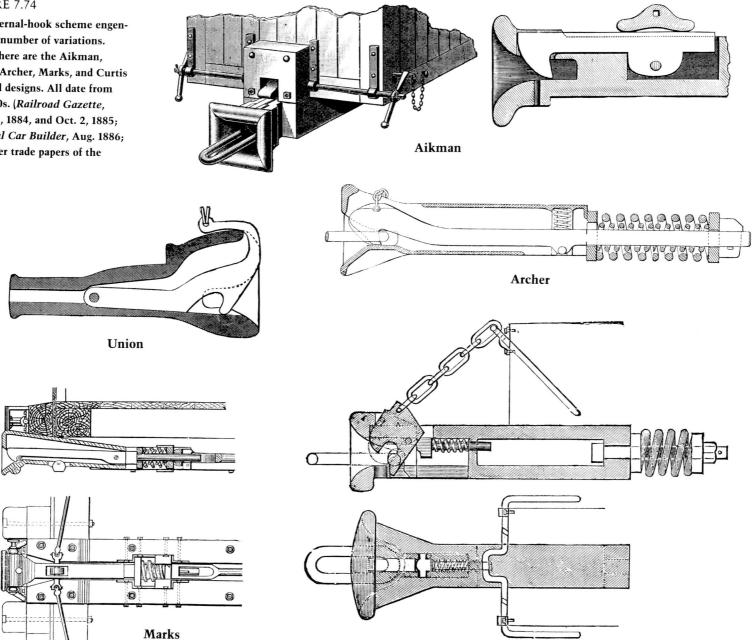

Aikman

Union

Archer

Marks

Curtis & Wood

A sizable number of patentees were convinced
that tinkering with the link-and-pin coupler was
fundamentally the wrong path. They felt that it
should be totally abandoned, and that an answer to
the coupler problem would be found only when
fresh and original thinking was applied. In the van-
guard of such bold innovators was Ezra Miller. Not
that Miller was the first to discard the old form for
some new pattern; he was simply the first to do
so with any success. Miller championed the
external-hook coupler whereby two giant iron
crochetlike hooks would grapple together to form
a union. Miller did not invent or originate the idea
of hook couplers when his first patent was issued
in 1863, but he did manage to work out the engineering details and successfully market a device

that the railroad industry would accept—that is,
accept for passenger equipment. The Miller coupler had become a national standard on passenger
cars by about 1875; however, it was only rarely
applied to freight equipment. Just why this was
true was never explained in the debates that surrounded the coupler question. Why didn't the railroads simply apply Miller hooks to the freight car
fleet? It was a known and tested mechanism.
Thousands were in daily service. There was nothing strange or unproved about it. Maybe it was the
cost; Miller's royalty fees were fairly stiff. Maybe
the buffer arrangement could not be made to fit
under the ends of a freight car unless platforms
were added. It is more likely that the Miller hook
was simply considered obsolete by 1885 and the in-

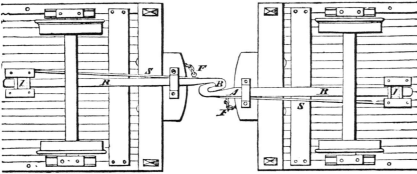

FIGURE 7.75

Luther Adams's hook coupler was patented in 1859. It predated the Miller hook by a few years, yet Miller's plan came to dominate the passenger car coupler market. In Adams's plan the release lever, marked *D* in the drawing, was worked from the roof. (*Railroad Record*, Jan. 19, 1860)

dustry was hoping to find some more suitable plan. Whatever the reason or reasons, it appears that only the Richmond & Allegheny Railroad applied Miller hooks to its freight cars.[168]

The railroads' coolness to the Miller couplers for freight service seems to have had little effect on the enthusiasm of patentees for the idea. Dozens, perhaps hundreds of patents were issued for external-hook couplers. Just a few years before the Miller hook appeared, Luther Adams of Blanchester, Ohio, patented a blunt or rounded-end hook coupler in December 1859 (Fig. 7.75). A leaf spring centered the couplers and kept them engaged. Uncoupling was achieved by a long lever that reached to the roof. Many years later Lucien Barnes of Rochester came forward with a double-spear hook that grappled a vertical pin supported at the top by

a conventional link-and-pin drawhead and at the bottom by a thick plate. This contrivance would surely have been more at home on a whaling ship than on a boxcar (Fig. 7.76). It was tried at the Buffalo coupler trials, but its only admirer was the local scrap-iron merchant. Even more monstrous creations were exhibited by Meadows & Mead of McMinnville, Tennessee, and B. B. Morgan of Ann Arbor (Fig. 7.76). Both of these contestants liked the vertical-hook plan. Neither worked well enough to warrant further testing, and at the end of the Buffalo trials they too found a welcome only in the scrapyard. Only one of the external-hook couplers achieved any measure of success, and this was the Hilliard drop-hook coupling. The inventor, Thomas J. Hilliard of Conway, Arkansas, patented his Mastodon-trap coupler in April 1882. It featured a heavy pawl shaped something like a massive paint scraper that would bump over the top of its opposing mate and clamp down to make the fastening (Fig. 7.76). The Hilliard coupler performed well enough at the Massachusetts Railroad Commission trials to be adopted, but it fared less well at the 1885 MCB Association contest and after a brief currency on a few New England roads, it too disappeared.

SPEARS AND OTHER DESIGNS

Hooks were really not the answer for some inventors. They preferred spears or harpoons as worthy replacements for links and pins. The novelty of this style of car coupling is unquestionable, but its utility is a matter of doubt. The oldest recorded automatic American car coupling was, in fact, on the harpoon style. The plan was created by Conrad H. Hunt and William Browne of Fredericksburg, Virginia. A patent was granted on December 26, 1837 (No. 538), but it is uncertain if the plan was ever tested, and it is even less certain that it was ever adopted for actual service. An iron spear, 1 inch thick by 3 to 5 inches deep, was attached to one car; the opposing car had to have parallel receiving plates ready to intercept the spear. The spear had a stud projecting out from either side that would slip into longitudinal slots cut into the receiving plates. Leaf springs pressed the receiving plates inward so as to grip the spear and hold the stud in the slots. Among the deficiencies of the Hunt and Browne design was the need to have all cars facing one way, since the spear and plates must always oppose one another. In addition, all cars must be nearly on an exact and uniform level for the coupling to work. Because neither of these prerequisites could be met in 1837 or at any other period of railroading, the fixed-spear scheme was forthwith abandoned.

About a decade after the Hunt and Browne failure, A. G. Heckrotte of Washington, D.C., promoted a design patented by William C. Bussey of Rock Grove, Illinois, on July 17, 1847 (No. 5,194).

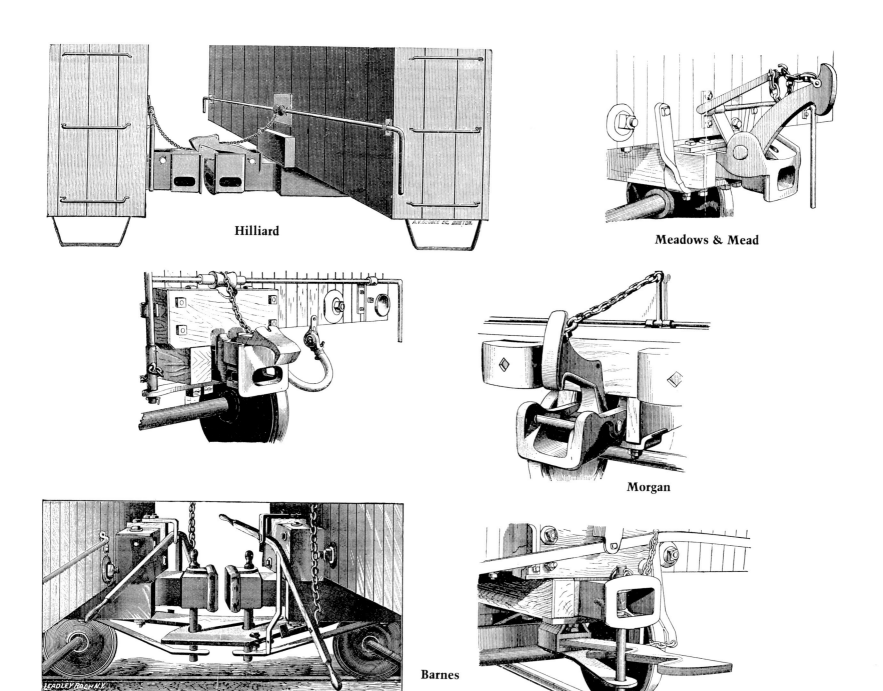

Hilliard

Meadows & Mead

Morgan

Barnes

FIGURE 7.76

Hooks were looked upon with considerable favor by some coupler designers. Shown here are the Hilliard, Meadows & Mead, Morgan, and Barnes designs. None of these schemes achieved more than occasional use except the Hilliard. (*Railway Age*, Dec. 18, 1884; *Railroad Gazette*, Jan. 2 and Sept. 25, 1885)

The Bussey/Heckrotte design incorporated the basic elements of nearly all subsequent harpoon-style couplers. The principal feature was a loose, double-ended spear. The spear replaced the link of the conventional coupler. The drawhead had a spring-loaded catch to hold the notched head of the spear. With near-perfect alignment, Heckrotte's device would probably have worked, but there is no record that any railroad adopted this design.

In June 1875 J. G. Rogers of Victoria, Missouri, patented a double-tipped spear that was round in cross-section. This formidable missile was intended to plunge into and between two spring-loaded levers that would grasp and hold it (Fig. 7.77). Le Roux & Van Aarle preferred a harpoon made of stock, square in cross-section. It had a

clumsy-looking center pin to prevent the spear from penetrating too far inside the drawhead (Fig. 7.77). Had this style of coupler become the new national standard, trainmen would have found a new challenge to their strength when attempting to lug several of Le Roux's spears from car to car. The Wapakoneta coupler must have been inspired by a visit to a textile factory, for the pointed link is a loom shuttle lookalike. The cast-steel link was held in place by a spring latch. Less lethal-appearing harpoon coupler links were promoted as well. Both the Gray and Murphy designs feature blunt, rounded-end links that were not at all war-like in appearance. A plan view of the Gray link is shown in Figure 7.77.

There were other coupler patentees whose ideas fit into none of the broad schemes already out-

FIGURE 7.77

Spears and harpoons were preferred by another group of would-
be coupler inventors. Shown here
are the Rogers, Le Roux & Van
Aarle, Wapakoneta, and Gray designs. Note the double-ended
spear link in the Wapakoneta
plan. All date from the 1880s except the Rogers design, which
was patented in June 1875. (U.S.
Patent drawing; *National Car
Builder*, July 1882; *Railroad
Gazette*, Sept. 25, 1885; and
Railway News, June 10, 1886)

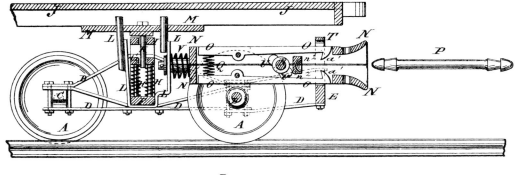

Rogers

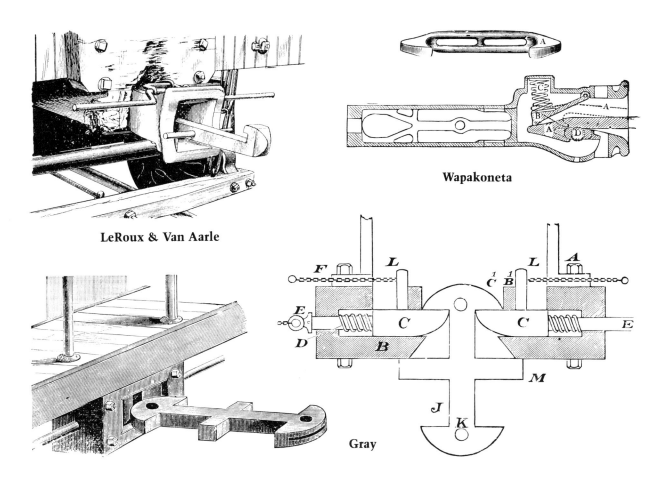

LeRoux & Van Aarle

Wapakoneta

Gray

lined. Their ideas are not worth examining in any
depth, but at least a few are worth mentioning just
to show the diversity of opinion on this important
problem. It is a factor in why the solution was so
long in coming. With so much advice and so much
of it different from the advice just offered, it is no
wonder that car builders were puzzled, annoyed,
and frustrated. In some cases, however, the choice
was not so difficult. No practical mechanic would
likely have selected any of the following examples.
In 1867 Ernst Von Jeinsen offered to the railroad industry the bizarre-looking creation pictured in Figure 7.78. Von Jeinsen had his priorities somewhat
reversed. His coupler was automatic only in uncoupling; it was necessary to handle the coupling
process manually. Little enthusiasm was exhibited for his invention.

Some years later an inventor named Hitchcock
overcame the basic defect of the Von Jeinsen device by producing a scheme no less strange in appearance but one that was capable, at least in theory, of self-coupling. The independence of mind
exhibited by the inventor is remarkable. Hitchcock devised a curved, inclined shelf that would
lift and guide a hook up and over its mate, allowing
it to fall and couple when in the correct position.
The hook heads apparently swiveled in the process of uncoupling (Fig. 7.78). In the purest democratic spirit Hitchcock was allowed to demonstrate his juggernaut at the 1885 Buffalo coupler
trials. It failed to couple on curves and so was rejected—to no one's surprise except the inventor's.

Hitchcock's ideas seem conventional compared
with one of the other contestants appearing at Buf-

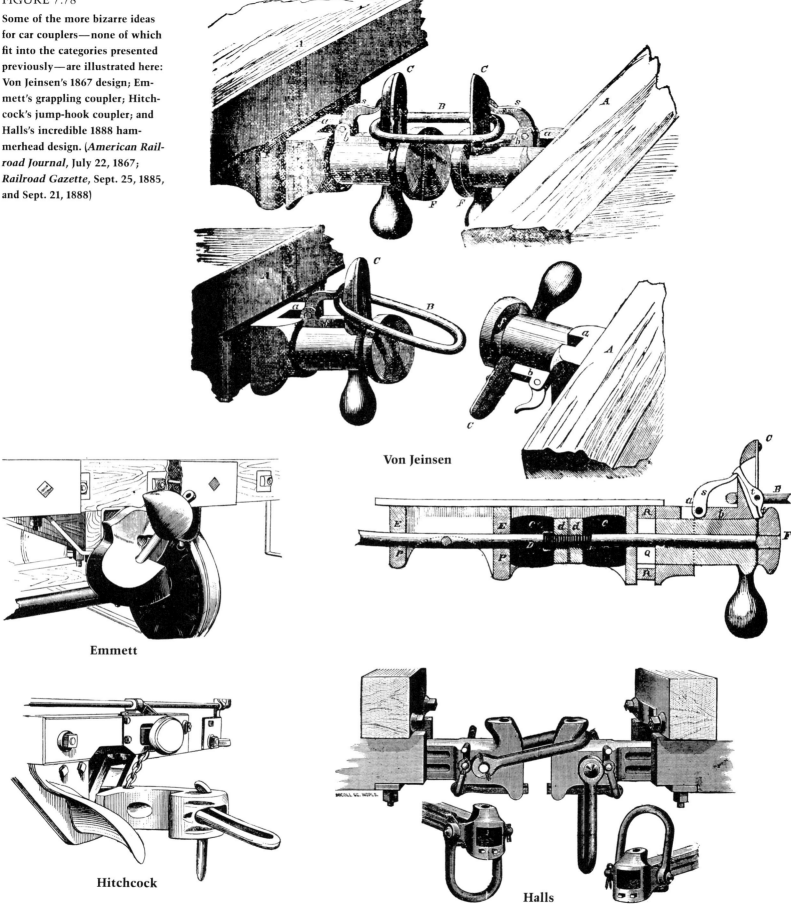

FIGURE 7.78

Some of the more bizarre ideas for car couplers—none of which fit into the categories presented previously—are illustrated here: Von Jeinsen's 1867 design; Emmett's grappling coupler; Hitchcock's jump-hook coupler; and Halls's incredible 1888 hammerhead design. (*American Railroad Journal*, July 22, 1867; *Railroad Gazette*, Sept. 25, 1885, and Sept. 21, 1888)

Von Jeinsen

Emmett

Hitchcock

Halls

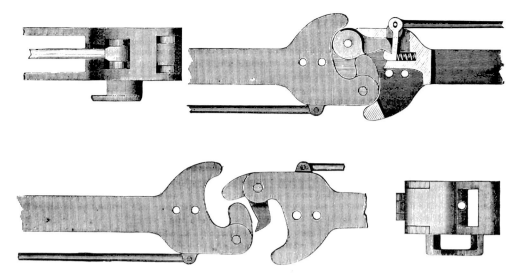

FIGURE 7.79

Eli Janney introduced the basic plan for the modern automatic coupler in his April 1873 patent. The knuckle or hinged-jaw principle is clearly shown in this engraving published the year after the patent was granted. (*National Car Builder***, Nov. 1874)**

falo in September 1885. No one had ever before seen anything quite like the device exhibited by William Emmett of Logansport, Indiana (Fig. 7.78). Just how it worked remains something of a mystery. A large screw ran through the center of its contorted body. The arm—projecting upward and fitted with a turnip-shaped mass—apparently grappled onto a like-shaped object on the opposing coupler. Once locked, the embrace of Emmett's brainchild could not be undone, and the car builders decided that it was best to retire Mr. Emmett and his coupler permanently to Indiana.

The creation of such monstrosities should logically have ended once the Janney coupler was selected, but the spirit of enterprise was not so easily squelched. Witness the labors of another Hoosier inventor, T. E. Halls of Indianapolis. A hinged link looped over the hammerhead end of the opposing drawbar to make the connection (Fig. 7.78). Striker pins worked a lever that flipped the link over the hammerhead when the drawbars were bumped together. Just how the uncoupling was accomplished is uncertain, but it would appear that the trainman had to step between the cars. Halls patented his design in March 1888 and persuaded a few local railroads to test it. His effort was, of course, futile in view of the success of the Janney-style coupler.

VERTICAL-PLANE JANNEY-STYLE COUPLERS

The final and major school of car couplers to be considered here was the winner of the coupler sweepstakes. It was the vertical-plane, swing-jaw, or knuckle coupler design that remains in use today and can be seen wherever railroads operate in North America. A major advantage of the vertical-plane design was its ability to mate even when the levels of the opposing cars were different. Most couplers, including the link-and-pin type, were not very forgiving when it came to being on a nearly equal level. A freight car's height above the rails varied depending on original construction, on whether it was loaded or empty, or on the condi-

tion of the springs or frame; the broad-faced knuckle coupler thus offered a definite advantage. This basic form of coupler appeared rather late in the furious hunt for a practical way to unite railroad cars. For all the ingenuity and effort expended, there was little measurable achievement after almost half a century of experimentation.

Thousands of ideas were advanced, as the previous examples illustrate, but most of them amounted to sideways rather than forward motion. Lost in all this activity were the efforts of an obscure ex-farmer and dry goods clerk from Alexandria, Virginia, named Eli H. Janney.[169] How did he come upon the idea for a revolving knuckle or jaw that would swing open to release or swing shut and lock to couple? Was it genius, divine inspiration, or simple luck? Whatever the source, Janney's achievement clearly involved a prolonged investigation and a dogged study of the problem. He began work on it in 1865, not long after leaving the Confederate Army. His first patent, issued in April 1868, was a dead end, but five years later he was awarded patent papers for what was to become the basic American railroad car coupler.

At the time Patent No. 138,405 was issued on April 29, 1873, no one in the engineering community recognized the profound consequences of Janney's improvement. Like his rivals, Janney was undoubtedly rebuffed more often than welcomed by the presiding master car builder, but he persisted, and by late 1874 a number of railroads were testing his coupler.[170] Southern roads like the Virginia Midland and the Atlanta & West Point were among those who tried his device, perhaps only because they did not want to turn away a former Confederate colonel (Fig. 7.79). Ironically, Janney's great patron was found in Yankeeland, for the Pittsburgh, Fort Wayne & Chicago took a decided liking to the colonel's work and was willing to undertake extensive road tests.[171] The shape of the knuckle's inner face as well as the configuration of the bifurcated head required refinements. The spring-loaded locking device was never really satisfactory, especially for heavier train, and was finally discarded altogether by 1882. These changes were, whenever possible, reflected in subsequent patents issued to Janney.

The inventor was supported at least in part by his original partners in Alexandria, but in 1877 he fell under the protective wing of William McConway and John J. Torley, founders and machinists in Pittsburgh. McConway and Torley started out in hardware manufacturing in 1869, specializing at first in harness fittings, but moved into railway supplies within a few years. The Pittsburgh partners were practical mechanics; they understood tooling, material strength, and just where to add metal or take it away. They could thus guide Janney in the bench-work improvements so necessary to make his coupler a workaday success. Janney

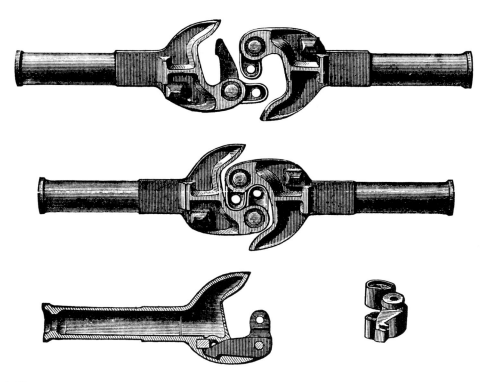

FIGURE 7.80

Janney's design evolved and was
strengthened through the tutelage
of ironworkers William Mc-
Conway and John J. Torley of
Pittsburgh, who took over the
manufacture of the Janney
coupler in 1877. The design
shown here, which dates from
1882, features the gravity-style
lock pin that replaced the earlier
spring latch. (*Engineering*,
July 13, 1883)

had solved the conceptual problem, but he needed expert assistance in perfecting the details for the marketplace.

Just a year before Janney became formally associated with his Pittsburgh mentors, the Pennsylvania Railroad began experimenting with the Virginian's apparatus, possibly encouraged to do so by the promising results reported by the line's western arm, the Pittsburgh, Fort Wayne & Chicago. The Pennsylvania was looking for an alternative to the Miller coupler, which most other domestic lines had already adopted for their passenger cars. Miller had somehow offended the officials of the nation's largest railroad. The Pennsylvania would not adopt the hook coupler but realized that it must find something better than the link and pin. Janney thus had a wonderful opportunity. With the help of McConway and Torley he could offer a workable coupler. The customer was eager to help as well, and in a rare display of patronage, the client railroad actually worked to assist the inventor and the manufacturer. In 1878 the PFW&C equipped all of its passenger cars with Janney couplers. The parent line and most of its subsidiaries adopted the design as standard in the following year. This endorsement by America's leading railroad represented a major victory for Janney.

The next order of business was to perfect a swing-jaw coupler suitable for freight service. An even more durable and trouble-free product was needed to withstand the rigors of the goods trade. In February 1882 Janney patented a new locking arrangement that worked by a heavy drop pin held in place by gravity rather than by a spring (Fig. 7.80). The Pennsylvania Railroad applied the new style of coupler to one hundred stock cars during

the following summer. In the fall of 1883 the Burlington, after trying Janney couplers on one hundred freight cars, was so satisfied that it began applying them in large numbers to the remainder of its fleet.[172]

Janney's invention, only a paper vision a decade earlier, was now a practical reality. It would soon grow into an item of major commercial value. McConway and his associates recognized that the invention's time had come and so organized the Janney Car-Coupling Company early in 1884.[173] McConway owned all but forty of the shares; Janney himself owned none, apparently having sold his rights to his former partners. Janney continued, however, to work on perfecting his design until a few months before his death. He lived on to see his coupler become a robust and dependable product well able to withstand the hard knocks of freight service. In a description printed in October 1888 a Janney coupler was described as having a malleable-iron head and barrel.[174] The barrel was 5 inches in diameter and ⅞ inch thick. The knuckle was forged iron or cast steel. The knuckle pin, 1⅜ inches in diameter, was made of steel. Fully assembled, the coupler measured 33¼ inches long and weighed 210 pounds (Fig. 7.81).

While Janney and his rivals labored to perfect their hardware, the car-building professionals vacillated on what to do. Finally, after years of procrastinating, the Master Car Builders Association decided to hold its own trials in Buffalo, New York. The Erie allowed its Hamburg Street yard to be used for a three-day test in September 1885. Contestants had to submit full-size working couplers; models and drawings alone would not be considered. M. N. Forney, the editor of the *Railroad Gazette* and a writer widely respected for his engineering sagacity, was to supervise the testing and prepare a record of the results.

It was to be the Olympics of car couplers, a jousting match with boxcars rather than armored knights coming into collision. The goal was not to unseat an opponent but to shackle up with an identical partner fastened to the opposing boxcar. Some clamped together without difficulty, while others failed to do so. Some indeed grappled so firmly that they could not be separated. Some worked well enough on the straight track but would not function on a curve. One broke its drawhead during the test. In all, there were forty-two contestants, most of which have been described earlier in this discussion. They came from all over the land. Three were Canadian. Some came from big cities like Boston, Chicago, and Pittsburgh. Others came from hamlets like Logansport, Indiana; Lyons, New York; and McMinnville, Tennessee.

By the time of the Buffalo trials coupler promoters had acquired a decided preference for the vertical-plane style, encouraged no doubt by Janney's success. Janney had in fact engendered several

FIGURE 7.81

The Janney and its various parts. The body of the coupler is *86*, the jaw or knuckle is *2A*, and the locking pin is *96 W.I.* Note that the knuckle had a large notch cut out and a hole bored through its face so that link-and-pin couplings could be made. (McConway & Torley catalog, 1899)

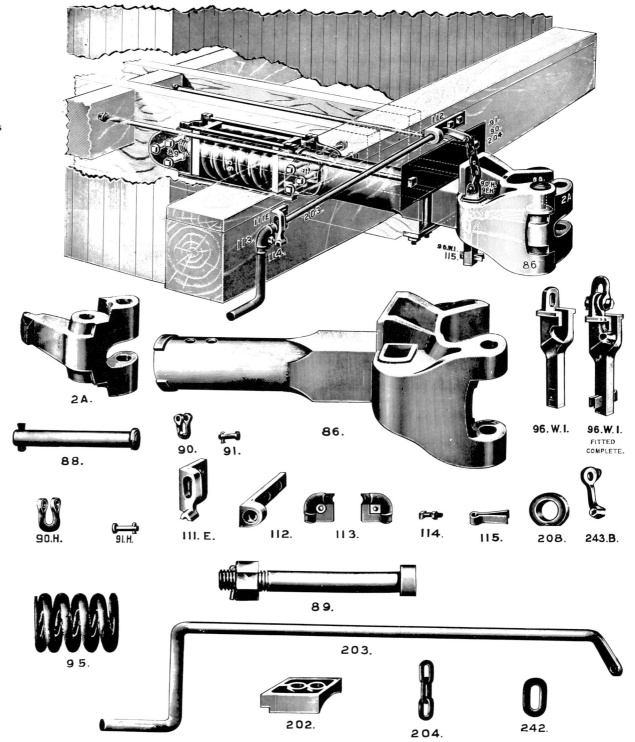

dozen imitators by the middle 1880s. Many more would pile onto the bandwagon once the trials were concluded. Most of the Janney copycats arriving to clang iron in Buffalo were nearly mirror images of their prototype. The Dowling, Thurmond, and Hien designs fit into this class. The last name in fact was a Janney patent licensee. But there were some strange and weird variants on the Janney plan; most are shown in Figure 7.82. Even those that look like the true Janneys incorporated different jaw, lock-pin, and head designs.

The Cowell vertical-plane coupler was one

standout in this group, for it really did not literally imitate Janney's design but was more of an articulated variation of a Miller hook coupler (Fig. 7.83). Renselaer A. Cowell of Cleveland, Ohio, obtained a number of coupler patents starting in 1868. Most were for the drop-pin variety, but by 1884 he had graduated to a swing-head plan. The Fitchburg and the Lehigh Valley railroads bought some of Cowell's apparatus. Their faith in the veteran Ohio mechanic was borne out by the coupler's good showing at the Buffalo trials, for it was one of twelve selected for more testing. Titus & Bossinger of

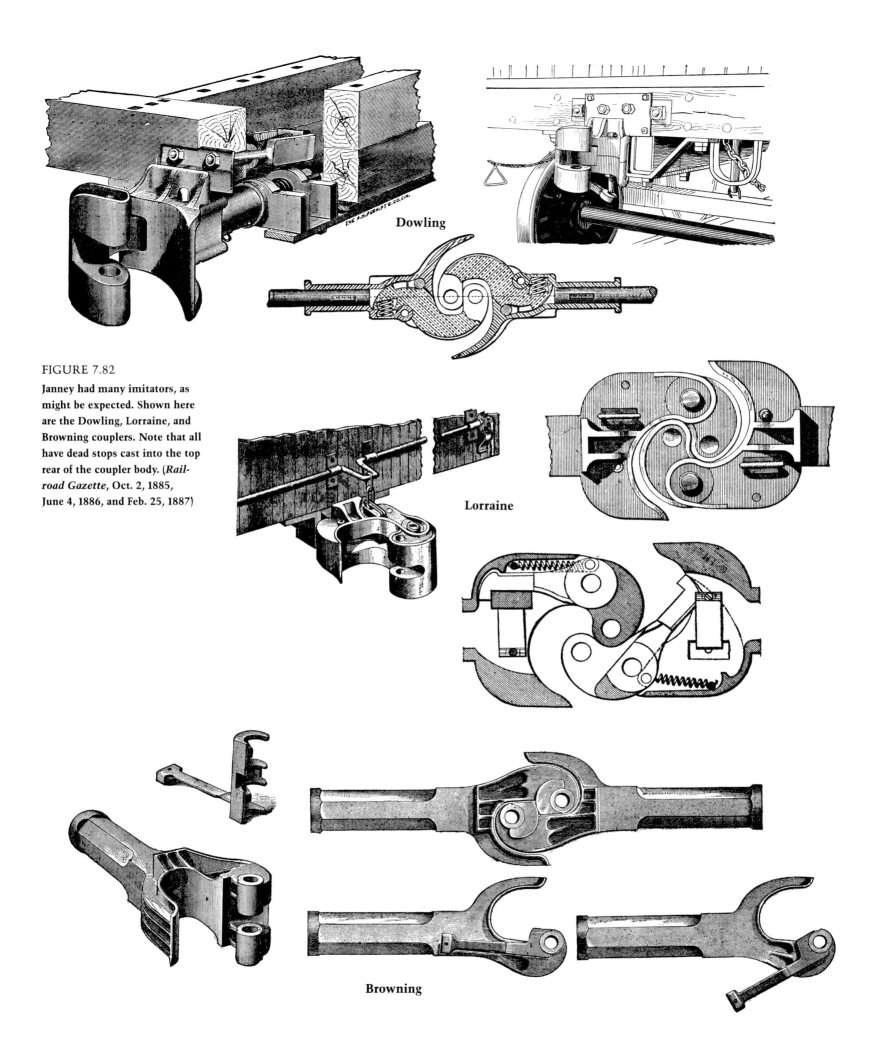

Dowling

Lorraine

Browning

FIGURE 7.82

Janney had many imitators, as might be expected. Shown here are the Dowling, Lorraine, and Browning couplers. Note that all have dead stops cast into the top rear of the coupler body. (*Railroad Gazette*, Oct. 2, 1885, June 4, 1886, and Feb. 25, 1887)

FIGURE 7.83

The top series of engravings shows the R. A. Cowell coupler, which was a swing-jaw design adopted by several railroads before the major shift to Janney's plan began in 1888. A similar but older design, by Titus & Bossinger, is shown in the bottom series of engravings. (*Railroad Gazette*, Sept. 26, 1884, and Jan. 2, 1885)

Cowell

Titus & Bossinger

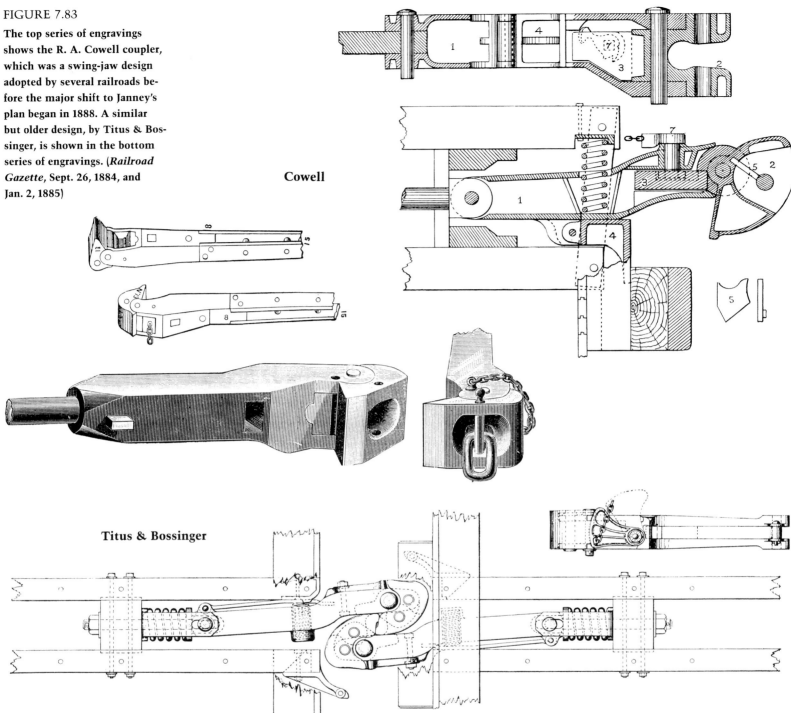

Huntington, West Virginia, had been working on a coupler very similar to Cowell's since 1876 (Fig. 7.83). It too functioned to the satisfaction of Forney and his fellow judges during the Buffalo contest and so was placed on the recommended list. Of the twelve finalists named below, half were vertical-plane couplers:

Ames, fixed-link internal hook (Fig. 7.31)
Archer, internal hook (Fig. 7.74)
Cowell, vertical plane (Fig. 7.83)
Dowling, vertical plane (Fig. 7.82)
Gifford, pin supporter (Fig. 7.72)

Hien, vertical plane (Fig. 7.84)
Janney, vertical plane (Fig. 7.80)
Marks, internal hook (Fig. 7.74)
McKeen, pin supporter (Fig. 7.70)
Perry, pin supporter (Fig. 7.71)
Thurmond, vertical plane (Fig. 7.84)
Titus & Bossinger, vertical plane (Fig. 7.83)

FORCING STANDARDIZATION

For all the attention given these trials, the tangible results were few. Nothing was really settled. No one coupler was selected, nor really was that expected. Only a preliminary report was ever rendered, but it seems clear that the Master Car Builders Association Executive Committee was strongly in favor of the Janney coupler. They could not openly endorse it, of course, because the association was prohibited from adopting patented or brand-name products. In fact, the curious term *vertical-plane coupler* was invented so that the membership might talk about the Janney design without mentioning its specific name. Meanwhile, the car builders became so involved in selecting some form of power braking for freight cars that the pesky coupler question was once again set aside.

Nothing happened in 1886, but at the following year's convention the MCB Association was once again confronted with the matter of selecting something to replace the ancient link and pin. The advent of air brakes spelled their doom because of the excessive slack present in the loose link-and-pin arrangement. The trunk lines wanted to force a decision before the various state railroad commissions expanded their often ill-advised lists of couplers legal within their boundaries. New York and Michigan, for example, did not include Janney in their approved lists. The MCB membership voted in favor of the vertical-plane coupler, but not by an overwhelming majority. The proposition just barely garnered the two-thirds majority required for passage. And even this weak endorsement had no real value because no one was required to do anything. It was little more than a "We think this is a good idea and hope you will adopt it" sort of requirement—a toothless law, but it at least got the conversion under way and established a direction to follow. To dissipate the notion that the association had adopted a specific patented design, the generic name for the new style of car connector was not Janney or even vertical-plane but rather the MCB coupler.

Early in 1888 the MCB Executive Committee met with McConway and Torley seeking a waiver of patent rights to the contour lines of the Janney coupler jaw as noted in claims 8 and 9 of Janney's patent of February 5, 1879. This was agreed to, and so much of the Janney design reverted to the public domain and could thus be adopted as a standard by the MCB Association. Every coupler manufacturer might make free use of this portion of Janney's design, and so it was assured that all vertical-plane couplers would mate successfully. The notion that any one inventor might reap a fortune in license fees was also quashed. Janney himself appears to have prospered very modestly from his landmark invention.[175] His estate consisted of no real property but only his furniture and other personal possessions plus some shares in the Janney coupler company. The estate was so small that no net worth was recorded in probate court records.

Now that the basic design issue had finally been settled, Janney's rivals rushed in to claim as much of the market as they could. There was little danger of a monopoly or a single supply source, for dozens of vertical-plane-coupler manufacturers clamored to satisfy the wants of the trade. Competitors like Smillie and Barnes willingly switched their once-firm allegiance to the hook or automated link-and-pin couplers and bounced onto the MCB coupler bandwagon. Within a few years they were followed by the Browning, the Burns, the Boston, the Buckeye, the California, the Drexel, the Lorraine—the list goes on and on to the end of the alphabet (Fig. 7.84). By 1891 there were nineteen major coupler manufacturers, and by the first year of the new century the list had grown to ninety.[176] Each had a differently shaped head and body; the release mechanisms were often substantially different, as was the placement of the release pin. Nothing was interchangeable or standard other than the inner contour of the knuckle. Railroads were thus forced to maintain a warehouse full of repair parts. One of the looked-for benefits of adopting the MCB coupler was thus dashed by this refusal to devise a specific standard and by a misguided willingness to throw the problem back into the marketplace for solution.

The misgivings of the opponents of the MCB coupler seemed at first to be substantiated by the less-than-perfect performance of the Janney coupler in actual service. It proved to be a weak link and was very prone to failure. In 1891 the Burlington reported 540 broken couplers, plus 2,400 failed knuckles.[177] The Boston & Albany, whose mechanical head was dead set against the Janney coupler, smugly noted 296 failures in just two months in 1890.[178] Defenders noted that such reports were overly critical and that the failure rates—1 percent for couplers and 3 percent for knuckles—were acceptable. They went on to insist that the Janney was perfect in concept, however imperfect it might be in its details.

It must be noted that the knuckles were greatly weakened by the slot and hole that had to be provided during the transition period when both Janney and link-and-pin couplers were being used. This hybrid arrangement posed other hazards to knuckles; they were regularly the target of rammings by links projecting out from a car still fitted with the old-style couplers. In time the couplers were made sturdier and more reliable by increasing the bulk of all parts and by adopting stronger materials. Malleable iron gave way to cast steel after a series of tests conducted at the Watertown (N.Y.) Arsenal in 1893 showed clearly that the more expensive material would more than pay for itself through longer service. In an effort to convince the car builders that their product was fully guaranteed, McConway and Torley promised to replace every broken coupler or part that was returned to the factory.

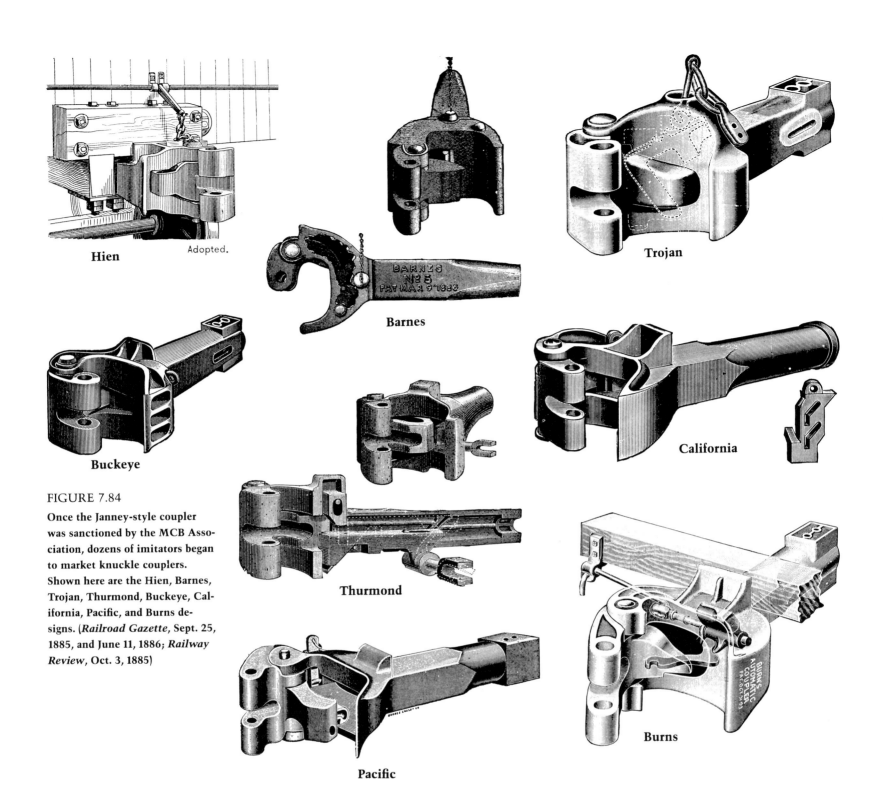

Hien Adopted.

Barnes

Trojan

Buckeye

California

Thurmond

Pacific

Burns

FIGURE 7.84

Once the Janney-style coupler was sanctioned by the MCB Association, dozens of imitators began to market knuckle couplers. Shown here are the Hien, Barnes, Trojan, Thurmond, Buckeye, California, Pacific, and Burns designs. (*Railroad Gazette*, Sept. 25, 1885, and June 11, 1886; *Railway Review*, Oct. 3, 1885)

Predictably, the conversion to the MCB standard proceeded at a languid pace. The process was expensive and inconvenient. There were more important or more attractive things to spend the car department's budget on. Most railroads needed more rolling stock and were thus more inclined to make such purchases a bigger budget item than a new style of coupler. Compassion for the trainman tended to fade once the sermons of Forney and other reformers slipped from memory. Most car builders were either simply unmotivated or downright opposed to the Janney coupler and its imita-

tors. Leander Garey, longtime president of the MCB Association, was candid about his indifference to the entire coupler question in remarks made at the 1885 MCB convention. Fitch D. Adams, a founder and former president of the MCB, had spoken several times about the need for a better car coupler, yet after the Janney style was adopted, he voiced his opposition to it at the spring 1890 meeting of the New England Railroad Club. John T. Chamberlain, master car builder for the Boston & Maine and later president of the MCB Association, was up-front about his hostility to the Janney-style coupler.[179] He preferred the link and pin as improved by the Safford drawhead. Not surprisingly, Chamberlain reported that his men were in complete agreement with his opinions. What is surprising are the remarks made before the New York State Railroad Commission in 1891 by Frank Sweeny, grand master of the Switchman's Mutual Aid Association. One would have expected Sweeny to plead for the immediate adoption of the MCB coupler to safeguard the welfare of his brethren, but instead he spoke out with considerable feeling against the new style of couplers. They were an imposition and a hazard, and he and his group wanted nothing more than the God-given right to spend their remaining days working with links and pins. Sweeny persuaded several switchmen to appear in support of his opinions.

This combination of indifference and opposition is manifest in the actual number of freight cars fitted with MCB couplers. In 1888 it amounted to less than 10 percent of the fleet. Within two years the number had risen by about 4 percent, to 114,364 cars.[180] At this rate the conversion would take not years but decades. In 1892 there was vague talk about having 20 percent of the cars equipped with safety couplers, but no one was very certain that even this modest goal would be met. Something was needed to get the industry moving, and one man thought he knew how to do it. That person was Lorenzo S. Coffin (1823–1915), a sometime preacher, farmer, and teacher born in New Hampshire but a resident of Iowa for most of his life. Coffin was a reformer and do-gooder who seemed to collect causes as he grew older. Temperance, homes for elderly workers, and education were among his interests, but he is best remembered for his devotion to improving the lot of railroad trainmen.

Coffin's attention was drawn to this problem in about 1881 while riding a freight train as a right-of-way agent. Being a kindly, gregarious man, he struck up a conversation with the rear brakeman while riding in the caboose. The brakeman was enough of a veteran to have already lost several fingers. While switching some cars at the end of the trip, the poor fellow's remaining fingers were crushed. Coffin was appalled. It was not just the man's suffering, but the fact that he was now unfit for any manual work. Coffin decided that something must be done to correct this terrible evil. That day the railroad industry lost a right-of-way agent and gained a critic of unparalleled passion and energy.

A skillful propagandist, Coffin began to speak and write on the subject. Church and civic groups, newspaper editors, magazine publishers, political figures—all heard from Father Coffin, as he came to be known. He rode freight trains to gather testimony from railroad workers. He made so much noise that he was appointed to the Iowa Railroad Commission. This body, like its counterparts elsewhere, was concerned mainly with rates, quality of service, and passenger safety; what happened to employees was of little concern. Coffin changed all of that. The Iowa commission suddenly became very concerned about railroad employee safety. In 1890 Iowa passed a safety appliance act, largely in reaction to Coffin's efforts. But Coffin wanted to protect workers not just in his area but across the entire United States. He spoke at the annual meetings of both the Master Car Builders and Master Mechanics associations. The reception was sympathetic but noncommittal. Coffin wanted action, not meaningless pleasantries. The establishment of the Interstate Commerce Commission made it clear that the nation's railroads were subject to some form of federal control, and so Coffin was inspired to seek passage of a national railway safety law.

The passage of such a bill would not be easy, for the railroad industry could be counted on to fight it head-on. It was a major industry. Its influence was considerable. Many politicians were dependent on its largess. And who was this man Coffin? A nobody. A little farmer from Iowa. A crank, a reformer, a nut. No one need pay attention to him. Or so some would think, but Coffin was persistent and persuasive—a crank, perhaps, but assuredly not a nut. He offered a reasoned and compassionate message that few could fault. The government had, after all, been active in transportation safety since the steamship inspection act of 1839; why shouldn't it deal with rail safety as well? Spokesmen for the industry argued that a mandated conversion to automatic couplers and power brakes, no matter how desirable, would bankrupt the industry. Estimates as high as $50 million were bantered about, but a more realistic estimate put it at $25 million to $30 million.[181] The trade press was basically loyal to the industry's stand that safety equipment was desirable but should be adopted on a voluntary basis, not legislated. The *Railroad Gazette* went further; after making a nod to trainman safety, it said that "Salvation Army methods" were the wrong approach to things mechanical.[182]

The safety legislation failed at first, but Coffin persisted and came back in 1892. This time he had found a powerful ally in President Benjamin Harrison. After more wrangling, the bill was finally passed and signed into law on March 2, 1893. The railroads were given five years to equip all their rolling stock with couplers that would couple automatically on impact. Now all options but one had been closed. Coffin could retire to Fort Dodge and devote his remaining years to the Anti-Saloon League. The railroads, no doubt sullen in defeat, began a halfhearted effort to comply with the Safety Appliance Act. In late 1897, with only a few months left to comply, the railroads reported that only 44.5 percent of their cars had MCB couplers.[183] Their excuse was the hard times following the 1893 panic. Traffic was down, revenues were depressed, there just was no money to buy all of those expensive brake and coupler sets. They wanted five more years. Father Coffin threatened to come out of retirement. The unions protested. Congress reacted by cutting the baby in half and granting a two-year extension. The railroads reacted by moving ahead at their own pace. They sought a second extension, which was granted, and so the conversion date was put off until August 1, 1900. After that time no cars could operate in interstate traffic unless they complied with the Safety Appliance Act.

Was the battle for self-acting couplers worth it? It was surely a long and costly war. In the first few years of the safety coupler, accidents actually went up. This was due mainly to dangers inherent in operating with a mix of coupler types. Lashing a link and pin to a Janney was more treacherous than making the connection with two old-style couplings. A major passenger train accident was in fact accountable to this makeshift method. The temporary hookup of a Cowell to a Miller led to the disastrous wreck at Hamburg, New York, in March 1890. The connection was an imperfect one, and the train broke apart twice. At the second parting the rear section overtook the front section of the train; when it crashed into the head-end portion, several passengers were killed or injured.

Even when two MCB-style couplers were successfully joined, trainmen were often confused about how to make them part. The release mechanisms were all different from make to make. Some were lifted from the top by a lever and chain. Others were pulled from below. Still others opened from the side, while one inspired manufacturer had a front-end release. Because trainmen were rarely graduate mechanical engineers, they would often step inside the cars to see firsthand how this latest iron puzzle was meant to operate. Such firsthand inspections were especially necessary at night—also the most dangerous time to be standing between two cars.

Coupler accidents peaked in 1893, when 11,710 incidents were reported by the Interstate Commerce Commission. That number fell over the next few years and began to show a noticeable improvement by about the time the conversion was finished. In 1900 coupler-related accidents were down to 4,198 or 9 percent of all employee accidents. Just twelve years earlier, couplers had accounted for 32 percent of railroad workers' injuries. By 1902, with the conversion complete and everyone connected with car handling familiar with the new-style coupler, accidents of this type dropped to 4 percent. The trainman's main hazard remained falling off the train. The terror of crushed fingers and smashed ribcages had been largely banished through the good work of Eli Jan-

ney, William McConway, Lorenzo Coffin, and a host of other caring men who helped develop and promote the knuckle coupler.

DRAFT GEARS

Draft gears fasten couplers to the car frame. The attachment is not a rigid one, however, for the draft gear serves a second office as a shock absorber. Lewis Sillcox, in his 1941 book *Mastering Momentum*, said that it was a machine designed to store and transform energy. It must be capable of cushioning and dissipating the impact of terrific blows. Demands on the draft gear are at their peak when the train starts and stops, but even when it is simply rolling along the track, the individual cars tend to travel at slightly varying speeds; hence they are continuously approaching and receding from one another. This travel is limited by the slack afforded by the draft gear. Under normal conditions the draft gear allows just enough flexibility that the cars and their consignments are not injured, but too violent a movement can cause damage to both and in extreme cases can lead to a derailment. A few inches of free travel between the cars in a long train allows the train to bunch up and creates slack so that the engine can start a heavy load by getting the forward end rolling before the rearmost cars add their burden. In the wood-car era an added boost was achieved by compressing the draft springs; they would then help push the train forward once the engine began to move.

Because most trains of the time were underpowered, serial starting was seen as a benefit of generous slack between the cars. In a fifty-car freight train there would be about 9 feet of slack. So much freedom, particularly on hand-braked trains, could cause major problems in both starting and stopping. When pulling out, couplers could be pulled loose, unless the engineer showed considerable skill. When starting, the engineer must not be too timid or the train would stall, yet if he was too vigorous, a break-in-two was almost assured. When stopping, the rear of the train began catching up with the locomotive. If the buff or running in of the draft gears was excessive, cars, ladings, and track would all suffer, for the dynamics of the ponderous train must somehow be controlled. When so much free rolling tonnage is involved, the mastery of momentum becomes a very large task indeed.

The cushioning problem was not serious in the beginning years of American railroads. Slow speeds and light rolling stock combined to forestall the worst effects of train dynamics. Leaf and coil springs addressed the draft-gear question satisfactorily between approximately 1835 and 1885. Fleeting experiments were made with air and rubber springs. The latter were fairly popular from the late 1840s to the end of the Civil War. As many as

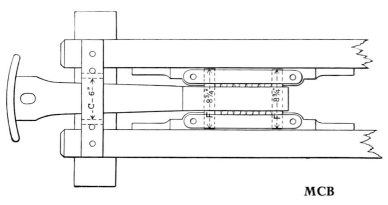

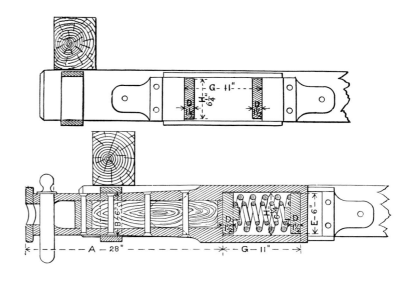

MCB

FIGURE 7.85

Conventional designs for draft gears tended to use a coil-within-a-coil style of draw spring. Shown here are the MCB standard adopted in 1879, a Norfolk & Western draft gear (note the key and tail bolt), and a Pennsylvania Railroad Graham-style draft gear, again with the old-fashioned key-and-tail-bolt connection. (*MCB Report*, 1886 and 1893; *Car Builders Dictionary*, 1888)

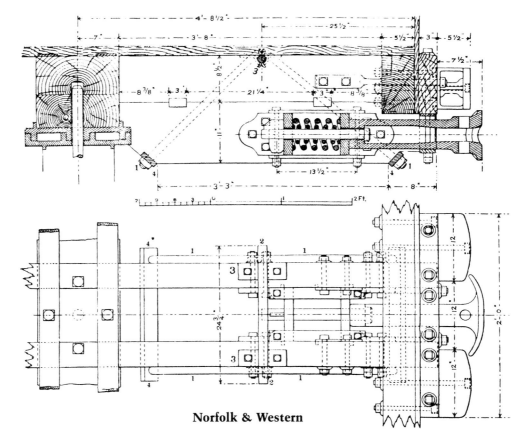

Norfolk & Western

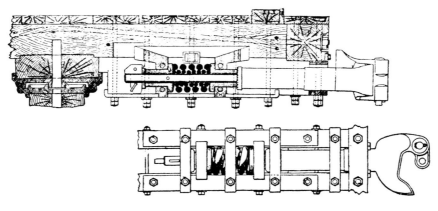

Pennsylvania Graham-style

nine rubber disc springs, separated by thick washers, were housed inside a cast-iron housing.[184] The failings of rubber, as already explained in the spring section of this chapter, disqualified gum blocks as being suitable for the couplers' tail stocks, but only after a stubborn effort to make rubber springs succeed. The steel springs of the period were imperfect, but at least they didn't freeze solid or pound flat after a few months' service. When greater capacity was needed, a larger spring was adopted. In 1879 the Master Car Builders Association adopted a coil spring with a 13,000-pound capacity as its standard (Fig. 7.85). The design was hardly revolutionary; it reflected standard practice going back a decade or more.

In 1884 the MCB Association boosted the recommended spring design to 18,000 pounds to handle the larger cars coming into daily use. Nine years later a somewhat larger spring was adopted. It measured 6¼ inches in diameter by 8 inches long and had a travel or compression distance of 2⅛ inches. The rating of 22,000 pounds was a trifle optimistic, and in actual use it was figured more accurately at only 19,000 pounds. The MCB Association favored double-coil springs made from tempered round rod, but a goodly number of car builders liked the volute spring for draft-gear work, and so with the usual independence and perversity of the car-building fraternity, most members chose to ignore the very standards so laboriously established by their own national association. An example of the volute draft spring appears earlier in this chapter (Fig. 7.66).

Two methods were used to attach the draft spring to the drawhead. The older idea was the stern or tail-bolt scheme whereby a large iron rod was cast into the rear end of the coupler or drawhead. A rectangular hole was made near the end of the rod for a tapered key. The spring was pushed over the rod; this was followed by a thick washer, and then the key was driven into place. Sometimes the rod was threaded and a nut was used in place of the key. Simple and cheap, the stern assembly was a natural favorite. It came apart easily when repairs were needed, which was seen as a decided advantage; but regrettably, it also came apart easily while in service. The problem was the key or nut. Should it come loose, fall out, or snap off, the entire coupler pulled out and the train was broken in two. Putting the entire load of the train on a single key was very poor engineering. For all its obvious faults, the tail-bolt fastening remained in favor well into the 1890s.[185] An example of this plan is shown by a drawing of a Union Tank Line coupler reproduced earlier in this chapter (Fig. 7.64; see also Fig. 7.85).

A more secure mounting was offered by the pocket or yoke-end design for couplers. An open rectangular box was formed from a heavy iron strap. One end was riveted to the rear end of the drawhead casting. The spring was placed inside the box or pocket opening. Brackets bolted to the draft timbers carried the load. The spring pushed against heavy iron crosspieces placed at each end of the draft-spring pocket, called the drawbar follower-plates. Should the drawhead be wrought rather than cast iron, the design was even stronger because the pocket could be an integral part of the main structure. An early example of this design is shown in the 1861 volute-spring draft gear shown in Figure 7.66.

Whatever plan was followed, draft gears remained a major source of trouble on American railroads. It was generally conceded that most of the cripples were resting idle on the side tracks because of broken draft gears, but the seriousness of the problem depended on whom you asked. In 1875 the MCB Association contended that 50 percent of all car running repairs were accountable to defective couplers and draft gears. Nearly a quarter-century later, the *Railroad Gazette* put the figure at 70 percent.[186] This notion was contested by a few car builders who felt that it was inflated and that the level might be more correctly figured at only 12 percent. Perhaps the trainmen on the 12 percent roads had developed an extremely light touch in handling their trains, but it seems strange that such a remarkable data spread should exist for this single item of repair costs. The denial of a problem is often good evidence that one exists, and the fact that so much effort was put into developing a better draft-gear design is another sure indicator that car builders were concerned.

Many solutions were offered to devise a more solid way of fastening the coupler to the car frame. The most solid and straightforward of these was the continuous drawbar. The couplers at both ends of the car were tied together by a heavy iron bar. The frame was thus entirely relieved of the buff and draw strains of train service. The idea could hardly be older; it was introduced in 1835 by Jacob Rupp, an employee of the Baltimore & Ohio. The B&O developed a resolute attachment to the plan and used it not only for boxcars but also for iron-body pot hoppers well into the 1880s. The stout, 2-inch-diameter rods were split at the center, where a nonthreaded connector that looked like a turnbuckle fastened the ends together by means of keys. Rupp's scheme was strong because it used so much iron, but it did not address with the cushioning requirement, which was seen, rightly, as the draft gear's second major duty. It offered no extra degree of slack because there were no springs, but perhaps all the slop offered by the links was enough to satisfy this need.

It hardly took cleverness to add a set of springs to the continuous-drawbar plan, and this appears to have been accomplished around 1875 by Samuel Griffith, master car builder for the Indianapolis,

FIGURE 7.86

Continuous drawbars offered a strong unitized form of construction. The top drawing shows a plan developed around 1875 by Samuel Griffith that was later exploited by the Continuous Draw Bar Co. The Central of Georgia scheme shown in the center view retained the traditional draft timbers. A Denver & Rio Grande design of 1889 is shown in the bottom view. (*MCB Report*, 1893; Voss, *Railway Car Construction*)

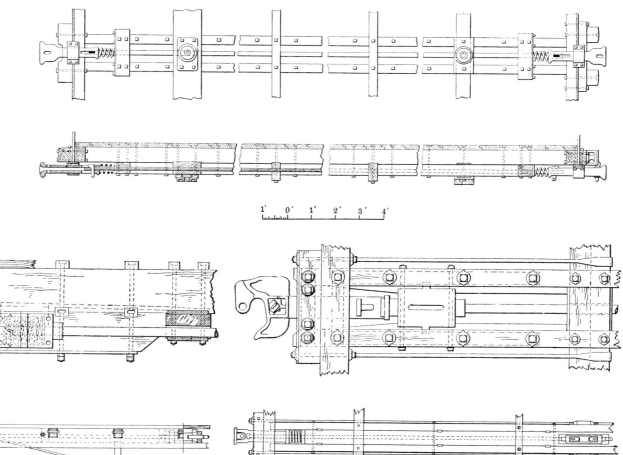

Cincinnati & Lafayette Railroad.[187] Griffith found some backers, and the Continuous Draw Bar Company was organized to exploit his invention, such as it was. A flat bar was used in place of Rupp's rods, springs were inserted, and the central connector was replaced by end unions placed just behind the draft springs. In 1879 Griffith's firm claimed that ten thousand cars on fifty railroads were using the device. About a year and a half later the number jumped to thirty thousand cars. Repair costs were down by 50 percent, or so it was claimed. In 1890 the Continuous Draw Bar Company inflated its repair cost savings claim to 90 percent. Even discounting the extravagant boosterism of Griffith's company, several major railroads were clearly enthusiastic about the general concept, if not necessarily of Griffith's specific design. The Baltimore & Ohio and its subsidiary, the Ohio & Mississippi, were most loyal to the Griffith design, while the Denver & Rio Grande and the Central of Georgia adopted similar but modified plans of their own (Fig. 7.86).[188] The Central of Georgia was quite outspoken in its praise for the continuous drawbar. The majority of American railroads, however, felt that it used too much iron and so was too costly.

The New York Central championed a semicontinuous drawbar devised around 1875 by one of its

car builders, David Hoit (1831–1885), superintendent of the West Albany car shop. Hoit's drawbar was a thick, round rod that ran from the drawhead spring back about one-third of the length of the frame. The rod passed through a cross-timber where a second draft spring was attached to the lengthy bar. In this way two draft springs for each coupler were in place to offer a greater cushion and so relieve some of the stress from the draft sills. An example of Hoit's plan is given in Chapter 5 (Fig. 5.12).

The majority of car builders eschewed the notions put forth by Griffith and Hoit, preferring instead short-stemmed coupler attachments. Once again, this general agreement on how best to do something by no means indicated a uniform approach to working out the details. Most car builders seemed content with a single draft spring. Spring pairs were also laid out in parallel, piggyback, and tandem patterns. Janney devised a secondary coupler stem placed above the main stem that worked against an auxiliary coil spring and was depended upon to absorb the heavier blows a car might sustain (Fig. 7.87). McConway and Torley promoted the plan through advertisements in railroad trade papers, especially during the 1890s, but Janney's double-stem draft gear never became an industry standard.

FIGURE 7.87

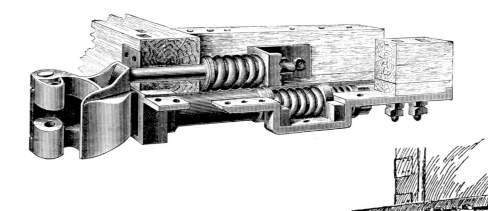

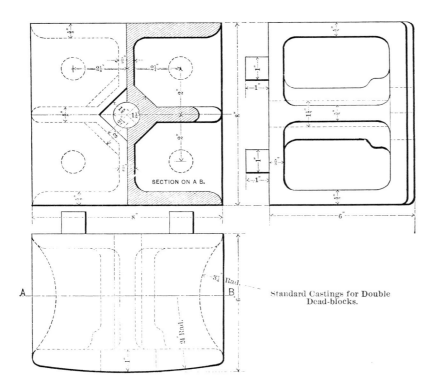

SECTION ON A B.

Standard Castings for Double Dead-blocks.

A — B

FIGURE 7.88

The deadly dead blocks meant to save the draft gear posed a grave danger to trainmen. On the left is the MCB standard adopted in 1882. On the right is an engraving published in 1892. (*MCB Report*, 1886; Thomas M. Cooley, Thomas C. Clarke, et al., *The American Railway* [New York, 1892])

For all the experimentation with multiple springage, there was an unshakable belief in dead blocks as the ultimate savior for draft gears. No less an authority than D. L. Barnes gave voice to this opinion, in a paper read before the New York Railroad Club in May 1891.[189] Barnes contended that draft springs could absorb only 4.5 percent of a 30-ton car coupling at 4 miles per hour. The speed was more than ideal but can hardly be counted as excessive. This meant that under fairly normal conditions the draft gear and the connecting draft timbers would be called upon to handle 95.5 percent of the forces involved when two cars coupled. In such an event, one or both elements might fail. Dead blocks were an imperative. Wood blocks had been favored for many years because they were cheap, were easily replaced, and offered a certain amount of cushioning effect. By the 1880s heavy cars and trains proved too much for plain timber blocks. A heavy blow would shatter them. Iron reinforcing bands were tried. Finally, cast-iron blocks were adopted. They were not made solid but were hollow, open-sided boxes with strengthening ribs (Figs. 7.88, 3.22). Iron was conserved and weight reduced by making them in this fashion. Malleable iron produced yet stronger blocks.

The major drawback to dead blocks was the hazard they posed to trainmen. To perform their office, the blocks must be placed close to the coupler; so positioned, they took direct aim at anyone standing between the cars. Switchmen must stand at this precise spot to work the couplers. The deadly dead blocks could crush and mangle a man even in the most gentle coupling. They were unforgiving should flesh and bone come into their ponderous grip. To avoid this hazard, some railroads adopted the dead-stop coupler. A lug or stop was cast into the upper back of the coupler head just behind the pinhole. This big stop would butt against a single wooden block attached to the end sill. Unlike the dead block, it did not have the trainman in its path and would offer no hazard in a normal coupling operation. Of course, anyone intent on losing a limb could do so by inserting the same between the dead stop and its buffing block while coupling. Several examples of dead-stop couplers appear in illustrations reproduced earlier in this chapter (Figs. 7.81, 7.84).

In the wood-car era the chief victim of draft-gear failures was the draft timber. These short wooden beams were fastened directly below the center sills at each end of the car. When cars came together too hard, the draft timbers tore and shattered. They looked like bits of a shipwreck washed up on

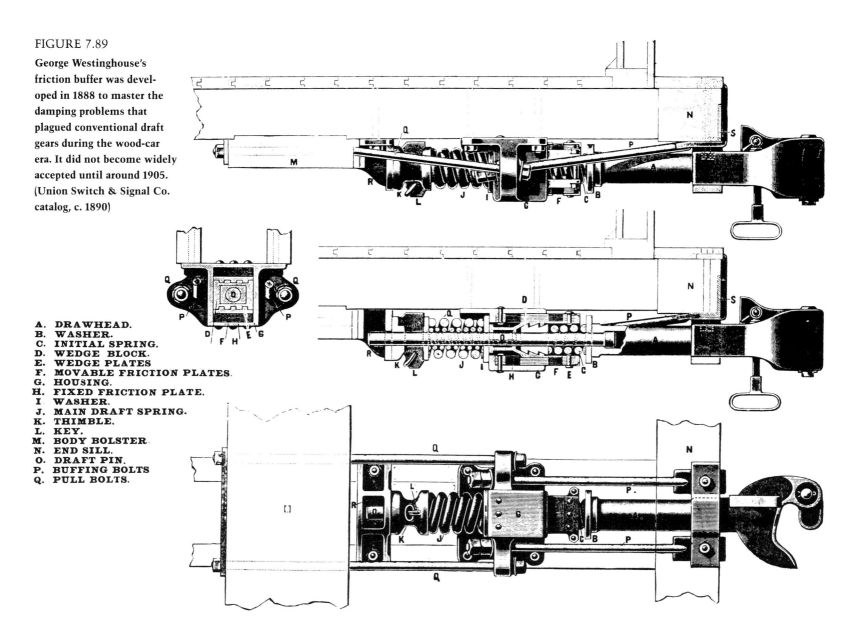

A. DRAWHEAD.
B. WASHER.
C. INITIAL SPRING.
D. WEDGE BLOCK.
E. WEDGE PLATES.
F. MOVABLE FRICTION PLATES.
G. HOUSING.
H. FIXED FRICTION PLATE.
I. WASHER.
J. MAIN DRAFT SPRING.
K. THIMBLE.
L. KEY.
M. BODY BOLSTER.
N. END SILL.
O. DRAFT PIN.
P. BUFFING BOLTS.
Q. PULL BOLTS.

the beach. Beefing up the dimension of the draft timbers would help them sustain the hard service they faced. As floor-frame members grew in size, so too did the draft timbers. By the middle 1890s draft timbers measured 5½ by 8 inches.[190] White oak was the favored material. Seven-eighth-inch bolts fastened the draft timbers to the main frame. Cast-iron key blocks were mortised between the timbers to relieve the bolts of some of the strain and to retard displacements of timbers. The key blocks were cast hollow about 2 inches square and extended the full width of the sills. The bolts passed through them as an added measure for stability.

These safeguards proved largely futile, for draft-timber failures grew worse as car size continued to grow. In 1892 the Southern Pacific began a study of draft-timber repairs.[191] It found in the first year of its survey that 20 percent of freight car draft timbers required replacement. In just three years that figure had jumped by 10 percent. By 1898 draft timbers could not be expected to last beyond three and

a half years, and as cars grew larger, their life expectancy was sure to go down. Car builders were forced to admit that the time of the draft timber was past. An era had ended, and all-metal mountings were a necessity. Pressed and cast steel were tried, but as in most composite structures, metal and timber tended to work loose and so rendered a rickety fabric. All-steel underframes at last solved the draft-gear attachment problem.

George Westinghouse was the only engineer of any national stature to get involved in draft-gear design.[192] He became interested in the problem during the brake trials at Burlington, Iowa (described in the next section of this chapter), where he noticed how badly fifty freight cars behaved when tied together. The excess slack, jerky motion, and general sloppiness of the couplings and draft gear impressed upon him that a better plan must be devised. He concluded that the dependence on common coil springs was the principal flaw in the system. Springs might effectively absorb shocks, but they must then recoil, which re-

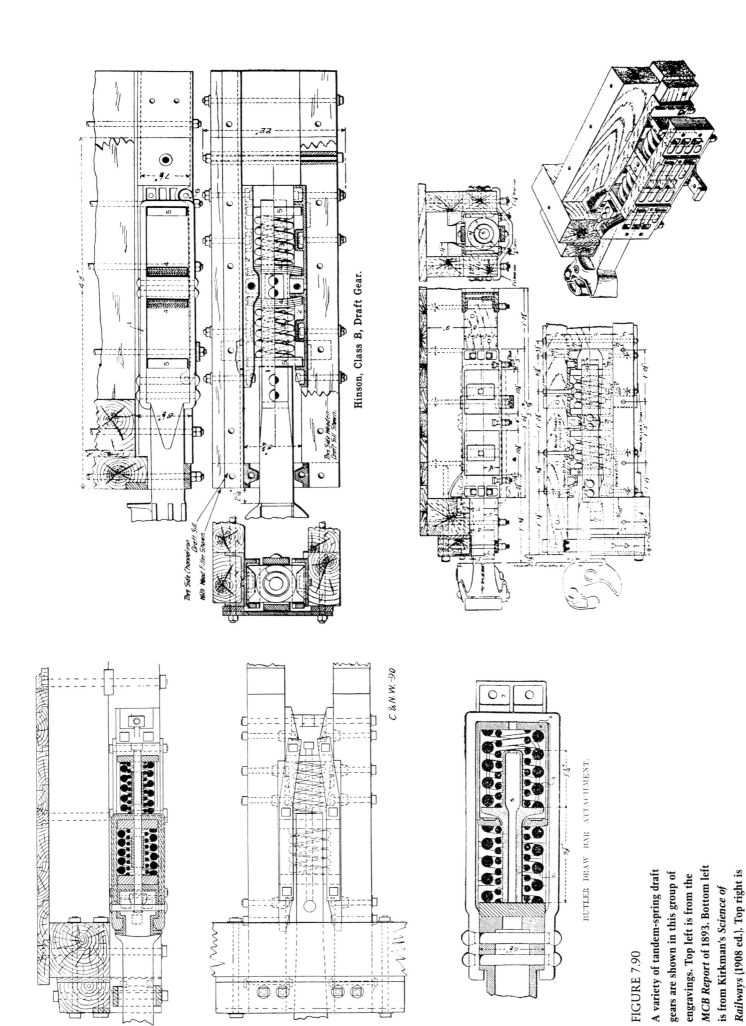

Hinson, Class B, Draft Gear.

C&N.W.-'90

BUTLER DRAW BAR ATTACHMENT.

FIGURE 7.90

A variety of tandem-spring draft gears are shown in this group of engravings. Top left is from the *MCB Report* of 1893. Bottom left is from Kirkman's *Science of Railways* (1908 ed.). Top right is from the *Railroad Gazette,* June 13, 1902. The bottom right engraving of Miner's draft gear is from F. J. Krueger's *Freight Car Equipment* (Detroit, 1910).

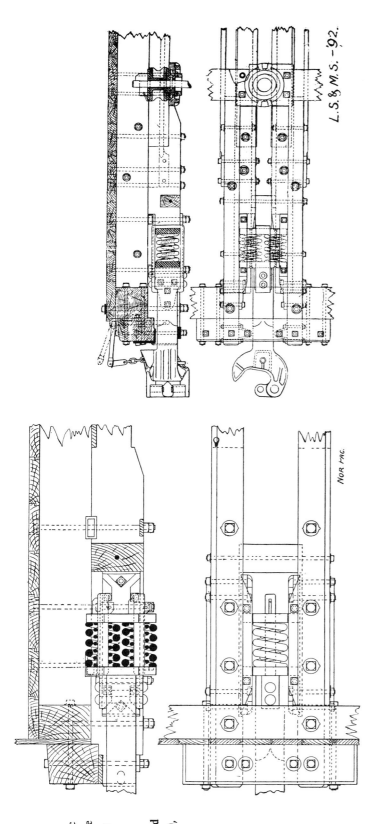

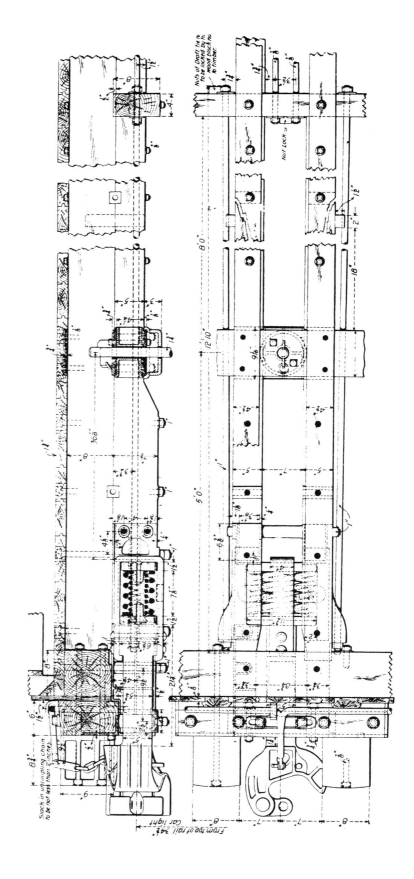

FIGURE 7.91
The parallel spring arrangement for draft gears is shown in these engravings. The Northern Pacific plan (*top left*) and the Lake Shore arrangement (*top right*) are both from the *MCB Report* of 1893. The bottom engraving shows a Gould-style draft gear as depicted in the *Railroad Gazette,* June 10, 1898.

sulted in an undesirable kickback. How to dampen or absorb the recoil became the focus of Westinghouse's research.

The solution lay in a set of interlocking friction plates set up as a triple wedge (Fig. 7.89). The plan was patented in October 1888, and sample units were soon placed on the market by Westinghouse's own Union Switch & Signal Company. It was rated at 30 tons of resistance but could develop as much as 50 tons. A set cost $30, which despite the merits of the system discouraged wide-scale application, at least in the first years after its introduction. Those who did not object to the cost complained that the friction gear was too stiff and caused coupler breakages. Defenders of Westinghouse countered that the friction gear's ability to absorb six to seven times the load of an ordinary draft spring could not be overlooked as the steel-car era progressed. By 1899 three thousand cars had

Westinghouse friction draft gear. As more steel cars rolled onto the main line, the tide turned more in Westinghouse's favor. By 1902 forty-one thousand cars carried his brand of draft gear. The number grew so rapidly over the next few years that 150,000 had been sold by 1905.

Inspired at least in part by Westinghouse's success, a sizable number of competitors sprang up. Draft-gear designs and patents blossomed by the hundreds. In 1903–1904 alone one hundred patents were issued for this category of invention. As usual, railroad mechanics were more confused then helped by such a surfeit of choices. A Graham or a Brian? Was it best to go with a known brand like Gould or Miner, or to take a chance with something clever but unknown like the Piper draft gear? Examples of the major and minor players in this specialized arena of railway equipment are shown in Figures 7.90, 7.91, and 7.92.

FIGURE 7.92

Many bizarre schemes were worked out by railway suppliers seeking to perfect the draft gear. Placing the springs sideways or perpendicular to one another was thought by some to achieve that goal. The Piper gear shown in the upper engraving is from the *Railroad Gazette*, June 13, 1902. The Republic gear shown in the bottom view was manufactured by the Western Railway Equipment Co. of St. Louis. It is reproduced from Krueger's *Freight Car Equipment*.

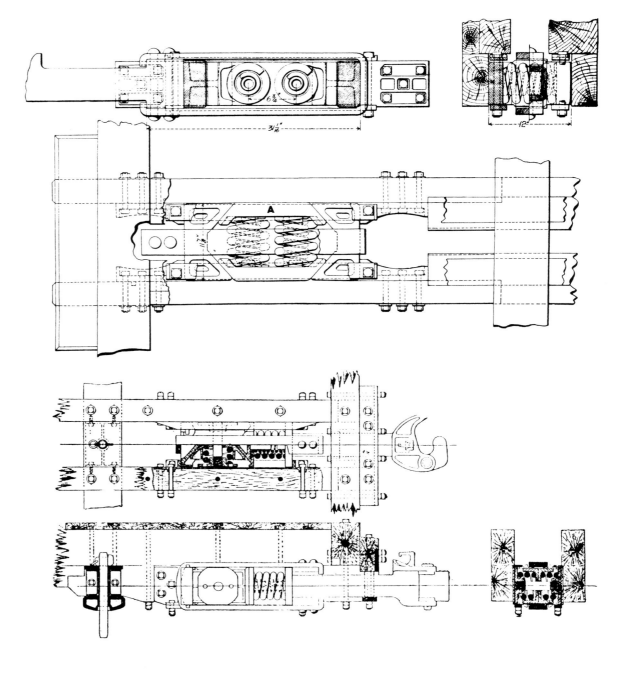

Brakes

If the same energy had been devoted to the means of stopping trains as was expended on finding ways to propel them, then an effective braking system would have been found a generation or more before the air brake was adopted. This is not to say that the issue of railway braking was ignored. It was a busy arena with countless schemers and inventors at work on the best way of stopping locomotives and cars. However, the interest and activity were concentrated outside the industry, whose managers seemed to look with either disdain or indifference upon the plans offered.

This conservatism was especially pronounced for freight equipment. Runaway passenger trains led to clamorous newspaper coverage, lawsuits, claims, and other nuisances, and so railway managers were more easily persuaded to adopt better braking for passenger trains. By 1875, in fact, just about all passenger trains had some form of power brakes. But such costly apparatus was hardly considered necessary for slow-moving goods trains. If the merchandise got scattered over a cornfield and a few trainmen died, so what? It was just another expected part of railroading. Spending big money on brakes for a bunch of boxcars hardly seemed justified. This overall viewpoint was never so bluntly articulated, but it was clearly suggested. The railroad trade press and the reports of the Master Car Builders Association continually voiced concern over the cost of outfitting freight cars with power brakes, or even double-acting manual brakes. This conservative attitude was most likely rooted in the poverty of the American railroad. Funds were especially tight during the first years of the steam railroad because of the scarcity of investment capital in the United States. Doing things on the cheap became a way of life for car builders, and this meant doing without such luxuries as power brakes.

Some railroads were ready to carry the vow of poverty to the extreme by employing no brakes of any type on their cars. Just how common such negligence was cannot be determined, but at least a few lines operated brakeless cars into the 1850s. This practice led to a major accident on the Western Railroad of Massachusetts in December 1840.[193] A long freight train estimated at thirty to forty cars was scheduled to stop at the crest of a big hill outside Springfield. According to normal operating plans, the train would be broken in two for the descent. A locomotive at the head of each section would make for a safer descent, but the weather was foul and before the crew knew it, they had passed over the crest with the entire train and only one engine. The whistle called for down brakes but to little avail, for there were only two or three brakemen on the train. Worse yet, the Boston & Worcester cars mixed into the train had no brakes whatever. The runaway resulted in a major smash-up at the bottom of the grade. Four crew members died in the wreckage. The adverse publicity prompted the B&W to retrofit its goods-carrying rolling stock with elementary manual braking apparatus. Both it and the connecting Western Railroad increased the number of brakemen assigned to their trains.

Elsewhere in Massachusetts, the Boston & Maine saw no reason to spend large sums on retarding its freight trains.[194] A brake van at the end of the train plus the tender brakes should do it. What was good enough for the motherland of railways was thought sufficient for lines in New England. The Old Colony agreed with its neighbor, the B&M, and was running brakeless cars into the 1850s.[195] Trainmen would thrust a stick or fence rail through the spokes of the wheels to hold them in place on sidings. New England had no monopoly on brakeless cars. A section of the Petersburg Railroad running out of Gaston, North Carolina, regularly ran fifteen-car trains that had brakes on only the first three or four cars.[196] The tender brakes were also applied when needed. Out west on the Michigan Central, tender brakes remained the sole means of stopping for both freight and passenger cars until at least the end of the 1840s.[197]

HAND BRAKES

Railroad brakes, whether mechanically or manually powered, traditionally depend on a shoe pressing against the wheel tread. The friction created converts the energy of the rolling train into heat and so stops the train. The shoe—being made from a material softer than the wheel—burns away, giving a distinctive odor to the process of stopping a train. Originally brake shoes were wooden. Cast iron became the norm after about 1860 and remained in favor until fairly recent times, when composite shoes came into fashion. The first brakes were operated by means of levers, possibly inspired by wagon and stagecoach designs. Some of these were foot-operated by brakemen stationed on top of the cars. The Camden & Amboy adopted this system for its tenders and threw in a gig top to protect the brakeman from the weather. A Baltimore & Ohio boxcar of the 1830s, illustrated in Chapter 2 (Fig. 2.14), shows a lever brake staff at each end of the car intended for arm rather than leg power. The Liverpool & Manchester four-wheel goods wagon of about 1830, also pictured in Chapter 2 (Fig. 2.9), shows a horizontal rather than a vertical brake lever, a plan followed in England for over a century. A rather complex rig for four-wheel cars was worked out in 1836 by Henry Waterman for the Mohawk & Hudson Railroad. It was an improvement over the British plan: all wheels were braked, and it could be worked from either end of the car. Waterman's rock-shaft brake was at the same time more com-

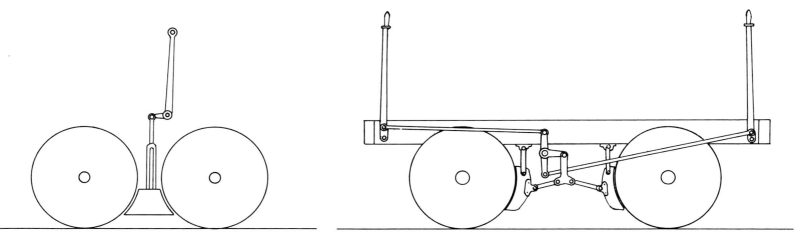

FIGURE 7.93

A plan for double-acting brakes devised in 1836 by Henry Waterman is shown on the right. A simpler scheme used by the Danville & Pottsville Railroad, also in the 1830s, is reproduced on the left. (After a drawing in S. M. Whipple, *Report on the Tanner Brake Case* [1874])

plex and costly than the less efficient Stephenson design. It is pictured in Figure 7.93.

By around 1850 most American railroads had abandoned levers for wheels. Some were mounted on the end of the car, while others were fastened to tall staffs that reached just above the roofline. The wheels themselves were always spoked. Most had rather plain, straight spokes made slightly oval in cross-section. Others had more ornamental curved spokes. Some were made with forked ends. Many examples can be seen in drawings and photographs reproduced throughout Chapters 2 through 6. One exceptional example is reproduced here to show the care that went into the design of so small and inconsequential a part of a freight car (Fig. 7.94). It would be difficult to imagine a more elaborate design for an object whose main purpose was utilitarian. The fact that the casting was malleable rather than cast iron permitted the thin cross-sections of casting.

The shape of the wheel or the type of shoe used meant little without the strong arms of a brakeman to work them. Most histories of braking show a bias against manual braking. It is generally pictured as inadequate, if not wholly unsafe, and yet for short trains, light cars, and slow speeds, manual braking was not terribly deficient. Gradients, obviously, could be a major factor, but most railroads other than mountainous ones rarely tolerated grades much over 1 percent. Henry G. Prout, in his 1921 *Life of George Westinghouse*, states that a stopping distance of 1,600 feet was normal for a manually braked train going 30 miles per hour. However, a test made in 1856 with a seven-car train going 28 mph stopped in only 376 feet. Surely the test was a staged stop with an alert crew, but even if they were only half as effective in normal service, the stopping distance was far better than Prout's estimate.

My own experience with a hand-braked train in September 1981 convinced me that manual brakes are perfectly satisfactory for very light rolling stock on relatively level track. The occasion was the operation of the *John Bull* locomotive and a

single coach on the 150th anniversary of the engine's first run.[198] The crew was totally inexperienced; who could have had firsthand knowledge of the operation of locomotives and cars dating from the 1830s? There was no one to guide or train us, and yet we found the hand brakes to be very responsive. With a little practice we could stop the train with dispatch. After the first day of operation we never felt that the train was out of control or posed more than a normal danger to us or any of the bystanders. It rained for two days during our run, but even so the brakes worked satisfactorily.

Although manual brakes must have been entirely satisfactory for short, light trains on level track, these conditions did not always prevail. And even when they did, many railroads simply employed too few brakemen for the job. In 1853 the New York legislature conducted an investigation into the cause of railroad accidents that recorded some rare statistics on just how many brakemen rode the freights.[199] On the Northern Railroad each train had one brakeman for every five cars. The Cayuga & Susquehanna had one man for every eight to fifteen cars. The Rochester & Syracuse ran fifteen- to thirty-car trains with three to four men. The little Oswego & Syracuse, a slow-speed line with short trains, was even more conservative; it had only one brakeman per train and apparently put inordinate faith in the fireman's ability to man the tender brakes. The ratio of one man for every five cars, which is the best coverage shown by this sample, seems niggardly even for trains grinding along at 10 miles per hour.

The bad reputation of hand-braking, then, may be more a matter of inadequate staffing than the failure of the manual system itself to function. A man with five or more cars to service can hardly be expected to make his way over and across his many charges to set the handwheels properly. To do his work correctly, he must apply each set of brakes with a light or medium pressure. Get too zealous at the wheel, and the shoes will clamp on so hard that the wheels will skid. Skidding wheels

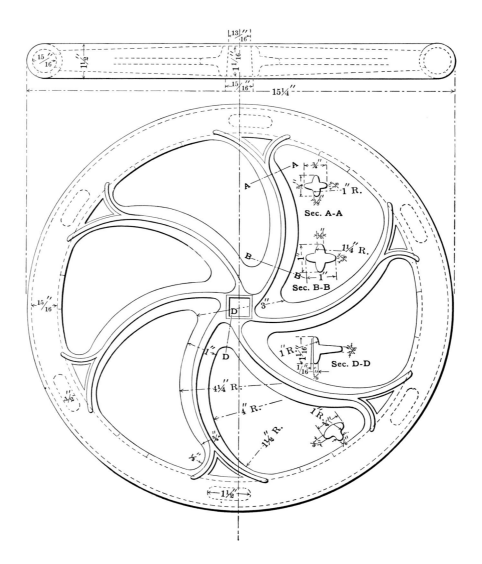

do not stop the train; they cause flat spots on the tread. Yet in the panic to stop, a brakeman will most likely bear down hard on the first few wheels he can reach. Lazy and indifferent men will do so even when there is no emergency. Assuming that it took about thirty seconds to set a brake and move on to the next one, a lone brakeman could probably not attend to more than three cars before the train stopped. Much of the train's stopping capacity was never applied simply because there was not enough staff to work it.

The brakeman's job was a poor one. It combined danger with low pay. It meant long hours on the road; sixteen hours was not unusual. It meant exposure to all extremes of weather. On a windy day one could be blown from the top of a car. Winter's chill and summer's heat were more noticeable in this exposed position high above the ground with no shade or shelter of any kind. Snow and sleet presented the greatest hazard to the occupants of the roof walks. Even on a dry day a misstep or a lurch of the cars could propel a man to his doom. Yet the trains ran every day, and the men clung to their brake wheels even in the worst days of winter (Fig. 7.95). On a fair day the men would stand tall on the roof boards as though on an excursion. The often-

reproduced picture of a Central Pacific freight going around Cape Horn in the mountains above Sacramento shows such a seemingly carefree scene.[200] Perhaps the train was stationary, posed for the photographer—we shall never know; but there is a feeling of movement to the picture intensified by the cavalier poses of the brakemen perched high above the rocks.

In general, the brakeman's job involved either too much or too little excitement. The trip could involve racing downgrade and jumping from car to car while trying to prevent a runaway. Or it could mean long hours wasted at a siding waiting for the eastbound train to pass. The shifts from bareback rider to idle pensioner and back were disturbing to many men who were looking for a steadier, more predictable kind of job.

There were other hazards to being a brakeman in those rawhide days of railroading. Low bridges and tunnels could wipe a man off a car should he fail to respond quickly to the telltales that hung over the tracks as a warning. Chewing cinders as the locomotive barked up a heavy grade was a disagreeable part of the job. Minor burns and eye injuries were other plagues. The working conditions were truly deplorable. And yet just after the air

FIGURE 7.95

The hardships and dangers of the brakeman's job are depicted in this engraving. (*Harper's Weekly*, March 10, 1877)

brake took over, old railroaders lamented that Westinghouse had taken the manhood out of railroading.[201] There was no ginger left in the job; modern brakemen were insipid, pale fellows more fit for the seminary than the robust job of old when the crew knew how to handle a brake wheel.

Actually, brakemen of the old sort did not disappear entirely with the adoption of the air brake. Edward Hungerford, writing in 1909, noted that most freight trains had no fewer than three active brakemen.[202] One was stationed in the locomotive cab at the head of the train, one at the end of the train in the caboose, and a third on top of the cars near the middle. They were carried as backups in case of an air-brake failure. In addition, some manual braking was routinely done to save wear and tear on the air brakes. Hand brakes never actually disappeared, of course, because even today it is necessary to secure idle cars in yards and on sidings. The most modern rolling stock is still adorned with manually powered wheels or ratchet levers that through their connections reach back and under the cars to pull the shoes against the wheels, just as it was done at the dawn of American railroading.

SINGLE-END BRAKING VERSUS FOUNDATION PLANS

In their zeal to keep everything cheap and simple, car builders determined that brakes on only one end of the car were entirely adequate. Hence, only four of eight wheels were called upon to help stop the train. In the 1830s four-wheel cars might have only 50 percent braking as well. The British

goods wagon illustrated in Chapter 2 (Fig. 2.9) demonstrates this scheme. While no one argued about the merits of all-wheel braking, few were willing to pay for it. Complex and costly linkages were required to connect the brake beams of both trucks to a single wheel at either end of the car. Car builders were generally not inclined to undertake the additional first cost and maintenance much before 1880. This was true even for big roads like the New York Central, though its water-level route may well have influenced this policy.[203] Granger lines like the Burlington and the Katy were more reasonable in espousing a single-end brake design, considering their lower traffic density.[204] The New Haven was building new cars at least as late as 1880 with brakes at one end only, perhaps just because of old-fashioned Yankee conservatism.[205]

More safety-conscious lines, particularly those with severe gradients, tended to be far more liberal when it came to the double-braking issue. The Baltimore & Ohio was hardly affluent in the 1850s, but it was convinced of the good sense of brake shoes for each and every wheel, even on the lowly freight car. The oldest known plan for a boxcar with eight-wheel braking depicts a B&O car. The engraving is reproduced in Chapter 2 as Figure 2.27. In 1878 the Erie and the Boston & Albany were both cited as early advocates of all-wheel braking, but unfortunately no specific dates in either case were given.[206] Those in favor of eight-wheel braking argued that their method not only allowed quicker, safer stops; it also reduced the wear and tear of stopping the train by spreading the work out over all of the wheels instead of just half. The brakeman's job was made easier as well, since he was not required to race over so many cars to perform his job. All-wheel braking could also help reduce stopping times. Improvements were certainly needed in that respect; in 1878 a big freight—forty-five to sixty cars—required one to one and a half minutes to stop, even though speeds rarely topped 10 miles per hour.[207]

The proponents of all-wheel braking won the debate on the matter of passenger cars; however, just how to arrange the mechanism for any style of rail vehicle remained an issue for generations. The essential problem was to devise a simple yet effective system of levers, chains, and rods to connect the brake beams on both sets of trucks to the brake staff at each end of the car. This network was called the brake-gear foundation plan. As might be expected, dozens of plans were proposed and tested. No one plan became universal, but a few like the Hodge, Stevens, and Tanner schemes tended to dominate. Whatever plan was used, it had to offer a way to reverse the thrust of the motion so that the left- and right-facing brake beams could be made to move against the wheels.[208] The Hodge system required six levers and seven rods to

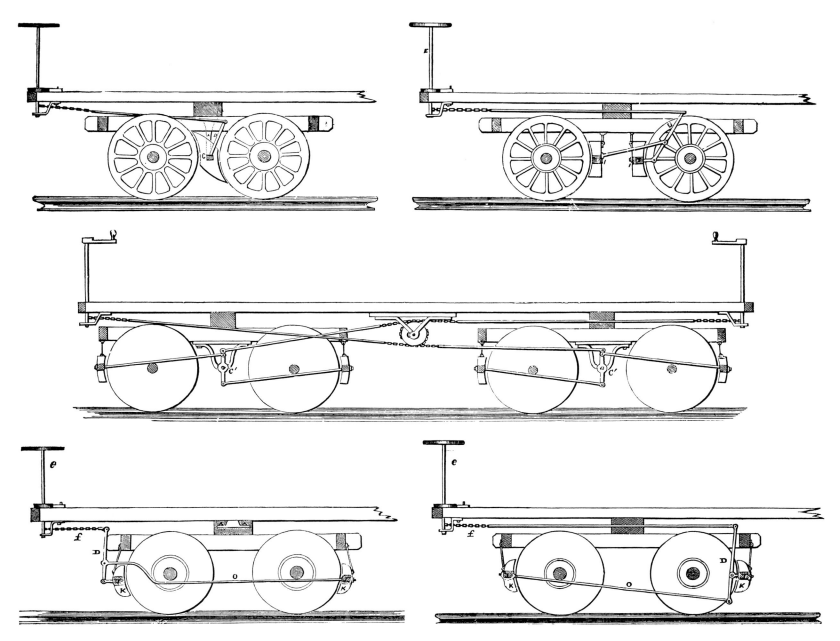

FIGURE 7.96

Some early brake-gear foundation plans are shown here. The upper two date from the 1830s, while the bottom pair are from about 1860. The middle engraving shows Millholland's roller-chain plan of 1843. (*Whipple's Railroad Reporter*, June 1861)

accomplish its mission. The Stevens foundation gear had only four levers and five rods, while the Tanner employed three levers and six rods. While the Hodge scheme would appear to have been the least attractive because of its numerous parts, it was in fact the favorite because it worked so well in combination with air brakes. The plan was, however, worked out many years before pneumatics were introduced for train stopping. Nehemiah Hodge of North Adams, Massachusetts, patented a foundation plan on October 2, 1849.

Just a few years earlier, James Millholland, master mechanic for the Baltimore & Susquehanna, had worked out an ingenious double-truck braking system that employed no levers whatever.[209] Instead, he used large-diameter drums or spools to reverse directions (Fig. 7.96). When the main rod/chain pulled in one direction a secondary chain and rod would move in the opposite direction. Simple and direct, the scheme illustrates Millhol-

land's unique approach to common engineering problems. In a patent taken out in 1843 for six-wheel running gears, he showed clasp brakes and so anticipated by more than half a century a fundamental improvement in railway braking. The patent drawing is reproduced in Chapter 2 (Fig. 2.16). Yet for all Millholland's inventiveness, no one seemed ready even to try his spool foundation plan other than the proprietary railroad out of Baltimore.

Novelty did have its day on the Pennsylvania Railroad with the large-scale installation of Wharton switches beginning in 1870.[210] How could this decision influence the form of brake-gear foundation plans? As it happened, part of the Wharton switch stuck up above the normal rail level and so could snag low-hanging parts of a conventional brake rigging. The problem was especially bad for tenders, nearly all of which had small 28- to 30-inch wheels. A solution was devised by an obscure designer named J. F. Elder at the road's Altoona

FREIGHT CAR TECHNOLOGY

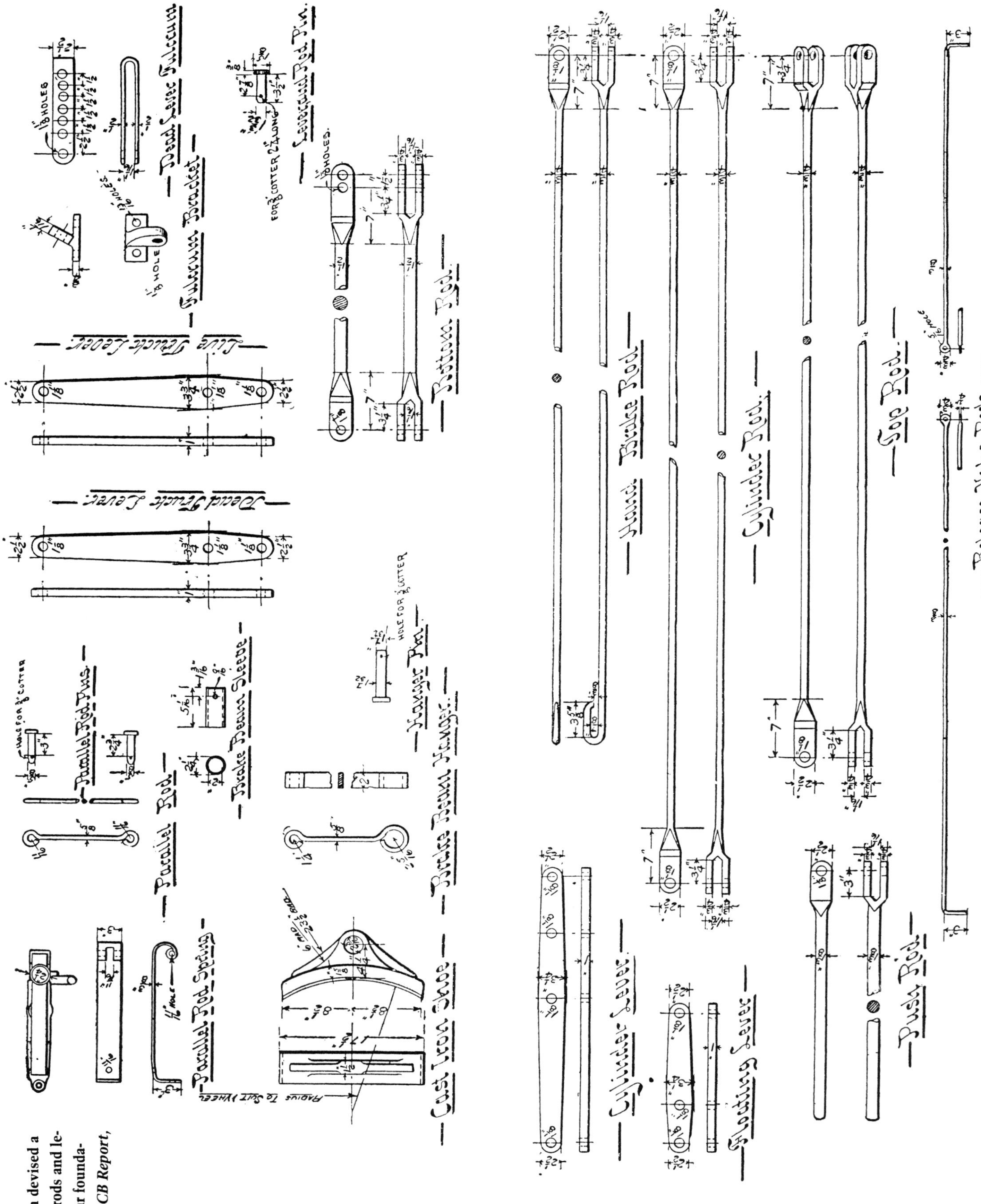

FIGURE 7.97

The MCB Association devised a standard plan for the rods and levers used in brake-gear foundation plans in 1889. (*MCB Report*, 1890)

THE AMERICAN RAILROAD FREIGHT CAR

shops. The plan can hardly be called simple, for it involved not only the usual levers and rods but also short chains and pulleys fixed on the ends of the levers. Whatever its failings, it was found acceptable by the mechanical moguls at Altoona, and Elder's plan was adopted as the system's standard.

The variety of foundation plans adopted are shown in the drawings reproduced in Chapters 2 through 6 and in all editions of the *Car Builders Dictionary*. Even with the coming of power brakes, all rail cars needed some form of hand-braking to direct the force of the air cylinder to the brake shoes. In addition, cars had to be parked securely when not in service, lest they wander away in disastrous journeys as runaways. A common complaint about foundation gears after the adoption of power-braking was the overly light proportions of the rods and levers. They would bend and deflect, requiring a greater effort on the part of the air cylinder piston to stop the train.[211] Car builders were urged to abandon the false economy of lightweight brake rigging and to follow the substantial standard designs offered by the Master Car Builders Association (Fig. 7.97).

MOMENTUM BRAKES

If all-wheel braking seemed like a futuristic dream, the ideal of continuous train brakes must have appeared even more remote. Continuous brakes consisted of a uniform and united system of braking whereby the brakes on all the cars could be set or released at once. The braking would be even and not dependent on the uncoordinated actions of half a dozen men scattered over the length of the train. Smooth, gentle stops would be but one benefit of continuous braking; repair costs would plummet as well because such gradual stops would not burn up shoes, flatten wheels, or shear pins. In the 1830s the Mauch Chunk and the Danville & Sunbury, small coal lines in Pennsylvania, fitted their cars with lever brakes that could be connected by ropes or wooden rods (Fig. 2.5). The Mauch Chunk operated trains of fourteen cars in this fashion—with all the brakes worked by one man on the rear car—but the scheme does not appear to have made any headway on main-line railroads.

Unfortunately the admirable idea of continuous braking first became associated with the momentum- or buffer-brake scheme, which proved itself such a lamentable failure. It was a scheme that attracted engineering luminaries like George Stephenson as well as mechanical nobodies like Stephen Tallman. Even Henry H. Westinghouse (1853–1933), George's younger brother, expressed his abiding faith in the momentum brake in a talk before the New York Railroad Club in January 1888. Ironically, Henry was a lifelong official of his brother's air-brake company, yet he remained a believer in the potential greatness of a competitive braking system because of the abundant "free power" inherent in every moving train. A freight of thirty-five to forty cars equals 800 tons when moving at just 20 miles per hour; it develops 10,686 foot-tons, an abundance of power to stop the train. The dilemma was, of course, how to harness this power for braking purposes. Everyone taken in by the illusion of buffer brakes, including Henry Westinghouse was doomed to disappointment.

Momentum braking was such a seductive scheme because it promised a free ride. All the inventor need do was to harness the enormous energy created by the forward motion of the train and reverse it via levers, chains, rods, or windup drums so that the rolling tonnage of engine and cars could be utilized to stop the train. It was like capturing the power of the surf, an endlessly attractive and endlessly frustrating exercise. Two basic schemes evolved. The buffer or compression brake was the most popular. It derived its power from the backward movement of the drawbars when the train's speed slackened. This was a normal consequence of the engineer shutting off steam and the fireman setting the tender hand brake—the slowing of the engine would cause the cars to bunch in behind it, and this compressing movement would automatically cause the buffer brakes to act. The more the cars crowded together, the harder the brakes would come on. The second plan involved a windup mechanism powered by the revolving car wheels or axles, usually through a friction drive. The windup drum pulled in a chain connected to the brake gear and so stopped the train. Its force would lessen as the train slowed. The windup action was triggered either by the motion of the buffer or, in a few later designs, by electric or air controls.

Most of these devices were a mechanical tour de force and looked like industrial-sized apple peelers, all gears, cams, levers, and ratchets. How any sensible mechanic could expect these complex devices to function long or well on the underside of a rough-and-tumble boxcar is difficult to imagine. Yet dozens of promoters were obviously optimistic, and the railroad industry, though a trifle skeptical, seemed ready to test any braking system that promised a cheaper alternative to the air brake. The long and discouraging search for a substitute for Westinghouse's product was largely motivated by the prevailing belief that air brakes were simply too costly for freight equipment.[212]

The efforts of John K. Smith of Port Clinton, Pennsylvania, provide us with the earliest record of an American momentum brake.[213] In October 1835 the Franklin Institute offered a brief report on Smith's self-acting brake. The car's center sill was made up of three parallel beams; the loose central one slid back and forth with the motion of the couplers. It was attached to the brake rigging and so was a buffer or compression brake in its simplest form. Smith had devised a way to connect the shifting center sill to horizontal lever brake staffs used on coal jimmies. The plan was said to have performed well for several months on the Little Schuylkill Railroad. Like most buffer brakes, it had to be disconnected when backing the cars. This caused considerable inconvenience and delay, because the crew had to walk the length of the train to disconnect the brake linkage. It was necessary to reverse the hookup procedure when the train was ready to go forward again. Smith received a patent in December 1835, but there is no evidence that his invention went beyond the testing stage.

It was just a few years later that George S. Griggs, master mechanic for the Boston & Providence Railroad, introduced the windup style of momentum brake into American practice.[214] A patent issued in December 1839 included a freight car drawing, indicating that the inventor had goods service in mind for his continuous-braking scheme (Fig. 7.98). Later reports show that passenger cars were included in this plan. A flat belt driven by an axle-mounted pulley ran over a second loose pulley under the car. A clutch controlled by a rope that went forward to the locomotive would, when engaged, turn a drum that reeled in a chain and so set the brakes on all eight wheels of the car. Griggs thus championed not only continuous braking but all-wheel braking.

The plan attracted some attention and praise and was adopted by several railroads in the Boston area. In tests on the Boston & Worcester, it stopped a sixteen-car freight train going down a 30-foot-to-the-mile grade at 15 miles per hour in just a little over 300 feet. The Griggs brake was declared equal to six brakemen and safer yet because unlike the brakemen, it would not abandon the train at the first sign of danger. It cost $25 per car, no trifling sum in 1839. But according to at least one critic this was no excuse for not adopting it; it would be simply parsimonious to hold back payment for a scheme that would save lives and property. Some of the Boston lines continued to use the Griggs brake until as late as 1848. Just why it was not more widely adopted is not explained, but we can assume with reasonable certainty that the inherent defects of the momentum brake, soon to be explained, prevented a more general use of this belt-driven windup brake.

Once Smith and Griggs had opened the momentum-brake era in the United States, a number of other inventors, predictably, followed their lead. Turner, Feger, Ambler, and a host of others patented and tested their designs in the 1840s through the 1860s. Most were only minor variations on the themes established by Smith and Griggs, but a few were decidedly novel. Lewis Kirk, for example, formerly mechanical head of the Philadelphia &

Reading, patented a hydrostatic-buffer brake on May 10, 1859 (No. 23,923), that featured a water cylinder and pump under each car. The drawbar movement actuated the pump, which delivered fluid to the cylinder. The corresponding motion of the piston set the brakes. Joseph Olmsted of Galesburg, Illinois, had even more advanced ideas.[215] In 1867 he began tests on the Burlington of a windup momentum brake with a clutch controlled by an electromagnet; by pushing a button, the engineer could apply the brakes along the whole train. A test train moving at 40 miles per hour was stopped in 672 feet. The cost of the apparatus—$50 per car—apparently discouraged further testing.

The most widely applied of the momentum brakes in this middle period of continuous braking, just before the air brake took hold, was the chain brake devised in 1855 by William Loughridge of Weverton, Maryland. A single windup drum was placed on the locomotive rather than individual mechanisms under each car. The friction wheel was forced against one of the driving-wheel tires when the crew wanted to stop the train. A long chain ran the length of the cars to activate the brakes. On the one hand Loughridge succeeded in simplifying the momentum brake, but on the other he introduced a new element for failure. The long chain required numerous couplings and compensating devices so that it might pass from car to car and yet maintain enough tension for the brakes to be set and released efficiently. Loughridge's brake was adopted by several major railroads for passenger trains but does not seem to have been used in freight service.

After several years of tinkering, Loughridge abandoned the chain brake and devoted his energies to steam and air brakes. An obscure inventor from Eastmanville, Michigan, named Albert F. Gue felt that Loughridge was ill-advised to give up so easily. In 1874 Gue took up the chain/momentum brake with the specific idea that it could be successfully developed for freight train service. Gue modified the Loughridge concept in several ways. First, he divided the system in half. The front of the train would be braked from the locomotive, while the rear would be controlled from the caboose. The windup winches would be powered through friction wheels by the forward motion of the train, as in all windup systems; however, the engagement clutches would be operated manually by the engineer at the front of the train and the conductor at the rear. Gue also realized that the longer the chain, the more troublesome the compensation problem became, so he proposed that only four or five cars at either end of the train would be hooked up. The remaining cars could either run brakeless or be serviced individually by the brakeman.

Gue obtained patents in May 1874 and February

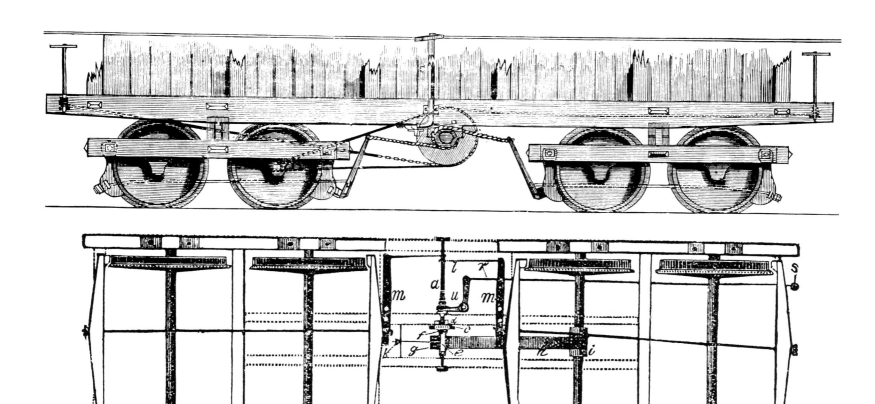

FIGURE 7.98

The Griggs windup momentum brake was introduced in 1830. The more important parts are *a*, countershaft; *c*, gear wheel; *d*, clutch; *f*, gear pinion; *h*, belt; *i*, axle pulley; *r*, control rod; and *s*, control lever. (*American Railroad Journal*, Sept. 15, 1841)

1875. These were assigned to George F. Field & Associates of Boston. Testing began on the Boston & Maine and the Old Colony in 1875.[216] Light chain was used at first, but Gue and Field found ½-inch wire rope more satisfactory. It was claimed that a train moving at 25 miles per hour could stop in the same distance as one equipped with air brakes, though admittedly the caboose wheels seemed to skid when stopping. The apparatus cost $50 on the locomotive, $35 on the caboose, and $10 on each car. A very optimistic report published in 1880—apparently written by the firm's press agent or the inventor's mother—claimed that the Gue and Field brake required little more than a few drops of oil once a month to keep it in tip-top order. The upbeat publicity attending the wire-cable train brake may not have resulted in many orders, but it does seem to have generated more interest in momentum brakes for train service.

William L. Card of Moberly, Missouri, did not believe in the chain and cable brake; he wanted to return to the basic notion of the old-fashioned buffer brake, whereby the action of the drawbar was directed to the brake rigging. In February 1878 Card obtained a patent—the first of four—that was meant to perfect the buffer brake in its final form.[217] Card's plans were ingenious, for he devised a sophisticated design that answered two of the major criticisms of this form of brake. A centrifugal ball-style governor mounted on the axle prevented engagement unless the train was moving at 6 miles per hour or faster. This eliminated

the annoying interference of the brakes during switching movements. A locking device prevented the automatic brake from engaging when the train backed up. While this helped in switching, it also meant that the train was once again dependent on manual braking when in reverse.

Though only a halfway solution, Card's design was seen as a triumph by its promoters, the American Brake Company of St. Louis. The drawing reproduced in Figure 7.99 explains the details of Card's design more clearly. The American buffer brake was in production by April 1881, and by early the next year some twenty railroads were using it. Most of these were small-lot test applications except for the Frisco, which became a patron of the Card brake. In 1882 the line had three hundred cars fitted with the apparatus. By 1885 this number had expanded to 2,000 on the Frisco and 650 on the Wabash.

Encouraging test results, all presumably generated by the manufacturer, were published in the trade press. Few editors seemed ready to offend an advertiser, and so almost any claim was printed. In a test made in October 1879 on a level track (train size not given) the train stopped in 466 feet at 20 miles per hour. Upping the speed only served to improve the brake's efficiency. At 30 mph the stop was made in only 306 feet. Several months later another test was made on the Frisco between Springfield and Rolla, 125 miles of steep grades and sharp curves. The 491-ton train consisted of a locomotive, tender, sixteen freight cars, a caboose,

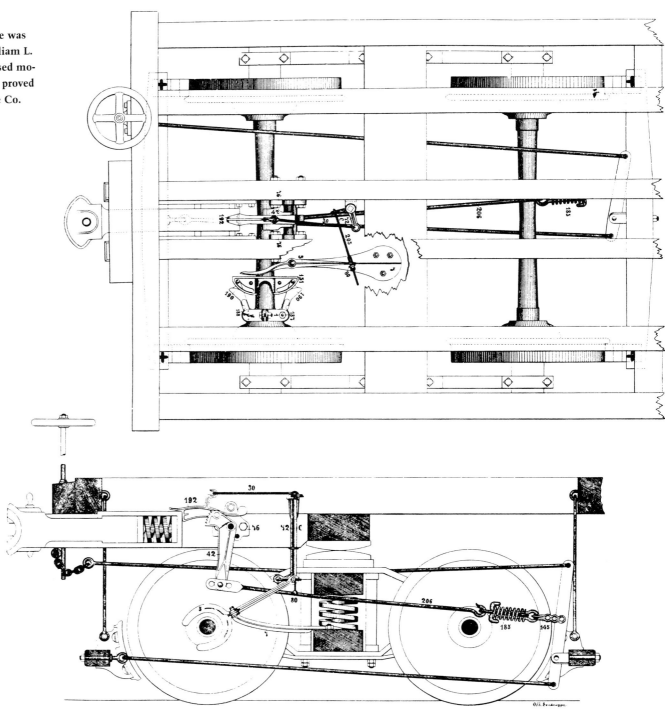

and a coach. Snow and bad weather added to the rigors of the test, as did grades as steep as 65 feet to the mile. Yet the stopping record was considered good: 1,230 feet in sixty-two seconds at 25 mph. In yet another test made near New Haven, a twenty-car coal train weighing 228 tons made a series of stops at an average speed of 26.5 mph. The stops averaged about 700 feet in thirty-seven seconds. In all tests, steam locomotive brakes, also manufactured by the American Company, were used in combination with the Card buffer brake. Exactly how much stopping the steam brake did as compared with the buffer rig cannot be determined, but the test results as reported in the trade press

were far different from those recorded at the brake trials conducted in 1886 at Burlington, Iowa, under the watchful eye of the Master Car Builders Association.

While the American brake was enjoying good press under hothouse conditions, Charles V. Rote of Lancaster, Pennsylvania, decided to imitate the example so ably set by the St. Louis firm. He obtained two patents on March 20, 1883, that were remarkably similar to the designs perfected by Card some five years earlier (Fig. 7.100). A firm was organized in Mansfield, Ohio, not long after the patents were issued, and testing began on the New York, Pennsylvania & Ohio.[218] Claims were made

FIGURE 7.100

Charles V. Rote's brake was mechanically very similar to Card's design. The motion of the coupler/draft gear worked the brake. (*Railroad Gazette*, Sept. 24, 1886)

that all cars on the line would be fitted with Rote's apparatus. A similar assertion was made about the New Haven; in all, however, it appears that only ten cars on the NYP&O were actually fitted with Rote's invention. The brake gear installed cost $31 per car and so presented an attractive saving over the Westinghouse brake, which cost about $50 per car.

Other inventors hoped to catch a ride on the popular wave for momentum brakes generated by the American Brake Company. In August 1879, just a little more than a year after Card began serious work on his designs, Stephen P. Tallman of New York City obtained the first of four patents for a windup style of momentum brake. Like the American brake, Tallman's creation seemed to perform very well, at least when the inventor was on hand to oversee its workings. In May 1880 a ten-car train on the Pennsylvania Railroad going downgrade was brought to a stop in 510 feet, yet only five cars in the trains were fitted with the windup brake.[219] In 1882 the Master Car Builders Association committee on brakes reported optimistically on the good showing made by Tallman on the New York & Harlem, the Rock Island,

and some of the cars belonging to the New York Live Stock Express Company.[220] The MCB membership seemed ready to take a hopeful view of any style of continuous brake that might prove cheaper than the air brake, but this rosy view of things would be dimmed forever once rigorous and objective testing began four years later in Burlington.

Just two years after the Tallman brake's debut, the Widdifield & Button brake was introduced. The chief architect of this windup style of train stopper was a Canadian named Watson P. Widdifield, a resident of Uxbridge, Ontario.[221] The inventor's main contribution to momentum-brake design was the introduction of a wearing friction drive. A large pulley and windup winch were mounted on a countershaft parallel to one of the axles. A small drive wheel made of fiber and soft metal was mounted on the axle (Fig. 7.101). The drive pulley was pushed against the soft, or wearable, wheel by the motion of the draft gear. The Lehigh Valley somehow became enamored of the Canadian design around 1884 and continued to run test trains for Widdifield as late as 1889, despite the brake's ignominious failure at the Bur-

FIGURE 7.101

The Widdifield & Button buffer brake used a windup chain powered by the motion of the car. The shaft pulley can be seen on the left side of the axle. The windup countershaft is directly behind the axle. (*Railroad Gazette*, Sept. 24, 1886)

lington trials some three years earlier.[222] Perhaps the Lehigh Valley's wish to believe was encouraged by the apparatus's low cost. At $10 per car, the Widdifield brake would surely have been a bargain—that is, if only it had worked as well as promised by its manufacturer.

For half a century there had been bright talk about the merits of the momentum brake. Dozens of schemes were put forward. Excellent test reports were published. Yet it always seemed to be the brake of the future, something obviously very good but not quite ready for the market; just a few more adjustments would make it perfect. This fantasy came to an abrupt end in July 1886, at the brake trials conducted under the hard and uncompromising eyes of the Master Car Builders Association brake committee. There was no sentiment in these trials. The judges were not made forgiving by a large luncheon and too much wine. This was a no-nonsense, no-smoke-screen style of testing, and accordingly the inherent defects of the momentum brake were exposed for all to see.

The Rote brake was found to be less effective than hand brakes, and it failed every test it was asked to pass. The Widdifield brake produced violent and jerky stops when put through the quick-stop test.[223] This defect was untypical, because most momentum brakes tended to brake hard at the front of the train and usually showed no braking power whatever at the rear. The Widdifield brake performed so abysmally with twenty-five to thirty cars that the fifty-car test was not even attempted. The American brake, the darling of momentum brakes, was exposed as an impostor. It worked fairly well with short trains of twenty-five

cars or less, but with fifty cars its stops were so violent and jerky that it was declared unsafe. The American failed the downhill test as well.

Curiously, the buffer-brake group did not feel beaten even after the humiliating results of the Burlington test were published. Widdifield rigged up an electromagnet to engage the friction-drive mechanism, thus eliminating the dependence on buffer control.[224] Card worked out a similar plan and actually entered it in the second Burlington brake trials of 1887.[225] But even electric controls could not salvage the momentum system, and after two failures, Card withdrew from the field. Electrical control was an old idea going back to the 1860s. It was revived in the early 1880s by the Waldumer Electric Brake Company of Cincinnati. But there were others living in the Queen City who felt that dependence on batteries and wires was not the right way to overcome the failings of the buffer brake. Sloan and Campbell developed a pneumatic control for the windup style of momentum brake.[226] An air line ran the length of the train to supply compressed air to cylinders under each car. The cylinder engaged or disengaged the windup apparatus as needed. The inventors claimed that their system used far less air than a conventional air brake because the forward motion of the train supplied the power. Normal air-braked cars could be mixed in with Sloan and Campbell cars. The Cincinnati Southern thought enough of the scheme to run a test train. But no one else appears to have shared the enthusiasm and the Sloan-Campbell brake joined the growing ranks of also-ran contenders who had sought to produce a cheap competitor to the Westinghouse brake.

While the car-building fraternity was being mesmerized by the antics of the momentum-brake hucksters, George Westinghouse was busy promoting his air brake for freight service. He had not, as is popularly supposed, literally invented the air brake; a number of English and American mechanics had patented or even demonstrated the idea years before Westinghouse took an interest in the subject.[227] His 1868–1869 creation was a lucky combination of tried or obvious elements necessary for a pneumatic train brake. The device was introduced at an auspicious time under the aegis of railway officials sympathetic to the young inventor's cause. The brake worked well enough and was adopted for passenger trains with a rapidity unusual for such a conservative industry. Because the number of cars involved was relatively modest, the investment was seen as not excessive, and it was surely a politic way of silencing the clamor for safer passenger service. Westinghouse dramatically improved the air brake in 1872–1873 and surpassed his predecessors by developing a fail-safe system that continued to function even if the train broke in two.

Yet for all Westinghouse's technical achievements, the industry showed little inclination to adopt pneumatic braking for anything beyond passenger trains. It was definitely a matter of numbers. Few trunk lines objected to paying $100 each for the few hundred coaches on their property, but it was another matter to make the same per-unit expenditure for fifty thousand boxcars. It went beyond the air tanks, cylinders, piping, and hoses needed for the cars (Fig. 7.102). The locomotive needed all of this rigging plus a fancy pump, which drove the cost up by $500. Figuring the brake sets at $50 each plus $42 for installation, the railroads rapidly lost their enthusiasm for a quick conversion to the Westinghouse system. The Burlington Railroad's management, for example was more open than most to the idea of air brakes for freight service, but the $1.38 million price tag was more than the line was ready to accept. It was estimated that for the industry as a whole, no less than $40 million would be required.[228] The buffer-brake contingent said that they could do the job for only $8 million, and it was this promise that kept their prospects alive for so long. But even if the cheaper alternative had worked, the capital expenditures were still very large, and prudent managers had to consider the most effective way to spend their capital. One historian, Albert Fishlow, is convinced that the money was better spent on steel rails than on power brakes.[229] Stronger track made for a safer, more productive railroad, and hence the reluctance to adopt air brakes was not entirely unreasonable.

In addition to the damper of the big numbers just given, car builders labored under a restraint voiced early in the power-brake era. It somehow became a fixed idea that no system of continuous brakes for freight service could cost more than $25 per car.[230] Just how this figure was established remains uncertain, but it was repeated often enough to become doctrine and so helped delay the acceptance of improved braking for American freight trains. Westinghouse's most serious competitors made an issue of the cost factor by offering cheaper apparatus. The Empire vacuum brake was offered at $45 per car, but the saving of just $5 did little to snare many customers. The Eames vacuum-brake company was more competitive and offered brake sets at $25 per car, the car builder's magic number. But even though the price was right, the vacuum brake was not really well suited for freight service, and so the search went on.

The debate over first cost was further complicated by considerations of repair costs. No one seemed to have very satisfactory figures, and the numbers offered were widely divergent.[231] Those friendly to the air brake said that it cost only a trifle over $17 per year to maintain the locomotive's air-brake apparatus, with most of the cost going to the ever-troublesome pump. Others put the cost at $87 per annum. The maintenance cost per car per year was figured at $4.50. None of these costs were thought unbearable when applied only to the passenger fleet, but they became gargantuan when multiplied by the number of freight cars.

Economics aside, the railroad mechanical professionals had some serious doubts about the air brake. These complaints came mainly from locomotive supervisors who found themselves saddled with a new and troublesome piece of apparatus to maintain. Traditionally, stopping the train had been the car department's problem, and now the motive power department was stuck with it. The art of locomotive maintenance had been refined to a high degree by 1870, and all was going along well until a fussy air pump was mounted on the boiler's side. It proved delicate and fragile.[232] It was built on too light a pattern and needed constant repairs and attention. Wilson Eddy, master mechanic for the Boston & Albany, complained at length about his problems with pumps. They could not remain idle for even a few days. Eddy said that two or three accidents were accountable to the Westinghouse brake because of defective pumps. His men preferred the more reliable Smith vacuum brake. J. H. Setchel, master mechanic for the Little Miami and a strong supporter of pneumatic braking, conceded that Eddy was at least partially correct. The pumps required careful adjusting and had to be well oiled. It was also true that the brakes tended to stick on when trying to spot the locomotive at water or wood stations. But Setchel insisted that these were small defects for such an effective preserver of life and property. Setchel suggested an axle-driven

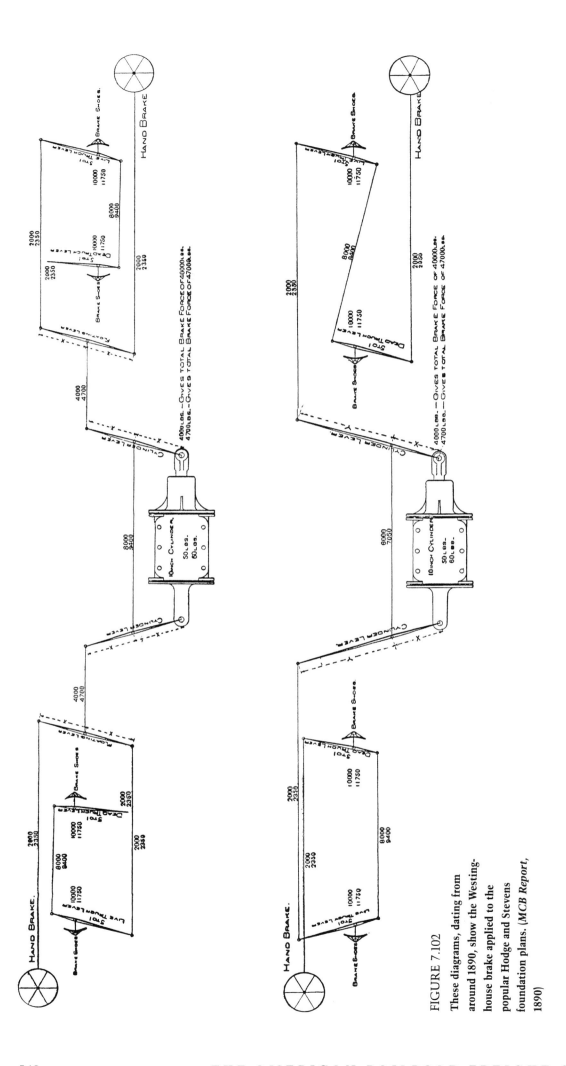

FIGURE 7.102

These diagrams, dating from around 1890, show the Westinghouse brake applied to the popular Hodge and Stevens foundation plans. (*MCB Report*, 1890)

pump as a cheaper and more reliable alternative to the Westinghouse steam pump.

Reuben Wells, another prominent master mechanic of the time and later superintendent of the Rogers Locomotive Works, offered a surprisingly negative opinion of the Westinghouse brake at the 1875 annual meeting of the Master Mechanics Association. He wondered in general about its utility for railroads other than the most crowded or the fastest. In his opinion, the majority of rail lines might well continue to rely on manual braking. Others went on to grumble at some length about the additional expenses of the "wheel account"; because of the heavy-handed application of the pneumatic, tires and treads were ground down in a deplorable fashion. Fast stops also caused flat spots that if very large would condemn the wheel to the scrap pile. It was also noted that some engineers were demanding more pay because of their new duty as the train stopper. While some lines had acceded to these demands, most felt that they offset the wage saving resulting from fewer brakemen. Despite the several complaints registered here, most of the master mechanics were ready to agree that the air brake's advantages outweighed its defects.

At the very time the master mechanics were finding fault with Westinghouse's creation, railway managers in England were examining the power-brake question in preparation for trials at Newark on the Midland Railway.[233] Air—a light, elastic fluid—was seen as far superior to, say, water, which was heavy, inelastic, and likely to freeze. From a theoretical viewpoint all agreed that compressed air was the ideal medium, but others, especially Francis W. Webb, locomotive superintendent for the London & North Western, said that air brakes were far too delicate and prone to leakage for the rigors of railway service. In his examination of the Westinghouse system he had counted 150 parts. That was 150 parts liable to fail, in Webb's estimation, and how could one expect an ordinary engine crew to master such complex mechanisms when they could barely operate a locomotive correctly? What was needed was something simple enough for a porter to understand and that, concluded Webb and his supporters, was the vacuum brake. And so Westinghouse was largely frozen out of the British market.

Westinghouse must have realized that selling American industry on freight air brakes would be about as difficult as entering the British market had been. He made no major converts until 1878, when the Denver & Rio Grande began applying pneumatics to their goods cars. Accordingly, the *Master Car Builders Report* for 1885 stated that the line had 6,806 cars outfitted, but because they were of the obsolete straight air pattern no one, not even Westinghouse, felt that they were a very effective demonstration of the benefits of the power brake for freight trains. Without the triple-valve or reserve air tanks, they lacked the fail-safe feature of the reformed Westinghouse system. A more substantial demonstration came about in 1883, when the Central Pacific announced its intention to adopt air brakes for freight cars.[234] By May 1884 the CP had 6,617 cars fitted and pushed the program with such vigor that within two years nearly all its stock—some twenty thousand cars—was running under the protection of air brakes. Other western trunk lines followed the CP's lead, notably the Santa Fe, the Union Pacific, and the Northern Pacific. In March 1885 the *Railroad Gazette* said that 35,855 freight cars had air brakes, most of them belonging to western lines, which reported wonderful improvements in their operations over the treacherous Rocky Mountain grades.[235]

At the same time, considerable progress was reported in the sphere of locomotive driver brakes.[236] Nearly half of the twenty-five thousand locomotives running in the United States had either steam or air brakes. While the driver-brake installation appears to have been rather evenly spread around the country, less progress was evident in the area of car equipage. The big eastern trunk lines may not have faced the gradients of the Rockies, but they did have far greater traffic densities and so needed power-braking as much as their western cousins. In 1883 Westinghouse offered a 20 percent discount on his freight car brake equipment, which was enough to encourage the western lines but not enough to generate any large sales east of the Mississippi. The Pennsylvania Railroad bought just enough sets to outfit its cattle cars, but no more.

While the eastern lines deferred, other forces were at work to push the industry into a wholesale acceptance of pneumatic braking. The annual growth in tonnage was a clear reminder that freight train speeds must be increased to handle the traffic on existing trackage. Increased speed called for an increase in braking efficiency; with—or in some cases even without—such improvement, railroads were generally increasing freight speeds from 10 and 15 to 20 and 25 miles per hour.[237] The other remedy commonly urged for the traffic problem was bigger cars. In 1884 it was said that one thousand 25-ton cars could move the same goods as seventeen hundred 15-ton cars.[238] Bigger cars would mean fewer cars and so a reduction in the number of expensive air-brake sets needed to move the same traffic.

As important as these arguments, perhaps even more so, was the emergence of Godfrey W. Rhodes, superintendent of motive power for the Burlington Lines, as a champion of the freight air brake. Just how Rhodes became so impassioned on the subject is unclear; his main responsibility was for the maintenance of locomotives at the Burlington's

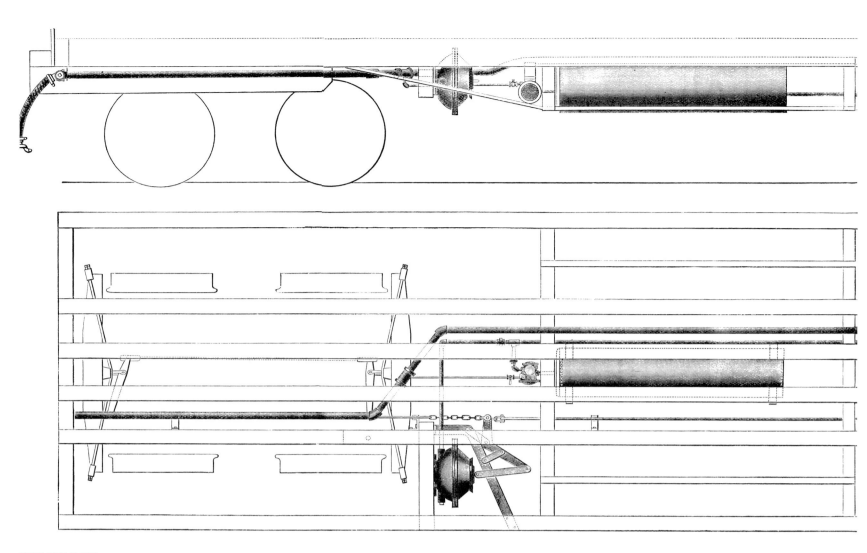

FIGURE 7.103

The Eames vacuum brake was a powerful competitor to the air brake in passenger car service but proved a failure for freight service. The two dish-shaped objects on either side of the frame are the vacuum cylinders. (*MCB Report*, 1887)

main shops in Aurora, Illinois. As a railroad mechanical man, he understood that the chief benefit of power brakes was that they put the train fully under the control of one man, the engineer. The slowing down or speeding up of the train's progress could be instantly effected by a little wrist action. Better time could be made over the road because of quicker stops, and accidents could be avoided or at least mitigated.

These matters had been discussed at the annual meetings of the Master Car Builders Association, which had maintained a committee on the subject of power brakes for a dozen years or more. But it talked in circles, and nothing of substance was ever effected. Rhodes asked to chair the committee, for he felt that the time was ripe to push freight braking technology. Rather than continue the aimless talking, Rhodes directed his committee to organize a field test near Burlington in December 1885. Naturally the Burlington Lines agreed to play host and made its local shop facilities and personnel available to the participants. It was more than an industry trade show or private showing of esoteric hardware; Rhodes intended to conduct a rigorous series of ten standard plus several special tests for every system of brake apparatus presented. The trials were very much like those held in Buffalo for couplers except that few contestants would appear.

As it happened, the trials could not be organized in time for the December schedule—perhaps just as well, because winter weather on the plains was not conducive to outdoor events, although it would have posed an added test for the apparatus. The trials began in July 1886 and with the prospect of a $40 million sale, more vendors than Westinghouse made an appearance. The momentum-brake people—Rote, Card, Widdifield et al.—failed and quit the field early on, as has already been mentioned. The only serious competitor to Westinghouse at the 1886 Burlington trials was the Eames Vacuum Brake Company of Watertown, New York.[239] The vacuum brake was well regarded by a number of U.S. railroads and proved remarkably effective on short trains starting in the early 1870s. As such, it was well suited for passenger work. The busy New York elevated lines operated trains on very short headways with both the Empire and Eames vacuum-brake systems. The millions of passengers safely carried over these lines for thirty years attested to their effectiveness. Vacuum brakes were simpler than air brakes because

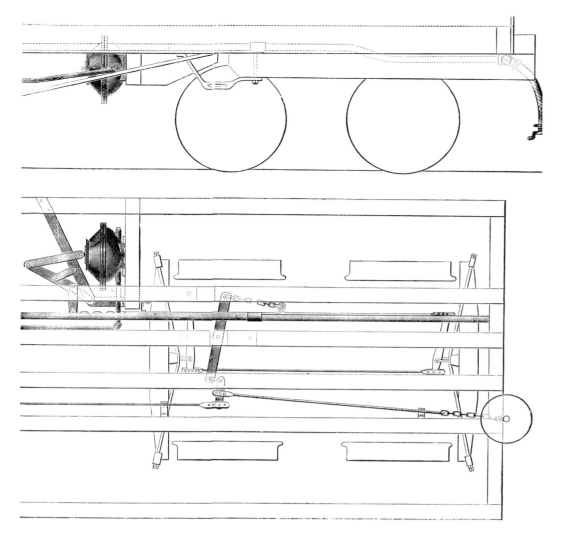

no expensive compressor was needed; hence a major objection to power braking was avoided. The vacuum was formed by a simple ejector mounted on the locomotive.

As early as 1872 George Glass, master mechanic for the Allegheny Valley Railroad, began to push the Smith vacuum brake for freight service.[240] He was impressed with the power of the Smith apparatus. It stopped one of his biggest engines with the throttle fully open. It also was capable of overcoming a major defect of the original Westinghouse brake—the total failure of the system should the train break in two. In the Smith system this was accomplished by placing a small rotary pump in the caboose that could, in an emergency situation, be powered by a hand crank or an axle drive. In a break-in-two, the conductor or rear brakeman could set the vacuum pump in motion and stop the runaway. Glass's enthusiastic report hardly engendered a stampede to Eames's Watertown factory. In fact, there appears to have been no large-scale purchases of vacuum-brake apparatus for freight cars, at least not until around 1880, when the Denver, South Park & Pacific outfitted seven hundred cars with Eames equipment.[241] One year later the Union Pacific purchased three

hundred sets, also for freight cars, from the Eames company.

The Eames firm hoped to break into the big time by matching or even outperforming Westinghouse at the Burlington trials. It was hoped that the simplicity and low cost of the vacuum system would bring victory. Eames developed an inexpensive kit costing $25 per car that consisted of two deep dishlike cast-iron cylinders covered with thick rubber diaphragms (Fig. 7.103). Each cylinder was about 15 inches in diameter. A vacuum reservoir tank, 15¾ inches in diameter by 70 inches long, was placed under the floor. Pipes and hoses connected it to the steam ejector on the locomotive. Steam passing through a venturi pipe inside the ejector evacuated the system. The atmospheric pressure would then cause the diaphragms to collapse inward into the dishlike cylinder. It was this motion that worked the brakes. A rod connected to an eye in the center of the diaphragm was attached to the foundation gear. Eames fashioned a rather complex floating fulcrum that was intended to distribute the power of the system to the weight of the car and would, when in working order, meet the varying power needs for loaded and empty cars. For all its preparations and optimism, the Eames firm failed to perform well during the trials. The vacuum brake was simply too weak for long trains, and one of the major goals of the industry was to operate freight trains of fifty cars and over.

While Eames's apparatus failed because of too little power, Westinghouse flunked for exactly the opposite reason. Actually, it was more a matter of uncontrollable power. The front of the train reacted calmly to the working of the air-brake valve, but the rear cars jumped and bucked so violently that some of them jumped off the track, couplers broke apart, wheels locked into a skid, and in a few cases the heavy truck king pins were sheared in half. Rhodes and some of his fellow judges riding at the back of the train were thrown the length of the car during one of these violent stops. The power to stop was obviously available, but unless it could be harnessed to make smooth, even stops, it would remain a worthless force much like that of a tornado.

Because no workable system was uncovered, Rhodes scheduled a second test in Burlington for May 1887.[242] Except for Card, the momentum-brake group had abandoned the field. However, Westinghouse was confronted by two new air-brake contestants: W. W. Hanscom of San Francisco and J. F. Carpenter of Germany. Eames returned but could show no greater muscle than the year before. Even at slow speeds the air brakes bested Eames by a ratio of 2 to 1. It was now a battle of the air brakes. Westinghouse was again embarrassed by rough stops. The problem was blamed on the inability to set all the brakes on at one time. The brakes came on serially, starting at the loco-

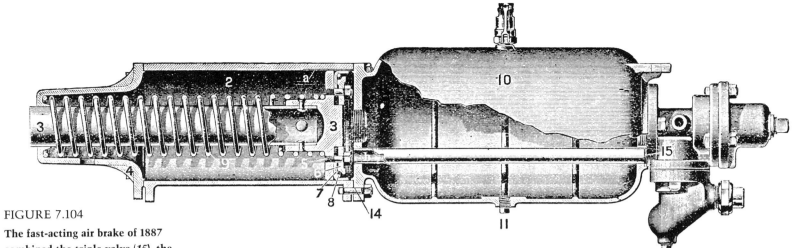

FIGURE 7.104

The fast-acting air brake of 1887 combined the triple valve (*15*), the air reservoir (*10*), and the cylinder (*9*) in a single bolt-together unit. The single-acting cylinder had a bore of 8 inches. (*MCB Report*, 1887)

FIGURE 7.105

The interior mechanism of the quick-acting triple valve is shown in this cutaway drawing. (*Engineer*, March 9, 1888)

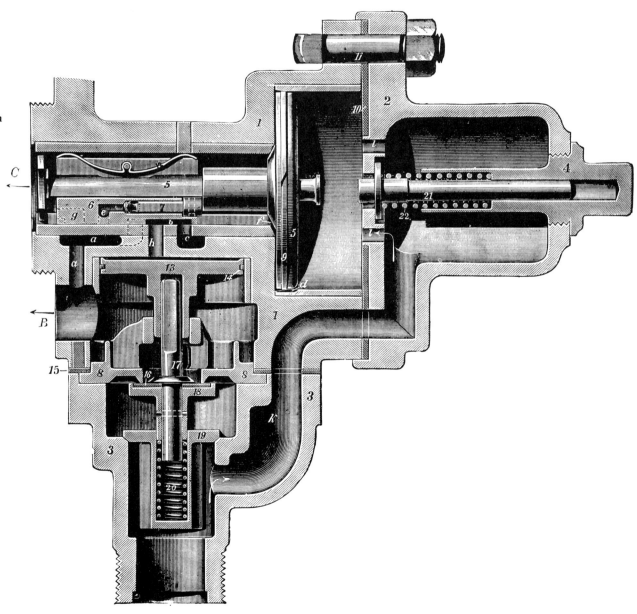

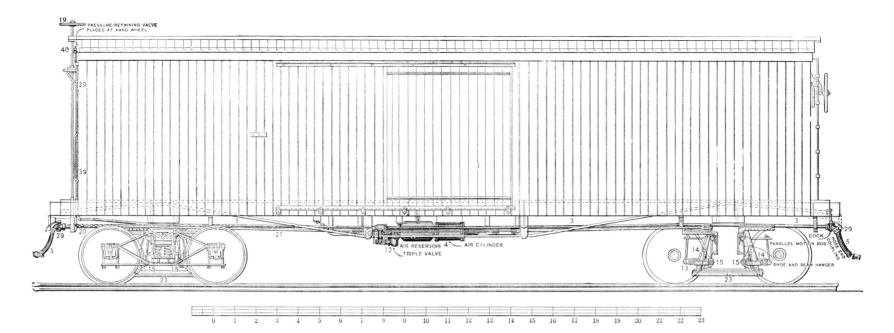

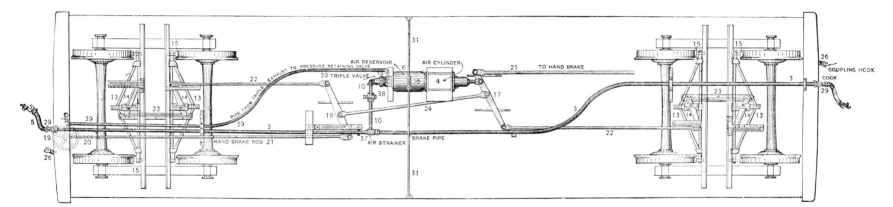

FIGURE 7.106

The general arrangement of the quick-acting brake is shown in this engraving. (Copied from M. N. Forney, *Catechism of the Locomotive* [New York, 1875; rpt. 1890])

motive and working toward the rear of the train. It took twenty seconds, which meant that the front of the train was nearly stopped before the rear brake even began to take hold. Carpenter seemed to have found a solution with his electric control.[243] The electro-pneumatic system had every brake on the train working within two seconds. The Eames electric control worked less well and was prone to burn out at its connections or otherwise be out of order.

The second trials ended inconclusively, with no single plan of power brake that really met the needs of the industry. Westinghouse went away only half-convinced that he must adopt electric controls if he wanted a brake that would work for fifty cars or more. This halfhearted conviction was apparently short-lived, for within a few weeks after leaving Iowa he announced a new nonelectric design that was purely pneumatic.[244] Brakes on the rear cars were operable within six seconds—not as good as the electro-pneumatic, but a grand improvement over the standard apparatus. By late June Westinghouse was back in Burlington and ran the course once again. This time it looked like he had solved the problem; the test gear worked su-

perbly with forty-five cars, but when five more were cut in, the old problems began to reappear. The jumps and crashes were not as severe as before, but the slideometer jumped by 30 inches during one test. It was back to the drawing board.

By the end of the summer of 1887, Westinghouse had worked out the defects and could confidently market a freight train brake that was both smooth and powerful. The changes to the basic design of the standard air brake were relatively modest. The train line was enlarged by ¼ inch, and the triple valve was redesigned. These and other minor changes speeded up the action of the system to two and a half seconds. For this reason the new style of brake was named the quick-acting brake. As in earlier designs of freight apparatus, the quick-acting was laid out as a bolt-together unit made up of cast-iron parts. The single-acting cylinder was on the left end (Fig. 7.104). A coil spring pushed the 8-inch-diameter piston back to the release position once the 50-pounds-per-square-inch air was vented. The piston was sealed by a large leather ring. The air reserve tank was in the center, and the triple valve was fastened to its right end (Figs. 7.105, 7.106).

A demonstrator train toured the nation to show the quick-acting brake in the field. Its technical perfection was unquestioned, but the industry continued its policy of reluctant acceptance. By October 1888 upwards of nineteen thousand sets had been sold and eight thousand more were on order, but this total represented only a small fraction of the freight cars in service.[245] In June of the following year 84,795 freight cars, or about 8 percent of the entire fleet, had air brakes. By 1891 the total had reached nearly 20 percent; however, the breakdown by railroad was very erratic.[246] The western roads made the best showing. Among the big trunk lines, the Northern Pacific had 75 percent of its freight cars equipped with air brakes, and the Santa Fe had 84 percent. The spunky little Colorado Midland was completely converted, as well it might be, considering its mountainous route. Curiously, the nearly level Atlantic Coast Line had 47 percent of its cars air-braked. Even more curious, that champion of air brakes, the Burlington Lines, had progressed no further than the 13 percent level. The eastern trunk lines made a poor showing: the Pennsylvania at 7 percent (Lines East), the B&O at 2 percent, and the Erie at 3 percent. Some of these lines could plead poverty, but such an excuse was hardly valid for the affluent Standard Railroad of the World.

With such an uneven spread of air-braked cars, there was no way to run solid trains of fully braked cars. It became the practice, therefore, to bunch the air-braked cars at the head of the train. This worked well enough for through trains but caused delays and considerable inconvenience on local freights, where cars were routinely picked up and set out. One other provisional practice followed on some roads during the conversion period was the mounting of dummy air lines on unconverted cars. A straight pipe with air hoses at each end was fastened to the cars' underframes.

The railroad industry showed no great enthusiasm for adopting air brakes for freight cars. Clearly, it would be a very long process if allowed to go on as a voluntary program. The railroads seemed ready to buy air brakes only for new cars. The pace began to quicken materially only after the Safety Appliance Act was passed by the U.S. Congress in 1893. The story of this major piece of federal safety legislation has already been outlined in the section dealing with couplers. But even the pressure of a national law could not hurry the railroads. Five more years passed, and the conversion was still not completed. It was another two years before air brakes were at last standard equipment and the American freight car crept reluctantly into the twentieth century.

EARLY IRON AND STEEL CARS

American railroading during the nineteenth century was an age of iron men and wooden cars. This assessment is true in general but not absolutely so, and those historians who contend that metal freight cars did not appear until 1897 are vastly oversimplifying the succession of the iron and steel cars over their wooden brethren. The story is as complex as Vatican politics and just about as notable for its false starts, dead ends, and misdirections. Just why did it take so long to convince the industry that metal was superior to wood for car construction? The superior strength and durability of metal make it the obvious choice, and yet a prolonged and often vitriolic debate ensued over the comparative strength, safety, cost, dead weight, convenience, and ease of repairs of the two competing substances.

The Pros and Cons of Metal Construction

In the one camp were the master car builders; in the other, the inventors and other advocates of the iron car. The builders were on the inside. They designed, built, or purchased the vast majority of freight cars used in the United States. They were in control, and they clearly sided with wooden construction. The old-line master car builders were almost to the man former carpenters, cabinetmakers, house builders, or sawmill operators. A reading of their obituaries in the reports of the Master Car Builders Association or in the trade journals of the period indicates that they were nearly all woodworkers. This surely was a major factor in their loyalty to wooden construction. It was familiar and tested. An attack on timber structures was in effect an attack on the conventional wisdom of the profession. The masters were comfortable with the sweet smell of sawdust and wood shavings. The scream of the power saw and the roar of the planer were their music. They were happy to have ironwork produced in the nearby foundry or smithy by tradesmen skilled in the art of metalworking.

The conservative stand of the car builders was based on more than a sentimental attachment to a traditional material. Their first obligation was to provide a supply of cheap, serviceable cars for the carriage of goods and passengers. Railroads are in the transportation business. They must have rugged, dependable equipment to move this traffic. Above all else, it must perform reliably and predictably in all seasons and under all operating conditions. A small mechanical failure could tie up the railroad, particularly the single-track lines so typical of nineteenth-century America, causing considerable annoyance, expense, and even danger. Railroads were not a testing laboratory but a workaday service industry. The men who operated them, accordingly, were not likely to be visionary or experimental in nature. Even if the presiding master car builder was given to an occasional innovation, his experiments were tempered by the trainmaster's complaints to the operating vice president that he was not being furnished with reliable equipment to move the traffic. No traffic, no revenue. Again, the builder was under great pressure to stand by only the tried and true.

Cost was another very real factor in car design and fabrication. Railroads had just so much capital to invest. They needed a given number of cars to move shipments, but they might not have funds to buy the very best cars available. Short-term solutions were often the only choice. Long-life, high-quality equipment would pay for itself in the long run, but many railroads were short of cash, and so of necessity they bought instead the cheapest wooden car available. Low cost also usually favored standard, off-the-shelf designs over more advanced plans. And so even the most open-minded car builder might be discouraged from testing iron cars because of an inadequate budget.

Of course, builders broke ranks and built and tested iron and steel freight cars. Some actually showed enthusiasm for this heretical form of construction. Beginning in the 1840s the Baltimore & Ohio and the Reading both began buying or building rather sizable fleets of iron-body coal cars. Before 1890, some seven years before the age of the metal freight car supposedly opened, we can account for over twenty thousand iron and steel freight cars in service in the United States.[1] This total suggests a path very different from that usually followed by a tradition-bound industry. The more normal course for radical innovation is early suggestion followed by rejection, then a few widely spaced test units, and then long years—often generations—later, a final grudging acceptance of the idea. The introduction of metal passenger cars followed this scenario. But the story of the metal freight car is less predictable. For certain tasks, such as the carriage of coal, iron-body cars were found satisfactory and were built in large numbers by a few railroads. But the practical success of these cars did not really influence the thinking of most builders.

They were talked about, to be sure, even within the pages of the Master Car Builders Association's reports, but what was good for the bituminous coal trade was not necessarily right for the Iowa grain harvest. And so while we can talk about over twenty thousand iron cars in the wooden era, they in fact constituted only a small portion of the overall freight car fleet. By 1890, for example, there were over 1 million freight cars in service in the United States; even if all twenty thousand plus iron cars built between 1840 and 1890 were running, they would represent a very small minority of the total.

Ironically, the friends of the iron freight car—those who pushed with greatest vigor and enthusiasm from outside the industry—were in effect its worst enemy. Car builders developed an early disdain for inventors' patented nostrums for the ills of car design. Radical solutions and extravagant claims did much to disenchant the practical mechanics in charge of America's rolling stock. The overblown talk of the advocates was found particularly offensive. Actually, they had a few solid points to make, but in their zeal to sell a new idea they resorted to flackery. In 1877 the *Railroad Gazette* reported on this weariness with unsupported assertions, with sanguine hopes, and most particularly with "talk" —"talk that . . . has seldom reached the dignity of a careful investigation of the subject, but has very generally been only an easy-going, gossipy kind of conversational allusion to it."[2] The talk dwelt on facts the car builders were almost always able to refute by practical example.

Fact: Iron and steel are stronger and more durable than wood. The strength of steel is rated at about 16,000 pounds per square inch for construction purposes, while yellow pine, a popular car-frame timber, was rated at only about 1,500 pounds per square inch. However, this benefit is offset by the weight of the materials; steel weighs about 490 pounds per cubic foot, yellow pine only 45 pounds.[3] Iron was said to be thirteen times stronger than the

best white oak. As wood dries, it becomes more brittle and hence weaker. Then again, iron also fatigues and weakens with age. The rotting of wood is equivalent to the rusting of iron—another balancing of disadvantages. Wood shrinks with age and weakens the structure because of loosened mortise joints, but then again, rivets and bolts tend to loosen with the vibration associated with freight-car service.

Metal cars, it was argued, would outlast wooden cars by many years. A wooden freighter had an average service life of about sixteen years. Most experienced car builders would have gone along with that figure; the interchange rules of the Master Car Builders Association confirmed it when it came to compensation for a wreck or lost car. Yet not everyone was ready to concede even this point to the friends of iron. John Kirby (1823–1915), longtime master car builder for the Michigan Southern & Northern Indiana, cited one hundred wooden cars purchased in 1856. After more than twenty years of hard service, they were ready for more use.[4] They had received roof, journal-bearing, and other routine repairs, but the structures were sound despite frequent overloading. Kirby claimed that there were hundreds of other veteran wooden cars operating well beyond their supposed life expectancy, and he felt that twenty years was a more realistic estimate. But even granting this estimate, a well-made iron or steel car could turn in between thirty and forty years of service. Long service is not necessarily an enormous advantage, however, because the car is often obsolete in design after only ten or fifteen years.

Fact: Metal cars can carry a greater payload than those of wooden construction. Here was a very basic issue indeed. It was one of the greatest challenges faced by the car designer—how to produce the greatest strength with the least dead weight. Operating costs were based directly on the weight of the train. The lighter the cars, the greater the profit realized, because the energy was employed more in moving the goods than in moving the weight of the train itself. This would mean not only fuel saving but the possibility of fewer and lighter locomotives, lighter bridges and track structure, faster schedules, and perhaps even fewer trains. And so the potential savings realized from cars with a greater capacity-to-weight ratio, or dead-weight ratio, was extremely important. The problem, of course, was not to sacrifice strength unduly to save weight. The car must be durable enough to handle its cargo safely. Metal advocates contended that because their material was so much stronger, they could produce equipment with a better dead-weight ratio—a logical enough argument until one considers the greater weight of the materials being compared.

There was actually a real advantage here when it came to comparing steel with wood, but not as great a difference as the earlier promoters of iron cars would claim. The inventor S. J. Seely of Brooklyn stated in 1862 that his patented iron car would weigh only half as much as an equivalent wooden vehicle.[5] But when sample cars were constructed, the results usually fell far short of the advertising. A report published sixteen years later regarding another iron car noted rather pointedly that while the inventor had promised a weight of 8 to 10 tons, the sample as delivered registered 12.75 tons.[6] When American lines were chided for not following Germany's progressive example in substituting iron for wood, the *Railroad Gazette* calculated the dead weights involved and offered the surprising report that the German cars had a dismal dead-weight ratio of 1.00 to 1.05, while the supposedly obsolete American wooden car had a ratio of 2 to 1.[7] This comparison was not completely fair, because the small four-wheel vehicles favored in Europe did not compare favorably with large double-truck cars, even when built of wood. However, as freight car size continued to grow, the dead-weight ratio began to favor metal construction because wooden floor sills grew out of proportion in larger and longer cars.

Fact: Metal cars are cheaper than wooden cars. This was an easy one to knock down. Certainly in first cost it was untrue. But it became less untrue as time went on; wood was becoming scarcer and dearer, iron and steel more plentiful and cheap. This is not to say that forests were totally depleted—in gross footage there was no decline in timber production—but what did begin to disappear along with the virgin forests was prime framing timber.[8] As floor sills went up in price, car builders were forced to consider alternatives. Until 1880 the United States imported the majority of its rails from Great Britain, even though a healthy and growing domestic iron industry had already been established. In that year over 3 million tons of rolled-iron products were produced in the United States.[9] Tonnage almost doubled during the next decade. Even in the 1860s there was an abundance of iron plate and rolled shapes to build railroad cars, but the cost difference was apparently too great to encourage such a movement.

Cost of materials, then, was a very real factor in deciding the form of construction to follow. Prices are not available so far as I can uncover for angles and plate in this early period, but it is probably safe to assume that the costs of comparable wooden and iron beams would be at great variance. Structural iron elements were, however, being used on a limited basis in France as early as 1786.[10] The practice spread to Britain and the United States in the 1830s. Fabricated beams made up from plate and angles could produce substantial iron members. Late in the following decade solid rolled I-beams appeared. The perfect material for car frames had been commercially available for nearly half a century before it saw any extended employment for that purpose.

Unfortunately, there appears to be no solid cost data for metal cars in the nineteenth century, but by the turn of the century some figures do appear. In 1908 the Master Car Builders Association valued a new wood-framed 40-foot boxcar at $440, less the trucks. A steel-framed boxcar was figured at $740, less the trucks. Note the comparison is not of equals, for the wooden car would likely have been no more than 40 tons and would have been fully depreciated in sixteen and a half years, while the steel car could be counted on for twenty-five years of service. Another way to illustrate the difference between wood and metal construction is to compare trucks. Those with wooden bolsters cost $215 per pair, those with metal bolsters commanded $315, and all-metal trucks for a 50-ton-capacity car cost $425. In February 1910 *Railroad Magazine* reported that steel-framed boxcars cost between $1,100 and $1,400, while an all-steel hopper car went for $2,000. These prices are for complete cars.

Costs involved other factors than just the outlay for materials. What about repairs and maintenance? Car builders voiced several concerns. How would you go about restoring an iron car after it had wrapped itself around a bridge pier? Accidents were common enough. The shops were ready to rebuild a wooden car, but they had no tools or experience when it came to iron bodies. If a car frame was bent from overloading, how could it ever be straightened out again? Just maintaining the painted surface would be a big enough headache. Freight cars are knocked about and treated roughly by the trainmen and laborers who load and move them. Without a good protective coating they would rust away rapidly. Electric and gas welding were not available until the twentieth century, and so holes must be patched with riveted plates— again labor-intensive and hence costly. And what about disposing of old cars? Wooden ones were set on fire; the iron fittings could be salvaged from the ashes. But dismantling an iron car by hand—a necessity before the days of the acetylene torch— cost more than the value of the salvaged iron. It would be cheaper to bury it in a landfill or swamp.

The cost of retooling the repair shops and retraining the men was a serious consideration as well. Tools were certainly available. Punches, shears, rolls, bulldozers, and press breaks were common among ship and locomotive builders. But they represented no small investment when railroad managers began to consider refitting all the division car repair shops on a given railroad. Most major railroads were divided into divisions 100 to 150 miles in length, meaning that upwards of two thousand car shops would require new tools. That alone was a good reason to stay with wooden cars for just a few more years.

FIGURE 8.1

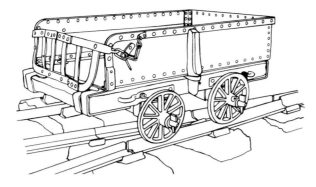

The builders' impatience with the metal-car advocates seems reasonable, at least before 1890. The talk of ultra-lightweight fire- and rotproof cars that could outcarry and outlast wooden cars was based in large part on false reasoning. Most of these early schemes were laid to rest in the cemetery of brokenhearted inventors. It was not a lack of original thinking that prevented the adoption of a basically sound technical idea. The metal freight car proved itself a splendid vehicle for commerce, but only after a lengthy development period. It was evolutionary, almost Darwinian in the sense of a stronger species overtaking a weaker one by natural selection. It came about not by inspiration or patents but through the plodding work and study of ordinary engineers who created a practical package for everyday use.

In some respects the metal car was nudged into favor by the very men who at first opposed it. Iron came to play an increasing role in wood-car design; by the 1890s composite construction was commonplace. Iron rods and plates had been used to strengthen car bodies and trucks since the earliest days of railroading. Ross Winans's patent of 1834 shows cast-iron body and truck bolsters. Wrought-iron body bolsters for freight cars were adopted in 1852 by the Illinois Central. Within another twenty years they had become rather common. Truss rods intended to strengthen the floor frame became common after 1870. Continuous draft gears—long rods that ran the entire length of the car and had a coupler at each end—removed the pulling load from the wooden underframe. But continuous drawbars were far from universal, and most repair shops were forever replacing draft timbers. One reporter commented on a big pile of broken and split draft timbers he saw at a western repair yard in 1890.[11] Almost none of it was rotten—it was sound timber. But bolts had pulled through holes, splitting the large wooden sills. The repair was a major one, meaning that the car was out of service for days. In 1896 two major railroads experimented with 10-inch steel channels bolted to center sills to strengthen the draft-gear connection.[12] The idea is so obvious that other lines must have tried the same expedient. And so steel came into freight car construction by both the front and back doors, one element at a time.

Economic changes are probably even more important than the working out of engineering details. By the 1890s railroad traffic had grown so huge that larger cars offered one solution for carriage over existing lines. Wooden cars were reaching a practical size limit. Cheaper steel coupled with the scarcity of framing timber had an effect on the decision to go with the metal car. The steel industry was looking for new markets and so encouraged the trend. Metalworking tools had become not only better but more common by this time. Pneumatic tools, particularly hand-held riveting guns, speeded car fabrication. Perhaps the story is best explained by the statement that the metal freight car finally triumphed simply because its time had come.

Early Iron Cars

The origins of the iron freight car are surprisingly early. Such cars were projected and actually built when steam-powered, public railways were still only a vision. By some miracle, a relic from the remote beginning of the iron-car era has survived.

OLD WORLD PROTOTYPES

A little iron tramway wagon stands near the center of the exhibit floor at the National Railway Museum in York, England (Fig. 8.1). Visitors pass it with hardly a glance. They are intent on seeing the Royal railway carriages and gaily painted steam locomotives displayed elsewhere in the gallery. And yet the unglamorous little car may be one of the most important transport specimens in the York collection. It may also be the oldest surviving artifact directly associated with two major innovations in freight-handling equipment. Its detachable body links it to containerization, now the apogee of modern shipping. The iron body itself is a precursor to the giant steel freight cars now so essential to the efficient movement of bulk cargoes.

So little solid information about the car exists that it is difficult to make any concrete statements with assurance, but even if the data could be found, the cleverest interpretive label would probably cause only a few visitors to stop and marvel over this weathered four-wheel vehicle. It is not prepossessing in size or appearance and looks rather like a badly designed contractor's dump car. The low-sided metal body rests on a wooden frame and has an estimated capacity of 2 to 3 tons. The car measures about 6 feet long by 5 feet wide by 3 feet tall. The wheels are flangeless, intended for the 50½-inch-gauge Peak Forest Tramway in the Manchester-Sheffield area.[13]

The horse-powered line opened in 1799 to carry lime and limestone. The car itself has been dated 1797, but its actual age is uncertain. Because the tramway operated as late as 1926, the preserved car could actually date from a far later period. There is, however, contemporary evidence for iron cars in England that date not long after the Peak Forest line began operations. Iron-frame bogie cars were used on the Pennydarren Tramway in South Wales in 1800. Examples are preserved by the Baltimore & Ohio Museum in Baltimore. The Oystermouth Railway, another British line, began operations in March 1807 with a passenger car constructed chiefly of iron.[14]

But more germane to the present discussion were the iron-bodied cars observed at work on the Stockton & Darlington Railway in 1826 and 1827 by two German engineers.[15] These cars were said to be very similar to the standard coal wagon, which measured 8 feet long over the frame and was carried on four 30-inch-diameter wheels. The body measured 6 feet 4 inches long by 4 feet 8 inches wide by 3 feet 4 inches high. The capacity was given as 53 hundredweight, or just under 3 tons. The metal-bodied cars were used for lime

FIGURE 8.2

The Stockton & Darlington Railway had some iron-bodied lime cars in service in the mid-1820s. (*Dictionnaire Technologique Chemin de Fer* [Paris, 1835]; metric measurements)

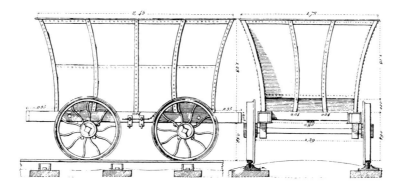
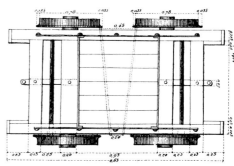

FIGURE 8.3

A Belgian boxcar of about 1870—a neatly executed design. (M. Ch. Couche, *Voie, Matériel Roulant et Exploitation Technique des Chemins de Fer* [Paris, 1873])

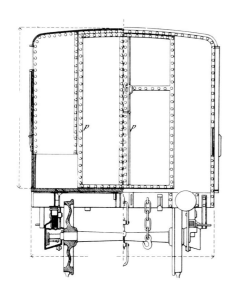
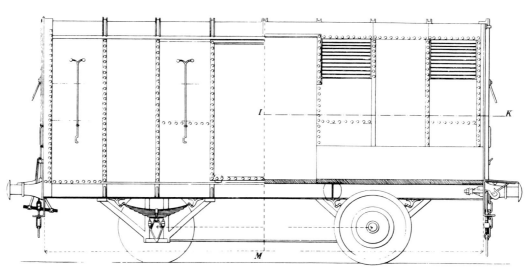
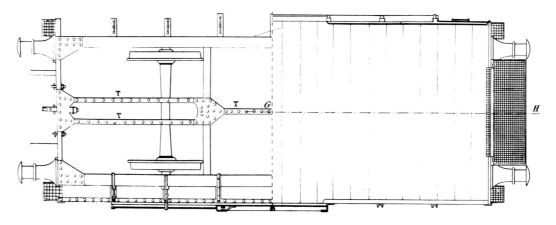

EARLY IRON AND STEEL CARS

FIGURE 8.4

Panama Railroad boxcar, c. 1860,
made by Harlan & Hollingsworth
of Wilmington, Del. Note the
iron frame. (Smithsonian
Neg. 81-8124)

FIGURE 8.5

The Portland Co. built this flatcar
for the Panama Railroad in 1866.
(Maine Historical Society,
Portland)

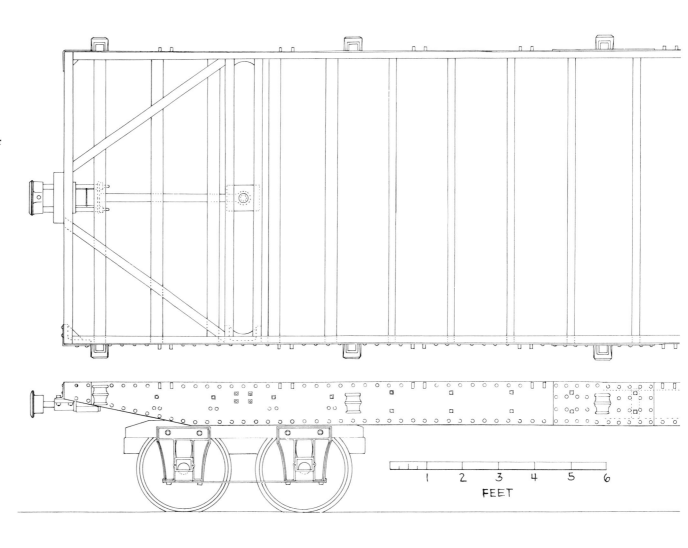

service and were so built because the lime was often so hot when loaded that it might have set fire to an all-wooden car. The German engineers do not say when the cars entered service, but they were in use at the time of their visit. The railway opened in 1825. Drawings of these cars were published in a French technical dictionary of 1835 and are reproduced here as Figure 8.2. Similar cars were also used on the Liverpool & Manchester Railway, which opened in 1830. One of these cars is shown attached behind the tender of the locomotive *Manchester* in a lithograph published in 1831.[16]

Just how widely these first-generation iron-body wooden-frame cars were used in Great Britain or elsewhere in Europe is not recorded in the literature available to me. But there is enough information to indicate that interest in developing such cars was continuous from that time forward. The Eastern Railway of France built an iron coal car in 1840. Within twenty years the railway was committed to composite iron and wood freight vehicles and in 1874 began to construct all-iron cars.[17] By 1895 it had over twenty thousand metal or metal-framed cars in service. The Belgian railways showed a similar enthusiasm.[18] They produced an all-iron boxcar in about 1845, probably the first built in the world. Within a few years they were committed to all-metal or composite construction

(Fig. 8.3). A few German lines followed the French and Belgians in the early 1860s and had over two thousand such cars in use by around 1867. Meanwhile, the British had not abandoned interest in the subject—drawings for both freight and passenger iron cars are given in D. K. Clark's *Railway Machinery* (London, 1855).

American commercial car shops seemed ready to capitalize on foreign interest in iron freight cars, especially for Latin American service. The Panama Railroad operated in a climate so unfavorable to wood that at least some of the early equipment was iron. The Portland Company of Maine produced some cars in the 1860s, as did Harlan & Hollingsworth of Wilmington, Delaware.[19] A surviving drawing for a boxcar produced by the Wilmington firm depicts an arch-roof vehicle with a body measuring 26 feet 6 inches long and 8 feet 6 inches high (Fig. 8.4). The design is notable for its metal floor frame. In 1866 the Portland Company built one or more iron flatcars for Panama (Fig. 8.5). The 30-foot-long cars measured 8 feet 6 inches in width. The main supports were the two deep iron plates (15 inches by ¼- or 5/16 inch thick) that served as the side sills. There were no secondary or intermediate sills. The heavy crosspieces in the frame were presumably wooden, as was the deck, which is not shown. The 40-inch-wheelbase trucks had side frames built up from plate and angle irons. The smallish wheels (30 inches) were used to lower the deck level. The coupler is of the two-bar variety with a cast-iron head. It is attached to the truck's king pin because there are no draft-gear timbers. We can offer two more examples of metal cars built by U.S. shops for export to Latin America: Billmeyer & Small of York, Pennsylvania, constructed an iron car for Cuba in 1877, and in 1890 the Middletown, Pennsylvania, plant fabricated four hundred metal cars for Brazil.[20]

IRON COAL CARS OF THE 1840S

What influence these overseas developments had on American car design is difficult to determine. Some happenstance exchange of ideas surely took place, but it was not until rather late in the nineteenth century that organizations like the International Railway Congress began to encourage the systematic study of railway engineering on an international basis. Whatever inspired interest in the subject, both the Baltimore & Ohio and the Philadelphia & Reading began to produce iron coal cars in the 1840s. The Reading started first, in 1843.[21] Within a year it had 856 small four-wheel iron-body wood-frame coal cars in operation. These were not all-metal cars; only the bodies and hardware were iron. No drawings from this early period exist, but visual material from the following decade has survived. A photograph in the collections of the George Eastman House, Rochester —taken around 1855 in the anthracite coal fields

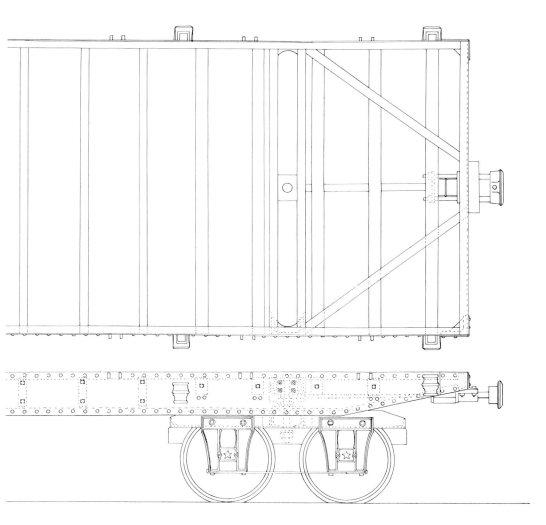

FIGURE 8.6

near Pottsville by the Philadelphia photographer F. Langenheim—shows several small, boxy open-top metal cars fitted with wheels outside the journals. Unfortunately, the image size is very small, but the vehicles are similar to the Reading car shown in Figure 8.6. Regrettably, the original drawing is torn and a critical piece containing the date is missing, but all other drawings in this group are from the 1850s and 1860s. The car in the drawing has 36-inch wheels, a 51-inch wheelbase, and a body measuring 10 feet 2 inches long by 4 feet 1 inch high by 6 feet wide. Sloping side walls and the ratchet wheel indicate a hopper-bottom opening for unloading. The long lever visible is for the primitive braking system. Cars of this type were said to weigh from 2.25 to 2.50 tons and to carry 5 tons, indicating a very good dead-weight ratio.[22] Wooden cars of the same weight were rated at only 4 tons, proving in at least this instance that the iron-car cranks were not always guilty of gross exaggeration when it came to reporting a superior weight-to-capacity ratio.

The officials of the Reading seemed satisfied with the composite car, for more were ordered; one estimate reports the line as having upwards of seven thousand in operation during their years of popularity.[23] Three eight-wheel iron-body coal hoppers were also produced by the Reading some-

time before 1868.[24] Until about 1870 iron coal cars outnumbered wooden ones, but by 1877 the road seems to have stopped building new ones because of high repair costs.[25] As early as 1865 complaints were being voiced about maintenance costs, and up to one-tenth of the fleet was out of use for servicing.[26] No details were offered on the exact nature of the defects, whether they involved problems with the iron bodies or with other components of the car, such as the running gear. And so this pioneer effort that started out with such promise wound down rather quickly. *Poor's Manual of Railroads* for 1888 reported that only 403 of the little iron four-wheelers were still listed on the active roster in 1887, and another 1,442 were in storage awaiting disposition. By 1896 only two were in service, hauling ashes from the Philadelphia engine terminal. The Reading had meanwhile adopted a wooden eight-wheel hopper car, but even it employed a very large quantity of iron and might be loosely termed a composite car because the sloping floor or hopper bottom was made of iron plate.[27] The floor area was sizable, measuring about 6 feet wide by 18 feet long. To be self-supporting, plates of this size would have to have been at least ¼ inch thick.

Just one year after the Reading began production, the Baltimore & Ohio began a program of

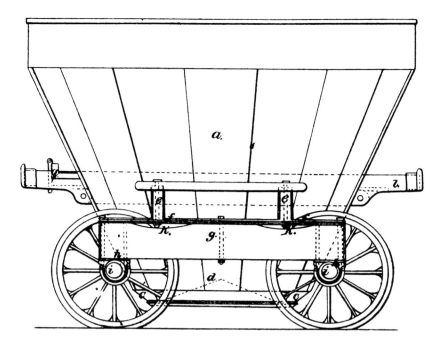

FIGURE 8.7

Patent drawing for a Baltimore & Ohio pot hopper. Only one car was made on this four-wheel plan. (R. Winans Patent, June 26, 1847)

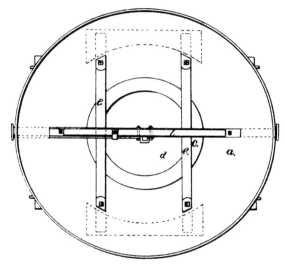

The bottom of the compartments was made cone-shaped with a hinged door or bottom plate so that the coal might be discharged by its own weight. The body was not beautiful; it looked like three vertical concrete mixers that had been welded together by a deranged blacksmith. But despite its awkward appearance, it was structurally rational and very strong.

About the only major design defect was the inaccessibility of the end hopper-bottom hatch doors. The latches were directly behind the truck side-pin bracket. Operating crews apparently managed to work around this inconvenient arrangement, for it was never changed. The trucks—which were really not trucks in the conventional sense—were as ingenious and original as the iron pot bodies. They had no bolster or center pin; rather, they were attached to the wooden main frame by pins fastened to a substantial cast-iron pedestal bracket (Fig. 8.9). Large rubber springs (coil springs were often used after about 1870) surrounded each pin, but in the drawing they are hidden from view by the supporting pedestals. The double-iron-bar side frames of the truck were able to swing a few inches to and fro on center pins, thus allowing the car to negotiate curves and switches. The axles remained parallel to each other because of the way the side frames were attached to the car. Note also how the journal boxes were attached to the side frame—by a single bolt. All the fits were made free and loose to add to the flexibility of the arrangement. At the very slow speeds—rarely more than 10 miles per hour—attempted by coal trains on the B&O during this period, these cars tracked very well. They were also said to be very agile on the rough yard trackage found around the coal mines of the Cumberland region.

The pot hopper is popularly credited to Ross Winans (1796–1877) of Baltimore. He was an ingenious and original mechanic without question, but one who has been wrongly credited with a number of important innovations, most notably the eight-wheel car.[28] There is also reason to question his claim to be sole originator of the pot hopper. Major confusion on this point stems from a patent issued to Winans on June 26, 1847 (No. 5,175), that does cover cars of this same type. Drawings are shown only for four- and six-wheel pot hoppers (Fig. 8.7). Actually, cars of this type appeared on the B&O three years before the patent was issued, and a contemporary memorandum dated May 1844 shows rough sketches and calculations for four-, six-, and eight-wheel cars.[29] This document specifically credits the six-wheel car to James Murray (1812–1895), then presiding master of machinery for the B&O and later an employee of Winans. Winans is credited with the eight-wheel design, but the first cars were built on Murray's plan. The memo calls for $\frac{1}{10}$- or $\frac{1}{8}$-inch iron plate—fine for saving weight, but surely too light

iron coal cars that would continue into the modern steel-car period. Once committed, this line never wavered from its faith in iron-body hoppers. They became a standard design that was modified only twice during the half-century these curious vehicles were in production. In appearance they were curious indeed, and, like so much else in the B&O equipment roster, wonderfully strange and original. The Grasshopper, Muddigger, and Camel locomotives—described as "alike peculiar"—had fitting workmates in the pot hoppers. These cars were built in four-, six-, and eight-wheel varieties. The eight-wheelers were the most numerous and will therefore represent the entire group for this general introductory description.

The main frame was wooden—basically two heavy timbers mounted outside an iron body composed of three intersecting cylinders. They overlapped so that only the end compartments were nearly a full circle; the center section was cut off materially at both ends by straight dividing plates.

EARLY IRON AND STEEL CARS

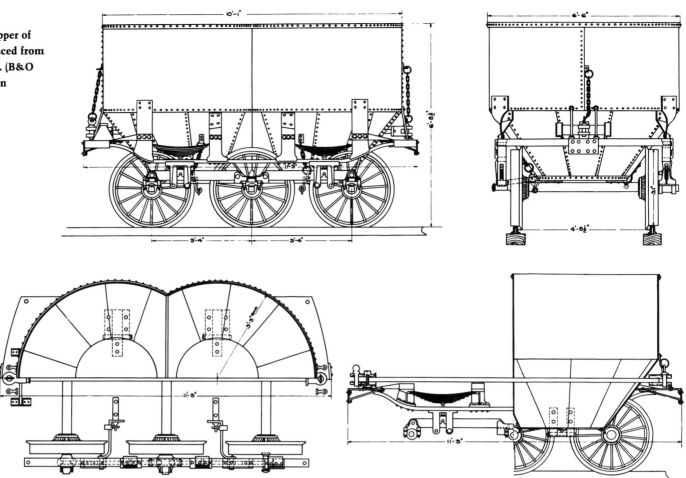

FIGURE 8.8

B&O six-wheel pot hopper of around 1845–1850. Traced from an old drawing in 1926. (B&O Photocopy, Smithsonian Neg. 64836)

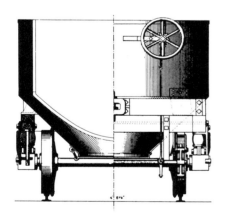

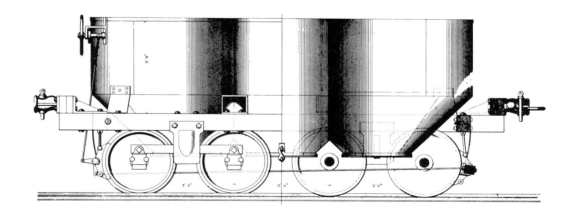

FIGURE 8.9

A German drawing—published in 1862 but based on data gathered in 1859—of a B&O 10-ton-capacity pot hopper. (A. Bendel, *Aufsatze Eisenbahnwesen in Nord-Amerika* [Berlin, 1862])

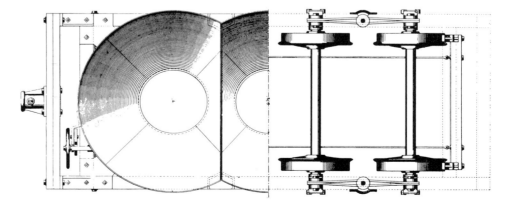

FIGURE 8.10

Pot hoppers were used by firms other than the B&O. This car was built in 1856 for the Lonaconing Coal Co. by Harlan & Hollingsworth. (Smithsonian Neg. 81-8127)

for a container meant to withstand the rigors of the coal trade. Late accounts give side-wall thickness as ¼ inch.

The first of Murray's six-wheelers was mentioned in the 1844 annual report of the railroad as an improved sheet-iron car of cylindrical form, which cost $340 and weighed only 2.5 tons yet offered the astonishing capacity of 7 tons. The next report mentioned that thirty iron cars were on order. The 1846 report listed 201 six-wheel iron hoppers in service. The number of six-wheelers peaked in 1855, when 254 were in service. An undated drawing of one of these curiosities was preserved by a draftsman, J. S. Bell, employed at the B&O main shops at Mt. Clare in the 1860s. A modern tracing was made by the railroad in 1926, reproduced here, on a much reduced scale, as Figure 8.8. Notice especially that the car is of all-iron construction and differs greatly from the wood-frame eight-wheel pot hopper described at the opening of this discussion. The Murray car was surely serviceable, but its capacity was limited. The western Maryland mines were shipping coal in ever-increasing amounts, and the B&O wanted a 10-ton hopper car. Winans's idea was adopted. An experimental eight-wheeler was produced in 1846, according to the annual report for that year. It weighed a sane 3.5 tons and carried 8.5 tons. Two more were produced in 1850. By the following year

seventy-three were on the property, and very soon the eight-wheelers would outnumber the now old-fashioned Murray cars. By 1855, according to the annual report, there were 774 eight-wheelers in use. The number of six-wheelers began to decline. Within ten years only ninety-six were running, and the eight-wheel fleet had grown to about eleven hundred. All of the Murray cars were retired over the next few years while the eight-wheel fleet continued to grow. There were never more than about twenty-five hundred pot hoppers in service at any one time on the B&O.

A very good picture of a fully developed 10-ton-capacity pot hopper is provided by a drawing in a German text on American railroads published in 1862 but based on data gathered three years earlier (Fig. 8.9).[30] The car measured 18 feet 4 inches long over the couplers. The pots were 7 feet in diameter and measured 63 inches high. The axles were on 42-inch centers. The drawing has two mistakes that should be noted—the continuous drawbar running the full length of the car between the couplers has been omitted for unknown reasons, and the engraver also erred in the end-view cross-section by making the plate too heavy.

If the B&O was pleased with the pot hoppers, so were many other lines. Most of these owners were in the same western Maryland coal fields served by the B&O. They included the Cumberland &

FIGURE 8.11

Several pot hoppers laden with Cumberland coal stand in the B&O yard at Martinsburg, W.Va. This photo dates from 1871. To the right stands a Winans Camel, which was capable of propelling forty-five pot hoppers over the mountainous line at 10 mph. (Smithsonian Neg. 48366)

Pennsylvania, the Georges Creek & Cumberland (later part of the C&P), and the West Virginia Central (a B&O subsidiary), as well as the following mining companies: the Allegheny, the Barton, and the Lonaconing (Fig. 8.10). Far to the south of the Cumberland mines, the Richmond & Danville became something of a convert to the pot hoppers. In 1849 the line ordered sixty six-wheelers from the Tredegar Iron Works in Richmond.[31] Of all the properties just mentioned, the Cumberland & Pennsylvania was probably the most devoted convert. In 1874 the C&P commented on the advantages of the pot hoppers, saying that they saved 7,500 pounds of dead weight over an equivalent wooden hopper car and cost $100 less to build.[32] They were easy to straighten out after wrecks. By the early 1880s the C&P had over 450 pot hoppers.[33] They continued in service on this line into the twentieth century.

During the 1870s and 1880s the B&O sought ways to increase the capacity of pot hoppers. The side sheets of some cars were raised from 12 to as much as 22 inches. The C&P followed suit, and two French engineers who prepared a study of American railroads at the time of the U.S. Centennial said that the capacity was raised from 8 to 15 tons.[34] In 1879 the B&O's mechanical chief, J. C. Davis (1818–1890), designed an enlarged pot hopper that weighed 7.3 tons and carried 13 tons.[35] The iron body was maintained at the old 14-foot length and 7-foot width, but the side walls were raised by 21 inches (Figs. 8.11, 8.12). A drawing dated June 1879 offers the following bill of material, less weights and parts for trucks:

Sheet iron, 2,575 pounds
Wrought iron, 1,321 pounds
Cast iron, 1,678 pounds
Rivets, 54 pounds
Bolts, 181 pounds
Nuts, 55 pounds
Gum springs, 87 pounds
Brass, 55 pounds
Lumber, 445 pounds
Total weight, 6,451 pounds

During this same period the B&O pressed pot hoppers into the grain trade.[36] Canvas covers were stretched over the open tops of the cars to protect the grain. Had Davis inadvertently created the ancestor of the covered hopper car that has become America's basic grain-carrying vehicle?

The pot hoppers must have proved themselves well to remain in favor through the reign of nearly half a dozen B&O master car builders. It is understandable that a mechanical anomaly might become the pet of an eccentric mechanical chief, particularly if it was his own invention. But the loyalty of such tough-minded mechanics as Thatcher Perkins, J. C. Davis, Jacob Shyrack, E. W. Grieves, Samuel Hayes, and Henry Tyson attests to their serviceability. In the spring of 1882 Loren Packard, then the presiding car chief, announced plans to build more pot hoppers. Even though the plan was antique, the cars performed well, and no cheaper substitute was available. New drawings were made for a super–pot hopper with a 20-ton capacity, or twice that of the old standard.

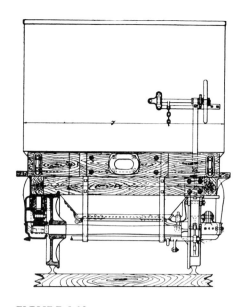

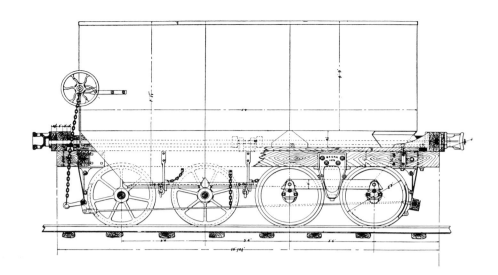

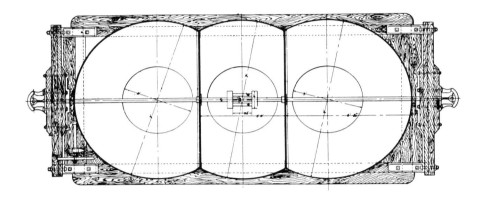

FIGURE 8.12

An enlarged pot-hopper design was developed in 1879 by J. C. Davis. Higher sides increased capacity to 13 tons. (*Locomotive Engineering*, April 1921)

FIGURE 8.13

Twenty-ton-capacity pot hoppers introduced in 1883 were the final development of this unique form of railroad freight car. (Smithsonian Neg. 49646)

EARLY IRON AND STEEL CARS

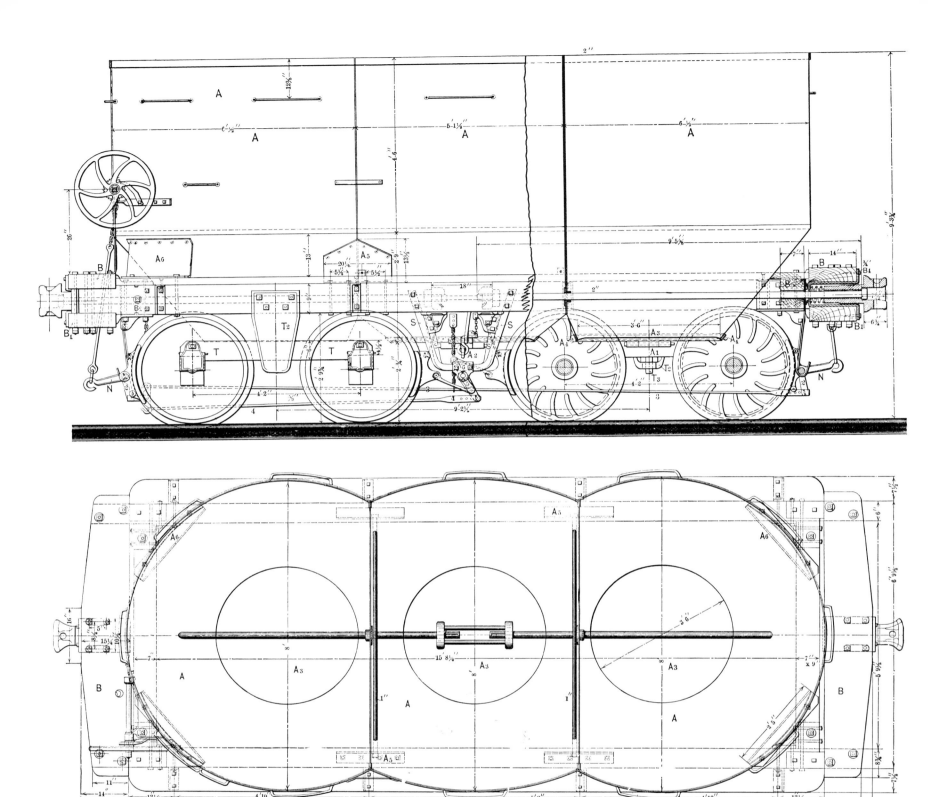

FIGURE 8.14
**Detailed mechanical drawing of
20-ton-capacity hoppers. (***National Car Builder***, April 1884)**

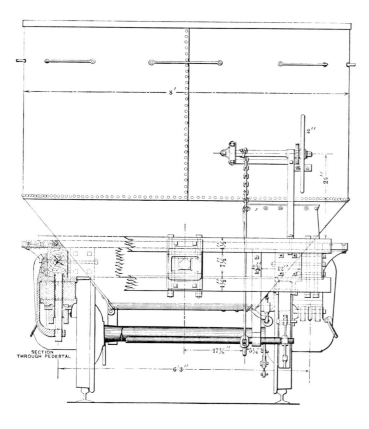

TABLE 8.1   Features of Baltimore & Ohio Class Q-1 Hopper

Length outside of Buffers		20 ft. 10½ in.	
" " End Sills		19 " 5 "	
" " Hopper		17 " 2½ "	
Width outside of Side Sills		6 " 9½ "	
" " Hopper		8 " 0 "	
" " Foot Boards		8 " ¼ "	
Diameter of Hopper Door		3 " 6 "	
Height from Rail to Top of Hopper		9 " 3¾ "	
" " " Angle		4 " 10 "	
" " " Door		2 " ¾ "	
" " " Top of Side Sill		3 " 7 "	
" " " Center of Buffer		3 " 2 "	

FINISHED SIZES OF TIMBER.

8 Side Sills, Georgia yellow pine	19 ft. × 6 × 9 in.
2 Foot Boards, "	17¼ " × 2 × 7½ "
2 End Sills, white oak	6½ " × 7 × 9 "
4 End Timbers, white oak	7¾ " × 4½ × 14 "
2 Brake Blocks, "	3½ × 6½ × 6½ "
4 End Coping, "	4 ft. 7 in. × 2 × 18 "
4 Side " "	2 ft. × 2 × 8 "
2 " " "	3 ft. 4½ in. × 2 × 8 "
4 Corner Pieces, "	2 " 3 " × 2 × 10 "
8 Dust Guards, poplar	8¾ × 6 × ½ "

IRON (IN PART.)

4 Pedestals, No. 26—cast	715 lbs
4 Saddles, " 30— "	180 "
8 Knees, Corners of Frame—cast	139 "
2 Draw-Bar Rods—wrought	210 "

Source: National Car Builder, April 1884, p. 46.

CONSTRUCTION.

The axles are M. C. B. standard, except center of axle, which is 4½ inches in diameter; the finished weight to average 390 lbs.

Spring casting is placed on lateral spring plates central between wheels, and held in position by one countersunk bolt; the spring resting on casting.

Pedestals are bolted to sill by four ¾-in. bolts 8½ in. long. Pedestal castings project under sill which rests on spring through which a 2-in. pin 20 in. long passes through pedestal spring, spring casting and bottom of pedestal, keyed at bottom of pedestals.

Cast-iron draw-bar lips project top and bottom 2 in. From lip to face of draw-bar 6¾ in. Draft-rod continuous with swivel in center and head on end at draw-bar. Draft-rod head 3¼ in. square, 2 in. thick. Turn-buckles 18 in. between cast ends, which are 2½ in. thick, the two rods which pass through castings 1¼ in. round riveted ends. Keys for draft-rods 3 × ⅜ in. thick, 4 in. at top.

Draw-bar springs 6 in. diameter, 2¼ in. hole, 2-coil round bar; capacity 16,000 lbs. with ⅛ motion lift. Material, best crucible steel. Pedestal springs of rubber, 10 in. diameter, 6 in. high, 2¼ in. hole, and to compress not more than ¾ in. under 16,000 lbs.

Side sills framed on end sills with double tenons. Foot-boards 2 × 7⅛ in. run on top of side sills between end timbers, secured to sills by four cast brackets each side.

Hopper pot of best No. 10 flange iron; sheets riveted with ¼ × ¾ rivets, each sheet to lap ⅝ in. from edge to center of hole. Hopper door 44½ in. diameter, 5⁄16 in. thick, hung to hopper by two hinges 3⅛ × ½, secured to hopper by 4 rivets through each hinge and sheet; door hinges 6 × ½ in. secured by 4 rivets ⅝ × 1¾ in. Hopper secured to frame by 4 side and 4 corner braces, with six ½ × 1½ in. rivets. Center brace secured to side sill by four ⅝ × 10½ bolts.

Top of hopper made of No. 11 refined iron sheets riveted together with 5⁄16 × ⅝ rivets, single lap; to have 12 long handles 18 in. long, and 2 short ones 10 in. long.

The wheels are 33-inch, double-plate, weighing not less than 565 lbs., and guaranteed for three years.

WEIGHT OF IRON.

Refined iron	1,739	lbs.
Flange "	1,810	"
Cast "	2,889½	"
Rivets	69	"
Nuts	60	"
Bolts	206¾	"
Washers	16½	"
Key bolts	18¼	"
Brasses	104	"
4 pair wheels	4,520	"
4 axles	1,560	"
Wrought iron	2,1?8	"
Nails	2	"
Draw-bar springs	33	"
Rubber springs	120	"

The first of the new class Q-1 hoppers was produced in September 1883 (Figs. 8.13, 8.14). Overall length was now 20 feet 1½ inches. The pots were 8 feet in diameter, or 1 foot greater than the older plans. The drawbars were dropped so as to be on center with the side sills—making for a stronger and more rational arrangement of the draft rigging. Old-fashioned rubber springs—great cylinders of gum rubber 10 inches in diameter—were readopted in place of the double coil spring used by Davis. Some of the leading mechanical features of the cars are given in Table 8.1.[37]

In May 1884 the Mt. Clare shops completed construction of three hundred Q-1 pot hoppers.[38] Another five hundred were produced by the Milton Car Works later the same year. It is possible that some were made as late as 1892; however, their end was very near at hand. In 1896 E. W. Grieves, addressing the Master Car Builders Association annual meeting, stated that the pot hoppers were the best iron coal cars ever built but that it was not practical to refit them for air brakes and Janney couplers as required by the federal Safety Appliance Act, and so they were being retired.[39] By late 1897 fifteen hundred had gone to the scrap line. Their end was also hastened by the large all-steel hopper cars then coming into service.[40] The two could not be operated on the same train successfully.

However, the pot hoppers survived for another decade on the Cumberland & Pennsylvania, which found the small 13-ton variety well suited for transferring coal to Chesapeake & Ohio canal boats for transhipment to Washington and points east.[41] In 1897, at the very time the B&O was retiring its old stock, the C&P was building new pot hoppers. A few must have lingered around for another decade or so, most likely in work train service, because the B&O produced three for its Centennial fair exhibition in 1927. The official catalog issued at the time of the exhibition mistakenly describes the cars as reproductions of an 1850 prototype. A recent inspection of the two surviving cars at the B&O Museum in Baltimore shows them to be genuine antiques, although it is possible that some of the wooden members were replaced in 1927. The size of the cars is too great to date from 1850, and the placement of the draft gear would make them B&O class Q-1. Accordingly, they cannot date earlier than 1883. But whatever their provenance, they are among the very few original American relics of the iron-car era.

IRON BOXCARS OF THE 1850S AND 1860S

The success of the iron hopper car on the Baltimore & Ohio would suggest the adoption of this form of construction for general merchandise traffic. This was done at least on an experimental basis. The Reading was almost purely a coal-hauling railroad; its experience with iron-body cars was not an entirely happy one, and so it made no further efforts in this direction. The industry in general agreed: the iron car was an interesting idea for someone else to study.

The B&O seemed immune to the prevailing fashion, however, and produced two cylindrical cars in 1846 that were said to look something like a modern tank car, except that they were far smaller and had access doors in the ends.[42] One was on four wheels and the other on eight. The smaller car weighed 2 tons and carried 2.5 tons, while the larger car weighed 6 tons and carried 8 tons. The eight-wheel car was at first listed for general merchandise traffic, but it was soon limited to the carriage of gunpowder like its small sister. No more cylinder cars were produced, but in 1852 Samuel J. Hayes (1816–1882), master mechanic for the B&O, began to produce iron boxcars along more conventional lines.[43] They were used for the carriage of gunpowder or barrels of flour. The railroad's annual report for 1855 lists forty-two in service. Some years later, in his new position with the Illinois Central, Hayes built a small group of swell-sided iron boxcars for the gunpowder trade. According to the *Railroad Gazette* of November 26, 1870, they had side walls of 14-gauge iron reinforced with angle iron. The floor was of 11-gauge plate, but it was covered with wood, presumably as a precaution against sparks that might ignite the powder. By 1872 the line had four Hayes gunpowder cars in service.

One of Hayes's successors at the B&O, Thatcher Perkins, reinstituted the fabrication of iron boxcars, and unlike those produced during the Hayes administration, the new series is fairly well documented, by both correspondence and a contemporaneous drawing formerly in the possession of L. W. Sagle of the B&O Public Relations Department (Fig. 8.16). In a letter to the road's president dated February 10, 1862, Perkins reported that two iron house cars were under construction at the Mt. Clare shops. They would cost $649 each. They could be produced for from $575 to $600 each in lots of one hundred, whereas a wooden boxcar cost from $450 to $475. Perkins felt that iron cars would be the better bargain in the long run and asked for permission to proceed. Permission was granted, and in February of the following year the last of the group was finished. The one hundred had cost $599.05 each, which was very close to the original estimate despite rising labor and material costs due to the war. Perkins contended that a contractor would ask $800 each.

Referring to Figure 8.16, we find a wooden frame, an arch roof, and an iron body of about the same configuration as an ordinary wooden boxcar of the period, except that the side walls bulged out slightly. The body measured 24 feet long, 8 feet 2 inches wide, and 7 feet 1 inch high. The side and roof sheets were ⅛ inch thick. The body side posts

and roof ribs or carlines were of 2- by 2-inch oak. The wooden frame was reinforced by a continuous drawbar and double iron truss rods at the bolsters. Curiously, Perkins, like the designers of the pot hoppers, overlooked the chief advantage of metal construction—a strong underframe that could both support a heavy payload and serve as a solid foundation for the couplers. The body might as well be wooden because it would serve for the life of a typical freight car. In the case of the hopper, there was more of an argument for a body material strong enough to withstand the constant abrasion of loading and unloading coal. But with boxcars intended for general freight, the body served only as a shelter. It would have made more sense to fix on composite construction, making the frame of metal and body of wood, but Perkins chose the opposite course.

Perkins was happy in his decision and produced more iron boxcars. By 1865 140 were on the roster, and production continued even after Perkins left his post at Mt. Clare, so that within a few years over three hundred class U boxcars were in service. These cars were intended for general freight, but they were not insulated and so became very hot in the summer, making them unsuitable for certain classes of merchandise.[44]

They were also said to sweat, which again damaged shipments as well as caused internal body rust. Ventilators were installed to promote air circulation, which relieved the sweating but

also admitted dust and sparks. A statement printed in 1872 said that the line continued to believe in iron boxcars and would extend the experiment, but it appears that the cars were basically a failure, for the line returned to wooden equipment.[45] The iron cars were gradually retired. Some ended up as office shanties or storage sheds. Two were resurrected in 1927 for exhibit at the Fair of the Iron Horse, which marked the B&O's one hundredth year. They are now in the B&O Museum collection with reproduction trucks of a later period (Fig. 8.17).

The Mt. Clare shops were not the lone center of iron-car design and construction in mid-nineteenth-century America. The action was widely scattered, very often with individual designers working on the same basic idea unaware that they were duplicating the efforts of their fellows. Some schemes were grandiose, such as the announced organization of a firm in Harlem, New York, in 1853 specifically for the manufacture of iron railroad cars.[46] It was claimed that many orders were in hand, and yet the plant never opened. Some schemes never went beyond the Patent Office. Richard Montgomery of New York City obtained a patent for a metal freight car on August 7, 1860 (No. 29,510), which was revised and reissued in 1862 and 1869. It used corrugated iron sheets and pressed-metal parts that bear a striking resemblance to modern Budd stainless steel passenger cars. He had an earlier patent, issued on May 17, 1853 (No. 9,738), for pressed-iron boilers, which ironically was used in later years to disallow patent applications for pressed-steel car ends.

Most early patents for iron freight cars remained entombed within the classical walls of the Patent Office, yet a few ideas were actually tested. The Pittsburgh, Fort Wayne & Chicago Railway exhibited a strong interest in iron passenger cars in 1860 and tested some boxcars during the same period.[47] In 1865 Charles S. Munn of Chicago was reported to be building similar test cars for the Michigan Central, the Chicago, Burlington & Quincy, and the American Express Company.[48] While technical details on most of these early efforts are paltry, we have a fine record of a very advanced design turned out in about 1864 for the Allegheny & Bald Eagle Railroad by Harlan & Hollingsworth (Fig. 8.18). The plan is a very progressive one and clearly foreshadows the modern hopper car with sloping bottom sheets and center drop doors. The chance survival of the drawing, which is the only record of this remarkably advanced design, would make anyone dealing with technical history wonder how many other missing links of this importance have disappeared without a trace.

Iron-body coal cars appeared rather unexpectedly on a tiny standard-gauge railway in Califor-

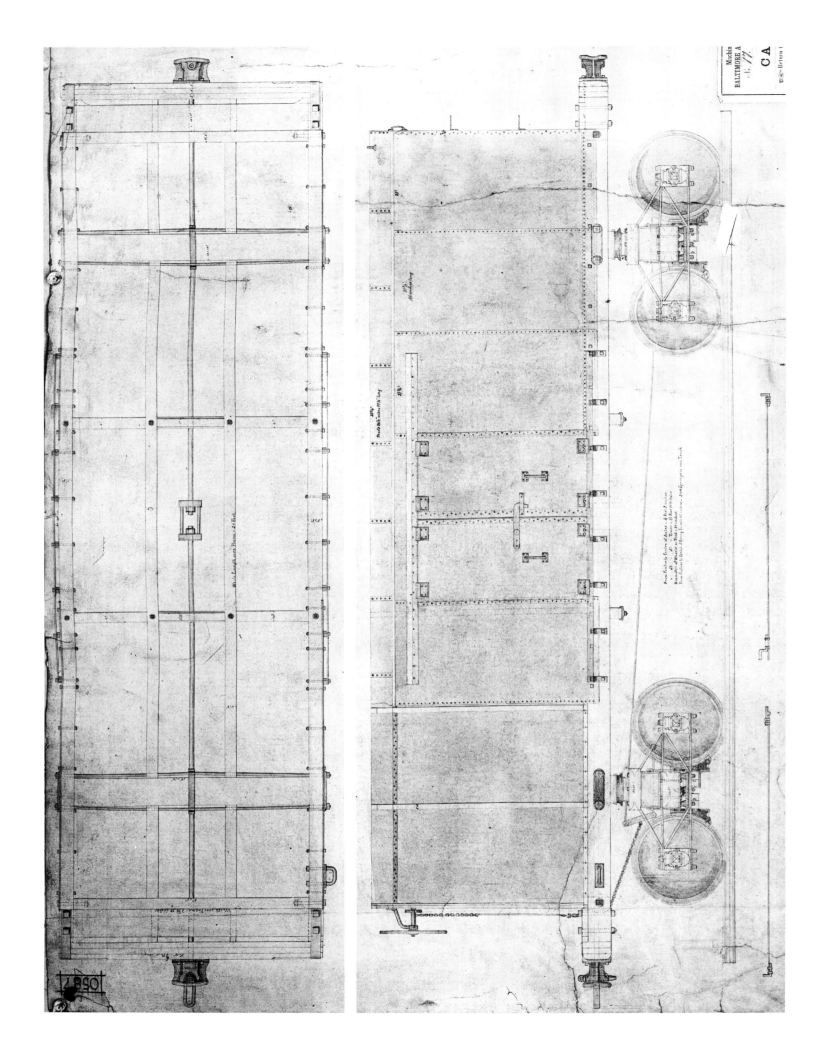

THE AMERICAN RAILROAD FREIGHT CAR

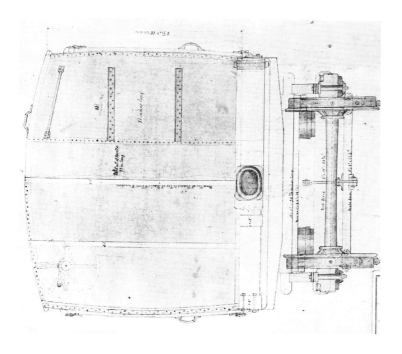

FIGURE 8.16

Photocopy of a Perkins iron boxcar of 1862. Regrettably, the original drawing was destroyed in a fire at the Mt. Clare shops around 1922. (Smithsonian Neg. 49655)

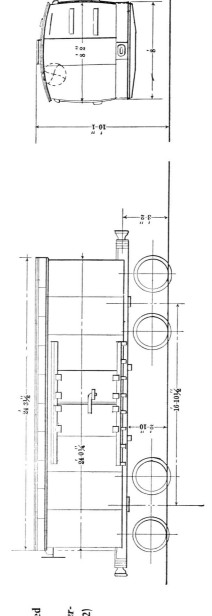

FIGURE 8.17

Diagram drawing of the preserved Perkins iron boxcar now at the B&O Museum, Baltimore. (*American Railroad Journal*, Dec. 1892)

EARLY IRON AND STEEL CARS

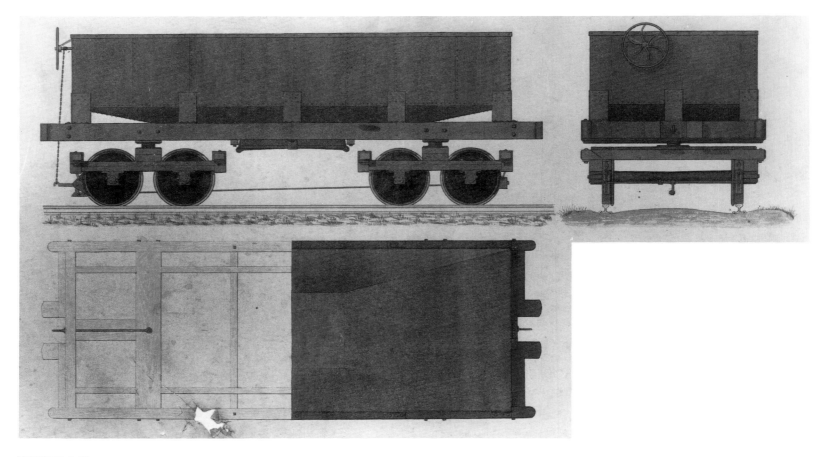

FIGURE 8.18

Harlan & Hollingsworth produced this drawing around 1864 for the Allegheny & Bald Eagle Railroad. It foreshadows the modern steel hopper car in its general plan. (Smithsonian Neg. 81-8102)

nia, and so the idea of metal freight equipment spread to the farthest reaches of the nation. The Pittsburg Railroad, located about thirty miles east of San Francisco, connected coal mines with barges on the San Joaquin River. The line opened in February 1866 and was stocked with two six-wheel tank engines and some thirty four-wheel coal cars. The locomotives were produced by the Union Iron Works of San Francisco, which was also most likely the fabricator of the line's other rolling stock. The cars were a decided novelty and looked something like a single-pot hopper (Fig. 8.19). The riveted iron body was elliptical in shape and curved down to form a hopper with a bottom drop door. The windup chain was operated by a windup shaft placed lengthwise rather than across the body, which was the usual position. The large and permanently mounted handwheel in place of a square shaft intended for a loose crank handle was another novel feature. The wooden frame and curious brake rig are also worthy of study.

The design for these odd little cars might well be traced back to the East, for the Union Iron Works draftsman and general manager, Irving M. Scott, began his mechanical career in Baltimore. For a time he worked at the machine shops of Murray & Hazelhurst, which built locomotives and cars for the Baltimore & Ohio. As a young man, Scott most likely had a direct hand in the planning and construction of pot hoppers. At least there appears to be a logical explanation for the appearance of mod-

ified pot hoppers 3,000 miles west of the Cumberland coal fields.

A few frustrated iron-car advocates saw the Civil War as an opportunity to demonstrate the merits of their ideas. The need for bulletproof cars was recognized early in the war. Just weeks after hostilities began, Rufus A. Wilder, mechanical head of the Mine Hill Railroad of Schuylkill Haven, Pennsylvania, hastily scratched out a rough plan for an armored car that could safely move along hostile tracks.[49] Such a vehicle could also be used on friendly lines to protect track workers from enemy snipers. A car on Wilder's plan, with ½-inch boiler plate, was fabricated by the Baldwin Locomotive Works using a wooden baggage car for the undercarriage. An engraving of the car was reproduced in the issue of *Leslie's Weekly Magazine* dated May 18, 1861. Later the following year one or more armored cars were being built for General Logan's use in Tennessee.[50] S. J. Seely of Brooklyn petitioned the War Department to adopt his scheme for an "imperishable" iron car made with curved ends and an arch roof.[51] Seely's car would be an advance over the composite cars of the B&O because of an angle and T-iron frame.

A rival of Seely and a figure to be discussed at greater length later in this chapter, Bernard J. La Mothe, had somewhat greater success with the authorities in Washington.[52] Two sample cars were actually delivered. The first was sizable for a

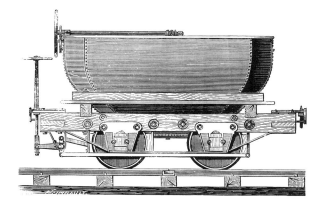

freight car of that period. Delivered in January 1862, it measured 44 feet long and featured a gas-pipe frame patented on September 24, 1861 (No. 33,350). But in the end La Mothe had no greater success than Seely or Wilder or the host of other enthusiasts who attempted vainly to broadcast the merits of their plans. Even the war emergency could not boost the fortunes of the iron car, whose future remained in question until the very end of the nineteenth century.

A major achievement made during the war years was the decision of the New York Central Railroad to adopt iron boxcars on a large scale. The NYC, like the B&O and the Reading, was a major rail-road. If these three prestigious companies were willing to employ iron cars, the remainder of the industry must follow their lead. But that belief mistakenly assumed that the success of such equipment was assured once a major test was un-der way. However, the results of these experiments were hardly conclusive. The economic and operat-ing merits of the iron car remained as controver-sial as ever. One source claims that the first NYC car was produced in 1859, while the earliest con-temporary evidence found to date is a brief note published in April 1861.[53] The cars were designed and built by John McB. Davidson of Albany. But like the date of construction, the style of framing is reported in conflicting ways as well. One source says that iron-clad oak sills were used; another says that riveted channels, six in number, formed the floor frames.[54] Because the cars were produced over several years, it is possible, of course, that both methods were used. The sides were of $\frac{1}{10}$-inch iron reinforced by 2- by 2- by $\frac{1}{4}$-inch angles. The railroad reported 121 iron boxcars in service in 1862.[55] The number steadily increased to a peak of 719 in 1864 and then gradually began to diminish as cars were wrecked or retired.

At the peak of production in 1863, Davidson was reported as building nine cars a week.[56] His major customer was the New York Central, but he was also producing a limited number for the American and Adams express companies. Production appar-ently ceased during the next year, but some of Davidson's handiwork remained in service for an-

other thirty years, and one was in use for sixty years. Even so, there was no agreement on the util-ity of this equipment. In 1883, while undergoing repairs, they were said to have "stood the wear and tear of service remarkably well."[57] A few years ear-lier, parties clearly friendly to the progress of metal construction berated critics of the Davidson cars, saying that they were as good as new after more than twenty years' service.[58]

But the men whose opinions really counted—the master car builders—voiced a different view. Eugene Chamberlain, superintendent of pool cars for the New York Central & Hudson River Rail-road, could only heap sarcasm on the memory of iron boxes.[59] He described them as things of "shreds and patches." As a young car repairer, he had resented being asked to lay aside his trusty saw, jack plane, and hammer to become an embryo boilermaker. He agreed with a fellow builder that they were destructive of every form of life con-tained therein. Chamberlain added that they had the same effect on the men required to repair them. They were in fact something like a boil—always in evidence. Quippishly he suggested that the last of the old iron cars be sent to Alaska for use as giant heating stoves.

In 1890 the New York Central's mechanical de-partment decided that the old Davidson cars were used up, and it began to retire them as rapidly as possible.[60] But one of the relics escaped the scrap line. A Davidson car was reportedly taken south by the Union Army in 1861. There it served with distinction under Generals Buell and Sherman as a munitions carrier. During one campaign it was dumped in a river, and there it reposed until Re-construction. It was pulled out of the water and mud and returned to service as a baggage car. It was retired in 1916 but put to work again four years later. On February 15, 1922, the old relic was blown to pieces when a case of dynamite accidentally dis-charged. One of Davidson's products had lasted long enough to witness its creator's vision mate-rialize: the metal car had become the standard freight car in America.

INNOVATIONS IN METAL UNDERFRAMING

It was not until the 1870s that promoters of the iron car made a determined effort to capitalize on the good foundation possible with metal under-framing. This was the principal advantage the new form offered over traditional wooden construc-tion. Success with metal underframes for locomo-tive tenders beginning in the 1860s naturally sug-gested the adoption of the idea for cars. One of the pioneers in metal tender-frame design was Ben-jamin W. Healey (c. 1821–1881), superintendent of the Rhode Island Locomotive Works. A patent was granted on June 7, 1870. In 1873 Healey built five flatcars that were elongated copies of his standard iron tender undercarriage for the Stonington Line.[61]

EARLY IRON AND STEEL CARS

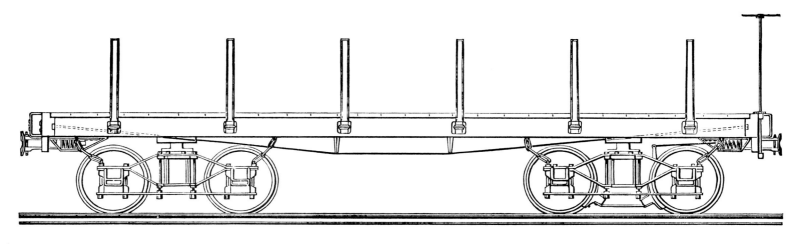

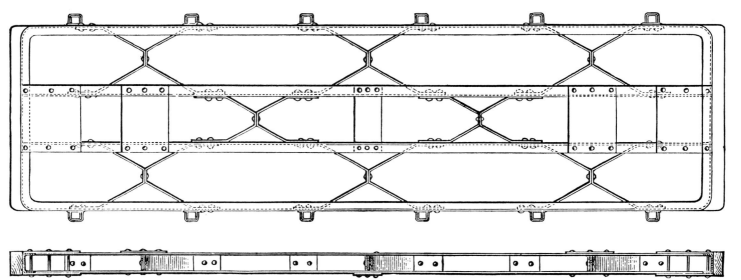

FIGURE 8.20

Benjamin Healey's iron-frame flatcar was based on a tender frame devised by Healey during his tenure as superintendent of the Rhode Island Locomotive Works in the 1860s. (*National Car Builder*, June 1879)

FIGURE 8.21

Samuel W. Murray's iron-frame boxcar was patented in 1877. (E. Lavoinne and E. Pontzen, *Les Chemins de Fer en Amérique* [Paris, 1882]; metric measurements)

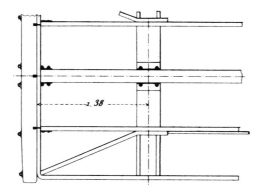

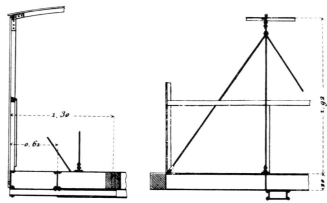

The frame was made up from 4½-inch channel iron cross-braced with flat bars in an X pattern. Truss rods in the conventional position helped stiffen the structure (Fig. 8.20). Wooden decking was used. The weight was given as 8.5 tons, the capacity as 25 tons. Healey claimed that the same general design, less the truss rods, had been used successfully on nearly two thousand locomotive tenders.

Healey's experiments may have prompted George Richards, master mechanic for the nearby Boston & Providence, to try his hand at iron-car construction.[62] He decided to make use of some old 57-pounds-to-the-yard T-rail stacked outside the repair shop. Rail is not a very efficient beam because so much of the material is concentrated in the head or wearing surface, but most railroad mechanics were forever trying to find new uses for secondhand rail. There were always great piles of it strewn around railroad property, and the trade journals frequently reported the latest ideas for putting it to some useful purpose—bridge construction was a favorite but rarely successful adaptive use. Richards used four rails per car frame, which he stiffened with truss rods. More than one design was used, because he reported car weights

FIGURE 8.22

The Lake Shore & Michigan
Southern acquired a few iron-
frame flats in the late 1870s. Note
the unusual truss rods of flat bar
stock. (Western Railroad Club
Proceedings, 1893)

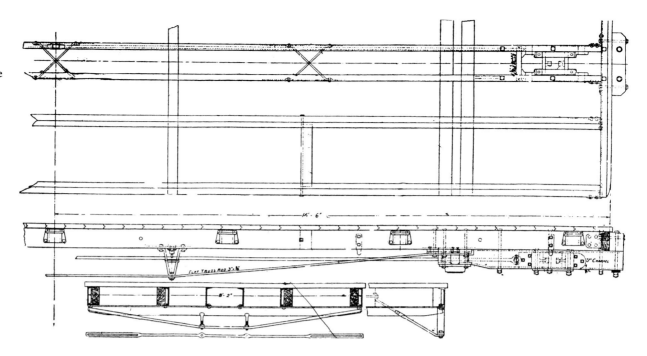

of 17,270 and 18,750 pounds with capacities of 34,630 and 43,550 pounds respectively. The frames were fitted with gondola bodies, and Richards contemplated building boxcars with rail frames. But the product of this idea was a very heavy car not worth the small saving resulting from the use of secondhand material.

Other mechanics were approaching the problem more scientifically. One of the better plans was patented on October 31, 1876 (No. 183,856), by Frederick J. Kimball of Philadelphia.[63] The Milton, Pennsylvania, car works produced a boxcar for the Empire Fast Freight Line in 1877. A sketchy drawing of some framing details was published several years later in a French work on American railways. It shows very light but deep section channels and I-beams braced by flat bars. The body bolster was a neatly fabricated box girder. Curiously, the center and end sills were wooden. The builder of Kimball's car was Murray, Dougal & Company of Milton, a firm well known for its iron tank cars. These oil carriers, already described in Chapter 5, had metal frames. The firm's main partner, Samuel W. Murray, decided to develop an iron box of his own plan not long after completion of the Kimball car. His design featured pressed-metal side panels as described in Patent No. 196,926 issued on November 6, 1877 (Fig. 8.21). It is uncertain if a full-size test car was produced; however, others were active in iron-car design during the 1870s. The combination box/oil tank cars of Joel F. Keeler and Horatio Brooks were described and illustrated in Chapter 5.

Another plan was tried by the Lake Shore & Michigan Southern for a group of flatcars built in the late 1870s.[64] Six channel beams, four wooden sills, and flat iron truss rods made up of ¾- by

5-inch stock, in place of the usual round rods, made up the floor-frame elements (Fig. 8.22). After fifteen years' service the cars were said to have performed well and cost little to maintain. And this report came from a railroad whose longtime master car builder, John Kirby, was generally hostile to iron construction. In 1895 Kirby built a larger steel-frame flatcar carried on twelve wheels. It measured 40 feet long and had a capacity of 50 or more tons. The 9-inch channel sills were supported by six truss rods.

The question of scientific iron-car construction would naturally attract the attention of railroad bridge builders. Cars are something like a bridge on wheels, and so the problems and calculations involved in designing the two are actually not very different. Two engineers associated with the Kellogg Bridge Works of Buffalo, Charles H. Kellogg and John W. Seaver, decided to solve this long-standing design problem (Fig. 8.23).[65] They felt that most inventors outside the industry were too ambitious in their efforts to better the dead-weight ratio of the wooden car in some spectacular way and so produced cars too lightly built to withstand the rigors of normal service. Orthodox car builders, on the other hand, tended to make a literal substitution of iron for wood and thus produced an overbuilt and overweight vehicle. Kellogg and Seaver felt that as experienced bridge builders they could formulate the perfect compromise. They promised to build a flatcar without old-fashioned truss rods but using I-beams sufficiently heavy to give stiffness in all directions, while holding the overall weight to a minimum (Fig. 8.24).

Kellogg and Seaver first offered an 8-ton flatcar with a capacity of 20 tons at $550, or just $100

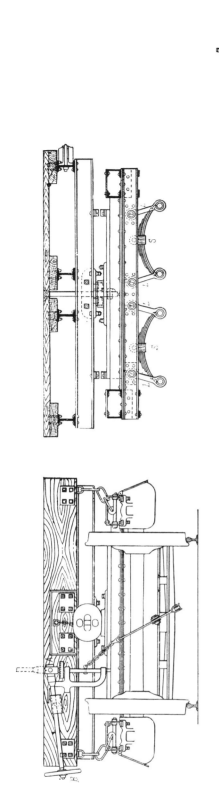

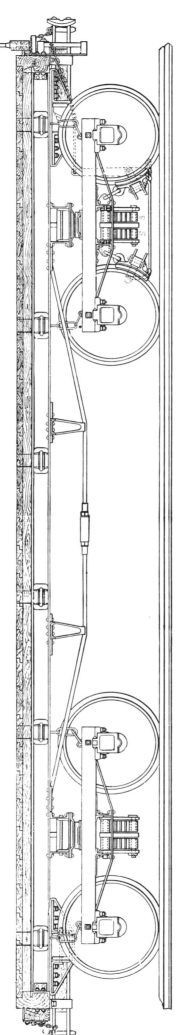

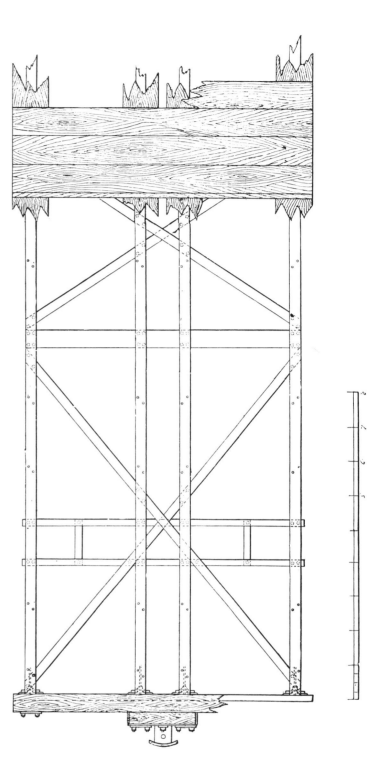

FIGURE 8.23

Charles H. Kellogg and John W. Seaver, bridge builders of Buffalo, claimed to have produced the best possible design by applying science to freight car planning. (*Railroad Gazette*, Jan. 24, 1879)

FIGURE 8.24

Kellogg and Seaver improved their initial plan by enlarging the girder size and eliminating the truss rods. (*Railway Age*, 1879)

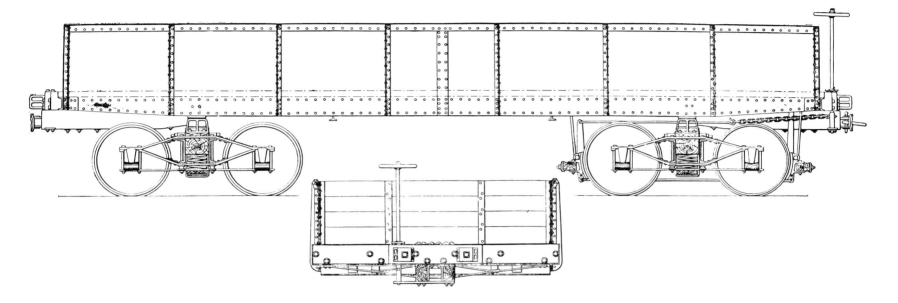

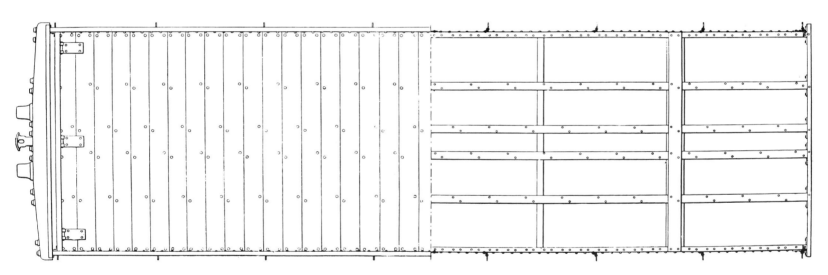

FIGURE 8.25

This modern-looking gondola was Kellogg and Seaver's last effort to break into the freight car market. (*Railroad Gazette*, Sept. 1, 1882)

more than a wooden car. Some flats were produced for the Buffalo & South Western in 1879.[66] Earlier in the same year they claimed to be building an all-iron coal car. Just how many cars Kellogg produced is not recorded, but the firm's interest in the subject continued for at least three years. A drawing of a modern-looking gondola car was published in 1882.[67] The 30-foot-long drop-end car was rated at 25 tons and was said to weigh only 9 tons (Fig. 8.25). The floor and end sills were wooden, as were the truck bolster and spring plank. The floor frame was composed of I-beams. Kellogg claimed that the gondola had a dead-weight ratio 36 percent better than an equivalent wooden car. Yet for all the

claims and real merits of Kellogg's design, production appears to have been very limited.

TUBE OR GAS-PIPE CARS

The rational designs of Healey and Kellogg ironically had far less impact and certainly received far less attention than a curious scheme patented in the previous decade by a French immigrant dentist named Bernard J. La Mothe. A very brief mention of La Mothe's 1861 patent was made earlier in this discussion, but it was not mentioned that La Mothe had been actively engaged in iron-car design since the early 1850s. At first he concentrated his energies on introducing a "life-saving" iron

FIGURE 8.26

La Mothe's gas-pipe scheme was patented in 1861. Efforts were made to apply it to both passenger and freight car construction. (*American Artisan*, April 8, 1868)

FIGURE 8.27

The gas-pipe boxcar was enlarged and strengthened in this later plan of the 1870s. (*National Car Builder*, June 1879)

FIGURE 8.28

Iron-tube freight cars were produced in impressive numbers by a variety of independent car manufacturers; however, most were purchased by private car line operators rather than by railroads. (*Master Mechanics Magazine*, June 1888)

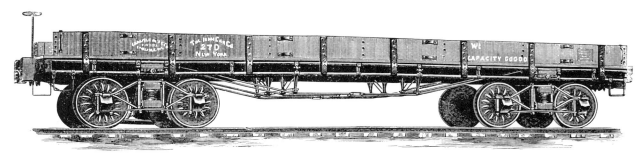

572   THE AMERICAN RAILROAD FREIGHT CAR

FIGURE 8.29

This high-sided hopper-bottom gondola was built for the Reading in about 1888 by the Carlisle Manufacturing Co. Note that the weight is not stenciled on the wooden body. (*National Car Builder*, June 1889)

passenger car, but when this endeavor proved hopeless, he turned to freight cars. His first plan was for a basketweave framework made up of riveted iron straps or flat bars. At about the time the passenger car venture was folding, he conceived of substituting iron rods or tubing for the framing elements. Tubing was—theoretically, at least—a very attractive material because it gave the greatest amount of strength for a given amount of material. A hundred pounds of iron formed into a tube will produce a shape stronger than a bar, channel, or I-beam. Gas pipe, even in rather large sizes, was commonly available in the 1860s because of the prevalence of gas lighting, and so there was a cheap and ready source of building material.

La Mothe seemed on solid ground in advocating tube or gas-pipe construction; however, the great flaw of his plan was the lack of a secure and practical way to join the pieces. Threaded joints weakened the structure because metal was cut away at the joints—already the weakest part of the structure. It also made for complicated assembly and disassembly. Gas or electric welding was the answer, but these techniques would not be developed for generations to come. But La Mothe thought that he could solve the problem with cast-malleable-iron clamps and U-bolts. The solution was clever but not totally sound, and the thousands of tube cars built over the next several decades all suffered from loose joints, among other complaints.

By the late 1860s the gas-pipe freight car was fully developed, and a handsome perspective drawing of a 28-foot-long boxcar was published in *American Artisan*.[68] The main floor sills were made up of three 1½-inch-diameter pipes stacked one on top of the other with a combination spacer and U-bolt clamping device (Fig. 8.26). The floor was reinforced by wire-rope truss rods. The upper body was framed with an intricate nest of small gas pipes. The interior and exterior were covered with sheet metal. Weight was estimated at 8 tons,

and the inventor talked rather uncertainly about a capacity of from 12 to 20 tons. It was this type of imprecise and exaggerated talk that so irritated the car builders.

There was a report in 1872 that the Detroit Car Works was building gas-pipe cars for La Mothe's New York Iron Car Company, but it doesn't appear that many were actually produced until the middle of the same decade.[69] By this time the National Tube Works had taken an interest in the project— it was such a good potential market for the company's gas pipe—and began producing freight cars to La Mothe's design at its plants at Boston and at McKeesport, Pennsylvania.[70] Other car plants, including an unnamed shop in Providence and the Osgood Bradly works in Worcester, were involved in producing La Mothe cars.[71] Flat-, box-, and even a few refrigerator cars were built during the 1870s for the Providence & Worcester, the Empire Fast Freight Line, the Pittsburgh branch of the B&O, the Boston & Albany, the Burlington, P. T. Barnum's Circus, the Atlantic Coast Line, and the Canada Southern.[72] In all, these orders represented about one hundred cars. The refrigerator cars were said to weigh 23,000 pounds, or 3 tons less than a wooden car. The designers hoped to shave off another ton by refining their plans (Fig. 8.27). The P&W felt safe in turning the pipe-frame cars over to shippers, like stone and iron merchants, who regularly overloaded their equipment. One of the 20-ton boxcars was loaded with 28 tons of ice and suffered no ill effects—at least according to a prospectus issued by the National Tube Works in 1877.[73]

The trade journals voiced some objections despite these positive reports. The June 1874 *National Car Builder* thought that the design was too much of an innovation to be adopted without a very long testing period. The September 1877 *Railway World* was disappointed by the great weight of the iron cars and questioned their ability to carry big loads, outside of a few staged trials, on a regular

EARLY IRON AND STEEL CARS

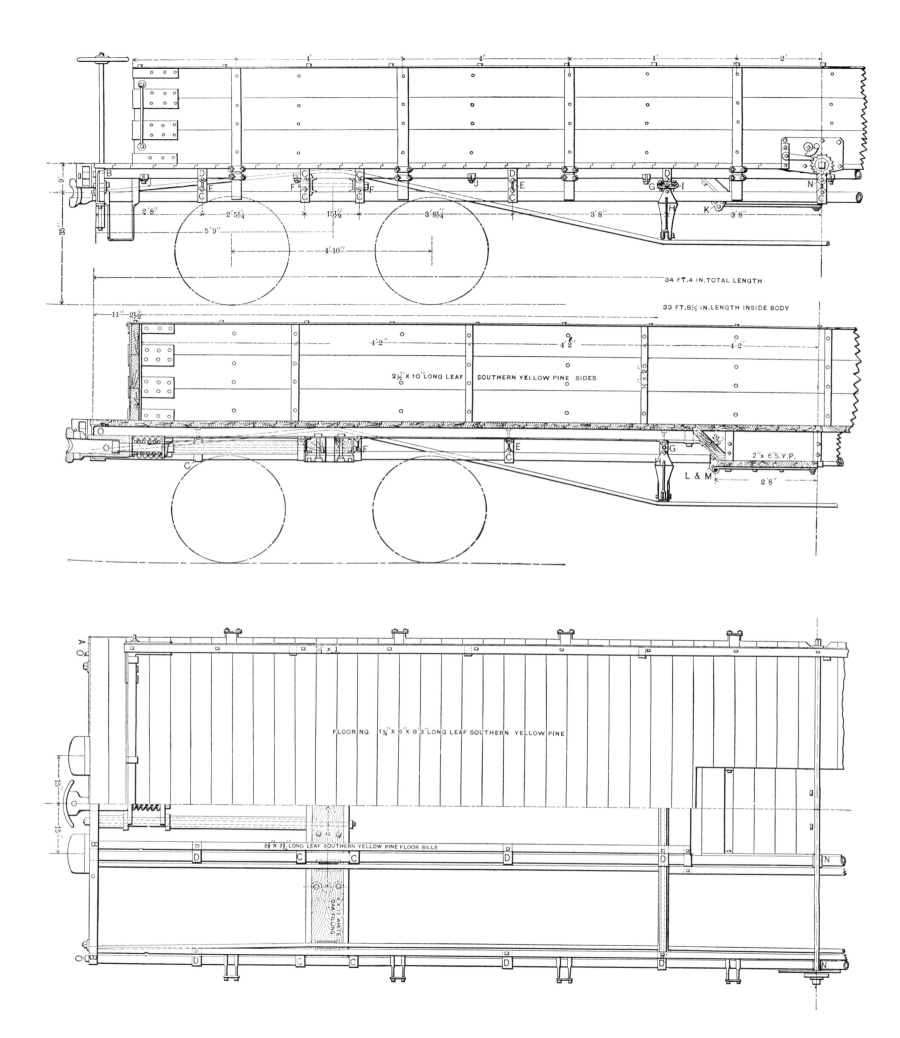

34 FT. 4 IN. TOTAL LENGTH

33 FT. 8½ IN. LENGTH INSIDE BODY

2½" X 10' LONG LEAF   SOUTHERN YELLOW PINE   SIDES

2" x 6" S.Y.P.

FLOORING 1¾" X 6" X 8'3" LONG LEAF SOUTHERN YELLOW PINE

2½" X 2½" LONG LEAF SOUTHERN YELLOW PINE FLOOR SILLS

6" X 12" WHITE OAK FILLING

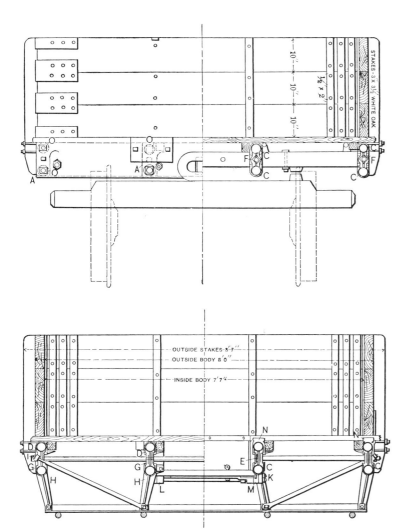

FIGURE 8.30

The complexity of the tube-frame car is well illustrated by these drawings. A sizable proportion of wood was incorporated into the design despite earlier efforts of La Mothe to interest the railroads in all-iron construction. (*American Journal of Railway Appliances*, June 30, 1884)

workaday basis. And what about the condensation and heating problem suffered by the B&O and New York Central boxcars of the previous generation? La Mothe tried to answer these complaints with design adjustments. The main-frame pipe size was increased from 1½ to 2 or even 2½ inches in diameter. The boxcars were insulated either with wood or combined wood and paper linings. By 1878 the boxcars had grown to 30 feet 9 inches long by 8 feet 9 inches wide by 7 feet 7 inches high.[74] Weight was given as 17,700 pounds and capacity as 35,000 pounds. If true, La Mothe had produced an extraordinary freight car.

There was enough truth in La Mothe's claims to keep the pipe-car project alive, but not enough to sustain the National Tube Works' support beyond a few years. The Panic of 1873 didn't help La Mothe's plans. The depression was unremitting, and new car orders in general were at a very low ebb for many years. If only standard cars were being purchased, the market for experimental ones was even more diminished. Early in 1882 La Mothe's original firm was succeeded by the United States Tube Rolling Stock Company.[75] As in the old firm, outside contractors were turned to for the fabrication of cars (Fig. 8.28). During the 1880s the

following firms produced pipe cars: the Gilbert Car Works (Troy, N.Y.); the Middletown (Pa.) Car Works; the Carlisle (Pa.) Manufacturing Company; the Milton (Pa.) Car Company; the Bloomsburg (Pa.) Car Works; the Harrisburg Car Company; the Lebanon (Pa.) Manufacturing Company; and the Allison Car Works (Philadelphia). To attract new customers the design was modified and improved. The end sills and bolster—formerly of timber—were made of channel iron. The floor-sill pipes were made with a camber to prevent buckling when overloaded.

Orders were received from two major railroads, the Pennsylvania and the Philadelphia & Reading.[76] The Pennsylvania received several 25-ton-capacity gondolas based on the dimensions of its class E, at the time one of the largest freight cars used in the United States. The wooden class E measured 35 feet 6 inches long and weighed 21,800 to 22,300 pounds (depending on the type of wood used). The tube cars produced by the Middletown Car Works were set to work hauling pig iron or coal. Some carried loads far in excess of their rated limit of 25 tons; they were reported to carry 33.5 tons, and yet the tare weight was given as 18,900 pounds, which represented a saving of 2,900 to 3,400 pounds over the wooden car. The P&R received some high-sided hopper-bottom gondolas or coal cars from the Carlisle Manufacturing Company. Advertisements with a fine line engraving of one of these cars appeared in railroad trade journals in the late 1880s (Fig. 8.29). A curious oversight (or was it a deliberate omission?) is the failure to stencil on the car's weight directly above the registered capacity of 30,000 pounds. An identical omission will be noticed in an engraving of a low-sided gondola used in an 1888 advertisement for the products of the Iron Car Company.[77]

The unexplained secrecy about tare weights did not prevent the Iron Car Company—successor to the short-lived U.S. Tube Rolling Stock Company—from seeking to publicize its wares. The National Exhibition of Railway Appliances, held in Chicago in 1883, was a product showcase intended for the railway supply trade. The Iron Car Company sent out a 35-foot-long gondola whose capacity, it claimed, was limited not by the strength of the frame but by the capacity of the axles.[78] The weight was given as 18,900 pounds. From this figure and the length, the car must have been the twin of those produced the year before for the Pennsylvania. The frame sills were made up of 2½-inch-diameter pipe. The car was loaded with 76,000 pounds of pig iron. An inspection made several days later showed no signs of a structural breakdown. A few years after this stationary test, the Iron Car Company said that one of its units carried 80,000 pounds in regular service.[79]

The firm reported with fiendish joy that one of its cars had survived the worst freight train wreck ever to occur in New England. The car was in the middle of the runaway train. When the debris was cleared away, the iron car emerged with only one of its tubes bent.[80] The managers also published letters from agents of two other major railroads attesting to the durability of tube cars involved in less spectacular derailments.[81] Even when heavily loaded, the cars sustained no major damage.

But the greatest show of strength took place in April 1888, when no fewer than 112 iron tube flats and gondolas were assembled on twenty parallel tracks to move the imposing Brighton Beach Hotel to a more favorable location 500 feet from its original foundation.[82] It wasn't the distance that posed the problem; it was the 5,000-ton wooden structure. Six locomotives were required to move the enormous load. The Iron Car Company reaped excellent news coverage; engravings and reports appeared in many national magazines and major newspapers. It was a publicity masterpiece, but in all fairness wooden cars could have done the job as well. Wooden flatcars of 20 and even 30 tons capacity were not exceptional by the late 1880s. Like the pipe-frame cars, they too could accept loads far greater than their rated capacities, particularly when static or required to move only at very slow speeds over a substantial track. John Kirby, a veteran car builder quoted earlier in this chapter, said that in all his years in the car repair shop—and thousands of cars passed through each season—he could recall only two wood-frame cars that actually broke down because of overloading. Even so, the hotel move was popularly credited to the superior strength of the gas-pipe cars.

Superficially, the good press seems to have improved the fortunes of the Iron Car Company. Orders were received from the Mobile & Ohio, the Central of Georgia, the Norfolk & Southern, the Old Colony, and several short lines. In October 1888 the company reported building twenty-five hundred cars over the previous eighteen months and was sending some sample cars to England.[83] Eleven months later Iron Car had taken over the Minnesota Car Company in Duluth and was reorganizing it to produce another fifteen hundred cars.[84] Within a matter of weeks it was reported that Iron Car had to date built 4,950 cars, work was under way on 3,400, and orders had been received for yet another 7,625.[85] In light of such statistics, critics could no longer dismiss the tube car as an experiment.

Or could they? Behind all the big numbers and flackery stood some very grave problems. Few of the new orders were actually from railroads. Most were from private car lines, which leased cars at mileage or per-diem rates. Such firms flourished when times were good and railroads were hard-pressed for equipment; they would rent anything that would help move the traffic.[86] The great majority of tube cars—estimates vary between eight thousand and ten thousand—were owned by the Southern Iron Car Line with headquarters in Atlanta, which leased them to various railroads.[87] This firm was headed by Eugene C. Spalding (1862–1902), a man prominent in railroad accounting circles and a longtime advocate of per-diem charges for interchanged freight cars. Spalding was good with figures and a master of the ledger book, but he was not a mechanic or an engineer. He accepted the Iron Car Company's sales pitch apparently without question. The purchasing decision in this case was directly in the hands of a manager who was not a technical expert in the area of freight car design, construction, or maintenance.

Despite Spalding's patronage, the Iron Car Company was in serious trouble. The firm, headed by Colonel John W. Post of New York City, had expanded too fast. In May 1890 the property was seized by creditors with claims of $800,000.[88] Chief among the claimants was the Huntingdon Car Works, which had built a majority of the cars. Efforts to reorganize were reported as late as July, but the financial support needed to rescue the company was apparently not forthcoming. Colonel Post organized a successor enterprise named the Steel Tubular Car Company. He announced plans for a manufacturing complex on the scale of Pullman Illinois, but nothing materialized and the tube-car scheme died at last, almost thirty years after having been conceived by an obscure French dentist.

The story of the pipe car does not end with the demise of Dr. La Mothe and Colonel Post's dream. A postmortem must be written about the thousands of pipe cars that rattled around the freight yards of North America for at least another decade. The Southern Iron Car Company had only two thousand cars in active service at the time of Spalding's death in 1902, and most of these were gone within two years.[89] The others had been retired and salvaged to pay the bills of the faltering firm. And there were now reports and candid admissions about the servicing problems attending the tube cars. To save weight the earliest design omitted a center sill, resulting in a poor connection for the draft gear and coupler (Fig. 8.30). Beginning in 1892 the Southern Line began to rebuild the older models with a 6- by 12-inch wooden center sill—one can only imagine the smug chuckles emanating from the old-line car builders on first seeing this alteration.[90] Despite all the claims about strength and rigidity, the structure tended to work loose and droop at the ends as well as sag in the middle. There were repeated complaints about the difficulties of repairs and disassembly; one critic claimed that not even a plumber could repair them.[91] Parts and tools were an obvious problem, though not a defect of the design itself, in shops staffed and equipped for wooden cars.

And finally, there was the costly and difficult

FIGURE 8.31

Two sample cars for the Furness Railway in Scotland were produced to the design of the Iron Car Co. in 1888 by the Huntingdon Car Works. (*Engineering*, Dec. 7, 1888)

task of refitting the tube cars with automatic couplers and air brakes as required by federal law— the Safety Appliance Act of 1893. It was necessary to raise the end sill level to apply the couplers, and this expense together with the cost of the new apparatus just wasn't worth the investment in cars already fifteen to twenty years old that had so many other problems. Most simply went to the scrap yards. A token number, however, were said to be in service on the Chicago & Eastern Illinois as late as perhaps 1920.[92] A relic of the gas-pipe era was uncovered in recent years in the Nevada desert, where rail enthusiasts found part of the frame of a Carson & Colorado boxcar. This western narrow-gauge line acquired a number of pipe-frame cars, presumably built by Allison in the early 1880s, that remained in service until the early

twentieth century. They were called "tin boxes" on the C&C.

At the very time when the tube car was on the decline in America, an effort was under way to introduce it in Europe and elsewhere.[93] In 1895 G. W. Ettenger (former general manager of the Iron Car Company) and George E. Church opened an office in London to promote the tube car. They secured British patents and began pushing the idea in Great Britain, Belgium, and India. Actually, there had been an earlier effort to introduce tube construction to the British market. In 1888 two gondolas were built for the Furness Railway in Scotland by the Huntingdon Car Works (Fig. 8.31). The designs appear to be stock U.S. plans right down to the arch-bar truck and link-and-pin couplers. The unaltered appearance did not seem to dampen the

EARLY IRON AND STEEL CARS

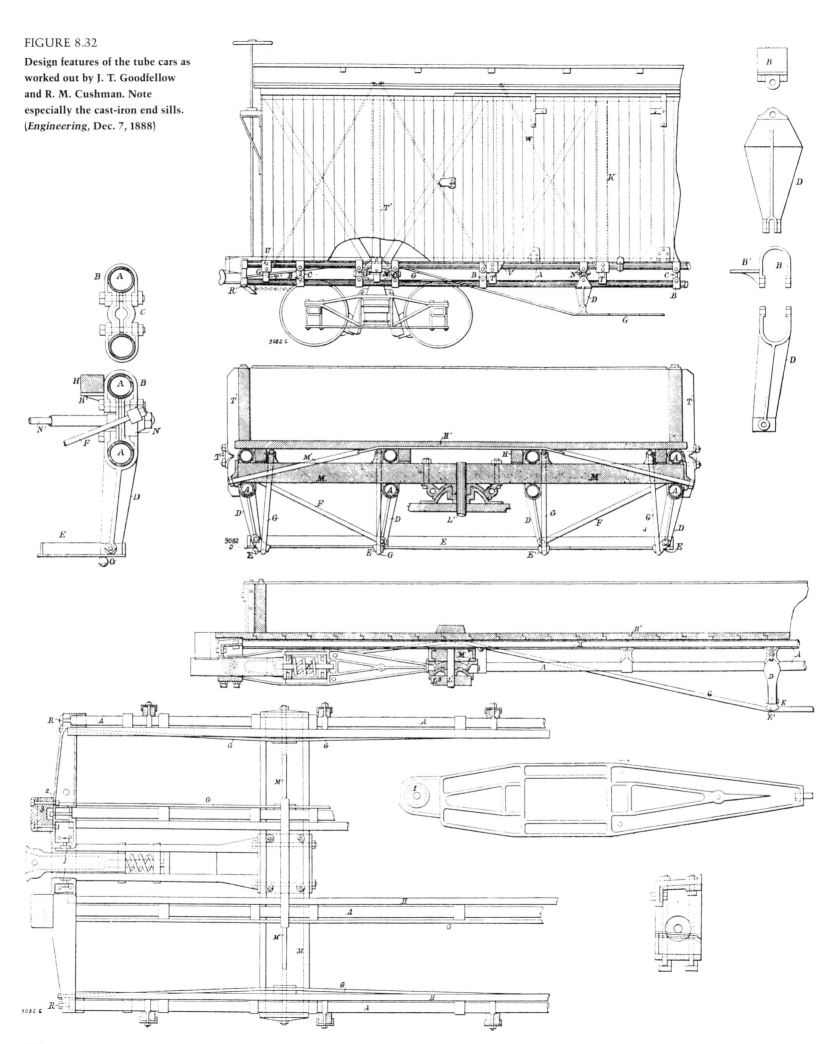

FIGURE 8.32

Design features of the tube cars as worked out by J. T. Goodfellow and R. M. Cushman. Note especially the cast-iron end sills. (*Engineering*, Dec. 7, 1888)

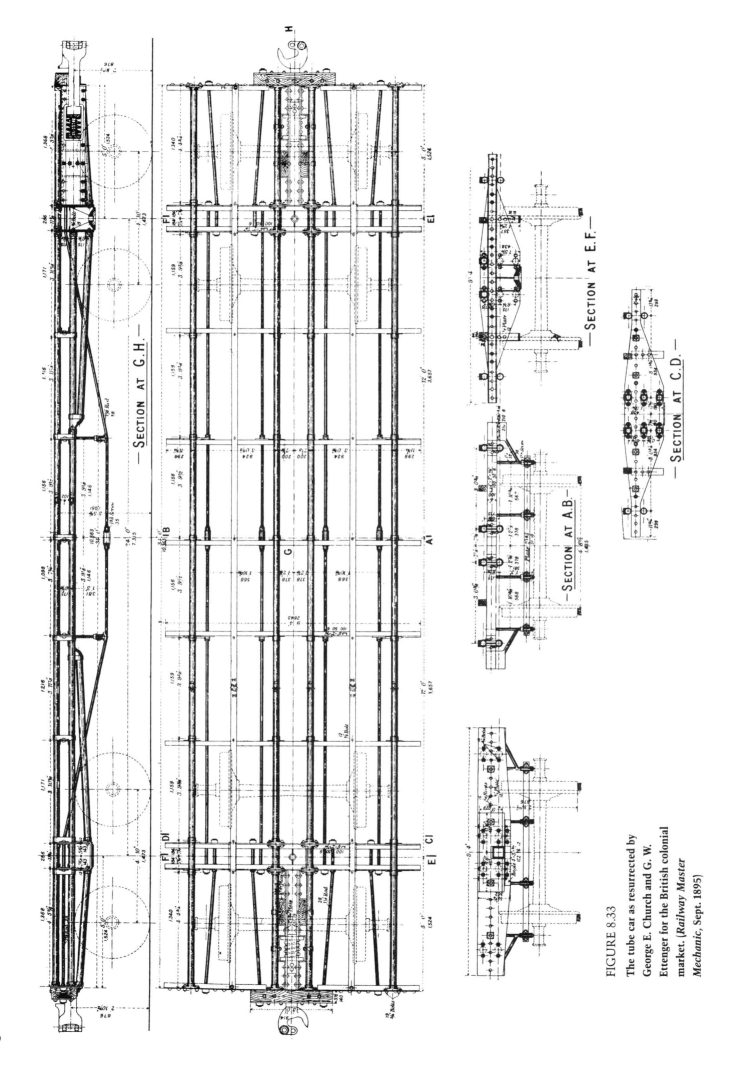

— Section at G.H. —

— Section at E.F. —

— Section at A.B. —

— Section at C.D. —

FIGURE 8.33

The tube car as resurrected by George E. Church and G. W. Ettenger for the British colonial market. (*Railway Master Mechanic*, Sept. 1895)

enthusiasm of the British trade press, which praised the Yankee tube cars for their size, capacity, and flexibility. One of the samples was made as a low-sided gondola. Its reported weight of 8.3 tons and capacity of 30 tons were considered astonishing. Its high-sided sister handled 18 tons of coke, which being a high-cube low-density cargo was not expected to establish any dramatic weight record.

The drawings pertaining to these cars, published in the December 7, 1888, issue of *Engineering*, reveal several interesting design elements (Fig. 8.32). The clamping brackets and assembly details of the pipe framing are worthy of study. The hollow cast-iron end sills are of interest as well; note the diamond-shaped openings visible in the perspective drawings of the complete cars. The massive cast-iron draft gear is an even more notable and innovative feature. Goodfellow and Cushman were not agents for the Iron Car Company in Britain; they were in fact designers employed by the Iron Car Company. R. M. Cushman had been active in tube-car design since the early 1870s, and J. T. Goodfellow had assigned a patent (June 1888) to the firm and was listed as its engineer in 1889.

Efforts to introduce the tube design beyond Scotland's Furness line appear to have been futile. Church and Ettenger met with a trifle more success (Fig. 8.33). They too failed to find much acceptance in the mother country, but British investment in foreign railways reached beyond their own colonies and was truly worldwide. The nitrate railways in Chile, for example, were receptive, and new tube-form cars built after the Church and Ettenger patents were supplied as late as 1905 by the Metropolitan Carriage & Wagon Company of Birmingham, England. Most of them were retired in the 1920s, but some were retained for light service into the next decade and perhaps even later.

The Steel-Car Revolution

There is no absolute division separating the history of the iron and steel cars. The iron car drifted into the age of steel via the tube car, but as a convenient if not very precise date, 1890 marks the beginning of the steel-car era. Any witnesses of the events of that year would report no change; the wooden freight car remained firmly on the throne. The proponents of metal construction remained out of favor and out of power. The talk about strength, dead weight, and durability continued as always, but no one seemed to listen very carefully. One gains the impression that most master car builders were not listening at all. They were too busy boasting about the progress made in wood-car design and capacity since 1870. Their achievements were amazing, and there was reason for honest pride. And yet the status quo and the stability of the moment were very deceptive. It was like Eu-

rope in the summer of 1914. An acute observer could sense that something was about to happen; a revolution was in the air. But almost no one knowledgeable about freight cars in 1890 would have predicted the overthrow of the wooden dynasty within seven short years. From the advantage of hindsight—our historical perspective—it is obvious that much had already changed by 1890.

THE INEVITABLE DEMISE OF WOOD

The mere passage of time was altering the makeup of the car-building profession. The majority of the pioneers—such as John Kirby, Leander Garey, D. H. Baker, and Fitch Adams—were either dead or about to die, retired or about to retire. They were replaced by a new generation of mechanics who were graduates not of the sawmill and cabinet shop, but of Georgia Tech and the Stevens Institute. These men were educated to handle the slide rule rather than the carpenter's square. They were exposed to a broader view of engineering, materials, and the techniques of fabrication. Their thinking was shaped more by science than by experience. They read textbooks and trade journals that described oceangoing ships, bridges, and tall buildings, all made from steel. It was this new material that permitted structures not possible with more traditional materials like stone and wood. By 1890 about the only technicians showing a loyalty to wood framing were car and house builders, who were hardly regarded as the new wave.

Steel's increasing popularity was enhanced by falling costs due to better production methods and the availability of a greater variety of stock shapes and sizes. Rolled beams once selling for $70 per ton had fallen to $40 per ton by 1891. Western riverboat builders, once loyal patrons of wooden construction because of its cheapness, began to adopt metal hulls when the price differential disappeared because of cheaper iron and steel.[94] Steel production and its integration into everyday products were boosted by a human dynamo named Andrew Carnegie. His interest and that of his advisers did much to promote the ascendancy of steel cars. Carnegie's involvement was worth more than the efforts of all the metal-car advocates of previous generations.

Steelworking tools had become more common by 1890. As they became less specialized, their cost fell and their operation became more familiar to a larger number of workmen. Pneumatic hand tools, drills, riveting hammers, and chisels were also rapidly coming into favor by 1890. They were powerful and fast, even when handled by semiskilled men, and did much to facilitate and lower the cost of steel fabrication. The advent of air-powered tools is one of several developments outside the railroad industry that had a direct effect on the decision to adopt steel freight cars; yet like

FIGURE 8.34

Very large capacity steel flatcars began to appear in small numbers on American railroads in the late 1880s. The example shown here was constructed by the Pennsylvania Railroad in 1888. (*Engineering News*, Jan. 30, 1892)

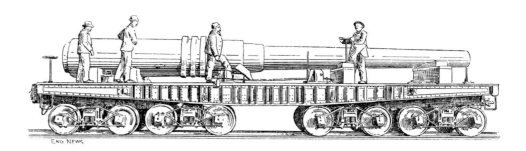

FIGURE 8.35

Steel crane cars are another example of metal construction for special-purpose cars. (*Rail-road Gazette*, Oct. 18, 1889)

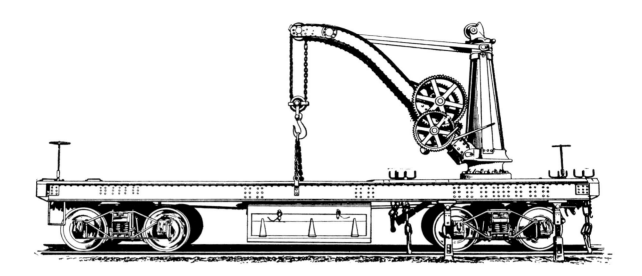

so many tangential subjects, they can only be mentioned here in passing.

The railroad industry was gaining experience with metal cars even while maintaining that they were unacceptable. Iron body bolsters and all-iron trucks were common by 1890, as already suggested. The most old-fashioned car builder acknowledged their superiority. Those open-minded enough to study overseas developments reported that metal underframes were universal in Germany and France and even found widespread use in Mexico.[95] Car builders willing to observe the activities of their fellow locomotive superintendents would find that about half of all new locomotives were pulling steel-framed tenders.[96] And there was no hiding the huge steel-framed flatcars introduced in the late 1880s. These special-service cars meant to carry naval guns or huge rolls of wire rope were the biggest freight cars in service (Fig. 8.34). The Altoona shops built a sixteen-wheel 39-foot flatcar in 1888 that weighed 51,800 pounds.[97] The car was limited to a 50-ton load because of limited track capacity, but it was felt that the car itself could safely handle 60 tons. Two twenty-four-wheel sisters followed in 1892, and during the following year a monstrous thirty-two-wheeler, which measured over 90 feet in length and carried loads approaching 150 tons, was outshopped by Altoona.[98]

The Pennsylvania also began building steel-framed crane cars during the same period (Fig.

8.35). Other railroads like the Buffalo, New York & Pennsylvania produced similar cars during the 1880s, as did the Industrial Works of Bay City, Michigan.[99]

The ministrations of Andrew Carnegie, the new generation of master car builders, a burgeoning steel industry, pneumatic tools, and a greater experience with metal framing were all factors in the successful transition from wood to steel construction. But there was an even more powerful argument for this radical change in car building, and it really wasn't a matter of one material or technique being superior to another. It was the economics of scale. It was part of that fundamental—surely the most fundamental—element in American freight car development: big cars versus small cars.

The simplest and most basic fact about American freight cars is that they have grown steadily from 1830 to the present time. Railroad managers could point to a dozen and more reasons to adopt ever-larger cars. Larger cars are more productive in moving a given quantity of goods, particularly bulk materials such as grain, coal, and sand. Bulk commodities have traditionally moved by rail and barge, and those managing their transport found at an early date that a few big containers moved goods more cheaply than many smaller units. A long list of potential savings can be offered. Because fewer big cars are needed in trains of equal capacity, the number of wearing parts—axles, wheels, brake shoes, couplers, draft gears—is reduced. Big cars result in shorter trains, meaning shorter sidings and the possibility of more trains on the main line since they are less likely to overlap blocks. Fewer cars per train means fewer locomotives, smaller operating crews, and lower labor costs. Because fewer wheels and axles are employed, friction is reduced, as is the time needed for oiling and inspection. There would be fewer empty-car movements and a saving in per-diem charges because these are levied per car on a time or mileage basis rather than by tonnage measure. Switching costs are reduced because fewer cars are involved in moving the same amount of traffic.

The big-car trend was dominant because it increased the railroad's capacity and efficiency for a relatively modest investment. Following are just a few specific examples of the benefits possible.[100] One comparison shows a weight saving of 315 tons from using the larger steel cars:

> Wood: Fifty 30-ton cars = 1,500 tons capacity; weight 825 tons
>
> Steel: Thirty 50-ton cars = 1,500 tons capacity; weight 510 tons

The assumed weight of 16.5 tons each for the wooden cars seems a trifle high; most 30-ton wooden cars weighed 15 tons (or less) each. Hence the total for the wood-car train should be more like 750 tons, yet the comparison is still favorable to steel. In another comparison the Pennsylvania's standard wooden hopper is contrasted with the Pittsburgh & Lake Erie's new steel hoppers:

> Wood: Ninety-six 20-ton cars = 1,920 tons capacity; dead weight 990 tons; train length 2,400 feet
>
> Steel: Thirty-five 55-ton cars = 1,920 tons capacity; dead weight 635 tons; train length 1,070 feet

Here we find the big steel cars producing a train less than half as long with a dead weight nearly 40 percent less, 488 fewer wheels, brake shoes, and journal bearings, and 122 fewer couplers and draft gears. This example is rather slanted in favor of the new steel cars. By the late 1890s 20-ton wooden hoppers were obsolete and had been superseded by larger 30-ton units. Likewise, most of the steel hoppers carried no more than 50 tons. Refiguring the balance, then, yields results less favorable to steel:

> Wood: Sixty 30-ton cars = 1,800 tons capacity; dead weight 765 tons; train length 1,800 feet
>
> Steel: Thirty-six 50-ton cars = 1,800 tons capacity; dead weight 620 tons; train length 1,120 feet

Repair cost considerations certainly favored big steel cars. Accountants couldn't seem to agree precisely on the cost of wood-car repairs; the figures ranged from $35.00 to $95.98 per year. But there was general agreement that they well exceeded repair costs for steel units, which would not be much over $15 per annum.[101] One railroad made an investigation of cost benefits and concluded that it should retire "4,000 coal and coke cars ranging from 9 to 22 years in age, and having from 40,000 to 60,000 lbs. capacity. It was shown that these cars, on the average, cost $95.98 a year for repairs, or 37.8 percent of the average value of the cars. It was shown conclusively that the company could buy 3,000 new steel cars having a total capacity 20 percent greater than that of . . . 4,600 wooden cars, and out of the amount it would cost to maintain the wooden cars for one year they could pay 6 percent interest on the cost price of the new steel cars and have remaining over $215,000."[102]

Indeed, the repair costs for wooden cars were becoming a burden. There were several reasons: developments in wood-car design, changes in the timber supply, even consequences of the railroad safety movement. In the twenty years since 1870, wood-car design had been vastly improved and enlarged. Length had increased from 28 feet to 36 feet and capacity from 10 tons to 30 tons; more impressive, dead weight had been reduced by as much as 37 percent.[103] Draftsmen had found ingenious ways to lay out high-capacity frames made from massive timbers clustered to form a heavy bridge. Master mechanics were justly proud of these achievements, but even a veteran wood-car man like David L. Barnes was ready to concede that conventional materials and solutions had been taken to their logical limits. Frames required the best grade of timber for maximum strength and durability, but by the late 1890s the virgin forests were about gone. Prime oak and yellow pine were scarce and costly; only Oregon fir remained in good supply, and it was not as strong.[104] The quality of the available wood was therefore producing weaker freight cars at the very time when operating crews were taking a more cavalier attitude toward the care and handling of their charges. Switching crews were becoming heavy-handed, now that their fellow workers no longer had to stand between cars to manipulate link-and-pin couplers. By 1900 these dangerous devices were fast disappearing in favor of the new Janney couplers, which required only a bump to lock together. But this bump too often became a timber-smashing crash. The repair lines grew, and sturdy steel-framed cars began to look even better.

In September 1896 *Locomotive Engineering* sensed a growing seriousness among the wood-car lobbyists and sought to calm their fears by reprinting a piece from another magazine that rarely dwelled upon car-building matters. The *Hard Wood Record* appealed to lumber suppliers who sought assurances about the future of the wooden age. The editors obliged with a series of real and contrived consolations. The supposed extra weight of metal cars would overburden track and bridges and so discourage their adoption. Greater braking capacity would be needed to stop steel-car trains. Metal cars would suffer damage from cargoes like sulphuric coal and chemicals. (This prediction was indeed partially fulfilled, for early steel hoppers did tend to rust out quickly before the introduction of copper-bearing steels; much later, rust-resistant alloys like Cor-Ten steel became something of a standard for coal hopper cars.) In an unaccustomed display of concern for the workingman's well-being, the editors also expressed dismay over the slipperiness of smooth steel-plate floors—a misplaced worry, because most steel cars retained wooden flooring.

In short, according to *Hard Wood*, the metal freight car had already proved itself a failure. Prussian railways had employed metal cars since 1870 and maintained careful cost records on them and on the wooden goods wagons then in service. The *Hard Wood* publishers insisted that these records showed wooden cars to be far cheaper to maintain, and that the Prussians had decided to return to wooden construction at least for the larger-size cars, even though timber costs were far higher there than in North America. It was contended that while minor damages to wooden cars could be fixed on the spot, such defects on an iron car required its return to the shop. The argument here is more than a little off the mark, for Prussian railways did not in fact return to wooden construction.

The *Hard Wood* people did have one real danger to point out concerning steel versus wood, and that was the monopoly danger the railroad industry might face. The wood-car industry was characterized by a multiplicity of suppliers. Dozens of vendors were available, and new ones easily entered the market when orders were strong. Anyone with a large woodworking shop could enter the car trade without great difficulty. But the fabrication of steel cars was a far trickier business. It took specialized tools and costly plant and equipment. In 1896 only about two firms were capable of producing steel cars. In addition, timber might be obtained from hundreds of suppliers, but steel could be obtained only from a handful of makers, most of whom were about to consolidate into a colossus named U.S. Steel. Yet even the monopoly danger turned out to be something of a red herring, because once the manufacture of steel cars got under way, a bevy of manufacturers entered the field and no one maker ever really dominated the business. *Hard Wood* was dead wrong in saying that its readers should not feel "alarmed by the publication of steel-car scare articles." Wood-car advocates had much to be alarmed about, as the events unfolding late in the 1890s would prove. Let us now examine what some of the steel-car pioneers were up to during this period.

OUTSIDERS: HARVEY AND OTHER PIONEERS

During the first years of the steel car it appeared that history was about to repeat itself. Again it appeared that outsiders would act as missionaries whose alien doctrines would be either ignored or resisted by the natives. The principal missionaries were again nearly all from outside the railroad industry. Many were obscure individuals who might wrongly or rightly be dismissed as cranks or visionaries, but at least a few were railway supply men of some prominence, like Samson Fox and Charles T. Schoen, and at least one was one of the world's greatest capitalists, Andrew Carnegie. But in fact history did not repeat itself, however much the first scenes recalled the old scenario.

The man who opened the story was the one least likely to impress a crusty old-line master car builder. He was R. P. Lamont, an engineering student whose 1890–1891 thesis project was to design an all-steel boxcar. While working on the drawings, this eaglet found work at the Michigan Central car shops in Detroit (Fig. 8.36). He was coached and to a degree encouraged by his fellow workers. The resulting design was reviewed favorably in the engineering trade press.[105]

Lamont planned for a 5½-inch channel-iron frame stiffened by truss rods. His drawing showed conventional arch-bar trucks, but he suggested the substitution of pressed steel. The body was to be fabricated from ribbed sheet metal. No cars appear to have been made on this plan. A less imaginative plan was executed at the same time (1892) by the Norfolk & Southern Railway's superintendent of motive power, G. R. Joughins. The resulting steel flatcar made of 8-inch channel beams carried a test load of 40 tons.[106] Meanwhile, a far more ambitious attempt to introduce steel freight equipment was under way in Chicago.

Ironically, the latest advocate for metal freight cars was a veteran lumberman named Turlington W. Harvey (1835–1909), who built a huge business consisting of sawmills, lumberyards, and forest properties scattered across the country.[107] Like so many other wealthy Chicago businessmen, Harvey wanted to emulate the region's most admired capitalist, George M. Pullman, by opening a model industrial town. Harvey picked up a large parcel of land 3 miles south of Pullman, Illinois, and began developing it in 1888. Among the first manufacturers proposed for the town of Harvey was the American Fire Proof Steel Car Company. The firm intended to market passenger cars based on the 1887 patent of William W. Green and James Murison.[108] Harvey sponsored the enterprise, but it soon became apparent that railroads would not patronize the offerings of Messrs. Green and Murison. George L. Harvey (d. 1923), a son of the lumber baron, felt that buyers could, however, be found for steel freight cars. Young Harvey argued

EARLY IRON AND STEEL CARS

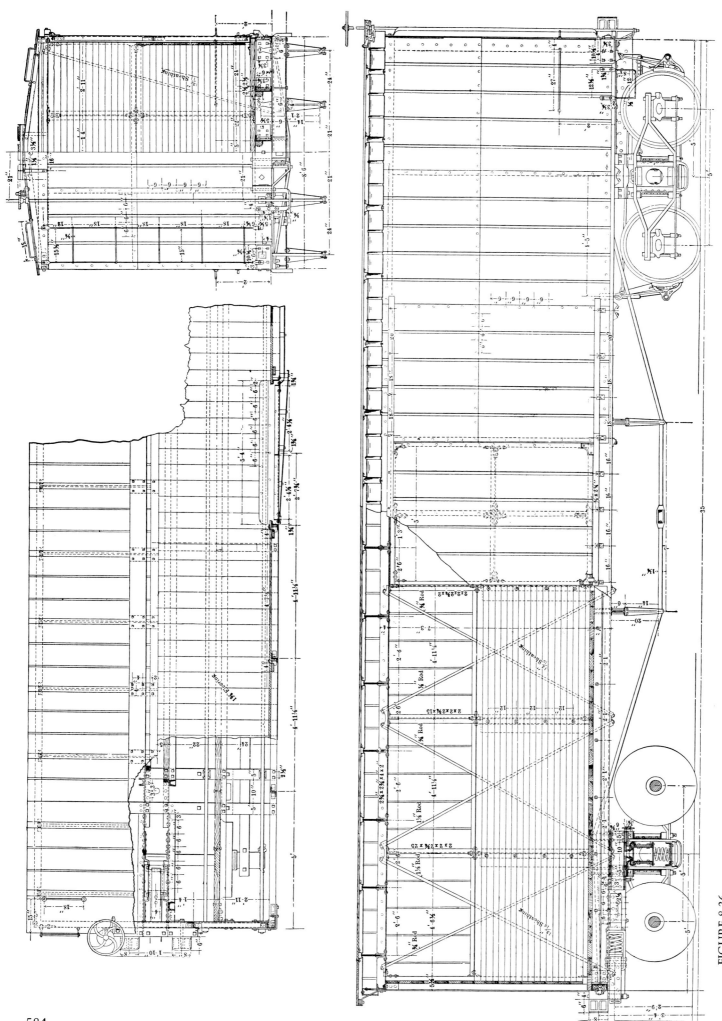

584

FIGURE 8.36

Traditional car design was
adapted for steel construction
by a young engineer named R. P.
Lamont, under the tutelage of
seasoned car builders. (Western
Railroad Club *Proceedings*, 1891)

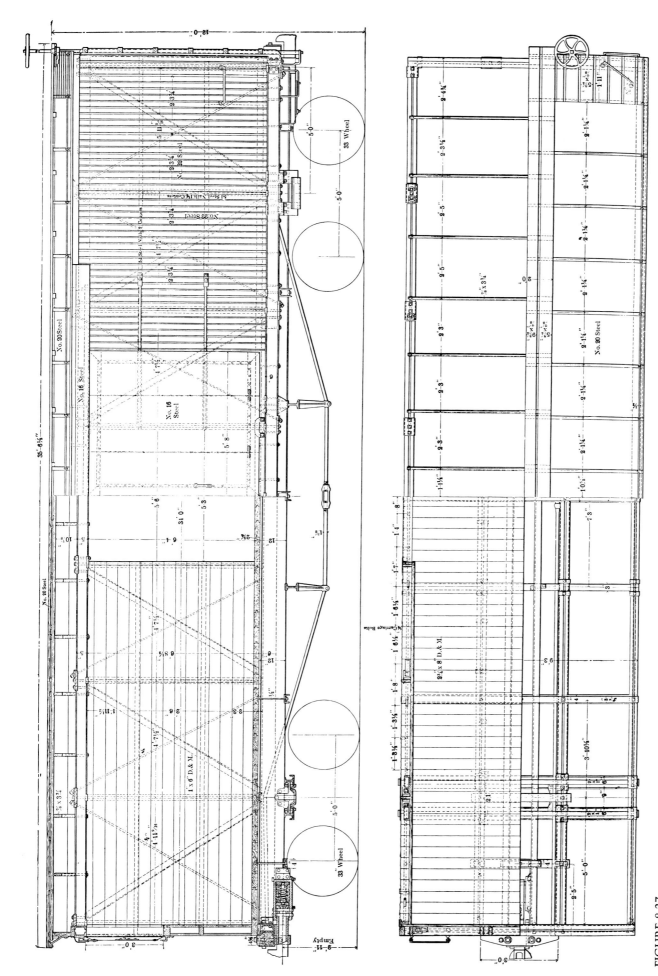

FIGURE 8.37

George L. Harvey's box was basically a steel-frame car with a wooden body covered over with light sheet metal. (*Railroad Gazette*, June 5, 1891)

EARLY IRON AND STEEL CARS

FIGURE 8.38

End view of Harvey's boxcar.
(*Railroad Gazette*, June 5, 1891)

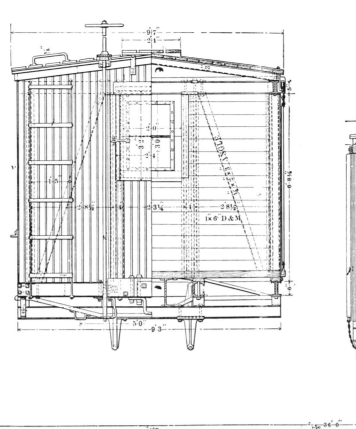

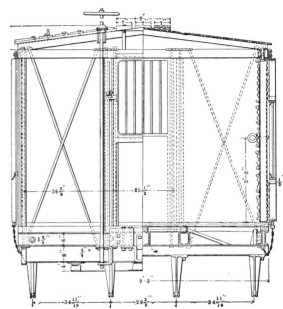

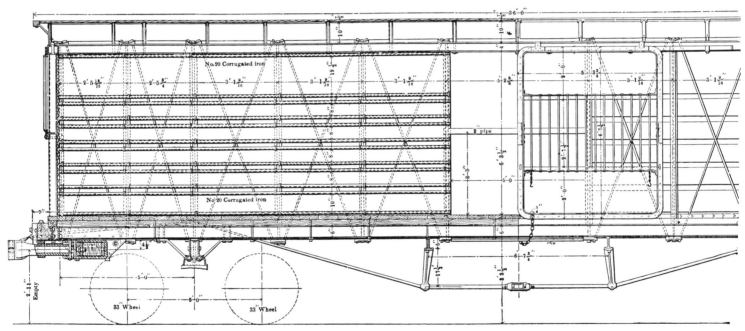

FIGURE 8.39

Harvey's original design for a steel stock car was nearly all steel except for the floor and roof sheathing. Later designs used more wood in the body structure.
(*Railway Review*, Dec. 13, 1890)

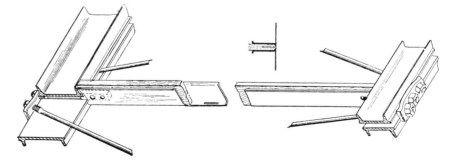

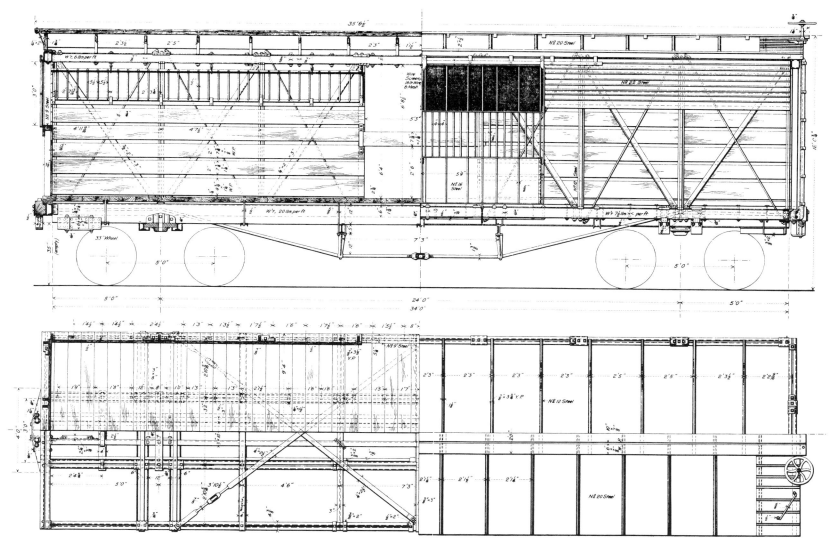

FIGURE 8.40

Harvey's mature stock car design shows a nearly total use of standard steel shapes for the framing members. (*Engineering News*, July 11, 1891)

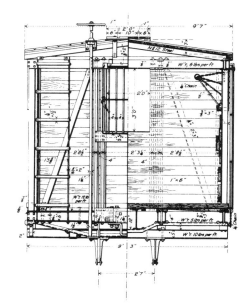

that such cars could be made from the standard offerings of rolling mills, such as I-beams and channel beams, without resorting to more specialized and expensive shapes. Even while the new plant was being erected, Harvey proceeded to have several sample cars built at the Burlington's shops at Aurora, Illinois.[109] The first unit, a gondola, was ready in October 1890. It was an unremarkable specimen that looked very much like a wood-frame car of the day, complete with truss rods. It measured 34 feet long and had low wooden sides. What made it special was its weight, a mere 23,500 pounds for a car rated at 60,000 pounds capacity. Aurora built two steel hopper cars in December at Harvey's request that again were very closely modeled on conventional wooden designs.

The car shop at Harvey opened in May or June 1891.[110] The main building intended for assembly as first projected was 450 feet long and wide enough for five tracks. Overhead cranes were available. An assembly-line process was spoken of, whereby assembly would progress step by step to delivery to the paint shop. Capacity was given at fifteen cars per day. But when the works opened, the assembly shop was scaled down to 300 by 100

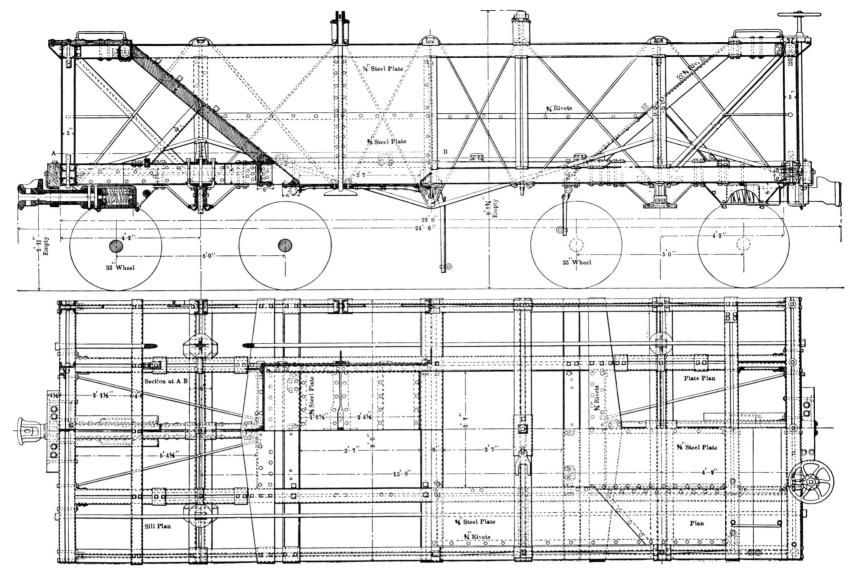

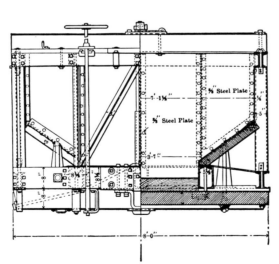

FIGURE 8.41

Harvey's ore car was an un-imaginative copy of wood-car design. (*Railway Master Mechanic*, June 1891)

feet, and only seventy-five men were employed. The plant was more than adequate for the business transacted as events developed over the next two years.

The younger Harvey had used the time available before his own plant opened to devise detailed plans for box-, ore-, and stock cars (Figs. 8.37–8.43). He followed the policy established with the sample cars and took care not to create something radical that might excite criticism among the car-building fraternity as it existed in 1890. His designs in general sided with the safe and the orthodox. Nor would he employ exotic pressed-steel shapes; he made do with standard 6- and 12-inch channels cut and bolted together with simple brackets and plates. He wanted an assembly that a common mechanic could understand and repair, and one that did not require the services of a skilled blacksmith.[111] Repair shop managers were promised that they could call upon any ordinary steel supply house for repair materials.

The boxcar measured 35 feet 6 inches in length and weighed 28,900 pounds (Figs. 8.37, 8.38). It had 12-inch channels for center sills, which weighed 20 pounds per foot. The roof planks were covered with sheet metal, and the ends were sheathed with corrugated sheets. In his initial design (1890–1891), Harvey planned for a steel body frame made up of U-posts, stiffened by angle iron and rod X-braces. The upper cord was a 5-inch channel.[112] The outside sheathing was to be 22-gauge corrugated steel. The door was also fabricated from sheet metal. The interior was lined with wood. But it appears that Harvey compromised after a time, for later drawings show greater use of wood in the body. Except for the body and floor framing, wood is the dominant material. It should also be noted that Harvey gave up the U-iron posts and carlines (roof beams) in favor of ordinary T-irons.

The 30-ton-capacity stock car was of a similar plan; however, less wood was used for the body (Figs. 8.39, 8.40). The body posts and carlines were U-shaped steel. In the first design, reproduced in the 1890 *Railway Review*, truss rods in an X pattern stiffened the body. The body slats were formed from corrugated metal, but according to sources published ten months later, the compression members of the body truss were made from 2-by 3-inch angles, and the tension members were rods.[113] Only the upper panel—something like a very broad letter board—was made of sheet metal. The lower slats were wooden.

Harvey used a standard frame for the gondola, box-, stock, and refrigerator cars, a scheme sure to win the approval of every practical car builder in the country, at least in theory. But the standard frame would not work for hopper or tank cars, and

Harvey never progressed very far with plans for the hopper. Drawings for an ore car indicate that his thinking in this area was not very fruitful (Fig. 8.41). The design was an unimaginative copy of a wooden car that made very poor use of the new material. In no way did the design capitalize on the skin or body unit strength of the hopper itself. Harvey did better when it came to tank design. The result was a clean-looking vehicle with a modern outline. The first tank car was nearly finished in December 1892.[114] A car of this type was exhibited at the Columbian Exposition during the following year (Fig. 8.42).

With working designs in hand, Harvey waited for the orders to roll in. The initial response was gratifying. The Milwaukee Road purchased twenty-five stock cars. Other orders came from the Burlington, the Chicago & Erie, the Lake Shore, and the Calumet & Blue Island railroads. The Hutchenson Refrigerator Car Line, the Illinois Steel Company, the American Cotton Seed Oil Company, and the Western Rolling Stock Company also bought from Harvey. Even Canada's Grand Trunk made a purchase. Business was so encouraging that the plant was enlarged in 1892.[115] But repeat orders appear to have been rare, and as new car orders drifted away Harvey opened a repair department to keep his men employed. The Panic of 1893 did not help matters. Even Harvey's father was undoubtedly hard-pressed by the business downturn, and the firm's benevolent angel may not have been able to save the son's pet enterprise. The plant closed amid reports that the managers of the car business spent too much time developing

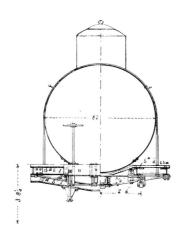

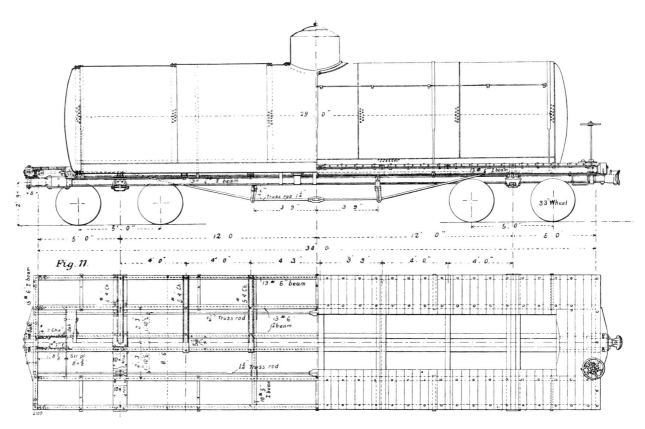

FIGURE 8.42

Harvey's tank car, less the truss rods, looks like a twentieth-century design. (James Dredge, *Record of the Transportation Exhibits at the World's Columbian Exposition of 1893* **[London, 1894])**

EARLY IRON AND STEEL CARS

FIGURE 8.43

Harvey's later designs made more
skillful use of standard steel
shapes. Note how a 5-inch I-beam
has been bent with a simple
bulldozer to form a needle beam
for the truss rod queen posts.
(Western Railroad Club
Proceedings, 1893)

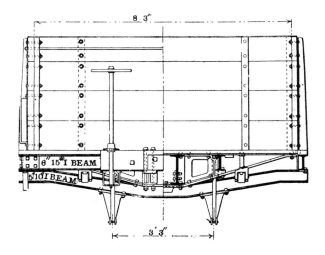

the town and not enough introducing a new technology to a tradition-bound industry.[116] By the time it closed, Harvey's firm had built around one hundred steel freight cars.

INSIDERS: TAYLOR, FOX, CARNEGIE, AND SCHOEN

At the very time Harvey's dream was evaporating in the dismal afterglow of the 1893 panic, the assistant to the president of Carnegie Steel, Charles L. Taylor (1857–1922), was traveling in Europe.[117] During his journey, Taylor was impressed by the thousands of metal-framed freight and passenger cars performing everyday service. They were not outlandish experimental units; many were veterans with decades of use. Taylor returned to Pittsburgh convinced that the time for steel railroad cars had already arrived and that American railroad managers must be made aware of their dilatoriness. Taylor's enthusiastic report was well received at Carnegie Steel, where orders for rails and other staples were on the decline. The plant managers were pessimistic in general about the prospects for the rail market, because the national railroad network was just about complete, and major new construction was not likely. They were seeking new markets.[118]

Taylor discovered that some efforts were already under way to promote steel freight cars, and if the railroads would not encourage these efforts, then Carnegie Steel would become the patron of the steel car. There had already been enough of talk, plans, and patents; what was needed was cars. Practical, working examples to demonstrate the good sense of the idea must be constructed. And if the mossback railroads wouldn't order them, then Carnegie Steel would. Taylor found that three car manufacturers were actively working toward the same goal. They were Willard Pennock, one of the owners of the Minerva, Ohio, car company; Charles Schoen, a manufacturer of pressed-steel car parts in Pittsburgh; and Samson Fox, an English steelmaker who had opened a branch plant in Joliet, Illinois, to produce freight car trucks. The last-named of the trio was already experienced in

the new art of steel construction and was surely one of the most interesting figures involved in this specialized area of technology.

Samson Fox (1838–1903) was born in Bowling, Yorkshire, the son of a poor weaver (Fig. 8.44).[119] At age ten he was sent to work in a textile mill, but he broke away from this drudgery within five years to become an apprentice toolmaker. He showed natural talent as a mechanic even as an adolescent, and his reputation grew rapidly. Fox opened his own toolmaking business when in his late twenties. He showed such management skills that in 1874 he was able to organize the Leeds Forge for the manufacture of corrugated firebox marine boilers of his own invention. They proved a wonderful success, and Fox was soon a wealthy man. He became an acquaintance of the Prince of Wales as well as a generous patron of music.

During his busy life Fox generated 150 patents, most of them dealing with the forming of steel plate. His fascination with pressed steel led to designs for metal railroad cars. He was inspired to devise a lightweight car after seeing a 200-pound man riding a 33-pound bicycle.[120] Fox wondered: Why can't railway cars have a weight-to-capacity ratio more on that scale? He also felt that wooden car frames had too many parts; a Great Western goods wagon, for instance, had sixty-eight individual pieces. Fox's neat, clean underframe had only thirteen parts and weighed 35 percent less.

In 1887 Fox began producing steel freight cars with $5/16$-inch-thick steel parts made by his hydraulic press. Customers were found in England, Belgium, Argentina, India, Spain, Japan, and Bengal.[121] North America, with the world's largest railway network, appeared to be such a natural market that Fox decided to build a branch plant in the center of the United States. In October 1888 the Joliet, Illinois, works opened. A number of lines purchased pressed-steel trucks and other freight car parts, but none would place an order for a complete car. By 1892 two hundred trucks had been sold to North American lines.[122] Carnegie Steel's order for seven flatcars in 1894 was apparently the first placed at Joliet. Delivery was made in the fall. The 34-foot-long cars weighed 25,150 pounds and were rated to carry 40 tons. The design, as worked out by the maker and Taylor, appears to have strayed considerably from Fox's plan, because the main frame was made from I-beams and not pressed-metal plates. They nonetheless proved successful and were still in daily service in 1903, according to a statement published that year.[123]

At about the same time that Carnegie was contracting with Fox, Taylor approached Willard Pennock. He and a brother had operated the Minerva Car Works since around 1875, and they had built thousands of garden-variety wooden cars of no particular distinction. About the time Harvey was

FIGURE 8.44

Samson Fox (1838–1903), a British pioneer of steel freight cars who also attempted their manufacture in the United States. (*Railway Age*, Oct. 30, 1903)

starting up his plant south of Chicago, Pennock decided to perfect his own idea for a steel car. In 1892 he published drawings for a gondola with a unitized style of construction, whereby the frame and body were combined as a single structure.[124] The scheme was not remarkably well executed, being a complex assemblage of bolt-together plates and tie rods (Fig. 8.45). Pennock worked up plans for flat- and boxcars and claimed that sample cars would be produced, but nothing appears to have actually been done. The car builder remained intrigued with the project and gladly accepted Taylor's order to build a steel flatcar.

The car, delivered in the summer of 1895, was much improved over the early plan. It had a bolt-together box-steel side sill, fish-bellied in shape, and 12-inch I-beams for center sills. It weighed 22,620 pounds complete with its Fox pressed-steel trucks.[125] Less trucks, the weight was 11,780 pounds. The car was conservatively rated for 30 tons, but when loaded with 59 tons of steel billets, it deflected only ⅝ inch at the center, and the test engineers felt that it could withstand up to 104 tons before failing. While its weight was only 1 or 2 tons less than that of an equivalent wooden car, its capacity was far greater.

Pennock hoped to expand his role in the introduction of the metal car by becoming associated with the Illinois Steel Company, which had taken over Harvey's plant. The works, reopened in 1896 under the name of the Universal Construction Company, announced that it would produce cars after the patented designs of Harvey and Pennock.[126] Illinois Steel planned to transform iron ore into steel, steel into sheets and shapes, and these products into finished cars—the perfect vertical monopoly. A flat was finished in June 1896 at the Harvey plant; again the buyer was Carnegie Steel. It was loaded with 70 tons, which deflected the frame by 1⅛ inches but caused no permanent set. A train of five steel gondolas was exhibited the following month at Chicago's Lake Shore station by the Universal Company. They were boldly lettered, "Indestructible Can Not Burn, Splinter, Shrink or Decay." A 40-ton ore car produced by the firm was displayed at the Master Car Builders Association annual meeting during the same summer. It was built to Pennock's drawings and registered a weight of 26,480 pounds.

Meanwhile, Taylor was testing the flatcars around the yard trackage at the steel plant. They worked very well. He then sent some of them on a cross-country trip carrying heavy steel plate to the West Coast. Again they performed admirably. Taylor now decided that some all-metal hopper cars were needed to demonstrate the merits of steel. He worked on the actual design with J. B. Hardie, an engineer in Carnegie's employ. He decided to use the Pennsylvania Railroad's class GG wooden hopper for a model.[127] The class GG was among the

newest and largest self-clearing hoppers in service and might be considered the most advanced wooden hopper car then in existence. These cars measured 30 feet long, weighed 35,200 pounds, and had a rated capacity of 35 tons. Hardie and Taylor designed a car of the same size—the sides were actually 3 inches higher than those of the class GG—but the steel hopper weighed 39,950 pounds, or 4,750 pounds more than its wooden rival. However, the steel hopper would easily carry 50 tons; hence the dead-weight ratio was much in its favor.

Taylor would not entrust his plans to either Fox or Pennock but had two sample cars fabricated at the Keystone Bridge Works, a Carnegie Steel subsidiary (Fig. 8.46). Fox did, however, furnish the trucks. The finished cars were tested with 125,000 pounds of wet sand, or 12.5 tons above their rated capacity. There was no noticeable deflection. The cars were shown at the Master Car Builders Association annual convention with no immediate effect. Even with Carnegie sponsorship, the steel car was still viewed as experimental—a good idea no doubt, something to consider, but a subject that surely needed more study. The car builders did publish a drawing of the Carnegie hopper in their report for 1896 but could make no definite recommendation on its merits or defects despite two more years of study. In 1899 the MCB Association didn't even bother to comment on this, the most topical subject facing the managers of the nation's car fleet.

Taylor and Carnegie came to understand that master car builders would not react even after a number of successful demonstrators were placed in service. Here was a repeat of the indifference or hostility promoters of metal cars had traditionally faced. But Carnegie was no ordinary promoter; he was a man of immense wealth and power. He in fact was assembling his own railroad, the Pittsburgh, Bessemer, & Lake Erie.[128] It might not be the equal of the New York Central or the Union Pacific, but it was a substantial trunk line that would move a large traffic—coal, limestone, and iron ore—from Lake Erie to Pittsburgh. For years Carnegie had complained about excessive freight rates; now he would have his very own line for the carriage of raw materials.

While the new southern end of the line was under construction (April 1896–October 1897), Taylor saw an opportunity to push his pet scheme. The new railroad would need rolling stock, and for nearly two years Taylor had been indoctrinating his boss about the vast potential of the steel-car market. Carnegie himself saw the PB&LE as a test laboratory for such exotic products as metal crossties, so why not steel freight cars as well? With this decision in hand, Taylor began searching for a supplier. The Keystone Bridge Works wasn't really set up for the task—a few experimental

FIGURE 8.45

Willard Pennock's idea for a strong car was a solid bed of inverted steel channels bolted together. Pennock did build some steel cars, but on a more economical plan than the one shown here. (*National Car Builder*, April 1892)

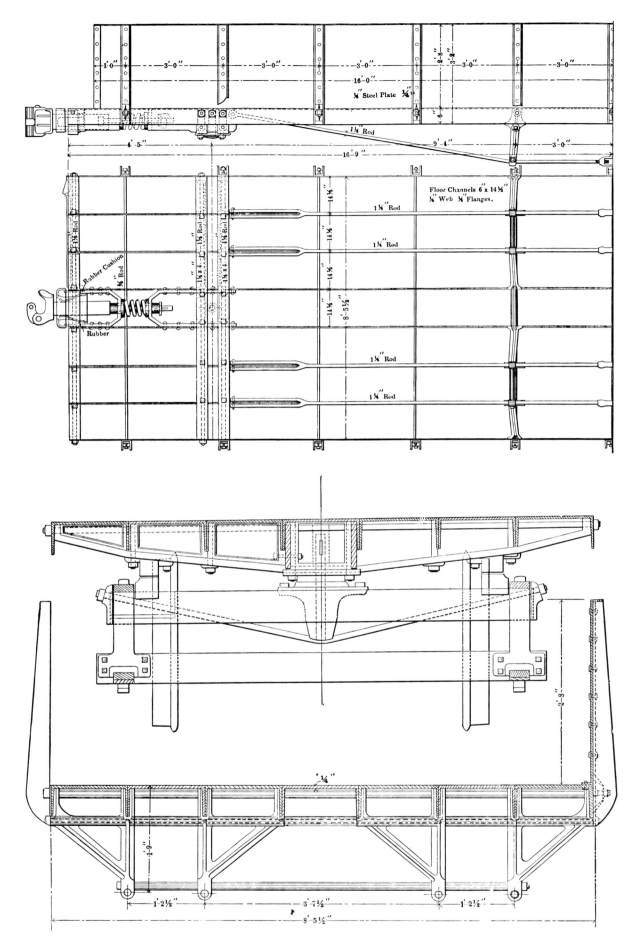

THE AMERICAN RAILROAD FREIGHT CAR

FIGURE 8.46

Charles L. Taylor, an official of Carnegie Steel, had several experimental cars built between 1894 and 1896. Pictured here is one of two hoppers built by the Keystone Bridge Works in 1896. The photo shows it after its retirement in 1928. (Bessemer & Lake Erie Photo)

FIGURE 8.47

Charles T. Schoen (1844–1917), the American counterpart to Samson Fox, was a pioneer of the commercially successful steel freight car. (*Railway Age*, Feb. 24, 1899)

units, perhaps, but not cars in quantity. The railroad needed six hundred hoppers just to begin service and would require hundreds more once the traffic really began to roll. Time was short; the railroad would open in just six months. Early in 1897 quotations for two hundred cars were sought from all interested steel fabricators.[129] Fox and Pennock failed to win the order. In April it was announced that a contract for six hundred, not two hundred, steel hoppers had been awarded to the Schoen Pressed Steel Company of Pittsburgh.

Charles T. Schoen (1844–1917), like Samson Fox and Andrew Carnegie, started life as a poor boy (Fig. 8.47).[130] Following his father's example, Schoen began his working life as a cooper in Wilmington, Delaware. He left the barrel-making trade and in 1865 moved to Philadelphia, where he learned to make springs. In later years, after mastering the art of metalworking, Schoen and a brother opened their own business. Springs led them into the production of other railway supplies, which in turn inspired a general investigation of freight car design. Schoen, still an obscure parts manufacturer, was soon bitten by the metal-car bug. In 1888 he began to market pressed-steel stake pockets (for flatcars), corner irons, and center plates, which he rightly claimed to be stronger and lighter than the cast-iron fittings they were intended to replace. In July 1890 he patented a general design for a steel car.

A few weeks before the patent was issued, the *Railroad Gazette* published a drawing prepared by the inventor for a steel gondola. It was much praised by journalists, not so much for its ingenuity as for its resemblance to a conventional wooden car.[131] This idea was reinforced by the truss rods, wooden floor deck, and arch-bar trucks (Fig. 8.48). Only the pressed-steel ends, side panels, and I-beam floor sills could possibly offend the most conservative master car builder. Schoen stated that sample cars would be made once his new factory in Pittsburgh was opened. But even after his removal to the Iron City in 1892, there is no evidence that his design was put into tangible form. It is likely that he pursued his idea only in the drafting office, until receiving a tender to bid from Taylor some five years later.

Taylor wanted cars built to a design worked out by himself, Hardie, and a young Keystone Bridge engineer named Guy M. Gray. The design was a modification of the 1896 test hoppers (Fig. 8.49). Schoen wanted to build after his own plans. A compromise was effected whereby four hundred of the order would be made on the Carnegie plan and the remainder on drawings worked up by Schoen and his assistants, George I. King, John M. Hansen, and Emil C. P. Swensson (Fig. 8.50).[132] The six hundred cars were to have a great variety of trucks— Schoen, Fox, Vogt, and Kindl.[133] Each hopper was to cost $1,000, or roughly twice the cost of a wooden hopper. But no wooden hopper, at any price, would carry 50 tons or last thirty years. Delivery was to begin in ninety days.

The drawings furnished by Taylor called for a vehicle fabricated from standard rolling-mill shapes (Fig. 8.50). The side frames were very deep but light channels. Fabrication was just a matter of cutting, punching, and riveting flat plate and rolled sections. Schoen and his designers felt this was all wrong. It was crude and unscientific, repre-

EARLY IRON AND STEEL CARS

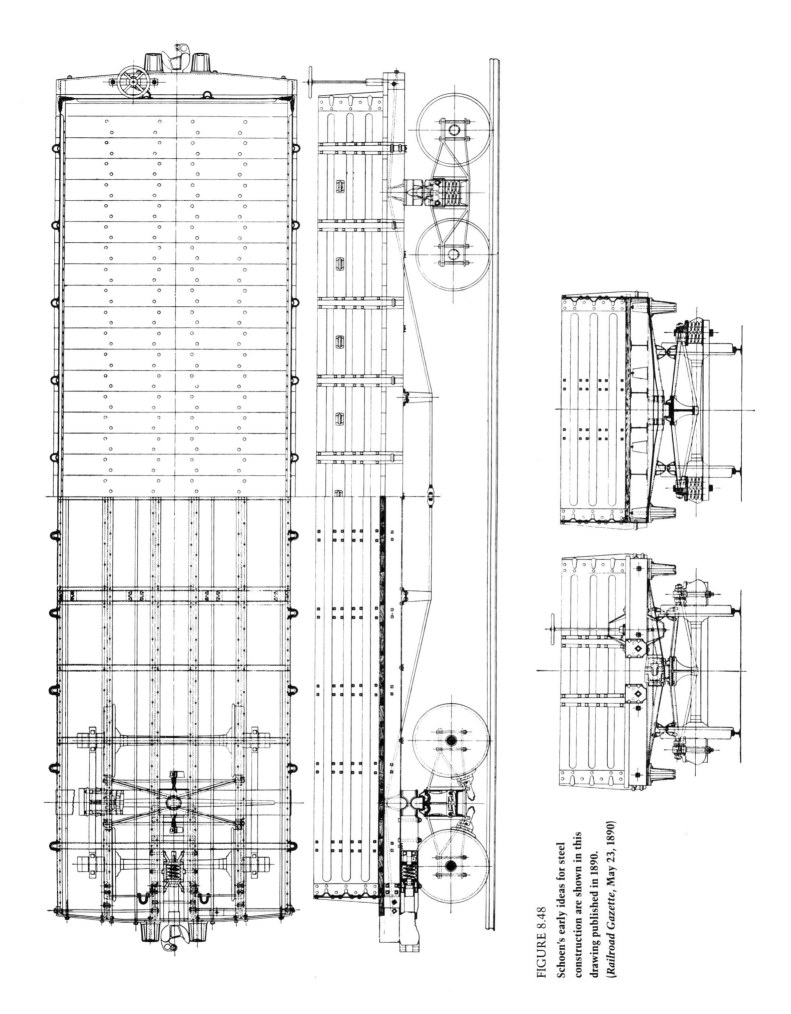

FIGURE 8.48

Schoen's early ideas for steel construction are shown in this drawing published in 1890. (*Railroad Gazette*, May 23, 1890)

FIGURE 8.49

Drawings of the first production-model all-steel freight cars built in the United States. Two hundred cars were produced on this plan by Schoen in 1897. (*American Railroad Journal*, May 1903)

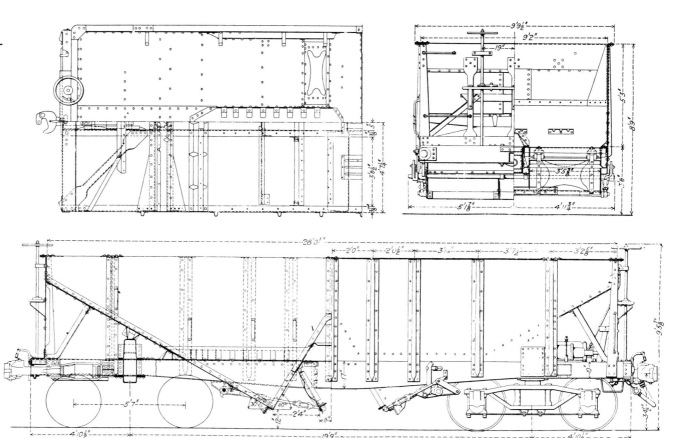

FIGURE 8.50

One of the Carnegie Steel hopper cars built by Schoen in 1898 that used standard shapes. It was carried on Vogt trucks. This photograph was made after the car's retirement in 1928. (Bessemer & Lake Erie Photo)

EARLY IRON AND STEEL CARS

senting a waste of material and needless extra weight. Each structural part of the car should instead be a custom shape designed for maximum strength and minimum weight. It should be strong where the strength was most needed. The side sill, for example, should be big in the middle and smaller at the ends. In addition, pressed-steel parts could be made in odd shapes that would eliminate the need for a multitude of small parts—which again would mean greater strength, fewer fastenings, and less weight. Stamped parts could also be made with ridges or depressions, which added stiffness yet required no more material.

In theory, pressed-steel construction made more economical use of materials and should result in weight saving of 10 to 11 percent.[134] The actual weight saving of the pressed-steel over the structural-steel cars was 2,800 pounds, or about 9 percent. But it was achieved at a price. Costly tooling was needed to produce the parts, and unless vast numbers were made, unit costs were high. Because car building tended to be a custom business with each buyer having his own ideas, standardization was more talked about than practiced. Stocking and shipping these special parts added to the expense and inconvenience of repairs. If a car could be repaired with a piece of standard channel or angle, almost any nearby supplier would suffice. In the end, most car builders decided to compromise—use a few pressed-steel parts where they rendered the maximum benefit and otherwise make do with standard shapes.

During its earliest years Schoen's plant was zealous in adhering to the absolute doctrine of just the right shape for each part. Drawings of the original PB&LE hoppers illustrate the details of the design (Fig. 8.49). They should be compared with the Carnegie structural cars (Fig. 8.50). Both sets of cars were rated at 50 tons capacity, and their weights were very similar—37,150 and 34,350 pounds, with the Schoen cars registering the lesser weight. These pioneers proved to be all their designers could hope for. They were instant celebrities. One journal correctly said that they would be more carefully watched than any freight cars ever built.[135] They ran until 1928. Three were preserved at this time, including one of the 1896 Taylor cars. They were among the most historic freight cars in America and deserved to be saved for future study, but misguided patriotism sent them to the scrap yard during World War II.

THE REVOLUTION IS COMPLETE

It had happened at last—a revolution in freight car construction. It represented a sharp break with past notions and practices. As expected, the Carnegie-controlled Pittsburgh, Bessemer & Lake Erie came forward with large repeat orders. In 1898 it purchased four hundred more hoppers from Schoen. During the next two years the PB&LE ac-

quired yet another seventeen hundred gondolas and hoppers.[136] In 1898 the Pennsylvania cautiously purchased five sample hoppers from Schoen (Fig. 8.51). By May of that year the line was convinced of their utility and ordered another one thousand. Within four years the Pennsylvania had ten thousand and in 1905 it bought twenty thousand. By late 1908 the line was a confirmed convert to steel, having a total of eighty-seven thousand steel and steel-framed freight cars.[137] During this same period other lines began to accept the steel car. The Erie bought a group of steel-framed boxcars from the Michigan-Peninsular Car Company of Detroit in 1897. The undercarriage was furnished by Fox.[138] Two years later the Michigan Southern ordered one thousand cars from Schoen.[139] In 1902 even the conservative Baltimore & Ohio adopted the steel hopper as its new standard.

The triumph of the steel car made Schoen a success. His business exploded as car orders poured in. For a brief time, at least, his firm monopolized the steel-car market. By June 1899 he had received orders for fifteen thousand cars.[140] A product that had been practically unsaleable two years earlier was now suddenly very much in demand. Schoen sought to satisfy the demand by increasing production. A great new plant was hastily erected on a 180-acre property in nearby McKees Rocks, Pennsylvania.[141] Four thousand more employees were hired. At the same time, Schoen arranged a consolidation of his firm and Fox's American branch works at Joliet as the Pressed Steel Car Company. The new organization was incorporated on January 13, 1899, with a capital of $25 million.[142] A story was published several years later in *The Engineer*, a respected British technical journal, claiming that Fox had demanded payment in gold for his valuable Joliet property, and that Schoen obligingly shipped between 8 and 9 tons of the precious metal to England.[143] The tale has recently been pronounced more colorful than true. No one has denied, however, that the flamboyant James B. Brady (Diamond Jim) was sales manager for Pressed Steel.

Even while Schoen was involved in the whirlwind of new orders and business mergers, he faced a nasty series of negotiations with Andrew Carnegie. The crafty steelmaker smelled huge profits in steel-car building and seriously considered entering the trade himself. Why should a nobody like Charles T. Schoen make profits that could as easily go to Carnegie Steel? By late 1898 Carnegie had decided to build his own plant, against the advice of some of his advisers, notably Charles M. Schwab.[144] It was argued that the steel company should be a supplier and not become more involved in the production of mechanical equipment. Schoen, of course, hoped to prevent the creation of such a powerful competitor. He offered to

FIGURE 8.51

Most major railroads began to buy steel freight cars soon after the initial PB&LE order. One of these 1898 pioneers, shown here, is now in the collection of the Pennsylvania State Railroad Museum in Strasburg. It was built by Schoen and is nearly identical to the PB&LE design. (Pennsylvania Railroad Photo)

TABLE 8.2   Freight Car Production, Steel versus Wood

Year	Steel & Steel-Framed	Wood
1900	29,800	112,000
1905	75,000	90,400
1910	162,600	13,700
1915	53,800	4,300
1920	60,900	0
1925	105,700	600

Source: American Railway Car Institute, *Statistics of Car Building* (New York, 1944); *Railroad Digest,* Jan. 1901, p. 48; and *Railroad Gazette,* Jan. 1, 1906, p. 26, and Feb. 2, 1906, p. 98.

Notes: Figures have been rounded off to the nearest 100. The figures do not include new cars built in railroad repair shops. This could add as much as 25 percent to the totals given, though few repair shops were capable of producing new steel cars much before 1910.

TABLE 8.3   Freight Cars in Service, Steel versus Wood

Year	Steel & Steel-Framed	Wood	Total
1910	555,000	1,580,000	2,135,000
1915	1,177,000	1,081,000	2,258,000
1920	1,516,000	806,000	2,322,000
1925	1,793,000	564,000	2,357,000
1930	1,972,000	380,000	2,352,000

Source: Interstate Commerce Commission, *Statistics of Railways in the United States.*

Note: By 1940 only 3.4 percent of U.S. freight cars were of wooden construction.

buy all of his steel—80,000 to 90,000 tons annually—from Carnegie if he would stay out of the car business. The offer was refused. Schoen then threatened to buy all the major car plants and totally boycott Carnegie. Schoen's imaginary monopoly wouldn't buy an ounce of steel from Carnegie, not a plate, rivet, axle, or washer. But Schoen didn't begin to have the capital (or the backers) to pull off this grandiose blackmail. In another communication between the adversaries, Carnegie demanded a "bonus" of $1 million to stay out of the car business.

In the end a curious settlement was reached, whereby Carnegie built the car plant and leased it to Schoen with the proviso that an annual bonus of $100,000 be tendered and all steel be purchased from Carnegie. The lease was good for only one year, so Schoen was very much under the steelmaker's control. The entire episode showed a less savory side of Carnegie's personality, according to the biographer of C. M. Schwab. It showed Carnegie as the hard trader, ready to manipulate and, if necessary, ruin his competitors. Of course, Carnegie's tough handling of Schoen ended in 1901 when U.S. Steel took over the property and Carnegie retired to devote himself full-time to his charities and fortune. Yet stormy times lay ahead for Schoen. He was forced out of the Pressed Steel Car Company in 1901 by a group of investors, apparently led by Frank N. Hoffstot (1861–1938).[145] Brady and Schoen had already had a falling out. Brady and Pressed Steel's chief designer, John Hansen (1873–1929) collaborated a year later to establish the Standard Steel Car Company in Butler, Pennsylvania.

TABLE 8.4   Steel and Steel-Framed Freight Cars, Class One Railroads Only

Year	Total Fleet	All-Steel	Steel-Framed	Steel & Steel Framed	Percentage
1905	1,731,000	—	—	150,000[a]	8.6
1910	2,135,000	311,000	241,000	552,000[b]	25.8
1915	2,258,000	501,000	676,000	1,177,000	51.8
1920	2,322,000	630,000	886,000	1,516,000	65.2
1925	2,357,000	757,000	1,036,000	1,793,000	76.0
1930	2,352,000	854,000	1,118,000	1,972,000	83.8
1935	1,835,000	809,000	908,000	1,717,000	93.5
1940	1,653,000	956,000	640,000	1,596,000	96.5

Source: Unless otherwise noted, figures are from annual issues of the Interstate Commerce Commission's *Statistics of Railways in the United States* and the Association of American Railways.

a. *Railroad Gazette,* Jan. 12, 1906, p. 2.

b. *Railway Age,* April 1, 1910, p. 891. Reprinted from International Railway Congress *Bulletin,* Feb. 1910; hence figures are more reflective of late 1909 and thus rather low for 1910.

By the opening of the twentieth century it was clear that the railroad industry had at last come to accept the metal freight car (Tabs. 8.2, 8.3, 8.4). In 1901 15 percent of new cars built were of steel. Four years later the figure jumped to 45 percent, and there were now about 150,000 of them in service. This represented 12 percent of the entire fleet—a major achievement for just seven years. It would be almost fifteen years before steel and steel-framed cars could claim a clear majority. The audience was still somewhat hostile, with some master car builders ordering wood-framed cars as late as 1925.[146] There were still supplies of cheap framing timber that could produce an inexpensive freight carrier. It should be noted that the numbers ordered after 1910 were relatively small, as shown in Table 8.2.

In some respects the adoption of the steel car was less revolutionary than it may appear, for it was really more a triumph of steel framing. Except for hopper cars and some gondolas, most so-called steel cars were actually of composite construction. The floor and body frames were all metal, but the body and floor themselves were wooden. Flatcars and even some gondolas traditionally had thick plank flooring. Many gondolas, particularly before 1940, had wooden sides as well. Until relatively modern times boxcars had wooden interiors, which were favored because they protected the cargoes and were easy to fasten to. The decision to pursue composite construction was purely rational, for wood made a perfect secondary material. It was cheap and easily replaced. The short life of freight equipment made expendable wood components ideal. They were largely nonstructural and performed the shelter role rather than the more demanding carrying function. Refrigerator cars were the last stronghold of composite construction, and this was true not simply for the reasons just given but because of insulation concerns. Gradually, less and less wood was used in car construction. The better class of boxcars were made with steel ends and roofs. By 1940 it was a universal practice, and most major roads had given up on exterior wood side sheathing; interiors, however, tended to remain wooden. Care, then, must be taken when speaking about the conversion from wood to steel construction.

There are clear and sensible reasons for why it took almost two generations to introduce the metal car to American railroads. This complex layering of technical, economic, manufacturing, and even psychological considerations has been outlined in some detail in the foregoing discussion. And yet perhaps the best explanation is the simplest one: it was an idea whose time had come. The American freight car had finally received a proper backbone.

BIOGRAPHICAL SKETCHES OF RAILROAD CAR BUILDERS

The selection of biographical sketches presented here does not pretend to be complete or even completely accurate. Some important individuals may have been inadvertently omitted. Some significant achievements of those included may have been overlooked as well. The reason for any such defects is the obscurity of the subject matter. Many of these men were hardly front-ranking individuals even in the world of engineering, much less in the society at large, and hence their lives are poorly recorded. We can find little more than birth and death dates plus sketchy outlines of their careers. Most car builders rated little more than a few sentences at the time of their passing. Details on their personalities, management skills, or inventions are notably lacking. Just who introduced innovations like truss rods, arch-bar trucks, or double roofing remains uncertain. And so the selection of lives presented here can hardly be called a definitive list of America's most important railroad car builders. It is at best a sample of the great and the not so great.

Fundamentally, the job of master car builder hardly called for greatness. The job essentially involved good housekeeping. The car builder was expected to keep a ready supply of cars in good repair. He was expected to keep things tidy and going and to make do with a small budget. Dependable, practical, sensible men were best suited for this work. Steadiness and attention to detail were far more important than brilliance, inspiration, or creative genius. There was hardly a need or even a place for many Mozarts or Leonardos in the car department. And because car building was a haven for ordinary men, strong on common sense and weak on glamour, none inspired any insightful biographies. Only one, George I. King, considered his career worthy of an autobiography, and it is a rather slim volume that speaks more to family history than to car building.

Most of the persons described here obtained only a minimal education, and a majority rose up from the mechanical trades. Carpenters, bridge builders, and blacksmiths are well represented, as might be expected, because this survey is limited largely to the nineteenth century when an advanced education was less common, nor was it considered as necessary as it is today. Native intelligence, energy, and motivation were considered the right ingredients for success in the wood-car era. There is also a sprinkling of lawyers, general traders, college graduates, and public benefactors. By 1890 more car builders tended to be graduates of engineering colleges such as MIT. Whenever possible, car builders representing major railroads or commercial car manufacturers have been selected. Others were picked because they were presidents of the Master Car Builders Association or served in the ranks for so many years.

The data presented here is drawn from such obvious sources as the railway trade press (*Railway Age*, etc.); the *Biographical Directory of Railway Officials*, 1st ed. 1885 (later titled *Who's Who in Railroading*); and obituary notices in MCB Association reports. A few car builders, especially those who became rich in the trade, were included in a variety of publications such as the *National Cyclopedia of American Biography*.

Sketches of other car builders not included here but sometimes associated with freight equipment appear in my earlier work, *The American Railroad Passenger Car* (Baltimore, 1978).

DAVID H. BAKER (1811–1894) was the veteran car builder and longtime general foreman of the Pennsylvania Railroad's big car shop at Jersey City. Born in Schenectady, N.Y., he became an apprentice cabinetmaker in that same place at age sixteen. Baker worked briefly in the car department of the Mohawk & Hudson before signing on with the Utica & Schenectady early in 1835 as a pattern maker. He quickly rose to manage the car shop for the U&S. In September 1849 he was appointed by the New Jersey Railroad & Transportation Co. as foreman of the East Newark car shop. Later he was named master car builder and transferred to the line's Jersey City facility. When the Pennsylvania leased the property in 1871, he maintained his old position but received the new title of general foreman. He remained on the job until his death in February 1894. Baker was one of the founding members of the MCB Association and was made an honorary member of that body in 1893 for his long and valuable service to the car-building interests of the nation.

JOHN C. BARBER (1844–1919) was a specialist in freight car truck design and held numerous patents in this field. At the time of his death it was estimated that 20 percent of the trucks in service were fitted with his lateral-play-limiting device. Barber was born in Lawrence County, N.Y., and began his railroad career at the Chicago & North Western Fond du Lac (Wis.) repair shops in 1864. He was a restless sort of mechanic and could not seem to stay put for more than a year or two at a time. He drifted from the Northern Pacific to the Katy to the Texas & Pacific, and then to the Missouri Pacific. In 1885 he retired from railway employment to become an independent inventor. The Northern Pacific lured him back for a few years, but then in 1896 he organized the Standard Car Truck Co. in Chicago and devoted the remainder of his life to the railway supply trade.

DAVID L. BARNES (1858–1896) was born near Providence, R.I., a descendant of thrifty and pious New England stock. When only fifteen, Barnes was apprenticed to a civil engineer and worked as a surveyor. He entered Brown College in 1876 and graduated with an M.A. He also attended MIT but earned no degree before entering the locomotive-building trade in 1879. He eventually rose to the position of chief draftsman at the Rhode Island Locomotive Works. Late in 1888 Barnes was hired by the *Railroad Gazette* and wrote extensively on car and locomotive design topics. Within a few years he became a strong advocate of steel freight cars and was in a position to broadcast his opinions. Despite a heavy writing schedule he continued to find time for consulting engineering assignments. One of these included design work for the Fox pressed-steel car truck.

JOSEPH W. BETTENDORF (1864–1933) promoted the cast-steel freight car trucks that came to replace the standard arch-bar truck. He was born in Leavenworth, Kans., and by age eighteen was working as a mechanic in a plow factory. By 1886 he had prospered enough to open a plant in Davenport, Iowa, to manufacture metal wheels in conjunction with his inventive older brother, W. P. Bettendorf (1857–1910). The agricultural machinery business prospered, and so the brothers decided to enter the railway equipment field. They began to produce steel car parts in the 1890s and were soon producing large components such as I-beam car bolsters. In 1902 the car part business grew so rapidly that they built a steel-car plant nearly a mile in length in a new town, named Bettendorf after the brothers, three miles outside Davenport. The shop could produce twenty-five cars a day. The rattle and hammering at the Bettendorf shop helped sound a finale to the wood-car era.

GEORGE D. BURTON (1855–1918) designed and marketed a palace cattle car starting in the early 1880s. Burton was not the first in this field, but for a time he was one of the more successful private stock-car operators. Actually, it was only one of the many interests of this inventive Yankee born in Temple, N.H. He was particularly active in electrical apparatus design, and by the time of his death he had been granted six hundred U.S. and foreign patents. More information on his involvement with the cattle car trade is given in Chapter 4.

FERDINAND E. CANDA (1842–1920) was both a gifted car designer and a high-level executive manager. Canda was apparently born in New York City into a prominent old French family; his father, for example, had been a general under Napoleon. During the 1870s Canda was in the Chicago area, where he built railroads as a general contractor. He also built freight cars and bridges in a shop located in Blue Island. By about 1881 he had become general manager of the Ensign Car Co. of Huntington, W.Va., but within a few years, though now a vice president with Ensign, he removed to New York City. Canda appears to have become an associate of the financier C. P. Huntington at this time. He continued his contacts with Ensign and became president in 1889. He remained active as a designer, and examples of his work can be seen elsewhere in this volume in sections devoted to cattle cars, flatcars, and the like. Perhaps his most notable designs were the 50-ton-capacity wood-frame cars built for the Southern Pacific in 1899. While Canda remained president of Ensign and a director of American Car & Foundry until at least 1905, his interests drifted into steelmaking, and he established the Chrome Steel Works at Carteret, N.J. Canda patented almost forty improvements for railroad cars, ranging from car wheels to door brackets.

GEORGE F. CHALENDER (1827–?) was an advocate of reinforcing wooden car components with iron. He succeeded in greatly stiffening wooden passenger car bodies with thin iron plates running the length of the body, below the windows. Of more consequence was his iron-bolster swing-motion truck described and pictured in Chapter 7. The truck became known by the name of his associate and partner, Hans Thielsen. Chalender was born in Bordentown, N.J. He began railway work in 1845 as a fireman on the Reading, was advanced to the rank of engineer, and then went west in

1851. After a few years in Chicago he found work on the Burlington & Missouri River as master mechanic in 1856. During his years at the Burlington (Iowa) shops he designed cars such as the ventilated boxcar illustrated in Chapter 4. In early 1875 he left the Burlington briefly for the Michigan Central but returned to become systemwide superintendent of the Mechanical Department. In 1881 he accepted the same position with the Atlantic & Pacific Railway and within a few years was assistant superintendent of that line. He retired in October 1885. No reports of his death have been uncovered.

JOHN T. CHAMBERLAIN (1849–1922) was born in England but came to this country as a child. At age eighteen he was apprenticed to the Atlantic & Great Western's Kent (Ohio) car shop. Two years later he transferred to the Boston & Albany shops in Allston, Mass. He rose to the rank of foreman and served in that capacity between 1878 and 1885. Over the next few years he worked for the Boston & Worcester and the Burton Stock Car Co. before settling in with the Boston & Maine in 1890. He would remain with the B&M for seventeen years. It was during this period that Chamberlain led a spirited attack against the adoption of automatic couplers. Despite this rather cranky campaign, Chamberlain retained the respect of his peers, who ironically elected him president of the MCB Association in 1901 just as the Janney-style coupler was becoming universal for freight cars. He retired from railway work in 1907.

JOHN W. CLOUD (1851–1936) was one of the better-trained persons associated with railroad cars during the wood-car era. Born in Woodbury, N.J., he graduated from the Lawrence Scientific School at Harvard before entering the employ of the Pennsylvania Railroad in 1876. For ten years he worked as engineer of tests at the Altoona shops. Wood, iron, paint, lubricating oil—all materials used in car construction and operations were subject to testing at the Pennsylvania's laboratories. Cloud left the "Standard Railroad of the World" in 1887 for a two-year tenure with the Erie as superintendent of motive power. In 1889 he joined the Westinghouse Air Brake Co. but found time to design independently his patented style of car truck. In 1899 Westinghouse sent him to England to manage its branch works. Cloud was made chairman of the board of the British Westinghouse Air Brake Co. in 1920. He retired eleven years later.

ISAAC H. CONGDON (1833–1899) was best known in railroad car circles for the ingenious cast-iron brake shoe made with wrought-iron inserts he introduced in 1876. Congdon was born in Granville, Mass., and began railway service in Cleveland as a machinist with the Cleveland, Co-lumbus & Cincinnati Railway. After a brief employment back home in New England, he returned to the CC&C as foreman of the machine shop. In 1866 he was selected as master mechanic by the Union Pacific and was stationed at the main shop in Omaha. Some years later a very critical report by a consulting engineer charged Congdon with failure to establish proper engineering standards, modern methods, and tools; this prompted the railroad's management to force him into an early retirement late in 1885. To soften this humiliation, a testimonial and a handsome bonus of $2,500 were tendered. Congdon apparently spent the next fourteen years in retirement and remained in Omaha. He was most likely a better mechanic than an administrator, for his massive smokestack was widely used on the Union Pacific and its subsidiaries. Congdon's snowplow also enjoyed fairly widespread employment as well.

GEORGE W. CUSHING (1833–1906) is remembered as a locomotive designer. He created some very decorative engines for the Chicago & North Western, but he was also very active in car design. Several examples are given in this volume, including the Missouri, Kansas & Texas boxcar illustrated in Chapter 3. He was born in Portland, Maine, and spent his early working years at Wilmarth's Union Works in Boston. In his late twenties he moved west and spent most of his life in that part of the country. He moved on every few years and spent time as mechanical head of the Wabash, the MKT, the Denver & Rio Grande, and the Northern Pacific. He had two terms with the last-named road, in 1871–1873 and in 1882–1887. The following two years were spent with the Reading, but in 1889 he returned west to take over the Union Pacific's mechanical department. What happened next may explain why Cushing moved about so often. He became popular with his employees by freely raising wages. This policy proved most unpopular with his supervisors, however. They were also unhappy with Cushing's administrative talents, or lack thereof. Evidence that he was involved in a shady deal with parts suppliers led to his dismissal in 1892. He found employment with a railway supply company but never again worked for a railroad.

THEODORE N. ELY (1846–1916) was born in Watertown, N.Y., in 1846 and graduated from Rensselaer Polytechnic Institute as a civil engineer in his twentieth year. He worked briefly for a foundry and then a coal mine before settling into a railroad career in 1868. His first railroad job was with the Fort Wayne branch of the Pennsylvania Railroad. His talent and energy soon won him a position with the parent road. In 1873 he was made superintendent of motive power. Within nine years he was made motive power chief for the entire Pennsylvania system. Ely started the chemical and physical testing of materials at the Altoona shops to ensure better and more economical rolling stock. He was a strong advocate of standard designs and larger freight cars (see Chapter 3). Ely retired in 1911 after more than forty years of service to the Pennsylvania's mechanical department, which was responsible for the largest freight car fleet in the nation.

CHARLES L. FREER (1856–1919) was born in Kingston, N.Y. As a young man he became an accountant and worked as paymaster for the Ulster & Delaware Railroad. He later held a similar position with a small railroad in Indiana. In about 1880 he joined the Peninsular Car Co. in Detroit and was made its secretary. He acquired an interest in the firm, and as its fortunes improved, Freer became a wealthy man. He retired when Peninsular was absorbed by American Car & Foundry in 1899. He spent the remainder of his life developing his art collection. The Freer Gallery in Washington, D.C., was founded by profits from the manufacture of freight cars. Freer was one of the few car builders who was also a major art patron.

WILLIAM GARSTANG (1851–1924) was born in Wigan, England, and came to North America with his family when about eight years of age. His education was rather meager; the only specialized training he received after public school was some night-school courses in drafting. He entered the work force as a water boy and at age twelve was apprenticed as a machinist in the Lake Shore's Cleveland shops. About six years later he became a machinist and foreman in the shops of the Atlantic & Great Western. In about 1880 he moved on to the Cleveland & Pittsburgh. Between 1888 and 1893 he was master mechanic on the Chesapeake & Ohio and then went on to take a similar position with the Big Four. He was active in car design; an example of a cattle car designed by Garstang appears in Chapter 4. He was chairman of the Master Car Builders Association standard car-wheel committee for several years before retiring from the railway mechanical field in 1913. Perhaps his greatest monument is the Beech Grove shops, near Indianapolis, Ind., which he designed and oversaw construction for. These shops now serve as Amtrak's main car repair facility.

HUGH GRAY (1807–1885), an early midwestern car builder, was a native of Scotland who emigrated to this country in 1835. He moved to Detroit and worked as a house builder until he found a job in the Michigan Central's car shop. In 1848 he was engaged by the fledgling Galena & Chicago Union and became master car builder for this tiny line that would one day blossom as the Chicago & North Western. Gray is credited with introducing the grain door, so that boxcars might be used as bulk grain carriers, eliminating the need to bag wheat, corn, and oats for shipment. It is also claimed that Gray was an innovator in the use of rooftop handwheels, though like all such claims, this one would be hard to prove. Gray was a vice president and early member of the Master Car Builders Association. He retired from railway work in 1881.

EDWARD W. GRIEVES (1843–1917) was born in the car-building town of Wilmington, Del. He apprenticed as a woodworker and graduated to the pattern shop at the Harlan & Hollingsworth car plant. In time he was promoted to foreman and then chief draftsman. In 1884 he left his hometown employer to become master car builder for the Baltimore & Ohio, staying on at the Mt. Clare shops until 1898. He was president of the MCB Association for two terms, 1893 and 1894. After leaving the B&O Grieves worked in the railway supply field until his retirement in 1915.

GEORGE HACKNEY (1826–?), active as a mechanical supervisor on several western lines, was a native of England and worked for the London, Brighton, & South Coast Railway from 1848 to 1861. He came to the United States in the latter year and subsequently worked for the Chicago & North Western and the Burlington before being appointed superintendent of machinery for the Santa Fe in 1878. The 1880 Santa Fe cattle car illustrated in Chapter 4 was designed during Hackney's time as head of the mechanical department. He resigned his position with the Santa Fe in June 1889. Two of his sons, Clem and Herbert, were also active in the railway mechanical field.

EDWIN M. HERR (1860–1932) was born in Lancaster, Pa., and graduated some eighteen years later from Yale's Scheffield Scientific School. He immediately entered railway service, working as a telegrapher and draftsman. By 1887 he had been appointed engineer of tests at the Burlington Aurora shops. He changed jobs about every two years and shifted between the mechanical departments of several large western roads, including the Milwaukee, the Chicago & North Western, and the Northern Pacific. Herr stayed with the NP for nearly seven years, starting in September 1898. An example of a car designed by him is shown in Figure 3.29. In 1905 Herr decided to leave railway service for the supply trade and went with Westinghouse Air Brake as general manager. He moved up in the Air Brake management to become president of the firm before retiring in 1929.

JOHN HICKEY (1845–1917) was a midwesterner who spent much of his career on far western lines. He was born in Painesville, Ohio, and benefited from some college education before apprenticing as a machinist with the Cuyahoga Steam Furnace Co. of Cleveland. He worked for the Chicago, Burlington & Quincy at Aurora, the Chicago & North Western, and several other midwestern lines. By 1882 he had reached the rank of master mechanic. Between 1891 and 1897 he was master mechanic and superintendent of rolling stock for the Northern Pacific. An interesting NP hopper car dating from Hickey's incumbency is shown in Chapter 5. Hickey hired on with the Rio Grande in 1897. He retired in 1905 as a consulting engineer.

DAVID HOIT (1831–1885) was born in Malone, N.Y. Soon after leaving school he grew disenchanted with life on the farm and took to the road with little more than a bundle and handful of change. After two years of wandering and several modest jobs, he moved to Massillon, Ohio, in 1852 and became apprenticed to a carriage maker. Four years later he went to Fort Wayne and became assistant foreman of the Pittsburgh, Fort Wayne & Chicago car shop. In 1868 he removed to Toledo to become master car builder for the Wabash. Hoit's mechanical and management abilities prompted the New York Central to appoint him head of its car shops in West Albany in 1874 upon the retirement of that facility's veteran superintendent, Joseph Jones. Ten years later Hoit left West Albany to take on the management of the Gilbert Car Works in Troy. His health broke down from overwork; he took a leave of absence and traveled to Bermuda in hope of restoring his health. This remedy failed, and Hoit died there on May 19, 1885. Hoit was something of a mechanical inventor and obtained several patents for improvements to tools and railway car appliances. None appears to have been very notable, but one, a special form of drawbar is illustrated in Figure 5.12.

GEORGE W. ILGENFRITZ (1821–1891) was a pioneer railroad car builder of York, Pa. Details on his early life are sketchy, but he was put to work at the age of twelve. He learned blacksmithing and coach building. By 1845 he was able to open his own shop for the production of agricultural machinery. Railroad car production began some years later. By 1867 he claimed to be making seventy-five to one hundred freight cars a month. He quit the field temporarily at about this time after a disagreement with his partner, Michael Schall. Doing nothing proved difficult for a man who had gone to work before entering his teens, and so Ilgenfritz reentered the car business and remained active until age and health forced him to retire. His son-in-law and grandson, Arthur and George King, were also active in the car-building trade.

ARTHUR KING (1841–1917) was born in Harpers Ferry, Va., the son of a master armorer at the Federal Arsenal in that place. His family removed to nearby Martinsburg, and there young Arthur was apprenticed as a machinist once his meager schooling was concluded. At the outbreak of the Civil War he moved to Philadelphia and worked for a rifle maker. In about 1867 he moved to York, Pa., and entered the employ of G. W. Ilgenfritz. This wise young man married his new boss's daughter and was soon made foreman of Ilgenfritz's car shop. In 1879 King joined Michael Schall in reopening the moribund Middletown Car Works. He became sole proprietor in 1891 after Schall's failure. Like most small car builders, King's career at Middletown was at best a mixed one. There was no solid record of profits, yet this shop turned out thousands of freight cars, including the gondola pictured in Figure 5.37.

GEORGE I. KING (1871–1953) followed his father and grandfather into the car-building trade. He was born in York, Pa., a small community that boasted no fewer than three railroad car builders within its boundaries. Unlike most of his ancestors, King was not forced to enter the job market upon completion of his elementary education. In 1889 he began studies at MIT but returned to the Middletown Car Works in 1891 to assist his father. He returned to MIT in 1896 to complete his engineering studies and then found employment as a draftsman with the Schoen Steel Car Co. in 1897. He and Schoen proved incompatible, and so he moved to the Michigan-Peninsular Car Co. in Detroit in 1898. This firm soon merged with American Car & Foundry, and King was named manager of the steel-car department. Family needs, however, prompted him to return to Middletown in 1901. King became intrigued with the details of steel freight car design and obtained a number of patents that saw at least some use during the early years of the twentieth century. A somewhat earlier design for hopper doors is shown in Figure 5.51. After the sale of Middletown, King apparently left the car-building field to devote his energies to consulting and lecturing.

PULASKI LEEDS (1845–1903) was born in Darien, Conn., and entered railway service when just sixteen years of age. He worked his way up from machinist apprentice to locomotive engineer. In 1879 he was made mechanical superintendent for the Boston & New York Air Line, but within three years he went west to take on similar duties with the Indianapolis, Decatur & Springfield. In 1889 Leeds was made master mechanic of the Louisville & Nashville, with headquarters in the Louisville shops. Locomotives were presumably Leeds's chief interest, but he was attracted to truck and car design as well. One example of his efforts in this area is shown by the coke car illustrated in Chapter 5 (Fig. 5.64). Most car builders died quietly at home surrounded by their families in the best Victorian tradition. Leeds was not so fortunate. A recently discharged employee confronted Leeds in his office. When Leeds refused to draft a favorable letter of reference, the workman shot him and then turned the gun on himself. Leeds had been an active member of the Master Car Builders Association and had served briefly as its president.

JOHN W. MARDEN (1841–1922) was active in the car-building trade for over half a century. This durable railroader was born in Concord, N.H., and entered railway service in 1861 as a journeyman with the Concord Railroad. He worked for other minor New England lines, gradually rising through the ranks to general foreman, then master car builder. In February 1881 he succeeded Enos Varney as master car builder on the Fitchburg, one of New England's larger railroads. In the following year he joined the Master Car Builders Association and proved to be more than a passive member. He worked hard on all committee assignments and repeatedly served on the Executive Committee. He was president of the organization in 1903. Marden also served as president of the New England Railroad Club. After the Boston & Maine leased the Fitchburg, Marden was promoted to the position of car department superintendent for the entire B&M system. The veteran car builder retired in January 1911. An example of his work can be seen in Figure 5.35.

ROBERT McKENNA (1825–1908) was born in Girvan, Scotland, and was first employed as an apprentice pattern maker at age fourteen. At twenty-three he came to realize the limitations of his homeland and emigrated to the United States. He worked as a journeyman pattern maker until 1853, when he entered the employ of the Hudson River Railroad's car department. He was quickly promoted to foreman and then general foreman. In 1870 he removed to Scranton, Pa., to become master car builder for the Lackawanna, whose main shops were there. McKenna continued his labors at the Scranton shops for nearly three decades before he retired in 1899.

JAMES McMILLIAN (1838–1902). The car-building trade created its share of millionaires. Among them was a young Canadian of Scottish ancestry, born in Hamilton, Ont. McMillian exhibited a flair for business and accounting at an early age and entered the hardware trade at fourteen years of age. In 1856 he moved to Detroit and began a rapid rise in the business circles of that developing lake port. He worked as purchasing agent for a local railroad for several years before entering the car-building trade in 1864. He and his partner, John S. Newberry, formed the Michigan Car Co., which within a decade had grown into one of the nation's major freight car producers. Its position in the field was considerably enhanced after its takeover of the Peninsular Car Co.—also located in Detroit—in 1891. The capacity of the combined plants was one hundred cars per day. McMillian controlled a large car-wheel foundry and an axle-forging shop as well as a branch car plant in Adrian, Mich. After amassing a large fortune McMillian decided to become more active in politics. For many years he served the Republican party at the state level; then in 1889 he ran for the U.S. Senate and won. He served a second term and was one of the few active car builders to attain such high political office.

HARVEY MIDDLETON (1852–1923) was born in Philadelphia and after completing high school served a four-year apprenticeship at Wm. Sellers & Co., machine tool builders. In 1876 he entered railway service as a machinist with the Philadelphia & Erie. Within two years he was promoted to assistant master mechanic. In 1880 he left the P&E and began a wandering career that led to one- and two-year tenures as mechanical chief with the St. Paul, Minnesota & Manitoba, the Louisville & Nashville, the Santa Fe, and the Union Pacific. Between 1891 and 1896 he worked at the Pullman Car Works, first as superintendent of construction and later as manager. By 1896 Middleton was restless once again and so accepted a job with the Baltimore & Ohio as superintendent of motive power. In 1900 he left the B&O to enter private practice as head of a large contracting company. He returned to the work force during World War I as a consulting engineer with the U.S. Army.

WILLIAM H. MINER (1862–1930) was a draft-gear expert who labored in several railroad car repair shops for over fifteen years before opening his own business. He was born in Juneau, Wis., and began railway service with the Wabash at sixteen years of age. Ten years later he was appointed superintendent of the Lafayette Car Works. He then worked briefly for the Michigan Central and a private freight line, the California Fruit Transport. In 1891 he developed a tandem draft gear whose commercial potential encouraged him to open a plant in Chicago for its manufacture. By 1898 Miner draft gears were used on fifteen thousand cars. In 1905 Miner established a testing laboratory devoted to perfecting draft-gear design. His firm developed and manufactured other freight car parts and became a major supplier in this field. Miner grew rich in the railway supply field and shared his wealth with several charities.

JAMES MURRAY (1812–1895) was a pioneer railway mechanical officer active in several areas of design other than car building. While master of machinery for the Baltimore & Ohio (1843–1846), he introduced the unique iron pot hopper cars described in Chapter 8. After leaving the B&O he operated a machine shop in Baltimore in partnership with Henry R. Hazelhurst. This firm built boilers, marine engines, locomotives, and railroad freight cars. Around 1855 he became an associate of Thomas Winans. Some twenty years later he was sent to Europe to represent Winans's interests there. Murray died in London at the advanced age of eighty-three. At the time of his death he was credited with building the first roundhouse in the United States.

SAMUEL W. MURRAY (1829–1909) was born in Lewisburg, Pa., into a family of early Scots settlers. After completing schooling at a local academy, Murray began a three-year apprentice term with the Portland Co., locomotive, car, and machinery builders of Portland, Maine. Now well trained in the mechanical trade, Murray worked for a number of machine shops, including the Lancaster Locomotive Works and M. W. Baldwin. In 1864, in partnership with several others, he opened a car plant in Milton, Pa. This firm became one of the largest freight car builders in Pennsylvania. It was specializing in iron tank cars by about 1875. A major fire leveled the plant in 1880, but it was rebuilt and continued on long after its founder's death. Murray's firm became part of American Car & Foundry in 1899 and continued in production until the present time.

JOHN S. NEWBERRY (1832–1887) was born in Waterville, N.Y., and moved to Michigan with his family at the age of five. He graduated first in his class from the University of Michigan at age eighteen and seemed destined for an academic career, but he became a civil engineer instead and worked in the construction department of the Michigan Central. He entered the legal profession for several years but then decided to join his friend James McMillian in the car-building business. Newberry was named president of the newly formed Michigan Car Co. in 1863. The rise and progress of this firm are related in the sketch of James McMillian.

LOREN PACKARD (1843–1895) was born in Northumberland, N.H., and became involved with railroad car work just as the Civil War was ending. He became shop foreman at the Wason Car Co. before taking over the New Haven Railroad's car shops in 1876. In May 1881 he succeeded J. S. Shryack as master car builder for the Baltimore & Ohio. His next job move, in March 1884, involved an even larger responsibility—master car builder for the New York Central's West Albany shops. He remained there until his death eleven years later. Two of Packard's car designs are represented in Figures 3.30 and 5.11.

ARIO PARDEE (1810–1892) was the principal partner in Pardee, Snyder & Co., car builders of Watsonville, Pa. He was born in Chatham, N.Y. At nineteen years of age he started work as a rodman with the engineering corps laying out the Delaware and Raritan Canal. He went on to survey railroads in Pennsylvania and became engaged in the development of coal fields in that state. One of his mines tapped into a large vein of anthracite, making Pardee a wealthy man. He invested largely in iron works, some as distant as Tennessee. Some of his earnings went into the car plant mentioned above, and while Pardee was probably not an active car builder, his role as a capitalist was essential to the operation of the car-building industries. Someone had to put up cash or credit for buildings, supplies, and tools before any cars could be produced. His partner, Henry F. Snyder, provided the mechanical and management skills needed to produce cars. Pardee became a major benefactor of Lafayette College (Easton, Pa.) starting in 1864. His help transformed this failed local institution into a major scientific college.

THATCHER PERKINS (1812–1883) is commonly associated exclusively with locomotive design, but his pioneering work in iron-car construction and his advocacy of wood-frame trucks make him at least a minor celebrity in the area of freight cars as well. Brief mention of his activities in these areas is made in Chapters 5 and 8, while his remodeling of the Louisville & Nashville repair shops is noted in Chapter 1. Perkins spent much of his career in the employ of the Baltimore & Ohio and the L&N as manager of rolling stock. During the 1850s he was a partner in the locomotive and car-building firm of Smith & Perkins (Alexandria, Va.). Perkins was born in North Berwick, Maine, and died in Baltimore.

JOHN PLAYER (1847–1914) was born in Woolwich, England, and learned the machinery trade in the famous arsenal located in his hometown. He emigrated to America in 1873 and became a machinist with the Iowa Central. By 1884 he was in charge of this road's mechanical department. In 1887 he moved on to the Wisconsin Central in the same position. By June 1890 Player had switched employers once again. This time he found work with a major trunk line, the Santa Fe, and became active again in both locomotive and car design. Player was a strong advocate of non-swing-motion freight car trucks. He also devised a cast-steel truck and a form of compound locomotive. Care should be taken not to confuse him with another John Player, long associated with the Brooks Locomotive Works.

GEORGE M. SARGENT (1830–1913), a prominent figure in the railway supply business, was born in Sedgewick, Maine. He enjoyed no particular measure of success until the late 1860s, when he entered the malleable-iron business. He became aware of the need for longer-wearing brake shoes and so bought a partial interest in a design patented in 1876 by I. H. Congdon. In Congdon's design, a cast-iron shoe was fitted with steel inserts. Sargent opened a foundry in Chicago for the manufacture of this device. By 1895 he was producing 30,000 brake shoes (500 tons' worth) per month. Sargent's firm became part of the American Brake Shoe & Foundry Co., which he helped organize in 1901.

CHARLES A. SCHROYER (1853–1931) was born in the car-building town of Milton, Pa. After completing his education in local public schools, Schroyer entered railway service on the Philadelphia & Erie in 1877. He remained there until 1885, when he worked briefly for the Ohio Falls Car Co. In 1886 he served as assistant superintendent of the Big Four's car department. Not long afterward he became superintendent of the Chicago & North Western's car department and remained in that position for over thirty years. Schroyer was very active in the Master Car Builders Association and served on many committees. He was president of the organization for two terms, 1899 and 1900. He also served as president of the Western Railway Club in 1898. Schroyer retired in September 1919 after a long and dedicated career in the car-building trade.

CHARLES F. SCOVILLE (1821–1890) was born in Torringsford, Conn. As a young man he learned the millwright's trade and so was well prepared for car building when he joined the American Car Co. of Seymour, Conn., in 1852. He soon left New England for Chicago, where he eventually joined the car department of the Illinois Central. When Wells, French & Co., bridge builders of Chicago, decided to open a freight car plant in 1871, Scoville was selected as superintendent and remained active in that position until a few months before his death. Wells, French & Co. grew into a major car-building establishment and became part of American Car & Foundry in 1899.

WILLIAM B. SNOW (1821–?) was master car builder for the Illinois Central for many years. He was born in Bellows Falls, Vt., and worked briefly for the Western Railroad and for Tracey & Fales, car builders of Hartford, Conn. He moved west in 1857 and found employment with the Illinois Central as foreman of the car shop in Chicago. He left to work for Pullman between 1872 and 1875 but then returned to the IC, assuming duties as master car builder. Snow remained active in this position until 1894. The dates of his retirement and death have not been uncovered, but examples of IC cars

built during his regime are shown in Figures 2.23 and 3.47.

HENRY F. SNYDER (1829–1886), a member of a prominent Pennsylvania family, was born in Selinsgrove, Pa. Because of a natural aptitude for mechanics, he became an apprentice in a Williamsport machine shop in 1850. After a few years he went east to develop his knowledge of steam engineering with such notable firms as the Allaire Works and the Rogers Locomotive Works. He then returned to Williamsport and opened a machine shop with two of his brothers as partners. The firm prospered until the Panic of 1873 caused it to fail. With the backing of a local capitalist, Ario Pardee, Snyder opened a car plant in Watsonville, Pa. This firm produced thousands of freight cars until around the 1890s.

RICHARD H. SOULE (1849–1908) was born in Boston. He entered railway service in 1875 as a draftsman at the Pennsylvania Railroad's Altoona shops. He worked in the mechanical departments of several of the Pennsylvania's subsidiaries until 1883, when he accepted a job with the New York, West Shore & Buffalo as superintendent of motive power. After two interim jobs Soule took charge of the rolling stock of the Norfolk & Western in June 1891. He became an agent for the Baldwin Locomotive Works six years later and spent the last years of his career as a consulting engineer. Soule was an active member of the Master Car Builders Association and read papers on topics ranging from air brakes to steel underframing. He was a member of numerous technical committees as well as the Executive Committee of the organization.

JOSEPH W. SPRAGUE (1831–?) managed the Ohio Falls Car Co. for over a quarter of a century. He was born in Salem, Mass., to a distinguished New England family. After graduating from Harvard in 1852 he set out to follow a career in civil engineering in the Midwest. While working in that region in 1866 he was asked by stockholders of the bankrupt Ohio Falls Car Co. to take over the firm's management. Sprague put the firm's affairs in order and built a new series of shops in 1872. He saw it through a second bankruptcy. When he retired in 1888 Ohio Falls was one of the largest and most profitable car builders in the nation. It became part of the American Car & Foundry merger scheme in 1899 and remained active until about 1945. An 1872 view of the Ohio Falls works is shown in Figure 1.69.

ISAAC W. VAN HOUTEN (1813–1900) was born in Mechanicsville, N.J. Like so many other car builders of his time, Van Houten started out as a house carpenter. In 1843 he joined the car department of the New Jersey Railroad & Transportation

Co. as a journeyman. He left the Jersey City shops in 1849 and joined the Pennsylvania Railroad's car department, whose main shops were then in Harrisburg rather than Altoona. Starting in 1850 he was foreman of the Pennsylvania's car shops at West Philadelphia. Van Houten remained on the job at West Philadelphia for forty-four years. When he retired at last, in 1894, Van Houten was indeed a veteran car builder, having spent over half a century in his trade. He was apparently well liked by his colleagues, for he was elected the first president of the Master Car Builders Association and held that office for five terms.

ENOS VARNEY (1813–1880) was born to a poor family in Rochester, N.H., the eldest of eleven children. Varney found employment in a brickyard as a boy but worked his way up in the blue-collar world as a carpenter. He emigrated to California in 1845. Not having found his fortune, he returned to New England the following year. He took a job in the car shops of the Fitchburg Railroad and by 1853 was named master car builder. He continued his duties faithfully until he was killed in a railroad accident at Littleton, N.H.

BENJAMIN K. VERBRYCK (1827–1891) was born in Tappan, N.Y., but left the family farm to become a carpenter. By 1852 he found work at the Erie's Piermont shops. By 1860 he had advanced to general foreman of the car shop. When the Pier-

mont shop closed three years later, he transferred to the Erie's new shops in Jersey City. In 1871 he left the East Coast to become master car builder for the Rock Island lines, where he remained until the end of his career. He was an active member of the Master Car Builders Association and was its president in 1886 and 1887. A car frame designed by Verbryck is shown in Figure 3.19.

ARTHUR M. WAITT (1858–1920) was an MIT graduate, born in Boston, who entered railway service in 1879 as a draftsman with the Chicago, Burlington & Quincy's mechanical department. Waitt began a series of career moves that covered the Eastern Railroad, the Boston & Maine, and several other roads. He did not remain in any of these jobs for more than three or four years. Even the great New York Central could not command his loyalty for more than four years. In 1903 Waitt began a study tour around the United States and Europe to investigate the latest advances in electric traction. At the close of his wanderings, he became a consulting engineer. From his job titles it would appear that Waitt was strictly a locomotive man, but there is evidence that he was equally interested in car design. Drawings for two open-top cars are shown in Figures 5.23 and 5.59. In addition, Waitt was very active in the discussions and committee reports of the Master Car Builders Association, where he seemed to have something to say about every topic under consideration.

CHRONOLOGY

1795	The Beacon Hill incline (Boston) wooden tramway is used to carry bricks. This is presumably the site of the first U.S. freight car operation.
1826	The Granite Railway (Quincy, Mass.) uses four-wheel winch cars to transport stone blocks. See 1834.
1827	The Mauch Chunk gravity railway in eastern Pennsylvania begins operations with four-wheel hopper cars of 1.5 tons capacity.
1828	Simpler four-wheel cars with very small iron wheels supersede the winch cars on the Granite Railway.
1829	The Delaware & Hudson Canal Co.'s gravity railway begins operations with four-wheel hopper cars of 2.5 tons capacity.
1830s	Most railroads charge 20 cents per ton-mile. In the next decade, rates drop rapidly.
1830–1835	Four-wheel open gondola-like wooden cars are the common form of American freight car. Capacity is generally 3 tons and length about 10 feet.
1830	The Pontchartrain Railroad imports several English freight cars. Other pioneer American lines study or copy British goods-wagon design. The Baltimore & Ohio Railroad carries President Adams's carriage on a flat to open piggyback operations in North America.
1832	A single horse pulls three cars, each loaded with 25 barrels of flour, on the B&O. The cargo weighs 8 tons. Speed is 3 to 5 mph, depending on the grade. House cars (boxcars) begin to appear on the B&O. Covered cars offer protection to goods being shipped, and by around 1840 the idea becomes standard for merchandise cars.
1834	The Granite Railway combines four cars to produce a sixteen-wheel flatcar for the carriage of very large stone blocks.

1835	The B&O receives the first double-truck (eight-wheel) 24-foot-long boxcars.
c. 1836	Container service begins on the Pennsylvania rail-canal system between Philadelphia and Pittsburgh.
1836–1838	Covered hoppers are used to haul grain on the Tonawanda and Mad River & Lake Erie railroads.
1838	Eight-wheel flatcars, 41 feet long, carry canal boats weighing 25 to 30 tons on the Morris Canal inclines.
1840–1865	Boxcar size and capacity are relatively fixed. The typical car carries 8 to 10 tons and measures 24 to 28 feet long.
1840	Eight-wheel freight cars are now standard on most U.S. railroads. Only four-wheel coal cars are built in large numbers after this date. There are about sixty-five hundred freight cars in service.
1842	Refrigerator cars, insulated and using ice, are in service on the Western Railroad (Mass.). The Reading opens its big coal dock at Port Richmond (Philadelphia). Milk car service begins on the Erie.
1843	The Reading introduces iron-body wood-frame coal cars.
1844	The B&O introduces iron pot hoppers for coal service.
1845	Crawford patents the air brake in Great Britain.
1846	The B&O builds two experimental iron boxcars.
1849	Berney patents a side-dump car.
1850	There are about 30,700 freight cars in service. There are forty-one railroad and private car builders in the United States. The B&O carries 132,000 tons of coal.
1851	Refrigerator cars are used on the Ogdensburg & Lake Champlain.
1852–1853	Kasson's Locomotive Express service begins; it uses very large flatcars.

1852 The B&O begins limited production of iron-body wood-frame boxcars.

Cutler and Rapp receive a U.S. patent for an air brake.

1853 Lyman begins operation of refrigerator cars in New York State.

1854 Cabooses are in service on the Erie; the origin of the first such cars is unrecorded.

1855 Four-wheel coal jimmies number in the thousands on eastern coal roads like the Readings, the Lackawanna, and the Central of New Jersey. They carry 5 to 6 tons each.

Spoked wheels have now largely been replaced by solid disc or plate-style cast-iron wheels.

The B&O has 774 iron pot hopper coal cars in service.

1857 The Reading carries 7,200 tons of coal per day.

1859 The Michigan Central is operating drovers' cabooses.

The Winslow double wood and sheet-iron car roof is patented. It becomes popular after 1870.

1860s Large shippers begin to demand rebates.

1860 There are approximately one hundred thousand freight cars in service.

Coal has become a major traffic on certain eastern railroads. The B&O derives one-third of its freight revenues from this source.

A humane cattle car is patented by Swearingen.

Most railroads charge 2.2 cents per ton-mile.

U.S. railroads carry 3.2 billion ton-miles.

1861 Only 53 percent of U.S. railroads are standard-gauge, making the interchange of cars difficult.

1862 The New York & Harlem opens a dairy depot in New York City.

1863 Seely patents a cylindrical iron oil tank car.

The Union Line for fast freight is established.

1864 The New York Central has over seven hundred iron boxcars in service.

The Union Transportation Co.—the beginning of large-scale private freight car operations—encourages the interchange system.

1865 The Red Line for fast freight is established.

The eastern railroad network is in place. Railroads begin to dominate the national freight business, leading to a decline in canal and riverboat traffic.

Densmore begins large-scale oil shipments in wooden tank cars.

Chicago's Union Stock Yard opens, encouraging the shipment of cattle by rail.

The Empire Line begins service.

1866 Very long flatcars are used by Jackson & Sharp to ship light cars.

1867 The B&O builds some high-cube cars for the carriage of empty barrels.

The Master Car Builders Association is organized.

The Central Pacific's Sacramento shops open. This facility will soon develop into the largest repair and construction shop in the Far West.

The Blue Line is established, starting service with four hundred cars.

The first U.S. patent for a refrigerator car is granted to Sutherland (see 1842 and 1851).

c. 1868 The Davis-Hammond refrigerator car enters service.

1868 The Railroad Car Trust (Philadelphia) opens a new era in bank-sponsored purchase of railroad cars.

Adjustable-gauge freight cars are introduced on the Grand Trunk for Chicago-to-Boston service.

1869 The Pennsylvania Railroad introduces its class NA four-wheel caboose.

The sixteen-wheel 40-ton-capacity gun car is running on the Pennsylvania Railroad.

The Iowa Pool is created to divide up traffic and revenue on certain western railroads' territories near Omaha.

Westinghouse patents but does not invent the air brake.

1869– The Pennsylvania Railroad opens a major
1879 new freight car shop at Altoona.

1870s– Rate wars rage as competing railroads
1880s seek to capture one another's traffic.

1870s Western railroads promote wheat farming.

Very long flatcars are now available for the long-distance delivery of streetcars and narrow-gauge cars.

Most cattle cars are 28 feet long or shorter, with a capacity of 10 tons.

1870 U.S. railroads carry 8 billion to 10 billion ton-miles.

There are approximately three hundred thousand freight cars in service.

Large-scale interchange of freight cars is under way.

Both the B&O and the Pennsylvania Railroad adopt standard freight car design classifications.

Few freight cars, except for flats, use truss rods before this time.

Arch-bar trucks replace the wood-beam variety as the U.S. standard.

Shipments of dressed beef by refrigerator cars begin.

1872   Roll on/roll off circus wagon train operations begin.

A federal law to regulate humane rail shipment of cattle is passed.

1873   The Blue Line makes express shipments of butter from the Midwest to eastern cities.

The MCB Association adopts the Sellers standard screw-thread system and the 3½- by 7-inch axle journal.

The financial panic of this year depresses freight traffic and orders for new cars until the end of the decade.

Rockefeller acquires control of the Union Tank Line.

1874   The MCB Association adopts a standard journal box.

1875   Empire Transportation has thirteen hundred tank cars in operation.

There are between fifty thousand and fifty-five thousand coal jimmies in service.

Iron body bolsters are now commonly used by most car builders, but some railroads continue to buy wooden-bolster cars for another twenty years.

1877   The Wicks and Tiffany refrigerator car patents are granted.

The American Humane Association is organized to lobby for humane shipment of cattle.

Railroad strikes cause serious disruption to freight service.

1878–   Twenty-ton-capacity boxcars are now
1879   more common.

1880s The long-distance transport of produce in refrigerator cars becomes more common.

Thirty-car freight trains weighing over 1,000 tons are typical.

1880   There are approximately 540,000 freight cars in service.

U.S. railroads carry 32.3 billion ton-miles.

Most railroads charge 0.75 cent per ton-mile.

Cabooses have at last become common on U.S. railroads.

Refrigerator cars are now in wide use; about thirteen hundred are in service.

Center-support (three-truck) freight cars are patented by Terry.

1881– Twenty-five-ton boxcars begin to appear.
1882

1881   The Pennsylvania Railroad averages only 37 miles per day per freight car.

1882   The New York Central adopts standard freight car designs systemwide.

The Eastman heater car is patented.

1883   Burton begins humane cattle car service.

1884   Mather enters the cattle-shipping business.

Jenkins patents a live-poultry car.

Grain shipments constitute nearly 50 percent of the New York Central's freight traffic.

The Reading has 3,134 20-ton hopper cars in service.

1885   The Arms Palace Horse Car Co. begins operations.

The Long Island Railroad begins piggyback service for farm produce wagons.

Cattle cars are typically 34 feet long with 20 tons capacity.

The first *Railway Equipment Register* is published, listing all cars in interchange service.

There are six thousand refrigerator cars in service.

Thirty-ton boxcars are being introduced.

Most freight cars now have four truss rods and turnbuckles.

1886   The Standard Code is adopted for railroad lantern, hand, and whistle signals.

Southern railroads convert to standard gauge, easing the problems of car interchange.

1887   Over one thousand Eastman heater cars are in service for potatoes and other goods that need protection from freezing.

A 55-ton twelve-wheel flatcar is built by the Lehigh Valley.

There are three thousand Tiffany refrigerator cars in service.

The Interstate Commerce Commission is created, essentially to deal with railroad freight rates.

1888   The iron gun car, with sixteen wheels and 60 tons capacity, is running on the Pennsylvania Railroad.

There are sixty-one hundred tank cars in the United States.

The Fox pressed-steel truck is introduced to the U.S. market.

1889   The progressive system of freight car manufacturing apparently begins at the Haskell & Barker Car Works in Michigan City, Ind.

1890s Pressed-steel trucks enjoy a brief popularity.

1890   There are approximately 1.1 million freight cars in service.

U.S. railroads carry 79.1 billion ton-miles.

Most major railroads offer special flatcars for heavy and large loads.

Pressed-steel body bolsters are introduced by Schoen.

The Erie purchases ore cars with three four-wheel trucks.

High-sided hoppers of 30 to 40 tons capacity are introduced.

Low-mileage and idle cars are a basic industry problem that won't go away. Only about 8 percent of the nation's cars are in motion at any one time.

Oil tank cars of 8,000-gallon capacity are common.

Fifty-car freight trains weighing 1,500 tons are typical.

Upwards of ten thousand gas-pipe-frame freight cars are in service.

1892   Average car mileage drops to 24.7 miles, down 22 percent since 1883.

The Venice Transportation Co. operates 60-foot-long low-level flats.

Harvey builds a steel-frame tank car for the Union Tank Line.

1893   High-cube boxcars with interior measurements of 9 feet 2 inches by 60 feet are built for light but bulky goods.

A federal law governing essentially brakes and couplers—the Safety Appliance Act—is passed.

Cast-steel body bolsters are introduced by the American Steel Foundry Co. of St. Louis.

1894   Tank cars for bulk shipments of molasses are introduced.

The Pennsylvania Railroad's class GG wooden 35-ton-capacity hopper is introduced.

c. 1895   Swing-motion freight car trucks decline in favor as heavier cars become more common.

The Brill Car Co. uses a 65-foot-long flat to deliver streetcars.

1895   About one-third of all mileage is for empty cars.

The Chesapeake & Ohio's cattle car, 36 feet long, weighs only 27,300 pounds.

Mather has three thousand stock cars in operation.

There are twenty-eight thousand refrigerator cars in service, up from six thousand in 1885.

1896   Slow speeds prevail for freight trains; 10 mph is typical. Few cars travel 1,000 miles in less than five days.

c. 1897   Cast-steel trucks are produced for freight car service. They become popular after 1905.

1897   An oyster tank car is built by Pullman after Stilwell's plan.

The first large order for all-steel freight cars is given to Schoen's Pressed Steel Car Co. by the Pittsburgh, Bessemer & Lake Erie.

The B&O has retired most of its pot hoppers.

1899   Schoen and Fox merge as the Pressed Steel Car Co.

The MCB Association has adopted four standard journal sizes.

The Lorain Steel Co. buys 66-foot-long gondolas of 40 tons capacity.

Most four-wheel coal jimmies are retired in favor of larger cars.

Canda designs and builds 50-ton-capacity wood-frame box- and hopper cars.

1900   Sixty-five commercial car builders employ thirty-three thousand men and produce 116,500 freight cars.

U.S. railroads are 93.5 percent single-track.

There are approximately 1.3 million freight cars in service.

U.S. railroads carry 141.1 billion ton-miles.

1902–   Railroad freight tonnage jumps by 46 per-
1907   cent.

1902   Only 3,300 of 200,000 miles of U.S. railroads have automatic signaling. Another 30,000 miles are governed by manual block signals.

The per-diem system for freight car interchange replaces the mileage system, except for private freight cars.

1903   The Elkins Act is passed to deal with rebates.

1905   About 45 percent of new freight cars are steel or steel-framed.

1906   The Hepburn Act is passed to deal basically with setting maximum freight rates.

1910   Over 90 percent of new freight cars are steel or steel-framed.

There are approximately 2.1 million freight cars in service.

FURTHER READING

Primary sources for American railroad freight cars are almost nonexistent. A few original specifications have been preserved. Trivial items can be found in manuscript collections such as the John W. Garrett Papers, held by the Maryland Historical Society, but there are no solid and extensive bodies of papers from any leading nineteenth-century master car builder. The papers of Octave Chanute are housed in the Library of Congress, and yet, despite his many years in the railway equipment field, the papers deal almost exclusively with aircraft. The basic sources of information for this volume were the railroad journals. The *American Railroad Journal*, the *Railroad Gazette*, and especially the *National Car Builder* are abundantly represented in the notes to each chapter. A description of these and other magazines appears on pp. 613–614 of this volume.

Many—in fact the majority—of the volumes consulted are not listed here because they are so numerous, and because they normally were used only to establish traffic or corporate details and did not offer much on the subject of railroad freight cars. Even so, my debt to authors such as John Stover, Alvin Harlow, and H. Roger Grant must at least be mentioned because they saved me from hundreds of hours of additional research. The list that follows is, I realize, distressingly short, but the literature directly bearing on the early American freight car is hardly voluminous.

BOOKS

Botkin, B. A., and Alvin F. Harlow. *A Treasury of Railroad Folklore*. New York, 1953.

Epstein, Ralph C. *GATX: A History of the General American Transportation Corporation 1898–1949*. New York, 1948.

Fishlow, Albert. *American Railroads and the Transformation of the Ante-Bellum Economy*. Cambridge, Mass., 1965.

Forney, Mathias N., ed. *The Car Builders Dictionary*. New York, 1879.

Grant, H. Roger. *Brownie the Boomer: The Life of Charles P. Brown, an American Railroader*. De Kalb, Ill., 1991.

_____. *The Corn Belt Route: A History of the Chicago Great Western Railroad Company*. De Kalb, Ill., 1984.

Harlow, Alvin F. *Road of the Century*. New York, 1947.

_____. *Steelways of New England*. New York, 1946.

Hitt, Rodney, ed. *The Car Builders Dictionary*. New York, 1903 and 1906.

Johnson, Emory R., and Grover G. Huebner. *Railroad Traffic and Rates*. 2 vols. New York, 1911.

Krueger, F. J. *Freight Car Equipment*. Detroit, 1910.

Lister, Francis E. *The Car Builders Dictionary*. New York, 1909.

Lucas, Walter A. *100 Years of Railroad Cars*. New York, 1958.

Perry, H. M. *Repairs of Railway Car Equipment*. Chicago, 1899.

Stimson, O. M. *Modern Freight Car Estimating*. Anniston, Ala., 1897.

Stover, John F. *American Railroads*. Chicago, 1961.

_____. *History of the Illinois Central Railroad*. New York, 1975.

_____. *The Life and Decline of the American Railroad*. New York, 1970.

Voss, William. *Railway Car Construction*. New York, 1892.

Wait, J. C., ed. *The Car Builders Dictionary*. New York, 1895.

Wellington, Arthur M. *The Car Builders Dictionary*. New York, 1888.

_____. *The Economic Theory of the Location of Railways*. New York, 1893.

PERIODICALS

American Journal of Railway Appliances. Vol. 1, 1883–vol. 21, 1901.

Space was devoted to rolling stock declines during 1890s; then the journal evolved into a commentary on machine tools and related matters. Title changed to *Railway Machinery* in 1901. Not a major source.

American Railroad Journal. Vol. 1, 1832–vol. 148, 1975.

The earliest years contain only scattered notes on cars, but more information was included after about 1890. A rich source from that time forward.

In 1913 the title changed to *Railway Mechanical Engineer.* In 1953 it changed again, to *Railway Locomotive and Cars.* This journal continued to be published under the original volume sequence until January 1975.

National Car Builder. Vol. 1, 1870–vol. 26, 1895.
The earliest years are incomplete; there were few illustrations until after 1880. The name was changed in 1886 to *National Car and Locomotive Builder.* It is the single most valuable source of information on the subject.

Railroad Car Journal. Vol. 1, 1890–vol. 12, 1902.
Started as *The Journal of Railroad Car Heating and Ventilating.* Changed to *Railroad Car Journal* in October 1891. Name again changed, to *Railroad Digest,* in 1901; the periodical expired the following year. Well illustrated.

Railroad Gazette. Vol. 1, 1870–vol. 44, 1908.
Always concerned with engineering matters, this paper devoted frequent, if not abundant, space to railroad car items. Ocassional drawings were published on the subject. The paper merged with *Railway Age* in 1908. M. N. Forney was the editor for many years.

Railway Age. Vol. 1, 1876–present.
A journal originally devoted to corporate matters; little space was given to rolling stock until about 1890. Then the subject was well covered until about 1955, when it was deemphasized.

NOTES

ABBREVIATIONS

Am. Rly Rev.   *American Railway Review*
Am. Rly Times   *American Railway Times*
ARRJ   *American Railroad Journal*
ASCE Trans.   American Society of Civil Engineers *Transactions*
ASME Trans.   American Society of Mechanical Engineers *Transactions*
DAB   *Dictionary of American Biography*
Eng. Mag.   *Engineering Magazine*
Eng. News   *Engineering News*
Loco. Eng.   *Locomotive Engineering*
Mast. Mech. Mag.   *Master Mechanics' Magazine*
MCB Report   *Master Car Builders Report*
Nat. Cyc. Am. Bio   *National Cyclopedia of American Biography*
NCB   *National Car Builder*
Poor's   *Poor's Manual of Railroads*
R&LHS Bull.   Railway and Locomotive Historical Society *Bulletin*
Rly. Age   *Railway Age*
Rly. Rev.   *Railway Review*
Rly. & Loco. Eng.   *Railway and Locomotive Engineering*
Rly. Mast. Mech.   *Railway Master Mechanic*
RR. Adv.   *Railroad Advocate*
RRCJ   *Railroad Car Journal*
RRG   *Railroad Gazette*
RR. Mag.   *Railroad Magazine*
RR. Reco.   *Railroad Record*

CHAPTER 1

1. Albert Fishlow, *American Railroads and the Transformation of the Ante-Bellum Economy* (Cambridge, Mass., 1965), p. 202.
2. E. R. Dewsnup, ed., *Railway Organization and Working* (Chicago, 1906).
3. *Eng. Mag.*, May 1895, p. 271.
4. Fishlow, *American Railroads*, p. 301.
5. James A. Ward, *Railroads and the Character of America, 1820–1887* (Knoxville, 1986), p. 78.
6. *Rly. Age*, March 28, 1902, p. 558.
7. George R. Taylor, *Transportation Revolution, 1815–1860* (New York, 1951), p. 2.
8. Fishlow, *American Railroads*, p. 297.
9. Arthur T. Hadley, *Railroad Transportation* (New York, 1885), p. 65.
10. Robert W. Fogel, *Railroads and American Economic Growth* (Baltimore, 1964).
11. Taylor, *Transportation Revolution*, p. 43.
12. A. D. Chandler, Jr., ed., *The Railroads: The Nation's First Big Business* (New York, 1965), p. 22.
13. Stuart R. Daggett, *Principles of Inland Transportation* (New York, 1928), p. 202.
14. *Eng. Mag.*, May 1895, p. 271.
15. *R&LHS Bull.* 21 (March 1930), table.
16. John Moody, *How to Analyze Railroad Reports* (New York, 1919), p. 81.
17. *Eng. Mag.*, April 1895, p. 8.
18. Slason Thompson, *Railway Statistics of the United States* (Chicago, 1925), p. 96.
19. Chandler, *Railroads*, p. 14.
20. Fishlow, *American Railroads*, p. 316.
21. *ASCE Trans.* 11 (1882), p. 367.
22. Emory R. Johnson and Grover G. Huebner, *Railroad Traffic and Rates* (New York, 1911), vol. 1, p. 159.
23. *RRCJ*, Jan. 1900, p. 4.
24. Johnson and Huebner, *Traffic and Rates*, p. 3.
25. Ibid., p. 51.
26. *Poor's*, 1888, p. 208.
27. Arthur M. Wellington, *The Economic Theory of the Location of Railways* (New York, 1893), p. 35.
28. Thomas M. Cooley, Thomas C. Clarke, et al., *The American Railway* (New York, 1892), p. 437.
29. E. H. Hungerford, *The Story of the Baltimore & Ohio Railroad, 1827–1927* (New York, 1928), p. 128.
30. John F. Stover, *History of the Baltimore & Ohio Railroad* (Lafayette, Ind., 1987), p. 89.
31. David Stevenson, *Civil Engineering in North America* (London, 1838), p. 272.
32. Canal History and Technology Symposium (Easton, Pa.) *Proceedings* 3 (1984), p. 74.
33. E. H. Mott, *The Story of the Erie* (New York, 1901), p. 483.
34. Robert L. Black, *The Little Miami Railroad* (Cincinnati, 1940), pp. 77–78; *RR. Reco.*, Jan. 18, 1855, p. 775.
35. *RRG*, May 4, 1895, p. 248.
36. Ibid., Feb. 10, 1882, p. 88; *Poor's*, 1882, p. 867.
37. Wellington, *Location*, p. 724; *RRG*, Jan. 10, 1890, p. 26.
38. J. E. Watkins, proof sheets for his *History of the Pennsylvania Railroad* (1896), vol. 2, n.p.
39. Richard C. Overton, *The Burlington Route* (New York, 1965), p. 102.
40. Henry M. Flint, *Railroads of the United States* (Philadelphia, 1868), p. 320; R. L. Frey, ed., *Railroads in the Nineteenth Century* (New York, 1988), p. 194.
41. *New Yorker*, March 6, 1989, p. 52.

42. Ralph W. Hidy et al., *The Great Northern Railway* (Boston, 1988), p. 99.
43. Ibid.; E. H. Hungerford, *The Modern Railroad* (Chicago, 1918), p. 360; *R.R. Mag.*, Dec. 1931, p. 55.
44. Johnson, *Traffic and Rates*, vol. 1, p. 322.
45. *Cassier's Magazine*, June 1894, p. 116.
46. Jules I. Bogen, *The Anthracite Railroads* (New York, 1927), pp. 23–29.
47. Fishlow, *American Railroads*, p. 88.
48. T. T. Taber, *The Delaware, Lackawanna, and Western Railroad in the Nineteenth Century* (Williamsport, Pa., 1977), p. 86.
49. Robert F. Archer, *A History of the Lehigh Valley Railroad* (Berkeley, 1977), pp. 50–108.
50. Carl W. Condit, *The Port of New York* (Chicago, 1980), vol. 1, p. 365.
51. Bogen, *Anthracite Railroads*, p. 188.
52. Archer, *Lehigh Valley*, p. 118.
53. Herbert H. Harwood, *Impossible Challenge: The Baltimore and Ohio Railroad in Maryland* (Baltimore, 1979), p. 43.
54. Charles W. Turner, *Chessie's Road* (Richmond, 1956), p. 107.
55. John F. Stover, *History of the Illinois Central Railroad* (New York, 1975), pp. 191 and 234.
56. Eli Bowen, *The Pictorial Sketchbook of Pennsylvania* (Philadelphia, 1852), p. 37.
57. Harwood, *Impossible Challenge*, pp. 53 and 110.
58. *Scientific American*, April 15, 1882, p. 226.
59. Hungerford, *Modern Railroad*, p. 340.
60. Fishlow, *American Railroads*, p. 163.
61. W. B. Catton, *John W. Garrett* (Ph.D. diss., Northwestern Univ., 1959), p. 541.
62. *RR. Reco.*, Sept. 13, 1860, p. 340.
63. George R. Taylor and Irene D. Neu, *The American Railroad Network, 1861–1890* (Cambridge, Mass., 1956), p. 67.
64. J. P. Maxwell, *Report on U.S. Railroads to the New Zealand General Assembly* (pamphlet, 1888), p. 19; *RRG*, May 23, 1884, p. 396.
65. Edwin A. Pratt, *American Railways* (New York, 1903), p. 39.
66. Taylor and Neu, *Railroad Network*, p. 74.
67. *RR. Mag.*, Oct. 1930, p. 433.
68. Catton, *Garrett*, p. 489.
69. Harwood, *Impossible Challenge*, p. 156.
70. John Grafton, *New York in the Nineteenth Century* (New York, 1977), p. 233.
71. *The Vanderbilt System* (1887), p. 13 (promotional booklet issued by the New York Central Railroad).
72. Newspaper reference courtesy of E. T. Francis of Livingston, N.J.
73. Walter A. Lucas, *From the Hills to the Hudson* (New York, 1944), p. 236.
74. John H. White, Jr., *The Great Yellow Fleet* (San Marino, Calif., 1986), p. 31.
75. *Railroad History* 159 (Autumn 1988), p. 48.
76. Stover, *Illinois Central*, pp. 190 and 234.
77. *RRG*, June 6, 1884, p. 437.
78. *Rly. Age*, July 22, 1887, p. 512.
79. Pratt, *American Railways*, p. 178.
80. Fishlow, *American Railroads*, pp. 68 and 80.
81. *RR. Reco.*, June 21, 1860, p. 210.
82. B&O annual report, 1858, p. 41.
83. Alex Groner, *The American Heritage History of American Business and Industry* (New York, 1972), p. 169. In this text the Hannibal & St. Joseph is cited, but because this line did not serve the Abilene area, it is assumed that the KP was the actual line involved.
84. J. C. Hoadley, *Report on the Transportation of Live Stock* (Massachusetts State Board of Health, 1875), p. 84; Pratt, *American Railways*, p. 44.
85. S. G. Reed, *A History of the Texas Railroads* (Houston, 1941), p. 744.
86. Johnson and Huebner, *Traffic and Rates*, vol. 1, p. 106.
87. Larry Barsness, *Heads, Hides, and Horns* (Fort Worth, Tex., 1985), pp. 124–139.
88. Brooke Hindle, ed., *Material Culture of the Wooden Age* (Tarrytown, N.Y., 1981).
89. Daggett, *Inland Transportation*, p. 217.
90. S. M. Derrick, *Centennial History of the South Carolina Railroad* (Columbia, S.C., 1930), pp. 119 and 211.
91. Reed, *Texas Railroads*, p. 733.
92. Daggett, *Inland Transportation*, p. 13.
93. *Official Guide to Railways*, 1893, p. 669, carries a full-page advertisement for the M&MT Co. See also Groner, *American Business and Industry*, p. 139.
94. *RRG*, Feb. 3, 1888, p. 71.
95. Ibid.
96. Alfred Marshall, in his masterful book *The Principles of Economics* (1890; rpt. London, 1930), pp. 285–286, developed the following list of factors that determined the rates for transport and other services:
   1. The service is essential and no substitute is available at a moderate price.
   2. The demand is constant and inelastic. The user will pay a greater price rather than do without the service.
   3. The service is a small part of the total cost of a product.
   4. A small decline in demand will cause a considerable fall in rates.
97. *Poor's*, 1870–1871, p. xxxiv.
98. Hadley, *Railroad Transportation*, p. 114.
99. Ibid., p. 129.
100. Chandler, *Railroads*, p. 159.
101. Johnson and Huebner, *Traffic and Rates*, vol. 1, p. 338.
102. Wellington, *Location*, p. 226.
103. Cooley, Clark, et al., *American Railway*, p. 436.
104. Wellington, *Location*, p. 753.
105. Johnson and Huebner, *Traffic and Rates*, vol. 1, p. 331.
106. Ibid., p. 352.
107. Hungerford, *Modern Railroad*, p. 336.
108. Reed, *Texas Railroads*, p. 609.
109. Derrick, *South Carolina Railroad*, opp. p. 84.
110. *ARRJ*, Aug. 12, 1848, p. 686; the tariff dated July 15, 1847, was published as an advertisement by the Georgia RR.
111. Lucas, *From the Hills to the Hudson*, p. 246.
112. The IC tariff is in the collection of the Museum of Science and Industry, Chicago.
113. Robert S. Henry, *This Fascinating Railroad Business* (Indianapolis, 1942), p. 342.
114. *Encyclopedia of American Business and Biography, Railroads in the Twentieth Century* (New York, 1988), p. 358.
115. Hungerford, *Modern Railroad*, p. 330.
116. *RR. Mag.*, July 1943, p. 78.
117. Taylor, *Transportation Revolution*, p. 135.
118. Hadley, *Railroad Transportation*, p. 17.
119. *Scribner's Magazine*, Jan. 1902, p. 19.
120. Taylor, *Transportation Revolution*, p. 133.
121. Frank W. Stevens, *Beginnings of the New York Central Railroad* (New York, 1926), p. 86; Henry Tanner, *Description of the Canals and Rail Roads*

of the United States (New York, 1840), p. 21.

122. Fishlow, *American Railroads*, p. 321.
123. Taylor, *Transportation Revolution*, p. 135.
124. *Cassier's Magazine*, June 1894, p. 114.
125. Slason Thompson, *A Short History of American Railways* (New York, 1925), p. 243.
126. *Rly. Rev.*, May 2, 1885, p. 212.
127. *Scribner's Magazine*, Jan. 1902, p. 19.
128. Wellington, *Location*, p. 753; *Encyclopedia of American Business History* (New York, 1988), p. 243.
129. *New York State Railroad Commissioner's Report*, 1855, p. 13.
130. *RRG*, April 15, 1898, p. 278.
131. *Rly. Rev.*, May 9, 1885, p. 217.
132. Stephen Salisbury, *The State, the Investor, and the Railroad* (Cambridge, Mass., 1967), pp. 206–222.
133. Catton, *Garrett*, pp. 147 and 363.
134. August Derleth, *The Milwaukee Road* (New York, 1948), p. 142.
135. Hadley, *Railroad Transportation*, p. 71.
136. Alvin F. Harlow, *Road of the Century* (New York, 1947), p. 233.
137. Fishlow, *American Railroads*, pp. 181 and 185.
138. Catton, *Garrett*, p. 563.
139. Stover, *B&O*, p. 123.
140. Harlow, *Road of the Century*, p. 325.
141. Johnson and Huebner, *Traffic and Rates*, vol. 1, p. 296.
142. Daggett, *Inland Transportation*, p. 371.
143. Overton, *Burlington Route*, p. 157.
144. Reed, *Texas Railroads*, pp. 544–551.
145. *RRG*, Dec. 1, 1882, p. 743.
146. Daggett, *Inland Transportation*, p. 371.
147. Johnson and Huebner, *Traffic and Rates*, vol. 1, p. 296.
148. James Ward, *J. Edgar Thomson: Master of the Pennsylvania* (Westport, Conn., 1980), pp. 118–122.
149. Harlow, *Road of the Century*, p. 98.
150. Overton, *Burlington Route*, pp. 108 and 197.
151. Peter C. Marzio, ed., *A Nation of Nations* (New York, 1976), p. 186.
152. Maury Klein, *Union Pacific: Birth of a Railroad* (Garden City, N.Y., 1987), p. 584.
153. Hungerford, *Modern Railroad*, p. 331.
154. Reed, *Texas Railroads*, p. 550.
155. *Encyclopedia of American Business History*, p. 48.
156. *RR. Mag.*, July 1943, pp. 28 and 33.
157. Daggett, *Inland Transportation*, p. 320.
158. Klein, *Union Pacific*, p. 503.
159. Harlow, *Road of the Century*, p. 387.
160. Bogen, *Anthracite Railroads*, p. 37.
161. *Railroad History* 145 (Autumn 1981), p. 77.
162. Daggett, *Inland Transportation*, p. 472; Derleth, *Milwaukee Road*, p. 100.
163. Taylor, *Transportation Revolution*, p. 58.
164. Derrick, *South Carolina Railroad*, pp. 113 and 287.
165. *ARRJ*, May 11, 1867, p. 437, and Sept. 14, 1889; *New York Times*, May 13, 1869.
166. Lewis H. Haney, *A Congressional History of Railways in the United States* (New York, 1908), vol. 2, pp. 249–255.
167. Just to mention a few of the books devoted to the ICC, readers are advised to consult the following studies: Ari and Olive Hoogenboom, *A Short History of the ICC* (New York, 1976); Albro Martin, *Enterprise Denied* (New York, 1971); and George W. Hilton, *The Transportation Act of 1958* (Bloomington, Ind., 1969).
168. Chandler, *The Railroads*, p. 103. Exactly when the divisional idea was adopted is unclear, but D. C. McCallum speaks of it as an established fact in his report on management structure for the Erie ("Superintendent's Report," in New York & Erie annual report, 1855).
169. Wellington, *Location*, p. 231.
170. *Rly. Age*, Sept. 1, 1881, p. 501.
171. *Eng. Mag.*, May 1895, p. 273.
172. *ARRJ*, June 25, 1853, p. 404.
173. Taylor and Neu, *Railroad Network*, p. 50.
174. Ibid., p. 31.
175. Rolland L. Maybee, *Railroad Competition and the Oil Trade 1855–1873* (Mount Pleasant, Mich., 1940), p. 81.
176. Catton, *Garrett*, pp. 380 and 475.
177. Taylor and Neu, *Railroad Network*, p. 9.
178. Ibid., p. 59.
179. *Poor's*, 1872–1873, p. 485.
180. *RRG*, Nov. 11, 1887, p. 733.
181. Stevens, *New York Central*, pp. 104 and 320.
182. Ibid., p. 327.
183. Salisbury, *State, Investor, and Railroad*, p. 84.
184. Harwood, *Impossible Challenge*, p. 34.
185. Dionysius Lardner, *Railway Economy* (New York, 1850), p. 140; Frederick S. Williams, *Our Iron Roads* (London, 1888), p. 312; *ARRJ*, Aug. 8, 1846, p. 501.
186. *RR. Reco.*, Sept. 7, 1854, p. 438.
187. Charles Ellet, Jr., *Remarks on the Gauge of the Covington and Ohio Railroad* (pamphlet, 1853), p. 6.
188. *Railroad History* 141 (Autumn 1979), p. 105.
189. *MCB Report*, 1874, p. 89; Taylor and Neu, *Railroad Network*, p. 72.
190. *ASME Trans.*, Dec. 1882, p. 386.
191. *RRG*, Nov. 29, 1873, p. 477.
192. Pratt, *American Railways*, p. 67.
193. *RRG*, March 7, 1902, p. 159.
194. Wellington, *Location*, p. 48.
195. *Rly. Age*, April 19, 1883, p. 221.
196. *RR. Mag.*, June 1910, p. 169.
197. Hungerford, *Modern Railroad*, p. 338.
198. *Scribner's Magazine*, May 1889, p. 568.
199. *Rly. Age*, March 18, 1892, p. 220, and March 25, 1892, p. 237.
200. Alvin F. Harlow, *Steelways of New England* (New York, 1946), p. 357.
201. *Daily News*, March 19, 1971.
202. *Rly. Age*, Dec. 25, 1891, p. 996; *Loco. Eng.*, March 1892, p. 197.
203. *RRG*, Nov. 29, 1873, p. 477.
204. *Eng. News*, Dec. 21, 1899, p. 403.
205. *Eng. Mag.*, May 1895, p. 273.
206. Dewsnup, *Railway Organization and Working*, p. 103.
207. Pratt, *American Railways*, p. 68.
208. Interview opinion offered to the author by George W. Hilton in August 1989.
209. Johnson and Huebner, *Traffic and Rates*, vol. 1, p. 196.
210. *ASCE Trans.*, 1883, p. 130.
211. Johnson and Huebner, *Traffic and Rates*, vol. 1, p. 203.
212. Ibid., p. 196.
213. *Poor's*, 1894, p. viii.
214. *ASME Trans.*, 1882, p. 382.
215. Ibid.
216. *Eng. Mag.*, May 1895, p. 273.
217. Pratt, *American Railways*, p. 67.
218. Wellington, *Location*, p. 168.
219. *NCB*, Sept. 1894, p. 137.
220. *RRG*, Aug. 20, 1886, p. 571.

221. J. A. Droege, *Freight Terminals and Trains* (New York, 1912), p. 18.
222. *Rly. Age*, 1902, p. 559.
223. Wellington, *Location*.
224. *ASCE Trans.*, 1883, p. 126.
225. *NCB*, Sept. 1894, p. 137.
226. *ARRJ*, June 25, 1853, p. 404.
227. *Official Guide of Railways*, June 1870, p. i.
228. Albert Fink, *Investigation into the Cost of Passenger Traffic* (1876), p. 2. This pamphlet touches on freight traffic as well, despite its title.
229. *RRG*, Feb. 15, 1895, pp. 97 and 99.
230. *ASCE Trans.*, 1882, p. 388.
231. Harold F. Williamson, *American Petroleum Industry* (Evanston, Ill., 1959), p. 537.
232. *RRG*, March 6, 1896, p. 159.
233. Hadley, *Railroad Transportation*, p. 107.
234. Henry, *Fascinating Business*, p. 292.
235. Droege, *Freight Terminals and Trains*, p. 1.
236. Johnson and Huebner, *Traffic and Rates*, vol. 1, p. 160.
237. *Rly. Age*, June 9, 1905, p. 873.
238. J. S. Bell, *Early Motive Power on the Baltimore and Ohio Railroad* (New York, 1912), p. 21.
239. *NCB*, June 1894, p. 89.
240. Harlow, *Steelways*, p. 126.
241. Baltimore & Susquehanna annual report, 1849, p. 15.
242. M. N. Forney, ed., *Catechism of the Locomotive* (New York, 1875), pp. 407–410.
243. *ARRJ*, June 25, 1853, p. 403.
244. John H. White, Jr., *American Locomotives: An Engineering History, 1830–1880* (Baltimore, 1968), p. 366.
245. Ibid., p. 438.
246. *Recent Locomotives* (New York, 1883), p. 11.
247. *Eng. News*, Jan. 9, 1896, p. 18.
248. Forney, *Catechism*, p. 408.
249. Wellington, *Location*, pp. 278 and 315.
250. *Eng. News*, Jan. 9, 1896, p. 18.
251. Harlow, *Steelways*, p. 135.
252. The 1840 Baldwin catalog was reprinted in about 1950 by Graham Hardy, a Virginia City (Nev.) book dealer. Location of the original is unknown.
253. F. A. Ritter Von Gerstner, *Die innern Communication der Vereinigten Staaten von Nord-America* (Vienna, 1842–1843), p. 251.
254. *Rly. Age*, Oct. 26, 1882, p. 592.
255. The Rogers Locomotive Works catalog for 1876 quotes a letter to *Am. Rly. Times*, Jan. 23, 1859.
256. Zerah Colburn, *The Locomotive Engine* (Philadelphia, 1851), p. 112.
257. *ARRJ*, Sept. 3, 1853, p. 571.
258. White, *American Locomotives, 1830–1880*, p. 72.
259. Wellington, *Location*, p. 566.
260. *ASCE Trans.*, 1883, p. 142.
261. Emory Edwards, *Modern American Locomotive Engines* (Philadelphia, 1883), p. 116.
262. *RRG*, July 25, 1879, p. 408.
263. Wellington, *Location*, p. 100.
264. *NCB*, Feb. 1883, p. 23; *RRG*, July 12, 1895, p. 463.
265. J. Luther Ringwalt, *Development of Transportation Systems in the United States* (Philadelphia, 1888), p. 383.
266. *ARRJ*, Feb. 1892, p. 52.
267. Wellington, *Location*, p. 566.
268. *RRG*, Aug. 12, 1898, p. 584.
269. Wellington, *Location*, p. 101.
270. U.S. Census, 1880; Thompson, *Railway Statistics*; *RRG*, July 12, 1895, p. 463.
271. Fishlow, *American Railroads*, p. 408; Chandler, *The Railroads*, p. 16; U.S. Census, 1880.
272. *Century Magazine*, Aug. 1901, p. 598.
273. *Eng. News*, Jan. 9, 1896, p. 18.
274. W. Fred Cottrell, *The Railroader* (Stanford, Calif., 1940), p. 83.
275. Mott, *Erie*, p. 426.
276. *Scribner's Magazine*, Nov. 1888, p. 546.
277. Richard Reinhardt, *Workin' on the Railroad* (Palo Alto, Calif., 1970), p. 90.
278. Ibid., p. 96.
279. B. A. Botkin and A. F. Harlow, *A Treasury of Railroad Folklore* (New York, 1953), p. 89. There are several books on the subject of hobos, including *Hard Travellin'* by Kenneth Allsop (New York, 1967).
280. *Eng. News*, Jan. 9, 1896, p. 18.
281. *Technology and Culture*, April 1982, p. 195.
282. Chandler, *The Railroads*, p. 134.
283. Johnson and Huebner, *Traffic and Rates*, vol. 1, p. 17.
284. *RR. Mag.*, Aug. 1912, p. 51.
285. Hungerford, *B&O*, vol. 1, opp. p. 168.
286. *RRCJ*, Dec. 1900, p. 342.
287. Wellington, *Location*, p. 206.
288. *NCB*, Nov. 1883, p. 130.
289. Botkin and Harlow, *Folklore*, p. 89.
290. *RRCJ*, Dec. 1900, p. 342.
291. Hidy et al. *Great Northern*, p. 42.
292. Reinhardt, *Workin'*, p. 94.
293. *Railroad History* 161 (Autumn 1989), p. 43.
294. Botkin and Harlow, *Folklore*, p. 170.
295. *ARRJ*, Dec. 23, 1854, p. 801.
296. Walter Licht, *Working for the Railroad* (Princeton, 1983), p. 179.
297. *ARRJ*, March 24, 1880, p. 290.
298. *Rly. Age*, May 6, 1880, p. 236.
299. Cooley, Clarke, et al., *American Railway*, p. 383.
300. Henry, *Fascinating Business*, p. 408; *Nat. Cyc. Am. Bio.*, vol. 17 (1927), p. 258.
301. Stover, *Illinois Central*, p. 230; *ARRJ*, Oct. 1885, p. 195, and Feb. 1886, p. 328.
302. *Encyclopedia Americana* (1959 ed.), vol. 29, p. 656.
303. Hungerford, *Modern Railroad*, pp. 418–431.
304. Stover, *Baltimore and Ohio*, pp. 201–203.
305. Chandler, *The Railroads*, p. 136.
306. Ibid.
307. Henry, *Fascinating Business*, p. 405.
308. *Encyclopedia of American Business: Nineteenth Century Railroads* (New York, 1988), p. 44.
309. *RR. Mag.*, Feb. 1936, p. 5; see also Gerald G. Eggert, *Railroad Labor Disputes* (Ann Arbor, Mich., 1967).
310. *RR. Mag.*, July 1909, p. 346; Joseph Taylor, *A Fast Life on the Modern Highway* (New York, 1874), p. 181.
311. *Century Magazine*, Aug. 1901, p. 598.
312. *Railroad History* 159 (Autumn 1988), p. 19.
313. Joseph F. Wall, *Andrew Carnegie* (New York, 1970), p. 124.
314. Harlow, *Steelways*, p. 365.
315. *RRG*, June 20, 1890, p. 438. The Newcastle & Frenchtown had an even earlier block system, but it was too early and too short-lived to exert much influence on subsequent signal installations. It was also more of a train locater than a system for governing train movements.
316. James L. Holton, *The Reading Railroad* (Laury's Station, Pa., 1989), p. 143.
317. *Rly. Rev.*, Feb. 17, 1906, p. 113.
318. Mary Brignano and Sid Navratil, *The Search for Safety* (New York, 1981), p. 81 (commissioned

by Union Switch & Signal Division, American Standard).

319. *ARRJ*, April 2, 1870, p. 360; *DAB*, vol. 8 (1932), p. 145.

320. Carl W. Condit, *The Railroad and the City* (Columbus, 1977), p. 127.

321. *RRG*, Jan. 11, 1901, p. 18.

322. Pratt, *American Railways*, p. 258; B. Adams and R. Hitt, *Railroad Signal Dictionary* (New York, 1908), unnumbered pages following the preface.

323. Condit, *Railroad and City*, p. 127.

324. Wellington, *Location*, p. 95.

325. *RR. Adv.*, Feb. 7, 1856, p. 8.

326. Hidy et al., *Great Northern*, p. 12.

327. Fink, *Cost of Passenger Traffic*.

328. *NCB*, Aug. 1872, p. 7.

329. *ASME Trans.*, 1882, p. 380.

330. *RRG*, Feb. 15, 1895, p. 99.

331. Harwood, *Impossible Challenge*, p. 67.

332. Wellington, *Location*, p. 102.

333. *Gaskell's Family Atlas* (Chicago, 1886), p. 60.

334. *Search for Safety*, p. 27; Oliver Jensen, *American Heritage History of Railroads in America* (New York, 1975), p. 192.

335. *RR. Mag.*, Aug. 1912, p. 520.

336. Catasauqua & Fogelsville Railroad employees timetable for Jan. 1879; *Rly. Age*, June 6, 1878.

337. Ibid.

338. Adams and Hitt, *Signal Dictionary*, fig. 152.

339. Wm. B. Sipes, *The Pennsylvania Railroad* (Philadelphia, 1875), p. 256.

340. *Technology and Culture*, April 1982, p. 195.

341. *RR. Mag.*, Aug. 1951, p. 12; W. M. Camp, *Notes on Track* (Chicago, 1903), p. 900.

342. *Traveler's Official Guide*, Oct. 1881.

343. *R&LHS Bull.* 50 (Oct. 1939), p. 42.

344. *Rly. Age*, Nov. 23, 1882, p. 659.

345. *RRG*, Jan. 3, 1890, p. 15.

346. Klein, *Union Pacific*, p. 498.

347. Camp, *Track*, p. 622.

348. Botkin and Harlow, *Folklore*, p. 98.

349. Ibid., p. 44.

350. A. K. McClure, *Old Time Notes of Pennsylvania* (Philadelphia, 1905), vol. 2, p. 158.

351. Edward H. Hungerford, *Men and Iron* (New York, 1938), p. 116.

352. Wall, *Carnegie*, p. 171.

353. *RRG*, Nov. 9, 1872, p. 485, and Feb. 15, 1895, p. 97.

354. Camp, *Track*, p. 622.

355. Ibid., p. 1187.

356. Flint, *Railroads of the U.S.*, pp. 314 and 424.

357. Wellington, *Location*, p. 119.

358. Hadley, *Railroad Transportation*, p. 106.

359. Camp, *Track*, p. 71.

360. *Journal of the Society for Industrial Archeology*, 1978, pp. 1–14.

361. *Eng. News*, Jan. 9, 1896, p. 18.

362. Wellington, *Location*, p. 278.

363. *Railroad History* 146 (Spring 1982), p. 37.

364. *RR. Adv.*, Sept. 8, 1855, p. 1.

365. *Great Railway Celebration* (Baltimore, 1857), p. 161 (company-sponsored history of the B&O and its associated lines); *ARRJ*, Sept. 1, 1855, p. 556.

366. *Railroad History* 129 (Autumn 1973), p. 47.

367. Condit, *Railroad and City*, p. 21.

368. *R&LHS Bull.* 105 (Oct. 1961), p. 10.

369. Keith L. Bryant, *History of the Atchison, Topeka, and Santa Fe Railway* (New York, 1974), p. 43; *Rly. Age*, June 26, 1879, p. 36.

370. *RRG*, Feb. 15, 1884, p. 138, and Feb. 22, 1884, p. 155.

371. National Railway Historical Society *Bulletin* 5 (1989), p. 12.

372. Harlow, *Steelways*, p. 355.

373. Reinhardt, *Workin'*, p. 177; William S. Kennedy, *Wonders and Curiosities of Railways* (Chicago, 1884), p. 82.

374. *RRG*, Feb. 3, 1888, p. 77.

375. Klein, *Union Pacific*, p. 285.

376. Gerald M. Best, *Snowplow* (Berkeley, 1966), p. 11.

377. Wm. B. Wilson, *History of the Pennsylvania Railroad* (Philadelphia, 1899), vol. 1, p. 309.

378. Best, *Snowplow*, p. 94.

379. Robert M. Vogel, *The Engineering Contributions of Wendell Bollman*, U.S. National Museum bulletin 240 (Washington, D.C., 1964), p. 89.

380. Paul Westhaeffer, *History of the Cumberland Valley Railroad* (Washington, D.C., 1979), p. 19.

381. According to Taylor and Neu, *Railroad Network*, p. 33, car ferry service began in 1867, but John Pixton's *The Marietta and Cincinnati Railroad* (University Park, Pa., 1966), p. 40, claims that it began in March 1860.

382. Stover, *Illinois Central*, p. 158.

383. *Railroad History* 161 (August 1989), p. 68.

384. George W. Hilton, *Great Lakes Car Ferries* (Berkeley, 1962).

385. Herman Haupt, *Reminiscences* (Milwaukee, 1901), p. 160.

386. *Nat. Cyc. Am. Bio.*, vol. 11 (1921), p. 20.

387. *RR. Mag.*, Sept. 1912, p. 620.

388. Harlow, *Steelways*, p. 176; Klein, *Union Pacific*, pp. 261, 267, 278.

389. Taylor and Neu, *Railroad Network*, p. 41.

390. Condit, *Railroad and City*, p. 59.

391. *Trains* magazine, April 1990, p. 34.

392. *Eng. News*, May 31, 1894, p. 443.

393. *Rly. Rev.*, May 13, 1893, p. 301.

394. *Eng. News*, Dec. 21, 1899, p. 403.

395. Pratt, *American Railways*, p. 105.

396. *RR. Mag.*, Aug. 1912, p. 513.

397. *Rly. Age*, July 12, 1889, p. 455.

398. Camp, *Track*, p. 452.

399. *Eng. News*, May 31, 1894, p. 444.

400. Pratt, *American Railways*, p. 111.

401. Lucius Beebe and Charles Clegg, *Hear the Train Blow* (New York, 1952), p. 140.

402. White, *American Locomotives, 1830–1880*, pp. 47 and 68; *A Century of Reading Motive Power* (Philadelphia, 1941), p. 46.

403. Claims for the first hump yard have been made for Speldorf, Germany, 1876 (Camp, *Track*, p. 457), and the Edge Hill yard, Manchester, England, 1873 (*Eng. News*, May 31, 1894, p. 445).

404. Wellington, *Location*, p. 818.

405. Condit, *Port of New York*, vol. 1, p. 357.

406. Johnson and Huebner, *Traffic and Rates*, vol. 1, p. 33.

407. Condit, *Port of New York*, vol. 1, pp. 51 and 166.

408. Wellington, *Location*, p. 823.

409. *RR. Mag.*, Aug. 1912, p. 516.

410. Daggett, *Inland Transportation*, p. 414.

411. *Rly. Age*, July 12, 1889, p. 455.

412. Ibid.

413. George Drury, *Train Watcher's Guide* (Milwaukee, 1984), p. 24.

414. *Rly. Age*, April 30, 1882, p. 176.

415. Ibid., May 6, 1892, p. 351.

416. Hungerford, *Modern Railroad*, p. 107.

417. Pratt, *American Railways*, p. 64.

418. Condit, *Railroad and City*, p. 16.

419. Salisbury, *State, Investor, and Railroad*, pp. 120 and 195.
420. Flint, *Railroads of the United States*, p. 212.
421. Harlow, *Road of the Century*, p. 190; Hungerford, *Modern Railroad*, p. 353.
422. *Rly. Rev.*, May 2, 1885, p. 205.
423. Wellington, *Location*, p. 818.
424. Taylor, *Transportation Revolution*, p. 138.
425. Harwood, *Impossible Challenge*, p. 67.
426. *RRG*, Nov. 28, 1874, p. 461.
427. *Rly. Age*, Dec. 22, 1893, p. 902.
428. Harwood, *Impossible Challenge*, p. 95.
429. J. J. Thomas, *Fifty Years on the Rail* (New York, 1912), p. 13.
430. *ASCE Trans.* 12 (1883), p. 151.
431. *Rly. Age*, May 23, 1896, p. 271.
432. *NCB*, April 1888, p. 61.
433. *RRG*, May 15, 1875, p. 202.
434. Botkin and Harlow, *Folklore*, p. 312.
435. Wellington, *Location*, p. 529.
436. Droege, *Freight Terminals and Trains*, p. 168.
437. *RRG*, May 9, 1879, p. 259.
438. *RRG*, Feb. 15, 1895, p. 97.
439. *RR. Mag.*, Oct. 1930, p. 478.
440. *Rly. Age*, March 9, 1894, p. 141.
441. *Rly. Age*, Oct. 9, 1896, p. 280.
442. Johnson and Huebner, *Traffic and Rates*, vol. 1, p. 254.
443. *RR. Reco.*, April 4, 1861, p. 77.
444. Fishlow, *American Railroads*, p. 421.
445. *NCB*, Feb. 1874, p. 21, based on data published in the Dec. 1873 *Railway Monitor*.
446. Ibid., Aug. 1874, p. 120.
447. Slason Thompson, *Railway Statistics of the United States 1906* (Chicago, 1907).
448. Ringwalt, *Development of Transportation Systems*, p. 210; *NCB*, Oct. 1879, p. 148.
449. *RRCJ*, April 1900, p. 98.
450. *MCB Report*, 1886, pp. 103 and 146.
451. *RRG*, Feb. 1878, p. 80; *Loco. Eng.*, June 1895, p. 359.
452. *RRG*, Dec. 23, 1887, p.824.
453. Private cars have been discussed generally in the following sources: *Business History Review*, Summer 1970; L. H. Weld, *Studies in History, Economics and Public Law* (Columbia University) 31, no. 1 (1908); J. W. Midgley's series in *Railway Age*, Oct. 1902–Nov. 1904; and White, *Great Yellow Fleet*. These sources have been drawn upon freely in the present text.
454. *Rly. Age*, May 6, 1880, p. 234.
455. *Railroad History* 141 (Autumn 1979), p. 98.
456. *Rly. Rev.*, June 28, 1884, p. 341.
457. Maybee, *Competition and Oil Trade*, p. 111.
458. George H. Burgess and M. C. Kennedy, *Centennial History of the Pennsylvania Railroad* (Philadelphia, 1949), p. 327.
459. Wilson, *Pennsylvania Railroad*, vol. 2, p. 69; *RRG*, Oct. 26, 1877, p. 475.
460. *DAB*, vol. 4, p. 271.
461. *RRCJ*, Nov. 1894, p. 244.
462. Maybee, *Competition and Oil Trade*, p. 131.
463. *ARRJ*, July 27, 1867, p. 703, and Feb. 1, 1868, p. 102.
464. Catton, *Garrett*, p. 429; *ARRJ*, Aug. 12, 1871, p. 877.
465. *ARRJ*, Feb. 5, 1870, p. 146.
466. John S. Sherman, ed., *Moody's Manual of Railroads* (New York, 1931), p. 1799.
467. Freeman Hubbard, *Encyclopedia of North American Railroading* (New York, 1981), p. 115.
468. *Moody's Manual of Railroads*, p. 1668.
469. Johnson and Huebner, *Traffic and Rates*, vol. 1, p. 225.
470. *RRCJ*, Feb. 1895, p. 42.
471. *Official Railway Equipment Register*, Oct. 1885, p. 62, and 1888, p. 183.
472. Data on the U.S. Rolling Stock Co. is drawn from *RRG*, Sept. 5, 1874, p. 349, and Feb. 27, 1875, p. 87; and *Rly. Age*, April 20, 1882, p. 212. See also *Poor's* for the same period.
473. *Rly. Age*, Dec. 9, 1904, p. 821.
474. Ralph C. Epstein, *GATX: A History of the General American Transportation Corp.* (New York, 1948), passim.
475. *Rly. Age*, Dec. 19, 1902, p. 676.
476. Rudolf A. Clemen, *American Live Stock and Meat Industry* (New York, 1923), p. 398.
477. Johnson and Huebner, *Traffic and Rates*, vol. 1, p. 223.
478. Maybee, *Competition and Oil Trade*, p. 268.
479. Daggett, *Inland Transportation*, p. 323.
480. *Rly. Age*, Oct. 16, 1903, p. 505.
481. Hadley, *Railroad Transportation*, p. 88.
482. Johnson and Huebner, *Traffic and Rates*, vol. 1, p. 246.
483. *Rly. Rev.*, March 14, 1891, p. 382.
484. *Rly. Age*, May 13, 1892, p. 382.
485. Maybee, *Competition and Oil Trade*, p. 115.
486. *RRCJ*, June 1898, p. 143.
487. William Voss, *Railway Car Construction* (New York, 1892), p. 1.
488. *Rly. Mast. Mech.*, Feb. 1903, p. 40.
489. *RRG*, July 6, 1900, p. 464.
490. *Rly. Age*, June 9, 1905, p. 873.
491. J. B. Jervis, *Railway Property* (Philadelphia, 1861), p. 195.
492. *NCB*, May 1889, p. 64.
493. *RRG*, Oct. 20, 1899, p. 730.
494. *Rly. Rev.*, Sept. 1891, p. 128; *Railroad History* 138 (Spring 1978), p. 6.
495. Ibid.
496. U.S. Census, 1900, pt. 4, pp. 263–289.
497. *Autobiography of George I. King* (privately printed, 1917), pp. 17–29.
498. *Railroad History* 138 (Spring 1978), p. 42.
499. *Rly. Rev.*, Jan 12, 1884, p. 23.
500. *RRG*, Nov. 1, 1889, p. 719.
501. *Rly. Rev.*, December 26, 1885, p. 641.
502. *NCB*, Dec. 1878, p. 178; *Poor's*, 1896, p. 1298.
503. *Rly. Rev.*, Oct. 22, 1881, p. 586; *Chicago Herald*, Nov. 12, 1892.
504. *Poor's*, 1901, p. 1101.
505. *Rly. Rev.*, Oct. 29, 1892, p. 688; *Rly. Age*, Aug. 26, 1898, p. 611.
506. Francis M. Caulkins, *History of Norwich* (1866), p. 623.
507. *Autobiography of King*, p. 16.
508. Stover, *Illinois Central*, p. 247.
509. *Rly. Rev.*, March 27, 1886, p. 145.
510. U.S. Census, 1870, p. 406. Commercial and railroad repair shops are apparently not separated in this count. Car repair plus streetcars are included in the numbers for 1870, which are hence somewhat inflated.
511. *NCB*, Oct. 1879, p. 148.
512. Listings of American car builders mentioned here are given in *Railroad History* 138 (Spring 1978) and 154 (Spring 1986).
513. Biographies of many car builders mentioned here are given in Appendix A of John H. White, Jr., *The American Railroad Passenger Car* (Baltimore,

1978). Other data is drawn from the railroad trade press and local histories.

514. *L&N Magazine*, July 1925, p. 10.
515. *NCB*, Oct. 1873, p. 247.
516. *RR. Adv.*, Sept. 20, 1856, p. 1.
517. *Railroad History* 138 (Spring 1978), p. 79.
518. Data on the Patten Car Co. is in the papers of the European and North American Railroad, Maine Historical Society, Portland.
519. *Rly. Rev.*, June 6, 1883, p. 12.
520. Gill pamphlet is in the collections of the Ohio Historical Society, Columbus.
521. *Rly. Rev.*, Aug. 31, 1889, p. 503.
522. *NCB*, Oct. 1872, p. 10, and Sept. 1873, p. 215.
523. James L. Bishop, *History of Manufacturers* (Philadelphia, 1866), p. 39.
524. *NCB*, Jan. 1871, p. 6.
525. *Rly. Age*, Jan. 18, 1882, p. 41.
526. Ibid., Dec. 31, 1885, p. 810.
527. Ibid., June 23, 1881, p. 341.
528. David L. Hay, "Before Ford: Assembly Line Techniques," paper given at 1985 SHOT meeting, Dearborn, Mich.
529. *RRG*, Nov. 4, 1871, p. 321.
530. *Rly. Rev.*, Aug. 20, 1881, p. 460.
531. *Lippincott's Magazine*, April 1873, p. 377.
532. *Rly. Age*, Sept. 10, 1885, p. 585.
533. *NCB*, March 1874, p. 34.
534. *Mast. Mech. Mag.*, May 1892, p. 71; *RRG*, July 4, 1884, p. 499.
535. *NCB*, April 1880, p. 62.
536. Stanley Buder, *Pullman: An Experiment in Industrial Order and Community Planning, 1880–1930* (New York, 1967), passim.
537. R. G. Dunn credit reports, Baker Library Manuscript Division, Harvard School of Business, Cambridge, Mass.
538. *The Manufactories and Manufacturers of Pennsylvania of the Nineteenth Century* (Philadelphia, 1875), p. 283.
539. *Lippincott's Magazine*, May 1873, p. 528.
540. *York Dispatch*, Dec. 24, 1879.
541. *RR. Adv.*, Sept. 20, 1856, p. 1.
542. *Rly. Rev.*, Aug. 31, 1889, p. 503.
543. *York Daily*, April 4, 1873.
544. *ARRJ*, Feb. 1899, p. 37.
545. *Rly. Mast. Mech.*, March 1907, p. 83.
546. *Car Builder*, March 1955, p. 13 (house organ of the Pullman Standard Co.).
547. *Rly. Age*, Jan. 1, 1949, p. 36.
548. *ARRJ*, Feb. 1908, p. 63.
549. *Standard*, Oct. 1917, p. 4 (house organ of the Pullman Car Works).
550. *Rly. Rev.*, April 16, 1892, p. 248.
551. *Manufacturers of Pennsylvania*, p. 282.
552. Hay, "Before Ford."
553. *Rly. Age*, April 10, 1879, p. 159.
554. *RRCJ*, May 1892, p. 132.
555. *Rly. Age*, Aug. 21, 1884, p. 256.
556. *RRG*, Sept. 1, 1882, p. 541.
557. U.S. Census, 1910, vol. 8, p. 475.
558. *RRG*, Oct. 4, 1873, p. 404.
559. Data on Altoona is drawn from ibid., Dec. 4, 1875, p. 502; Dredge, *Pennsylvania Railroad*, p. 90; *NCB*, Sept. 1881, p. 102; and *ARRJ*, Jan. 1897, p. 2.
560. *NCB*, July 1890, p. 102.
561. Data on the L&N shops is from *RR. Reco.*, June 22, 1854, p. 261; *RRG*, Sept. 30, 1871, p. 299; *NCB*, June 1883, p. 65; and *ARRJ*, June 1906, p. 209.
562. Data on the Sacramento shops is from *Western Railroad Gazette*, Sept. 4, 1869, p. 1; *NCB*, May

1880, p. 79; *Rly. Rev.*, Aug. 11, 1888, p. 466; and *Mast. Mech. Mag.*, March 1892, p. 40.

CHAPTER 2

1. Data on the Granite Railway is drawn from several sources: C. B. Stuart's *Lives and Works of Civil and Military Engineers of America* (New York, 1871); *Winans vs. Eastern RR Evidence for the Respondents* (Boston, 1854); *The First Railroad in America* (Boston, 1926); and A. R. Telle's *History of Milton, Mass.* (1887).
2. A copy of the Granite Railway print is in the Worcester (Mass.) Art Museum.
3. *Boston Mechanic* 4 (1835), pp. 6–9.
4. *R&LHS Bull.* 110 (April 1964), p. 59; John N. Hoffman, *Anthracite in the Lehigh Region of Pennsylvania* (Washington, D.C., 1968), pp. 99–107. The Mauch Chunk line was rebuilt in 1844–1845 with incline planes and a new up line. It stopped carrying coal in about 1870. It operated as a tourist line until 1933 and was scrapped four years later.
5. F. A. Ritter Von Gerstner, *Die innern Communication der Vereinigten Staaten von Nord-America* (Vienna, 1842–1843). A. Bendel, *Aufsatze Eisenbahnwesen in Nord-Amerika* (Berlin, 1862), text p. 20 and atlas plate 3.
6. Walter A. Lucas, ed., *Railroadians of America Book 3* (New York, 1941), p. 115.
7. Newcomen Society (England) *Transactions*, 1978–1979, p. 120. Allen's English diary is reproduced in *R&LHS Bull.* 89 (Nov. 1953), pp. 97–138.
8. Nicholas Wood, *A Practical Treatise on Rail-Roads* (London, 1832), p. 528.
9. Von Gerstner, p. 330.
10. *R&LHS Bull.* 82 (April 1951); Jim Shaughnessy, *The Delaware and Hudson* (Berkeley, 1967); Edward Steers, "The Delaware and Hudson Canal Company's Gravity Railroad," Canal History and Technology Symposium (Easton, Pa.) *Proceedings* 2 (1983).
11. *Journal of the Franklin Institute*, April 1829, p. 235. See also John H. White, Jr., *The American Railroad Passenger Car* (Baltimore, 1978), p. 514. There were another dozen patents issued on railroad cars before 1835, but I cannot find one that received any extensive application.
12. Thomas Earle, *A Treatise on Rail-Roads* (Philadelphia, 1830).
13. *R&LHS Bull.* 139 (Autumn 1978), pp. 65–77, discusses American railroad engineers who visited England before 1840.
14. *Shipper and Carrier* (B&M magazine), Aug. 1925, p. 17. The Stephenson drawing is reproduced in this trade journal. It is presently located in the collections of the California State Railroad Museum, Sacramento.
15. "Early Days of Railroading," paper read before the Lowell Historical Society on March 2, 1909, and printed by the Lowell Historical Society in 1913.
16. The so-called Petersburg scrapbook is in the possession of the Division of Mechanical and Civil Engineering, National Museum of American History, Smithsonian Institution.
17. *ARRJ*, Sept. 8, 1832, p. 579.
18. Golda M. McArdle, "The Pontchartrain Railroad" (M.A. thesis, Tulane Univ. 1931), p. 44. See also *Rly. Age*, Aug. 13, 1885, p. 518.
19. *Winans vs. the New York & Erie Railway*, published extracts from a patent suit printed in New York in 1859, p. 327.

20. Von Gerstner, pp. 250–252.
21. Petersburg Railroad annual report, 1835, pp. 14 and 15.
22. Eugene Ferguson, ed., *Early Engineering Reminiscences (1815–1840) of George Escol Sellers* (Washington, D.C., 1965), p. 155.
23. Ibid.
24. *Winans vs. Eastern RR.* (New York, 1853), p. 1251.
25. Von Gerstner, p. 88.
26. Ibid., pp. 297 and 312.
27. Jonathan Knight and Benj. H. Latrobe, *Report upon the Locomotive Engines . . . of the Principal Rail Roads in the Northern and Middle States* (Baltimore, 1838), p. 21.
28. *Rly. Rev.*, Sept. 24, 1892, p. 596.
29. *Winans vs. Eastern RR.* (1853), p. 464.
30. *Winans vs. the New York & Erie* (1859), p. 317.
31. *Winans vs. Eastern RR.* (1853), p. 771.
32. Details of Winans's eight-wheel car case are given in White, *American Railroad Passenger Car*, pp. 18–20.
33. *ARRJ*, Dec. 8, 1855, p. 777, reporting data collected in April–May 1844. The age for the cars involved in the test is given in this report.
34. Emile With (b. 1816), *Nouveau Manuel Complet: Construction des Chemins de Fer* (Paris, 1857), atlas plate 11.
35. M. W. Baldwin letters, Historical Society of Pennsylvania, Philadelphia.
36. H. T. Gause, *Semi Centennial Memoir of the Harlan and Hollingsworth Company* (Wilmington, Del., 1886).
37. Von Gerstner, pp. 88, 238, 278.
38. *Winans vs. Eastern RR.* (1853), pp. 997 and 1026.
39. Baltimore & Susquehanna Railroad annual report, 1849, p. 14.
40. R. H. G. Thomas, *The Liverpool and Manchester Railway* (London, 1980), p. 184. See also Bertram Baxter, *Stone Blocks and Iron Rails* (Newton Abbot, Devon, 1966).
41. Baggage crates were in use on the Camden & Amboy before December 1833. A lawsuit reported in the New York *Journal of Commerce*, May 28, 1834, noted the loss of theatrical goods when a baggage crate was accidentally dropped in the water at Bordentown, N.J., while a C&A steamer was being unloaded.
42. Data on the sectional canal boats comes principally from the following sources: *Reminiscences of Sellers*; *Baldwin Record of Recent Construction*, no. 97 (Philadelphia, 1920); *Rly. Rev.*, Nov. 27, 1886, p. 638; Von Gerstner, pp. 129–130; and *Pennsylvania Magazine of History and Biography*, Oct. 1945, pp. 294–313.
43. The O'Connor invoice is copied on Smithsonian Neg. 79-13729.
44. See White, *American Railroad Passenger Car*, pp. 454–455, for more data on the baggage crate system.
45. Von Gerstner, p. 353.
46. *NCB*, Dec. 1870, p. 5.
47. R. E. Carlson, *The Liverpool and Manchester Railway Project 1821–1831* (New York, 1969), p. 218; Thomas, *Liverpool and Manchester Railway*, p. 184.
48. *RRCJ*, June 1898, p. 143. A car of this same general type is illustrated by an engraving in the 1879 *Car Builders Dictionary*, p. 203.
49. *Am. Rly. Rev.*, Aug. 4, 1859, p. 8; *ARRJ*, Oct. 15, 1859, p. 659.
50. *ARRJ*, May 22, 1858, p. 321.

51. Alvin F. Harlow, *Steelways of New England* (New York, 1946), recounts the recollections of A. V. H. Carpenter, a trainman on the Vermont Central.
52. *Am. Rly. Rev.*, Dec. 29, 1859, p. 1, and Feb. 7, 1861, p. 67.
53. *R&LHS Bull.* 88 (May 1953), p. 129.
54. *Am. Rly. Rev.*, Nov. 24, 1859, p. 7.
55. *ARRJ*, June 14, 1851, p. 371.
56. *Am. Rly. Rev.*, May 10, 1860, p. 277.
57. *ARRJ*, June 12, 1845, p. 382.
58. *Chart of the Boston and Worcester and Western RR* (Boston, 1847), an illustrated booklet in the collection of the American Antiquarian Society, Worcester, Mass.
59. *Richmond Business Directory*, 1855, p. 60.
60. *RRCJ*, Aug. 1894, p. 167.
61. *Rly. Mast. Mech.*, June 1911, p. 218.
62. Archives, Museum of American History, Smithsonian Institution.
63. *Rly. & Loco. Eng.*, May 1921, p. 134. The same drawing is reproduced in the 1858 supplement atlas to Douglas Galton's *Report on the Railways of the U.S.* (London, 1857–1858).
64. *RRCJ*, June 1898, p. 143.
65. The B&O boxcars of 1862 are generally similar to the 1856 car depicted in this volume except for the floor frames and trucks. Photocopies of the drawing are available on Smithsonian Negs. 64839 and 64841.
66. *Rly. Mast. Mech.*, June 1911, p. 218.
67. *NCB*, Feb. 1884, p. 19; the note speaks of C&A cars built twenty years earlier.
68. *RRG*, April 11, 1879, p. 196, speaks of cars built in 1856–1869.
69. William Voss, *Railway Car Construction* (New York, 1892).
70. *NCB*, July 1883, p. 77; iron body bolsters are credited to a Mr. French of the Illinois Central.
71. Ibid., Aug. 1874, p. 117.
72. Garrett papers, Maryland Historical Society, Baltimore.
73. *Rly. Mast. Mech.*, June 1911, p. 218.
74. *The Shoreliner*, 1980, p. 10, published by the New Haven Railroad Historical and Technical Society, reproduces an ambrotype from the collection of the George Eastman House, Rochester, N.Y.
75. White, *American Railroad Passenger Car*, p. 466.
76. *Journal of the Franklin Institute*, Feb. 1840, p. 99.
77. Archives, Museum of American History, Smithsonian Institution.
78. *Rly. Mast. Mech.*, Jan. 1899, p. 12.
79. *Loco. Eng.*, Aug. 1896, p. 681.
80. *RRCJ*, June 1898, p. 143.
81. Peter Pollack, *Picture History of Photography* (New York, 1970), p. 126. The photo is owned by the Chicago Historical Society.
82. Barry B. Combs, *Westward to Promontory* (Palo Alto, Calif., 1969); Lucius Beebe, *The Central Pacific and Southern Pacific Railroads* (Berkeley, 1963).
83. All by E. P. Alexander: *The Pennsylvania RR: A Pictorial History* (New York, 1947); *On the Main Line: The Pennsylvania Railroad in the Nineteenth Century* (New York, 1971); and *Down at the Depot* (New York, 1970).

CHAPTER 3

1. *NCB*, March 1892, p. 39.
2. *RRCJ*, June 1897, p. 137.
3. *RRG*, July 12, 1882, p. 442.

4. Ibid., Oct. 26, 1883, p. 702.
5. Ibid., July 1, 1871, p. 160.
6. *NCB*, April 1876, p. 59.
7. Albro Martin, *James J. Hill and the Opening of the Northwest* (New York, 1976), p. 102.
8. *NCB*, Nov. 1885, p. 141.
9. *Engineer*, Dec. 28, 1883, p. 498.
10. *NCB*, May 1882, p. 59.
11. *RRG*, March 25, 1887, p. 192.
12. *Rly. & Loco. Eng.*, April 1924, p. 120.
13. *RRG*, Jan. 11, 1884, p. 29; *ARRJ*, June 1902, p. 174.
14. *NCB*, Oct. 1879, p. 446; *Rly. Age*, Aug. 28, 1879, p. 409.
15. Photographs of B&M four-wheel boxcars are in the Von Name Collection, Library of Congress; the B&M station photos are in the Division of Mechanical and Civil Engineering, National Museum of American History, Smithsonian Institution.
16. *Rly. Rev.*, March 26, 1881, p. 165.
17. *RRG*, Feb. 15, 1895, p. 97.
18. Ibid., Oct. 20, 1899, p. 730.
19. *ARRJ*, Nov. 1892, p. 500.
20. *NCB*, Nov. 1885, p. 149.
21. *RRG*, Jan. 27, 1893, p. 68.
22. *ARRJ*, Sept. 1896, p. 209; *RRCJ*, June 1897, p. 12.
23. *RRCJ*, Dec. 1893, p. 293.
24. *Engineer*, Dec. 28, 1883, p. 498.
25. *NCB*, Jan. 1873, p. 11, and Dec. 1876, p. 184.
26. *RRG*, March 25, 1887, p. 192.
27. *RRG*, July 21, 1893, p. 539.
28. *ARRJ*, April 1899, p. 112.
29. "The Narrow Gauge Fallacy," *Railroad History*, no. 141 (Autumn 1979), pp. 77–97. See also George W. Hilton, *American Narrow Gauge Railroads* (Stanford, Calif., 1990).
30. *Engineering*, Dec. 11, 1871, p. 441.
31. *NCB*, Aug. 1884, p. 105.
32. *RRG*, Oct. 26, 1883, p. 702.
33. *RRCJ*, July 1893, p. 258.
34. *RRG*, Oct. 9, 1893, p. 711.
35. *ARRJ*, Feb. 1896, p. 21.
36. *MCB Report*, 1880, p. 42.
37. *RRG*, Jan. 27, 1899, p. 58; *NCB*, Nov. 1895, p. 170.
38. *RRG*, March 25, 1887, p. 192; *Rly. Rev.*, Oct. 11, 1890, p. 598.
39. *RRG*, Jan. 27, 1899, p. 58.
40. Marshall M. Kirkman, *Science of Railways* (Chicago, 1908), unnumbered volume entitled *Cars, Construction, Handling, and Suspension*, p. 20.
41. *NCB*, Nov. 1895, p. 170.
42. William Voss, *Railway Car Construction* (New York, 1892), p. 27.
43. *RRG*, March 6, 1885, p. 145.
44. *RRCJ*, June 1897, p. 137.
45. *NCB*, April 1881, p. 41.
46. Ibid., April 1884, p. 40.
47. *MCB Report*, 1878, p. 138.
48. *NCB*, April 1892, p. 57.
49. *MCB Report*, 1878, p. 138; *NCB*, Nov. 1883, p. 131.
50. *RRG*, Dec. 9, 1910, p. 1133.
51. Ibid., Feb. 12, 1892, p. 115.
52. Ibid., Dec. 30, 1892, p. 977.
53. See John H. White, Jr., *The American Railroad Passenger Car* (Baltimore, 1978), pp. 42–44, for more data on car lumber.
54. *Rly. Rev.*, April 2, 1892, p. 207.
55. *RRG*, June 9, 1899, p. 395.
56. Ibid., June 8, 1888, p. 362.
57. Sherry Olsen, *The Depletion Myth* (Baltimore, 1971), p. 207.
58. *NCB*, April 1881, p. 41.
59. *RRG*, June 21, 1889, p. 405; design of car actually dates from Aug. 1886.
60. *NCB*, Aug. 1877, p. 117, says wood-and-iron-frame cars were tested starting in Nov. 1875.
61. *Rly. Rev.*, Feb. 29, 1896, p. 117.
62. *RRG*, June 10, 1898, p. 412; *ARRJ*, April 1899, p. 112.
63. *Eng. News*, June 14, 1890, p. 568.
64. *Mast. Mech. Mag.*, Dec. 1904, p. 513.
65. O. M. Stimson, *Modern Freight Car Estimating* (Anniston, Ala., 1897), p. 458.
66. Ibid., p. 459.
67. *NCB*, May 1872, p. 7, and May 1876, p. 68; *RRCJ*, Dec. 1897, p. 384; *Restoration Feasibility Investigation on Nine Selected Passenger and Freight Cars* (report to Nevada State Museum by Short Line Enterprises, June 1981).
68. *Rly. Age*, Nov. 16, 1900, p. 388.
69. *New England Railroad Club*, Oct. 11, 1898, p. 46; *ARRJ*, Sept. 1896, p. 214.
70. *RRCJ*, March 1897, p. 77, and Oct. 1897, p. 293.
71. *Rly. Age*, Nov. 16, 1900, p. 388.
72. Voss, *Railway Car Construction*, p. 47.
73. Affiliation for B&O System Historical Research Handbook Insert 5120.16, Aug. 1983, by E. W. and J. W. Barnard and D. Jones, Columbus, Ohio.
74. *Restoration Feasibility Investigation*, p. 239.
75. More repair cost data can undoubtedly be uncovered. My search was hardly exhaustive. It might be found in both published and manuscript sources that I have not explored, but it is a task I am content to leave to an economic historian.
76. B&O annual report, 1847, pp. 46 and 47.
77. *Rly. Rev.*, Oct. 26, 1889, p. 620; *RRCJ*, June 1895, p. 104; *Scribner's Magazine*, May 1889, p. 584.
78. Institution of Civil Engineers *Proceedings* 28 (1869), p. 360, Zerah Colburn's report on U.S. rolling stock.
79. *NCB*, April 1879, p. 54.
80. *Poor's*, 1884, p. 362.
81. *Eng. Mag.*, April 1895, p. 12.
82. Arthur M. Wellington, *The Economic Theory of the Location of Railroads* (New York, 1893), p. 167.

CHAPTER 4

1. The literature on American agriculture and its dependence on transportation is extensive. I have drawn on only a sampling of this scholarly outpouring and have not attempted to offer more than a few of the more obvious ideas here. Especially useful were Rudolf A. Clemen's *American Live Stock and Meat Industry* (New York, 1923) and Mary Yeager's *Competition and Regulation: The Development of Oligopoly in the Meat Packing Industry* (Greenwich, Conn., 1981). Data from these valuable studies is used throughout this chapter without specific reference.
2. *Rly. Rev.*, Sept. 10, 1881, p. 510.
3. Ibid., Jan. 29, 1887, p. 62.
4. *Rly. Age*, Oct. 16, 1879, p. 502; Edwin A. Pratt, *American Railways* (New York, 1903), p. 45.
5. *Rly. Age*, Nov. 2, 1882, p. 45.
6. Ibid., Oct. 16, 1884, p. 651; *RRG*, Aug. 12, 1892, p. 601.
7. *RRG*, Dec. 10, 1909, p. 1153.
8. *NCB*, Aug. 1874, p. 120.
9. J. A. Droege, *Freight Terminals and Trains* (New York, 1912), p. 187.
10. *RRG*, Aug. 12, 1892, p. 601.
11. *NCB*, March 1880, p. 40.
12. *RRG*, Dec. 10, 1909, p. 1154.

13. *Car Builders Dictionary*, 1888, figs. 136–138.
14. *RRG*, May 26, 1882, p. 313.
15. *NCB*, Nov. 1880, p. 169.
16. Ibid., March 1880, p. 40.
17. Ibid., Sept. 1881, p. 108.
18. *Mast. Mech. Mag.*, Oct. 1889, p. 172.
19. *ARRJ*, April 1899, p. 112.
20. *Rly. Age*, Oct. 16, 1879, p. 502, and Sept. 16, 1880, p. 492; *NCB*, March 1885, p. 38. Henry Bergh and the ASPCA do not appear to have been active in the cattle humane movement.
21. *Am. Rly. Times*, Dec. 3, 1870, p. 357.
22. *Rly. Rev.*, Sept. 9, 1882, p. 515.
23. *Rly. Age*, July 1, 1887, p. 456.
24. *RRG*, Dec. 10, 1909, p. 1153.
25. *American Artisan*, Feb. 21, 1866, p. 250; *Rly. Age*, Sept. 16, 1880, p. 492.
26. *Pittsburgh Commercial*, Aug. 29, 1873.
27. Lewis H. Haney, *A Congressional History of Railways in the United States* (New York, 1908), vol. 2, p. 260.
28. Ibid.; *RRG*, June 14, 1873, p. 236.
29. Data on Humane Association is drawn from *Rly. Age*, Nov. 11, 1880, p. 585; Feb. 17, 1881, p. 98; and Feb. 2, 1882, p. 68; *RRG*, Aug. 27, 1880, p. 459, and Apr. 25, 1884, p. 317.
30. *Loco. Eng.*, Nov. 1892, p. 393.
31. *Am Rly. Times*, Dec. 3, 1870, p. 357.
32. *Rly. Age*, Oct. 16, 1879, p. 502.
33. *Pittsburgh Commercial*, Aug. 29, 1873.
34. *NCB*, June 1874, p. 90; *Rly. Age*, Nov. 2, 1882, p. 618; *Rly. Rev.*, Sept. 2, 1882, p. 499.
35. *ARRJ*, June 1883, p. 98, and Aug. 1883, p. 188; *Rly. Rev.*, Sept. 30, 1882. For biographical data on Burton, see *Nat. Cyc. Am. Bio.*, vol. 12 (1904), p. 467, and *Who Was Who*, vol. 1, p. 174.
36. *RRG*, Sept. 9, 1887, p. 584.
37. Ibid., March 22, 1889, p. 195.
38. Ibid., Dec. 7, 1888, p. 807.
39. *ARRJ*, Dec. 1883, p. 386; *Rly. Rev.*, May 31, 1890, p. 306.
40. *Rly. Rev.*, Aug. 16, 1884, p. 430.
41. *RRG*, March 2, 1888, p. 134.
42. Ibid., July 7, 1893, p. 500.
43. *Rly. Rev.*, May 10, 1890, p. 269.
44. *Nat. Cyc. Am. Bio.*, vol. C (1930), p. 126; *Rly. Age*, Oct. 19, 1903, p. 505.
45. *Rly. Age*, July 28, 1881, p. 433.
46. *Journal of Railway Appliances*, March 1889, p. 40.
47. *Rly. Rev.*, June 12, 1886, p. 304.
48. *RRG*, Aug. 12, 1892, p. 601.
49. *Rly. Age*, Feb. 28, 1902, p. 267.
50. Ibid., Sept. 16, 1880, p. 492; *ARRJ*, Aug. 20, 1881, p. 953.
51. *Rly. Age*, Feb. 28, 1902, p. 267.
52. *Rly. Rev.*, Oct. 26, 1889, p. 620.
53. *NCB*, Sept. 1877, p. 134.
54. *RRG*, Dec. 27, 1895, p. 850.
55. *Rly. Age*, July 13, 1917, p. 85. See also U.S. Patent No. 308,808 (Dec. 2, 1884).
56. *Rly. Age*, Dec. 19, 1902, p. 676.
57. *RRG*, June 20, 1884, p. 469.
58. James Dredge, *Record of the Transportation Exhibits at the World's Columbian Exposition of 1893* (London, 1894), p. 474; *RRCJ*, Jan. 1893, p. 65.
59. *Rly. Rev.*, July 19, 1890, p. 425.
60. *Rly. Age*, April 26, 1883, p. 239; *Cincinnati Commercial*, Nov. 1, 1882. See also the Clemen and Yeager works cited in note 1.
61. *ARRJ*, June 15, 1842, pp. 364–366.
62. Oscar Edward Anderson, *Refrigeration in America* (Princeton, 1953), p. 30.
63. *RRG*, May 14, 1886, p. 325.
64. Ibid.; *Rly. Age*, July 5, 1895, p. 326; *Vermont Quarterly*, June 1946, pp. 121–134.
65. *RR. Reco.*, Aug. 18, 1853, p. 387.
66. *Scientific American*, Nov. 7, 1857, p. 70; *ARRJ*, May 12, 1860, p. 406; *Am. Rly. Rev.*, May 24, 1860, p. 312; *American Artisan*, Sept. 4, 1867, p. 131.
67. *The Station Agent*, Dec. 1889, p. 155; rpt. in *Railroad History*, no. 141 (Autumn 1979), p. 98.
68. *Western Railroad Gazette*, April 19, 1869, p. 1; *Illinois Central Magazine*, Oct. 1928, p. 3; C. J. Corliss, *Main Line of Mid America* (New York, 1950), pp. 295–296.
69. Rudolf A. Clemen, "George H. Hammond," Newcomen Society in North America pamphlet, 1946. Similar data is given in *DAB*, vol. 8 (1932), pp. 204–205.
70. *Western Railroad Gazette*, July 24, 1869, p. 1; *American Artisan*, Sept. 8, 1869, p. 149; *NCB*, Sept. 1872, p. 8; *Ice and Refrigeration*, Nov. 1901, p. 207.
71. Institution of Civil Engineers *Proceedings*, 1878, p. 33.
72. *NCB*, Dec. 1878, p. 180.
73. *RRG*, May 9, 1879, p. 259.
74. *NCB*, May 1880, p. 73.
75. *RRG*, Feb. 6, 1880, p. 82; *NCB*, Oct. 1881, p. 115.
76. *Rly. Age*, July 12, 1883, p. 419; *Rly. Rev.*, Jan. 29, 1887, p. 68.
77. *MCB*, 1883, p. 68.
78. *Rly. Rev.*, March 31, 1888, p. 169.
79. *Railway Equipment Register*, 1897, p. 229, and 1900, pp. 221 and 228.
80. *Car Builders Dictionary*, 1895, pp. 148–149.
81. *RRCJ*, Nov. 1895, p. 256.
82. *RRG*, Feb. 4, 1887, p. 73.
83. Ibid., April 22, 1898, p. 293.
84. *NCB*, Oct. 1876, p. 150.
85. Institution of Civil Engineers *Proceedings*, 1878, pp. 33–35.
86. *Rly. Rev.*, July 3, 1880, pp. 334 and 337.
87. *RRG*, Aug. 15, 1884, p. 598.
88. *Rly. Rev.*, Sept. 2, 1882, p. 504.
89. Ibid.
90. *Rly. Rev.*, March 15, 1884, p. 144, and Jan. 21, 1888, p. 40.
91. *RRG*, Sept. 21, 1894, p. 644.
92. Ibid., Oct. 30, 1891, p. 758.
93. Ibid., Sept. 4, 1897, p. 507.
94. Milk and dairy traffic data was drawn from Ralph Selitzer's *The Dairy Industry in America* (New York, 1976); E. H. Mott's *Between the Ocean and the Lake* (New York, 1901), pp. 407–409; *ARRJ*, July 1843, p. 220 and Dec. 20 and 27, 1867, pp. 987 and 1027; *Rly. Age*, Oct. 3, 1902, p. 339, and Nov. 3, 1903, p. 626; and International Railway Congress *Bulletin*, Dec. 1909, p. 1682.
95. *NCB*, April 1877, p. 54.
96. *Rly. Rev.*, Feb. 17, 1900, p. 88.
97. *Car Builders Dictionary*, 1895, figs. 168 and 596–597. The same drawings are repeated in the 1898 edition, which was reprinted in 1977 by Newton K. Gregg, Novato, Calif.
98. Ibid., figs. 267–270.
99. *Rly. Age*, Oct. 3, 1902, p. 339.
100. Selitzer, *Dairy Industry in America*, p. 193.
101. *RRG*, Aug. 16, 1873, p. 333.
102. *Rly. Age*, Nov. 6, 1903, p. 626.
103. Data on Eastman is drawn from *RRG*, Feb. 24, 1882, p. 115, and Feb. 8, 1884, p. 105; and *Rly. Age*,

Feb. 25, 1887, p. 139. See also U.S. Patent No. 269,189 (Dec. 19, 1882).

104. Information on fruit traffic is drawn from G. M. Best, *Iron Horses to Promontory* (San Marino, Calif., 1969), p. 143; *Rly. Rev.*, June 23, 1888, p. 364; *Mast. Mech. Mag.*, Dec. 1894, p. 199; and E. Lavoinne and E. Pontzen, *Les Chemins de Fer en Amérique* (Paris, 1882), vol. 2, p. 82.
105. *NCB*, April 1871, p. 5.
106. *RRG*, May 24, 1878, p. 261.
107. *Restoration Feasibility Investigation on Nine Selected Passenger and Freight Cars* (report to Nevada State Museum by Short Line Enterprises, June 1981), p. 149.
108. Lavoinne and Pontzen, text vol. 2, p. 79, and atlas plate 7.
109. There are abundant drawings and photographs to document the unchanging nature of the ventilated car. Examples will be found in *NCB*, Feb. 1883 (Louisville & Nashville) and Jan. 1884 (Central Pacific); see also John O'Connell's *Railroad Album* (Chicago, 1954), pp. 40 and 45.
110. Lucius Beebe, *The Central Pacific and Southern Pacific Railroads* (Berkeley, Calif., 1963), p. 150.
111. *NCB*, June 1884, p. 68.
112. *Rly. Rev.*, Sept. 19, 1891, p. 612.
113. Ibid., July 28, 1894, p. 428; *RRG*, May 25, 1894, p. 366.
114. *Technology and Culture*, Jan. 1969, p. 35.
115. *RRG*, March 22, 1895, p. 183.
116. The Glen Summit car is pictured in a Murray, Dougal & Co. catalog of around 1896 owned by Thomas T. Taber of Muncy, Pa.
117. *ARRJ*, Oct. 27, 1866, p. 1019.
118. *RRCJ*, Feb. 1898, p. 43.

CHAPTER 5

1. *NCB*, Dec. 1891, p. 181; *RRG*, July 25, 1879, p. 408.
2. Exact statistics, as usual, are difficult to uncover, but numbers for individual roads found in *Poor's* and published annual reports produce this total. *RRG*, Oct. 4, 1873, p. 400, claims that 58,355 four-wheel cars were operating on U.S. railroads; most of these were almost certainly jimmies.
3. Drawings of a fully developed British coal wagon are included in William Strickland's *Report on Canals, Railways, Roads, and Other Subjects* (Philadelphia, 1826). Other illustrations will be found in Charles E. Lee's *The Evolution of Railways* (London, 1943).
4. *RRG*, Sept. 8, 1876, p. 390. The drawings were produced in the 1879 and 1888 editions of the *Car Builders Dictionary*.
5. James Dredge, *The Pennsylvania Railroad* (London, 1879), p. 157.
6. *RRCJ*, no. 3 (Spring 1972), pp. 5–6.
7. Smith Neg. 69, Stevens Institute of Technology, Hoboken, N.J., shows a coal dock at Hoboken, November 1887, with a string of DL&W jimmies unloading. See also T. T. Taber's *The Delaware, Lackawanna, and Western in the Twentieth Century* (Williamsport, Pa., 1981), part 2, p. 704.
8. *NCB*, Jan. 1876, p. 7.
9. Ibid., March 1879, p. 35.
10. Ibid., Jan. 1878, p. 5; *MCB Report*, 1878, p. 120; *RRG*, May 10, 1878, p. 230.
11. *NCB*, Dec. 1884, p. 154.
12. *Mast. Mech. Mag.*, April 1888, p. 51.
13. *Eng. News*, April 5, 1890, p. 327.
14. *NCB*, March 1890, p. 33.
15. *RRG*, May 19, 1899, p. 350; *RRCJ*, Dec. 1899, p. 370.
16. *Rly. Rev.*, Feb. 4, 1899, p. 65.
17. *RRG*, June 9, 1882, p. 342.
18. *NCB*, Jan. 1880, p. 5.
19. *RRG*, April 6, 1888, p. 217.
20. *ARRJ*, Nov. 1891, p. 505.
21. *RRG*, March 6, 1896, p. 157.
22. *Railway Equipment Register*, April 1897, p. 204. The 33100 was one of 230 cars in the 33000–33249 series.
23. Ibid., p. 140. The N&W had around sixty-five hundred hopper-bottom gondolas as well. See also *Car Builders Dictionary*, 1898, figs. 302–304.
24. *RRG*, June 15, 1900, p. 385.
25. Ibid., April 28, 1899, p. 294.
26. *Car Builders Dictionary*, 1898, fig. 231; James Dredge, *Record of the Transportation Exhibits at the World's Columbian Exposition in 1893* (London, 1894), p. 450.
27. Dredge, *Pennsylvania Railroad*, pp. 154–155.
28. *NCB*, May 1880, p. 76.
29. *RRG*, April 17, 1896, p. 264.
30. Ibid., Sept. 14, 1888, p. 600; *Rly. Master Mech.*, Sept. 1889, p. 154.
31. *RRG*, Dec. 16, 1887, p. 808.
32. *Rly. Rev.*, Aug. 9, 1890, p. 460; transcript of Pullman's Chicago car plant for 1890–1891.
33. Data on Reading number series was provided by George M. Hart, former director of the Pennsylvania State Railroad Museum, Strasburg.
34. *NCB*, Nov. 1892, p. 166.
35. *Loco. Eng.*, March 1897, p. 232.
36. *NCB*, March 1884, p. 31; *RRG*, July 18, 1884, p. 532.
37. *Rly. Age*, Oct. 24, 1902, p. 428.
38. *RRG*, May 2, 1902, p. 326. In a sketchy history of Lehigh Valley coal cars, 1877 is given as the date 20-ton-capacity eight-wheel cars were introduced.
39. *NCB*, Jan. 1885, p. 9.
40. Ibid., Nov. 1884, p. 163.
41. *RRG*, Jan. 11, 1884, p. 29.
42. *NCB*, June 1888, p. 86.
43. *Rly. Rev.*, Jan. 26, 1889, p. 45.
44. *RRCJ*, May 1897, p. 112.
45. *RRG*, June 8, 1894, p. 399.
46. *ARRJ*, March 1896, p. 34; *RRG*, Feb. 4, 1898, p. 75.
47. *RRCJ*, Aug. 1894, p. 184; *ARRJ*, Oct. 1903, p. 352.
48. *RRG*, June 9, 1899, p. 397; *Rly. Age*, Dec. 22, 1899, p. 957.
49. *RRG*, June 8, 1894, p. 399.
50. R. J. Wayner, *Cars of the Pennsylvania Railroad* (New York, 1977), p. 41, and *New York Central Cars* (New York, n.d.), p. 34.
51. *NCB*, May 1880, p. 73, and Aug. 1880, p. xii; *RRG*, July 29, 1887, p. 495; *Poor's*, 1884, p. 14.
52. *RRG*, July 29, 1887, p. 496.
53. *RRCJ*, Nov. 1892, p. 35, and Feb. 1894, p. 24.
54. *Car Builders Dictionary*, 1898, fig. 333.
55. *NCB*, Oct. 1871, p. 5.
56. Society for Industrial Archaeology *Newsletter*, Summer 1981.
57. *R&LHS Bull.* 111 (Oct. 1964).
58. *R&LHS Bull.* 45 (Jan. 1938), pp. 17–18; Ted Wurm and Harre Demoro, *Silver Short Line* (Glendale, Calif., 1983), p. 35.
59. E. Lavoinne and E. Pontzen, *Les Chemins de Fer en Amérique* (Paris, 1882), atlas plate 11, fig. 11.
60. Jim Shaughnessy, *The Delaware and Hudson: An Illustrated History* (Berkeley, 1967), p. 136.
61. *NCB*, Oct. 1881, p. 116.
62. Peter Temin, *Iron and Steel in Nineteenth-Century America* (Cambridge, Mass., 1964), p. 195.

63. *NCB*, Nov. 1886, p. 149.
64. *Rly. Rev.*, Oct. 17, 1891, p. 673.
65. *Eng. News*, Oct. 26, 1893, p. 348; *RRCJ*, Dec. 1892, p. 58.
66. *Rly. Age*, Feb. 18, 1898, p. 121.
67. *Eng. News*, April 19, 1890, p. 369.
68. *RRCJ*, Feb. 1894, p. 24.
69. *Rly. Rev.*, March 14, 1896, p. 145.
70. Ibid., Feb. 7, 1891, p. 88.
71. Lucius Beebe and Charles Clegg, *Virginia and Truckee* (Oakland, 1949), p. 153.
72. A Murray, Dougal & Co. catalog of about 1896 shows a picture of the Glen Summit mineral water car on p. 62.
73. *RRCJ*, May 1894, p. 76.
74. The following works were consulted and the information found has been used without specific credit in the text:
    A. H. Z. Carr, *John D. Rockefeller's Secret Weapon* (New York, 1962).
    W. C. Darrah, *Pithole, the Vanished City* (Gettysburg, Pa., 1972).
    P. H. Giddens, *Early Days of Oil: A Picture History* (Princeton, 1948) and *Pennsylvania Petroleum 1750–1872, a Documentary History* (Titusville, Pa., 1947).
    J. T. Henry, *Early and Later History of Petroleum* (Philadelphia, 1873).
    R. L. Maybee, *Railroad Competition and the Oil Trade 1855–1873* (Mt. Pleasant, Mich., 1940).
    H. T. Rosenberger, *Philadelphia and Erie Railroad* (Potomac, Md., 1975).
    H. F. Williamson and A. R. Daum, *The American Petroleum Industry* (Evanston, Ill., 1959), vol. 1.
    See also *Century Magazine*, July 1883, and *Harper's Magazine*, Dec. 1864, for general articles on the oil industry.
75. *Am. Rly. Rev.*, March 21, 1861, p. 165.
76. The Densmores became financial backers of Christopher L. Sholes, a machinist from Milwaukee, in about 1869. James Densmore offered more than money to the struggling inventor; he pushed and bullied Sholes into perfecting a practical typewriter. So driven, Sholes succeeded where others had failed. The Remington Arms Company began making the first production units in 1874. *Nat. Cyc. Am. Bio.*, vol. 3 (1891), pp. 316–317, contains biographies of the Densmore brothers, George W. N. Yost, and others associated with early typewriters. James died in 1886 and Amos in 1893.
77. Michael H. Adler, *The Writing Machine: A History of the Typewriter* (London, 1973), p. 142.
78. *Scientific American*, Nov. 1, 1862, p. 280.
79. Seely may have had relatives active in the oil fields, for an S. M. Seely of New York City bought one of the best oil lands in Oil Creek in 1864, and a C. A. Seely reported on the oil industry in *Scientific American*, Sept. 1, 1866, p. 144.
80. *Scientific American*, Oct. 21, 1865, p. 257. Keeler's patent model is in the collection of the Smithsonian Institution. The model and patent drawing are for a four-wheel vehicle, but it is most likely that the prototype car was on eight wheels, considering its length of 28 feet.
81. *R&LHS Bull.* 97 (Oct. 1957), pp. 68–72, contains a brief article on the ill-fated cylinder cars. Original patent models for the 1879 Sly's cylinder car patent and the Myers 1851 patent are housed in the Smithsonian and the B&O Museum (Baltimore), respectively.
82. A replica of a Densmore car was built in 1959 by the Union Tank Line as part of the Drake Well Centennial. Regrettably, the proportions of the car are rather off the mark. It is much too short, a defect exaggerated by the long-wheelbase trucks. The car frame is a most inexact reproduction for a wooden car of any period and seems to follow no documented plan. The Drake Well Museum (Titusville, Pa.), which exhibits the car on its grounds, plans to reconstruct it in a more convincing manner when time and funds are available.
83. *Engineering*, Feb. 16, 1872, p. 102. Warden is recorded as having two patents for oil-carrying ships issued in his name during 1871, but no record has yet been found for a railroad tank car patent bearing his name during this period.
84. Lavoinne and Pontzen, atlas plate 11; *Asher and Adams' New Columbian Railroad Atlas* (New York, 1879), p. 252.
85. Copies of these Oil City views are in the Drake Well Museum collection. They were published by Frank, Robbins in about 1875–1880.
86. Advertisements illustrated with a Murray, Dougal & Co. car of this type first appear in *Poor's* for 1894. The same line cut is used in a Murray, Dougal & Co. advertisement in the New York Railroad Club's *Proceedings* for 1900.
87. *The Keystone*, March 1982, p. 8. This journal is published by the Pennsylvania Railroad Technical and Historical Society. See also *Railway World*, June 9, 1877.
88. *RRG*, Jan. 6, 1893, p. 6; *Engineer*, April 20, 1894, p. 336; see also Chapter 8 for more on Harvey.
89. *Asher and Adams' Railroad Atlas*, p. 252.
90. The 1896 Milton Car Works catalog is owned by T. T. Taber of Muncy, Pa. The twin-tank car is illustrated on p. 26.
91. *RRG*, April 5, 1873, p. 144.
92. Carr, *Rockefeller's Secret Weapon*, p. 86.
93. Data on the Brooks car is from *RRG*, Sept. 22, 1876, p. 413; and *Rly. Age*, Aug. 29, 1902, p. 207, and Sept. 19, 1902, p. 281.
94. Milton Car Works catalog, 1896, p. 61.
95. Williamson and Daum, *American Petroleum Industry*, vol. 1, p. 183.
96. Ida M. Tarbell, *History of the Standard Oil Company* (New York, 1904), p. 24; Carr, *Rockefeller's Secret Weapon*, p. 21.
97. *RRG*, Oct. 26, 1877, p. 475.

CHAPTER 6

1. David Stevenson, *Civil Engineering in North America* (London, 1838), plate 6; *Civil Engineers and Architects Journal*, March 1842, p. 105.
2. Data on Kasson is drawn from *RR. Reco.*, Nov. 17, 1853, p. 597; and *ARRJ*, July 30, 1853, p. 487, and Nov. 5, 1853, p. 719.
3. Basic data on Lovell was kindly furnished by Professor John Lozier of Bethany College from the P. I. Perrin notebooks at the Old Colony Historical Society, Taunton, Mass. More data was found in Boston city directories.
4. See reference to Perrin notebooks in the preceding note.
5. *Railroad History*, no. 149 (Autumn 1983), pp. 56 and 65; George B. Abdill, *A Locomotive Engineer's Album* (Seattle, 1965), p. 158.
6. Data on the V&T car was generously provided by Stephen Drew of the California State Railroad Museum, Sacramento.

7. *Railroad History*, no. 141 (Autumn 1979), p. 90.

8. Data on Venice is based on listings in the *Railway Equipment Registers* for 1897, 1910, 1918, and 1920, together with a brief biographical notice of its long-time general manager, H. V. Gehm, in the 1901 *Railway Officials of America*, p. 198. The firm appears to have expired in 1935.

9. *RRG*, April 27, 1894, p. 298, and June 7, 1895, p. 363. Drawings are reproduced in James Dredge, *Record of the Transportation Exhibits at the World's Exposition of 1893* (London, 1894), plate 120.

10. E. Lavoinne and E. Pontzen, *Les Chemins de Fer en Amérique* (Paris, 1882), p. 86 and plate 8. See also J. E. Watkins, *History of the Pennsylvania Railroad* (proof sheets, 1896), vol. 2, pp. 325, 329–330.

11. James D. Horan, *Timothy O'Sullivan, America's Forgotten Photographer* (New York, 1966), p. 142, has two views of the Civil War gun car.

12. *Rly. Mast. Mech.*, Oct. 1895, p. 156. The same drawings were reproduced a few months later in both *Rly. Rev.* and *RRCJ*.

13. *RRG*, Aug. 16, 1889, p. 536; *RRCJ*, Aug. 1898, p. 230.

14. *RRCJ*, June 1895, p. 111.

15. *RRG*, June 15, 1894, p. 421.

16. *RRCJ*, Feb. 1898, p. 34, and May 1898, p. 118.

17. *RRG*, March 31, 1899, p. 226; *Railroad Digest*, Feb. 1901, p. 60.

18. Information on Von Baader was kindly provided by Ludwig Schetzbaum, curator of the Rail Section, Deutsches Museum, Munich, Germany.

19. Engravings of 1830-vintage carriages on flatcars were published in *The Life of C. B. Vignoles* by his son O. J. Vignoles (London, 1889), and on the masthead of the earliest issue of *ARRJ*.

20. *Mast. Mech. Mag.*, Oct. 1897, p. 142. The trip took place in Dec. 1847.

21. George Measom, *Illustrated Guide to the Great Northern Railway* (London, 1861), p. 544.

22. *The Locomotive*, Dec. 15, 1936, p. 402.

23. *Distribution Age*, Feb. 1967, p. 36, refers to lift-van service since 1901.

24. Charles Dollfus, *Histoire de la Locomotion Terrestre* (Paris, 1935), p. 92; *ARRJ*, April 11, 1846, p. 229.

25. *Baltimore Gazette*, Dec. 17, 1830.

26. Barry B. Combs, *Westward to Promontory* (Palo Alto, Calif., 1969), p. 22.

27. *RRG*, April 20, 1877, p. 181.

28. Tom Parkinson and Charles P. Fox, *The Circus Moves by Rail* (Boulder, 1978); and a four-part series of articles appearing in *Bandwagon*, the journal of the Circus Historical Society, between Nov./Dec. 1983 and May/June 1984.

29. *Harper's Weekly*, Jan. 31, 1885; *Scribner's Magazine*, May 1889, p. 583; *Boston and Maine Magazine*, May/June 1955, p. 8; Vincent F. Seyfried, *The Long Island Rail Road*, part 6 (Garden City, N.Y., 1975), pp. 182–190.

30. *Railroad History*, no. 151 (Autumn 1984), p. 31; *Trains* magazine, Jan. 1960, p. 27.

31. *National Geographic Magazine*, Oct. 1923, p. 372.

32. Eric Rath, *Container Systems* (New York, 1973), p. 6.

33. For early British and German rail/canal containers, see M. J. T. Lewis, *Early Wooden Railways* (London, 1970), pp. 59, 290, and 339.

34. James Anderson, *Recreations in Agriculture, Natural History, Arts, and Miscellaneous Literature*, vol. 4 (London, 1801).

35. R. H. G. Thomas, *Liverpool and Manchester Railway* (London, 1980), p. 184. The engraving of the L&M goods station is reproduced in *Rly. Gaz.*, Sept. 16, 1938, p. 64.

36. *Rly. Age*, April 9, 1932, p. 614, mentions containers on the Canterbury line in July 1833.

37. *ARRJ*, April 11, 1846, p. 229.

38. *Illustrated London News*, June 29, 1846; rpt. in *Rly. Age*, Oct. 29, 1927, p. 857.

39. Canal History and Technology Symposium (Easton, Pa.) *Proceedings* 2 (1983), p. 109.

40. John Lippitt's patent Aug. 14, 1877, No. 194,097.

41. *ARRJ*, March 1901, p. 93.

42. Data on Eaton's 1877 patent was provided by Professor Fritz Lehmann, University of British Columbia, Vancouver, B.C., Canada.

43. *Rly. Age*, Feb. 8, 1901, p. 101.

44. *NCB*, Oct. 1882, p. 117.

45. *Rly. Rev.*, Oct. 9, 1889, p. 647.

46. *RRG*, Feb. 8, 1895, p. 84; *ARRJ*, Jan. 1896, p. 6; *RRCJ*, April 1898, p. 100.

47. *Rly. Rev.*, May 5, 1900, p. 241; *Rly. Age*, July 21, 1901, p. 22.

48. *Car Review*, Aug. 1909, p. 21; *Rly. Age*, July 1, 1904, p. 30, and March 24, 1905, p. 41; *Rly. Mast. Mech.*, July 1904, p. 274.

49. *ARRJ*, Nov. 1908, p. 424.

50. *RRG*, Nov. 3, 1905, p. 420.

51. *ARRJ*, March 1901, p. 93.

52. Ibid.; *Rly. Age*, Feb. 8, 1901, p. 101 (Mowry is misspelled as Mowery).

53. *Rly. Rev.*, Aug. 24, 1889, p. 489.

54. Ibid., Jan. 26, 1889, p. 44.

55. Ibid., Jan. 4, 1896, p. 7.

56. *Rly. Age*, March 28, 1977, p. 9.

57. *Rly. Rev.*, March 15, 1884, p. 133.

58. Ibid.

59. *RRG*, Oct. 19, 1883, p. 690.

60. *MCB Report*, 1879, p. 94.

61. *ASCE Trans.*, 1882, p. 306.

62. Ibid., pp. 301–307.

63. *RRG*, May 23, 1879, p. 284.

64. *ARRJ*, June 1900, p. 171.

65. *Rly. Mast. Mech.*, June 1897, p. 80.

66. *RRG*, Oct. 4, 1878, p. 483.

67. Ibid., March 1, 1895, p. 136.

68. *ARRJ*, Dec. 1901, pp. 371 and 389; June 1902, p. 182.

69. *RRG*, July 13, 1883, p. 464.

70. *ASME Trans.*, 1882, pp. 293 and 299.

71. *RRG*, March 17, 1876, p. 119, and Dec. 20, 1878, p. 610. Both are illustrated in Chapter 3.

72. Ibid., Feb. 27, 1880, p. 121; April 30, 1880, p. 232; and April 14, 1882, p. 220.

73. Ibid., Jan. 9, 1903, p. 32.

74. *ARRJ*, Sept. 1895, p. 400.

75. Ibid., June 1902, p. 168; *RRG*, May 1, 1908, p. 610.

76. *ARRJ*, May 1905, p. 153; June 1905, p. 206; and July 1905, p. 254.

77. *RRCJ*, Nov. 1896, p. 286, and Feb. 1897, p. 32.

78. Ibid., Jan. 1901, p. 48.

79. *RRG*, March 13, 1896, p. 176.

80. *Rly. Age*, Dec. 8, 1899, p. 906.

81. General information on cabooses has been freely drawn from William Knapke's *The Railroad Caboose* (San Marino, Calif., 1968) and other secondary works such as Robert S. Henry's *This Fascinating Railroad Business* (Indianapolis, 1942), often without specific reference in this text. See also *RRCJ*, Dec. 1900, p. 342.

82. John H. White, Jr., *American Locomotives: An En-*

gineering History, 1830–1880 (Baltimore, 1968), pp. 221–222.

83. Rly. & Loco. Eng., March 1901, p. 110.

84. RR. Adv., March 14, 1855, p. 3; Loco. Eng., May 1897, p. 359.

85. Am. Rly. Rev., Nov. 24, 1859, p. 7; ARRJ, Oct. 15, 1859, p. 659.

86. ARRJ, Aug. 23, 1862, p. 645.

87. Am. Rly. Rev., Oct. 24, 1861, p. 21.

88. The most convenient source for U.S. Military Railroad photographs is George Abdill's picture book Civil War Railroads (Seattle, 1961).

89. Both Russell photographs are reproduced in Combs, Westward to Promontory, pp. 54 and 55, and G. M. Best, Iron Horses to Promontory (San Marino, Calif., 1969), p. 49.

90. Poor's, 1881, p. 633.

91. Omer Lavallee to John H. White, letter dated Jan. 8, 1985, in correspondence files of Division of Transportation, Smithsonian Institution. Keefer's 1860 report is summarized in R&LHS Bull. 56 (October 1941).

92. Lavallee to White, personal letter 1985.

93. Iron Horse News (Colorado Railroad Museum newsletter), April 1, 1986.

94. Thomas C. Cochran, Railroad Leaders (New York, 1965), p. 439.

95. Rly. Rev., Aug. 29, 1903, p. 640.

96. Loco. Eng., May 1897, p. 359; Car Builders Dictionary, 1895, figs. 81–82, illustrates an Erie four-wheel caboose.

97. Data on the Pennsylvania caboose is drawn from James Dredge, Pennsylvania Railroad (London, 1879); photographs in the Smithsonian Chaney negative collection; ARRJ, June 1904, p. 209; and Robert J. Wayner, ed., Pennsy Car Plans (New York, 1969).

98. SVRR caboose photo in files of the N&W Railway, Roanoke, Va.

99. Caboose Data Book No. 2: Cabin Cars of the Pennsylvania and Long Island Railroads (Hicksville, N.Y., 1982); Rly. Rev., June 14, 1902, p. 448.

100. NCB, Aug. 1883, p. 90.

101. RRG, June 30, 1905, p. 766.

102. Rly. Age, Aug. 9, 1901, p. 102.

103. April/May 1985 newsletter issued by the Friends of B&O System Research.

104. Smithsonian Chaney Neg. 15508.

105. NCB, Dec. 1874, p. 179.

106. RRG, Sept. 7, 1894, p. 606.

107. The Portland Co. records are preserved by the Maine Historical Society, Portland.

108. Knapke, Railroad Caboose, p. 30; John O'Connell, Railroad Album (Chicago, 1954), p. 33.

109. NCB, Aug. 1881, p. 90, and Nov. 1881, p. 138.

110. RRG, May 4, 1883, p. 278.

111. National Model Railroad Association Bulletin, June 1979, p. 39.

112. Railroad Stories, Oct. 1937, p. 48, reprints a newspaper account dated Dec. 9, 1898.

113. Am. Rly Rev., Oct. 24, 1861, p. 21.

114. Frank W. Stevens, Beginnings of the New York Central Railroad (New York, 1926), p. 348. The engraving reproduced by Stevens is from Henry O'Reilly's History of Rochester (1838).

115. A. Bendel, Aufsatze Eisenbahnwesen in Nord-Amerika (Berlin, 1862), plate 13, fig. 15.

116. Rly. Age, Oct. 26, 1882, p. 589.

117. George Abdill, Rails West (Seattle, 1960), p. 33; R. C. Reed, Train Wrecks (Seattle, 1968), p. 142.

118. ARRJ, Aug. 1899, p. 271.

119. Gilbert A. Lathrop, Little Engines and Big Men (1954).

120. Master Mechanics Report, 1875, p. 114.

121. NCB, Nov. 1881, p. 138.

122. Ibid., May 1880, p. 73.

123. Restoration Feasibility Investigation on Nine Selected Passenger and Freight Cars (report to Nevada State Museum by Short Line Enterprises, June 1981; Railroad History, no. 138 (Spring 1978), p. 77.

124. D. P. Holbrook and S. D. Lorenz, Waycars of the CB&Q (Danvers, Mass., n.d. [c. 1978]), p. 12.

125. ARRJ, Oct. 15, 1859, p. 659; R&LHS Bull. 19 (Sept. 1929), p. 18.

126. Am. Rly. Rev., Nov. 24, 1859, p. 7.

127. Drawing of 1872 Central Pacific drovers' caboose is in the collections of the California State Railroad Museum, Sacramento.

128. RRG, Aug. 21, 1908, p. 752.

129. Ibid., April 22, 1898, p. 289.

130. RW&O tintype is in the collection of the Smithsonian Division of Photographic History.

CHAPTER 7

1. Readers should consult John H. White, Jr., The American Railroad Passenger Car (Baltimore, 1978), for more details on trucks, wheels, axles, and other matters discussed in this chapter.

2. Ibid., pp. 59 and 60.

3. ARRJ, Sept. 16, 1848, plate opp. p. 600.

4. James Dredge, The Pennsylvania Railroad (London, 1879), plate 54.

5. USMRR 1865 truck is pictured on Smithsonian Neg. 81-8119.

6. The Merrilees Collection is now housed in the Canadian National Archives, Ottawa, Ont.

7. Several McMurray photos are reproduced in L. W. Sagle's B&O Power (Medina, Ohio, 1964); see pp. 24, 208, and 210.

8. NCB, March 1873, p. 59.

9. Ibid., Jan. 1876, p. 4.

10. E. Lavoinne and E. Pontzen, Les Chemins de Fer en Amérique (Paris, 1882), text p. 76 and atlas plate 7; RRG, Sept. 30, 1871, p. 299.

11. John H. White, Jr., American Locomotives: An Engineering History, 1830–1880 (Baltimore, 1968), p. 47.

12. White, American Railroad Passenger Car, p. 504.

13. Drawing of the live-spring truck is copied on Smithsonian Neg. 81-8109.

14. Scientific American, Sept. 11, 1845.

15. Rly. Rev., Nov. 24, 1888, p. 677.

16. William Voss, Railway Car Construction (New York, 1892), p. 71.

17. NCB, April 1874, p. 53.

18. Ibid., Oct. 1880, p. 157.

19. RRG, Nov. 5, 1886, p. 753.

20. Master Mechanics Report, 1893, p. 227.

21. Angus Sinclair, Development of the Locomotive Engine (1907), p. 257.

22. J. E. Watkins's proof sheets for his projected History of the Pennsylvania Railroad (1896), vol. 2, n.p.

23. RRCJ, May 1899, p. 114.

24. RRG, March 17, 1876, p. 117.

25. Ibid., May 12, 1882, p. 282.

26. NCB, June 1885, p. 72.

27. RRG, Sept. 14, 1888, p. 601.

28. Watkins, Pennsylvania Railroad, vol. 2, n.p.

29. ARRJ, Aug. 7, 1869, p. 875, and Nov. 27, 1869, p. 1321; Engineering, Nov. 19, 1869, p. 334.

30. Tisdale's patents were issued on March 10, 1863 (No. 37,889), June 14, 1864 (No. 43,163), and May 25, 1869 (No. 90,410).
31. *Van Nostrand's Engineering Magazine*, Aug. 1870, p. 215.
32. *ARRJ*, April 11, 1846, p. 229.
33. Snow's patents were issued on Aug. 23, 1870 (No. 106,737), and Aug. 15, 1871 (No. 118,065).
34. *NCB*, Nov. 1874, p. 163, reported new hopper cars built by Ohio Falls as having Cleveland trucks. The same is true for gondolas built by the Missouri Car Works; *Transactions of the Institution of Engineers and Shipbuilders in Scotland* 26 (1884–1885), plate 20.
35. Voss, *Railway Car Construction*, p. 63; the text was perhaps a trifle out of date and probably reflects a division of opinion that was already less equal and more in favor of rigid trucks.
36. *RRG*, Dec. 5, 1890, p. 842.
37. *NCB*, Oct. 1890, p. 147, and Feb. 1891, p. 30.
38. *Rly. Rev.*, June 11, 1892, p. 368.
39. White, *American Railroad Passenger Car*, p. 498.
40. *NCB*, Jan. 1874, pp. 5 and xi.
41. Ibid., June 1884, p. 75.
42. Voss, *Railway Car Construction*, p. 67 and fig. 86.
43. *NCB*, Oct. 1890, p. 147.
44. Data on the Thielsen truck is from *NCB*, June 1873, p. 137; Sept. 1878, p. 132; and Sept. 1881, p. 111; *RRG*, May 4, 1877, p. 198, and June 15, 1877, p. 272; and Richard C. Overton, *Burlington Route* (New York, 1965), p. 38.
45. *RRG*, May 12, 1882, p. 282.
46. *RRCJ*, May 1899, p. 117; T. Bute and A. Borries, *Das Amerikanache Eisenbahnwesen* (Wiesbaden, 1892), plate 35.
47. Data on Barber is from *RRG*, Feb. 20, 1891, p. 124; James Dredge, *Record of the Transportation Exhibits at the World's Columbian Exposition of 1893* (London, 1894), plate 95; *RRCJ*, May 1899, p. 154; *Rly. & Loco. Eng.*, April 1903, p. 12; and *Railway Mechanical Engineer*, Jan. 1920, p. 59.
48. *NCB*, Feb. 1891, p. 30.
49. Ibid., Oct. 1890, p. 147.
50. Ibid.
51. *ARRJ*, Sept. 1896, p. 217.
52. Data on Terry's car is from *Rly. Age*, July 22, 1880, p. 393; *Rly. Rev.*, July 19, 1882, p. 398; and *NCB*, May 1882, p. 55.
53. Data on Finlay is from *RRG*, May 23, 1884, p. 399; July 11, 1884, p. 154; Aug. 28, 1885, p. 556; Sept. 17, 1886, p. 641; and Sept. 1, 1893, p. 650; and *NCB*, Oct. 1889, p. 154.
54. *Rly. Age*, April 12, 1901, p. 426.
55. Ibid., Aug. 1985, p. 38.
56. *NCB*, Feb. 1875, p. 21.
57. Ibid., Feb. 1874, p. 21, and Feb. 1884, p. 20.
58. Dredge, *World's Columbian Exposition*, plate 131.
59. *MCB Report*, 1885, pp. 46–55.
60. Ibid., 1886, p. 183.
61. *RRCJ*, Dec. 1894, p. 261.
62. *NCB*, Oct. 1889, p. 154.
63. Charles Paine, *The Elements of Railroading* (New York, 1895), p. 118.
64. *NCB*, Oct. 1889, p. 154.
65. *Rly. Age*, Dec. 1, 1911, p. 1127.
66. Ibid.
67. *MCB Report*, 1883, p. 15.
68. *RRCJ*, June 1898, p. 153.
69. *MCB Report*, 1885, p. 53.
70. *NCB*, Dec. 1891, p. 185.

71. Howard Fleming, *Narrow Gauge Railways in America* (New York, 1875), pp. 54–56.
72. *RRG*, May 24, 1895, p. 329.
73. F. J. Krueger, *Freight Car Equipment* (Detroit, 1910), p. 186.
74. *NCB*, Feb. 21, 1891, p. 21.
75. *RRG*, Jan. 25, 1895, p. 50.
76. *NCB*, Jan. 1889, p. 3.
77. *RRG*, Nov. 5, 1886, p. 753.
78. Ibid., April 18, 1902, p. 289; *Rly. & Loco. Eng.*, March 1896, p. 249.
79. *NCB*, Jan. 1889, p. 3.
80. *RRG*, July 9, 1897, p. 494.
81. Ibid., Jan. 25, 1895, p. 50, and July 26, 1895, p. 495.
82. *RRCJ*, Sept. 1899, p. 157.
83. *RRG*, June 10, 1898, p. 403.
84. Roy V. Wright, ed., *Car Builders Cyclopedia* (New York, 1928), pp. 824, 825, and 830.
85. *RRG*, Oct. 6, 1893, p. 733, and May 25, 1895, p. 329; *NCB*, April 1893, p. 54.
86. *RRCJ*, May 1899, p. 116.
87. *Loco. Eng.*, April 1896, p. 358; *ARRJ*, May 1898, p. 153.
88. *Rly. & Loco. Eng.*, June 1906, p. 254.
89. *Rly. Age*, Jan. 9, 1903, p. 43.
90. *RRCJ*, Nov. 1898, p. 335.
91. *Car Builders Dictionary*, 1919, p. 65.
92. *Rly. Age*, July 6, 1929, p. 73.
93. *Railway Mechanical Engineer*, May 1934, p. 175.
94. Ibid.
95. *Rly. Age*, July 1, 1939, p. 6.
96. Institution of Mechanical Engineers *Proceedings*, 1903, p. 919; *RRG*, Oct. 10, 1903, p. 769.
97. *RRG*, Jan. 10, 1890, p. 31.
98. Claims for the Fox truck appeared in *RRG*, March 13, 1891, p. 186; *Rly. Rev.*, March 19, 1892, p. 175; and *MCB Report*, 1896, p. 65.
99. Edward B. Dorsey, *English and American Railroads Compared* (New York, 1887).
100. Parker Morell, *Diamond Jim* (New York, 1934), p. 44.
101. *RRG*, Nov. 27, 1891, p. 846.
102. *Rly. Mast. Mech.*, Feb. 1892, p. 28.
103. *Loco. Eng.*, May 1896, p. 423.
104. *Western Railroad Club Proceedings*, 1893, p. 99.
105. *MCB Report*, 1896, p. 65.
106. *RRG*, July 26, 1895, p. 495.
107. *New York Railroad Club Proceedings*, April 19, 1900, p. 3.
108. *RRCJ*, June 1898, p. vi; *Car Builders Dictionary*, 1906 (rear advertising section), p. 16.
109. *Rly. Mast. Mech.*, June 1897, p. 82.
110. *Loco. Eng.*, March 1898, p. 145.
111. *RRCJ*, July 1900, p. 210.
112. ACF catalog "B," 1903; *Car Builders Dictionary*, 1903, p. 293.
113. *RRCJ*, Sept. 1899, p. 259; O. M. Stimson, *Modern Freight Car Estimating* (Anniston, Ala., 1897), p. 29.
114. *ARRJ*, Feb. 1898, p. 34.
115. *MCB Report*, 1884, pp. 36 and 108; see also plates 1 and 2 at the rear of the volume.
116. *RRG*, July 30, 1897, p. 139, and Jan. 13, 1899, p. 22; *RRCJ*, March 1898, p. 74. Shickle, Harrison & Howard was a beginning source of many prominent railway suppliers. The firm itself was gradually bought up by ACF between 1899 and 1902. Gallagher left in 1899 to join Harry Scullin in forming Scullin Steel. Clarence Howard left somewhat later to take over the Commonwealth Steel Co. This data was furnished by Mark J. Cedeck of the Mercantile Library Assoc. and F. Travers Burgess, a St.

Louis attorney, active for many years with the General Steel Castings Corp., successor to Commonwealth Steel.

117. *Car Builders Dictionary*, 1903, fig. 3740. Andrew D. Young and Eugene F. Provenzo, *History of St. Louis Car Co.* (Berkeley, 1978), p. 82, reproduces a photo of an Illinois Central flatcar carried on Ajax trucks.

118. *RRG*, March 2, 1900, p. 139, and July 28, 1906, p. 74.

119. Fred W. Schneider III and Stephen P. Carlson, *PCC—The Car That Fought Back* (Glendale, Calif., 1980).

120. *ARRJ*, Aug. 1904, p. 319; *Nat. Cyc. Am. Bio.*, vol. 21, p. 260.

121. *RRCJ*, Jan. 1899, p. 4.

122. Ibid.

123. Ibid.

124. *Car Builders Dictionary*, 1888, fig. 1930.

125. *Cassier's Magazine*, March 1910, p. 440.

126. Much of the discussion of wheels presented here is based on amply documented material appearing in White, *American Railroad Passenger Car*, starting on p. 526.

127. Robert H. Ramage, *Introducing the Tough Guy*, a booklet published in 1947 by the Association of Chilled Car Wheel Manufacturers, Chicago, pp. 12 and 32.

128. Data on broad wheels is from George R. Taylor and Irene D. Neu, *The American Railroad Network, 1861–1890* (Cambridge, Mass., 1956), p. 59; *NCB*, Sept. and Nov. 1877, pp. 131 and 169; March 1879, p. 42; and *Car Builders Dictionary*, 1888, p. 28.

129. *ARRJ*, June 26, 1847, p. 408.

130. Ibid., June 25, 1853, p. 403.

131. *RRG*, April 20, 1883, p. 248. An article in *Engineer*, Dec. 28, 1883, p. 499, claims that only twenty-eight thousand were condemned each year but does not reflect on how many failed in service.

132. Ramapo Wheel & Foundry Co. catalog, c. 1887, p. 14.

133. *ARRJ*, June 26, 1847, p. 408.

134. *RR Mag.*, March 1969, p. 18.

135. More data on axles, journals, and springs will be found in White, *American Railroad Passenger Car*, starting on p. 510.

136. *RRG*, June 7, 1873, p. 229.

137. Paine, *Elements of Railroading*, p. 113; *RR Adv.*, Sept. 20, 1856, p. 2.

138. Joseph Anthony, *Forms of Railroad Car Axles*, a booklet published for the author in Greenbush, N.Y., 1868.

139. *NCB*, Aug. 1886, p. ix.

140. *Railway Mechanical Engineer*, Nov. 1915, p. 569.

141. *RRCJ*, June 1899, p. 155.

142. Data on the Fletcher lid is from *MCB Report*, 1889, p. 45; *NCB*, July 1890, p. 97; and *Car Builders Dictionary*, 1888, fig. 2019.

143. *NCB*, May 1873, p. 109; *Scientific American*, July 17, 1875.

144. *ARRJ*, Jan. 5, 1850, p. 11.

145. Ibid., Aug. 16, 1856, p. 521; *RR. Adv.*, July 19, 1856, p. 4.

146. *ARRJ*, Oct. 1896, p. 251.

147. General information on couplers has been freely drawn from White, *American Railroad Passenger Car*, and C. H. Clark's coupler article in *Technology and Culture*, April 1972, pp. 170–208.

148. *RR. Mag.*, Jan. 1930, p. 235.

149. *NCB*, June 1876, p. 85; *MCB Report*, 1885, p. 36.

150. *RRG*, July 18, 1884, p. 534.

151. Undated catalog issued around 1880 by Wilson,

152. *R&LHS Bull.* 6 (1923), p. 76.

153. The *ARRJ* engraving is reproduced in White, *American Railroad Passenger Car*, p. 576.

154. *RRG*, Sept. 19, 1884, p. 680.

155. Ibid., Oct. 4, 1873, p. 402; March 21, 1874, p. 102; July 20, 1877, p. 328; and July 18, 1879, p. 388.

156. Ibid., Oct. 4, 1873, p. 402.

157. *Technology and Culture*, April 1972, p. 179.

158. Richard Reinhardt, *Workin' on the Railroad* (Palo Alto, Calif., 1970, rpt. 1971), p. 280.

159. Ibid., p. 86. Reinhardt's source for this story was Herbert E. Hamblen (1849–1901), *General Managers Story* (New York, 1898). The story is fiction and its subject is never identified, but it may be semi-autobiographical, because the author worked as a locomotive engineer between 1880 and 1894.

160. *Technology and Culture*, Jan. 1982, p. 42.

161. *RRG*, Sept. 19, 1884, p. 686; Oct. 3, 1884, p. 714; and Jan. 23, 1885, p. 60; *Rly. Rev.*, Oct. 11, 1884, p. 532.

162. *RRG*, May 7, 1886, p. 309, and July 9, 1886, p. 471.

163. *MCB Report*, 1884, pp. 131–151; *RRG*, Oct. 2, 1885, p. 630.

164. *ARRJ*, Aug. 1884, p. 129; *NCB*, Aug. 1891, p. 120.

165. There is no direct evidence for the concern about failure. It was suggested by W. L. Withuhn in his as yet unpublished study of locomotive superheating and compounding where the risk of failure convinced master mechanics to choose the cheaper alternative.

166. *RRG*, Oct. 4, 1873, p. 402, and July 4, 1884, p. 496; *RR. Mag.*, Jan. 1930, p. 236.

167. *NCB*, Dec. 1876, p. xi; *Car Builders Dictionary*, 1879, p. 30 (rear advertisement), and 1888, figs. 453–459; *Rly. Age*, July 17, 1879.

168. *NCB*, Aug. 1882, p. 92.

169. *DAB*, vol. 9 (1932), p. 609, offers a sketchy biography of Janney.

170. *NCB*, Nov. 1874, p. 165.

171. *Technology and Culture*, April 1972, p. 188.

172. *RRG*, Oct. 5, 1883, p. 658, and March 7, 1884, p. 189.

173. *Rly. Rev.*, Feb. 16, 1884, p. 91.

174. Ibid., Oct. 20, 1888, p. 603.

175. City of Alexandria Will Book no. 1 (1871–1917), p. 191.

176. *RRG*, Nov. 13, 1891, p. 794; *RRCJ*, Sept. 1900, p. 247.

177. *RRG*, March 4, 1892, p. 171.

178. *Eng. News*, May 3, 1890, p. 422.

179. *NCB*, Dec. 1891, p. 184.

180. *RRG*, March 4, 1892, p. 171.

181. *RRG*, Nov. 13, 1891, p. 798; *Rly. Rev.*, April 1, 1893, p. 197.

182. *RRG*, Dec. 25, 1891, p. 916.

183. Ibid., Dec. 10, 1897, p. 864.

184. *ARRJ*, Sept. 16, 1848, opp. p. 600.

185. Ibid., March 1896, p. 38.

186. *RRG*, Nov. 17, 1899, p. 794.

187. Data on the Continuous Draw Bar Co. is drawn from *NCB*, Aug. 1875, p. 123; Aug. 1879, p. ii; and Jan. 1881, p. ii; and *Poor's*, 1890, p. 1304.

188. *RRCJ*, Aug. 1893, p. 305, and July 1893, p. 262.

189. *NCB*, June 1892, p. 90.

190. *ARRJ*, March 1896, p. 38.

191. Ibid., Oct. 1899, p. 332.

192. Westinghouse draft-gear data is drawn from White, *American Railroad Passenger Car*, pp. 575–578, with additional data from *Eng. News*, March 1,

Walker & Co., in the collections of the DeGolyer Foundation Library, Southern Methodist University, Dallas, Tex.

1890, p. 207; *Rly. Rev.*, July 11, 1903, p. 550; and *RRG*, Oct. 13, 1899, p. 705, and Jan. 13, 1905, p. 25.

193. Massachusetts Committee on Railways and Canals, *Senate Document 55*, Jan. 17, 1842.

194. Francis B. C. Bradlee, *The Boston and Maine Railroad* (Salem, 1921), p. 39.

195. *R&LHS Bull.* 6 (1923), p. 77.

196. F. A. Ritter Von Gerstner, *Die innern Communication der Vereinigten Staaten von Nord-America* (Vienna, 1842–1843), p. 255.

197. *NCB*, April 1890, p. 58.

198. John H. White, Jr., *The John Bull: 150 Years a Locomotive* (Washington, D.C., 1981).

199. *Causes of Railroad Accidents*, a report to the New York Legislature (Albany, 1853).

200. Lucius Beebe, *The Central Pacific and Southern Pacific Railroads* (Berkeley, Calif., 1963), p. 52.

201. *RR. Mag.*, Aug. 1909, p. 387.

202. *Outing Magazine*, June 1909, p. 273.

203. *Car Builders Dictionary*, 1879, fig. 61.

204. *NCB*, Sept. 1879, p. 131; *RRG*, May 24, 1878, p. 261.

205. *NCB*, Oct. 1880, p. 157.

206. *MCB Report*, 1878, p. 41.

207. Ibid., p. 13.

208. Brake-gear diagrams are given in every early edition of the *Car Builders Dictionary* as well as in White, *American Railroad Passenger Car*, p. 542.

209. S. M. Whipple, *The Tanner Brake Case*, prepared for the Eastern Railroad Assoc., 1871. See also John H. White, Jr., "James Millholland," *U.S. National Museum Bulletin* 252, paper 69 (Washington, D.C., 1967).

210. *Rly. & Loco. Eng.*, Jan. 1903, p. 27. See also *Car Builders Dictionary*, 1879, fig. 645, which shows a diagram of Elder's brake rig.

211. *RRCJ*, June 1895, p. 116.

212. *MCB Report*, 1875, p. 9.

213. *Journal of the Franklin Institute*, Oct. 1835, p. 227; *ARRJ*, April 2, 1836, p. 193.

214. *ARRJ*, Sept. 15, 1841, p. 168, and Oct. 28, 1848, p. 691; S. A. Howland, *Steamship Disasters and Railroad Accidents in the United States* (Worcester, 1840), p. 262.

215. *Western Railroad Gazette*, Oct. 30, 1869, p. 1.

216. *MCB Report*, 1875, p. 10; *NCB*, June 1880, p. 97.

217. Data on the Card brake is from *Rly. Rev.*, April 17, 1880, p. 181; Jan. 7, 1882, p. 14; and April 28, 1883, p. 235; and *NCB*, April 1881, p. 45.

218. Data on Rote is from *Rly. Rev.*, Nov. 8, 1884, p. 578, and Dec. 6, 1884, p. 636; *ARRJ*, June 1885, p. 86; and *RRG*, Sept. 24, 1886, folding plate.

219. *Rly. Rev.*, July 31, 1880, p. 374.

220. *MCB Report*, 1882, p. 10.

221. *Rly. Rev.*, Sept. 19, 1885, p. 454.

222. Ibid., Jan. 19, 1889, p. 30.

223. *MCB Report*, 1887, p. 103.

224. *Rly. Rev.*, Jan. 19, 1889, p. 30.

225. *MCB Report*, 1887, pp. 119 and 134.

226. *Rly. Rev.*, Feb. 13, 1886, p. 82.

227. *Trains* magazine, 1978, p. 48. Much has been written about Westinghouse, including two book-length biographies. See also White, *American Railroad Passenger Car*, and a lengthy article in *Business History Review*, Spring 1984.

228. *MCB Report*, 1885, p. 75.

229. Albert Fishlow, *American Railroads and the Transformation of the Ante-Bellum Economy* (Cambridge, Mass., 1965).

230. *NCB*, Dec. 1876, p. 184.

231. *Master Mechanics Report*, 1875, p. 126.

232. Ibid., 1872, pp. 113–133, offered a lengthy report and discussion on the pros and cons of air brakes. See also reports for 1874 and 1875.

233. *History of Technology Annual* 11 (1986) contains an article on early power brakes in Great Britain.

234. Central Pacific Railway annual reports, 1883, p. 34, and 1884, p. 42; *MCB Report*, 1885, p. 75.

235. *RRG*, March 27, 1885, p. 204.

236. *ASCE Trans.*, 1885, p. 471.

237. Ibid., p. 411; *MCB Report*, 1885, p. 81.

238. *RRG*, Jan. 11, 1884, p. 29.

239. White, *American Railroad Passenger Car*, pp. 548–551, contains a discussion of vacuum brakes.

240. *Master Mechanics Report*, 1872, p. 132.

241. *Rly. Rev.*, April 2, 1881, p. 179.

242. *MCB Report*, 1887, contains a lengthy description of the Burlington trials starting on p. 25.

243. *School of Mines Quarterly* (Columbia University), Jan. 1884, p. 132.

244. *RRG*, June 10, 1887, p. 380; *Rly. Rev.*, July 2, 1887, p. 392.

245. *RRG*, Oct. 12, 1888, p. 671.

246. *NCB*, Dec. 1892, p. xxx; *Eng. News*, May 2, 1891, p. 430.

CHAPTER 8

1. Pot hoppers (mainly B&O) 3,000–4,000; Reading coal cars 6,000–7,000; New York Central iron boxcars 700; iron pipe cars 10,000 plus; misc. 500 plus or minus: total 20,200–22,200. Documentation for these figures is given later in the text and notes.

2. *RRG*, July 6, 1877, p. 304.

3. Ibid., Sept. 28, 1900, p. 637; Lionel S. Marks, ed., *The Mechanical Engineer's Handbook* (New York, 1930), pp. 422 and 426.

4. *RRG*, April 11, 1879, p. 196. During the wood-car era Rule 17 of the MCB Association stated that average freight car life was sixteen years; a car was thus depreciated at 6 percent per year.

5. *ARRJ*, Oct. 25, 1862, p. 839.

6. *NCB*, Feb. 1878, p. 24.

7. *RRG*, Dec. 27, 1889, p. 854.

8. *MCB Report*, 1880, p. 41. See also John H. White, Jr., *The American Railroad Passenger Car* (Baltimore, 1978), pp. 42–44, for more on wood supply.

9. Peter Temin, *Iron and Steel in Nineteenth-Century America* (Cambridge, Mass., 1964), pp. 274–275.

10. *Smithsonian Journal of History*, Fall 1968, pp. 41–76; C. E. Peterson, ed., *Building in Early America* (Radnor, Pa., 1976), pp. 96–118.

11. *RRG*, Nov. 7, 1890, p. 771.

12. *Rly. Rev.*, Feb. 29, 1896, p. 117; *RRG*, Feb. 28, 1896, p. 144.

13. D. Ripley, *The Peak Forest Tramway*, booklet reprinted 1972, Oakwood Press. National Railway Museum catalog dates the car as 1797, but the museum staff agree that the car's precise age is in question.

14. Bertram Baxter, *Stone Blocks and Iron Rails* (Newton Abbot, Devon, 1966), p. 86.

15. C. Von Oeynhausen and H. Von Dechen, *Railways in England* (1826–1827; German text trans. and rpt. by Newcomen Society, London, 1971), pp. 31 and 32.

16. C. F. D. Marshall, *Centennial History of the Liverpool and Manchester Railway* (London, 1930), p. 88.

17. *ARRJ*, Aug. 1896, pp. 171–177; *RRG*, Jan. 24, 1908, p. 127.

18. *RRG*, June 16, 1876, p. 263.

19. *American Railway Times*, May 31, 1862, p. 175.
20. *Rly. Rev.*, June 21, 1890, p. 368.
21. *ARRJ*, Feb. 1844, p. 85.
22. A. Bendel, *Aufsatze Eisenbahnwesen in Nord-Amerika* (Berlin, 1862), pp. 42–43; data gathered 1859.
23. *Loco. Eng.*, April 1896, p. 307.
24. *Poor's*, 1868, p. 261.
25. *MCB Report*, 1877, p. 44. Curiously, the *Engineer*, June 15, 1877, p. 407, reported the Reading as building fifty new iron coal cars.
26. *Scientific American*, Dec. 9, 1865, p. 371.
27. *NCB*, March 1884, p. 31.
28. White, *American Railroad Passenger Car*, pp. 18–20.
29. B&O annual report, 1844, p. 20. The Winans-Murray memo of May 1844 was uncovered in the B&O Museum archives (Baltimore) by John Hankey, curator of the museum.
30. Bendel, plate 17.
31. M. C. Clark, "The Richmond and Danville Railroad" (M.A. thesis, George Washington University, 1959), p. 27.
32. *NCB*, Oct. 1874, p. 147.
33. *R&LHS Bull.* 65 (March 1945), p. 43.
34. E. Lavoinne and E. Pontzen, *Les Chemins de Fer en Amérique* (Paris, 1882), p. 71; data gathered in 1876.
35. *Loco. Eng.*, April 1921, p. 103.
36. *MCB Report*, 1877, p. 43.
37. *NCB*, April 1884, p. 46.
38. *RRG*, May 2, 1884, p. 342.
39. *MCB Report*, 1895, p. 270.
40. *Rly. Mast. Mech.*, Nov. 1897, p. 157.
41. *RRG*, Sept. 5, 1902, p. 689; *ARRJ*, May 1907, p. 161. There is a photograph of a pot hopper reportedly built in 1897 for the C&P in the Coleman Collection, Ohio State Museum, Columbus.
42. B&O annual report, 1846, p. 46; *ARRJ*, May 1907, p. 160.
43. *RR. Adv.*, Sept. 12, 1857, p. 73; *Western Railroad Gazette*, Oct. 17, 1857, p. 4.
44. *ARRJ*, May 1907, p. 160.
45. *Industrial Monthly*, Feb. 1872, p. 52.
46. *RR. Rec.*, July 28, 1853, p. 341.
47. White, *American Railroad Passenger Car*, pp. 121–122.
48. Ibid., p. 124.
49. *Railroad History*, no. 130 (Spring 1974), p. 51.
50. *ARRJ*, Oct. 18, 1862, p. 819.
51. Ibid., Oct. 25, 1862, p. 839.
52. White, *American Railroad Passenger Car*, p. 117.
53. *RR. Mag.*, June 1939, p. 115; *Am. Rly. Rev.*, April 4, 1861, p. 197.
54. *NCB*, Nov. 1883, p. 121; *RRG*, Feb. 21, 1879, p. 93.
55. *Poor's*, 1868–1869, p. 36.
56. *ARRJ*, March 7, 1863, p. 233.
57. *NCB*, Nov. 1883, p. 121.
58. *RRG*, Feb. 21, 1879, p. 93; *NCB*, April 1882, p. 39.
59. New York Railroad Club *Proceedings*, Jan. 17, 1908, p. 966.
60. *Rly. Mast. Mech.*, Oct. 1890, p. 167.
61. *NCB*, June 1879, p. 85.
62. *Railway World*, Nov. 24, 1877, p. 1117.
63. Ibid., May 12, 1877, p. 439; Lavoinne and Pontzen, atlas vol. 2, plate 8.
64. Western Railroad Club Report, 1893, p. 106; *RRCJ*, June 1895, p. 111.
65. *RRG*, Jan. 24, 1879, p. 43, and March 14, 1879, p. 139.
66. *NCB*, April 1879, p. 54.
67. *RRG*, Sept. 1, 1882, p. 537.
68. *American Artisan*, April 8, 1868, p. 193.
69. *NCB*, Aug. 1872, p. 13.
70. Ibid., Feb. 1878, p. 24. Production began two years before the article.
71. Ibid., Dec. 1876, p. 182.
72. Refer to notes 70 and 71 and *NCB*, June 1879, p. 87; *ARRJ*, March 16, 1878, p. 289; *RRG*, May 9, 1879, p. 259.
73. NTW prospectus booklet, 1877; original in De-Golyer Foundation Library, Southern Methodist University, Dallas, Tex.
74. *RRG*, Nov. 22, 1878, p. 568.
75. *NCB*, Feb. 1882, p. 23, and April 1882, p. 39.
76. *Rly. Rev.*, June 17, 1882, p. 351, and Oct. 21, 1882, p. 604.
77. *Rly. Mast. Mech.*, Jan. 1888, p. 199.
78. *Rly. Rev.*, June 23, 1883, p. 359.
79. *Rly. Mast. Mech.*, Dec. 1888, p. 189.
80. Ibid., March 1889, p. 41. The wreck occurred on the Fitchburg in Jan. 1889.
81. *Rly. Rev.*, Nov. 10, 1888, p. 652.
82. *NCB*, May 1888, pp. 65 and 77.
83. *RRG*, Oct. 12, 1888, p. xxiii.
84. Ibid., Sept. 13, 1889, p. 603.
85. Ibid., Oct. 11, 1889, p. 667.
86. Ibid., May 1, 1891, p. 307.
87. Ibid., April 20, 1894, p. 284; *Rly. Rev.*, Dec. 24, 1904, p. 907.
88. *Iron Age*, July 3, 1890, p. 15; *RRG*, July 11, 1890, p. 498; *Rly. Rev.*, July 12, 1890, p. 413.
89. *Rly. Age*, Dec. 19, 1902, p. 676; *Rly. Rev.*, Dec. 24, 1904, p. 907.
90. *RRG*, April 20, 1894.
91. Ibid., May 1, 1891, p. 307; *NCB*, March 1894, p. 40; *RRCJ*, Feb. 1895, p. 29.
92. T. H. Russell, *Railway Library Car Shop Practice* (Chicago, 1925), vol. 1, p. 68. The tube cars were said to be in service until "recent years," presumably meaning 1920. Data on the C&C tin boxcars is from Stephen Drew, curator of the California State Railroad Museum, Sacramento.
93. *RRCJ*, Feb. 1895, p. 28; *Rly. Mast. Mech.*, Sept. 1895, p. 140; *The Locomotive* (London), May 14, 1932, p. 164.
94. Western Railroad Club *Proceedings*, Nov. 1895, p. 95; *Cassier's Magazine*, Feb. 1897, p. 266.
95. Western Railroad Club *Proceedings*, Nov. 1895, p. 93.
96. *NCB*, March 1894, p. 32.
97. *RRG*, May 11, 1888, p. 298.
98. *NCB*, May 1893, p. 70.
99. Ibid., May 1884, p. 54; *Rly. Rev.*, April 6, 1889, p. 184.
100. *Loco. Eng.*, March 1898, p. 163; *ARRJ*, May 1903, p. 172.
101. *RRG*, Sept. 28, 1900, p. 637, and Jan. 24, 1908, p. 129. J. D. McIlwain claimed that wood-car maintenance equaled $50 per year; see Western Railroad Club *Proceedings*, 1893–1894, p. 114.
102. *RRG*, Jan. 24, 1908, p. 129.
103. Western Railroad Club *Proceedings*, Nov. 1891, p. 91.
104. *RRG*, Jan. 24, 1908, p. 128.
105. *Rly. Mast. Mech.*, Feb. 1892, p. 27. Lamont's design dates from about 1890–1891.
106. *NCB*, March 1894, p. 40; *ARRJ*, Feb. 1894, p. 91.
107. *Encyclopedic Biography of Illinois*, vol. 1 (1892), p. 171.
108. White, *American Railroad Passenger Car*, p. 130.
109. *RRG*, Oct. 24, 1890, p. 741, and Dec. 19, p. 882.
110. Ibid., Feb. 28, 1890, p. 148; *Rly. Rev.*, Feb. 1, 1890, p. 60.
111. *NCB*, Dec. 1893, p. 188.

112. *RRG*, June 5, 1891, p. 387; *NCB*, March 1894, p. 49.
113. *Rly. Rev.*, Dec. 13, 1890, p. 747; *RRG*, Sept. 18, 1891, p. 650; *Eng. News*, July 11, 1891, fold-in plate.
114. *RRG*, Dec. 30, 1892, p. 995; *Rly. Rev.*, Jan. 7, 1893, p. 6.
115. *Eng. News*, May 19, 1892, p. 499.
116. *Rly. Rev.*, Jan. 4, p. 14, and May 30, 1896, p. 296.
117. *ARRJ*, May 1903, p. 168. Taylor was a graduate engineer (Lehigh Univ.) who worked as a chemist and supervisor at several steel companies before becoming Carnegie's assistant in 1893. See *Who Was Who* Historical, vol. 1 (1897–1942), p. 1218.
118. Robert Hessen, *Steel Titan: The Life of Charles M. Schwab* (New York, 1975), p. 86.
119. Data on Fox comes from several sources: *Engineering*, Oct. 30, 1903, p. 604; *Dictionary of National Biography*, vol. 1 (1901–1911), p. 52; Geo. A. Newby, *Behind the Fire Doors: Fox's Corrugated Furnace and the High-Pressure Steamship* (published by the author in Harrogate, England, 1979).
120. *ARRJ*, June 1894, p. 261.
121. *Iron* (London), Dec. 6, 1889, p. 483.
122. *Rly. Mast. Mech.*, Feb. 1892, p. 28.
123. *ARRJ*, May 1903, p. 168.
124. *NCB*, April 1892, p. 63.
125. *Rly. Mast. Mech.*, Dec. 1895, p. 186.
126. *RRG*, June 12, 1896, p. 411; *Rly. Rev.*, June 13, 1896, p. 325, and July 18, 1896, p. 396.
127. *ARRJ*, Oct. 1903, p. 352.
128. Roy C. Beaver, *The Bessemer and Lake Erie Railroad* (San Marino, Calif., 1969). The line was created by taking over the existing Pittsburgh, Shenango & Lake Erie and adding a 40-mile southern extension to bring it into the Pittsburgh area. The line was first named the BP&LE and became the B&LE in 1900.
129. *RRG*, April 2, 1897, p. 233.
130. Biographical data on Schoen is drawn chiefly from *Nat. Cyc. Am. Bio.*, vol. 25, p. 398; *Rly. Age*, Feb. 24, 1899, p. 141; and *RRG*, Nov. 24, 1899, p. 305.
131. *RRG*, May 23, 1890, p. 352.
132. All four men claim a major role in the design of the pressed-steel version of the 1897 PB&LE order. King—son of the former owner of the Middletown, Pa., car works——worked for Schoen but had a falling out with his employer and left in 1901, according to a statement published in his privately printed autobiography of 1917. Hansen was Schoen's engineer until he too left in 1902 to help organize the Standard Steel Car Co. Swensson, a native of Denmark and president of the ASCE, also worked in Schoen's drafting office. His claim appears in *Nat. Cyc. Am. Bio.*, vol. 14, p. 277.
133. *ARRJ*, May 1903, p. 170.
134. *RRG*, Sept. 28, 1900, p. 637.
135. *Rly. Mast. Mech.*, July 1897, p. 98.
136. Beaver, *B&LE*, p. 174.
137. *Eng. News*, Jan. 7, 1904, pp. 3–5; *ARRJ*, March 1909, p. 424.
138. *Rly. Rev.*, June 5, 1897, p. 322.
139. *RRG*, July 14, 1899, p. 510.
140. *RRCJ*, June 1899, p. 144.
141. *RRG*, July 14, 1899, p. 503.
142. Ibid., Jan. 20, 1899, p. 48.
143. *Engineer*, Oct. 30, 1903, p. 431; *The Harrogate Advertiser* (English newspaper), March 12, 1977.
144. The Carnegie-Schoen controversy is covered in Hessen's *Steel Titan* and J. F. Wall's *Andrew Carnegie* (New York, 1970).
145. *Rly. Age*, Jan. 7, 1939, p. 105.
146. Ibid., Jan. 4, 1930, p. 75.

INDEX